August 13–17, 2012
Helsinki, Finland

Association for
Computing Machinery

Advancing Computing as a Science & Profession

SIGCOMM'12

Proceedings of the ACM SIGCOMM 2012 Conference

Applications, Technologies, Architectures, and Protocols for Computer Communication

Sponsored by:

SIGCOMM

Association for
Computing Machinery

Advancing Computing as a Science & Profession

The Association for Computing Machinery
2 Penn Plaza, Suite 701
New York, New York 10121-0701

Notice to Past Authors of ACM-Published Articles
ACM intends to create a complete electronic archive of all articles and/or other material previously published by ACM. If you have written a work that has been previously published by ACM in any journal or conference proceedings prior to 1978, or any SIG Newsletter at any time, and you do NOT want this work to appear in the ACM Digital Library, please inform permissions@acm.org, stating the title of the work, the author(s), and where and when published.

ISBN: 978-1-4503-1419-0

Additional copies may be ordered prepaid from:

ACM Order Department
PO Box 30777
New York, NY 10087-0777, USA

Phone: 1-800-342-6626 (USA and Canada)
+1-212-626-0500 (Global)
Fax: +1-212-944-1318
E-mail: acmhelp@acm.org
Hours of Operation: 8:30 am – 4:30 pm ET

Printed in the USA

SIGCOMM'12 General Chairs' Welcome Message

Welcome to ACM SIGCOMM 2012 in Helsinki! As in previous years, SIGCOMM is being organized as a week full of research events: the main conference is both preceded and followed by eight workshops on current topics in and around networking, and four tutorials that are looking beyond.

We are happy that we were able to maintain registration fees at roughly the same level as the last time SIGCOMM was held in Europe in 2009. We were even able to reduce some of the student registration fees compared to 2009. Given the high costs of living, services, and goods in Finland and economical unknowns in Europe (including the US Dollar exchange rate), creating a balanced budget has been a challenging task.

Holding SIGCOMM 2012 in Helsinki was only made possible through the generous financial support from our many (industrial) supporters of the conference. We are particularly indebted to our two hosts, **Nokia** and **Aalto University**, whose extremely substantial financial contributions provided an early, solid basis for organizing a successful event. It would simply not have been possible to organize SIGCOMM in Helsinki without them.

It is our privilege to acknowledge the generous support from our Silver supporters: **Cisco Systems**, **Ericsson**, **NetApp**, **NEC** and the **City of Helsinki**; and Bronze supporters: **Akamai**, **AT&T**, **Alcatel-Lucent Bell Labs**, **Comcast**, **Conviva**, **Facebook**, **Google**, **HP**, **Huawei**, the **Internet Society**, **Microsoft Research**, **Riverbed**, **Telefonica** and **Verisign Labs**.

SIGCOMM has a long tradition of financially supporting student participation. We have continued this tradition by keeping the cost of student registrations as low as possible, by organizing a student dinner for students to meet with leaders of the research community, and by continuing to provide a substantial number of student travel grants. We are thankful for student travel support provided by the **US National Science Foundation**, **ACM SIGCOMM**, the **Internet Society**, and **Internet-Science.eu**.

Organizing a conference such as SIGCOMM requires the involvement and dedication of a large and enthusiastic group of volunteers. We have enjoyed working with an outstanding group of colleagues who served on the organization committee and have selflessly and generously contributed their time and energy for more than a year. The conference would be impossible to organize without their volunteer spirit.

We thank all those who contributed to the organization of the technical program of the main conference and the workshops: Program committee chairs **Venkat Padmanabhan** and **George Varghese**, workshop chairs **Dina Katabi** and **Jitendra Padhye**, the chairs of the eight selected workshops, and the chairs of the Demo and Poster program, **Kevin Almeroth** and **Colin Perkins**.

We also thank all those who ensured that everything ran smoothly behind the scenes: Treasurer and registration chair **Joe Touch**; publication chair **Andrea Passarella**; publicity chair **Ryuji Wakikawa**; web chair **Liang Wang**; travel grant chairs **Marcelo Bagnulo**, **Polly Huang**, and **Sharon Goldberg**; and, for the overall local organization and taking care of all special needs, local chairs **Jussi Kangasharju** and **Jukka Manner**.

We want to thank the ACM SIGCOMM chair **Bruce Davie** and conference coordinator **Jaudelice de Oliveira** for their active role and help in all stages of preparation of the conference. Last, but not least, we gratefully acknowledge the help from ACM Headquarters and Sheridan Printing COmpany with the preparation of the conference and workshop proceedings.

We hope that you will find SIGCOMM 2012 to be a stimulating and scientifically enriching experience.

We are sure that you will find the time to enjoy the beautiful and vibrant city of Helsinki, World Design Capital 2012, with its long hours of daylight.

Lars Eggert	**Jörg Ott**
SIGCOMM 2012 General Co-Chair	*SIGCOMM 2012 General Co-Chair*
NetApp, Germany	*Aalto University, Finland*

Message from the PC Chairs

Welcome to the 2012 ACM SIGCOMM Conference in Helsinki, Finland, which is a wonderful location and incidentally is the farthest north that the conference has ventured to in its long history! It is our pleasure to introduce the technical program of the conference. This year's program continues SIGCOMM's tradition of presenting outstanding work that suggests new areas of research in networking. The program committee accepted 31 papers selected from 235 submissions. We felt it was important to have a broad view of networking; this allowed us to choose papers on such diverse topics as on-chip networking, key-value stores, and measuring click spam. While this year's program continues to reflect a recent renaissance in data centers and wireless, traditional topics such as routing, QoS, and measurement are also well represented.

The papers came from a rich mix of authors from both academia and industry across the globe. Authors from organizations in around 25 countries submitted papers, and the final program includes authors from over 30 organizations in 8 countries. It was a delight to work with a terrific technical program committee of 49 members from all over the world and including distinguished academic and industrial researchers. We chose the PC with care, trying to balance experience with youth, and recognizing authors who have made consistent technical contributions in SIGCOMM and other top conferences, while ensuring broad technical area coverage and a wide geographic representation. Members of the PC were drawn from 9 countries in 4 continents, with a quarter of the PC comprising first-time members, both relatively young networking researchers and experienced researchers from sister areas such as operating systems and mobile systems. We are grateful for the insight and care that our PC displayed, working hard and responding to our every request. We are also delighted to report that the first-time PC members were a wonderful addition, enriching the PC with their insights and exhibiting a high degree of diligence and professionalism.

We continued the tradition of a two-round review process, with a first round of three reviews for each paper. Around half the submitted papers were chosen for the second round, which then generated three more reviews. After the second round and a subsequent online discussion phase, around 60 papers were chosen for discussion at the PC meeting. As chairs we went less by numerical score in order to decide which paper to advance than by the actual reviews, realizing that numerical scores vary widely with reviewers. In a few cases, we sought external reviewers to obtain an expert opinion. The PC meeting was held in San Jose, CA, USA, over a day and a half. While all of the papers in the discussion set were discussed and had tentative decisions made on day 1, the half-day on day 2 allowed plenty of time for the PC to revisit, and in some cases reverse, the decisions made on day 1, and even bring up papers for discussion papers that hadn't been shortlisted for discussion on day 1.

Despite following the traditional process, this year's review process was marked by two new mechanisms. First, we reintroduced an author response phase, which had last been tried in SIGCOMM over a decade ago. To make the author response effective, we formulated a careful set of instructions, including illustrative examples, to help make the author responses more factual and effective. The author response phase was intended partly to reduce the possibility of mistaken reviews, but also to promote an approximation to the healthy dialog between authors and reviewers inherent in the classical journal process. We were delighted with the quality of the author responses we saw even with (eventually) rejected papers: respectful, factual, and insightful. The responses were all considered as part of the PC discussion, and some indeed changed the reviewers' minds. Second, the entire PC made an attempt to provide constructive reviews, emphasizing the positive, and suggesting where the paper could be improved. To make explicit the value we placed on good reviews and recognize exceptional contributions in this regard by PC members, we instituted 3 Outstanding Reviewer Awards, which will be presented at the conference.

The selection of the award papers started with nominations made by PC members. We as the PC chairs shortlisted 5 of the nominated papers and passed these on to an awards committee chaired by David Wetherall and also comprising T.V. Lakshman, Dave Levin, and Ratul Mahajan, all members of the PC. We are grateful to the members of the awards committee for conducting a thorough process and sending us their selections in quick time. It is a pleasure to present the Best Paper Award for ACM SIGCOMM 2012 to the paper titled "Multi-Resource Fair Queueing for Packet Processing" by Ali Ghodsi, Vyas Sekar, Matei Zaharia, and Ion Stoica, and the Best Student Paper Award to the paper titled "Picasso: Flexible RF and Spectrum Slicing" by Steven Hong, Jeff Mehlman, and Sachin Katti. For the Outstanding Reviewer Awards, we, the PC chairs, made the selections based on the quality and timeliness of reviewing, participation in the online and PC meeting discussions, and the willingness to go the extra mile when needed. It was challenging for us to make a selection from among the many PC members who contributed so strongly, but after careful thought, it is our pleasure to recognize Aditya Akella, Matt Caesar, and Dave Levin with the Outstanding Reviewer Awards for SIGCOMM 2012. Congratulations to all of the award winners, authors as well as reviewers.

As the PC chairs, we are grateful to our entire PC that worked so hard and often went the "extra mile" at short notice while trying to "do the right thing". We thank our external reviewers and Terry Lam from UCSD, who maintained the HotCRP site for us and was always prompt in attending to our requests. We thank Geoff Voelker for his help with Banal and KK Ramakrishnan who handled several papers on which both the PC chairs had conflicts and was a de facto, "third chair". We are grateful to the SIGCOMM 2012 General Chairs, Lars Eggert and Jörg Ott, who made our job easy, to Andrea Passarella for being prompt and ever-helpful with publications-related matters, to Liang Wang for always being responsive on web-related matters, and to Joe Touch for his help with the PC meeting logistics. We thank our industry sponsors and the SIGCOMM Technical Steering Committee Members, who were always ready with advice.

We hope you will enjoy the program and also take the opportunity to interact with the remarkable group of people assembled in Helsinki. To our mind, networking continues to be a vibrant and exciting field. We hope that the program will inspire you to create and explore new ideas for the future of networking.

Sincerely,

Venkat Padmanabhan
Microsoft Research India

George Varghese
UC San Diego

ACM SIGCOMM 2012 PC Chairs

Table of Contents

Session 4: Measuring Networks

Session Chair: Sergey Gorinsky *(Institute IMDEA Networks)*

Session 5: Data Centers: Resource Management

Session Chair: Emin Gün Sirer *(Cornell University)*

Session 6: Wireless and Mobile Networking

Session Chair: Srinivasan Seshan *(Carnegie Mellon University)*

Posters & Demonstrations II

Session 7: Network Formalism and Algorithmics

Session Chair: Matt Caesar *(University of Illinois)*

Session 8: Streaming and Content Networking

Session Chair: Fabián Bustamante *(Northwestern University)*

Session 9: Routing

Session Chair: Craig Partridge *(Raytheon BBN Technologies)*

Session 10: Data Centers: Network Resilience

Session Chair: Bruce Maggs *(Duke University & Akamai Technologies)*

SIGCOMM 2012 Conference Organization

General Chairs: Lars Eggert *(NetApp, Germany)*
Jörg Ott *(Aalto University, Finland)*

Program Chairs: Venkat Padmanabhan *(Microsoft Research, India)*
George Varghese *(UCSD, USA)*

Workshop Chairs: Dina Katabi *(MIT, USA)*
Jitendra Padhye *(Microsoft Research, USA)*

Poster and Demo Chairs: Kevin Almeroth *(University of California, Santa Barbara, USA)*
Colin Perkins *(University of Glasgow, UK)*

Publication Chair: Andrea Passarella *(CNR IIT, Italy)*

Local Arrangements Chairs: Jussi Kangasharju *(University of Helsinki, Finland)*
Jukka Manner *(Aalto University, Finland)*

Publicity Chair: Ryuji Wakikawa *(Toyota ITC, USA/Japan)*

Treasurer & Registration Chair: Joe Touch *(USC/ISI, USA)*

Travel Grant Chairs: Marcelo Bagnulo *(Universidad Carlos III de Madrid, Spain)*
Polly Huang *(National Taiwan University, Taiwan)*
Sharon Goldberg *(Boston University, USA)*

Web Chair: Liang Wang *(University of Helsinki, Finland)*

Conference Coordinator: Jaudelice de Oliveira *(Drexel University, USA)*

Program Committee: Aditya Akella *(University of Wisconsin, USA)*
Virgilio Almeida *(UFMG, Brazil)*
Mostafa Ammar *(Georgia Tech, USA)*
Sujata Banerjee *(HP Labs, USA)*
Fabián E. Bustamante *(Northwestern University, USA)*
Matt Caesar *(University of Illinois, USA)*
Krishna Chintalapudi *(Microsoft Research India)*
Dah Ming Chiu *(Chinese University of Hong Kong)*
Romit Roy Choudhury *(Duke University, USA)*
Mark Crovella *(Boston University, USA)*
Jon Crowcroft *(Cambridge University, UK)*
Bruce Davie *(Nicira Networks, USA)*
Jeff Dean *(Google, USA)*
Kevin Fall *(Qualcomm, USA)*

Program Committee (continued):

Nick Feamster *(University of Maryland, USA)*
Sergey Gorinsky *(Institute IMDEA Networks, Spain)*
Krishna Gummadi *(MPI-SWS, Germany)*
Kyle Jamieson *(University College London, UK)*
Brad Karp *(University College London, UK)*
Dina Katabi *(MIT, USA)*
Sachin Katti *(Stanford University, USA)*
Ramana Kompella *(Purdue University, USA)*
Arvind Krishnamurthy *(University of Washington, USA)*
Craig Labovitz *(DeepField Networks, USA)*
T. V. Lakshman *(Bell Labs, Alcatel-Lucent, USA)*
Kirill Levchenko *(UC San Diego, USA)*
Dave Levin *(HP Labs, USA)*
Bruce Maggs *(Duke University and Akamai Technologies, USA)*
Ratul Mahajan *(Microsoft Research Redmond, USA)*
Dave Maltz *(Microsoft, USA)*
Z. Morley Mao *(University of Michigan, USA)*
Nick McKeown *(Stanford University, USA)*
Mike Mitzenmacher *(Harvard University, USA)*
Craig Partridge *(Raytheon BBN Technologies, USA)*
Lili Qiu *(UT Austin, USA)*
K. K. Ramakrishnan *(AT&T Labs Research, USA)*
Bhaskaran Raman *(IIT Bombay, India)*
Pablo Rodriguez *(Telefonica Research, Spain)*
Ant Rowstron *(Microsoft Research Cambridge, UK)*
Srinivasan Seshan *(Carnegie Mellon University, USA)*
Hovav Shacham *(UC San Diego, USA)*
Emin Gün Sirer *(Cornell University, USA)*
Ion Stoica *(UC Berkeley, USA)*
Lakshmi Subramanian *(New York University, USA)*
Renata Teixeira *(CNRS and UMPC Sorbonne Universities, France)*
Jon Turner *(Washington University, USA)*
Kobus Van der Merwe *(AT&T Labs - Research, USA)*
David Wetherall *(University of Washington, USA)*
Haifeng Yu *(National University of Singapore)*

Additional reviewers:

Hari Adiseshu *(Bell Labs, Alcatel-Lucent, USA)*
Deniz Altınbüken *(Cornell University, USA)*
Michela Becchi *(University of Missouri, USA)*
Dan Blumenthal *(UC Santa Barbara, USA)*
Sem Borst *(Bell Labs and Eindhoven University of Technology, Netherlands)*
Stephen Checkoway *(Johns Hopkins, USA)*
George Danezis *(Microsoft Research, UK)*
Amogh Dhamdhere *(CAIDA, USA)*
Robert Escriva *(Cornell University, USA)*
Murali Kodialam *(Bell Labs, Alcatel-Lucent, USA)*
Elisavet Kozyri *(Cornell University, USA)*
Kunal Punera *(Yahoo! Research, USA)*
Krishna Puttaswamy *(Bell Labs, Alcatel-Lucent, USA)*
Michael Schapira *(Hebrew University of Jerusalem, Israel)*
Prashant Shenoy *(UMass Amherst, USA)*
Tim Sherwood *(UC Santa Barbara, USA)*
Ji-Yong Shin *(Cornell University, USA)*
Vijay Sivaraman *(University of New South Wales, Australia)*
Bill Thies *(Microsoft Research, India)*
Patrick Thiran *(EPFL, Switzerland)*
Philip Watts *(University College London, UK)*
Walter Willinger *(AT&T Labs Research, USA)*
Tilman Wolf *(UMass Amherst, USA)*

SIGCOMM 2012 Sponsors & Supporters

Sponsor:

acm sigcomm

Hosts:

NOKIA Connecting People

Aalto University

Silver Supporters:

CISCO

City of Helsinki

ERICSSON

NEC

NetApp

Bronze Supporters:

Akamai

at&t

Alcatel·Lucent

comcast

CONVIVA

facebook

Google

hp

HUAWEI

Internet Society 20 YEARS

Microsoft Research

riverbed

Telefónica Telefónica Investigación y Desarrollo

VERISIGN

Multi-Resource Fair Queueing for Packet Processing

Ali Ghodsi[†,‡], Vyas Sekar[◇], Matei Zaharia[†], Ion Stoica[†]
[†] University of California, Berkeley [◇] Intel ISTC [‡] KTH/Royal Institute of Technology
{alig, matei, istoica}@cs.berkeley.edu, vyas.sekar@gmail.com

ABSTRACT

Middleboxes are ubiquitous in today's networks and perform a variety of important functions, including IDS, VPN, firewalling, and WAN optimization. These functions differ vastly in their requirements for hardware resources (*e.g.,* CPU cycles and memory bandwidth). Thus, depending on the functions they go through, different flows can consume different amounts of a middlebox's resources. While there is much literature on weighted fair sharing of link bandwidth to isolate flows, it is unclear how to schedule *multiple* resources in a middlebox to achieve similar guarantees. In this paper, we analyze several natural packet scheduling algorithms for multiple resources and show that they have undesirable properties. We propose a new algorithm, Dominant Resource Fair Queuing (DRFQ), that retains the attractive properties that fair sharing provides for one resource. In doing so, we generalize the concept of virtual time in classical fair queuing to multi-resource settings. The resulting algorithm is also applicable in other contexts where several resources need to be multiplexed in the time domain.

Categories and Subject Descriptors: C.2.6 [Computer-Communication Networks]: Internetworking

Keywords: Fair Queueing, Middleboxes, Scheduling

1. INTRODUCTION

Middleboxes today are omnipresent. Surveys show that the number of middleboxes in companies is on par with the number of routers and switches [28]. These middleboxes perform a variety of functions, ranging from firewalling and IDS to WAN optimization and HTTP caching. Moreover, the boundary between routers and middleboxes is blurring, with more middlebox functions being incorporated into hardware and software routers [2, 6, 1, 27].

Given that the volume of traffic through middleboxes is increasing [20, 32] and that middlebox processing functions are often expensive, it is important to schedule the hardware resources in these devices to provide predictable isolation across flows. While packet scheduling has been studied extensively in routers to allocate link bandwidth [24, 10, 29], middleboxes complicate the scheduling problem because they have *multiple* resources that can be congested. Different middlebox processing functions consume vastly different amounts of these resources. For example, intrusion detection functionality is often CPU-bound [13], software routers bottleneck on memory bandwidth when processing small packets [8], and forwarding of large packets with little processing can bottleneck on link bandwidth. Thus, depending on the processing needs of the flows going through it, a middlebox will need to make scheduling decisions across multiple resources. This becomes more important as middlebox resource diversity increases (*e.g.,* GPUs [30] and specialized hardware acceleration [23, 5]).

Traditionally, for a single resource, weighted fair sharing [10] ensures that flows are isolated from each other by making *share guarantees* on how much bandwidth each flow gets [24]. Furthermore, fair sharing is *strategy-proof*, in that flows cannot get better service by artificially inflating their resource consumption. Many algorithms, such as WFQ [10], GPS [24], DRR [29], and SFQ [18], have been proposed to approximate fair sharing through discrete packet scheduling decisions, but they all retain the properties of share guarantees and strategy-proofness. We would like a multi-resource scheduler to also provide these properties.

Share guarantees and strategy-proofness, while almost trivial for one resource, turn out to be non-trivial for multiple resources [16]. We first analyze two natural scheduling schemes and show that they lack these properties. The first scheme is to monitor the resource usage of the system, determine which resource is currently the bottleneck, and divide it fairly between the flows [14]. Unfortunately, this approach lacks both desired properties. First, it is not strategy-proof; a flow can manipulate the scheduler to get better service by artificially *increasing* the amount of resources its packets use. For example, a flow can use smaller packets, which increase the CPU usage of the middlebox, to shift the bottleneck resource from network bandwidth to CPU. We show that this can double the manipulative flow's throughput while hurting other flows. Second, when multiple resources can simultaneously be bottlenecked, this solution can lead to oscillations that substantially lower the total throughput and keep some flows below their guaranteed share.

A second natural scheme, which can happen by default in software router designs, is to perform fair sharing independently at each resource. For example, packets might first be processed by the CPU, which is shared via stride scheduling [31], and then go into an output link buffer served via fair queuing. Surprisingly, we show that even though fair sharing for a single resource is strategy-proof, composing per-resource fair schedulers this way is not.

Recently, a multi-resource allocation scheme that provides share guarantees and strategy-proofness, called Dominant Resource Fairness (DRF) [16], was proposed. We design a fair queueing algorithm for multiple resources that achieves DRF allocations. The main challenge we address is that existing algorithms for DRF provide fair sharing in *space*; given a cluster with much larger number

SIGCOMM'12, August 13–17, 2012, Helsinki, Finland.

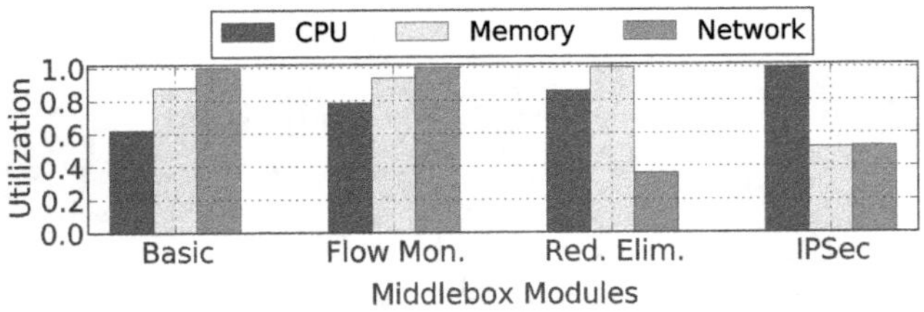

Figure 1: Normalized resource usage of four middlebox functions implemented in Click: basic forwarding, flow monitoring, redundancy elimination, and IPSec encryption.

of servers than users, they decide how many resources each user should get on each server. In contrast, middleboxes require sharing in *time*; given a small number of resources (*e.g.*, NICs or CPUs) that can each process only one packet at a time, the scheduler must interleave packets to achieve the right resource shares over time. Achieving DRF allocations in time is challenging, especially doing so in a memoryless manner, *i.e.*, a flow should not be penalized for having had a high resource share in the past when fewer flows were active [24]. This memoryless property is key to guaranteeing that flows cannot be starved in a work-conserving system.

We design a new queuing algorithm called Dominant Resource Fair Queuing (DRFQ), which generalizes the concept of virtual time from classical fair queuing [10, 24] to multiple resources that are consumed at different rates over time. We evaluate DRFQ using a Click [22] implementation and simulations, and we show that it provides better isolation and throughput than existing schemes.

To summarize, our contributions in this work are three-fold:

1. We identify the problem of multi-resource *fair queueing*, which is a generalization of traditional single-resource fair queueing.
2. We provide the first analysis of two natural packet scheduling schemes—bottleneck fairness and per-resource fairness—and show that they suffer from problems including poor isolation, oscillations, and manipulation.
3. We propose the first multi-resource queuing algorithm that provides both share guarantees and strategy-proofness: Dominant Resource Fair Queuing (DRFQ). DRFQ implements DRF allocations in the time domain.

2. MOTIVATION

Others have observed that middleboxes and software routers can bottleneck on any of CPU, memory bandwidth, and link bandwidth, depending on the processing requirements of the traffic. Dreger *et al.* report that CPU can be a bottleneck in the Bro intrusion detection system [13]. They demonstrated that, at times, the CPU can be overloaded to the extent that each second of incoming traffic requires 2.5 seconds of CPU processing. Argyraki *et al.* [8] found that memory bandwidth can be a bottleneck in software routers, especially when processing small packets. Finally, link bandwidth can clearly be a bottleneck for flows that need no processing. For example, many middleboxes let encrypted SSL flows pass through without processing.

To confirm and quantify these observations, we measured the resource footprints of several canonical middlebox applications implemented in Click [22]. We developed a trace generator that takes in real traces with full payloads [4] and analyzes the resource consumption of Click modules using the Intel(R) Performance Counter Monitor API [3]. Figure 1 shows the results for four applications. Each application's maximum resource consumption was normalized to 1. We see that the resource consumption varies across modules: basic forwarding uses a higher relative fraction of link bandwidth than of other resources, redundancy elimination bottlenecks on memory bandwidth, and IPSec encryption is CPU-bound.

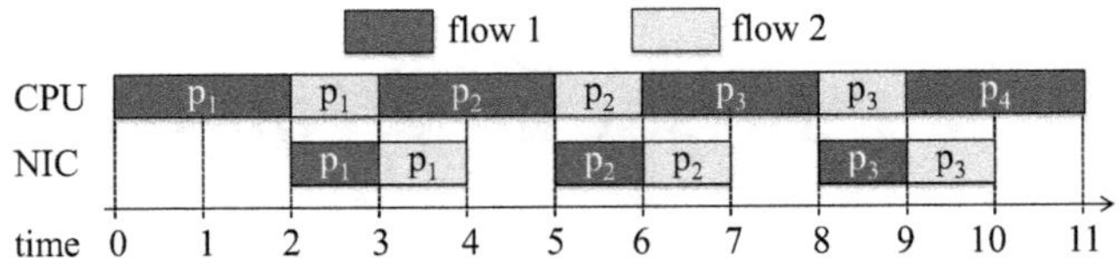

Figure 2: Performing fair sharing based on a single resource (NIC) fails to meet the share guarantee. In the steady-state period from time 2–11, flow 2 only gets a third of each resource.

Many middleboxes already perform different functions for different traffic (*e.g.*, HTTP caching for some flows and basic forwarding for others), and future software-defined middlebox proposals suggest consolidating more functions onto the same device [28, 27]. Moreover, further functionality is being incorporated into hardware accelerators [30, 23, 5], increasing the resource diversity of middleboxes. Thus, packet schedulers for middleboxes will need to take into account flows' consumption across multiple resources.

Finally, we believe multi-resource scheduling to be important in other contexts too. One such example is multi-tenant scheduling in deep software stacks. For example, a distributed key-value store might be layered on top of a distributed file system, which in turn runs over the OS file system. Different layers in this stack can bottleneck on different resources, and it is desirable to isolate the resource consumption of different tenants' requests. Another example is virtual machine (VM) scheduling inside a hypervisor. Different VMs might consume different resources, so it is desirable to fairly multiplex their access to physical resources.

3. BACKGROUND

Designing a packet scheduler for multiple resources turns out to be non-trivial due to several problems that do not occur with one resource [16]. In this section, we review these problems and provide background on the allocation scheme we ultimately build on, DRF. In addition, given that our goal is to design a packet queuing algorithm that achieves DRF, we cover background on fair queuing.

3.1 Challenges in Multi-Resource Scheduling

Previous work on DRF identifies several problems that can occur in multi-resource scheduling and shows that several simple scheduling schemes lack key properties [16].

Share Guarantee: The essential property of fair queuing is *isolation*. Fair queuing ensures that each of n flows can get a guaranteed $\frac{1}{n}$ fraction of a resource (*e.g.*, link bandwidth), regardless of the demand of other flows [24].[1] Weighted fair queuing generalizes this concept by assigning a weight w_i to each flow and guaranteeing that flow i can get at least $\frac{w_i}{\sum_{j \in W} w_j}$ of the sole resource, where W is the set of active flows.

We generalize this guarantee to multiple resources as follows:

Share Guarantee. A backlogged flow with weight w_i should get at least $\frac{w_i}{\sum_{j \in W} w_j}$ fraction of one of the resources it uses.

[1] By "flow," we mean a set of packets defined by a subset of header fields. Administrators can choose which fields to use based on organizational policies, *e.g.*, to enforce weighted fair shares across users (based on IP addresses) or applications (based on ports).

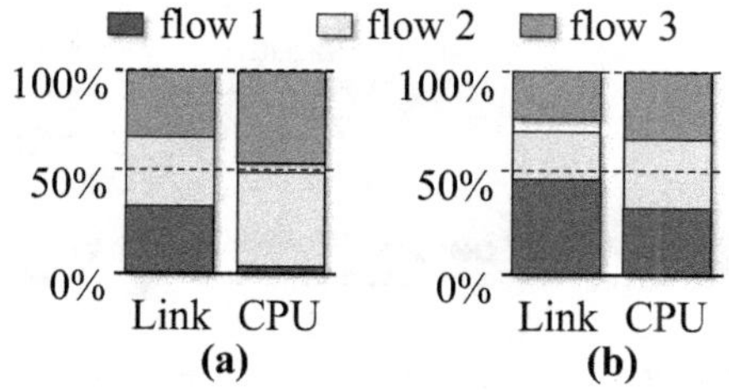

Figure 3: Bottleneck fairness can be manipulated by users. In (b), flow 1 increases its CPU usage per packet to shift the bottleneck to CPU, and thereby gets more bandwidth too.

Surprisingly, this property is not met by some natural schedulers. As a strawman, consider a scheduler that only performs fair queueing based on one specific resource. This may lead to some flows receiving less than $\frac{1}{n}$ of all resources, where n is the total number of flows. As an example, assume that there are two resources, CPU and link bandwidth, and that each packet first goes through a module that uses the CPU, and thereafter is sent to the NIC. Assume we have two flows with *resource profiles* $\langle 2, 1\rangle$ and $\langle 1, 1\rangle$; that is, packets from flow 1 each take 2 time units to be processed by the CPU and 1 time unit to be sent on the link, while packets from flow 2 take 1 unit of both resources. If the system implements fair queuing based on only link bandwidth, it will alternate sending one packet from each flow, resulting in equal allocation of link bandwidth to the flows (both flows use one time unit of link bandwidth). However, since there is more overall demand for the CPU, the CPU will be fully utilized, while the network link will be underutilized at times. As a result (see Figure 2), the first flow receives $\frac{2}{3}$ and $\frac{1}{3}$ of the two resources, respectively. But the second flow only gets $\frac{1}{3}$ on both resources, violating the share guarantee.

Strategy-Proofness: The multi-resource setting is vulnerable to a new type of *manipulation*. Flows can manipulate the scheduler to receive better service by artificially inflating their demand for resources they do not need.

For example, a flow might increase the CPU time required to process it by sending smaller packets. Depending on the scheduler, such manipulation can increase the flow's allocation across *all* resources. We later show that in several natural schedulers, greedy flows can as much as double their share at the cost of other flows.

These types of manipulations were not possible in single-resource settings, and therefore received no attention in past literature. It is important for multi-resource schedulers to be resistant to them, as a system vulnerable to manipulation can incentivize users to *waste* resources, ultimately leading to lower total goodput.

The following property discourages the above manipulations:

Strategy-proofness. A flow should not be able to finish faster by increasing the amount of resources required to process it.

As a concrete example, consider the scheduling scheme proposed by Egi *et al.* [14], in which the middlebox determines which resource is bottlenecked and divides that resource evenly between the flows. We refer to this approach as *bottleneck fairness*. Figure 3 shows how a flow can manipulate its share by wasting resources. In (a), there are three flows with resource profiles $\langle 10, 1\rangle$, $\langle 10, 14\rangle$ and $\langle 10, 14\rangle$ respectively. The bottleneck is the first resource (link bandwidth), so it is divided fairly, resulting in each flow getting one third of it. In (b), flow 1 increases its resource profile from $\langle 10, 1\rangle$ to $\langle 10, 7\rangle$. This shifts the bottleneck to the CPU, so the system starts to schedule packets to equalize the flows' CPU usage. However, this gives flow 1 a higher share of bandwidth as well, up from $\frac{1}{3}$ to almost $\frac{1}{2}$. In similar examples with more flows, flow 1 can almost double its share.

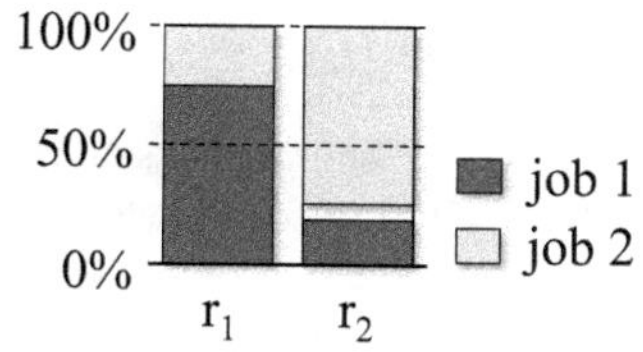

Figure 4: DRF allocation for jobs with resource profiles $\langle 4, 1\rangle$ and $\langle 1, 3\rangle$ in a system with equal amounts of both resources. Both jobs get $\frac{3}{4}$ of their dominant resource.

We believe the networking domain to be particularly prone to these types of manipulations, as peer-to-peer applications already employ various techniques to increase their resource share [26]. Such an application could, for instance, dynamically adapt outgoing packet sizes based on throughput gain, affecting the CPU consumption of congested middleboxes.

3.2 Dominant Resource Fairness (DRF)

The recently proposed DRF allocation policy [16] achieves both strategy-proofness and share guarantees.

DRF was designed for the datacenter environment, which we briefly recapitulate. In this setting, the equivalent of a flow is a job, and the equivalent of a packet is a job's task, executing on a single machine. DRF defines the *dominant resource* of a job to be the resource that it currently has the biggest share of. For example, if a job has 20 CPUs and 10 GB of memory in a cluster with 100 CPUs and 40 GB of memory, the job's dominant resource is memory, as it is allocated $\frac{1}{4}$ of it (compared to $\frac{1}{5}$ for CPU). A job's *dominant share* is simply its share of its dominant resource, *e.g.*, $\frac{1}{4}$ in this example. Informally, DRF provides the allocation that "equalizes" the dominant shares of different users. More precisely, DRF is the max-min fair allocation of dominant shares.

Figure 4 shows an example, where two jobs run tasks with resource profiles $\langle$4 CPUs, 1 GB$\rangle$ and $\langle$1 CPU, 3 GB$\rangle$ in a cluster with 2000 CPUs and 2000 GB of memory. In this case, job 1's dominant resource is CPU, and job 2's dominant resource is memory. DRF allocates $\langle$1500 CPUs, 375 GB$\rangle$ of resources to job 1 and $\langle$500 CPUs, 1500 GB$\rangle$ to job 2. This equalizes job 1's and job 2's dominant shares while maximizing the allocations.

We have described the DRF *allocation*. Ghodsi *et al.* [16] provide a simple *algorithm* to achieve DRF allocations in space (*i.e.*, given a cluster of machines, compute which resources on which machines to assign to each user). We seek an algorithm that achieves DRF allocations in *time*, multiplexing resources across incoming packets. In Section 5, we describe this problem and provide a queuing algorithm for DRF. The algorithm builds on concepts from fair queuing, which we review next.

3.3 Fair Queuing in Routers

Fair Queuing (FQ) aims to implement max-min fair allocation of a single resource using a fluid-flow model, in which the link capacity is infinitesimally divided across the backlogged flows [10, 24]. In particular, FQ schedules packets in the order in which they would finish in the fluid-flow system.

Virtual clock [33] was one of the first schemes using a fluid-flow model. It, however, suffers from the problem that it can punish a flow that in the past got better service when fewer flows were active. Thus, it violates the following key property:

Memoryless scheduling. A flow's current share of resources should be independent of its share in the past.

In the absence of this memoryless property, flows may experience starvation. For example, with virtual clock, if one flow uses a link at full rate for one minute, and a second flow becomes active, then only the second flow is serviced for the next minute, until their virtual clocks equalize. Thus, the first flow starves for a minute.[2]

The concept of *virtual time* was proposed to address this pitfall [24]. Instead of measuring real time, virtual time measures the amount of work performed by the system. Informally, a virtual time unit is the time it takes to send *one* bit of a *unit*-weight flow in the fluid-flow system. Thus, it takes l virtual time units to send a packet of length l. Thus, virtual time progresses faster than real-time when fewer flows are active. In general, assuming a flow with weight w, it takes l/w virtual time units to send the packet in the fluid-flow system.

Virtual time turns out to be expensive to compute exactly, so a variety of algorithms have been proposed to implement FQ efficiently by approximating it [10, 24, 18, 29, 9]. One of the main algorithmic challenges we address in our work is to extend this concept to multiple resources that are consumed at different rates over time.

4. ANALYSIS OF EXISTING POLICIES

We initially explored two natural scheduling algorithms for middleboxes. The first solution, called *bottleneck fairness*, turns out to lack both strategy-proofness and the sharing guarantee. The second, called *per-resource fairness*, performs fair sharing independently at each resource. This would happen naturally in routers that queue packets as they pass between different resources and serve each queue via fair sharing. We initially pursued per-resource fairness but soon discovered that it is not strategy-proof.

4.1 Bottleneck Fairness

In early work on resource scheduling for software routers, Egi *et al.* [14] point out that most of the time, only one resource is congested. They therefore suggest that the system should dynamically determine which resource is congested and perform fair sharing on that resource. For example, a middlebox might place new packets from each flow into a separate queue and serve these queues based on the packets' estimated CPU usage if CPU is a bottleneck, their memory bandwidth usage if memory is a bottleneck, etc.

This approach has several disadvantages. First, it is not strategy-proof. As we showed in Section 3.1, a flow can nearly double its share by artificially increasing its resource consumption to shift the bottleneck.

Second, when neither resource is a clear bottleneck, bottleneck fairness can rapidly oscillate, affecting the throughput of all flows and keeping some flows below their share guarantee. This can happen readily in middleboxes where some flows require expensive processing and some do not. For example, consider a middlebox with two resources, CPU and link bandwidth, that applies IPsec encryption to flows within a corporate VPN but forwards other traffic to the Internet. Suppose that an external flow has a resource profile of $\langle 1, 6\rangle$ (bottlenecking on bandwidth), while an internal flow has $\langle 7, 1\rangle$. If both flows are backlogged, it is unclear which resource should be considered the bottleneck.

Indeed, assume the system decides that the first resource is the bottleneck and tries to divide it evenly between the flows. As a result, the first resource will process seven packets of flow 1 for every single packet of flow 2. Unfortunately, this will congest the second resource right away, since processing seven packets of flow 1 and one packet of flow 2 will generate a higher demand for resource 2 than resource 1, *i.e.,* $7\langle 1, 6\rangle + \langle 7, 1\rangle = \langle 14, 43\rangle$. Once

[2]A workaround for this problem would be for the first flow to never use more than half of the link capacity. However, this leads to inefficient resource utilization during the first minute.

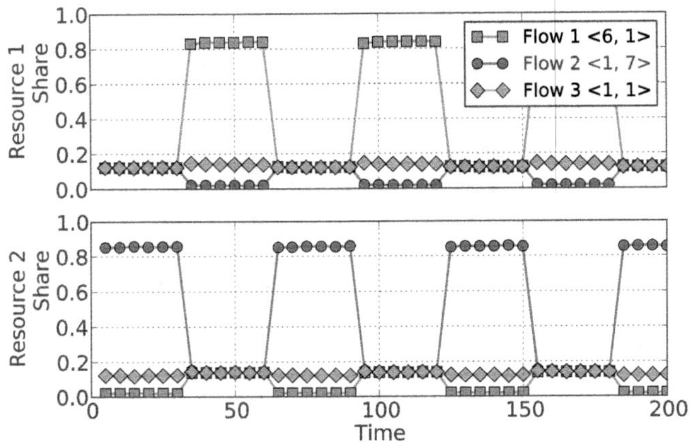

Figure 5: Example of oscillation in Bottleneck Fairness [14]. Note that flow 3 stays below $\frac{1}{3}$ share of both resources.

resource 2 becomes the bottleneck, the system will try to divide this resource equally. As a result, resource 2 will process six packets of flow 2 for each packet of flow 1, which yields an overall demand of $\langle 1, 6\rangle + 6\langle 7, 1\rangle = \langle 43, 12\rangle$. This will now congest resource 1, and the process will repeat.

Such oscillation is a problem for TCP traffic, where fast changes in available bandwidth leads to bursts of losses and low throughput. However, bottleneck fairness also fails to meet share guarantees for non-TCP flows. For example, if we add a third flow with resource profile $\langle 1, 1\rangle$, bottleneck fairness always keeps its share of *both* resources below $\frac{1}{3}$, as shown in Figure 5. This is because there is no way, while scheduling based on one resource, to increase all the flows' share of that resource to $\frac{1}{3}$ before the other gets congested.

4.2 Per-Resource Fairness (PF)

A second intuitive approach is to perform fair sharing independently at each resource. For example, suppose that incoming packets pass through two resources: a CPU, which processes these packets and then an output link. Then one could first schedule packets to pass through the CPU in a way that equalizes flows' CPU shares, by performing fair queuing based on packets' processing times, and then place the packets into buffers in front of the output link that get served based on fair sharing of bandwidth.

Although this approach is simple, we found that it is not strategy-proof. For example, Figure 6(a) shows two flows with resource profiles $\langle 4, 1\rangle$ and $\langle 1, 2\rangle$ that share two resources. The labels of the packets show when each packet uses each resource. For simplicity, we assume that the resources are perfectly divisible, so both flows can use a resource simultaneously. Furthermore, we assume that the second resource can start processing a packet only after the first one has finished it, and that there is only a 1-packet buffer for each flow between the resources. As shown in Figure 6(a), after the initial start, a periodic pattern with a length of 7 time units emerges. As a result, flow 1 gets resource shares $\langle \frac{4}{7}, \frac{1}{7}\rangle$, *i.e.,* it gets $\frac{4}{7}$ of the first resource and $\frac{1}{7}$ of the second resource. Meanwhile, flow 2 gets resource shares $\langle \frac{3}{7}, \frac{6}{7}\rangle$.

Suppose flow 1 artificially increases its resource consumption to $\langle 4, 2\rangle$. Then per-resource fair queuing gives the allocation in Figure 6(b), where flow 1's share is $\langle \frac{2}{3}, \frac{1}{3}\rangle$ and flow 2's share is $\langle \frac{1}{3}, \frac{2}{3}\rangle$. Flow 1 has thus increased its share of the first resource by 16%, while decreasing flow 2's share of this resource by 22%.

This behavior surprised us, because fair queuing for a single resource *is* strategy-proof. Intuitively, flow 1 "crowds out" flow 2 at the second resource, which is the primary resource that flow 2

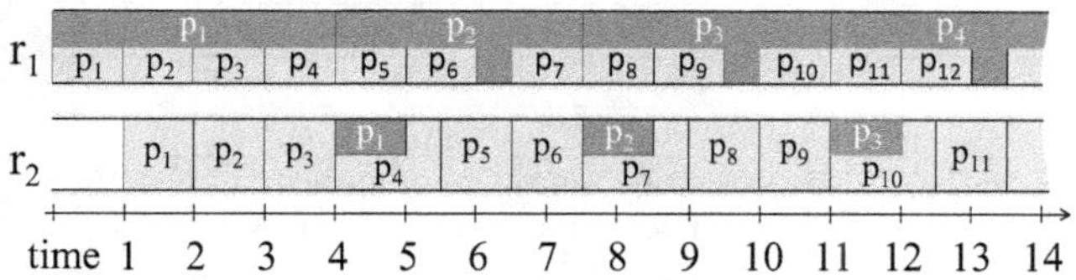

(a) Allocation with resource profiles $\langle 4, 1\rangle$ and $\langle 1, 2\rangle$.

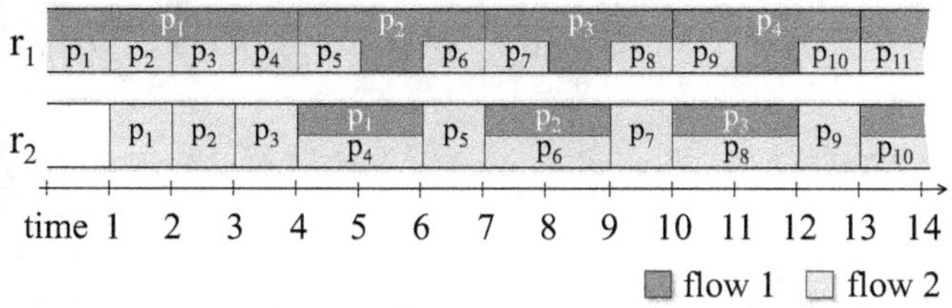

(b) Allocation with resource profiles $\langle 4, 2\rangle$ and $\langle 1, 2\rangle$.

Figure 6: Example of how flows can manipulate per-resource fairness. A shaded box shows the consumption of one packet on one resource. In (b), flow 1 increases per-packet resource use from $\langle 4, 1\rangle$ to $\langle 4, 2\rangle$ to get a higher share of resource 1 ($\frac{2}{3}$ as opposed to $\frac{4}{7}$).

needs, by increasing its share, and this causes the buffer for flow 2's packets between the two resources to be full more of the time. This leaves more time for flow 1 at the first resource.

We found that the amount by which flows can raise their share with per-resource fairness is as high as $2\times$, which provides substantial incentive for applications to manipulate the scheduler. We discuss an example in Section 8.2.1. We have also simulated other models of per-resource fairness, including ones with bigger buffers and ones that let multiple packets be processed in parallel (*e.g.*, as on a multicore CPU), but found that they give the same shares over time and can be manipulated in the same way.

Finally, from a practical viewpoint, per-resource fairness is hard to implement in systems where there is no buffer between the resources, *e.g.*, a system where scheduling decisions are only taken at an input queue, or a processing function that consumes CPU and memory bandwidth in parallel (while executing CPU instructions). Our proposal, DRFQ, is directly applicable in these settings.

5. DOMINANT RESOURCE FAIR QUEUING

The goal of this section is to develop queuing mechanisms that multiplex packets to achieve DRF allocations.

Achieving DRF allocations in middleboxes turns out to be algorithmically more challenging than in the datacenter context. In datacenters, there are many more resources (machines, CPUs, etc) than active jobs, and one can simply divide the resources across the jobs according to their DRF allocations at a given time. In a packet system, this is not possible because the number of packets in service at a given time is usually much smaller than the number of backlogged flows. For example, on a communication link, at most one packet is transmitted at a time, and on a CPU, at most one packet is (typically) processed per core. Thus, the only way to achieve DRF allocation is to share the resources in time instead of space, *i.e.*, multiplex packets from different flows to achieve the DRF allocation over a longer time interval.

This challenge should come as no surprise to networking researchers, as scheduling a single resource (link bandwidth) in time was a research challenge receiving considerable attention for years. Efforts to address this challenge started with idealized models (*e.g.*, fluid flow [10, 24]), followed by a plethora of algorithms to accurately and efficiently approximate these models [29, 18, 9].

We begin by describing a unifying model that accounts for resource consumption across different resources.

5.1 Packet Processing Time

The mechanisms we develop generalize the fair queueing concepts of virtual start and finish times to multiple resources, such that these times can be used to schedule packets.

To do this, we first find a metric that lets us compare virtual times across resources. Defining the unit of virtual time as the time it takes to process one bit does not work for all resource types. For example, the CPU does not consume the same amount of time to process each bit; it usually takes longer to process a packet's header than its payload. Furthermore, packets with the same size, but belonging to different flows, may consume different amounts of resources based on the processing functions they go through. For example, a packet that gets handled by the IPSec encryption module will consume more CPU time than a packet that does not.

To circumvent these challenges, we introduce the concept of *packet processing time*. Denote the k:th packet of flow i as p_i^k. The processing time of the packet p_i^k at resource j, denoted $s_{i,j}^k$, is the time consumed by resource j to process packet p_i^k, normalized to the resource's processing capacity. Note that processing time is not always equal to packet service time.[3] Consider a CPU with four cores, and assume it takes a single core 10 μs to process a packet. Since the CPU can process four such packets in parallel, the normalized time consumed by the CPU to process the packet (*i.e.*, the processing time) is 2.5 μs. However, the packet's service time is 10 μs. In general, the processing time is the inverse of the throughput. In the above example, the CPU throughput is 400,000 packets/sec.

We define a unit of virtual time as one μsec of processing time for the packet of a flow with weight one. Thus, by definition, the processing time, and by extension, the virtual time, do not depend on the resource type. Also, similarly to FQ, the unit of virtual time does not depend on number of flows backlogged. Here, we assume that the time consumed by a resource to process a packet does not depend on how many other packets, if any, it processes in parallel.

We return to processing time estimation in §5.7 and §7.1.

5.2 Dove-Tailing vs. Memoryless Scheduling

The multi-resource scenario introduces another challenge. Specifically, there is a tradeoff between *dove-tailing* and *memoryless scheduling*, which we explain next.

Different packets from the same flow may have different processing time requirements, *e.g.*, a TCP SYN packet usually requires more processing time than later packets. Consider a flow that sends a total of 10 packets, alternating in processing time requirements $\langle 2, 1\rangle$ and $\langle 1, 2\rangle$, respectively. It is desirable that the system treats this flow the same as a flow that sends 5 packets, all with processing time $\langle 3, 3\rangle$. We refer to this as the *dove-tailing* requirement.[4]

Our dove-tailing requirement is a natural extension of fair queuing for one resource. Indeed, past research in network fair queuing attempted to normalize the processing time of packets of different length. For example, a flow with 5 packets of length 1 KB should be treated the same as a flow with 10 packets of length 0.5 KB.

[3]Packet service time is the interval between (1) the time the packet starts being processed and (2) the time at which its processing ends.

[4]Dove-tailing can occur in practice in two ways. First, if there is a buffer between two resources (*e.g.*, the CPU and a link), this will allow packets with complementary resource demands to overlap in time. Second, if two resources need to be consumed in parallel (*e.g.*, CPU cycles and memory bandwidth), there can still be dove-tailing from processing multiple packets in parallel (*e.g.*, multiple cores can be working on packets with complementary needs).

At the same time, it is desirable for a queuing discipline to be *memoryless*; that is, a flow's current share of resources should not depend on its past share. Limiting memory is important to prevent starving flows when new flows enter the system, as discussed in Section 3.3.

Unfortunately, the memoryless and dove-tailing properties cannot both be *fully* achieved at the same time. Dove-tailing requires that a flow's relative overconsumption of a resource be compensated by its *past* relative underconsumption of a resource, *e.g.*, packets with profile $\langle 1, 2\rangle$ and $\langle 2, 1\rangle$. Thus, it requires the scheduler to have memory of past processing time given to a flow.

Memoryless and dove-tailing are at the extreme ends of a spectrum. We gradually develop DRFQ, starting with a simple algorithm that is fully memoryless, but does not provide dove-tailing. We thereafter extend that algorithm to provide full dove-tailing but without being memoryless. Finally, we show a final extension in which the amount of dove-tailing and memory is configurable. The latter algorithm is referred to as DRFQ, as the former two are special cases. Before explaining the algorithms, we briefly review Start-time Fair Queuing (SFQ) [18], which our work builds on.

5.3 Review: Start-time Fair Queuing (SFQ)

SFQ builds on the notion of virtual time. Recall from Section 3.3 that a virtual time unit is the time it takes to send *one* bit of a *unit-weight* flow in the fluid-flow system that fair queuing approximates. Thus, it takes l virtual time units to send a packet of length l with weight 1, or, in general, l/w units to send a packet with weight w. Note that the virtual time to send a packet is always the same regardless of the number of flows; thus, virtual time progresses slower in real-time when more flows are active.

Let p_i^k be the k-th packet of flow i. Upon p_i^k's arrival, all virtual time based schedulers assign it a start and a finish time $S(p_i^k)$ and $F(p_i^k)$, respectively, such that

$$F(p_i^k) = S(p_i^k) + \frac{L(p_i^k)}{w_i}, \qquad (1)$$

where $L(p_i^k)$ is the length of packet p_i^k in bits, and w_i is the weight of flow i. Intuitively, functions S and F approximate the virtual times when the packet would have been transmitted in the fluid flow system.

In turn, the virtual start time of the packet is:

$$S(p_i^k) = \max\left(A(p_i^k), F(p_i^{k-1})\right), \qquad (2)$$

where $A(p_i^k)$ is the virtual arrival time of p_i^k. In particular, let a_i^k be the (real) arrival time of packet p_i^k. Then, the $A(p_i^k)$ is simply the virtual time at real time a_i^k, *i.e.*, $V(a_i^k)$.

Fair queueing algorithms usually differ in (1) how $V(t)$ (virtual time) is computed and (2) which packet gets scheduled next.

While there are many possibilities for both choices, SFQ proceeds by (1) assigning each packet a virtual time equal to the start time of the packet currently in service (that is, $V(t)$ is the start time of the packet in service at real time t) and (2) always scheduling the packet with the lowest virtual start time. We discuss why these choices are attractive in middleboxes in Section 5.7.

5.4 Memoryless DRFQ

In many workloads, packets within the same flow have similar resource requirements. For such workloads, a memoryless DRFQ scheduler closely approximates DRF allocations.

Assume a set of n flows that share a set of m resources j, $(1 \leq j \leq m)$, and assume flow i is given weight w_i, $(1 \leq i \leq n)$.

Throughout, we will use the notation introduced in Table 1.

Notation	Explanation
p_i^k	k-th packet of flow i
a_i^k	arrival time of packet p_i^k
$s_{i,j}^k$	processing time of p_i^k at resource j
$S(p)$	virtual start time of packet p in system
$F(p)$	virtual finish time of packet p in system
$V(t)$	system virtual time at time t
$V(t, j)$	system virtual time at time t at resource j
$S(p, j)$	virtual start time of packet p at resource j
$F(p, j)$	virtual finish time of packet p at resource j

Table 1: Main notation used in the DRFQ algorithm.

Achieving a DRF allocation requires that two backlogged flows receive the same processing time on their respective dominant resources, *i.e.,* on the resources they respectively require the most processing time on. Given our unified model of processing time, we can achieve this by using the maximum processing time of each packet when computing the packet's virtual finish time, *i.e.,* using $\max_j\{s_{i,j}^k\} \times \frac{1}{w_i}$ for the kth packet of flow i with weight w_i.

For each packet we record its virtual start time and virtual finish time as follows:

$$S(p_i^k) = \max\left(V(a_i^k), F(p_i^{k-1})\right), \qquad (3)$$

$$F(p_i^k) = S(p_i^k) + \frac{\max_j\{s_{i,j}^k\}}{w_i}. \qquad (4)$$

Thus, the finish time is equal to the virtual start time plus the processing time on the dominant resource. For a non-backlogged flow, the start time is the virtual time at the packet's arrival. For a backlogged flow, the max operator in the equation ensures that a packet's virtual start time will be the virtual finish time of the previous packet in the flow.

Finally, we have to define the virtual time function, V. Computing the virtual time function exactly is generally expensive [29] and even more so for DRF allocations. We therefore compute it as follows:

$$V(t) = \begin{cases} \max_j\{S(p, j) | p \in P(t)\} & \text{if } P(t) \neq \emptyset \\ 0 & \text{if } P(t) = \emptyset \end{cases} \qquad (5)$$

where $P(t)$ are the packets currently in service at time t. Hence, virtual time is the maximum start time of any packet p that is currently being serviced.

Note that in the case of a single link where there is at most one packet in service at a time, this reduces to setting the virtual time at time t to the start time of the packet being serviced at t. This is exactly the way virtual start time is computed in SFQ. While there are many other possible computations of the virtual time, such as the average between the start and the finish times of the packets in service, in this paper we only consider an SFQ-like computation. In Section 5.7 we discuss why an SFQ-like algorithm is particularly attractive for middleboxes.

Table 2 shows two flows with process times $\langle 4, 1\rangle$ and $\langle 1, 3\rangle$, respectively, on all their packets. The first flow is backlogged throughout the example, with packets arriving much faster than they can be processed. The second flow is backlogged in two bursts (packets p_2^0 to p_2^3 and p_2^4 to p_2^7). In the time interval 0 to 3, both flows are backlogged, so virtual start times are simply equal to the previous packet's virtual finish time. At time 10, the second flow's second burst starts with p_2^4. Assume that the middlebox is then processing p_1^5, which has virtual start time 20. Thus, $V(a_2^4) = 20$, making the virtual start time 20, instead of the previous packet's virtual finish time 12. Thereafter, the inflow of packets from the two flows is

Flow 1 with processing times $\langle 4, 1\rangle$								
Packet	p_1^0	p_1^1	p_1^2	p_1^3	p_1^4	p_1^5	p_1^6	p_1^7
Real arrival time	0	1	2	3	4	5	6	7
Virt. start/finish	0/4	4/8	8/12	12/16	16/20	20/24	24/28	28/32
Flow 2 with processing times $\langle 1, 3\rangle$								
Packet	p_2^0	p_2^1	p_2^2	p_2^3	p_2^4	p_2^5	p_2^6	p_2^7
Real arrival time	0	1	2	3	10	11	12	13
Virt. start/finish	0/3	3/6	6/9	9/12	20/23	23/26	26/29	29/32
Scheduling Order								
Order	1	2	3	4	5	6	7	8
Packet	p_1^0	p_2^0	p_2^1	p_1^1	p_2^2	p_1^2	p_2^3	p_1^3
Order	9	10	11	12	13	14	15	16
Packet	p_1^4	p_1^5	p_2^4	p_2^5	p_1^6	p_2^6	p_1^7	p_2^7

Table 2: Basic example of how memoryless DRFQ works with two flows. The first flow is continuously backlogged. The second flow is backlogged in two bursts.

Flow 1 alternating $\langle 1, 2\rangle$ and $\langle 2, 1\rangle$						
Packet	p_1^0	p_1^1	p_1^2	p_1^3	p_1^4	p_1^5
Real start time	0	1	2	3	4	5
Virtual start time	0	2	4	6	8	10
Virtual finish time	2	4	6	8	10	12
Flow 2 with processing times $\langle 3, 3\rangle$						
Packet	p_2^0	p_2^1	p_2^2	p_2^3	p_2^4	p_2^5
Real start time	0	1	2	3	4	5
Virtual start time	0	3	6	9	12	15
Virtual finish time	3	6	9	12	15	18
Scheduling Order						
Order	1	2	3	4	5	6
Packet	p_1^0	p_2^0	p_1^1	p_2^1	p_1^2	p_2^2
Order	7	8	9	10	11	12
Packet	p_1^3	p_1^4	p_2^3	p_1^5	p_2^4	p_2^5

Table 3: Why dove-tailing fails with memoryless DRFQ.

again faster than the service time, leading to start times equal to the finish time of the previous packet.

Table 3 shows why dove-tailing fails with this memoryless DRFQ algorithm. One flow's packets have alternating process times $\langle 1, 2\rangle$ and $\langle 2, 1\rangle$, while the second flow's packets have process times $\langle 3, 3\rangle$. Both flows are continuously backlogged with higher inrate than service rate. With perfect dove-tailing, the virtual finish time of p_1^3 should be the same as that of p_2^1. Instead, flow 1's virtual times progress faster, making it receive poorer service.

5.5 Dove-tailing DRFQ

To provide dove-tailing, we modify the memoryless DRFQ mechanism to keep track of the start and finish times of packets on a per-resource basis. The memoryless algorithm scheduled the packet with the smallest start time. Since a packet will now have multiple start times (one per resource), we need to decide which of them to use when making scheduling decisions. Given that we want to schedule based on dominant resources, we will schedule the packet whose *maximum* per-resource start time is smallest across all flows. This is because the packet's maximum start time is its start time on its flow's dominant resource. If two packet's have the same start time, we lexicographically compare their next largest start times.

More formally, we will compute the start and finish times of a packet at each resource j as:

$$S(p_i^k, j) = \max\left(V(a_i^k, j), F(p_i^{k-1}, j)\right), \quad (6)$$

$$F(p_i^k, j) = S(p_i^k, j) + \frac{s_{i,j}^k}{w_i}. \quad (7)$$

Flow 1 alternating $\langle 1, 2\rangle$ and $\langle 2, 1\rangle$						
Packet	p_1^0	p_1^1	p_1^2	p_1^3	p_1^4	p_1^5
Real start time	0	1	2	3	4	5
Virtual start time R_1	0	1	3	4	6	7
Virtual finish time R_1	1	3	4	6	7	9
Virtual start time R_2	0	2	3	5	6	8
Virtual finish time R_2	2	3	5	6	8	9
Flow 2 with processing times $\langle 3, 3\rangle$						
Packet	p_2^0	p_2^1	p_2^2	p_2^3	p_2^4	p_2^5
Real start time	0	1	2	3	4	5
Virtual start time R_1	0	3	6	9	12	15
Virtual finish time R_1	3	6	9	12	15	18
Virtual start time R_2	0	3	6	9	12	15
Virtual finish time R_2	3	6	9	12	15	18
Scheduling Order						
Order	1	2	3	4	5	6
Packet	p_1^0	p_2^0	p_1^1	p_2^1	p_1^2	p_1^3
Order	7	8	9	10	11	12
Packet	p_2^2	p_1^4	p_1^5	p_2^3	p_2^4	p_2^5

Table 4: How dove-tailing DRFQ satisfies dove-tailing.

As mentioned above, the scheduling decisions should be made based on the maximum start or finish times of the packets across all resources, *i.e.*, $S(p_i^k)$ and $F(p_i^k)$ where

$$S(p_i^k) = \max_j\{S(p_i^k, j)\}, \quad (8)$$

$$F(p_i^k) = \max_j\{F(p_i^k, j)\}. \quad (9)$$

In the rest of this section, we refer to $S(p_i^k)$ and $F(p_i^k)$ as simply the start and finish times of packet p_i^k.

Finally, we now track virtual time per resource, *i.e.*, $V(t, j)$ at time t for resource j. We compute this virtual time independently at each resource:

$$V(t, j) = \begin{cases} \max_{p \in P(t)}\{S(p, j)\} & \text{if } P(t) \neq \emptyset \\ 0 & \text{if } P(t) = \emptyset \end{cases} \quad (10)$$

Table 4 shows how dove-tailing DRFQ schedules the same set of incoming packets as the example given in Table 3. Now virtual start and finish times are provided for both resources, R_1 and R_2. When comparing the two scheduling orders it is evident that dove-tailing DRFQ improves the service given to the first flow. For example, p_1^3 is now scheduled before p_2^2, rather than after as with memoryless DRFQ. Though real processing time and virtual start times are different, the virtual finish times clearly show that two packets of flow 1 "virtually" finish for every packet of flow 2. As start times are based on finish times of the previous packet, the schedule will reflect this ordering.

Table 5 shows how dove-tailing DRFQ is not memoryless. Flow 1 initially has processing time $\langle 2, 1\rangle$ for packets p_1^0 through p_1^2. But packets p_1^3 through p_1^5 instead have processing time $\langle 0.2, 1\rangle$. Flow 2's packets all have processing time $\langle 2, 1\rangle$. As can be seen, once flow 1's processing time switches, it gets scheduled twice in a row (p_1^4 and p_1^5). This example can in be extended to have an arbitrary number of flow 1's packets scheduled consecutively, increasing flow 2's delay arbitrarily.

5.6 Δ–Bounded DRFQ

We have explored two algorithms that trade off sharply between memoryless scheduling and dove-tailing. We now provide an algorithm whose degree of memory and dove-tailing can be controlled through a parameter Δ.

Flow 1 p_1^0–p_1^2 require $\langle 2,1\rangle$, and p_1^3–p_1^5 require $\langle 0.2,1\rangle$						
Packet	p_1^0	p_1^1	p_1^2	p_1^3	p_1^4	p_1^5
Real start time	0	1	2	3	4	5
Virtual start time R_1	0	2	4	6	6.2	6.4
Virtual finish time R_1	2	4	6	6.2	6.4	6.6
Virtual start time R_2	0	1	2	3	4	5
Virtual finish time R_2	1	2	3	4	5	6
Flow 2 with processing times $\langle 2,1\rangle$						
Packet	p_2^0	p_2^1	p_2^2	p_2^3	p_2^4	p_2^5
Real start time	0	1	2	3	4	5
Virtual start time R_1	0	2	4	6	8	10
Virtual finish time R_1	2	4	6	8	10	12
Virtual start time R_2	0	1	2	3	4	5
Virtual finish time R_2	1	2	3	4	5	6
Scheduling Order						
Order	1	2	3	4	5	6
Packet	p_1^0	p_2^0	p_1^1	p_2^1	p_1^2	p_2^2
Order	7	8	9	10	11	12
Packet	p_1^3	p_2^3	p_1^4	p_1^5	p_2^4	p_2^5

Table 5: Example of dove-tailing DRF not being memoryless. As of packet p_1^3, flow 1's processing time switches from $\langle 2,1\rangle$ to $\langle 0.1,1\rangle$.

Such customization is important because in practice it is not desirable to provide unlimited dove-tailing. If a flow alternates sending packets with processing times $\langle 1,2\rangle$ and $\langle 2,1\rangle$, then the system can buffer packets and multiplex resources so that in real time, a pair of such packets take time equivalent to a $\langle 3,3\rangle$ packet. Contrast this with the flow first sending a long burst of 1000 packets with processing time $\langle 1,2\rangle$ and thereafter a long burst of 1000 packets with processing time $\langle 2,1\rangle$. After the first burst, the system is completely done processing most of those packets, and the fact that the processing times of the two bursts dove-tail does not yield any time savings. Hence, it is desirable to bound the dove-tailing to match the length of buffers and have the system be memoryless beyond that limit.

Δ–Bounded DRFQ is similar to dove-tailing DRFQ (§5.5), except that the virtual start and finish are computed differently. We replace the virtual start time, Eq. (6), with:

$$S(p_i^k, j) = \max\left(V(a_i^k, j), B_1(p_i^{k-1}, j)\right) \quad (11)$$

$$B_1(p, j) = \max\left(F(p, j), \max_{j' \neq j}\{F(p, j')\} - \Delta\right) \quad (12)$$

Thus, the start time of a packet on each resource can never differ by more than Δ from the maximum finish time of its flow's previous packet on any resource. This allows each flow to "save" up to Δ processing time for dove-tailing.

We similarly update the virtual time function (Eq. 10) to achieve the same bounding effect:

$$V(t, j) = \begin{cases} \max_{p \in P(t)}\{B_2(p, j)\} & \text{if } P(t) \neq \emptyset \\ 0 & \text{if } P(t) = \emptyset \end{cases} \quad (13)$$

$$B_2(p, j) = \max\left(S(p, j), \max_{j' \neq j}\{S(p, j')\} - \Delta\right) \quad (14)$$

Dove-tailing DRFQ (§5.5) and memoryless DRFQ (§5.4) are thus special cases of Δ–bounded DRFQ. In particular, when $\Delta = \infty$, the functions B_1 and B_2 reduce to functions F and S as in the previous section, and Δ–bounded DRFQ becomes equivalent to dove-tailing DRFQ. Similarly, if $\Delta = 0$, then B_1 and B_2 reduce to the maximum per-resource start and finish time of the flow's previous packet, respectively. Thus, Δ–bounded DRFQ becomes memoryless DRFQ. For these reasons, we simply refer to Δ–bounded DRFQ as DRFQ in the rest of the paper.

5.7 Discussion

The main reason we chose an SFQ-like algorithm to approximate DRFQ is that SFQ does not need to know the processing times of the packets before scheduling them. This is desirable in middleboxes because the CPU and memory bandwidth costs of processing a packet may not be known until after it has passed through the system. For example, different packets may pass through different processing modules (*e.g.,* HTTP caching) based on their contents.

Like SFQ, DRFQ schedules packets based on their virtual start times. As shown in Eq. (13), the virtual start time of packet p_i^k depends only on the start times of the packets in service and on the finish time of the previous packet, $F(p_i^{k-1})$. This allows us to delay computing $S(p_i^k)$ until just after p_i^{k-1} has finished, at which point we can use the *measured values* of packet p_i^{k-1}'s processing times, $s_{i,j}^{k-1}$, to compute its virtual finish time.

Although the use of a SFQ-like algorithm allows us to defer computing the processing time of each packet until after it has been processed (*e.g.,* after we have seen which middlebox modules it went through), there is still a question of *how* to measure the consumption. Unfortunately, measuring the exact CPU and memory consumption of *each* packet (*e.g.,* using CPU counters [3]) is expensive. However, in our implementation, we found that we could estimate consumption quite accurately based on the packet size and the set of modules it flowed through. Indeed, linear models fit the resource consumption with $R^2 > 0.97$ for many processing functions. In addition, DRFQ is robust to misestimation—that is, flows' shares might differ from the true DRF allocation, but each flow will still get a reasonable share as long as the estimates are not far off. We discuss these issues further in Section 7.1.

6. DRFQ PROPERTIES

In this section, we discuss two key properties of Δ-bounded DRFQ. Lemma 6.1 bounds the unfairness between two backlogged flows over a given time interval. This bound is independent of the length of the interval. Lemma 6.2 bounds the delay of a packet that arrives when the flow is idle. These properties parallel the corresponding properties in SFQ [18].

Due to space concern, we defer proofs to our technical report [15].

A flow is called *dominant-resource monotonic* if, during any of its backlogged periods, its dominant resource does not change. A flow in which all packets have the same dominant resource is trivially a dominant-resource monotonic flow. In this section, $s_{i,r}^{\uparrow}$ denotes $\max_k s_{i,r}^k$.

Consider a dominant-resource monotonic flow i, and let r be the dominant share of i. Then the virtual start times of i's packets at resource r do not depend on Δ. This follows trivially from Eq. (11), as $B_1(p_i, r)$ is equal to $F(p_i, j)$ for any packet p_i of i. For this reason, the bound in the next lemma does not depend on Δ.

THEOREM 6.1. *Consider two dominant-resource monotonic flows i and j, both backlogged during the interval $[t_1, t_2)$. Let $W_i(t_1, t_2)$ and $W_j(t_1, t_2)$ be the total processing times consumed by flows i and j, respectively, on their dominant resource during interval $[t_1, t_2)$. Then, we have*

$$\left|\frac{W_i(t_1, t_2)}{w_i} - \frac{W_j(t_1, t_2)}{w_j}\right| < \frac{s_{i,d_i}^{\uparrow}}{w_i} + \frac{s_{j,d_j}^{\uparrow}}{w_j}, \quad (15)$$

where $s_{q,d_q}^{\uparrow}$ represents the maximum processing time of a packet of flow q on its dominant resource d_q.

The next result does not assume that flows are dominant-resource monotonic but assumes that each packet has a non-zero demand for every resource.

THEOREM 6.2. *Assume packet p_i^k of flow i arrives at time t, and assume flow i is idle at time t. Assume all packets have non-zero demand on every resource. Let n be the total number of flows in the system. Then the maximum delay to start serving packet p_i^k, $D(p_i^k)$, is bounded above by*

$$D(p_i^k) \leq \max_r \left(\sum_{j=1, j \neq i}^{n} s_{j,r}^{\uparrow} \right), \quad (16)$$

where $s_{j,d_j}^{\uparrow}$ represents the maximum processing time of a packet of flow j on its dominant resource d_j.

7. IMPLEMENTATION

We prototyped DRFQ in the Click modular router [22]. Our implementation adds a new DRFQ-Queue module to Click, consisting of roughly 600 lines of code. This module takes as input a *class specification* file identifying the types of traffic that the middlebox processes (based on port numbers and IP prefixes) and a model for estimating the packet processing times for each middlebox function. We use a model to estimate the packets' processing times because measuring the exact CPU and memory usage of a single packet is expensive.

The main difference from a traditional Queue is that DRFQ-Queue maintains a per-flow buffer with a fixed capacity *Cap* per-flow and also tracks the last virtual completion time for each flow. As each packet arrives, it is assigned a virtual start time and is added to the buffer corresponding to its flow. If the specific buffer is full, the packet is dropped, but the virtual completion time for the flow is not incremented. On every call to dequeue a packet, DRFQ-Queue looks at the head of each per-flow queue and returns the packet with the lowest virtual start time.

To decide when to dequeue packets to downstream modules, we use a separate token bucket for each resource; this ensures that we do not oversaturate any particular resource. On each dequeue, we pull out a number of tokens from each resource corresponding to the packet's estimated processing time on that resource. In addition, we periodically check the true utilization rate of each resource, and we scale down the rate at which we dequeue packets if we find that we have been estimating processing times incorrectly. This ensures that we do not overload the hardware.

7.1 Estimating Packets' Resource Usage

To implement any multi-resource scheduler, one needs to know the consumption of each packet for each resource. This was simple when the only resource scheduled was link bandwidth, because the size of each packet is known. Consumption of other resources, such as CPU and memory bandwidth, is harder to capture at a fine granularity. Although CPU counters [3] can provide this data, querying the counters for each packet adds overhead. Fortunately, it is possible to estimate the consumption accurately. We used a two-step approach similar to the one taken in [14]: (i) Determine which modules each packet passed through. Fortunately, with DRFQ it is necessary to know this only *after* the packet has been processed (*c.f.,* §5.7), (ii) Use a model of each module's resource consumption as a function of packet size to estimate the packet's consumption.

For the second step, we show that modules' resource consumption can be estimated accurately using a simple linear model based on packet size. Specifically, for a given module m and resource r, we find parameters $\alpha_{m,r}$ and $\beta_{m,r}$ such that the resource consumption of a packet of size x is $\alpha_{m,r}x + \beta_{m,r}$. We have fit such linear models to four Click modules and show that they predict consumption well, fitting with $R^2 \geq 0.97$ in most cases. For example, Figure 7 shows the CPU and memory bandwidth consumption of a redundancy elimination module. We list the R^2 values for other modules in Table 6. The only cases where R^2 is lower than 0.97 are for the CPU consumptions of basic forwarding and statistical monitoring, which are nearly constant but have jumps at certain packet sizes, as shown in Figure 8. We believe this to be due to CPU caching effects.

Further refinements can be made based on the function of the module. For example, if a module takes more time to process the

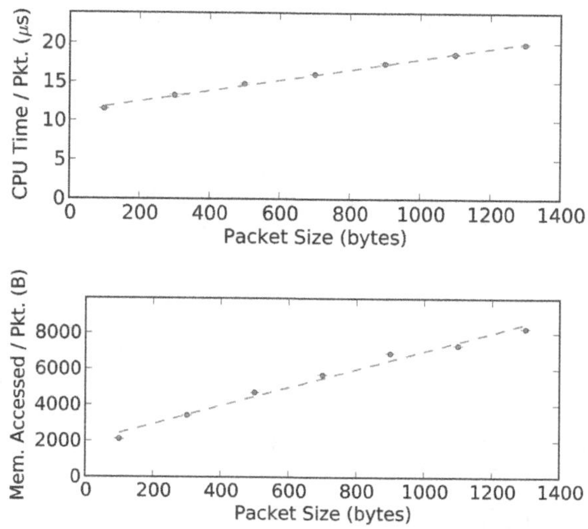

Figure 7: Per-packet CPU and memory b/w consumption of the redundancy elimination. Results are averaged over five runs that measure the consumption of processing 10,000 packets each. Error bars show max and min values. As seen, for both memory and CPU, linear models fit well with $R^2 > 0.97$.

Module	R^2 for CPU	R^2 for Memory
Basic Forwarding	0.921	0.994
Redundancy Elim.	0.997	0.978
IPSec Encryption	0.996	0.985
Stat. Monitoring	0.843	0.992

Table 6: R^2 values for fitting a linear model to estimate the CPU and memory bandwidth use of various modules.

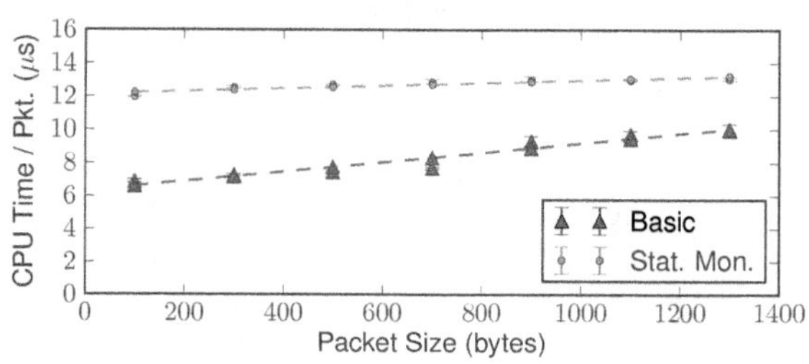

Figure 8: CPU usage vs. packet size for basic forwarding and statistical monitoring.

first few packets of a flow (for some one-time work), one can use a separate linear model for them.

Finally, if the estimation is wrong, DRFQ will still run, but flows' shares may be off by the ratio to which processing times have been misestimated. One can also imagine dynamically recomputing each flow's usage, but we chose not to explore estimation further in this paper as it is orthogonal to our main focus of defining a suitable allocation policy.

8. EVALUATION

We evaluated DRFQ using both our Click implementation and packet-level simulations. We use Click to show the basic functioning of the algorithm and simulations to compare it in more detail against other schedulers. Our workload is mostly dominant-resource monotonic, so we used the $\Delta = 0$ configuration by default, unless otherwise stated.

8.1 Implementation Results

We ran a Click-based multi-function middlebox in usermode on an Intel(R) Xeon(R) CPU 2.8GHz X5560 machine with a 1Gbps Ethernet link. We connected this machine to a traffic generator that uses Click to send packets from multiple flows. We configured the middlebox to apply three different processing functions to these flows based on their port number: basic forwarding, per-flow statistical monitoring, and IPSec encryption. Because our machine only had one 1 Gbps link, we throttled its outgoing bandwidth to 200 Mbps to emulate a congested link, and throttled the fraction of CPU time that the DRFQ module is allowed to use for processing to 20% so that the CPU can also be a bottleneck at this rate.

8.1.1 Dynamic Allocation

We begin by generating three flows that each send 25,000 1300-byte UDP packets per second to exceed the total outgoing bandwidth capacity. We configured the flows such that: (i) Flow 1 only undergoes basic forwarding, which is link bandwidth bound, (ii) Flow 2 undergoes IPSec, which is CPU-bound, (iii) Flow 3 requires statistical monitoring, which is bandwidth-bound but uses slightly more CPU than basic forwarding.

Figure 9 shows the resource shares of the flows over time (measured using timing instrumentation we added in Click), as we start and end them at different points. We see that Flow 1 initially has a complete share of the network, but only 20% of the CPU since it only requires lightweight processing. When Flow 2 arrives, Flow 1's CPU and network share expectedly decreases. Note, however that Flow 1's network share is more than 2× higher that Flow 2 because Flow 2 has a different dominant resource, namely CPU. Also, the two flows' resource demands dovetail, and their dominant shares are equalized. Finally when Flow 3 arrives, we observe that the network shares of Flow 1 and Flow 3 are equalized (as the dominant resource is link bandwidth for both), and Flow 2's share decreases further to equalize the dominant shares.

8.1.2 Isolation of Small Flows

Next, we extend the above setup to analyze the impact of DRFQ on short flows. As before, Flow 1 and Flow 2 require basic and IPSec processing respectively, and they are set to send 40,000 packets/second each to exceed the outgoing bandwidth. We then add two new flows, Flow 3 and Flow 4, both using only basic processing, but sending packets at a much lower rate of 1 packet/second and 0.5 packets/second, respectively. Ideally, we want these low-rate flows to have no backlog and not be impacted by the larger queues from the high-rate flows. Figure 10 confirms that this happens in practice, showing the steady-state latency of the four flows:

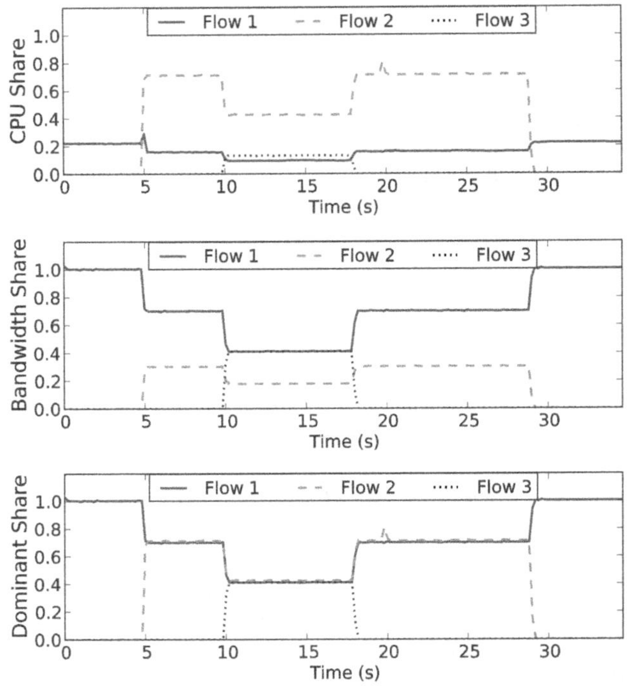

Figure 9: Shares of three competing flows arriving at Click at different times. Flow 1, 2, and 3 respectively undergo basic forwarding, IPSec, and statistical monitoring.

both low-rate flows see more than an order of magnitude lower per-packet latency than the larger ones. We also notice that the high-rate IPSec flow has a higher latency than the high-rate basic flow because it has a smaller bandwidth share but the same queue size.

8.1.3 Comparison with Bottleneck Fairness

We also implemented bottleneck fairness [14] in Click to test whether the oscillations that occur when two resources are in demand affect performance. For these experiments, we used TCP flows, and added 20 ms of network latency to get realistic behavior for wide-area flows. We made bottleneck fairness check for a new bottleneck every 300 ms. We ran two TCP flows for 30 seconds each: one that only undergoes basic processing, and one that undergoes CPU-intensive redundancy elimination as well.

Table 7 shows the throughputs of both flows running separately (one at a time), together under bottleneck fairness, and together

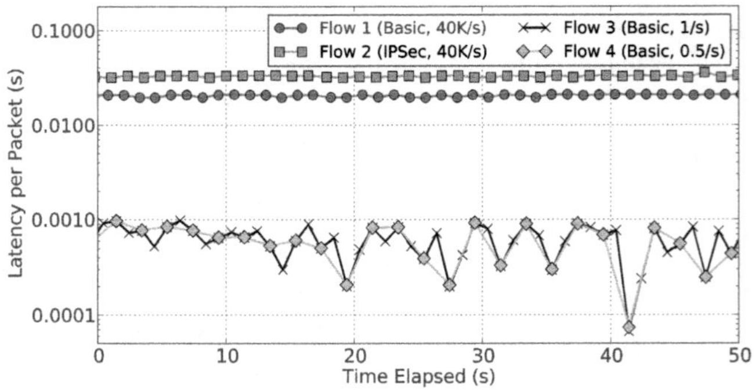

Figure 10: Latencies of DRFQ scheduling in Click when two bottlenecked flows (sending 40,000 packets/s each) and two low-rate flows (sending 0.5-1 packets/s) compete.

Scenario	Flow 1 (BW-bound)	Flow 2 (CPU-bound)
Running alone	191 Mbps	33 Mbps
Bottleneck	75 Mbps	32 Mbps
DRFQ	160 Mbps	28 Mbps

Table 7: Throughput of bandwidth and CPU intensive flows alone and under bottleneck fairness and DRFQ.

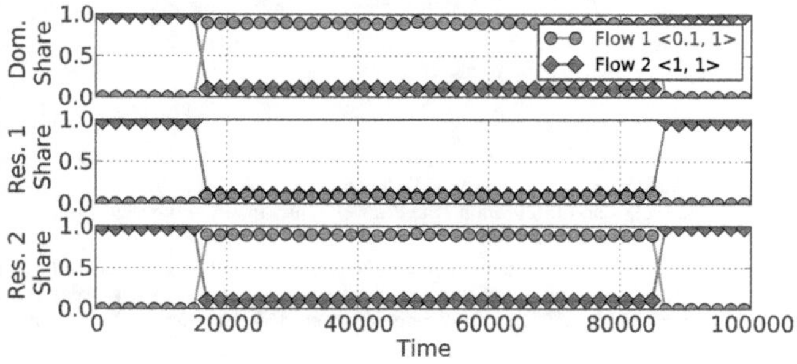

Figure 11: Fair queuing applied to only the first resource violates the share guarantee for flow 2.

under DRFQ. With bottleneck fairness, the oscillations in available bandwidth for flow 2 cause it to lose packets, back off, and get less than half the share it had running alone (*i.e.*, less than its share guarantee). This does not happen for the second flow because its rate is smaller so its queue in the middlebox does not overflow. In contrast, however, DRFQ provides a high throughput for both flows, letting both use about 83% of the bandwidth it would have alone because their demands dove-tail.

8.2 Simulation Results

We compare DRFQ with alternative solutions using per-packet simulations. The results are based on a discrete-event simulator that assumes resources are being used serially. It implements different queuing principles, including DRFQ, by looking at an input queue of packets and selecting which flow's packet should get processed next. It uses Poisson arrivals and normally distributed resource consumption. The packet processing times have means according to each flow's provided resource profile and standard deviation set to a tenth of the mean.

8.2.1 Comparison With Alternative Schedulers

Single-resource Fair Queuing: The first approach we test applies fair queuing on just one resource (*e.g.,* link bandwidth). This is the allocation that would result if traditional weighted fair queuing were used, ignoring the multi-resource consumption of packets. Figure 11 shows the simulation of a scenario in which one flow uses an equal amount of two resources, *i.e.,* $\langle 1, 1\rangle$. Another flow, with profile $\langle 0.1, 1\rangle$, starts and ends at times 15,000 and 85,000, respectively. Fair queuing is only applied to the first resource. We see that the share guarantee is violated; the $\langle 1, 1\rangle$ flow gets only 10% of each resource when the other flow is active.

Bottleneck Fairness: Figure 5 from Section 4.1 shows how Bottleneck Fairness behaves when multiple resources are bottlenecked. We let two flows have resource profiles $\langle 6, 1\rangle$ and $\langle 1, 7\rangle$. Then, we let one flow use equal amounts of two resources, $\langle 1, 1\rangle$. Bottleneck queuing was configured to dynamically switch to the current bottleneck (every 30 time units). As can be seen in Figure 5, oscillations occur when the bottleneck shifts. As a result, the first flow only gets 10% of either resource, far less than its share guarantee of $\frac{1}{3}$.

Per-Resource Fairness: Figure 12 investigates how a flow can manipulate per-resource fair queuing by changing its demands (*e.g.,* by changing packet sizes) to receive better service. It simulates a scenario with ten flows. The first flow has resource profile $\langle 20, 1\rangle$, whereas the last nine have $\langle 10, 11\rangle$. At time 25,000 the first flow artificially changes its demand to roughly $\langle 20, 11\rangle$, leading it to double its share under per-resource fair queuing. Meanwhile, the same change under DRFQ has no effect on the shares.

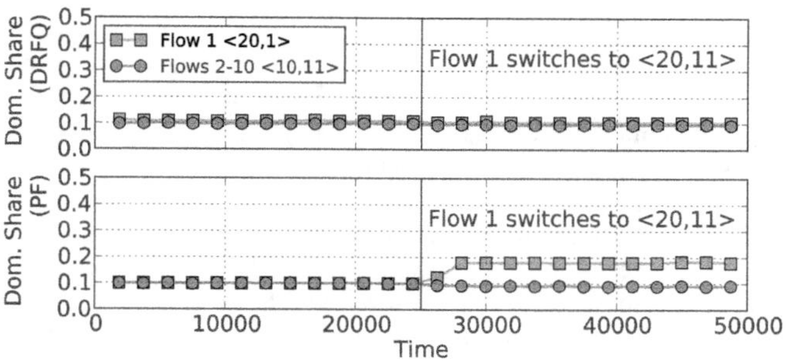

Figure 12: A flow manipulating PF to double its share.

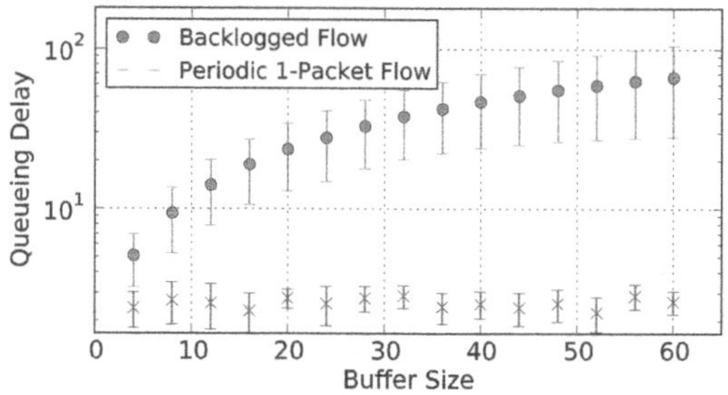

Figure 13: Per-packet delay of a backlogged flow compared to a periodic single-packet flow under DRFQ.

8.2.2 Isolation Under DRFQ

Next, we investigate packet delays under DRFQ. Figure 13 shows two different flows. The first is constantly backlogged to the level that it overflows all buffers and suffers from packet drops. The second flow periodically sends single packets, spread far apart in time. For both flows we measure the queuing delay for every packet and plot the mean and standard deviation. The x-axis shows the same simulation for various buffer sizes. As the buffer size is increased, the delay on the backlogged flow increases, as the incoming rate of packets is much higher than what the system can handle. The periodic flow, however, is unaffected by the backlogged flow and receives constant delay, irrespective of the buffer length.

8.3 Overhead

To evaluate the overhead of our Click implementation of DRFQ, we used the aforementioned trace generator to create a synthetic 350 MB workload from actual traces [4]. We ran the workload through two applications: flow monitor and intrusion detection system (IDS). For each application, we measured the overhead with and without DRFQ. Flow monitor's overhead was 4%, whereas IDS' overhead was 2%. While this is already low, we believe the overhead can be further reduced. First, DRFQ requires per-flow queues that are currently implemented in software. Many software routers and middleboxes already have support for in-hardware queues. Second, the overhead can be reduced using fair queueing per-class or per-aggregate basis, rather than per-flow basis.

9. RELATED WORK

Our work builds on WFQ [10, 24] as it, in similarity with many GPS approximations [18, 9, 17], uses the notion of virtual time. In particular, we approximate virtual time using start times as in SFQ [18], as it helps us avoid knowing in advance what middlebox modules a packet will traverse. As our evaluation shows, naively performing fair queuing on a single resource provides poor isolation for flows, violating the share-guarantee. Our attempt to extend WFQ by doing per-resource fair queuing (§4.2) turned out to violate strategy-proofness. Thus, DRFQ generalizes WFQ to multiple resources while providing isolation and strategy-proofness.

In the context of middleboxes, Egi *et al.* [14] proposed bottleneck fairness for software routers. We share their motivation for multi-resource fairness. However, we showed (§4.1) that their mechanism can not only provide poor isolation, but it can lead to heavy oscillations that severely degrade system performance. Dreger *et al.* [12] suggest measuring resource consumption of modules in NIDS and shutting off modules that overconsume resources. This approach is infeasible as some modules must run at all times, *e.g.,* a VPN module. Moreover, shutting down modules does not provide isolation between flows. With our approach, the flows that overconsume resources will fill buffers, eventually leading to modules not processing them, but each flow is sure to at least get its share guarantee of service.

In the context of active networks, Alexander *et al.* [7] propose a scheduling architecture called RCANE. This approach is akin to Per-Resource Fairness and therefore violates strategy-proofness.

Multi-resource fairness has also been investigated in the context of micro-economic theory. Ghodsi *et al.* [16] provide an overview and compare with the method preferred by economists, Competitive Equilibrium from Equal Incomes (CEEI). They show that CEEI is not strategy-proof and has several other undesirable properties. Dolev *et al.* [11] proposed an alternative to DRF. It too fails to be strategy-proof and is also computationally expensive to compute.

Our focus in this paper has been on achieving DRF allocations in the time domain. Others have analyzed how DRF allocations can be computed [19] and extended [21, 25] in the space domain.

10. CONCLUSION

Middleboxes apply complex processing functions to an increasing volume of traffic. Their performance characteristics are different from traditional routers; different processing functions have different demands across multiple resources, including CPU, memory bandwidth, and link bandwidth. Traditional single resource fair queuing schedulers therefore provide poor isolation guarantees between flows. Worse, in systems with multiple resources, flows can shift their demand to manipulate schedulers to get better service, thereby wasting resources. We have analyzed two schemes that are natural in the middlebox setting—bottleneck fairness and per-resource fairness—and shown that they have undesirable properties. In light of this, we have designed a new algorithm, DRFQ, for multi-resource fair queueing. We show through a Click implementation and extensive simulations that, unlike other approaches, our solution does not suffer from oscillations, provides flow isolation, and is strategy-proof. For future research directions, we believe DRFQ is applicable in many other multi-resource fair queueing contexts, such as VM scheduling in hypervisors.

11. ACKNOWLEDGEMENTS

We thank Adam Oliner, Ganesh Ananthanarayanan, and Patrick Wendell for useful feedback on earlier drafts of this paper. This research is supported in part by NSF FIA Award #CNS-1038695, NSF CISE Expeditions award CCF-1139158, a Google PhD Fellowship, gifts from Amazon Web Services, Google, SAP, Blue Goji, Cisco, Cloudera, Ericsson, General Electric, Hewlett Packard, Huawei, Intel, MarkLogic, Microsoft, NetApp, Oracle, Quanta, Splunk, VMware and by DARPA (contract #FA8650-11-C-7136).

12. REFERENCES

[1] Crossbeam network consolidation. `http://www.crossbeam.com/why-crossbeam/consolidation/`, June 2012.

[2] F5 Networks products. `http://www.f5.com/products/big-ip/`, June 2012.

[3] Intel perf. counter mon. `http://software.intel.com/en-us/articles/intel-performance-counter-monitor/`, June 2012.

[4] M57 network traffic traces. `https://domex.nps.edu/corp/scenarios/2009-m57/net/`, Feb. 2012.

[5] Palo alto networks. `http://www.paloaltonetworks.com/`, June 2012.

[6] Vyatta Software Middlebox. `http://www.vyatta.com`, June 2012.

[7] D. S. Alexander, P. B. Menage, A. D. Keromytis, W. A. Arbaugh, K. G. Anagnostakis, and J. M. Smith. The Price of Safety in An Active Network. *JCN*, 3(1):4–18, March 2001.

[8] K. Argyraki, K. Fall, G. Iannaccone, A. Knies, M. Manesh, and S. Ratnasamy. Understanding the packet forwarding capability of general-purpose processors. Technical Report IRB-TR-08-44, Intel Research Berkeley, May 2008.

[9] J. Bennett and H. Zhang. WF^2Q: Worst-case fair weighted fair queueing. In *INFOCOM*, 1996.

[10] A. Demers, S. Keshav, and S. Shenker. Analysis and simulation of a fair queueing algorithm. In *SIGCOMM*, pages 1–12, 1989.

[11] D. Dolev, D. G. Feitelson, J. Y. Halpern, R. Kupferman, and N. Linial. No justified complaints: on fair sharing of multiple resources. In *ITCS*, pages 68–75, 2012.

[12] H. Dreger, A. Feldmann, V. Paxson, and R. Sommer. Operational experiences with high-volume network intrusion detection. In *ACM Conference on Computer and Communications Security*, pages 2–11, 2004.

[13] H. Dreger, A. Feldmann, V. Paxson, and R. Sommer. Predicting the resource consumption of network intrusion detection systems. In *RAID*, 2008.

[14] N. Egi, A. Greenhalgh, M. Handley, G. Iannaccone, M. Manesh, L. Mathy, and S. Ratnasamy. Improved forwarding architecture and resource management for multi-core software routers. In *NPC*, pages 117–124, 2009.

[15] A. Ghodsi, V. Sekar, M. Zaharia, and I. Stoica. Multi-resource fair queueing for packet processing. Technical Report UCB/EECS-2012-166, EECS Department, University of California, Berkeley, June 2012.

[16] A. Ghodsi, M. Zaharia, B. Hindman, A. Konwinski, I. Stoica, and S. Shenker. Dominant resource fairness: Fair allocation of multiple resource types. In *NSDI*, 2011.

[17] S. J. Golestani. A self-clocked fair queueing scheme for broadband applications. In *INFOCOM*, pages 636–646, 1994.

[18] P. Goyal, H. Vin, and H. Cheng. Start-time fair queuing: A scheduling algorithm for integrated services packet switching networks. *ACM Transactions on Networking*, 5(5):690–704, Oct. 1997.

[19] A. Gutman and N. Nisan. Fair Allocation Without Trade. In *AAMAS*, June 2012.

[20] M. Honda, Y. Nishida, C. Raiciu, A. Greenhalgh, M. Handley, and H. Tokuda. Is it still possible to extend TCP? In *Proc. IMC*, 2011.

[21] C. Joe-Wong, S. Sen, T. Lan, and M. Chiang. Multi-resource allocation: Fairness-efficiency tradeoffs in a unifying framework. In *INFOCOM*, 2012.

[22] E. Kohler, R. Morris, B. Chen, J. Jannotti, and M. F. Kaashoek. The Click modular router. *ACM Trans. Comput. Syst.*, 18, August 2000.

[23] M. Kounavis, X. Kang, K. Grewal, M. Eszenyi, S. Gueron, and D. Durham. Encrypting the Internet. In *Proc. SIGCOMM*, 2010.

[24] A. Parekh and R. Gallager. A generalized processor sharing approach to flow control - the single node case. *ACM Transactions on Networking*, 1(3):344–357, June 1993.

[25] D. C. Parkes, A. D. Procaccia, and N. Shah. Beyond Dominant Resource Fairness: Extensions, Limitations, and Indivisibilities. In *ACM Conference on Electronic Commerce*, 2012.

[26] M. Piatek, T. Isdal, T. Anderson, A. Krishnamurthy, and A. Venkataramani. Do incentives build robustness in bittorrent. In *NSDI'07*, 2007.

[27] V. Sekar, N. Egi, S. Ratnasamy, M. Reiter, and G. Shi. Design and implementation of a consolidated middlebox architecture. In *NSDI*, 2012.

[28] V. Sekar, S. Ratnasamy, M. Reiter, N. Egi, and G. Shi. The Middlebox Manifesto: Enabling Innovation in Middlebox Deployments. In *HotNets 2011*, Oct. 2011.

[29] M. Shreedhar and G. Varghese. Efficient fair queuing using deficit round robin. *ACM Transactions on Networking*, 4(3):375–385, 1996.

[30] R. Smith, N. Goyal, J. Ormont, K. Sankaralingam, and C. Estan. Signature Matching in Network Processing Using SIMD/GPU Architectures. In *Int. Symp. on Performance Analysis of Systems and Software*, 2009.

[31] C. A. Waldspurger. *Lottery and Stride Scheduling: Flexible Proportional Share Resource Management.* PhD thesis, MIT, Laboratory of Computer Science, Sept. 1995. MIT/LCS/TR-667.

[32] Z. Wang, Z. Qian, Q. Xu, Z. M. Mao, and M. Zhang. An Untold Story of Middleboxes in Cellular Networks. In *SIGCOMM*, 2011.

[33] L. Zhang. Virtual clock: a new traffic control algorithm for packet switching networks. *SIGCOMM CCR*, 20:19–29, August 1990.

Making Middleboxes Someone Else's Problem: Network Processing as a Cloud Service

Justine Sherry
UC Berkeley

Shaddi Hasan
UC Berkeley

Colin Scott
UC Berkeley

Arvind Krishnamurthy
University of Washington

Sylvia Ratnasamy
UC Berkeley

Vyas Sekar
Intel Labs

ABSTRACT

Modern enterprises almost ubiquitously deploy middlebox processing services to improve security and performance in their networks. Despite this, we find that today's middlebox infrastructure is expensive, complex to manage, and creates new failure modes for the networks that use them. Given the promise of cloud computing to decrease costs, ease management, and provide elasticity and fault-tolerance, we argue that middlebox processing can benefit from outsourcing the cloud. Arriving at a feasible implementation, however, is challenging due to the need to achieve functional equivalence with traditional middlebox deployments without sacrificing performance or increasing network complexity.

In this paper, we motivate, design, and implement APLOMB, a practical service for outsourcing enterprise middlebox processing to the cloud. Our discussion of APLOMB is data-driven, guided by a survey of 57 enterprise networks, the first large-scale academic study of middlebox deployment. We show that APLOMB solves real problems faced by network administrators, can outsource over 90% of middlebox hardware in a typical large enterprise network, and, in a case study of a real enterprise, imposes an average latency penalty of 1.1ms and median bandwidth inflation of 3.8%.

Categories and Subject Descriptors
C.2.0 [**Computer-Communication Networks**]: General—*Security and firewalls*; C.2.1 [**Network Architecture and Design**]: [Distributed applications]; C.2.3 [**Network Operations**]: [Network management]

General Terms
Design, Management, Measurement

Keywords
Middlebox, Cloud, Outsourcing

1. INTRODUCTION

Today's enterprise networks rely on a wide spectrum of specialized appliances or *middleboxes*. Trends such as the proliferation of smartphones and wireless video are set to further expand the range of middlebox applications [19]. Middleboxes offer valuable benefits, such as improved security (*e.g.*, firewalls and intrusion detection systems), improved performance (*e.g.*, proxies) and reduced bandwidth costs (*e.g.*, WAN optimizers). However, as we show in §2, middleboxes come with high infrastructure and management costs, which result from their complex and specialized processing, variations in management tools across devices and vendors, and the need to consider policy interactions between these appliance and other network infrastructure.

The above shortcomings mirror the concerns that motivated enterprises to transition their in-house IT infrastructures to managed cloud services. Inspired by this trend, we ask whether the promised benefits of cloud computing—reduced expenditure for infrastructure, personnel and management, pay-by-use, the flexibility to try new services without sunk costs, *etc.*—can be brought to middlebox infrastructure. Beyond improving the status quo, cloud-based middlebox services would also make the security and performance benefits of middleboxes available to users such as small businesses and home and mobile users who cannot otherwise afford the associated costs and complexity.

We envision enterprises outsourcing the processing of their traffic to third-party *middlebox service providers* running in the cloud. Our proposal represents a significant change to enterprise networks, and hence we first validate that this exercise is worthwhile by examining what kind of a burden middleboxes impose on enterprises. The research literature, however, offers surprisingly few real-world studies; the closest study presents anecdotal evidence from a single large enterprise [42]. We thus start with a study of 57 enterprise networks, aimed at understanding (1) the nature of real-world middlebox deployments (*e.g.*, types and numbers of middleboxes), (2) "pain points" for network administrators, and (3) failure modes. Our study reveals that middleboxes do impose significant infrastructure and management overhead across a spectrum of enterprise networks and that the typical number of middleboxes in an enterprise is comparable to its traditional L2/L3 infrastructure!

Our study establishes the costs associated with middlebox deployments and the potential benefits of outsourcing them. We then examine different options for architecting cloud-based middlebox services. To be viable, such an architecture must meet three challenges:

(1) Functional equivalence. A cloud-based middlebox must offer functionality and semantics equivalent to that of an on-site middlebox – *i.e.*, a firewall must drop packets correctly, an intrusion detection system (IDS) must trigger identical alarms, *etc.* In contrast to traditional endpoint applications, this is challenging because middlebox functionality may be topology dependent. For example, traffic compression must be implemented *before* traffic leaves the enterprise access link, and an IDS that requires stateful processing must see *all* packets in *both* directions of a flow. Today, these requirements are met by deliberately placing middleboxes 'on path' at network choke points within the enterprise – options that are not readily available in a cloud-based architecture. As we shall see, these topological constraints complicate our ability to outsource middlebox processing.

SIGCOMM'12, August 13–17, 2012, Helsinki, Finland.

(2) *Low complexity at the enterprise.* As we shall see, an outsourced middlebox architecture still requires some supporting functionality at the enterprise. We aim for a cloud-based middlebox architecture that minimizes the complexity of this enterprise-side functionality: failing to do so would detract from our motivation for outsourcing in the first place.
(3) *Low performance overhead.* Middleboxes today are located on the direct path between two communicating endpoints. Under our proposed architecture, traffic is instead sent on a detour through the cloud leading to a potential increase in packet latency and bandwidth consumption. We aim for system designs that minimize this performance penalty.

We explore points in a design space defined by three dimensions: the redirection options available to enterprises, the footprint of the cloud provider, and the complexity of the outsourcing mechanism. We find that all options have natural tradeoffs across the above requirements and settle on a design that we argue is the sweet spot in this design space, which we term APLOMB, the Appliance for Outsourcing Middleboxes. We implement APLOMB and evaluate our system on EC2 using real end-user traffic and an analysis of traffic traces from a large enterprise network. In our enterprise evaluation, APLOMB imposes an average latency increase of only 1 ms and a median bandwidth inflation of 3.8%.

To summarize, our key contributions are:

- A study of costs and concerns in 57 real-world middlebox deployments, across a range of enterprise scenarios.
- A systematic exploration of the requirements and design space for outsourcing middleboxes.
- The design, implementation, and evaluation of the APLOMB architecture.
- A case study of how our system would impact the middlebox deployment of a large enterprise.

A core question in network design is where network functionality should be embedded. A wealth of research has explored this question for various network functionality, such as endpoints *vs.* routers for congestion control [20, 35, 29] and on-path routers *vs.* off-path controllers for routing control plane functions [28, 39]. Our work follows in this vein: the functionality we focus on is advanced traffic processing (an increasingly important piece of the network *data* plane) and we weigh the relative benefits of embedding such processing in the cloud *vs.* on-path middleboxes, under the conjecture that the advent of cloud computing offers new, perhaps better, options for supporting middlebox functionality.

Roadmap: We present our study of enterprise middlebox deployments in §2. In §3 we explore the design space for outsourcing middleboxes; we present the design and evaluation of the APLOMB architecture in §4 and §5 respectively. We discuss outstanding issues in §6 and related work in §7 before concluding in §8.

2. MIDDLEBOXES TODAY

Before discussing outsourcing designs, we draw on two datasets to discuss typical middlebox deployments in enterprise networks and why their challenges might be solved by the cloud. We conducted a survey of 57 enterprise network administrators, including the number of middleboxes deployed, personnel dedicated to them, and challenges faced in administering them. To the best of our knowledge, this is the first large-scale survey of middlebox deployments in the research community. Our dataset includes 19 small (fewer than 1k hosts) networks, 18 medium (1k-10k hosts) networks, 11 large (10k-100k hosts) networks, and 7 very large (more than 100k hosts) networks.

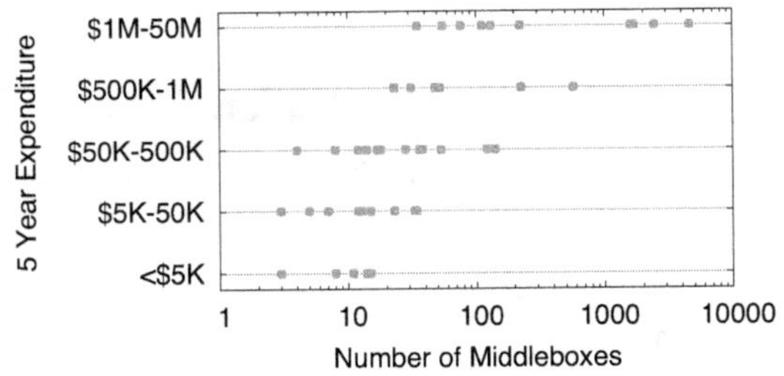

Figure 2: Administrator-estimated spending on middlebox hardware per network.

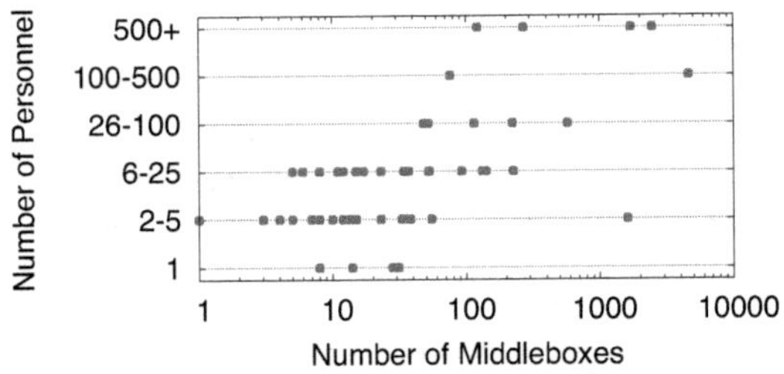

Figure 3: Administrator-estimated number of personnel per network.

We augment our analysis with measurements from a large enterprise with approximately 600 middleboxes and tens of international sites; we elaborate on this dataset in §5.3.

Our analysis highlights several key challenges that enterprise administrators face with middlebox deployments: large deployments with high capital expenses and operating costs (§2.1), complex management requirements (§2.2), and the need for overprovisioning to react to failure and overload scenarios (§2.3). We argue these factors parallel common arguments for cloud computation, and thus make middleboxes good candidates for the cloud.

2.1 Middlebox Deployments

Our data illustrates that typical enterprise networks are a complex ecosystem of firewalls, IDSes, web proxies, and other devices. Figure 1 shows a box plot of the number of middleboxes deployed in networks of all sizes, as well as the number of routers and switches for comparison. Across all network sizes, the number of middleboxes is on par with the number of routers in a network! The average very large network in our data set hosts 2850 L3 routers, and 1946 total middleboxes; the average small network in our data set hosts 7.3 L3 routers and 10.2 total middleboxes.[1]

These deployments are not only large, but are also costly, requiring high up-front investment in hardware: thousands to millions of dollars in physical equipment. Figure 2 displays five year expenditures on middlebox hardware against the number of actively deployed middleboxes in the network. All of our surveyed very large networks had spent over a million dollars on middlebox hardware in the last five years; the median small network spent between $5,000-50,000 dollars, and the top third of the small networks spent over $50,000.

Paralleling arguments for cloud computing, outsourcing middlebox processing can reduce hardware costs: outsourcing eliminates most of the infrastructure at the enterprise, and a cloud provider can provide the same resources at lower cost due to economies of scale.

[1]Even 7.3 routers and 10.2 middleboxes represents a network of a substantial size. Our data was primarily surveyed from the NANOG network operators group, and thus does not include many of the very smallest networks (*e.g.* homes and very small businesses with only tens of hosts).

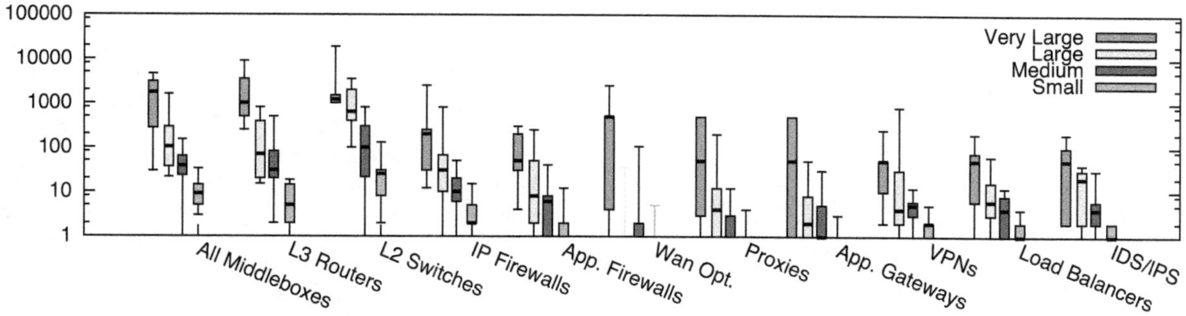

Figure 1: Box plot of middlebox deployments for small (fewer than 1k hosts), medium (1k-10k hosts), large (10k-100k hosts), and very large (more than 100k hosts) enterprise networks. Y-axis is in log scale.

2.2 Complexity in Management

Figure 1 also shows that middleboxes deployments are diverse. Of the eight middlebox categories we present in Figure 1, the median very large network deployed seven categories of middleboxes, and the median small network deployed middleboxes from four. Our categories are coarse-grained (*e.g.* Application Gateways include smartphone proxies and VoIP gateways), so these figures represent a *lower bound* on the number of distinct device types in the network.

Managing many heterogeneous devices requires broad expertise and consequently a large management team. Figure 3 correlates the number of middleboxes against the number of networking personnel. Even small networks with only tens of middleboxes typically required a management team of 6-25 personnel. Thus, middlebox deployments incur substantial operational expenses in addition to hardware costs.

Understanding the administrative tasks involved further illuminates why large administrative staffs are needed. We break down the management tasks related to middleboxes below.

Upgrades and Vendor Interaction. Deploying new features in the network entails deploying new hardware infrastructure. From our survey, network operators upgrade in the median case every four years. Each time they negotiate a new deployment, they must select between several offerings, weighing the capabilities of devices offered by numerous vendors – an average network in our dataset contracted with 4.9 vendors. This four-year cycle is at the same time both too frequent and too infrequent. Upgrades are too frequent in that every four years, administrators must evaluate, select, purchase, install, and train to maintain new appliances. Upgrades are too infrequent in that administrators are 'locked in' to hardware upgrades to obtain new features. Quoting one administrator:

> Upgradability is very important to me. I do not like it when vendors force me to buy new equipment when a software upgrade could give me additional features.

Cloud computing eliminates the upgrade problem: enterprises sign up for a middlebox *service*; how the cloud provider chooses to upgrade hardware is orthogonal to the service offered.

Monitoring and Diagnostics. To make managing tens or hundreds of devices feasible, enterprises deploy network management tools (e.g., [15, 9]) to aggregate exported monitoring data, *e.g.* SNMP. However, with a cloud solution, the cloud provider monitors utilization and failures of specific devices, and only exposes a middlebox *service* to the enterprise administrators, simplifying management at the enterprise.

Configuration. Configuring middleboxes requires two tasks. *Appliance configuration* includes, for example, allocating IP addresses, installing upgrades, and configuring caches. *Policy configuration* is customizing the device to enforce specific enterprise-wide policy goals (*e.g.* a HTTP application filter may block social network sites). Cloud-based deployments obviate the need for enterprise administrators to focus on the low-level mechanisms for appliance configuration and focus only on policy configuration.

	Misconfig.	Overload	Physical/Electric
Firewalls	67.3%	16.3%	16.3%
Proxies	63.2%	15.7%	21.1%
IDS	54.5%	11.4%	34%

Table 1: Fraction of network administrators who estimated misconfiguration, overload, or physical/electrical failure as the most common cause of middlebox failure.

Training. New appliances require new training for administrators to manage them. One administrator even stated that existing training and expertise was a key question in purchasing decisions:

> Do we have the expertise necessary to use the product, or would we have to invest significant resources to use it?

Another administrator reports that a lack of training limits the benefits from use of middleboxes:

> They [middleboxes] could provide more benefit if there was better management, and allocation of training and lab resources for network devices.

Outsourcing diminishes the training problem by offloading many administrative tasks to the cloud provider, reducing the set of tasks an administrator must be able perform. In summary, for each management task, outsourcing eliminates or greatly simplifies management complexity.

2.3 Overload and Failures

Most administrators who described their role as engineering estimated spending between one and five hours per week dealing with middlebox failures; 9% spent between six and ten hours per week. Table 1 shows the fraction of network administrators who labeled misconfiguration, overload, and physical/electrical failures as the most common cause of failures in their deployments of three types of middleboxes. Note that this table is *not* the fraction of failures caused by these issues; it is the fraction of administrators who estimate each issue to be the *most common* cause of failure. A majority of administrators stated misconfiguration as the most common cause of failure; in the previous subsection we highlight management complexity which likely contributes to this figure.

On the other hand, many administrators saw overload and physical/electrical problems as the most common causes of errors. For example, roughly 16% of administrators said that overload was the most common cause of IDS and proxy failure, and 20% said that physical failures were the most common cause for proxies.

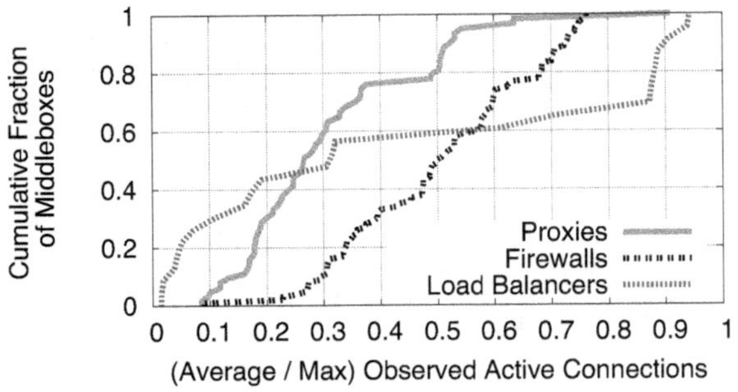

Figure 4: Ratio of average to peak active connections for all proxies, firewalls, and load balancers in the very large enterprise dataset.

A cloud-based capability to elastically provision resources avoids overload by enabling on-demand scaling and resolves failure with standby devices – without the need for expensive overprovisioning.

2.4 Discussion

To recap, our survey across 57 enterprises illuminates several middlebox-specific challenges that cloud outsourcing can solve: large deployments with high capital and operating expenses, complex management requirements inflating operation expenses, and failures from physical infrastructure and overload. Cloud outsourcing can cut costs by leveraging economies of scale, simplify management for enterprise administrators, and can provide elastic scaling to limit failures.

Outsourcing to the cloud not only solves challenges in existing deployments, but also presents new opportunities. For example, resource elasticity not only allows usage to scale *up*, but also to scale *down*. Figure 4 shows the distribution of average-to-max utilization (in terms of active connections) for three devices across one large enterprise. We see that most devices operate at moderate to low utilization; e.g., 20% of Load Balancers run at <5% utilization. Today, however, enterprises must invest resources for peak utilization. With a cloud solution, an enterprise can lease a large load balancer only at peak hours and a smaller, cheaper instance otherwise. Furthermore, a pay-per-use model democratizes access to middlebox services and enables even small networks who cannot afford up-front costs to benefit from middlebox processing.

These arguments parallel familiar arguments for the move to cloud computation [23]. This parallel, we believe, only bolsters the case.

3. DESIGN SPACE

Having established the potential benefits of outsourcing middleboxes to the cloud, we now consider how this might be achieved. To understand the challenges involved in such outsourcing, it is useful to reflect on how middleboxes are deployed today within an enterprise. Consider a middlebox m that serves traffic between endpoints a and b. Our proposal changes the *placement* of m – moving m from the enterprise to the cloud. This eliminates three key properties of today's middlebox placement:

1. *on-path*: m lies on the direct IP path between a and b
2. *choke point*: all paths between a and b traverse m
3. *local*: m is located inside the enterprise.

The challenges in outsourcing middleboxes arise as a result of losing these three properties. First, being on-path and at a choke point makes it easy for a middlebox to obtain the traffic it must process and specifically ensures that the middlebox sees both directions of traffic flow between two endpoints. (Bidirectional visibility is critical since most middleboxes operate at the session level.) Second, being on-path implies that a middlebox introduces no additional latency into the path. In contrast, sending traffic on a detour through the cloud could increase path latency. Third, middleboxes such as proxies and WAN optimizers, which we dub *location dependent* boxes, rely on being physically local to reduce latency and bandwidth costs. For example, proxies effectively terminate communication from an enterprise host a to an external host b thus reducing communication latency from that of path a-m-b to that of a-m. Similarly, by using redundancy elimination techniques, WAN optimizers avoid transmitting data over the wide area.

This raises three natural questions that together define the overall design space for outsourcing middleboxes to the cloud:

1. What is the effective *complexity* of the network architecture at the enterprise after outsourcing – *e.g.*, what types of middleboxes can be outsourced and what enterprise-side functionality is needed to achieve such outsourcing?
2. What *redirection* architecture is required to retain the functional equivalence and low latency operation; *e.g.*, ensuring that both directions of traffic between a and b via the cloud consistently traverse the same cloud PoP?
3. What type of *provider footprint* is needed for low latency operation; *e.g.*, is an Amazon-style footprint with a few distributed PoPs sufficient or do we need a larger, Akamai-scale footprint?

At a minimum, we need some generic device to redirect the enterprise's traffic to the cloud; we call this an *Appliance for Outsourcing Middleboxes* or *APLOMB*. We explore options for redirection in §3.1 and discuss strategies for low latency operation and evaluate the impact of the provider footprint in §3.2. In addition, to retain the savings in bandwidth consumption that local proxies and WAN optimizers offer, we consider extending APLOMB to have compression capabilities (which we call APLOMB+) in §3.3.

3.1 Redirection

We consider three natural approaches for redirecting the traffic to the cloud for middlebox processing and analyze their latency vs. complexity tradeoffs.

Bounce Redirection: In the simplest case, the APLOMB gateway at the enterprise tunnels both ingress and egress traffic to the cloud, as shown in Figure 5(a). Incoming traffic is bounced to the cloud PoP (1), processed by middleboxes, and then sent back to the enterprise (2,3). Outgoing traffic is similarly redirected (4-6). An immediate drawback of this architecture is the increase in end-to-end latency due to an extra round trip to and from the cloud PoP for each packet. Nevertheless, this design is feasible – especially if cloud providers have a large footprint such that the RTT to/from the cloud is small – and provides an attractive benefit in the simplicity of the architecture. Only the APLOMB gateway needs to be cloud-aware and no modification is required to existing enterprise applications. Naming and addressing for enterprise hosts are unchanged. Furthermore, bounce redirection requires minimal functionality and configuration at the gateway – a few static rules to redirect traffic to a cloud PoP.

IP Redirection: To avoid the extra round-trips in Figure 5(a), we can route traffic directly to/from the cloud, such that traffic goes from an Internet user directly to the cloud, and then to the enterprise. One possible design to achieve this is to have the cloud provider announce IP prefix P on the enterprise's behalf. Hosts communicating with the enterprise direct their traffic to P and thus their enterprise-bound traffic is received by the cloud provider. The cloud provider, after processing the traffic, tunnels the traffic to the

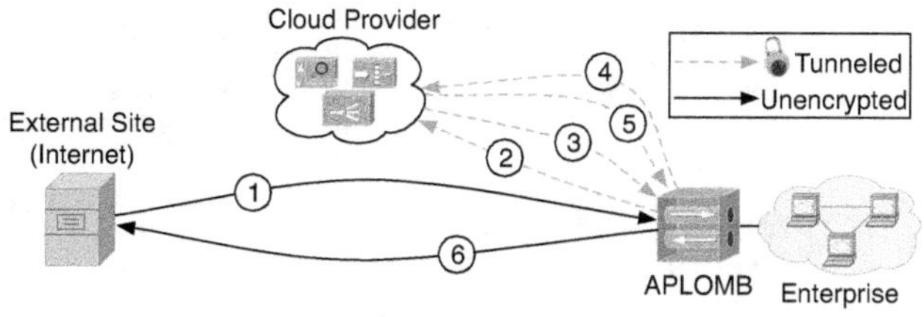

(a) "Bounce" redirection inflates latency.

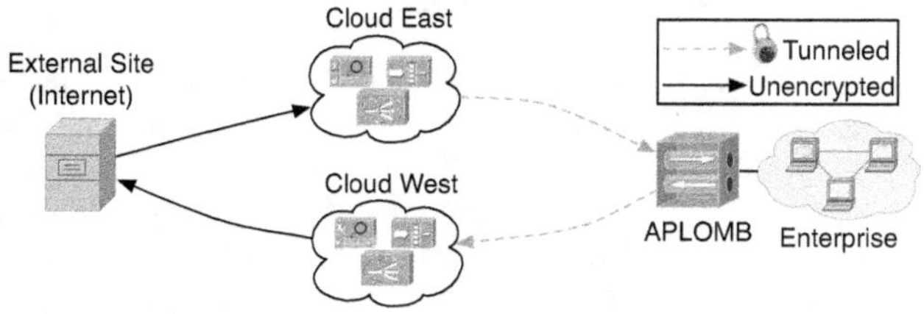

(b) Direct IP redirection in multi-PoP deployments cannot ensure that bidirectional traffic traverses the same PoP.

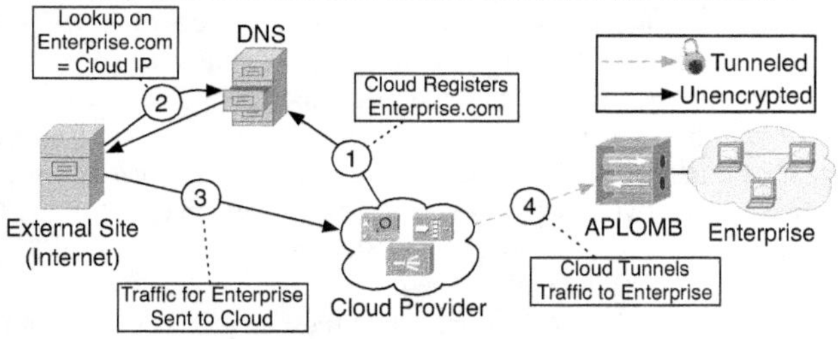

(c) DNS-based redirection minimizes latency and allows providers to control PoP selection for each request.

Figure 5: Comparison of redirection architectures.

enterprise gateways. However, in practice enterprises may want to redirect through several of the cloud provider's datacenters for improved latency, load distribution, and fault tolerance. (We'll discuss this "multi-pop" selection in depth in the following subsection.) For this, the cloud provider might advertise P from multiple PoPs so that client traffic is effectively anycasted to the closest PoP. Unfortunately, IP-based redirection in a multi-PoP scenario will break the semantics of stateful middleboxes as it cannot ensure that traffic from a client a to enterprise b will be routed to the same cloud PoP as that from b to a, as shown in Figure 5(b). Furthermore, because traffic is redirected at the network layer based on BGP path selection criteria (*e.g.*, AS hops), the enterprise or the cloud provider has little control over which PoP is selected and cannot, for example, pick PoPs to optimize end-to-end latency.

DNS Redirection: To allow redirection using multiple cloud PoPs, we can rely on DNS-based redirection similar to its use in CDNs (as shown in Figure 5(c)). Here, the cloud provider runs DNS resolution on the enterprise's behalf and registers DNS names for the client's external services, in this example '`MyEnterprise.com`' (step 1) [2]. A user accessing `MyEnterprise.com` (step 2) is directed to the cloud PoP via DNS (step 3). The traffic is then processed by relevant middleboxes and tunneled to the enterprise (step 4). Outbound return traffic from the enterprise (and traffic initiated by enterprise hosts) is also easy to control; the APLOMB gateway device uses a simple lookup to send traffic to the same PoP the inbound client would receive from DNS. This ensures that traffic between the enterprise and an external host will always traverse the same cloud PoP, even when network-level routing changes. However, DNS-based redirection introduces a challenge in outsourcing traffic for legacy applications which provide external clients with IP addresses rather than DNS names.

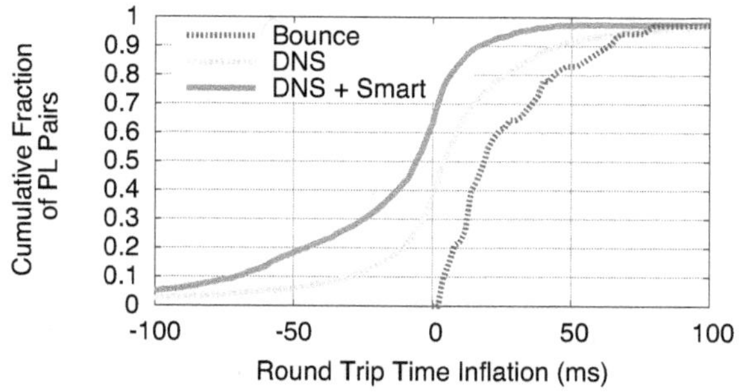

Figure 6: Round Trip Time (RTT) inflation when redirecting traffic between US PlanetLab nodes through Amazon PoPs.

In our system architecture, we choose DNS redirection because it avoids the latency penalty of bounce redirection and provides more control over redirection than pure IP redirection.

3.2 Minimizing Latency

DNS redirection can force bidirectional traffic through the same cloud provider PoP, thus bringing traffic to a 'choke point' where network processing can occur. However, any redirection has the potential to inflate end to end latency between an outsourcing enterprise and external clients – even if one avoids obvious pitfalls like the 'bounce' architecture. We now consider a DNS-based outsourcing architecture where the cloud provider has multiple PoPs, and discuss latency-optimized PoP selection strategies as well as how large a footprint a cloud provider should possess in order to provide redirection services that do not excessively inflate latency.

3.2.1 Smart Redirection

We quantify the latency impact of the bounce, IP, and DNS redirection options in Figure 6 using measurements from over 300 PlanetLab nodes and twenty Amazon CloudFront locations. We consider an enterprise "site" located at one of fifty US-based PlanetLab sites while the other PlanetLab nodes emulate "clients". For each site e, we pick the closest Amazon CloudFront PoP $P_e^* = \arg\min_P Latency(P, e)$ and measure the impact of tunneling traffic to/from this PoP.

Figure 6 shows that the bounce redirection can increase the end-to-end RTT by more than 50ms for 20% of inter-PlanetLab paths. The basic DNS-based redirection, where the enterprise tunnels traffic to and from the cloud PoP to which it has the minimum RTT, reduces the 80th percentile of latency inflation $2\times$ compared to bounce redirection. In fact, for more than 30% of the pairwise measurements, the latency is actually lower than the direct IP path. This is because of well-known triangle inequality violations in inter-domain routing and the fact that cloud providers are well connected to tier-1/2 ISPs [31].

To reduce latency further, we redirect traffic not through the cloud PoP with the minimum RTT to and from the enterprise, but redirect traffic on a per-destination basis through the PoP that minimizes end-to-end latency. That is, instead of using a single fixed PoP P_e^* for each enterprise site e, we choose the optimal PoP for each client and site c, e combination (i.e., $\arg\min_P Latency(P, c) + Latency(P, e)$). We quantify the inflation with this redirection using the earlier setup with Amazon CloudFront sites as PoPs and PlanetLab nodes as enterprise sites. Figure 6 shows that with "Smart Redirection", more than 70% of the cases have zero or negative inflation and 90% of all traffic has less than 10ms inflation.

Smart redirection requires that the APLOMB appliance redirect traffic to different PoPs based on the client's IP and maintain persistent tunnels to multiple PoPs instead of just one tunnel to its clos-

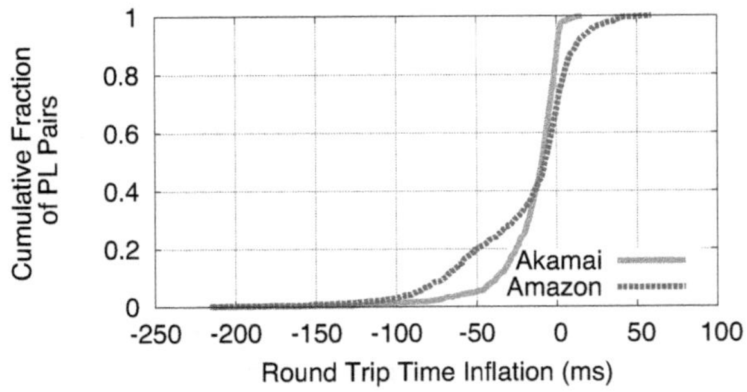

Figure 7: PlanetLab-to-PlanetLab RTTs with APLOMB redirection through Amazon and Akamai.

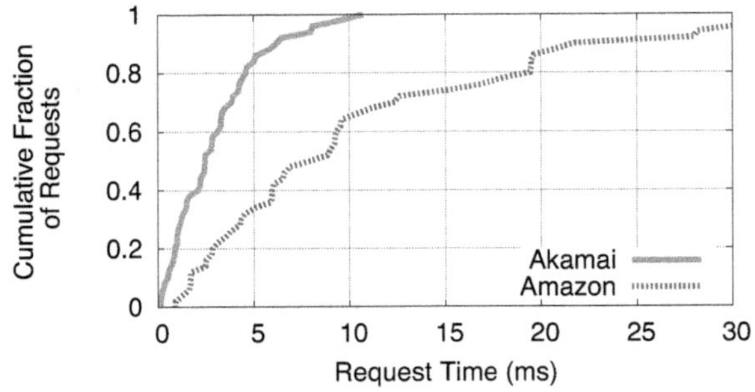

Figure 8: Direct RTTs from PlanetLab to nearest Akamai or Amazon redirection node.

est PoP. This requirement is modest: mappings for PoP selection can be computed at the cloud provider and pushed to APLOMB appliances, and today's commodity gateways can already support hundreds of persistent tunneled connections.

3.2.2 Provider Footprint

Next, we analyze how the middlebox provider's choice of geographic footprint impacts latency. Today's clouds have a few tens of global PoPs and expand as new demand arises [4]. For greater coverage, we could envision a middlebox provider with a footprint comparable to CDNs such as Akamai with thousands of vantage points [45]. While it is clear that a larger footprint provides lower latency, it is not clear how large a footprint is required in the context of outsourcing middleboxes.

To understand the implications of the provider's footprint, we extend our measurements to consider a cloud provider with an Akamai-like footprint using IP addresses of over 20,000 Akamai hosts [26]. First, we repeat the the end-to-end latency analysis for paths between US PlanetLab nodes and see that a larger, edge-concentrated Akamai footprint reduces tail latency, but the overall changes are marginal compared to a smaller but well connected Amazon-like footprint. End-to-end latency is the metric of interest when outsourcing most middleboxes – all except for 'location dependent' appliances. Because roughly 70% of inter-PlanetLab node paths actually experience *improved* latency, these results suggest that a middlebox provider can service most customers with most types of middleboxes (*e.g.*, NIDS, firewalls) with an Amazon-like footprint of a few tens of PoPs.

Some 'location dependent' middleboxes such as cache/proxies and protocol accelerators, are best served by cloud PoPs that are close to enterprise (minimizing latency to and from the middlebox). To evaluate whether we can outsource even these location dependent middleboxes without a high latency penalty (we discuss bandwidth penalties in §3.3), we look at the RTT between each PlanetLab node and its closest Akamai node in Figure 8. In this case, we see a more dramatic impact of Akamai's footprint as it provides sub-millisecond latencies to 20% of sites, and less than 5 ms latencies to almost 90% of sites. An Amazon-like footprint provides only 30% of sites with an RTT <5 ms. Our results suggest that an Amazon-like footprint can provide low latency only for a limited portion of US clients; to provide low latency service for a nation-wide client base, an Akamai-like footprint is necessary.

Type of Middlebox	Enterprise Device	Cloud Footprint
IP Firewalls	Basic APLOMB	Multi-PoP
Application Firewalls	Basic APLOMB	Multi-PoP
VPN Gateways	Basic APLOMB	Multi-PoP
Load Balancers	Basic APLOMB	Multi-PoP
IDS/IPS	Basic APLOMB	Multi-PoP
WAN optimizers	APLOMB+	CDN
Proxies	APLOMB+	CDN

Table 2: Complexity of design and cloud footprint required to outsource different types of middleboxes.

3.3 Location Dependent Services

As discussed earlier, location dependent appliances optimize both latency and bandwidth consumption. The above results show that, with an appropriate provider footprint, these appliances can still offer significant latency savings. We now consider whether we can retain bandwidth savings. Unfortunately, this is a harder problem since bandwidth optimizations must fundamentally be implemented before the enterprise access link in order to be useful. We discuss three possible options next.

The first is to simply not outsource these appliances. From the enterprises we surveyed and Figure 1, we see that WAN optimizers and proxies are currently only deployed in large enterprises and that APLOMB is of significant value even if it doesn't cover proxies and WAN optimizers. Nevertheless, we would like to do better and hence ask whether full-fledged middleboxes are really needed or whether we could achieve much of their benefit with a more minimal design.

The second option we consider is to embed *general-purpose* traffic compression capabilities into the APLOMB appliance – we call this augmented appliance APLOMB+. We evaluate APLOMB+ against traditional WAN optimizers in §5.3 using measurements from a large enterprise and show that protocol-agnostic compression [21] can provide similar bandwidth savings.

We cannot, however, claim that such a minimal capability exists for every conceivable middlebox (*e.g.*, consider an appliance that encodes outgoing traffic for packet loss protection), nor that APLOMB+ can fully replicate the behavior of dedicated appliances. Thus, our third option considers more general support for such functions at the APLOMB gateway. For this, we envision a more "active" appliance architecture that can run specialized software modules (*e.g.*, a FEC encoder) that are dynamically installed by the cloud provider or the enterprise administrator. Although more general, this increases device and configuration complexity, and we do not consider this option in this paper.

3.4 Discussion

In the following section, we describe our APLOMB design, which implements all of the above design choices. We now reflect on our original goal—outsourcing as many middleboxes as possible—and investigate how well our architecture achieves this goal.

Table 2 identifies the design option (and hence its associated complexity) that is needed to retain the functional equivalence of the middleboxes observed in our survey.[2] We consider "outsource-

[2]A subtle issue is whether load balancers really need to be physically close to backend servers; *e.g.*, for identifying load imbalances

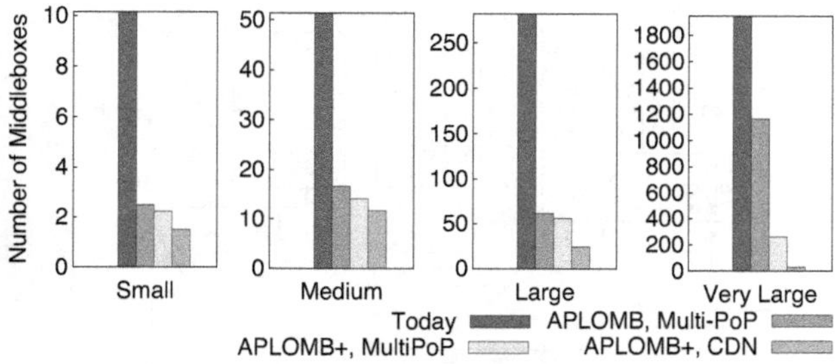

Figure 9: Average number of middleboxes remaining in enterprise under different outsourcing options.

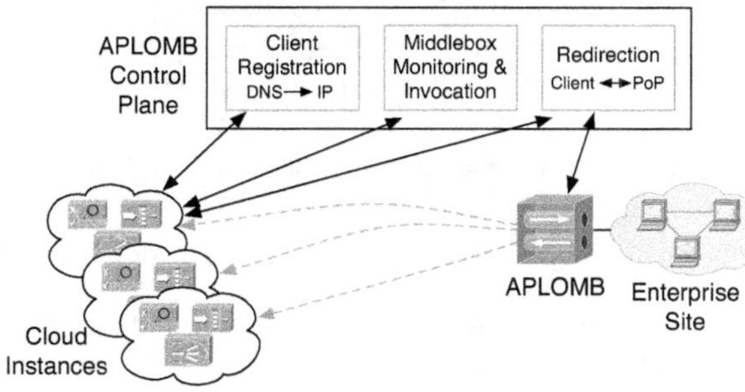

Figure 10: Architectural components of APLOMB.

able" middleboxes to be those whose functionality can be performed in the cloud with little (<5ms) or no latency inflation for at least 80% of PoPs. Most middleboxes can be outsourced with a 'Basic APLOMB' device (tunneling to and from the cloud using smart redirection, but without compression) and only an Amazon-like 'Multi-PoP' footprint. The only exceptions are WAN optimizers and proxies; these require an APLOMB+ (compression-performing) gateway and may require an Akamai-like 'CDN' cloud footprint for more remote enterprise locations. We also note that with an Akamai-like footprint, the bounce redirection becomes a feasible and potentially much simpler outsourcing solution.

Based on this analysis, Figure 9 shows the number of middleboxes that remain in an average small, medium, and large enterprise under different outsourcing deployment options. This suggests that small and medium enterprises can achieve almost all outsourcing benefits with a basic APLOMB architecture using today's cloud providers. (We revisit the remaining middleboxes, which are largely 'internal firewalls', in §5.3.) The same basic architecture can outsource close to 50% of the appliances in very large enterprise networks; using APLOMB+ increases the percentage of outsourced appliances to close to 90%.

4. APLOMB: DETAILED DESIGN

In describing the detailed design of the APLOMB architecture, we focus on three key components as shown in Figure 10: (1) a APLOMB gateway to redirect enterprise traffic, (2) the corresponding functions and middlebox capabilities at the cloud provider, and (3) a control plane which is responsible for managing and configuring these components.

4.1 Enterprise Configuration

Redirecting traffic from the enterprise client to the cloud middlebox provider is simple: an APLOMB gateway is co-located with the enterprise's gateway router, and enterprise administrators supply the cloud provider with a manifest of their address allocations. APLOMB changes neither routing nor switching, and end hosts require no new configuration.

4.1.1 Registration

APLOMB involves an initial registration step in which administrators provide the cloud provider with an *address manifest*. These manifests list the enterprise network's address blocks in its *private* address space and associates each address or prefix with one of three types of address records:

Protected services: Most private IP addresses are registered as protected services. These address records contain an IP address or prefix and the public IP address of the APLOMB device at the gateway to the registered address(es). This registration allows inter-site enterprise traffic to traverse the cloud infrastructure (*e.g.* a host at site A with address 10.2.3.4 can communicate with a host at site B with address 10.4.5.6, and the cloud provider knows that the internal address 10.4.5.6 maps to the APLOMB gateway at site B). The cloud provider allocates no permanent public IP address for hosts with 'protected services' addresses; Internet-destined connections instead undergo traditional NAPT.

DNS services: For hosts which accept incoming traffic, such as web servers, a publicly routeable address must direct incoming traffic to the appropriate cloud PoP. For these IP addresses, the administrator requests DNS service in the address manifest, listing the private IP address of the service, the relevant APLOMB gateway, and a DNS name. The cloud provider then manages the DNS records for this address on the enterprise client's behalf. When a DNS request for this service arrives, the cloud provider (dynamically) assigns a public IP from its own pool of IP addresses and directs this request to the appropriate cloud PoP and subsequent APLOMB gateway.

Legacy IP services: While DNS-based services are the common case, enterprise may require legacy services that require fixed IP addresses. For these services, the enterprise registers the internal IP address and corresponding APLOMB gateway, and the cloud provider allocates a static public IP address at a single PoP for the IP service. For this type of service, we fall back to the single-PoP Cloud-IP solution rather than DNS redirection discussed in §3.

4.1.2 APLOMB gateway

The APLOMB gateway is logically co-located with the enterprise's gateway router and has two key functions: (1) maintaining persistent tunnels to multiple cloud PoPs and (2) steering the outgoing traffic to the appropriate cloud PoP. The gateway registers itself with the cloud controller (§4.3), which supplies it with a list of cloud tunnel endpoints in each PoP and forwarding rules (5-tuple $\rightarrow$ cloud PoP Identifier) for redirection. (The gateway router blocks all IP traffic into the network that is not tunneled to a APLOMB gateway.) For security reasons, we use encrypted tunnels (e.g., using OpenVPN) and for reducing bandwidth costs, we enable protocol-agnostic redundancy elimination [21]. Note that the functionality required of the APLOMB gateway is simple enough to be bundled with the egress router itself or built using commodity hardware.

For scalability and fault tolerance, we rely on traditional load balancing techniques. For example, to load balance traffic across multiple APLOMB gateways, the enterprise's private address space can be split to direct traffic to, *e.g.* 10.1.0.0/17 to one gateway, and 10.1.128.0/17 to another. To handle gateway failures, we envision APLOMB hardware with fail-open NICs configured to direct the packets to a APLOMB replica under failure. Since each APLOMB box keeps almost no per-flow state, the replica receiving

at sub-millisecond granularity. Our conversations with administrators suggest that this is not a typical requirement and thus that load balancers can be outsourced.

traffic from the failed device can start forwarding the new traffic without interruption to existing flows.

4.2 Cloud Functionality

The cloud provider has three main tasks: (1) publicly addressable IP addresses to the appropriate enterprise customer and internal private address, (2) apply middlebox processing services to the customers' traffic according to their *policies* (§4.3), and (3) tunnel traffic to and from the appropriate APLOMB gateways at enterprise sites. Thus, the core components at the cloud PoP are:

- *Tunnel Endpoints* to encapsulate/decapsulate traffic from the enterprise (and to encrypt/decrypt and compress/decompress if enabled)
- *Middlebox Instances* to process the customers' traffic
- *NAT Devices* to translate between publicly visible IP addresses and the clients' internal addresses. NAT devices manage statically configured IP to IP mappings for DNS and Legacy IP services, and generate IP and port mappings for Protected Services (§4.1).
- *Policy switching* logic to steer packets between the above components.

Fortunately, it is possible to realize each of these components with existing solutions and there are many research and commercial solutions to provide these features (*e.g.* [34, 25, 8]). These solutions differ along two key dimensions depending on whether the middlebox services are: (1) provided by the cloud infrastructure provider (e.g., Amazon) or by third-party cloud service providers running within these infrastructure providers (e.g., [17]), and (2) realized using hardware- (e.g., [14, 11]) or software-based middleboxes (e.g., [41, 46, 13, 42]. Our architecture is agnostic to these choices and accommodates a broad range of deployment scenarios as long as there is some feasible path to implement the four components described above. The specific implementation we present in this paper runs as a third-party service using software-based middleboxes over an existing infrastructure provider.

4.3 Control Plane

A driving design principle for APLOMB is to keep the new components introduced by our architecture that are on the critical path – *i.e.*, the APLOMB gateway device and the cloud terminal endpoint – as simple and as stateless as possible. This not only reduces the enterprise's administrative overhead but also enables seamless transition in the presence of hardware and network failures. To this end, the APLOMB Control Plane manages the relevant network state representing APLOMB gateways, cloud PoPs, middlebox instances, and tunnel endpoints. It is responsible for determining optimal redirection strategies between communicating parties, managing and pushing middlebox policy configurations, and dynamically scaling cloud middlebox capacity to meet demands.

In practice, the control plane is realized in a *cloud controller*, which manages every APLOMB gateway, middlebox, tunneling end point, and the internals of the cloud switching policy.[3] Each entity (APLOMB device, middlebox, *etc.*) registers itself with the controller. The controller sends periodic 'heartbeat' health checks to each device to verify its continued activity. In addition, the controller gathers RTTs from each PoP to every prefix on the Internet (for PoP selection) and utilization statistics from each middlebox (for adaptive scaling). Below we discuss the redirection optimiza-

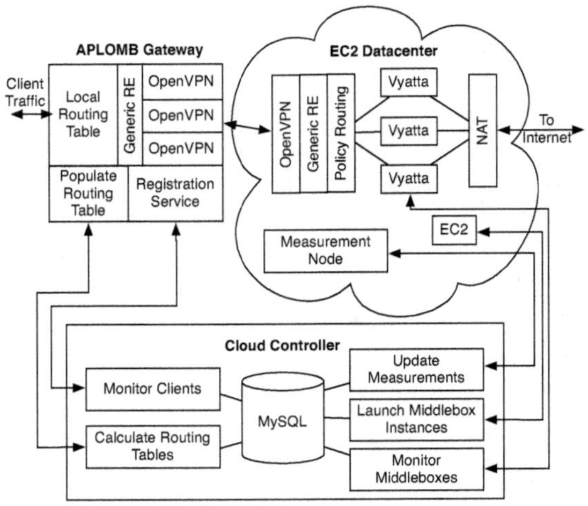

Figure 11: Software architecture of APLOMB.

tion, policy management, and middlebox scaling performed by the cloud controller.

Redirection Optimization. Using measurement data from the cloud PoPs, the cloud controller pushes the current best (as discussed in §3.2) tunnel selection strategies to the APLOMB gateways at the enterprise and mappings in the DNS. To deal with transient routing issues or performance instability, the cloud controller periodically updates these tunneling configurations based on the newest measurements from each cloud PoP.

Policy Configuration. The cloud controller is also responsible for implementing enterprise- and middlebox-specific *policies*. Thus, the cloud provider provides a rich policy configuration interface that exports the available types of middlebox processing to enterprise administrators and also implements a programmatic interface to specify the types of middlebox processing required [33]. Enterprise administrators can specify different *policy chains* of middlebox processing for each class of traffic specified using the traditional 5-tuple categorization of flows (i.e., source and destination IPs, port values and the protocol). For example, an enterprise could require all egress traffic to go through a firewall → exfiltration engine → proxy. and require that all ingress traffic traverse a firewall → IDS, and all traffic to internal web services further go through an application-level firewall. If appropriate, the provider may also export certain device-specific configuration parameters that the enterprise administrator can tune.

Middlebox Scaling. APLOMB providers have a great deal of flexibility in how they actually implement the desired middlebox processing. In particular, as utilization increases on a particular middlebox, the APLOMB provider simply increases the number of instances of that middlebox being utilized for a client's traffic.

Using data from heartbeat health checks on all middleboxes, the cloud controller detects changes in utilization. When utilization is high, the cloud controller launches new middleboxes and updates the policy switching framework; when utilization is low, the cloud controller deactivates excess instances. While scaling is simpler if all middlebox processing is performed in software on standard virtual machines, providers using hardware middleboxes could achieve the same result using policy switching alone. Techniques for dynamic scaling under load are well-known for cloud computing applications like web servers [12]; as such we do not go into detail here.

4.4 Implementation

We built a prototype system for cloud middlebox processing using middlebox processing services running on EC2 and APLOMB

[3]While the cloud controller may be in reality a replicated or federated set of controllers, for simplicity this discussion refers to a single logically centralized controller.

endpoints in our lab and at the authors' homes. We consciously choose to use off-the-shelf components that run on existing cloud providers and end host systems. This makes our system easy to deploy and use and demonstrates that the barriers to adoption are minimal. Our APLOMB endpoint software can be deployed on a stand-alone software router or as a tunneling layer on an end host; installing and running the end host software is as simple as connecting to a VPN.

Figure 11 is a software architecture diagram of our implementation. We implement a cloud controller on a server in our lab and use geographically distributed EC2 datacenters as cloud PoPs. Our cloud controller employs a MySQL database to store data on middlebox nodes, RTTs to and from cloud PoPs, and registered clients. The cloud controller monitors APLOMB devices, calculates and pushes routing tables to the APLOMB devices, requests measurements from the cloud PoPs, monitors middlebox instances, and scales middlebox instances up or down as demand varies.

At the enterprise or the end host, the APLOMB gateway maintains several concurrent VPN tunnels, one to a remote APLOMB at each cloud PoP. On startup, the APLOMB software contacts the cloud controller and registers itself, fetches remote tunnel endpoints for each cloud PoP, and requests a set of initial tunnel redirection mappings. A simple tunnel selection layer, populated by the cloud controller, directs traffic to the appropriate endpoint tunnel, and a redundancy elimination encoding module compresses all outgoing traffic. When run on a software router, ingress traffic comes from an attached hosts for whom the router serves as their default gateway. Running on a laptop or end host, static routes in the kernel direct application traffic to the appropriate egress VPN tunnel.

EC2 datacenters host tunnel endpoints, redundancy elimination decoders, middlebox routers, and NATs, each with an inter-device switching layer and controller registration and monitoring service. For tunneling, we use OpenVPN [10], a widely-deployed VPN solution with packages for all major operating systems. We use a Click [36] implementation of the redundancy elimination technique described by Anand et al [21]. For middlebox processing, we use Vyatta [16], a customizable software middlebox. Our default Vyatta installation performs firewalling, intrusion detection, caching, and application-layer web filtering. Policy configurations (§4.3) are translated into Vyatta configurations such that each client can have a unique Vyatta configuration dependent on their needs. Finally, each cloud PoP also hosts one 'measurement node', which periodically issues ping measurements for RTT estimation to assist in PoP selection.

5. EVALUATION

We now evaluate APLOMB. First, we present performance benchmarks for three common applications running over our implementation (§5.1).We then demonstrate APLOMB's dynamic scaling capability and its resilience to failure (§5.2). Having shown that APLOMB is practical, we return to our goal of outsourcing all middlebox functionality in an enterprise with a trace-driven evaluation of middlebox outsourcing using APLOMB, applied to data from a middlebox deployment in a large enterprise (§5.3).

5.1 Application Performance

We first demonstrate that APLOMB's architecture is practical for enterprise use with performance benchmarks for common applications using our APLOMB implementation.

HTTP Page Loads: In Figure 12, we plot page load times (fetching the front page and all embedded content) from a university network for the Alexa top 1,000 most popular web pages with and without APLOMB processing. We performed this experiment with

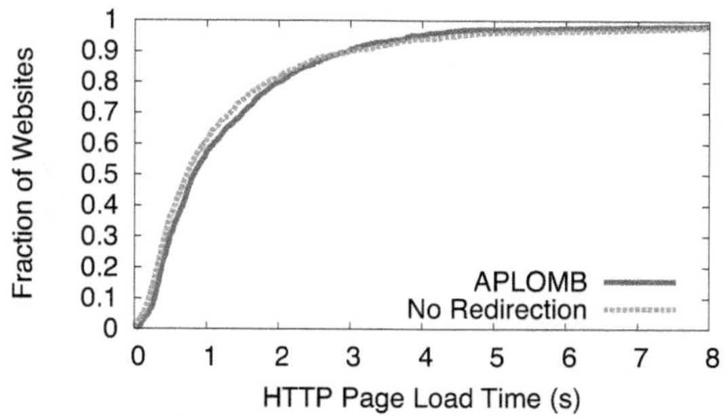

Figure 12: CDF of HTTP Page Load times for Alexa top 1,000 sites with and without APLOMB.

a vacant cache. For pages at the 50th percentile, page loads without APLOMB took 0.72 seconds, while page loads with took 0.82 seconds. For pages at the 95th percentile, using APLOMB results in shorter page load times: 3.85 seconds versus 4.53 seconds.

BitTorrent: While we don't expect BitTorrent to be a major component of enterprise traffic, we chose to experiment with Bit Torrent because it allowed us to observe a bulk transfer over a long period of time, to observe many connections over our infrastructure simultaneously, and to establish connections to non-commercial endpoints. We downloaded a 698MB public domain film over BitTorrent with and without APLOMB from both a university network and from a residential network, five times repeatedly. The average residential download took 294 seconds without APLOMB, with APLOMB the download speed increased 2.8% to 302 seconds. The average university download took 156 seconds without APLOMB, with APLOMB the average download took 165 seconds, a 5.5% increase.

Voice over IP: Voice over IP (VoIP) is a common enterprise application, but unlike the previously explored applications, VoIP performance depends not only on low latency and high bandwidth, but on low *jitter*, or variance in latency. APLOMB easily accommodates this third demand: we ran VoIP calls over APLOMB and for each call logged the jitter estimator, a running estimate of packet interarrival variance developed for RTP. Industry experts cite 30ms of one-way jitter as a target for maximum acceptable jitter [7]. In the first call, to a residential network, median inbound/outbound jitter with APLOMB was 2.49 ms/2.46 ms and without was 2.3 ms/1.03 ms. In the second, to a public WiFi hotspot, the median inbound/outbound jitter with APLOMB was 13.21 ms/14.49 ms and without was 4.41 ms/4.04 ms.

In summary, these three common applications suffer little or no penalty when their traffic is redirected through APLOMB.

5.2 Scaling and Failover

To evaluate APLOMB's dynamic scaling, we measured traffic from a single client to the APLOMB cloud. Figure 13 shows capacity adapting to increased network load over a 10-minute period. The client workload involved simultaneously streaming a video, repeatedly requesting large files over HTTP, and downloading several large files via BitTorrent. The resulting network load varied significantly over the course of the experiment, providing an opportunity for capacity scaling. The controller tracks CPU utilization of each middlebox instance and adds additional capacity when existing instances exceed a utilization threshold for one minute.

While middlebox capacity lags changes in demand, this is primarily an artifact of the low sampling resolution of the monitoring infrastructure provided by our cloud provider. Once a new middlebox instance has been allocated and initialized, actual switchover time to begin routing traffic through it is less than 100ms. To handle failed middlebox instances, the cloud controller checks for reacha-

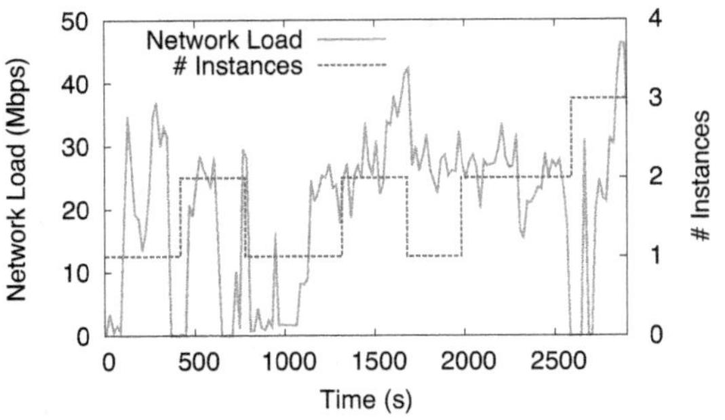

Figure 13: Network load (Y_1) and number of software middlebox instances (Y_2) under load. Experiment used low-capacity instances to highlight scaling dynamics.

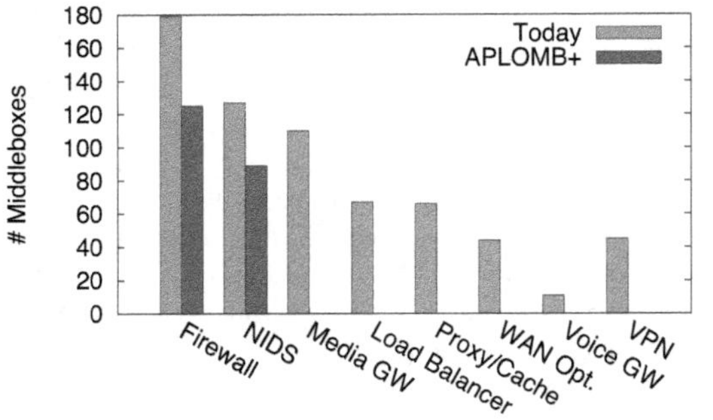

Figure 14: Number of middleboxes in the enterprise with and without APLOMB+. The enterprise has an atypical number of 'internal' firewalls and NIDS.

bility between itself and individual middlebox instances every second; when an instance becomes unreachable, APLOMB ceases routing traffic through it within 100ms. Using the same mechanism, the enterprise APLOMB can cope with failure of a remote APLOMB, re-routing traffic to another remote APLOMB in the same or even different cloud PoP, providing fault-tolerance against loss of an entire datacenter.

5.3 Case Study

We set out with the goal of outsourcing as many middleboxes as possible and reducing enterprise costs, all the while without increasing bandwidth utilization or latency. We revisit this using the data from the very large enterprise to determine:

- How many middleboxes can the enterprise outsource?
- What are the gains from elastic scaling?
- What latency penalty will inter-site traffic suffer?
- How much does the enterprise's bandwidth costs increase?

Middleboxes Outsourced: Figure 14 shows that the large enterprise can outsource close to 60% of the middleboxes under a CDN footprint with APLOMB+.

This high fraction of outsourceability comes despite an atypically high deployment of "internal" firewalls and NIDS at this enterprise. Internal firewalls protect a host or subnetwork not only from Internet-originated traffic, but from traffic originated within the enterprise; the most common reason we found for these deployments was PCI compliance for managing credit card data. While the average enterprise of this size deploys 27.7 unoutsourceable internal firewalls, this enterprise deploys over 100 internal firewalls. From discussions with the network's administrators, we learned these were installed in the past to protect internal servers against worms that preferentially scanned internal prefixes, *e.g.* CodeRed and Nimda. As more IT infrastructure moves to the cloud (see §6), many internal firewalls will be able to move to the cloud as well.

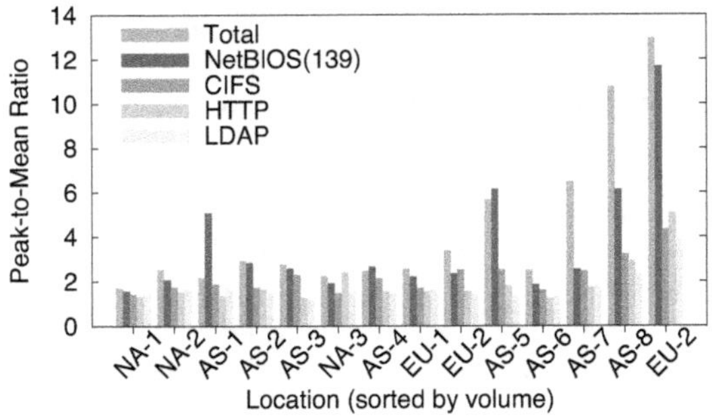

Figure 15: Ratio of peak traffic volume to average traffic volume, divided by protocol.

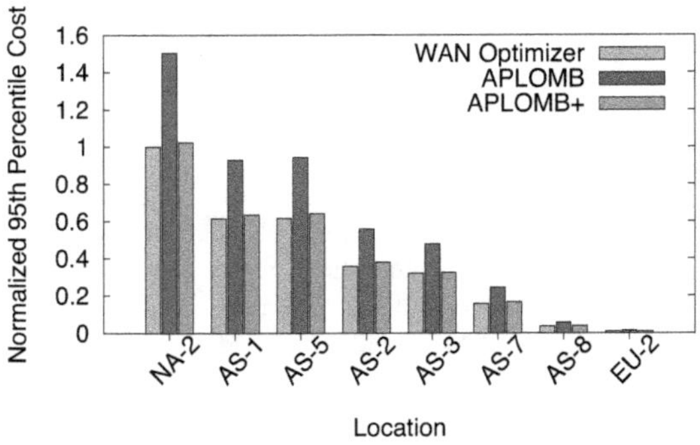

Figure 16: 95th percentile bandwidth without APLOMB, with APLOMB, and with APLOMB+.

Cost Reduction: To evaluate benefits from elastic scaling, in Figure 15 we focus on each site of the enterprise and show the ratio of peak-to-average volumes for total inter-site traffic. We use sites across three continents: North America (NA-x), Asia (AS-x), and Europe (EU-x). The peak represents a conservative estimate of the traffic volume the enterprise has provisioned at the site, while the average is the typical utilization; we see that most sites are provisioned over $2\times$ their typical load, and some of the smaller sites as much as $12\times$! In addition, we show peak-to-average values for the top four protocols in use. The per-protocol numbers are indicative of elasticity savings per middlebox, as different protocols are likely to traverse different middleboxes.

Latency: We measured redirection latency for inter-site traffic between the top eleven sites of the enterprise through the APLOMB infrastructure by pinging hosts at each site from within EC2. We found that for more than 60% of inter-site pairs, the latency with redirection is almost identical to the direct RTT. We found that most sites with inflated latency were in Asia, where EC2 does not have a wide footprint.

We also calculated a weighted inflation value, weighted by traffic volume and found that in expectation a typical redirected packet experiences only 1.13 ms of inflation. This results from the fact that the inter-site pairs with high traffic volume actually have negative inflation, by virtue of one or both endpoints being in the US or Europe, where EC2's footprint and connectivity is high.

Bandwidth: Last, we evaluate bandwidth inflation. We ran a traffic trace with full packet payloads collected at a different small enterprise [38] through our APLOMB prototype with and without generic redundancy elimination. Without Generic RE, the bandwidth utilization increased by 6.2% due to encryption and encapsulation overhead. With Generic RE, the bandwidth utilization reduced by 28%, giving APLOMB+ a 32% improvement over basic APLOMB.

As we observed in §2, many larger enterprises already compress their inter-site traffic using WAN optimizers. To evaluate the impact of switching compression for inter-site traffic from a traditional WAN optimizer solution to APLOMB+, we compared our observed benefits to those provided by WAN optimizers at eight of the large enterprise sites. In Figure 16, we measure the bandwidth cost of a given site in terms of the 95th percentile of the total traffic volume with a WAN Optimizer, with APLOMB, and with APLOMB+. With APLOMB, the worst case inflation is 52% in the median case and at most 58%; APLOMB+ improves this to a median case of 3.8% inflation and a worst case of 8.1%.

6. DISCUSSION

Before concluding, we mention some final thoughts on the future of "hybrid" enterprise/cloud architectures, potential cost models for bandwidth, and security challenges that continue to face APLOMB and cloud computing.

IT Outsourcing and Hybrid Clouds: APLOMB complements the ongoing move by enterprises from locally-hosted and managed infrastructure to outsourced cloud infrastructure. A network administrator at one large enterprise we surveyed reported their company's management had issued a broad mandate to moving a significant portion of their IT infrastructure to the cloud. Federal government agencies are also rapidly moving their IT infrastructure to the cloud, in compliance with a mandate to adopt a "cloud first" policy for new services and to reduce the number of existing federal datacenters by 800 before 2015 [37]. As these services move to the cloud, the middleboxes protecting them (including internal firewalls, which APLOMB itself cannot outsource) will move to the cloud as well.

Nevertheless, many enterprises plan to keep at least some local infrastructure, citing security and performance concerns for applications currently deployed locally [18]. Further, user-facing devices such as laptops, desktops, smartphones, and printers will always remain within the enterprise – and the majority of middlebox services benefit these devices rather than servers. With some end hosts moving to the cloud, and the majority remaining behind in the enterprise, multiple vendors now offer services for integrating public cloud services with enterprises' existing infrastructure [1, 3], facilitating so-called "hybrid clouds" [32]. APLOMB allows administrators to evade the middlebox-related complexity in this hybrid model by consolidating middleboxes in only one deployment setting.

Bandwidth Costs: APLOMB reduces the cost of middlebox infrastructure, but it may increase bandwidth costs due to current cloud business models. Today, tunneling traffic to a cloud provider necessitates paying for bandwidth twice – once for the enterprise network's access link, and again at the cloud provider. Nevertheless, this does not mean that APLOMB will double bandwidth costs for an enterprise. We observed earlier that redundancy elimination and compression can reduce bandwidth demands at the enterprise access link by roughly 30%. This optimization is not possible without redirection through a cloud PoP, and could allow a lower capacity, less expensive access link to satisfy an enterprise's needs.

The largest factor in the cost of APLOMB for an enterprise is the bandwidth cost model used by a cloud provider. Today, cloud providers price bandwidth purely by volume; for example, Amazon EC2 charges between \$0.05-\$0.12 per GB of outgoing traffic, decreasing as volume increases (all incoming traffic is free). On the other hand, a dedicated APLOMB service provider would be able to take advantage of wholesale bandwidth, which is priced by transfer rate. We convert between the two pricing strategies (per-GB and per-Mbps) with the rough conversion factor of 1Mbps sustained monthly throughput equaling 300GB per month. This is in comparison with "wholesale" bandwidth prices of \$3-\$5 per Mbps for high-volume customers. As a result, though current pricing strategies are not well-suited for APLOMB, a dedicated APLOMB provider could offer substantially lower prices. Indeed, Amazon offers a bulk-priced bandwidth service, "DirectConnect", which offers substantially lower per-GB costs for high-volume customers [1]. Table 3 provides a comparison of the bandwidth costs for a hypothetical enterprise which transfers 500TB of traffic per month to and from a cloud service provider under each of these models. These charges a minimal compared to expected savings in hardware, personnel, and other management costs.

Pricing Model	Total Cost	\$/GB	\$/Mbps
Standard EC2	30003.20	0.0586	17.58
Amazon DirectConnect	11882.50	0.0232	6.96
Wholesale Bandwidth	6826.70	0.0133	4.00

Table 3: Cost comparison of different cloud bandwidth pricing models given an enterprise with a monthly transfer volume of 500TB (an overestimate as compared to the very large enterprise in our study); assumes conversion rate of 1Mbps of sustained transfer equals 300GB over the course of a month.

Security Challenges: Adopting APLOMB brings with it the same security questions as have challenged cloud computing. These challenges have not stopped widespread adoption of cloud computing services, nor the willingness of security certification standards to certify cloud services (for example, services on Amazon EC2 can achieve PCI-1 compliance, the highest level of certification for storing credit card data). However, these challenges remain concerns for APLOMB and cloud computing in general. Just as cloud storage services have raised questions about providing a cloud provider unencrypted access to data, cloud middlebox services give the cloud provider unencrypted access to traffic flows. Although VMs and other isolation techniques aim to protect customers of a cloud service from other, malicious, customers, some have demonstrated in the cloud computing context information leakage, *e.g.* through side-channels [40]. APLOMB encrypts tunneled traffic to and from the enterprise to protect against man-in-the-middle attacks, and allocates each client it's own set of VMs for middlebox processing, but ultimately it will not appeal to companies whose security policies restrict them from cloud computing in general.

7. RELATED WORK

Our work contributes to and draws inspiration from a rich corpus of work in cloud computing, redirection services, and network management.

Cloud Computing: The motivation for APLOMB parallels traditional arguments in favor of cloud computing, many of which are discussed by Armbrust et al. [23]. APLOMB also adapts techniques from traditional cloud solutions, *e.g.* utilization monitoring and dynamic scaling [12], and DNS-based redirection to datacenters with optimal performance for the customer [44].

Middlebox Management: Others have tackled middlebox management challenges within the enterprise [33, 34, 24, 27, 42]. Their solutions offer insights we can apply for managing middleboxes within the cloud – *e.g.*, the policy-routing switch of Joseph et al. [34], the management plane of Ballani et al. [24], and the consolidated appliance of Sekar et al. [42]. None of these proposals consider moving middlebox management out of the enterprise entirely, as we do. Like us, ETTM [27] proposes removing middleboxes from the enterprise network but, where we advocate moving them to the cloud, ETTM proposes the opposite: pushing middlebox processing to enterprise end hosts. As such, ETTM still retains the problem of middlebox management in the enterprise. Sekar et al [42] report on the middlebox deployment of a single large enterprise; our survey is broader in scope (covering a range of management and failure concerns) and covers 57 networks of various scales. They also propose a consolidated middlebox architecture that aims to ameliorate some of the administrative burden associated with middlebox management, but they do not go so far as to propose removing middleboxes from the enterprise network entirely.

Redirection Services: Traffic redirection infrastructures have been explored in prior work [22, 43, 47] but in the context of improving Internet or overlay routing architectures as opposed to APLOMB's goal of enabling middlebox processing in the cloud. RON showed how routing via an intermediary might improve latency; we report similar findings using cloud PoPs as intermediaries. Walfish et al. [47] propose a clean-slate architecture, DOA, by which end hosts explicitly address middleboxes. Gibb et al. [30] develop a service model for middleboxes that focuses on service-aware routers that redirect traffic to middleboxes that can be in the local network or Internet.

Cloud Networking: Using virtual middlebox appliances [16] reduces the physical hardware cost of middlebox ownership, but cannot match the performance of hardware solutions and does little to improve configuration complexity. Some startups and security companies have cloud-based offerings for specific middlebox services: Aryaka [5] offers protocol acceleration; ZScalar [17] performs intrusion detection; and Barracuda Flex [6] offers web security. To some extent, our work can be viewed as an extreme extrapolation of their services and we provide a comprehensive exploration and evaluation of such a trend. CloudNaaS [25] and startup Embrane [8] aim at providing complete middlebox solutions for enterprise services that are *already* in the cloud.

8. CONCLUSION

Outsourcing middlebox processing to the cloud relieves enterprises of major problems caused by today's enterprise middlebox infrastructure: cost, management complexity, capacity rigidity, and failures. Our survey of 57 enterprise network managers guides the design of APLOMB, a practical system for middlebox processing in the cloud. APLOMB succeeds in outsourcing the vast majority of middleboxes from a typical enterprise network without impacting performance, making scalable, affordable middlebox processing accessible to enterprise networks of every size.

Acknowledgements

We thank the anonymous reviewers for their helpful feedback; our shepherd, Jacobus Van der Merwe, for his guidance in developing the final draft of our paper; the NANOG operators for their enthusiastic participation in our survey; Neil Doran, Zhixin Tang, and Mark Poepping for helping us refine initial versions of our survey; and Ashok Anand for sharing the WAN optimization modules.

9. REFERENCES

[1] Amazon Direct Connect. http://aws.amazon.com/directconnect/.
[2] Amazon Route 53. http://aws.amazon.com/route53.
[3] Amazon Virtual Private Cloud. http://aws.amazon.com/vpc/.
[4] Amazon Web Services launches Brazil datacenters for its cloud computing platform. http://phx.corporate-ir.net/phoenix.zhtml?c=176060&p=irol-newsArticle&ID=1639908.
[5] Aryaka WAN Optimization. http://www.aryaka.com.
[6] Barracuda Web Security Flex. http://www.barracudanetworks.com/ns/products/web_security_flex_overview.php.
[7] Cisco: Quality of Service Design Overview. http://www.ciscopress.com/articles/article.asp?p=357102.
[8] Embrane. http://www.embrane.com/.
[9] Network Monitoring Tools. http://slac.stanford.edu/xorg/nmtf/nmtf-tools.html.
[10] OpenVPN. http://www.openvpn.com.
[11] Palo Alto Networks. http://www.paloaltonetworks.com/.
[12] Rightscale Cloud management. http://www.rightscale.com/.
[13] Riverbed Virtual Steelhead. http://www.riverbed.com/us/products/steelhead_appliance/virtual_steelhead.php.
[14] Symantec: Data Loss Protection. http://www.vontu.com.
[15] Tivoli Monitoring Software. http://ibm.com/software/tivoli/products/monitor.
[16] Vyatta Software Middlebox. http://www.vyatta.com.
[17] ZScaler Cloud Security. http://www.zscaler.com.
[18] Cloud computing - 31 companies describe their experiences. http://www.ipanematech.com/information-center/download.php?link=white-papers/White%20Book_2011-Cloud_Computing_OBS_Ipanema_Technologies_EBG.pdf, 2011.
[19] Enterprise Network and Data Security Spending Shows Remarkable Resilience. http://www.abiresearch.com/press/3591, 2011.
[20] M. Allman and V. Paxson. TCP congestion control. RFC 5681.
[21] A. Anand, A. Gupta, A. Akella, S. Seshan, and S. Shenker. Packet Caches on Routers: The Implications of Universal Redundant Traffic Elimination. In *Proc. of SIGCOMM*, 2008.
[22] D. Andersen, H. Balakrishnan, F. Kaashoek, and R. Morris. Resilient overlay networks. In *SOSP*, 2001.
[23] M. Armbrust et al. A view of cloud computing. *Commun. ACM*, April 2010.
[24] H. Ballani and P. Francis. CONMan: a step towards network manageability. In *SIGCOMM*, 2007.
[25] T. Benson, A. Akella, A. Shaikh, and S. Sahu. Cloudnaas: a cloud networking platform for enterprise applications. In *Proc. SOCC*, 2011.
[26] D. R. Choffnes and F. E. Bustamante. Taming the torrent: a practical approach to reducing cross-isp traffic in peer-to-peer systems. In *SIGCOMM*, 2008.
[27] C. Dixon, H. Uppal, V. Brajkovic, D. Brandon, T. Anderson, and A. Krishnamurthy. ETTM: a scalable fault tolerant network manager. In *NSDI*, 2011.
[28] N. Dukkipati and N. McKeown. Why flow-completion time is the right metric for congestion control. *CCR*, January 2006.
[29] S. Floyd. HighSpeed TCP for large congestion windows. RFC 3649.
[30] G. Gibb, H. Zeng, and N. McKeown. Outsourcing network functionality. In *HotSDN*, 2012.
[31] K. P. Gummadi, H. V. Madhyastha, S. D. Gribble, H. M. Levy, and D. Wetherall. Improving the reliability of Internet paths with One-hop Source Routing. In *Proc. OSDI*, 2004.
[32] M. Hajjat, X. Sun, Y.-W. E. Sung, D. A. Maltz, S. Rao, K. Sripanidkulchai, and M. Tawarmalani. Cloudward bound: Planning for beneficial migration of enterprise applications to the cloud. In *SIGCOMM*, 2012.
[33] D. Joseph and I. Stoica. Modeling middleboxes. *Network, IEEE*, 22(5), 2008.
[34] D. A. Joseph, A. Tavakoli, and I. Stoica. A policy-aware switching layer for data centers. In *SIGCOMM*, 2008.
[35] D. Katabi, M. Handley, and C. Rohrs. Congestion control for high bandwidth-delay product networks. In *SIGCOMM*, 2002.
[36] E. Kohler, R. Morris, B. Chen, J. Jannotti, and M. F. Kaashoek. The Click modular router. *ACM ToCS*, August 2000.
[37] V. Kundra. 25 Point Implementation Plan to Reform Federal Information Technology Management. Technical report, US CIO, 2010.
[38] M57 packet traces. https://domex.nps.edu/corp/scenarios/2009-m57/net/.
[39] N. McKeown et al. OpenFlow: enabling innovation in campus networks. *CCR*, March 2008.
[40] T. Ristenpart, E. Tromer, H. Shacham, and S. Savage. Hey, you, get off of my cloud: exploring information leakage in third-party compute clouds. In *CCS*, 2009.
[41] M. Roesch. Snort - Lightweight Intrusion Detection for Networks. In *LISA*, 1999.
[42] V. Sekar, S. Ratnasamy, M. K. Reiter, N. Egi, and G. Shi. The middlebox manifesto: enabling innovation in middlebox deployment. In *HotNets*, 2011.
[43] I. Stoica et al. Internet indirection infrastructure. *ToN*, April 2004.
[44] A. Su, D. Choffnes, A. Kuzmanovic, and F. Bustamante. Drafting behind Akamai (Travelocity-based detouring). In *SIGCOMM*, 2006.
[45] V. Valancius, N. Laoutaris, L. Massouli'e, C. Diot, and P. Rodriguez. Greening the internet with nano data centers. In *Proc. CoNEXT*, 2009.
[46] Visolve. Transparent caching using Squid. http://www.visolve.com/squid/whitepapers/trans_caching.pdf, 2006.
[47] M. Walfish, J. Stribling, M. Krohn, H. Balakrishnan, R. Morris, and S. Shenker. Middleboxes no longer considered harmful. In *OSDI*, 2004.

HyperDex: A Distributed, Searchable Key-Value Store

Robert Escriva
Computer Science Department
Cornell University
escriva@cs.cornell.edu

Bernard Wong
Cheriton School of Computer Science
University of Waterloo
bernard@uwaterloo.ca

Emin Gün Sirer
Computer Science Department
Cornell University
egs@systems.cs.cornell.edu

ABSTRACT

Distributed key-value stores are now a standard component of high-performance web services and cloud computing applications. While key-value stores offer significant performance and scalability advantages compared to traditional databases, they achieve these properties through a restricted API that limits object retrieval—an object can only be retrieved by the (primary and only) key under which it was inserted. This paper presents HyperDex, a novel distributed key-value store that provides a unique `search` primitive that enables queries on secondary attributes. The key insight behind HyperDex is the concept of *hyperspace hashing* in which objects with multiple attributes are mapped into a multidimensional hyperspace. This mapping leads to efficient implementations not only for retrieval by primary key, but also for partially-specified secondary attribute searches and range queries. A novel chaining protocol enables the system to achieve strong consistency, maintain availability and guarantee fault tolerance. An evaluation of the full system shows that HyperDex is 12-13× faster than Cassandra and MongoDB for finding partially specified objects. Additionally, HyperDex achieves 2-4× higher throughput for `get`/`put` operations.

Categories and Subject Descriptors

D.4.7 [**Operating Systems**]: Organization and Design

Keywords

Key-Value Store, NoSQL, Fault-Tolerance, Strong Consistency, Performance

1. INTRODUCTION

Modern distributed applications are reshaping the landscape of storage systems. Recently emerging distributed key-value stores such as BigTable [11], Cassandra [32] and Dynamo [19] form the backbone of large commercial applications because they offer scalability and availability properties that traditional database systems simply cannot provide. Yet these properties come at a substantial cost: the data retrieval API is narrow and restrictive, permitting an object to be retrieved using only the key under which it was stored, and the consistency guarantees are often quite weak. Queries based on secondary attributes are either not supported, utilize costly secondary indexing schemes or enumerate all objects of a given type.

This paper introduces HyperDex, a high-performance, scalable, consistent and distributed key-value store that provides a new `search` primitive for retrieving objects by secondary attributes. HyperDex achieves this extended functionality by organizing its data using a novel technique called *hyperspace hashing*. Similar to other hashing techniques [23, 29, 36, 47], hyperspace hashing deterministically maps objects to servers to enable efficient object insertion and retrieval. But it differs from these techniques because it takes into account the secondary attributes of an object when determining the mapping for an object. Specifically, it maps objects to coordinates in a multi-dimensional Euclidean space – a *hyperspace* – which has axes defined by the objects' attributes. Each server in the system is mapped onto a region of the same hyperspace, and owns the objects that fall within its region. Clients use this mapping to deterministically insert, remove, and search for objects.

Hyperspace hashing facilitates efficient search by significantly reducing the number of servers to contact for each partially-specified `search`. The construction of the hyperspace mapping guarantees that objects with matching attribute values will reside on the same server. Through geometric reasoning, clients can restrict the search space for a partially-specified query to a subset of servers in the system, thereby improving search efficiency. Specificity in searches works to the clients' advantage: a fully-specified search contacts exactly one server.

A naive hyperspace construction, however, may suffer from a well-known problem with multi-attribute data known as "curse of dimensionality [6]." With each additional secondary attribute, the hyperspace increases in volume exponentially. If constructed in this fashion, each server would be responsible for a large volume of the resulting hyperspace, which would in turn force search operations to contact a large number of servers, counteracting the benefits of hyperspace hashing. HyperDex addresses this problem by partitioning the data into smaller, limited size *subspaces* of fewer dimensions.

Failures are inevitable in all large-scale deployments. The standard approaches for providing fault tolerance store objects on a fixed set of replicas determined by a primary key. These techniques, whether they employ a consensus algorithm among the replicas and provide strong consis-

SIGCOMM'12, August 13–17, 2012, Helsinki, Finland.

tency [18,46], or spray the updates to the replicas and only achieve eventual consistency [19,32,45,49], assume that the replica sets remain fixed even as the objects are updated. Such techniques are not immediately suitable in our setting because, in hyperspace hashing, object attributes determine the set of servers on which an object resides, and consequently, each update may implicitly alter the replica set. Providing strong consistency guarantees with low overhead is difficult, and more so when replica sets change dynamically and frequently. HyperDex utilizes a novel replication protocol called *value-dependent chaining* to simultaneously achieve fault tolerance, high performance and strong consistency. Value-dependent chaining replicates an object to withstand f faults (which may span server crashes and network partitions) and ensures linearizability, even as replica sets are updated. Thus, HyperDex's replication protocol guarantees that all `get` operations will immediately see the result of the last completed `put` operation – a stronger consistency guarantee than those offered by the current generation of NoSQL data stores.

Overall, this paper describes the architecture of a new key-value store whose API is one step closer to that of traditional RDBMSs while offering strong consistency guarantees, fault-tolerance for failures affecting a threshold of servers and high performance, and makes three contributions. First, it describes a new hashing technique for mapping structured data to servers. This hashing technique enables efficient retrieval of multiple objects even when the search query specifies both equality and range predicates. Second, it describes a fault-tolerant, strongly-consistent replication scheme that accommodates object relocation. Finally, it reports from a full implementation of the system and deployment in a data center setting consisting of 64 servers, and demonstrates that HyperDex provides performance that is comparable to or better than Cassandra and MongoDB, two current state-of-the-art cloud storage systems, as measured using the industry-standard YCSB [15] benchmark. More specifically, HyperDex achieves 12-13× higher throughput for search workloads than the other systems, and consistently achieves 2-4× higher throughput for traditional key-value workloads.

The rest of this paper is structured as follows: Section 2 describes hyperspace hashing. Section 3 specifies how to deal with spaces of high dimensionality through data partitioning. Section 4 specifies the value-dependent replication protocol used in HyperDex. Section 5 outlines our full implementation of HyperDex. Section 6 evaluates HyperDex under a variety of workloads. Section 7 discusses related work for hyperspace hashing and HyperDex. We discuss how our system relates to the CAP Theorem in Section 8 and conclude with a summary of our contributions.

2. APPROACH

In this section, we describe the data model used in HyperDex, outline hyperspace hashing, and sketch the high-level organization and operation of the system.

2.1 Data Model and API

HyperDex stores objects that consist of a key and zero or more secondary attributes. As in a relational database, HyperDex objects match an application-provided schema that defines the typed attributes of the object and are persisted in tables. This organization permits straightforward migration from existing key-value stores and database systems.

HyperDex provides a rich API that supports a variety of datastructures and a wide range of operations. The system natively supports primitive types, such as `strings`, `integers` and `floats`, as well as composite types, such as `lists`, `sets` or `maps` constructed from primitive types. The dozens of operations that HyperDex provides on these datatypes fall into three categories. First, *basic operations*, consisting of `get`, `put`, and `delete`, enable a user to retrieve, update, and destroy an object identified by its key. Second, the `search` operation enables a user to specify zero or more ranges for secondary attributes and retrieve the objects whose attributes fall within the specified, potentially singleton, ranges. Finally, a large set of atomic operations, such as `cas` and `atomic-inc`, enable applications to safely perform concurrent updates on objects identified by their keys. Since composite types and atomic operations are beyond the scope of this paper, we focus our discussion on the `get`, `put`, `delete`, and `search` operations that form the core of HyperDex.

2.2 Hyperspace Hashing

HyperDex represents each table as an independent multi-dimensional space, where the dimensional axes correspond directly to the attributes of the table. HyperDex assigns every object a corresponding coordinate based on the object's attribute values. An object is mapped to a deterministic coordinate in space by hashing each of its attribute values to a location along the corresponding axis.

Consider, for the following discussion, a table containing user information that has the attributes **first-name**, **last-name**, and **telephone-number**. For this schema, HyperDex would create a three dimensional space where the first-name attribute comprises the x-axis, the last-name attribute comprises the y-axis, and the telephone-number attribute comprises the z-axis. Figure 1 illustrates this mapping.

Hyperspace hashing determines the object to server mapping by tessellating the hyperspace into a grid of non-overlapping regions. Each server is assigned, and is responsible for, a specific region. Objects whose coordinates fall within a region are stored on the corresponding server. Thus, the hyperspace tessalation serves like a multi-dimensional hash bucket, mapping each object to a unique server.

The tessalation of the hyperspace into regions (called the *hyperspace mapping*), as well as the assignment of the regions to servers, is performed by a fault-tolerant coordinator. The primary function of the coordinator is to maintain the hyperspace mapping and to disseminate it to both servers and clients. The hyperspace mapping is initially created by dividing the hyperspace into hyperrectangular regions and assigning each region to a virtual server. The coordinator is then responsible for maintaining this mapping as servers fail and new servers are introduced into the system.

The geometric properties of the hyperspace make object insertion and deletion simple. To insert an object, a client computes the coordinate for the object by hashing each of the object's attributes, uses the hyperspace mapping to determine the region in which the object lies, and contacts that server to store the object. The hyperspace mapping obviates the need for server-to-server routing.

2.3 Search Queries

The hyperspace mapping described in the preceding sections facilitates a geometric approach to resolving `search`

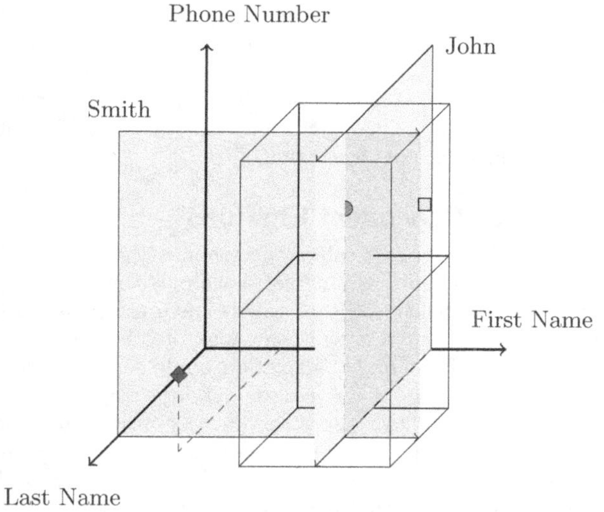

Figure 1: Simple hyperspace hashing in three dimensions. Each plane represents a query on a single attribute. The plane orthogonal to the axis for "Last Name" passes through all points for `last_name = 'Smith'`, while the other plane passes through all points for `first_name = 'John'`. Together they represent a line formed by the intersection of the two search conditions; that is, all phone numbers for people named "John Smith". The two cubes show regions of the space assigned to two different servers. The query for "John Smith" needs to contact only these servers.

operations. In HyperDex, a `search` specifies a set of attributes and the values that they must match (or, in the case of numeric values, a range they must fall between). HyperDex returns objects which match the `search`. Each `search` operation uniquely maps to a *hyperplane* in the hyperspace mapping. A `search` with one attribute specified maps to a hyperplane that intersects that attribute's axis in exactly one location and intersects all other axes at every point. Alternatively, a `search` that specifies all attributes maps to exactly one point in hyperspace. The hyperspace mapping ensures that each additional search term potentially reduces the number of servers to contact and guarantees that additional search terms will not increase `search` complexity.

Clients maintain a copy of the hyperspace mapping, and use it to deterministically execute `search` operations. A client first maps the `search` into the hyperspace using the mapping. It then determines which servers' regions intersect the resulting hyperplane, and issues the `search` request to only those servers. The client may then collect matching results from the servers. Because the hyperspace mapping maps objects and servers into the same hyperspace, it is never necessary to contact any server whose region does not intersect the `search` hyperplane.

Range queries correspond to extruded hyperplanes. When an attribute of a `search` specifies a range of values, the corresponding hyperplane will intersect the attribute's axis at every point that falls between the lower and upper bounds of the range. Note that for such a scheme to work, objects' relative orders for the attribute must be preserved when mapped onto the hyperspace axis.

Figure 1 illustrates a query for `first_name = 'John'` and `last_name = 'Smith'`. The query for `first_name = 'John'` corresponds to a two-dimensional plane which intercepts the `first_name` axis at the hash of `'John'`. Similarly, the query for `last_name = 'Smith'` creates another plane which intersects the `last_name` axis. The intersection of the two planes is the line along which all phone numbers for John Smith reside. Since a search for John Smith in a particular area code defines a line segment, a HyperDex search needs to contact only those nodes whose regions intersect that segment.

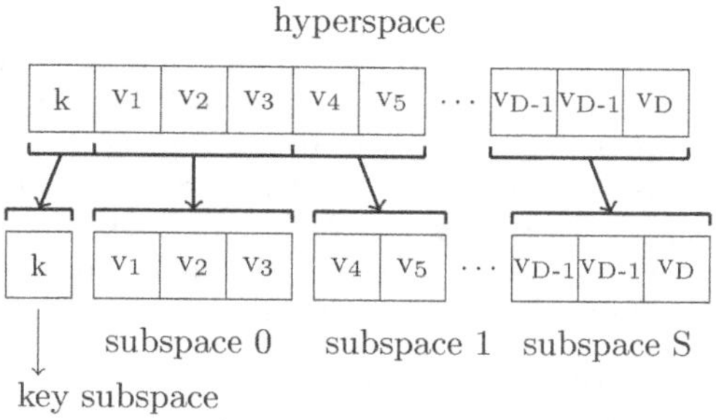

Figure 2: HyperDex partitions a high-dimensional hyperspace into multiple low-dimension subspaces.

3. DATA PARTITIONING

HyperDex's Euclidean space construction significantly reduces the set of servers that must be contacted to find matching objects.

However, the drawback of coupling the dimensionality of hyperspace with the number of searchable attributes is that, for tables with many searchable attributes, the hyperspace can be very large since its volume grows exponentially with the number of dimensions. Covering large spaces with a grid of servers may not be feasible even for large data-center deployments. For example, a table with 9 secondary attributes may require 2^9 regions or more to support efficient searches. In general, a D dimensional hyperspace will need $O(2^D)$ servers.

HyperDex avoids the problems associated with high-dimensionality by partitioning tables with many attributes into multiple lower-dimensional hyperspaces called *subspaces*. Each of these subspaces uses a subset of object attributes as the dimensional axes for an independent hyperspace. Figure 2 shows how HyperDex can partition a table with D attributes into multiple independent subspaces. When performing a `search` on a table, clients select the subspace that contacts the fewest servers, and will issue the `search` to servers in exactly one subspace.

Data partitioning increases the efficiency of a `search` by reducing the dimensionality of the underlying hyperspace. In a 9-dimensional hyperspace, a `search` over 3 attributes would need to contact 64 regions of the hyperspace (and thus, 64 servers). If, instead, the same table were partitioned into 3 subspaces of 3 attributes each, the `search` will never contact more than 8 servers in the worst case, and exactly one server in the best case. By partitioning the table, HyperDex reduces the worst case behavior, decreases the number of servers necessary to maintain a table, and increases the likelihood that a `search` is run on exactly one server.

Data partitioning forces a trade-off between search generality and efficiency. On the one hand, a single hyperspace can accommodate arbitrary searches over its associated attributes. On the other hand, a hyperspace which is too large will always require that partially-specified queries contact many servers. Since applications often exhibit search locality, HyperDex applications can tune search efficiency by creating corresponding subspaces. As the number of subspaces grows, so, too, do the costs associated with maintaining data consistency across subspaces. Section 4 details how HyperDex efficiently maintains consistency across subspaces while maintaining a predictably low overhead.

3.1 Key Subspace

The basic hyperspace mapping, as described so far, does not distinguish the key of an object from its secondary attributes. This leads to two significant problems when implementing a practical key-value store. First, key lookups would be equivalent to single attribute searches. Although HyperDex provides efficient search, a single attribute search in a multi-dimensional space would likely involve contacting more than one server. In this hypothetical scenario, key operations would be strictly more costly than key operations in traditional key-value stores.

HyperDex provides efficient key-based operations by creating a one-dimensional subspace dedicated to the key. This subspace, called the *key subspace*, ensures that each object will map to exactly one region in the resulting hyperspace. Further, this region will not change as the object changes because keys are immutable. To maintain the uniqueness invariant, `put` operations are applied to the key subspace before the remaining subspaces.

3.2 Object Distribution Over Subspaces

Subspace partitioning exposes a design choice in how objects are distributed and stored on servers. One possible design choice is to keep data in normalized form, where every subspace retains, for each object, only those object attributes that serve as the subspace's dimensional axes. While this approach would minimize storage requirements per server, as attributes are not duplicated across subspaces, it would lead to more expensive search and object retrieval operations since reconstituting the object requires cross-server cooperation. In contrast, an alternative design choice is to store a full copy of each object in each subspace, which leads to faster search and retrieval operations at the expense of additional storage requirements per server.

Hyperspace hashing supports both of these object distribution techniques. The HyperDex implementation, however, relies upon the latter approach to implement the replication scheme described in Section 4.

3.3 Heterogeneous Objects

In a real deployment, the key-value store will likely be used to hold disparate objects with different schema. HyperDex supports this through the table abstraction. Each table has a separate set of attributes which make up the objects within, and these attributes are partitioned into subspaces independent of all other tables. As a result, HyperDex manages multiple independent hyperspaces.

4. CONSISTENCY AND REPLICATION

Because hyperspace hashing maps each object to multiple servers, maintaining a consistent view of objects poses a challenge. HyperDex employs a novel technique called *value-dependent chaining* to provide strong consistency and fault tolerance in the presence of concurrent updates.

For clarity, we first describe value-dependent chaining without concern for fault tolerance. Under this scheme, a single failure leaves portions of the hyperspace unavailable for updates and searches. We then describe how value-dependent chaining can be extended such that the system can tolerate up to f failures in any one region.

4.1 Value Dependent Chaining

Because hyperspace hashing determines the location of an object by its contents, and subspace partitioning creates many object replicas, objects will be mapped to multiple servers and these servers will change as the objects are updated. Change in an object's location would cause problems if implemented naively. For example, if object updates were to be implemented by simply sending the object to all affected servers, there would be no guarantees associated with subsequent operations on that object. Such a scheme would at best provide eventual consistency because servers may receive updates out-of-order, with no sensible means of resolving concurrent updates.

HyperDex orders updates by arranging an object's replicas into a value-dependent chain whose members are deterministically chosen based upon an object's hyperspace coordinate. The head of the chain is called the *point leader*, and is determined by hashing the object in the key subspace. Subsequent servers in the chain are determined by hashing attribute values for each of the remaining subspaces.

This construction of value-dependent chains enables efficient, deterministic propagation of updates. The point leader for an object is in a position to dictate the total order on all updates to that object. Each update flows from the point leader through the chain, and remains pending until an acknowledgement of that update is received from the next server in the chain. When an update reaches the tail, the tail sends an acknowledgement back through the chain in reverse so that all other servers may commit the pending update and clean up transient state. When the acknowledgement reaches the point leader, the client is notified that the operation is complete. In Figure 3, the update u_1 illustrates an object insertion which passes through h_1, h_2, h_3, where h_1 is the point leader.

Updates to preexisting objects are more complicated because a change in an attribute value might require relocating an object to a different region of the hyperspace. Value-dependent chains address this by incorporating the servers assigned to regions for both the old and new versions of the object. Chains are constructed such that servers are ordered by subspace and the servers corresponding to the old version of the object immediately precede the servers corresponding to the new version. This guarantees that there is no instant during an update where the object may disappear from the data store. For example, in Figure 3, the update u_2 modifies the object in a way that changes its mapping in Subspace 0 such that the object no longer maps to h_2 and instead maps to h_5. The value-dependent chain for update u_2 is h_1, h_2, h_5, h_3. The update will result in the object being stored at h_5, and subsequently removed from h_2, as acknowledgments propagate in reverse.

Successive updates to an object will construct chains which overlap in each subspace. Consequently, concurrent updates may arrive out of order at each of these points of overlap. For example, consider the update u_3 in Figure 3. The value-dependent chain for this update is h_1, h_5, h_3. Notice that it

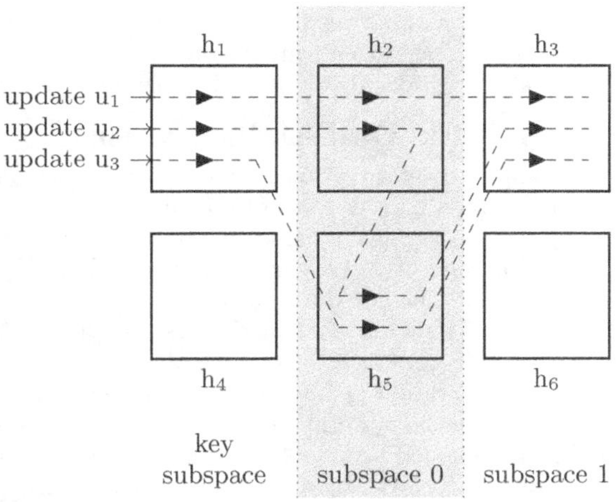

Figure 3: HyperDex's replication protocol propagates along value-dependent chains. Each update has a value-dependent chain that is determined solely by objects' current and previous values and the hyperspace mapping.

is possible for u_3 to arrive at h_5 before u_2. If handled improperly, such races could lead to inconsistent, out-of-order updates. Value-dependent chains efficiently handle this case by dictating that the point leader embed, in each update, dependency information which specifies the order in which updates are applied. Specifically, the point leader embeds a version number for the update, and the version number, hash and type of the previous update. For instance, u_3 will have a version number of 3 and depend upon update u_2 with version number 2 and type `put`. Servers which receive u_3 before u_2 will know to delay processing of u_3 until u_2 is also processed.

By design, HyperDex supports destructive operations that remove all state pertaining to deleted objects. Examples of destructive operations include `delete` and the cleanup associated with object relocation. Such operations must be carefully managed to ensure that subsequent operations get applied correctly. For instance, consider a `del` followed by a `put`. Since we would like a `del` to remove all state, the `put` must be applied on servers with no state. Yet, if another `del`/`put` pair were processed concurrently, servers which had processed either `del` would not be able to properly order the `put` operations. Value-dependent chains ensure that concurrently issued destructive operations are correctly ordered on all servers. Each server independently delays operations which depend upon a destructive operation until the destructive operation, and all that came before it, are acknowledged. This ensures that at most one destructive operation may be in-flight at any one time and guarantees that they will be ordered correctly. The delay for each message is bounded by the length of chains, and the number of concurrent operations.

4.2 Fault Tolerance

To guard against server failures, HyperDex provides additional replication within each region. The replicas acquire and maintain their state by being incorporated into value-dependent chains. In particular, each region has $f+1$ replicas which appear as a block in the value-dependent chain. For example, we can extend the layout of Figure 3 to tolerate one failure by introducing additional hosts h_1' through h_6'. As with regular chain replication [57], new replicas are introduced at the tail of the region's chain, and servers are bumped forward in the chain as other servers fail. For example, the first update in Figure 3 has the value-dependent chain $h_1, h_1', h_2, h_2', h_3, h_3'$. If h_2 were to fail, the resulting chain would be $h_1, h_1', h_2', h_2'', h_3, h_3'$. This transition will be performed without compromising strong consistency.

Point leader failures do not allow clients to observe an inconsistency. For instance, if h_1, the point leader, were to fail in our previous example, h_1' will take over the role of point leader. When a client detects a point leader failure, it will notify the application and preserve at-most-once semantics. Further, because all client acknowledgments are generated by the point leader, the client will only see a response after an object is fully fault-tolerant.

In our implementation, HyperDex uses TCP to transmit data which ensures that messages need only be retransmitted when servers fail and are removed from the value-dependent chain. In response to failures, HyperDex servers retransmit messages along the chains to make progress. Upon chain reconfiguration, servers will no longer accept messages from failed servers, ensuring that all future messages traverse the new, valid chain.

4.3 Server and Configuration Management

HyperDex utilizes a logically centralized coordinator to manage system-wide global state. The coordinator encapsulates the global state in the form of a *configuration* which consists of the hyperspace mapping between servers and regions and information about server failures. The coordinator assigns to each configuration an epoch identifier, a strictly increasing number that identifies the configuration. The coordinator distributes configurations to servers in order by epoch, and updates the configuration as necessary. HyperDex's coordinator holds no state pertaining to the objects themselves; only the mapping and servers.

HyperDex ensures that no server processes late-arriving messages from failed or out-of-date servers. Each HyperDex server process (an *instance*) is uniquely identified by its IP address, port and *instance id*. Instance ids are assigned by the coordinator and are globally unique, such that servers can distinguish between two instances (e.g. a failed process and a new one) that reuse the same IP and port number. HyperDex embeds into messages the instance ids, the regions of the hyperspace and indices into the chain for both the sender and recipient. A recipient acting upon a message can validate the sender and recipient against its most recent configuration. If the message contains a stale mapping, it will not be acted upon. If, instead, the mapping is valid, the host processes the message accordingly.

Each host changes its configuration in an all-or-nothing fashion which appears instantaneous to threads handling network communication. This is accomplished on each host by creating state relevant to the new configuration, pausing network traffic, swapping pointers to make the new state visible, and unpausing network traffic. This operation completes in sub-millisecond time on each host.

The logically centralized coordinator does not impose a bottleneck. Configurations are small in practice, and proportional to the size of the cluster and number of tables active on the cluster. Furthermore, in cases where bandwidth from the coordinator becomes a bottleneck, the coordinator need only distribute deltas to the configuration. Clients maintain a complete copy of the configuration in memory and perform local computation on the hyperspace mapping.

4.4 Consistency Guarantees

Overall, the preceding protocol ensures that HyperDex provides strong guarantees for applications. The specific guarantees made by HyperDex are:

Key Consistency All actions which operate on a specific key (e.g., **get** and **put**) are linearizable [26] with all operations on all keys. This guarantees that all clients of HyperDex will observe updates in the same order.

Search Consistency HyperDex guarantees that a search will return all objects that were committed at the time of search. An application whose put succeeds is guaranteed to see the object in a future search. In the presence of concurrent updates, a search may return both the committed version, and the newly updated version of an object matching the search.

HyperDex provides the strongest form of consistency for key operations, and a conservative and predictable consistency guarantees for search operations.

5. IMPLEMENTATION

HyperDex is fully implemented to support all the features described in this paper. The implementation is nearly 44,000 lines of code. The HyperDex software distribution contains an embeddable storage layer called HyperDisk, a hyperspace hashing library, the HyperDex server, the client library and the HyperDex coordinator, as well as full client bindings for C, C++, and Python, and partial bindings for Java, Node.JS and Ruby.

This section describes key aspects of the HyperDex implementation.

5.1 HyperDex Server

High throughput and low latency access are essential for any key-value store. The HyperDex server achieves high performance through three techniques; namely, edge-triggered, event-driven I/O; pervasive use of lock-free datastructures and sharded locks to maximize concurrency; and careful attention to constant overheads. The HyperDex server is structured around a concurrent, edge-triggered event loop wherein multiple worker threads receive network messages and directly process client requests. Whereas common design patterns would distinguish between I/O threads and worker threads, HyperDex combines these two functions to avoid internal queuing and communication delays. Because the event-loop is edge-triggered, unnecessary interaction with the kernel is minimized. Socket buffer management ensures that threads never block in the kernel when sending a message and consumes a constant amount of memory per client.

HyperDex makes extensive use of lock-sharding and lock-free datastructures to reduce the probability of contention whenever possible. Per-key datastructures are protected by an array of locks. Although nothing prevents two threads from contending for the same lock to protect different keys, the ratio of locks to threads is high enough to reduce this occurrence to 1 in 10^6. Global datastructures, such as lookup tables for the per-key datastructures, are concurrent through the use of lock-free hash tables. Our implementation ensures that background threads may safely iterate over global state while worker threads insert and remove pointers to per-key state.

Finally, the use of cache-conscious, constant-time data structures reduces the overheads associated with common operations such as linked-list and hash-table management.

5.2 HyperDisk: On-Disk Data Storage

A key component of server performance for any key-value store is the storage back end used to organize data on disk. Since hyperspace hashing is agnostic to the choice of the back end, a number of design options are available. At one extreme, we could have used a traditional database to store all the objects of a server in a single, large, undifferentiated pool. While this approach would have been the simplest from an implementation perspective, it would make HyperDex dependent on the performance of a traditional database engine, require manual tuning of numerous parameters, and subject the system to the vagaries of a query optimizer.

Instead, HyperDex recursively leverages the hyperspace hashing technique to organize the data stored internally on a server. Called HyperDisk, this approach partitions the region associated with a server into smaller non-overlapping sub-regions, where a sub-region represents a file on disk, and objects are located in the file that contains their coordinate. Each file is arranged as a log, where insertions and deletions operate on the tail, and where a search operation linearly scans through the whole file.

HyperDisk's hashing scheme differs from the standard hyperspace hashing in two ways: first, HyperDisk partitions only the region assigned to a HyperDex server; and second, HyperDisk may alter the mapping dynamically to accommodate the number of objects stored within a region. Overall, recursive application of hyperspace hashing enables HyperDex to take advantage of the geometric structure of the data space at every system layer.

5.3 Distributed Coordination

In HyperDex, the hyperspace mapping is created and managed by a logically centralized coordinator. Since a physically centralized coordinator would limit scalability and pose a single point of failure, the HyperDex coordinator is implemented as a replicated state machine. It relies on a coordination service [2, 9, 27] to replicate the coordinator on multiple physical servers. The coordinator implementation ensures that servers may migrate between coordinators so that no coordinator failure leads to correlated failures in the system. The coordinator directs all failure recovery actions. Servers may report observed failures to the coordinator, or the coordinator may directly observe failures through periodic failure detection pings to servers.

Overall, the replicated state machine implementation ensures that the coordinator acts as a single, coherent component with well-defined state transitions, even though it is comprised of fault-tolerant, distributed components.

6. EVALUATION

We deployed HyperDex on both a small and medium-size computational cluster and evaluated the performance of each deployment using the Yahoo! Cloud Serving Benchmark (YCSB) [15], an industry-standard benchmark for cloud storage performance. Our evaluation also examines the performance of HyperDex's basic operations, specifically, **get**, **put**, and **search**, using targeted micro-benchmarks. These micro-benchmarks isolate specific components and help expose the performance impact of design decisions. For both

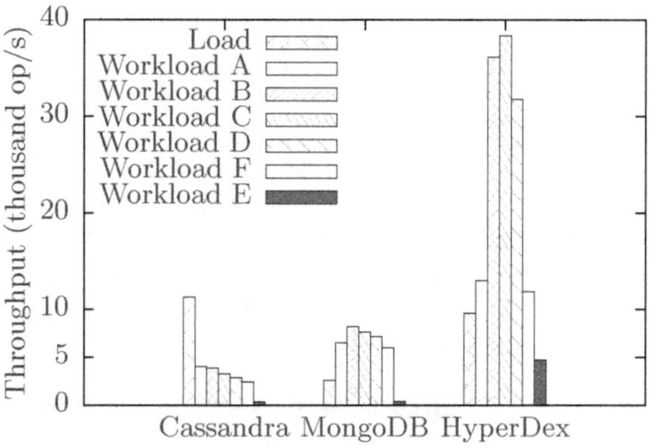

Figure 4: Average throughput for a variety of real-world workloads specified by the Yahoo! Cloud Serving Benchmark. HyperDex is 3-13 times faster than Cassandra and 2-12 times faster than MongoDB. Workload E is a search-heavy workload, where HyperDex outperforms other systems by more than an order of magnitude.

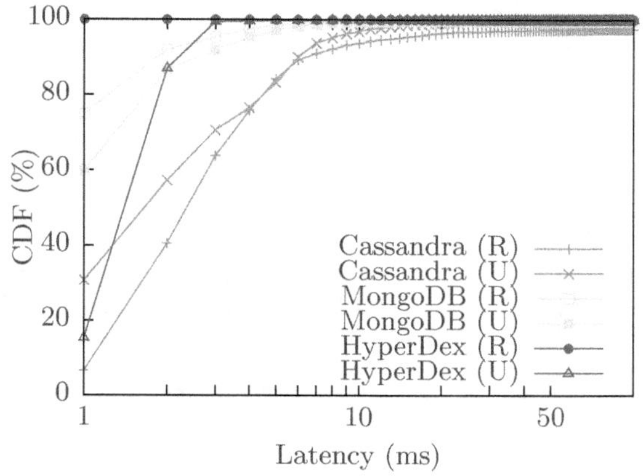

Figure 5: GET/PUT performance. Latency distribution for Workload A (50% reads, 50% updates, Zipf distribution).

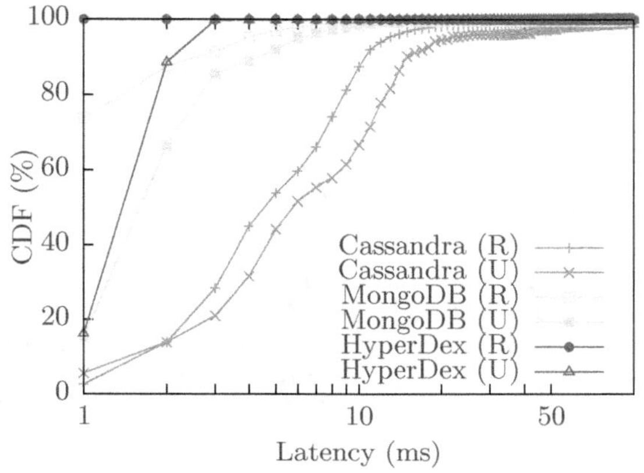

Figure 6: GET/PUT performance. Latency distribution for Workload B (95% reads, 5% updates, Zipf distribution). HyperDex maintains low latency for reads and writes.

YCSB and the micro-benchmarks, we compare HyperDex with Cassandra [32], a popular key-value store for Web 2.0 applications, and MongoDB [37], a distributed document database.

The performance benchmarks are executed on our small, dedicated lab-size cluster in order to avoid confounding issues arising from sharing a virtualized platform, while the scalability benchmarks are executed on the VICCI [42] testbed. Our dedicated cluster consists of fourteen nodes, each of which is equipped with two Intel Xeon 2.5 GHz E5420 processors, 16 GB of RAM, and a 500 GB SATA 3.0 Gbit/s hard disk operating at 7200 RPM. All nodes are running 64-bit Debian 6 with the Linux 2.6.32 kernel. A single gigabit Ethernet switch connects all fourteen machines. On each of the machines, we deployed Cassandra version 0.7.3, MongoDB version 2.0.0, and HyperDex.

For all tests, the storage systems are configured to provide sufficient replication to tolerate one node failure. Each system was configured to use its default consistency settings. Specifically, both Cassandra and MongoDB provide weak consistency and fault-tolerance guarantees; because acknowledgments are generated without full replication, small numbers of failures can lead to data loss. In contrast, HyperDex utilizes value-depending chaining and, as a result, always provides clients with strong consistency and fault-tolerance, even in the presence of failures. Since MongoDB allocates replicas in pairs, we allocate twelve machines for the storage nodes, one machine for the clients, and, where applicable, one node for the coordinator. HyperDex is configured with two subspaces in addition to the key subspace to accommodate all ten attributes in the YCSB dataset.

6.1 Get/Put Performance

High `get`/`put` performance is paramount to any cloud-based storage system. YCSB provides six different workloads that exercise the storage system with a mixture of request types and object distributions resembling real-world applications (Table 1). In all YCSB experiments, the database is preloaded with 10,000,000 objects and each operation selects the object and operation type in accordance with the workload specification. Figure 4 shows the throughput achieved by each system across the YCSB workloads. HyperDex provides throughput that is between a factor of two to thirteen higher than the other systems. The largest performance gains come from improvements in `search` performance. Significant improvements in `get`/`put` performance is attributable mostly to the efficient handling of `get` operations in HyperDex. Our implementation demonstrates that the hyperspace construction and maintenance can be realized efficiently.

In order to gain insight into the performance of the system, we examine the request latency distributions of the different systems under all read/write workloads. HyperDex's performance is predictable: all reads complete in under 1 ms, while a majority of writes complete in under 3 ms. Cassandra's latency distributions follow a similar trend for workloads B, C, D and F and show a slightly different trend for workload A. MongoDB, on the other hand, exhibits lower latency than Cassandra for all operations. For all workloads, HyperDex completes 99% of operations sooner than either Cassandra and MongoDB. Figures 5 and 6 show the latency distributions for workloads A and B respectively.

For completeness, we present the performance of all three systems on a write-heavy workload. Figure 7 shows the latency distribution for inserting 10,000,000 objects to set up the YCSB tests. Consistency guarantees have a significant effect on the `put` latency. MongoDB's default behavior considers a `put` complete when the request has been queued in

Name	Workload	Key Distribution	Sample Application
A	50% Read/50% Update	Zipf	Session Store
B	95% Read/5% Update	Zipf	Photo Tagging
C	100% Read	Zipf	User Profile Cache
D	95% Read/5% Insert	Temporal	User Status Updates
E	95% Scan/5% Insert	Zipf	Threaded Conversations
F	50% Read/50% Read-Modify-Write	Zipf	User Database

Table 1: The six workloads specified by the Yahoo! Cloud Serving Benchmark. These workloads model several applications found at Yahoo! Each workload was tested using the same YCSB driver program and system-specific Java bindings. Each object has ten attributes which total 1 kB in size.

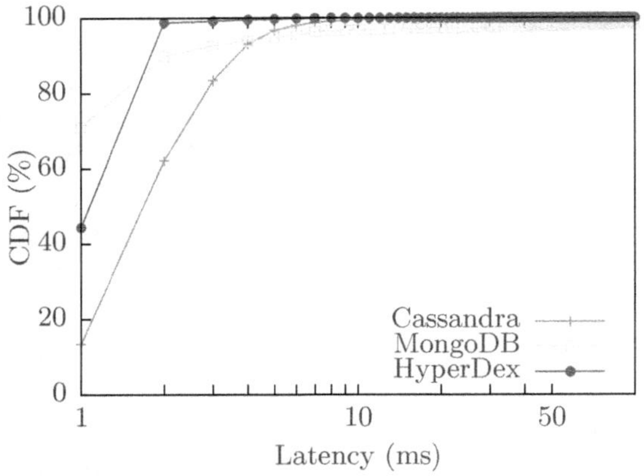

Figure 7: `PUT` performance. Latency distribution for 10,000,000 operations consisting of 100% insertions. Each operation created a new object under a unique, previously unused key. Although fast, HyperDex's minimum latency is bounded by the length of the value-dependent chains.

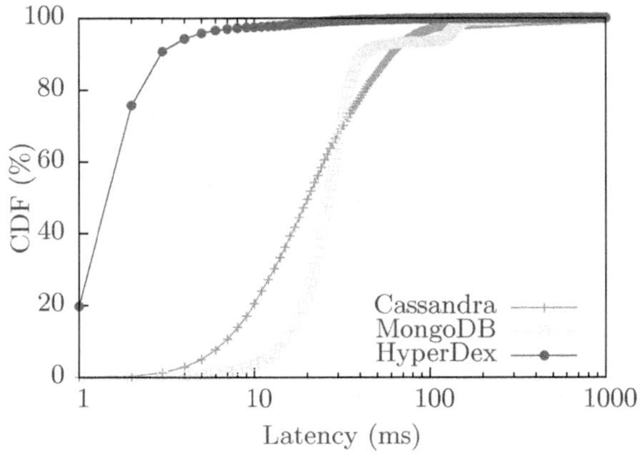

Figure 8: `SEARCH` performance. Latency distribution for 10,000 operations consisting of 95% range queries and 5% inserts with keys selected from a Zipf distribution. HyperDex is able to offer significantly lower latency for non-primary key range queries than the other systems are able to offer for primary-key range queries.

a client's send buffer, even before it has been seen by any server. Cassandra's default behavior considers a `put` complete when it is queued in the filesystem cache of just one replica. Unlike these systems, the latency of a HyperDex `put` operation includes the time taken to fully replicate the object on $f + 1$ servers. Because MongoDB does not wait for full fault tolerance, it is able to complete a majority of operations in less than 1 ms; however, it exhibits a long-tail (Figure 7) that adversely impacts average throughput. Similarly, Cassandra completes most operations in less than 2 ms. Despite its stronger fault-tolerance guarantees, HyperDex completes 99% of its operations in less than 2 ms.

6.2 Search vs. Scan

Unlike existing key-value stores, HyperDex is architected from the ground-up to perform `search` operations efficiently. Current applications that rely on existing key-value stores emulate search functionality by embedding additional information about other attributes in the key itself. For example, applications typically group logically related objects by using a shared prefix in the key of each object, and then rely upon the key-value store to locate keys with the common prefix. In Cassandra, this operation is efficient because keys are stored in sorted order, and returning all logically grouped keys is an efficient linear scan. Fittingly, the YCSB benchmark calls this a `scan` operation. HyperDex's `search` functionality is a strict superset of the `scan` operation. Rather than using a shared prefix to support scans, HyperDex stores, for each object, the prefix and suffix of the key as two secondary attributes. Scans are then implemented as a multi-attribute search that exactly matches a provided prefix value and a provided range of suffix values. Thus, all YCSB benchmarks involving a `scan` operation operate on *secondary attributes* in HyperDex, but operate on the *key* for other systems.

Despite operating on secondary attributes instead of the key, HyperDex outperforms the other systems by an order of magnitude for `scan` operations (Figure 8). Seventy five percent of search operations complete in less than 2 ms, and nearly all complete in less than 6 ms. Cassandra sorts data according to the primary key and is therefore able to retrieve matching items relatively quickly. Although one could alter YCSB to use Cassandra's secondary indexing schemes instead of key-based operations, the result would be strictly worse than what is reported for primary key operations. MongoDB's sharding maintains an index; consequently, `scan` operations in MongoDB are relatively fast. The `search` performance of HyperDex is not attributable to our efficient implementation as `search` is more than an order of magnitude faster in HyperDex, which eclipses the 2-4× performance advantage observed for `get`/`put` throughput. Hyperspace hashing in HyperDex ensures that search results are located on a small number of servers; this enables effective pruning of the search space and allows each search to complete by contacting exactly one host in our experiments.

An additional benefit of HyperDex's aggressive search pruning is the relatively low latency overhead associated with `search` operations. Figure 9 shows the average latency of a

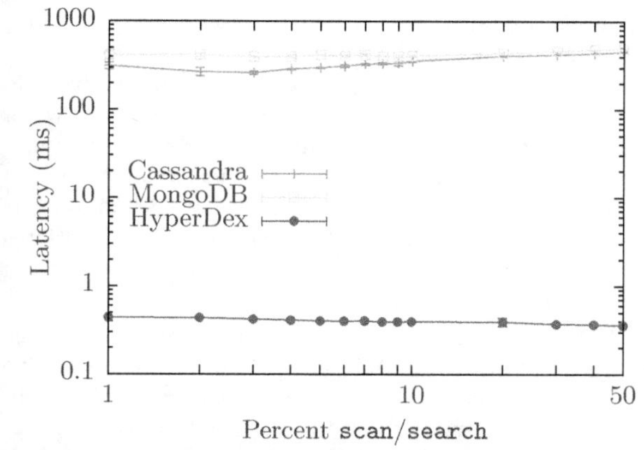

Figure 9: The effect of an increasing scan workload on latency. HyperDex performs significantly better than the other systems even as the scan workload begins to approach 50%.

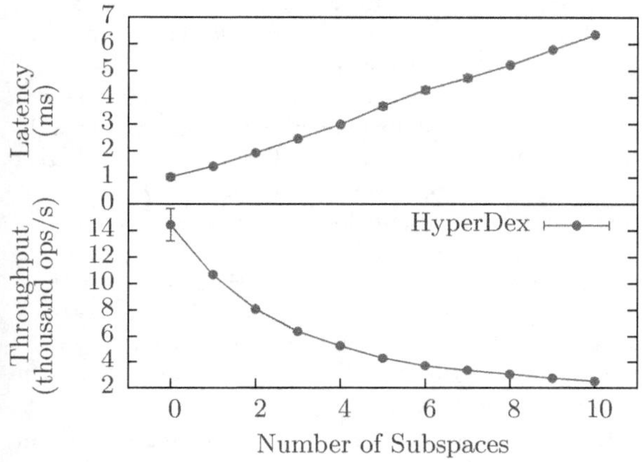

Figure 10: Latency and throughput for put operations as a function of non-key subspaces The error bars indicate standard deviation from 10 experiments. Latency increases linearly in the length of the chain, while throughput decreases proportionally. In applications we have built with HyperDex, all tables have three or fewer subspaces.

single `scan` operation as the total number of `scan` operations performed increases. In this test, searches were constructed by choosing the lower bound of the range uniformly at random from the set of possible values, as opposed to workload E which uses a Zipf distribution to select objects. Using a uniform distribution ensures random access, and mitigates the effects of object caching. HyperDex consistently offers low latency for `search`-heavy workloads.

A critical parameter that affects HyperDex's search performance is the number of subspaces in a HyperDex table. Increasing the number of subspaces leads to additional opportunities for pruning the search space for *search* operations, but simultaneously requires longer value-dependent chains that result in higher `put` latencies. In Figure 10, we explore the tradeoff using between zero and ten additional subspaces beyond the mandatory key subspace. Note that adding ten additional subspaces increases the value-dependent chain to be at least 22 nodes long. As expected, HyperDex's `put` latency increases linearly with each additional subspace.

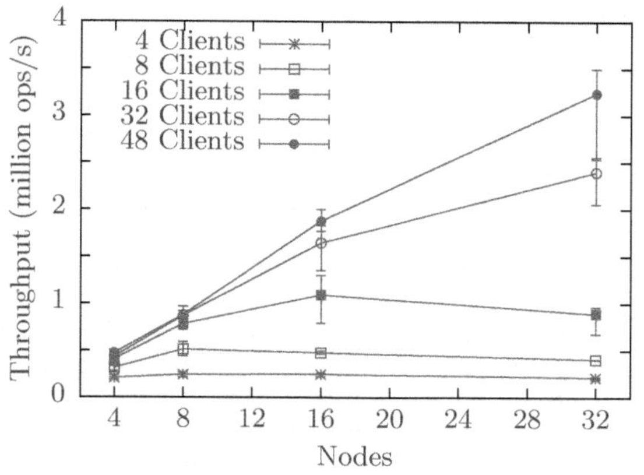

Figure 11: HyperDex scales horizontally. As more servers are added, aggregate throughput increases linearly. Each point represents the average throughput of the system in steady state over 30 second windows. The error bars show the 5th and 95th percentiles.

6.3 Scalability

We have deployed HyperDex on the VICCI [42] testbed to evaluate its performance in an environment representative of the cloud. Each VICCI cluster has 70 Dell R410 PowerEdge servers, each of which has 2 Intel Xeon X5650 CPUs, 48 GB of RAM, three 1 TB hard drives, and two 1 Gbit ethernet ports. Users are provided with an isolated virtual machine for conducting experiments. Each virtual machine comes preinstalled with Fedora 12 and runs the 2.6.32 Linux kernel.

We examined the performance of a HyperDex cluster as the cluster increases in size. Increasing the number of servers in the cluster provides HyperDex with additional resources and leads to a proportional increase in throughput. In Figure 11, we explore the change in system throughput as resources are added to the cluster. As expected, HyperDex scales linearly as resources are added to the cluster. Each point in the graph represents the average throughput observed over a 30 second window and the error bars show the 5th and 95th percentiles observed over any 1-second window. At its peak, HyperDex is able to average 3.2 million operations per second.

The workload for Figure 11 is a 95% read, 5% write workload operating on 8 B keys and 64 B values. The measurements reported are taken in steady state, with clients randomly generating requests according to the workload. This workload and measurement style reflects the workload likely to be encountered in a web application. The reported measurements exclude the warm-up time for the system. In all experiments, 15 seconds was sufficient to achieve steady state. Clients operate in parallel, and are run on separate machines from the servers in all but the largest configurations. Clients issue requests in parallel, and each client maintains an average of 1,000 outstanding requests per server. Increasing the number of clients does not significantly impact the achievable average throughput.

This experiment shows that a medium-sized HyperDex cluster is able to achieve high throughput for realistically sized deployments [3]. Additional resources allow the cluster to provide proportionally better throughput.

7. RELATED WORK

Database system Storage systems that organize their data in high-dimensional spaces were pioneered by the database community more than thirty years ago [5,7,24,31,39,41,51]. These systems, collectively known as Multi-Dimensional Databases (MDB), leverage multi-dimensional data structures to improve the performance of data warehousing and online analytical processing applications. However, unlike hyperspaces in HyperDex, the data structures in MDBs are designed for organizing data on a single machine and are not directly applicable to large-scale distributed storage systems. Alternatively, more recent database systems [1, 17] have begun exploring efficient mechanisms for building and maintaining large-scale, tree-based distributed indices. In contrast, the mapping HyperDex constructs is not an index. Indices must be maintained and updated on object insertion. Hyperspace hashing, on other hand, is a mapping that does not change as objects are inserted and removed.

Peer-to-peer systems Past work in peer-to-peer systems has explored multi-dimensional network overlays to facilitate decentralized data storage and retrieval. CAN [47] is a distributed hash-table that, much like HyperDex, organizes peers in a multi-dimensional space. However, CAN only provides key inserts and lookups; the purpose of CAN's multi-dimensional peer configuration is to limit a CAN node's peer-set size to provide efficient overlay routing. HyperDex provides search, and does not do routing in the space.

MURK [21], SkipIndex [58], and SWAM-V [30] dynamically partition the multi-dimensional space into kd-trees, skip graphs, and Voronoi diagrams respectively to provide multi-dimensional range lookups. Although conceptually similar to HyperDex, providing coordination and management of nodes for these dynamic space partitioning schemes is significantly more complex and error-prone than for HyperDex's static space partitioning and require additional operational overhead. These systems also do not address several critical and inherent problems associated with mapping structured data into a multi-dimensional space and providing reliable data storage. Specifically, they are not efficient for high dimensional data due to the curse-of-dimensionality, and they either lack data replication or provide only eventually consistent operations on replicated data. Addressing these problems by augmenting dynamic space partitioning schemes with subspaces and value-dependent chaining would further increase the complexity and overhead of node coordination and management. Mercury [8] builds on top of a Chord [55] ring, and uses consistent hashing [29] on each attribute as secondary indexes. Although Mercury's implementation is unavailable, Cassandra uses a similar design and its performance illustrates how a multi-ring-based system would perform. Arpeggio [13] provides search over multiple attributes by enumerating and creating an index of all $\binom{k}{x}$ fixed-size subsets of attributes using a Chord ring. Both of these approaches insert redundant pointers into rings without concern for consistency.

Space-filling curves A common approach to providing multi-attribute search uses space-filling curves to partition multi-dimensional data across the storage nodes. This approach uses the space filling curve to map the multi-dimensional data into a single dimension, which then enables the use of traditional peer-to-peer techniques for performing searches. SCRAP [21], Squid [52] and ZNet [53] are examples of this approach with each node responsible for data in a contiguous range of values. Similarly, MAAN [10] performs the same mapping, but uses uniform locality preserving hashing. Space-filling curves do not scale well when the dimensionality is high, because a single search query may be partitioned into many one-dimensional ranges of considerably varying size. Furthermore, unlike in HyperDex, fully-qualified searches, where values for all attributes are specified, may involve contacting more than one node in space-filling curve-based systems.

NoSQL Storage A new class of scalable storage systems, collectively dubbed "NoSQL", have recently emerged with the defining characteristic that they depart from the traditional architecture and SQL interface of relational databases. It is common practice for NoSQL systems to make explicit tradeoffs with respect to desirable properties. For instance, many NoSQL systems explicitly sacrifice consistency – even in the common case without failures – to achieve availability under extreme failure scenarios. NoSQL systems include document databases [16, 37] that offer a schema-free data model, in-memory cache solutions [36, 43, 48] that accelerate application performance, and graph databases [38] which model interconnected elements and key-value stores which offer predictable performance. Still, some systems do not fit neatly into these categories. For example, Yahoo!'s PNUTS [14] system supports the traditional SQL selection and projection, functions but does not support joins.

HyperDex explores a new point in the NoSQL design space. The rich HyperDex API provides qualitatively new functionality not offered by other NoSQL systems, HyperDex sets a new bar for future NoSQL systems by combining strong consistency properties with fault-tolerance guarantees, a rich API and high performance.

Key-Value Stores Modern key-value stores have roots in work on Distributed Data Structures [23,36] and distributed hash tables [29,47,50,55]. Most open source key-value stores draw heavily from the ring-based architecture of Dynamo [19] and the tablet-based architecture of BigTable [11]. For instance, Voldemort [45] and Riak [49] are heavily influenced by Dynamo's design. Other systems like HBase [25] and HyperTable [28] are open source implementations of BigTable. Cassandra [32] is unique in that it is influenced by BigTable's API and Dynamo's ring structure. Like HyperDex, all of these systems are designed to run in a datacenter environment on many machines.

Recent work on key-value stores largely focuses on improving performance by exploiting underlying hardware or manipulating consistency guarantees for a performance advantage. Fawn KV [4] builds a key-value store on underpowered hardware to improve the throughput-to-power-draw ratio. SILT [33] eliminates read amplification to maximize read bandwidth in a datastore backed by solid-state disk. RAMCloud [40] stores data in RAM and utilizes fast network connections to rapidly restore failed replicas. TSSL [54] utilizes a multi-tier storage hierarchy to exploit cache-oblivious algorithms in the storage layer. Masstree [35] uses concatenated B^+ trees to service millions of queries per second. COPS [34] provides a high-performance geo-replicated key-value store that provides causal+ consistency. Other systems [12, 19] trade consistency for other desirable properties, such as performance. Spanner [18] uses Paxos in the wide area to provide strong consistency. Spinnaker [46] uses Paxos to build a strongly consistent store that performs nearly as well as those that are only eventually consistent [19,44,56].

Each of these existing systems improves the performance, availability or consistency of key value stores while retaining the same basic structure: a hash table. In HyperDex, we take a complementary approach which expands the key-value interface to support, among other operations, search over secondary attributes.

8. DISCUSSION

Surprisingly, the open-source release of the HyperDex system [20] uncovered various misunderstandings surrounding the CAP Theorem [22]. The popular CAP refrain ("C, A, P: pick any two") causes the subtleties in the definitions of C, A and P to be lost, even on otherwise savvy developers. There exists no conflict between claims in our paper and the CAP Theorem. The failure model used in the CAP Theorem is unconstrained; the system can be subject to partitions and node failures that affect any number of servers and network links. No system, including HyperDex, can simultaneously offer consistency and availability guarantees using such weak assumptions. HyperDex makes a stronger assumption that limits failures to affect at most a threshold of servers and is thus able to provide seemingly impossible guarantees.

9. CONCLUSIONS

This paper described HyperDex, a second-generation NoSQL storage system that combines strong consistency guarantees with high availability in the presence of failures and partitions affecting up to a threshold of servers. In addition, HyperDex provides an efficient search primitive for retrieving objects through their secondary attributes. It achieves this extended functionality through *hyperspace hashing*, in which multi-attribute objects are deterministically mapped to coordinates in a low dimension Euclidean space. This mapping leads to efficient implementations for key-based retrieval, partially-specified searches and range-queries. HyperDex's novel replication protocol enables the system to provide strong consistency without sacrificing performance. Industry-standard benchmarks show that the system is practical and efficient.

The recent trend toward NoSQL data stores has been fueled by scalability and performance concerns at the cost of functionality. HyperDex bridges this gap by providing additional functionality without sacrificing scalability or performance.

10. ACKNOWLEDGMENTS

We would like to thank Pawel Loj for his contributions on cluster stop/restart, Deniz Altınbüken for her ConCoord Paxos implementation, and members of the HyperDex open source community for their code contributions, and the VICCI team for granting us time on their cluster. This material is based upon work supported by National Science Foundation under Grants No. CNS-1111698 and CCF-0424422 and by the National Science and Engineering Research Council of Canada.

11. REFERENCES

[1] M. K. Aguilera, W. M. Golab, and M. A. Shah. A Practical Scalable Distributed B-Tree. In *PVLDB,* 1(1), 2008.

[2] D. Altınbüken and E. G. Sirer. Commodifying Replicated State Machines with OpenReplica. Computing and Information Science, Cornell University, Technical Report 1813-29009, 2012.

[3] S. Anand. `http://www.infoq.com/presentations/NoSQL-Netflix/`.

[4] D. G. Andersen, J. Franklin, M. Kaminsky, A. Phanishayee, L. Tan, and V. Vasudevan. FAWN: A Fast Array of Wimpy Nodes. In Proc. of *SOSP,* Big Sky, MT, Oct. 2009.

[5] R. Bayer. The Universal B-Tree for Multidimensional Indexing: General Concepts. In Proc. of *WWCA,* Tsukuba, Japan, Mar. 1997.

[6] R. E. Bellman. *Dynamic Programming.* Princeton University Press, 1957.

[7] J. L. Bentley. Multidimensional Binary Search Trees Used for Associative Searching. In *CACM,* 18(9), 1975.

[8] A. R. Bharambe, M. Agrawal, and S. Seshan. Mercury: Supporting Scalable Multi-Attribute Range Queries. In Proc. of *SIGCOMM,* Portland, OR, Aug. 2004.

[9] M. Burrows. The Chubby Lock Service for Loosely-Coupled Distributed Systems. In Proc. of *OSDI,* Seattle, WA, Nov. 2006.

[10] M. Cai, M. R. Frank, J. Chen, and P. A. Szekely. MAAN: A Multi-Attribute Addressable Network for Grid Information Services. In Proc. of *GRID Workshop,* Phoenix, AZ, Nov. 2003.

[11] F. Chang, J. Dean, S. Ghemawat, W. C. Hsieh, D. A. Wallach, M. Burrows, T. Chandra, A. Fikes, and R. Gruber. BigTable: A Distributed Storage System for Structured Data. In Proc. of *OSDI,* Seattle, WA, Nov. 2006.

[12] J. Cipar, G. R. Ganger, K. Keeton, C. B. M. III, C. A. N. Soules, and A. C. Veitch. LazyBase: Trading Freshness for Performance in a Scalable Database. In Proc. of *EuroSys,* Bern, Switzerland, Apr. 2012.

[13] A. T. Clements, D. R. K. Ports, and D. R. Karger. Arpeggio: Metadata Searching and Content Sharing with Chord. In Proc. of *IPTPS Workshop,* La Jolla, CA, Feb. 2005.

[14] B. F. Cooper, R. Ramakrishnan, U. Srivastava, A. Silberstein, P. Bohannon, H.-A. Jacobsen, N. Puz, D. Weaver, and R. Yerneni. PNUTS: Yahoo!'s Hosted Data Serving Platform. In *PVLDB,* 1(2), 2008.

[15] B. F. Cooper, A. Silberstein, E. Tam, R. Ramakrishnan, and R. Sears. Benchmarking Cloud Serving Systems with YCSB. In Proc. of *SoCC,* Indianapolis, IN, June 2010.

[16] CouchDB. `http://couchdb.apache.org/`.

[17] A. Crainiceanu, P. Linga, J. Gehrke, and J. Shanmugasundaram. Querying Peer-to-Peer Networks Using P-Trees. In Proc. of *WebDB Workshop,* Paris, France, June 2004.

[18] J. Dean. Designs, Lessons, and Advice from Building Large Distributed Systems. Keynote. In Proc. of *LADIS,* Big Sky, MT, Oct. 2009.

[19] G. DeCandia, D. Hastorun, M. Jampani, G. Kakulapati, A. Lakshman, A. Pilchin, S. Sivasubramanian, P. Vosshall, and W. Vogels. Dynamo: Amazon's Highly Available Key-Value Store. In Proc. of *SOSP,* Stevenson, WA, Oct. 2007.

[20] R. Escriva, B. Wong, and E. G. Sirer. Hyperdex. http://hyperdex.org/.

[21] P. Ganesan, B. Yang, and H. Garcia-Molina. One Torus to Rule Them All: Multidimensional Queries in P2P Systems. In Proc. of *WebDB Workshop,* Paris, France, June 2004.

[22] S. Gilbert and N. A. Lynch. Brewer's Conjecture and the Feasibility of Consistent, Available, Partition-Tolerant Web Services. In *SIGACT News,* 33(2), 2002.

[23] S. D. Gribble. A Design Framework and a Scalable Storage Platform to Simplify Internet Service Construction. U.C. Berkeley, 2000.

[24] A. Guttman. R-Trees: A Dynamic Index Structure for Spatial Searching. In Proc. of *SIGMOD,* 1984.

[25] HBase. `http://hbase.apache.org/`.

[26] M. Herlihy and J. M. Wing. Linearizability: A Correctness Condition for Concurrent Objects. In *ACM ToPLaS,* 12(3), 1990.

[27] P. Hunt, M. Konar, F. P. Junqueira, and B. Reed. ZooKeeper: Wait-Free Coordination for Internet-Scale Systems. In Proc. of *USENIX,* Boston, MA, June 2010.

[28] Hypertable. `http://http://hypertable.org/`.

[29] D. R. Karger, E. Lehman, F. T. Leighton, R. Panigrahy, M. S. Levine, and D. Lewin. Consistent Hashing and Random Trees: Distributed Caching Protocols for Relieving Hot Spots on the World Wide Web. In Proc. of *STOC,* El Paso, TX, May 1997.

[30] F. B. Kashani and C. Shahabi. SWAM: A Family of Access Methods for Similarity-Search in Peer-to-Peer Data Networks. In Proc. of *CIKM,* Washington, D.C., Nov. 2004.

[31] A. Klinger. Patterns and Search Statistics. In *Optimizing Methods in Statistics,* Academic Press, 1971.

[32] A. Lakshman and P. Malik. Cassandra: A Decentralized Structured Storage System. In Proc. of *LADIS,* Big Sky, MT, Oct. 2009.

[33] H. Lim, B. Fan, D. G. Andersen, and M. Kaminsky. SILT: A Memory-Efficient, High-Performance Key-Value Store. In Proc. of *SOSP,* Cascais, Portugal, Oct. 2011.

[34] W. Lloyd, M. J. Freedman, M. Kaminsky, and D. G. Andersen. Don't Settle for Eventual: Scalable Causal Consistency for Wide-Area Storage with COPS. In Proc. of *SOSP,* Cascais, Portugal, Oct. 2011.

[35] Y. Mao, E. Kohler, and R. T. Morris. Cache Craftiness for Fast Multicore Key-Value Storage. In Proc. of *EuroSys,* Bern, Switzerland, Apr. 2012.

[36] Memcached. `http://memcached.org/`.

[37] MongoDB. `http://www.mongodb.org/`.

[38] Neo4j. `http://neo4j.org/`.

[39] J. Nievergelt, H. Hinterberger, and K. C. Sevcik. The Grid File: An Adaptable, Symmetric Multikey File Structure. In *ACM ToDS,* 9(1), 1984.

[40] D. Ongaro, S. M. Rumble, R. Stutsman, J. K. Ousterhout, and M. Rosenblum. Fast Crash Recovery in RAMCloud. In Proc. of *SOSP,* Cascais, Portugal, Oct. 2011.

[41] J. A. Orenstein and T. H. Merrett. A Class of Data Structures for Associative Searching. In Proc. of *PODS,* 1984.

[42] L. Peterson, A. Bavier, and S. Bhatia. VICCI: A Programmable Cloud-Computing Research Testbed. Princeton, Technical Report TR-912-11, 2011.

[43] D. R. K. Ports, A. T. Clements, I. Zhang, S. Madden, and B. Liskov. Transactional Consistency and Automatic Management in an Application Data Cache. In Proc. of *OSDI,* Vancouver, Canada, Oct. 2010.

[44] D. Pritchett. BASE: An ACID Alternative. In *ACM Queue,* 6(3), 2008.

[45] Project Voldemort. `http://project-voldemort.com/`.

[46] J. Rao, E. J. Shekita, and S. Tata. Using Paxos to Build a Scalable, Consistent, and Highly Available Datastore. In *PVLDB,* 4(4), 2011.

[47] S. Ratnasamy, P. Francis, M. Handley, R. M. Karp, and S. Shenker. A Scalable Content-Addressable Network. In Proc. of *SIGCOMM,* San Diego, CA, Aug. 2001.

[48] Redis. `http://redis.io/`.

[49] Riak. `http://basho.com/`.

[50] A. I. T. Rowstron and P. Druschel. Pastry: Scalable, Decentralized Object Location, and Routing for Large-Scale Peer-to-Peer Systems. In Proc. of *ICDSP,* volume 2218, 2001.

[51] H. Samet. Spatial Data Structures. In *Modern Database Systems: The Object Model, Interoperability, and Beyond,* Addison Wesley/ACM Press, 1995.

[52] C. Schmidt and M. Parashar. Enabling Flexible Queries with Guarantees in P2P Systems. In *Internet Computing Journal,* 2004.

[53] Y. Shu, B. C. Ooi, K.-L. Tan, and A. Zhou. Supporting Multi-Dimensional Range Queries in Peer-to-Peer Systems. In Proc. of *IEEE International Conference on Peer-to-Peer Computing,* Konstanz, Germany, Aug. 2005.

[54] R. P. Spillane, P. J. Shetty, E. Zadok, S. Dixit, and S. Archak. An Efficient Multi-Tier Tablet Server Storage Architecture. In Proc. of *SoCC,* 2011.

[55] I. Stoica, R. Morris, D. R. Karger, M. F. Kaashoek, and H. Balakrishnan. Chord: A Scalable Peer-to-Peer Lookup Service for Internet Applications. In Proc. of *SIGCOMM,* San Diego, CA, Aug. 2001.

[56] D. B. Terry, M. Theimer, K. Petersen, A. J. Demers, M. Spreitzer, and C. Hauser. Managing Update Conflicts in Bayou, a Weakly Connected Replicated Storage System. In Proc. of *SOSP,* Copper Mountain, CO, Dec. 1995.

[57] R. van Renesse and F. B. Schneider. Chain Replication for Supporting High Throughput and Availability. In Proc. of *OSDI,* San Francisco, CA, Dec. 2004.

[58] C. Zhang, A. Krishnamurthy, and R. Y. Wang. SkipIndex: Towards a Scalable Peer-to-Peer Index Service for High Dimensional Data. Princeton, Technical Report TR-703-04, 2004.

Picasso: Flexible RF and Spectrum Slicing

Steven Hong
Stanford University
hsiying@stanford.edu

Jeffrey Mehlman
Stanford University
jmehlman@stanford.edu

Sachin Katti
Stanford University
skatti@stanford.edu

ABSTRACT

This paper presents the design, implementation and evaluation of Picasso, a novel radio design that allows simultaneous transmission and reception on separate and arbitrary spectrum fragments using a single RF front end and antenna. Picasso leverages this capability to flexibly partition fragmented spectrum into multiple slices that share the RF front end and antenna, yet operate concurrent and independent PHY/MAC protocols. We show how this capability provides a general and clean abstraction to exploit fragmented spectrum in WiFi networks and handle coexistence in dense deployments. We prototype Picasso, and demonstrate experimentally that a Picasso radio partitioned into four slices, each concurrently operating four standard WiFi OFDM PHY and CSMA MAC stacks, can achieve the same sum throughput as four physically separate radios individually configured to operate on the spectrum fragments. We also demonstrate experimentally how Picasso's slicing abstraction provides a clean mechanism to enable multiple diverse networks to coexist and achieve higher throughput, better video quality and latency than the best known state of the art approaches.

Categories and Subject Descriptors: C2.1 [**Computer Systems Organization**]: Computer Communication Networks
General Terms: Algorithms, Performance, Design
Keywords: Radio Virtualization, Interference Cancellation

1. INTRODUCTION

Many applications can benefit from the capability to simultaneously and independently use arbitrarily sized but separate spectrum fragments with a single radio and antenna. By this capability we mean that the radio can simultaneously transmit, simultaneously receive, or simultaneously transmit (TX) and receive (RX) on arbitrary but separate spectrum fragments. For example, we can use it for spectrum aggregation in fragmented ISM bands as shown in Fig. 1(A). A WiFi AP can run independent OFDM PHY and CSMA MAC protocols on two WiFi channels to simultaneously serve two legacy WiFi clients assigned to different channels and achieve significantly higher throughput than a legacy AP that is restricted to the use of only one channel at a time. Similarly, a WiFi client radio with such a capability can simultaneously connect to multiple WiFi APs on different channels and obtain a much higher aggregate throughput than current radios that can transmit or receive on only one channel at a time.

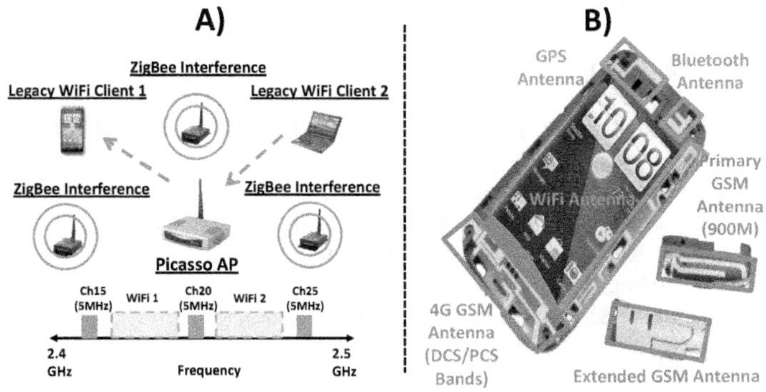

Figure 1: A) Fragmented Spectrum is common in ISM band: Picasso AP can aggregate multiple fragments and simultaneously service multiple clients on different channels. **B) Samsung Galaxy Printed Antenna Layout:** Instead of having an antenna per protocol, Picasso enables multiple protocols to use the same radio and antenna, saving space on the device.

Such a capability could also be used for radio sharing and coexistence. Portable consumer devices such as smartphones must accommodate a growing list of separate ISM band protocols such as WiFi, WiFi-Direct, Zigbee, NFC, and Bluetooth. Current practice is to use a separate radio and antenna for each protocol, shown in Fig. 1(B), but as the number of radios increases it becomes difficult to find enough space to separately place all the antennas these radios would need (e.g., the iPhone 4 "antenna-gate" was caused by antennas placed too closely [30]). Instead, a radio with a single antenna that allows simultaneous TX/RX on arbitrary spectrum fragments could be shared among all the above protocols. WiFi would use one fragment, Zigbee would use another, and so on, saving valuable real estate on space-constrained devices. They can operate their own independent PHY/MAC protocols on the shared radio without interfering with each other.

However, it is non-trivial to design such a radio. The key obstacle is that current radios cannot simultaneously transmit and receive on different arbitrary spectrum fragments with a single, shared Radio Frequency (RF) front end and antenna. The reason for this is that the transmitted signal causes high-powered self-interference, which saturates the RX chain and Analog-to-Digital Converter (ADC), consequently nulling the received signal. While the standard solution is to utilize static, analog RF filters to eliminate the self-interference, such an option is infeasible because spectrum fragmentation is dynamic–available spectrum in the ISM band varies in space and time, depending on the presence of other wireless networks. Consequently, if a radio wants to leverage all the available spectrum and be able to simultaneously transmit and receive on different fragments, the shared analog front end would need programmable analog filters that can be dynamically configured to let only the received signals through and filter out the self-interference. Analog filters however are typically statically configured and programmable analog filters that can be changed dynamically are expensive and impractical to deploy in current radios [18].

In this paper, we present the design and implementation of Picasso, a novel full duplex circuit design that sufficiently cancels (instead of filters) the self-interference in analog and prevents RX front end and ADC saturation, enabling the radio to cleanly recover the received signal. Our key contribution here is a circuit design that (1) isolates TX and RX signals at a single antenna by incorporating a circulator [23], and (2) exploits the fact that the self-interference signal travels through the fixed, known circulator channel to design a passive self-interference cancellation circuit. This allows a *radio to simultaneously transmit and receive on arbitrary spectrum fragments even while using a single RF front end and antenna.*

Our design improves on all prior related work on full duplex [9, 14, 5] wireless since they require at least two antennas (one for TX, one for RX), and these need to be separated by 15–20cm, which is untenable for small personal gadgets such as smartphones or tablets.

Picasso leverages this full duplex capability to build an abstraction that allows one to flexibly slice a single radio and antenna into separate independent slices operating on different spectrum fragments. Each slice is associated with a specific spectrum fragment in the ISM band (whose width/position can be programmatically specified). The key property is that the operation of each slice is decoupled from the other slices, i.e., the slice is free to run whatever narrowband PHY and MAC protocols it chooses, and the protocol behavior is not impacted by any other slice that may be present on the shared radio and antenna. Thus, in the above scenarios, the AP would have two slices corresponding to the two spectrum fragments and run two independent WiFi OFDM/CSMA protocols on the two slices in parallel. Similarly, a radio could be shared amongst multiple protocols (e.g. WiFi, Bluetooth, and NFC) by assigning independent slices to the appropriate spectrum fragment and running the corresponding protocol.

Further, to ensure that each slice can use existing, well-engineered narrowband PHY and MAC protocols on each slice, Picasso includes a reconfigurable filter engine that transparently takes signals spread over different spectrum fragments, and efficiently filters and resamples them so that the higher layers just see a simple sample stream consisting of narrowband digital samples. The higher layers are then free to process these samples with any narrowband PHY technique they choose, and schedule access to the slice with a MAC protocol of their choice. The slice, for all intents and purposes, appears as their own piece of spectrum centered at zero, operating on their own radio. Picasso thus completely abstracts out the complexity of spectrum fragmentation.

We design and implement a prototype of Picasso on Xilinx Virtex-5 FPGA-based software radios [32]. Our implementation consists of both the radio design that provides the slicing abstraction, as well as a WiFi-like contiguous OFDM PHY and CSMA MAC to operate on top of the slices. We show that Picasso's implementation of the slicing abstraction provides strong decoupled operation, i.e. there's no SNR loss because of either programmable filtering or simultaneous TX/RX. In other words, a Picasso radio achieves the same throughput as one would have achieved by using several independent radios statically configured to operate on individual slices.

To demonstrate the generality and applicability of Picasso, we demonstrate that Picasso provides clean abstractions to solve coexistence problems in dense deployments by evaluating a scenario where a number of diverse networks have to coexist with each other while sharing radios and fragmented spectrum (shown later in Fig. 16).

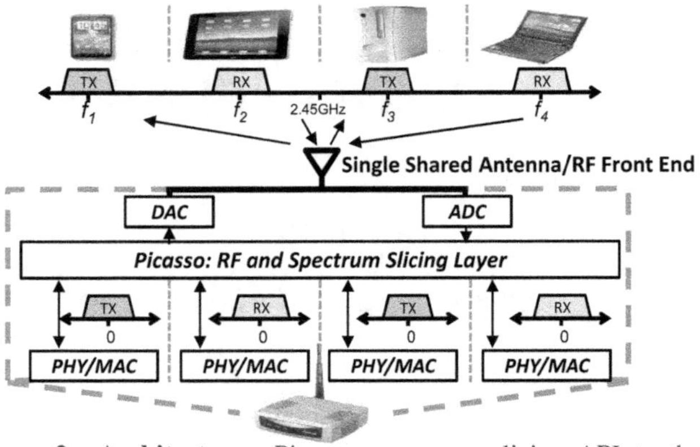

Figure 2: Architecture: Picasso exposes a slicing API to the PHY/MACs of the different slices and allows them to independently operate.

Each coexisting network has different objectives: the WiFi network wants to maximize throughput, the gameplay network wants to maximize video quality, and the game controller network wants to minimize latency. We compare Picasso with a state-of-the-art technique for exploiting fragmented spectrum, SWIFT [24], and show that Picasso achieves nearly $2\times$ higher throughput, a 10dB higher PSNR in video quality, and lowers latency by a factor of $2\times$. Further, the improvements are achieved with simple narrowband OFDM PHY and CSMA MAC implementations running on top of each slice, leading to a modular design that reuses existing, well-engineered components.

Picasso builds on our recent workshop paper [12] that sketched out the basic approach. In this paper, we present a novel full duplex design, as well as present a full design and prototype implementation. Finally, Picasso's focus is the mechanism to implement slicing of spectrum and RF front ends. An equally interesting question is given the set of coexisting networks and available fragmented spectrum, what is the best allocation of slices, and the distributed signaling algorithm needed to inform all devices their slice allocations? While this question is orthogonal to Picasso's design, we show in Sec. 7 that Picasso simplifies this problem considerably compared to a design that uses traditional radios which are incapable of slicing.

2. OVERVIEW

Picasso's key contribution is that it allows one to flexibly slice a radio into multiple slices each operating on different, arbitrary spectrum fragments with their own independent PHY/MAC protocols. Fig. 2 shows how Picasso differs from current designs–between the PHY/MAC layers and the actual transmission of RF signals on the antenna, there is an additional slicing layer. Picasso exposes an API to the PHY/MAC layers which consists of streams of complex digital baseband samples flowing between the slicing and the PHY layers for the slices. Each stream carries an internal header which includes a tuple specifying the spectrum slice on which those digital complex samples are transmitted or received. Picasso thus makes the architecture more amenable to evolution. By decoupling how spectrum and radios are sliced from how packets are processed and scheduled, the PHY and MAC can evolve independently and innovations can be easily integrated without having to change the radio. To realize this abstraction, Picasso needs to overcome two key challenges.

1. RF Isolation: First, Picasso must provide complete isolation between the operation of multiple slices. Slice isolation is relatively easy in the scenario where the device is simultaneously transmitting (or receiving) on all adjacent slices. But when the radio is transmitting on one spectrum slice while receiving in another, isolation is significantly harder. In order to provide this functionality in practice, the radio has to cope with high-powered self-interference introduced by its own transmitted signal at the antenna and the RX chain, which causes *receiver saturation* [14]. Since our design utilizes a single antenna, the transmitted signal directly leaks into the RX chain.

The problem is that the self-interference is billions of times stronger than the received signal (e.g., for WiFi the self-interference would be nearly 80+dB stronger). As we will show in Section 3, the dynamic range of practical ADCs is not large enough to acquire the received signal in the face of such large self-interference, so the receiver saturates and the received signal is lost in quantization. Note that this is a problem even though the transmitted signal is on a different band than the received signal. Spectrum slices over which a radio might transmit or receive are not fixed in advance and change over time and space, so we must be able to receive over the entire band and have no choice but to allow all signals across the ISM band through.

A naive way to prevent receiver saturation is to utilize multiple, parallel receive RF chains. However, supporting multiple RF chains on mobile devices is expensive [11] in terms of space and power, and does not scale well with an increasing number of slices. Another plausible solution to prevent receiver saturation is to use statically configured analog filters to filter out the transmitted signal and only pass the received signal. But again, since spectrum slices assigned vary over time and space, the analog filters would have to be dynamically reconfigurable. However, analog filters are typically optimized for static operation [18]. Hence in order to support dynamic configurations, we would require a bank of static analog filters of varying widths, with at least as many filters as the combinatorial number of spectrum slices, possible slice widths, and center frequencies.

2. Programmable Abstraction: Picasso's second challenge is to provide the abstraction of a single narrowband radio on each allocated spectrum slice to the higher layer protocols. To do so, Picasso must ensure that the digital samples being sent down from the PHY for transmission occupy a bandwidth equal to the slice's spectrum allocation w, and are centered at DC. Picasso must also ensure that the samples are shifted to the correct center frequency f_c that is associated with that slice, and manage this abstraction across multiple MAC/PHY instances. The opposite is performed on reception.

Picasso's goal is to design a radio with slicing capability using only one TX and RX RF chain, one antenna, and one simple wideband analog filter. Solving these challenges in aggregate enables Picasso to virtualize the radio, allowing multiple independent MAC/PHY protocols to share the same RF front end. The next two sections elaborate on how to solve these challenges.

3. Picasso: RF FRONT END SLICING

In this section, we describe how Picasso slices a single RF frontend. We focus on the challenging case where the radio is simultaneously transmitting and receiving over different slices. Picasso's key insight is to cancel the self-interference in analog instead of filtering it. Picasso subtracts the self-interference signal from the received signal so that its effect is eliminated and receiver saturation does not occur. We describe the design of Picasso's novel self interference cancellation technique below, beginning with a discussion of how much cancellation is needed for Picasso to prevent receiver saturation.

How much self-interference cancellation is needed? The amount of cancellation required depends on two factors:

1. Dynamic range/resolution of ADC
2. Range of the expected signal strengths

To provide specific numbers, we focus on the requirements for an 802.11 WiFi system, since it is the protocol with the highest transmit power in the ISM band. Dynamic range (DR) is defined as the

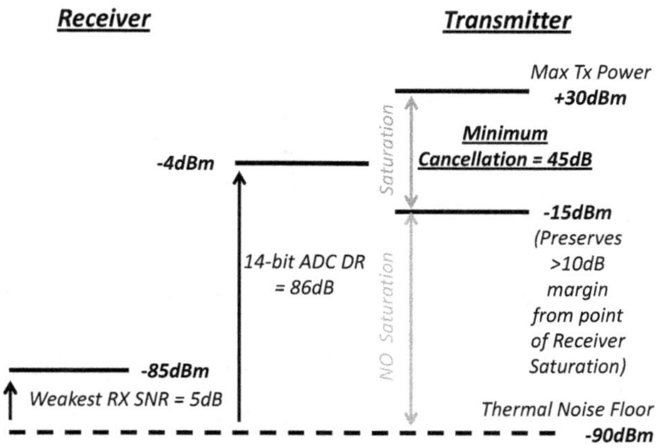

Figure 3: Minimum self-interference cancellation required is 45dB: The maximum transmit power for WiFi is 23 dBm, with a PAPR of 7dB, and so the maximum peak power is 30dBm. The dynamic range of a 14-bit ADC ranges from -90 to -4 dBm, but to preserve >10dB to receive the weakest detectable signal (5dB is the minimum but we preserve at least 10dB to provide some buffer), the maximum power of the canceled transmit signal must be less than -15dBm.

ratio between largest and smallest acceptable values of a variable of interest. At the transmitter, the dynamic range of the Digital-to-Analog Converter (DAC) determines the maximum ratio between the powers of the strongest and weakest transmissions. At the receiver, the ADC's dynamic range defines the maximum ratio between the strongest and weakest received signal power. When the dynamic range is exceeded, the converter's quantization noise buries the weaker signals. The dynamic range of a DAC or ADC is calculated as follows [20]

$$\text{DR (dB)} = 6.02 \times n + 1.76\text{dB} \qquad (1)$$

where n is the number of bits in the DAC/ADC Resolution. For example, Picasso uses 14-bit ADCs, which provide about 86dB of dynamic range.

At the transmitter, the maximum ratio between transmit powers over different slices will rarely exceed 30dB, so DAC dynamic range is usually not a concern. On the other hand, if the transmitter is operating while the system attempts to receive, the dynamic range of the receiver ADC is critical because the transmitted signal is much stronger than the received signal. To determine the required ADC dynamic range, we calculate the second variable –the range of expected signal strengths. 802.11 is built to operate at SNRs as low as 5dB. Because the typical thermal noise-floor for WiFi systems is approximately -90dBm, the power of the weakest decodable signal is -85dBm. On the other side of the spectrum, the maximum output from a WiFi 2.4 GHz antenna is 23 dBm. Further, OFDM exhibits relatively high PAPR (Peak to Average Power Ratio) of around 7dB [3], so the highest transmit power a signal will have is around 30dBm.

Commodity 14-bit ADCs [33] that can operate at a rate of 200 MS/s (enough to to cover the 100MHz 2.4GHz ISM band) are commercially available. A 14-bit ADC has 86dB of dynamic range. However, with a noise floor of -90dBm and a max transmit power of 30dBm, we would need 120dB of dynamic range, far beyond the capabilities of commodity ADCs. Hence in order to ensure receiver saturation does not occur, Picasso must ensure that self-interfering transmitted signal power is reduced by at least 45dB. Fig. 3 demonstrates the required interference power reduction in order to comfortably prevent receiver saturation.

Note that the above analysis is for the *worst-case* scenario with maximum transmit power and PAPR for OFDM signals. In practice, transmit powers used are smaller than 23dBm and a PAPR of

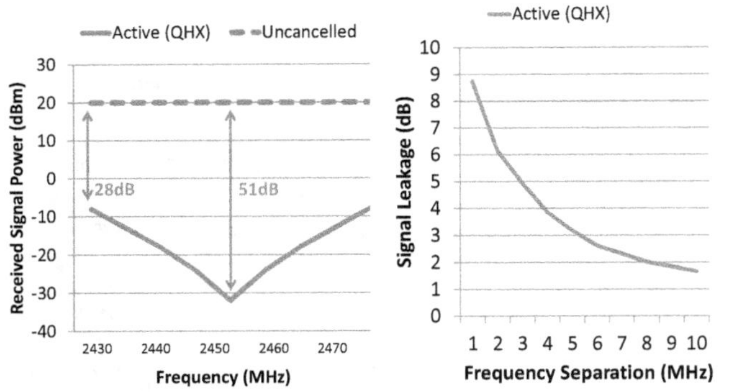

Figure 4: A)In Band Cancellation - QHX-Balun Cancellation achieves sufficient in-band cancellation; **B) Out of Band Cancellation Leakage** - QHX-Balun cancellation leaks a significant amount of signal power into adjacent bands, raising the noise floor by as much as 9dB.

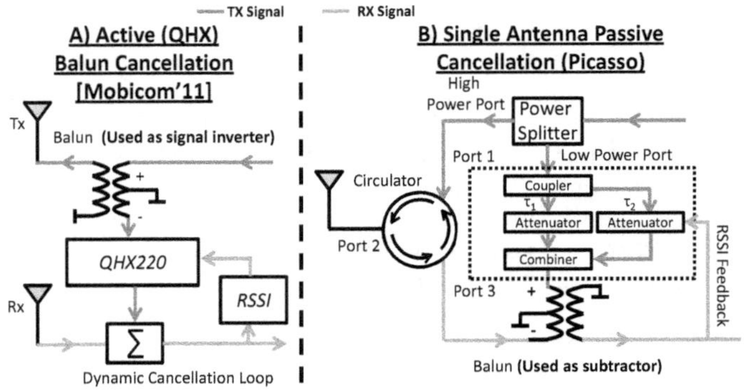

Figure 5: A)QHX-Balun Cancellation Diagram; B)Picasso Single Antenna, Passive Cancellation: Picasso's cancellation design is *passive*, preventing interference leakage into adjacent bands, and *requires only a single antenna.*

7dB is quite high [3]. Further, other ISM PHYs such as Zigbee and Bluetooth do not have high PAPRs. Hence, if Picasso can design a technique that provides 45dB of self interference reduction, it should be more than sufficient.

3.1 Prior Full Duplex Techniques

There are three recent full duplex designs [14, 9, 15] that provide sufficient self interference cancellation (45dB). However, none of this work is directly applicable for Picasso because of the following reasons:

- First, all prior designs use two or more antennas to achieve full duplex, i.e. they use TX and RX antennas that are separated by distances ranging from 15–20 cm. However, in a mobile device such as a smartphone or a small tablet where circuit real estate is at a premium, such multiple and distant antenna placement is infeasible (e.g., the iPhone 4 "antenna-gate"). Picasso's design must use a single antenna for both TX/RX to ensure it is applicable to small consumer devices.
- Second, as we show below, while prior work provides sufficient in-band cancellation, they do not provide isolation across slices. Specifically, since the interference cancellation technique uses active circuits that leak interference to neighboring spectrum, they harm the received signal's SNR and hurt performance.

To cancel self-interference, the best performing prior design [14] utilized separate TX/RX antennas separated by 15cm, so the transmitted signal is received after some variable attenuation and delay at the RX antenna. A cancellation circuit tries to obtain the *inverse* of the transmitted signal using a a balanced/unbalanced (balun) transformer, dynamically adjusts the attenuation and delay of the inverse signal to match the over-the-air self interference, and combines it with the signal received on the RX antenna. The prior design used active QHX220 chips to mimic programmable attenuation and delay, thus we call this technique *QHX-balun cancellation*. We implemented and evaluated the amount of cancellation one could get with QHX-balun cancellation, with TX/RX antennas spaced 15cm apart. We found that the QHX-balun cancellation techniques provide a significant amount of cancellation, 35dB on average for a 40MHz transmitted signal (Fig. 4(A)) which, when combined with the signal attenuation of 20dB due to the TX/RX antenna separation, is comfortably sufficient for our receivers that use 14-bit ADCs. The number is consistent with the prior reported results.

However, there is a catch. Even though the technique manages to cancel the self-interference, it ends up leaking a lot of interference into adjacent spectrum where there is no transmitted signal. We plot the amount of interference leakage in adjacent spectrum as a function of the spectrum distance from the edge of the transmit band. As Fig. 4(B) shows, the QHX-balun based technique leaks a significant amount of interference, as high as 10dB in adjacent bands that are separated by 1 MHz.

Apart from the fact that it requires separate TX and RX antennas, *interference leakage in adjacent spectrum* is a show stopper for Picasso since the adjacent spectrum may be used to receive concurrently, and the received signal's SNR will be reduced. To put the impact of interference leakage in perspective, a 10dB reduction in SNR can result in an order of magnitude decrease in throughput. The culprit is the active QHX chip used in the prior design. Even though it does provide self-interference cancellation in the transmit spectrum fragment, it distorts and leaks the transmitted signal power into neighboring spectrum fragments.

3.2 Picasso's Self-Interference Cancellation Technique

Picasso designs a novel self-interference cancellation technique that makes two significant contributions over all prior full duplex designs: (1) Single antenna full duplex and (2) Passive, leakage-free cancellation. We describe each in more detail below.

Single Antenna Full Duplex: Fig. 5(B) shows the block diagram of Picasso's single antenna, full duplex circuit. There are two key differences. First, the two separate antennas are replaced by a single antenna and a circulator (which we describe below). Second, instead of the active programmable QHX-220 component, there is a passive cancellation circuit which consists of two fixed delay lines and programmable, resistive switched attenuators. The key takeaway is that all the components in Picasso's design are ***passive***, whereas prior approaches used active components. We briefly describe the circuit components and how they work together to provide the required 45dB of cancellation.

A circulator [23] is a three-port device–labeled 1, 2 and 3 (Fig. 5(B))–that provides limited isolation between any two ports, and has historically been used in ranging systems such as radar [23]. A common circulator design contains an arrangement of conductor striplines and ferrite blocks. When the ferrites are biased, the circulator produces magnetic fields which enhance or oppose any RF energy flowing across the striplines. This biasing causes directional RF signal flow, routing the energy in a circular fashion from port 1 to 2, 2 to 3, and 3 to 1, but not in the reverse direction. Thus the power scattering matrix which represents the power flow allowed by the circulator between any pair of ports, S, for an ideal circulator that has perfect isolation quantified by the factor $\Gamma = 0$ can be written as

$$S = \begin{pmatrix} \Gamma & 1-\Gamma^2 & \Gamma \\ \Gamma & \Gamma & 1-\Gamma^2 \\ 1-\Gamma^2 & \Gamma & \Gamma \end{pmatrix} \quad (2)$$

where the row indicates the source port and the column indicates the destination port. Microwave circulators are available at ISM band

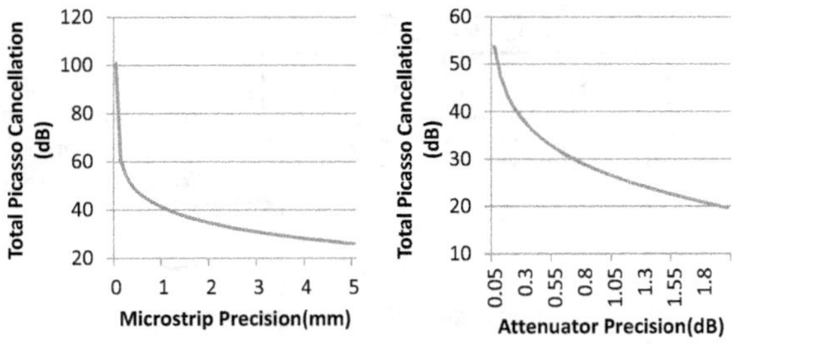

Figure 6: A)Picasso Cancellation vs. Delay Spread For delay mismatches larger than those introduced by 1mm copper traces($\approx$ 4.76ns), Picasso will not be able to achieve sufficient cancellation. **B)Picasso Cancellation vs. Attenuation Spread:** Picasso requires attenuation spreads of less than .5dB.

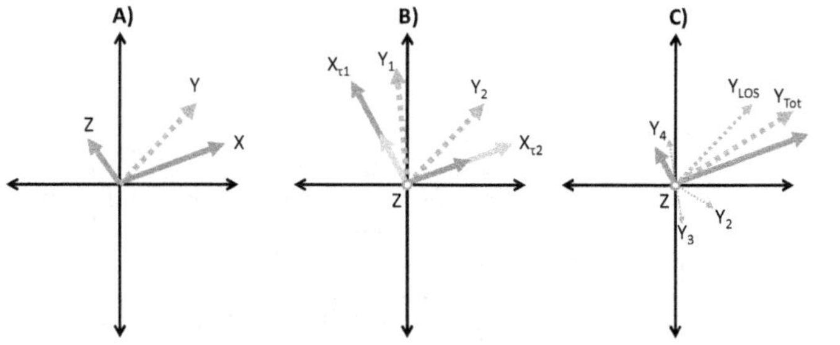

Figure 7: Reference Signal shown in blue (X), interference signal shown in green (Y), residual signal shown in red (Z). **A) Effect of delay mismatch on signal cancellation:** Delay offsets result in imperfect cancellation. **B) 2 fixed delay paths with tunable attenuators:** Different attenuator values compensate for different delay values: **C) Helps mitigate effects of multipath due to reflections:** Picasso is tuned to cancel the aggregate response (Y_{Tot}), not just the LOS component.

frequencies off the shelf [34] and their operation is similar in principle. Our design utilizes a 3-port circulator to concurrently route the TX and RX signals to and from the antenna as shown in Fig. 5(B). The TX signal is first passed through a 8dB power splitter, which splits the input signal into a high-power and low-power output. The high-power output has 85% of the input power and is connected to port 1 of the circulator. Hence, very little transmit signal power is wasted. The low-power output is used as a reference signal for self interference cancellation, which we describe after the next paragraph.

The antenna is connected to port 2 and the RX chain is connected to port 3. An ideal circulator would ensure that the signal on port 1 (the TX signal) would never leak to port 3 (the RX signal). However, in practice, the impedance of the circulator input ports are not perfectly matched with the outputs resulting in less than perfect isolation, $\Gamma \neq 0$, and there is leakage from port 1 to port 3. Despite this mismatch, we show in Section 6 that the circulator can still provide around 15dB of isolation between the transmit and receive chains, i.e. the self-interference we see on the receive port is 15dB lower than the transmitted signal power. However, remember that we need to provide 45dB of self-interference reduction for Picasso–15dB is not enough. Further, the cancellation should not leak any additional interference to neighboring spectrum bands.

Leakage-Free, Passive Cancellation: Our cancellation technique builds on two key observations. First, the main cause of interference leakage is the active QHX220 chip which serves as the programmable attenuation/delay component. Hence, to avoid interference leakage, we have to use passive components which do not introduce such distortion. However, even though passive and programmable attenuators are available off-the-shelf, passive delay lines are not.

Our key insight here is that the transmitted signal is going through static circuit components–it never travels through the air. Hence, the variance in the expected delay is significantly smaller, and Picasso can measure this limited range via a simple calibration procedure (these devices are designed to provide predictable behavior for a given frequency band). Picasso discards the active components required to amplify the signal to compensate for the larger possible range of delays and attenuation.

While the circulator limits the possible range of variation, environmental factors such as temperature and multipath reflections still cause some spread in the delay and attenuation of the self-interference. To simulate what effect these distributions have on cancellation, Fig. 6 shows a simulation, where we send a 40MHz OFDM signal at 2.4GHz carrier, and pass it through a single fixed delay and attenuation using measurements from our prototype. Next, we simulate varying delay and attenuation spreads by varying the wire length and attenuation components. As expected, we see that cancellation declines with increasing variance in the spread of delay and attenuation. While the circuit can achieve 30dB+ of cancellation when the length and attenuation deviations are small (less than 1mm in length and .5dB in attenuation spread), the performance suffers at larger mismatches.

To get a sense for what reasonable spreads in delay and attenuation can be, we look at what happens when a hand touches the antenna, a situation which commonly occurs in a mobile handset application. The hand changes the impedance of the antenna, which in turn causes reflections through the circulator to manifest at different delays. Our Vector Network Analyzer measurements showed that touching a standard WiFi Dipole antenna resulted in delay and attenuation spreads as large as 20ns and 3dB respectively. In copper traces, this would correspond to delay offsets of more than 4.2mm. As predicted in Fig. 6, we saw the cancellation drop precipitously from 40dB+ to less than 20dB. These possible perturbations underscore the need for a certain degree of tunability in order to ensure that the system is robust. While tunable delay elements are not readily available, tunable attenuators are. We leverage these devices in order to provide a margin for disturbances.

Our cancellation circuit utilizes fixed delay lines and programmable attenuators to compensate for a range of disturbances, including those which result in variable delay spreads seen in the example above. The trick is to realize that a change in delay manifests itself as a change in phase in the frequency domain. Analog signals of a particular frequency can be represented by a phasor/vector [19] so to aid our illustration, we sketch the vector representation of signal cancellation with delay mismatch ($\hat{\tau}$) in Fig. 7A. As you can see, the greater the delay mismatch, the larger the residual signal becomes.

A simple solution here would be to simply change the phase of the reference signal as well. The QHX220 does exactly this, but unfortunately requires integrated LNAs to inject additional power in order to change the phase by controlling the I and Q voltages. As we showed, these active components leak significant power into adjacent bands and thus should be avoided if possible.

We observe that we can achieve the desired phase shift if we simply utilize two delay paths instead of one. Recall that while programmable delay elements are not readily available, programmable attenuators are. By provisioning a programmable attenuator on each of our delay paths and then taking the aggregate response of both paths, we are able to mimic phase offset and adjust to variable delay spreads caused by perturbations. Again, let's take a look at the vector representation. As you can see in Fig. 7B, the aggregate response of two delay lines with programmable attenuators enables us to change the phase of the signal. Notice that in a single path, a programmable attenuator would not help at all with delay spread matching; the addition of the additional degree of freedom provided by the second path is essential.

Lastly, this approach enables us to combat internal reflections within the circulator and multipath in the surrounding environment. Because of imperfect isolation, the output signal from the circulator is actually spread in time. For instance, the power from port 1 leaking

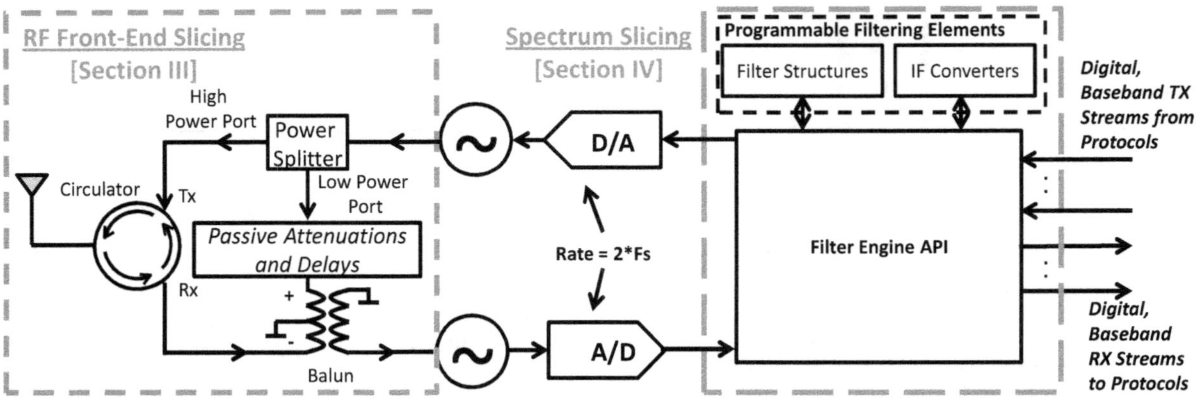

Figure 8: Picasso Block Diagram: The self-interference cancellation setup uses passive components for matching attenuation and delay. The filter engine setup processes digital baseband streams and appropriately resamples, filters and maps each stream to the appropriate fragment.

into 3 which we are trying to cancel, actually leaks to 2 as well. It then subsequently leaks back into 1 and ends up at our receive port again, albeit at a lower power (45dB lower with our circulator). If we simply match for the first delay path, we would be limiting our cancellation at 45dB due to this reflected path. Similarly, the multipath from the environment is typically on the order of 30-50dB lower as well. As you can see in Fig. 7C, the multipath components slightly affect the LOS component. With our cancellation circuit, the attenuation values can be fine-tuned to cancel the aggregate signal and achieve higher cancellation.

In summary, our cancellation circuit consists of two programmable attenuators and fixed delay line lengths which are statically configured to roughly straddle the delay the transmitted signal experiences through the circulator and any connecting wires (e.g., one delay line has less delay while the other has a larger delay than the transmitted signal). The attenuator values for each delay path are adaptively adjusted once every millisecond using a simple gradient descent algorithm [14] in order to compensate for changing environmental factors. Note that despite the fact that the attenuator values are adaptively changed, the attenuators themselves as well as the delay lines are passive. Thus, Picasso does not leak interference into adjacent bands as the QHX220 nonlinear active components do [14].

Reducing Transmit Insertion Loss: A final wrinkle is that instead of using the balun as a signal inverter, we use it as a subtractor circuit. Fig. 5(A) shows how a balun in the prior design [14] takes an input signal on the unbalanced tap and produces two output signals which are inverses of each other on the balanced taps. The drawback of this approach is that the input power is split between the two inverses, and thus half the power is spent in obtaining the reference signal for cancellation. Our insight is that the balun's operation can be modeled in reverse as taking two inputs on the balanced side and producing the subtraction of the input signals as the signal on the unbalanced side, shown in Fig. 5(B). Hence if the two inputs are replicas of each other, the output will be a zero signal. By using the balun to subtract rather than obtain the inverse, unlike prior designs, Picasso saves nearly half the transmit power from going to waste.

Picasso passes the output of the cancellation circuit to one of the balun inputs, and the other input is from port 3 of the circulator. We show in Sec. 6 that coupled with the circulator, this technique provides sufficient cancellation (>45dB).

Picasso's design thus eliminates the need for separate TX and RX antennas, preserving valuable real-estate in mobile devices. Further, unlike all prior work [14] which need to accommodate the large variance of the wireless channel between the TX and RX antennas with active cancellation circuits, Picasso's technique *requires no active components*. Because the circulator scattering characteristics are relatively fixed, once Picasso is calibrated at production time only small adjustments are required. Since the tuning scale is relatively small, all the components used in the design can be passive, enabling Picasso to guarantee that there is no interference leakage.

4. Picasso: SPECTRUM SLICING

Picasso provides transparent spectrum slicing, i.e., it allows each slice to have its own separate PHY and MAC protocol. The higher layer protocols are presented with the abstraction of a single narrowband radio that is exclusively for their use on the allocated spectrum slice. Picasso designs a digital filter engine that provides such spectrum slicing. Specifically, recall that digital samples being sent down from the PHY for transmission occupy a bandwidth equal to the slice's spectrum allocation w, and are centered at zero. The filter engine's job is to ensure that the samples are shifted to the correct center frequency f_c that is associated with that slice. On the receive side, the filter engine is doing the reverse,

The filtering engine contains an ADC and DAC, both of which are capable of operating at 200MS/s (the required Nyquist rate to create signals that span the entire 100MHz ISM band). Further, on the analog RF side, there is a single oscillator at 2.45GHz which upconverts the signal to the ISM band. The filter engine performs three high level tasks to shape the signals for transmission (to receive a signal, the steps are reversed):

1. **Resampling:** Since the DAC expects an input signal at 200MS/s, first Picasso upsamples the 40MS/s streams to 200MS/s. To accomplish this, the upsampler will interpolate (insert extra samples) to reach the 200MS/s mark.
2. **Filtering:** Upsampling creates aliases [20] that can cause interference. Hence the filter engine must low-pass filter both upsampled streams to remove any undesirable aliasing effects generated by the first step and retain only the upsampled-baseband version of each stream. To accomplish this, Picasso designs a reconfigurable filter bank, which consists of banks of programmable filters consisting of FIR, IIR, and resampling filter building blocks. These filters will be configured and sequenced to both resample and remove the aliasing effects.
3. **Mapping to the appropriate spectrum slices:** After the first two steps, the filtering engine has two 200MS/s streams each occupying 20MHz at the center frequency. The final step is to move the 20MHz occupancies to the specified fragments in the 100MHz band, i.e. to -38MHz and 22MHz, respectively (corresponding to 2.412GHz and 2.472GHz at a center frequency of 2.45GHz). For this Picasso uses intermediate frequency converters, which map the signal from incoming digital baseband to a specified digital intermediate frequency.

Finally, these streams are added together and sent to the DAC, then upconverted to the 2.45GHz carrier frequency and transmitted. Fig. 8, the right hand side shows the block diagram of the current design of Picasso's filter engine.

Note that the filtering also automatically takes care of any residual self interference. The analog self-interference cancellation prevents the receiver ADC from saturating, but by itself, it is insufficient to fully cancel out the self-interference. But now with the ADC dynamic range no longer saturated, Picasso's filtering engine can digitally remove the remaining self-interference. To summarize, Picasso's design consists of 2 main components, the self-interference cancellation block and the filtering engine. Individually, each component is inadequate but together, they enable Picasso to provide the full slicing abstraction.

5. IMPLEMENTATION

Picasso (Fig. 9) is implemented in two parts – the self interference cancellation is realized as a discrete board level prototype (shown in yellow) and the spectrum slicing portion is realized using a Virtex-5 LX30 FPGA based software radio from National Instruments[32] (shown in red). The FPGA is connected to an NI 5781 Baseband Transceiver which uses the USRP XCVR2450 as the radio front end. The radio front ends are then connected to Picasso's self-interference cancellation board.

5.1 Self-Interference Cancellation

ADC resolution is an important specification because it determines the receiver's dynamic range. The NI 5781 baseband transceiver features quadrature 200MS/s 14-bit ADCs and 200MS/s 16-bit DACs and provide approximately 86dB of dynamic range and coverage over the entire 100MHz ISM band.

Analog components have an active power range of operation over which they act linearly and predictably. Beyond these points, performance is often uncharacterized and starts to break down. Let's take a look at the XCVR2450, an off-the-shelf radio front end utilized in Picasso, to see which components we need to keep an eye on and make sure are not saturated.

One of the XCVR2450's active components we have to watch out for is the low noise amplifier (LNA). The figure of merit for the LNA is the input 1dB compression point, which represents the maximum power that this device can handle before it saturates. The XCVR2450 LNA [36] has a 1dB compression point of 0dBm. As long as we maintain a sufficient level (6dB is a good rule of thumb) below this point, the LNA performance will not suffer. With a maximum self-interference signal of 30dBm including PAPR, Picasso must provide at least 36dB worth of cancellation. Another component in the XCVR2450 to keep an eye on is the mixer [37], which has a Third Order Intercept (IIP3) of 11.5dBm. The IIP3 is a measure of the nonlinearity of the component and specifies the maximum handleable power before nonlinearities are no longer negligible. Because it comes after the LNA which has a maximum gain of 17dBm, we require 41.5dB of cancellation to limit the input power at the LNA (11.5dBm-17dB-6dB = −11.5dBm), allowing the gain control access to its full range. Fortunately, we've already obtained enough cancellation for the ADC Dynamic Range which turns out to be the most sensitive part of our system. Both of these cancellation numbers (36dB and 41.5dB) are lower than the 45dB needed for the ADC, so as long as Picasso can achieve 45dB of self-interference cancellation, RF performance should not suffer.

5.2 Filter Engine

The filter engine is implemented on two main processing modules in the software radio, a Virtex-5 95T FPGA and a real-time processor (NI PXIe-8133 RT Module). The Virtex-5 95T FPGA has 14,720 slices and nearly 9Mb of random access memory (RAM). The slices are used for almost every operation on the FPGA, while the RAM use is limited to storing data while averaging and providing temporary storage when restructuring the output of the definition of FFTs from bit order to natural order. For the filter structures, we utilize DSP48E slices available on the Xilinx LX30 [38]. Each of these slices is a highly configurable arithmetic logic unit which feature pipelined multiplier, adder and accumulator stages and can be clocked at up to 550MHz. Slices can be individually programmed and/or cascaded to implement FIR, IIR, and resampling filtering with relative ease.

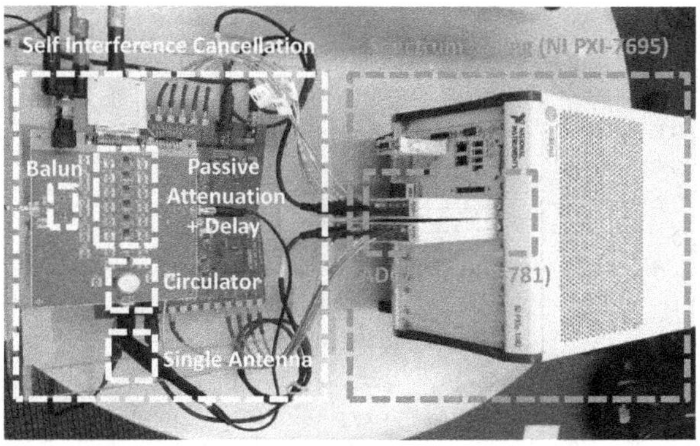

Figure 9: Picasso Radio Physical Setup: Built with off the shelf components and software radios from National Instruments.

5.3 PHY/MAC Implementation

We have also implemented a WiFi-style OFDM PHY and a CSMA MAC split across the FPGA and the realtime OS. The OFDM PHY can be configured to operate over different bandwidths (from 20MHz to 5MHz), and runs in realtime on the Software Defined Radio platform. It supports all the WiFi constellations (from BPSK to 64-QAM) as well as channel coding rates (1/2, 2/3, and 3/4 convolutional coding). We have not implemented an automatic rate adaptation algorithm, but in the experiments we empirically pick the combination of constellation size and channel coding rate that maximizes the link throughput. Thus, our prototype implements all the relevant features from a WiFi PHY/MAC. Note that we can run multiple instances of the WiFi PHY and MAC, one for each slice on the device.

6. EXPERIMENTAL EVALUATION

In this section we evaluate the performance of Picasso's prototype in an indoor testbed of NI Virtex-5 software radios that implement Picasso. As discussed before, we have implemented a WiFi style OFDM PHY and multiple instances (up to 4 due to FPGA hardware constraints) of these can be run on top of the slices in a single Picasso radio node. We benchmark Picasso's overall throughput performance, the efficacy of its self interference cancellation and programmable filtering components.

Compared Approach: Our main claim is that Picasso can slice a single RF front end and available fragmented spectrum into multiple slices, each of which can operate independent and decoupled PHY protocols. Hence the natural comparison is to a system that uses *multiple, separate radios*, one for each slice, with each radio configured to work on the corresponding specific spectrum slice, using its own PHY protocol. Unless stated otherwise, our comparison metric is total throughput.

An artifact of our current prototype is that our NI FPGA software radios use the USRP XCVR2400 as the RF front end. The XCVR's widest analog filter is only 40MHz, hence even though our ADCs and DACs have enough speed to cover the entire 100MHz ISM band, the RF front end limits us to 40MHz. Hence, in all our experiments the maximum spectrum that is available, either contiguous or the sum of the fragments, is 40MHz. However, if a wider front end were available, all of the results would carry over proportionally since none of our techniques are bandwidth-limited.

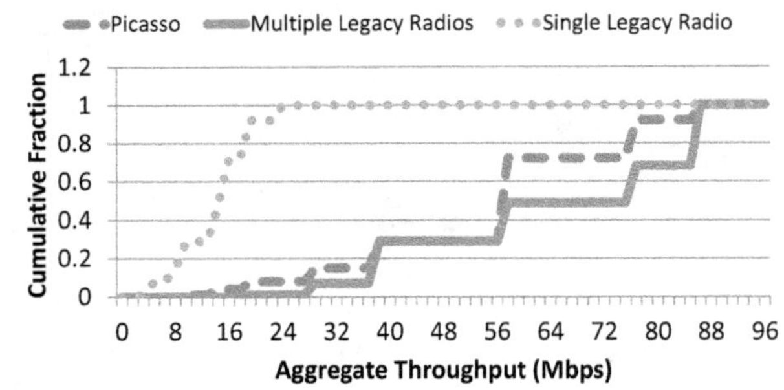

Figure 10: Aggregate Throughput for 4 Links: Picasso provides nearly the same throughput over 4 links as 4 independent radios would, and significantly outperforms the throughput a single legacy radio could provide over the same set of spectrum fragmented.

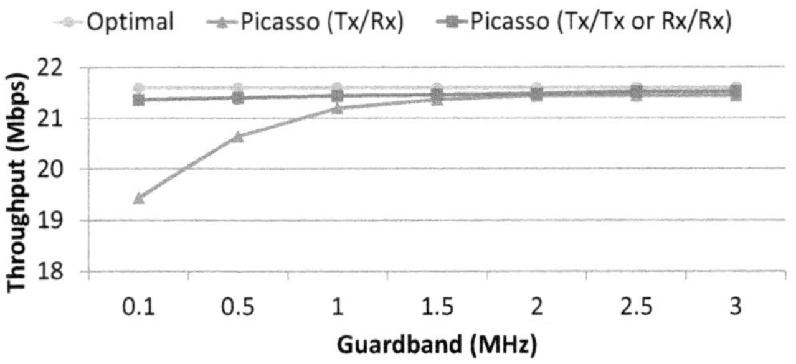

Figure 11: Picasso Throughput for Varying Guardbands: For guardband separations as small as 500 KHz, Picasso throughput does not suffer. At smaller separations, the Rx/Rx case maintains its performance but interference leakage causes a 20% degradation in performance for the Tx/Rx case.

6.1 Picasso: Overall Performance

Method: We statically place one Picasso node in the testbed and allow it to operate over four spectrum slices with separate WiFi PHY instances for each slice. For each of the four spectrum slices, we create a software radio client which is implementing a contiguous narrowband WiFi PHY on top of the spectrum slice. The four clients are placed at random locations in the testbed. The experiments are run in a 40MHz piece of spectrum, which is divided into four non-contiguous fragments randomly. The fragment sizes vary from 5-10MHz. We then measure the maximum possible PHY throughput by choosing the bitrate and coding combination for each of the spectrum fragments and measure the PHY throughputs achieved. To compare against the multiple radio setup, we replace the Picasso node with four physically separate nodes, each of them running a single narrowband contiguous PHY instance configured to work on one of the spectrum slices. Once again we measure PHY throughputs by picking the highest bitrate and coding combination. We also compute the throughput achieved by legacy radios which can only transmit on one spectrum fragment at a time, hence they always use the largest fragment. The experiment is repeated for each location and each setting of spectrum fragmentation. We plot the CDF of these throughputs in Fig. 10.

Analysis: Picasso achieves almost the same throughput with one radio as the physically independent setup with four physically separate radios, validating the claim that slicing allows decoupled operation of each slice. As expected, it outperforms the legacy radio approach by nearly a factor of $4\times$.

At the higher throughputs, there is a 10% dip in Picasso's performance relative to the multiple radio configuration. The reason is that at these throughputs, the transmission is likely using a very dense constellation and high coding rates (e.g. 64 QAM and 3/4 rate convolutional coding). When there are two adjacent slices very close to each other, separated only by the minimum default value of 500KHz, there is a small amount of signal power leakage from Picasso's self interference cancellation and programmable filtering. Hence if the Picasso node is transmitting while it is simultaneously receiving a densely coded stream, the small leakage can cause a slight performance drop, which we see in the above experiment. The loss is relatively small however, around 10%, and can be completely avoided by increasing the guard band between slices a bit higher than the default value of 500KHz. We leave that tradeoff as a tunable parameter which the network designer can choose depending on his needs. The next experiment sheds more light on this tradeoff.

6.1.1 *Impact of Slice Spectrum Separation*

Method: In this experiment we use Picasso nodes that are using 10MHz spectrum fragments each, but the separation between the fragments is an experimental parameter. We conduct two sets of experiments, one where a Picasso node is using two spectrum slices, one for TX and one for RX; and another where the Picasso node is either receiving or transmitting on both spectrum fragments. The first configuration tests whether Picasso's self interference cancellation technique leaks interference, and the second tests of Picasso's programmable filtering component leaks interference. We place two Picasso nodes in the testbed such that they can barely use the densest constellation and highest coding rate, since this is the configuration which suffers the most from interference leakage. The separation between the two spectrum slices is varied in both experiments and the PHY throughput achieved is measured as before. Fig. 11 plots the throughput achieved vs. the spectrum slice separation for Picasso along with the results for filtering and self interference cancellation individually.

Analysis: Picasso's throughput doesn't suffer much even for separations as small as 1MHz for either experiment. At 500KHz (the minimum default slice spectrum separation in Picasso), the simultaneous TX/RX case starts to see a small throughput loss of around $5-10\%$ which progressively increases to 20% for separations of 100KHz. The two RX case suffers relatively minor throughput loss. The reason is that Picasso's self-interference cancellation, despite using passive analog components to ensure frequency flat behavior and minimize leakage, still leaks a small amount of interference to the close neighboring band. Perfectly flat, non-leaking analog components are hard to obtain off-the-shelf, we suspect performance can be further improved with better analog components.

6.1.2 *Impact of Relative Signal Powers at Different Slices*

Another practical concern is whether the throughput achieved in each slice depends on the relative powers of the transmissions and receptions on other slices. For example, suppose a Picasso radio is operating two slices. While one slice is transmitting at 23dBm, the other slice is receiving a signal which after attenuation has a power of -85dBm. In other words there is a gap of 108dB between the two signals. One might wonder if throughput is affected because of any dynamic range issues with Picasso. We evaluate this scenario experimentally below.

Method: Here we use a Picasso node that is utilizing two spectrum slices of width 10MHz each, and the two slices are connected to two separate client radios configured to work on the respective slices. The Picasso radio as well as the clients are configured to transmit with a power of 23dBm on each slice. We vary the locations of the two radios, and for each location measure the difference in signal power at the two slices. If the Picasso radio is transmitting on one slice and receiving on the other, then the difference is between the transmit power of 23dBm and the received signal power on the other slice. If the Picasso radio is receiving on both slices, the difference is between the received signal powers on both slices. We plot the

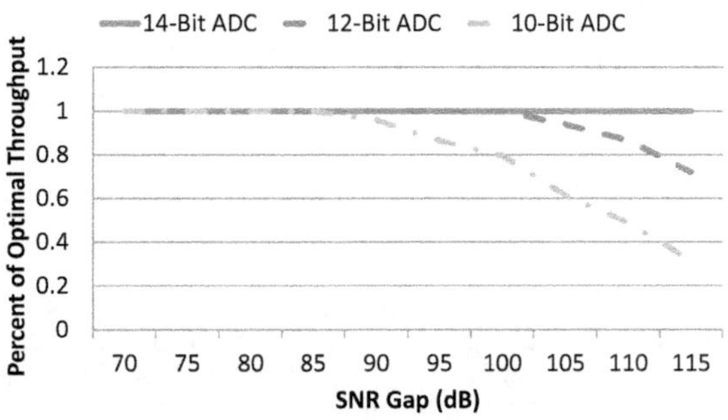

Figure 12: Impact of Relative SNRs of Different Slices: 14-Bit ADC provides sufficient dynamic range for all possible SNR gaps. 12-Bit ADC may suffice in certain scenarios if performance loss at high SNR gaps are acceptable.

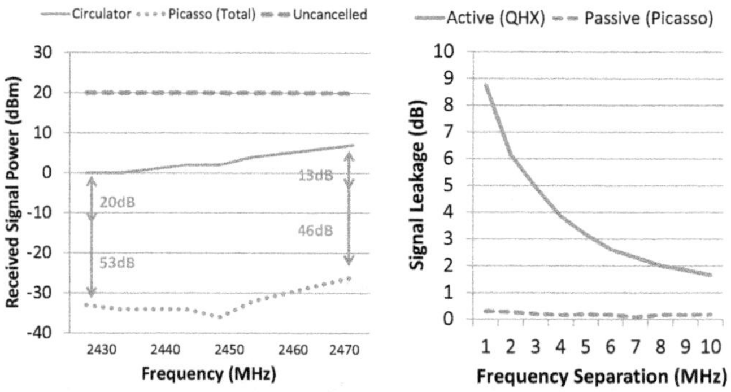

Figure 13: A)In Band Cancellation - Picasso achieves sufficient in-band cancellation (> 45dB) to prevent receiver saturation; **B) Out of Band Cancellation Leakage** - While QHX-Balun cancellation leaks significant interference into adjacent sidebands, Picasso cancellation does not.

throughput achieved by each slice when both slices are active simultaneously normalized by the throughput when only that slice is active vs. the difference in signal powers between the two slices at the Picasso radio in Fig. 12.

Analysis: As we can see, with Picasso's current prototype, even very high signal power differences between slices have no impact on throughput. The reason is that Picasso's self-interference cancellation coupled with a 14-bit ADC provides sufficient dynamic range so that throughput does not suffer in spite of large gaps between signal powers in concurrently active slices. To check the impact of the ADC resolution, we repeat the same experiment with 12-bit and 10-bit ADCs. Recall from Sec. 3 that lower resolution ADCs have lower dynamic range and hence are more likely to get saturated if the self-interference is too strong. As Fig. 12 shows, there is a slight loss of performance with 12-bit ADCs, and more with 10-bit ADCs. This is expected, since the lower dynamic range leaves less ADC resolution for the received signal and consequently introduces higher quantization noise when the difference in signal powers is very high. Nonetheless, the performance is still reasonable, and suggests even a 12-bit ADC might suffice for Picasso in most cases.

6.2 Picasso's Self Interference Cancellation

We experimentally evaluate the performance of Picasso's cancellation technique. From a single Picasso node, we transmit a 40 MHz OFDM signal and measure the amount of cancellation Picasso's technique achieves. We plot the cancellation achieved vs. the bandwidth of the transmitted signal Fig. 13(A).

As we can see the cancellation achieved on average is around 45-50dB, which is sufficient for Picasso's purposes. Further, note the almost complete lack of interference leakage as Fig. 13(B) shows. Even when the adjacent spectrum fragment is as close as 1MHz, there is almost no interference leakage from the transmitted signal. At the minimum slice guard band value of 500KHz there is a small leakage which shows up as the slight loss of throughput we discussed in the throughput experiment in Sec. 6.1.1. And note that even though the experiment was run for several minutes and across different days, we could use the same statically configured setting to consistently obtain the above cancellation. Prior techniques need to adapt on a per packet basis.

Throughput Impact of using Passive Components: In Section. 3.1 we discussed why the prior QHX-balun based self-interference cancellation techniques used for full duplex are not a good fit for Picasso. Here we evaluate experimentally the throughput impact if the QHX-balun based cancellation technique is used compared to Picasso's technique.

Method: We run an experiment with two nodes which have their transmit and receive antennas spaced 15 centimeters apart and implementing QHX based balun cancellation. Each radio has two slices of 10MHz width, and the spectrum separation between the slices is varied. For each slice spectrum separation setting, we apply both cancellation techniques separately using the same power and gain settings. We once again measure PHY throughput for both techniques and plot it vs. the separation between the transmit and receive spectrum fragments in Fig. 14.

Analysis: Fig. 14 quantifies the throughput impact of the out-of-band interference leakage the prior QHX-balun cancellation technique introduces. Picasso's technique, because it does not leak self-interference, achieves a throughput that is nearly $5\times$ better when the separation is around 500KHz.

6.3 Programmable Filter Engine

In this section, we discuss the efficacy of Picasso's programmable filtering. We focus on two metrics that can be traded off against each other: out-of-band interference rejection and the complexity of the filtering implementation. Complexity is measured as the number of coefficients used in the linear digital filter. The logic utilization for the filter engine is a function of the filter length, coefficient precision, coefficient symmetry, and input data precision. While the filter engine can be configured to utilize any parameters, Picasso utilizes 16-bit symmetric coefficients with an input sample precision of 16 bits. Thus, the only remaining variable parameter is the filter length. The complexity of the filter must also be weighed against the configuration speed as the number of clock cycles required to reconfigure the filter engine is directly proportional to the number of coefficients used in the filter.

6.3.1 Complexity vs. Interference Rejection

To determine how different filter lengths impact performance, we conduct the following experiment using 2 Picasso nodes: node 1 transmits two 10 MHz OFDM signals in two spectrum slices which node 2 must decode. We vary the separation between the two spectrum fragments. Performance is measured in terms of the achievable PHY throughput between nodes 1 and 2. We plot the throughput achieved for different filter complexities and varying separation between spectrum fragments.

Analysis: From Fig. 15, we can see that as filter lengths increase, the throughput between nodes 1 and 2 increase. This is expected, a longer filter length does a much better job of rejecting out-of-band interference since its frequency response starts approximating the ideal square shape and roll-off is reduced. In Picasso's implementation, we use 20'th order filters, which consume modest resources and exhibit very good interference rejection even when the adjacent slice is as close as 500KHz.

6.3.2 Complexity vs. Reconfiguration Speed

The filter engine facilitates on the fly filter configuration by reloading new coefficient vectors. The limitation is that the filtering operation must be halted while the new coefficient values are loaded

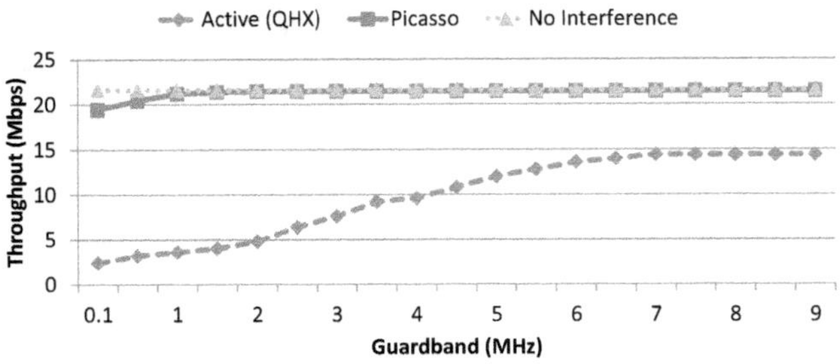

Figure 14: Throughput Comparisons with different self interference cancellation schemes: Active cancellation introduces sideband leakage which reduces system throughput. Picasso's passive cancellation does not, which allows for near optimal throughput.

and some of the internal data structures are subsequently initialized. While longer filters invariably take a longer time to load, they provide better performance as shown in Fig. 15. The worst-case load time for the longest filter is less than 4-5μs, which is less than one WiFi slot length.

7. APPLICATIONS

Picasso's slicing capability is a general primitive that can be used for a variety of applications. In this paper we show how Picasso can help diverse networks coexist gracefully in dense deployments and dynamic fragmented spectrum. More broadly though, Picasso can be applied to designing WiFi APs, multi-channel mesh networks, and even adaptive duplexers in phones. We discuss these applications briefly in Sec. 9.

We demonstrate how Picasso helps tackle coexistence by using the example deployment shown in Fig. 16. The deployment paints a typical home network: a WiFi AP aims to provide Internet access to a laptop, a tablet, an Xbox and a smartphone. Concurrently, there is a multiplayer game going on, and the Xbox wants to stream high definition individual gameplay video to the smartphone and tablet, and common gameplay to a TV. The smartphone and tablet also double up as wireless gameplay controllers. Further, with the emergence of P2P technologies such as WiFi-Direct [31] (a new WiFi standard for directly connecting two devices without incurring the overhead of an AP), except for the WiFi network, all the other connections will likely be connected via independent WiFi-Direct links.

As we can see, the scenario has a number of diverse networked applications that have to coexist while sharing a single radio on each device in the fragmented ISM band spectrum. Further, their requirements are diverse. The WiFi network wants to maximize throughput. The gameplay streaming group (Xbox to smartphone, tablet and TV) wants to maximize video quality. The gameplay controller group have relative low throughput requirements but low latency is a critical requirement for seamless gameplay.

Picasso provides a clean architecture for multiple such networked applications to coexist by allowing the applications to decouple their operation from each other using slicing. Specifically, Picasso enables us to slice each device's radio into three slices. The first slice would be assigned to the WiFi network, which is now free to operate its own optimized PHY and MAC protocols to maximize throughput. The second slice would be for the gameplay video streaming application (Xbox, smartphone, tablet and TV), and if it so chooses, would operate independent PHY and MAC protocols optimized to maximize video quality. Finally, a third slice is for the gameplay controller application (Xbox, smartphone, tablet), with an independent PHY/MAC optimized for low latency. Thus Picasso *decouples the operation of the networked applications from each other*, and enables them to gracefully coexist yet independently design their protocols to optimize their respective objectives.

Compared Approaches: We compare the performance of Picasso with the following approaches:

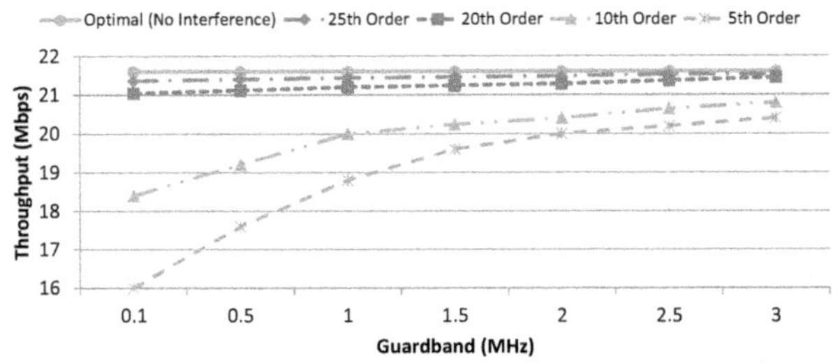

Figure 15: Guardband separation between adjacent fragments vs. achievable throughput for varying filter lengths: Out of band rejection performance increases for higher order filters.

1. **Multiple Radios (MR) [1]**: This is the optimal scheme which uses separate dedicated radios for each network group. In other words each device would have multiple radios, one for each networked application.
2. **SWIFT [24]**: This is the state-of-the-art practical technique that can weave multiple spectrum fragments into a single logical connection via a modified OFDM PHY. Medium access is based on CSMA. However, standard CSMA is not optimized to provide QoS, since it uses the same backoff counter for all nodes. Hence we modify SWIFT's CSMA to use 802.11e's MAC protocol [6], a modification to the standard 802.11 MAC protocol for providing QoS. Specifically, WiFi network traffic is classified as best effort, and uses a congestion window range of $31 - 1023$. Video streaming traffic has higher priority and when it is contending, it uses a congestion window range of $15 - 31$, and finally $7 - 15$ for the latency-sensitive traffic which has the highest priority. These are standard parameters from the 802.11e specification. Finally, note that a SWIFT radio has to either transmit or receive on all fragments at the same time, it cannot operate in full duplex mode.

Traffic Model: In order to emulate a realistic network, each device's traffic pattern is modeled using empirical trace based models [10]. Specifically for the WiFi network group, we use a traffic model that models Internet use behavior on WiFi networks [10]. For the video streaming group, we use a traffic model that a video encoder would generate in order to guarantee 720P quality video. Note that contrary to perception, guaranteeing a consistent video quality does not imply a consistent data rate, due to variations in the video frames the actual data rate required to maintain a fixed video quality varies over time. Finally for the gameplay controller group, we assume that short packets are generated at random intervals with an average period of 50 milliseconds between them [21], which is slightly lower than the median human response times of 200 milliseconds.

Slice Allocation: A natural question is how are slices allocated for Picasso as well as the multiple radio approach. We use a simple greedy algorithm. For the latency-sensitive group, we allocate a small spectrum slice of 1MHz that is sufficient for the low rate latency-sensitive traffic. The next slice is allocated to the video streaming group, with sufficient spectrum to provide 720p streaming (which requires roughly 3Mbps per stream, and thus nearly 10Mbps). Leftover spectrum is allocated to the WiFi network group. Note that this implies that the WiFi network might be allocated multiple slices corresponding to whatever spectrum fragments are leftover. The network will run separate narrowband OFDM PHY and CSMA MAC instances for these slices. We assume that a higher layer mechanism such as multi-path TCP [27] (MPTCP) is used to aggregate the throughput.

An important point is that this is just one allocation; a different policy might use a different allocation (e.g. a parent may prioritize WiFi network throughput over the gameplay streaming!). However, the key point is that Picasso decouples policy from mechanism, and enables a wide range of policies.

Another issue is how slice allocation decisions are coordinated and

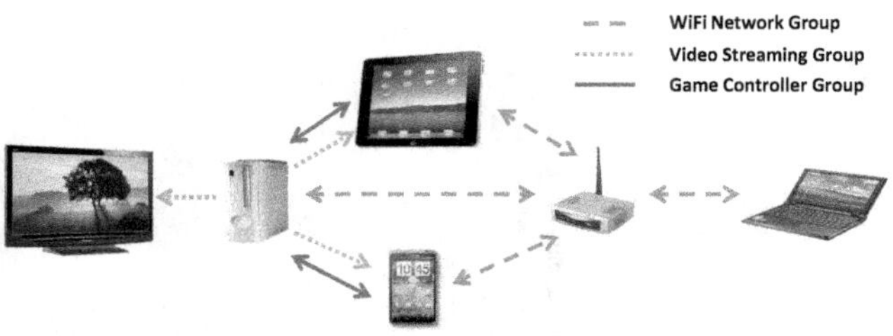

Figure 16: Coexistence in Dense Environments: Picasso enables devices operating in multiple networks to flexibly share fragmented spectrum and RF front ends, yet operate in a decoupled manner.

communicated to all the gadgets. To achieve this, we allocate a slice of 1MHz as a control channel, that all devices access to announce their QoS needs, and coordinate to figure out the best slice allocation. Note that such a simple decoupled control channel functionality is also because of the slicing capability, non-Picasso radios would have to co-ordinate in band using mechanisms such as RTS/CTS.

The SWIFT approach of course does not have any slice allocation issues. All devices operate over the entire available spectrum, including fragmented spectrum.

Optimized PHY/MACs for the individual slices: Each networked application uses an optimized PHY/MAC for their slices. The WiFi network group has two choices, it could run CSMA with OFDM where all the devices contend, or it could use FDMA like prior work has suggested [28] by taking advantage of Picasso's slicing capability to allocate slices for each individual device in the group. We chose the latter approach since FDMA has much less overhead compared to CSMA. For the video streaming slice, we use a simple OFDM PHY and CSMA MAC, since all the transmissions originate from the Xbox and thus CSMA causes no contention overhead. For the gameplay controller group, we use CSMA since that provides the lowest latency access.

Experimental Method: We begin by evaluating the performance of Picasso along three axes - WiFi Network's Aggregate Throughput, Video Streaming Group PSNR, and Gaming Group Average Latency - and compare it with the performance of the multiple radio and SWIFT approaches.

We randomly vary two independent parameters in our network. First, we vary the relative channel conditions between the devices. Because the varying channel conditions inevitably lead to a disparity in absolute performance, we normalize all three axes relative to the values achieved by the MR approach to demonstrate the relative performance gains of implementing Picasso over SWIFT independent of the channel conditions. Second, we vary the fragmentation pattern of the network. We assume that there are other narrowband interferers in the ISM band (e.g. Zigbee devices) which are transmitting on random frequencies, and hence the spectrum available to the Networks is fragmented. The Picasso and MR approaches use the slice allocations from the algorithm described above. SWIFT devices of course always use all the available spectrum.

Analysis: Fig. 17(A) plots the first tradeoff between latency and throughput. Picasso provides nearly 2× higher throughput and 2× lower latency than SWIFT. With SWIFT, there is a positive correlation between the two variables, the higher the throughput achieved by the WiFi network group, the higher the latency for the controller group. The reason behind this effect is that higher throughput implies latency-sensitive packets have to wait longer to obtain channel access and in turn incur higher latency using SWIFT-CSMA. On the other hand, Picasso has none of these issues, since the latency-sensitive traffic has a separate slice, they obtain a consistent low latency. Further, note that Picasso is able to obtain consistently higher throughput than SWIFT. The reason is that since SWIFT mediates access to the channels in time, there is significant overhead associated with idle slots, collisions, and backoffs. Because Picasso slices in frequency, the only overhead it pays is in the lengths of the guardbands, which enables it to make more efficient use of the spectrum.

Fig. 17(B) plots the second tradeoff between latency and PSNR. Picasso provides nearly 10dB higher median PSNR than SWIFT for the video streams. Once again, there is a slight positive correlation here between PSNR and Latency for SWIFT with higher video qualities corresponding to higher latencies. The reason is the same as above: the video traffic occupies a significant fraction of the channel time to maintain high video quality, with the latency-sensitive traffic starved for channel access in spite of the aggressive backoff counters it employs. Picasso again is able to maintain slice isolation and thus can guarantee that the latency-sensitive traffic obtains consistently low latency, and that the video streams also obtain a predictable rate to maintain a high video quality. However, sometimes SWIFT achieves higher PSNRs than Picasso, which is expected since SWIFT could allocate a majority of the channel access for video transmissions depending on the channel access. That increases video quality, but comes at the expense of latency and throughput for the controller and WiFi groups. Further, note that SWIFT cannot guarantee a predictable high rate, the video quality varies significantly across time, which causes a loss of video fidelity.

Fig. 17(C) plots the third tradeoff between video quality and throughput. Picasso, as expected, maintains consistent video quality and normalized throughput. SWIFT cannot guarantee consistent throughput for the video streams, hence video encoding adapts to a lower quality. SWIFT does not achieve higher throughput either, since as discussed above, SWIFT CSMA adds significant overhead due to contention and limits the actual useful throughput that can be achieved. With Picasso, in the WiFi network group, we can assign slices for each individual device that is part of the group and operate FDMA for medium access, and thus eliminate contention overhead. Consequently, Picasso achieves both higher throughput and video quality.

8. RELATED WORK

Picasso bridges and builds upon research in two previously unrelated areas of wireless: in-band full duplex and transceiver design for fragmented spectrum.

Picasso distinguishes itself from prior single-channel full-duplex cancellation work [5, 14, 9, 15] by achieving self-interference cancellation using a single antenna and passive attenuation/delay components which do not leak interference into adjacent spectrum. Furthermore, unlike [14], Picasso does not require constant tuning and as such would be much easier to manufacture on a chip.

The RF front end and spectrum slicing abstraction that Picasso provides differentiate it from prior transceiver designs for fragmented spectrum because it provides strong decoupled operation between slices. Because prior works [24, 7, 22, 26] are unable to transmit while simultaneously receiving, the usage and performance on all spectrum fragments are inevitably tied together, greatly complicating the design of higher-layer coexistence protocols.

The virtue of Picasso is that because it enables higher layers to systematically exploit each fragment of spectrum independently, coexistence protocols can be greatly simplified. Thus, Picasso's slicing capability provides a key building block for coexistence protocols [25, 2, 28], facilitating the allocation of available spectrum fragments amongst competing radios. Picasso in turn builds on top of prior spectrum sensing frameworks such as [13, 16, 28], which provide information about the RF environment and inform Picasso regarding availability of spectrum slices.

9. DISCUSSION & CONCLUSION

Picasso's ability to simultaneously use arbitrary spectrum fragments with a single radio/antenna allows it to virtualize a single radio

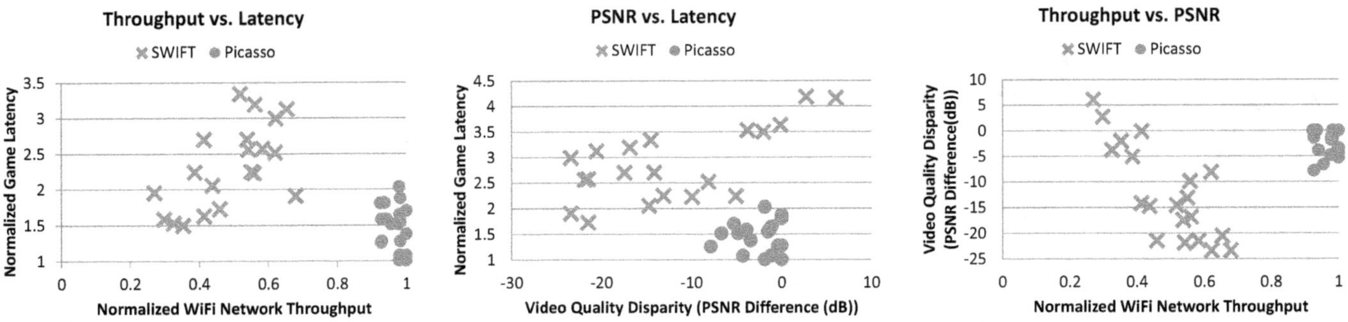

Figure 17: A) Throughput vs. Latency: Picasso eliminates the contention latency incurred by the gaming system when competing with larger WiFi Network packets;**B) PSNR vs. Latency:** Picasso can provide latency and deadline guarantees whereas SWIFT cannot; **C) Throughput vs. PSNR:** Picasso achieves higher aggregate throughput and PSNR by eliminating the idle slots inherent in SWIFT CSMA due to contention and random access.

into separate independent slices operating on different spectrum fragments. This slicing abstraction provides a clean solution for many wireless applications, of which fragmented spectrum exploitation, heterogeneous network coexistence, and radio sharing stand out.

We believe that Picasso is a general architectural solution that is not just limited to operation in the unlicensed bands. For instance, cellular spectrum fragmentation is likely to remain an issue globally because of short-sighted regulatory planning. The problem is compounded by the fact that even the same service providers own different fragments of spectrum in different regions, forcing mobile chipsets to accommodate a wide frequency range of operation in order to support roaming. Not only would a system like Picasso enable handset manufacturers to save costs by replacing the disparate chipsets with a single integrated solution, it would also facilitate global roaming and liberate consumers to more easily switch network operators, potentially driving improved quality of service due to increased competition between service providers.

Implementing Picasso within a mobile chipset would be challenging, but not impossible. Isolation devices such as circulators [40], are readily available in miniaturized form, but provide slightly less isolation. Consequently, due to the limited power handling capability of miniaturized microwave components, Picasso would likely require additional self-interference cancellation on the order of 55dB or so in order to ensure that the LNA, mixers, and lower resolution ADCs operate within the specified power ranges. We believe that such performance is feasible and will be realized in the near term.

Acknowledgements: We would like to thank Dave Levin, Ratul Mahajan, Asaf Cidon, and the anonymous SIGCOMM reviewers for their insightful comments. This research was made possible through the financial support of the NSF POMI Expedition Grant, as well as NDSEG and NSF Graduate Research Fellowships.

10. REFERENCES

[1] P. Bahl, A. Adya, J. Padhye, A. Walman, "Reconsidering Wireless Systems With Multiple Radios," In *ACM SIGCOMM CCR*, 2004

[2] P. Bahl, et. al "KNOWS: Kognitiv Networking Over White Spaces." In *IEEE DySPAN*, 2007

[3] M. Briggs, J. Martinez, D. Bare, "Power Measurements of OFDM Signals," In *IEEE Symposium on Electromagnetic Compatability*, 2004.

[4] R. Chandra, R. Mahajan, T. Moscibroda, R. Raghavendra, P. Bahl, "A Case for Adapting Channel Width in Wireless Networks," In *ACM SIGCOMM*, 2008

[5] J. Choi, M. Jain, K Srinivasan, P. Levis, S. Katti, "Achieving Single Channel, Full Duplex Wireless Communication", In *ACM MOBICOM*, 2010

[6] S. Choi, J. Prado, N. Shankar, S. Mangold, "IEEE 802.11 e contention-based channel access (EDCF) performance evaluation", In *IEEE ICC*, 2003.

[7] E. Coffman, et. al, "Channel Fragmentation in Dynamic Spectrum Access Systems - a Theoretical Study," In *ACM SIGMETRICS*, 2010

[8] FCC Table of Frequency Allocations, *http://transition.fcc.gov/oet/spectrum/table/fcctable.pdf*, May 6, 2011.

[9] E. Everett, M. Duarte, C. Dick. A. Sbharwal, "Empowering Full-Duplex Wireless Communication by Exploiting Directional Diversity," *IEEE Asilomar*, 2011.

[10] A. Gember, A. Anand, A. Akella, "A Comparative Study of Handheld and Non-Handheld TraffiñÄc in Campus Wi-Fi Networks," *Passive and Active Measurement Conf*, 2011.

[11] T. Gill, "RF performance of mobile terminals - a challenge for the industry", *http://www.cambridgewireless.co.uk/Presentation/Trevor%20Gill%20-%20Radio%20SIG%20310311.pdf*, 2010.

[12] S. Hong, J. Mehlman, S. Katti, "Picasso: Full Duplex Signal Shaping to Exploit Fragmented Spectrum." In *ACM HotNets*, 2011.

[13] S. Hong, S. Katti, "DOF: A Local Wireless Information Plane", In *ACM SIGCOMM*, 2011

[14] M. Jain, J. Choi, T. Kim, D. Bharadia, K. Srinivasan, P. Levis, S. Katti, P. Sinha, S. Seth, "Practical Real-Time Full Duplex Wireless", In *ACM MOBICOM*, 2011.

[15] M. Khojastepour, et. al, "The Case for Antenna Cancellation for Scalable Full Duplex Wireless Communications" In *ACM HOTNETS*, 2011.

[16] K. Lakshminarayanan, S. Sapra, S. Seshan, and P. Steenkiste. "Rfdump: An Architecture For Monitoring the Wireless Ether." In *ACM CoNEXT*, 2009.

[17] T. Lamprecht, et. al, "Passive alignment of optical elements in a printed circuit board", In *Electronic Components and Technology Conference*, 2006.

[18] G. Matthaei, et. al, "Microwave filters, impedance-matching networks, and coupling structures", *Dedham, Mass: Artech House Books*, 1980.

[19] T. Moon, W. Stirling, "Mathematical Methods and Algorithms for Signal Processing," *Prentice Hall*, 1999.

[20] A. Oppenheim, R. Schafer, "Discrete Time Signal Processing: 3rd Edition," *Prentice Hall Inc*, 2009

[21] C. Palazzi, S. Ferretti, S. Cacciaguerra, M. Roccetti, "A RIO-Like Technique for Interactivity Loss-Avoidance in Fast-Paced Multiplayer Online Games," In *ACM Computers in Entertainment Magazine*, 2005.

[22] J. Poston, W. Horne, "Discontiguous OFDM Considerations for Dynamic Spectrum Access in Idle TV Channels." In *IEEE DySPAN*, 2005.

[23] D. Pozar, "Microwave Engineering: 3rd Edition", *Wiley*, 2004.

[24] H. Rahul, N. Kushman, D. Katabi, F. Edalat, and C. Sodini. "Learning to Share: Narrowband friendly wideband radios." In *ACM SIGCOMM*, 2008

[25] K. Tan, J. Fang, Y. Zhang, S. Chen, L. Shi, J. Zhang, Y. Zhang, "Fine Grained Channel Access in Wireless LAN." In *ACM SIGCOMM*, 2010.

[26] K. Tan, et. al, "Spectrum Virtualization Layer." *MSR Tech Report*, http://research.microsoft.com/apps/pubs/default.aspx?id=154410, 2011.

[27] D. Wischick, C. Raiciu, A. Greenhalgh, M. Handley, "Design, Implementation and Evaluation of Congestion Control for Multipath TCP," In *USENIX NSDI*, 2011

[28] L. Yang, W. Hou, L. Cao, B. Zhao, H. Zheng, "Supporting Demanding Wireless Applications with Frequency-agile Radios", In *USENIX NSDI*, 2010

[29] L. Yang, B. Zhao, H. Zheng, "The Spaces Between Us: Setting and Maintaining Boundaries in Wireless Spectrum Access", In *ACM MOBICOM*, 2010

[30] Gizmodo, "IPhone 4 Antenna-Gate", *http://gizmodo.com/5846638/giz-explains-whats-so-smart-about-the-iphone-4ss-antenna*,2011.

[31] WiFi Direct Industry White Paper, *http://www.wi-fi.org/discover-and-learn/wi-fi-direct*, 2010.

[32] NI PXIe-8133 User Manual, *http://www.ni.com/pdf/manuals/372870b.pdf*, 2011.

[33] NI5781 Datasheet, *http://sine.ni.com/ds/app/doc/p/id/ds-212/lang/en*, 2011.

[34] MECA Circulator, *http://www.e-meca.com/rf-circulator-isolator.php*

[35] UHD Daughterboard Application Notes, *http://files.ettus.com/uhd_docs/manual/html/dboards.html*, 2011.

[36] Avago MGA-82563 17dBm GaAs MMIC Amplifier *www.avagotech.com/docs/AV02-1985EN*.

[37] Avago AD8347 Direct Conversion Quadrature Demodulator, *www.analog.com/static/imported-files/data_sheets/AD8347.pdf*.

[38] UG193: XtremeDSP User Guide, *http://www.xilinx.com/support/documentation/user_guides/ug193.pdf*, June 1, 2010.

[39] DS249: LogiCORE IP CORDIC v4.0 Data Sheet, *http://www.xilinx.com/support/documentation/ip_documentation/cordic_ds249.pdf*, March 1, 2011.

[40] 3.2 mm Miniaturized Circulator Data Sheets, *http://www.hitachimetals.com/product/rf_microwave/isolators/small/3_2mmSmallIsolators.html*, 2012.

Spinal Codes

Jonathan Perry, Peter A. Iannucci, Kermin Fleming, Hari Balakrishnan, Devavrat Shah
Massachusetts Institute of Technology
Cambridge, Mass., USA
{yonch,iannucci,kfleming,hari,devavrat}@mit.edu

ABSTRACT

Spinal codes are a new class of rateless codes that enable wireless networks to cope with time-varying channel conditions in a natural way, without requiring any explicit bit rate selection. The key idea in the code is the sequential application of a pseudo-random hash function to the message bits to produce a sequence of coded symbols for transmission. This encoding ensures that two input messages that differ in even one bit lead to very different coded sequences after the point at which they differ, providing good resilience to noise and bit errors. To decode spinal codes, this paper develops an approximate maximum-likelihood decoder, called the *bubble decoder*, which runs in time polynomial in the message size and achieves the Shannon capacity over both additive white Gaussian noise (AWGN) and binary symmetric channel (BSC) models. Experimental results obtained from a software implementation of a linear-time decoder show that spinal codes achieve higher throughput than fixed-rate LDPC codes [11], rateless Raptor codes [33], and the layered rateless coding approach [8] of Strider [12], across a range of channel conditions and message sizes. An early hardware prototype that can decode at 10 Mbits/s in FPGA demonstrates that spinal codes are a practical construction.

Categories and Subject Descriptors: C.2.1 [Network Architecture and Design]: Wireless communication

General Terms: Algorithms, Design, Performance

Keywords: Wireless, rateless, channel code, capacity, spinal code, encoder, decoder

1. INTRODUCTION

Signal attenuation, noise, multipath fading, and interference all make it difficult to achieve high throughput over wireless networks. Achieving high throughput is challenging even when the channel is characterized by a single invariant parameter such as the noise variance or the bit-error rate, but in practice, mobility and interference cause conditions to vary over multiple time-scales. Currently deployed solutions to this problem in wireless LANs and cellular networks are *reactive*; they measure the channel to dynamically select a "bit rate"—i.e., a modulation scheme, channel code, and code rate—from a set of pre-defined choices.

SIGCOMM'12, August 13–17, 2012, Helsinki, Finland.

An alternate approach is to use a *rateless code* between the sender and receiver [8, 12, 27]. With a rateless code, the sender encodes the message bits so that every achievable higher rate is a prefix of achievable lower rates. The sender keeps transmitting coded data until the receiver informs the sender that it has correctly decoded all the data (or the sender gives up). An ideal rateless code can be decoded correctly with modest computational complexity as soon as the effective rate drops below the capacity of the channel; the prefix property of such a code eliminates the need for the heuristics used in explicit bit rate selection.

This paper presents the encoding, decoding, and performance of *spinal codes*, a new class of rateless codes. Spinal codes can encode message bits directly to constellation symbols (which is somewhat unusual), or they can produce a sequence of coded bits to be transmitted using any pre-existing symbol set. The first approach is preferable because the spinal decoder can extract information from the raw received symbols without a demapping step, permitting the use of the same dense constellation at all signal-to-noise ratios (SNR). Even without control over the physical layer, spinal codes may be useful over an existing physical layer modulation to improve throughput and error resilience.

Spinal codes apply a *hash function* sequentially to successive portions of the original message bits to produce a pseudo-random mapping from message bits to coded bits, and then use a *constellation mapping function* to produce a sequence of symbols for transmission. Thanks to the sequential application of the hash function, two input messages differing in even a single bit result in independent, seemingly random symbols after the point at which they differ: any difference between messages is magnified in the encoded symbols. This property makes the code robust to noise and errors; it achieves reasonable rates even at SNR as low as −5 dB.

No previous code uses a hash function because good hash functions, by design, do not possess a simple, invertible structure. Therefore, unlike previous codes such as low-density parity check (LDPC) [11], LT [21], or Reed-Solomon [42] codes whose decoders exploit graphical or algebraic properties, spinal decoding requires a different strategy.

So how does one decode spinal codes? Our solution draws its strength from the sequential structure of the encoding. The idea is to search over a tree, breadth first, with edges weighted by the difference between the actual received symbols and the result of *replaying the encoder* for different potential message bit sequences. The shortest path is an exact maximum-likelihood (ML) decoding, but the size of the tree is exponential in the message length. We introduce a polynomial-time method, called the *bubble decoder*, which prunes the tree to produce an approximate ML estimate of the message. Theoretically, the bubble decoder achieves capacity. To the best of our knowledge, spinal codes are the first rateless codes with an efficient encoder and decoder that achieve capacity over both AWGN and BSC (bit-flip) models.

Our experiments conducted using a *linear-time* spinal decoder compare spinal codes to Raptor [33, 26] over a dense QAM-256 constellation, Strider [12], and LDPC codes [11]. We present four principal experimental results. First, the rate improvements of spinal codes over Raptor and Strider (both rateless codes) for AWGN channels are significant:

SNR	Raptor/QAM-256	Strider
High (> 20 dB)	21%	40%
Medium (10 – 20 dB)	12%	25%
Low (< 10 dB)	20%	32%

Second, spinal codes outperform the best envelope of the 802.11n family of LDPC codes, because of a "hedging" effect that we identify. Rateless codes over modest code block lengths are able to take advantage of channel variations that are inevitable even when the SNR is fixed, unlike fixed-rate codes.

Third, on a Rayleigh fading channel model, spinal codes outperform Strider by between 19% and 28% at an SNR of 10 dB and by between 28% and 33% at an SNR of 20 dB. We augmented Strider with a puncturing strategy, which improved Strider's performance. We also found that spinal codes perform reasonably well even when the decoder does not have detailed or accurate fading information, unlike Strider.

Fourth, for small packets, typical of Internet telephony or gaming applications, spinal codes outperform Raptor by between 14% – 20%, and Strider by between 2.5× and 10×.

To demonstrate that spinal codes are indeed practical to implement, we have built a hardware prototype that achieves over-the-air rates comparable to equivalent software simulations for SNR between 2 dB and 15 dB. The FPGA decoder runs at 10 Mbits/s. Using the appropriate tools, we estimate that a silicon implementation of the design would operate at about 50 Mbits/s.

We believe these experimental results show that spinal codes are a promising and practical advance for wireless networks. Moreover, from a conceptual standpoint, spinal codes present a framework for making Shannon's random coding ideas, which form the basis of many fundamental capacity proofs, practical.

2. RELATED WORK

Current wireless networks, including 802.11 and various wide-area cellular wireless standards, provide a large number of physical layer (PHY) configurations, including various codes, various parameters for these codes, several choices of symbol sets (i.e., constellations) over which to modulate bits, and a way to map groups of bits to symbols (e.g., Gray code). These networks implement *explicit, reactive* bit rate adaptation policies to dynamically select and configure the discrete choices and parameters provided by the PHY [15, 18, 41, 30, 43, 7].

In recent years, there has been strong interest in rateless codes over both erasure (packet loss) and wireless (AWGN and BSC) channel models. By "rateless", we mean a code where the sequence of coded bits (or symbols) when the code achieves a higher rate is a *prefix* of the sequence when the code achieves a lower rate [8]. This prefix property allows a decoder to process coded data incrementally until successful decoding is possible. Shannon's random codebook approach used in the proofs of capacity is inherently rateless, achieving capacity for channels characterized by a single noise or error parameter [32, 34]. Unfortunately, it is computationally intractable.

The desire for computationally efficient, capacity-achieving rateless codes led to Shokrollahi's work on Raptor codes [33, 9], which are built on Luby's LT codes [21]. They achieve capacity for the Binary Erasure Channel where packets are lost with some probability. On AWGN and BSC models (which model wireless better), not much is known about how close Raptor codes come to capacity. There have, however, been several attempts made to extend Raptor codes to the AWGN channel [26, 35, 4]; we compare spinal codes with an extension of the method of Palanki and Yedidia [26].

Erez et al. recently proposed a "layered approach" to design rateless codes for the AWGN channel [8]. This approach combines existing fixed-rate base codes to produce symbols in a rateless manner. By carefully selecting linear combinations of symbols generated by the base codes, they show that the resulting rateless code can achieve capacity as the number of layers increases, provided the fixed-rate base code achieves capacity at some fixed SNR. Strider [12] uses this layered approach, with a base turbo-code [6].

In contrast, spinal codes are not layered codes; they do not rely on existing fixed-rate base codes. Unlike Strider, which takes an existing fixed-rate code and symbol set system and makes modifications to the lowest PHY procedures to achieve linear combinations of symbols, the construction of spinal codes provides a single (arguably simpler) mechanism to overcome channel impediments. Spinal codes also naturally extend to the BSC case, whereas the layered approach does not. We compare spinal codes with Strider in §8.

Unlike most existing practical codes, spinal codes are nonlinear; i.e., the coded symbols (bits) are not linear combinations of input message bits. Using hash functions, it produces good coded sequences without requiring operations such as multiplying message bits by a random matrix or using complex graph structures. Raptor and LT codes use a pseudo-random number generator (through choice of the graph structure) to choose which bits to XOR together, whereas spinal codes use a pseudo-random number generator directly to produce symbols.

The M-algorithm [1, 17, 28] is a method to efficiently decode over a tree for random convolutional codes [40]. Our bubble decoder may be viewed as a generalization of the classical sequential decoding algorithm as well as the M-algorithm, as explained in §4.3.

We also note a superficial similarity between the direct coding to symbols used in spinal codes and Trellis Coded Modulation (TCM) [38, 37]. TCM was crafted specifically to achieve high minimum distance between codewords under a sparse constellation for convolutional codes, whereas spinal codes aim to attain higher mean distance, obviating the need for sparse constellations. TCM is not rateless, does not achieve capacity for AWGN, is not workable (in any obvious way) for BSC, and is specific to convolutional codes.

Among fixed-rate codes for communication applications, convolutional codes [40], LDPC [11] and turbo-codes [6] are the most widely used. Because LDPC and turbo-codes perform well, much simulation work has been done on puncturing these codes and combining them with incremental redundancy in an attempt to emulate rateless operation [23, 20, 13, 31]. We compare spinal codes to LDPC codes decoded using a strong belief propagation decoder.

3. ENCODING SPINAL CODES

This section describes the encoder for spinal codes. We describe it in the context of a system that has full control of the physical layer, so the encoder produces a sequence of symbols for transmission and the decoder operates on the received symbol sequence to produce an estimate of the original message bits. By slightly modifying the encoder and decoder, it is straightforward to apply the code to a system that has an existing mapping from (coded) bits to symbols.

The encoding procedure takes the input message bits, $M = m_1 m_2 \dots m_n$, and produces a sequence of symbols on the *I*-*Q* plane. At the receiver, the PHY receives a stream of symbols on the *I*-*Q* plane. The decoder processes this stream sequentially, continuing

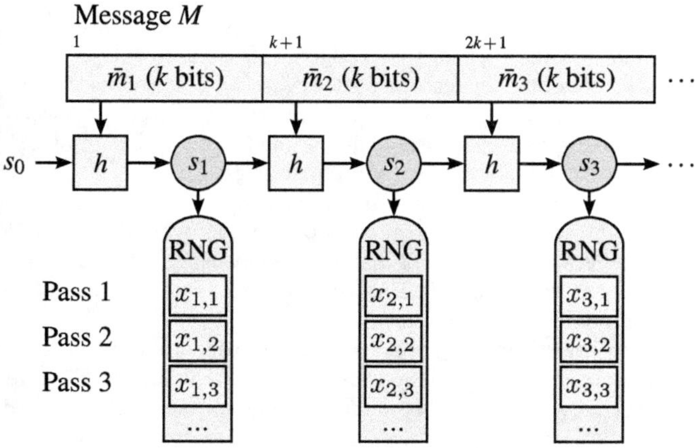

Figure 1: Encoding process. Start with a hash function, h. Compute spine values $s_i = h(s_{i-1}, \bar{m}_i)$. Seed RNG with s_i. For pass ℓ, map c bits from RNG to symbol $x_{i,\ell}$.

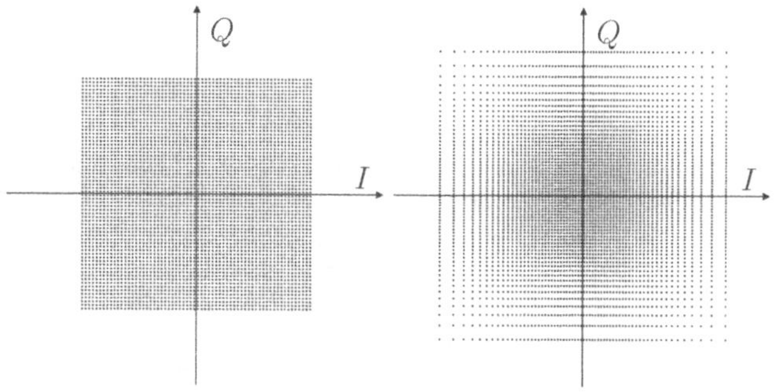

Figure 2: Uniform (left) and truncated Gaussian (right) constellation mapping functions. Same average power; $c = 6$; truncated Gaussian with $\beta = 2$.

until the message is successfully decoded, or until the sender (or receiver) gives up, causing the sender to proceed to the next message. In practice, a single link-layer frame might comprise multiple coded messages, as explained in §6.

When used in rateless mode, spinal encoder can produce as many symbols as necessary from a given sequence of message bits. The sequence of coded bits or symbols generated at a higher code rate is a prefix of that generated at all lower code rates.

3.1 Spine Construction

At the core of the spinal code is a hash function, h, and a pseudo-random number generator, RNG, known to both the transmitter and receiver. h takes two inputs: (1) a ν-bit state and (2) k message bits. It returns a new ν-bit state. That is,

$$h : \{0,1\}^{\nu} \times \{0,1\}^{k} \to \{0,1\}^{\nu}.$$

The initial value, s_0, of the ν-bit state is known to both the encoder and decoder, and may be considered (for now) to be the string 0^{ν} without loss of generality.

As shown in Figure 1, the idea is to build a *spine* of ν-bit states by sequentially hashing together groups of k bits from the input message. We denote bits $m_{ki+1} \dots m_{k(i+1)}$ as $\bar{m}_i$, so the sequence of states is simply

$$s_i = h(s_{i-1}, \bar{m}_i),\ s_0 = 0^{\nu}.$$

Each of these n/k states, or *spine values* (n being the number of bits in the input message), is used to seed a random number generator, RNG. Each RNG generates a sequence of pseudo-random c-bit numbers, which are converted into output symbols using a constellation mapping function (§3.3). RNG is a deterministic function from a ν-bit seed and an index to a c-bit number:

$$\text{RNG} : \{0,1\}^{\nu} \times \mathbb{N} \to \{0,1\}^{c}.$$

The sequence of states computed by repeatedly applying h is superficially similar to a traditional convolutional encoding, but there are three key differences. First, the hash function has a richer pseudo-random (and generally nonlinear) structure and operates on a significantly larger ν-bit state, where ν is on the order of 32. (Hash collisions are a potential concern; §8.5 shows that they can be made extremely rare.) Traditional convolutional codes update their state according to a linear (exclusive-or) function. The larger state space of the spinal encoder gives rise to the second major difference: the "constraint length" of this encoding goes all the way back to the start of the message, because the state at the end of any stage depends on all the input message bits in the message until that point. The third key difference is that, whereas a convolutional encoder has a constant ratio of the number of input to output bits (i.e., a fixed rate), the spinal code is rateless because one can generate as many transmission symbols as desired using the RNG. h and RNG together allow the spinal encoding to not only achieve good separation between codewords, but also ratelessness.

3.2 Hash Function and RNG

We choose h uniformly using a random seed from a *pairwise independent* family of hash functions $\mathscr{H}$ [24]. This property guarantees that for two distinct hash inputs x and y, every pair of output values a and b is equally likely. This property is standard and attainable in practice. The encoder and decoder both know h, RNG, and the initial value s_0; if s_0 is chosen pseudo-randomly, the resulting symbol sequence is pseudo-random, providing resilience against "bad" or adversarial input message sequences (one may view the use of a pseudo-random s_0 as analogous to a scrambler).

Because our requirements for RNG are similar to those for h, one suitable choice for RNG is to combine h with a ν-to-c-bit shift register.

3.3 Rateless Symbol Generation

The output of the encoder is delivered in a series of *passes* of n/k symbols each, as depicted in Figure 1. The encoder produces symbols $x_{i,1}$ for the first pass, where $x_{i,1}$ is the output of a deterministic *constellation mapping function* acting on the first c-bit number generated by the ith RNG (seeded by s_i). It produces symbols $x_{i,\ell}$ for subsequent passes by generating additional outputs from each of the random number generators. The encoder continues to loop back and generate additional symbols until the receiver manages to decode the message or the sender or receiver decides to give up on the message.

Let b be a single c-bit input to the constellation mapping function. For the BSC, the constellation mapping is trivial: $c = 1$, and the sender transmits b. For the AWGN channel (with or without fading), the encoder needs to generate I and Q under an average power constraint. The constellation mapping function generates I and Q independently from two separate RNG outputs of c bits each.

We examine the two constellation mappings shown in Figure 2. The first is *uniform*, and the second produces a *truncated Gaussian* via the standard normal CDF, Φ. In terms of the average power P,

$$\begin{aligned}&\text{Uniform: } b \to (u - 1/2)\sqrt{6P}\\ &\text{Gaussian: } b \to \Phi^{-1}(\gamma + (1-2\gamma)u)\sqrt{P/2}\end{aligned} \qquad u = \frac{b + 1/2}{2^c}$$

where $\gamma \equiv \Phi(-\beta)$ limits the Gaussian's range to $\pm\beta\sqrt{P/2}$. β controls the truncation width. Very small corrections to P are omitted.

4. DECODING SPINAL CODES

In this section, we present an efficient *bubble decoder* for spinal codes. This is an approximate ML decoder whose time and space complexity are polynomial in the number of bits in the message being recovered. Later in this section and in Appendix 9, we show that the polynomial-time approximation of ML decoding achieves capacity over both the AWGN and BSC models. In §8, we show experimentally that a linear-time bubble decoder achieves throughput close to the Shannon limit, outperforming state-of-the-art rated codes (LDPC) and recent rateless codes (Raptor and Strider).

4.1 The Problem

The central concept in ML spinal decoding is to search for the encoded message that differs least from the received signal. Given a vector of observations $\bar{y}$ and an encoder function $\bar{x}(M)$ that yields the vector of transmitted symbols for a message M, the ML rule for the AWGN channel is then

$$\hat{M} \in \operatorname*{arg\,min}_{M' \in \{0,1\}^n} \|\bar{y} - \bar{x}(M')\|^2. \tag{1}$$

That is, the receiver's estimated message $\hat{M} \in \{0,1\}^n$ is the one that produces an encoded vector $\bar{x}(\hat{M})$ closest to $\bar{y}$ in ℓ_2 distance. For the BSC, the only change is to replace the ℓ_2 Euclidean distance with Hamming distance.

Because spinal codes are rateless, the lengths of vectors $\bar{x}$ and $\bar{y}$ increase as the transmitter sends more symbols through the channel. Suppose that the transmitter has sent N symbols up to the present. The set of all transmitted words $\{\bar{x}(M') \text{ for all } M' \in \{0,1\}^n\}$ forms a dense cloud in N-dimensional space. Under Gaussian noise, nearby points in the cloud are initially indistinguishable at the receiver. As N increases, however, the average separation of the points increases. Eventually, n/N (whose dimensions are bits/symbol) drops below the Shannon capacity. The separation of the points then becomes great enough that the correct message is the arg min, and decoding terminates.

The brute-force approach to ML decoding is to conduct an exhaustive search over all 2^n possible messages. Thus, the key question is whether it is possible to develop a practical and implementable spinal decoder. Fortunately, the sequential structure of spinal codes and the powerful mixing effect of h and RNG enable efficient decoding, as explained next.

4.2 Decoding over a Tree

Because the spinal encoder applies the hash function sequentially, input messages with a common prefix will also have a common spine prefix, and the symbols produced by the RNG from the shared spine values will be identical. The key to exploiting this structure is to decompose the total distance in (1) into a sum over spine values. If we break $\bar{y}$ into sub-vectors $\bar{y}_1, \ldots, \bar{y}_{n/k}$ containing symbols from spine values s_i of the correct message, and we break $\bar{x}(M')$ for the candidate message into n/k vectors $\bar{x}_i(s'_i)$, the cost function becomes:

$$\|\bar{y} - \bar{x}(M')\|^2 = \sum_{i=1}^{n/k} \|\bar{y}_i - \bar{x}_i(s'_i)\|^2. \tag{2}$$

A summand $\|\bar{y}_i - \bar{x}_i(s_i)\|^2$ only needs to be computed once for all messages that share the same spine value s_i. The following construction takes advantage of this property.

Ignoring hash function collisions (discussed in §8.5), decoding can be recast as a search over a tree of message prefixes. The root of this *decoding tree* is s_0, and corresponds to the zero-length message. Each node at depth d corresponds to a prefix of length kd bits, and is labeled with the final spine value s_d of that prefix. Every node has 2^k children, connected by edges $e = (s_d, s_{d+1})$ representing a choice of k message bits $\bar{m}_e$. As in the encoder, s_{d+1} is $h(s_d, \bar{m}_e)$. By walking back up the tree to the root and reading k bits from each edge, we can find the message prefix for a given node.

To the edge incident on node s_d, we assign a *branch cost* $\|\bar{y}_d - \bar{x}_d(s_d)\|^2$. Summing branch costs on the path from the root to a node gives the *path cost* of that node, equivalent to the sum in (2). The ML decoder finds the leaf with the lowest cost, and returns the corresponding complete message.

The sender continues to send successive passes until the receiver determines that the message has been decoded correctly. The receiver stores all the symbols it receives until the message is decoded correctly. For instance, if six symbols have been received so far for each spine value, then the vectors $\bar{x}_d$ in the cost computation above have six components.

4.3 Bubble Decoding: Pruning the Tree

Suppose that M and M' differ only in the ith bit. Comparing $\bar{x}(M)$ with $\bar{x}(M')$, we find that symbols from spine values $\lfloor i/k \rfloor, \ldots, n/k$ in the two transmissions are completely dissimilar. If M is the correct decoding, then M' will have a larger path cost than M. The gap will be largest when i is small. This observation suggests a helpful intuition, which can be derived more formally from the proof of the theorem in the appendix: alternate codewords with comparable path costs to the ML decoding *differ only in the last $O(\log n)$ bits.*

Building on this idea, suppose that we have constructed the entire ML decoding tree and computed the path costs for all of the leaves. If we pick the best 100 leaves and trace them back through the tree, we expect to find that within a few times $\log n$ steps, they all converge to a small number of common ancestors.

Consequently, our proposed bubble decoder accepts two parameters: the *depth d*, and the *beam width B*. Instead of searching the entire decoding tree, we maintain B common ancestors, termed the *beam*, and a partial decoding tree rooted on each ancestor. The pseudo-code for the bubble decoder is:

```
Let T_0 be the tree of nodes out to depth d from root.
beam ← {T_0}   # set of rooted trees
for i = 1 to n/k − d do
  candidates ← [ ]   # list of (tree,cost) tuples
  for T ∈ beam do
    for T' ∈ subtrees(root(T)) do
      Expand T' from depth d − 1 to depth d.
      Compute and store path_cost in expanded nodes.
      cost ← min{path_cost(x) | x ∈ leaves(T')}
      candidates.append((T', cost))
  # get B lowest cost candidates, breaking ties arbitrarily
  beam ← best(candidates, B)
return best(candidates, 1)
```

These steps are depicted in Figure 3.

When $d = 1$, this method is the classical beam search (from AI), also termed the M-algorithm in communication systems [1]. When $d = n/k - \log_k B$, we recover the full ML decoder without any tree pruning.

4.4 Tail Symbols

The bubble decoder ends up with a list of B messages, from which it should produce the best one. One approach is to reconstruct and validate all B messages, but the cost of doing so may be too high when B is large. An alternative is to send multiple symbols from $s_{n/k}$ in each pass (*tail symbols*), enough that if the correct candidate is in the beam, it has the lowest path cost. Then, only the lowest path

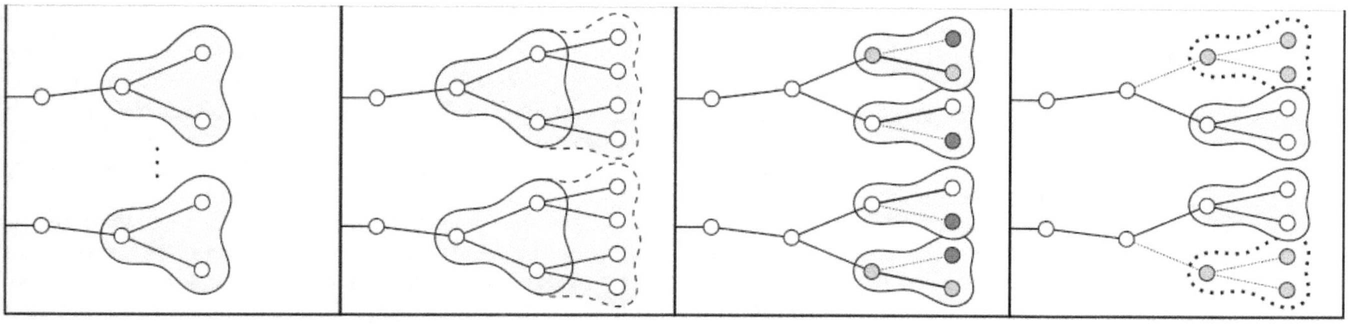

(a) After step $i-1$ (b) Expand (c) Score (d) Pop

Figure 3: Sequential decoding process using the bubble decoder with $B=2$, $d=2$, $k=1$. (a) At the beginning of step i, the partial trees have depth $d-1$. (b) Grow them out to depth d. (c) Propagate the smallest path costs back through the tree. (d) Select the B best children, pruning the rest. Advance to the next step and repeat.

cost message needs to be validated. We find that adding even just one tail symbol works well.

4.5 Decoding Time and Space Complexity

A single decoding attempt requires $n/k-d$ steps. Each step explores $B2^{kd}$ nodes at a cost of L RNG evaluations each, where L is the number of passes. Each step selects the best B candidates in $O(B2^k)$ comparisons using the selection algorithm. The overall cost is $O(\frac{n}{k}BL2^{kd})$ hashes and $O(\frac{n}{k}B2^k)$ comparisons.

Storage requirements are $O(B2^{kd}(k+v))$ for the beam and the partial trees, plus $O(\frac{n}{k}B(k+\log B))$ for message prefixes.

If B is polynomial in n, or if B is constant and $d=O(\log n)$, the total number of states maintained and the time complexity of operations remains polynomial in n. If both B and d are constant, the complexity of the bubble decoder is *linear* in n. Our experimental results are for such linear-time configurations, with B maximized subject to a compute budget (§8.5).

In comparison, LDPC and Raptor decoders use several iterations of belief propagation (a global operation involving the entire message). Turbo decoders also require many full runs of the BCJR [2] or Viterbi algorithm [40]. All told, LDPC, Raptor, and turbo decoders perform several tens to thousands of operations per bit.

A spinal decoder with an appropriate choice of parameters performs a comparable number of operations per bit to these codes, achieves competitive throughput (§8), and is parallelizable (§7.2). The spinal decoder has the additional advantage that the decoder can run as symbols arrive because it operates sequentially over the received data, potentially reducing decode latency.

4.6 Capacity Results

For the AWGN channel with the uniform constellation mapping, we establish that a polynomial-time decoder achieves rates within a small constant (≈ 0.25 bits/symbol) of capacity. The proof appears in the appendix. A recent companion paper [3] states and establishes capacity results for the AWGN channel with the Gaussian constellation, and for the BSC: *the spinal decoder achieves capacity under these settings.*

THEOREM 1 (AWGN CHANNEL PERFORMANCE). *Let $\mathsf{C}_{\mathsf{awgn}}(\mathsf{SNR})$ be the AWGN channel capacity per channel use. With the uniform constellation, a bubble decoder polynomial in n achieves BER $\to 0$ as $n\to\infty$ for any number of passes L such that*

$$L\Big[\mathsf{C}_{\mathsf{awgn}}(\mathsf{SNR})-\delta\Big] > k,$$

with the degree of the polynomial inversely proportional to $(\mathsf{C}_{\mathsf{awgn}}(\mathsf{SNR})-\delta-k/L)$ and

$$\delta \equiv \delta(c,P^*,\mathsf{SNR}) \approx 3(1+\mathsf{SNR})2^{-c}+\frac{1}{2}\log\left(\frac{\pi e}{6}\right). \quad (3)$$

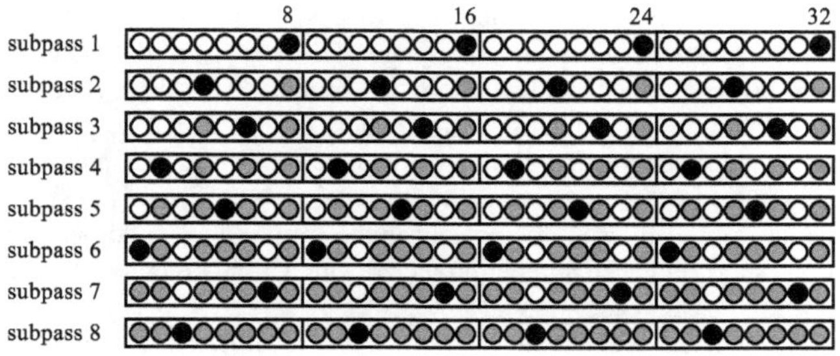

Figure 4: Puncturing schedule. In each subpass, the sender transmits symbols for spine values marked by dark circles; shaded circles represent spine values that have already been sent in a previous subpass.

This result suggests that with the uniform constellation mapping, by selecting a large enough $c=\Omega(\log(1+\mathsf{SNR}))$, it is possible to achieve rates within $\frac{1}{2}\log(\pi e/6)\approx 0.25$ of $\mathsf{C}_{\mathsf{awgn}}(\mathsf{SNR})$. As mentioned above, it is possible to close this theoretical gap with an appropriately chosen Gaussian constellation mapping. In simulation with finite n, however, we do not see significant performance differences between the Gaussian and uniform mappings.

5. PUNCTURING

In §3, the sender transmits one symbol per spine value per pass. If it takes ℓ passes to decode the message, the rate achieved is k/ℓ bits per symbol, with a maximum of k. Moreover, at moderate SNR, when ℓ is a small integer, quantization introduces plateaus in the throughput. Because the decoding cost is exponential in k, we cannot simply increase k to overcome these disadvantages.

Spinal codes may be *punctured* to achieve both high and finer-grained rates, without increasing the cost of decoding. Rather than sending one symbol per spine value per pass, the sender skips some spine values and, if required, fills them in subsequent "subpasses" before starting the next pass.

Figure 4 shows transmission schedule we implemented (others are possible). Each pass is divided into eight subpasses (rows in the figure). Within a subpass, only the spine values corresponding to dark circles are transmitted. Decoding may terminate after any subpass, producing a fine-grained set of achievable rates. This schedule nominally permits rates as high as $8k$ bits per symbol.

Puncturing does not change the decoder algorithm. For any missing spine value in a subpass, the associated branch costs are treated as 0, and the children are computed as before (all the children of a given parent will have the same score). If the correct candidate falls out of the beam, decoding will indeed fail in this subpass. If B is large enough, the correct candidate may remain in the beam until the next non-omitted spine value arrives. In our experiments, we find that $B = 256$ exhibits the positive benefits of puncturing; as computing becomes cheaper, increasing B further will cause the benefits of puncturing to be even more pronounced.

6. LINK-LAYER FRAMING

To use spinal codes, two changes to the traditional (e.g., 802.11) link-layer protocol are useful. First, because the code is rateless, the encoder and decoder must maintain some state across distinct frame transmissions, and use the *cumulative* knowledge of transmitted symbols to decode a message. The amount of data kept is small (on the order of a few kilobytes), similar to H-ARQ receivers implementing incremental redundancy (e.g., 3G, LTE). To prevent an erased frame transmission (e.g., the receiver fails to lock on to the preamble) from de-synchronizing the receiver (which needs to know which spine values are being sent in the frame), the sender should use a short sequence number protected with a highly redundant code (cf. the PLCP header in 802.11).

Second, it is useful to divide a single link-layer frame into multiple *code blocks*, each encoded separately. This use of code blocks is unusual, but not unique (cf. 802.11n with LDPC). The reason we use it is that, for a fixed compute budget at the decoder, shorter coded messages come closer to the Shannon capacity (§8). Each code block has a maximum length, n (1024 bits in our experiments).

At the link layer, the sender takes a datagram from the network layer and divides it into one or more code blocks of size not exceeding n bits. It computes and inserts a 16-bit CRC at the end of each block to construct a link-layer frame. This frame is handed to the encoder, which encodes each code block independently to produce symbols.

The sender transmits a certain number of symbols, and then pauses for feedback from the receiver. An important concern for any rateless code over half-duplex radios is that the receiver cannot send feedback when the sender is still transmitting, and the sender may not know when to pause for feedback. To achieve high throughput, a good algorithm is required for determining pause points. We have addressed this problem in more recent work [16].

At the receiver, the decoder processes the received symbols for each code block. If any block gets decoded successfully (the CRC passes), the next link-layer ACK indicates that fact. The ACK timing is similar to 802.11, but the ACK contains one bit per code block.

7. IMPLEMENTATION

This section describes implementation considerations for spinal codes. After describing general implementation principles, we describe the salient features of our hardware prototype.

The first goal when implementing a communication system is selecting the range of conditions under which we would like the system to perform well. For wireless networks, we expect the maximum SNR observed in practice to be 30 to 35 dB [14, 10]. At the lower end, we would like to support as low an SNR as possible, to work in challenging conditions with high external interference.

An implementation of spinal codes should pre-select (one or more) hash functions, RNGs, and values of k, perhaps even at the time of protocol standardization. Of course, these could be selected dynamically from a set of possibilities, but the sender and receiver need to agree on them. k determines the maximum possible rate (with our puncturing schedule, it is $8k$ bits/symbol), but the complexity of decoding is exponential in k, so smaller values have lower decoding cost. In §8.5, we find that $k = 4$ provides performance close to capacity for SNRs as high 35 dB.

An attractive property of spinal codes is that, given a value of k, the rate achieved under any given set of channel conditions depends *only* on the decoder's computational capabilities. The *same* encoded transmission can achieve a higher rate at a decoder that invests a greater amount of computation. With bubble decoding, each receiver can pick a B and d independently (so, for instance, a base station might pick values larger than a phone, and mobile devices could pick different values). The transmitter requires no knowledge about the receiver's capabilities, which avoids to need to negotiate supported modulation and coding schemes (bit rates) on every association.

7.1 Implementation Decisions

Choosing h. Spinal codes rely on the "mixing" ability of the hash function to provide pairwise independence. We initially used Salsa20 [5], a cryptographic-strength function with demonstrated mixing properties. On each use, Salsa20 requires 320 XORs, 320 additions and 320 rotations on 32-bit words. With these results in hand, we compared code performance with two other much cheaper hash functions developed by Jenkins, one-at-a-time and lookup3.[1] The one-at-a-time hash requires just 6 XORs, 15 bit shifts and 10 additions per application. Our simulations showed no discernible difference in performance between these three hash functions. We used one-at-a-time in our implementation and experiments.

RNG. We implemented RNG using one-at-a-time; to get the tth output symbol, the encoder and decoder call $h(s_i, t)$. This method has the desirable property that not every output symbol has to be generated in sequence: if some frames containing symbols are not recovered, the decoder need not generate the missing symbols.

Other parameters. We find that $c = 6$, $B = 256$, $k = 4$, $d = 1$ are good choices of parameters; see §8.5 for supporting results.

PHY and link layers. The hardware implementation runs atop an OFDM PHY. It uses code block sizes of up to 1024 bits with a 16-bit CRC, dividing a longer packet into multiple 1024-bit code blocks.

Decoder details. The bubble decoder may be invoked multiple times on the same message with different numbers of input symbols. At first glance, it would seem like a good idea to cache explored nodes in the decoding tree between decoder runs, so in subsequent runs the scores would only need to be incrementally updated rather than recomputed. However, until enough symbols have arrived to successfully decode the message, the new symbols end up changing pruning choices to the extent that caching turns out to be unhelpful. Instead, the decoder stores the received symbols, and uses them to rebuild the tree in each run.

7.2 Hardware Implementation

For high-speed, low-power wireless operation, a code must be feasible in hardware. To demonstrate feasibility, we implemented a prototype spinal encoder and $d = 1$ bubble decoder using the Airblue [25] platform. We incorporated spinal codes into Airblue's 802.11 OFDM stack to create 802.11-like 20 MHz and 10 MHz OFDM transceivers (Figure 5).

[1] `http://en.wikipedia.org/wiki/Jenkins_hash_function`

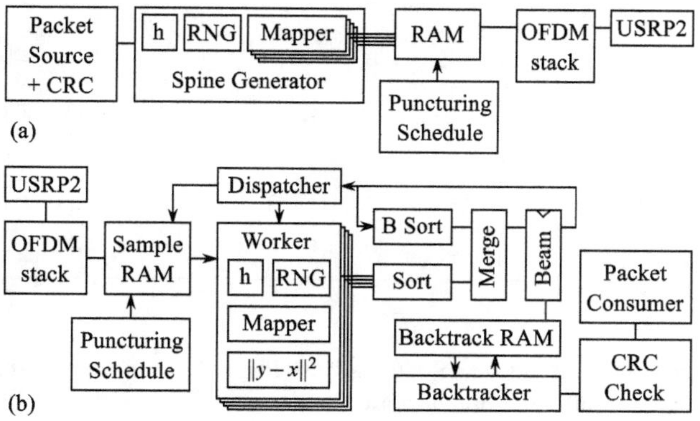

Figure 5: Hardware implementation block diagram, showing the spinal transmitter (a) and receiver (b).

Spinal codes are attractive to implement in hardware because of the high parallelism and low latency made possible by their tree-like structure. These properties contrast with existing high-performance codes like turbo and LDPC, which have limited parallelism and longer latency due to their iterative structure. Although the spinal encoder is a straightforward sequence of hashes, RNG evaluations, and constellation mappings, the decoder requires careful design to take advantage of parallelism.

As samples arrive from the OFDM stack, they are written into an SRAM in unpunctured order, with passes for a given spine value located at adjacent addresses for batch reads. When a decode attempt starts, a dispatch unit instructs M identical worker units to explore all possible decodings of the first k bits, starting from state s_0. Each worker has a certain number of hash units, which serve double duty for computing h and RNG. A worker explores a node by computing several hashes per cycle until it has mapped, subtracted, squared, and accumulated the branch cost over all available passes.

Over the course of several cycles, the dispatcher and the workers will deliver $B2^k$ scored candidate nodes to the selection unit. This stage, corresponding to the two inner loops in the algorithm in §4.3, is highly parallelizable: the work accomplished per cycle is linear in the number of workers.

These candidates stream into a selection unit, which identifies the best B of them. The selection unit sorts the M candidates delivered in a given cycle, selecting the best B. The candidates from prior cycles will have already been winnowed down to the best B, so the system merges those with the B from this cycle. The result is B items in bitonic (not sorted) order. The system stores this list in a register, and on the next cycle finishes sorting these B in parallel with sorting the new M.

Once all $B2^k$ nodes have been scored and selected, the best B become the new beam, and are copied to the backtrack memory. This step advances the outer loop of the algorithm. On the last iteration, the system fully sorts the B candidates and picks the best one, then follows backtrack pointers to recover the message, and checks its CRC.

This prototype has a throughput of up to 10 Mbps in FPGA technology. Synthesized using the Synopsis Design Compiler for the TSMC 65 nm process, the design can sustain 50 Mbps. This decoder is competitive with algorithms like Viterbi decoding in terms of logic area (.60 mm^2 versus .12 mm^2), which is encouraging considering the decades of research and development devoted to Viterbi decoding. In more recent work, we have developed a hardware spinal decoder (with some refinements and generalizations to the above approach) that is competitive with turbo decoding.

8. EVALUATION

Our goal is to evaluate spinal codes under various conditions and compare it to the following codes:

LDPC. We use the same combinations of code rates and modulations for our LDPC implementation as in 802.11n [19], using soft demapped information. The code block size $n = 648$ bits. We implemented a belief propagation decoder that uses forty full iterations with a floating point representation [39]. To mimic a good bit rate adaptation strategy such as SoftRate [41] working atop the LDPC codes, we plot the *best envelope* of LDPC codes in our results; i.e., for each SNR, we report the highest rate achieved by the entire family of LDPC codes.

Raptor code. We follow a similar construction optimized for the AWGN channel to Yedidia & Palanki [26], with an inner LT code generated using the degree distribution in the Raptor RFC [22], and an outer LDPC code as suggested by Shokrollahi [33] with a forty-iteration belief propagation decoder. The outer code rate is 0.95 with a regular left degree of 4 and a binomial right degree. We experimented with different symbol sets, and report results for the dense QAM-256 constellation as well as QAM-64. We calculate the soft information between each received symbol and the other symbols, a process that takes time exponential in the number of constellation points: QAM-2^α requires time $\Theta(2^{\alpha/2})$.

Strider. Our Strider implementation is a C++ port of the Matlab software from Gudipati [12]. We use the recommended 33 data blocks (layers), a rate-$1/5$ base turbo code with QPSK modulation, and up to 27 passes. Unless mentioned otherwise, we use the recommended code block size of 50490 bits. A significant enhancement we added to Strider is *puncturing*, to enable it to achieve a finer-grained set of rates than in the original work (denoted by "Strider+").

8.1 Experimental Setup and Metrics

Software platform. To evaluate the different codes under the same conditions, we integrated all codes into a single framework, built with no sharing of information between the transmitter and receiver components. A generic *rateless execution engine* regulates the streaming of symbols across processing elements from the encoder, through the mapper, channel simulator, and demapper, to the decoder, and collects performance statistics. All codes run through the same engine. In most cases, we measure performance across an SNR range from −5 dB to 35 dB, stepping by 1 dB at a time.

Hardware experiments. We use on-air hardware experiments to cross-validate the software platform results and to demonstrate that spinal codes perform well in hardware under real-world, wide-band conditions. We use high-speed transceivers constructed using Airblue [25], which is built out of Xilinx XUPV5 FPGA boards and USRP2 radio front-ends. All on-air experiments were conducted in the 2.4 GHz ISM band in an unshielded laboratory at MIT CSAIL. We tested spinal codes with both 20 MHz and 10 MHz waveforms.

Metrics. We evaluate two metrics: the *rate* and the *gap to capacity*. We measure the rate in bits per symbol, so multiplying that number by the channel bandwidth (in Hz), and subtracting OFDM overheads, would give the throughput in bits per second.

The gap to capacity is often a more instructive metric than the rate because it allows us to compare how close different codes are to the Shannon limit. The "gap to capacity" of a code, $\mathscr{C}$, at a given SNR, is defined as how much more noise a capacity-achieving code can handle and still provide the same throughput as $\mathscr{C}$. For example, say a code achieves a rate of 3 bits/symbol at an SNR of 12 dB. Because the Shannon capacity is 3 bits/symbol at 8.45 dB, the gap to capacity is $8.45 - 12 = -3.55$ dB.

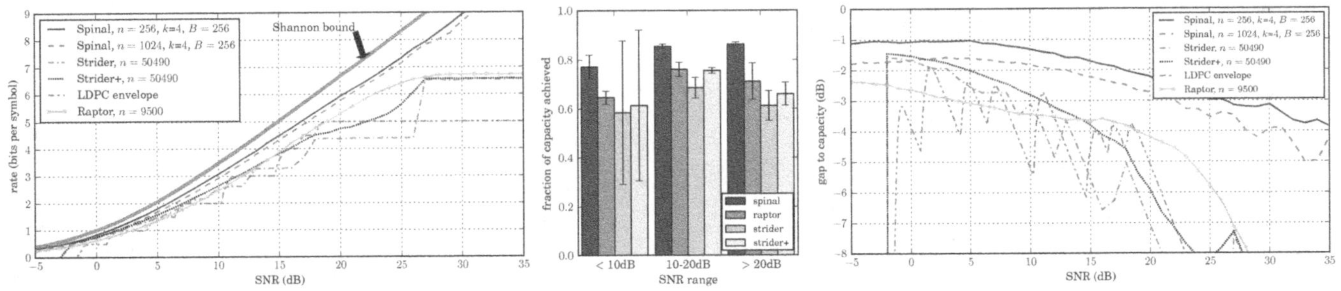

Figure 6: Rates achieved by spinal code with $k = 4$, $B = 256$, $d = 1$, and the other codes (Strider+ is Strider with our puncturing enhancement). Experiments at each SNR average Raptor performance over 100-300 kbits of data, Strider over 5-20 Mbits, LDPC over 2 Mbits, and spinal codes over 0.6 to 3 Mbits.

8.2 AWGN Channel Performance

Figure 6 shows three charts comparing Raptor codes, Strider, and LDPC codes to spinal codes from experiments run on the standard code parameters for each code. The first two charts show the rates as a function of SNR, while the third shows the gap to capacity. The two spinal code curves (256 and 1024 bits) both come closer to Shannon capacity than any of the other codes across all SNR values from -5 dB to 35 dB. The gap-to-capacity curves show that spinal codes consistently maintain a smaller gap than all the other codes.

We aggregate by SNR to summarize gains under different conditions. Above an SNR of 20 dB, spinal codes obtain a rate 21% higher than Raptor/QAM-256, 40% higher than Strider, and 54% higher than the LDPC envelope. Between 10 and 20 dB, spinal codes achieve a rate 25% higher than Strider and 12% higher than Raptor/QAM-256. At SNRs below 10 dB, spinal codes achieve a rate 20% higher than Raptor/QAM-256 and 32% higher than Strider.

Strider. Strider uses 33 parallel rate-1/5 turbo codes with QPSK modulation, so without puncturing, the rates it achieves track the expression $(2/5) \cdot 33/\ell$ bits/symbol, where ℓ is the number of passes required for successful decoding. In the tested SNR range, Strider needs at least $\ell = 2$ passes to decode, for a maximum rate of 6.6 bits/symbol. The puncturing enhancement we added (Strider+) produces the more graded set of achieved rates shown in Figure 6. At low SNR, we find that Strider is unable to successfully decode as many messages as spinal codes. Another source of inefficiency in Strider is that the underlying rate-1/5 turbo code has a non-negligible gap to capacity. The results (without puncturing) are generally consistent with Figure 4a in the Strider paper [12]; it is important to note that the "omniscient" scheme discussed in that paper is constrained to modulation and coding schemes in 802.11a/g, and as such has a significant gap to the Shannon capacity.

Raptor. We are unaware of any previously reported Raptor result for the AWGN channel that achieves rates as high as those shown in our implementation [26]. We believe that one reason for the good performance is that we have a careful demapping scheme that attempts to preserve as much soft information as possible. That said, spinal codes still perform 12%–21% better across the entire SNR range, with the greatest gains at low and high SNRs. There are two reasons for better performance: first, spinal codes naturally incorporate soft information, while Raptor (and also Strider) loses information in the mapping/demapping steps, and second, the LT code used in Raptor has some information loss. We experimented with Raptor/QAM-64 as well, finding that it performs a little better at low-to-medium SNR (16% worse than spinal codes, rather than 20%), but does much worse (54%) at high SNR. The dense QAM-256 constellation does entail a significantly higher decoding cost for Raptor, whereas spinal codes naturally support dense constellations.

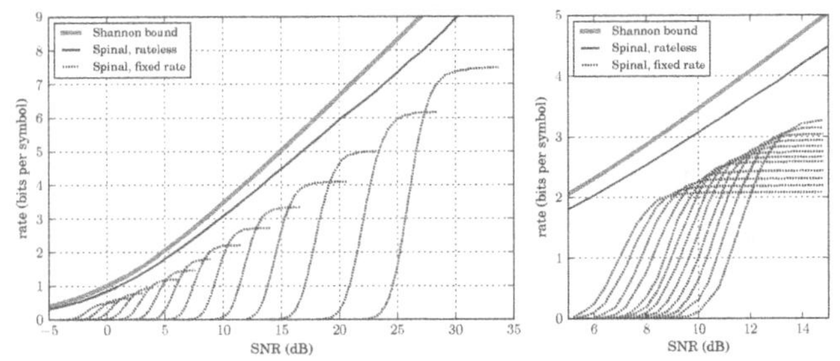

Figure 7: Throughput of the rateless spinal code compared to various rated versions of the spinal code.

LDPC. The primary reason why spinal codes do better than the best envelope of LDPC codes has to do with the ability of rateless codes to take advantage of "lucky" channel conditions. We term this the *hedging effect*. Intuitively, hedging is the ability to decode in less time when the noise is low, without sacrificing reliability. This property is more general than the LDPC comparison. In particular, Figure 7 demonstrates that the *rateless* spinal code outperforms every *rated* version of the spinal code at every SNR.

Constant SNR means that the distribution of the noise does not vary, but the realized noise does vary substantially over time. Because rated codes cannot adapt to realized noise, they must be risk-averse to ensure a high probability of decoding. Hence, they tend to occupy the channel for longer than strictly necessary. By contrast, rateless codes can use the channel for less time when the realized noise is small and thus achieve higher rates. Due to the law of large numbers (precisely, concentration), this effect diminishes with increasing message length. For the same reason, rated codes approach capacity only for long message sizes.

Small code block sizes. The results presented above picked favorable code block (message) sizes for each code. For many Internet applications, including audio and games, the natural packet size is in the 64-256-byte range, rather than tens of thousands of bits. Understanding the performance of different codes in this regime would help us evaluate their effectiveness for such applications.

Figure 8 shows the rates achieved by spinal codes, Raptor, and Strider at three small packet sizes: 1024, 2048, and 3072 bits. Each column shows the results obtained for data transfers in the SNR range 5 to 25 dB. In this range, spinal codes outperform Raptor by between 14% and 20% for these packet sizes.

The gains over Strider are substantial (2.5× to 10×) even when puncturing is used. To handle small packets in Strider, we used the same number of layers and reduced the number of symbols per

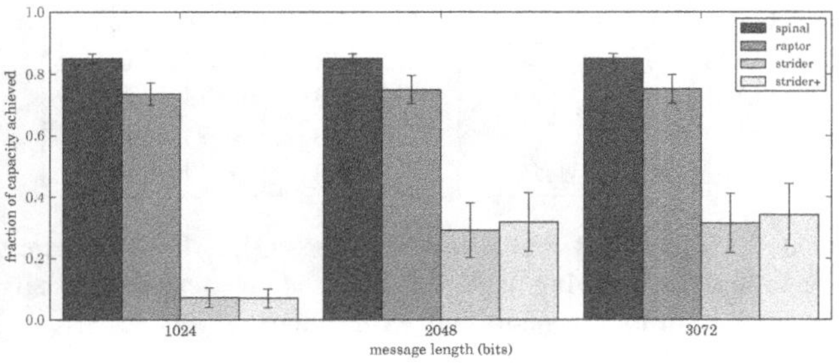

Figure 8: Average fraction of capacity in range 5-20 dB for spinal codes, Raptor and Strider at different message sizes.

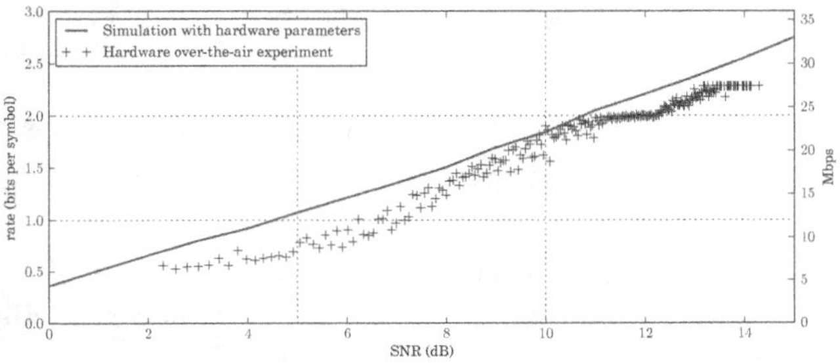

Figure 9: Rates achieved by the hardware over-the-air experiment, compared to a software simulation with similar parameters. Throughput (right axis) shows equivalent link rate for a 20 MHz 802.11a/g channel.

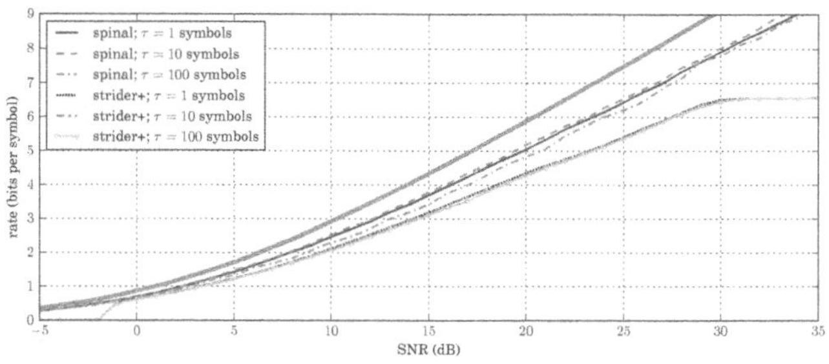

Figure 10: Performance of spinal codes and strider in a simulation model of a Rayleigh fading environment. The decoders are given exact fading channel parameters.

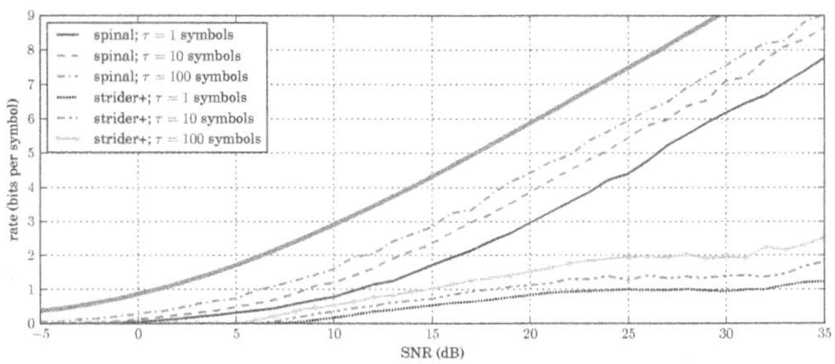

Figure 11: Performance of the AWGN decoders on the Rayleigh simulation. This experiment examines the decoders' resilience to varying or inaccurate channel information.

layer, which is a reasonable method. It is possible that reducing the number of layers might help, but it is unclear how best to handle smaller sizes in Strider.

8.3 Hardware Over-the-air Results

Figure 9 shows the results obtained by measuring the Airblue spinal code implementation in over-the-air experiments for $n = 192$ bits, $k = 4$, $c = 7$, $d = 1$, and $B = 4$. Each + sign in the figure is the rate measured by transmitting at least 20 messages over a 10 MHz band. The measured on-air decoding performance closely tracks the results of a similarly configured software simulator across a large SNR range (the range achievable using commodity USRP2/RFX2400 radio frontends), providing important real-world validation of the code's performance. Differences include effects of fixed-point precision, but should not affect the take-away point: a reasonable implementation is both achievable and operational.

8.4 Fading Channel Performance

This section describes experiments with spinal codes and Strider+ over a fading channel model [36]. The model is a Rayleigh fading environment with two parameters (σ^2, τ). The transmitted signal x is transformed according to $y = hx + n$, where y is the received signal, n is Gaussian noise of power σ^2, and h is a complex fading coefficient randomized every τ symbols to a complex number with uniform phase and Rayleigh magnitude.

The first experiment shows the performance of the codes on fading channels, with both codes incorporating detailed fading information. In the second experiment, neither decoder is given fading information. As such, this experiment evaluates the robustness of the codes to varying conditions and to inaccurate estimates of channel parameters, as might occur in practice.

Figures 10 and 11 show the results of both experiments for three different coherence times, specified as multiples of one symbol time. In both graphs, the top curve is the capacity of the *fading* channel. It is noteworthy that spinal codes perform roughly similarly at all the measured coherence times when fading information is available. Compared to Strider+, at 10 dB, the improvement is between 11% and 20% (without puncturing in Strider, the gains are between 19% and 28%). At an SNR of 20 dB, the gains are between 13% and 20% (without puncturing, between 28% and 33%). When no fading information is available, spinal codes achieve much higher rates than Strider+ (Figure 11).

These results show that spinal codes perform well across a wide range of time-varying conditions, and that spinal decoding is robust even when the decoder does not have accurate fading information.

8.5 Exploration of Spinal Code Parameters

Collision probability. Spines for two distinct messages can converge when there is a hash collision, i.e., $h(s_i, \bar{m}_i) = h(s'_i, \bar{m}'_i)$. Collisions degrade the decoder's ability to discriminate between candidate messages with different prefixes. The probability that colliding messages exist can be made exponentially small in the message length n by choosing $\nu \geq 3n$ (cf. the Birthday Paradox).

In practice, it is not necessary to eliminate all collisions to achieve high performance. A collision potentially reduces performance if it occurs between the correct message and another candidate in the beam. Each iteration explores $B2^{kd}$ nodes. In a decode attempt, a node collides with the correct one with probability $\sim (n/k)2^{-\nu}B2^{kd}$, so these events are rare if $\nu \gg \log(B) + \log(n) + kd$. For example, with $n = 256$, $k = 4$, $B = 256$, $d = 1$, and $\nu = 32$, a collision occurs only once per 2^{14} decodes on average.

Picking k and B. Figure 12 shows that $k = 4$ performs well across a range of compute budgets (the x axis is proportional to $B2^k/k$). Smaller values of k under-perform at higher SNRs; larger values of k don't do well at low compute budgets. Each decoder can use a value

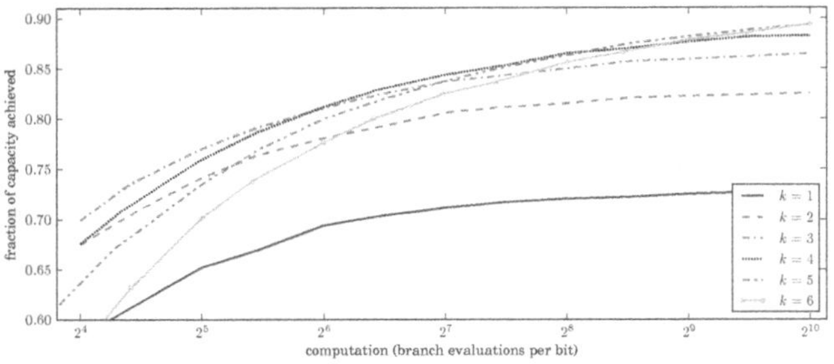

Figure 12: How compute budget per bit ($B2^k/k$) affects performance in the SNR range 2-24 dB, for different k. A choice of $k = 4$ yields codes that perform well over the entire range of budgets. This graph also shows that $B = 256$ is a good choice.

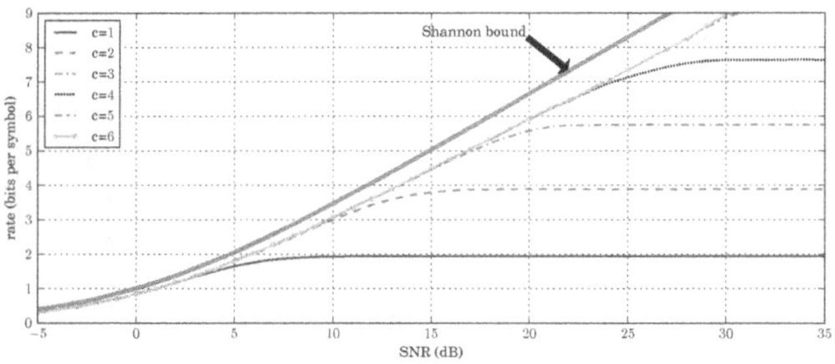

Figure 13: Throughput with different densities of output symbols. $c = 6$ is a good choice for this range of SNRs.

of B according to its price/performance ratio. As computation becomes cheaper, increasingly higher budgets can be used, translating to higher B, to get better performance. From this graph, we conclude that $k = 4$ is a good choice for the SNR range we are targeting. For our experimental compute budgets, $B = 256$ is a reasonable choice.

Picking c. The number of output bits, c, limits the maximum throughput. When c is small, even if the channel's SNR can support a high rate, there are simply too few bits transmitted to decode with high throughput. Figure 13 shows that $c = 6$ is the right choice for the range of SNR values we are concerned with.

Peak-to-average power ratio (PAPR). A practical modulation scheme should have a modest PAPR, defined in terms of the output waveform $y(t)$ as $10 \cdot \log_{10} \frac{\max |y(t)|^2}{\text{mean}|y(t)|^2}$. High PAPR is a problem because the linearity of radio components degrades when waveforms have large peaks. In a non-OFDM wireless system, dense constellations usually have a high PAPR: for QAM-4 it is 0 dB, while for QAM-∞ it is 4.77 dB.

These results, however, do not carry over to the 802.11a/g OFDM stack, which our PHY uses. For such OFDM systems using scrambling, PAPR is typically 5-12 dB [29], depending on the transmitted symbols. As shown in Table 1, OFDM obscures all but negligible differences between the PAPRs of dense constellations and standard WiFi constellations.

Code block length. A strength of the spinal code is good memory in the encoding, so bad information from a burst of noise can be corrected by all following symbols if necessary. But this memory also has a price: once a path is pruned out, the probability of the decoder resynchronizing to a useful path is low. The decoder has to receive more symbols until the true path is not pruned. However small this probability, for fixed k and B, a longer code block means more opportunities for the true path to be lost. Hence, longer code

Constellation	Mean PAPR	99.99% below
QAM-4	7.34 dB	11.31 dB
QAM-64	7.31 dB	11.41 dB
QAM-2^{20}	7.31 dB	11.43 dB
Trunc. Gaussian, $\beta = 2$	7.29 dB	11.47 dB

Table 1: Empirical PAPR for 802.11a/g OFDM with various constellations, showing negligible effect of constellation density. Each row summarizes 5 million experiments.

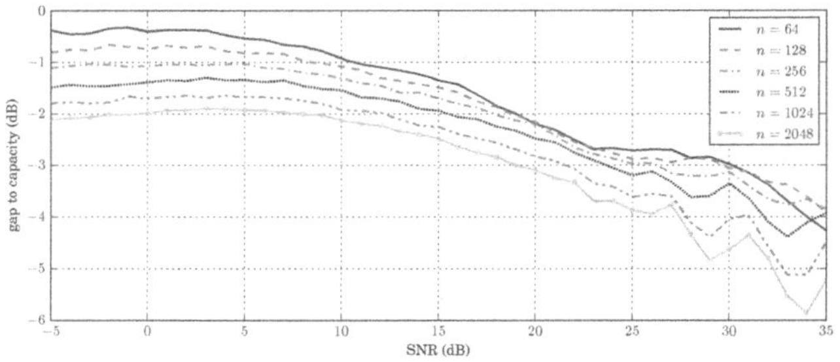

Figure 14: Effect of code block length on performance ($k = 4$, $B = 256$). Some puncturing artifacts can be seen above 25 dB, where less than one pass is transmitted on average.

blocks require either more symbols per bit or a larger B in order to decode, even with the same SNR, as reflected in Figure 14.

9. CONCLUSION

This paper described the design, implementation, and evaluation of rateless spinal codes. The key idea in spinal codes is the sequential application of a random hash function to the message bits to produce a sequence of coded bits and symbols for transmission. We described a novel, efficient, capacity-achieving bubble decoder for spinal codes. Our experimental results show that spinal codes out-perform Raptor, Strider, and the best envelope of 802.11n LDPC codes by significant amounts over a range of channel conditions and code block sizes. Our hardware prototype in Airblue [25] runs at 10 Mbits/s on FPGA hardware, and we estimate it can run at 50 Mbits/s in silicon.

This paper opens up several avenues for future work. First, developing a wireless network architecture atop spinal codes that provides a different wireless link abstraction from today: a link is that is always reliable at all SNR above some well-defined threshold, but which produces outages below the threshold, eliminating highly variable packet delays. Second, developing a good link-layer protocol for rateless codes to deal with the issues raised in §6. Third, investigating the joint-decoding properties of codes that use hash functions. And last but not least, the ideas presented in this paper may provide a constructive framework for de-randomizing, and realizing in practice, a variety of random-coding arguments widely used in information-theoretic proofs.

ACKNOWLEDGMENTS

We thank Joseph Lynch for helping us collect hardware results and Aditya Gudipati for support in implementing Strider. We thank David Andersen, Nick Feamster, Daniel Halperin, Mark Handley, Kyle Jamieson, Henry Pfister, Tom Richardson, Pramod Viswanath, Lizhong Zheng, and the SIGCOMM reviewers for helpful comments. Three generous graduate fellowships supported this work: the Irwin and Joan Jacobs Presidential Fellowship (Perry and Iannucci); the Claude E. Shannon Assistantship (Perry), and the Intel Fellowship (Fleming). Intel also partially supported the Airblue platform. We thank the members of the MIT Center for Wireless Networks and

Mobile Computing, including Amazon.com, Cisco, Intel, Mediatek, Microsoft, and ST Microelectronics, for their interest and support.

REFERENCES

[1] J. Anderson and S. Mohan. Sequential coding algorithms: A survey and cost analysis. *IEEE Trans. on Comm.*, 32(2):169–176, 1984.

[2] L. Bahl, J. Cocke, F. Jelinek, and J. Raviv. Optimal Decoding of Linear Codes for Minimizing Symbol Error Rate (Corresp.). *IEEE Trans. Info. Theory*, 20(2):284–287, 1974.

[3] H. Balakrishnan, P. Iannucci, J. Perry, and D. Shah. De-randomizing Shannon: The Design and Analysis of a Capacity-Achieving Rateless Code. *arXiv:1206.0418*, June 2012.

[4] R. Barron, C. Lo, and J. Shapiro. Global design methods for raptor codes using binary and higher-order modulations. In *MILCOM*, 2009.

[5] D. Bernstein. The Salsa20 Family of Stream Ciphers. *Lecture Notes in Computer Science*, 4986:84–97, 2008.

[6] C. Berrou, A. Glavieux, and P. Thitimajshima. Near Shannon limit error-correcting coding and decoding: Turbo-codes (1). In *ICC*, 1993.

[7] J. Bicket. Bit-Rate Selection in Wireless Networks. Master's thesis, Massachusetts Institute of Technology, Feb. 2005.

[8] Erez, U. and Trott, M. and Wornell, G. Rateless Coding for Gaussian Channels. *IEEE Trans. Info. Theory*, 58(2):530–547, 2012.

[9] O. Etesami, M. Molkaraie, and A. Shokrollahi. Raptor codes on symmetric channels. In *ISIT*, 2005.

[10] J. Frigon and B. Daneshrad. Field measurements of an indoor high-speed QAM wireless system using decision feedback equalization and smart antenna array. *IEEE Trans. Wireless Comm.*, 1(1):134–144, 2002.

[11] R. Gallager. Low-density parity-check codes. *IRE Trans. Information Theory*, 8(1):21–28, 1962.

[12] A. Gudipati and S. Katti. Strider: Automatic rate adaptation and collision handling. In *SIGCOMM*, 2011.

[13] J. Ha, J. Kim, and S. McLaughlin. Rate-compatible puncturing of low-density parity-check codes. *IEEE Trans. Info. Theory*, 2004.

[14] D. Halperin, W. Hu, A. Sheth, and D. Wetherall. Predictable 802.11 packet delivery from wireless channel measurements. In *SIGCOMM*, 2010.

[15] G. Holland, N. Vaidya, and P. Bahl. A Rate-Adaptive MAC Protocol for Multihop Wireless Networks. In *MobiCom*, 2001.

[16] P. Iannucci, J. Perry, H. Balakrishnan, and D. Shah. No Symbol Left Behind: A Link-Layer Protocol for Rateless Codes. In *MobiCom*, 2012.

[17] F. Jelinek. Fast sequential decoding algorithm using a stack. *IBM Journal of Research and Development*, 13(6):675–685, 1969.

[18] G. Judd, X. Wang, and P. Steenkiste. Efficient Channel-aware Rate Adaptation in Dynamic Environments. In *MobiSys*, June 2008.

[19] IEEE Std 802.11n-2009: Enhancements for Higher Throughput.

[20] J. Li and K. Narayanan. Rate-compatible low density parity check codes for capacity-approaching ARQ scheme in packet data communications. In *Int. Conf. on Comm., Internet, and Info. Tech.*, 2002.

[21] M. Luby. LT codes. In *FOCS*, 2003.

[22] M. Luby, A. Shokrollahi, M. Watson, and T. Stockhammer. Raptor Forward Error Correction Scheme for Object Delivery. RFC 5053 (Proposed Standard), Oct. 2007.

[23] R. Mantha and F. Kschischang. A capacity-approaching hybrid ARQ scheme using turbo codes. In *GLOBECOM*, 1999.

[24] M. Mitzenmacher and E. Upfal. *Probability and computing: Randomized algorithms and probabilistic analysis*, chapter 13, pages 321–326. Cambridge University Press, 2005.

[25] M. C. Ng, K. E. Fleming, M. Vutukuru, S. Gross, Arvind, and H. Balakrishnan. Airblue: A System for Cross-Layer Wireless Protocol Development. In *ANCS*, Oct. 2010.

[26] R. Palanki and J. Yedidia. Rateless codes on noisy channels. In *ISIT*, 2005.

[27] J. Perry, H. Balakrishnan, and D. Shah. Rateless Spinal Codes. In *HotNets-X*, Oct. 2011.

[28] G. Pottie and D. Taylor. A comparison of reduced complexity decoding algorithms for trellis codes. *JSAC*, 7(9):1369–1380, 1989.

[29] C. Schurgers and M. B. Srivastava. A Systematic Approach to Peak-to-Average Power Ratio in OFDM. In *SPIE's 47th Meeting*, 2001.

[30] S. Sen, N. Santhapuri, R. Choudhury, and S. Nelakuditi. AccuRate: Constellation-based rate estimation in wireless networks. *NSDI*, 2010.

[31] S. Sesia, G. Caire, and G. Vivier. Incremental redundancy hybrid ARQ schemes based on low-density parity-check codes. *IEEE Trans. on Comm.*, 52(8):1311–1321, 2004.

[32] C. Shannon. Communication in the presence of noise. *Proc. of the IRE*, 37(1):10–21, 1949.

[33] A. Shokrollahi. Raptor codes. *IEEE Trans. Info. Theory*, 52(6), 2006.

[34] N. Shulman. *Universal channel coding*. PhD thesis, Tel-Aviv University, 2004.

[35] E. Soljanin, N. Varnica, and P. Whiting. Incremental redundancy hybrid ARQ with LDPC and Raptor codes. *IEEE Trans. Info. Theory*, 2005.

[36] E. Telatar. Capacity of Multi-Antenna Gaussian Channels. *European Trans. on Telecom.*, 10(6):585–595, 1999.

[37] G. Ungerboeck. Channel coding with multilevel/phase signals. *IEEE Trans. Info. Theory*, IT-28(1):55–67, Jan. 1982.

[38] G. Ungerboeck and I. Csajka. On improving data-link performance by increasing the channel alphabet and introducing sequence coding. In *ISIT*, 1976.

[39] A. Vila Casado, M. Griot, and R. Wesel. Informed dynamic scheduling for Belief-Propagation decoding of LDPC codes. In *IEEE ICC*, 2007.

[40] A. Viterbi. Error bounds for convolutional codes and an asymptotically optimum decoding algorithm. *IEEE Trans. Info. Theory*, 13(2):260–269, 1967.

[41] M. Vutukuru, H. Balakrishnan, and K. Jamieson. Cross-Layer Wireless Bit Rate Adaptation. In *SIGCOMM*, 2009.

[42] S. Wicker and V. Bhargava. *Reed-Solomon codes and their applications*. Wiley-IEEE Press, 1999.

[43] S. Wong, H. Yang, S. Lu, and V. Bharghavan. Robust Rate Adaptation for 802.11 Wireless Networks. In *MobiCom*, 2006.

APPENDIX: PROOF SKETCH OF THEOREM 1

We establish the theorem for the real-valued AWGN channel; the complex channel simply doubles the capacity. We only cover the uniform constellation mapping here. This argument applies to a bubble decoder with depth $d = 1$ and width B polynomial in n (exponent to be determined). The proof for $d' \neq 1$, $B' \geq \max(1, 2^{-kd}B)$ follows similarly and is omitted for space.

Inputs to the constellation mapping function are uniformly distributed, thanks to the mixing properties of the hash function and RNG. Under the uniform mapping, these inputs are mapped directly to the range $[-\sqrt{3P/2}, \sqrt{3P/2}]$, giving an average symbol power (variance) slightly less than $P^* \triangleq P/2$; the difference vanishes as $c \to \infty$. The Shannon capacity [32] of the AWGN channel with average power constraint P^* is

$$\mathsf{C}_{\mathsf{awgn}}(P^*) = {}^1\!/_2 \log_2(1 + \mathsf{SNR}) \text{ bits/symbol}, \tag{4}$$

where $\mathsf{SNR} = \frac{P^*}{\sigma^2}$ denotes the signal-to-noise ratio.

The rate of the spinal code after L passes is k/L. Reliable decoding is not possible until L passes have been received such that k/L is less than $\mathsf{C}_{\mathsf{awgn}}(P^*)$. We shall show that for essentially the smallest L that satisfies this inequality, our polynomial-time decoder will produce the correct message with high probability. In the remainder of this section, we assume L is the smallest value such that $k/L < \mathsf{C}_{\mathsf{awgn}}(P^*) - \delta(c, \mathsf{SNR}, P^*)$. Without puncturing, a single-pass transmission of the code results in rate $R_{max} = n/(n/k) = k$. That is, the rate after L passes, k/L, is R_{max}/L.

The spinal encoder (§3) generates $x_{i,\ell}$, the ith symbol of the ℓth pass, from the spine value s_i, $1 \leq i \leq n/k$, $1 \leq \ell \leq L$. Now s_i depends on s_{i-1} and the k bits $\bar{m}_i$, s_{i-1} in turn depends on s_{i-2} and $\bar{m}_{i-1}$, and so on. Choosing ν large enough to avoid collisions with extremely high probability, and using the pairwise independence properties of h and RNG, $x_{i,\ell} = f(h_{i,\ell}(s_0, \bar{m}_1, \ldots, \bar{m}_i))$, where $h_{i,\ell}$ is a random hash function satisfying pairwise independence, and f is the deterministic constellation mapping. That is, the first i symbols of any pass ℓ depend only on the first ik bits, denoted by $M_i = (m_1, \ldots, m_{ik})$. Therefore, two codewords that differ only in the bits indexed higher than ik get mapped to the same first i symbols (in each pass), but all subsequent symbols generated for them are independent of each other and entirely random.

After receiving symbols for L passes, the bubble decoder sequentially expands the decoding tree, maintaining up to B candidate message prefixes $M_i^1, \ldots, M_i^B$ at each stage i with the hope that for $i = n/k$, they will be closest messages $M^1, \ldots, M^B \in \{0,1\}^n$ to the received $\bar{y}$: i.e., that $\bar{x}(M^1), \ldots, \bar{x}(M^B)$ minimize $\|\bar{y} - \bar{x}(M')\|$ over all $M' \in \{0,1\}^n$.

For convenience, the theorem modifies the bubble decoder slightly, replacing $\bar{y}$ with $\alpha\bar{y}$ where $\alpha = P^*/(P^* + \sigma^2)$. (This modification effectively is a linear MMSE decoder rather than ML, and is weaker than ML.) A capacity proof for this modified weaker bubble decoder implies that the result holds for the original.

Our goal is to show that, with high probability, at every stage of decoding, the bubble decoder has the prefix M_i of the correct message M in its beam of B candidates. We shall additionally show that in this situation, the B candidate messages differ from each other (and hence from the correct message) only in the last $O(\log n)$ bits. If the decoder declares any one of the B candidates as the decoded message, the bit error rate (BER) is $O(\frac{\log n}{n}) \to 0$. That is, the BER goes to 0 as desired.

Intuitively, good messages have low path costs and bad messages have high path costs. Provided we are δ away from the Shannon bound, the difference between good and bad is large enough to make a reliable hard-decoding decision. We establish two invariants, which together give the desired result. The first is an upper bound on the path cost of the correct message, independent of constellation map. The second is a lower bound on the path cost of an incorrect message, and depends on the constellation map.

Invariant 1. Given message $M \in \{0,1\}^n$, the encoder transmits symbols $\bar{x}(M)$. Let $\bar{x}_i(M) = (x_{i,1}(M), \ldots, x_{i,L}(M))$ denote the first L symbols generated from the ith spine value of M and let $\bar{y}_i$ be their noisy versions at the receiver. Then, for small enough $\varepsilon > 0$, with probability $1 - O(\exp(-\Theta(\varepsilon^2 iL)))$,

$$\sum_{j=1}^{i}\sum_{\ell=1}^{L}(\alpha y_{j,\ell} - x_{j,\ell}(M))^2 \le (1+\varepsilon)\frac{iLP^*}{1+\mathsf{SNR}}, \tag{5}$$

for all $1 \le i \le L$. To see why, consider the following: for each j, ℓ, under the AWGN channel,

$$y_{j,\ell} = x_{j,\ell}(M) + z_{j,\ell}, \tag{6}$$

where $z_{j,\ell}$ is independent Gaussian noise (mean 0, variance σ^2). Therefore,

$$\begin{aligned}\left(\alpha y_{j,\ell} - x_{j,\ell}(M)\right)^2 &= \alpha^2 z_{j,\ell}^2 + (1-\alpha)^2 x_{j,\ell}(M)^2\\ &\quad - 2\alpha(1-\alpha)z_{j,\ell}x_{j,\ell}(M).\end{aligned} \tag{7}$$

By the independence of $x_{j,\ell}(M)$ and $z_{j,\ell}$, and because $\mathbb{E}[x_{j,\ell}(M)^2] \approx P^*$, the mean of the RHS of (7) is

$$\begin{aligned}\alpha^2\sigma^2 + (1-\alpha)^2 P^* &= \frac{(P^*)^2\sigma^2}{(P^*+\sigma^2)^2} + \frac{P^*\sigma^4}{(P^*+\sigma^2)^2}\\ &= \frac{P^*\sigma^2}{(P^*+\sigma^2)} = \frac{P^*}{1+\mathsf{SNR}}.\end{aligned} \tag{8}$$

Because all the summands on the LHS of (5) are independent and identically distributed (i.i.d.), and there are iL summands in total, the mean of the LHS is precisely $iLP^*/(1+\mathsf{SNR})$. Now the LHS of (5) can be written as three summations, each having iL terms, one each corresponding to the terms on the RHS of (7). Because each of these is a summation of i.i.d. random variables with exponentially decaying tails (and $x_{j,\ell}^2$ is bounded by $3P/2$), applying a Chernoff-style bound implies a concentration of these terms around their means within error $\varepsilon iLP^*/(1+\mathsf{SNR})$), with probability decaying as $O\big(\exp\big(-\Theta(\varepsilon^2 iL)\big)\big)$ for small enough $\varepsilon > 0$. This argument completes the justification of (5).

For $iL = \Omega(\varepsilon^{-2}\log n)$, the bound holds with probability at least $1 - 1/n^4$. Hence, by the union bound ("the probability that at least one of the events happens is no greater than the sum of the probabilities of the individual events"), it holds simultaneously for all $i = \Omega(\varepsilon^{-2}\log n)$ with probability $\ge 1 - O(1/n^3)$. We need this bound for any contiguous set of indices $(q, q+1, \ldots, q+i)$ with $i = \Omega(\varepsilon^{-2}\log n)$. Since there are $O(n)$ such intervals, by another application of the union bound, this claim holds true with probability at least $1 - O(1/n^2)$.

Invariant 2. Consider $M' = (m_1', \ldots, m_n')$ with $m_1' \ne m_1$, i.e., M' and $M = (m_1, \ldots, m_n)$ differ at least in the first bit. All the coded symbols of M and M' are mapped independently and at random. That is, $\bar{x}(M')$ is independent of $\bar{y}$. We'll focus, for the time being, on the first iL symbols. Now for the uniform constellation, one way to obtain $\bar{x}(M')$ is to sample a point uniformly at random in the cube $[-\sqrt{3P/2}, \sqrt{3P/2}]^{iL}$, and then map the co-ordinates in each of the iL dimensions to the nearest quantized value (at granularity $2^{-c}\sqrt{6P}$). Therefore, the probability of $\bar{x}(M')$ falling within squared distance $(1+\varepsilon)iLP^*/(1+\mathsf{SNR})$ of $\alpha\bar{y}$ is bounded above by the probability that a uniformly sampled point in $[-\sqrt{3P/2}, \sqrt{3P/2}]^{iL}$ falls within $r^2 \equiv (1+\varepsilon+\delta_1)iLP^*/(1+\mathsf{SNR})$ of $\alpha\bar{y}$, with $\delta_1 = 6(1+\mathsf{SNR})2^{-c}$. For the uniform distribution, this is merely the ratio of the volume of a ball of radius r and the volume of the cube $[-\sqrt{3P/2}, \sqrt{3P/2}]^{iL}$: using Stirling's approximation that $\ln K! \sim K\ln K - K$,

$$\begin{aligned}\frac{1}{(iL/2)!}\left(\frac{\pi r^2}{6P}\right)^{iL/2} &= \frac{1}{(iL/2)!}\left(\frac{\pi(1+\varepsilon+\delta_1)iL}{12(1+\mathsf{SNR})}\right)^{iL/2}\\ &\approx \left(\frac{\pi e(1+\varepsilon+\delta_1)}{6(1+\mathsf{SNR})}\right)^{iL/2}\\ &\approx 2^{-iL(\mathsf{C}_{\mathsf{awgn}}(P^*)-\Delta)},\end{aligned} \tag{9}$$

where $\Delta \approx \frac{1}{2}\big(\varepsilon + \delta_1 + \log(\pi e/6)\big)$.

Completing the proof using Invariants 1 and 2. Consider a bubble decoder at stage i trying to estimate M_i using $\bar{y}^i$. From Invariants 1 and 2, conditional on event (5) happening (which happens with high probability for $i = \Omega(\varepsilon^{-2}\log n)$), the chances of an M', differing from M in the first bit, having $\bar{x}^i(M')$ closer to $\alpha\bar{y}^i$ compared to $\bar{x}^i(M)$, is given by (9). There are at most 2^{ik-1} such messages. By the union bound, the chance for such an event is at most

$$P(i) = 2^{-iL\left(\mathsf{C}_{\mathsf{awgn}}(P^*)-\Delta-\frac{k}{L}\right)}. \tag{10}$$

Thus, if $k/L < \mathsf{C}_{\mathsf{awgn}}(P^*) - \frac{1}{2}(\delta_1 + \log(\pi e/6))$, then choosing $\varepsilon = \frac{1}{2}\left(\mathsf{C}_{\mathsf{awgn}}(P^*) - \frac{1}{2}(\delta_1 + \log(\pi e/6)) - \frac{k}{L}\right)$ makes the exponent negative, i.e., $P(i)$ will decay exponentially in i. Given $M' \ne M$, let $q(M, M') = \min\{p : m_p \ne m_p'\}$ be the first bit index where they differ. The above argument has $q(M, M') = 1$. In general, while decoding at stage i, at most 2^{jk} distinct M' have $q(M, M') \in \{(i-j)k+1, \ldots, (i-j+1)k-1\}$. In this scenario, the chance that any such M' has $\bar{x}^i(M')$ closer to $\alpha\bar{y}^i$ compared to $\bar{x}^i(M)$ is the same as (10), with i replaced by j. This is because, given that M' and M have the same first $(i-j)k$ bits, their codewords $\bar{x}^{i-j}(M) = \bar{x}^{i-j}(M')$. Thus, the probability of the decoder finding $\bar{x}^i(M')$ closer to $\alpha\bar{y}^i$ is at most $P(j)$. Since Invariant 1 holds for $i = \Omega(\varepsilon^{-2}\log n)$, at any decoding stage i, the chance that any $M' \ne M$ with $i - q(M, M') = \Omega(\varepsilon^{-2}\log n)$ has $\bar{x}^i(M')$ closer to $\alpha\bar{y}^i$ compared to $\bar{x}^i(M)$ is bounded by

$$\sum_{j=1}^{i-\Omega(\varepsilon^{-2}\log n)} P(j) = O\left(\frac{1}{n^2}\right), \tag{11}$$

for an appropriately large constant in the $\Omega(\cdot)$ term above. Thus, at any stage of decoding, only messages M' that differ from M in only the most recent $O(\varepsilon^{-2}\log n)$ bits can be closer to $\alpha\bar{y}^i$. But the number of such messages is polynomial in n, with degree depending on ε^{-2}, and $\varepsilon = \frac{1}{2}\left(\mathsf{C}_{\mathsf{awgn}}(P^*) - \frac{1}{2}(\delta_1 + \log(\pi e/6)) - \frac{k}{L}\right)$ (half the *gap*). Therefore, choosing B for the decoder to be this polynomial value, we can ensure that at each stage, the correct message M is one of the candidates. When decoding ends, the remaining candidates are only those that differ from M in the last $O(\varepsilon^{-2}\log n)$ bits.

Efficient and Reliable Low-Power Backscatter Networks

Jue Wang Haitham Hassanieh Dina Katabi Piotr Indyk
Massachusetts Institute of Technology
{jue_w, haithamh, dk, indyk}@mit.edu

Abstract – There is a long-standing vision of embedding backscatter nodes like RFIDs into everyday objects to build ultra-low power ubiquitous networks. A major problem that has challenged this vision is that backscatter communication is neither reliable nor efficient. Backscatter nodes cannot sense each other, and hence tend to suffer from colliding transmissions. Further, they are ineffective at adapting the bit rate to channel conditions, and thus miss opportunities to increase throughput, or transmit above capacity causing errors.

This paper introduces a new approach to backscatter communication. The key idea is to treat all nodes as if they were a single virtual sender. One can then view collisions as a code across the bits transmitted by the nodes. By ensuring only a few nodes collide at any time, we make collisions act as a *sparse code* and decode them using a new customized compressive sensing algorithm. Further, we can make these collisions act as a *rateless code* to automatically adapt the bit rate to channel quality –i.e., nodes can keep colliding until the base station has collected enough collisions to decode. Results from a network of backscatter nodes communicating with a USRP backscatter base station demonstrate that the new design produces a 3.5× throughput gain, and due to its rateless code, reduces message loss rate in challenging scenarios from 50% to zero.

Categories and Subject Descriptors C.2.2 [**Computer Systems Organization**]: Computer-Communications Networks

General Terms Algorithms, Design, Performance

Keywords Backscatter, RFID, Wireless, Compressive Sensing

1. INTRODUCTION

Backscatter devices – like RFIDs – differ from other low-power communication technologies (UWB, ZigBee, Bluetooth, etc.) in that they do not expend their own energy on data transmission [16]. In backscatter systems, a device called *the reader* transmits a high power RF signal. A nearby backscatter node can modulate the reader's signal by changing the impedance match on its own antenna in order to convey a message of zeros and ones back to the reader. This behavior allows a node to communicate at almost zero power, by capturing the energy in the reader's RF signal [24]. Further, new generations of backscatter nodes are equipped with various sensing capabilities and can communicate over multiple meters [52]. They may also come equipped with a small onboard battery for sensing and computation but use backscatter to reduce communication cost [42]. These capabilities make them ideal for ultra-low power ubiquitous networks. In fact, there is a growing interest in deploying both RFID and integrated RFID-sensor networks [19, 45]. A typical application involves tagging all items in a store with RFIDs, so that a customer can pay for her purchases simply by pushing her shopping cart through the checkout line [6]. Other applications include monitoring temperature, pressure, humidity, etc. [18, 45]. Recent reports show that industries such as healthcare, retailing, oil and aviation are moving towards deploying such backscatter networks for object tracking, asset monitoring, and emerging Machine-to-Machine (M2M) applications [18, 19, 38].

There are two main challenges, however, in deploying reliable and efficient backscatter networks:

- *(a) Node Identification Overhead:* Backscatter nodes cannot hear each other's transmissions [8]. Thus, they rely on the reader to schedule their medium access. The reader, however, does not know a priori which nodes want to transmit (e.g., which items are in a customer's shopping cart). Thus, existing backscatter protocols [14] go through an identification phase before transmission, which consumes a significant fraction of the total communication time (30%-60% in common modes of RFID EPC Gen-2 standard [14]). At a high level, the reader divides time into slots. A node that wants to transmit picks a random temporary id and sends it in a random slot. Since the nodes cannot hear each other, some of them may collide. To increase reliability one needs to eliminate collisions. Eliminating collisions however requires making the total number of time slots fairly large, which reduces efficiency as many empty slots are produced.
- *(b) Ineffective Bit Rate Adaptation:* Ultra-low power technologies, including backscatter, use very simple modulations that transmit only 1 bit/symbol (e.g., ON-OFF keying or BPSK) [48, 16]. Denser modulation schemes (e.g., 4QAM, 16QAM, or 64QAM) are avoided because they require power hungry linear amplifiers [35]. This leaves backscatter nodes unable to leverage a good channel to transmit more bits per symbol and increase bandwidth efficiency. It is also difficult for them to reduce the bit rate to increase robustness to bad channels [7]. Such adaptation requires the reader to send feedback to each node so it can adjust the redundancy in its messages. However, such feedback is not cost effective because backscatter messages are short and often bursty [14].

This paper introduces Buzz, a communication system for backscatter networks that improves both reliability and efficiency. The key idea underlying Buzz is to treat backscatter nodes as if they were a single virtual sender. Collisions can then be viewed as a code across the bits transmitted by different nodes. However, simply letting nodes repeatedly collide leads to multiple copies of the same collision (i.e., the same codeword), which is not effective for decoding [46]. Instead, Buzz makes each node transmit only in a small random subset of the collisions, and does so without imposing overhead on the nodes themselves. The resulting code is sparse and also rateless (i.e., nodes can keep colliding until the reader collects enough collisions to decode). Buzz then leverages the sparsity

SIGCOMM'12, August 13–17, 2012, Helsinki, Finland.

and rateless nature of the code to enable fast identification and distributed rate adaptation, as described below.

Node Identification: We would like to identify the K nodes that want to transmit in a network of N nodes, where $K \ll N$ (e.g., 20 items in a customer's shopping cart among one million items in a Wal-Mart store). We can model the scenario as an N-element *sparse* binary vector $\mathbf{x}$ that is zero everywhere except in K locations. Let all K nodes that have data transmit concurrently, where a node i transmits a known binary pattern A_i. The signal received by the reader, $\mathbf{y}$, can be represented as:

$$\begin{aligned} \mathbf{y} &= [A_1 \ldots A_i \ldots A_N]\mathbf{x} \\ &= \mathbf{Ax} \end{aligned} \quad (1)$$

Eq. 1 is a standard compressive sensing problem, where one wants to retrieve a *sparse* vector $\mathbf{x}$ using linear combinations $\mathbf{y} = \mathbf{Ax}$. Thus, we can use a compressive sensing solution to efficiently recover $\mathbf{x}$ with a small number of linear combinations.

However, traditional compressive sensing algorithms are computationally infeasible to apply in this setting because $\mathbf{A}$ has as many columns as the number of nodes in the network (e.g., one million items in a Wal-Mart store). To develop a practical solution, we exploit the sparsity in $\mathbf{x}$ to eliminate large chunks of columns in $\mathbf{A}$. In particular, we hash the elements of $\mathbf{x}$ into buckets. All ids that hash to empty buckets can be eliminated while ids that hash to non-empty buckets can be disambiguated using a much smaller scale compressive sensing solution. In §5, we incorporate the effect of the wireless channel and extend this idea to form a full-fledged identification protocol for backscatter networks.

Distributed Bit Rate Adaption: In contrast to traditional rate adaptation, which focuses on point-to-point communication, Buzz looks at the network as a whole and adapts the aggregate bit rate of all backscatter nodes to channel conditions. Fig. 1(a) shows the existing system where backscatter nodes transmit sequentially. In this design, each node's share of the medium is the same. A node with a good channel probably does not need the amount of share it gets, while another node with a bad channel would not be able to deliver its data within its share. In contrast, backscatter nodes in Buzz randomly collide in different time slots and keep doing so until the reader signals that it has correctly received their data, as shown in Fig. 1(b). These collisions act as a rateless code *across* nodes in the network and allow us to implicitly redistribute the mismatched network resources.

However, decoding these collisions to recover the transmitted bits is an expensive joint optimization problem [51]. To address this issue, in Buzz each node contributes to only a small random subset of the collisions so that the resulting code is sparse, i.e., has a low density. Similar to LDPC codes, such low density codes can be decoded using a linear time decoder based on belief propagation. However, in contrast to LDPC codes, which are centralized block codes, Buzz's code is both distributed (i.e., it operates across the bits of many nodes) and rateless (i.e., the reader collects collisions until it has enough to decode). Using these properties, Buzz provides automatic bit rate adaptation to backscatter networks.

Summary of Results: We built a prototype of Buzz and evaluated it in a testbed of 16 computational UHF RFIDs and a USRP backscatter reader. We compared Buzz with TDMA and CDMA based schemes in a wide range of channel conditions, which leads to the following findings:

- Averaged across experiments with different numbers of concurrent tags and over 600 traces in different channel conditions, Buzz improved the overall communication efficiency of backscatter networks by 3.5×. This gain is the combination of two factors: First, Buzz reduced the time spent in the identification phase by 5.5×, compared to the Framed Slotted Aloha protocol used in the EPC Gen-2 standard; Second, during the data transmission phase, Buzz's bit rate adaptation on average delivered a throughput gain of 2× over TDMA and CDMA.
- Buzz enables backscatter networks to work in far more challenging channel conditions than previously possible. In challenging conditions, TDMA and CDMA systems experienced a message loss rate as high as 50% and 100% respectively whereas Buzz's loss rate was zero due to its rateless code.

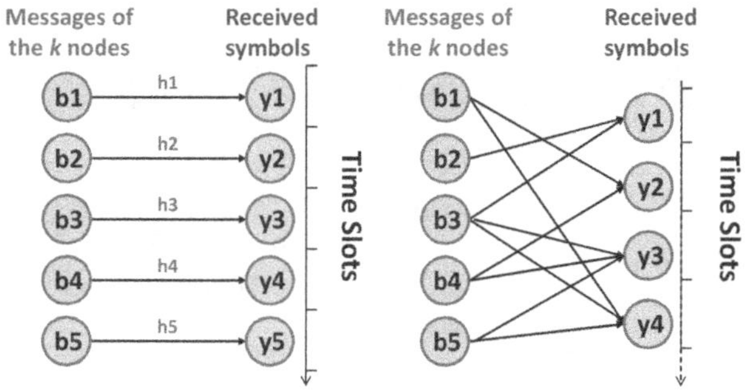

(a) Avoiding Collisions (b) Allowing Collisions

Figure 1—(a) Current systems try to avoid collisions by having nodes sequentially transmit. (b) In Buzz a random subset of nodes collide in each time slot, forming a distributed rateless code.

Contributions: This paper makes the following contributions:

- It introduces the concept of using randomized collisions as a distributed code across the bits of multiple transmitters.
- It presents a new low-complexity compressive sensing algorithm that enables fast backscatter node identification.
- It presents the first automatic rate adaptation protocol that adapts the collective bit rate of multiple transmitters, which do not individually change their transmission bit rate.
- It demonstrates a working system that provides a severalfold improvement to the efficiency and reliability of backscatter networks.

2. BACKSCATTER COMMUNICATION

In backscatter networks, the reader transmits a high power continuous waveform. Backscatter nodes transmit their signal by reflecting back the continuous waveform using ON-OFF keying. The nodes transmit a "1" bit by changing the impedance on their antennas to reflect the reader's signal and a "0" bit by remaining in their initial silent state [16].

There are four main distinctions between backscatter networks and the more familiar WiFi networks.

- There is no carrier frequency offset Δf_c between different nodes' transmissions since nodes do not generate their own RF signal but rather reflect the reader's signal [16].
- Backscatter nodes transmit and receive in a narrow bandwidth due to their power limitation [14]. As a result, the multipath effect of wireless communication is negligible and the system can be modeled as a single tap channel (one complex number) [46].
- Nodes are naturally synchronized by the reader's query and small synchronization errors do not matter since they transmit at very low bit rates (tens to hundreds of kbps) [14]. In §8.1, we present measurement results for commercial passive RFID tags and computational RFID tags.

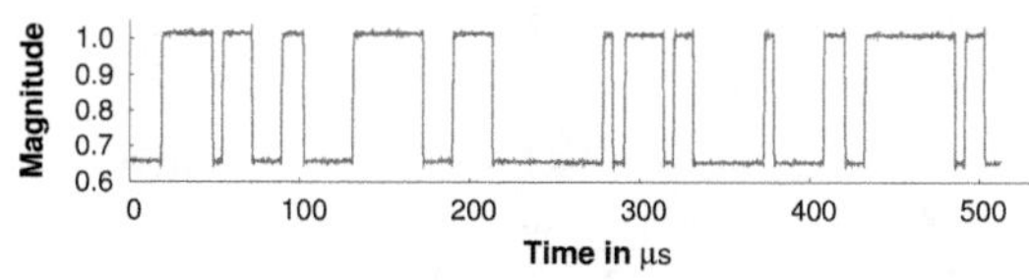

(a) A Single Node Transmission

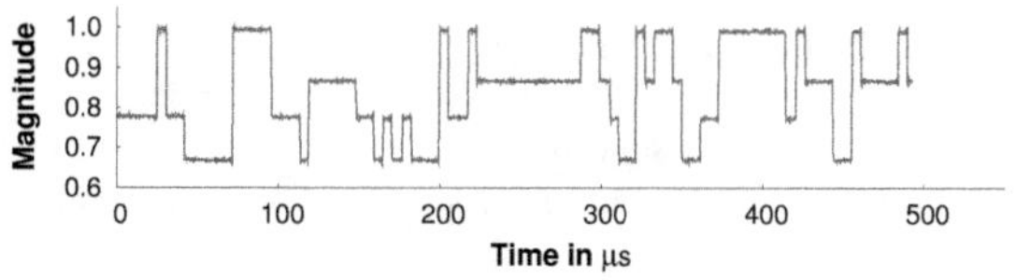

(b) Collision of Two Nodes' Transmissions

Figure 2—(a) When a single node transmits, the received signal has two levels which distinguish a "0" bit and a "1" bit. (b) When two nodes' signals collide, the received signal has four levels corresponding to "00", "01", "10" and "11".

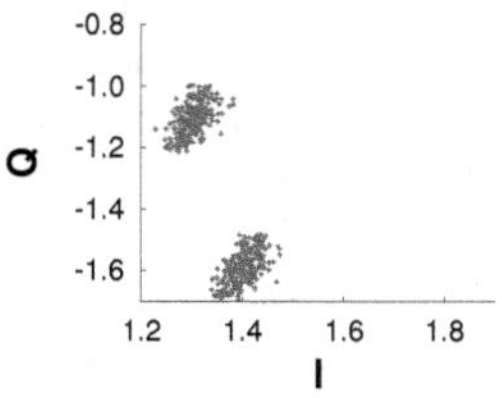

(a) Single Node Constellation

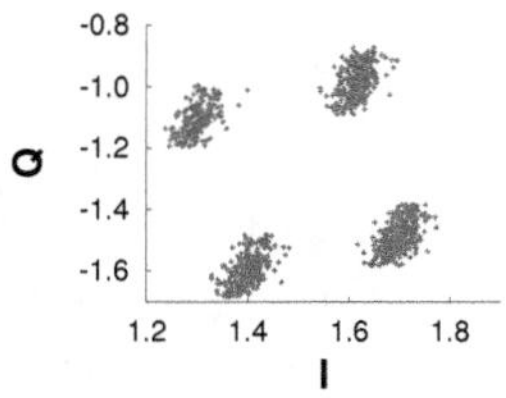

(b) Two Node Constellation

Figure 3—(a) A single node's transmission has two constellation points. (b) Concurrent transmissions of two nodes lead to four constellation points. Decoding a denser constellation increases the throughput of the network.

- The reader is a single power-full device and decoding complexity can be delegated to it while keeping the backscatter nodes simple and power efficient [27].

3. ILLUSTRATIVE EXAMPLES

We start with two simple examples that illustrate the benefits of collisions and provide some intuition into how one may harvest these opportunities.

3.1 Collisions Enable Distributed Rate Adaptation

As mentioned in §1, ultra-low power devices typically use simple modulations that transmit only one bit per symbol (e.g., ON-OFF keying, BPSK, or binary ASK) [48]. Such a choice of modulation is conservative, i.e., it assumes bad channel conditions. In many cases, the wireless channel can support denser modulations that transmit multiple bits per symbol, e.g., 4QAM, 16QAM or higher. Today's low power devices cannot harvest these opportunities to transmit at higher bit rates and increase their throughput.

Collisions can address the issue. Fig. 2 shows backscatter transmission signals received at a USRP-based reader. When a single node transmits (Fig 2(a)), the signal exhibits two levels, which can be used to distinguish a "1" bit from a "0" bit. In contrast, when transmissions from two backscatter nodes collide (Fig. 2(b)), the received signal exhibits four levels that correspond to the values of the two colliding bits: "00", "01", "10", and "11". Knowing the channel coefficients of the nodes, a backscatter reader that receives the signal in Fig. 2(b) can easily distinguish these four levels, and decode both nodes from a single collision, obtaining a bit rate of 2 bits/symbol. Hence, by letting these two nodes collide, one can double the aggregate bit rate.

In fact, letting two backscatter nodes collide is fairly similar to having one virtual node transmit using a denser modulation. This can be more easily seen by comparing Fig. 3(a), which plots the constellation of one backscatter node transmitting alone, and Fig. 3(b), which plots the denser constellation with two nodes transmitting together. Allowing two nodes to collide produces a 4-point constellation. The difference between the two constellations in Fig. 3 resembles the difference between BPSK, a 1 bit/symbol modulation, and 4QAM, a 2 bits/symbol modulation. Allowing more nodes to collide leads to even denser constellations (e.g., three colliding nodes produce an 8-point constellation).

The above example reveals an opportunity for increasing network throughput via collisions. Harvesting the opportunity however requires more than simply letting nodes collide. Decoding is possible in Fig. 3(b) because the four constellation points are distinguishable. If the channels however were noisier or the spacing of the constellation were less ideal, the different levels in Fig. 2(b) could be confused, leading to decoding errors. Without an additional mechanism, the backscatter nodes would need to know the channel's signal-to-noise ratio (SNR) a priori to decide whether it can support 2 bits/symbol. To address this issue, we design a rateless code to enable automatic rate adaptation. In contrast to conventional rateless codes [25, 39], the new code operates across multiple nodes to adapt their aggregate bit rate. The encoding at each node involves only transmitting its original message, or remaining silent. In §6, we describe the code and its corresponding decoder.

TX	Slot 1	Slot 2	Slot 3
Pattern 1	0	1	1
Pattern 2	1	0	0
Pattern 3	1	0	1
Pattern 4	1	1	1

Table 1—Transmit Patterns

Pattern	011	100	101	111
011	~~022~~	111	112	122
100	111	~~200~~	201	211
101	112	201	~~202~~	212
111	122	211	212	~~222~~

Table 2—Collision Patterns

3.2 Collisions Facilitate Assigning Unique IDs

It might seem that collisions make it harder to distinguish different nodes. However, in this section we will show that designing for collisions can improve the system's ability to assign unique temporary ids to the nodes. Consider the following toy example, where two backscatter nodes need to obtain unique ids. The reader divides time into slots and the two nodes transmit in these slots. Suppose that the total time is three slots. We will compare the probabilities that the two nodes will obtain unique ids in two designs:

- **Option 1:** Each node picks a time slot at random and transmits in it. If the two nodes pick different time slots, they can use their time slots as their unique ids.
- **Option 2:** Each node transmits in all three time slots, but the transmission pattern is randomly chosen from the following set of patterns: 011, 100, 101 and 111. It then transmits that pattern over the period of the three time slots, as shown in Table 1. In this case, the pattern serves as the node's id.

In option 1, the two nodes will fail to obtain unique ids if they transmit in the same time slot , which occurs with probability 1/3. In option 2, the nodes will collide in the three slots. For simplicity, let us assume that the channels of the two nodes are the same. The reader will receive one of the collision patterns in Table 2. The reader can then map the received pattern to identify which two patterns were picked by the two nodes. Since there are four possible patterns, the two nodes will be indistinguishable with probability 1/4. Note that while the two options have the same overhead (i.e., three time slots), option 2 reduces the probability that the two nodes end up with indistinguishable ids from 1/3 to 1/4. This toy example shows that designing for collisions can improve the distinguishability of the nodes and reduce the probability of failing to assign them different ids. In §5 we propose a new compressive sensing algorithm which generalizes this basic intuition to any number of nodes and practical settings.

4. PROBLEM DOMAIN

Buzz is designed to increase communication efficiency and reliability in backscatter networks. We consider a deployment of backscatter nodes that capture some measurements of interest (e.g., temperature readings, object tracking). The setup is fairly similar to that of a wireless LAN where a backscatter reader plays the role of the access point. Both uplink and downlink communications have to go through the reader.

Before we delve into the details of our design, it is important to distinguish two modes of operation for backscatter networks:

(a) Event-Driven Communication: This is the more common mode of operation where a backscatter network is used to detect certain events of interest. The classic example is the shopping cart application, in which backscatter nodes are attached to items in a store [6]. A customer pays for her purchases simply by pushing her cart past the scanning area, where a backscatter reader queries the items in the cart for their ids and reports the ids to an application that looks up their prices and presents the charge to the customer. Other event-driven applications include inventory management, object tracking, and event detection [18, 38].

Communication in event-driven networks can be divided into three phases. We describe them at a high level and refer the reader to the EPC Gen-2 standard for more details:

- *Identification:* First the reader has to identify the nodes that have data to transmit. The identification phase uses temporary ids. This is because the *globally* unique id is often long, while the *temporary* id can be short since the uniqueness needs to hold only for the nodes that want to transmit around the same time. For example, the EPC Gen-2 standard uses 16-bit temporary ids during the identification phase [14].
- *Resource Allocation:* In the next step the reader schedules node transmissions. In the EPC Gen-2 standard, the reader assigns the nodes different time slots [14]. In a CDMA or FDMA system, the reader assigns the nodes different CDMA codes or different frequency bands. In these schemes, the reader has to transmit to each node its corresponding assignment.
- *Uplink Data Transmission:* Finally, the nodes transmit their data to the reader using the allocation in the previous phase. The data could be the node's global id or other context information.

Event-driven networks can use Buzz as follows: To assign unique temporary ids, they can use Buzz's compressive sensing identification protocol. As for the second phase, Buzz eliminates the need for the reader to allocate resources to the nodes because it enables the nodes to use a randomized approach to access the medium and transmit their data. Hence, in Buzz the nodes can immediately proceed to the third phase and transmit their data using Buzz's rate adaptation protocol.

(b) Periodic Backscatter Networks: Here, backscatter nodes sense some process and periodically report their measurements to the reader. An example application of such networks is creating real-time heat maps of a data center, where low-power backscatter nodes periodically sense and report the temperature [45].

In periodic backscatter networks, the set of nodes that have data is known a priori. Hence, the network can adopt a static schedule for accessing the medium which eliminates the need for an identification phase. Therefore, periodic backscatter networks can directly use Buzz's rateless bit rate adaptation protocol to maximize the efficiency and reliability of their communication.

5. NODE IDENTIFICATION USING COMPRESSIVE SENSING

We have a backscatter network with N nodes, but only $K \ll N$ have data. We would like to identify these K nodes. We can model this problem as follows: consider a binary vector $\mathbf{x}_{N\times 1}$ where each element x_i corresponds to the backscatter node with id i and $x_i = 1$ if node i has data to transmit. Our aim is to estimate $\mathbf{x}$. One option for solving the problem is to allocate N time slots that correspond to the N elements of $\mathbf{x}$, and have each of the K nodes transmit in its own time slot (i.e., send a "1" bit). This option addresses the collision problem since each node sends in a unique time slot, but it is very wasteful because $K \ll N$.

Protocol: In our scheme, the reader triggers the nodes to start transmitting. Each node that has data (i.e., $x_i = 1$) then uses its id i as a seed to a pseudorandom binary number generator.[1] For each time slot, the node generates a random bit ("0" or "1") and transmits if the random bit is "1". Therefore in each time slot, the reader receives one wireless symbol, which is the collision of the transmissions from a subset of the K nodes. The nodes continue generating a random bit and transmitting it until the reader recovers the vector $\mathbf{x}$. At this point the reader triggers the nodes to stop transmitting which it can do by terminating its RF signal [16].

Corresponding Code: Let M be the length of the binary string that each node has transmitted before the reader terminates the process. Let $\mathbf{A}$ be the $M \times N$ random binary matrix where each column of $\mathbf{A}$ corresponds to the string transmitted by a particular node and each row of $\mathbf{A}$ corresponds to a time slot. Thus, $\mathbf{A}_{j,i} = 1$ if backscatter node i transmits in time slot j. The reader receives a vector $\mathbf{y}$ of M symbols where each symbol is obtained in a time slot. The vector $\mathbf{y}$ can be written as:

$$\mathbf{y} = \mathbf{A}_{M\times N}\mathbf{x}_{N\times 1} \tag{2}$$

Since $\mathbf{x}$ is sparse, i.e., it has only $K \ll N$ non-zero entries, compressive sensing theory tells us that we can efficiently estimate $\mathbf{x}$ with high accuracy given only a few symbols, specifically $M \approx K\log(N/K)$ [13, 9].

The above argument ignores the channels of the backscatter nodes. However, recall that backscatter nodes transmit in a narrow band ($\leq$ 640 KHz) [14] and hence their channels can be modeled as a single complex number (i.e., a single tap channel) [46]. Incorporating the channels, Eq. 2 can be rewritten as:

$$\mathbf{y} = \mathbf{A}\begin{bmatrix} h_1x_1 \\ \vdots \\ h_Nx_N \end{bmatrix} = \mathbf{A}\begin{bmatrix} h_1 & & \\ & \ddots & \\ & & h_N \end{bmatrix}\mathbf{x} = \mathbf{AHx} = \mathbf{Az} \tag{3}$$

where $\mathbf{H}_{N\times N}$ is a diagonal channel matrix and $\mathbf{H}_{i,i} = h_i$ is the complex channel coefficient of the ith node. Since the vector $\mathbf{z} = \mathbf{Hx}$ is sparse and has only K non-zero entries, this is again a compressive sensing problem. In this case, the compressive sensing algorithm will estimate the sparse vector $\mathbf{z} = \mathbf{Hx}$, where $z_i = h_i$ if node i has data and 0 otherwise. This allows us to identify the nodes with data to transmit and estimate their channel coefficients.

5.1 Optimizing Performance

While the formulation of the core problem in Eq. 3 is efficient in terms of the number of time slots, the overhead on the reader is

[1]Generating random numbers is a routine operation carried out in today's RFID systems for node identification as required by the EPC Gen-2 standard [14], and does not introduce extra overhead on the nodes [3].

likely to be prohibitive. Specifically, to estimate $\mathbf{z} = \mathbf{Hx}$, compressive sensing requires the matrix $\mathbf{A}$ to be known at the decoder. In theory, the reader can generate this matrix by using the same pseudorandom number generator used by the nodes and feeding it with all ids in the network. However, in practice, this is computationally infeasible since the number of columns in $\mathbf{A}$ is equal to the number of nodes in the network, which can be huge. Also, even if the reader generates the matrix offline and stores it, it still cannot operate on it because of the computational complexity and the associated increase in the running time. Note that this is not an issue for the backscatter nodes because each node only needs to know the column of $\mathbf{A}$ corresponding to its own id, which it can generate using its id as a seed. Nodes that do not have data do not need to generate their columns in the matrix because the model multiplies those columns by zero. The reader however does not know a priori which nodes have data. It will obtain this information only after solving the matrix equation.

Thus, we transform the above compressive sensing problem into an equivalent compressive sensing problem, but with significantly reduced computational complexity. In doing so, we aim to make both the number of transmitted symbols and the computational complexity functions of K, the number of nodes with data, and independent of N, the total number of nodes in the network.

We start by focusing on a temporary id space. Each of the K nodes that want to transmit randomly picks a temporary id and uses it as its identifier for subsequent communication in the data transmission phase. The temporary id space can be much smaller than N since it is only used to distinguish between the K nodes that want to transmit. However, without knowing K we might define a space too small in which case some of the K nodes will pick the same temporary id and become indistinguishable, or an unnecessarily large space where the computational complexity is still high. Therefore, we first estimate K to properly set the size of the temporary id space. Then, we further reduce the scale of the problem by quickly eliminating large chunks of this temporary id space. Finally, we use standard compressive sensing to solve a small sparse recovery problem. In the rare event where the estimate of K causes an error, the reader starts over as is the case in today's RFID systems.

A. First Stage: Estimating K.

Inspired by prior work that uses streaming algorithms to efficiently count the number of active flows on high speed links [15], here we use a simple streaming technique to quickly estimate the number of backscatter nodes with data to transmit. Again we divide time into short slots that correspond to the transmission of one bit, i.e., one wireless symbol. Every s time slots constitute a single step. In the first step, and for each of the s time slots, each node with data transmits a "1" bit with probability 0.5 and stays silent with probability 0.5; in the second step and for each of the s time slots, each node will transmit "1" with probability 0.25; and so on. In step j and for each of the s time slots in that step, each node transmits a "1" bit with probability $p_j = 2^{-j}$.

The reader only distinguishes between two states of the time slots, "occupied" and "empty". This can be done based on the power level. The expected number of empty slots in each step is $s \times (1-p_j)^K$. The percentage of empty slots is $E_j = (1-p_j)^K$. Thus one can calculate an estimate of K as:[2]

$$\widehat{K} = \frac{\log E_{j^*}}{\log(1-p_{j^*})} \tag{4}$$

where j^* is the step that E_j exceeds a predefined threshold and the algorithm terminates. Once we reach a step in which the probability p_j is very low, the slots will become mostly empty. By setting the threshold for the percentage of empty slots close to 1, we can get a fairly tight estimate of K in $\log K$ steps. One can prove the following:

LEMMA 5.1. *There exists an absolute constant $C > 1$ such that for any $\epsilon, \delta \in (0,1)$, if the number of slots per step s is equal to $C\log(1/\delta)/\epsilon^2$, then with probability at least $1 - O(\log k \cdot \delta)$, we have $\widehat{K} = (1 \pm \epsilon)K$ and $j^* = \log K + O(1)$.*

We note the proof is similar in nature to the standard analysis of streaming and estimation algorithms as seen in [15, 32, 29]. Finally, we note that Buzz does not need an exact estimate of K. An approximate estimate with the guarantees in the above lemma is sufficient for achieving the performance of our identification protocol.[3]

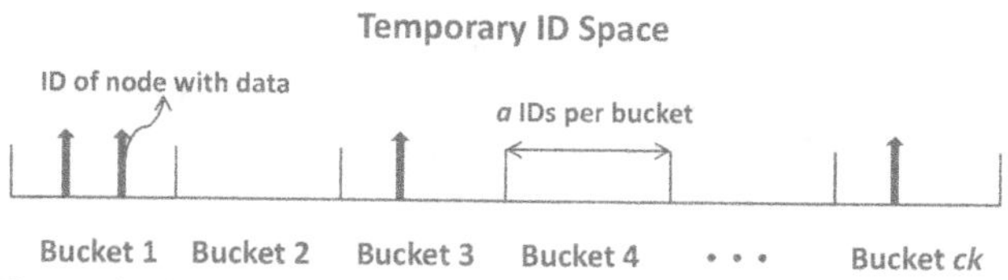

Figure 4—The temporary id space of size $a \times c \times K$ is hashed into cK buckets, each containing a ids. At most K buckets will have ids of the nodes of interest. Temporary ids hashed to empty buckets can be ruled out (e.g., ids hashed to buckets 2 and 4).

B. Second Stage: Reducing the Scale of Compressive Sensing.

Once we have an estimate of K, we can reduce the scale of the problem from the size of the entire node population N to the size of a temporary id space, which is a function of K. Specifically, we set the size of this temporary id space to $a \times c \times K$. The parameters a and c will be used in the following sections to balance the tradeoff between transmission time and decoding complexity.

Only K of the $a \times c \times K$ temporary ids correspond to nodes that want to transmit. To further reduce the scale of this sparse recovery problem, we hash the $a \times c \times K$ possible ids into cK buckets. Each bucket will contain a ids as shown in Fig. 4. We represent these cK buckets by allocating cK time slots; each slot's duration is of the length of one bit. A backscatter node with data to transmit will transmit "1" in time slot j if its temporary id hashes to bucket j. The reader then checks the power in each time slot. We denote a bucket as empty if no power is detected in its corresponding time slot. Temporary ids that hash to empty buckets can be eliminated. Given that there are only K transmitting nodes, at most K buckets will be non-empty. Since each bucket contains a ids, at the end of this stage, at most aK ids remain as candidate node ids.

C. Third Stage: Compressive Sensing Decoding.

Now that we have reduced the scale of the compressive sensing problem to recovering K temporary ids out of only aK possible ids, we can apply the protocol described at the beginning of §5. Let $\mathbf{A}'$ be a reduced version of the matrix $\mathbf{A}$ that keeps only the columns corresponding to the remaining aK possible temporary ids; and $\mathbf{x}'$ and $\mathbf{H}'$ are the similarly reduced forms of $\mathbf{x}$ and $\mathbf{H}$. The reader only needs to regenerate $\mathbf{A}'$, as opposed to $\mathbf{A}$, and solve the system:

$$\begin{aligned} \mathbf{y} &= \mathbf{A}'\mathbf{H}'\mathbf{x} \\ &= \mathbf{A}'\mathbf{z}' \end{aligned} \tag{5}$$

[2]The numerator of Eq. 4 can be written as $\log \min(E_{j^*}, 1-1/s)$ to handle the case where all slots are empty i.e. $E_{j^*} = 1$.

[3]Having an estimate of the number of nodes is also beneficial to existing protocols. In §10, we feed our estimate of K to the protocol specified in the EPC Gen-2 standard, and show that Buzz still significantly outperforms it.

To decode, the reader uses compressive sensing to estimate the elements of vector $\mathbf{z}'$. In compressive sensing, estimating $\mathbf{z}'$ is formulated as an optimization problem [13, 9]:

$$\begin{aligned} \underset{z'}{\text{minimize}} \quad & \|\mathbf{z}'\|_1 \\ \text{subject to} \quad & \mathbf{A}'\mathbf{z}' = \mathbf{y}, \end{aligned} \tag{6}$$

where $\|.\|_1$ is the L1 norm.

Since the space of the problem is now aK as opposed to N, compressive sensing tells us that for a sparsity level $a = \omega(1)$ we only need to receive about $K\log(aK/K) = K\log(a)$ symbols to decode $\mathbf{z}'$ if elements in $\mathbf{A}'$ are randomly generated binary numbers [13].

D. Communication Cost and Decoding Complexity.

In terms of the communication cost, the first and second stages of Buzz's identification algorithm require $s \times j^* = s\log(K)$ and cK time slots (i.e., wireless symbols) respectively. The compressive sensing stage requires $M \approx K\log a$ time slots. Thus, the communication complexity of Buzz is $O(s\log(K) + cK + K\log a)$ bits.

In terms of decoding complexity, the running time of the first two stages is $s\log(K) + cK$. The third stage involves solving a compressive sensing problem for an $M \times aK$ matrix $\mathbf{A}'$, where $M \approx K\log a$. There is a large body of algorithms in compressive sensing that use linear programming to efficiently solve the problem in Eq. 6 [11, 10]. In our implementation, we use an algorithm based on the interior-point method [22][4], which has the running time of $T = O(I \cdot T_{A'})$, where $I \approx \sqrt{aK}$ is the number of iterations in the algorithm, and $T_{A'}$ is the cost of a matrix-vector multiplication involving $\mathbf{A}'$ [4]. For efficiency, we use sparse binary matrices $\mathbf{A}'$ for which $T_{A'} = O(aK\log(aK/K)) = O(aK\log a)$ [4]. This yields

$$T = O(I \cdot T_{A'}) = O(\sqrt{aK} \cdot aK\log a) = O((aK)^{3/2}\log a).$$

Therefore, the total decoding complexity of the three stages is equal to $O(s\log(K) + cK + (aK)^{3/2}\log a)$.

In our implementation of the K estimation stage, we set $s = 4$ and the termination threshold to 0.75, and use the estimated $\widehat{K}$ to set parameters for the latter stages. The latter two stages of the customized compressive sensing approach provide a trade-off between the decoding complexity and the communication cost. One should choose the parameters a and c properly based on system capabilities, constraints and requirements. In our experiments, we set c to 10 and $a = K$.

6. DISTRIBUTED RATE ADAPTATION

We describe a protocol that enables backscatter networks to adapt the uplink bit rate to the channel quality. In contrast to traditional work on rate adaptation [47, 25], which adapts the rate of individual senders, this protocol adapts the aggregate rate of the backscatter nodes that want to transmit. Said differently, we model the system as a single virtual sender that is realized by the collective transmissions of the backscatter nodes. We then design a protocol that optimizes the bit rate of the virtual sender given the channel conditions. The protocol operates in a distributed fashion and the reader does not schedule the transmissions/collisions of nodes.

(a) Protocol: In Buzz's rate adaptation protocol, the reader initiates the data transmission phase by sending a single command to all backscatter nodes. Upon decoding this command, each node starts a random binary number generator which is seeded by its own temporary id and the current time slot[5] and returns a random bit. In each time slot, if the random bit generated is "1" the node will transmit its message, otherwise the node remains silent. Hence, depending on the results of the random binary number generator, random subsets of the nodes will collide together. The nodes repeat this process (generating a random bit and transmitting the message if the bit is "1") until the reader signals them all to stop.

The reader receives the colliding messages and tries to decode them (as described below). Once it has decoded all K messages and the decoded messages pass the CRC check, the reader triggers the backscatter nodes to stop transmitting the current messages (and if applicable move on to the new message). The reader can do this by terminating its RF signal, which naturally terminates all backscatter transmissions [16]. Note that the reader does not send signals to indicate the beginning or end of a time slot. The nodes automatically move on to the next time slot until there is no RF power to reflect.

(b) Encoder: The result of the above protocol is a rateless code. For clarity, we first express the code assuming that each of the K messages is one bit. We use the vector $\mathbf{b}$ to denote the K bits that the K nodes transmit, where $b_i \in (0, 1)$ is the bit of node i. When transmitted according to the above protocol, these bits will collide randomly on the channel and the reader will receive the vector $\mathbf{y}$:

$$\mathbf{y}_{L\times 1} = \mathbf{D}_{L\times K}\mathbf{H}_{K\times K}\mathbf{b}_{K\times 1} \tag{7}$$

where $\mathbf{H}$ is a diagonal matrix whose diagonal element $h_{i,i}$ represents the channel of backscatter node i; $\mathbf{D}$ is a binary matrix, where element $d_{j,i}$ is the jth random number used by node i in the above protocol; L is the number of time slots until the reader decodes all K bits and triggers the nodes to stop. Note that while $\mathbf{b}$ is a *binary* vector that corresponds to the K bits transmitted by the backscatter nodes, $\mathbf{y}$ is a vector of complex numbers that refer to the L wireless symbols received at the reader.

We generalize the above to P-bit messages as follows:

$$\mathbf{Y}_{L\times P} = \mathbf{D}_{L\times K}\mathbf{H}_{K\times K}\mathbf{B}_{K\times P}. \tag{8}$$

The only difference from Eq. 7 is that now both the received and the transmitted signals are expressed as matrices.

(c) Belief Propagation Decoder: The reader knows $\mathbf{H}$ from the node identification step. It can generate $\mathbf{D}$ because it also knows the ids of the nodes, i.e., seeds, and has the same random number generator they use. It knows $\mathbf{Y}$ because this is the collision signal that it receives. The task is to retrieve the original messages $\mathbf{B}$.

One approach to decode is inverting the encoding matrix $\mathbf{M}_{L\times K} = \mathbf{D} \times \mathbf{H}$ (if $L \neq K$ one can use the pseudo inverse) [46]. This solution is undesirable because it does not take into account the reader's knowledge that the matrix $\mathbf{B}$ is binary. It is also highly expensive from a computational standpoint. To provide the rateless property, the reader has to invert the matrix for each value $L' = 1, \ldots, L$ until it finds the right L' at which all messages pass the CRC check. Thus, instead of inverting the matrix, Buzz uses a belief propagation algorithm that decodes incrementally and also embeds the knowledge that the messages are binary strings.

We will explain how Buzz decodes the jth bit transmitted by all nodes. The same procedure follows for decoding all the remaining bits in the message. The jth bit of each node collides only with the jth bits of other nodes and can be decoded separately from the other bits in the message. Hence, we will decode using Eq. 7.

Eq. 7 can be modeled as a bipartite graph shown in Fig. 5 where the K vertices on the left represent the original bits $\mathbf{b}$, and the vertices on the right represent the received symbols $\mathbf{y}$. There is an edge from b_i to y_j if $d_{j,i} = 1$, the weight of which is $h_{i,i}$. Hence, the set

[4] Faster algorithms exist in the literature to solve the standard compressive sensing problem such as [5].

[5] Unlike previous stages, in the uplink data transmission case, the duration of each time slot is the length of the nodes' messages. For simplicity we assume it is fixed.

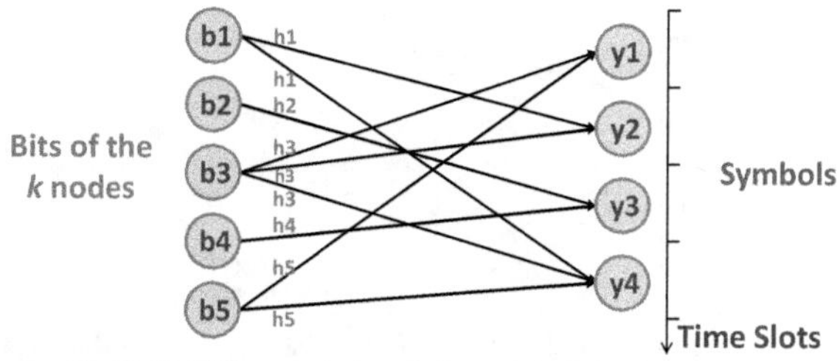

Figure 5—Belief Propagation Bipartite Graph: On the left are the bits transmitted by the backscatter nodes. On the right are the received symbols. An edge represents a node's contribution to a received symbol, i.e., the node participates in the collision on that symbol. If b_i is flipped in the belief propagation decoding process, only the gains of bits that have collided with i will be updated.

1 Pseudocode for jointly decoding one bit of all nodes

INPUT:
$\mathbf{y}, \mathbf{H}, \mathbf{D}$
OUTPUT:
Decoded bits $\hat{\mathbf{b}}$
PROCEDURE:
Initialization:
$\hat{\mathbf{b}} \leftarrow K$ independent random bits
Belief propagation decoding:
Calculate gains $G_1, G_2, \ldots, G_K$ using Eq. 9
repeat
  Find i s.t. $G_i = \max G_1, G_2, \ldots, G_K$
  $\hat{b}_i \leftarrow \hat{b}_i \bigoplus 1$
  Update G_i and the gains of node i's neighbors' neighbors
  if $G_1, G_2, \ldots, G_k \leq 0$ **then**
    Done
until Done

of incoming edges to y_j identifies all bits whose collision creates y_j weighted by their channels. The set of outgoing edges from a bit b_i identifies all collision symbols to which it has contributed.

The objective of the belief propagation algorithm is to find a vector $\hat{\mathbf{b}}$ that could have produced the received signal $\mathbf{y}$, or more concretely to find $\hat{\mathbf{b}}$ that minimizes the error:

$$\min_{\hat{\mathbf{b}}} \|\mathbf{DH}\hat{\mathbf{b}} - \mathbf{y}\|_2.$$

The algorithm works as follows. Initialize $\hat{\mathbf{b}}$ to a random binary vector. In each iteration, flip one bit in $\hat{\mathbf{b}}$ that best minimizes the remaining error $\|\mathbf{DH}\hat{\mathbf{b}} - \mathbf{y}\|_2$ and keep flipping until the error cannot be reduced any further by flipping any single bit, i.e., reaching an optimum.

To quickly find the bit, which once flipped would yield the biggest gain, Buzz maintains for each bit i in $\hat{\mathbf{b}}$ a variable G_i, representing the gain achieved by flipping bit i, i.e., the reduction in error. G_i is calculated as follows:

$$\begin{aligned} \tilde{\mathbf{b}} &= \hat{\mathbf{b}} \\ \tilde{b}_i &= \hat{b}_i \oplus 1 \\ G_i &= \|\mathbf{DH}\hat{\mathbf{b}} - \mathbf{y}\|_2 - \|\mathbf{DH}\tilde{\mathbf{b}} - \mathbf{y}\|_2 \end{aligned} \tag{9}$$

At the beginning of the first iteration, G_i is calculated based on the randomly initialized $\hat{\mathbf{b}}$. Then after flipping bit i of $\hat{\mathbf{b}}$, Buzz updates G_i and the gains of all nodes which have collided with node i, i.e., update G_l if the ith column and l^{th} column of $\mathbf{D}$ have 1 in at least one same row. This could be more clearly seen in the bipartite graph. After flipping bit i of $\hat{\mathbf{b}}$, only the gains of the bits that are neighbors of neighbors of the vertex $\hat{\mathbf{b}}_i$ in the bipartite graph need to be updated, as other bits are not affected. For example, in Fig. 5 if bit 4 was flipped, we only update the gains of bits 2 and 4. If bit 1 was flipped, we update the gains of bits 1, 3 and 5. A pseudocode for decoding one bit is shown in Alg. 1.

The algorithm finishes its decoding of the jth bits in the messages once all gains are negative. It then moves on to the bits at the $j+1$ position. After decoding all positions, the algorithm checks whether the resulting messages pass the CRC check. If any message passes the CRC check, the values of the bits in the message will be fixed for all subsequent decoding. The reader keeps collecting collisions until all decoded messages pass the CRC check.

(d) Discussion: A few points are worth elaborating on:

- We choose the matrix $\mathbf{D}$ to be sparse, i.e., each row of $\mathbf{D}$ has only a few ones but is mostly zero. A sparse $\mathbf{D}$ reduces the decoding time because only a few gains need to be updated when a bit is flipped. It also allows the reader to decode the messages from fewer collisions because it reduces the number of local minima of the error function. This argument is similar to why belief propagation works well with low density parity check code [20]. Hence, Buzz uses a low density code by making the matrix D sparse. The sparsity of D is related to K, the number of nodes that want to transmit. The reader communicates its estimate of K to the nodes, which they give as an input to the random binary number generator to bias its output.
- Due to the near-far effect, different nodes may have very different channel coefficients. Unlike in CDMA, the near-far effect is mitigated in Buzz because: first, the code is sparse (only a few nodes collide at a time); second, each time, a node's signal combines with a different subset of nodes, providing diversity. The results in §9 and §10 are in the presence of near-far effect. Further, this property can be leveraged to improve decoding and reduce the overall time. Specifically, bits of nodes with good channels will have the largest gains and will directly converge to the right binary values in the first few iterations. Once they are decoded and pass the CRC check, we set their gains to be negative infinite so that flipping the bits of nodes with worse channels in later iterations will not flip them back to a wrong value. This effectively cancels their contributions to the collisions.
- If a backscatter node runs out of power in the middle of the data collection phase, its impact on the other nodes will be minimal. Specifically, the nodes which have already been decoded will not be affected. This node's influence on the system translates to additional noise to the collisions it participated in, i.e., it will take more collisions to decode the remaining nodes.
- Buzz achieves its goal of delivering a rateless code that adapts to channel conditions. Specifically, if the decoding process finishes in L time slots with zero bit error, it means $K \times P$ data bits are successfully delivered using $L \times P$ symbols. Hence, the rate of the system is $\frac{K}{L}$ bits/symbol. When the channels are good, then we have $L < K$ and the system delivers multiple bits per symbol. In contrast, in current backscatter systems, nodes transmit sequentially, thus making the highest rate 1 bit/symbol regardless of the quality of the channel [14]. Furthermore, in current systems if the channel of some node is bad and cannot support 1 bit/symbol, decoding will fail. In contrast, with Buzz, a node with a bad channel will just take more collisions to decode. This rateless design eliminates the need for frequent feedback as required in today's RFID systems, reducing the overhead on both the nodes and the reader.

7. IMPLEMENTATION

We build a prototype of Buzz using USRP software radio [28] as the reader and the UMass Moo computational RFIDs [50] as backscatter nodes in the network.

Reader: We adopt a USRP implementation of an EPC Gen-2 RFID reader developed in [8] and customize it to incorporate Buzz's node

(a) UMASS Moo Computational RFID

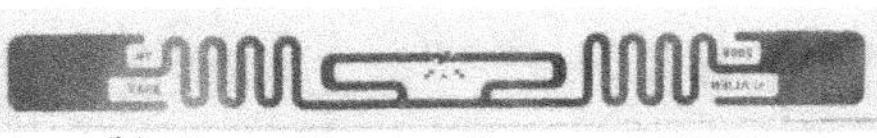

(b) General Purpose Commercial RFID

Figure 6—We experiment with two types of RFID tags: (a) The Moo computational RFIDs with programmable microcontrollers and (b) the Alien Squiggle General Purpose UHF RFID Tags.

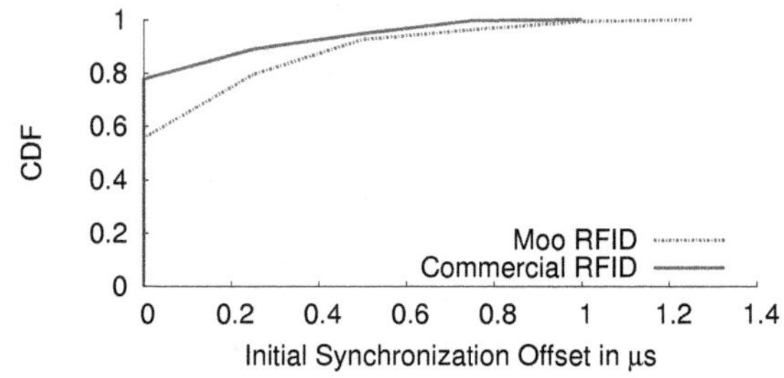

Figure 7—CDF of the initial synchronization offset between different RFID tags: Both commercial and computational RFIDs have an initial offset of less than 1μsec which has negligible impact on the performance of Buzz.

identification and distributed rate adaptation algorithms described in §5 and §6. This mainly includes changing the reader commands to allow backscatter nodes to distinguish which algorithm to run. Due to the overhead of user mode processing in software radios, we do not run the decoders in real time. Instead, we collect the traces from the USRP receiver and process them offline.

UHF backscatter systems operate between 860 MHz and 960 MHz [14]. We use RFX900 USRP daughterboards and Cushcraft 900 MHz RFID antennas which operate in this range [34]. We run all experiments at a carrier frequency of 925 MHz. The USRP reader transmits its queries at a data rate of 27 kbps and captures measurements of the tags' signals at 4 MHz.

Backscatter Nodes: The Moo tag, shown in Fig 6(a) is a passive computational RFID that operates in the UHF band [50]. It has an onboard temperature sensor, accelerometer, and a programmable MSP430F2618 microcontroller [50]. The original code base implemented in the Moo already supports most parts of the EPC Gen-2 standard and has all components necessary for Buzz, including decoding the reader's commands, reading data from memory and transmitting it in a random time slot, and changing the impedance of the antenna to transmit zeros and ones [50]. Since Buzz's algorithms require randomization in many stages, we randomize the encoding process offline and program 20 different randomly encoded traces into each Moo tag in advance. This allows the tags to transmit different data in each run without being reprogrammed.

In addition to the Moo, we also experiment with commercial RFIDs (The Alien Squiggle Inlay [1]) shown in Fig. 6(b). We use them to study the level of synchronization in general purpose RFIDs as we will discuss in §8.1.

Setup: We deploy the tags on a movable plastic cart placed on a 1.5m×3m table, where the USRP reader also sits. Distances between tags and reader vary in a range of [0.5,6] feet. The maximum distance is limited by the capabilities of the Moo, whose typical operating range is 2 feet [50].

8. MICROBENCHMARK

We start with a few experiments that provide insight into the working of the system.

8.1 Synchronization

Buzz assumes that transmissions of backscatter nodes are synchronized. In this section, we present measurement findings for both commercial RFIDs and computational RFIDs, and show that the synchronization accuracy in practice is sufficient for Buzz's performance. In particular, synchronization in backscatter networks has two aspects:

- *The starting offset of the nodes' transmissions:* Backscatter nodes are naturally synchronized because they are triggered by the reader's signal. However, the jitter in detecting the reader's signal can lead to initial offsets.
- *Drift in the nodes' digital clocks:* Backscatter nodes use a digital clock to time their operations. Different clocks have different drifts, which can be corrected by having each node estimate its clock drift relative to a virtual clock maintained by the reader and compensate for the difference. This is done by counting the number of clock ticks between two pulses from the reader separated by a fixed time interval.

Initial Offset: We measure the offset using two types of off-the-shelf passive UHF RFID tags: Alien Squiggle commercial RFIDs [1] and the Moo computational RFIDs [50]. We let the tags concurrently reply to a reader's query and measure the offset between the starting points of their transmissions. We use 20 different tags of each type and vary the number of concurrent tags in each run between 2 and 8.

Fig. 7 shows the CDF for the measured offsets. The offset's 90th percentile is 0.3 μs for the commercial tags and 0.5 μs for the Moo, and the maximum is less than 1 μs. The default bit rate used in practical settings and recommended by manufacturers is 64 kbps [2, 27]. At such rates, a 1 μs offset is about 6.5% of the bit length (16 μs). We show in §9 that its impact on performance is negligible.

The EPC Gen-2 standard allows for a maximum bit rate of 640 kbps [14]. Although at such maximum rate the offset can be 65% of the bit length, the reader can sample the backscatter signal at a much higher rate and obtain many samples per bit. The reader can then use the middle samples of each bit to increase robustness to synchronization errors, which affect the samples at the boundary of a bit. For instance, with a sampling rate of 20 MHz (which is simple with today's hardware capability [28]), the reader can get 10 synchronized samples of each bit when the tags transmit at the maximum bit rate of 640 kbps. Alternatively, one can also design the circuit of the backscatter nodes to further improve synchronization accuracy [12, 23, 41].

Drift Correction: To test how well backscatter nodes can correct for their clock drift, we program the Moo tags to use a reader's virtual clock and insert extra clock cycles to correct for the drift. We let two tags transmit the same data stream at 80 kbps and capture the trace on the USRP reader with and without correcting for the clock drift. Fig. 8 shows a trace for which the initial synchronization offset is zero to help visualize the effect of clock drift over time. After 2 ms ($\approx$ 160 bits) in Fig. 8(a), the clock drift causes the bits of the two tags to be misaligned by 50% of the symbol length. Fig. 8(b) shows that after correcting for the clock drift, the bits of the two tags are still aligned after 2 ms. The results reported in this paper use this method for drift correction. We note that the clock drift of each tag is fairly stable [12, 50]. We let each Moo tag compute its clock drift

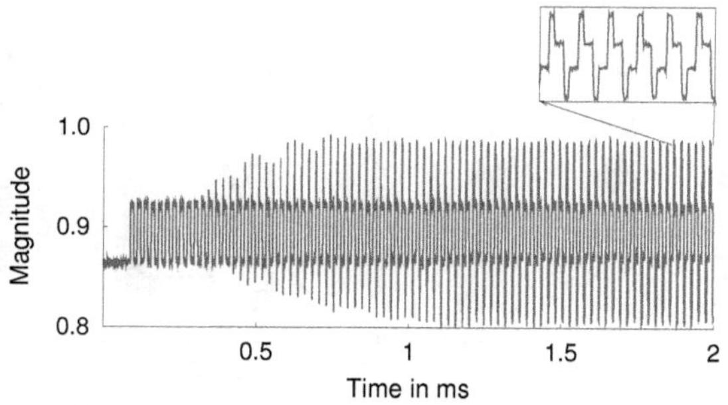

(a) Collision Without Correction for the Clock Drift

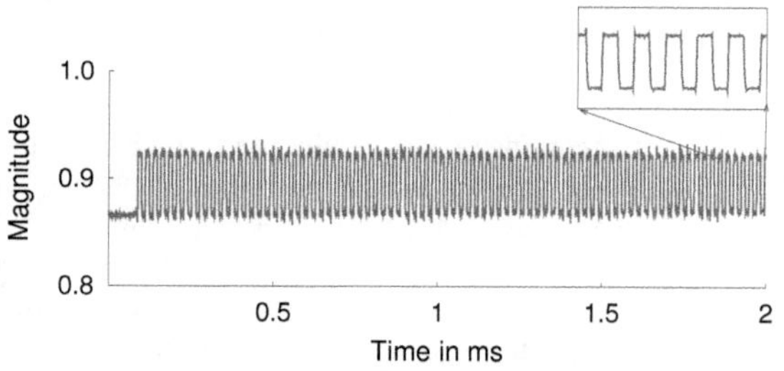

(b) Collision after Correcting for the Clock Drift

Figure 8—Two tags concurrently transmit the same data: (a) Without correcting for the clock drift, the bits of the two tags are misaligned by 50% of the symbol length after 2 ms. (b) When the tags correct for their clock drifts, their bits remain aligned.

once and used the same estimate throughout our experiments over a period of four months.

8.2 Understanding the Belief Propagation Decoder

To better understand how Buzz's belief propagation algorithm decodes collisions to achieve distributed rate adaptation, let us zoom in on one example where 14 Moo tags had messages to send and the messages were decoded in ten slots. In line with the EPC Gen-2 standard, each tag transmitted a 96-bit message at a bit rate of 80 kbps [14].

Fig. 9 shows how the average bit rate adapted over time. For each time slot, the figure shows the number of tag messages already decoded in previous slots as a dark blue bar and the number of newly decoded tag messages upon the arrival of the current time slot as a light blue bar. The decoding process started off strong and quickly reached its peak. Eleven tag messages were decoded within the first four slots, resulting in a peak bit rate of 2.75 bits/symbol. The decoding of these eleven tags can generally be categorized into two types. Certain tags had a very good channel and were therefore immediately decoded from the first collision they participated in, such as the two tags decoded in the very first time slot. In contrast, other tags' messages were uncovered as the belief on their data gradually built up when the tags they collided with were decoded. This process resembled a ripple effect, resulting in the high decoding efficiency in slot 3 and 4.

After slot 4, it took Buzz another six slots to decode the three remaining tags (in the same experiment, TDMA and CDMA consistently failed to decode the last two tags' messages). Notably, the last tag participated in four collisions to finally get decoded. Because of this particular tag, the average bit rate was dragged down and other tags had to transmit an extra number of times. By the end of the transfer, the aggregate bit rate was 1.4 bits/symbol.

This experiment also shows a limitation of Buzz. Namely, some nodes may transmit after they have been decoded. This is due to the fact that Buzz treats all nodes with data to transmit as a single distributed sender. An alternative design choice could have the

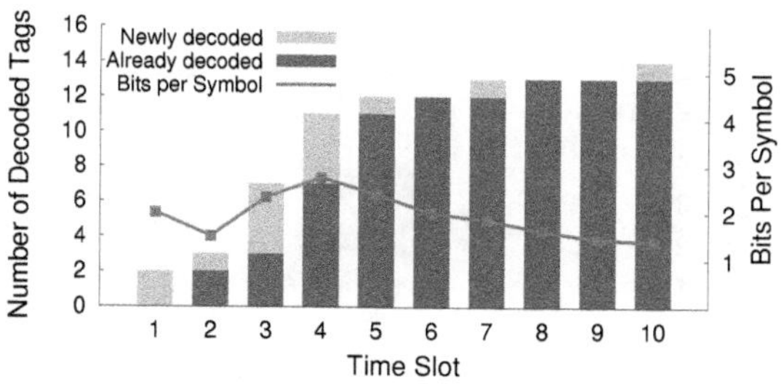

Figure 9—The progress of Buzz's decoding of 14 tags during 10 time slots: Buzz quickly decoded 11 tags within the first 4 slots, and then gradually adapted the aggregate bit rate to channel conditions to decode all remaining tags.

reader ACK the tags whose messages have been decoded, so that they remain silent in subsequent time slots. To silence the tags, the reader would at least need to ACK by echoing the temporary id of each decoded tag. A back-of-the-envelope calculation based on the EPC Gen-2 standard suggests that a 75% overhead on top of the uplink transmission time is necessary to silence these 14 tags [14]. Therefore, the cost of silencing individual tags seems to outweigh the benefits, and so Buzz avoids such a design.

9. RESULTS FOR UPLINK TRANSMISSION

In this section, we focus on the efficiency and reliability of the uplink transmission phase in backscatter networks. We assume that the backscatter reader has already performed node identification and resource allocation (evaluated in §10). We compare Buzz with the following two baseline schemes:

- **TDMA:** Tags sequentially transmit one after another. In line with the EPC Gen-2 standard, each tag uses a bit rate of 80 kbps for both TDMA and Buzz. Miller-4 code is used in TDMA to increase its robustness [14].
- **CDMA:** We compare Buzz with synchronous CDMA using Walsh codes, because asynchronous CDMA severely suffers from the near-far problem [30].[6] Walsh codes are the most commonly used orthogonal codes in CDMA systems. For K tags, we use a set of Walsh codes with a spreading factor of K. We let CDMA use the same symbol rate as Buzz, 80k symbols per second, and adopt a standard CDMA decoder for orthogonal codes [30] to decode the received CDMA signal.

We run the experiments for $K = 4, 8, 12$, and 16 Moo tags. For each run of all experiments, the tags decode which scheme to use from the reader signal (TDMA, CDMA, or Buzz), and transmit a 32-bit message with a 5-bit cyclic-redundancy check. We repeat the experiments at ten different locations for each value of K. At each location, we run the three schemes to collect five traces each, immediately one after another without changing the environment, i.e., moving the tags or reader antennas.

Transmission Efficiency: Fig. 10 shows the time it takes the three schemes to complete data transfer. TDMA and CDMA have a fixed aggregate bit rate, and therefore always finish transmission in a fixed amount of time for each value of K. In contrast, Buzz only completes the data transmission phase when the reader has successfully decoded the messages of all tags. Hence the total transfer time for Buzz is variable. As we can see, Buzz finishes in approximately half the time of TDMA and CDMA on average,[7] translating

[6]In cellular CDMA systems, the near-far problem is addressed by adjusting the power level at the transmitters [30], which is not feasible for backscatter nodes.

[7]The bump on the CDMA curve at $K = 12$ is due to the lack of

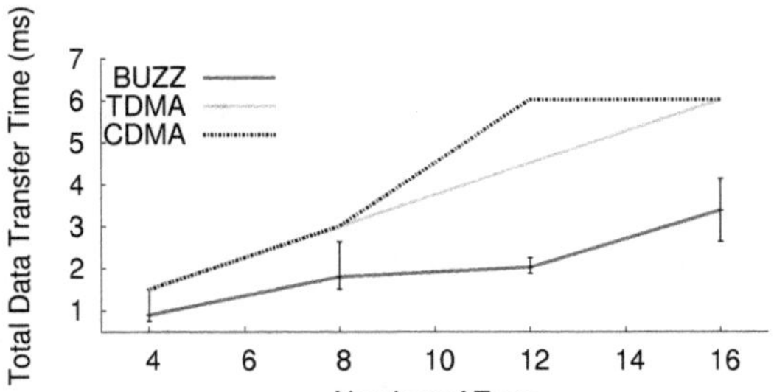

Figure 10—Total data transfer time: Buzz achieves 2× improvement in total transfer time over TDMA and CDMA. Unlike TDMA and CDMA which have a fixed rate of 1 bit/symbol, the rateless nature of Buzz allows it to adapt to an average rate of 2 bits/symbol.

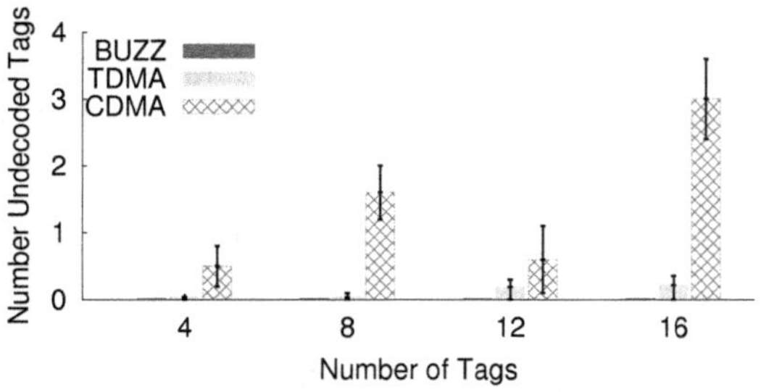

Figure 11—Message errors: CDMA has the lowest reliability. TDMA has few errors because it uses Miller-4 code. Buzz's bars are not visible because it has zero errors due to its rateless code.

to a 2× increase in aggregate bit rate of the network. However, this statistic understates the efficiency gain Buzz brings about. Unlike Buzz, TDMA and CDMA do not adapt their rates, and therefore might have decoding errors upon the completion of all the nodes' transmissions. These decoding errors lead to retransmissions, further reducing their efficiency.

Message Reliability: Commercial backscatter systems use a checksum to decide whether a message is decoded correctly and send a NAK if the checksum does not pass [14]. Thus, the number of messages decoded incorrectly represents a lower bound of retransmissions needed. Fig. 11 shows the number of tags whose messages were not correctly decoded for the same traces in Fig. 10.

Due to its automatic rate adaptation, Buzz decodes all messages correctly. TDMA has very few errors because it uses Miller-4 code which increases its robustness to bad channels. CDMA is less reliable, and as the number of tags in the network increases, its reliability quickly degrades. This is mainly because all tags get equal shares of the channel in CDMA, despite the fact that there is generally a disparity in channel quality between them. Note that in Fig. 11, CDMA for $K = 12$ has fewer undecoded messages than both $K = 8$ and $K = 16$. The reason for this is that we let each of the twelve tags use unique Walsh code of 16 bits, as no Walsh code of 12 bits is available [30]. This implies that one could use a longer code to increase reliability in CDMA, however, this comes at the expense of efficiency.

Next, we focus on more challenging channel conditions to further investigate the reliability of Buzz and TDMA. In this experiment, we use four tags and keep moving the tags further and further away from the reader to worsen the channels of all tags. Fig. 12 shows the comparison between Buzz and TDMA. When the channel quality in the network is good, Buzz decodes all four tags' messages correctly in a single time slot (4 bits/symbol), while TDMA requires 4 time slots to decode. As the channel worsens, TDMA starts failing to

a set of orthogonal Walsh codes of length 12 [30]. So instead, for $K = 12$ we use the orthogonal Walsh codes of length 16.

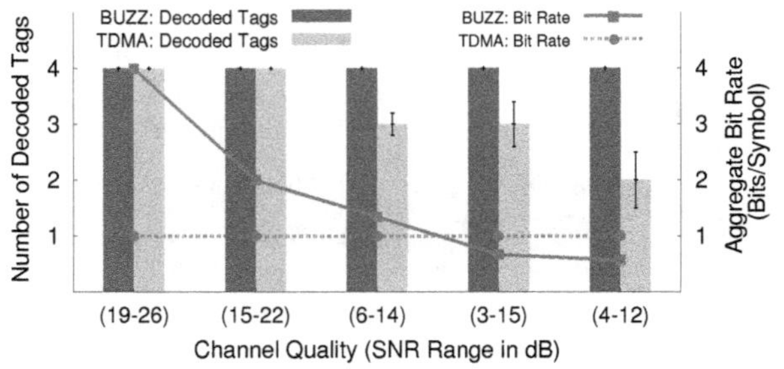

Figure 12—Under challenging conditions where TDMA experiences 50% (median) of message loss, Buzz adapts the bit rate to below 1 bit/symbol to match the channel quality and is able to decode with zero errors.

decode while Buzz gradually adapts the aggregate bit rate to below 1 bit/symbol. Notably, in the most challenging channel conditions, TDMA decodes two tags' messages incorrectly out of four, experiencing a 50% message loss rate. It is worth noting that for the same tag placement, CDMA has a message loss rate of 100%. In contrast, Buzz adapts the aggregate bit rate of the network to 0.57 bits/symbol, taking seven time slots to correctly decode all tags' messages. This demonstrates Buzz's ability to adapt to a bit rate of below 1 bit/symbol under challenging channel conditions, which is important for improving reliability of backscatter systems.

Power Efficiency: We investigate Buzz's impact on energy consumption of the backscatter nodes, and compare it to TDMA and CDMA. Backscatter energy consumption however is ultra-low, and therefore if we measure it after individual message transmission, the measurement noise may potentially exceed the consumed energy. Thus, to obtain robust measurements, we make each scheme reply to a large number of queries from the reader and measure the total energy consumption. The Moo has a small capacitor, leaving us unable to measure this accumulative energy consumption. As a workaround, we attach a large capacitor ($C = 0.1F$) to the Moo so that it can store enough energy in advance for our experiment.

In each experiment, the USRP reader sends a sequence of 8800 queries spaced by 10.2 ms to $K = 8$ tags.[8] Upon receiving each query, each tag wakes up and replies its message using the three schemes. We measure the energy consumed by the transfer as:

$$E_{consumed} = \frac{1}{2}CV_0^2 - \frac{1}{2}CV_f^2, \tag{10}$$

where V_0 is the voltage on the backscatter node's capacitor before the transfer and V_f is the voltage at the end of the transfer. We measure $E_{consumed}$ for three different values of V_0 since the amount of energy spent by a node depends on its initial voltage. For each value of V_0, we repeat the experiment for ten different positions of the tags and reader. At each position, we run the schemes one after another without changing the environment.

Fig. 13 shows the average net energy consumption for a single run (averaged by 8800) across tags and positions. As expected, each tag transmits for a much longer time in CDMA, and hence consumes more energy. In contrast, independent of the starting voltage V_0, Buzz does not consume much more energy than TDMA. There are two primary reasons for this. First, the rateless code in Buzz is sparse. Each node only transmits its message for very few times. Second, although in TDMA each node transmits its message

[8]Backscatter nodes continue to harvest energy from the reader's RF signal even while they drain energy performing the operations of the three schemes. To ensure a fair comparison, we keep the duration of the reader's continuous waveform the same between queries for all three schemes.

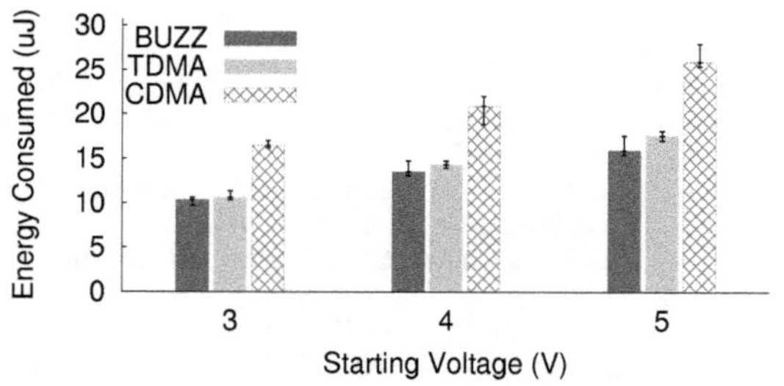

Figure 13—Energy consumption: Buzz does not consume much more energy than TDMA. Both Buzz and TDMA are significantly more energy efficient than CDMA.

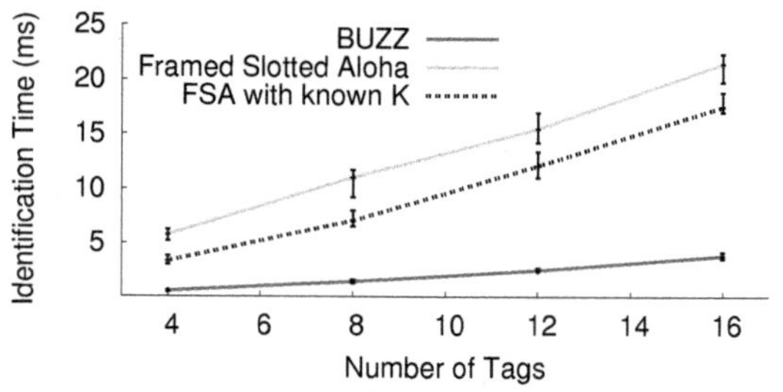

Figure 14—Identification time: Compared to Framed Slotted Aloha used in commercial RFID solutions, Buzz is significantly more efficient in identifying the nodes that want to transmit. Using the estimate of K from Stage 1 of Buzz's identification protocol improves the efficiency of Framed Slotted Aloha by 20%-40%.

only once, TDMA uses Miller-4 codes to protect the messages as recommended and hence has to switch the impedance on the antenna at 8 times of the data rate, which consumes more energy. If TDMA did not use Miller-4, it would lose its robustness and experience higher error rates [7]. Consequently, Buzz does not incur additional energy overhead to improve the efficiency and reliability of backscatter systems.

10. IDENTIFYING THE BACKSCATTER NODES

Recall that Buzz's node identification protocol consists of three stages: 1) estimating the number of backscatter nodes with data, 2) reducing the scale of the compressive sensing problem, and 3) compressive sensing decoding. In this section, we compare the efficiency of Buzz's three-step protocol with two baselines.

- **Framed Slotted Aloha (FSA):** This is the scheme adopted in the EPC Gen-2 standard [14]. Here, the reader initiates the identification phase by sending a query and allocating 2^Q time slots. Each tag picks a random slot and transmits a 16-bit random number (RN16), which serves as its temporary id. If a tag's id is correctly decoded (i.e., no collision), the reader will ACK this 16-bit temporary id. The reader adjusts Q to better accommodate the remaining active tag population, and repeats the procedure until all tags are identified. We follow the Q adjustment algorithm described in the standard [14], which initially sets $Q = 4$. It increases Q to $Q + C$ when a collision is detected, and reduces Q to $Q - C$ when no tag replies. We use $C = 0.3$ as recommended by the standard.
- **FSA Augmented with Estimated K:** We feed $\widehat{K}$, the estimate of K derived from Stage 1 of Buzz's identification protocol to FSA. Once FSA knows $\widehat{K}$, it sets $Q = log_2\widehat{K}$, i.e., allocating $\widehat{K}$ slots, instead of $Q = 4$ as the initial value. The maximum throughput for FSA is known to be $\frac{1}{e} = 36.8\%$, and is achieved when the number of allocated slots is equal to the number of nodes [8, 16]. Hence, setting $Q = log_2\widehat{K}$ improves the efficiency of FSA. Further, knowing $\widehat{K}$ can help FSA work with a smaller temporary id space as in Buzz. Specifically, we let each tag in FSA transmit a shorter random temporary id in the slot it picks instead of the RN16. This reduces both the uplink and downlink transmission time in FSA.

We run the experiment for $K = 4$, 8, 12 and 16 Moo tags, and repeat for ten locations each. Fig. 14 shows the total amount of time spent on identifying the tags, including the overhead for ACKing the tags. While in Buzz, the reader only sends a single signal to tell all tags to stop, in the compared schemes, each tag needs to be ACKed individually.

For 16 tags, using Buzz's compressive sensing scheme achieves a 5.5× reduction in identification time over original FSA, and is 4.5× more efficient than FSA with estimated K. Also, as we can see, using the estimate of K from Stage 1 of Buzz reduces the identification time in FSA by 20%-40%. This is mainly because the size of the temporary id space is reduced to a function of K, instead of 2^{16}. In conclusion, by compressing the sparse identity space and reducing the reader feedback overhead, Buzz reaches the goal of significantly speeding up the node identification phase in backscatter systems. Combined with the 2× throughput gain in data transmission, Buzz reduces the overall communication time by 3.5×.

11. RELATED WORK

Various protocols have been proposed to improve the performance of backscatter communication [31]. Most of them are based on TDMA, such as Framed Slotted Aloha adopted by the EPC Gen-2 RFID protocol [14] and the binary search tree algorithm [31]. Researchers have also studied the use of CDMA in this scenario [37]. FDMA and SDMA based schemes are proposed in [36, 49]. The common problems with these anti-collision protocols are twofold. First, since collisions (over time, frequency, code or space) can never be efficiently eliminated, these systems still end up wasting resources over collisions. Second, they divide the medium evenly among all backscatter nodes which typically have diverse channels.

In the context of ultra-low power systems (e.g., passive RFIDs), [43] is the only piece of work similar to ours in the sense that it also aims to decode collisions. However, it does not reduce identification overhead, or address the rate adaptation problem, but instead focuses on decoding repeated collisions in a low frequency proximity card environment. Further, the heavy machine learning decoders employed in [43] are not practical in high throughput systems.

There has recently been a lot of work on decoding collisions in WiFi networks to increase throughput [26, 21]. However, applying these techniques to backscatter networks is challenging. Successive interference cancellation requires exponential differences in the power levels of colliding nodes [26]. ZigZag decoding requires an offset between colliding messages [21], which is inefficient given the short messages in backscatter systems.

In the area of compressive sensing, the closest to our work is an algorithm in [40] which eliminates chunks of the space to improve running time. However, it uses complex deterministic codes which are not applicable to low power backscatter networks. [17] provides a theoretical analysis of using compressive sensing for sparse detection in on-off random access channels. It requires nodes to transmit arbitrary values chosen from a Gaussian distribution, which is infeasible since backscatter nodes can only transmit binary values.

Belief propagation in a sparse scenario is best-known for its use in LDPC codes [20]. Our belief propagation algorithm is similar to a bit flipping algorithm introduced in [33]. However, the algorithmic novelty of Buzz lies in distributedly coding the bits of mutiple transmitters, on the air. Accordingly, Buzz's bit flipping algorithm is devised to effectively decode such a rateless code on a complex constellation graph representing multiple sources, unlike bit flipping decoding in the modulo-2 domain often employed by the coding theory community [33, 44].

Lastly, the rate adaptation problem in wireless networks has been extensively studied. Our design is inspired by recent advances in automatic rate adaptation using rateless codes, such as spinal and strider codes [39, 25], but differs in that we adapt the aggregate bit rate of the network through the design of a sparse distributed rateless code that is easy to decode using belief propagation.

12. CONCLUSION

For ultra-low power backscatter networks to reach the stage of widespread deployment, issues of reliability and efficiency must be addressed. This paper addresses these issues by modeling uplink transmissions in backscatter networks as if they were performed by a single virtual sender and treating collisions as a *sparse rateless* code across the nodes. We introduce a novel compressive sensing algorithm which significantly speeds up backscatter node identification and a belief propagation algorithm which enables distributed rate adaptation in data transfer. Evaluation of the new design shows a large improvement in both reliability and efficiency of backscatter communication.

Acknowledgments: We thank Nate Kushman, Ben Ransford, Shane Clark, Fadel Adib, Arthur Berger, Yu-Chih Tung, Phillip Nadeau, Shyamnath Gollakota, Hong Zhang, the reviewers and our shepherd, Michael Mitzenmacher for their insightful comments. We also thank Omid Aryan for helping with the experimental setup. This research is funded by NSF and the Interconnect Focus Center. We thank the members of the MIT Center for Wireless Networks and Mobile Computing, including Amazon.com, Cisco, Intel, Mediatek, Microsoft, and ST Microelectronics, for their interest and support.

13. REFERENCES

[1] Alien Technology Inc. ALN-9640 Squiggle Inlay. www.alientechnology.com.

[2] Alien Technology Inc. Common RFID Implementation Issues. Tech. Report. http://www.alientechnology.com/docs/.

[3] P. Bardell, W. McAnney, and J. Savir. *Built-In Test for VLSI: Pseudorandom Techniques*. John Wiley & Sons, 1987.

[4] R. Berinde, A. Gilbert, P. Indyk, H. Karloff, and M. Strauss. Combining geometry and combinatorics: a unified approach to sparse signal recovery. *Allerton Conference*, 2008.

[5] R. Berinde and P. Indyk. Sequential sparse matching pursuit. *Allerton Conference*, 2009.

[6] M. Brazeal. *RFID: Improving the Customer Experience*. Paramount Market Publishing, 2009.

[7] M. Buettner and D. Wetherall. A Gen 2. RFID Monitor Based on the USRP. *SIGCOMM Communication Review*, 2010.

[8] M. Buettner and D. Wetherall. A Software Radio-based UHF RFID Reader for PHY/MAC Experimentation. *IEEE RFID*, 2011.

[9] E. Candès, J. Romberg, and T. Tao. Stable signal recovery incomplete and inaccurate measurements. *Comm. Pure Appl. Math.*, 2006.

[10] E. Candes and T. Tao. Near-optimal signal recovery from random projections and universal encoding strategies. *IEEE Transactions on Information Theory*, November 2004.

[11] I. Carron. Compressive sensing: Section 4 sparse recovery solvers. *http://sites.google.com/site/igorcarron2/cs*, 2012.

[12] N. Cho, S.-J. Song, S. Kim, S. Kim, and H.-J. Yoo. A 5.1-μw UHF RFID tag chip integrated with sensors for wireless environmental monitoring. In *ESSCIRC*, 2005.

[13] D. Donoho. Compressed sensing. *IEEE Trans. on Info. Theory*, 2006.

[14] EPCglobal Inc. EPCglobal Class 1 Generation 2 V. 1.2.0. http://www.gs1.org/gsmp/kc/epcglobal/uhfc1g2.

[15] C. Estan, G. Varghese, and M. Fisk. Bitmap algorithms for counting active flows on high speed links. In *IMC*, 2003.

[16] K. Finkenzeller. *RFID Handbook*. John Wiley & Sons, 2010.

[17] A. Fletcher, V. Goyal, and S. Rangan. A sparsity detection framework for on-off random access channels. In *ISIT*, 2009.

[18] Frost & Sullivan. Global RFID healthcare and pharmaceutical market. Industry Report, 2011.

[19] Frost & Sullivan. Global RFID market. Industry Report, 2011.

[20] R. Gallager. Low-density parity-check codes. *IEEE Transactions on Information Theory*, 1962.

[21] S. Gollakota and D. Katabi. ZigZag decoding: Combating hidden terminals in wireless networks. In *SIGCOMM*, 2008.

[22] M. Grant, S. Boyd, and Y. Ye. CVX: Matlab software for disciplined convex programming. http://cvxr.com/cvx.

[23] R. E. Greeff, F. W. Smith, and D. K. Ovard. RFID device time synchronization. Patent US7889083, 2006.

[24] J. Griffin and G. Durgin. Complete link budgets for backscatter-radio and RFID systems. *IEEE Antennas and Propagation Magazine*, 2009.

[25] A. Gudipati and S. Katti. Strider: Automatic rate adaptation and collision handling. In *ACM SIGCOMM*, 2011.

[26] D. Halperin, T. Anderson, and D. Wetherall. Taking the sting out of carrier sense: Interference cancellation for wireless lans. In *ACM MobiCom*, 2008.

[27] Impinj Speedway. R420 RFID reader. www.impinj.com.

[28] E. Inc. Universal Software Radio Peripheral. http://ettus.com.

[29] T. Jayram and D. Woodruf. Optimal bounds for Johnson-Lindenstrauss transforms and streaming problems with sub-constant error. In *SODA*, 2011.

[30] M. Karim and M. Sarraf. *W-CDMA and CDMA2000 for 3G mobile networks*. McGraw-Hill, 2002.

[31] D. Klair, K.-W. Chin, and R. Raad. A survey and tutorial of RFID anti-collision protocols. *IEEE Comm. Surveys*, 2010.

[32] M. Kodialam and T. Nandagopal. Fast and reliable estimation schemes in RFID systems. *MobiCom*, 2006.

[33] Y. Kou, S. Lin, and M. Fossorier. Low-density parity-check codes based on finite geometries: a rediscovery and new results. *Transactions on Information Theory*, 2001.

[34] Laird Technologies. Crushcraft S9028PCRW RFID antenna. http://www.arcadianinc.com/.

[35] T. Lee. *The Design of CMOS Radio-Frequency Integrated Circuits*. Cambridge University Press, 1998.

[36] H.-C. Liu and J.-P. Ciou. Performance analysis of multi-carrier RFID systems. In *SPECTS*, 2009.

[37] C. Mutti and C. Floerkemeier. CDMA-based RFID systems in dense scenarios: Concepts and challenges. In *IEEE Int. Conf. RFID*, 2008.

[38] M. Pelino, C. Mines, J. Warner, and S. Musto. M2M connectivity helps telcos offset declining traditional services. Forrester Research, 2011.

[39] J. Perry, H. Balakrishnan, and D. Shah. Rateless spinal codes. In *HotNets-X*, 2011.

[40] E. Porat and M. Strauss. Sublinear time, measurement optimal, sparse recovery for all. In *SODA*, 2012.

[41] J. Posamentier. RFID tag clock synchronization. Patent US20070205871, 2006.

[42] PowerID. Battery assisted passive RFID tags read at 160+ feet. The RFID Network, 2012. www.rfid.net.

[43] D. Shen, G. Woo, A. Lippman, D. Reed, and J. Wang. Separation of multiple passive RFID signals using software defined radio. In *IEEE Int. Conference on RFID*, 2009.

[44] M. Sipser and D. Spielman. Expander codes. *IEEE Transactions on Information Theory*, 42:1710–1722, 1996.

[45] C. Swedberg. Visual data center combines RFID with 3-d thermal imaging. *RFID Journal*, July 2010.

[46] D. Tse and P. Vishwanath. *Fundamentals of Wireless Communications*. Cambridge University Press, 2005.

[47] M. Vutukuru, H. Balakrishnan, and K. Jamieson. Cross-layer wireless bit rate adaptation. In *ACM SIGCOMM*, 2009.

[48] A. Wang, S. Cho, C. Sodini, and A. Chandrakasan. Energy efficient modulation and mac for asymmetric RF microsensor systems. In *Int. Symposium on Low Power Electronics*, 2001.

[49] J. Yu, K. Liu, and G. Yan. A novel RFID anti-collision algorithm based on sdma. In *WiCOM*, 2008.

[50] H. Zhang, J. Gummeson, B. Ransford, and K. Fu. Moo: A batteryless computational RFID and sensing platform. Tech Report UMASS, 2011. http://spqr.cs.umass.edu/moo/.

[51] Y. Zhang, H.-H. Chen, and M. Guizani. *Cooperative Wireless Communications*. CRC Press, 2009.

[52] T. Zimmerman. Assessing the capabilities of RFID technologies. Gartner, 2009.

Network-Aware Service Placement in a Distributed Cloud Environment

Moritz Steiner, Bob Gaglianello, Vijay Gurbani,
Volker Hilt, W. D. Roome, Michael Scharf, and Thomas Voith
Bell Labs, Alcatel-Lucent
{*first.last*}@alcatel-lucent.com

ABSTRACT

We consider a system of compute and storage resources geographically distributed over a large number of locations connected via a wide-area network. By distributing the resources, latency to users can be decreased, bandwidth costs reduced and availablility increased. The challenge is to distribute services with varying characteristics among the data centers optimally. Some services are very latency sensitive, others need vast amounts of storage, and yet others are computationally complex but do not require hard deadlines on execution. We propose efficient algorithms for the placement of services to get the maximum benefit from a distributed cloud systems. The algorithms need input on the status of the network, compute resources and data resources, which are matched to application requirements.

This demonstration shows how a network-aware cloud can combine all three resource types - computation, storage, and network connectivity - in distributed cloud environments. Our dynamic service placement algorithm monitors the network and data center resources in real-time. Our prototype uses the information gathered to place or migrate services to provide the best user experience for a service.

Categories and Subject Descriptors

C.2.1 [**Computer-Communication Networks**]: Network Architecture and Design—*Network operating systems*

Keywords

Cloud, service placement

1. INTRODUCTION

Cloud computing has seen sizeable growth in the past few years and analysts predict the trend is going to accelerate [3]. Today, public cloud providers operate a small number of large data centers, offering compute and storage resources. The cloud operators do not own nor do they operate the network that connects their data centers to the public Internet or to their customer's sites. Network operators, on the other hand, operate the network but are not aware of the cloud services using that network. We argue that many new opportunities arise by combining the data center and network operations and by adding compute and storage resources at multiple locations within the network.

One such opportunity is to run services closer to users, thereby running a service at many locations in parallel. This facilitates offering of latency-sensitive services from the cloud, and provide performance as good – or better – than users currently receive from their local computers. Examples include remote storage, virtual desktops, or action games. Integrated network and cloud providers can offer services in high quality to end users, since they control all resources involved in the delivery of the service. Finally, bandwidth costs can be reduced by running services on data-centers close to the users.

SIGCOMM'12, August 13–17, 2012, Helsinki, Finland.
ACM 978-1-4503-1419-0/12/08.

2. RESEARCH CHALLENGES

To obtain the maximum benefit from a distributed cloud system, we need efficient service-placement algorithms that can accomodate different, sometimes contradicting, goals [1]. A service provider may, for example, want to minimize communication, compute and storage costs, maximize link and data-center utilization and guarantee the user's quality of experience. This is a nontrivial, multi-dimensional optimization problem.

The placement and resource allocation algorithms need input from the data centers and from the network describing available resources and some normalized notion of the cost of accessing that resource. A challenge is to summarize and abstract this data to a level that is meaningful for solving the optimization and without revealing too many details about the communication network and data centers. Besides information about the resources, the algorithms need to know the application's profile: is it latency sensitive? does it require a large amount of storage? how do the different components of an application communicate? which components are user-facing? are there any legal restrictions involved in executing services for a given user outside a defined country, etc.? There are, of course, existing research and standardization activities that can be leveraged, but regardless, some extensions are needed for distributed cloud environments.

A further challenge is to offer an environment that enables service developers to rapidly deploy new services that are distributed, elastic, close to the users, and resilient. This requires a management system optimized for distributed clouds.

3. DEMONSTRATION

We have implemented a solution for network-aware service placement in a distributed cloud, which will be shown during the demonstration. The solution has several components. First, we utilize CloudStack [2], an open source cloud management system. By using such a system, we show that our solution does not depend on proprietary protocols. Second, we implemented our own Web-based frontend "dashboard" that communicates with the other components, and provides a central interface for monitoring and configuring the system. Finally, we use the ALTO protocol [4], developed within the Internet Engineering Task Force (IETF) as a standard in-

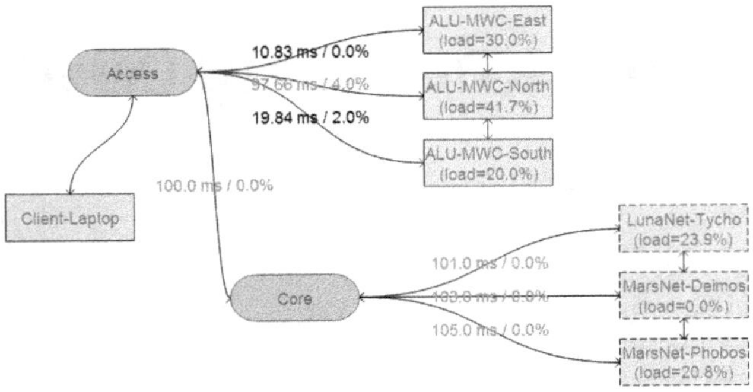

Figure 1: The demonstration network topology, including live measurement data.

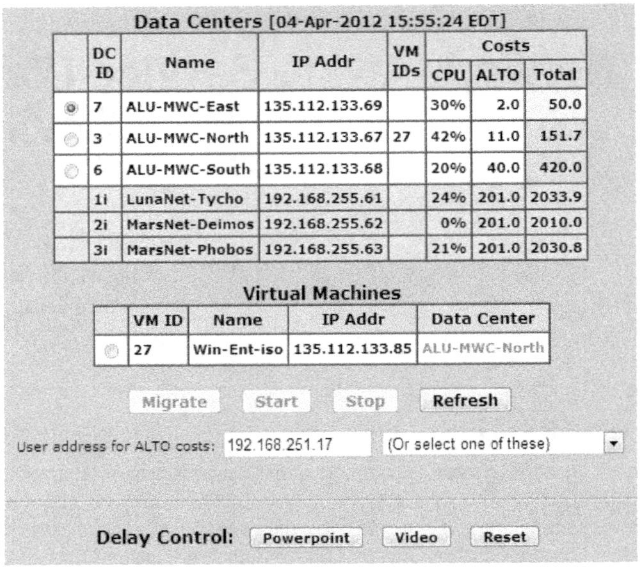

Figure 2: The cloud control dashboard showing the data centers and the virtual machines.

terface to obtain information about the communications network. We developed our own ALTO server that collects network topology data and provides an abstract view of that topology and costs. The networking information used by the ALTO server is gathered from live measurements in the demonstration network and includes dynamic loss and delay information. The linux commands "tc" and "netem", included in current distributions, are used to introduce latency and loss into the system.

The overall function of the system is as follows: The cloud dashboard queries the ALTO server for the network metrics and the cloud management for the data center load. Using this input and the user location, the dashboard runs a placement decision algorithm, which computes a score for each data center and recommends the data center with the best overall score.

The demonstration utilizes four laptops and an Ethernet switch to simulate multiple data centers connected to and controlled by a cloud management system. Each of three laptops form a mini data center. The fourth laptop hosts the cloud management software, the dashboard, the measurement logic and the ALTO server. An additional laptop runs the remote desktop client, and another laptop or tablet displays the network topology and the dashboard of the cloud operations center.

4. SCENARIOS

Our demonstration uses a virtual desktop as a sample application. Access to a remote desktop is a promising use case for outsourcing enterprise infrastructure. It uses the standard RDP protocol [5] to connect the user to the host of the virtual desktop. Because it is very latency sensitive, the virtual desktop service highlights the advantages of having many distributed data centers over having one or a few data centers.

The demonstration starts by running a high definition video on the laptop serving as a client. Note that the virtual desktop, running in the remote data-center, renders the video and writes the video bitstream to the virtual screen. RDP sends those screen updates over the network to the client computer, and the client displays them on the user's physical screen. Initially the video plays smoothly, and the audio is synchronized with the video. This is the experience the user expects.

Then we degrade the user's experience by introducing latency and loss between the client computer and the data-center hosting the virtual desktop. In the real world, this could happen on an overloaded link, or when the user disconnects from the network and resumes after reconnecting from a different location. The video becomes choppy and out of sync with the audio. In the demonstration, we allow the user to "look behind the curtain" to see the cause of the performance problems: we display the network topology and live measurements of link and data center load (Figure 1). Provisioning of both network and compute performance highlights that selecting the right data center is a non-trivial optimization problem.

Next we show how our placement algorithm uses this data to calculate an overall score for each data center and suggest a datacenter to host the user's virtual desktop. Then we use our dashboard to migrate the virtual desktop to the new data-center (Figure 2). This is a manual step in our demonstration; in reality, the system would do this automatically without user intervention. Finally, we show how the video quality – the user's experience – is restored to what it was at the beginning.

We further show a user working on a Powerpoint presentation with the virtual desktop. At first, Powerpoint responds smoothly to the user's edits. But as we increase the network latency, the user finds it difficult, and finally impossible, to work effectively. Again, we use our dashboard and network monitoring tools to select a better data-center and migrate the desktop to it, thus restoring the user's productivity.

In summary, our demonstration shows the tight coupling of the network with a distributed cloud solution using a system that integrates several of our own software components addressing individual challenges. We demonstrate that in case of network or data center problems migrating a service to another data center can mitigate the performance issues.

5. REFERENCES

[1] M. Alicherry and T. V. Lakshman. Network Aware Resource Allocation in Distributed Clouds. In *INFOCOM*, 2012.

[2] cloudstack - open source cloud computing. `http://www.cloudstack.org/`.

[3] Gartner Cloud Market Prediction. `http://tiny.cc/x696bw`.

[4] IETF ALTO working group. `http://datatracker.ietf.org/wg/alto/`.

[5] Windows Remote Desktop Services. `http://tiny.cc/zt05bw`.

SP4: Scalable Programmable Packet Processing Platform

Harjot Gill
University of Pennsylvania
Philadelphia, USA

Dong Lin
University of Pennsylvania
Philadelphia, USA

Lohit Sarna
University of Pennsylvania
Philadelphia, USA

Robert Mead
University of Pennsylvania
Philadelphia, USA

Kenton C.T. Lee
University of Pennsylvania
Philadelphia, USA

Boon Thau Loo
University of Pennsylvania
Philadelphia, USA

ABSTRACT

We propose the demonstration of SP4, a software-based programmable packet processing platform that supports (1) stateful packet processing useful for analyzing traffic flows with session semantics, (2) uses a task-stealing architecture that automatically leverages multi-core processing capabilities in a load-balanced manner without the need for explicit performance profiling, and (3) a declarative language for rapidly specifying and composing new packet processing functionalities from reusable modules. Our demonstration showcases the use of SP4 for performing high-throughput analysis of traffic traces for a variety of applications, such as filtering out unwanted traffic and detection of DDoS attacks using machine learning based analysis.

Categories and Subject Descriptors

C.2.1 [**Computer-Communication Networks**]: Network Architecture and Design

Keywords

Packet analysis, multicore, declarative networking

1. INTRODUCTION

As network applications become increasingly complex and heterogeneous, there is an increasing need for extensibility at the data plane, in order to carry out sophisticated functionalities such as traffic flow management, filtering out unwanted traffic, and content-based networking. Emerging software-defined networking platforms such as OpenFlow [6] provides extensibility by enabling packet processing at the software level. However, enabling such functionality in software requires significantly high bandwidth and compute capacity from servers.

One promising direction proposed recently is the use of multi-core machines to enable high-performance programmable software routers [3, 11], that can scale as the number of cores/machines increases. However, existing solutions either lack programming tools to rapidly customize different data plane functionalities, require explicit performance profiling to allocate cores to processing functionalities, or are not capable of performing complex operations beyond stateless per-packet processing, such as basic IP routing and packet encryption. To address these limitations, we present SP4 (stands for *Scalable Programmable Packet Processing Platform*), a software-based router platform that aims to provide data plane customizability and rapid deployment at high throughput and low latency.

We have developed a prototype of SP4 based on Rapidnet declarative networking engine [8], enhanced with task-stealing model of parallelism, and uses component-based dataflow architecture that allows for rapid assembly of packet processing functionalities. SP4 uses SP4LOG, a declarative language inspired by declarative networking [5] for constructing dataflows by composing existing data processing components. SP4LOG increases customizability through the use of high-level programming abstractions for composition. For example, our SMTP, VoIP and DDoS dataflows can be specified in only 7, 10 and 5 rules respectively.

SP4 integrates a threading library based on the task-stealing model [9], that enables us to enable fine-grained parallelism at the level of individual components in the dataflow, hence achieving both automatic load-balancing and high throughput processing. As opposed to naïve per-packet parallelism [3] which is only suitable for stateless workload (e.g. packet-routing, IPSec etc.), SP4 can process stateful multi-packet protocols (e.g. SIP) with high degree of parallelism. To ensure correct packet ordering in the presence of parallel processing, SP4 allows packets to be ordered based on its specified *context* attribute, defined in terms of application-specific semantics (e.g., a SIP session).

2. SYSTEM OVERVIEW

We provide an overview of SP4 by describing its dataflow pipeline. Refer to [4] for more details on SP4LOG.

SP4 uses a component-based dataflow architecture [7] that allows for rapid assembly of packet processing functionalities from reusable components into dataflow pipelines. These component modules are typically wrappers over existing code, that perform functionalities for various operations such as aggregating traffic statistics, or performing deep-packet inspections into specific application traffic.

The dataflow pipeline consists of several components connected in a linear chain, which are directly compiled from SP4LOG programs. Figure 1 shows an example dataflow pipeline that illustrates the execution model of SP4. SP4 supports three types of reusable components (e.g. elements), whose types are specified when the dataflow is constructed:

- **Serial.** Packets are processed in strict FIFO order. This is done for operations where total order is essential. In our example, `Packet Capture` is a serial component.
- **Parallel.** Incoming packets to a parallel component can be processed by several concurrent threads in a manner where ordering does not matter. For instance, once SMTP messages are assembled, regular expression matches on individual email mes-

SIGCOMM'12, August 13–17, 2012, Helsinki, Finland.
ACM 978-1-4503-1419-0/12/08.

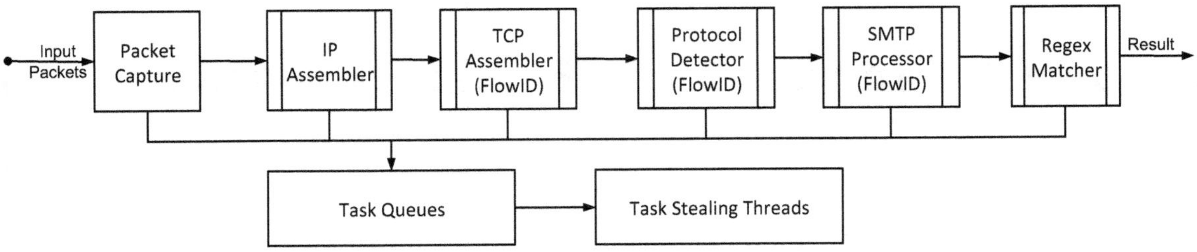

Figure 1: SP4 Dataflow pipeline example based on SMTP analysis. Parallel components have double lines, and context-ordered elements additionally have their context-keys shown in (...).

sages can be parallelized and processed in arbitrary order. `IP Assembler` and `Regex Match` are parallel components.

- **Parallel context-ordered.** These components are processed in partial order. A *context-key* is specified, in which all packets with the same key should be processed in order of their arrival into the system. But the ordering of packets with different keys is not required. For instance, `TCP Assembler` have to assemble message in a partial order (based on TCP flows). `Protocol Detector` and `SMTP Processor` are also context-ordered components.

3. DEMONSTRATION PLAN

Our demonstration will showcase dataflow parallelization techniques used in SP4 to support automatic load-balancing and fine-grained parallelism, and present concrete use-cases of the system for high-throughput analysis of network traces. Our demonstration consists of the following use cases:

SMTP analysis. As our first use-case, we will demonstrate a naive regex based data-leak detection system that monitors SMTP traffic. The dataflow is shown in Figure 1. The setup consists of SMTP protocol based email traffic generator, a network tap and a SP4 node. To emulate real network traffic, we will use an email generator written in JavaMail API that sends emails with preconfigured content to a mail server. A SP4 node deployed on a linux machine will execute a dataflow for analyzing the SMTP traffic. Specifically, this dataflow will assemble TCP flows, detect SMTP protocol, assemble SMTP messages and red-flag if message matches customizable regex based rules.

VoIP call interception. As our second use-case, we will demonstrate VoIP traffic capture of a live SIP call. Our experimental setup consists of SIP clients (running on Android smartphones), a network tap and a SP4 node. To emulate real network traffic load, we will use SIPp traffic generator tool[10] in addition to real SIP clients. The workflow will monitor and track all active SIP sessions based on preconfigured interception rules. The RTP stream interceptor component will decode and dump the RTP stream to local disk if the corresponding SIP session is matched by the SIP transaction processor component. Finally the dumped stream will be available for playback as sound media, post session termination. SP4 will maintain call-states while dynamically load-balancing the processing across threads.

DDoS attack detection. Our third use-case evaluates the performance and reliability of SP4 in detecting Distributed Denial of Service Attack (DDoS) patterns by using computationally intensive Support Vector Machine algorithm (SVM), a popular general-purpose algorithm used for machine learning. The dataflow includes two additional components for feature extraction and SVM classification. Our experimental setup consists of a DDoS traffic generator and a SP4 node. The SVM Classification module is implemented using libsvm [1], and is trained outside of SP4 on DARPA intrusion detection dataset [2] and PREDICT dataset. In our demonstration, SP4 node will tap network traffic generated by a DDoS trace-driven traffic generator, extract and update features that are significant for machine learning for each active connection, and red-flag potential DDoS incidents if the set of feature value is classified as positive by SVM classification component.

In all use cases, we will demonstrate the speedup and throughput obtained by SP4 by running the system on multiple quad-core laptops deployed at the conference venue. If network connectivity allows, we will also connect to powerful desktop machines at Penn with multiple cores (and hyperthreading enabled) to demonstrate throughput and speedup capabilities of SP4.

Acknowledgements. This project is supported in part by NSF grants CCF-0820208, IIS-0812270, CNS-0845552, and CNS-1040-672, and the DARPA SAFER award N66001-C-4020.

4. REFERENCES

[1] C.-C. Chang and C.-J. Lin. LIBSVM: A library for support vector machines. *ACM Transactions on Intelligent Systems and Technology*, 2:27:1–27:27, 2011. Software available at `http://www.csie.ntu.edu.tw/~cjlin/libsvm`.

[2] DARPA Intrusion Detection Data Sets. `https://www.predict.org/`.

[3] M. Dobrescu, N. Egi, K. Argyraki, B.-G. Chun, K. Fall, et al. Routebricks: Exploiting parallelism to scale software routers. In *SOSP*, Oct 2009.

[4] H. Gill, D. Lin, T. Kothari, and B. T. Loo. Declarative multicore programming of software-based stateful packet processing. In *Declarative Aspects and Applications of Multicore Programming*. `netdb.cis.upenn.edu/papers/sp4_damp12.pdf`.

[5] B. T. Loo, T. Condie, M. Garofalakis, D. E. Gay, J. M. Hellerstein, P. Maniatis, R. Ramakrishnan, T. Roscoe, and I. Stoica. Declarative Networking. *CACM*, 2009.

[6] N. McKeown, T. Anderson, H. Balakrishnan, G. Parulkar, L. Peterson, and J. Rexford. Openflow: Enabling innovation in campus networks. In *ACM SIGCOMM*, April 2008.

[7] R. Morris, E. Kohler, J. Jannotti, and M. F. Kaashoek. The click modular router. In *ACM Transactions on Computer Systems*, Aug 2000.

[8] RapidNet: A Declarative Toolkit for Rapid Network Simulation and Experimentation. `http://netdb.cis.upenn.edu/rapidnet/`.

[9] J. Reinders. Intel thread building blocks. In *OâĂŹReilly Associates*, 2007.

[10] SIPp Open Source test tool / traffic generator for the SIP protocol. `http://sipp.sourceforge.net/`.

[11] T. Wolf, N. Weng, and C.-H. Tai. Runtime support for multicore packet processing systems. *Network, IEEE*, 21(4):29 –37, July-August 2007.

Distributed Content Storage for Just-in-Time Streaming

Sourav Kumar Dandapat
Department of CSE,
IIT Kharagpur, India
sdandapat@cse.iitkgp.ernet.in

Sanyam Jain
Department of CSE,
IIT Kharagpur, India
jain.sanyam@gmail.com

Niloy Ganguly
Department of CSE,
IIT Kharagpur, India
niloy@cse.iitkgp.ernet.in

Romit Roy Choudhury
Department of ECE and CS,
Duke University, USA
romit@ee.duke.edu

ABSTRACT

We propose a content distribution strategy over municipal WiFi networks where Access Points (APs) collaboratively cache popular multimedia content, and disseminate them in a manner that each mobile device has the portion of the content *just-in-time* for playback. If successful, we envision that a child will be able to seamlessly watch a movie in a car, as her tablet downloads different parts of the movie over different WiFi APs at different times.

Categories and Subject Descriptors

C.2.4 [**Distributed System**]: Distributed Applications

General Terms

Design, Performance

Keywords

Content Distribution, Municipal WiFi Network, Distributed Content Storage

1. INTRODUCTION

Video on Demand (VoD) has gained immense popularity in recent years whereby almost 50% of the entire Internet traffic is due to video. Video distribution becomes a challenging task especially when we consider distribution over wireless networks (which carries 18% of overall traffic). The challenges arise from low wireless data rates, high packet loss probability, link disconnections due to mobility, etc. Hence attention needs to be paid towards designing strategies to distribute content over wireless network [1] to fulfill the dual objectives of upholding user-experience and congestion control.

The increasingly large number of WiFi APs deployed across cities opens up new opportunities for seamless distribution of such content. A feasible strategy can be to cache popular content in such APs, whereby a traveler moving (say, by car) past those APs can download content from them. Specifically, a traveler during her journey might be interested in a movie, so she starts the process of simultaneous download and playback. However, it is almost infeasible to download a complete movie (assumed to be large in size) from a single AP due to the short association duration of the mobile device. So the download needs to be performed through successive APs and hence (different parts of) the same movie needs to be available in other APs in order to support just-in-time download. Caching the entire (movie) file in *every* AP reduces the total number of movies available to users, and is undesirable. We propose a content distribution strategy that ensures minimum redundancy storage while also offering uninterrupted playback to all users, regardless of their travel path.

SIGCOMM'12, August 13–17, 2012, Helsinki, Finland.
ACM 978-1-4503-1419-0/12/08.

2. CONTENT DISTRIBUTION

Let us assume that the total number of APs in a municipal WiFi network is n and every AP is capable of storing m files (each of same size). So the described network can cache up to $n \times m$ number of unique files.

To allow seamless download on the move, one trivial solution is to keep same set of m files across all n APs. However, it would limit the total number of cached file to m from $n \times m$. Almost similar efficiency can be achieved if we break each file into set of chunks of identical size and place 1st chunk at the 1st AP that client comes across, 2nd chunk at the 2nd AP and so on. This solution, while requiring less storage, would not be useful for another client (say) C, which travels the same path in the reverse direction. In this case, C would still download all the chunks but in reverse order, hence has to wait until the end (referred as *idle* time) before it can start playback.

Chunk Frequency: An intuition which can be derived from the above mentioned situation is that to eliminate *idle* time, the 1st chunk of the file must be available at every AP given that a user can start her journey from any point. If we assume that a car normally moves at 40 km/hour and the range of an AP is 50 m (radius), then the car would stay associated with an AP for 9 seconds. If a file can be downloaded at 1 Mbps rate then at most 9 Mb can be downloaded from an AP, which would then require approximately 61 seconds to play (assuming 80 minutes is needed to watch a 700 MB movie). By this time of 61 seconds, a user may encounter other APs, depending on the density of WiFi deployment. Assuming, a WiFi AP every $100m$, the vehicle may encounter 6 APs. Thus, the 2nd and 3rd chunks of the file must be available at least in one of the next 6 and 12 APs respectively, starting from the source where the user starts moving. To generalize, the qth chunk must be available within $(q-1) \times p$ APs, while $q \geq 2$ or frequency distribution of qth chunk f_q is $\geq 1/((q-1) \times p)$, where p is the number of APs a user encounters while playing a video chunk.

Chunk Distribution Strategy: We assume that each traveler follows the shortest path from the source to her destination. To determine which chunks must be cached at each AP, we need to consider all shortest paths between pairs of locations. We assume X_1, X_2, ...X_n represent n APs of the network. We also assume that a binary variable z_i^j is associated with every X_i AP, where $z_i^j = 0$ signifies the absence of the jth chunk at node X_i and $z_i^j = 1$ signifies presence. To ensure that the qth chunk is present at least once in a path (X_i, ..., $X_{i+(q-1)\times p}$) of length $(q-1) \times p$, the following condition must be satisfied: $z_i^q + z_{i+1}^q + ... + z_{i+(q-1)\times p}^q \geq 1$. We assume that file has been divided into κ chunks and each AP can host at most k ($\leq \kappa$) chunks.
With these constraints in mind we minimize the objective function f_{obj} that represents total storage required for the file caching. We use a Integer Linear Programming (ILP) solver (LPSOLVE) – the ILP package outputs the assignments of chunks to AP.

Random Distribution Strategy (RDS): Since the ILP package also provides the frequency of different chunks in the entire network for the proposed solution, a random distribution strategy can be to distribute identical number of chunks randomly among APs. We use the RDS as a baseline scheme for comparison.

3. EXPERIMENT

We have considered a region of the IIT Delhi campus (rectangular area with dimension $1km \times 1km$) for our case study [2]. 83 APs are assumed to be placed in the entire region. The movie file distributed is of size 45 Mb and divided into 5 equal sized chunks.
We use network simulator (NS3) to simulate and evaluate the performance of our proposed solution. "The ONE" [3] simulator is used to generate shortest path mobility model trace of clients. We place chunks in an AP cache as obtained from the ILP solution stated above. A client sends request to the associated AP for the desired file. In response, AP sends available chunks of requested file. The client selects a particular chunk for downloading depending on chunk selection strategy. We have done experiments for different *storage constraints* (percentage of AP memory allocated to storing a given file).

4. RESULTS AND DISCUSSION

Metrics: The performance of the proposed algorithm is evaluated using the following metrics:

(a) **Interruption Fraction(%)($\mathcal{IF}$):** Lets assume that a client takes t_1 unit time in her total journey and was idle (could not watch movie) for t_2 unit time. $\mathcal{IF}$ of this client can be formally expressed as $\frac{t_2 \times 100}{t_1}$.

(b) **Download Efficiency($\mathcal{DE}$):** It signifies average number of chunks downloaded from each AP. If a device crosses p APs and collects q chunks then $\mathcal{DE}$ of the device is $\frac{q}{p}$.

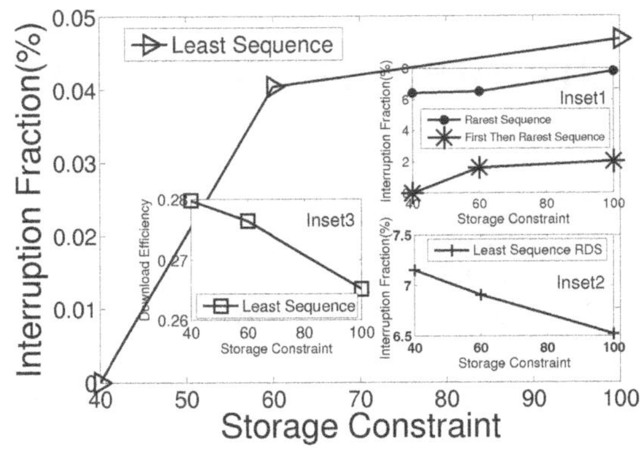

Figure 1: $\mathcal{IF}$ and $\mathcal{DE}$ of proposed and random schemes w.r.t chunk selection strategies.

Discussion: Fig. 1 shows variation of $\mathcal{IF}$ with different storage constraint of the proposed algorithm while client always downloads the chunk with Least Sequence First (LSF). Result shows that the $\mathcal{IF}$ is almost negligible in this strategy. Inset1 of Fig. 1 shows the variation of $\mathcal{IF}$ while the chunk selection strategies are first then rarest (priority of first chunk is highest then it selects rarest chunk) and rarest respectively. Rarest selection strategy has significantly higher $\mathcal{IF}$ as client needs to wait long for the initial chunks. Inset2 of Fig. 1 shows the variation of $\mathcal{IF}$ of RDS while chunk selection strategy is LSF. $\mathcal{IF}$ of RDS is significantly high compared to the proposed algorithm. An interesting observation is that in our algorithm the performance degrades (though very slowly) with loosening of storage constraint while it has opposite effect in case of RDS. This is because with loosening of constraint, rare pieces have a tendency of clustering, which sometimes makes them unavailable, whereas the average pieces are relatively easier to find, thus improving performance of RDS. The same tendency is noticed when we capture the $\mathcal{DE}$ (Inset3).

5. CONCLUSION & FUTURE DIRECTION

The key contribution of this poster is in showing that chunks of a (video) file can be scattered in distributed caches (in WiFi APs) in a manner that a mobile user may be able to download the needed chunk just in time for playback. This may allow the user to watch streaming content without interruption, thereby offloading cellular network traffic. Our simulation results present early promise, motivating a full scale implementation and evaluation on a real end-to-end system.

6. ACKNOWLEDGEMENT

Authors would like to thank Vodafone for their partial financial support.

7. REFERENCES

[1] J. Eriksson, H. Balakrishnan, S. Madden, *Cabernet: Vehicular Content Delivery Using WiFi*, Mobicom 2008

[2] http://www.openstreetmap.org/

[3] The Opportunistic Network Environment simulator www.netlab.tkk.fi/tutkimus/dtn/theone/

Blockmon: A High-Performance Composable Network Traffic Measurement System

Felipe Huici
NEC Europe Ltd

Andrea di Pietro
CNIT

Brian Trammell
ETH Zurich

Jose Maria Hidalgo
OPTENET

Daniel Martinez Ruiz
OPTENET

Nico d'Heureuse
NEC Europe Ltd

ABSTRACT

Passive network monitoring and data analysis, crucial to the correct operation of networks and the systems that rely on them, has become an increasingly difficult task given continued growth and diversification of the Internet. In this demo we present Blockmon, a novel composable measurement system with the flexibility to allow for a wide range of traffic monitoring and data analysis, as well as the necessary mechanisms to yield high performance on today's modern multi-core hardware. In this demo we use Blockmon's GUI to show how to easily create Blockmon applications and display data exported by them. We present a simple flow meter application and a more involved VoIP anomaly detection one.

Categories and Subject Descriptors

C.2 [**Computer-Communication Networks**]: General

Keywords

Network monitoring, data processing, high performance

1. INTRODUCTION

Two salient trends have dominated Internet-scale measurement over the past decade: the continued growth of the Internet, both in terms of attached nodes and total data transferred; and the diversification of devices attached to and applications running over the network.

These challenges point to the need for a high-performance, yet easily-extensible solution. To this end we demonstrate Blockmon, a system for supporting high-performance *composable measurement*: building fast network measurement applications out of small, discrete blocks.

Systems work in network measurement is by no means a new field. The modular principles in Blockmon were inspired by the Click modular router [4], though Click is oriented towards packet-based processing. Other programmable tools such as CoMo [3] or PF_RING [2] are either less flexible than or do not perform as well as Blockmon.

Blockmon contributes some key innovations to this wide base of work, focusing on applying programmable measurement to today's traffic loads and diversity while maximizing the use of modern, multi-core commodity hardware to ensure high performance. Blockmon is available as open-source software at `http://blockmon.github.com/blockmon`.

SIGCOMM'12, August 13–17, 2012, Helsinki, Finland.
ACM 978-1-4503-1419-0/12/08.

2. SYSTEM DESCRIPTION

At a high-level, Blockmon provides a set of units called *blocks*, each of which performs a certain discrete processing action, for instance parsing a DNS response, or counting the number of distinct VoIP users on a link. The blocks communicate with each other by passing *messages* via *gates*; one block's output gates are connected to the input gates of other blocks, which allows runtime indirection of messages. A set of inter-connected blocks implementing a measurement application is called a *composition*; a generic composition is shown in figure 1.

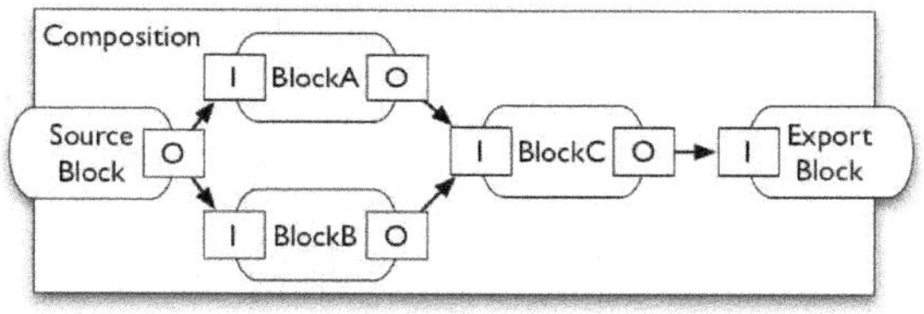

Figure 1: Example composition

The Message class provides a basic interface for identifying message types, and for supporting import and export of messages in order to connect compositions across nodes. Messages are constant in order to ensure that they can be shared without contention among multiple blocks concurrently, and provide a tagging mechanism to allow Blocks to add small bits of data to a message in a thread-safe manner. This latter mechanism allows blocks to process high volume messages (e.g., packets) in a pipeline without incurring the overhead of having to allocate new message objects.

The assignment of activities to threads and threads to CPU cores can have a large impact on performance [1]. To leverage this, Blockmon schedules work in thread pools. Each block is assigned to a pool via the composition, and pools can be pinned to specific cores. This model allows flexibility in terms of which block is executed on which CPU core.

Input gates are implemented either via direct invocation, where messages are passed via method call; or indirect invocation, where messages are passed via a lock-free rotating queue. Direct invocation causes the downstream block to run within the thread of the upstream block, while indirect invocation separates threads from each other without lock contention. Whether blocks are invoked directly or indirectly is another method for tuning performance.

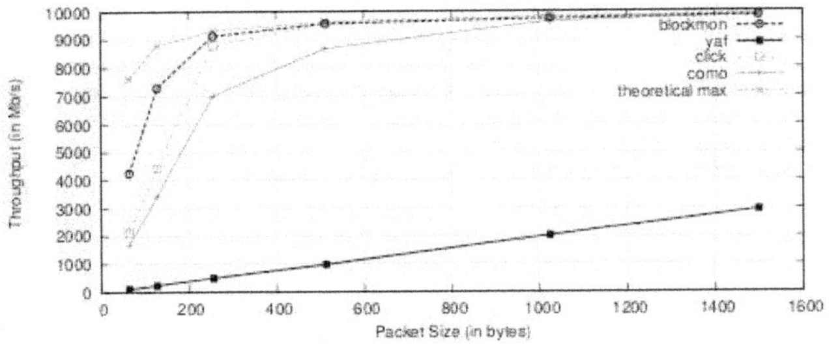

Figure 2: Performance comparison between Blockmon and other existing systems.

Additional performance improvements include efficient message passing via C++11 shared pointers and object-move semantics, batch allocation of memory for messages, and support for fast capture blocks using novel packet capture engines such as PFQ [5]. These improvements combine to produce the results in figure 2, which show Blockmon outperforming other systems when they all run a simple application that keeps per-flow statistics (byte and packet counts).

3. DEMONSTRATION DESCRIPTION

We will demonstrate two applications built on top of Blockmon: a simple monitoring application that keeps per-flow statistics such as byte and packet counts, and a more complex application that can detect anomalies or abuse in VoIP CDR (call data record) traces. The demonstration will focus on the detection of telemarketing activity. We will show the operation of Blockmon together with a Python-based back-end daemon and a web GUI front-end implemented in PHP and Javascript.

The Blockmon GUI consists of two views. The first is the composition view, which allows users to graphically design a composition and run it in the system. This view connects to the Blockmon back-end via a JSON-based protocol and retrieves a list of available blocks, as well as information about them (e.g., what they do, what parameters and gates they have, etc). The user then drags and drops blocks onto the main canvas, configures their parameters and connects them. In addition, the GUI allows blocks to be mapped to threadpools, and threadpools to be mapped to the available CPUs (see figure 3).

Once this is all done, the GUI sends a command to the back-end to start the composition running. While running, the user can inspect internal variables of blocks. For instance, the *PacketCounter* block has a variable *pktcnt* which returns the number of packets the block has seen so far; writing to the *reset* variable resets this count.

The GUI's second view is a dashboard, which is in charge of displaying the actual data produced by the running Blockmon compositions. The dashboard provides a number of display types based on the Google Chart Tools[1] and Highcharts JS[2] (e.g., line graphs, bar and pie charts, etc.) which are constantly updated according to the output of the compositions. The user can configure the most appropriate graphs to monitor the output of a composition, which is temporarily stored in an internal buffer to allow the dashboard graph to read and update the results on the screen.

The simple flow statistics application is intended as a "Hello, World" demonstration of Blockmon and the concepts

[1] `https://google-developers.appspot.com/chart/`
[2] `http://www.highcharts.com/`

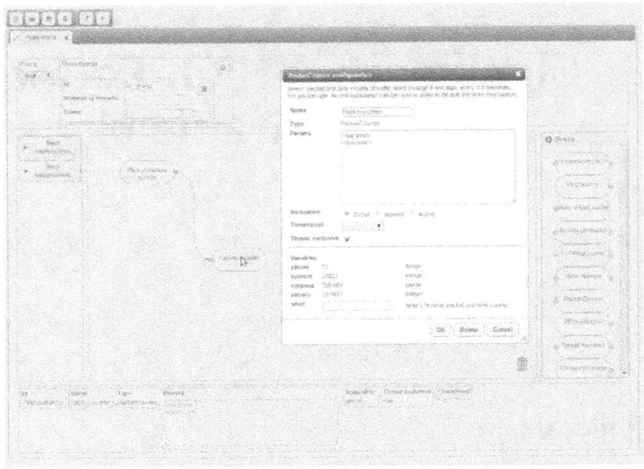

Figure 3: Blockmon GUI showing a composition.

of composable measurement. To demonstrate it, we draw its composition in the GUI, then start it. The composition keeps track of flows, periodically exporting data in IPFIX format about the N flows with the largest byte and packet counts. We then show how to set up the dashboard to display a graph showing the moving byte rates of the top N flows. We finally run the drawn composition and show the tracked flows statistics in the configured dashboard graphs.

We then turn to a VoIP anomaly detection application, intended as a demonstration of the ease of porting an existing non-trivial application to Blockmon. This application detects anomalies such as telemarketers based on CDR (call data record) traces. Once again we show how to draw up its composition and set up the dashboard in the GUI. This time the dashboard displays a listing of the most anomalous users, including time-varying anomaly detection scores for the system's various anomaly detection algorithms. We will also show how to use the composition view to inspect the value of a block's variable by displaying the rate at which the block *CDRSource* is processing CDRs.

4. REFERENCES

[1] Egi, N., Greenhalgh, A., Handley, M., Hoerdt, M., Huici, F., and Mathy, L. Towards high performance virtual routers on commodity hardware. In *Proceedings of ACM CoNEXT 2008* (Madrid, Spain, December 2008).

[2] Fusco, F., and Deri, L. High speed network traffic analysis with commodity multi-core systems. In *Proceedings of the 10th annual conference on Internet measurement* (2010), IMC '10, ACM, pp. 218–224.

[3] Iannaccone, G. Fast prototyping of network data mining applications. In *Passive and Active Measurement Conference 2006* (Adelaide, Australia, March 2006).

[4] Kohler, E., Morris, R., Chen, B., Jahnotti, J., and Kasshoek, M. F. The click modular router. *ACM Transaction on Computer Systems 18*, 3 (2000), 263–297.

[5] N.Bonelli, Pietro, A. D., Giordano, S., and Procissi, G. On multi–gigabit packet capturing with multi–core commodity hardware. In *Passive and Active Measurement conference (PAM)* (2012).

PaDIS Emulator: An Emulator to Evaluate CDN-ISP Collaboration

Ingmar Poese
T-Labs/TU Berlin
ingmar@net.t-labs.tu-berlin.de

Benjamin Frank
T-Labs/TU Berlin
bfrank@net.t-labs.tu-berlin.de

Simon Knight
University of Adelaide
simon.knight@adelaide.edu.au

Niklas Semmler
TU Berlin
nsemmler@net.t-labs.tu-berlin.de

Georgios Smaragdakis
T-Labs/TU Berlin
georgios@net.t-labs.tu-berlin.de

Abstract

We present PaDIS Emulator, a fully automated platform to evaluate CDN-ISP collaboration for better content delivery, traffic engineering, and cost reduction. The PaDIS Emulator enables researchers as well as CDN and ISP operators to evaluate the benefits of collaboration using their own operational networks, configuration, and cost functions.

The PaDIS Emulator consists of three components: the network emulation, the collaboration mechanism, and the performance monitor. These layers provide scalable emulation of the interaction between an ISP or a number of ISPs with multiple CDNs and vice versa. PaDIS Emulator design is flexible in order to implement a wide range of collaboration mechanisms on virtualized or real hardware, and evaluate them before introduction to operational networks.

Categories and Subject Descriptors

C.2.1 [**Computer-Communication Networks**]: Network Architecture; C.2.3 [**Network Operations**]: Network Management

General Terms

Performance, Measurement.

Keywords

Traffic Engineering, Content Distribution, Network Optimization.

1. BACKGROUND

Recent studies [5, 6] show that a large fraction of Internet traffic originates from a small number of Content Delivery Networks (CDNs). To cope with increasing demand for content, CDNs have deployed massively distributed hosting infrastructures. These infrastructure pose challenges for CDNs as they must dynamically map end-users to servers without full awareness of network conditions within an ISP and end-user locations [1]. ISPs also struggle to cope with the rapid traffic shifts caused by the dynamic server selection policies of CDNs. Their traffic engineering is based on the presumption that traffic is predictable. It adapts the routing and operates on time scales of several hours [3], thus, it is too slow to react to rapid traffic changes caused by CDNs.

The challenges faced separately by CDNs and ISPs can become an opportunity for collaboration. This will improve operation, reduce costs, and enhance end-user experience. In previous work, we designed and implemented a recommendation system called Provider-aided Distance Information System (PaDIS), run by an ISP to enable CDN-ISP collaboration [9]. PaDIS offers an interface for CDN-ISP collaboration without the need to share any operational information from either party and therefore avoids the confidentiality concerns about leaking operational secrets. Our previous work showed significant performance gains for ISPs, CDNs, and end-users [4]. Also, CDNs' mapping improved and ISPs gained another tool in their traffic engineering suite.

SIGCOMM'12, August 13–17, 2012, Helsinki, Finland.
ACM 978-1-4503-1419-0/12/08.

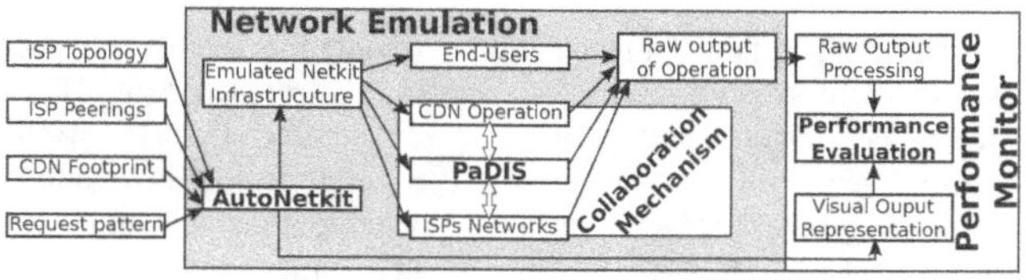

Figure 1: PaDIS Emulator Workflow.

In this demo we present a fully-automated framework that emulates the topology, operation, and policies of both ISPs and CDNs. Our emulation framework, PaDIS Emulator, can be used to quantify the benefits of CDN-ISP collaboration. It also helps identify the best practices to achieve the maximum benefit through collaboration. Our architecture is scalable to emulate a large number of CDNs and ISPs. Thus, it can be used as a collaboration and decision making tool of both ISPs and CDNs.

2. ARCHITECTURE

PaDIS Emulator has been designed with three-layers: (1) Network Emulation, (2) the Collaboration Mechanism, and (3) a Performance Monitor, see Figure 1. Only the first layer uses emulation; the other two can run either on an emulated or hardware-based infrastructure. This logical separation allows us to use real router and network service software in an emulated environment for evaluation, but with the opportunity to easily apply it to a real deployment.

2.1 Network Emulation

In the Network Emulation, both the ISP topologies and CDN footprints are created using virtual machines and links. To simplify the emulated setting, a graphical editor, Yed, is used for modeling. Alternatively, the configuration can be converted directly from operational data, such as IGP, BGP.

For the basic setup, PaDIS Emulator can run on high-level representation of ISPs and their interconnection, the deployment foot-

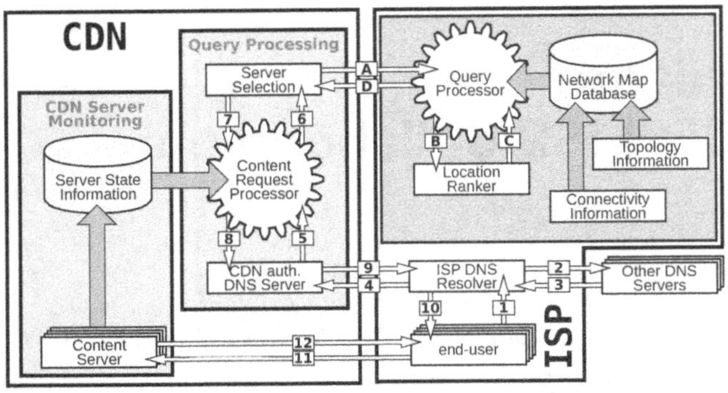

Figure 2: PaDIS Collaboration Scheme.

print of CDNs and the demand patterns of users to CDNs. However, additional information, such as specific DNS deployment, BGP policies, peering costs, etc., can easily be integrated. Once the specifications have been completed, we use AutoNetkit [8], an auto-configuration tool for Netkit, to allow the entire virtual setup to be fully automated. Netkit runs the Quagga routing daemon inside User-Mode Linux virtual machines to emulate ISP networks. Furthermore, it can run any additionally needed software. AutoNetkit allows for a fully automated setup of emulated networks, including IP addressing, routing, peering and DNS configuration. We added extra features to support PaDIS Emulator including auto-deploying software, and configuring link properties and meta information on peering links.

The footprint of a CDN is installed into the PaDIS Emulator by deploying custom or generic CDN software. Today, the CDN deployment is very complex [7]. Some CDNs, e.g., Limelight and Google, rely on large datacenters and connect to ISPs either through private peering or Internet Exchange Points. Others, such as Akamai, deploy their infrastructure deep inside an ISP's network. PaDIS Emulator can configure both of these CDN strategies, either individually, joined or as hybrids. In fact, PaDIS Emulator allows multiple independent CDNs, with different combinations of deployment and strategies to be run inside the same emulation.

A crucial task in CDN operation is the mapping of users to the server. This is mostly done through DNS and/or HTTP redirection. PaDIS Emulator contains a fully-functional DNS hierarchy. The positioning of DNS servers and resolvers can either be done automatically or through a user-specified custom configuration. This feature is important: the performance of content delivery relies heavily on the deployment and operation of ISP-operated DNS [9, 1]. Finally, request patterns and profiles of ISP users for content hosted in the CDNs can be pre-defined or modelled as statistical function.

2.2 Collaboration Scheme

The Collaboration Scheme layer is the heart of PaDIS Emulator. ISPs run PaDIS, while CDNs are enabled to utilize this novel service. PaDIS uses its network monitoring to collect *Connectivity* and *Topology Information* and keeps an up-to-date network database. This information is used to aid the CDN in its server selection process by ranking candidate servers in relation to the network state, requested content, and the desired performance metric. On the CDN side, operation works as it does today (see Figure 2, steps 1-10), with the option of obtaining ISP recommendation when users are assigned to servers, (steps A-D). To quantify the efficiency of PaDIS, the collaboration scheme can be dynamically enabled or disabled on a per-CDN basis.

When the collaboration mechanism is enabled, the set of servers, determined by the Server Selection, able to satisfy the content request is sent to PaDIS. The CDN, instead of performing its own server selection, utilizes PaDIS to obtain ISP guidance. If needed, the CDN can also specify the performance metrics relevant for the mapping, e.g., delay, bandwidth, etc., or any combination. The CDN also sends the location of the request: either the DNS server that issued the request, or, if the DNS client extension is supported [2] the subnet/IP address of the end-user. When PaDIS receives the request, it ranks the candidate servers based on the end-user location, the performance metrics and the state of the network. Finally, the ranked list is returned to the CDN which, in turn, compiles the final answer for the end-user.

2.3 Performance Monitor

The Performance Monitor layer consists of a number of network and content delivery performance metrics which are measured and reported. The monitor will show the layout of the network used in the *Network Emulation*, annotated with regularly updated statistics, e.g., link utilization, traffic load on peering points and estimated costs. Furthermore, CDN performance statistics are shown, including the server load, delivering cost, and the user-server mapping efficiency.

3. DEMONSTRATION

Our demonstration runs on virtualized hardware, but the general concept is not limited to it. CDNs and ISPs are able to evaluate the benefits of collaboration through PaDIS Emulator while taking into account CDN peerings, capacity and mapping strategies as well as ISP topology, routing, peering agreements and policies. PaDIS Emulator allows the utilization of different optimization functions by CDNs and ISPs. To visualize the results, PaDIS Emulator collects and reports the utilization of each ISP link, the load on the CDNs, the stress on the peering links and the end-user download time. Furthermore, to quantify the effect of CDN-ISP collaboration, it allows the user to enable and disable the collaboration at run time, and reports the gains due to collaboration.

During the demo session we present to the audience a preconfigured network in PaDIS Emulator and how the reported performance metrics change when modifying e.g., the mapping strategy, the optimization function or even completely disabling the collaborative optimization. We also encourage everyone to try their own settings/modifications in PaDIS Emulator and observe the impact on the emulated network and the reported performance values.

4. REFERENCES

[1] B. Ager, W. Mühlbauer, G. Smaragdakis, and S. Uhlig. Comparing DNS Resolvers in the Wild. In *ACM IMC*, 2010.

[2] C. Contavalli, W. van der Gaast, S. Leach, and D. Rodden. IETF draft: Client IP Information in DNS Requests, 2011.

[3] B. Fortz and M. Thorup. Optimizing OSPF/IS-IS Weights in a Changing World. *IEEE JSAC*, 20(4), 2002.

[4] B. Frank, I. Poese, G. Smaragdakis, S. Uhlig, and A. Feldmann. Content-aware Traffic Engineering. In *ACM SIGMETRICS*, 2012.

[5] A. Gerber and R. Doverspike. Traffic Types and Growth in Backbone Networks. In *OFC/NFOEC*, 2011.

[6] C. Labovitz, S. Lekel-Johnson, D. McPherson, J. Oberheide, and F. Jahanian. Internet Inter-Domain Traffic. In *ACM SIGCOMM*, 2010.

[7] T. Leighton. Improving Performance on the Internet. *CACM*, 2009.

[8] H. Nguyen, M. Roughan, S. Knight, N. Falkner, O. Maennel, and R. Bush. How to Build Complex, Large-Scale Emulated Networks. In *TridentCom*, 2010.

[9] I. Poese, B. Frank, B. Ager, G. Smaragdakis, and A. Feldmann. Improving Content Delivery using Provider-Aided Distance Information. In *ACM IMC*, 2010.

Signposts: End-to-End Networking in a World of Middleboxes

Andrius Aucinas, Amir Chaudhry, Jon Crowcroft, Sebastian Probst Eide, Steve Hand, Anil Madhavapeddy, Andrew W. Moore, Charalampos Rotsos, Narseo Vallina-Rodriguez
University of Cambridge, UK
first.last@cl.cam.ac.uk

Richard Mortier
University of Nottingham, UK
richard.mortier@nottingham.ac.uk

ABSTRACT

This demo presents Signposts, a system to provide users with a secure, simple mechanism to establish and maintain communication channels between their personal cloud of named devices. Signpost names exist in the DNSSEC hierarchy, and resolve to secure end-points when accessed by existing DNS clients. Signpost clients intercept user connection intentions while adding privacy and multipath support. Signpost servers co-ordinate clients to dynamically discover routes and overcome the middleboxes that pervade modern edge networks. The demo will show a simple scenario where an individual's personal devices (phone, laptop) are interconnected via Signposts while sitting on different networks behind various middleboxes. As a result they will be able to fetch and push data between each other, demonstrated by, e.g., simple web browsing, even as the network configuration changes.

Categories and Subject Descriptors

C.2.1 [**Network Architecture and Design**]: Distributed networks; C.2.2 [**Network Protocols**]: Routing protocols

General Terms

Design, Experimentation, Measurement

Keywords

Naming, DNS, Middlebox, User-centered, Edge network

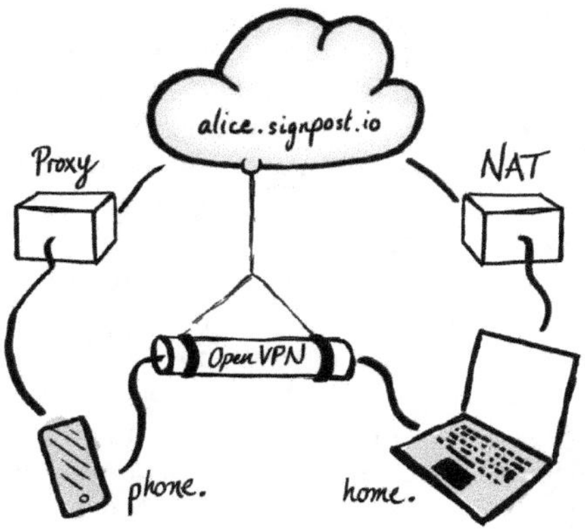

Figure 1: Schematic of local channel creation via a cloud-based Signpost server. Alice's machine `home` attempts to resolve `phone.alice.signpost.io`, which results in a route being created between the two devices (e.g. OpenVPN).

1. SIGNPOSTS AND EFFECTFUL NAMES

The modern Internet has broadly divided into two halves: a performant, global, core network, and an edge network through which end-users access services. Devices in the core network typically route to each other freely, and major content providers form significant networks in their own right (e.g., Facebook and Google), often referred to as "clouds".

However, activities that are simple in the cloud – such as establishing peer-to-peer links – are disproportionately complicated in the edge. A typical home network is obscured from the cloud by firewalls, Network Address Translators (NATs), and proxies, all of which complicate the creation of incoming connections. As a result, when a user wants their devices to communicate the only practical option is to route all content and communication via the cloud. There are substantial downsides to routing via the cloud as compared to ad-hoc channels over local networks – e.g., inability to operate without an Internet connection, increased potential for privacy violation, the relative lack of bandwidth and higher latency, as well as energy and financial costs [6, 2].

These factors all motivate the need for *Signposts*, our network service that enables user-centric control of connectivity in the edge network, both within a personal cloud of devices and across to the devices of others. Such a service must account for the need to operate when disconnected from the global Internet, whether due to intermittent connectivity or the interposition of middleboxes that break the all-pairs connectivity model of the traditional Internet.

Signposts automate the process of establishing and maintaining routes across the modern Internet by exploiting DNS as a ubiquitous signalling channel (§3). Every user registers a unique domain name (e.g., `alice.signpost.io`) and binds stable names to their devices as a one-off process (e.g., `phone` and `home`). DNS responses for these domains are served by authoritative Signpost servers running in the cloud and on

SIGCOMM'12, August 13–17, 2012, Helsinki, Finland.
ACM 978-1-4503-1419-0/12/08.

their home networks. This infrastructure takes care of resolving names into network addresses and setting up routes between devices. Since middleboxes make it impossible to predict a network's capabilities ahead of time, the act of DNS name resolution triggers actions that automatically setup appropriate NAT traversal, VPNs and proxies between clients and servers: the act of resolving a device's name has side-effects in the network, a scheme reminiscent of circuit-switched networks [5] and illustrated in Figure 1.

2. DEMO: ALICE'S PERSONAL CLOUD

Signposts are concerned solely with establishing communication channels between devices, not with the transport of data over those channels. Consider the simple demo scenario depicted in Figure 1. Alice has a Signpost server running in a globally visible location in the public cloud. This serves her public key and zone, `alice.signpost.io`, with DNSSEC providing a chain of signed attestations back to the DNS root that the record has not been tampered with en route. Alice has two network-connected devices to which she has bound the concrete names `phone` and `home` and for which the server will coordinate name resolution.

Automatically establishing a connection between `phone` and `home` via the Signpost then becomes relatively straightforward. One client, say `home`, initiates the process by attempting a DNS resolution of `phone.alice.signpost.io`. Normal DNS mechanisms cause this query to reach Alice's Signpost, which has been delegated the zone `alice.signpost.io`. The Signpost resolves the name `phone` by probing and establishing routes using a bidirectional signalling channel established between the Signpost and Alice's devices. The Signpost dynamically probes a variety of channels to determine whether multiple routes can be established, and create them if so. The result is that various channels (e.g., VPN, TOR, etc) are created between Alice's phone and computer, and at least one valid IP address endpoint returned in the DNS query if any viable route exists.

Without Signposts, Alice would have had to manually configure a port-forward at her NAT box, or run VPN software on her phone. Not normally visible to users, the demo incorporates a lightweight mechanism to display path existence and properties.

3. DISCUSSION

Why DNS? Most devices used to access the Internet are essentially anonymous from a network perspective, with only transient names if they have names at all. The Signpost network assigns secure, stable names to each device, and provides a way of resolving these names into concrete network addresses. Signposts extend the DNS protocol [7] for this purpose, for the following reasons:

- *Ubiquity.* DNS is among the most widely deployed services on the Internet. Effectively every Internet-connected client supports name resolution, and has access to the DNS when connected.
- *Reach.* As such a critical part of the Internet's infrastructure, and unlike TCP, HTTP and similar protocols, DNS tends not to be manipulated by middleboxes other than modified DNS servers themselves [3, 4].
- *Security.* The DNSSEC security extensions have recently been deployed on the live root servers [1]. DNSSEC provides origin authentication and integrity protection for DNS records, and (along with SSL) represents one of the two global public key infrastructures.

Identity. The Signpost system requires individuals to register their own domain. Thus, by running a Signpost server which has been delegated their zone, an individual has an authenticated public identity on the Internet via their public-private key-pair, using standard DNSSEC records. This pushes user security credentials into the naming fabric of the Internet, meaning that any other network service which can perform name lookups can efficiently verify a user key.

Multipath. In addition, Signposts solves the problem of multipath route discovery as name resolution results in dynamic probing. This integrates well into end-point upgrades such as multipath TCP.

4. SUMMARY

A feature of the development of the Internet has been the gradual erosion of the end-to-end principle. In a world of mobile gateways, NATs and other middlebox impositions, it is less clear what it means for the "ends" to connect. Signposts is a system which allows us to evolve past this current, rather unsatisfactory, state of affairs. It provides a control plane that enables *edge-to-edge* connectivity, between the many heterogeneous and often mobile edge networks that have now sprung up around the Internet.

More information and source code is available via `http://signpost.io`

5. ACKNOWLEDGEMENTS

This work was partially funded by the RCUK Horizon Digital Economy Research Hub grant, EP/G065802/1.

6. REFERENCES

[1] R. Arends, R. Austein, M. Larson, D. Massey, and S. Rose. DNS Security Introduction and Requirements. RFC 4033, IETF, March 2005.

[2] J. Baliga, R. Ayre, K. Hinton, and R. Tucker. Green cloud computing: Balancing energy in processing, storage, and transport. *Proceedings of the IEEE*, 99(1):149 –167, January 2011.

[3] B. Carpenter and S. Brim. Middleboxes: Taxonomy and Issues. RFC 3234, IETF, February 2002.

[4] M. Honda, Y. Nishida, C. Raiciu, A. Greenhalgh, M. Handley, and H. Tokuda. Is it still possible to extend TCP? In *Proceedings of the 11th ACM SIGCOMM Internet Measurement Conference*, IMC '11, pages 181–194, Berlin, Germany, November 2011. ACM.

[5] I. M. Leslie. Extending the local area network. Technical Report UCAM-CL-TR-43, University of Cambridge, Computer Laboratory, 15 JJ Thomson Avenue, Cambridge CB3 0FD, United Kingdom.

[6] A. Li, X. Yang, S. Kandula, and M. Zhang. Cloudcmp: comparing public cloud providers. In *Proceedings of the 10th ACM SIGCOMM Internet Measurement Conference*, IMC '10, pages 1–14, Melbourne, Australia, November 2010. ACM.

[7] P. Mockapetris. Domain names – concepts and facilities. RFC 1034, IETF, November 1987.

Towards SmartFlow: Case Studies on Enhanced Programmable Forwarding in OpenFlow Switches

Felicián Németh, Ádám Stipkovits, Balázs Sonkoly and András Gulyás*
HSN Lab, Department of Telecommunications and Media Informatics
Budapest University of Technology and Economics
{nemethf,stipkovits,sonkoly,gulyas}@tmit.bme.hu

ABSTRACT

The limited capabilities of the switches renders the implementation of unorthodox routing and forwarding mechanisms as a hard task in OpenFlow. Our high level goal is therefore to inspect the possibilities of slightly smartening up the OpenFlow switches. As a first step in this direction we demonstrate (with Bloom filters, greedy routing and network coding) that a very limited computational capability enables us to natively support experimental technologies while preserving performance. We distribute the demos[1] in source files and as a ready-to-experiment VM image to promote further improvements and evaluations.

Categories and Subject Descriptors

C.2.6 [**Computer-Communication Networks**]: Internetworking—*routers*

Keywords

SDN, OpenFlow, Bloom filters, Greedy routing, Network Coding

1. INTRODUCTION

Since its first proposition, OpenFlow (OF) [1] has become the key enabler and driver of innovation in networks. By now, numerous innovative applications have appeared on top of OF, targeting mostly the well-known issues of packet switched networks. The spectrum is spanning over protocols, routing, traffic engineering technologies and security. These sets of applications rely on the standard philosophy of OF, namely to execute forwarding-related tasks in the switches and move the control-specific operations completely into the controller. While such a scheme is satisfactory for a whole bunch of applications, we demonstrate that this view is somewhat limited when the support for unorthodox network technologies are considered.

In our demo we present OF testbeds for experimental technologies with non-negligible future potential. These testbeds will require increasingly more functionality from the OF switches. First we show that *Bloom filter* based forwarding [2] can be implemented in a *standard manner* over OF version 1.1[2], if the flow entries of the switches are interpreted in an *unorthodox* way. Secondly, we introduce the OF implementation of *"greedy" or geographical* [3] routing and show that although its implementation can be done in a standard fashion (using OF controllers), the beneficial properties of this basically distributed technology can *only be exploited by adding some computational capability to the switches.* Finally we present a testbed for inter session *network coding* [4] and demonstrate that if we want to apply this technology in OF then the usage of *smarter switches is imperative.* Our case studies support that introducing only lightweight and low-level *enhancements*, numerous networking operations may become available in OF. For the universal treatment of such enhancements we envision a generic language for switch computations similar to the vertex shading languages of video cards.

*MTA-BME Future Internet Research Group

[1]Available from http://github.com/nemethf/sigcomm2012

SIGCOMM'12, August 13–17, 2012, Helsinki, Finland.
ACM 978-1-4503-1419-0/12/08.

2. CASE STUDIES

Stateless multicast with Bloom filters: We assign Bloom IDs to switch ports and put a Bloom filter containing the packet's path into the packet's destination Ethernet address field. The Bloom filter is generated by OR-ing the Bloom IDs of the ports that the packet needs to pass through.

Our testbed shows that a switch can be "forced" to follow a Bloom filter based forwarding mechanism relying solely on standard OF 1.1 protocol[3], however we need to slightly deviate from the regular usage of flow entries: normally, flow entries correspond to specific set of *flows*, whereas we proactively configure a separate flow table for each *port* as shown in Table 1. The first entry matches if the packet's Bloom filter contains the port's Bloom ID. In case of a match the switch forwards the packet via the corresponding port and independently of the matching result, the packet is matched against the next port's flow table hereby providing the stateless multicast switching capability.

Fix sized flow tables with Greedy routing: Greedy routing relies on embedding of the switches by assigning them coordinates in a predefined metric space. A switch then forwards packets towards its neighbor closest to the packets' destination[4]. Knowing the topology and the coordinates of each node, the controller can configure the flow tables of the switches (see Table 2). Afterwards, routing operates without controller supervision.

[2]Note that this is not true for OF 1.0

[3]Currently, commercial OF switches support ver. 1.0 only.

[4]We store 2-dimensional coordinates in Ethernet addresses.

Table 1: Bloom filter in OF. Flow entries in the ith flow table.

match condition	associated actions
addr & BloomID$_i$ =BloomID$_i$	output(port$_i$) goto-table(i+1)
(always match)	goto-table(i+1)

Table 2: A single flow entry implementing greedy routing. address-n1...nN stands for the coordinates of the switch's neighbors.

match condition	associated actions
(always match)	write-metadata(max. number, invalid port) update-distance(address-n1, port-to-n1) update-distance(address-n2, port-to-n2) ... update-distance(address-nN, port-to-nN) output-by-metadata

For enabling greedy routing we need only two simple extra actions. These experimenter actions assume that the metadata register holds two values: a distance and a port number. The action *update-distance* calculates the distance between its first argument and the (geographical) destination address of the packet being processed. If the distance is smaller than the one stored in the metadata register, then it saves the distance and its second argument in the metadata register. Our second general purpose action, *output-by-metadata* forwards packet to the port stored in the metadata.

Obviously, the routing decision can be implemented as a controller application without any need for protocol extensions. However, controller based greedy routing results in slower connection setup and error recovery, and more importantly the flow table size is proportional to the actual number of flows. On the other hand, by using enhanced switches, the table size is proportional only to the number of ports and the forwarding speed is independent of the traffic load.

200% link utilization with Network Coding: Network coding (NC) may mix or encode packets before forwarding, hereby improving resource efficiency and robustness [4]. However, implementing NC as an OF controller logic is practically impossible since every packet would have to be sent back and forth to the controller for en-/decoding. We extended the OF protocol with three simple actions to support XOR-based mixing of two flows. NC metadata is sent in three MPLS labels following the format proposed in [5] (Fig. 1). The outermost label is a flow identifier that can be used as a matching condition in the flow table. Sequence numbers can be prepended to packets by the action *set-mpls-label-from-counter* and two flows can be combined with the *XOR-encode* action. XOR-encode does not alter the packet in the OF processing pipeline, it creates a copy of the packet and puts it in the encoding queue or if there is a pending packet from the other flow, then it mixes them, replaces the MPLS flow-id, and sends the resulting packet through the pipeline. This final step ensures general applicability because further processing of the newly created packet can be specified with standard flow table entires. The *XOR-decode* action works similarly, but it searches matching sequence numbers when it dequeues a packet from the decoding queue.

MPLS label: flow-id	MPLS label: seq.no 1	MPLS label: seq.no 2	data

Figure 1: NC packet format. If seq. numbers are both non-zero, then the original packets are *xor*ed in data.

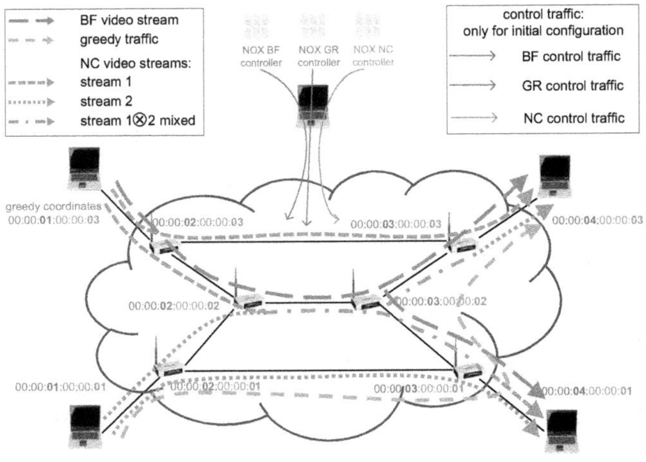

Figure 2: Demonstrated topology.

These three new OF actions are quite simple, but in the meantime they are general enough to support several NC use-cases defined in [4], e.g., intersession multicasting, delay or energy minimalization. To the best of our knowledge this will be the first NC implementation in OF and we hope that our case study will trigger further developments and discussions about the NC-OF interworking.

3. DEMO SCENARIO

For demonstration we have ported our extensions into OpenWRT firmware and built a testbed with TP-Link devices.[5] The demo topology shown in Fig. 2 can accommodate all of our testbeds. Video streams are sent using different forwarding schemes in a configurable manner. Changing the active controller results in switching among the three network operations.

4. REFERENCES

[1] N. McKeown et al. OpenFlow: enabling innovation in campus networks. *ACM SIGCOMM CCR*, 38(2), 2008.

[2] P. Jokela et al. LIPSIN: Line speed publish/subscribe inter-networking. *ACM SIGCOMM CCR*, 39(4), 2009.

[3] M. Boguna, F. Papadopoulos, and D. Krioukov. Sustaining the internet with hyperbolic mapping. *Nature Communications*, 1(6), 2010.

[4] P. A. Chou and Y. Wu. Network coding for the internet and wireless networks. *IEEE Signal Processing Magazine*, 24(5), 2007.

[5] T. Biermann, A. Schwabe, and H. Karl. Creating butterflies in the core – a network coding extension for MPLS/RSVP-TE. In *Proc. of Networking*, 2009.

[5]Our implementation extends OF 1.1 software switch by TrafficLab, Ericsson Research.

An OpenFlow-Based Energy-Efficient Data Center Approach

Michael Jarschel, Rastin Pries
University of Würzburg, Institute of Computer Science
Würzburg, Germany
{michael.jarschel|pries}@informatik.uni-wuerzburg.de

Categories and Subject Descriptors

C.2.1 [**COMPUTER-COMMUNICATION NETWORKS**]: Network Architecture and Design—*Network communications, Network topology*; C.4 [**PERFORMANCE OF SYSTEMS**]: Reliability, availability, and serviceability

Keywords

Data Center, OpenFlow, Energy-Efficiency

1. INTRODUCTION

Infrastructure as a service (IaaS) is one of the prevalent business models in cloud computing and has generated much customer interest over the past few years. An IaaS provider offers the temporary deployment and maintenance of a custom virtual host and network infrastructure to its customers on which arbitrary applications can be run and/or hosted. Providers of such a service face several challenges in their data centers. One of the main issues is the inherent heterogeneity of systems and applications from different customers. As a result, a variety of different load and traffic patterns has to be handled by the same data center infrastructure. There are two ways to remedy this. The first is the classical approach of overprovisioning to sustain a constant service quality. This is usually very inefficient and only open to companies with a huge budget and a lot of resources. The second one is smart resource management. An IaaS provider has to find a good balance between the various customer application requirements and the efficient use of the available resources in the data center. The ECDC (Energy effiCient Data Center) approach we showcase in this demo is such a smart mechanism for finding this balance. It leverages monitoring information from machines as well as network devices and environmental data to create a coherent view of the current situation in a data center. This alone enables a single network operator to react to system changes as soon as they happen. However, the monitoring data is also used by a smart control application to react in certain situations by redistributing virtual machines, traffic flows and VLANs as well as powering devices up and down.

SIGCOMM'12, August 13–17, 2012, Helsinki, Finland.

2. ARCHITECTURE

Figure 1 shows the ECDC architecture in a simple data center scenario. On the right we see the servers hosting the customers' virtual infrastructures. They are organized in racks with a top of the rack switch each. The entry point into the network from the service side are the virtual switches integrated into the hypervisor of each server. On the left we see two types of customers that are connecting to the data center network from the customer side, i.e. the Internet. The type A customer is a private home user who wants to use an entertainment service, e.g. video streaming, hosted by a service provider as a virtual infrastructure in rack B and C in the data center. The type B customer is a business user who uses a business application set up by his company in the data center, e.g. a virtual desktop infrastructure (VDI). As both types of customers have different demands and requirements, their traffic is kept in separate VLANs in the data center network to be able to manage them independently. The connectivity for type A customers is represented by red lines, for type B green lines are used. Packets are tagged at the entry switch into the data center network. The network itself is OpenFlow-enabled [3]. Forwarding decisions for all network elements, i.e. access, edge, and hypervisor switches are handled by a central entity - the OpenFlow controller (OFC). The control connection for each network element is established via a physically isolated management network. This network also connects the controller, OpenFlow switches, physical servers, and environmental sensors to the central data center management entity - the management station. The management station queries monitoring information on CPU-, network-, and memory-load, as well as power consumption from the connected devices via SNMP. Armed with this host of information, the management station generates the appropriate network policy for the OFC, distributes virtual ma-

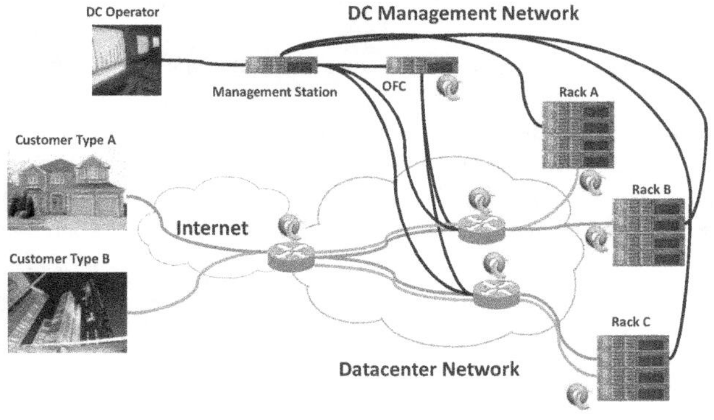

Figure 1: ECDC Architecture

chines across the servers, and powers down unused devices in order to ensure an efficient utilization of all resources while maintaining a good service quality for the customer at all times. The management station achieves this by observing a number of configured thresholds and timeouts for each service class. If a monitored parameter, e.g. CPU load, falls below or rises above a threshold for a certain amount of time as defined by a timeout, the management station will take action, e.g. by consolidating multiple virtual machines to one host or in the opposite case by spreading them over multiple hosts. Once such an action is triggered, the network, i.e. the OpenFlow controller, is immediately notified and can adapt the flow rules in the switches according to the new situation with little delay, minimizing the impact on the service. The gathered information is also presented to the data center operator through a graphical user interface as illustrated in Figure 2, which displays the current topology of the network and a time series of monitored values as configured by the operator. The operator is then able to facilitate changes in the operation of the data center, if this is necessary, e.g. through the introduction of a new service class.

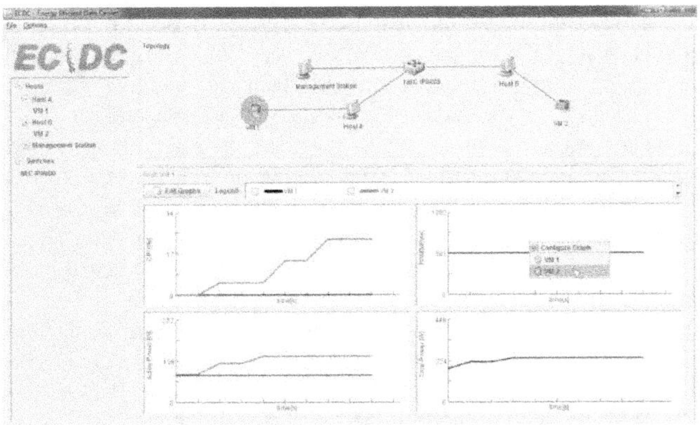

Figure 2: ECDC Operator GUI

3. DEMONSTRATION

The demonstration testbed is hosted on the German-Lab [6] (G-Lab) facility in Wuerzburg, Germany. Four rack servers are used as computing nodes running OpenNebula [1] and KVM [2] as hypervisor using the Open vSwitch [4] as virtual switch. The management station is hosted on a fifth server running the OpenNebula management software as well as our Java-based data center management software. The management network is a legacy IP network realized by the Cisco top of rack switch of the G-Lab facility. As OpenFlow controller, we use BigSwitch's Floodlight hosted in a G-Lab virtual machine. The OpenFlow data center network is represented by a Pronto 3290 OpenFlow switch [5]. The demonstration will show the operation of the ECDC-enabled data center over the course of a business day. Using our own traffic generator, we emulate the behavior of the two types of users introduced in Section 2. For demonstration purposes we condensed the emulated "day" to a short cycle. During the progression of this cycle, we show the changes in the system as an operator would perceive them using our ECDC GUI as our software adapts the resource allocation according to the demand. In the topology section of the GUI (cf. Figure 2) topology changes caused by the migration of virtual machines as well as the powering up and down of physical hosts will be displayed. In the monitoring section the collected information reflecting load changes in servers is illustrated by line graphs. If an OpenFlow switch is selected in the topology display, the monitoring section changes to show the switch's flow table entries. By selecting an entry, the path of the corresponding flow through the network is displayed in the topology section.

4. CONCLUSION

In this demonstration we showcase our smart data center management software ECDC. It allows for an integrated adaption of computing and network resources according to the required capacity to ensure a smooth operation of services in an IaaS scenario. At the same time the software aims to minimize the carbon footprint of the data center in question by consolidating capacities and powering down those not needed. To achieve this, we leverage the modern software defined networking approach as well as proven open-source cloud management software. Challenges we are facing here are the integration of network and server virtualization by using the OpenVSwitch in combination with OpenNebula. We believe there is a lot of potential for optimization in data centers and this approach might also be applicable to a larger scale scenario.

5. ACKNOWLEDGMENTS

The authors would gratefully thank Thomas Höhn, Christopher Metter, and Phuoc Tran-Gia from the University of Würzburg for the work and fruitful discussions. This work was funded by the international bureau of the Federal Ministry of Education and Research (Förderkennzeichen VNM10/073).

6. REFERENCES

[1] J. Fontán, T. Vázquez, L. Gonzalez, R. Montero, and I. Llorente. Opennebula: The open source virtual machine manager for cluster computing. In *Open Source Grid and Cluster Software Conference*, 2008.

[2] A. Kivity, Y. Kamay, D. Laor, U. Lublin, and A. Liguori. kvm: the linux virtual machine monitor. In *Proceedings of the Linux Symposium*, volume 1, pages 225–230, 2007.

[3] N. McKeown, T. Anderson, H. Balakrishnan, G. Parulkar, L. Peterson, J. Rexford, S. Shenker, and J. Turner. Openflow: enabling innovation in campus networks. *ACM SIGCOMM Computer Communication Review*, 38(2):69–74, 2008.

[4] B. Pfaff, J. Pettit, T. Koponen, K. Amidon, M. Casado, and S. Shenker. Extending networking into the virtualization layer. *Proc. HotNets (October 2009)*, 2009.

[5] Pica8. Pronto 3290 OpenFlow Switch. http://www.pica8.org/products/p3290.php.

[6] D. Schwerdel, D. Günther, R. Henjes, B. Reuther, and P. Müller. German-lab experimental facility. *Future Internet-FIS 2010*, pages 1–10, 2010.

Reduction-based Analysis of BGP Systems with BGPVerif

Anduo Wang* Alexander J.T. Gurney* Xianglong Han* Jinyan Cao*
Carolyn Talcott‡ Boon Thau Loo* Andre Scedrov*
*University of Pennsylvania †SRI International
{anduo,jinyan,hanxiang,agurney,boonloo}@cis.upenn.edu, clt@csl.sri.com,
scedrov@math.upenn.edu

ABSTRACT

Today's inter-domain routing protocol, the Border Gateway Protocol (BGP), is increasingly complicated and fragile due to policy misconfiguration by individual autonomous systems (ASes). Existing configuration analysis techniques are either manual and tedious, or do not scale beyond a small number of nodes due to the state explosion problem. To aid the diagnosis of misconfigurations in real-world large BGP systems, this paper presents *BGPVerif* , a reduction based analysis toolkit. The key idea is to reduce BGP system size prior to analysis while preserving crucial correctness properties. BGPVerif consists of two components, *NetReducer* that simplifies BGP configurations, and *NetAnalyzer* that automatically detects routing oscillation. BGPVerif accepts a wide range of BGP configuration inputs ranging from real-world traces (Rocketfuel network topologies), randomly generated BGP networks (GT-ITM), Cisco configuration guidelines, as well as arbitrary user-defined networks. BGPVerif illustrates the applicability, efficiency, and benefits of the reduction technique, it also introduces an infrastructure that enables networking researchers to interact with advanced formal method tool.

Categories and Subject Descriptors

C.2.2 [**Computer-Communication Networks**]: Network Protocols

General Terms

Verification, Management

Keywords

Border gateway protocol, reduction, formal analysis

1. INTRODUCTION

The Internet today runs on a complex routing protocol called the *Border Gateway Protocol* or *BGP* for short. BGP enables autonomous systems (ASes) worldwide to achieve global connectivity, subject to each system's local policy (what paths are allowed, and the route preference used to select best paths). The convergence behavior of the global Internet depends on how each ISP configures its policy. Prior work has shown that policy misconfigurations can lead to route oscillation and slow convergence [2].

While our understanding of BGP safety has improved dramatically in the past decade, there is a lack of automated tools that can aid in the safety analysis process. Analysis today either consists of manual proofs and counter-examples on small instances, or custom simulators that are unable to explore the full protocol state space.

In this demonstration, we present *BGPVerif*, a formal analysis toolkit that enables networking researchers to study and analyze large BGP systems in a sound and automatic fashion. BGPVerif consists of two components. *NetAnalyzer* [10] is an automated analysis toolkit based on the Maude rewriting logic engine [4]. In NetAnalyzer, the BGP system is encoded as a transition system driven by rewriting rules. Maude's high-performance rewriting engine is used to analyze BGP instances automatically, by simulating execution runs, as well as exhaustively exploring all execution runs for possible divergence.

While automated, NetAnalyzer is susceptible to the state explosion problem as the number of nodes increases. The *NetReducer* component applies a novel reduction theory [9, 8] for reducing BGP network size while preserving convergence properties. In combination, NetAnalyzer and NetReducer enable BGPVerif to perform safety analysis efficiently, often completing the analysis on large networks that would otherwise not be possible to study within reasonable time.

2. OVERVIEW

As shown in Figure 1, BGPVerif takes as input the configuration of a BGP system, which includes the network topology and routing policies for all routers. Policies include routing filters and preferences based on path attributes. BGPVerif converts this configuration to an *Extended Path Digraph* (EPD) [8] instance, an internal representation used by BGPVerif to perform the reduction. The EPD structure includes an explicit representation of both the network topology and the preference-induced dependencies among paths. EPD extends prior models [1, 6], and provides a basis for reducing instances prior to analysis [9].

Given the EPD representation of the BGP system, the NetReducer component first performs a reduction to output the reduced EPD. The basic idea of reduction is straightforward: to simplify the BGP system of interest, one repeatedly merges all pairs of nodes where the merger does not affect the convergence behavior of the entire network. There are two rules for identifying such node pairs, and each of them only requires checking of local policy details. (1) *Duplicate* reduction merges two nodes whose routing configurations agree upon what paths and route preferences they learned from their neighbors. (2) *Complementary* reduction merges two nodes whose neighbors agree upon the paths and preferences learned from them.

In our demonstration prototype, reduction can perform reduction in two modes: (1) In *batch* mode, BGPVerif repeatedly merges all reducible nodes and outputs the final reduced BGP system in the

SIGCOMM'12, August 13–17, 2012, Helsinki, Finland.
ACM 978-1-4503-1419-0/12/08.

form of a EPD; (2) Alternatively, in the *interactive* mode, BGPVerif allows the user to select which nodes to try to reduce.

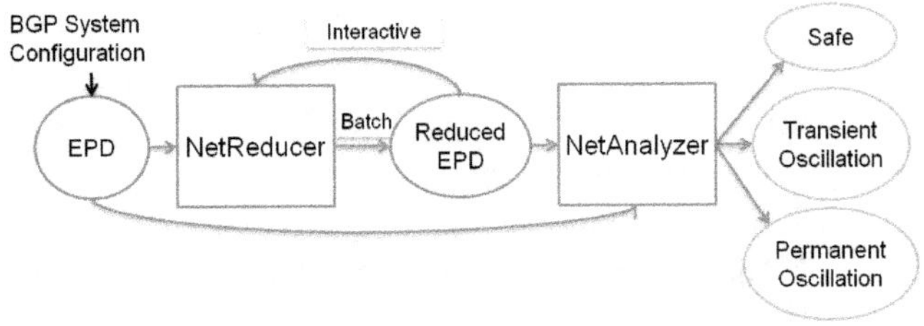

Figure 1: BGPVerif System Overview.

Finally, given a BGP configuration's EPD (either before or after reduction), the NetAnalyzer component uses the Maude analyzer [10, 9, 3] to classify the configuration as either convergent or oscillatory, and the oscillation as either transient or permanent.

3. DETAILS OF DEMONSTRATION

BGPVerif provides a graphical user interface for the user to specify BGP configuration inputs, and invoke reduction or analysis. The interface embeds a visualizer that demonstrates the dynamic reduction process as a transformation of the network topology augmented with routing policies. In the back-end, BGPVerif includes the reduction engine NetReducer and the analysis engine NetAnalyzer based on Maude. NetReducer is responsible for computing the internal EPD representation, and performing the actual reduction, whereas NetAnalyzer extends Maude with additional library support for detecting route oscillations in BGP systems.

Specify Configuration.

In the BGPVerif GUI, the user inputs the topology and policy of the BGP configuration of interest, or selects a configuration already provided. The user can edit these pre-specified configurations as well. The built-in network topologies include real world Rocketfuel data [7], and randomly generated networks using GT-ITM [5], which is controlled by the input parameters. The built-in routing policies include standard iBGP configurations such as full mesh, route reflection and confederations [11].

Once the BGP system configuration is loaded, BGPVerif generates the internal EPD representation by running BGP in a special mode that deduces all possible paths and their relative preferences. The user can now choose `batch` or `interactive reduce` commands from NetReducer, or the `analyze` command of NetAnalyzer.

NetReducer: Simplify BGP Systems.

In batch mode, NetReducer will try all combinations of nodes to see which are reducible. This makes the batch mode useful as an automated step prior to analysis. In the fine-grained interactive mode, the user can specify either one or two nodes that she wants to merge. The NetReducer can check whether a given pair of nodes is indeed reducible; or, given only a single node, can search for a second node to form a reducible pair. The interactive mode provides insights into convergence behavior even without a following analysis operation.

Consider the scenario where a network operation would like to expand her BGP system by adding new routers, without changing the convergence behavior. Rather than analyzing the two BGP systems before and after migration separately, she can instead incrementally grow the BGP network size while preserving convergence throughout. At each step, the operator adds a node, and by interactive reduction, she can decide if it can be merged with any of the existing nodes. She can also add a new router in a chosen place by deliberately making it duplicate or complementary with an existing node.

NetAnalyzer: Detect Route Oscillations.

NetAnalyzer provides an interactive graphical interface for networking researchers to make use of the formal analyses implemented in Maude [10, 9, 3]. It allows the user to analyze BGP configurations using reduction techniques that accelerate analysis by orders of magnitude, enabling automated analysis of configurations that could not previously be handled.

NetAnalyzer utilizes Maude's built-in exhaustive search ability and adds library support for detecting BGP oscillations. Our library allows us to exhaustively search all execution runs to detect route oscillation. On termination, NetAnalyzer can correctly classify a BGP configuration as either safe (no oscillations in any protocol runs), transient oscillation (route oscillation happens in some execution traces), or permanent oscillation (convergence is impossible).

4. ACKNOWLEDGMENT

This research is funded in part by NSF grants (CCF-0820208, CNS-0830949, CNS-0845552, CNS-1040672, TC-0905607 and CPS-0932397), AFOSR grant FA9550-08-1-0352, and ONR grant N00014-11-1-0555.

5. REFERENCES

[1] GRIFFIN, T. G., AND WILFONG, G. An analysis of BGP convergence properties. In *SIGCOMM* (1999).

[2] LABOVITZ, C., MALAN, G., AND JAHANIAN, F. Internet Routing Instability. *TON* (1998).

[3] MASON, I. A., AND TALCOTT, C. L. Iop: The interoperability platform & imaude: An interactive extension of maude. *Electron. Notes Theor. Comput. Sci. 117* (Jan. 2005), 315–333.

[4] MAUDE. `http://maude.cs.uiuc.edu/`.

[5] MODELING TOPOLOGY OF LARGE INTERNETWORKS. `http://www.cc.gatech.edu/projects/gtitm/`.

[6] SOBRINHO, J. Network routing with path vector protocols: theory and applications. In *SIGCOMM* (2003).

[7] SPRING, N., MAHAJAN, R., AND WETHERALL, D. Measuring ISP topologies with Rocketfuel. In *ACM SIGCOMM* (2002).

[8] WANG, A., GURNEY, A. J. T., TALCOTT, C., LOO, B. T., AND SCEDROV, A. A calculus of policy-based routing systems. In *PODC* (2012). Brief announcement.

[9] WANG, A., TALCOTT, C., GURNEY, A. J. T., LOO, B. T., AND SCEDROV, A. Reduction-based formal analysis of BGP instances. 18th International Conference on Tools and Algorithms for the Construction and Analysis of Systems (TACAS 2012).

[10] WANG, A., TALCOTT, C., JIA, L., LOO, B. T., AND SCEDROV, A. Analyzing BGP instances in Maude. In *13th IFIP International Conference on Formal Methods for Open Object-based Distributed Systems. 31th IFIP International Conference on FORmal TEchniques for Networked and Distributed Systems. (FMOODS-FORTE 2011).*

[11] ZHANG, R., AND BARTELL, M. *BGP Design and Implementation.* Cisco Press, 2003.

Route Shepherd: Stability Hints for the Control Plane

Alexander J. T. Gurney
University of Pennsylvania
Philadelphia, USA

Xianglong Han
University of Pennsylvania
Philadelphia, USA

Yang Li
University of Pennsylvania
Philadelphia, USA

Boon Thau Loo
University of Pennsylvania
Philadelphia, USA

ABSTRACT

The Route Shepherd tool demonstrates applications of choosing between routing protocol configurations on the basis of rigorously-supported theory. Splitting the configuration space into equivalence classes allows the identification of which parameter combinations lead to protocol stability, and which do not. This ahead-of-time analysis generates a predicate, in the form of a combination of linear integer inequalities, which can be used in several complementary ways by downstream applications. Examples presented include warning operators about errors in advance, recovery from protocol oscillation, plotting a series of safe parameter changes, and understanding the dynamics of the routing system.

Categories and Subject Descriptors

C.2.2 [**Computer-Communication Networks**]: Network Protocols—*Routing protocols*; C.2.3 [**Computer-Communication Networks**]: Network Operations—*Network management*

General Terms

Management, Reliability

Keywords

Border Gateway Protocol, routing policy, stable path problems, partial specification

1. INTRODUCTION

We present the Route Shepherd toolkit, which aims to provide analysis and generation of safe routing configurations based on well-established routing theory [4]. Network operators often have to reconfigure their routing setups, in order to perform traffic engineering, update ISP policies, or perform a live migration of one protocol to another [2, 5, 7, 8]. A significant challenge facing operators is the large array of configuration choices that could be applied. Many of these have non-obvious interactions: the chosen paths are computed in a complex fashion, depending on the values of these settings and the timing of network events. Sometimes, particular combinations of parameters can result in persistent oscillation, or loss of connectivity. These situations are dependent on the precise numeric values of the parameters (e.g. link weights) involved—a very small change can have huge effects—and it can be hard to see how to salvage a proposed network design once such problems occur.

SIGCOMM'12, August 13–17, 2012, Helsinki, Finland.
ACM 978-1-4503-1419-0/12/08.

As an example, operators might want to adjust link weights for their internal network, so that performance metrics such as congestion can be minimized [5, 7]. However, it is hard to make such changes with confidence, since one has to take multi-protocol and cross-domain interaction into account. Individual changes can be simulated, one at a time, but this technique can only sample a few points out of an enormous configuration space. A misstep can be costly, since not all anomalies disappear after parameters are reverted [3]. Moreover, one cannot revert a problematic change made in somebody else's network.

To address the above challenges, the Route Shepherd toolkit provides a mechanism to explore many routing configurations at once, and summarize the results for the benefit of operators. Route Shepherd aims to be able to deal with configurations not one-by-one, but in equivalence classes, giving a higher-level perspective. In this way, unsafe configurations can be avoided en masse, as the operator is presented with the information at the earliest opportunity. In the demo, we will show how this can be done in the case of changes to a single parameter at a time, by highlighting ahead of time the *safe and unsafe ranges* for that value. That is, based on the current state of the rest of the network, we can tell which changes to this parameter will lead to oscillation and which will not. This is done *without* the need to simulate all of the different possibilities.

Based on the same data, we can also plot *safe courses* in the configuration space—going from one complete parameter set to another, while avoiding transient anomalies, or recovering from an unsafe configuration by making a sequence of changes to bring the network back into compliance.

2. SYSTEM OVERVIEW

The goal of our system is to allow network operators to understand the properties of a wide range of configuration possibilities, in a lightweight fashion. The essence of our tool is a pipeline (see Figure 1) for identifying *representative* configurations and analyzing only these.

From the original network configuration (1), we build a combinatorial model of the possible routing outcomes (2). This intermediate representation is based on the idea of *partial configuration*, where some parameters are treated as variables [4]. From this, we identify representative configurations, each one associated with a set of parameter bounds (3). The basic idea here is that only certain parameter changes affect the routing outcome, and so equivalent changes can be grouped together. Our underlying theory gives us confidence that these representatives do indeed cover only safe configurations. However, not all of the representatives deemed unsafe will actually diverge, since our theory is not precise enough to include all nuances of protocol behavior. We therefore introduce two further filtering steps (4): numerically impossible configura-

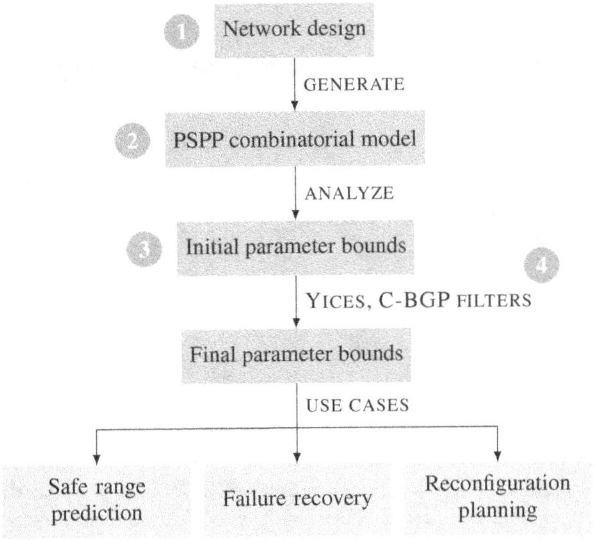

Figure 1: Tool pipeline

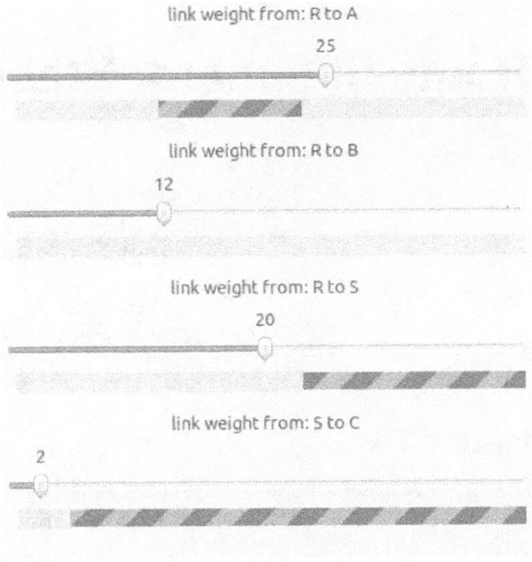

Figure 2: Link weight controls and predictions

tions are rejected by the Yices SMT solver [1], and any remaining configurations are screened with the C-BGP [6] solver, for the final stability test. Although we end up using simulation with C-BGP, the context of our earlier theory work means that we only have to simulate a few different examples, in order to achieve coverage of the configuration space.

We have validated the results of this tool against networks of Quagga software routers, to confirm its predictions against the actual network behavior. The demo uses several Quagga instances on a single computer, each running in a user-mode Linux virtual machine.

3. DEMONSTRATION PLAN

We demonstrate our tool in two modes: the normal interactive mode, allowing users to tweak the running system and see the consequences, and a replay mode, where human interaction is not required. We have developed a GUI that illustrates the action of the simulated network control plane, and allows users to alter certain chosen parameters. Additionally, the controls are decorated with annotations indicating *safe ranges* for each parameter, based on the current routing state; see Figure 2. These indicate which changes are potentially unsafe (red slashed range), and which are guaranteed not to result in oscillation (green range).

The interface also shows the network map, and highlights the chosen routes as they change over time. The program can be configured with a choice of different network topologies. These include simple 'toy' examples, where it is easier to comprehend why given states are safe or otherwise, and also more realistic and complex scenarios, showing why our tool is useful for these situations that are more difficult to understand without automated support.

We present the following additional use cases:

Recovery. We will show how a persistent oscillatory state can be halted by means of a minimal number of automatically chosen parameter changes. This is achieved by using the Yices solver in MAX-SMT mode.

Reconfiguration. In addition to the ability to interactively change parameters one-by-one, we show how the system can plot a series of safe changes, from one complete parameter set to another, without causing any transient routing anomalies.

Video

A demonstration video is available online at http://netdb.cis.upenn.edu/routeshepherd/.

Acknowledgements

This research is partly supported by the NSF Expeditions in Computer Augmented Program Engineering (ExCAPE) project, ITR-1138996. Further support was provided by the NSF grants CCF-0820208, IIS-0812270, CNS-0845552 and CNS-1040672.

References

[1] B. Duterte and L. de Moura. A fast linear arithmetic solver for DPLL(T)*. In *Proc. Conference on Computer Aided Verification*, 2006.

[2] P. François, M. Shand, and O. Bonaventure. Disruption-free topology reconfiguration in OSPF networks. In *Proceedings of IEEE INFOCOM*, 2007.

[3] T. Griffin and G. Huston. RFC 4264: BGP wedgies, 2005.

[4] A. J. T. Gurney, L. Jia, A. Wang, and B. T. Loo. Partial specification of routing configurations. In *Workshop on Rigorous Protocol Engineering (WRiPE)*, 2012.

[5] A. M. C. A. Koster and X. Muñoz, editors. *Graphs and algorithms in communication networks.* Springer, 2010.

[6] B. Quoitin and S. Uhlig. Modeling the routing of an Autonomous System with C-BGP. *IEEE Network*, 19(6), 2005.

[7] S. Raza, Y. Zhu, and C.-N. Chuah. Graceful network state migrations. *IEEE/ACM Transactions on Networking*, 19(4), 2011.

[8] L. Vanbever, S. Vissicchio, C. Pelsser, P. François, and O. Bonaventure. Seamless network-wide IGP migrations. In *Proceedings of ACM SIGCOMM*, 2011.

Efficiently Migrating Stateful Middleboxes

Vladimir Olteanu
Universitatea Politehnica Bucuresti
vladimir.olteanu@cs.pub.ro

Costin Raiciu
Universitatea Politehnica Bucuresti
costin.raiciu@cs.pub.ro

Categories and Subject Descriptors

C.2.1 [**Computer-Communication Networks**]: Network Architecture and Design—*packet-switching networks*

Keywords

Middlebox, Migration

1. INTRODUCTION

Middleboxes that hold per-flow state and perform Layer 4+ processing are widely deployed in the Internet today: a recent study shows their presence on at least a third of the studied paths [5]. These middleboxes provide functionality ranging from security to performance optimization, and are becoming ubiquitous with time.

To reduce costs and enable fast functionality updates there is an ongoing trend of migrating away from specialized hardware implementations of middleboxes to software running on commodity servers [9].

Programmable switches (such as OpenFlow) coupled with x86 machines have been proposed as the natural architecture to create scalable middleboxes that are also easy to deploy and update [3]. The basic recipe is very simple: a collection of x86 servers are connected to an OpenFlow switch, which is in turn "on-path" for the traffic. The servers implement distributedly the functionality of a single middlebox (possibly in virtual machines), such as carrier-grade NAT or firewall. The programmable switch is a key ingredient, splitting load between the machines.

To make such distributed middleboxes scalable, we need ways to seamlessly move processing and its associated flow state between local or remote servers. This would allow the platform to deal with load surges by adding servers, and to efficiently scale down by shutting machines off. Processing could even be migrated to different middleboxes in other parts of the world to optimize other aspects such as user-perceived delay.

2. EXISTING APPROACHES

Flow processing middleboxes keep most of their state in memory to reduce packet-level delays. Hence, migrating flow processing boils down to efficient memory migration. To this end, process migration algorithms [7] are the obvious starting point.

SIGCOMM'12, August 13–17, 2012, Helsinki, Finland.
ACM 978-1-4503-1419-0/12/08.

There are three basic techniques used for process migration: *Pre-copy:* Processing is performed by the original host and the memory pages are copied to the destination machine, in succesive rounds. The first round copies all active pages, while subsequent rounds only copy "dirty" pages. *Stop-and-copy:* The source machine stops processing; the memory pages are copied and processing resumes on the destination machine. *Demand migration:* The destination machine starts processing and faults unavailable pages across as they are accessed.

Process migration algorithms typically use two of the three techniques above and work reasonably well because of the principle of locality: if a page is accessed, it is likely that it will be accessed again soon. That is why the combination of a few pre-copy rounds followed by a quick stop-and-copy phase also works well for virtual machine migration [2].

Our focus is, however, on network processing that keeps per-flow state. Say we're using the same algorithm to migrate flow-states. Here is what will happen between two successive pre-copy iterations: a significant number of states will die off, because most flows are short-lived [8], a significant number of new states will be established and a seemingly random subset of the existing states will be updated, depending on which packets hit the middlebox in the meantime.

The principle of locality does not apply to flow-states and this makes pre-copying ineffective. Migration causes a long period of downtime due to the stop-and-copy phase which has to copy all the state across.

3. STATE MIGRATION

Stateful network processing allocates state for each flow (e.g. 5 tuple), and memory accesses are made to this flow-state when packets belonging to the flow arrive at the middlebox. Such processing also maintains global state (e.g. state about flow aggregates), but this is typically much smaller in size.

Our focus is on efficiently migrating per-flow state. The basic idea is to allow short flows to die and their corresponding state to be deleted before any flow-state transfer takes place. To achieve this goal, we run in parrallel both the source and the destination processing for a short period of time. When migration begins, a small set of pages is copied to B containing code and invariant global state. After this, the algorithm consists of two overlapping phases: the idle phase and the iterative freeze-and-copy phase.

In the *idle* phase, all traffic is initially forwarded to A. A stops establishing new state for flows it hasn't seen yet. All

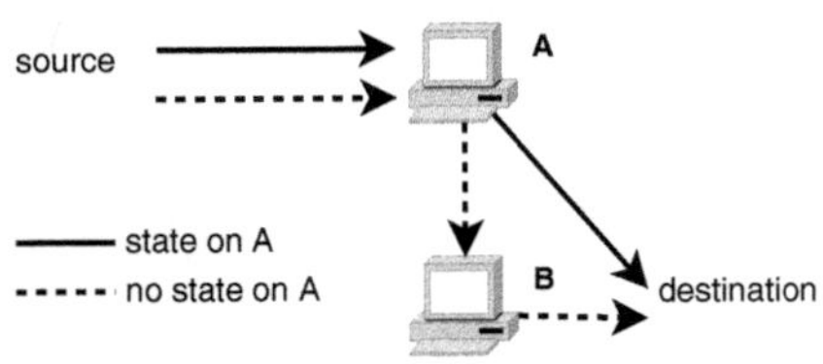

(a) Daisy Chaining

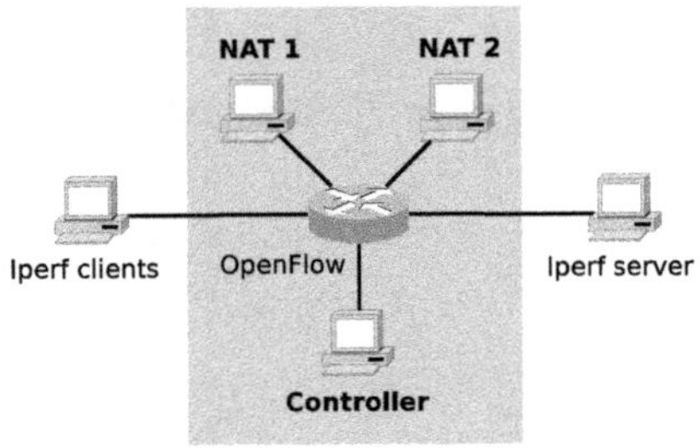

(b) Experimental Setup

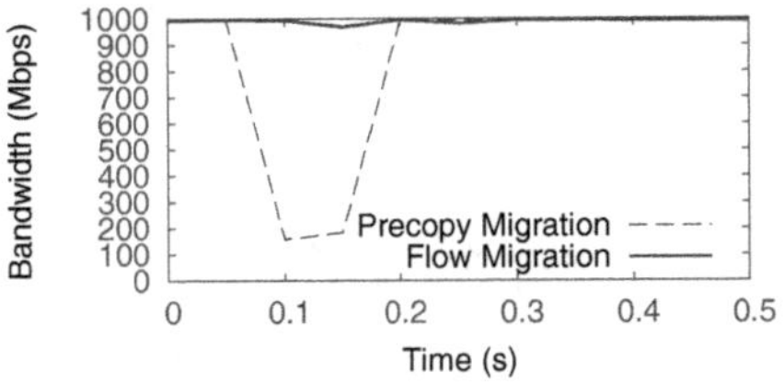

(c) Impact of migration on bandwidth

packets that hit A but don't have any matching state are forwarded to B, who processes them normally.

The *idle* phase takes advantage of the fact that most flows are short-lived. Rather than transfer those states, they can be kept on the original machine until they expire while, at the same time, allowing states for new flows to be established on the other machine. The two machines are daisy-chained (packets will always hit A first, regardless of direction; see figure 1(a)).

The *freeze-and-copy* phase is iterative and only starts when the rate at which states expire on A falls below a certain threshold; the remaining states are likely long-lived and need to be transfered. A freezes some of its remaining states and begins transferring them to B. All packets that would modify a frozen state are buffered by A. When the transfer finishes, all frozen states are deleted and the buffered packets are forwarded to B. The process is repeated until all states are copied to B. This phase only deals with long-lived flows and freezing aims to reduce the number of dropped packets. Daisy-chaining continues until the *freeze-and-copy* state finishes. At this point, there is no flow state left on machine A and traffic is redirected in the OF switch to hit machine B directly.

4. APPLICATION: CARRIER-GRADE NAT

We have implemented in Click [6] a scalable carrier-grade NAT with reactive load-balancing as a proof of concept for our flow state migration algorithm. Current best practices [4] along with OpenFlow's limitations [1] have driven our design of the NAT. Internal (IP, port) tuples are dynamically mapped onto their external counterparts; all connections corresponding to one such mapping constitute a flow whose state can be independently frozen and moved. The smallest migration goal is made up of all flows that share a certain external IP.

We have run experiments with a Gigabit Ethernet link saturated by 100 running instances of iperf, using the experimental setup shown in Fig.1(b). NAT box 1 is initially active, and it also has state for 65,000 fictive connections, emulating an external IP with the port space fully used. Each state is 24 bytes in size, bringing the total to roughly 1.5 MB. We migrate the flow states to NAT 2 both via our flow migration algorithm and via a single stop-and-copy round. Figure 1(c) shows the impact of the migration process on the bandwidth. While freeze-and-copy incurred only some minor disruption due to minor RTT inflation caused by the freeze time, the stop-and-copy round caused packet loss disrupting the iperf flows.

The downtime for stop-and-copy migration increases linearly with the number of flows and the size of the flow state. The NAT per flow state is small, but migrating more connections will bring proportionally more downtime which is unacceptable for production use.

5. NEXT STEPS

An open question is how effective precopy is with real traffic traces. Our simulations[1] with zipf-like flows sizes show that precopy has limited effect, however we need to experiment with real traffic traces to get a full picture.

Ideally, migration should be provided by the OS without changing the apps, and this is our end goal. It would allow us to migrate existing applications such as Bro or Snort without rewriting them, and giving our techniques wider applicability.

Achieving this goal is challenging. Running both source and destination processing in parallel is tricky as it requires efficient synchronization of accesses to global state. Done ineficiently, such synchronization can undo the performance benefits of flow-migration. Further, to implement daisy chaining, the OS must know which flows have state associated and which not; acquiring this knowledge without application support seems feasible only if both flow definitions and state expiration mechanisms are known to the OS.

6. REFERENCES

[1] Openflow switch specification, version 1.0.0.

[2] C. Clark and et al. Live migration of virtual machines. In *Proc. NSDI*, 2005.

[3] A. Greenhalgh and et al. Flow processing and the rise of commodity network hardware. *SIGCOMM CCR*, 39(2):20–26, 2009.

[4] S. Guha, K. Biswas, B. Ford, S. Sivakumar, and P. Srisuresh. NAT Behavioral Requirements for TCP. RFC 5382, Oct. 2008.

[5] M. Honda, Y. Nishida, C. Raiciu, A. Greenhalgh, M. Handley, and H. Tokuda. Is it still possible to extend tcp? In *ACM IMC*, 2011.

[6] E. Kohler, R. Morris, B. Chen, J. Jahnotti, and M. F. Kasshoek. The click modular router. *ACM Transaction on Computer Systems*, 18(3):263–297, 2000.

[7] R. Lawrence. A survey of process migration mechanisms.

[8] L. Peterson. Inter-as traffic patterns and their implications. In *Proc. IEEE GLOBECOM*, 1999.

[9] V. Sekar and et al. The design and implementation of a consolidated middlebox architecture. In *Proc. NSDI*, 2012.

[1]Not shown here; these results led us to only consider stop-and-copy as the alternative migration solution

A Demonstration of Ultra-Low-Latency Data Center Optical Circuit Switching

Nathan Farrington, George Porter, Pang-Chen Sun, Alex Forencich
Joseph Ford, Yeshaiahu Fainman, George Papen, Amin Vahdat

UC San Diego
http://mordia.net

ABSTRACT

We designed and constructed a 24×24-port optical circuit switch (OCS) prototype with a programming time of 68.5 μs, a switching time of 2.8 μs, and a receiver electronics initialization time of 8.7 μs [1]. We demonstrate the operation of this prototype switch in a data center testbed under various workloads.

Categories and Subject Descriptors

C.2.1 [**Network Architecture and Design**]: Circuit-switching Networks

Keywords

data center networks, optical circuit switching

1. INTRODUCTION

Data center networks are crucial to the scalability and performance of data center applications, yet are often under provisioned due to their high CAPEX and OPEX [2]. Recent work combines both traditional electronic packet switches (EPS) as well as optical circuit switches (OCS) [3]–[5] to reduce cost. Unfortunately, the relatively slow OCS switching times of 12 ms make them suitable for a limited class of workloads such as rack-to-rack backup or virtual machine migration. We designed and constructed a 24×24-port optical circuit switch (OCS) prototype, called Mordia, for use in data center networks, with the specific goal of supporting the much more common class of workloads that exhibit all-to-all communication patterns, such as MapReduce and on-line webpage rendering using distributed memory caches.

2. DESIGN & IMPLEMENTATION

The prototype is separated into a data plane (Figure 1) and a control plane (Figure 2). The data plane is constructed as an optical ring of six stations. Each station contains all of the optical components to switch four input/output port pairs. We use fixed-wavelength laser transmitters and tunable wavelength-selective switch (WSS)-based receivers.

The control plane is responsible for programming the WSS modules and synchronizing with all devices connected to the OCS. The controller maintains a round-robin schedule of input-output port mappings. Every 80 μs, the controller programs the six WSS modules with the next input-output port mapping in the list. The round-robin schedule allows the OCS to support all-to-all communication patterns at very high speeds and provides throughput and latency fairness.

Both before and after reconfiguring the WSS modules, the controller broadcasts a synchronization packet to connected devices. This allows the connected devices to learn the current input-output port mapping and when it is safe to begin transmitting.

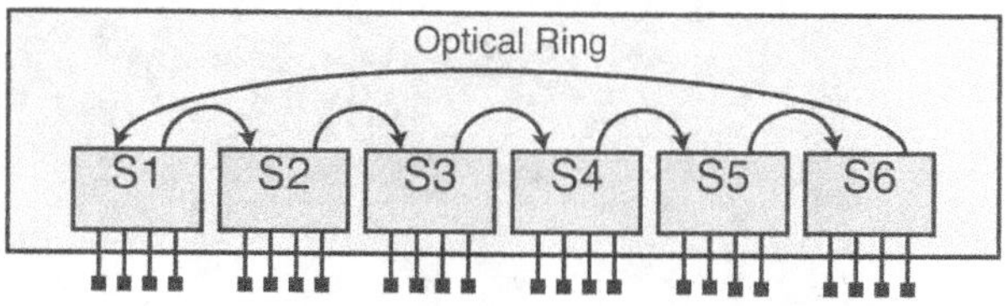

Figure 1. Data plane

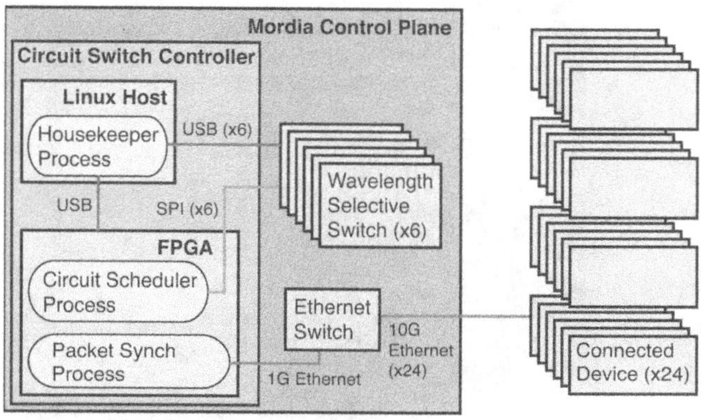

Figure 2. Control plane

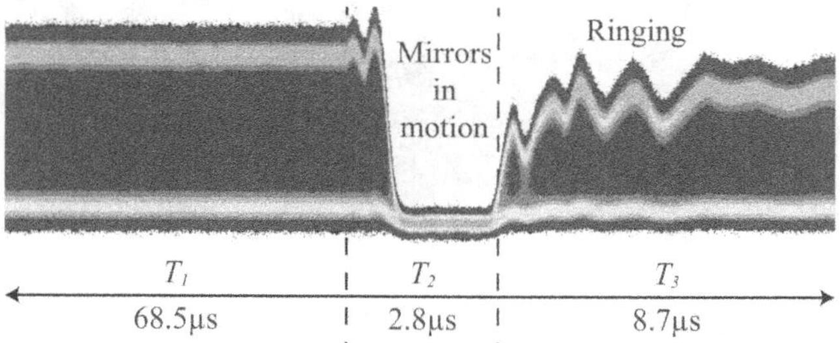

Figure 4. Measurements of switching time

Figure 3 shows that the implementation of the prototype occupies an entire datacenter rack. Each of the bottom three sliding trays contains the components for two stations. The FPGA placement is challenging because it must connect to all six WSS modules with short ribbon cables.

3. MEASUREMENTS

Figure 4 shows physical measurements of the OCS. We use the notation from [6]. The mirrors are only in motion for a small fraction of the total time. The majority of the loss-of-light time is due to ringing (T_3). Table 1 compares the switching speed of the Mordia prototype to Helios [3, 6]. Figure 5 shows measurements of the loss-of-light time ($T_2 + T_3$) as seen from the connected devices. The resulting minimum duty cycle is approximately 85%.

SIGCOMM'12, August 13–17, 2012, Helsinki, Finland.
ACM 978-1-4503-1419-0/12/06.

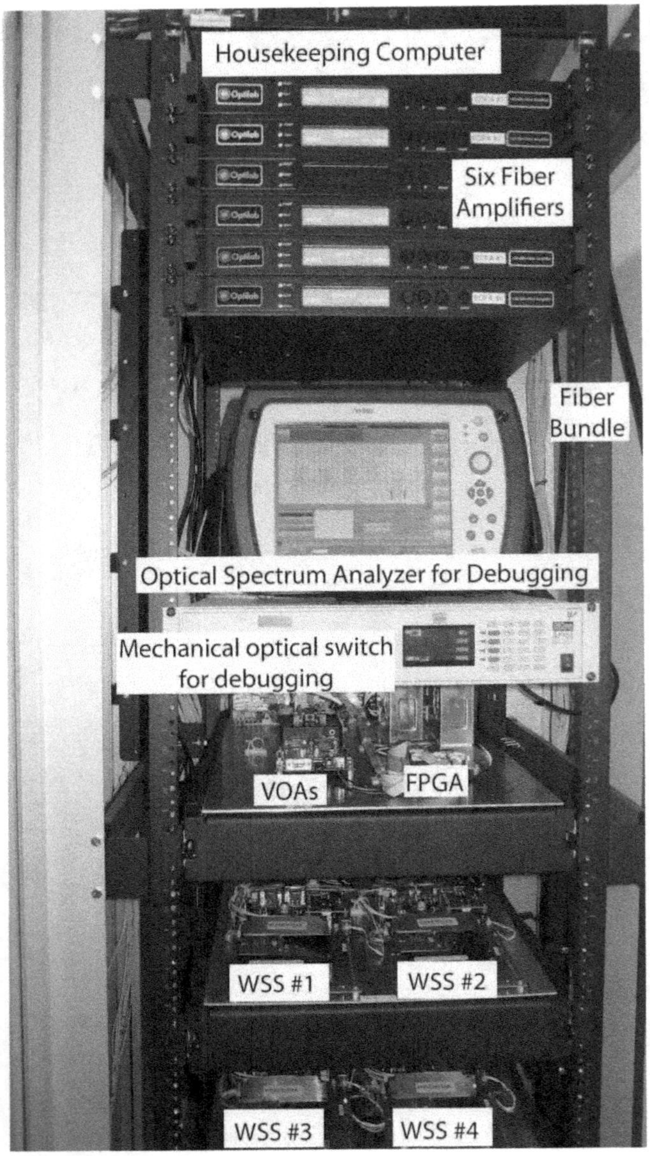

Figure 3. Rack containing Mordia OCS prototype

These measurements show that Mordia is three orders of magnitude faster than Helios and that the performance problems associated with the original Helios prototype have largely been solved in the Mordia prototype.

4. WORKLOADS

The demonstration consists of various workloads on a 24-node compute cluster and a visualization of the communication patterns in real time. Each workload can be run on either a traditional EPS or the Mordia OCS so as to compare the performance directly.

For example, one workload is a synthetic traffic generator called UDP Blaster where each host sends constant bitrate UDP traffic to each other host in the cluster. Good performance for this workload requires that the connected devices synchronize with the Mordia OCS so that they can transmit at the right time.

Another workload is TritonSort, the world's fastest large data sorting system [7]. TritonSort also has an all-to-all communication pattern, but uses TCP instead of UDP, and is rate limited by the disk I/O bandwidth. A successful demonstration of TritonSort on the

Table 1. Performance of Helios and Mordia

	Helios	Mordia	Speedup
T_1	5 ms	68.5 μs	73x
T_2	12 ms	2.8 μs	4,286x
T_3	15 ms	8.7 μs	1,724x

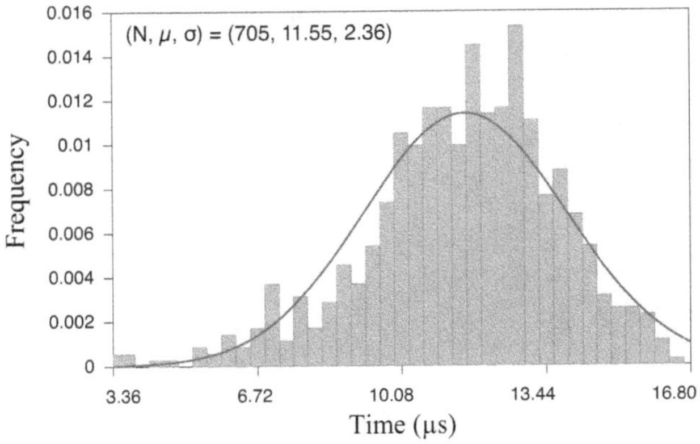

Figure 5. Measurements of loss-of-light time (T_2+T_3)

Mordia OCS prototype will may require careful analysis and possible modification of TCP.

5. ACKNOWLEDGMENTS

We would like to thank the NSF Center for Integrated Access Networks (#0812072) and gifts from Cisco Systems and Google, Inc. We would also like to acknowledge technical assistance from Mod Marathe at Cisco, Haw-Jyh Liaw of NetLogic, and Patrick Geoffray at Myricom, Inc.

REFERENCES

[1] Farrington, N., Porter, G., Sun, P.-C., Forencich, A., Ford, J., Fainman, Y., Papen, G., and Vahdat, A. The Design and Implementation of a Fast, Scalable Data Center Optical Circuit Switch. *Under review.*

[2] Al-Fares, M., Loukissas, A., and Vahdat, A. A Scalable, Commodity Data Center Network Architecture. In *SIGCOMM '08*.

[3] Farrington, N., Porter, G., Radhakrishnan, S., Bazzaz, H.H., Subramanya, V., Fainman, Y., Papen, G., and Vahdat, A. Helios: A Hybrid Electrical/Optical Switch Architecture for Modular Data Centers. In *SIGCOMM '10.*

[4] Wang, G., Andersen, D.G., Kaminsky, M., Papagiannaki, K., Ng, T.S.E., Kozuch, M., and Ryan, M. c-Through: Part-time Optics in Data Centers. In *SIGCOMM '10.*

[5] Chen, K., Singla, A., Singh, A., Ramachandran, K., Xu, L., Zhang, Y., Wen, X., and Chen, Y. OSA: An Optical Switching Architecture for Data Center Networks with Unprecedented Flexibility. In *NSDI '12.*

[6] Farrington, N., Fainman, Y., Liu, H., Papen, G., and Vahdat, A. Hardware Requirements for Optical Circuit Switched Data Center Networks. In *Optical Fiber Conference (OFC/NFOEC) '11.*

[7] Rasmussen, A., Porter, G., Conley, M., Madhyastha, H., Mysore, R.N., Pucher, A., Vahdat, A. TritonSort: A Balanced Large-Scale Sorting System. In *NSDI '11.*

AutoNetkit: Simplifying Large Scale, Open-Source Network Experimentation

Simon Knight
University of Adelaide,
Australia

Askar Jaboldinov
Loughborough University,
United Kingdom

Olaf Maennel
Loughborough University,
United Kingdom

Iain Phillips
Loughborough University,
United Kingdom

Matthew Roughan
University of Adelaide,
Australia

ABSTRACT

We present a methodology that brings simplicity to large and complex test labs by using abstraction. The networking community has appreciated the value of large scale test labs to explore complex network interactions, as seen in projects such as PlanetLab, GENI, DETER, Emulab, and SecSI. Virtualization has enabled the creation of many more such labs. However, one problem remains: it is time consuming, tedious and error prone to setup and configure large scale test networks. Separate devices need to be configured in a coordinated way, even in a virtual lab.

AutoNetkit, an open source tool, uses abstractions and defaults to achieve both configuration and deployment and create such large-scale virtual labs. This allows researchers and operators to explore new protocols, create complex models of networks and predict consequences of configuration changes. However, our abstractions could also allow the discussion of the broader configuration management problem. Abstractions that currently configure networks in a test lab can, in the future, be employed in configuration management tools for real networks.

Categories and Subject Descriptors

C.2.3 [**Computer-Communication Networks**]: Network Operations—*Network Management*

General Terms

Design, Experimentation, Management

1. INTRODUCTION

Emulated networks, which run a real router operating systems inside virtual machines, offer realistic yet inexpensive network experimentation. However they are time-consuming to configure, which limits their use in network research. AutoNetkit solves this problem, providing an integrated platform that allows the experimenter to visually describe their network experiment at a high-level, with resource allocation, device configuration handled automatically. The platform (shown in Figure 1) is Open-Source and generates router configurations from a network-level abstract representation, allocates appropriate resources, and deploys the network on emulated devices. This combination of simple specification with automated deployment and data collection allows for rapid iteration of network design leading to efficient evaluation of "what-if" scenarios, and providing a platform for research into complex policy and protocol interactions.

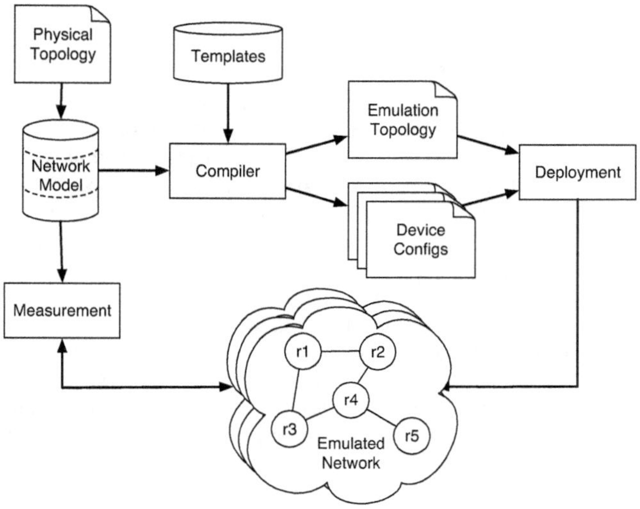

Figure 1: AutoNetkit: High-level descriptions are transformed into device-level configurations for deployment to an emulated network.

There are a number of platforms which can be used to conduct a network experiment on. Physical hardware networks are the most realistic, but are expensive, with a large scale testbed being beyond the budget of most researchers, and it is time consuming to manually configure and connect the physical routers. When available, hardware networks are typically small scale, limiting their usefulness in the study of large scale network behaviour.

Simulations, such as C-BGP [7], typically focus on one aspect of network behaviour, such as the routing decision process. This allows for large-scale experiments to be conducted on modest computer hardware, but at limited realism: modern networks are a complex interaction of multiple protocols, which cannot be easily simulated. Furthermore, interpretations of a standard (such as an RFC) can differ between vendor implementations, and bugs can exist in vendor software releases. To be relevant to real-world networks, the research community needs to study the behaviour of real vendor equipment, inclusive of RFC interpretations and bugs.

SIGCOMM'12, August 13–17, 2012, Helsinki, Finland.
ACM 978-1-4503-1419-0/12/08.

Emulated networks sit between these two extremes: by running the same operating system image as physical routers, they display the same protocol interactions and decision processes. Using virtual machines cuts costs dramatically with many virtual routers running on a commodity server.

However, emulated networks retain a main problem of hardware networks: the routers require configuration. For small networks, an experimenter may manually enter the configurations, but this approach doesn't scale well. While we may not need to plug cables into the router, the connection topology still needs to be created. In addition, IP addresses must be allocated and assigned to interfaces, routing protocols configured with appropriate AS numbers, IP addresses, and hostnames. This is time-consuming, tedious, and error-prone, and must be conducted before the actual network experiment itself is created. If the experiment is to be conducted on a different emulation platform (such as when comparing vendors) or if the network topology needs to be changed, the whole configuration process must be repeated.

2. AUTONETKIT

Emulation offers many advantages to network research. An automated configuration process will reduce the chance of errors, and free researchers to spend time on the experiments instead of setup. A standard platform reduces the space required in a paper to describe the experiment setup, and simplifies reproducibility of experiments. AutoNetkit addresses this configuration problem and simplifies the process of network experimentation.

The current version builds upon previous work [5], with a number of key new features. We allow the user to specify the network topology at a high-level, in a standard graph exchange format. This allows the use of a GUI-based graph editor, such as yED (`www.yworks.com/products/yed`) a freeware graph-drawing tool similar to Visio or OmniGraffle. In these graphs nodes correspond to routers, with edges indicating their connectivity. Node and edge attributes allow hostnames, AS numbers, and link weights to be specified. From these IGP and BGP topologies can be inferred (e.g. an eBGP session is created for a link that spans two Autonomous Systems). Additional attributes allow for advanced iBGP topologies and DNS databases to be configured. Autonetkit supports both OSPF and IS-IS IGPs, iBGP design patterns, and a clean abstraction for inter-domain BGP policy.

AutoNetkit supports Netkit [6], and can create and deploy to Junosphere [2], Dynamips [1], and C-BGP, and provides a tool to run measurement commands (such as traceroute or viewing the routing table) across the emulated network. It is easy to deploy the same experiment to multiple platforms: only the emulation server needs to be changed.

3. PERFORMANCE

As an example, we used AutoNetkit to setup a *bad gadget* [3]. We were able to setup and deploy this case-study in less than ten minutes to each of our supported emulated platforms. We verfied the expected oscillatory behaviour through automated traceroutes. On Cisco IOS and Juniper Junos routers the IGP tie-break worked as expected. However, the routing did not oscillate on Quagga and C-BGP — due to a difference in one of the final steps of the BGP best-path selection process. This example illustrates both the importance of testing behaviour on real router platforms, and the ease of doing so with AutoNetkit.

While AutoNetkit allows rapid design and testing of small case studies, it is especially useful for large networks. Our combination of simple visual network design and multi-AS support complete with BGP policy allows for realistic large-scale network experiments.

In another example, we have created a 89-node 132-link network topology, based on example networks from The Internet Topology Zoo [4]. AutoNetkit generated the configuration files for this network in less than 3 seconds on a 2007 MacBook Pro. The time taken to deploy these configuration files to a number of emulation platforms are shown in Table 1.

Platform	Memory (GB)	Startup Time (sec)
Netkit	12.6	402
Dynagen/GNS3	20.9	55
Junosphere	136.8	1011
C-BGP	0.020	< 0.2

Table 1: Resource use for 89 router example network.

4. SUMMARY

Our work provides an environment for network researchers to play and experiment. The tool takes a high-level abstract representation of a network and compiles it to a choice of emulated environments. The result is a level of scalability and flexibility that is hard to achieve in a hardware testbed, but with substantially more realism than a simulation.

AutoNetkit is fast and is capable of building network configurations in seconds, even for large networks. By using a diagram-based, high-level abstraction we allow a fast capture of the network design, enabling the examination and investigation of complex "what-if" scenarios.

AutoNetkit is currently used by industry, in teaching, and the networking research community. It is open-source and written in Python, providing a multi-platform tool that is freely available online at `www.autonetkit.org`.

5. ACKNOWLEDGMENTS

This project was supported by the Australian Government through an Australian Postgraduate Award, Australian Research Council Discovery Grants DP110103505 and DP0985063, and Cisco Grant 2011-89493(3696). We also would like to thank Hung Nguyen, Nick Falkner and Michael Rumsewicz for their helpful input.

6. REFERENCES

[1] Dynamips: Cisco 7200 simulator.

[2] Junosphere. `http://www.juniper.net/as/en/products-services/software/junos-platform/junosphere/`.

[3] T. Griffin, F. B. Shepherd, and G. Wilfong. The stable paths problem and interdomain routing. *IEEE/ACM Trans. on Networking*, 2002.

[4] S. Knight, H. Nguyen, N. Falkner, R. Bowden, and M. Roughan. The Internet Topology Zoo. *Selected Areas in Communications, IEEE Journal on*, 29(9):1765–1775, 2011.

[5] H. Nguyen, M. Roughan, S. Knight, N. Falkner, R. Bush, and O. Maennel. How to build complex, large-scale emulated networks. In *TridentCom*, Berlin, Germany, May 2010.

[6] M. Pizzonia and M. Rimondini. Netkit: easy emulation of complex networks on inexpensive hardware. In *TridentCom*, 2008.

[7] B. Quoitin. C-BGP, an efficient BGP simulator. http://cbgp.info.ucl.ac.be/, 2003.

Bulk of Interest: Performance Measurement of Content-Centric Routing

Matthias Wählisch
Freie Universität Berlin
waehlisch@ieee.org

Thomas C. Schmidt
HAW Hamburg, Informatik
t.schmidt@ieee.org

Markus Vahlenkamp
HAW Hamburg, Informatik
markus@vahlenkamp.net

ABSTRACT

The paradigm of information-centric networking subsumes recent approaches to integrate content replication services into a future Internet layer. Current concepts foster either a dynamic mapping that directs content requests to a nearby copy, or an immediate routing on content identifiers. In this paper, we evaluate in practical experiments the performance of content routing, which we analyze with a focus on conceptual aspects. Our findings indicate that the performance of the content distribution system is threatened by a heavy management of states that arise from the strong coupling of the control to the data plane in the underlying routing infrastructure.

Categories and Subject Descriptors

C.2.1 [**Computer-Comm. Networks**]: Network Architecture and Design

General Terms

Measurement, Security

Keywords

Performance, Experimental Evaluation, Routing

1. INTRODUCTION

One major dedication of today's Internet is the global distribution of content in huge amounts. Content distribution networks (CDNs) facilitate an efficient, wide-area replication of static data for selected content providers, whereas the end-to-end design of TCP/IP does not foresee implicit replication and in-network storage. There is no openly available standard solution for the asynchronous, global replication of popular content in the current Internet. Currently, *Information-Centric Networks* (ICN) [1] propose to fill this gap.

Essentially two approaches to routing exist in current ICN proposals, an evolutionary path that routes on IP, and 'clean slate' concepts that route on content identities. NetInf extends the current Internet by a resolution service that maps content names to topological IDs like IP addresses. TRIAD, DONA and NDN perform content retrieval by routing on names. Route responses and the data itself are then forwarded along reverse paths (RPF), either by using IP as a lower layer, or without IP but by dedicated RPF states.

Operating on the content itself forces network infrastructure into a content awareness. A mapping service is not only required to resolve names to source locations, but must advising a nearby replica, the existence of which it learned from the data distribution system. Content routers need to rely on names in its interface tables and – for RPF-based forwarding schemes – a reverse state for every data unit. This control information is highly dynamic and requires regular updates from the data plane. The ICN paradigm thereby opens up the control plane to continuous modifications from the data plane. This is in contrast to the current Internet, where DNS and routing states remain unaltered by data-driven events such as transmitted file names or data locations.

In this poster, we briefly present our study [2] of joint control and data plane behaviour under varying data conditions for the example of NDN [3] (§ 2). We are in particular interested in the response to load of the ICN infrastructure. Experiments are performed in test networks running PARC's CCNx software and reveal a flaw of performance (by concept, not implementation) with increasing demands to the routing system (§ 3).

2. BASIC MEASUREMENT SETUP

For our measurement study, we use the CCNx implementation version 0.5.1 [4], i.e., the client library to announce content interests, the content repository to store data, and the `ccnd` to forward subscription and data. The following analysis focuses on the effects on the router side. Even though the measurements relate mainly to the `ccnd`, we do *not* evaluate the implementation but use it as one real-world instance of the information-centric network deployment to illustrate protocol mechanisms. To gain a fine-grained view, we concentrate on the local system, as well as inter-router dependencies.

The basic network topology is represented by a chain of two CCNx routers directly interlinked at 100 Mbit/s, one end connects the content consumer, the other the content repository. We leave default values for all CCNx parameters. CCNx routers communicate via TCP.

3. RESULTS

To analyze the performance of content consumption in our experiment, the content receiver initiates parallel download of multiple 10 Mbit files over a constant time. We consider

SIGCOMM'12, August 13–17, 2012, Helsinki, Finland.
ACM 978-1-4503-1419-0/12/08.

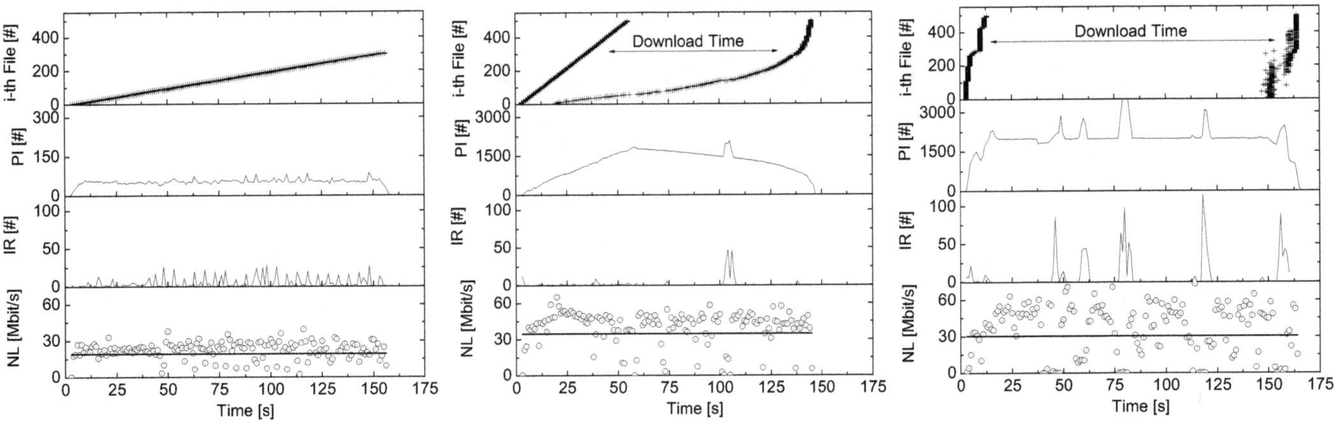

(a) 2 files per second

(b) 10 files per second

(c) 100 files per second

Figure 1: Parallel download of 10 Mbit files: Start and stop time of the download per file at the receiver & resource consumption at its designated router [Pending Interests (PI), Interest Retransmits (IR), and Network Load (NL) including the mean goodput]

three extremes, the request of 2 files, 10 files, and 100 files per second. Fig. 1 shows the start and completion time of the download per file (top graph), the Pending Interest Table (PIT) size, the effective number of Interest retransmissions, and the traffic load including the mean goodput at the first hop. For visibility reasons, we rescaled one y-axis (PI) in Fig. 1(a).

With an increasing number of parallel downloads, not only the download time increases significantly, but also the interval of the request and receive phase grows in the scenarios of overload. While the download time is almost constant for two files per second (cf., Fig. 1(a)), the stop time diverges non-linearly from the beginning of the download in the cases of excessive parallelism (cf., Fig. 1(b),(c)). 150 *s* are needed to download *each* single file in the worst case (Fig. 1(c)), while the link capacity would permit to retrieve *all* files in about 10 *s*.

The reason for this performance flaw is visualized in the subjacent graphs. A higher download frequency leads to an increasing number of simultaneous PIT entries, which require coordination with the data plane. Each file request will be split into the request of multiple chunks, in which the generation of corresponding interest messages will be pipelined. As soon as the content traverses, Interest states dissolve and thus release memory. These operations cause a continuous maintenance of states triggered by data traversal, and a simultaneous burst in CPU load (CPU exhaustion not shown). Finally it results in growing Interest retransmits after droppings or timeouts (shown in second lowest graphs), which again leads to retransmissions of data chunks. As an overall net effect, the network utilization fluctuates significantly, but does not adapt to actual user demands: Even though data requests could fill the links easily, the average load remains about constant at 30 %.

4. CONCLUSIONS & DISCUSSIONS

We have analyzed content-centric routing performance under varying loads and closely examined the implications of data-driven state management. The exhaustion of memory and processing resources following excessive state allocations was identified as one major reason for service degradation. It is worth noting that these effects are not shortcomings of the CCNx implementation, but driven by concept. Resource-intensive state management opens the floor widely to DoS attacks. Threats arrive from *resource exhaustion*, *state decorrelation*, and *path and name infiltration* [2].

An obvious approach to mitigate the resource exhaustion problem is to limit the rates of state injection into the network. Applying restrictions per user, though, will require addressing and tracking of end nodes, and lead to traffic shaping and bandwidth restrictions. As content states will accumulate in the network, and inter-provider deployment almost surely will lead to a heterogeneous, unbalanced network transitions, rate limiting may milden, but cannot effectively prevent the resource exhaustion problems discussed in this paper. Conceptual seem necessary to make content centric networking scalable, robust, and resistant to infrastructure attacks.

Acknowledgment

This work has been supported by the German Federal Ministry of Education and Research (BMBF) within the project HAMcast (see http://hamcast.realmv6.org).

5. REFERENCES

[1] B. Ahlgren, C. Dannewitz, C. Imbrenda, D. Kutscher, and B. Ohlmann, "A Survey of Information-Centric Networking (Draft)," Dagstuhl Seminar Proceedings, Tech. Rep. 10492, 2011.

[2] M. Wählisch, T. C. Schmidt, and M. Vahlenkamp, "Backscatter from the Data Plane — Threats to Stability and Security in Information-Centric Networking," Open Archive: arXiv.org, Technical Report arXiv:1205.4778v1, 2012. [Online]. Available: http://arxiv.org/abs/1205.4778v1

[3] V. Jacobson, D. K. Smetters, J. D. Thornton, and M. F. Plass, "Networking Named Content," in *Proc. of the 5th Int. Conf. on emerging Networking EXperiments and Technologies (ACM CoNEXT'09)*. New York, NY, USA: ACM, Dec. 2009, pp. 1–12.

[4] PARC, "The CCNx Homepage," www.ccnx.org, 2012.

User-level Data Center Tomography

Neil Alexander Twigg
University of Stirling
nat@cs.stir.ac.uk

Marwan Fayed
University of Stirling
mmf@cs.stir.ac.uk

Colin Perkins
University of Glasgow
csp@csperkins.org

Dimitrios Pezaros
University of Glasgow
dp@dcs.gla.ac.uk

Posco Tso
University of Glasgow
posco@dcs.gla.ac.uk

ABSTRACT

Measurement and inference in data centers present a set of opportunities and challenges distinct from the Internet domain. Existing toolsets may be perturbed or be mislead by issues related to virtualization. Yet, while equally confronted by scale, data centers are relatively homogenous and symmetric. We believe these may be attributes to be exploited. However, data is required to better evaluate our hypotheses. Therefore, we introduce our efforts to gather data using a single framework from which we can launch tests of our choosing. Our observations reinforce recent claims, but indicate changes in the network. They also reveal additional obfuscations stemming from virtualization.

Categories and Subject Descriptors

C.2.3 [**Computer-Communication Networks**]: Network Operations; C.4 [**Performance of Systems**]: Measurement Techniques

Keywords

Data Centers, Network Measurement, Tomography

1. INTRODUCTION

Data center network performance and behaviour is poorly understood in the public sphere. A better understanding facilitates research in network management, planning, design of protocols, and the ability to recognise abnormal events.

From inside the data center, operators have a direct line to network equipment and statistics [2]. In this context, there exists tools and projects that seek to measure, monitor, and diagnose data center network-related events. While informative, the output of these efforts must necessarily be kept confidential to maintain competitive advantage. When results can be shared, the experiments are irreproducible for the same reason. This poses challenges when attempting to characterize or improve upon network-level performance and security.

In this poster we present our efforts to provide a client-level perspective. We seek to identify the type and resolution of data available to the paying customer, as well as the inferences that might be made with this data. We take the position that, from a user-perspective, the only level of service that matters is the one that the user sees. For example, metrics such as bisection bandwidth describe network-wide characteristics that a single user may never see.

The task of measuring data center networks from a user perspective is challenging, and is only partially analogous to Internet measurement and tomography. Although issues of scale are present in both contexts, measurement in data center networks changes in two ways. The first is arguably to the benefit of any measurement infrastructure: Relative to the Internet, data centers are homogeneous and symmetric. This suggests that it may be possible to infer network characteristics without the need to evaluate complete sets of end-to-end paths. Alternatively, it suggests that characteristics inferred for a path of length x may hold for all internal paths of length x. The second difference is the possible perturbation caused by virtualization. Virtualization is known to lead to misleading or adverse data in existing toolsets [3]. For reasons different from previous work, this is confirmed by our own observations.

The client-level perspective is motivated by a series of questions in our own work:

- What is the effect of the data center on time-sensitive applications, and can shortcomings be predicted or compensated?
- Can a client detect and react to anomalies in their own 'slice' of the network?
- How do we obtain, and subsequently provide to the wider research community, useful data sets for the purpose of simulation and modeling?

Answers to such questions are impossible without traces and representative models of data center characteristics. Anecdotally, one of our own virtual machines was found to be participating in a DoS attack without our knowledge, after relaxing firewall rules for just a few hours. The opacity of the network only hindered our efforts to diagnose and repair.

2. FRAMEWORK

We are working to fill these gaps by constructing a flexible framework in which measurements can be selected and executed in a dynamic fashion. Our framework was built and tested on Amazon's AMI linux for EC2. It uses only standard C libraries and may be easily ported to other platforms.

When launched in the data center, clients first retrieve and build the latest source code from a git repository. Upon completion, clients then communicate with an off-site server to retrieve the list of expected measurements, and learn of the private IP addresses in use by other clients as they come

SIGCOMM'12, August 13–17, 2012, Helsinki, Finland.
ACM 978-1-4503-1419-0/12/08.

Figure 1: Sample topology from node at root (blue) to 19 other nodes (yellow).

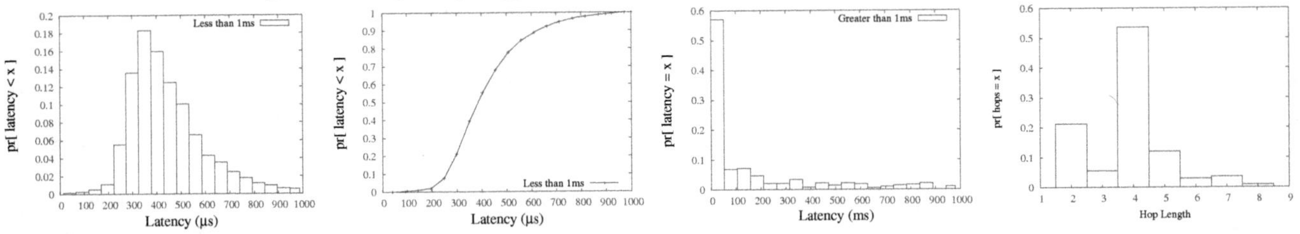

(a) edf of RTTs $< 1ms$.
(b) cdf of RTTs $< 1ms$.
(c) edf of RTTs $> 1ms$.
(d) edf of path lengths.

Figure 2: Summary of initial measurements

online. With a single point of distribution, experiments may be created, modified, and distributed quickly.

Current modules consist of ping- and traceroute-like measurements. In order to minimize timing issues, as well as embed relevant tags, the current modules build custom ICMP headers and use raw sockets. In the next section we validate the measurements taken using this framework by comparing against the work in [3]. We are also in talks with Amazon to determine accuracy. In future we will focus on bottleneck bandwidth estimation, as well as understanding the relationship between logical and physical paths.

3. INITIAL RESULTS

Initial tests were selected to address two concerns. First, the impact of layer-2 routing in the network with respect to its opacity at the IP layer. A multiple-hop layer-2 path between endpoints is presented to the user (at the IP layer) as a single hop. This makes inference of bottleneck links, for example, challenging. Second, fast round-trip times risk being too small to be reliably recorded given the granularity of the OS. Initial tests are positive, and reveal some truth in these assumptions, as discussed in following sections.

Network Maps

From the data gathered we can construct maps of the topology. A subset of a complete map appears in Figure 1 showing the topology from the view of the VM at the root to 19 other VMs in the network (leafs). The rich maps suggest there is sufficient IP-level detail irrespective of layer-2 routing. One interesting observation is that the host machine never responds to expired TTL. We have also been able to infer the practice of assigning to each physical machine a /24 network address by comparing VM with gateway addresses, and reconciling every VM always being reachable.

Round Trip Times

Our observations reinforce the high variability seen in [3] but differ in remaining respects. In that work it was observed that the first few RTT measurements in every new set of pings between a pair of nodes were abnormally high. Once those RTTs were removed, their observations revealed all RTTs to be less than $200\mu s$. We witnessed no such initial behaviour, as well as an increase in RTTs. Observations are summarized in Figures 2a to 2c. For clarity we have separated RTTs less than 1ms (the edf and cdf in Figures 2a and 2b respectively) from those more than $1ms$. The ratio of packets in the two categories is approximately 5:1. Interestingly, RTTs greater than $1ms$ appear to be fairly constant and persistent over spans of several seconds. We were also surprised by the appearance of negative RTT values in our measurements. We believe this to be a consequence of different timing as different cores revealed to the VMs.

Path Length

The distribution of path lengths also differs from the work in [3], which observes all path lengths to be either 3 or 4 hops, Figure 2d shows this no longer to be true. Most paths are 2, 4, or 5 hops in length, with a non-trivial number spanning as many as 8 hops.

Two years apart, our observations suggest that the means of the underlying distributions are similar, but that the variances are quite different. We see this as initial evidence of changes in the network and levels of service that need to be better understood. Observations also reveal negative pings and hidden nodes, both as a consequence of virtualization.

Network maps and measurements are viewable online at [1]. Members of the community will be encouraged to participate by injecting their own measurement modules. Recognizing that measurement is often a community effort, source code is expected to soon be available.

4. REFERENCES

[1] `http://d253108.cs.stir.ac.uk:89`.

[2] P. Gill, N. Jain, and N. Nagappan. Understanding network failures in data centers: Measurement, analysis, and implications. In *Proceedings of ACM SIGCOMM*, August 2011.

[3] G. Wang and T. E. Ng. The impact of virtualization on network performance of Amazon EC2 data center. In *Proceedings of the IEEE INFOCOM*, March 2010.

Towards Detecting BGP Route Hijacking using the RPKI

Matthias Wählisch
Freie Universität Berlin
waehlisch@ieee.org

Olaf Maennel
Loughborough University
olaf@maennel.net

Thomas C. Schmidt
HAW Hamburg
t.schmidt@ieee.org

ABSTRACT

Prefix hijacking has always been a big concern in the Internet. Some events made it into the international world-news, but most of them remain unreported or even unnoticed. The scale of the problem can only be estimated.

The Resource Publication Infrastructure (RPKI) is an effort by the IETF to secure the inter-domain routing system. It includes a formally verifiable way of identifying who owns legitimately which portion of the IP address space. The RPKI has been standardized and prototype implementations are tested by Internet Service Providers (ISPs). Currently the system holds already about 2% of the Internet routing table.

Therefore, in theory, it should be easy to detect hijacking of prefixes within that address space. We take an early look at BGP update data and check those updates against the RPKI—in the same way a router would do, once the system goes operational. We find many interesting dynamics, not all can be easily explained as hijacking, but a significant number are likely operational testing or misconfigurations.

Categories and Subject Descriptors

C.2.2 [**Computer-Communication Networks**]: Network Protocols—*Routing Protocols*

Keywords

BGP, RPKI, secure inter-domain routing, deployment

1. INTRODUCTION

The Border Gateway Protocol (BGP) was designed with next to no security in mind and is therefore quite vulnerable to attacks such as prefix hijacks, protocol attacks, and accidental misconfiguration [1, 2].

This lack of security made it easy to announce address space that does belong to someone else—without the owners permission. "Evil guys" use this to inject spam, to launch attacks, or perform other illegal activities. Given the critical importance of the Internet, and the widespread use of this hijacking, there has been a lot of concern around that topic [3].

The IETF Secure Inter-Domain Routing (SIDR) working group has taken up the task to secure the BGP routing system. It builds upon ideas to cryptographically verify BGP update messages [2]. At first, this must include a mechanism of certifying who owns what address space. RPKI certificate hierarchy [4] forms a hierarchical relationship that follows the address space allocation, e.g., IANA gives address space to ARIN, ARIN to ISPs, ISPs to customers. In this way Route Origin Authorizations (ROAs) can be cryptographically signed and published in repositories. On the other end, relying party tools download that information, verify and upload it to the router. The router, now, receives BGP update messages and is able to check them against those validated ROAs. This is called *BGP Origin Validation*. If a ROA is found following a longest common prefix match, the prefix update can either be *valid* or *invalid*. It is up to the ISP to decide about the consequences, but, the intended meaning of *invalid* is that the legitimate owner did *not* agree to the announcement.

In this poster, we aim at first steps in understanding to what extent BGP hijacking actually occurs in the Internet. To achieve this we are downloading ROAs in the same way an ISP is downloading from the public repositories. We then look at BGP tables and updates from RIPE's RRC00 [5] and RouteViews [6] collector, and match these real-world BGP data against the validated ROA information. We expect that the *invalid* messages fall into two categories: (1) hijacks or similar illegitimate activities or (2) system testings or misconfiguration, including operators not understanding the RPKI system. To our surprise it is challenging to distinguish between those two categories.

The RPKI implements the concept of positive attestation. Consequently, all legitimate origin ASes of an RPKI-enabled address block must be included in signed ROAs. An operator may easily miss some more-specific prefixes from customers or additional (e.g., sibling) ASes. An invalid BGP update, thus, is by no means obviously a hijack and the properties of system testing or misconfigurations are hardly distinguishable from prefix hijacks. Currently, up to 20% of the verifiable routing table are *invalid*, which we will analyze in detail in the next section.

2. INVALID ≠ INVALID?

We look at one month, April 2012, of BGP update data and try to develop a methodology that estimates the occurence of real BGP hijacking. On April 1st, 2012, the RRC00 table dump contained 432,864 prefixes of which 6,843 (1.6%) prefixes were *valid*, but also 1,665 (0.4%) prefixes were *invalid*. We now try to understand better what causes invalid prefixes. Prefixes are *invalid*, if (i) the BGP origin AS does not comply with the ROA origin AS (20% of all *invalid* cases), (ii) the announced prefix-length is longer than the specified max-length in the ROA (70%), or (iii) both (10%). RouteViews exhibit similar results.

When looking at those prefixes, which turned invalid because the origin AS differed, we noticed that in over 40% of the cases we found the valid origin AS as the first upstream AS (second last

SIGCOMM'12, August 13–17, 2012, Helsinki, Finland.
ACM 978-1-4503-1419-0/12/08.

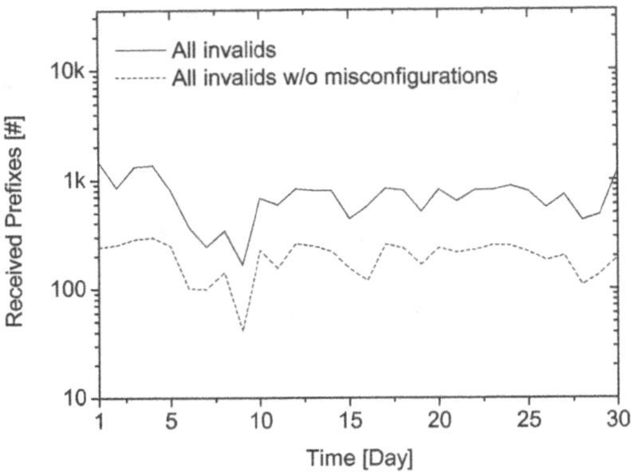

Figure 1: Number of invalid announcements for distinct prefixes per day

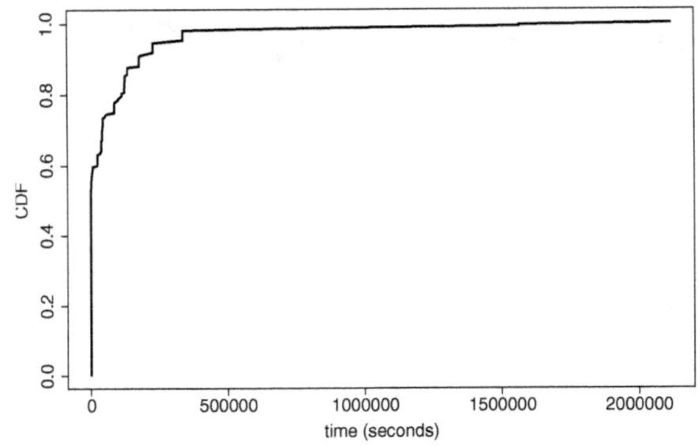

Figure 2: Duration of *invalid* in table.

AS on the path). This might lead to the speculation that operators of the upstream AS have "forgotten" to add at least some of their customers to the RPKI system. We also noticed this number has dropped significantly since January 2012, which might be due to increased training in how the RPKI works and how it secures the routing system.

Those considerations lead us to a set of assumptions, which we would like to understand better in order to distinguish misconfigurations from real illegitimate activity:

Misconfigurations are likely to cause (a) an *invalid* appearing on an AS path where the first upstream AS is one of the covering ROAs, (b) an *invalid* prefix that matches the ROA's origin AS, but is more specific compared to the maximum length defined in the ROA, or (c) an *invalid* announcement in which the ROA origin does not match the BGP origin AS, but both ASes are siblings.

Prefix Hijacking is indicated by (d) an *invalid* more-specific appearing temporarily and then disappearing again while a *valid* super-block is in the table, or (e) a *valid* prefix that is temporarily replaced by an *invalid* (at some observation points), while the corresponding ASes are not siblings.

Two points should be noted: (1) the proposed system for BGP origin validation is not intended to secure the routing system against attackers, who are faking their origin AS. For this reason alone, we do not consider BGP prefixes, where the origin AS is valid, but only the max-prefix length criteria is invalidating the announcement. We also believe that most of those cases are due to operators, who are currently in the process of adding their prefixes into the RPKI and have not done so for their whole network yet. (2) It should also be mentioned that hijacking often occurs in unallocated address space. However, unallocated address space is out of the scope of this work, as it is not secured by the RPKI.

We now look at BGP update dynamics for the whole month of April 2012. Figure 1 shows the number of invalid announcements for distinct prefixes per day, in which we exclude misconfigurations. Analyzing those in detail, only one *invalid* prefix appear as a more-specific of a *valid*, and 32 *invalid* prefixes occur, which temporarily replace an otherwise *valid* prefix (at some observation points). Figure 2 shows the duration of *invalids* in the table. While we would expect real hijacking to only last in the order of hours, we noticed that some "potential candidate prefixes" are still in the table for several days. We currently believe that many of those are still very likely operational tests and that our methodology needs further refinement. As a consequence this also implies, operators should be careful before turning on real filtering based on the RPKI.

The one prefix, which according to our methodology is most likely to be real hijcking, was in the table for 37.8 hours. While one prefix does does not sound too much, one should consider that the RPKI covers currently only 2% of the global routing table. If this would be a representative sample one might expect real hijacking in the order of 50 events per month.

3. CONCLUSION & FUTURE WORK

Securing the routing system is an important task for the Internet. Great progress has been made by the IETF SIDR working group, and prototype implementation of the system are being tested. In this poster we intended to look at BGP hijacking, but it turned out that we found many other interesting properties including how ISPs are starting to use the system. We argue that it is important to understand observable properties now, to support the efforts of the RPKI. For example, if we could detect common mistakes that operators are likely to make, this research could lead to improved training or to automated tools that watch-out for such misconfigurations.

In future work we intend to continue our investigation, and to improve our heuristics. We also intend to validate our findings by discussing with operators at forums such as NANOG and RIPE.

Acknowledgments

We would like to thank the anonymous reviewers for their valuable comments. This work was partially financed by a gift from Cisco University Research Program Fund, grant 2011-89493(3696), and partially supported by the German BMBF within the project Peeroskop.

4. REFERENCES

[1] K. Butler, T. R. Farley, P. McDaniel, and J. Rexford, "A Survey of BGP Security Issues and Solutions," *Proceedings of the IEEE*, vol. 98, no. 1, pp. 100–122, January 2010.

[2] G. Huston, M. Rossi, and G. Armitage, "Securing BGP - A Literature Survey," *Communications Surveys Tutorials, IEEE*, vol. 13, no. 2, pp. 199 –222, 2011.

[3] X. Hu and Z. Mao, "Accurate Real-time Identification of IP Prefix Hijacking," in *Security and Privacy, 2007. SP '07. IEEE Symposium on*, may 2007, pp. 3 –17.

[4] M. Lepinski and S. Kent, "An Infrastructure to Support Secure Internet Routing," 2012, RFC 6480.

[5] "RIPE's Routing Information Service," http://data.ris.ripe.net/.

[6] "University of Oregon RouteViews project," http://www.routeviews.org/.

Choice as a Principle in Network Architecture*

Tilman Wolf
ECE Department
University of Massachusetts
Amherst, MA, USA
wolf@ecs.umass.edu

Jim Griffioen, Ken Calvert
CS Department
University of Kentucky
Lexington, KY, USA
{griff,calvert}@netlab.uky.edu

Rudra Dutta, George Rouskas
CS Department
North Carolina State Univ.
Raleigh, NC, USA
{dutta,rouskas}@csc.ncsu.edu

Ilia Baldine
Renaissance Computing Inst.
University of North Carolina
Chapel Hill, NC, USA
ibaldin@renci.org

Anna Nagurney
Finance and Operations Mgnt.
University of Massachusetts
Amherst, MA, USA
nagurney@isenberg.umass.edu

ABSTRACT

There has been a great interest in defining a new network architecture that can meet the needs of a future Internet. One of the main challenges in this context is how to realize the many different technical solutions that have developed in recent years in a single coherent architecture. In addition, it is necessary to consider how to ensure economic viability of architecture solutions. In this work, we discuss how to design a network architecture where choices at different layers of the protocol stack are explicitly exposed to users. This approach ensures that innovative technical solutions can be used and rewarded, which is essential to encourage wide deployment of this architecture.

Categories and Subject Descriptors

C.2.6 [**Computer-Communication Networks**]: Internetworking; C.2.2 [**Computer-Communication Networks**]: Network Protocols—*Protocol architecture*

General Terms

Design

Keywords

network architecture, innovation, economics

1. INTRODUCTION

Much of the Internet' success comes from its ability to support a wide range of service at the edge of the network. However, the Internet offers little choice of service inside the network. It is widely agreed that this limitation inhibits the development and deployment of new networking services, protocols, security designs, management frameworks, and other components that are essential to support the increasingly diverse systems, applications, and communication paradigms of the next-generation Internet

*This material is based upon work supported by the National Science Foundation under Grant Nos. 1111040, 1111088, 1111256, 1111276.

SIGCOMM'12, August 13–17, 2012, Helsinki, Finland.
ACM 978-1-4503-1419-0/12/08.

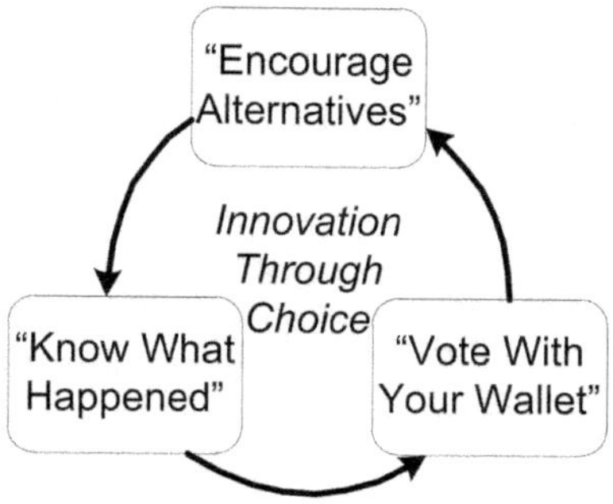

Figure 1: Foundational principles and their dependencies.

We believe that *choice* is the key aspect of a network architecture, and that it can drive innovations necessary for future networks. Choice implies that the entities using the network can select from a range of alternative services that may differ in functionality, performance, and cost. Choice can only arise if the network architecture supports the technology necessary for dynamic introduction of new alternatives. It is also necessary to put in place suitable economic processes to ensure that incentives trigger innovation and users can "vote with their wallets."

In this poster, we describe the outline of a network architecture that can provide choice as a core principle.

2. CHOICE INSIDE THE NETWORK

Choices can appear at different layers in the protocol stack of a network, ranging from different communication paths to different protocols and application-layer services. Our system for a market-driven competition of networking functionality is based on three key principles (illustrated in Figure 1):

- **Principle 1: Encourage Alternatives.** The underlying network infrastructure must provide the building blocks to create different types of services and to create alternative services of the same type. Support for alternatives allows users to select the service(s) that best meets their needs, and provides the best performance for their application. Implicit in this principle is the idea that—in contrast to the present Internet, where competition exists only at the application layer,

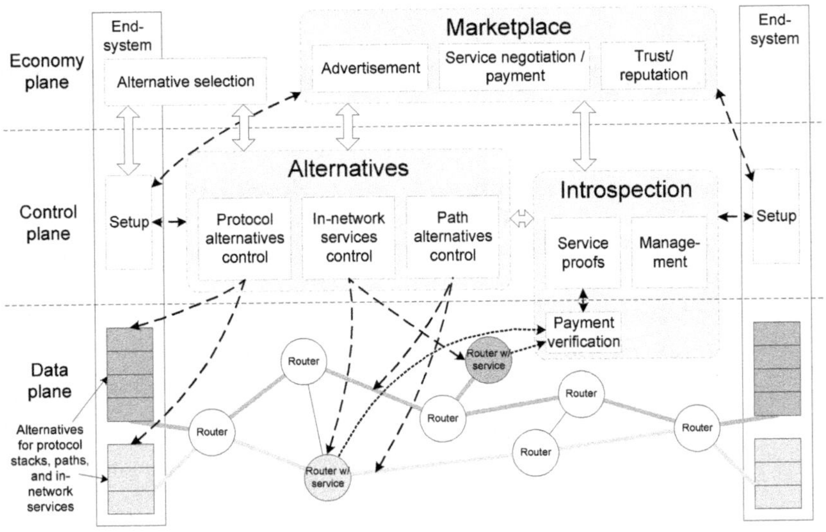

Figure 2: Overview of ChoiceNet architecture.

if at all—users must not be "stuck" when the service they receive is not consistent with their expectations. Rather, they must be able to choose a different service provider, to better meet their expectations.

- **Principle 2: Vote With Your Wallet.** The underlying network infrastructure must provide building blocks so that users can financially encourage providers that offer superior (and often innovative) services, while discouraging providers that offer inferior services and fail to innovate. In other words, the "money protocols" needed for users to vote with their wallet (i.e., pay for good services) must be designed into the network. We believe that incentives and competition are crucial for the long-term health of the network—winning ideas will succeed, grow, and promote additional competition, while losing ideas fade away.

- **Principle 3: Know What Happened.** Distinguishing services and providers that perform well from those that do not is crucial to enabling robust competition. In a complex system like the Internet, determining what happened (i.e., whom to blame) when an end-to-end service does not meet user expectations can be a challenging proposition because providers may be operating at different layers and in several locations along an end-to-end path. The network must provide building blocks that allow users and providers to determine, and exchange information about, the performance they experience. Such "introspection" capability of the network also enables innovative network management and monitoring tools, which themselves can evolve over time.

These principles are interdependent and mutually rely on each other. For example, if there are no other alternatives, there is no need to vote with one's wallet, nor does it help to know that the current alternative is not a good one.

3. NETWORK ARCHITECTURE

Embracing the three principles discussed above requires new features and mechanisms not present in the current Internet. Figure 2 illustrates schematically how the principles interact within a network—which we dub *ChoiceNet*—and the new features they support.

As a starting point, one needs support for alternatives, including the ability to create alternatives and select among them. To support alternatives, ChoiceNet offers new control-plane mechanisms. More importantly, ChoiceNet introduces a new *economy plane* that is responsible for supporting the types of business relationships and incentives that drive innovation and change in the real world. The economy plane supports advertisement of choices to users, as well as selection from among a set of alternatives.

Economy plane protocols enable users to negotiate desired levels of service, and pay (compensate) specific providers for services. It also includes mechanisms to help establish the identity, level of trust, and reputation of the parties involved in a business relationship.

4. USE SCENARIO

To illustrate the ability of ChoiceNet to support innovation in the network, we briefly describe one use scenario, highlighting the benefits of choice as a fundamental feature of the architecture. For this scenario, consider the selection of connection quality for video distribution.

A user may be offered different connection services for the purpose of watching streaming video. These connections may differ in their technical implementation (e.g., quality-of-service, use of caching, etc.). The user selects a service (i.e., a complete package of end-to-end connection and related services) and pays for its use. Depending on the user's satisfaction with the video experience, they continue to use the chosen service or switch to another (i.e., vote with their wallet).

Note that users does not require to have technical expertise in judging low-level networking metrics. They simply make choices based on their overall experience. Since economic rewards are explicitly represented in the architecture, network service providers are incentivized to create innovative offerings. This approach can therefore drive continued innovation throughout the network.

A Frequency Adjustment Architecture for Energy Efficient Router

Wenliang Fu, Tian Song
Beijing Laboratory of Intelligent Information Technology, School of Computer Science
Beijing Institute of Technology, Beijing, China, 100081
{fuwenl, songtian}@bit.edu.cn

ABSTRACT

With the rapid expansion of customer population and link bandwidth, energy expenditures of the Internet have been rising dramatically. To gain energy efficiency, we propose a novel router architecture, which allows each of its modules to adjust frequency according to traffic loads. Several modulation strategies are also discussed to ensure dwell time on low energy states and reduce blind switches. Our preliminary results show that the frequency adjustment router could save up to 40% of the total energy consumption.

Categories and Subject Descriptors

C.2.1 [**Network Architecture and Design**]: Network communications

Keywords

Energy Efficient Router, Frequency Adjustment

1. INTRODUCTION

Nowadays, much surplus and redundancy are introduced during network planning for rarely appeared traffic peak hours and device failures, wasting energy most of the time [1]. To cope with this situation, energy efficient technologies, such as smart port sleeping, dynamic buffer adapting and green routing, have been proposed to trade performance for energy efficiency in idle or low traffic time.

Smart port sleeping explores a sleep-wake mechanism between active and sleep modes for energy savings in the physical layer, and later leads to the establishment of standard IEEE 802.3az. However, due to considerable time and energy costs for transition, this method is only effective while port utilization rate is under 10% [2].

A. Vishwanath *et al.* explore buffer usage of routers and develop a buffer adapting algorithm for letting SRAM and DRAM buffers sleep while not being used [3]. Results show that up to 10% of the total energy consumption could be eliminated with negligible traffic congestion and latency.

From the network level, green routing exploits limiting traffic forwarding in a subset of network links. The essence of this idea is to optimize traffic modes for more energy efficient behaviors of network devices.

In this poster, we propose a frequency adjustment router architecture which allows each of its functional module to adopt different frequencies, offering various energy efficiencies, traffic capacities and processing latencies. Our frequency adjustment router has more potential for energy saving in the following aspects. First, aimed at functional modules which accounts for about 50% of a router's overall energy, our method has more energy saving potentials than other methods (ports and memory chips account for 10% and 11% of the total energy consumption respectively) [4]; Second, frequencies of each energy state could be customized for various traffic environments. Third, modulations between multiple frequencies are more flexible for strategic switching than simple sleep and active switching. In addition, our proposed frequency adjustment scheme is a essential supplement, not a replacement, for current energy efficient technologies, and the comprehensive utilization of those methods could certainly achieves higher energy efficiency.

SIGCOMM'12, August 13–17, 2012, Helsinki, Finland.
ACM 978-1-4503-1419-0/12/08.

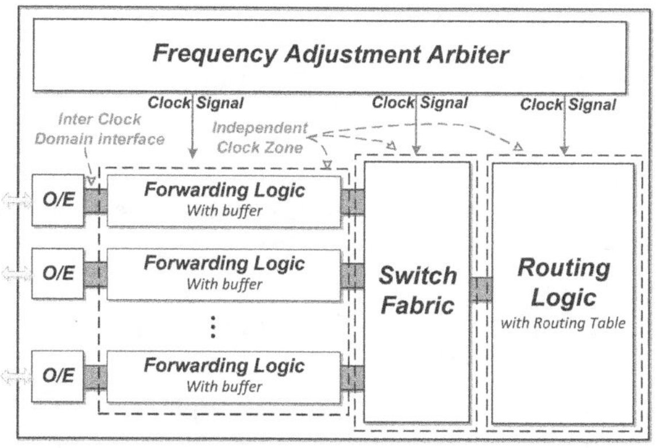

Figure 1: Block Diagram of Frequency Adjustment Router Architecture.

2. FREQUENCY ADJUSTMENT ROUTER ARCHITECTURE

Our proposed architecture is shown in Fig.1. We treat functional modules as basic adjustment units, namely frequency adjustment zones. To support free frequencies modulation between modules, those units are connected with inter clock domain interfaces for handling asynchronous clock problems. Generally, the interface is an asynchronous FIFO based one-way data path, which collects signal sequences and stores them in FIFO according to one clock, and reads the signals from FIFO and replays them with the pace of another clock. In order to modulate frequencies of those units cooperatively, a frequency adjustment arbiter is introduced

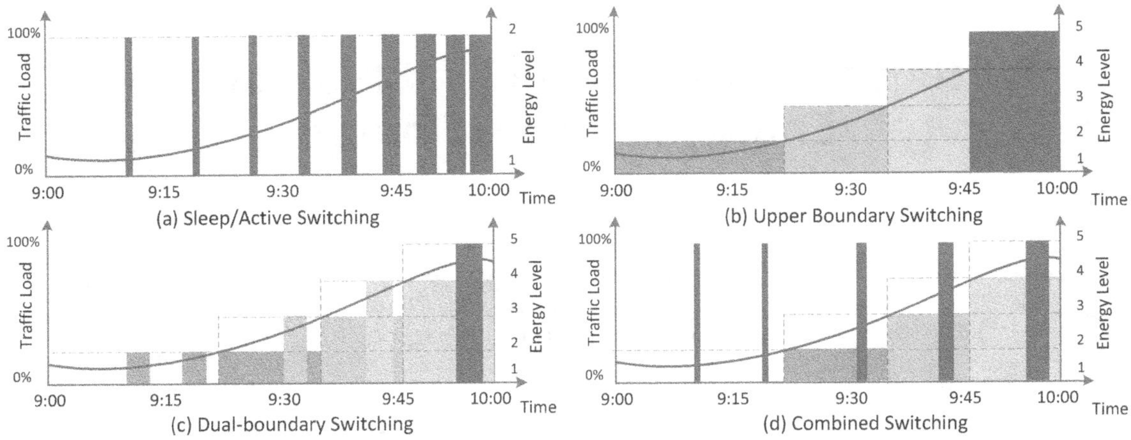

Figure 2: Energy State Switching Strategies.

to monitor system information, such as FIFO occupancy, port utilization and current energy status of each modules, and make frequency adjustment decisions.

3. DYNAMIC FREQUENCY ADJUSTMENT STRATEGIES

We developed several frequency switching strategies, distinguished by buffer occupancy threshold and cold down time between transitions, to guarantee stay time in low energy states and reduce blind switches, described as follow:

Sleep/active switching: Triggered by fixed buffer occupancy threshold, we propose a two-stage switch strategy as shown in Fig.2 (a), which allows modules to operate either in sleep mode for energy savings, or in active mode (marked as red) for processing packets in buffer.

Upper Boundary switching: This scheme permits multi-stage switching, and always adopts an upper frequency level that barely offering more capacity for handling the traffic. As shown in Fig.2 (b), energy states with different frequencies are distinguished by colors.

Dual-boundary switching: Fig.2 (c) describes a optimized strategy of upper boundary idea. By introducing a lower boundary, this scheme further reduces stay time in upper frequency levels with the cost of limited switches.

Combined switching: This strategy combines ideas of sleep/active switching and boundary switching, as shown in Fig.2 (d), treating low traffic time (such as nights) with sleeping strategy, and high traffic time with the aid of lower boundary for reducing frequent switches.

4. EXPERIMENT AND RESULT

We developed a router simulator which offers 10 frequencies range from 125 MHz to 12.5 MHz, corresponding to 1000Mbps to 100Mbps respectively in traffic capacity. The experiments were based on real traces captured bidirectionally from a office network of 1000 users, spinning over a period of more than three hours and the loss rate was kept under 1%. The preliminary results are shown in Fig.3, by limiting switches with preset cold down time between transitions and processing packets in buffer at the maximum speed, the combined strategy achieved the most significant savings than other methods.

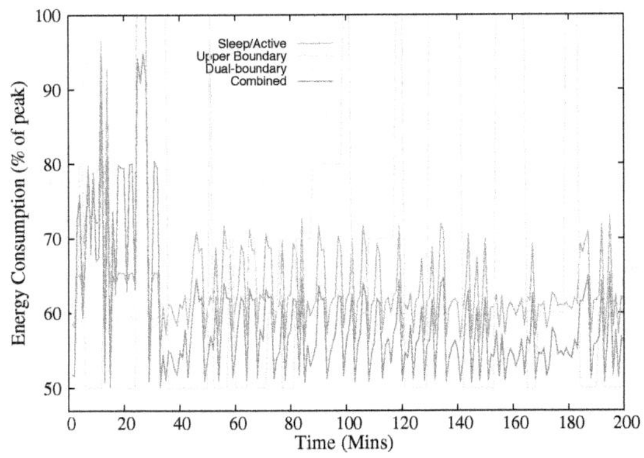

Figure 3: Frequency Adjustment for Energy Efficiency.

5. CONCLUSION

In this poster, we proposed a novel frequency adjustment router architecture, which allows each of its modules to adopt multiple and different frequencies. In addition, several frequency adjustment strategies are also described for tuning various network environments. Simulations on real traces show that our frequency adjustment router could save up to 40% of the total energy consumption.

6. REFERENCES

[1] The Applied Research Group. http://ipmon.sprint.com/.

[2] P. Reviriego, K. Christensen, J. Rabanillo, and J. Maestro. An initial evaluation of energy efficient ethernet. *Communications Letters, IEEE*, (99):1–3, 2011.

[3] A. Vishwanath, V. Sivaraman, Z. Zhao, C. Russell, and M. Thottan. Adapting router buffers for energy efficiency. In *Proceedings of the Seventh COnference on emerging Networking EXperiments and Technologies*, page 19. ACM, 2011.

[4] R.S. Tucker, R. Parthiban, J. Baliga, K. Hinton, R.W.A. Ayre, and W.V. Sorin. Evolution of wdm optical ip networks: A cost and energy perspective. *Journal of Lightwave Technology*, 27(3):243–252, 2009.

Detecting Third-party Addresses in Traceroute IP Paths

Pietro Marchetta, Walter de Donato, Antonio Pescapé
University of Napoli Federico II
Napoli, Italy
{pietro.marchetta,walter.dedonato,pescape}@unina.it

ABSTRACT

Traceroute is probably the most famous computer networks diagnostic tool, widely adopted for both performance troubleshooting and research. Unfortunately, traceroute is not free of inaccuracies.

In this poster, we present our ongoing work to address the inaccuracy caused by *third-party addresses*. We discuss the impact of third-party addresses on traceroute applications and present a novel active probing technique able to identify such addresses in traceroute traces. Finally, we detail preliminary results suggesting how this phenomenon has been largely underestimated.

Categories and Subject Descriptors

C.2.1 [**Computer-communication networks**]: Network Architecture and Design—*Network topology*

General Terms

Measurement

Keywords

Traceroute, Internet topology, AS-level path.

1. MOTIVATION AND SUMMARY

In the last decade many research works have used traceroute (or its variants) to infer network topological properties [5,8], and, more in general, in active monitoring approaches for anomaly detection, performance analysis, and geolocation. Traceroute is known to be affected by several inaccuracies that can produce errors or wrong assumptions when using its results to infer other information [7]: third-party (TP) addresses have been pointed out as an uncommon case mostly appearing at the border of multi-homed Autonomous Systems (ASes) [6].

To the best of our knowledge, the occurrence of TP addresses in IP paths has not been properly quantified and we believe that its impact has been largely underestimated. In this poster, to shed light on this topic, we propose a novel active probing technique based on the IP prespecified timestamp option: it allows to directly detect the presence of TP addresses in traceroute IP paths. We present a preliminary analysis conducted on about 13 K traceroute traces collected from a single vantage point. The analysis confirms the potentiality of our technique and unveils an unexpected amount of TP addresses. Finally, by performing IP-to-AS mapping on our dataset, we quantify the amount of potentially fake AS links due to TP addresses.

SIGCOMM'12, August 13–17, 2012, Helsinki, Finland.
ACM 978-1-4503-1419-0/12/08.

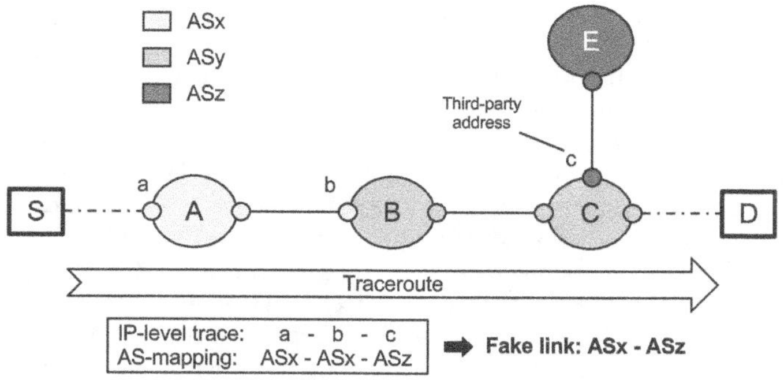

Figure 1: TP addresses inducing the inference of false AS links

2. TP ADDRESSES AND THEIR IMPACT

The RFC1812 [2] states that the source address of an ICMP reply packet should correspond to the outgoing interface of the reply, rather than the interface on which the packet triggering the reply was received. This behavior can cause a traceroute IP path to include addresses associated to interfaces not included in the actual traversed path. Let's see an example (Fig. 1): in the trace from S to D containing the sequence (a, b, c) of IP addresses (hereafter IPs), where a and b represent the incoming interfaces of routers A and B respectively, and c is the interface used by router C to send ICMP replies to the traceroute originator, the last IP is a TP address (being associated - in this specific trace - to an interface not effectively traversed by packets from S to D).

The occurrence of TP addresses can have a significant impact on some traceroute applications. A first example are the issues occurring when mapping traced IP addresses to AS numbers, in order to discover AS level links. In this case, as shown in previous works [6], TP addresses may cause the inference of false AS links. Considering again Fig. 1: if the IP address associated to interface b belongs to ASx, and the one associated to c belongs to the ASz addressing space, then the traceroute conversion will produce a false AS link, i.e. $ASx - ASz$. Another example is related to the identification of the subnets traversed by packets along the path toward a destination, as performed by the Tracenet tool [10]: the presence of TP addresses forces the tool to exploit several heuristics to state if a detected subnet is actually on the traversed path.

3. PROPOSED APPROACH

Our approach exploits the IP prespecified Timestamp (TS) option to identify TP addresses. The TS option allows to prespecify in a single packet up to four IP addresses from which a timestamp is requested. When processing this option, most devices insert their own timestamp only if the packet passes through the interface as-

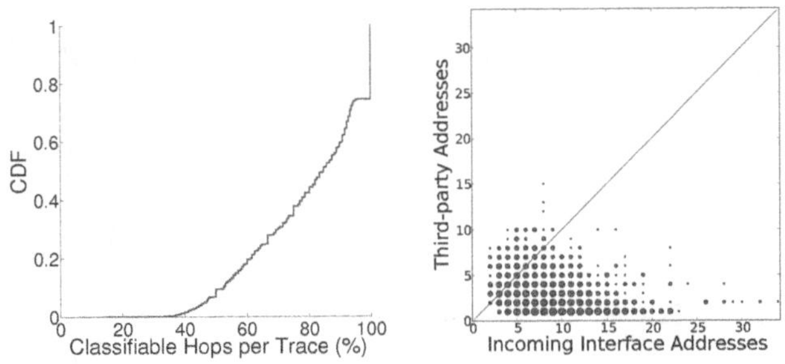

(a) Approach applicability. (b) Per-path classification.

Figure 2: Preliminary results.

sociated to the prespecified address [4]. Suppose IPx owned by one of these devices and belonging to a traceroute path. In order to understand if IPx is a TP address, we target the traceroute destination with an UDP probe[1] equipped with a TS option, in which IPx is prespecified four times: if the TS option brought back into the payload of the ICMP PORT UNREACH message contains at least one timestamp, according to the behavior described above, IPx belongs to the actual IP path[2], otherwise IPx is a TP address. To be able to distinguish between routers reporting a TP address and ones not implementing the TS option, after collecting a traceroute path, we first directly probe each discovered hop IPy with an ICMP ECHO REQUEST message in which IPy is prespecified: this step allows to identify the set of IP addresses classifiable by our approach. The proposed approach would allow to avoid complex heuristics for subnet positioning in Tracenet [10] and to take into account path fluctuations affecting the Palmtree [9] alias resolution technique.

4. PRELIMINARY RESULTS

To evaluate the proposed approach, we randomly selected 50 K destinations among the ones showing stable responsiveness according to the PREDICT project [1]. About 13 K destinations replied to UDP probes carrying the TS option. Focusing our attention on this subset, we launched a traceroute campaign, from our laboratory at University of Napoli, collecting a final dataset of 32 K IPs. Fig. 2(a) shows the percentage of hops classifiable in each trace: all the hops resulted classifiable in 26% of traces. In general, the proposed approach showed a promising level of applicability: on average and in the worst case, we were able to classify 80% and 38% of hops per trace, respectively. Furthermore, it classified 2.5 K IPs (7.8%) as TP addresses in at least one trace. Such value appears surprisingly high, suggesting that TP addresses are not so uncommon as previously hypothesized [6]. Fig. 2(b) compares, for each trace, the number of addresses classified as TP or associated to incoming interfaces (IN) and shows how in 324 traces (2.43% of the total) we found more TP addresses than IN IPs.

As shown in Fig. 3, a single TP address may affect several traces. The two most common TP addresses appeared respectively in 7.3 K and 5.2 K traces, essentially because they were located close to the vantage point: on average, a TP address appeared in 7.8 traces.

Finally, mapping each address to the owner AS [3], we found that our dataset covers 1.9 K distinct ASes. By considering two consecutive hops in a traceroute trace mapped to distinct ASes as a potential AS link, we extracted about 2 K AS links. Among these, 527 (26%) involved TP addresses, thus being potential source of inaccuracies according to the scenario depicted in Fig. 1. Although not necessarily all these links were false, such high percentage suggests that TP addresses may represent a significant cause of AS maps distortion. Our approach allows to identify the subset of links to carefully consider when the objective is to draw the AS level map exploiting the information obtained with traceroute.

[1] UDP probes allow to avoid ambiguities caused by the reverse path.
[2] In this case, IPx is probably associated to an incoming interface.

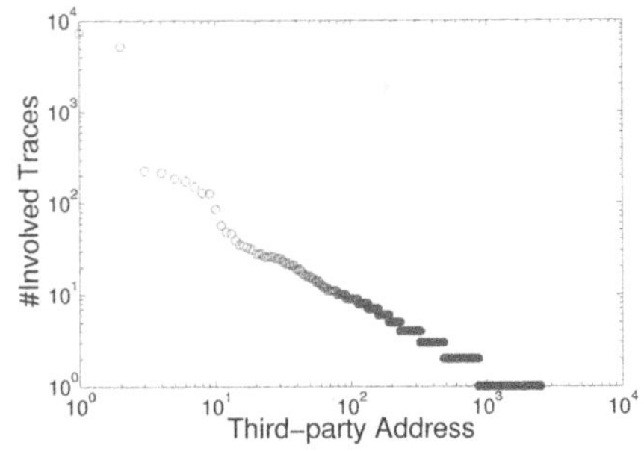

Figure 3: Third-party addresses over traces.

5. ONGOING AND FUTURE WORK

This analysis represents a first snapshot of an ongoing research work. We are extending our study in different ways: (i) adopting different probes to obtain a wider support; (ii) using a larger dataset, to better evaluate the applicability of our technique; (iii) exploiting multiple vantage points, to detect when IPs act as incoming and TP addresses. (iv) deepening the TP addresses impact on AS level topology exploiting ground truth. We plan to implement and release to the research community a modified traceroute version able to automatically discover the IP path and label IPs as TP address.

6. REFERENCES

[1] Scrambled internet trace measurement dataset. *PREDICT ID USC-LANDER / internet-address-hitlist, 2009-12-23 2012-02-14.*

[2] F. Baker. Ietf rfc1812: Requirements for ip version 4 routers. *http://ds.internic.net/rfc/rfc1812.txt.*

[3] T. Cymru. Ip to asn mapping. *http://www.team-cymru.org/Services/ip-to-asn.html*, 2012.

[4] W. de Donato, P. Marchetta, and A. Pescapé. A hands-on look at active probing using the ip prespecified timestamp option. In *PAM'12, Vienna, Austria*, 2012.

[5] B. Donnet and T. Friedman. Internet topology discovery: a survey. *IEEE Communications Surveys and Tutorials*, 9(4), 2007.

[6] Y. Hyun, A. Broido, and K. Claffy. On third-party addresses in traceroute paths. In *PAM'03, San Diego, CA, USA*, 2003.

[7] Y. Hyun, A. Broido, and K. Claffy. Traceroute and bgp as path incongruities. Technical report, CAIDA, University of California, San Diego, 2003.

[8] P. Marchetta, P. Mérindol, B. Donnet, A. Pescapè, and J.-J. Pansiot. Topology discovery at the router level: A new hybrid tool targeting isp networks. *JSAC*, 29(9):1776–1787, 2011.

[9] M. Tozal and K. Sarac. Palmtree: An ip alias resolution algorithm with linear probing complexity. *Computer Communications*, 2010.

[10] M. Tozal and K. Sarac. Tracenet: an internet topology data collector. In *Proc. of IMC'10*. ACM, 2010.

Reviving Delay-based TCP for Data Centers

Changhyun Lee
KAIST
Daejeon, Korea
chlee@an.kaist.ac.kr

Keon Jang
Microsoft Research
Cambridge, UK
keonjang@microsoft.com

Sue Moon
KAIST
Daejeon, Korea
sbmoon@kaist.edu

ABSTRACT

With the rapid growth of data centers, minimizing the queueing delay at network switches has been one of the key challenges. In this work, we analyze the shortcomings of the current TCP algorithm when used in data center networks, and we propose to use latency-based congestion detection and rate-based transfer to achieve ultra-low queueing delay in data centers.

Categories and Subject Descriptors

C.2.2 [**Computer-Communication Networks**]: Network Protocols

General Terms

Performance

Keywords

Data centers, Latency, TCP

1. INTRODUCTION

The recent growth of data centers has brought new challenges to network research. Among the challenges, minimizing network delay within a data center is one of the key concerns for service operators. Previous work [1] has well pointed out that user experience is badly affected in data center applications when even a single flow suffers from a large latency.

In today's data centers, the largest part of network delay comes from the queueing delay at switches since the propagation delay within a data center can be almost negligible; ideally, 100 meters of network cabling between two nodes adds only $0.5\mu s$ of propagation delay, while a single 1,500 byte packet queued at a 10Gbps switch already costs $1.2\mu s$. Recently, much effort has been put to reduce the queueing delay at switches. DCTCP [2] uses ECN marking to slow down flows before the queue becomes full. HULL [3] takes a further step and give up a little bandwidth to have even lower latency than DCTCP. Our work is also motivated to reduce the queueing delay in data centers, but we focus on the end-to-end congestion control not on queue management at switches. The final goal is to keep the queue occupancy low and fully utilize the link bandwidth at the same time.

In this work, we demonstrate why the current TCP cannot keep latency low when operating in data centers even with AQM policies and lay out directions for overcoming the hurdles. First, we present the problems in loss-based congestion detection and window-based congestion control. Second, we propose to use a delay-based congestion detection and rate-based transfer, and show why it could be a feasible solution for data center environment with simulation results. Finally, we present future directions and challenges of our approach.

SIGCOMM'12, August 13–17, 2012, Helsinki, Finland.
ACM 978-1-4503-1419-0/12/08.

2. ROOT SOURCES OF LARGE LATENCY

Loss-based congestion detection

How to detect congestion in network links and notify the sender is an important part of the congestion control mechanism. The most commonly used TCP version in the Internet, TCP Reno, relies on the packet loss event; a sender does not stop until one of the queues in the network path gets full and drops packets. This mechanism is born to keep the queues occupied. Queue management schemes using ECN may alleviate the situation, but they still suffer from the same dilemma; you need to fill up the queue first in order to empty the queue. As queueing delay is the major source of network delay in data centers, full queues are the last thing that data centers would like to have.

Window-based congestion control

TCP Reno controls the rate of a flow by adjusting the window size, and packets in the same window are sent back-to-back. Therefore packets arriving at a switch have a high level of burstiness and easily fill the queue. As minimizing the queue length is the key issue in data center networks, window-based transmission is not a proper scheme to be used.

The burstiness may fade out as the size of a window decreases, but there still is a limitation in window-based transfer. When network congestion is severe, the size of a window can reduce to one in low latency networks. We run a simulation for this scenario and show in Figure 1. We use a dumbbell topology where data nodes are connected through a single 10Gbps switch, and end-to-end propagation delay is $10\mu s$. Here we gradually increase the number of TCP flows from 0 to 50 until 2.1 second. The congestion window size of each flow is fixed to one during the simulation, so each flow repeatedly sends one data packet, waits for an ACK packet, and then sends another data packet; we call it pingpong transfer. This pingpong transfer is the slowest way of sending data in the current TCP, but we see that as soon as the number of flows exceeds 19, the queue is starting to grow and experience increasing delay. In this scenario, there is no way of doing throughput control on these flows other than stopping them for a retransmission timeout after a loss occurs. The wide-area Internet does not have to worry about this case because of large RTT, but it is a practical concern in networks of small propagation delay like data centers.

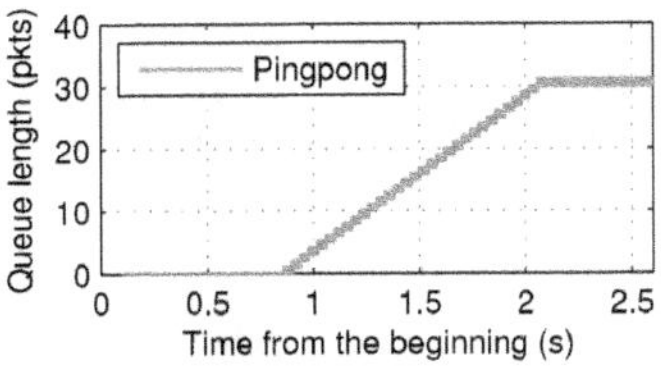

Figure 1: Minimum sending rate simulation of TCP

3. FUTURE DIRECTIONS

Based on the limitations of current TCP in data center networks, we present a preliminary design of our proposed congestion control mechanism. The primary goal is to achieve low queueing delay without adding any features to switches.

Latency-based congestion detection

To overcome the shortcomings of loss-based congestion detection, many TCP variants have been proposed in the literature. One of the solutions designed for the Internet is XCP [4]. It adds a congestion header to a packet which is frequently updated by specialized XCP routers. While the queueing delay reduction of XCP is notable, using it for data centers requires modification of all switches and routers in the networks, which is not truly practical. Recently proposed techniques for data centers such as DCTCP [2] and HULL [3] also require some level of modification at switches.

As a better way of congestion detection in an end-to-end manner, we propose to use a latency-based algorithm from TCP Vegas. Although TCP Vegas is not the most popular protocol in the Internet, it has much potential in data centers. First, network topology and routing in data centers are much more static than in the wide-area Internet. Stability in routing ensures that latency change occurs only from network congestion while delay measurement in the Internet suffers from various type of noises. Second, the portion of propagation delay and transmission delay in total end-to-end delay is much smaller than that of queueing delay. In this environment, one additional packet queueing greatly increases the total network delay unlike the Internet. Provided that accurate delay measurement in microseconds is feasible and affordable, delay is a good source to determine the congestion level in data center networks.

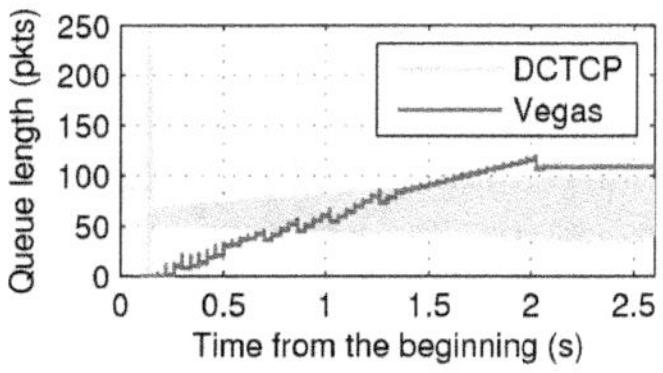

Figure 2: Queue length of TCP Vegas and DCTCP simulation

To explore the potentials of TCP Vegas in data center networks, we run a basic simulation and compare the result with DCTCP; we use the ns-2 code and sample simulation script of DCTCP available online. The same dumbbell topology is used as in the previous section, but the end-to-end propagation delay is changed to $50\mu s$. The maximum queue size at the switch is 250 packets. During the simulation, we increase the number of flows from 0 to 50 until 2.1 second. In Figure 2, we find that latency-based congestion detection achieves comparable queue length reduction with DCTCP and exhibits much less queue length variation. The queue length of TCP Vegas, unlike DCTCP, is proportional to the number of flows in the simulation result, and how to maintain the queue small regardless of the number of flows will be the primary goal of our future work.

Rate-based transfer using inter-packet gap

We have shown that window-based transfer is not an efficient way for low latency networks. To achieve fine-grained throughput control, we propose to use rate-based transfer similar to TFRC protocol [5]. According to the difference between the base RTT and the current RTT, we decide the time gap between two consecutive packets to be sent. When a network path is congested, the time gap is increased to slow down a flow. In this approach, we can micro-manage the rate of a flow by adjusting the time gap a bit by bit. When a network is too congested, a flow can be sent even slower than the pingpong manner by setting the time gap larger than RTT. The micro-grained time gap can be gradually reduced or increased to the point where network bandwidth is fully utilized and queue occupancy remains low.

Challenges

There are several remaining challenges in our new algorithm design and implementation. First, measuring the base RTT for a flow can be easily affected by background flows. The ideal base RTT value will be the sum of propagation delay and transmission delay of a network path, but getting an accurate base RTT in presence of background traffic is not an easy problem. We currently resort to the minimum value of observed RTTs since the beginning of a flow, but we consider creating a high priority channel only for base RTT measurement that is separated from data transfer channel. Second, sending packets with the exact inter-packet gap may not be easy in real world implementation. To fully benefit from rate-based transfer, inter-packet gap in practice can be as low as sub-microseconds, and precisely scheduling packets at operating system level can be hard to be done. Therefore we plan to offload TCP packet scheduling to a specialized hardware that helps accurate timing control. We are currently revising our algorithm design to have better performance and planning to do more simulations and experiments with various network topologies and traffic pattern scenarios.

4. ACKNOWLEDGMENTS

This research was supported by the KCC (Korea Communications Commission), Korea, under the R&D program supervised by the KCA (Korea Communications Agency) (KCA-2011-08913-05002).

5. REFERENCES

[1] C. Wilson, H. Ballani, T. Karagiannis, and A. Rowtron. Better Never than Late: Meeting Deadlines in Datacenter Networks. In *Proceedings of the ACM SIGCOMM conference*, 2011.

[2] M. Alizadeh, A. Greenberg, D. A. Maltz, J. Padhye, P. Patel, B. Prabhakar, S. Sengupta, and M. Sridharan. Data Center TCP (DCTCP). In *Proceedings of the ACM SIGCOMM conference*, 2010.

[3] M. Alizadeh, A. Kabbani, T. Edsall, B. Prabhakar, A. Vahdat, and M. Yasuda. Less Is More: Trading a Little Bandwidth for Ultra-Low Latency in the Data Center. In *Proceedings of USENIX NSDI conference*, 2012.

[4] D. Katabi, M. Handley, and C. Rohrs. Congestion Control for High bandwidth-delay Product Networks. In *Proceedings of the ACM SIGCOMM conference*, 2002.

[5] S. Floyd, M. Handley, J. Padhye, and J. Widmer. TCP Friendly Rate Control (TFRC): Protocol Specification. RFC 5348 (Proposed Standard), Sept. 2008.

FaaS: Filtering IP Spoofing Traffic as a Service

Bingyang Liu[*‡] Jun Bi[*‡] Xiaowei Yang[†]
[*]Network Research Center, Tsinghua University
[‡]Tsinghua National Laboratory for Information Science and Technology
[†]Department of Computer Science, Duke University
liuby@netarchlab.tsinghua.edu.cn junbi@tsinghua.edu.cn xwy@cs.duke.edu

Categories and Subject Descriptors

C.2.0 [**Computer Communication Networks**]: General—*Security and protection*

General Terms

Design

Keywords

IP Spoofing, Ingress Filtering, Economics

1. INTRODUCTION

IP spoofing weakens network security and accountability. Although a lot of techniques have been proposed to prevent IP spoofing, most of them are not implemented by device vendors, and the only available anti-spoofing tool in practice is ingress filtering [3].

However, ISPs do not have incentives to deploy ingress filtering as it only prohibits a deployer from sending spoofing traffic, but it hardly protects the deployer from receiving spoofing traffic. Recent research shows that the deployment of ingress filtering hasn't been improved in four years, and the Internet is still vulnerable to IP spoofing [2].

The reason that ingress filtering lacks deployment incentive can also be explained from an economic perspective, i.e. deploying ingress filtering is not profitable, because it introduces extra operational cost without generating sufficient revenue for the deployers. In this paper, we seek to increase the deployers' economic revenue to incentivize the deployment of ingress filtering.

Our proposal is Filtering as a Service (FaaS), a mechanism that creates a market for the IP spoofing prevention service. As shown in Figure 1, an Autonomous System (AS) in need of IP spoofing prevention (FaaS customer, or *f*-customer) can purchase the service from any AS who has deployed ingress filtering (*f*-provider). An *f*-customer benefits from its *f*-providers, and rewards them with service payments. An *f*-provider deploys ingress filtering, and opens a source of revenue by providing filtering service. Different from traditional ingress filtering, which drops any detected spoofing packet, an *f*-provider only filters the spoofing packets for its *f*-customers (specifically the spoofing packets whose dstIP or srcIP belong to its *f*-customers) to prevent free riders.

A deployer gains more revenue with FaaS than it does on the current Internet. On the current Internet, an AS deploying ingress filtering benefits all the ASes on the Internet, but it can only receive rewards from its customer ASes (e.g. the customers may prefer or even pay more for securer service). With FaaS, the deployer can

SIGCOMM'12, August 13–17, 2012, Helsinki, Finland.
ACM 978-1-4503-1419-0/12/08.

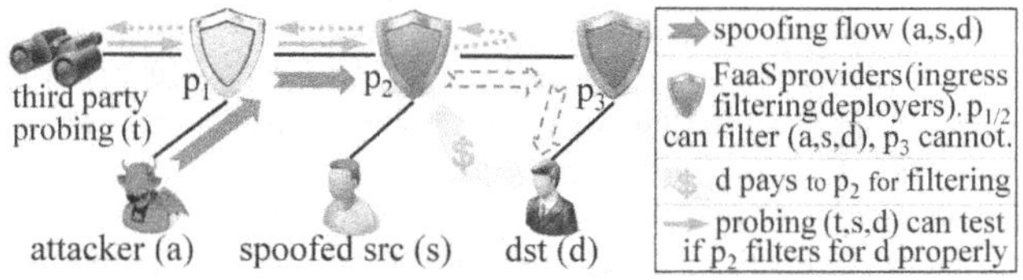

Figure 1: FaaS on a sample AS-level topology. (a, s, d) is a spoofing flow sent by a to d with s as the source address.

receive rewards from any AS who is concerned about IP spoofing and would like to purchase its service. Since the deployers can gain higher revenue, it can be expected that more ASes will be motivated to deploy ingress filtering, and the spoofing traffic on the Internet can be further mitigated.

2. PRELIMINARY DESIGN

In this section we present a preliminary design of FaaS, which consists of three components, an online trading platform, an accounting system and a consulting system.

2.1 Online Trading Platform

Since ingress filtering only works well near network edges, an *f*-customer has to purchase the filtering service from many *f*-providers, most of whom are geographically distant, which makes offline trading too expensive. Thus an online trading platform is needed to lower the trading cost. Different from today's C2C e-commerce platforms like eBay, a trader must prove that he or she is the legal representative of a particular AS who owns the the AS number and the IP address blocks. RPKI can be used as this proof, since it is standardized by IETF and steadily being deployed.

2.2 Accounting System

Without an accounting system, a malicious *f*-provider may cheat its *f*-customers, i.e. charges them but doesn't filter for them. This cheating is feasible because, when a *f*-customer is under a spoofing attack, it cannot tell which *f*-provider should be responsible for the attack. Thus, we design a third party distributed probing system, which periodically generates a modest amount of spoofing probes to explore the malicious *f*-providers. As in the example shown in Figure 1, t tests whether p_2 filters for d properly by emitting spoofing probes (t, s, d), whose TTL expires after passing p_2. If t receives any ICMP TTL Exceeded messages, it knows that (t, s, d) is not filtered by p_2, and p_2 could be a malicious *f*-provider. A concrete design of the probing system is our future work.

2.3 Consulting System

The consulting system helps an *f*-customer to choose an optimal set of *f*-providers to maximize its benefit (protection), and evaluates

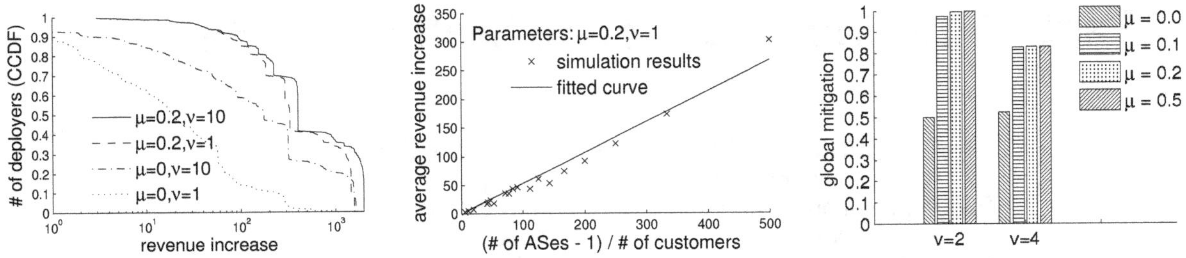

(a) Deployers' revenue increase (b) Average revenue increase (c) Global spoofing mitigation

Figure 2: Simulation results

the revenue increase for an f-provider. This section gives a brief introduction to the consulting system. A complete design can be found here *http://netarchlab.tsinghua.edu.cn/~liuby/pub/faas-tr.pdf*.

Assuming that filtering a spoofing flow (a, s, d) has a benefit of $w(a, s, d)$, filtering the spoofing flows whose dstIP or srcIP is c makes the benefit of c, which c wants to maximize. Intuitively, c can purchase the filtering service from all the f-providers to maximize its benefit. However, as shown in Figure 1, since a spoofing flow can be filtered by multiple f-providers, purchasing the service from all the f-providers implies that c pays for filtering a same flow for more than once, and c wants to keep this redundant payment rate under a threshold μ. Thus c's goal is to maximize its benefit under the redundant payment rate constraint. The consulting system solves this maximization problem by finding an optimal set of f-providers for c.

An f-provider p's f-customers are simply the ASes who choose p as their f-provider. Given the deterministic algorithm of ingress filtering, the benefit that p offers to its f-customers can also be computed, denoted by B_p. Similarly, we can compute the benefit that p offers to its customers with the traditional deployment on the Internet, denoted by B'_p. Assuming that the reward an AS pays to p is a constant fraction of the benefit that it gains from p, we can finaly quantify that FaaS increases p's revenue by B_p/B'_p times.

3. SIMULATION

In this section, we investigate the effectiveness of FaaS in increasing the deployers' revenue and mitigating global spoofing traffic by simulation. In the simulation, we assume that an AS will deploy ingress filtering if its revenue is increased by ν times; otherwise it won't deploy. We also assume that every AS concerns IP spoofing. The simulation is based on an AS-level topology, as same as the one used in [4]. We assign the spoofing flows with uniform and variant[1] filtering benefit respectively. The results show high similarity between the two assignments, and we only show the results of the uniform assignment due to the lack of space.

The deployers' revenue increase is shown in Figure 2a. When μ (the constraint on f-customers' redundant payment rate) is 0, the f-providers' revenue is lower because the f-customers are very conservative in choosing f-providers to avoid any extra payment. When $\mu = 0.2$, the revenue increase is higher, and *about 90% of deployers' revenue is increased by more than 100 times*. We believe that 0.2 redundant payment is affordable for the f-customers, because the Return On Investment is strong enough[2]. The figure also shows that the higher ν (deployers' expectation of the revenue increase) is, the higher the deployers' revenue is, because there are fewer f-providers and each f-provider can have more f-customers.

Recall that FaaS increases a deployer's revenue by allowing it to get rewards from more ASes. If an AS p, who has m customers, can attract all the other ASes to purchase its service, we can roughly estimate that its revenue can be increased by $\frac{n-1}{m}$ times. This speculation is verified in Figure 2b, which shows that the average revenue increase across the ASes with the same number of customers has a linear correlation with $\frac{n-1}{m}$ (correlation coefficient is 0.9898). The fitted arverage revenue increase is $ARI = 0.5347\frac{n-1}{m}$, which can be used as a simplified estimation of revenue increase in a new topology. The slope is less than 1 becuase a deployer transits more traffic and offers more benefit for its customers than other ASes, and thus it can receive more payments from its customers. The reason that we set $\mu = 0.2$ has been explained in the last paragraph, and we set $\nu = 1$ for a cross comparison with the current Internet. Most f-providers can have $n - 1$ f-customers under this setting.

Finally we evaluate how well FaaS can mitigate the global spoofing traffic. As shown in Figure 2c, when $\mu = 0.2$, *99.55% of the global spoofing traffic can be filtered while each deployer's revenue at least doubled* ($\nu = 2$), and the mitigation is 82.86% when each deployer's revenue is increased by at least 4 times ($\nu = 4$). The marginal improvement in mitigation is trivial when $\mu > 0.2$, which indicates that *0.2 redundant payment is enough for the* f*-customers to choose sufficient* f*-providers to filter most spoofing traffic.*

4. CONCLUSION

From the deployment progress of BGPSec, DNSSec and new anti-spoofing methods, we learn the lesson that deploying new techniques on the Internet infrastructure can be very challenging. In this paper, we make use of the existing technique, ingress filtering, to mitigate IP spoofing. We solve the deployment incentive problem of ingress filtering from a new, economic perspective. The simulation results show that our proposal can significantly increase the ISPs' deployment incentive and mitigate the global spoofing traffic.

5. ACKNOWLEDGEMENTS

This work is supported by National Science Foundation of China under Grant 61073172, and Program for New Century Excellent Talents in University. The authors also thank Qiang Cao, Yang Chen, Yu Chen and Ang Li for their help and comments.

6. REFERENCES

[1] The business value of ddos protection. Arbor Networks, 2010.

[2] R. Beverly, A. Berger, Y. Hyun, and k claffy. Understanding the efficacy of deployed internet source address validation filtering. In *IMC*, 2009.

[3] T. Ehrenkranz and J. Li. On the state of ip spoofing defense. *ACM Transactions on Internet Technology (TOIT)*, 9(2):6:1–6:29, 2009.

[4] X. Liu, A. Li, X. Yang, and D. Wetherall. Passport: Secure and adoptable source authentication. In *NSDI*, 2008.

[1] $w(a, s, d) = w(a)w(s)w(d)$ where $w(v)$ is v's IP address span.

[2] The ROI in DDoS protection is 281% [1]. 0.2 extra payment results in the new ROI of 218%, which is still at a high level.

Deadline-Aware Datacenter TCP (D^{2}TCP)

Balajee Vamanan
Purdue University
bvamanan@ecn.purdue.edu

Jahangir Hasan
Google Inc.
jahangir@google.com

T. N. Vijaykumar
Purdue University
vijay@ecn.purdue.edu

ABSTRACT

An important class of datacenter applications, called Online Data-Intensive (OLDI) applications, includes Web search, online retail, and advertisement. To achieve good user experience, OLDI applications operate under soft-real-time constraints (e.g., 300 ms latency) which imply deadlines for network communication within the applications. Further, OLDI applications typically employ tree-based algorithms which, in the common case, result in bursts of children-to-parent traffic with tight deadlines. Recent work on datacenter network protocols is either deadline-agnostic (DCTCP) or is deadline-aware (D^3) but suffers under bursts due to race conditions. Further, D^3 has the practical drawbacks of requiring changes to the switch hardware and not being able to coexist with legacy TCP.

We propose Deadline-Aware Datacenter TCP (D^{2}TCP), a novel transport protocol, which handles bursts, is deadline-aware, and is readily deployable. In designing D^{2}TCP, we make two contributions: (1) D^{2}TCP uses a distributed and reactive approach for bandwidth allocation which fundamentally enables D^{2}TCP's properties. (2) D^{2}TCP employs a novel congestion avoidance algorithm, which uses ECN feedback and deadlines to modulate the congestion window via a gamma-correction function. Using a small-scale implementation and at-scale simulations, we show that D^{2}TCP reduces the fraction of missed deadlines compared to DCTCP and D^3 by 75% and 50%, respectively.

Categories and Subject Descriptors

C.2.2 [**Computer Communication Networks**]: Network Protocols.

General Terms

Algorithms, Design, Performance.

Keywords

Datacenter, Deadline, SLA, TCP, OLDI, ECN, Cloud Services.

1. INTRODUCTION

Datacenters are emerging as critical computing platforms for ever-growing, high-revenue, online services such as Web search, online retail, and advertisement. These services employ Online Data Intensive applications (OLDI) [18] which have two distinguishing properties: (1) Because application latency affects user experience, and hence revenue [13], OLDI applications operate under soft-real-time constraints (e.g., 300 ms latency). (2) The applications employ tree-based, divide-and-conquer algorithms where every query operates on data spanning thousands of servers [12].

An OLDI query's overall time budget gets divided among the nodes of the algorithm's tree (e.g., 40 ms for a parent-to-leaf RPC). To avoid missing its deadline, a parent node sends out an incomplete response without waiting for slow children that have missed their deadlines. Because such incomplete responses adversely affect response quality, and hence revenue, achieving fewer missed deadlines is important. While a node's response time includes both computational and network latencies, we focus on reducing the network delay. We note that in addition to achieving fewer missed deadlines, a network protocol that allows tighter network budgets is invaluable as it allows more time for computation, thus producing higher-quality responses.

A key reason for increased network delay is that all the children of a parent node face the same deadline and are likely to respond nearly at the same time, causing a fan-in burst at the parent in the common case [10][1][25]. Further, because typical data centers host multiple applications at the same time to enable flexible use and high utilization of the datacenter resources, OLDI flows with different deadlines and background flows with no deadlines (e.g., Web index update) share the network. As such, multiple such bursts coinciding in time at a switch may lead to congestive packet drops and TCP retransmits, which frequently result in missed deadlines. We emphasize that the fan-in bursts are fundamental to OLDI applications and are not artifacts of the network. The shallow- and shared-buffer nature of datacenter switches, combined with the buffer-filling nature of long-lived TCP flows, precludes absorbing these bursts in packet buffers [1]. Current datacenters alleviate this problem by a combination of two approaches: (1) over-provision the network link bandwidths to absorb the bursts, and (2) increase the network's time budget (e.g., equal to, say, the 99th percentile of the observed network delay) leaving less time for computation. While the former incurs high cost, the latter either degrades response quality (e.g., less time for Page Rank in Web search), or requires more machines to compensate for less computation per machine and may worsen fan-in bursts by increasing the fan-in degree (i.e., more children per parent).

Recent work on datacenter networks either reduces the tail-end network latency or proposes deadline-aware networks. DCTCP [1] is an elegant proposal that targets the tail-end latency by gracefully throttling flows in proportion to the extent of congestion, thereby reducing queuing delays and congestive packet drops and, hence, also retransmits. DCTCP reduces the 99th-percentile of the network latency by 29%. Unfortunately, DCTCP is a deadline-agnostic protocol that equally throttles all flows, irrespective of whether their deadlines are near or far. D^3 shows that as much as 7% of flows may miss their deadlines with DCTCP [25]. D^3 tackles missed deadlines by pioneering the idea of incorporating deadline awareness into the network. In D^3, the

SIGCOMM'12, August 13–17, 2012, Helsinki, Finland.

switches grant a sender's request for the required bandwidth based on the flow size and deadline.

While D^3 improves upon DCTCP, D^3 has significant performance and practical shortcomings, which we cover in detail in Sections 2.4.2 and 2.4.3. On the performance side, D^3 employs a centralized and pro-active approach in which the switches allocate bandwidth greedily on a first-come-first-served basis. Such a greedy approach may allocate bandwidth to far-deadline requests arriving slightly ahead of near-deadline requests. We show in Section 4.2.2 that, due to this race condition, D^3 inverts the priority of 24%-33% requests, thus contributing to missed deadlines. Fixing these priority inversions is a hard problem as we explain in Section 2.4.2. On the practical side, D^3 has some major drawbacks as well. D^3 requires custom switching chips to handle requests at line rates. Such custom hardware would incur high cost and long turn-around time that hinders near-term deployment. Further, because legacy TCP flows do not recognize D^3's honor-based bandwidth allocation, D^3 cannot coexist with TCP. Such lack of protocol interoperability prevents incremental deployment necessary for introducing new technologies and for high availability during upgrades. As such, any datacenter protocol must be able to coexist with legacy TCP to be deployable in the real world.

In summary, DCTCP improves tail latency but is not deadline-aware whereas D^3 is deadline-aware but it has both performance and practical shortcomings. We stipulate that a datacenter network protocol should:

- meet OLDI deadlines, especially in fan-in-burst-induced congestion;
- achieve high bandwidth for background flows;
- work with existing switch hardware; and
- be able to coexist with legacy TCP.

We propose Deadline-Aware Datacenter TCP (D^2TCP), a novel transport protocol based on TCP that meets the above requirements. In designing D^2TCP, we make two contributions.

Our first contribution is D^2TCP's distributed and reactive approach. Because having global up-to-date information for all flows in a datacenter is technically infeasible given the rapid arrival rate and latency of flows, any network scheduling scheme must work with incomplete information. D^2TCP approaches this challenge by inheriting TCP's distributed and reactive nature, and adding deadline awareness to it. In contrast to D^3's centralized bandwidth allocation at the switches, which rules out per-flow state, D^2TCP's distributed approach allows the hosts to maintain the needed state without changing the switch hardware. In contrast to D^3's pro-active approach, which does not allow for correcting the decisions resulting from inaccurate information, D^2TCP's reactive approach allows senders to correct any temporary and small over-subscription of the network which can be absorbed by typical packet buffers.

Our second contribution is D^2TCP's novel congestion avoidance algorithm which uses ECN feedback and deadline information to modulate the congestion window size via a gamma-correction function [2]. The key idea behind the algorithm is that far-deadline flows back off aggressively and near-deadline flows back off only a little or not at all. Our algorithm simultaneously satisfies the four conditions stipulated above so that D^2TCP (1) achieves deadline-based prioritization in the presence of fan-in-burst-induced congestion so that flows with nearer deadlines are prioritized, while ensuring that congestion does not worsen; (2) achieves high bandwidth for background flows even as the short-lived D^2TCP flows come and go; (3) requires no changes to the switch hardware, so that deployment amounts to merely upgrading the TCP and RPC stacks; and (4) coexists with legacy TCP, allowing incremental deployment.

We demonstrate a real implementation of D^2TCP on a small 16-server testbed, and show that even at such small scale where fan-in-burst-induced congestion is much less severe than at real scale, D^2TCP reduces the fraction of missed deadlines by 20% compared to DCTCP, while requiring fewer than 100 additional lines of kernel code.

We perform further evaluations using at-scale simulations to show that D^2TCP

- reduces the fraction of missed deadlines compared to DCTCP and D^3 by 75% and 50%, respectively;
- achieves nearly as high bandwidth as TCP for background flows without degrading OLDI performance;
- meets deadlines that are 35-55% tighter than those achieved by D^3 for a reasonable 5% of missed deadlines, giving OLDIs more time for actual computation; and
- coexists with TCP flows without degrading their performance.

Note that we do not present formal proofs for D^2TCP's fairness and stability.

The remainder of the paper is organized as follows. In Section 2, we discuss the nature of OLDI applications and pinpoint the issues with the previous proposals. We describe the details of D^2TCP in Section 3. In Section 4, we describe our experimental methodology and present our experimental results. We discuss some related work in Section 5, and conclude in Section 6.

2. OLDIs AND DATACENTER NETWORK PROTOCOLS

We describe the nature of today's datacenter applications, and how this nature interacts with the previous proposals for datacenter network protocols.

2.1 Online Data Intensive Applications

As the name suggests, there are two defining properties of Online Data Intensive (OLDI) applications: *online* and *data-intensive* [18]. *Online* implies an interactive nature, wherein a user typically inputs a query via a browser and expects an immediate response. Consequently, OLDI applications are designed to respond within a short deadline (e.g., 300 milliseconds). *Data-intensive* means that the applications consult large data sets (e.g., entire index of the web) for computing the response. The large volume of data is typically distributed over thousands of servers, and each query hits every server.

The two properties combined lead to tree-based, divide-and-conquer algorithms for OLDI applications [12], as shown in Figure 1. The specific example we show is a two-level tree, however, the properties we describe hold for shallower and deeper trees as well. The user query arrives at the root, which broadcasts the query down to the leaves across which the data is partitioned. Each leaf sends its response to its parent which aggregates the results from all the leaves and sends a response to the root. The root, in turn, aggregates all the parents' results and ships off the final response to the user. This overall application architecture is called *scatter-gather* or *partition-aggregate*. The propagation of the request down to leaves and of the responses back up to the root must complete within the deadline. That overall budget gets divided among the levels of the tree. For example, a leaf may have to respond to its parent within 30 milliseconds (Figure 1). When this deadline expires, the parent aggregates the available results and ships off the final response. Any leaf that misses its deadline

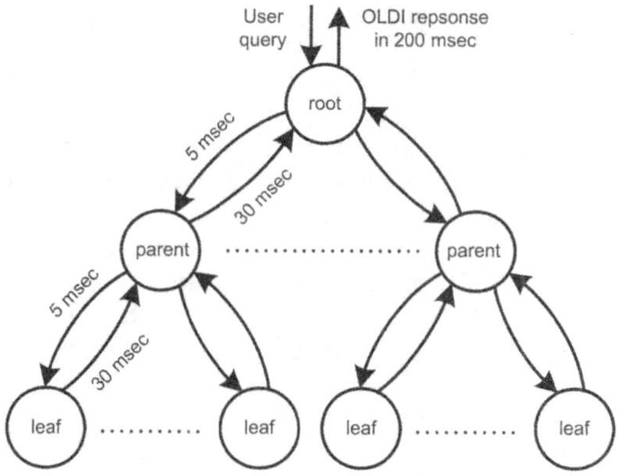

Figure 1: OLDI architecture

fails to contribute to the final response. Such missed deadlines result in incomplete, lower-quality responses.

It is instructive to note that each leaf's budget gets divided into two parts: the computation time on the leaf, and the communication latency between the leaf and its parent. This division balances two competing demands. A generous budget for communication ensures fewer missed deadlines by the network, but also means less time for computation, thereby penalizing the quality of results (e.g., less time for Page Rank in Web search). The upshot is that not only is it desirable to have fewer missed deadlines, but there is also great value in tightening the communication budget to give more time for computation.

2.2 OLDI Fan-in Congestion

Consider a parent's subtree when the parent sends a query to its leaves. Because all leaves receive the query at nearly the same time, and because they all face the same deadline, the leaves are likely respond around the same time. The result is a burst of responses fanning in to the parent in the common case [10] [1]. While such bursts can be smoothed by inserting jitter, doing so increases the tail-end latency [9], and is therefore of limited use. Further, because datacenters run multiple applications at the same time to improve utilization, multiple fan-in bursts (from same or different applications) may coincide in time at the same switch.

In addition to the bursty fan-in traffic described above, datacenter networks also carry background traffic consisting of long-lived flows. These flows push new control information and data to the nodes of the OLDI applications. While this traffic is usually not constrained by tight deadlines, it does involve large data transfers. Given the nature of TCP, these long flows tend to exercise the switch buffers to high utilization [1].

The combined network traffic described above often results in fan-in-burst-induced congestions which cause tail drops as the packet buffers fill up. Absorbing these bursts in larger packet buffers is precluded by two factors. First, datacenter switches employ network ASICs with on-chip packet buffer memory. Given the limitations of die size, on-chip packet buffers are naturally shallow. Switch designs with larger off-chip packet buffers are significantly more expensive and complex [14], and reserved for high-end core routers that must buffer for Internet scale RTTs. Second, the nature of long-lived TCP flows is to fill up larger buffers, which may lead to longer queuing delays and still cause missed deadlines for OLDI traffic [1].

While fan-in-burst-induced congestions are a fundamental characteristic of OLDI applications, the manner in which today's datacenter networks handle such fan-in bursts contribute to missed deadlines in two ways.

First, a TCP/IP network uses packet drops as a feedback to inform the senders about on-going congestion[1]. Under this mechanism, the sender must wait for a timeout to detect packet loss even as the deadline expires. Furthermore, the sender also halves its transmission rate to alleviate the congestion. The net result is that the leaves involved in a fan-in-burst-induced congestion are likely to miss their deadlines. Current datacenters address this problem by a combination of two approaches: (1) increasing the network link bandwidths and (2) increasing the network time budget to be greater than, say, the 99th percentile network latency. While the former incurs high cost, the latter reduces the time budget for computation, which either penalizes the response quality or requires more machines to compensate for less computation per machine and may worsen fan-in bursts by increasing the fan-in degree (i.e., more children per parent).

Second, TCP treats all congested flows equally which is sub-optimal when the congested flows' deadlines are different. Ideally, a network should prioritize the flows that are about to miss their deadlines while throttling the flows that can afford to wait. However, TCP is a *fair-share*[2] protocol which lacks such deadline-based distinction. Because fan-in bursts are common in OLDIs and fair-share protocols are deadline-agnostic, such protocols are not well-suited to datacenters [25].

2.3 Datacenter TCP (DCTCP)

Explicit Congestion Notification (ECN) [21] is an extension to the TCP/IP protocol that enables congestion feedback without using packet drops as the feedback mechanism. ECN relies on Active Queue Management (AQM) schemes like RED [8] to track congestion at a switch. When a switch encounters sustained congestion, it marks the Congestion Encountered (CE) bit in the IP header, thereby informing the endpoints about the congestion. The endpoints observe this CE bit feedback and reduce their transmission rate.

DCTCP [1] shows that ECN does not suffice to solve OLDI's fan-in burst problem. In datacenters, the number of congested flows is small enough that their congestion windows tend to be synchronized with each other. Furthermore, the traffic is bursty in nature. Therefore, halving the flows' windows in response to the ECN feedback causes the flows to thrash instead of gracefully converging to the available bandwidth. In summary, one bit of ECN feedback merely indicates the *presence* of congestion, but it carries no information about the *extent* of congestion.

To address this problem, DCTCP elegantly aggregates the one-bit ECN feedback from multiple packets and multiple RTTs to form a multiple-bit, weighted-average metric for sizing the window. Using this metric, the senders modulate their window sizes in a graceful manner, without thrashing. As a result, DCTCP reduces the 99th percentile of network latency in OLDIs by 29%. Thus, DCTCP frees up more time for computation in OLDIs.

2.4 Deadline Driven Delivery (D^3) Protocol

While DCTCP addresses the first issue mentioned in Section 2.2 (packet drops causing time-outs), DCTCP, being a fair-share protocol, does not address the second issue (lack of deadline awareness). As such, a later work called Deadline Driven Delivery (D^3) shows that DCTCP may cause up to 7% of the deadlines to

[1] The use of RED [8] without ECN merely triggers this feedback earlier than the onset of full congestion, but the feedback mechanism remains the same -- a packet drop.

[2] Strictly speaking TCP has an RTT-bias but we use the term fair-share as TCP *tries* to treat flows equally and the bias is an undesired side-effect.

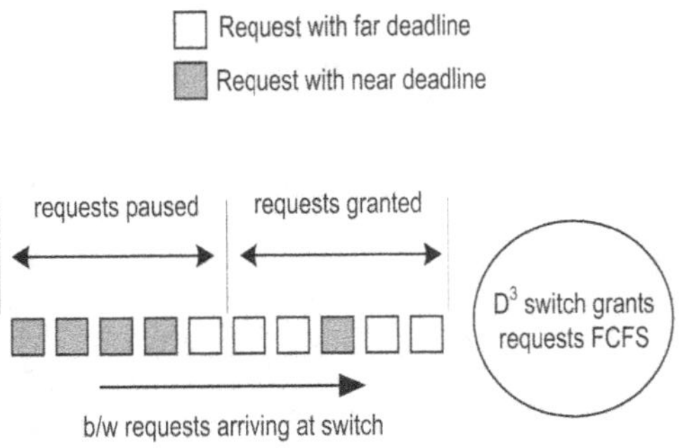

Figure 2: D3's priority inversion

be missed [25]. Citing TCP's problems, D^3's authors argue that TCP is fundamentally ill-suited for OLDIs and opt for an altogether new protocol.

D^3 pioneered the idea of incorporating deadline awareness into the network. To that end, D^3 proactively allocates bandwidth based on the idea of bandwidth reservation before data transmission. As applications know the deadline for a message, they pass this information to the transport layer in the request to send. Based on their deadlines and the amount of remaining data, senders must request bandwidth every round trip time (RTT) and send only the corresponding amount of data. Switches receive these requests, compute a decision, and place the decisions into the packet header. Thus, the senders learn how much data to transmit in the next RTT. Because requests are made in a deadline-aware manner, D^3 dramatically reduces the fraction of missed deadlines compared to DCTCP's [25].

2.4.1 D^3 operation

For every RTT, a sender computes the needed bandwidth as the amount of remaining data divided by the time until the deadline. The sender places the request into the D^3 packet header. The very first request in a flow uses a *SYN*-like packet carrying the request and no data payload. In subsequent RTTs, the bandwidth request for the next RTT is piggybacked on the data packets of the current RTT.

The switches receive these packets and extract the bandwidth request. The switches also maintain state for the bandwidth already allocated (and hence the remaining bandwidth) along the path in question. Based on this state, each switch makes a greedy choice and tries to grant as much of the request as possible. The switch places its grant-response in the packet header and forwards the packet. Each switch along a packet's path performs the same action, creating a vector of bandwidth grants in the header. The receiver copies this vector into the ACK packet back to the sender. The sender extracts the vector and chooses the minimum of all the grants to decide how much data to transmit for the next RTT.

2.4.2 Challenges in centralized and proactive scheduling

D^3 employs a centralized and pro-active approach to scheduling the network. To maximize network utilization, the switches allocate bandwidth greedily on a first-come-first-served basis. Because fan-in-burst-induced congestion is common in OLDIs, near- and far-deadline traffic often competes for bandwidth. Unfortunately, D^3's greedy approach may allocate bandwidth to far-deadline requests arriving slightly ahead of near-deadline requests (see Figure 2). Due to this race condition, D^3 causes frequent priority inversions, which contribute to missed deadlines. Indeed, our results in Section 4.2.2 show that D^3 inverts the priority of 24%-33% requests.

Fixing these priority inversions within a centralized, proactive design space is hard for multiple reasons. First, given the rapid arrival rate and tight deadlines of flows, it is not feasible for a switch to wait and gather information on all flows before making decisions. Instead the switch must make decisions on-the-fly without knowledge of near-future requests. Second, because maintaining detailed per-flow state at the switches is prohibitively complex, D^3 has no memory of which flows recently encountered priority inversions. If requests continue to arrive in an unfavorable order, the same priority inversions may happen again in subsequent RTTs. Third, maintaining some spare bandwidth in anticipation of future bursts is also hampered by the lack of detailed per flow state. Allocation of such spare bandwidth requires a priority list, so the switch can decide whether a given request should be granted out of the spare pool or be denied. Such a priority list again requires tracking detailed per-flow state. Further, a large spare may underutilize the network whereas a small spare may be insufficient to absorb bursts. A scheme that dynamically adapts the amount of the spare bandwidth, must balance the conflicting needs of maximizing utilization and accommodating near-deadline bursts, which is an open and complex research problem.

The above analysis assumes that D^3's bandwidth requests contain deadline information, even though as proposed in [25], D^3's bandwidth requests do not contain deadline information.

2.4.3 Challenges in practical deployment

To handle requests at line rates, D^3 requires custom switch ASICs. Unfortunately, such OLDI-specific, custom silicon would not only incur high cost due to low volumes, but also incur long turn-around time hindering near-term deployment.

Further, D^3, as proposed, cannot coexist with TCP. TCP flows passing through the same switches as D^3 flows would not recognize D^3's honor-based bandwidth allocation. Placing TCP and D^3 flows in separate QoS classes would provide isolation but would also require partitioning the bandwidth among the QoS classes. Such partitioning has far-reaching consequences and raises the hard issue of optimizing multiple applications' bandwidth shares, fairness, and network utilization. Tunneling TCP traffic inside D^3 is problematic because it results in two nested flow control loops with conflicting behavior. For example, TCP may increment its window size, while D^3 is decrementing it, and the surplus segments would incur TCP time-outs, back-off and retransmits. Protocol interoperability requirements in datacenters, though less stringent than those in the Internet, cannot be ignored completely as suggested by D^3. Due to the lack of interoperability, the upgrade of switches and applications to D^3 would all have to occur in an atomic manner. For such large, invasive changes to be atomic, the datacenter would have to be unavailable for a long enough time that the changes are all but implausible. Indeed, datacenter infrastructure upgrades are almost always incremental for this reason, so that old and new technologies coexist for some time. Finally, it may not be reasonable to expect *all* application writers to abandon TCP and develop D^3 versions just because *some* applications desire the use of D^3. As such, any datacenter protocol must be able to coexist with legacy TCP to be deployable in the real world.

3. Deadline-Aware Datacenter TCP (D^2TCP)

We now describe the design of Deadline-Aware Datacenter TCP (D^2TCP), a novel protocol for datacenter networks. We set out with the explicit goals of not requiring custom hardware and supporting coexistence with legacy TCP. The basic idea behind D^2TCP is to modulate the congestion window size based on both deadline information and the extent of congestion. In designing

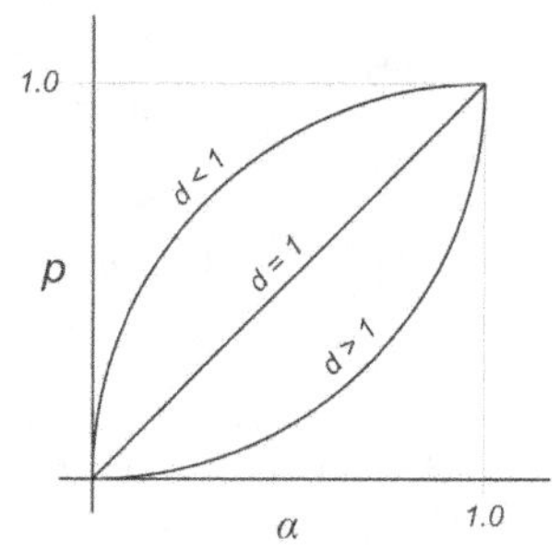

Figure 3: Gamma-correction function for congestion avoidance ($p = \alpha^d$)

D[2]TCP, we make two contributions: D[2]TCP's distributed and reactive approach for allocating bandwidth and D[2]TCP's deadline-aware congestion avoidance algorithm.

3.1 Distributed and Reactive Allocation

Meeting deadlines in fan-in-burst-induced congestion requires knowledge of the flows' deadlines and the extent of congestion. However, having complete and up-to-date global information for all flows in a datacenter is technically infeasible given the rapid arrival rate and latency of flows. Therefore, any network scheduling scheme must make decisions with incomplete information, and the challenge is to choose a compromise that is well suited to OLDIs in datacenters.

D[3]'s centralized and proactive approach compromises the handling of future bursts in order to maximize utilization. Because OLDI traffic is bursty, D[3] suffers from frequent priority inversions, making D[3]'s compromise not well suited to datacenters. In contrast, D[2]TCP inherits TCP's distributed and reactive nature, and adds deadline awareness to it. While D[2]TCP, like D[3], also makes decisions without complete and accurate information, D[2]TCP's compromises are better suited for datacenters.

D[2]TCP modulates the window size based on the deadlines and the extent of congestion. Each D[2]TCP sender sizes its window without knowing how many other flows are congested and by how much other flows will back off. The risk here is that multiple congested senders with tight deadlines may refuse to back off and over-subscribe to bandwidth at the congestion point. Fortunately, due to their stateful nature, the senders can correct this over-subscription by reacting to future congestion feedback in a careful and calculated manner, ensuring that the oversubscription is only temporary and small. Networks are equipped to deal with temporary oversubscriptions by virtue of packet buffers. We note that while D[2]TCP inherits it's distributed and reactive nature from TCP, our contributions are in adding deadline awareness without abandoning TCP's time-tested distributed and reactive approach; and in identifying and analyzing the fundamental difference between D[3] and D[2]TCP, and explaining why those differences matter in the context of datacenter network protocols.

To summarize, in contrast to D[3]'s centralized bandwidth allocation at the switches, which rules out per-flow state, D[2]TCP's distributed approach allows the hosts to maintain the needed state without changing the switch hardware. In contrast to D[3]'s pro-active approach, which does not allow for correcting the decisions resulting from inaccurate information, D[2]TCP's reactive approach allows senders to correct any temporary and small over-subscription of the network. Thus, the compromises made by a distributed, reactive scheme are better suited for datacenters.

3.2 Deadline-aware Congestion Avoidance

Our second contribution is D[2]TCP's novel congestion avoidance algorithm. Like D[3], we assume that applications pass the deadline information to the transport layer in the request to send data. This information then enables D[2]TCP to modulate the congestion window size in a deadline-aware manner. When congestion occurs, far-deadline flows back off aggressively, while near-deadline flows back off only a little or not at all. With such deadline-aware congestion management, not only can the number of missed deadlines be reduced, but also tighter deadlines can be met because the network adapts to the time budget. D[2]TCP requires no changes to the switch hardware, and only requires that the switches support ECN, which is true of today's datacenter switches. Therefore, D[2]TCP deployment amounts to merely upgrading the TCP and RPC stacks.

3.2.1 *Congestion avoidance algorithm*

The easiest way to explain the D[2]TCP congestion avoidance algorithm is to start with DCTCP and build deadline awareness on top of it. We expect that the switches are ECN-capable and are configured to mark CE bits when the packet buffer occupancy exceeds a certain threshold. Like DCTCP, we maintain α, a weighted average that quantitatively measures the extent of congestion:

$$\alpha = (1 - g) \times \alpha + g \times f$$

Here f is the fraction of packets that were marked with CE bits in the most recent window, and g is the weight given to new samples. We now define d as the deadline imminence factor, and explain its derivation later in Section 3.2.3. For now it suffices to know that a larger d implies a closer deadline. Based on α and d we compute p, the penalty function applied to the window size, as follows:

$$p = \alpha^d$$

This function was originally proposed for color correction in graphics [2], and was dubbed gamma-correction because the original paper uses γ as the exponent. Note that being a fraction, $\alpha \leq 1$ and therefore, $p \leq 1$. After determining p, we resize the congestion window W as follows:

$$\begin{aligned} W &= W \times \left(1 - \frac{p}{2}\right), & if\, p > 0 \\ &= W + 1, & if\, p = 0 \end{aligned}$$

In the case where α is zero (i.e., no CE-marked packets, indicating absence of congestion) and therefore p is zero, the window size is grown by one segment similar to TCP. And when all packets are CE-marked, $\alpha = 1$ and therefore $p = 1$, then the window size gets halved similar to TCP. For α between 0 and 1 the window size is modulated by p.

Figure 3 plots p, with a number of different curves for various values of d. The straight line in the middle shows $d = 1$, and hence $p = \alpha$. The curves below the straight line are for $d > 1$ (i.e., near-deadline flows incur lower penalty), and the curves above are for $d < 1$ (i.e., far-deadline flows incur higher penalty). Note that when $p = \alpha$ the behavior matches DCTCP. Essentially, DCTCP is a special case of D[2]TCP where $d = 1$ for all flows all the time, so they back off equally upon congestion irrespective of their deadlines. Accordingly, in D[2]TCP we use $d = 1$ for long flows that do not specify deadlines (i.e., D[2]TCP behaves like DCTCP).

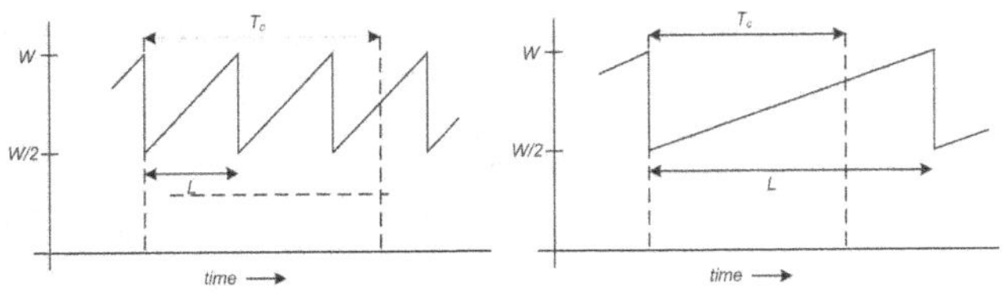

Figure 4: Sawtooth waves for deadline-agnostic behavior (a) $T_c > L$ and (b) $T_c < L$

3.2.2 *Impact of d on congestion behavior*

Looking at Figure 3 we see that when $d > 1$ (i.e., near-deadline flows), p increases only slowly in response to increases in α until α gets close to 1, at which point p rapidly converges to 1 as well. In other words, minor and mild congestions do not penalize near-deadline flows by much. However, severe congestions cause a full backoff similar to TCP and DCTCP, That is, when $\alpha = 1, p = 1$ the window size gets halved just like regular TCP.

When $d > 1$ (i.e., far-deadline flows), p increases rapidly even with small increases in α, and approaches 1 as α catches up to 1. Thus, even minor congestions cause rapid reduction in far-deadline flows' window sizes, but severe congestions do not penalize the flows any more than regular TCP or DCTCP would.

The combination of $d < 1$ and $d > 1$ behaviors complement each other under congestion situations. Far-deadline flows relinquish bandwidth so that near-deadline flows can have greater short-term share in order to meet their deadlines. Furthermore, if congestion continues to worsen after far-deadline flows have backed off, then two possible scenarios are at play: (1) there are many near-deadline flows not reducing their share and (2) there may be regular TCP flows consuming bandwidth in a deadline-agnostic manner. In both scenarios, the $d > 1$ condition ensures that as α grows even near-deadline flows will throttle themselves, allowing other near-deadline and regular TCP flows to make progress. In fact, if the near-deadline flows have different deadlines and hence different d, then the flows' back-off behavior will diverge as the congestion worsens. As a result, only the flows with the most imminent deadlines will win.

In summary, the gamma-correction function provides iterative feedback to near-deadline flows so that they do not drive the network to congestive collapse.

3.2.3 *Determining d based on deadlines*

We now explain how we determine d based on a given deadline value. The value of d should be such that the resulting congestion behavior allows the flow to safely complete within its deadline. We use deadline-agnostic congestion behavior ($p = \alpha$ and $W = \frac{W}{2}$ upon full congestion) as a starting point. Say T_c is the time needed for a flow to complete transmitting all its data under deadline-agnostic behavior; and D is the time remaining until its deadline expires. Now if the flow can just meet its deadline under deadline-agnostic congestion behavior (i.e., $T_c \cong D$) then $d = 1$ is appropriate. It also follows that if $T_c > D$ then we should set $d > 1$ to indicate a tight deadline, and vice versa. Therefore, we compute d as:

$$d = \frac{T_c}{D}$$

Because d appears in the exponent of the gamma-correction function, extreme values for d can result in p behaving like a binary value for the useful mid-range of α. In other words, the mere *presence* of congestion would determine window sizing behavior, instead of the *extent* of congestion. Essentially, we want the gamma-correction function to yield gentle curves about the straight line of $d = 1$. Therefore, we propose capping d to be within a desired range (Section 4.1.1). We explore the effects of varying this cap in Section 4.2.4.

We now compute T_c using Figure 4 which shows the sawtooth wave for deadline-agnostic congestion behavior. We pessimistically assume that the flow encounters full-on congestion (i.e., $\alpha = 1$ and $p = 1$); if there is less congestion then the flow will complete sooner, making T_c an upper bound for time to completion. We assume that a flow's current window size is W and it has B bytes remaining to transmit. Further, we make the following simplifying assumptions that congestion occurs in a repeating pattern: (1) Full-on congestion occurs when the window size is W; (2) consequently, the window size is halved in the next RTT eliminating the congestion; and (3) in response, the window size is increased by one segment every RTT until the window size reaches W again at which point full-on congestion occurs again. The resultant sawtooth wave has a time period L.

While T_c can be computed using a precise analysis, we found that a reasonable approximation suffices for achieving successful deadline-aware congestion avoidance. In Figure 4(a), the average window size over the duration of T_c is $\frac{3}{4}W$. Therefore, T_c can be approximated as:

$$T_c = B/(\frac{3}{4}W)$$

Though the average window size is different in the case of Figure 4(b), we still use this approximation for T_c in all our evaluations. A precise analysis of Figure 4(a) yields the following expression:

$$B = \left(\frac{W}{2} + \frac{W}{2} + 1 + \frac{W}{2} + 2 + \cdots \frac{W}{2} + L - 1\right) * \frac{T_c}{L}$$

Solving for T_c, we get:

$$T_c = \frac{B}{\frac{3W}{4} - \frac{1}{2}} \qquad \text{if } T_c > L$$

Similarly, an analysis of Figure 4(b) yields:

$$B = \frac{W}{2} + \frac{W}{2} + 1 + \frac{W}{2} + 2 + \cdots + \frac{W}{2} + T_c - 1$$

Solving for T_c, we get:

$$T_c = -\frac{W-1}{2} + \sqrt{1/4(W-1)^2 + 2B} \qquad \text{if } T_c < L$$

We found that D^2TCP performs similarly with precise and approximate expressions. Therefore, we do not present any evaluations using the precise expressions.

3.2.4 *Stability and convergence*

As stated above, our computation of d makes some approximations. Fortunately, D^2TCP has some in-built self-correction. If the algorithm under-estimates a flow's d, the flow may back off too much. However, in subsequent RTTs, the flow's D will decrease more than a commensurate decrease in its B. As a result d will increase, causing the flow to increase its transmission rate. The same is true for over-estimations of d. The flow will transmit faster than needed, and its B will decrease faster than needed. Consequently, subsequent RTTs will yield a lower d, thus correcting the over-estimation. Furthermore, if the aggressive transmission leads to severe congestion, then the natural response of the gamma function will dial up the value of p and throttle the flow as well.

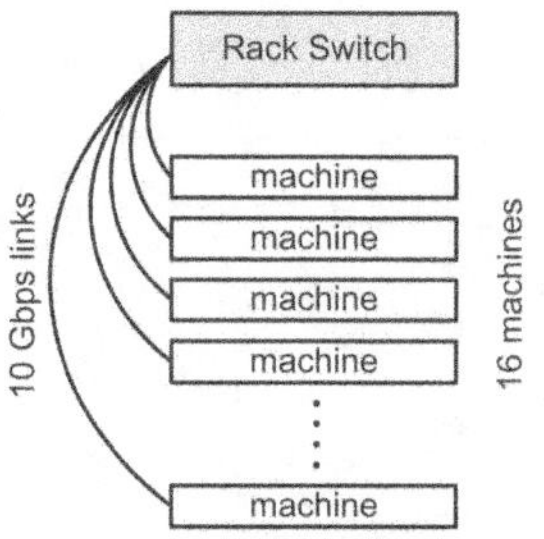

Figure 5: Real implementation testbed

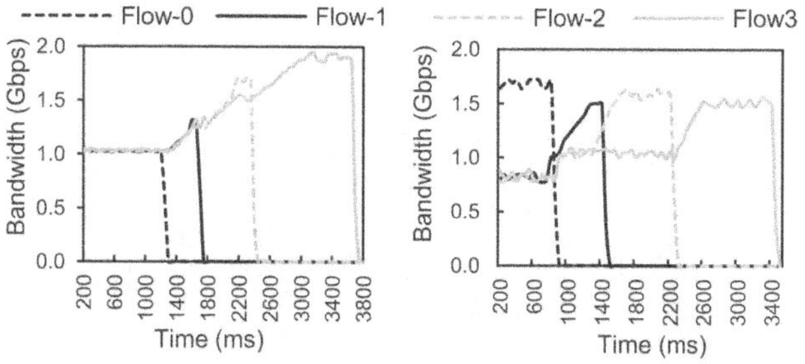

Figure 6: Throughput for DCTCP (left) vs. D^{2}TCP (right)

Another issue to consider with deadline-aware scheduling is the possibility of inadvertently driving the network into congestive collapse. Imagine if all flows were to demand unreasonably tight deadlines, then their resulting congestion avoidance behavior would be to refuse downsizing their windows, effectively overloading the network. However, D^{2}TCP's gamma-correction function guards against such overload at two levels: (1) The effect of tighter deadlines is captured by the value of d which, in turn, determines the window size (hence sending rate). Because we cap the maximum value of d at 2.0, we limit how aggressively near-deadline flows can send. Therefore, deadlines that are tighter beyond a certain limit get rounded down so applications will not drive the network into congestive collapse. (2) At extreme congestion, as α and p approach 1, D^{2}TCP defaults to TCP. Therefore, D^{2}TCP's worst case stability is similar to that of TCP.

3.2.5 D^{2}TCP summary

The D^{2}TCP algorithm is as follows: A sender that does not encounter CE-marked packets increases its window size by one segment; otherwise, the sender computes α, T_c, and then d. The sender then uses the gamma-correction function to obtain p and resizes the congestion window based on p.

We do not change other aspects of TCP, such as slow start, and retransmission and timeout when there is a packet loss.

Note that when flows do not have a deadline (e.g., background flows) we use $d = 1$ so that D^{2}TCP defaults to DCTCP. The stability of DCTCP for long flows is examined in [1].

4. EVALUATION

We evaluate D^{2}TCP using both simulations and a real implementation. We use the real implementation to run a set of microbenchmarks on a small testbed, and to validate our simulator. We rely on simulations to evaluate production-like workloads at the scale of thousands of servers. Furthermore, because D^3 requires custom hardware, we limited our comparisons against D^3 to simulations.

4.1 Small-scale Real Implementation

We first present evaluations on a small-scale real implementation of DCTCP and D^{2}TCP. We use a set of microbenchmarks that examine the basic functionality of D^{2}TCP as a deadline-aware network protocol.

4.1.1 Implementation methodology

We started with the publicly available DCTCP source code from [4]. We then built our D^{2}TCP implementation on top of DCTCP, which amounted to around 100 additional lines of code. We instrumented D^{2}TCP to operate over only a select range of TCP ports, thus allowing us to use the same kernel for different protocols. We deployed this implementation on the testbed depicted in Figure 5. The testbed consists of a Top-of-Rack switch connected to 16 server machines. The switch is based on a Broadcom Scorpion ASIC with 24x 10Gbps ports and a 4MB shared packet buffer, and the servers connect to the switch via 10Gbps links.

We set the key parameters of DCTCP and D^{2}TCP to match those in [1]: (1) g, the weighted averaging factor is 1/16; and (2) K, the buffer occupancy threshold for marking CE-bits, is 20 for 1Gbps links, and 65 for 10Gbps links. For D^{2}TCP we cap d, the deadline imminence factor, to be between 0.5 and 2.0 (except in Section 4.2.4, where we explore the effects of varying this cap). We set RTO_{min} for all the protocols to be 20 ms.

4.1.2 Deadline awareness

We begin by examining the fundamental ability of D^{2}TCP to schedule the network in a deadline aware manner via deadline-aware congestion avoidance. In this experiment we have four hosts transmit flows to a fifth "root" host. We choose flow sizes and deadlines to illustrate the impact of a deadline-aware protocol. We set the flow sizes as 150, 220, 350 and 500 MB, with respective deadlines of 1000, 1500, 2500 and 4000 ms. Note that the flow sizes and deadlines in this synthetic test are not intended to model a real workload.

In Figure 6 we show the throughput achieved by the various flows over time, for both DCTCP and D^{2}TCP. The difference between DCTCP and D^{2}TCP is most noteworthy in the 0-2200 ms timeframe. As expected, DCTCP grants all flows equal bandwidth, and consequently the near-deadline flows miss their deadlines. In contrast, D^{2}TCP's deadline-aware congestion avoidance allows the near deadline flows to take a larger share of the available bandwidth, and the far-deadline flows commensurately relinquish bandwidth. Consequently, the flow completion times for flow #0 and #1 under D^{2}TCP are significantly shorter than those under DCTCP. As the number of active flows decrease, the opportunity for deadline-aware scheduling among them also decreases. Consequently, flow #2 and #3 have similar completion times under both schemes. These results establish the utility of a deadline aware network protocol.

4.1.3 Mixing deadline and non-deadline traffic

Recall that when flows have no deadlines we set $d = 1$ which causes D^{2}TCP to behave identical to DCTCP. DCTCP's stability and convergence for traffic patterns that consist entirely of such long non-deadline flows are examined in [1] in detail. We do not evaluate such traffic mixes in our paper, as the results for D^{2}TCP would be identical to that for DCTCP.

Instead, we examine traffic patterns that consist of a mix of deadline and non-deadline traffic. To that end, we set up a small-scale OLDI application with one root and the number of leaves varying between 20 and 40. Because the testbed has only 16 server machines, we run multiple leaves on each physical machine. The root periodically sends a query to all the leaves, which in turn, idle for a fixed "computation time" and then respond to the query. The replies from the leaves are sized 100-

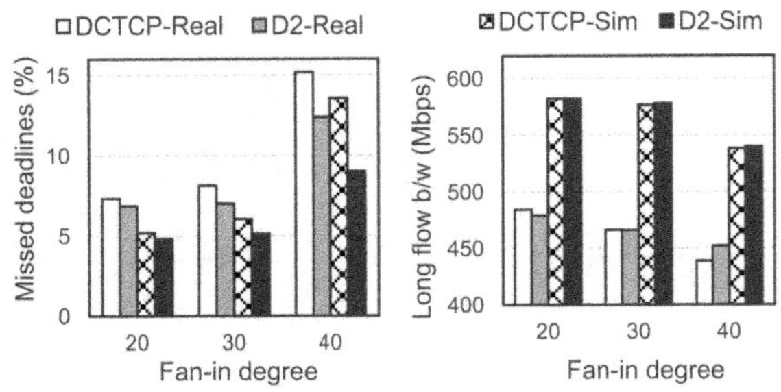

Figure 7: Missed deadlines (left) & bandwidth of long flows (right) in real implementation vs. simulation

500 KB, and have a deadline of 5-25 ms. In addition to this OLDI traffic, we have two servers initiate long-lived flows of 10MB to the root, once every 80 ms. The experiment lasts for the duration of 1000 OLDI queries. The aggregate traffic results in a network utilization of around 10% which is a realistic load for datacenters, and not an artificial overload scenario. Our goal in this experiment is to compare DCTCP and D^2TCP in the real implementation.

In Figure 7(left) we show the percentage of missed deadlines in the real implementation and in simulation of DCTCP and D^2TCP (shown as DCTCP-Real, D2-Real, DCTCP-Sim, and D2-Sim), while varying the fan-in between 20 and 40. We discuss DCTCP-Sim and D2TCP-Sim results in Section 4.1.4. We see that, across the board, D2-Real misses fewer deadlines than DCTCP-Real, with the difference increasing as we increase the fan-in. At a fan-in of 40, DCTCP-Real misses 15.2% of deadlines, whereas D2-Real misses 12.3% of deadlines.

In Figure 7(right) we show the throughput achieved by the non-deadline long flows for varying fan-in degrees. We see that at a fan-in degree of 40, the throughput achieved by D2-Real is 451 Mbps, which compares favorably to the throughput of 438 Mbps for DCTCP-Real. Therefore, the performance of deadline flows is not improved by degrading the throughput of non-deadline flows.

4.1.4 Simulator validation

To validate our simulator, we compare the real implementation and simulation results in Figure 7. We notice that the absolute numbers from the simulation are slightly off from the real implementation. For example, the percentage of missed deadlines for both DCTCP and D^2TCP under simulation is lower than those under real implementation. Such discrepancy is expected as simulations do not capture all the details and nuances of a real system, such as burst-smoothing jitter caused by unpredictable system events, interrupt coalescing, Large Segment Offload (LSO), and other TCP quirks.

Nevertheless, the relative performance difference between DCTCP and D^2TCP, and the trend in that difference, are similar across simulation and real implementation. At a fan-in of 30 in Figure 7, D^2TCP achieves 16% reduction in missed deadlines over DCTCP under simulations, whereas under real implementation the reduction is 15%. As we increase the fan-in degree, the relative performance difference between DCTCP and D^2TCP increases under both simulation and real implementation. Because of these key similarities, we believe that our at-scale simulation results are trustworthy. In addition, we also ensure that our simulation results closely match published DCTCP and D^3 results [25], as we show in the next section.

4.2 At-Scale Simulations

We now present our at-scale simulations. We model the network topology and the traffic after typical production deployments. Recall from Section 1 that D^2TCP's goal is to

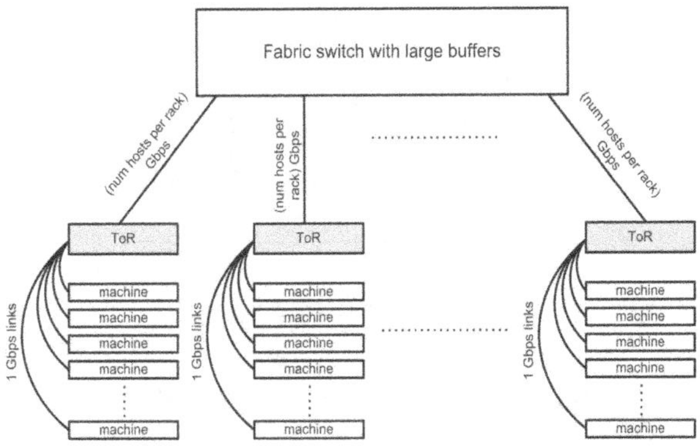

Figure 8: Simulated network

reduce the percentage of missed deadlines without degrading throughput for long-lived flows. Therefore, we focus on these metrics in our at-scale simulations.

4.2.1 Simulation methodology

We implemented DCTCP and D^2TCP on top of ns-3's TCP New Reno protocol [19], and enabled the marking of CE bits in the switch model of ns-3. For D^3, we wrote both the end-host protocol and the switch logic, based on the details in [25]. We set D^3's base rate to be one segment per RTT. Further, we use the same RRQ packet format described in [25] including the 8-bit bytes-per-microsecond field. All DCTCP and D^2TCP parameters match those in Section 4.1.1.

We performed our simulations on the network depicted in Figure 8, which uses a fat-tree topology typical of datacenter networks. There are 25 racks with each rack having up to 40 end-host machines. Thus, our simulations capture the behavior of a 1000-machine deployment. Each end-host connects to the top-of-rack (ToR) switch via a 1 Gbps link. Because the bottleneck in datacenter networks is usually the ToR switch [10] [1] [25], we abstracted away the rest of the fat-tree topology, replacing it with one large fabric switch with large buffers. Each ToR is connected to the fabric switch via a single link with a line rate equal to 1 * *number-of-hosts-in-a-rack* Gbps. We sized the packet buffers in the ToR switches to match typical buffer sizes of shallow-buffered switches in real data centers (4MB) [1]. We set the link latencies to 20 μs, achieving an average of RTT of 200 μs, which is representative of datacenter network RTTs.

We ran a set of five synthetic OLDI applications on the network, equally dividing the total number of end-hosts among the applications. The assignment of an application node to a physical end-host is random to capture the effects of (1) applications dynamically requesting and relinquishing virtual machines in the data center, and (2) virtual to physical machine assignment being completely fluid. Each application consists of a set of five identical OLDI trees, each with one parent and *n* leaves, which have the same settings for leaf-to-parent message size and for deadlines. These settings are different across the five applications. We varied *n*, the number of leaves per parent, in the trees to explore varying degree of fan-in-bursts.

The distributions of message sizes and of deadlines in real OLDIs are publicly available [1]. However, details such as the exact deadline for given leaf-to-parent ratio and message size, are not publicly available. Therefore, like [25], we chose semi-synthetic values for deadlines using the aforementioned distributions.

Whether a message misses its deadline or not depends entirely on the precise deadline assigned to a message size. As

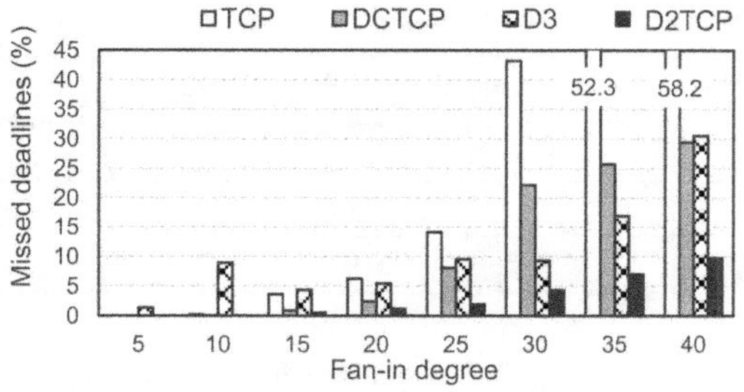

Figure 9: Missed deadlines for low variance (10%)

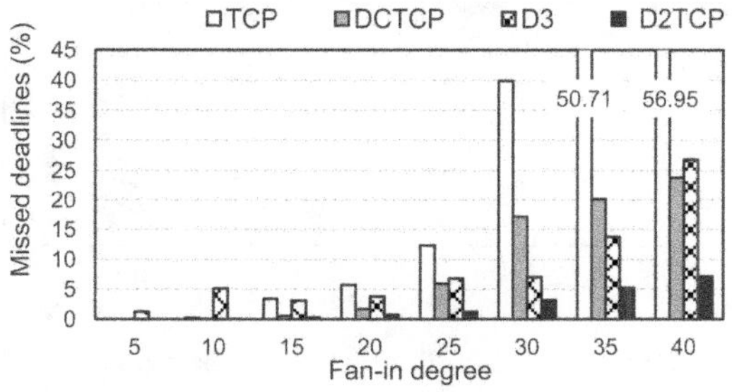

Figure 10: Missed deadlines for med. variance (50%)

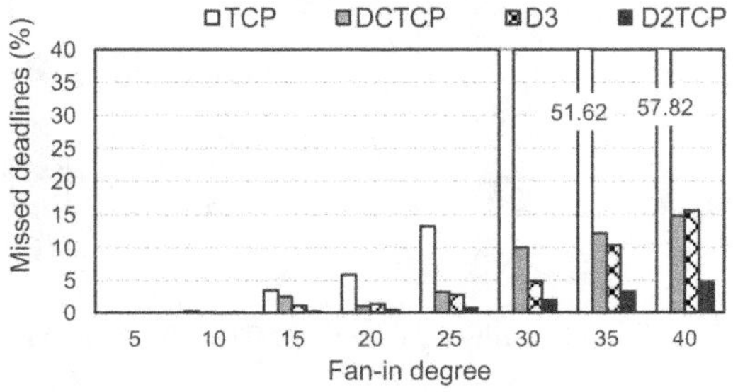

Figure 11: Missed deadlines for high variance (Exp)

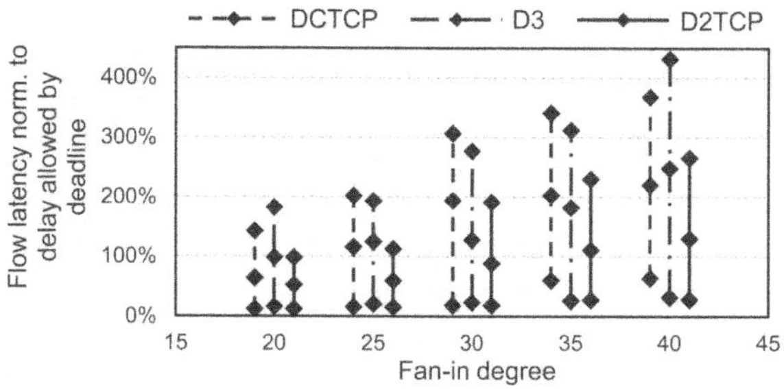

Figure 12: Normalized latencies at the 50th, 90th, and 99th percentiles

such, the choices of the deadlines directly impact the results of any experiment. Therefore, we carefully calibrated our message sizes and deadlines such that the number of missed deadlines for DCTCP and D^3 are in line with the results in the D^3 paper [25]. This corroboration is discussed in detail in Section 4.2.2. We set the five OLDI applications' message sizes to 2, 6, 10, 14, and 18 KB and deadlines to 20, 30, 35, 40 and 45 milliseconds, respectively. We used these calibrated message sizes and deadlines in all our experimental results, except in Section 4.2.6. Note that the message sizes (fixed size of few KBs), long flow sizes (a few MB), and number of concurrent long-lived connections (1 connection) are chosen to match the characteristics of production workloads [1]. In all our experiments, the network utilization is 10-20% which is a realistic load for datacenters, and not an artificial overload scenario.

4.2.2 OLDI performance

Recall that in OLDIs the parent-to-leaf RPC has an overall budget that gets divided into computation and network parts. If all leaves were to finish computation at exactly the same instant for every single query, then the network deadline would be a hard and fixed value. In reality, the computation time can vary across leaves if the work for a query is not evenly balanced. While the computation budget forms a hard upper bound, some leaves may respond sooner resulting in a slightly looser *effective* network deadline. In addition, some applications attempt to smooth fan-in bursts via user-injected jitter [9]. As such, the exact nature and architecture of an OLDI application can affect the distribution of effective deadlines. We stipulate that a robust network protocol should work across a spectrum of deadline distributions. To that end, we evaluate three deadline distributions which use the same base deadlines as specified in Section 4.2.1, to which each adds a different variation. Our low variance case models a 10% uniform-random variation added to the base deadline. The medium variance case adds a 50% uniform-random variation, and the high variance case models a one-sided exponential distribution with mean equal to the base deadline. Our high variance case matches the deadline distribution used in [25].

We now compare D^2TCP against DCTCP and D^3 in terms of the fraction of missed deadlines for our benchmark. Figure 9, Figure 10, and Figure 11 correspond to the low, medium and high variance deadline cases. In all three graphs, the Y axis shows the fraction of missed deadlines for TCP, DCTCP, D^3, and D^2TCP as we vary the degree of burstiness on the X axis by increasing the fan-in degree (i.e., number of leaves per parent) from 5 to 40. Typical OLDI applications' fan-in degrees fall in this range [1] [25].

Across all three graphs, we see that TCP misses a rapidly increasing number of deadlines as we increase the fan-in degree (i.e., burstiness). While DCTCP and D^3 improve over TCP, they still miss a significant fraction of deadlines. For the medium variance case both DCTCP and D^3 miss around 25% of the deadlines at the fan-in degree is 40. In comparison, D^2TCP keeps the fraction of missed deadlines under 7% at a fan-in degree of 40. Because this trend holds true for all three deadline variation cases, we argue that D^2TCP is robust enough to handle a wide spectrum of deadline distributions. For the remainder of the results section, we use only the medium variance case for all remaining experiments.

All the schemes perform better with the higher-variance deadlines. Such behavior is expected because higher variance smoothes out fan-in-burst-induced congestion. Note that for the high variance case, the fraction of missed deadlines for D^3 fall in the range of 0-15% which is close to that of 0-9% reported in the D^3 paper [25], confirming that our D^3 implementation is reasonable.

To explain the above results, we show another set of data for the medium variance case. Figure 12 shows the 50th, 90th, and 99th percentile latencies for DCTCP, D^3 and D^2TCP normalized to the delay allowed by the deadline. On each line, the three points from bottom to top correspond to the 50th, 90th, and 99th percentile latencies, respectively. As expected, D^2TCP's latencies are significantly lower than those of DCTCP and D^3, resulting in fewer missed deadlines. Overall, the latencies for all the schemes closely track the fraction of missed deadlines in Figure 10.

We now examine our earlier claims about D^3's shortcomings. Recall that D^3's greedy approach may allocate bandwidth to far-deadline requests arriving slightly ahead of near-deadline

Table 1: Priority inversion for D^3

Fan-in Degree	Low-Var.	Med. Var.	Hi. Var.
20	31.9	26.3	24.1
25	33.2	28.7	24.6
30	35.7	30.8	28.6
35	41.9	33.4	31.5
40	48.6	40.5	33.1

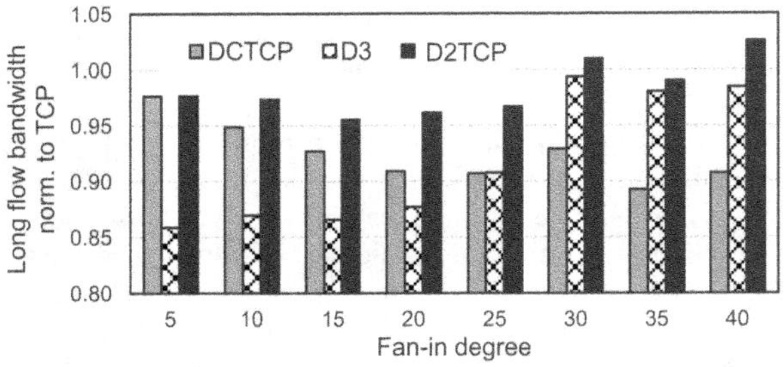

Figure 13: Bandwidth of background flows

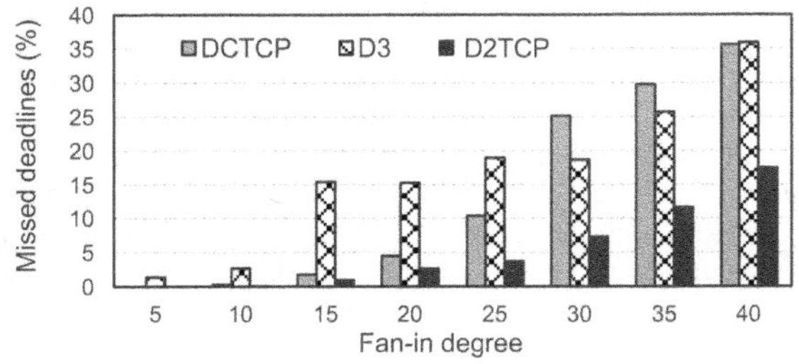

Figure 14: Missed deadlines under background flows

requests. This race condition causes D^3 to frequently invert priorities of congested flows. To confirm these claims, we compute the percentage of requests that are denied while later-deadline requests have been granted. This percentage is a measure of priority inversion in D^3. Table 1 shows this percentage for D^3 under all three variance cases, for various fan-in degrees. From the table, we see that even in a favorable setting (high variance deadlines with a fan-in degree of 20), D^3 incurs priority inversion for nearly 25% of all flows. Also note that the priority inversions worsen both as the fan-in degree increases, and as the variance in deadlines gets tighter causing more burstiness.

4.2.3 *Background flows*

To test whether long-lived, non-OLDI flows achieve high bandwidth even as short-lived OLDI flows come and go, we replace one leaf-to-parent flow in each OLDI tree with a long-lived background flow. This background flow has an exponential arrival with mean of 300 ms and sends 1 MB of data.

In Figure 13, we show the background flows' bandwidth for DCTCP, D^3, and D^2TCP, normalized to that for TCP on the Y axis as we vary the fan-in degree on the X axis. Background flows give up bandwidth to OLDI flows only for the short duration of fan-in-burst-induced congestion under all the schemes. Consequently, all the schemes perform well, achieving 85% or more of the bandwidth achieved by TCP. D^2TCP is slightly better than DCTCP which throttles background flows to make room unnecessarily for far-deadline, OLDI flows. Overall, D^2TCP achieves 95% or more of the bandwidth achieved by TCP.

To confirm that the background flows do not take bandwidth away from the OLDI flows, we show in Figure 14 the fraction of

Table 2: Long-flow b/w when D^2TCP & TCP coexist

Fan-in degree	Long flow bandwidth (Mbps)		
	All TCP	Mix #1	Mix #2
15	90	90	90
20	86	86	86

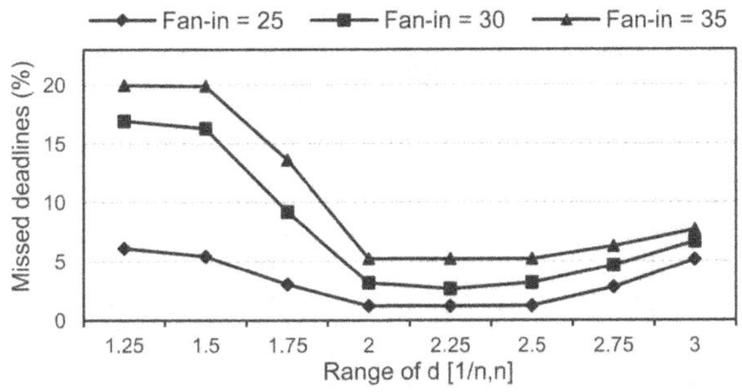

Figure 15: Missed deadlines while varying cap on d

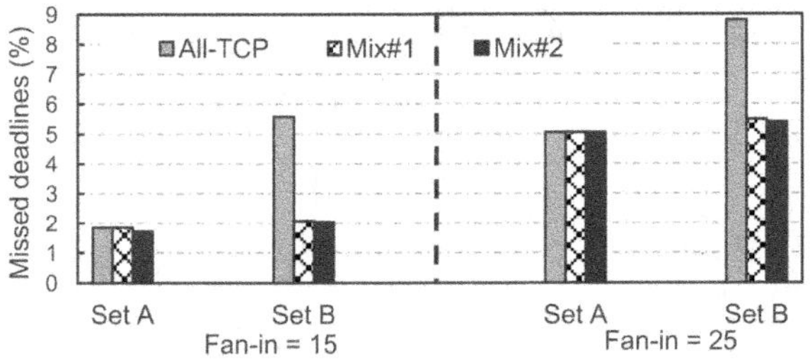

Figure 16: Missed deadlines when D^2TCP & TCP coexist

missed deadlines for the OLDI flows in the presence of the background flows. For all the schemes, the fraction remains similar to that in the absence of the background flows (Figure 10).

4.2.4 *Varying the cap on d – deadline imminence*

Recall from Section 3.2.3, that because d appears in the exponent of the gamma-correction function, extremely low and high values may cause undesirable behavior. Therefore we cap the value of d to be within the range $(n, 1/n)$. In all our experiments we set $n = 2.0$. In this section we evaluate the effects of varying n. In Figure 15, we show D^2TCP's percentage of missed deadlines for various fan-ins as we vary n between 1.25 and 3.0. As expected, when n is close to 1.0, D^2TCP's behavior matches DCTCP and the fraction of missed deadlines is high. As n increases to 2.0, the fraction drops dramatically but then levels off. At $n = 3.0$ and beyond we see that the fraction starts to increase slowly as the larger n allows near-deadline flows to increasingly ignore congestion feedback.

4.2.5 *Coexisting with TCP*

To demonstrate that D^2TCP can coexist with TCP without hurting bandwidth or deadlines, we use the same production benchmark but use a mix of D^2TCP and TCP for the various network traffic. Because we are specifically interested in TCP performance here, we restrict this experiment to fan-in degrees where TCP performance is acceptable in Figure 10 (i.e., missed deadline fraction of 5% or less, and fan-in degree of 15 and 20).

We run three experiments gradually adding D^2TCP traffic to a TCP datacenter. Imagine that the set of five OLDI applications in our workload are divided into two sets: set A consist of three OLDIs, and set B consists of the other two OLDIs. We start with

Table 3: Deadlines achieved by D^3 and D^2TCP for similar fraction of missed deadlines

Fan-in degree	D3's missed deadlines (%)	D^2TCP's missed deadlines (%)	D^2TCP's tighter deadline (%)
10	0.71	0.84	55
15	3.61	3.49	45
20	4.7	4.88	35

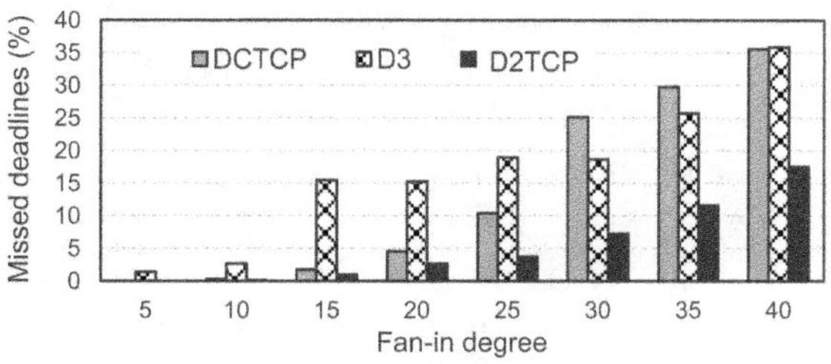

Figure 17: Missed deadlines under tighter deadlines

all five OLDIs and all long flows running on TCP. We first "upgrade" set B to D^2TCP, while set A continues to run on TCP. Next, we upgrade the background long-flow traffic to D^2TCP as well. In summary, the three setups are:

- All-TCP: 5x OLDIs + long flows; no D^2TCP.
- Mix#1 TCP: 3x OLDIs + long flows; D2TCP: 2x OLDIs.
- Mix#2 TCP: 3x OLDIs; D^2TCP: 2x OLDI + long flows.

In Figure 16, we show the fraction of missed deadlines (Y axis) for sets A and B in the three runs as we vary the fan-in degree (X axis). We separately analyze long flows below. For set A, comparing All-TCP and Mix#1 shows that TCP's (i.e., set A's) missed-deadline fraction is not worsened by sharing the network with D^2TCP traffic (set B in Mix#1). Furthermore, set B sees a reduction in missed deadlines upon migrating from TCP (All-TCP) to D^2TCP (Mix#1). Going back to set A, comparing Mix#1 and Mix#2 we see that TCP's (i.e., set A's) missed-deadline fraction is not hurt by background flows upgrading to D^2TCP. Similarly, set B's) missed-deadline fraction stays the same between Mix#1 and Mix#2, showing that background flows using D^2TCP do not hurt the OLDIs using D^2TCP.

In Table 2, we show the long-flow throughput achieved in the three runs. Going from All-TCP to Mix#1, the throughput does not degrade, showing that upgrading some OLDIs to D^2TCP does not hurt long TCP flows. Comparing Mix#1 and Mix#2 shows that D^2TCP's long-flow throughput is similar to that of TCP.

4.2.6 Tighter deadlines

To show that D^2TCP performs well over a range of deadlines, we evaluate D^2TCP under tighter deadlines than our default (Section 4.2.1). Because we found that deadlines tightened by 10% or 20% lead to similar behavior, we show results only for the 20% case in Figure 17. As expected, the tighter deadlines here result in more deadlines being missed under all the schemes than those missed in Figure 10. Nevertheless, D^2TCP maintains its advantage over DCTCP and D^3 under the tighter deadlines.

While the above results show the fractions of missed deadlines under tighter deadlines, it may be important to determine the inverse (i.e., how much tighter can the deadlines be for a target fraction of missed deadlines). For instance, this question is typically of interest to datacenter operators who would like to maintain the fraction of missed deadlines within an acceptable threshold and wish to know how much the communication deadlines can be tightened to allow more time for computation and improve response quality. In Table 3, we show the tightness of deadlines supported by D^2TCP as compared to D^3 for a target fraction of missed deadlines. We limit the study to a reasonable fraction of missed deadlines (i.e., 5%). From the table, we see that D^2TCP achieves deadlines that are tighter by 35-55%, which would make sizable room for computation.

5. RELATED WORK

There is an abundance of past work that deals with the subjects of congestion control, network scheduling, and reducing latencies. There are many schemes that build on top of TCP, while others are novel protocols altogether. A comprehensive review of all such work is beyond the scope of this paper, but we summarize some of the most relevant work here.

Earliest Deadline First (EDF) [17] is one of the earliest packet scheduling algorithms and is provably optimal when deadlines are associated with individual packets. When deadlines are associated with flows, however, applying EDF to individual packets as they arrive at the switch is not only suboptimal but can worsen the congestion in the network [26].

In [24], the authors show that having finer grain system clocks allow for a faster response to TCP timeouts, and help in reducing the net latency of TCP flows.

Rate Control Protocol (RCP) [5] can achieve 10-fold improvement in the completion times of small- to medium-sized flows in the Internet, particularly downloads representative of typical web browsing. RCP is provably optimal when minimizing overall completion times is the metric. RCP replaces TCP's slow start phase with an allocation equal to the fair share available at the bottleneck route. Like D^3, RCP requires hardware modification to the routers.

Live multimedia traffic also has a soft-real-time nature, and both proactive bandwidth reservation [6] and reactive [23] [16] schemes exist. TCP-RTM [16] observes that TCP always favors reliability over timeliness, and proposes extensions that improve performance of multimedia applications by allowing minimal amount of packet re-ordering and loss in the TCP stack.

Active Queue Management schemes like RED [8] and E-TCP [11] inject early warnings of congestion to TCP endhosts by randomly dropping packets when switch buffer occupancy is high. Because senders back-off before full on congestion, these schemes allow TCP to operate in the high throughput, fast-retransmit mode, instead of degrading to full back-off.

High-speed TCP [7], CUBIC [22], and XCP [15] all successfully improve the performance of TCP in high bandwidth-delay-product networks. They exploit the large degree of statistical multiplexing present, and also mitigate TCP's drastic reaction to packet losses. XCP shares some common design details with D^3 in that senders request bandwidth via a congestion header, and the switches populate their responses in this header.

Re-feedback [3] addresses the problem of fairness and stability in the Internet when untrusted senders may act selfishly in the face of congestion. Re-feedback incentivizes senders to populate packet headers with honest information about congestion situation so the network may schedule accordingly.

QCN [20] proposes to improve Ethernet performance in datacenters via multibit feedback from the switches to endhosts. By utilizing smarter switches and hardware-based reaction logic in the endhost NICs, QCN dramatically reduces recovery time during congestions, thus improving flow completion times. However, QCN cannot span beyond L2 domains limiting its scope of application. VCP [27] is another similar scheme that relies on ECN-like feedback via elaborate processing at the switches.

6. CONCLUSION

Online, data-intensive applications (OLDI) in datacenters (e.g., Web search, online retail and advertisement) achieve good user experience by controlling latency using soft-real-time constraints which translate to deadlines for network communication within the applications. Further, OLDI applications typically employ tree-based algorithms which, in the common case, result in fan-in burst of children-to-parent traffic with tight deadlines. Previous work on datacenter network protocols is either deadline-agnostic, or is deadline-aware but suffers under bursts due to race conditions.

We proposed Deadline-Aware DataCenter TCP (D^2TCP) which:

- prioritizes near-deadline flows over far-deadline flows in the presence of fan-in-burst-induced congestion;
- achieves high bandwidth for background flows even as the short-lived OLDI flows come and go;
- requires no changes to the switch hardware; and
- coexists with legacy TCP.

D^2TCP uses a distributed and reactive approach for bandwidth allocation that fundamentally enables D^2TCP's properties. D^2TCP's key mechanism is a novel congestion avoidance algorithm, which uses ECN feedback and deadlines to modulate the congestion window via a gamma-correction function.

Using small-scale real implementation, and at-scale simulations, we showed that D^2TCP

- reduces the fraction of missed deadlines compared to both DCTCP and D^3 by 75% and 50%, respectively;
- achieves nearly as high bandwidth as TCP for background flows without degrading OLDI performance;
- meets deadlines that are 35-55% tighter than those achieved by D^3 for the same reasonable fraction of missed deadlines (i.e., 5%), giving OLDIs more time for actual computation; and
- coexists with TCP flows without degrading their performance.

D^2TCP has significant performance and practical advantages. On the performance side, by reducing the number of missed deadlines, D^2TCP improves OLDI applications' response quality, and hence user experience. Further, by meeting tighter deadlines, D^2TCP allows more time for computation in OLDI applications and thereby further enhances OLDI response quality and user experience. Given that OLDI applications are likely to scale up in size to accommodate ever-growing data on the Web, D^2TCP's tighter deadlines may fundamentally enable this scale-up without degrading OLDI response quality. On the practical side, by requiring no changes to the switch hardware, D^2TCP can be deployed by merely upgrading the TCP and RPC stacks. Our prototype implementation of D^2TCP amounted to only 100 lines of kernel code. Finally, by being able to coexist with TCP, D^2TCP is amenable to incremental deployment, a key requirement for datacenter network protocols in the real world. The growing importance of OLDI applications implies that these significant advantages make D^2TCP an important ingredient for datacenters.

ACKNOWLEDGMENTS

We thank the SIGCOMM reviewers, and our shepherd David Maltz, for their insightful comments which helped us significantly improve the paper. We also thank Sridhar Raman and Abdul Kabbani for their help with real implementations of DCTCP and D^2TCP, and Gwendolyn Voskuilen for reviewing the paper.

REFERENCES

[1] M. Alizadeh, A. G. Greenberg, D. A. Maltz, J. Padhye, P. Patel, B. Prabhakar, S. Sengupta, and M. Sridharan. Data center TCP (DCTCP). In Proc. *SIGCOMM*, 2010.

[2] Charles A. Poynton (2003). Digital Video and HDTV: Algorithms and Interfaces. Morgan Kaufmann. pp. 260, 630. ISBN 1558607927.

[3] B. Briscoe et. al. Policing Congestion Response in an Internetwork using Re-feedback. In Proc. *SIGCOMM* 2005.

[4] Datacenter TCP, http://www.stanford.edu/~alizade/Site/DCTCP.htm

[5] Nandita Dukkipati. RCP: Congestion Control to Make Flows Complete Quickly. PhD Thesis, Department of Electrical Engineering, Stanford University, October 2006.

[6] D. Ferrari, A. Banerjea, and H. Zhang. Network support for multimedia: A discussion of the tenet approach. In Proc. *Computer Networks and ISDN Systems*, 1994.

[7] S. Floyd. RFC 3649: HighSpeed TCP for large congestion windows.

[8] S. Floyd and V. Jacobson. Random early detection gateways for congestion avoidance. *IEEE/ACM Transactions on Networking*, 1(4):397–413, 1993.

[9] S. Floyd and V. Jacobson. The synchronization of periodic routing messages. *IEEE/ACM Transactions on Networking*, 2(2):122-136, 1994.

[10] R. Griffith, Y. Chen, J. Liu, A. Joseph, and R. Katz. Understanding TCP incast throughput collapse in datacenter networks. In *WREN Workshop*, 2009.

[11] Y. Gu, D. Towsley, C. Hollot, and H. Zhang. Congestion control for small buffer high bandwidth networks. In Proc. *INFOCOM*, 2007.

[12] Urs Hoelzle, Jeffrey Dean, and Luiz André Barroso. Web Search for A Planet: The Architecture of the Google Cluster, In *IEEE Micro Magazine*, April 2003.

[13] T. Hoff. Latency is Everywhere and it Costs You Sales - How to Crush it, July 2009. http://highscalability.com/blog/2009/7/25/latency-iseverywhere-and-it-costs-you-sales-how-to-crush-it.html.

[14] S. Iyer et. al. Analysis of a memory architecture for fast packet buffers. In IEEE *HPSR* Workshop, 2001.

[15] D. Katabi, M. Handley, and C. Rohrs. Congestion Control for High Bandwidth-Delay Product Networks. In Proc. *SIGCOMM*, 2002.

[16] Sam Liang and David Cheriton. TCP-RTM: Using TCP for Real Time Applications. In Proc. *ICNP*, 2002.

[17] C. L. Liu and J. W. Layland. Scheduling Algorithms for Multiprogramming in a Hard-Real-Time Environment. *Journal of the ACM*, 20(1), 1973.

[18] D. Meisner, C. M. Sadler, L. A. Barroso, W. Weber and T. F. Wenisch. Power Management of Online Data-Intensive Services. In Proc. *ISCA*, June 2011.

[19] The ns-3 discrete-event network simulator. http://www.nsnam.org/

[20] R. Pan, B. Prabhakar, and A. Laxmikantha. QCN: Quantized congestion notification an overview. http://www.ieee802.org/1/files/public/docs2007/au_prabhakar_qcn_overview_geneva.pdf

[21] K. Ramakrishnan, S. Floyd, and D. Black. RFC 3168: The addition of explicit congestion notification (ECN) to IP.

[22] I. R. Sangtae Ha and L. Xu. Cubic: A new TCP-friendly high-speed TCP variant. In Proc. *SIGOPS-OSR*, 2008.

[23] V. Tsaoussidis and C. Zhang. 2002. TCP-Real: receiver-oriented congestion control. *The International Journal of Computer and Telecommunications Networking*. 40(4), 2002.

[24] V. Vasudevan et al. Safe and effective fine-grained TCP retransmissions for datacenter communication. In Proc. *SIGCOMM*, 2009.

[25] C. Wilson, H. Ballani, T. Karagiannis, A. Rowstron. Better Never Than Late: Meeting Deadlines in Datacenter Networks. In Proc. *SIGCOMM*, 2011.

[26] C. Wilson, H. Ballani, T. Karagiannis, and A. Rowstron. Better never than late: Meeting deadlines in datacenter networks. Technical Report MSR-TR-2011-66, Microsoft Research, May 2011.

[27] Y. Xia, L. Subramanian, I. Stoica, and S. Kalyanaraman. One more bit is enough. In Proc. *SIGCOMM*, 2005.

Finishing Flows Quickly with Preemptive Scheduling

Chi-Yao Hong
UIUC
cyhong@illinois.edu

Matthew Caesar
UIUC
caesar@illinois.edu

P. Brighten Godfrey
UIUC
pbg@illinois.edu

ABSTRACT

Today's data centers face extreme challenges in providing low latency. However, fair sharing, a principle commonly adopted in current congestion control protocols, is far from optimal for satisfying latency requirements.

We propose Preemptive Distributed Quick (**PDQ**) flow scheduling, a protocol designed to complete flows quickly and meet flow deadlines. PDQ enables flow preemption to approximate a range of scheduling disciplines. For example, PDQ can emulate a shortest job first algorithm to give priority to the short flows by pausing the contending flows. PDQ borrows ideas from centralized scheduling disciplines and implements them in a fully distributed manner, making it scalable to today's data centers. Further, we develop a multipath version of PDQ to exploit path diversity.

Through extensive packet-level and flow-level simulation, we demonstrate that PDQ significantly outperforms TCP, RCP and D^3 in data center environments. We further show that PDQ is stable, resilient to packet loss, and preserves nearly all its performance gains even given inaccurate flow information.

Categories and Subject Descriptors: C.2.2 [Computer-Communication Networks]: Network Protocols
General Terms: Algorithms, Design, Performance
Keywords: Data center, Flow scheduling, Deadline

1. INTRODUCTION

Data centers are now used as the underlying infrastructure of many modern commercial operations, including web services, cloud computing, and some of the world's largest databases and storage services. Data center applications including financial services, social networking, recommendation systems, and web search often have very demanding latency requirements. For example, even fractions of a second make a quantifiable difference in user experience for web services [7]. And a service that aggregates results from many back-end servers has even more stringent requirements on completion time of the back-end flows, since the service must often wait for the *last* of these flows to finish or else reduce the quality of the final results.[1] Minimizing delays from network congestion, or meeting soft-real-time deadlines with high probability, is therefore important.

Unfortunately, current transport protocols neither minimize flow completion time nor meet deadlines. TCP, RCP [10], ICTCP [22], and DCTCP [3] approximate *fair sharing*, dividing link bandwidth equally among flows. Fair sharing is known to be far from optimal in terms of minimizing flow completion time [4] and the number of deadline-missing flows [5]. As a result, a study of three production data centers [20] showed that a significant fraction $(7-25\%)$ of flow deadlines were missed, resulting in degradation of application response quality, waste of network bandwidth, and ultimately loss of operator revenue [3].

This paper introduces **Preemptive Distributed Quick (PDQ)** flow scheduling, a protocol designed to complete flows quickly and meet flow deadlines. PDQ builds on traditional real-time scheduling techniques: when processing a queue of tasks, scheduling in order of Earliest Deadline First (EDF) is known to minimize the number of late tasks, while Shortest Job First (SJF) minimizes mean flow completion time. However, applying these policies to scheduling data center flows introduces several new challenges.

First, EDF and SJF assume a centralized scheduler which knows the global state of the system; this would impede our goal of low latency in a large data center. To perform dynamic decentralized scheduling, PDQ provides a distributed algorithm to allow a set of switches to collaboratively gather information about flow workloads and converge to a stable agreement on allocation decisions. Second, unlike "fair sharing" protocols, EDF and SJF rely on the ability to *preempt* existing tasks, to ensure a newly arriving task with a smaller deadline can be completed before a currently-scheduled task. To support this functionality in distributed environments, PDQ provides the ability to perform distributed preemption of existing flow traffic, in a manner that enables fast switchover and is guaranteed to never deadlock.

Thus, PDQ provides a distributed flow scheduling layer which is *lightweight*, using only FIFO tail-drop queues, and *flexible*, in that it can approximate a range of scheduling disciplines based on relative priority of flows. We use this primitive to implement two scheduling disciplines: EDF to minimize mean flow completion time, and SJF to minimize the number of deadline-missing flows.

Through an extensive simulation study using real datacenter workloads, we find that PDQ provides strong benefits over existing datacenter transport mechanisms. PDQ is

[1]See discussion in [3], §2.1.

SIGCOMM'12, August 13–17, 2012, Helsinki, Finland.

most closely related to D^3 [20], which also tries to meet flow deadlines. Unlike D^3, which is a "first-come first-reserve" algorithm, PDQ proactively and preemptively gives network resources to the most critical flows. For deadline-constrained flows, our evaluation shows PDQ supports 3 times more concurrent senders than [20] while satisfying their flow deadlines. When flows have no deadlines, we show PDQ can reduce mean flow completion times by ~30% or more compared with TCP, RCP, and D^3.

The key contributions of this paper are:

- We design and implement PDQ, a distributed flow scheduling layer for data centers which can approximate a range of scheduling disciplines.
- We build on PDQ to implement flow scheduling disciplines that minimize mean flow completion time and the number of deadline-missing flows.
- We demonstrate PDQ can save ~30% average flow completion time compared with TCP, RCP and D^3; and can support 3× as many concurrent senders as D^3 while meeting flow deadlines.
- We show that PDQ is stable, resilient to packet loss, and preserves nearly all its performance gains even given inaccurate flow information.
- We develop and evaluate a multipath version of PDQ, showing further performance and reliability gains.

2. OVERVIEW

We start by presenting an example to demonstrate potential benefits of PDQ over existing approaches (§2.1). We then give a description of key challenges that PDQ must address (§2.2).

2.1 Example of Benefits

Consider the scenario shown in Figure 1, where three concurrent flows (f_A, f_B, and f_C) arrive simultaneously.

Deadline-unconstrained Case: Suppose that the flows have no deadlines, and our objective is to minimize the average flow completion time. Assuming a fluid traffic model (infinitesimal units of transmission), the result given by fair sharing is shown in Figure 1b: [f_A,f_B,f_C] finish at time [3,5,6], and the average flow completion time is $\frac{3+5+6}{3} = 4.67$. If we schedule the flows by SJF (one by one according to flow size), as shown in Figure 1c, the completion time becomes $\frac{1+3+6}{3} = 3.33$, a savings of ~29% compared to fair sharing. Moreover, for every individual flow, the flow completion time in SJF is no larger than that given by fair sharing.

Deadline-constrained Case: Suppose now that the flows have deadlines, as specified in Figure 1a. The objective becomes minimizing the number of tardy flows, i.e., maximizing the number of flows that meet their deadlines. For fair sharing, both flow f_A and f_B fail to meet their deadlines, as shown in Figure 1b. If we schedule the flows by EDF (one by one according to deadline), as shown in Figure 1c, every flow can finish before its deadline.

Now we consider D^3, a recently proposed deadline-aware protocol for data center networks [20]. When the network is congested, D^3 satisfies as many flows as possible according to the flow request rate in the order of their arrival. In particular, each flow will request a rate $r = \frac{s}{d}$, where s is the flow's size and d is the time until its deadline. Therefore, the

Flow ID	Size	Deadline
f_A	1	1
f_B	2	4
f_C	3	6

(a)

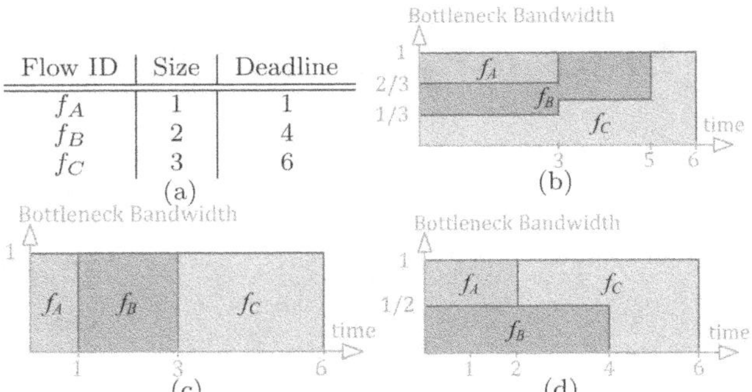

Figure 1: ***Motivating Example. (a) Three concurrent flows competing for a single bottleneck link; (b) Fair sharing; (c) SJF/EDF; (d) D^3 for flow arrival order $f_B \leadsto f_A \leadsto f_C$.***

result of D^3 depends highly on flow arrival order. Assuming flows arrive in the order $f_B \leadsto f_A \leadsto f_C$, the result of D^3 is shown in Figure 1d. Flow f_B will send with rate $\frac{2}{4} = 0.5$ and will finish right before its deadline. However, flow f_A, which arrives later than f_B, will fail to meet its deadline using the remaining bandwidth, as evident in Figure 1d. In fact, out of $3! = 6$ possible permutations of flow arrival order, D^3 will fail to satisfy some of the deadlines for 5 cases, the only exception being the order $f_A \leadsto f_B \leadsto f_C$, which is the order EDF chooses. Although D^3 also allows senders to terminate flows that fail to meet their deadlines (to save bandwidth), termination does not help in this scenario and is not presented in Figure 1d.

2.2 Design Challenges

Although attractive performance gains are seen from the example, many design challenges remain to realize the expected benefits.

Decentralizing Scheduling Disciplines: Scheduling disciplines like EDF or SJF are centralized algorithms that require global knowledge of flow information, introducing a single point of failure and significant overhead for senders to interact with the centralized coordinator. For example, a centralized scheduler introduces considerable flow initialization overhead, while becoming a congestive hot-spot. This problem is especially severe in data center workloads where the majority of flows are mice. A scheduler maintaining only elephant flows like DevoFlow [8] seems unlikely to succeed in congestion control as deadline constraints are usually associated with mice. The need to address the above limitations leads to PDQ, a fully distributed solution where switches collaboratively control flow schedules.

Switching Between Flows Seamlessly: The example of §2.1 idealistically assumed we can start a new flow immediately after a previous one terminates, enabling all the transmission schedules (Figure 1b, 1c and 1d) to fully utilize the bottleneck bandwidth and thus complete flows as quickly as possible. However, achieving high utilization during flow switching in practice requires precise control of flow transmission time. One could simplify this problem by assuming synchronized time among both switches and senders, but that introduces additional cost and effort to coordinate clocks. PDQ addresses this problem by starting the next set of waiting flows slightly before the current sending flows finish.

Prioritizing Flows using FIFO Tail-drop Queues: One

could implement priority queues in switches to approximate flow scheduling by enforcing packet priority. Ideally, this requires that each of the concurrent flows has a unique priority class. However, a data center switch can have several thousand active flows within a one second bin [6], while modern switches support only ∼10 priority classes [20]. Therefore, for today's data center switches, the number of priority classes per port is far below the requirements of such an approach, and it is unclear whether modifying switches to support a larger number of priority classes can be cost-effective. To solve this, PDQ explicitly controls the flow sending rate to regulate flow traffic and retain packets from low-priority flows at senders. With this flow pausing strategy, PDQ only requires per-link FIFO tail-drop queues at switches.

3. PROTOCOL

We first present an overview of our design. We then describe the design details implemented at the sender (§3.1), receiver (§3.2) and switches (§3.3). This section assumes each flow uses a single path. In §6, we will show how PDQ can be extended to support multipath forwarding.

Centralized Algorithm: To clarify our approach, we start by presenting it as an idealized *centralized* scheduler with complete visibility of the network, able to communicate with devices in the network with *zero delay*. To simplify exposition, the centralized scheduler assumes that flows have no deadlines, and our only goal is to optimize flow completion time. We will later relax these assumptions.

We define the *expected flow transmission time*, denoted by $\mathcal{T}_i$ for any flow i, to be the remaining flow size divided by its *maximal sending rate* $\mathcal{R}_i^{\max}$. The maximal sending rate $\mathcal{R}_i^{\max}$ is the minimum of the sender NIC rate, the switch link rates, and the rate that receiver can process and receive. Whenever network workload changes (a new flow arrives, or an existing flow terminates), the centralized scheduler recomputes the flow transmission schedule as follows:

1. B_e = available bandwidth of link e, initialized to e's line rate.
2. For each flow i, in increasing order of $\mathcal{T}_i$:
   (a) Let P_i be flow i's path.
   (b) Send flow i with rate $\mathcal{R}_i^{\text{sch}} = \min_{\forall e \in P_i}(\mathcal{R}_i^{\max}, B_e)$.
   (c) $B_e \leftarrow B_e - \mathcal{R}_i^{\text{sch}}$ for each $e \in P_i$.

Distributed Algorithm: We eliminate the unrealistic assumptions we made in the centralized algorithm to construct a fully distributed realization of our design. To distribute its operation, PDQ switches propagate flow information via explicit feedback in packet headers. PDQ senders maintain a set of flow-related variables such as flow sending rate and flow size and communicate the flow information to the intermediate switches via a scheduling header added to the transport layer of each data packet. When the feedback reaches the receiver, it is returned to the sender in an ACK packet. PDQ switches monitor the incoming traffic rate of each of their output queues and inform the sender to send data with a specific rate ($\mathcal{R}$>0) or to pause ($\mathcal{R}$=0) by annotating the scheduling header of data/ACK packets. We present the details of this distributed realization in the following sections.

3.1 PDQ Sender

Like many transport protocols, a PDQ sender sends a SYN packet for flow initialization and a TERM packet for flow termination, and resends a packet after a timeout. The sender maintains standard data structures for reliable transmission, including estimated round-trip time and states (e.g., timer) for in-flight packets. The PDQ sender maintains several state variables: its current sending rate ($\mathcal{R}_S$, initialized to zero), the ID of the switch (if any) who has paused the flow ($\mathcal{P}_S$, initialized to ø), flow deadline ($\mathcal{D}_S$, which is optional), the expected flow transmission time ($\mathcal{T}_S$, initialized to the flow size divided by sender NIC rate), the inter-probing time ($\mathcal{I}_S$, initialized to ø), and the measured RTT (RTT_S, estimated by an exponential decay).

The sender sends packets with rate $\mathcal{R}_S$. If the rate is zero, the sender sends a *probe* packet every $\mathcal{I}_S$ RTTs to get rate information from the switches. A probe packet is a packet with a scheduling header but no data content.

On packet departure, the sender attaches a scheduling header to the packet, containing fields set based on the values of each of the sender's state variables above. $\mathcal{R}_H$ is always set to the maximal sending rate $\mathcal{R}_S^{\max}$, while the remaining fields in the scheduling header are set to its current maintained variables. Note that the subscript H refers to a field in the scheduling header; the subscript S refers to a variable maintained by the sender; the subscript i refers to a variable related to the ith flow in the switch's flow list.

Whenever an ACK packet arrives, the sender updates its flow sending rate based on the feedback: $\mathcal{T}_S$ is updated based on the remaining flow size, RTT_S is updated based on the packet arrival time, and the remaining variables are set to the fields in the scheduling header.

Early Termination: For deadline-constrained flows, when the incoming flow demand exceeds the network capacity, there might not exist a feasible schedule for all flows to meet their deadlines. In this case, it is desirable to discard a minimal number of flows while satisfying the deadline of the remaining flows. Unfortunately, minimizing the number of tardy flows in a dynamic setting is an $\mathcal{NP}$-complete problem.[2]

Therefore, we use a simple heuristic, called *Early Termination*, to terminate a flow when it cannot meet its deadline. Here, the sender sends a TERM packet whenever *any* of the following conditions happen:

1. Deadline is past (Time > $\mathcal{D}_S$).
2. The remaining flow transmission time is larger than the time to deadline (Time + $\mathcal{T}_S > \mathcal{D}_S$).
3. The flow is paused ($\mathcal{R}_S = 0$), and the time to deadline is smaller than an RTT (Time + $RTT_S > \mathcal{D}_S$).

3.2 PDQ Receiver

A PDQ receiver copies the scheduling header from each data packet to its corresponding ACK. Moreover, to avoid the sender overrunning the receiver's buffer, the PDQ receiver reduces $\mathcal{R}_H$ if it exceeds the maximal rate that receiver can process and receive.

3.3 PDQ Switch

The high-level objective of a PDQ switch is to let the most *critical* flow complete as soon as possible. To this end, switches share a common flow comparator, which decides

[2]Consider a subproblem where a set of concurrent flows that share a bottleneck link all have the same deadline. This subproblem of minimizing the number of tardy flows is exactly the $\mathcal{NP}$-complete subset sum problem [11].

flow criticality, to approximate a range of scheduling disciplines. In this study, we implement two disciplines, EDF and SJF, while we give higher priority to EDF. In particular, we say a flow is more critical than another one if it has smaller deadline (emulating EDF to minimize the number of deadline-missing flows). When there is a tie or flows have no deadline, we break it by giving priority to the flow with smaller expected transmission time (emulating SJF to minimize mean flow completion time). If a tie remains, we break it by flow ID. If desired, the operator could easily override the comparator to approximate other scheduling disciplines. For example, we also evaluate another scheduling discipline incorporating flow waiting time in §7.

The switch's purpose is to resolve flow contention: flows can preempt less critical flows to achieve the highest possible sending rate. To achieve this goal, the switches maintain state about flows on each link (§3.3.1) and exchange information by tagging the scheduling header. To compute the rate feedback ($\mathcal{R}_H$), the switch uses a flow controller (controlling which flows to send; §3.3.2) and a rate controller (computing the aggregate flow sending rate; §3.3.3).

3.3.1 Switch State

In order to resolve flow contention, the switch maintains state about flows on each link. Specifically, it remembers the most recent variables ($<\mathcal{R}_i$, $\mathcal{P}_i$, $\mathcal{D}_i$, $\mathcal{T}_i$, $RTT_i>$) obtained from observed packet headers for flow i, which it uses to decide at any moment the correct sending rate for the flows. However, we do not have to keep this state for *all* flows. Specifically, PDQ switches only store the most critical 2κ flows, where κ is the number of *sending* flows (i.e., flows with sending rate $\mathcal{R}_S > 0$). Since PDQ allocates as much link bandwidth as possible to the most critical flows until the link is fully utilized, the κ most critical flows fully utilize the link's bandwidth; we store state for 2κ flows in order to have sufficient information immediately available to unpause another flow if one of the sending flows completes. The remaining flows are not remembered by the switch, until they become sufficiently critical.

The amount of state maintained at the switch thus depends on how many flows are needed to fill up a link. In most practical cases, this value will be very small because (i) PDQ allows critical flows to send with their highest possible rates, and (ii) switch-to-switch links are typically only $1-10\times$ faster than server-to-switch links, e.g., current data center networks mostly use 1 Gbps server links and 10 Gbps switch links[3], and the next generation will likely be 10 Gbps server links and 40 or 100 Gbps switch links. However, if a flow's rate is limited to something less than its NIC rate (e.g., due to processing or disk bottlenecks), switches may need to store more flows.

Greenberg et al. [12] demonstrated that, under a production data center of a large scale cloud service, the number of concurrent flows going in and out of a machine is almost never more than 100. Under a pessimistic scenario where *every* server concurrently sends or receives 100 flows, we have an average of 12,000 active flows at each switch in a VL2 network (assuming flow-level equal-cost multi-path forwarding and 24 10-Gbps Ethernet ports for each switch, the same as done in [12]). Today's switches are typically equipped with $4-16$ MByte of shared high-speed memory[4], while storing all these flows requires 0.23 MByte, only 5.72% of a 4 MByte shared memory. Indeed, in our simulation using the trace from [6], the maximum memory consumption was merely 9.3 KByte.

Still, suppose our memory imposes a hard upper limit M on the number of flows the switch can store. PDQ, as described so far, will cause under-utilization when $\kappa > M$ and there are paused flows wanting to send. In this underutilized case, we run an RCP [10] rate controller—which does not require per-flow state—alongside PDQ. We inform RCP that its maximum link capacity is the amount of capacity not used by PDQ, and we use RCP only for the less critical flows (outside the M most critical) that are not paused by any other switches. RCP will let *all* these flows run simultaneously using the leftover bandwidth. Thus, even in this case of large κ (which we expect to be rare), the result is simply a partial shift away from optimizing completion time and towards traditional fair sharing.

3.3.2 The Flow Controller

The flow controller performs Algorithm 1 and 3 whenever it receives a data packet and an ACK packet, respectively. The flow controller's objective is to *accept* or *pause* the flow. A flow is accepted if *all* switches along the path accept it. However, a flow is paused if *any* switch pauses it. This difference leads to the need for different actions:

Pausing: If a switch decides to pause a flow, it simply updates the "pauseby" field in the header ($\mathcal{P}_H$) to its ID. This is used to inform other switches and the sender that the flow should be paused. Whenever a switch notices that a flow is paused by another switch, it removes the flow information from its state. This can help the switch to decide whether it wants to accept other flows.

Acceptance: To reach consensus across switches, flow acceptance takes two phases: (i) in the forward path (from source to destination), the switch computes the available bandwidth based on flow criticality (Algorithm 2) and updates the rate and pauseby fields in the scheduling header; (ii) in the reverse path, if a switch sees an empty pauseby field in the header, it updates the *global* decision of acceptance to its state ($\mathcal{P}_i$ and $\mathcal{R}_i$).

We now propose several optimizations to refine our design:

Early Start: Given a set of flows that are not paused by other switches, the switch accepts flows according to their criticality until the link bandwidth is fully utilized and the remaining flows are paused. Although this ensures that the more critical flows can preempt other flows to fully utilize the link bandwidth, this can lead to *low link utilization when switching between flows.* To understand why, consider two flows, A and B, competing for a link's bandwidth. Assume that flow A is more critical than flow B. Therefore, flow A is accepted to occupy the entire link's bandwidth, while flow B is paused and sends only probe packets, e.g., one per its RTT. By the time flow A sends its last packet (TERM), the sender of flow B does not know it should start sending data because of the feedback loop delay. In fact, it could take one to two RTTs before flow B can start sending data.

[3]For example, the NEC PF5240 switch supports 48×1 Gbps ports, along with 2×10 Gbps ports; Pronto 3290 switch provides 48×1 Gbps ports and 4×10 Gbps ports.

[4]For example, the "deep-buffered" switches like Cisco Catalyst 4500, 4700, 4900 and 4948 series have 16 MByte shared memory, while shallow-buffered switches like Broadcom Triumph and Scorpion have 4 MByte shared memory [3].

Although the RTT in data center networks is typically very small (e.g., ~150 μs), the high-bandwidth short-flow nature makes this problem non-negligible. In the worst case where all the flows are short control messages (<10 KByte) that could finish in just one RTT, links could be idle more than half the time.

To solve this, we propose a simple concept, called *Early Start*, to provide seamless flow switching. The idea is to start the next set of flows slightly before the current sending flows finish. Given a set of flows that are not paused by other switches, a PDQ switch classifies a currently sending flow as nearly completed if the flow will finish sending in K RTTs (i.e., $\mathcal{T}_i < K \times RTT_i$), for some small constant K. We let the switch additionally accept as many nearly-completed flows as possible according to their criticality and subject to the resource constraint: aggregated flow transmission time (in terms of its estimated RTT) of the accepted nearly-completed flows ($\sum_i \mathcal{T}_i/RTT_i$) is no larger than K. The threshold K determines how early and how many flows will be considered as nearly-completed. Setting K to 0 will prevent concurrent flows completely, resulting in low link utilization. Setting K to a large number will result in congested links, increased queue sizes, and increased completion times of the most critical flows. Any value of K between 1 and 2 is reasonable, as the control loop delay is one RTT and the inter probing time is another RTT. In our current implementation we set $K = 2$ to maximize the link utilization, and we use the rate controller to drain the queue. Algorithm 2 describes this in pseudocode, and we will show that Early Start provides seamless flow switching (§5).

Dampening: When a more critical flow arrives at a switch, PDQ will pause the current flow and switch to the new flow. However, bursts of flows that arrive concurrently are common in data center networks, and can potentially cause frequent flow switching, resulting in temporary instability in the switch state.[5] To suppress this, we use dampening: after a switch has accepted a flow, it can only accept other paused flows after a given small period of time, as shown in Algorithm 1.

Suppressed Probing: One could let a paused sender send one probe per RTT. However, this can introduce significant bandwidth overhead because of the small RTTs in data center networks. For example, assume a 1-Gbps network where flows have an RTT of 150 μs. A paused flow that sends a 40-byte probe packet per RTT consumes $\frac{40\text{ Byte}}{150\ \mu s}/1$ Gbps $\approx$ 2.13% of the total bandwidth. The problem becomes more severe with larger numbers of concurrent flows.

To address this, we propose a simple concept, called *Suppressed Probing* to reduce the probing frequency. We make an observation that only the paused flows that are about to start have a need to send probes frequently. Therefore, it is desirable to control probing frequency based on the flow criticality and the network load. To control probing frequency, one would need to estimate *flow waiting time* (i.e., how long does it take until the flow can start sending). Although it is considered hard to predict future traffic workloads in data centers, switches can easily estimate a lower bound of the flow waiting time by checking their flow list. Assuming each flow requires at least X RTTs to finish, a PDQ switch estimates that a flow's waiting time is at least $X \times \max_{\forall \ell}\{\text{Index}(\ell)\}$ RTTs, where Index(ℓ) is the flow index in the list on link ℓ. The switch sets the inter-probing time field ($\mathcal{I}_H$) to $\max\{\mathcal{I}_H, X \times \text{Index}(\ell)\}$ in the scheduling header to control the sender probing rate ($\mathcal{I}_S$), as shown in Algorithm 3. The expected per-RTT probing overhead is significantly reduced from $O(n)$ (n flows, each of which sends one probe per RTT) to $\frac{1}{X}\sum_{k=1}^{n} 1/k = O(\log n)$. In our current implementation, we conservatively set X to 0.2 RTT_{avg}.

[5]We later show that PDQ can quickly converge to the equilibrium state when the workload is stable (§4).

Algorithm 1: PDQ Receiving Data Packet

if $\mathcal{P}_H$ = *other switch* **then**
    Remove the flow from the list if it is in the list;
    return;
if *the flow is not in the list* **then**
    if *the list is not full or the flow criticality is higher than the least critical flow in the list* **then**
        Add the flow into the list with rate $\mathcal{R}_i = 0$.
        Remove the least critical flow from the list whenever the list has more than κ flows.
    else
        Set $\mathcal{R}_H$ to RCP fair share rate;
        if $\mathcal{R}_H = 0$ **then** $\mathcal{P}_H$ = myID;
        return;
Let i be the flow index in the list; Update the flow information: $<\mathcal{D}_i,\mathcal{T}_i,RTT_i> = <\mathcal{D}_H,\mathcal{T}_H,RTT_H>$;
if W=min(*Availbw*(i),$\mathcal{R}_H$)>0 **then**
    if *the flow is not sending* ($\mathcal{P}_i \neq \emptyset$), *and the switch just accepted another non-sending flow* **then**
        $\mathcal{P}_H$ = myID; $\mathcal{P}_i$ = myID; // Pause it
    else $\mathcal{P}_H = \emptyset$; $\mathcal{R}_H$=W; // Accept it
else $\mathcal{P}_H$ = myID; $\mathcal{P}_i$ = myID; // Pause it

Algorithm 2: Availbw(j)

X=0; A=0;
for ($i = 0$; $i < j$; $i = i + 1$) **do**
    if $\mathcal{T}_i/RTT_i < K$ *and* $X < K$ **then**
        $X = X + \mathcal{T}_i/RTT_i$;
    else
        $A = A + \mathcal{R}_i$;
    if $A \geq C$ **then return** 0;
return $C - A$;

Algorithm 3: PDQ Receiving ACK

if $\mathcal{P}_H$ = *other switch* **then**
    Remove the flow from the list if it is in the list;
if $\mathcal{P}_H \neq \emptyset$ **then**
    $\mathcal{R}_H = 0$; // Flow is paused
if *the flow is in the list with index i* **then**
    $\mathcal{P}_i = \mathcal{P}_H$; $\mathcal{I}_H = \max\{\mathcal{I}_H, X \times i\}$; $\mathcal{R}_i = \mathcal{R}_H$;

3.3.3 The Rate Controller

The rate controller's objective is to control the aggregated flow sending rate of the flows accepted by the flow controller based on the queue size and the measured aggregate traffic. The rate adjustment serves the following purposes. First, whenever the queue builds up due to the use of Early Start, it helps drain the queue right after flow switching. Second, it helps tolerate the congestion caused by transient inconsistency. For example, if a packet carrying the pausing information gets lost, the corresponding sender that is supposed to stop will still be sending, and the rate controller can reduce the flow sending rate to react to the congestion. Finally, this allows PDQ to be friendly to other transport protocols in a multi-protocol network.

The rate controller maintains a single variable C to control the aggregated flow sending rate. This variable will be used to compute the sending rate field ($\mathcal{R}_H$) in the scheduling header, as shown in Algorithm 2.

The rate controller updates C every 2 RTTs because of the feedback-loop delay: we need about one RTT latency for the adjusted rate to take effect, and one additional RTT is used to measure the link congestion with that newly adjusted sending rate.

The rate controller updates C to $\max\{0, r_{\text{PDQ}} - q/(2 \times \text{RTT})\}$, where q is the instantaneous queue size and r_{PDQ} is the per-link aggregated rate for PDQ flows. If all traffic is transported using PDQ, one can configure the r_{PDQ} to be equal to the link rate. This allows PDQ flows to send with its highest possible rate. Otherwise, the network administrator can decide their priority by setting r_{PDQ} accordingly. For example, one could give preference to other protocols by periodically updating r_{PDQ} to the difference between the link rate and the measured aggregated traffic of the other protocols. Alternatively, one could set it based on the per-protocol traffic amount to achieve fairness across protocols.

4. FORMAL PROPERTIES

In this section, we present two formal properties of PDQ — deadlock-freedom and finite convergence time.

Assumptions: Without loss of generality, we assume there is no packet loss. Similarly, we assume flows will not be paused due to the use of flow dampening. Because PDQ flows periodically send probes, the properties we discuss in this section will hold with additional latency when the above assumptions are violated. For simplicity, we also assume the link rate C is equal to the maximal sending rate $\mathcal{R}_S^{\max}$ (i.e., $\mathcal{R}_S^{\text{sch}} = 0$ or C). Thus, each link accepts only one flow at a time.

Definitions: We say a flow is *competing* with another flow if and only if they share at least one common link. Moreover, we say a flow F_1 is a *precedential* flow of flow F_2 if and only if they are competing with each other and flow F_1 is more critical than flow F_2. We say a flow F is a *driver* if and only if (i) flow F is more critical than any other competing flow, or (ii) all the competing flows of flow F that are more critical than flow F are non-drivers.

Results (proof is in [14]): We verify that PDQ has no *deadlock*, which is a situation where two or more competing flows are paused and are each waiting for the other to finish (and therefore neither ever does). We further prove that PDQ will converge to the *equilibrium* in $P_{\max} + 1$ RTTs for stable workloads, where $P_{\max}$ is the maximal number of precedential flows of any flow. Given a collection of active flows, the equilibrium is defined as a state where the drivers are accepted while the remaining flows are paused.

5. PDQ PERFORMANCE

In this section, we evaluate PDQ's performance through comprehensive simulations. We first describe our evaluation setting (§5.1). Under a "query aggregation" scenario, PDQ achieves near-optimal performance and greatly outperforms D^3, RCP and TCP (§5.2). We then demonstrate that PDQ retains its performance gains under different workloads, including two realistic data center workloads from measurement studies (§5.3), followed by two scenarios to demonstrate that PDQ does not compromise on traditional congestion control performance metrics (§5.4). Moreover, PDQ retains its performance benefits on a variety of data center topologies (Fat-Tree, BCube and Jellyfish) and provides clear performance benefits at all scales that we evaluated (§5.5). Further, we show that PDQ is highly resilient to inaccurate flow information and packet loss (§5.6).

5.1 Evaluation Setting

Our evaluation considers two classes of flows:

Deadline-constrained Flows are time sensitive flows that have specific deadline requirements to meet. The flow size is drawn from the interval [2 KByte, 198 KByte] using a uniform distribution, as done in a prior study [20]. This represents query traffic (2 to 20 KByte in size) and delay sensitive short messages (>100 KByte) in data center networks [3]. The flow deadline is drawn from an exponential distribution with mean 20 ms, as suggested by [20]. However, some flows could have tiny deadlines that are unrealistic in real network applications. To address this, we impose a lower bound on deadlines, and we set it to 3 ms in our experiments. We use *Application Throughput*, the percentage of flows that meet their deadlines, as the performance metric of deadline-constrained flows.

Deadline-unconstrained Flows are flows that have no specific deadlines, but it is desirable that they finish early. For example, Dryad jobs that move file partitions across machines. Similarly, we assume the flow size is drawn uniformly from an interval with a mean of 100/1000 KByte. We use the average flow completion time as the performance metric.

We have developed our own event-driven packet-level simulator written in C++. The simulator models the following schemes:

PDQ: We consider different variants of PDQ. We use PDQ(Full) to refer to the complete version of PDQ, including Early Start (ES), Early Termination (ET) and Suppressed Probing (SP). Likewise, we refer to the partial version of PDQ which excludes the above three algorithms as PDQ(Basic). To better understand the performance contribution of each algorithm, we further extend PDQ(Basic) to PDQ(ES) and PDQ(ES+ET).

D^3: We implemented a complete version of D^3 [20], including the rate request processing procedure, the rate adaptation algorithm (with the suggested parameters $\alpha = 0.1$ and $\beta = 1$), and the quenching algorithm. In the original algorithm when the total demand exceeds the switch capacity, the fair share rate becomes negative. We found this can cause a flow to return the allocated bandwidth it already reserved, resulting in unnecessarily missed deadlines. Therefore, we add a constraint to enforce the fair share bandwidth fs to always be non-negative, which improves D^3's performance.

RCP: We implement RCP [10] and optimize it by counting the exact number of flows at switches. We found this improves the performance by converging to the fair share rate more quickly, significantly reducing the number of packet drops when encountering a sudden large influx of new flows [9].

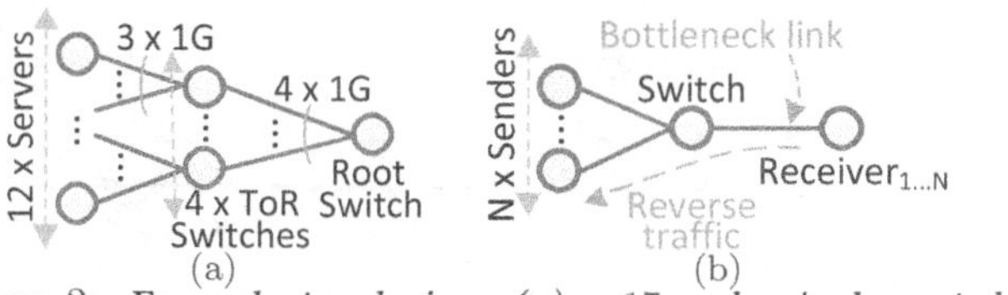

Figure 2: ***Example topologies: (a) a*** 17***-node single-rooted tree topology; (b) a single-bottleneck topology: sending servers associated with different flows are connected via a single switch to the same receiving server. Both topologies use*** 1 ***Gbps links, a switch buffer of*** 4 ***MByte, and FIFO tail-drop queues. Per-hop transmission/propagation/processing delay is set to*** 11/0.1/25 μs.

This is exactly equivalent to D^3 when flows have no deadlines.

TCP: We implement TCP Reno and optimize it by setting a small $RTO_{\min}$ to alleviate the TCP Incast problem, as suggested by previous studies [3, 19].

Unless otherwise stated, we use single-rooted tree, a commonly used data center topology for evaluating transport protocols [3, 19, 20, 22]. In particular, our default topology is a two-level 12-server single-rooted tree topology with 1 Gbps link rate (Figure 2a), the same as used in D^3. We vary the traffic workload and topology in §5.3 and §5.5.

5.2 Query Aggregation

In this section, we consider a scenario called *query aggregation*: a number of senders initiate flows at the same time to the same receiver (the aggregator). This is a very common application scenario in data center networks and has been adopted by a number of previous works [22, 20, 3]. We evaluate the protocols in both the deadline-constrained case (§5.2.1) and the deadline-unconstrained case (§5.2.2).

5.2.1 Deadline-constrained Flows

Impact of Number of Flows: We start by varying the number of flows.[6] To understand bounds on performance, we also simulate an *optimal* solution, where an omniscient scheduler can control the transmission of any flow with no delay. It first sorts the flows by EDF, and then uses a dynamic programming algorithm to discard the minimum number of flows that cannot meet their deadlines (Algorithm 3.3.1 in [16]). We observe that PDQ has near-optimal application throughput across a wide range of loads (Figure 3a).

Figure 3a demonstrates that Early Start is very effective for short flows. By contrast, PDQ(Basic) has much lower application throughput, especially during heavy system load because of the long down time between flow switching. Early Termination further improves performance by discarding flows that cannot meet their deadline. Moreover, Figure 3a demonstrates that, as the number of concurrent flows increases, the application throughput of D^3, RCP and TCP decreases significantly.

Impact of Flow Size: We fix the number of concurrent flows at 3 and study the impact of increased flow size on the application throughput. Figure 3b shows that as the flow size increases, the performance of deadline-agnostic schemes (TCP and RCP) degrades considerably, while PDQ remains

[6]We randomly assign f flows to n senders while ensuring each sender has either $\lfloor f/n \rfloor$ or $\lceil f/n \rceil$ flows.

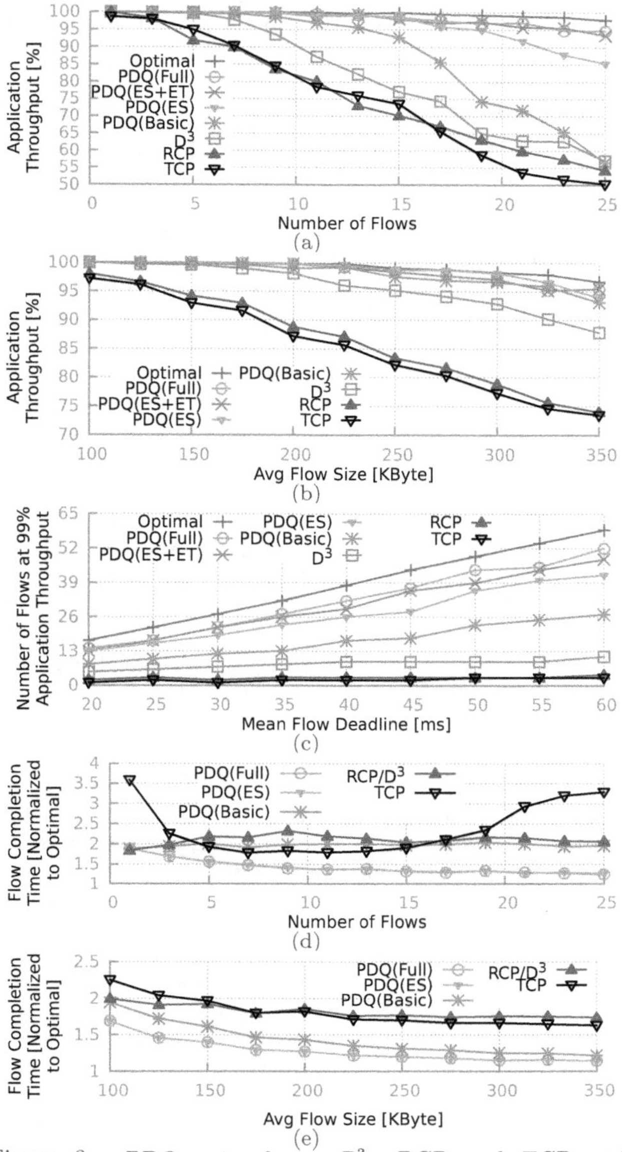

Figure 3: ***PDQ outperforms D^3, RCP and TCP and achieves near-optimal performance. Top three figures: deadline-constrained flows; bottom two figures: deadline-unconstrained flows.***

very close to optimal regardless of the flow size. However, Early Start and Early Termination provide fewer benefits in this scenario because of the small number of flows.

Impact of Flow Deadline: Data center operators are particularly interested in the operating regime where the network can satisfy almost every flow deadline. To this end, we attempt to find, using a binary search procedure, the maximal number of flows a protocol can support while ensuring 99% application throughput. We also vary the flow deadline, which is drawn from an exponential distribution, to observe the system performance with regard to different flow deadlines with mean between 20 ms to 60 ms. Figure 3c demonstrates that, compared with D^3, PDQ can support >3 times more concurrent flows at 99% application throughput, and this ratio becomes larger as the mean flow deadline increases. Moreover, Figure 3c shows that Suppressed Probing becomes more useful as the number of concurrent flows increases.

5.2.2 Deadline-unconstrained Flows

Impact of Flow Number: For deadline-unconstrained case, we first measure the impact of the number of flows on the average flow completion time. Overall, Figure 3d demonstrates that PDQ can effectively approximate the optimal flow completion time. The largest gap between PDQ and optimal happens when there exists only one flow and is due to flow initialization latency. RCP has a similar performance for the single-flow case. However, its flow completion time becomes relatively large as the number of flows increases. TCP displays a large flow completion time when the number of flows is small due to the inefficiency of slow start. When the number of concurrent flows is large, TCP also has an increased flow completion time due to the TCP incast problem [19].

Impact of Flow Size: We fix the number of flows at 3, and Figure 3e shows the flow completion time as the flow size increases. We demonstrate that PDQ can better approximate optimal flow completion time as flow size increases. The reason is intuitive: the adverse impact of PDQ inefficiency (e.g., flow initialization latency) on flow completion time becomes relatively insignificant as flow size increases.

5.3 Impact of Traffic Workload

Impact of Sending Pattern: We study the impact of the following sending patterns: **(i)** *Aggregation*: multiple servers send to the same aggregator, as done in the prior experiment. **(ii)** *Stride*(i): a server with index x sends to the host with index $(x+i)$ mod N, where N is the total number of servers; **(iii)** *Staggered Prob*(p): a server sends to another server under the same top-of-rack switch with probability p, and to any other server with probability $1-p$; **(iv)** *Random Permutation*: a server sends to another randomly-selected server, with a constraint that each server receives traffic from exactly one server (i.e., 1-to-1 mapping).

Figure 4 shows that PDQ reaps its benefits across all the sending patterns under consideration. The worst pattern for PDQ is the Staggered Prob(0.7) due to the fact that the variance of the flow RTTs is considerably larger. In this sending pattern, the non-local flows that pass through the core network could have RTTs $3-5$ times larger than the local flow RTTs. Thus, the PDQ rate controller, whose update frequency is based on a measurement of *average* flow RTTs, could slightly overreact (or underreact) to flows with relatively large (or small) RTTs. However, even in such a case, PDQ still outperforms the other schemes considerably.

Impact of Traffic Type: We consider two workloads collected from real data centers. First, we use a workload with flow sizes following the distribution from a large-scale commercial data center measured by Greenberg et al. [12]. It represents a mixture of long and short flows: Most flows are small, and most of the delivered bits are contributed by long flows. In the experiment, we assume that the short flows (with a size of <40 KByte) are deadline-constrained. We conduct these experiments with random permutation traffic.

Figure 5a demonstrates that, under this particular workload, PDQ outperforms the other protocols by supporting a significantly higher flow arrival rate. We observed that, in

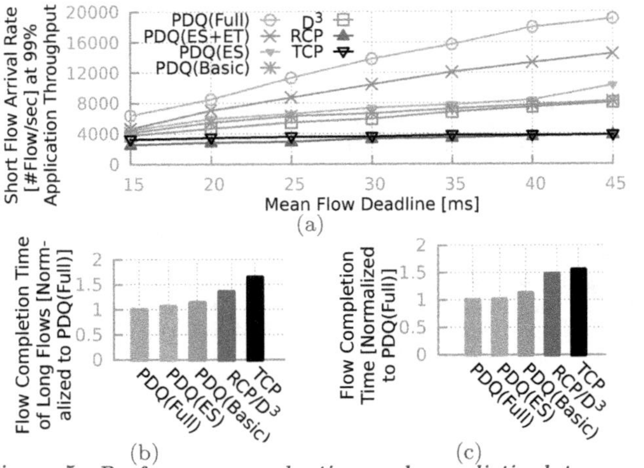

Figure 5: ***Performance evaluation under realistic data center workloads, collected from* (a, b) *a production data center of a large commercial cloud service [12] and* (c) *a university data center located in Midwestern United States (EDU1 in [6]).***

this scenario, PDQ(Full) considerably outperforms PDQ(ES+ET). This suggests that Suppressed Probing plays an important role in minimizing the probing overhead especially when there exists a large collection of paused flows. Figure 5b shows that PDQ has lower flow completion time for long flows: a 26% reduction compared with RCP, and a 39% reduction compared with TCP.

We also evaluate performance using a workload collected from a university data center with 500 servers [6]. In particular, we first convert the packet trace, which lasts 10 minutes, to flow-level summaries using Bro [1], then we fed it to the simulator. Likewise, PDQ outperforms other schemes in this regime (Figure 5c).

5.4 Dynamics of PDQ

Next, we show PDQ's performance over time through two scenarios, each with varying traffic dynamics:

Scenario #1 (Convergence Dynamics): Figure 6 shows that PDQ provides seamless flow switching. We assume five flows that start at time 0. The flows have no deadlines, and each flow has a size of ~1 MByte. The flow size is perturbed slightly such that a flow with smaller index is more critical. Ideally, the five flows together take 40 ms to finish because each flow requires a raw processing time of $\frac{1 \text{ MByte}}{1 \text{ Gbps}} = 8$ ms. With seamless flow switching, PDQ completes at ~42 ms due to protocol overhead (~3% bandwidth loss due to TCP/IP header) and first-flow initialization time (two-RTT latency loss; one RTT latency for the sender to receive the SYN-ACK, and an additional RTT for the sender to receive the first DATA-ACK). We observe that PDQ can converge to equilibrium quickly at flow switching time, resulting in a near perfect (100%) bottleneck link utilization (Figure 6b). Although an alternative (naive) approach to achieve such high link utilization is to let every flow send with fastest rate, this causes the rapid growth of the queue and potentially leads to congestive packet loss. Unlike this approach, PDQ exhibits a very small queue size[7] and has no packet drops (Figure 6c).

[7]The non-integer values on the y axis comes from the small probing packets.

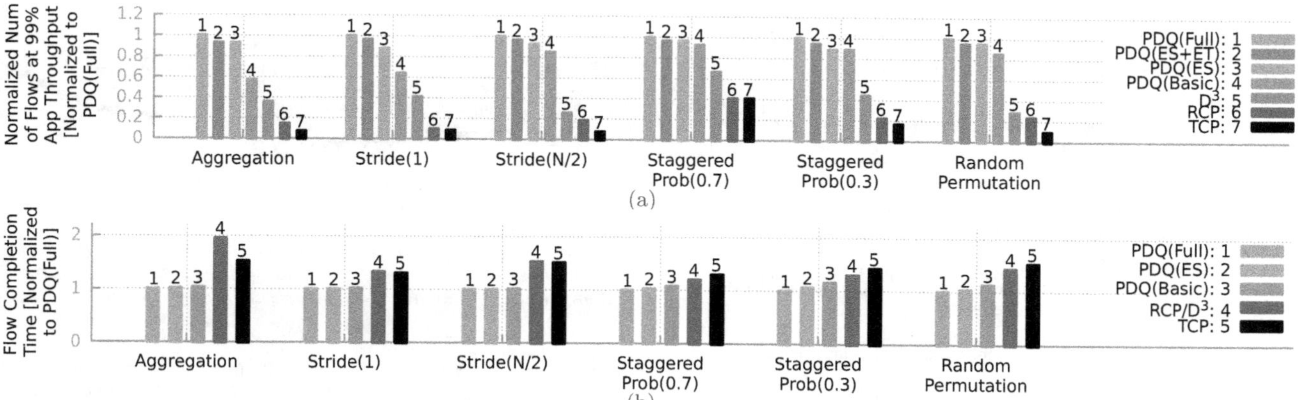

Figure 4: ***PDQ outperforms*** D^3***, RCP and TCP across traffic patterns.*** *(a) Deadline-constrained flows; (b) Deadline-unconstrained flows.*

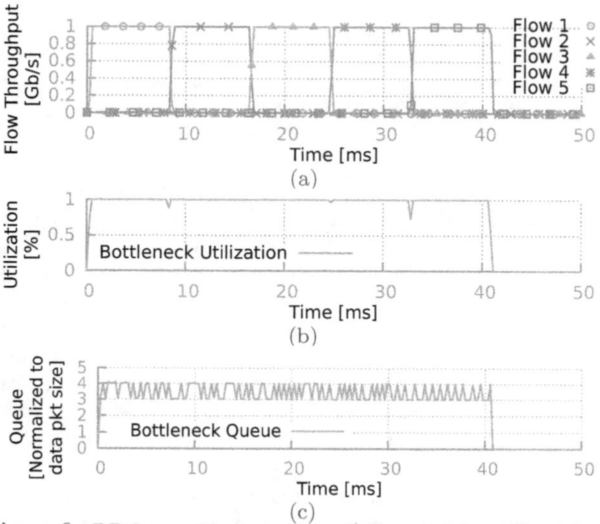

Figure 6: ***PDQ provides seamless flow switching. It achieves high link utilization at flow switching time, maintains small queue, and converges to the equilibrium quickly.***

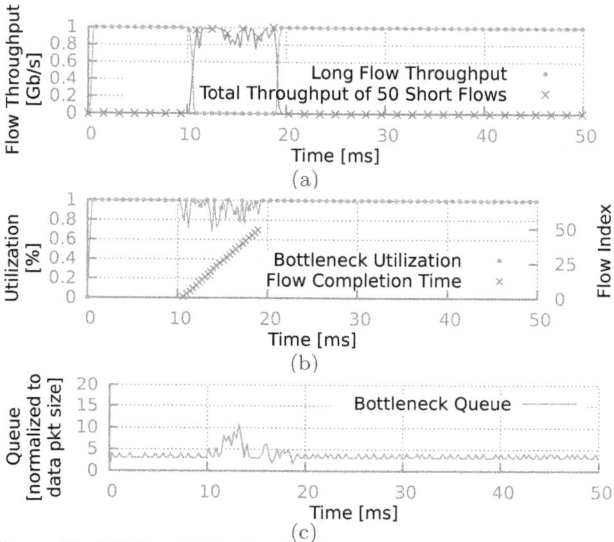

Figure 7: ***PDQ exhibits high robustness to bursty workload. We use a workload of*** 50 ***concurrent short flows all start at time*** 1 ***ms, and preempting a long-lived flow.***

Scenario #2 (Robustness to Bursty Traffic): Figure 7 shows that PDQ provides high robustness to bursty workloads. We assume a long-lived flow that starts at time 0, and 50 short flows that all start at 10 ms. The short flow sizes are set to 20 KByte with small random perturbation. Figure 7a shows that PDQ adapts quickly to sudden bursts of flow arrivals. Because the required delivery time of each short flow is very small ($\frac{20 \text{ KByte}}{1 \text{ Gbps}} \approx 153\ \mu s$), the system never reaches stable state during the preemption period (between 10 and 19 ms). Figure 7b shows PDQ adapts quickly to the burst of flows while maintaining high utilization: the average link utilization during the preemption period is 91.7%. Figure 7c suggests that PDQ does not compromise the queue length by having only 5 to 10 packets in the queue, which is about an order of magnitude smaller than what today's data center switches can store. By contrast, XCP in a similar environment results in a queue of ~60 packets (Figure 11(b) in [15]).

5.5 Impact of Network Scale

Today's data centers have many thousands of servers, and it remains unclear whether PDQ will retain its successes at large scales. Unfortunately, our packet-level simulator, which is optimized for high processing speeds, does not scale to large-scale data center topology within reasonable processing time. To study these protocols at large scales, we construct a *flow-level simulator* for PDQ, D^3 and RCP. In particular, we use an iterative approach to find the equilibrium flow sending rates with a time scale of 1 ms. The flow-level simulator also considers protocol inefficiencies like flow initialization time and packet header overhead. Although the flow-level simulator does not deal with packet-level dynamics such as timeouts or packet loss, Figure 8 shows that, by comparing with the results from packet-level simulation, the flow-level simulation does not compromise the accuracy significantly.

We evaluate three scalable data center topologies: (1) Fat-tree [2], a structured 2-stage Clos network; (2) BCube [13], a server-centric modular network; (3) Jellyfish [18], an unstructured high-bandwidth network using random regular graphs. Figure 8 demonstrates that PDQ scales well to large scale, regardless of the topologies we tested. Figure 8e shows that about 40% of flow completion times under PDQ are reduced by at least 50% compared to RCP. Only 5 – 15% of

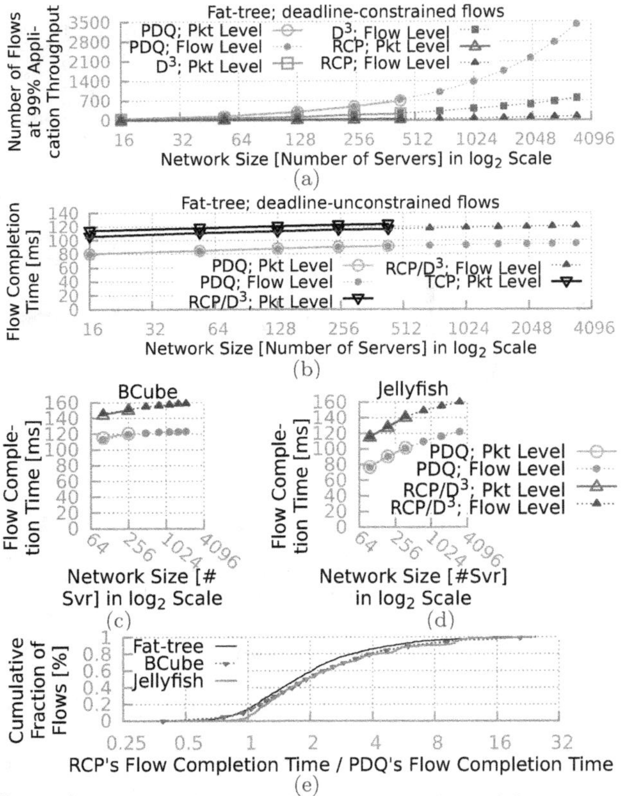

Figure 8: ***PDQ performs well across a variety of data center topologies. (a,b) Fat-tree; (c) BCube with dual-port servers; (d) Jellyfish with* 24*-port switches, using a* 2:1 *ratio of network port count to server port count. (e) For network flows, the ratio of the flow completion time under PDQ to the flow completion time under RCP (flow-level simulation; # servers is* ∼128*). All experiments are carried out using random permutation traffic; top figure: deadline-constrained flows; bottom four figures: deadline-unconstrained flows with* 10 *sending flows per server.***

the flows have a larger completion time, and no more than 0.9% of the flows have 2× completion time.

5.6 PDQ Resilience

Resilience to Packet Loss: Next, to evaluate PDQ's performance in the presence of packet loss, we randomly drop packets at the bottleneck link, in both the forward (data) and reverse (acknowledgment) direction. Figure 9 demonstrates that PDQ is even more resilient than TCP to packet loss. When packet loss happens, the PDQ rate controller detects anomalous high/low link load quickly and compensates for it with explicit rate control. Thus, packet loss does not significantly affect its performance. For a heavily lossy channel where the packet loss rate is 3% in both directions (i.e., a round-trip packet loss rate of $1-(1-0.03)^2 \approx 5.9\%$), as shown in Figure 9b, the flow completion time of PDQ has increased by 11.4%, while that of TCP has significantly increased by 44.7%.

Resilience to Inaccurate Flow Information: For many data center applications (e.g., web search, key-value stores, data processing), previous studies have shown that flow size can be precisely known at flow initiation time.[8] Even for

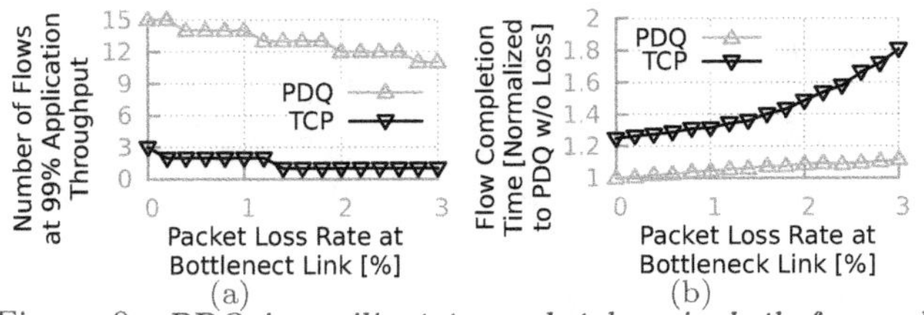

Figure 9: ***PDQ is resilient to packet loss in both forward and reverse directions: (a) deadline-constrained and (b) deadline-unconstrained cases. Query aggregation workload.***

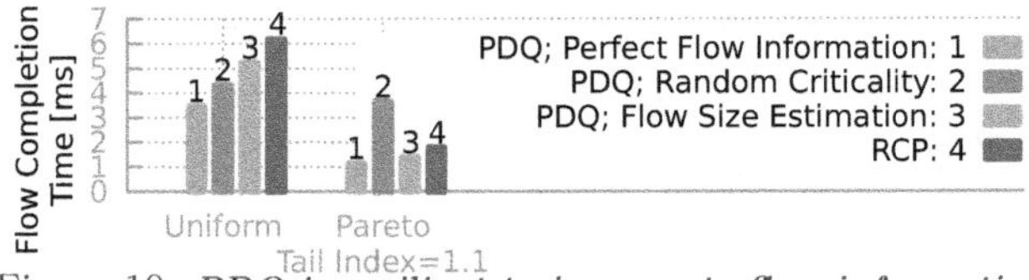

Figure 10: ***PDQ is resilient to inaccurate flow information. For PDQ without flow size information, the flow criticality is updated for every* 50 *KByte it sends. Query aggregation workload,* 10 *deadline-unconstrained flows with a mean size of* 100 *KByte. Flow-level simulation.***

applications without such knowledge, PDQ is resilient to inaccurate flow size information. To demonstrate this, we consider the following two flow-size-unaware schemes. *Random*: the sender randomly chooses a flow criticality at flow initialization time and uses it consistently. *Flow Size Estimation*: the sender estimates the flow size based on the amount of data sent already, and a flow is more critical than another one if it has smaller estimated size. To avoid excessive switching among flows, the sender does not change the flow criticality for every packet it sends. Instead, the sender updates the flow criticality only for every 50 KByte it sends. Figure 10 demonstrates two important results: (i) PDQ does require a reasonable estimate of flow size as random criticality can lead to large mean flow completion time in heavy-tailed flow size distribution. (ii) With a simple estimation scheme, PDQ still compares favorably against RCP in both uniform and heavy-tailed flow size distributions.

6. MULTIPATH PDQ

Several recent works [17, 21] show the benefits of multipath TCP, ranging from improved reliability to higher network utilization. Motivated by this work, we propose Multipath PDQ (M-PDQ), which enables a single flow to be striped across multiple network paths.

When a flow arrives, the M-PDQ sender splits the flow into multiple subflows, and sends a SYN packet for each subflow. To minimize the flow completion time, the M-PDQ sender periodically shifts the load from the paused subflows to the sending one with the minimal remaining load. To support M-PDQ, the switch uses flow-level Equal-Cost Multi-Path (ECMP) to assign subflows to paths. The PDQ switch requires no additional modification except ECMP. The M-PDQ receiver maintains a single shared buffer for a multipath flow to resequence out-of-order packet arrivals, as done in Multipath TCP [21].

We illustrate the performance gains of M-PDQ using BCube [13] a data center topology that allows M-PDQ to exploit the path diversity between hosts. We implement BCube address-based routing to derive multiple parallel paths. Using ran-

[8]See the discussion in §2.1 of [20].

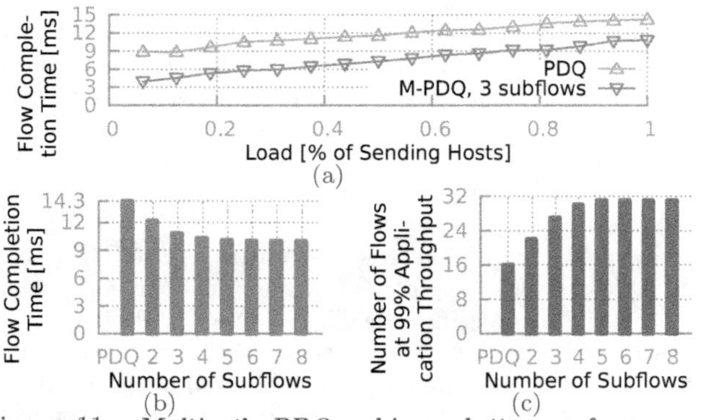

Figure 11: *Multipath PDQ achieves better performance. BCube(2,3) with random permutation traffic. (a, b) deadline-unconstrained, (c) deadline-constrained flows.*

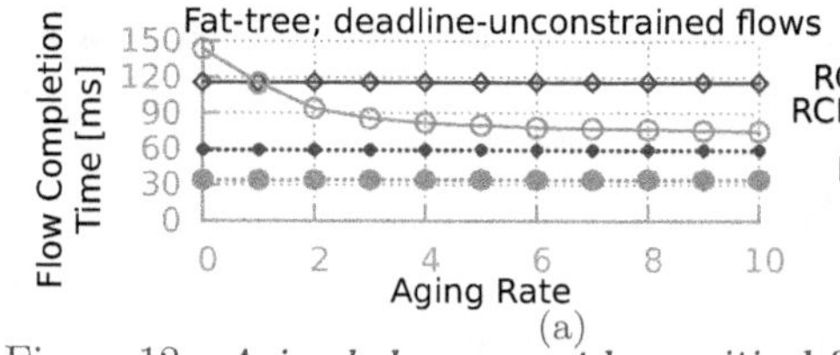

Figure 12: *Aging helps prevent less critical flows from starvation and shortens their completion time. The PDQ sender increases flow criticality by reducing $\mathcal{T}_H$ by a factor of $2^{\alpha t}$, where α is a parameter that controls the aging rate, and t is the flow waiting time (in terms of* 100 *ms). Flow-level simulation;* 128*-server fat-tree topology; random permutation traffic.*

dom permutation traffic, Figure 11a demonstrates the impact of the system load on flow completion time of M-PDQ. Here, we split a flow into 3 M-PDQ subflows. Under light loads, M-PDQ can reduce flow completion time by a factor of two. This happens because M-PDQ exploits more links that are underutilized or idle than single-path PDQ. As load increases, these advantages are reduced, since even single-path PDQ can saturate the bandwidth of nearly all links. However, as shown in Figure 11a, M-PDQ still retains its benefits because M-PDQ allows a critical flow to have higher sending rate by utilizing multiple parallel paths. Finally, we fix the workload at 100% to stress the network (Figures 11b and 11c). We observe that M-PDQ needs about 4 subflows to reach 97% of its full potential. By allowing servers to use all four interfaces (whereas single-path PDQ can use only one), M-PDQ provides a significant performance improvement.

7. DISCUSSION

Fairness. One could argue the performance gains of PDQ over other protocols stem from the fact that PDQ unfairly penalizes less critical flows. Perhaps counter-intuitively, the performance gain of SJF over fair sharing does not usually come at the expense of long jobs. An analysis [4] shows that at least 99% of jobs have a smaller completion time under SJF than under fair sharing, and this percentage increases further when the traffic load is less than half.[9] Our results further demonstrate that, even in complex data center networks with thousands of concurrent flows and multiple bottlenecks, 85 − 95% of PDQ's flows have a smaller completion time than RCP, and the worst PDQ flow suffers an inflation factor of only 2.57 as compared with RCP (Figure 8e). Moreover, unfairness might not be a primary concern in data center networks where the network is owned by a single entity that has full control of flow criticality. However, if desired, the operator can easily override the flow comparator to achieve a wide range of goals, including fairness. For example, to prevent starvation, the operator could gradually increase the criticality of a flow based on its waiting time. Using a fat-tree topology with 256 servers, Figure 12 demonstrates that this "flow aging" scheme is effective, reducing the worst flow completion time by ∼48%, while the mean flow completion time increases only 1.7%.

[9]Assuming a M/G/1 queueing model with heavy-tailed flow distributions; see [4].

When flow completion time is not the priority. Flow completion time is not the best metric for some protocols. For example, real-time audio and video may require the ability to *stream*, or provide a number of flows with a fixed fraction of network capacity. For these applications, protocols designed for streaming transport may be a better fit. One can configure the rate controller (§3.3.3) to slice the network into PDQ-traffic and non-PDQ-traffic, and use some other transport protocol for non-PDQ-traffic. In addition, there are also applications where the receiver may not be able to process incoming data at the full line rate. In such cases, sending any rate faster than what receiver can process does not offer substantial benefits. Assuming the receiver buffers are reasonably small, PDQ will back off and allocate remaining bandwidth to another flow.

Does preemption in PDQ require rewriting applications? A preempted flow is paused (briefly), not terminated. From the application's perspective, it is equivalent to TCP being slow momentarily; the transport connection stays open. Applications do not need to be rewritten since preemption is hidden in the transport layer.

Incentive to game the system. Users are rational and may have an incentive to improve the completion time of their own flows by splitting each flow into small flows. While a similar issue happens to D^3, TCP and RCP[10], users in PDQ may have an even greater incentive, since PDQ does preemption. It seems plausible to penalize users for having a large number of short flows by reducing their flows' criticality. Developing a specific scheme remains as future work.

Deployment. On end hosts, one can implement PDQ by inserting a shim layer between the IP and the transport layers. In particular, the sender maintains a set of PDQ variables, intercepts all calls between IP and transport layer, attaches and strips off the PDQ scheduling header[11], and passes the packet segment to IP/transport layer accordingly. Additionally, the shim layer could provide an API that al-

[10]In TCP/RCP, users may achieve higher aggregated throughput by splitting a flow into smaller flows; in D^3, users may request a higher rate than the flow actually needs.

[11]The 16-byte scheduling header consists of 4 fields, each occupying 4 bytes: $\mathcal{R}_H$, $\mathcal{P}_H$, $\mathcal{D}_H$, and $\mathcal{T}_H$. The PDQ receiver adds $\mathcal{I}_S$ and RTT_S to the header by reusing the fields used by $\mathcal{D}_H$ and $\mathcal{T}_H$. This is feasible because $\mathcal{D}_H$ and $\mathcal{T}_H$ are used only in the forward path, while $\mathcal{I}_S$ and RTT_S are used only in the reverse path. Any reasonable hashing that maps switch ID to 4-byte $\mathcal{P}_H$ should provide negligible collision probability.

lows applications to specify the deadline and flow size, or it could avoid the API by estimating flow sizes (§5.6). The PDQ sender can easily override TCP's congestion window size to control the flow sending rate. We note that PDQ requires only a few more variables per flow on end hosts. On switches, similar to previous proposals such as D^3, a vendor can implement PDQ by making modifications to the switch's hardware and software. Per-packet operations like modifying header fields are already implemented on most vendors' hardware (e.g., ASICs), which can be directly used by our design. The more complex operations like computing the aggregated flow rate and sorting/updating the flow list can be implemented in software. We note that PDQ's per-packet running time is $O(\kappa)$ for the top κ flows and $O(1)$ for the rest of the flows, where κ is a small number of flows with the highest criticality and can be bounded as in §3.3.1. The majority of the sending flows' scheduling headers would remain unmodified[12] by switches.

8. RELATED WORK

D^3: While D^3 [20] is a deadline-aware protocol that also employs explicit rate control like PDQ, it neither resequences flow transmission order nor preempts flows, resulting in a substantially different flow schedule which serves flows according to the order of their arrival. Unfortunately, this allows flows with large deadlines to hog the bottleneck bandwidth, blocking short flows that arrived later.

Fair Sharing: TCP, RCP [10] and DCTCP [3] all emulate fair sharing, which leads to suboptimal flow completion time.

TCP/RCP with Priority Queueing: One could use priority queuing at switches and assigning different priority levels to flows based on their deadlines. Previous studies [20] showed that, using two-level priorities, TCP/RCP with priority queueing suffers from losses and falls behind D^3, and increasing the priority classes to four does not significantly improve performance. This is because flows can have very different deadlines and require a large number of priority classes, while switches nowadays provide only a small number of classes, mostly no more than ten.

ATM: One could use ATM to achieve QoS priority control. However, ATM's CLP classifies traffic into only two priority levels, while PDQ gives each flow a unique priority. Moreover, ATM is unable to preempt flows (i.e., new flows cannot affect existing ones).

DeTail: In a recent (Oct 2011) technical report, Zats et al. propose DeTail [23], an in-network multipath-aware congestion management mechanism that reduces the flow completion time "tail" in datacenter networks. However, it targets neither mean flow completion time nor the number of deadline-missing flows. Unlike DeTail which removes the tail, PDQ can save ~30% flow completion time on average (compared with TCP and RCP), reducing the completion time of almost every flow (e.g., 85% − 95% of the flows, Figure 8e). We have not attempted a direct comparison due to the very different focus and the recency of this work.

9. CONCLUSION

We proposed PDQ, a flow scheduling protocol designed to complete flows quickly and meet flow deadlines. PDQ provides a distributed algorithm to approximate a range of scheduling disciplines based on relative priority of flows, minimizing mean flow completion time and the number of deadline-missing flows. We perform extensive packet-level and flow-level simulation of PDQ and several related works, leveraging real datacenter workloads and a variety of traffic patterns, network topologies, and network sizes. We find that PDQ provides significant advantages over existing schemes. In particular, PDQ can reduce by ~30% the average flow completion time as compared with TCP, RCP and D^3; and can support 3× as many concurrent senders as D^3 while meeting flow deadlines. We also design a multipath variant of PDQ by splitting a single flow into multiple subflows, and demonstrate that M-PDQ achieves further performance and reliability gains under a variety of settings.

[12]Until, of course, the flow is preempted or terminated.

10. REFERENCES

[1] Bro network security monitor. http://www.bro-ids.org.

[2] M. Al-Fares, A. Loukissas, and A. Vahdat. A scalable, commodity data center network architecture. In *SIGCOMM*, 2008.

[3] M. Alizadeh, A. Greenberg, D. A. Maltz, J. Padhye, P. Patel, B. Prabhakar, S. Sengupta, and M. Sridharan. Data center TCP (DCTCP). In *SIGCOMM*, 2010.

[4] N. Bansal and M. Harchol-Balter. Analysis of SRPT scheduling: Investigating unfairness. In *SIGMETRICS*, 2001.

[5] N. Bansal and M. Harchol-Balter. End-to-end statistical delay service under GPS and EDF scheduling: A comparison study. In *INFOCOM*, 2001.

[6] T. Benson, A. Akella, and D. A. Maltz. Network traffic characteristics of data centers in the wild. In *IMC*, 2010.

[7] J. Brutlag. Speed matters for Google web search, 2009.

[8] A. R. Curtis, J. C. Mogul, J. Tourrilhes, P. Yalagandula, P. Sharma, and S. Banerjee. DevoFlow: Scaling flow management for high-performance networks. In *SIGCOMM*, 2011.

[9] N. Dukkipati, Y. Ganjali, and R. Zhang-Shen. Typical versus worst case design in networking. In *HotNets*, 2005.

[10] N. Dukkipati and N. McKeown. Why flow-completion time is the right metric for congestion control. *SIGCOMM Comput. Commun. Rev.*, 2006.

[11] M. R. Garey and D. S. Johnson. *Computers and Intractability; A Guide to the Theory of NP-Completeness.* W. H. Freeman & Co., 1990.

[12] A. Greenberg, J. R. Hamilton, N. Jain, S. Kandula, C. Kim, P. Lahiri, D. A. Maltz, P. Patel, and S. Sengupta. VL2: A scalable and flexible data center network. In *SIGCOMM*, 2009.

[13] C. Guo, G. Lu, D. Li, H. Wu, X. Zhang, Y. Shi, C. Tian, Y. Zhang, and S. Lu. BCube: A high performance, server-centric network architecture for modular data centers. In *SIGCOMM*, 2009.

[14] C.-Y. Hong, M. Caesar, and P. B. Godfrey. Finishing flows quickly with preemptive scheduling. Technical report. http://arxiv.org/abs/1206.2057, 2012.

[15] D. Katabi, M. Handley, and C. Rohrs. Congestion control for high bandwidth-delay product networks. In *SIGCOMM*, 2002.

[16] M. L. Pinedo. *Scheduling: Theory, Algorithms, and Systems.* Springer, 2nd edition, 2002.

[17] C. Raiciu, S. Barre, C. Pluntke, A. Greenhalgh, D. Wischik, and M. Handley. Improving datacenter performance and robustness with multipath TCP. In *SIGCOMM*, 2011.

[18] A. Singla, C.-Y. Hong, L. Popa, and P. B. Godfrey. Jellyfish: Networking data centers randomly. In *NSDI*, 2012.

[19] V. Vasudevan, A. Phanishayee, H. Shah, E. Krevat, D. G. Andersen, G. R. Ganger, G. A. Gibson, and B. Mueller. Safe and effective fine-grained TCP retransmissions for datacenter communication. In *SIGCOMM*, 2009.

[20] C. Wilson, H. Ballani, T. Karagiannis, and A. Rowstron. Better never than late: Meeting deadlines in datacenter networks. In *SIGCOMM*, 2011.

[21] D. Wischik, C. Raiciu, A. Greenhalgh, and M. Handley. Design, implementation and evaluation of congestion control for multipath TCP. In *NSDI*, 2011.

[22] H. Wu, Z. Feng, C. Guo, and Y. Zhang. ICTCP: Incast congestion control for TCP in data center networks. In *CoNEXT*, 2010.

[23] D. Zats, T. Das, and R. H. Katz. DeTail: Reducing the flow completion time tail in datacenter networks. Technical report, Oct 2011.

DeTail: Reducing the Flow Completion Time Tail in Datacenter Networks

David Zats∓, Tathagata Das∓, Prashanth Mohan∓, Dhruba Borthakur◇, Randy Katz∓
∓ University of California, Berkeley ◇ Facebook
{dzats, tdas, prmohan, randy}@cs.berkeley.edu, dhruba@fb.com

ABSTRACT

Web applications have now become so sophisticated that rendering a typical page may require hundreds of intra-datacenter flows. At the same time, web sites must meet strict page creation deadlines of 200-300ms to satisfy user demands for interactivity. Long-tailed flow completion times make it challenging for web sites to meet these constraints. They are forced to choose between rendering a subset of the complex page, or delay its rendering, thus missing deadlines and sacrificing either quality or responsiveness. Either option leads to potential financial loss.

In this paper, we present a new cross-layer network stack aimed at reducing the long tail of flow completion times. The approach exploits cross-layer information to reduce packet drops, prioritize latency-sensitive flows, and evenly distribute network load, effectively reducing the long tail of flow completion times. We evaluate our approach through NS-3 based simulation and Click-based implementation demonstrating our ability to consistently reduce the tail across a wide range of workloads. We often achieve reductions of over 50% in 99.9th percentile flow completion times.

Categories and Subject Descriptors

C.2.2 [**Computer-Communication Networks**]: Network Protocols

Keywords

Datacenter network, Flow statistics, Multi-path

1. INTRODUCTION

Web sites have grown complex in their quest to provide increasingly rich and dynamic content. A typical Facebook page consists of a timeline-organized "wall" that is writeable by the user and her friends, a real-time cascade of friend event notifications, a chat application listing friends currently on-line, and of course, advertisements selected by displayed content. Modern web pages such as these are made up of many components, generated by independent subsystems and "mixed" together to provide a rich presentation of information.

Building such systems is not easy. They exploit high-level parallelism to assemble the independent page parts in a timely fashion, and present these incrementally, subject to deadlines to provide an interactive response. The final mixing system must wait for all subsystems to deliver some of their content, potentially sacrificing responsiveness if a small number of subsystems are delayed. Alternatively, it must present what it has at the deadline, sacrificing page quality and wasting resources consumed in creating parts of a page that a user never sees.

In this paper, we investigate how the network complicates such application construction, because of the high variation in performance of the network flows underlying their distributed workflows. By improving the statistics of network flow completion, in particular *by reducing the long flow completion tail*, the application gains better worst-case performance from the network. Applying the end-to-end principle, while the mixer software must still deal with subsystems that fail to respond by the deadline, an underlying network that yields better flow statistics reduces the conservativeness of time-outs while reducing the frequency with which they are triggered. The ultimate application-layer result is better quality and responsiveness of the presented pages.

Deadlines are an essential constraint on how these systems are constructed. Experiments at Amazon [26] demonstrated that failing to provide a highly interactive web site leads to significant financial loss. Increasing page presentation times by as little as 100ms significantly reduces user satisfaction. To meet these demands, web sites seek to meet deadlines of 200-300ms 99.9% of the time [12, 33].

Highly variable flow completion times complicate the meeting of interactivity deadlines. Application workflows that span the network depend on the performance of the underlying network flows. Packet arrival pacing is dictated by round-trip-times (RTTs) and congestion can significantly affect performance. While datacenter network RTTs can be as low as 250μs, in the presence of congestion, these times can grow by two orders of magnitude, forming a long tail distribution [12]. Average RTTs of hundreds of microseconds can occasionally take tens of milliseconds, with implications for how long a mixer application must wait before timing-out on receiving results from its subsystems.

Flash congestion is the culprit and it cannot be managed through conventional transport-layer means. Traffic bursts commonly cause packet losses and retransmissions [12]. Uneven load balancing often causes a subset of flows to experience unnecessarily high congestion [10]. The absence of traffic prioritization causes latency-sensitive foreground flows to wait behind latency-insensitive background flows [33]. Each contributes to increasing the long tail of flow completion, *especially for the latency-sensitive short flows critical for page creation.* While partial solutions exist [10, 12, 29,

33], no existing approach solves the whole problem. Fortunately, datacenter networks already contain the key enabling technology to reduce the long flow completion tail. They employ high-speed links and a scaled-out network topology, providing multiple paths between every source and destination [9, 23, 24].

Flash congestion can be reduced if it can be detected and if network-layer alternatives can be exploited quickly enough. We address this challenge by constructing a *cross-layer network stack that quickly detects congestion at lower network layers, to drive upper layer routing decisions, to find alternative lower-congestion paths to destinations.*

In this paper, we present the implementation and experimental evaluation of *DeTail*. DeTail is a cross-layer network stack design to reduce long-tailed flow completions in datacenter networks. It provides an effective network foundation for enabling mixer applications to assemble their complex content more completely and within responsiveness time constraints. The key contributions of this work are:

- Quantification of the impact of long-tailed flow completion times on different datacenter workflows;
- Assessment of the causes of long-tailed flow completion times;
- A cross-layer network stack that addresses them;
- Implementation-validated simulations demonstrating DeTail's reduction of 99.9^{th} percentile flow completion times by over 50% for many workloads without significantly increasing the median

In the following section, we analyze how long-tailed flow completion times affect workflows' interactive deadlines. In Section 3, we describe the causes of long-tailed flow completion times and the inadequacy of partial solutions. In Section 4, we introduce the cross-layer network-based approach DeTail uses to overcome these issues. In Section 5, we describe the NS-3-based simulation [6] and Click-based implementation [27] with which we evaluate DeTail. The evaluation of DeTail in Section 6 demonstrates reduced flow completion times for a wide range of workloads. We discuss various aspects of DeTail in Section 7. We describe how DeTail compares with prior work in Section 8 and conclude in Section 9.

2. IMPACT OF THE LONG TAIL

In this section, we begin by analyzing datacenter network traffic measurements, describing the phenomenon of the long tail. Next, we present two workflows commonly used by page creation subsystems and quantify the impact of the long flow completion time tail on their ability to provide rich, interactive content. We compare this with the performance that could be achieved with shorter-tailed distributions. We conclude this section with a discussion of how to quantify the long tail.

2.1 Traffic Measurements

Recently, Microsoft researchers [12] published datacenter traffic measurements for production networks performing services like web search. These traces captured three traffic types: (i) soft real-time queries, (ii) urgent short messages, and (iii) large deadline-insensitive background updates. Figure 1 reproduces graphs from [12], showing the distribution of measured round-trip-times (RTTs) from worker nodes to aggregators. The former typically communicate with mid-level aggregators (MLAs) located on the same rack. This graph represents the distribution of intra-rack RTTs.

Figure 1 shows that while the measured intra-rack RTTs are typically low, congestion causes them to vary by two orders of magnitude, forming a long-tail distribution. In this particular environment, intra-rack RTTs take as little as $61\mu s$ and have a median duration of $334\mu s$. But, in 10% of the cases, RTTs take over $1ms$. In fact, RTTs can be as high as $14ms$. These RTTs are the measured time between the transmission of a TCP packet and the receipt of its acknowledgement. Since switch queue size distributions match this behavior [11], *the variation in RTTs is caused primarily by congestion.*

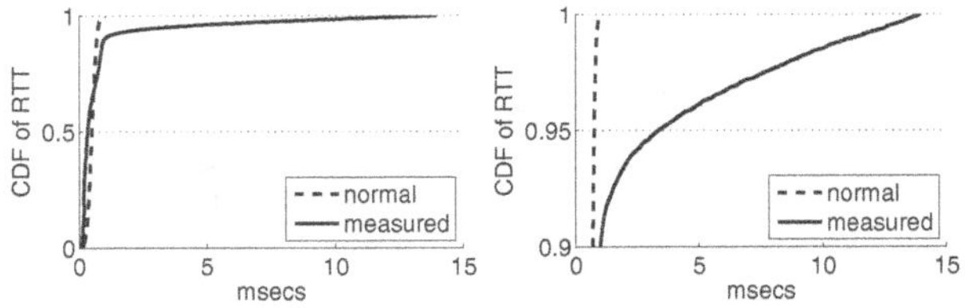

(a) Complete Distribution (b) 90th-100th Percentile

Figure 1: CDF of RTTs from the worker to the aggregator. We compare Microsoft's measured distribution [12] with a synthetic normal one having a 50% larger median.

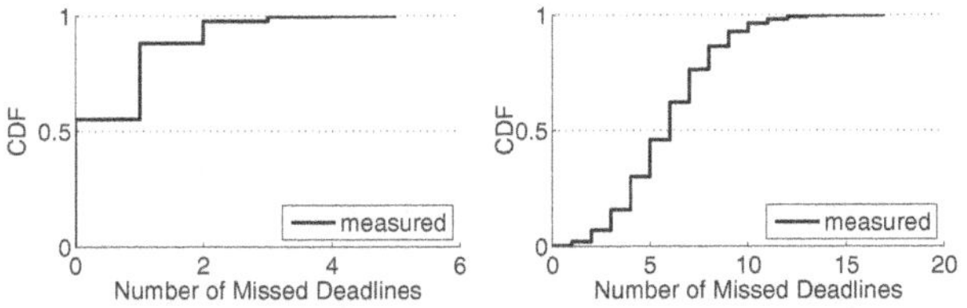

(a) 40 workers (b) 400 workers

Figure 2: Probability that a workflow will have a certain number of workers miss their $10ms$ deadline. All workers would meet their deadlines if RTTs followed the normal distribution.

For comparison, Figure 1 includes a synthetic distribution of RTTs following a normal distribution. While we set this distribution to have a median value that is 50% higher than that of the measured one, it has a much shorter tail.

As a measured distribution of datacenter flow completion times is unavailable, we conservatively assume each flow takes one RTT.

2.2 Impact on Workflows

Here we introduce the partition-aggregate and sequential workflows commonly used by page creation subsystems. For both workflows, we compare the impact of the long-tailed measured distribution with a shorter-tailed one. For this comparison, we focus on 99.9^{th} percentile performance as this is the common metric used for page creation [12, 33]. We see that a long-tailed distribution performs significantly worse than a shorter-tailed distribution, even when the latter has a higher median. We conclude this analysis with the key takeaways.

2.2.1 Partition-Aggregate

Partition-aggregate workflows are used by subsystems such as web search. Top-level aggregators (TLAs) receive requests. They divide (partition) the computation required to perform the request across multiple mid-level aggregators (MLAs), who further partition computation across worker nodes. Worker nodes perform the computation in parallel and send their results back to their MLA. Each MLA combines the results it receives and forwards them on to the TLA.

To ensure that the response is provided in a timely manner, it is common practice to give worker nodes as little as $10ms$ to perform their computation and deliver their result [12]. If a worker

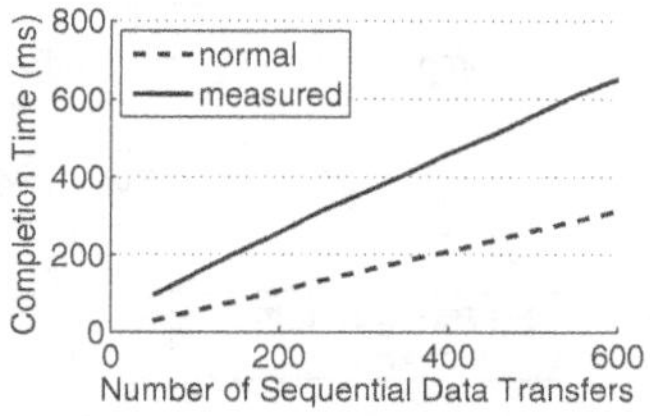

Figure 3: 99.9^{th} percentile completion times of sequential workflows. Web sites could use twice as many sequential requests per page under a shorter-tailed distribution.

node does not meet its deadline, its results are typically discarded, ultimately degrading the quality of the response.

To assess the impact of the measured RTT distribution (in Figure 1) on partition-aggregate workers meeting such deadlines, we analyze two hypothetical workflows. One has 40 workers while the other has 400. In Figure 2, we show the probability that a workflow will have a certain number of workers miss their deadlines. We assigned completion times to each worker by sampling from the measured RTT distribution. Those with completion times greater than $10ms$ were considered to have missed their deadlines. We performed this calculation 10000 times. Under the measured distribution, at the 99.9^{th} percentile, a 40-worker workflow has 4 workers (10%) miss their deadlines, while a 400-worker workflow has 14 (3.50%) miss theirs. Had RTTs followed the normal distribution, no workers would have missed their deadlines. This is despite the distribution having a 50% higher median than the measured one. This shows the hazard of designing for the median rather than long-tail performance.

These results assume that worker nodes do not spend any time computing the result they transmit. As the pressure for workers to perform more computation increases, the fraction of workers missing their deadlines will increase as well.

2.2.2 *Sequential*

In sequential workflows, a single front-end server fetches data from back-end servers (datastores) for every page creation. Future requests depend on the results of previous ones.

To quantify the impact of the long tail, we generated sequential workflows with varying numbers of data retrievals. We assumed that each data retrieval would use one flow and obtained values for retrievals by sampling from the appropriate distribution in Figure 1. We took the completion time of sequential workflows to be the sum of the randomly generated data retrieval times. We performed this calculation 10000 times.

In Figure 3, we report 99.9^{th} percentile completion times for different RTT distributions. Under the measured RTT distribution, to meet $200ms$ page creation deadlines, web sites must have less than 150 sequential data retrievals per page creation. Had RTTs followed the normal distribution, web sites could employ more than 350 sequential data retrievals per page. This is despite the distribution having a 50% higher median than the measured one. Again, designing for the median rather than long-tail performance is a mistake.

2.2.3 *Takeaways*

Long-tailed RTT distributions make it challenging for workflows used by page creation subsystems to meet interactivity deadlines. *While events at the long tail occur rarely, workflows use so many flows, that it is likely that several will experience long delays for every page creation.* Hitting the long tail is so significant that workflows actually perform better under distributions that have higher medians but shorter tails.

The impact is likely to be even greater than that presented here. Our analysis does not capture packet losses and retransmissions that are likely to cause more flows to hit the long tail.

Facebook engineers tell us that the long tail of flow completions forces their applications to choose between two poor options. They can set tight data retrieval timeouts for retrying requests. While this increases the likelihood that they will render complete pages, long tail flows generate non-productive requests that increase server load. Alternatively, they can use conservative timeouts that avoid unnecessary requests, but limit complete web page rendering by waiting too long for retrievals that never arrive. *A network that reduces the flow completion time tail allows such applications to use tighter timeouts to render more complete pages without increasing server load.*

2.3 Quantifying the Tail

Median flow completion time is an insufficient indicator of workflow performance. However, determining the right metric is challenging. Workflows only requiring 10 flows are much less likely to be affected by 99.9^{th} percentile flow completion times versus those with 1000 flows. To capture the effect of the long tail on a range of different workflow sizes, we report both 99th and 99.9^{th} percentile flow completion times.

3. CAUSES OF LONG TAILS

Section 2 showed how the long tail of flow completion times impacts page creation workflows. As mentioned earlier, flash congestion aggravates three problems that lead to long-tailed flow completion times: packet losses and retransmissions, absence of prioritization, and uneven load balancing. Here we describe these problems and how they affect the latency-sensitive short flows critical to page creation. We then discuss why current solutions fall short.

3.1 Packet Losses and Retransmissions

[12, 16, 31] study the effect of packet losses and retransmissions on network performance in datacenters. Packet losses often lead to flow timeouts, particularly in short flows where window sizes are not large enough to perform fast recovery. In datacenters, these timeouts are typically set to $10ms$ [12, 31]. Since datacenter RTTs are commonly of the order of $250\mu s$, just one timeout guarantees that the short flow will hit the long tail. It will complete too late, making it unusable for page creation. Using shorter timeouts may mitigate this problem, but it increases the likelihood of spurious retransmissions that increase network and server load.

Additionally, partition-aggregate workflows increase the likelihood of incast breakdown [12, 33]. Workers performing computation typically respond simultaneously to the same aggregator, sending it short flows. This sometimes leads to correlated losses that cause many flows to timeout and hit the long tail.

3.2 Absence of Prioritization

Datacenter networks represent a shared environment where many flows have different sizes and timeliness requirements. The traces from Section 2 show us that datacenters must support both latency-sensitive and latency-insensitive flows, with sizes that typically range from 2*KB* to 100*MB* [12].

During periods of flash congestion, short latency-sensitive flows can become enqueued behind long latency-insensitive flows. This increases the likelihood that latency-sensitive flows will hit the long tail and miss their deadlines. Approaches that do not consider different flow requirements can harm latency-sensitive flows.

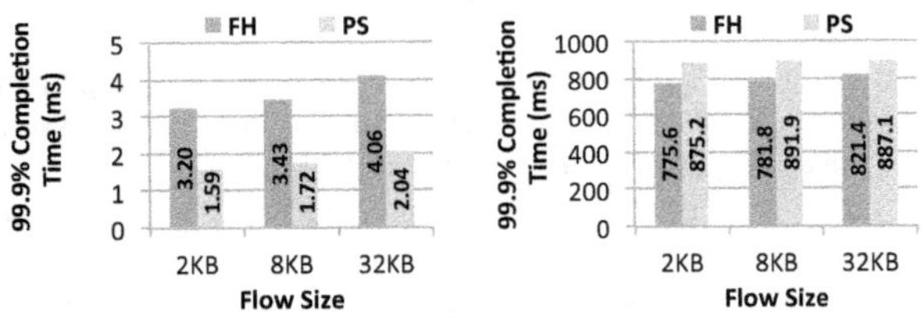

(a) Regular Topology (b) Degraded Link

Figure 4: Simulated 99.9^{th} percentile flow completion times of flow hashing (FH) and packet scatter (PS)

3.3 Uneven Load Balancing

Modern datacenter networks have scaled out, creating many paths between every source and destination [9, 23, 24]. Flow hashing is typically used to spread load across these paths while maintaining the single-path assumption commonly employed by transport protocols. Imperfect hashing, as well as varying flow sizes often lead to uneven flow assignments. Some flows are unnecessarily assigned to a more congested path, despite the availability of less congested ones. This increases the likelihood that they will hit the long tail.

This phenomena has been observed before for large flow sizes [10, 29]. Here we show that it is also a problem for the short flows common in page creation. We present a simulation on a 128-server FatTree topology with a moderate oversubscription factor of four (two from top-of-rack to aggregate switches and two from aggregate to core switches). For this experiment, we ran an all-to-all workload consisting solely of high-priority, uniformly chosen $2KB$, $8KB$, and $32KB$ flows. These sizes span the range of latency-sensitive flows common in datacenter networks [12].

In Figure 4(a), we compare the performance of flow hashing and a simple multipath approach: *packet scatter*. Packet scatter randomly picks the output port on which to send packets when multiple shortest paths are available. To factor out transport-layer effects, we used infinitely large switch buffers and also disabled rate-limiting and packet retransmission mechanisms. We see that packet scatter significantly outperforms traditional flow hashing, cutting 99.9^{th} percentile flow completion times by half. As we have removed transport-layer effects, these results show that *single path approaches reliant of flow hashing significantly under-perform multipath ones.*

Multipath approaches that do not dynamically respond to congestion, like packet scatter, may perform significantly worse than flow hashing for topological asymmetries. Consider a common type of failure, where a $1Gbps$ link between a core and aggregate switch has been degraded and now operates at $100Mbps$ [29]. Figure 4(b) shows that for the same workload, packet scatter can perform 12% worse than flow hashing. As we will see in Section 6, flow hashing itself performs poorly.

Topological asymmetries occur for a variety of reasons. Datacenter network failures are common [18]. Asymmetries can be caused by incremental deployments or network reconfigurations. Both static approaches (packet scatter and flow hashing) are unaware of the different capabilities of different paths and cannot adjust to these environments. *An adaptive multipath approach would be able to manage such asymmetries.*

3.4 Current Solutions Insufficient

DCTCP, D^3, and HULL [12, 13, 33] are single path solutions recently proposed to reduce the completion times of latency-sensitive flows. Single-path fairness and congestion control protocols have also been developed through the datacenter bridging effort [2]. These reduce packet losses and prioritize latency-sensitive flows. But they do not address the uneven load balancing caused by flow hashing, and hence still suffer the performance loss illustrated in Figure 4(a).

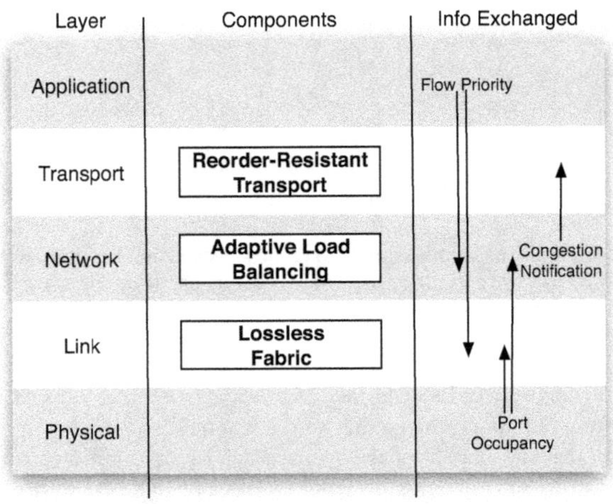

Figure 5: The DeTail network stack uses cross-layer information to address sources of long tails in flow completion times.

Recently two solutions have been proposed to more evenly balance flows across multiple paths. Hedera [10] monitors link state and periodically remaps flows to alleviate hotspots. Since Hedera remaps flows every five seconds and focuses on flows taking more than 10% of link capacity, it cannot improve performance for the short flows common in page creation.

The other solution is MPTCP [29]. MPTCP launches multiple TCP subflows and balances traffic across them based on congestion. MPTCP uses standard TCP congestion detection mechanisms that have been shown by DCTCP to be insufficient for preventing packet drops and retransmissions [12]. Also, while MPTCP is effective for flow sizes larger than $70KB$, it is worse than TCP for flows with less than 10 packets [29]. As small flows typically complete in just a few RTTs, host-based solutions do not have sufficient time to react to congested links and rebalance their load. *Current multipath-aware solutions cannot support the short flows common in page creation workflows.*

Most of the solutions discussed here seek to minimize in-network functionality. Instead they opt for host-based or controller-based approaches. Quick response times are needed to support the short, latency-sensitive flows common in page creation. In the following section, we present our network-oriented, cross-layer approach to meeting this goal.

4. DETAIL

In this section, we first provide an overview of DeTail's functionality and discuss how it addresses the causes of long-tailed flow completion times. We then describe the mechanisms DeTail uses to achieve this functionality and their parameterization.

4.1 Overview

DeTail is a cross-layer network-based approach for reducing the long flow completion time tail. Figure 5 depicts the components of the DeTail stack and the cross-layer information exchanged.

At the link layer, DeTail uses port buffer occupancies to construct a *lossless fabric* [2]. By responding quickly, lossless fabrics ensure that packets are never dropped due to flash congestion. They are only dropped due to hardware errors and/or failures. Preventing

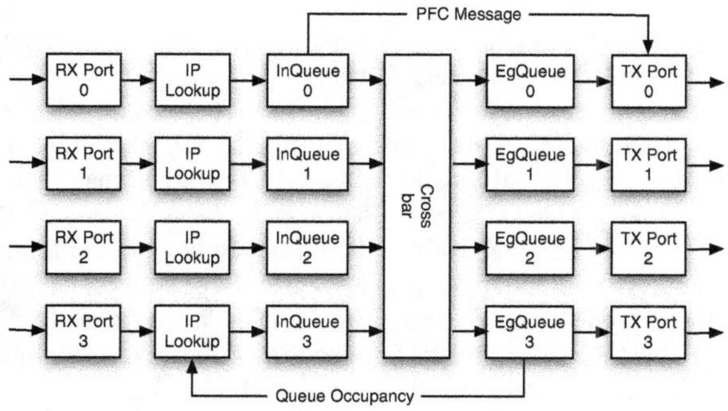

Figure 6: Assumed CIOQ Switch Architecture

congestion-related losses reduces the number of flows that experience long completion times.

At the network layer, DeTail performs per-packet adaptive load balancing of packet routes. At every hop, switches use the congestion information obtained from port buffer occupancies to dynamically pick a packet's next hop. This approach evenly smooths network load across available paths, reducing the likelihood of encountering a congested portion of the network. Since it is adaptive, it performs well even given topologic asymmetries.

DeTail's choices at the link and network layers have implications for transport. Since packets are no longer lost due to congestion, our transport protocol relies upon congestion notifications derived from port buffer occupancies. Since routes are load balanced one packet at a time, out-of-order packet delivery cannot be used as an early indication of congestion to the transport layer.

Finally, DeTail allows applications to specify flow priorities. Applications typically know which flows are latency-sensitive foreground flows and which are latency-insensitive background flows. By allowing applications to set these priorities, and responding to them at the link and network layers, DeTail ensures that high-priority packets do not get stuck behind low-priority ones. This assumes that applications are trusted, capable of specifying which flows are high priority. We believe that this assumption is appropriate for the kind of environment targeted by DeTail.

4.2 DeTail's Details

Now we discuss the detailed mechanisms DeTail uses to realize the functionality presented earlier. We begin by describing our assumed switch architecture. Then we go up the stack, discussing what DeTail does at every layer. We conclude by discussing the benefits of our cross-layer stack.

4.2.1 Assumed Switch Architecture

In Figure 6, we depict a four-port representation of a Combined Input/Output Queue (CIOQ) Switch. The CIOQ architecture is commonly used in today's switches [1,28]. We discuss DeTail's mechanisms in the context of this architecture and postpone discussion of others until Section 7. This architecture employs both ingress and egress queues, which we denote as InQueue and EgQueue, respectively. A crossbar moves packets between these queues.

When a packet arrives at an input port (e.g., RX Port 0), it is passed to the forwarding engine (IP Lookup). The forwarding engine determines on which output port (e.g., TX Port 2) the packet should be sent. Once the output port has been determined, the packet is stored in the ingress queue (i.e., InQueue 0) until the crossbar becomes available. When this happens, the packet is passed from the ingress queue to the egress queue corresponding to the desired output port (i.e., InQueue 0 to EgQueue 2). Finally, when the packet reaches the head of the egress queue, it is transmitted on the corresponding output port (i.e., TX Port 2).

To ensure that high-priority packets do not wait behind those with low-priority, the switch's ingress and egress queues perform strict priority queueing. Switches are typically capable of performing strict priority queueing between eight different priorities [4]. We use strict prioritization at both ingress and egress queues.

DeTail requires that the switch provide per-priority ingress and egress queue occupancies to higher layers in the stack. Each queue maintains a *drain bytes* counter per priority. This is the number of bytes of equal or higher priority in front of a newly arriving packet. The switch maintains this value by incrementing/decrementing the counters for each arriving/departing packet.

Having higher layers continuously poll the counter values of each queue may be prohibitively expensive. To address this issue, the switch associates a signal with each counter. Whenever the value of the counter is below a pre-defined threshold, the switch asserts the associated signal. These signals enable higher layers to quickly select queues without having to obtain the counter values from each. When multiple thresholds are used, a signal per threshold is associated with each counter. We describe how these thresholds are set in Section 4.3.2.

4.2.2 Link Layer

At the link layer, DeTail employs flow control to create a lossless fabric. While many variants of flow control exist [8], we chose to use the one that recently became part of the Ethernet standard: Priority Flow Control (PFC) [7]. PFC has already been adopted by vendors and is available on newer Ethernet switches [4].

The switch monitors ingress queue occupancy to detect congestion. When the drain byte counters of an ingress queue pass a threshold, the switch reacts by sending a Pause message informing the previous hop that it should stop transmitting packets with the specified priorities. When the drain byte counters reduce, it sends an Unpause message to the previous hop asking it to resume transmission of packets with the selected priorities[1].

During periods of persistent congestion, buffers at the previous hop fill, forcing it to generate its own Pause message. In this way, flow control messages can propagate back, quenching the source.

We chose to generate Pause/Unpause messages based on ingress queue occupancies because packets stored in these queues are attributed to the port on which they arrived. By sending Pause messages to the corresponding port when an ingress queue fills, DeTail ensures that the correct source postpones transmission.

Our choice of using PFC is based on the fact that packets in lossless fabrics can experience head-of-line blocking. With traditional flow control mechanisms, when the previous hop receives a Pause message, it must stop transmitting all packets on the link, not just those contributing to congestion. As a result, packets at the previous hop that are not contributing to congestion may be unnecessarily delayed. By allowing eight different priorities to be paused individually, PFC reduces the likelihood that low-priority packets will delay high priority ones. We describe how packet priorities are set in Section 4.2.5.

4.2.3 Network Layer

At the network layer, DeTail makes congestion-based load balancing decisions. Since datacenter networks have many paths between the source and destination, multiple shortest path options exist. When a packet arrives at a switch, it is forwarded on to the shortest path that is least congested.

[1]PFC messages specify the duration for which packet transmissions should be delayed. We use them here in an on/off fashion.

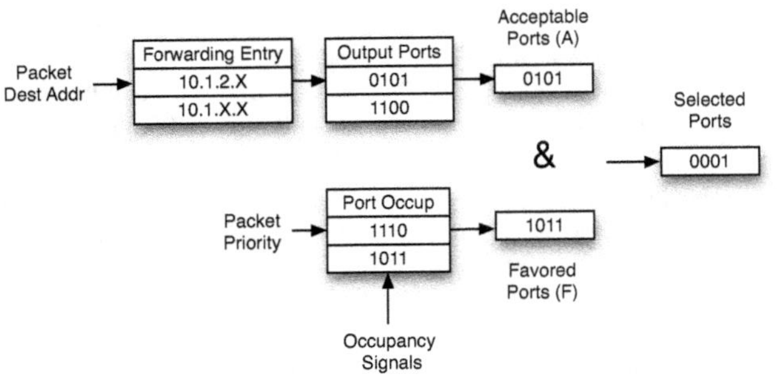

Figure 7: **Performing Adaptive Load Balancing** - A packet's destination IP address is used to determine the bitmap of *acceptable ports* (A). The packet's priority and port buffer occupancy signals are used to find the bitmap of the lightly loaded *favored ports* (F). A bitwise AND (&) of these two bitmaps gives the set of *selected ports* from which one is chosen.

DeTail monitors the egress queue occupancies described in Section 4.2.1 to make congestion-based decisions. Unlike traditional Ethernet, egress queue occupancies provide an indication of the congestion being experienced downstream. As congestion increases, flow control messages are propagated towards the source, causing the queues at each of the switches in the path to fill. By reacting to local egress queue occupancies we make globally-aware hop-by-hop decisions without additional control messages.

We would like to react by picking an acceptable port with the smallest drain byte counter at its egress queue for every packet. However, with the large number of ports in today's switches, the computational cost of doing so is prohibitively high. We leverage the threshold-based signals described in Section 4.2.1. By concatenating all the signals for a given priority, we obtain a bitmap of the favored ports, which are lightly loaded.

DeTail relies on forwarding engines to obtain the set of available shortest paths to a destination. We assume that associated with each forwarding entry is a bitmap of acceptable ports that lead to shortest paths for matching packets[2].

As shown in Figure 7, when a packet arrives, DeTail sends its destination IP address to the forwarding engine to determine which entry it belongs to and obtains the associated bitmap of acceptable ports (A). It then performs a bitwise *AND* (&) of this bitmap and the bitmap of favored ports (F) matching the packet's priority, to obtain the set of lightly loaded ports that the packet can use. DeTail randomly chooses from one of these ports and forwards the packet[3].

During periods of high congestion, the set of favored ports may be empty. In this case, DeTail performs the same operation with a second, larger threshold. If that does not yield results either, DeTail randomly picks a port from the bitmap. We describe how to set these thresholds in Section 4.3.2.

4.2.4 Transport Layer

A transport-layer protocol must address two issues to run on our load-balanced, lossless fabric. It must be resistant to packet reordering and it cannot depend on packet loss for congestion notification.

Our lossless fabric simplifies developing a transport protocol that is robust to out-of-order packet delivery. The lossless fabric ensures that packets will only be lost due to relatively infrequent hardware errors/failures. As packet drops are now much less frequent, it is not necessary that the transport protocol respond agilely to them. We simply need to disable the monitoring and reaction to out-of-order packet delivery. For TCP NewReno, we do this by disabling fast recovery and fast retransmit. While this leads to increased buffering at the end host, this is an acceptable tradeoff given the large amount of memory available on modern servers.

Obtaining congestion information from a lossless fabric is more difficult. Traditionally, transport protocols monitor packet drops to determine congestion information. As packet drops no longer happen due to congestion, we need another approach. To enable TCP NewReno to operate effectively with DeTail, we monitor the drain byte counters at all output queues. Low priority packets enqueued when the appropriate counter is above a threshold have their ECN flag set. This forces the low priority, deadline-insensitive TCP flow contributing to congestion to reduce its rate.

These types of modifications often raise concerns about performance and fairness across different transports. As the vast majority of datacenter flows are TCP [12] and operators can specify the transports used, we do not perform a cross-transport study here.

4.2.5 Application Layer

DeTail depends upon applications to properly specify flow priorities based on how latency-sensitive they are. Applications express these priorities to DeTail through the sockets interface. They set each flow (and hence the packets belonging to it) to have one of eight different priorities. As the priorities are relative, applications need not use all of them. In our evaluation, we only use two.

Applications must also react to extreme congestion events where the source has been quenched for a long time (Section 4.2.2). They need to determine how to reduce network load while minimally impacting the user.

4.2.6 Benefits of the Stack

DeTail's layers are designed to complement each other, overcoming limitations while preserving their advantages.

As mentioned earlier, link-layer flow control can cause head-of-line blocking. In addition to using priority, we mitigate this by employing adaptive load balancing and ECN. Adaptive load balancing allows alternate paths to be used when one is blocked and ECN handles the persistent congestion that aggravates head-of-line blocking.

DeTail's per-packet adaptive load balancing greatly benefits from the decisions made at the link and transport layers. Recall that using flow control at the link layer provides the adaptive load balancer with global congestion information, allowing it to make better decisions. And the transport layer's ability to handle out-of-order packet delivery allows the adaptive load balancer more flexibility in making decisions.

4.3 Choice of Settings

Now that we have described the mechanisms employed by DeTail, we discuss how to choose their parameters. We also assess how end-host parameters should be chosen when running DeTail.

4.3.1 Link Layer Flow Control

A key parameter is the threshold for triggering PFC messages. Pausing a link early allows congestion information to be propagated more quickly, making DeTail's adaptive load balancing more agile. At the same time, it increases the number of control messages. As PFC messages take time to be sent and responded to, setting the Unpause threshold too low can lead to buffer underflow, reducing link utilization.

To strike a balance between these competing concerns, we must first calculate the time to generate PFC messages. We use the same approach described in [7] to obtain this value.

[2]Bitmaps can be obtained with the TCAM and RAM approach as described in [9].

[3]Round-robin selection can be used if random selection is costly

For $1GigE$, it may take up to $36.456\mu s$ for a PFC message to take effect[4]. $4557B$ (bytes) may arrive after a switch generates a PFC message. As we pause every priority individually, this can happen for all eight priorities. We must leave $4557B \times 8 = 36456B$ of buffer space for receiving packets after PFC generation. Assuming 128*KB* buffers, this implies a maximum Pause threshold of $(131072B - 36456B)/8 = 11827$ *Drain Bytes* per priority. Setting the threshold any higher can lead to packet loss.

Calculating the Unpause threshold is challenging because the specifics of congestion cause queues to drain at different rates. Our calculations simply assume a drain rate of 1*Gbps*, requiring an Unpause threshold of at least $4557B$ to ensure the ingress queues do not overflow. However, ingress queues may drain faster or slower than $1Gbps$. If they drain slower, additional control messages may have to be sent, re-pausing the priority. If they drain faster, our egress queues reduce the likelihood of link underutilization.

These calculations establish the minimum and maximum threshold values to prevent packet loss and buffer underflow. Between the desire for agility and reduced control message overhead, we set the Unpause threshold to the minimum value of 4557 *Drain Bytes* and the Pause threshold to 8192 *Drain Bytes* (halfway between the minimum and the maximum). When fewer priorities are used, the Pause threshold can be raised without suffering packet loss. Given the desire for agile response to congestion, we leave it unmodified.

The tradeoffs discussed here depend on link speeds and buffer sizes. Analysis of how these tradeoffs change is left for future work.

4.3.2 Adaptive Load Balancing

When performing threshold-based adaptive-load balancing, we must determine how many thresholds to have for a given priority (i.e., most favored, favored, and least favored ports) as well as what these thresholds should be. Clearly, increasing the number of thresholds increases complexity, so the benefits of each additional threshold must outweigh the complexity cost.

Through a simulation-based exploration of the design space with the other parameters as described above, we determined that having two thresholds of 16*KB* and 64*KB* yields favorable results.

4.3.3 Explicit Congestion Notification

The threshold for setting ECN flags represents a tradeoff. Setting it too low reduces the likelihood of head-of-line blocking but increases the chance that low-priority flows will back off too much, underutilizing the link. Setting it too high has the opposite effect. Through experiments, we determined that a threshold of 64*KB* drain bytes appropriately makes this tradeoff.

4.3.4 End-Host Timers

Setting the timeout duration (i.e., RTO_{min} in TCP) of end host timers too low may lead to spurious retransmissions that waste network resources. Setting them too high leads to long response times when packets are dropped.

Traditionally, transport-layer protocols recover from packet drops caused by congestion and hardware failures. Congestion occurs frequently, so responding quickly to packet drops is important for achieving high throughput. However, DeTail ensures that packet drops only occur due to relatively infrequent hardware errors/failures. Therefore, it is more important for the timeout duration to be larger to avoid spurious retransmissions.

To determine a robust timeout duration for DeTail, we simulated all-to-all incast 25 times with varying numbers of servers (connected to a single switch) and different values of RTO_{min}. During every incast event, one server receives a total of 1*MB* from the remaining servers. We saw that values of $10ms$ and higher effectively avoid spurious retransmissions.

Unlike this simulation, datacenter topologies typically have multiple hops. Hence, we use $200ms$ as RTO_{min} for DeTail in our evaluations to accommodate the higher RTTs.

5. EXPERIMENTAL SETUP

Here we describe the NS-3 based simulator [6] and Click-based implementation [27] we use to evaluate DeTail.

5.1 NS-3 Simulation

Our NS-3 based simulation closely follows the switch design depicted in Figure 6. Datacenter switches typically have 128-256*KB* buffers per port [12]. To meet this constraint, we chose per-port ingress and egress queues of 128*KB*.

Network simulators typically assume that nodes are infinitely fast at processing packets, this is inadequate for evaluating DeTail. We extended NS-3 to include real-world processing delays. Switch delays of $25\mu s$ are common in datacenter networks [12]. We rely upon published specifications to break-down this delay as follows, providing explanations where possible:

- $12.24\mu s$ transmission delay of a full-size $1530B$ Ethernet frame on a 1*GigE* link.
- $3.06\mu s$ crossbar delay when using a speedup of 4. Crossbar speedups of 4 are commonly used to reduce head of line blocking [28].
- $0.476\mu s$ propagation delay on a copper link [7].
- $5\mu s$ transceiver delay (both ends of the link) [7].
- $4.224\mu s$ forwarding engine delay (the remainder of the $25\mu s$ budget).

We incorporate the transceiver delay into the propagation delay. The other delays are implemented individually, including the response time to PFC messages.

Packet-level simulators are known to have scalability issues, in terms of topology size and simulation duration [29]. We evaluated the feasibility of also developing a flow-level simulator, but concluded that it would be unable to shed light on the packet-level dynamics that are the focus of this paper.

NS-3's TCP model lacks support for ECN. Hence, our simulations do not evaluate explicit congestion notification (as discussed in Section 4.2.4). As we will show, even without ECN-based throttling of low priority flows our simulations demonstrate impressive results.

5.2 Click-based Implementation

To validate our approach, we implemented DeTail in Click [27]. Overall, our implementation mirrors the design decisions specified in Section 4 and portrayed in Figure 6. Here we describe the salient differences and analyze the impact they have on our parameters.

5.2.1 Design Differences

Unlike hardware switches, software routers typically do not emulate a CIOQ switch architecture. Instead, the forwarding engine places packets directly into the output queue. This output-queued approach is poorly suited to DeTail because we rely on ingress queues to determine when to send PFC messages.

[4]We do not consider jumbo frames. Also, PFC is only defined for 10*GigE*. We use 1*GigE* for manageable simulation times. We base PFC response times on the time specified for Pause Frames. This is appropriate since 10*GigE* links are given the same amount of time to respond to PFC messages are they are to Pause Frames.

To address this difference, we modified Click to have both ingress and egress queues. When packets arrive, the forwarding engine simply annotates them with the desired output port and places them in the ingress queue corresponding to the port on which they arrived. Crossbar elements then pull packets from the ingress queue to the appropriate egress queue. Finally, when the output port becomes free, it pulls packets from its egress queue.

Software routers also typically do not have direct control over the underlying hardware. For example, when Click *sends* a packet, it is actually enqueued in the driver's ring buffer. The packet is then DMAed to the NIC where it waits in another buffer until it is transmitted. In Linux, the driver's ring buffer alone can contain hundreds of packets. It is difficult for the software router to asses how congested the output link is when performing load balancing. Also, hundreds of packets may be transmitted between the time when the software router receives a PFC message and it takes effect.

To address this issue, we add rate limiters in Click before every output port. They clock out packets based on the link's bandwidth. This reduces packet buildup in the driver's and NIC's buffers, instead keeping those packets in Click's queues for a longer duration.

5.2.2 Parameter Modifications

The limitations of our software router impact our parameter choices. As it lacks hardware support for PFC messages, it takes more time both generate and respond to them.

Also, our rate limiter allows batching up to 6*KB* of data to ensure efficient DMA use. This may cause PFC messages to be enqueued for longer before they are placed on the wire and additional data may be transmitted before a PFC message takes effect. This also hurts high-priority packets. High priority packets will suffer additional delays if they arrive just after a batch of low priority packets has been passed to the driver.

To address these limitations, we increased our Pause / Unpause thresholds. However, instead of increasing ingress queue sizes, we opted to ensure that only two priorities were used at a time. This approach allows us to provide a better assessment of the advantages of DeTail in datacenter networks.

6. EXPERIMENTAL RESULTS

In this section, we evaluate DeTail through extensive simulation and implementation, demonstrating its ability to reduce the flow completion time tail for a wide range of workloads. We begin with an overview describing our traffic workloads and touch on key results. Next, we compare simulation and implementation results, validating our simulator. Later, we subject DeTail to a wide range of workloads under a larger topology than permitted by the implementation and investigate its scaled-up performance.

6.1 Overview

To evaluate DeTail's ability to reduce the flow completion time tail, we compare the following approaches:

Flow Hashing (*FH*): Switches employ flow-level hashing. This is the status quo and is our baseline for comparing the performance of DeTail.

Lossless Packet Scatter (*LPS*): Switches employ packet scatter (as already explained in Section 3) along with Priority Flow Control (PFC). While not industry standard, *LPS* is a naive multipath approach that can be deployed in current datacenters. The performance difference between *LPS* and DeTail highlights the advantages of Adaptive Load Balancing (ALB).

DeTail: As already explained in previous sections, switches employ PFC and ALB.

All three cases use strict priority queueing and use TCP NewReno as the transport layer protocol. For *FH*, we use a TCP RTO_{min} of $10ms$, as suggested by prior work [12, 31]. Since *LPS* and DeTail use PFC to avoid packet losses, we use the standard value of $200ms$ (as discussed in Section 4.3.4). Also, we use reorder buffers at the end-hosts to deal with out-of-order packet delivery.

We evaluate DeTail against *LPS* only in Section 6.4. For all other workloads, *LPS* shows similar improvements as DeTail and has been omitted for space constraints.

Traffic Model: Our traffic model consists primarily of high-priority data retrievals. For each retrieval, a server sends a 10-byte request to another server and obtains a variable sized response (i.e., data) from it. The size of the data (henceforth referred to as *retrieval data size*) is randomly chosen to be 2*KB*, 8*KB*, or 32*KB*, with equal probability. We chose discrete data sizes for more effective analysis of 99th and 99.9^{th} percentile performance. The rate of generation of these data retrievals (henceforth called *retrieval rate*) and the selection of servers for the retrievals are defined by the traffic workload. In most cases, we assumed the inter-arrival times of retrievals to be exponentially distributed (that is, a Poisson process). We also evaluated against more bursty traffic models having lognormal distributions with varying sigma (σ) values. Where specified, we also run low-priority, long background data transfers.

Key results: Throughout our evaluation, we focus on 99th and 99.9^{th} percentile completion times of data retrievals to assess DeTail's effectiveness. We use the percentage reduction in the completion times provided by DeTail over Flow Hashing as the metric of improvement. Our key results are:

- DeTail completely avoids congestion-related losses, reducing 99.9^{th} percentile completion times of data retrievals in all-to-all workloads by up to 84% over Flow Hashing.
- DeTail effectively moves packets away from congestion hotspots that may arise due to disconnected links, reducing 99.9^{th} percentile completion times by up to 89% over Flow Hashing. *LPS* does not do as well and actually performs worse than *FH* for degraded links.
- Reductions in individual data retrievals translate into improvements for sequential and partition-aggregate workflows, reducing their 99.9^{th} percentile completion times by 54% and 78%, respectively.

6.2 Simulator Verification

To validate our simulator, we ran our Click-based implementation on Deter [14]. We constructed a 36-node, 16-server FatTree topology. Over-subscription is common in datacenter networks [3]. To model the effect of a moderate over-subscription factor of four, we rate-limited the ToR-to-aggregate links to 500*Mbps* and the aggregate-to-core links to 250*Mbps*.

We designated half of the servers to be front-end (web-facing) servers and half to be back-end servers. Each front-end server continuously selects a back-end server and issues a high-priority data retrieval to it. The data retrievals are according to a Poisson process and their rate is varied from 100 to 1500 retrievals/second.

We simulated the same workload and topology, with parameters matched with that of the implementation. Figure 8 compares the simulation results with the implementation measurements. For rates ranging from 500 to 1500 retrievals/sec, the percentage reduction in completion time predicted by the simulator is closely matched by implementation measurements, with the difference in

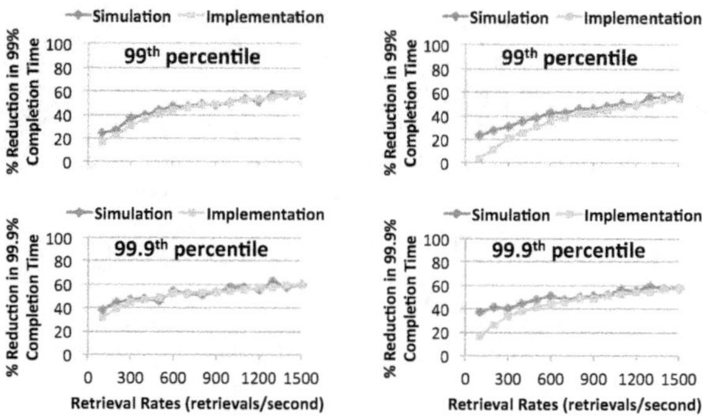

(a) $2KB$ (b) $8KB$

Figure 8: **Comparison of simulation and implementation results** - Reduction by DeTail over *FH* in 99th and 99.9^{th} percentile completion times of $2KB$ and $8KB$ data retrievals

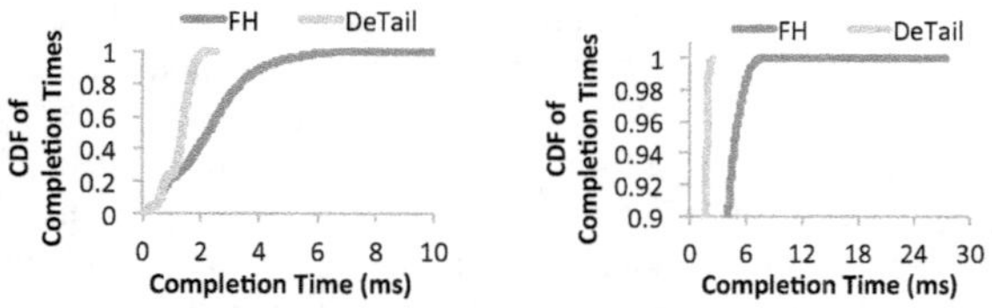

(a) Complete distribution (b) 90th-100th percentile

Figure 9: CDF of completion times of $8KB$ data retrievals under all-to-all workload of 2000 retrievals/second

the percentage being within 8% (results for $32KB$ data retrievals and *LPS* are similar and have been omitted for space constraints). Note that this difference increases for lower rates. We hypothesize that this is due to end-host processing delays that are present only in the implementation (i.e., not captured by simulation) dominating completion times during light traffic loads.

We similarly verified our simulator for lognormal distributions of data retrievals having a $\sigma = 1$. The simulation and implementation results continue to match, with the difference in the percentage growing slightly to 12%.This demonstrates that our simulator is a good predictor of performance that one may expect in a real implementation. Next, we use this simulator to evaluate larger topologies and a wider range of workloads.

6.3 Microbenchmarks

We evaluate the performance of DeTail on a larger FatTree topology with 128 servers. The servers are distributed into four pods having four ToR switches and four aggregate switches each. The four pods are connected to eight core switches. This gives an oversubscription factor of four in the network (two from top-of-rack to aggregate switches and two from aggregate to core switches). We evaluate two traffic patterns:

- **All-to-all**: Each server randomly selects another server and retrieves data from it. All 128 servers engage in issuing and serving data retrievals.
- **Front-end / Back-end**: Each server in first three pods (i.e, front-end server) retrieves data from a randomly selected server in the fourth pod (i.e., back-end server).

The data retrievals follow a Poisson process unless mentioned otherwise. In addition, each server is engaged in, on average, one $1MB$ low-priority background flow. Using a wide range of workloads, we illustrate how ALB and PFC employed in DeTail reduce the tail of completion times as compared to *FH*.

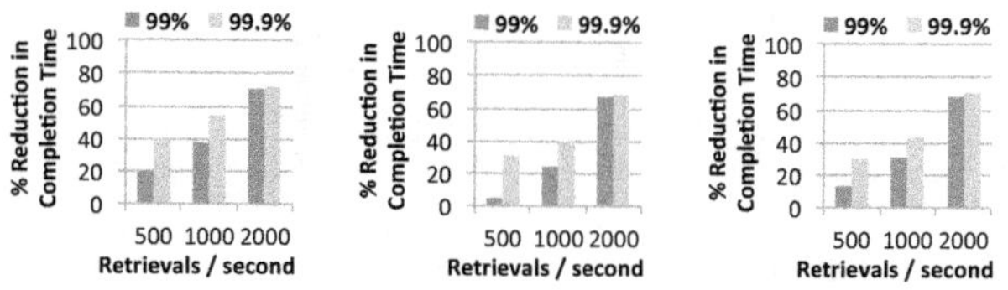

(a) $2KB$ (b) $8KB$ (c) $32KB$

Figure 10: **All-to-all Workload** - Reduction by DeTail over *FH* in 99th and 99.9^{th} percentile completion times of $2KB$, $8KB$ and $32KB$ retrievals

σ	0.5			1			2		
size (KB)	2	8	32	2	8	32	2	8	32
500 (r/s)	40%	20%	26%	38%	26%	26%	31%	30%	31%
1000 (r/s)	43%	30%	35%	46%	35%	37%	36%	23%	33%
2000 (r/s)	67%	62%	65%	68%	66%	67%	84%	76%	73%

Table 1: **All-to-all Workload with Lognormal Distributions** - Reduction in 99.9^{th} percentile completion time of retrievals under lognormal arrivals

All-to-all Workload: Each server generates retrievals at rates ranging from 500 to 2000 retrievals/second, which corresponds to load factors[5] of approximately 0.17 to 0.67. Figure 9 illustrates the effectiveness of DeTail in reducing the tail, by presenting the cumulative distribution of completion times of $8KB$ data retrievals under a rate of 2000 retrievals/second. While the 99th and 99.9^{th} percentile completion times under *FH* were $6.3ms$ and $7.3ms$, respectively, DeTail reduced them to $2.1ms$ and $2.3ms$; a reduction of about 67% in both cases. Even the median completion time improved by about 40%, from $2.2ms$ to $1.3ms$. Furthermore, the worst case completion time was $28ms$ under *FH* compared to $2.6ms$, which demonstrates the phenomenon discussed in Section 2. Flow completion times can increase by an order of magnitude due to congestion and mechanisms employed by DeTail are essential for ensuring tighter bounds on network performance.

Figure 10 presents the reductions in completion times for three data sizes at three retrieval rates. DeTail provided up to 70% reduction at the 99th percentile (71% at 99.9^{th} percentile) completion times. Specifically, the 99.9^{th} percentile completion times for all sizes were within $3.6ms$ compared to $11.9ms$ under *FH*. Within each data size, higher rates have greater improvement. The higher traffic load at these rates exacerbates the uneven load balancing caused by *FH*, which ALB addresses.

We also evaluate DeTail under more bursty traffic using lognormally distributed inter-arrival times. While keeping the same mean query rate (i.e., same load factors) as before, we vary the distribution's parameter σ from 0.5 to 2. Higher values of σ lead to more bursty traffic. Table 1 shows that DeTail achieves between 20% and 84% reductions at the 99.9^{th} percentile. Note that even at low load (500 r/s), for highly bursty ($\sigma = 2$) traffic DeTail achieves reductions greater than 30%.

Front-end / Back-end Workload: Each front-end server (i.e., servers in the first three pods) retrieves data from randomly selected back-end servers (i.e., servers in the fourth pod) at rates ranging from 125 to 500 retrievals/second, which correspond to load factors of approximately 0.17 to 0.67 on the aggregate-to-core links of the

[5]load factor is the approximate utilization of the aggregate-to-core links by high-priority traffic only

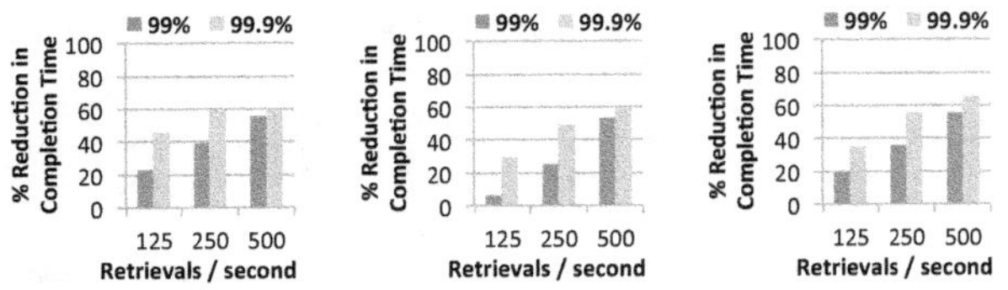

(a) $2KB$ (b) $8KB$ (c) $32KB$

Figure 11: **Front-end / Back-end Workload** - Reduction by DeTail over FH in 99th and 99.9^{th} percentile completion times of $2KB$, $8KB$ and $32KB$ data retrievals

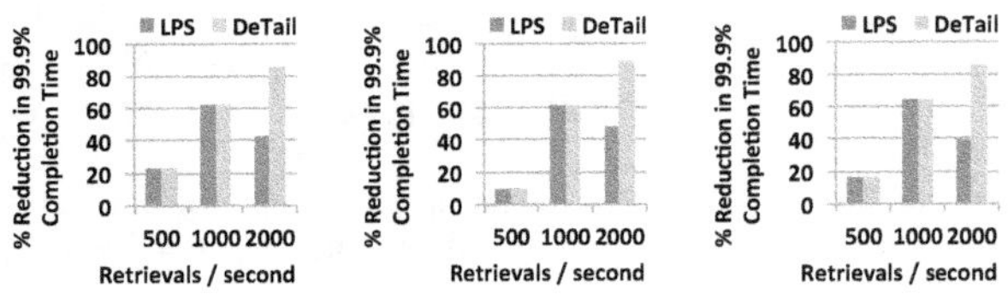

(a) $2KB$ (b) $8KB$ (c) $32KB$

Figure 12: **Disconnected Link** - Reduction by *LPS* and DeTail over FH in 99.9^{th} percentile completion times of $2KB$, $8KB$ and $32KB$ retrievals

fourth pod. Figure 11 shows that DeTail achieves 30% to 65% reduction in the completion times of data retrievals at the 99.9^{th} percentile. This illustrates that DeTail can perform well even under the persistent hotspot caused by this workload.

Long Background Flows: DeTail's approach to improving data retrievals (i.e., high-priority, short flows) does not sacrifice background flow performance. Due to NS-3's lack of ECN support, we evaluate the performance of background flows using the 16-server implementation presented earlier. We use the same half front-end servers and half-backend servers setup, and apply a retrieval rate 300 retrievals/second. Additionally, front-end servers are also continuously engaged in low-priority background flows with randomly selected back-end servers. The background flows are long; each flow is randomly chosen to be one of $1MB$, $16MB$ or $64MB$ with equal probability. Figure 14 shows that DeTail provides a 38% to 60% reduction over FH in the average completion time and a 58% to 71% reduction in the 99th percentile. Thus, DeTail significantly improves the performance of long flows. A detailed evaluation of DeTail's impact on long flows is left for future work.

6.4 Topological Asymmetries

As discussed in Section 3.3, a multipath approach must be robust enough to handle topological asymmetries due to network component failures or reconfigurations. We consider two types of asymmetries: disconnected links and degraded links. These asymmetries lead to load imbalance, even with packet scatter. In this section, we show how ALB can adapt to the varying traffic demands and overcome the limitations of packet-level scattering. Besides FH, we evaluate DeTail against *LPS* to highlight the strength of ALB over packet scatter (used in *LPS*). We assume that the routing protocol used in the network has detected the asymmetry and converged to provide stable multiple routes.

Disconnected Link: We evaluated the all-to-all workload with Poisson data retrievals on the same topology described in the previous subsection, but with the assumption of one disconnected aggregate-to-core link. Figure 12 presents the reduction in 99.9^{th} percentile completion times for both *LPS* and DeTail (we do not present 99th percentile for space constraints). DeTail provided 10% to 89% reduction, almost an order of magnitude improvement ($18ms$ under DeTail compared to $159ms$ under FH for $8KB$ retrievals at 2000 retrievals/second). *LPS's* inability to match DeTail's improvement at higher retrieval rates highlights the effectiveness of ALB at evenly distributing load despite asymmetries in available paths.

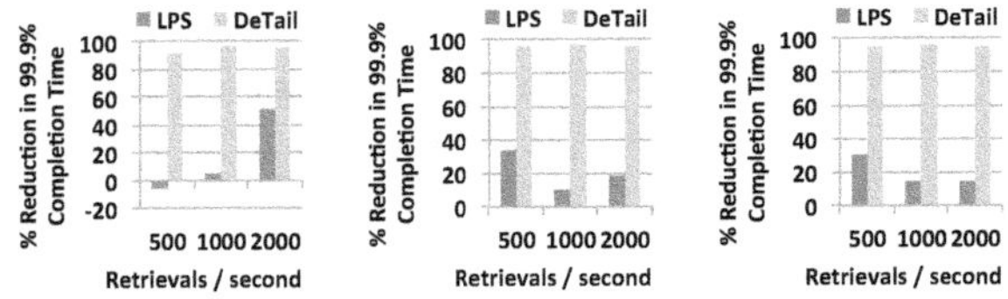

(a) $2KB$ (b) $8KB$ (c) $32KB$

Figure 13: **Degraded Link** - Reduction by *LPS* and DeTail over FH in 99.9^{th} percentile completion times of $2KB$, $8KB$ and $32KB$ data retrievals

Degraded Link: Instead of disconnecting, links can occasionally be downgraded from $1Gbps$ to $100Mbps$. Figure 13 presents the results for the same workload with a degraded core-to-agg link. DeTail provided more than 91% reduction compared to FH. This dramatic improvement is due to ALB's inherent capability to route around congestion hotspots (i.e., switches connected to the degraded link) by redirecting traffic to alternate paths. While the 99.9^{th} percentile completion time for $8KB$ at 2000 retrievals/second (refer to Figure 13(b)) under FH and *LPS* was more than $755ms$, it was $37ms$ under DeTail. In certain cases, *LPS* actually performs worse than FH (i.e., for $2KB$, 500 retrievals/second).

In both fault types, the improvement in the tail comes at the cost of increased median completion times. As we have argued earlier, this trade off between median and 99.9^{th} percentile performance is appropriate for consistently meeting deadlines.

6.5 Web Workloads

Next, we evaluate how the improvements in individual data retrievals translate to improvements in the sequential and partition-aggregate workflows used in page creation. Here we randomly assign half the servers to be front-end servers and half to be back-end servers. The front-end servers initiate the workflows to retrieve data from randomly chosen back-end servers. We present the reduction in the 99.9^{th} percentile completion times of these workflows.

Sequential Workflows: Each sequential workflow initiated by a front-end server consists of 10 data retrievals of size $2KB$, $4KB$, $8KB$, $16KB$, and $32KB$ (randomly chosen with equal probability). As described in the Section 2, these retrievals need to be performed one after another. Workflows arrive according to a Poisson process at an average rate of 350 workflows/second. Figure 15 shows that DeTail provides 71% to 76% reduction in the 99.9^{th} percentile completion times of individual data retrievals. In total, there is a 54% improvement in the 99.9^{th} percentile completion time of the sequential workflows – from $38ms$ to $18ms$.

Partition-Aggregate Workflows: In each partition-aggregate workflow, a front-end server retrieves data in parallel from 10, 20, or 40 (randomly chosen with equal probability) back-end servers. As characterized in [12], the size of individual data retrievals is set to $2KB$. These workflows arrive according to a Poisson process

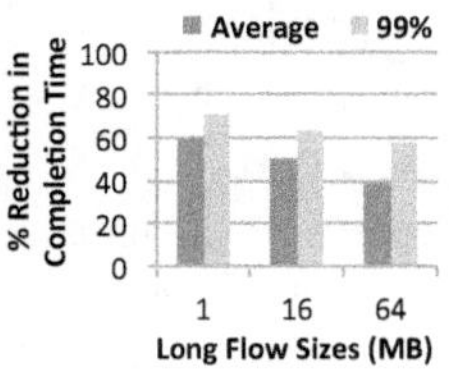

Figure 14: **Long Flows** - Reduction by DeTail in completion times of long, low-priority flows

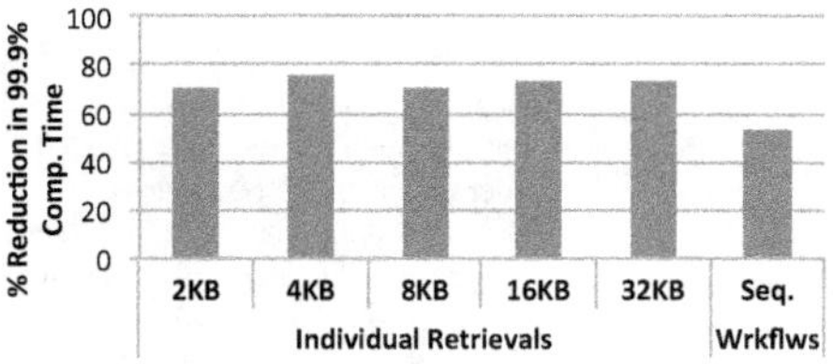

Figure 15: **Sequential Workflows** - Reduction by DeTail over *FH* in 99.9^{th} percentile completion times of sequential workflows and their individual data retrievals

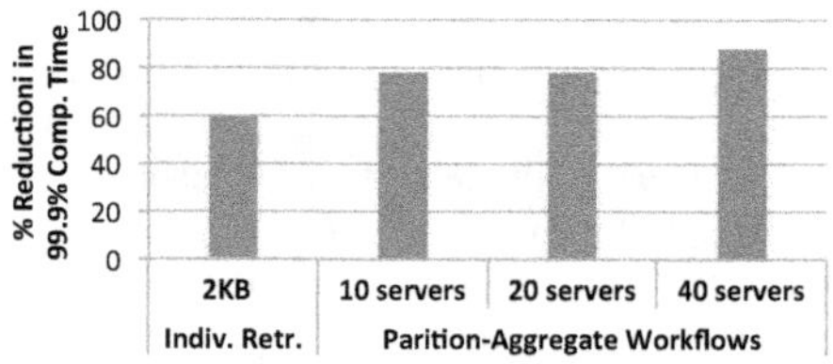

Figure 16: **Partition-Aggregate Workflows** - Reduction by DeTail over *FH* in 99.9^{th} percentile completion times of partition-aggregate workflows and their individual retrievals

at an average rate of 600 workflows/second. Figure 16 shows that Detail provides 78% to 88% reduction in 99.9^{th} percentile completion times of the workflows. Specifically, the 99.9^{th} percentile completion time of workflows with 40 servers was $17ms$ under DeTail, compared to $143ms$ under *FH*. This dramatic improvement is achieved by preventing the timeouts that were experienced by over 3% of the individual data retrievals under *FH*.

These results demonstrate that DeTail effectively manages network congestion, providing significant improvements in the performance of distributed page creation workflows.

7. DISCUSSION

We first describe how DeTail can be applied to other switch architectures. Next we present initial ideas about a DeTail-aware transport protocol.

7.1 Alternate Switch Architectures

Modern datacenters increasingly employ shared-memory top-of-rack switches [12]. In these switches, arriving packets are added to the output queue when the forwarding decision is made. They do not wait in input queues until the crossbar becomes available. This makes it difficult to determine which links contribute to congestion.

We address this by associating a bitmap with every input port. When an arriving packet is enqueued on a congested output queue, the bit corresponding to that port is set. When the output queue empties, the corresponding bits in the input ports are cleared. As input ports with any bits set in their bitmaps are contributing to congestion, this determines when we send Pause/Unpause messages. To handle multiple priorities, we use a per-port bitmap for each.

We have output queues only report congestion for a priority if its drain bytes have exceeded the thresholds specified earlier and if total queue occupancy is greater than $128KB$. This reduces the likelihood of underflow in the same way that the $128KB$ output queues do in the CIOQ architecture (see Section 4).

To evaluate this approach, we re-ran the Poisson all-to-all microbenchmark presented in Section 6. As before, we assume our switches have $256KB$ per-port. Shared-memory architectures dynamically set queue occupancy thresholds. We simulated a simple model that optimizes per-port fairness. When a switch's memory is exhausted, it drops packets from the queue with the highest occupancy. Arriving packets may only be dropped if they are destined for the most occupied queue. Priority is used to decide which of an output queue's packets to drop. We believe this is an idealized model of the performance a shared-memory switch with the same optimization strategy can achieve.

In Table 2, we present the reduction in 99.9^{th} percentile data retrieval times. Due to space constraints, we do not present 99th percentile results. With up to 66% reduction in completion time,

	500			1000			2000		
size (KB)	2	8	32	2	8	32	2	8	32
reduction	17%	10%	14%	38%	34%	35%	66%	64%	66%

Table 2: **Shared Memory** - Reduction by DeTail in 99.9^{th} percentile completion times for all-to-all workloads of exponentially distributed retrievals

these results show that DeTail's approach is beneficial for shared memory switches as well. We leave a thorough evaluation of DeTail's performance with shared-memory switches for future work.

7.2 DeTail-aware transport

The transport layer protocol presented in this paper is a retrofit of TCP NewReno. Delay-based protocols, such as TCP Vegas [15], may be better suited in these environments. Instead of waiting for packet drops that do not occur, they monitor increases in delay. Increased delay is precisely the behavior our lossless interconnect exhibits as congestion rises. We plan to investigate this approach further in the future.

8. RELATED WORK

In this section, we discuss prior work and how it relates to DeTail in three areas: Internet protocols, datacenters, and HPC interconnects, discussing each in turn.

8.1 Internet Protocols

The Internet was initially designed as a series of independent layers [17] with a focus on placing functionality at the end-hosts [30]. This approach explicitly sacrificed performance for generality. Improvements to this design, in terms of TCP modifications such as NewReno, Vegas, and SACK [15, 20, 25] and in terms of buffer management such as RED and Fair Queuing [19, 21] were proposed. All of these approaches focused on improving the notification and response of end-hosts. Consequently, they operate at coarse-grained timescales inappropriate for our workload.

DeTail differs from this work by taking a more agile in-network approach that breaks the single path assumption to reduce the flow completion time tail.

8.2 Datacenter Networks

Relevant datacenter work has focused on two areas: topologies and traffic management protocols. Topologies such as FatTrees, VL2, BCube, and DCell [9, 22–24] sought to increase bisection bandwidth. Doing so necessitated increasing the number of paths between the source and destination because increasing link speeds was seen as impossible or prohibitively expensive.

Prior work has also focused on traffic management protocols for datacenters. DCTCP and HULL proposed mechanisms to improve flow completion time by reducing buffer occupancies [12, 13]. D^3 sought to allocate flow resources based on application-specified deadlines [33]. And, the recent industrial effort known as Datacenter Bridging extends Ethernet to support traffic from other protocols that have different link layer assumptions [2]. All of these approaches focus on single-path mechanisms that are bound by the performance of flow hashing.

Datacenter protocols focused on spreading load across multiple paths have been proposed. Hedera performs periodic flow remapping of elephant flows [10]. MPTCP takes a step further, making TCP aware of multiple paths [29]. While these approaches provide multipath support, they operate at timescales that are too coarse-grained to improve the short flow completion time tail.

8.3 HPC Interconnects

DeTail borrows some ideas from HPC interconnects. Credit-based flow control has been extensively studied and is often deployed to create lossless fabrics [8]. Adaptive load balancing algorithms such as UGAL and PAR have also been proposed [8]. To the best of our knowledge, these mechanisms have not been evaluated for web-facing datacenter networks focused on reducing the flow completion tail.

A commodity HPC interconnect, Infiniband, has made its way into datacenter networks [5]. While Infiniband provides a priority-aware lossless interconnect, it does not perform Adaptive Load Balancing (ALB). Without ALB, hotspots can occur, leading a subset of flows to hit the long tail. Host-based approaches to performing load-balancing, such as [32] have been proposed. But these approaches are limited because they are not sufficiently agile.

9. CONCLUSION

In this paper, we presented DeTail, an approach for reducing the tail of completion times of the short, latency-sensitive flows critical for page creation. DeTail employs cross-layer, in-network mechanisms to reduce packet losses and retransmissions, prioritize latency-sensitive flows, and evenly balance traffic across multiple paths. By making its flow completion statistics robust to congestion, DeTail can reduce 99.9^{th} percentile flow completion times by over 50% for many workloads.

DeTail's approach will likely achieve significant improvements in the tail of flow completion times for the foreseeable future. Increases in network bandwidth are unlikely to be sufficient. Buffers will drain faster, but they will also fill up more quickly, ultimately causing the packet losses and retransmissions that lead to long tails. Prioritization will continue to be important as background flows will likely remain the dominant fraction of traffic. And load imbalances due to topological asymmetries will continue to create hotspots. By addressing these issues, DeTail enables web sites to deliver richer content while still meeting interactivity deadlines.

10. ACKNOWLEDGEMENTS

This work is supported by MuSyC: "Multi-Scale Systems Center", MARCO, Award #2009-BT-2052 and AmpLab: "AMPLab: Scalable Hybrid Data Systems Integrating Algorithms, Machines and People", DARPA, Award #031362. We thank Ganesh Ananthanarayanan, David Culler, Jon Kuroda, Sylvia Ratnasamy, Scott Shenker, and our shepherd Jon Crowcroft for their insightful comments and suggestions. We also thank Mohammad Alizadeh and David Maltz for helping us understand the DCTCP workloads.

11. REFERENCES

[1] Cisco nexus 5000 series architecture. http://www.cisco.com/en/US/prod/collateral/switches/ps9441/ps9670/white_paper_c11-462176.html.

[2] Data center bridging. http://www.cisco.com/en/US/solutions/collateral/ns340/ns517/ns224/ns783/at_a_glance_c45-460907.pdf.

[3] Datacenter networks are in my way. http://mvdirona.com/jrh/TalksAndPapers/JamesHamilton_CleanSlateCTO2009.pdf.

[4] Fulcrum focalpoint 6000 series. http://www.fulcrummicro.com/product_library/ FM6000_Product_Brief.pdf.

[5] Infiniband architecture specification release 1.2.1. http://infinibandta.org/.

[6] Ns3. http://www.nsnam.org/.

[7] Priority flow control: Build reliable layer 2 infrastructure. http://www.cisco.com/en/US/prod/collateral/switches/ps9441/ps9670/white_paper_c11-542809.pdf.

[8] ABTS, D., AND KIM, J. High performance datacenter networks: Architectures, algorithms, and opportunities. *Synthesis Lectures on Computer Architecture 6*, 1 (2011).

[9] AL-FARES, M., LOUKISSAS, A., AND VAHDAT, A. A scalable, commodity data center network architecture. In *SIGCOMM* (2008).

[10] AL-FARES, M., RADHAKRISHNAN, S., RAGHAVAN, B., HUANG, N., AND VAHDAT, A. Hedera: Dynamic flow scheduling for data center networks. In *NSDI* (2010).

[11] ALIZADEH, M. Personal communication, 2012.

[12] ALIZADEH, M., GREENBERG, A., MALTZ, D. A., PADHYE, J., PATEL, P., PRABHAKAR, B., SENGUPTA, S., AND SRIDHARAN, M. Data center tcp (dctcp). In *SIGCOMM* (2010).

[13] ALIZADEH, M., KABBANI, A., EDSALL, T., PRABHAKAR, B., VAHDAT, A., AND YASUDA, M. Less is more: Trading a little bandwidth for ultra-low latency in the data center. In *NSDI* (2012).

[14] BENZEL, T., BRADEN, R., KIM, D., NEUMAN, C., JOSEPH, A., SKLOWER, K., OSTRENGA, R., AND SCHWAB, S. Experience with deter: a testbed for security research. In *TRIDENTCOM* (2006).

[15] BRAKMO, L. S., O'MALLEY, S. W., AND PETERSON, L. L. Tcp vegas: new techniques for congestion detection and avoidance. In *SIGCOMM* (1994).

[16] CHEN, Y., GRIFFITH, R., LIU, J., KATZ, R. H., AND JOSEPH, A. D. Understanding tcp incast throughput collapse in datacenter networks. In *WREN* (2009).

[17] CLARK, D. The design philosophy of the darpa internet protocols. In *SIGCOMM* (1988).

[18] DEAN, J. Software engineering advice from building large-scale distributed systems. http://research.google.com/people/jeff/stanford-295-talk.pdf.

[19] DEMERS, A., KESHAV, S., AND SHENKER, S. Analysis and simulation of a fair queueing algorithm. In *SIGCOMM* (1989).

[20] FLOYD, S., AND HENDERSON, T. The newreno modification to tcp's fast recovery algorithm, 1999.

[21] FLOYD, S., AND JACOBSON, V. Random early detection gateways for congestion avoidance. *IEEE/ACM Trans. Netw. 1* (August 1993).

[22] GREENBERG, A., HAMILTON, J. R., JAIN, N., KANDULA, S., KIM, C., LAHIRI, P., MALTZ, D. A., PATEL, P., AND SENGUPTA, S. Vl2: a scalable and flexible data center network. In *SIGCOMM* (2009).

[23] GUO, C., LU, G., LI, D., WU, H., ZHANG, X., SHI, Y., TIAN, C., ZHANG, Y., AND LU, S. Bcube: A high performance, server-centric network architecture for modular data centers. In *SIGCOMM* (2009).

[24] GUO, C., WU, H., TAN, K., SHI, L., ZHANG, Y., AND LU, S. Dcell: a scalable and fault-tolerant network structure for data centers. In *SIGCOMM* (2008).

[25] JACOBSON, V., AND BRADEN, R. T. Tcp extensions for long-delay paths, 1988.

[26] KOHAVI, R., AND LONGBOTHAM, R. Online experiments: Lessons learned, September 2007. http://exp-platform.com/Documents/IEEEComputer2007OnlineExperiments.pdf.

[27] KOHLER, E., MORRIS, R., CHEN, B., JANNOTTI, J., AND KAASHOEK, M. F. The click modular router. *ACM Trans. Comput. Syst. 18* (August 2000).

[28] MCKEOWN, N. White paper: A fast switched backplane for a gigabit switched router. http://www-2.cs.cmu.edu/ srini/15-744/readings/McK97.pdf.

[29] RAICIU, C., BARRE, S., PLUNTKE, C., GREENHALGH, A., WISCHIK, D., AND HANDLEY, M. Improving datacenter performance and robustness with multipath tcp. In *SIGCOMM* (2011).

[30] SALTZER, J. H., REED, D. P., AND CLARK, D. D. End-to-end arguments in system design. *ACM Trans. Comput. Syst. 2* (November 1984).

[31] VASUDEVAN, V., PHANISHAYEE, A., SHAH, H., KREVAT, E., ANDERSEN, D. G., GANGER, G. R., GIBSON, G. A., AND MUELLER, B. Safe and effective fine-grained TCP retransmissions for datacenter communication. In *SIGCOMM* (2009).

[32] VISHNU, A., KOOP, M., MOODY, A., MAMIDALA, A. R., NARRAVULA, S., AND PANDA, D. K. Hot-spot avoidance with multi-pathing over infiniband: An mpi perspective. In *CCGRID* (2007).

[33] WILSON, C., BALLANI, H., KARAGIANNIS, T., AND ROWTRON, A. Better never than late: meeting deadlines in datacenter networks. In *SIGCOMM* (2011).

Inferring Visibility: Who's (Not) Talking to Whom?

Gonca Gürsun, Natali Ruchansky, Evimaria Terzi, and Mark Crovella
Department of Computer Science
Boston University

ABSTRACT

Consider this simple question: how can a network operator identify the set of routes that pass through its network? Answering this question is surprisingly hard: BGP only informs an operator about a limited set of routes. By observing traffic, an operator can only conclude that a particular route passes through its network – but not that a route does *not* pass through its network. We approach this problem as one of statistical inference, bringing varying levels of additional information to bear: (1) the existence of traffic, and (2) the limited set of publicly available routing tables. We show that the difficulty depends critically on the position of the network in the overall Internet topology, and that the operators with the greatest incentive to solve this problem are those for which the problem is hardest. Nonetheless, we show that suitable application of nonparametric inference techniques can solve this problem quite accurately. For certain networks, traffic existence information yields good accuracy, while for other networks an accurate approach uses the 'distance' between prefixes, according to a new network distance metric that we define. We then show how solving this problem leads to improved solutions for a particular application: traffic matrix completion.

Categories and Subject Descriptors: C.2.3 [Computer-Communication Networks]: Network Operations.

Keywords: BGP, matrix completion.

1. INTRODUCTION

The global state of the Internet's interdomain routing system at any instant is an almost unknowable mystery. Yet, absent this knowledge, a network operator cannot answer this very simple question: 'Which routes pass through my network?'

Answering this simple question can be useful in many ways. For example, it can inform traffic engineering and capacity planning; when traffic loads change, operators would like to know whether the cause is a change in demand, or a routing change. From a security standpoint, knowing whether a route passes through one's network can be used to identify packets with spoofed source addresses.

Knowing what routes pass through one's network can also provide business intelligence, by shedding light on the routing behavior of customers and competitors. As a concrete example, consider a network A with competitor B. Suppose B has customer C and A would like to make a bid for C's business. The price offered should depend on the proportion of C's traffic that is already passing through A. A simple way to estimate this is to see what fraction of C's routes are passing through A.

From a scientific standpoint, this question has basic appeal. It is equivalent to asking: 'If I capture a traffic matrix from a particular network, and see a zero value in the (i, j) position, what does that mean?'

In this paper we develop methods to answer this visibility question, taking the standpoint of a single network operator. To discover whether the route from i to j passes through the operator's network (i.e., is *visible* in the network), the operator has available two kinds of potentially useful information: observed traffic, and known BGP routes. While BGP provides knowledge of only a limited set of routes to each destination, observed traffic may potentially provide information about any source-destination pair. Hence our methods concentrate on the knowledge obtained from observed traffic. (More discussion of this issue appears in Section 8.)

Unfortunately, traffic only provides *positive* information about visibility: if traffic is observed flowing from i to j then (i, j) is known to be visible; but if no traffic is observed, then it is not possible to conclude anything about the visibility status of (i, j); i.e., traffic does not provide *negative* information about visibility. Thus the key question comes down to asking: if no traffic is observed from i to j, is it because the path from i to j does not pass through the observer, or because i is simply not sending traffic to j? Like Sherlock Holmes, we are interested in 'the dog that didn't bark.'

Our approach starts from simple intuition about how routes spread through the Internet; based on this understanding we develop a family of nonparametric classifiers. We show that different kinds of networks (essentially, networks in different tiers) require different instantiations of this classifier. We then show how to configure these classifiers for different kinds of networks, and demonstrate that we can answer our original question with a surprisingly high level of accuracy. We conclude with an example showing the utility of our classifier when applied to the traffic matrix completion problem.

SIGCOMM'12, August 13–17, 2012, Helsinki, Finland.

2. OVERVIEW

Our central notion is *visibility:* we say that a source-destination pair (i, j) is visible to AS x if, were i to send traffic to j, the traffic would pass through x. We refer to the AS whose owner wishes to infer visibility as the *observer* AS.[1] We use the term *path* to refer to the sequence of ASes taken along the route from a source (traffic generator) to a destination (traffic recipient). We naturally group addresses into prefixes as used in BGP, and associate prefixes with the ASes that appear at the end of their BGP routes. Hence, in what follows 'source' will mean 'source AS' or 'source prefix' (and likewise for 'destination.')

For any given observer AS, we define the corresponding *visibility matrix* T, which has sources on the rows and destinations on the columns, such that $T(i, j) = 1$ iff the source-destination pair (i, j) is visible to the observer, and zero otherwise. In our work, we assume that T is constant during the period of study; in practice we work with snapshots of T.

Of course, the observer does not actually have access to T. Instead the observer can measure traffic flowing through the network. We assume that traffic is measured over some fixed duration, and we organize measured traffic volumes into a matrix V, with the same rows and columns as T. For example, the value of $V(i, j)$ may be the number of bytes sent from i to j during the measurement interval. We refer to V as the *traffic matrix* or *volume matrix*. The (positive) visibility information contained in V defines the *observed* visibility matrix M, with $M(i, j) = 1$ iff $V(i, j) > 0$, and zero otherwise. Thus, M encodes the incomplete visibility information that is learned from observing traffic.

Hence, M is an approximation to T, with the property that if $M(i, j) = 1$, then $T(i, j) = 1$. However, if $M(i, j) = 0$, then $T(i, j)$ may be 0 or 1. Therefore, there are two types of zeros in the observed visibility matrix M: *true zeros* and *false zeros*. A true zero occurs when (i, j) is truly not visible; a false zero occurs when (i, j) is actually visible, but not communicating. In notation: $M(i, j)$ is a *true zero* if $M(i, j) = 0$ and $T(i, j) = 0$; it is a *false zero* if $M(i, j) = 0$ and $T(i, j) = 1$.

2.1 The VISIBILITY-INFERENCE problem

Our goal is to design a mechanism that outputs a matrix $\widehat{T}$ that is a better approximation of T than M is. We call $\widehat{T}$ the *predicted visibility matrix*, and we call the problem of inferring $\widehat{T}$ the VISIBILITY-INFERENCE problem. This is a classification problem that focuses on the zero-valued elements of M, which we denote by Z, (i.e., $Z = \{(i, j) \mid M(i, j) = 0\}$). For every $(i, j) \in Z$, the classifier decides whether (i, j) corresponds to a true zero or a false zero and sets $\widehat{T}(i, j) = 0$ or 1 respectively. In other words, our goal is to set the values of $\widehat{T}(Z)$ (i.e., the subset of the elements of $\widehat{T}$ that correspond to the indices in Z) in such a way that they agree with the corresponding values of $T(Z)$.

To fix terminology, we consider a classification of $(i, j) \in Z$ that outputs $\widehat{T}(i, j) = 0$ to be a *positive* classification – the element is predicted to be, in fact, a true zero. Correspondingly, $\widehat{T}(i, j) = 1$ is a *negative* classification – the element is predicted to be a false zero.

The performance of a classifier can then be expressed in

[1]We work at the granularity of ASes, rather than ISPs; some implications of this are discussed in Section 8.

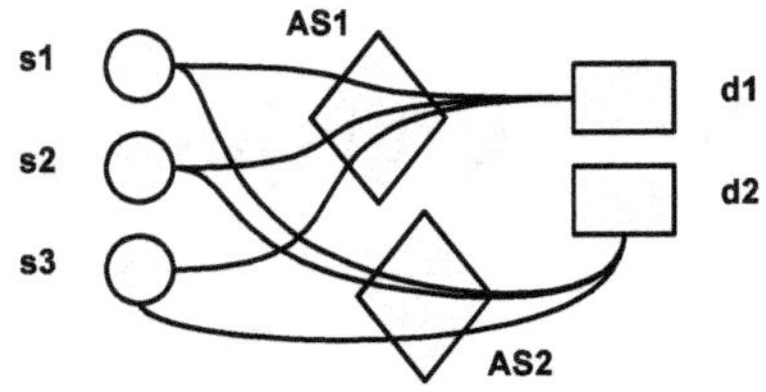

Figure 1: Groups of similar paths

terms of *True Positive Rate* (TPR) and the *False Positive Rate* (FPR) of the predicted matrix $\widehat{T}$. TPR is the classifier's accuracy on true zeros:

$$\text{TPR} = \frac{\text{number of correctly classified true zeros}}{\text{total number of true zeros}}$$

while 1−FPR is the classifier's accuracy on false zeros:

$$\text{FPR} = \frac{\text{number of incorrectly classified false zeros}}{\text{total number of false zeros}}$$

Thus we can formally describe the VISIBILITY-INFERENCE problem as the task of forming a predicted matrix $\widehat{T}$ such that TPR is as close to 1 as possible and FPR is as close to 0 as possible.

The classifiers we develop will allow tradeoffs between TPR and FPR. To examine this tradeoff and thus assess the overall performance of the classifier, we use ROC (Receiver Operating Characteristic) curves [7]. The ROC curve of $\widehat{T}$ plots TPR versus FPR for different settings of the classifier. The ideal classifier operates at the upper-left point $(0, 1)$. In order to describe the performance of the classifier over its entire range of settings with a single number, we use the *area under the ROC curve of* $\widehat{T}$ (AUC) which takes values between 0 and 1. An ideal classifier (one which can operate at the $(0, 1)$ point) has an AUC of 1, and in general, the closer to 1, the better the prediction of $\widehat{T}$.

2.2 Our approach

Having defined our problem, we now discuss the basic intuition behind our approach. We start from the observation that at any given point in the Internet, there are groups of (i, j) pairs that are routed similarly. This general idea is shown in Figure 1. For example, in some region of the Internet (say, a set of nearby ASes AS_1, AS_2, ...) one may often find a group of paths that all pass through AS_1, and yet none passes through AS_2. Another group may all pass through AS_2 but none through AS_1. At the highest level, our strategy is to identify path groups, and if traffic is observed on some paths in the group, infer that all members of the group are likely to be visible.

More concretely, for any given source-destination pair (i, j), we look for other pairs that we believe to have a similar visibility status as (i, j). We then use any knowledge we have about the other pairs to form a visibility estimate for (i, j). Essentially, this is nearest-neighbor classification (with incomplete information). The principal challenge then becomes to properly identify which source-destination pairs are 'neighbors' to (i, j).

On a high level, our classifier takes a specific form: given some $(i, j) \in Z$,

1. Select an (i, j)*-descriptive submatrix* from M. This

submatrix, denoted by $M(S_i, D_j)$, is defined by the set of sources S_i and destinations D_j.

2. Compute the value of some *structural property* of $M(S_i, D_j)$. We call this the *descriptive value* π_{ij}.

3. If the descriptive value is above a *threshold*, predict $\widehat{T}(i,j) = 1$; otherwise, predict $\widehat{T}(i,j) = 0$.

This approach performs classification of each $(i,j) \in Z$ independently. Although more complicated mechanisms could classify pairs jointly, our experimental evaluation indicates that the independent approach yields very accurate predictions.

The goal of the first step is to find a submatrix $M(S_i, D_j)$ whose elements have a similar visibility status with respect to the observer. In practice, we simplify this; we only require that sources in S_i are similar in some way, and that destinations in D_j are similar in some way, with respect to the observer. Of course, different kinds of similarity lead to different submatrices. We use two types of similarity, based on different information. The first is based on the *visibility information* – as encoded in the observed visibility matrix. The second is based on *proximity information* – as encoded in publicly available BGP tables. These two strategies are described in Sections 4 and 5, respectively.

The second step evaluates different *structural properties* of $M(S_i, D_j)$. These properties are discussed in detail in Sections 4 and 5; here we just note that these are simple properties like the number of nonzero elements in $M(S_i, D_j)$, the size of $M(S_i, D_j)$ or its density.

The third step classifies (i,j) based on whether its descriptive value π_{ij} is above a threshold. By varying this threshold, one can trade off TPR and FPR.

In practice, it is necessary to choose a threshold value to configure the classifier. If the classifier is robust, there will be a reasonably wide range of threshold values that yield good results, so its accuracy will not be highly sensitive to threshold settings. We show that this is the case for the classifiers we develop in Sections 4 and 5. In Section 6, we show how to actually choose a threshold value based on observable data, and we show that such automatically-chosen threshold results in a highly-accurate classifier.

3. DATASETS

To evaluate our methods, we need a large survey of ground-truth visibility matrices T. To obtain these matrices, we use a collection of BGP tables (collected on midnight UTC on December 6th, 2011) obtained from the Routeviews [14] and RIPE [12] repositories.

The full dataset is collected from 359 unique monitors; it consists of over 48 million AS paths, and contains 454,804 unique destination prefixes. (Note that not all BGP tables show paths to all prefixes.) Because these paths are the *active* paths at the time of collection, each path represents the sequence of ASes that traffic flows through when going from the particular monitor to the path's destination prefix.

Visibility Matrices. Using this data we construct visibility matrices for every AS appearing in the dataset – 23,511 ASes. In order for visibility matrices to be comparable across ASes, each matrix must be indexed by the same set of rows and columns. From the full dataset, we select a subset of monitors and a subset of prefixes such that new dataset contains the AS path from *every* monitor to *every* prefix. This results in 38 monitor ASes and 135,369 prefixes. For each of the 23,511 ASes, we construct a 0-1 visibility matrix of size 38 × 135,369. Thus, each visibility matrix has about 5.2 million entries; each matrix entry (i,j) records the visibility status of the pair (AS i, prefix j).

In a given observer AS's visibility matrix T, an entry (i,j) is 1 if the path from AS i to prefix j contains the observer AS, and 0 otherwise. Since we have all (5.2 million) active paths from every monitor AS to every prefix, we know with certainty the 0-1 value of every entry in T for every observer AS. Of course, these visibility matrices are only a portion of the complete visibility matrix for each observer, but each provides millions of ground-truth values for validating our methods.

For each ground-truth matrix T, we also need an observed visibility matrix M, which differs from T by having some 1s turned into (false) zeros. The number of false zeros in M is affected by the duration over which traffic is observed. Longer traffic observation periods result in smaller numbers of false zeros, as additional source-destination pairs generate observable traffic. Hence we study a range of false zeros, expressed as a fraction of visible elements (1s) in T: from 10% false zeros (corresponding to a long measurement period), to 95% false zeros (corresponding to very short measurement period). We generate false zeros by randomly flipping 1s in T to zeros; the result becomes the observed visibility matrix M. We call the fraction of 1s flipped in M the *flipping percentage*. When the flipping percentage is small (10%) classifiers have more information to work with; when the flipping percentage rises to 95% classifiers have very little information to work with and the classification problem is quite challenging.

Our bit-flipping strategy models the case where the traffic of each source-destination pair is independent of the others. While this reflects the basic unpredictability of traffic patterns, it is possible that correlations in traffic patterns could affect the accuracy of our classifier. For that reason, we also use another strategy: *destination-based* flipping. The goal is to model the situation where a particular destination prefix receives no traffic from any source – such as when there are no hosts provisioned with addresses from the prefix. In destination-based flipping, *all* of the 1s in an *entire* column are flipped to zero.

Observer AS Types. An important fact is that the arrangement of 1s in the visibility matrix of an AS shows distinct patterns that depend on the AS's topological location in the AS graph. For instance, in the visibility matrix of an AS that sits in the core of the graph, the 1-valued entries are scattered relatively uniformly; in contrast, for an AS that sits at the edge of the graph, the 1-valued entries are clustered in a small set of rows and columns. This is a natural consequence of the routing structure of the Internet.

Hence, as we will show in Sections 4 and 5, the topological location of an AS affects the relative performance of different classifiers. To distinguish ASes in different locations, we use two metrics: *node degree* and *k-shell number*. We compute these metrics using our BGP dataset; although these metrics are sensitive to missing edges in the BGP graph, we only use them for ranking ASes, not for quantitative comparisons.

Degree is the number of observed neighbors in the BGP graph derived from our set of paths. K-shell number measures the centrality of a node in a graph [1]. It is computed using the *k-core decomposition*, which separates the nodes of

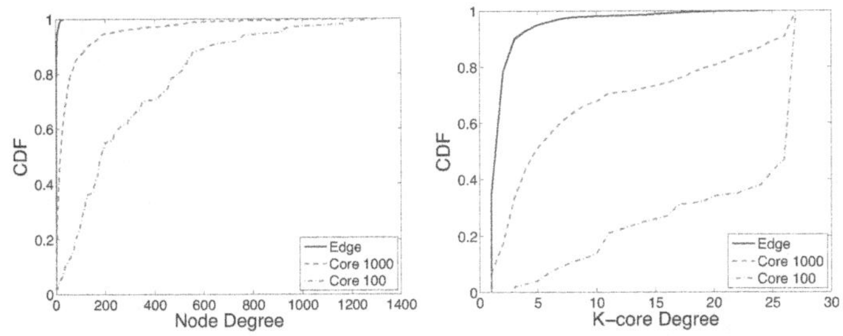

Figure 2: Properties of AS Sets Studied.

a graph into nested sets called 'shells.' As described in [5], this is a parameter-free way of characterizing the AS graph, and it corresponds to a natural notion of centrality.

Using these metrics, we define three sets of ASes to represent different topological positions. First, the `Core-100` set consists of the 100 ASes with highest k-shell number. Second, the `Core-1000` set consists of the 1000 ASes with the highest k-shell number (and so contains the `Core-100` set). Finally, the `Edge` set consists of 1000 ASes that have low degree and k-shell number. This set contains mainly stub ASes and ASes that are topologically close to stubs. This set is representative of 'typical' ASes in that almost 95% of the ASes in our dataset are stubs. The `Edge` set was constructed by randomly sampling among all the ASes not contained in the `Core-1000` set. Figure 2 shows the distribution of degree and k-shell numbers for these three sets. The Figure shows that the `Core-1000` set is intermediate in both respects between the `Edge` and `Core-100` sets, while the other two sets reflect extremes: boundary and center of the AS graph.

Traffic Matrix. Some of the experiments we report use knowledge of source-destination traffic volumes, i.e., the traffic matrix V. Traffic matrices are generally hard to obtain at the fine grain we work with in this paper (AS-prefix traffic volumes). However we were able to obtain **`netflow`** data from a Tier-1 provider (a member of the `Core-100` set) suitable for our purpose, and collected in the same timeframe as the BGP data. This consists of traffic volumes measured in bytes, over a duration of one day. We organized the flow data according to source AS and destination prefix, using the same row and column indices as our visibility matrices, into a traffic matrix V.

4. THE VISIBILITY-BASED METHOD

We start our exploration of problem solutions by examining the most straightforward approach to inferring visibility of unknown entries, namely, making use of the observed visibility matrix. We term this the *visibility-based* method. We first describe the visibility-based method by showing how it instantiates the general strategy described in Section 2.2; then, we examine the method's accuracy and applicability.

4.1 Method description

As described in Section 2.2, our general strategy seeks to find collections of source-destination pairs whose visibility status is likely to be similar to the target pair (i, j). In this section we use the positive information in M directly, and ask "How many sources and destinations have (positive) visibility that is similar to the target (i, j)?" To ask this question, we instantiate the generic method presented in Section 2.2 as follows:

Submatrix selection: Given $(i, j) \in Z$, the visibility-based method selects the (i, j)-descriptive submatrix as:

$$\begin{aligned} S_i &= \{i\} \cup \{i' \mid M(i', j) = 1\} \quad \text{and} \\ D_j &= \{j\} \cup \{j' \mid M(i, j') = 1\}. \end{aligned}$$

That is, the set S_i consists of i as well as all the sources that have have been observed to send traffic to destination j. Similarly, the set D_j contains j as well as all the destinations that have been observed to receive traffic from source i. Then S_i and D_j define the subtmatrix $M(S_i, D_j)$ that will be used to predict whether $M(i, j)$ is a true or a false zero.

The intuition behind this method is as follows. Referring back to Figure 1, imagine a set of sources $\tilde{S} = \{s_1, s_2, ...\}$ that have similar behavior with respect to the observer. That is, paths from sources $\tilde{S}$ to an arbitrary destination d all either go through the observer, or not. Then the pattern of 1s on the rows $s \in \tilde{S}$ of M will be similar. Likewise, if there are destinations $\tilde{D}$ with similar behavior, the patterns of 1's in the $\tilde{D}$ columns will be similar. Hence, by choosing S_i and D_j according to the above rules, we select a submatrix which will tend to be large and contain many 1s when the target (i, j) is a false zero.

Structural properties: There are a number of ways we might test the (i, j)-descriptive submatrix $M(S_i, D_j)$ to see whether it is large and contains a large number of 1s. These correspond to the following structural properties of $M(S_i, D_j)$:

- SIZE: If $|S_i| = s$ and $|D_j| = d$, the SIZE of $M(S_i, D_j)$ is simply $s \times d$.
- SUM: The SUM of $M(S_i, D_j)$ is the number of ones that appear in $M(S_i, D_j)$.
- DENSITY: The DENSITY of $M(S_i, D_j)$ is the ratio of the SUM to the SIZE of $M(S_i, D_j)$.

Classification criterion: The intuition behind this method suggests that when the (i, j)-descriptive submatrix is small, or contains few 1s, then (i, j) is likely a true zero; otherwise, it is likely a false zero. In practice we set a threshold β and our classifier becomes:

$$\widehat{T}(i, j) = \begin{cases} 1 \text{ if } \pi_{ij} > \beta & \text{(False Zero)} \\ 0 \text{ if } \pi_{ij} \leq \beta & \text{(True Zero).} \end{cases} \tag{1}$$

As already noted, in order for the classifier to be robust, there should be a significant region of β values over which TPR is close to 1 and FPR is close to 0.

4.2 Experimental results

Our experimental evaluation starts with the ground truth visibility matrices as described in Section 3. We test the classifier in each case on an equal number of true and false zeros, randomly selected – we test as many as possible while keeping the numbers equal, up to a limit of 2000 of each type. This balancing of test cases allows us to examine both TPR and FPR at the same resolution, and avoids bias stemming from the much larger set of true zeros than false zeros.

Descriptive power of the SUM property. To interpret the performance of the visibility-based method, it is helpful

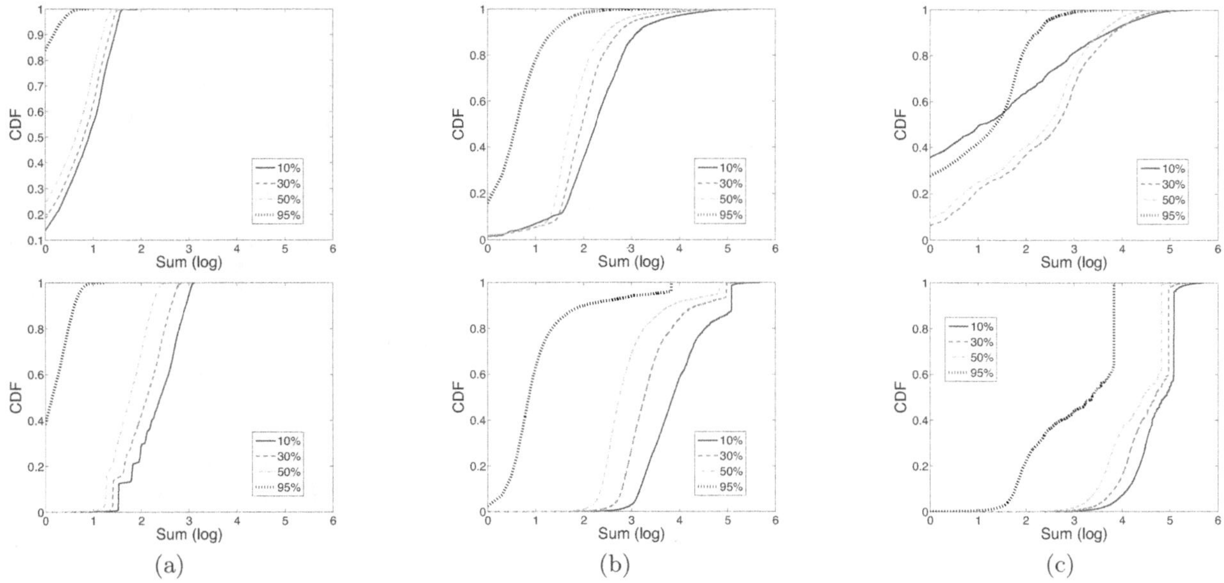

Figure 3: Submatrix Sum distribution for (a) **Edge**, (b) **Core-1000**, (c) **Core-100**. Upper: True Zeros; Lower: False Zeros. Scale on x axis is base 10 log.

to start by looking at the Sum values for submatrices, comparing the case for true and false zeros. (We concentrate on the Sum metric for reasons explained below.) Here we use knowledge of the true and false zeros, obtained from ground truth, in order to gain insight on the method (of course, actual performance results do not use this knowledge).

Figure 3 shows the CDF of Sum values on a log (base 10) scale for true zeros (upper) and false zeros (lower). To interpret these figures in the context of our classification problem, one can visualize a classification threshold drawn as a single vertical line through both an upper and lower plot. On the upper plot, a good classifier will place the majority of the distribution to the left of the line (yielding low FPR); on the lower plot, the majority should be to the right of the line (yielding high TPR).

The figure shows that the best case is for the **Edge** networks, shown on the left. Here it can be seen than there is a significant range of thresholds – values between about 10 and 50 – which almost completely separate the true and false zeros, in all cases except the extreme 95% case. Even in the 95% case, there is a significant opportunity for separating the two classes (at a different threshold).

The **Core-1000** networks are also generally separable at certain thresholds, but the **Core-100** networks less so. Examining the reasons for the difference between the **Edge** and **Core** networks, we find the following. Networks in the **Edge** class are typically 'stubs' in the AS graph (or close to stubs). The paths passing through a stub network are typically only those with sources or destinations in the set of prefixes that are announced by the network. Likewise, paths passing through a network that is near the edge, but not a stub, are primarily those that have sources or destinations announced by an AS in the network's customer cone. These situations correspond to a visibility matrix that is quite sparse, but having a few rows and columns that are almost completely filled.

In such cases, the Sum and Size properties behave similarly: for a true zero, both return very small values (Figure 3(a) upper) while for a false zero, both capture the 1s in the same row and column as (i, j) resulting in large values (Figure 3(a) lower). This explains why we only show results for Sum – the results for Size are very similar, and the results for Density are poor because the density is nearly constant.

In contrast, for the highest-tier networks (**Core-100**, Figure 3(c)) the arrangement of 1s in M is much more complex because of the spreading of paths through the network core. In such an AS, many pairs are visibile, so the likelihood of unrelated pairs having similar visibility is much higher than for **Edge** networks.

Classification accuracy. The differences between **Edge** and **Core** networks are reflected in the performance of the classifier. Figure 4 shows an 'aggregate ROC' curve across all networks in each set. The aggregate ROC is a composite constructed by collecting results for all (i, j) pairs tested in all networks. While this curve does not reflect the performance of any actual network (those results are next), it serves to give an overall sense of the classifier's performance.

The results show that it is possible to achieve excellent performance in the case of the **Edge** networks. Figure 4 shows a TPR of over 90% when FPR is zero, and an FPR of about 20% when TPR is 100%, for moderate flipping percentages. Thus, either the true or the false zeros can be labeled essentially perfectly, with small error for the other class. However, the picture is not so good for **Core-1000** and **Core-100** networks. For those networks, achieving TPR greater than about 95% requires a relatively high FPR, as high as 50% or more, even for moderate flipping percentages.

To give better insight into how the classifier performs in individual networks, we turn to Figure 5. In the Figure the AUC metric is used to judge the classifier's performance on each network, and the resulting CDF of AUC across all networks is presented. The Figure shows that for moderate flipping percentages the AUC for **Edge** networks is almost

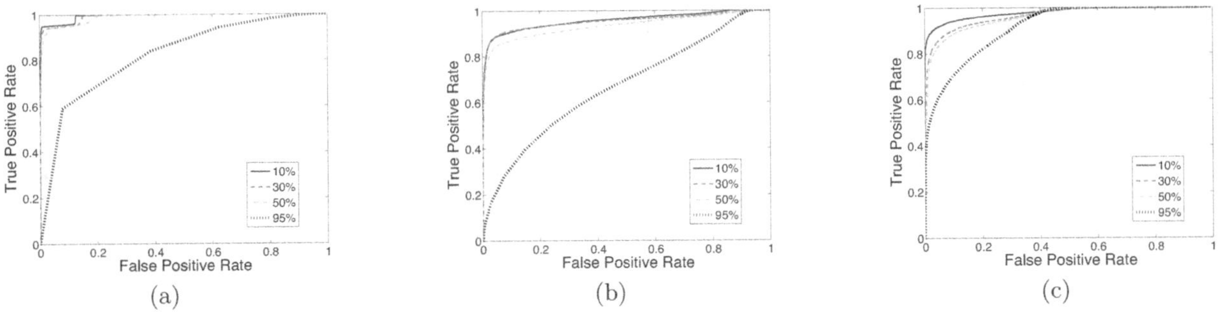

Figure 4: Aggregate ROC Curves for (a) `Edge`, (b) `Core-1000`, (c) `Core-100`.

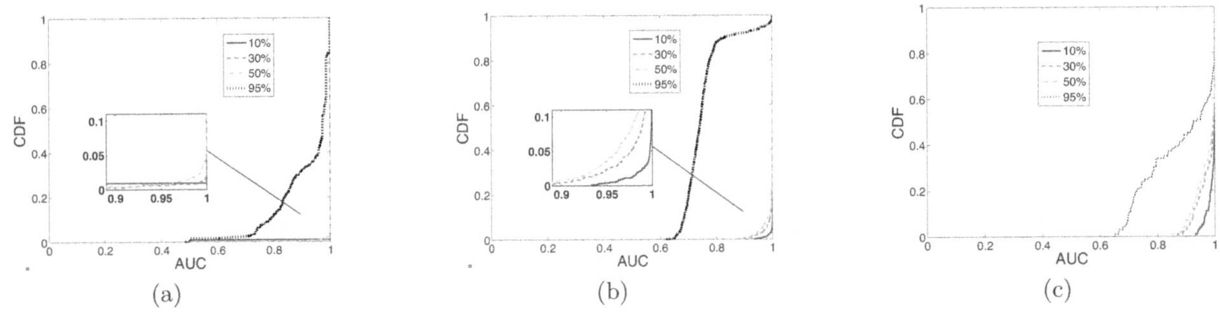

Figure 5: CDFs of AUC for (a) `Edge`, (b) `Core-1000`, (c) `Core-100`.

always very close to 1, while for `Core-1000` and `Core-100` networks, up to half of the AUCs are less than 1.

These results show that for `Edge` networks, the visibility method is quite accurate. However, the same is not true for `Core` networks; such networks apparently require a more sophisticated approach. We develop such an approach in the next section.

5. THE PROXIMITY-BASED METHOD

The challenge presented by `Core` networks is that observed visibility patterns are not sufficiently helpful in finding source-destination pairs that are routed similarly to the (i, j) target. In this section we develop a new distance metric for prefixes (sources and destinations) that helps infer visibility in such networks. We compute this metric using the limited amount of BGP state observable in publicly available datasets (as described in Section 3). We refer to this approach as the *proximity* method.

5.1 Routing-state distance

Intuitively, we seek a measure of distance (or dissimilarity) between prefixes. Our intent is that if the distance between prefixes is low, then we expect that paths to or from those prefixes have a similar visibility status for an arbitrary observer.

A natural measure of distance in this setting would be *hop distance.* Let prefixes p_1 and p_2 be announced by AS_1 and AS_2 respectively. Then the hop distance between p_1 and p_2 is simply the length of the path (e.g., shortest path) between AS_1 and AS_2 in the AS graph.

Unfortunately, this is a poor metric to use. One way to see this is simply to note that most ASes are very close in hop distance. For example, Figure 6(a) shows the distribution of hop distances for a random sample of 1000 AS-AS pairs in our data. Almost half of the hop distances are 1, 2, or 3. Such a metric clearly hides significant routing differences across prefix pairs.

What is needed is a metric that measures whether two prefixes are 'routed similarly in general.' To that end we define *routing-state distance (RSD).* First we define RSD formally, then we describe how we compute it in practice.

RSD: Definition. We start with a connected graph $G = (V, E)$. We make the following assumption: for each source-destination pair $(x_1, x_2) \in \{V \times V\}$ we assume that there is a unique node $x_3 = nexthop(x_1, x_2)$, and that by following the $nexthop(\cdot, x_2)$ function recursively, one will eventually reach x_2. We assume as well that $nexthop(x, x) = x$. Thus, $nexthop(x_1, x_2)$ is the next node on the unique path from x_1 to x_2; and that path stops at x_2.

We then define:

$$routestate(x) = \langle nexthop(x_1, x), nexthop(x_2, x), ..., nexthop(x_{|V|}, x)\rangle$$

and:

$$rsd(x_1, x_2) = \#\{x_i \mid nexthop(x_i, x_1) \neq nexthop(x_i, x_2)\}$$

Intuitively, $routestate(x)$ tells us what the 'direction' is to x from each node in the graph. Further, $rsd(x_1, x_2)$ is the number of positions where the vectors $routestate(x_1)$ and $routestate(x_2)$ differ. Thus, two nodes which appear to most other nodes to be in the same 'direction' are considered close under the *rsd* metric.

Referring again to Figure 1, assume that the next hop from d_1 to each s_i is AS1, and that the next hop from d_2 to each s_i is AS2. Then the nodes s_1, s_2, s_3 would be considered close to each other under the *rsd* metric. That is because d_1 considers all the s_i to be in the same direction, as does d_2.

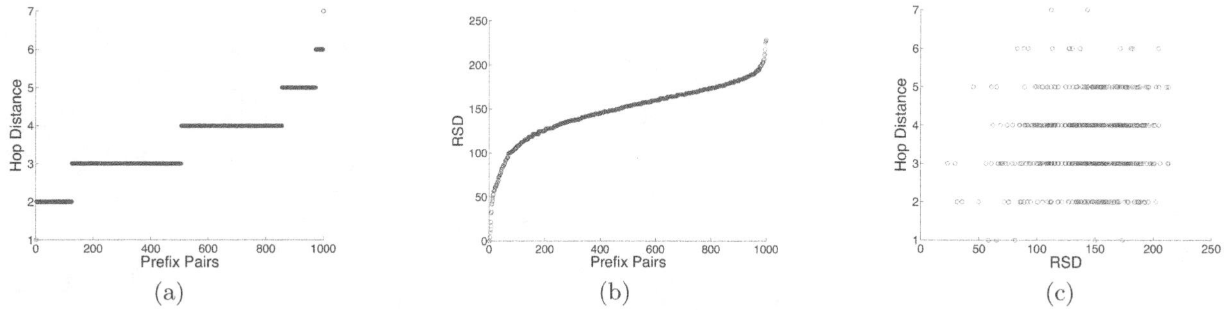

Figure 6: Distance distribution: (a) hop distance (b) $\widehat{rsd}$ (c) Comparison.

RSD has a number of good properties. It is in fact a metric (it is symmetric, positive definite, and obeys the triangle inequality). Further, it can be applied to many different situations, as long as the notion of a unique next hop is satisfied (as for example when using shortest paths, or for many routing situations). And, unlike hop-distance, it is fine-grained – taking on a wide range of values (from 0 to $|V|$).

RSD and BGP. We will compute the RSD between pairs of ASes, and then extend RSD to prefixes by noting the AS that announces the prefix. The general idea is to use the (comparatively small) set of publicly observable BGP paths to compute RSD. However, computing RSD from BGP data raises some implementation considerations. There are two main issues: (1) we cannot observe the *nexthop* function for every AS pair in the Internet; and (2) for some AS pairs the *nexthop* function is not uniquely defined.

The first issue concerns the fact that the publicly available BGP data consists (essentially) of paths from a set of monitor ASes to a large collection of prefixes. For any given AS pair (x_1, x_2), these paths may not contain information about $nexthop(x_1, x_2)$. We address this by *approximating* RSD. We make the following observation: some ASes have a much larger impact on RSD than others. For example, a stub AS has a highly uniform contribution to each *routestate* vector. If the stub AS x has a small set of providers, then for many AS pairs (x_1, x_2), $nexthop(x, x_1) = nexthop(x, x_2)$. Hence *most ASes contribute little information to RSD.*

Thus, we conclude that we can approximate RSD using a subset of all ASes, in particular those ASes with many neighbors; these ASes contribute the most information to RSD. We call these ASes the *basis set.* We select the basis set by identifying 77 ASes with the largest number of neighbors in our data. Happily, such ASes tend to appear on many paths in the publicly available data; hence we can often find $nexthop(x_1, x_2)$ when x_1 is in the basis set.

To address the case when $nexthop(x_1, x_2)$ is not available (for AS x_1 in the basis set), we performed extensive studies of the effect of missing *nexthop* information on RSD. Space does not permit a full summary of our findings; we note only that *proportional approximation* yields results that work well in practice. This approximation extends the *routestate* vector to include 'don't know' elements. We then define $\widehat{rsd}(x_1, x_2)$ as the fraction of *known* positions in which $routestate(x_1)$ and $routestate(x_2)$ differ, times the number of ASes in the basis set. This normalizes $\widehat{rsd}$ so that it always ranges between zero and the size of the basis set.

The second issue is that for some AS pairs there is more than one next hop. This happens when an AS uses detailed routing policies (such as hot-potato routing) that are not strictly per-neighbor. That is, traffic destined for the same prefix may take different next hops depending on where it enters the network. We address this problem the same way as in [10], i.e., by introducing the notion of 'quasi-routers.' Using the algorithm in [10] we divide each AS in the basis set into a minimal set of quasi-routers such that for each (quasi-router, prefix) pair there is a unique next hop AS. This expands the size of the basis set from 77 to 243.

Having addressed these two issues, we can compute $\widehat{rsd}$ for each pair of prefixes in our dataset. In Figure 6(b) we show the RSD values for the same set of prefix pairs as in Figure 6(a) (sorted in increasing order in both cases). The steep slope on the left of the figure shows that RSD can make fine distinctions between prefix pairs. Furthermore, it is clear that RSD is not simply another measure of hop distance: Figure 6(c) shows a scatterplot of RSD versus hop distance, indicating that there is little relationship between the two metrics. A last observation about RSD applied to prefixes is that it can be interpreted as a generalization of the notion of BGP atoms [3]. A BGP atom is a set of prefixes which are routed the same way everywhere in the Internet; so a BGP atom is a prefix set in which each prefix pair has an RSD of zero.

5.2 Method description

Using RSD, we again instantiate the generic method presented in Section 2.2.

Submatrix selection: Given $(i, j) \in Z$ and threshold τ, the proximity method selects the (i, j)-descriptive submatrix by selecting rows and columns from M that correspond to sources and destinations that are close to i and j respectively. That is,

$$\begin{aligned} S_i &= \{i\} \cup \{i' \mid \widehat{rsd}(i', i) \leq \tau\} \quad \text{and} \\ D_j &= \{j\} \cup \{j' \mid \widehat{rsd}(j', j) \leq \tau\}. \end{aligned}$$

We explored various threshold values τ. Space does not permit reviewing those results, but we found that $\tau = 50$ works well in practice. As can be seen in Figure 6(b), this means that we are placing a relatively small subset of rows and columns into $M(S_i, D_j)$.

Structural properties: As in the visibility-based approach we again compute the SIZE, SUM and the DENSITY of the $M(S_i, D_j)$ as input to the classifier. In this case, SUM and DENSITY prove to be equally effective discriminating features; we report only results for SUM.

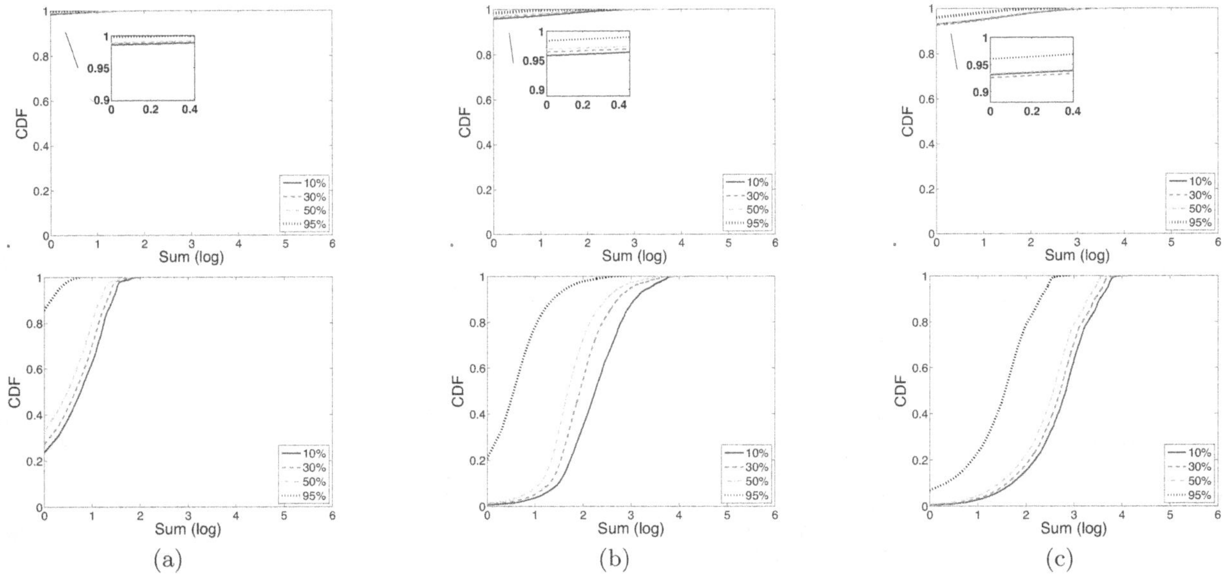

Figure 7: Submatrix SUM distribution for (a) **Edge**, (b) **Core-1000**, (c) **Core-100**. Upper: True Zeros; Lower: False Zeros.

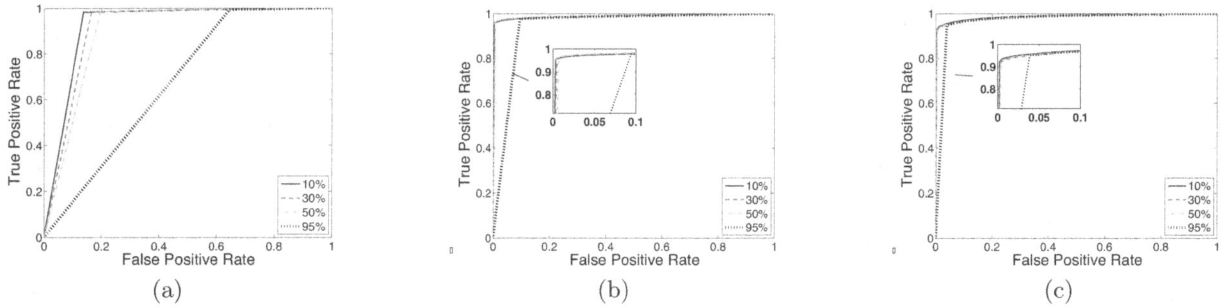

Figure 8: Aggregate ROC Curves for: (a) **Edge**, (b) **Core-1000**, (c) **Core-100**.

Classification criterion: The intuition behind the proximity method suggests that the target element (i, j) should have similar visibility to the paths captured in the $M(S_i, D_j)$ submatrix. Both SUM and DENSITY are large when the paths captured in $M(S_i, D_j)$ are visible; hence we use the same classification criterion as for the visibility method (Equation (1)).

5.3 Experimental results

We again start by examining the distribution of SUM values for submatrices $M(S_i, D_j)$, which are shown in Figure 7. The upper row of plots shows that in each set of ASes, the SUM metric is almost always zero when (i, j) is a true zero (i.e., more than 95% of the time). This makes sense, since the paths captured in $M(S_i, D_j)$ are similar to the path (i, j) in terms of routing behavior. For the case of false zeros, $M(S_i, D_j)$ is typically nonzero, but with different behavior for the **Edge** versus **Core** networks. The **Core-1000** and **Core-100** true-zero submatrices are easily separated from the false-zero submatrices. For moderate flipping levels, a threshold value anywhere in the range 1-10 is effective. However for the **Edge** networks, there is no similarly good threshold.

This can be understood as follows. The proximity method is not effective for **Edge** networks because it typically selects only a small set of sources and destinations for inclusion in $M(S_i, D_j)$ – those that are close to i and j in terms of $\widehat{rsd}$. Since **Edge** networks have sparse visibility matrices in general, the small size of $M(S_i, D_j)$ makes it possible that most or all 1s are flipped to zeros, leading to false zeros that are misclassified as true zeros. On the other hand, comparison of the top and bottom of Figure 7(b) and (c) shows that RSD is able to pick out those prefixes that are similar in routing behavior when the network is in the core and has denser visibility matrices, and this is what is needed for accurate classification.

The accuracy of classification in terms of aggregate ROC curves is shown in Figure 8. The figure shows that classification in general is very accurate for **Core** networks: for moderate flipping levels, accuracy is as high as 95% TPR with zero FPR. For **Edge** networks, as expected, classification accuracy is generally poorer, with an unavoidably nonzero FPR.

Figure 9 summarizes AUCs over all networks. For **Edge** networks, a significant number of AUCs are in the 0.5 range. An AUC of 0.5 is typically characteristic of a straight-line ROC curve from the origin to the upper-right corner. This happens when there is essentially no discriminatory power in

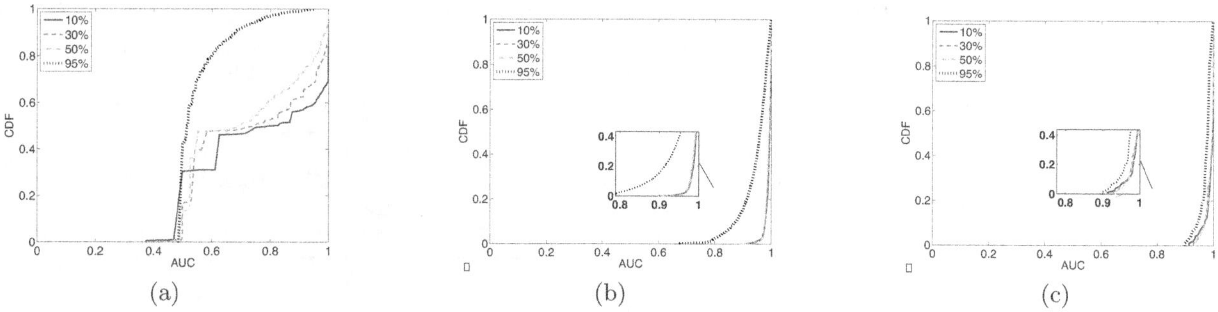

Figure 9: CDFs of AUC for (a) `Edge`, (b) `Core-1000`, (c) `Core-100`.

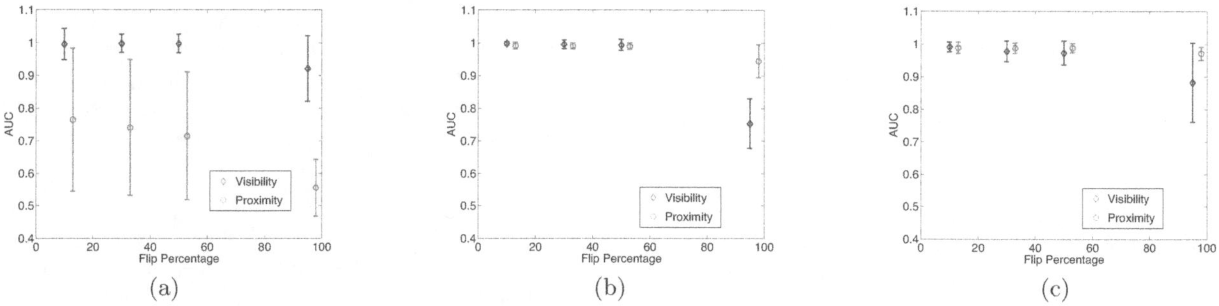

Figure 10: AUC Means for (a) `Edge`, (b) `Core-1000`, (c) `Core-100`.

the classifier. On the other hand, for `Core` networks, AUCs are generally quite high, typically 0.95 or higher.

5.4 Comparison

The AUC means for both methods are shown in Figure 10. Figure 10(a) shows that for `Edge` networks, the visibility-based method is vastly preferable, and maintains an accurate classification rate even for flipping percentages as high as 95%. However, Figures 10(b) and (c) show that the visibility-based method breaks down, especially for high flipping percentages, for `Core` networks. Since the VISIBILITY-INFERENCE problem is likely to be of more interest to `Core` networks in general, this motivates the effort to develop a more refined method targeted specifically to `Core` networks.

Figure 10 shows that this more refined method (proximity-based) is in fact clearly effective in both `Core-1000` and `Core-100` networks. For those networks, even at very high flip percentages, the proximity-based method using RSD is capable of AUCs that are nearly always close to 1.

6. CLASSIFICATION

The results in Sections 4 and 5 demonstrated both methods have high TPR and low FPR for large range of threshold values. Nonetheless, in practice we need to choose a threshold value (without access to ground-truth data). Here, we show that such a value can be picked automatically, by exploiting the information hidden in the 1s of the observed visibility matrix M.

The key idea is that, intuitively, we expect the descriptive submatrices of the 1-valued entries of M to be similar to the descriptive submatrices of the false-zero entries of M. Therefore, we can use the distribution of the characteristic values of the former, in order to choose a threshold that will correctly classify the latter.

To demonstrate the effectiveness of this strategy, we perform the following experiment. For each network in `Edge`, with visibility matrix M, we choose entries $M(i, j) = 1$ and form their corresponding (i, j)-descriptive submatrices using the methods described in Section 4.1. Then, we set the classification threshold to the low end of the distribution of SUM values obtained. Based on experience with the classifier, we selected the 3rd percentile of the distribution as a reasonable choice.

The leftmost part of Table 1 shows the the average (and standard deviation) of TPR and FPR of the visibility-based classifier across all the networks in `Edge` and for all four flipping percentages. The results show that the 3rd-percentile rule for picking the threshold value leads to average TPR of almost 1 and very small FPR for flip percentages up to 50%. Even for 95% flipping percentage, the average TPR drops only to 0.85 and the average FPR increases only to 0.17.

The last four columns of Table 1 show the same results for the `Core-1000` and `Core-100` networks. These results are obtained by performing the same experiment in those networks, with the only difference that the (i, j)-descriptive submatrices of the 1-valued entries of M were obtained using the proximity-based method described in Section 5.2. We use the proximity-based method since it was identified to outperform visibility-based classification for these networks.

The results show again that a threshold set to the 3rd-percentile of the sum of the descriptive submatrices of the 1-valued entries yields excellent average TPR and FPR for all flip percentages. Overall, these experiments confirm our intuition that the descriptive submatrices of the 1-valued entries of M are similar to the submatrices of the false zeros, and that using this information to select the classifier threshold leads to excellent prediction accuracy.

The results above are for the random bit-flipping strat-

Table 1: Mean and standard deviation of TPR and FPR for **Edge**, **Core-1000** and **Core-100**; 3rd-percentile SUM threshold.

	Edge		Core-1000		Core-100	
Flip %	TPR	FPR	TPR	FPR	TPR	FPR
10%	0.99 (0.062)	0.032 (0.15)	0.98 (0.033)	0.03 (0.04)	0.95 (0.067)	0.027 (0.022)
30%	0.99 (0.061)	0.045 (0.15)	0.98 (0.035)	0.03 (0.04)	0.95 (0.071)	0.028 (0.021)
50%	0.99 (0.067)	0.061 (0.17)	0.98 (0.033)	0.03 (0.05)	0.95 (0.064)	0.034 (0.025)
95%	0.85 (0.18)	0.08 (0.18)	0.98 (0.027)	0.21 (0.18)	0.96 (0.054)	0.069 (0.046)

Table 2: Mean and standard deviation of TPR and FPR for destination-based flipping on **Edge** and **Core-100**.

Edge		Core-100	
TPR	FPR	TPR	FPR
1.0 (0)	0.98 (0.11)	0.78 (0.30)	0.027 (0.17)

egy, corresponding to the case where traffic for each source-destination pair is independent. As described in Section 3, we also consider the case of correlated bit-flipping, in which all of the 1s for a specific destination (column) are flipped to zero. For each AS in the **Edge** and **Core-100** datasets, we applied this stategy 1000 times to a randomly chosen column. We report results for only the zeros in the chosen column, because classification accuracy of zeros outside of the chosen column is not significantly changed.

Table 2 shows the resulting classification accuracy, again using the visibility-based method for **Edge** networks and the proximity-based method for **Core-100** networks. It shows that for **Edge** networks, essentially all zeros are classified as true zeros. This occurs because of the nature of the visibility-based method. In that method, the 1s in the same column and rows as the zero $(i.j)$ determine the size of the submatrix. If the entire column is set to zero, the submatrix is necessarily too small to identify any false zeros. Note that classification accuracy for the rest of the matrix is not strongly affected, however.

The situation is quite different for the proximity-based method operating on **Core-100** networks. Here the classifier performs quite well, with a high TPR and a FPR of almost zero. That is, even when traffic patterns are such that a destination prefix receives no traffic at all, the traffic patterns in similarly-routed prefixes are sufficient to accurately separate true and false zeros.

7. APPLICATION: TM COMPLETION

In Section 1, we described a number of reasons one might seek a solution to the VISIBILITY-INFERENCE problem. Here we demonstrate an example in depth: improving the accuracy of *traffic matrix completion*.

The problem. Traffic matrix (TM) completion refers to estimating traffic volumes that are not directly measurable, whether due to missing or dropped data [16], lack of visibility [2], or other reasons. The general notion builds on work in statistical signal processing [4] that has identified sufficient conditions and appropriate algorithms for estimating missing elements of a partially-observed matrix.

TM completion starts with a partially-observed traffic matrix V. The *known* (i.e., observed) entries of V are the set Ω. Using Ω one seeks to estimate the remaining, unknown elements of V. If ground truth is not available, TM completion accuracy can be assessed through cross-validation, i.e., by holding a subset $R \subset \Omega$ out as a validation set. Then, $\Omega \setminus R$ is given as input to the matrix-completion method, which predicts the entire matrix $\widehat{V}$, including elements R. Accuracy is measured by the Normalized Mean Absolute Error (NMAE) on the validation set R:

$$\text{NMAE}(R) = \frac{\sum_{(i,j)\in R} | V(i,j) - \widehat{V}(i,j) |}{\sum_{(i,j)\in R} V(i,j)}.$$

Prior work on TM completion has been forced to treat zeros in V as missing values (e.g., [2]). This is because, as we have discussed throughout this paper, lack of observed traffic for an element (i,j) is ambiguous: it may or may not reflect a valid traffic measurement. Thus, a large amount of information – all false zeros in V, which represent valid traffic measurements – is thrown away.[2] We demonstrate the significance of this loss, and the improvement possible with our classification methods, using **LMaFit** [15], a well-known matrix-completion algorithm.

Data. We use the traffic matrix V described in Section 3; it comes from a large Tier-1 provider that is a member of the **Core-100** set. We also use the corresponding ground-truth matrix T for the same network, for validation (only). The only preprocessing we perform is to remove any rows and columns that are fully-zero in V, since it is well understood that matrix completion cannot estimate elements in such rows and columns. After this filtering step, we wind up with versions of V and T having 28 rows (source ASes) and 6198 columns (destination prefixes); V has a density of about 4%.

Improving TM completion accuracy. To evaluate the benefit of solving VISIBILITY-INFERENCE in this setting, we consider four possible cases for TM completion:

Ground Truth (GT): TM completion using perfect knowledge of false zeros (valid zero-valued traffic measurements). So for GT the set of known entries given to **LMaFit** consists of $\Omega \setminus R$ plus the known false zero entries as given by T.

Visibility (VIS): TM completion after labeling false zeros via the visibility-based method. So for VIS the set of known entries given to **LMaFit** consists of $\Omega \setminus R$ plus the false zeros identified by the visibility-based method.

Proximity (PROX): TM completion after labeling false zeros via the proximity-based method. Input to **LMaFit**

[2] We extend the notion of true and false zeros to V by identifying them with the true and false zeros of M. This means that a false zero in V is actually a *valid* traffic measurement; a true zero is not.

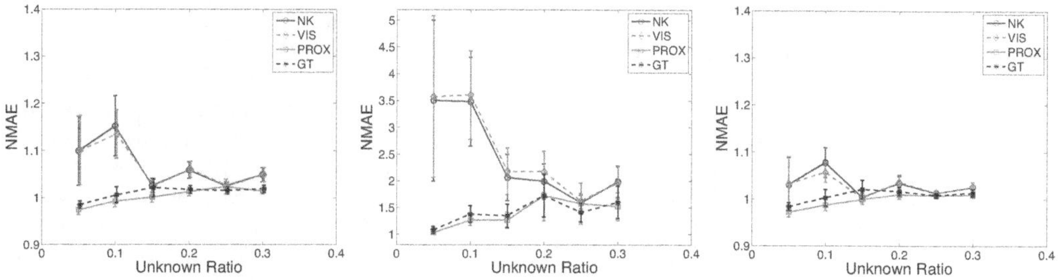

Figure 11: Accuracy of matrix completion for GT, VIS, PROX and NK for (a) all, (b) small, (c) large unknown elements in R.

is $\Omega \setminus R$ plus the false zeros identified by the proximity-based method.

No Knowledge (NK): TM completion with no knowledge of false zeros, i.e., as it has been done previously in the literature. In this case, the set of known entries given to `LMaFit` consists only of $\Omega \setminus R$.

Figure 11 shows the results, in which the fraction of Ω that is held out in R is varied from 0.05 to 0.3 (denoted on the plots as the *unknown ratio*). All results are averages over 20 different cross-validation sets R. We expect the results to depend on the size of the TM elements being estimated; additional knowledge of zero-values has a bigger influence on estimation of small elements. Hence, in Figure 11(a) we report results for all values in R, while in Figure 11(b) we show results for small values (entries smaller than the median) and in Figure 11(c) we show results for large values.

The results show that knowledge of false zeros can significantly improve the accuracy of TM completion. Comparing NK and GT one sees that accuracy can be improved as much as a factor of 3.5 when predicting small values. With respect to our classifiers, as we expect, PROX outperforms VIS since this is a `Core-100` AS. Most significantly, we observe that labeling false zeros using PROX yields improvements that are essentially the same as those obtained using ground truth.

8. DISCUSSION AND RELATED WORK

To the best of our knowledge, we are the first to define and address the VISIBILITY-INFERENCE problem, whose goal is inferring visibility from traffic. Therefore, we are not aware of any existing methods that solve exactly the same problem. Despite the unique question prompting our work, our methods have connections to much previous work. Here, we discuss those connections as well as the limitations of our approach.

BGP paths. The analysis of the properties of observed BGP paths has a long history, initially enabled by the Routeviews project [14]. Our work draws on the notion of BGP Atoms [3], which resulted from the observation that some sets of prefixes are routed identically everywhere in the Internet. The specifics of how routing policies affect observed BGP paths have been studied extensively [10]; these studies informed our work and guided our adaptation of RSD to BGP data.

Data Limitations. It is well-known that publicly available BGP tables provide an incomplete view of the AS graph. This has been reported widely, and recently reviewed in [13]. In this regard, it is important to note that our results are not based on the AS graph; in fact, the way we use BGP data does not introduce ambiguities due to missing links. As we describe in Section 3, we construct visibility matrices over only the set of sources and destinations for which we have every active path between a source and a destination. Thus, we omit source-destination pairs for which our BGP data may be missing links, and we can have high confidence in the 0-1 status of each element of each ground-truth T matrix. While we cannot rule out inaccuracies due to configuration errors, false BGP advertisements, or path changes that have not yet reached the monitors, we believe that these issues have negligible effect on our results. In the few cases where we discuss the AS graph (Section 3) we only take from it qualitative, not quantitative observations.

The experiments we perform make the assumption that paths visible in BGP match those that are actually taken by traffic. It is known that this is not always the case [8]. However, as [8] notes, "the two AS paths usually match." That paper shows that mismatches are rare — a few percent at most. Hence we don't expect that these mismatches dramatically change our results.

Scaling Considerations. As noted in Section 2, we consider observers to be ASes rather than ISPs. We don't believe this has a major impact on our results. While ISPs can certainly merge visibility and traffic information obtained from multiple ASes that they operate, such merging does not seem to fundamentally change the nature or difficulty of the VISIBILITY-INFERENCE problem. On the other hand, studying visibility at the AS level is natural because the AS is the granularity at which BGP selects routes.

Another issue concerns the scale of the full problem. The matrices we work with, although they have over 5 million elements, are small compared to a full Internet-wide matrix. A conservative estimate of of the size of a full AS-prefix matrix is on the order of 5 billion elements. For our Tier-1 provider, the fraction of nonzero traffic elements in V is about 0.1%. On the other hand, that provider's data was based on a sampling rate of 1/1000; hence very many small elements of V were lost due to the sampling process. So for the case of this provider, we can estimate that a day's traffic occupies on the order of 5 million nonzero elements in V, and probably much more.

It is not clear the extent to which routing changes affect the estimation process. The notion of visibility itself can be attributed to an instant in time, during which routing changes are assumed to be negligible. However, the need to collect traffic to populate M introduces sensitivity to routing changes. The effect of route changes is to inflate the number of 1s in M. The sensitivity of the inference process to this inflation needs more study; we only note here that a small fraction of prefixes are responsible for most route changes; and those prefixes receive comparatively little traffic [11].

Additional Approaches. Finally, we ask whether there are other ways ISPs could attack this problem.

One direction, as mentioned in Section 2, is for ISPs to make note of the routes that they learn through BGP. How many entries in $\widehat{T}$ could the observer fill in using BGP-learned routes? From BGP, the observer learns a path to each destination from each of its neighbors. Each path as well contains sub-paths. This implies that the number of $\widehat{T}$ elements that can be learned in this way for any given destination is equal to the number of unique ASes found in all neighbor paths to the destination. Experience with BGP suggests that this is not usually a large number, except for the small set of ASes with very high degree. For example, for each of the Routeviews monitors, the average number of elements per destination that can be learned by such a method is ~40. In comparison, for ASes in the `Core-1000` set, the average number of elements per destination potentially visible through traffic observation is ~500. Thus, we believe that BGP-learned paths are likely to add only incremental improvement to estimation of $\widehat{T}$.

Another approach would be for ISPs to use information about the *volume* of traffic observed. The general idea is to assume a distributional model for traffic – one which includes a nonzero probability that a traffic element is zero. One then models the observed data as a mixture of the values taken from two sources: the assumed traffic distribution (which generates false zeros), and an additional source of (true) zeros. Such 'zero-inflated' models are used, for example in ecology, in settings loosely analogous to the VISIBILITY-INFERENCE problem [9]. Taking this approach requires imposition of modeling assumptions, with all the difficulties that accompany it: addressing the model selection problem, estimating parameters, and assessing confidence in the results. However the potential exists to estimate the *number* of false zeros via this sort of method. While this approach presents many hurdles, a considerable amount of theory has been developed around how to do this in various settings, and some issues relevant to traffic models are being addressed [6].

9. CONCLUSIONS

We started from the simple question 'what routes pass through a network?' We showed that using the network's traffic data to answer this question is equivalent to identifying source-destination pairs that are *not* communicating at a given time.

Answering these questions prompted us to look for ways to identify sets of source-destination paths that are *routed similarly in general.* This can be thought of as an considerable generalization of the notion of BGP atoms: rather than groups of prefixes that are routed *identically*, we look for groups of paths that are routed *similarly*. Surprisingly, we show that such groups of paths can be identified in a large set of representative locations in the Internet.

A key enabling idea has been the definition of a new distance metric for network prefixes: routing-state distance. Using it we were able to extrapolate from the relatively small amount of information available in publicly accessible BGP tables to estimate routing similarity between any two prefix pairs in the Internet.

We realized these ideas in the form of a family of classifiers; applying these classifiers to traffic measurements we showed that one can generally answer our motivating question with a high degree of accuracy.

While a number of challenges remain and considerable additional work is warranted, we are hopeful that the methods we develop here can improve knowledge of route visibility and further inform both operators and researchers.

Acknowledgements. This work was supported by NSF grants CNS-1017529, CNS-0905565, CNS-1018266, CNS-1012910, CNS-1117039, by a GAANN Fellowship, and by grants from Microsoft, Yahoo!, and Google. The authors thank the SIGCOMM referees and shepherd for their help in improving the paper.

10. REFERENCES

[1] J. I. Alvarez-Hamelin, L. Dall'Asta, A. Barrat, and A. Vespignani. K-core decomposition: a tool for the visualization of large scale networks. Technical report, Arxiv, 2005.

[2] V. Bharti, P. Kankar, L. Setia, G. Gürsun, A. Lakhina, and M. Crovella. Inferring invisible traffic. In *CoNEXT*, Philadelphia, PA, 2010.

[3] A. Broido and k. claffy. Analysis of RouteViews BGP data: policy atoms. In *NRDM*, 2001.

[4] E. J. Candès and B. Recht. Exact matrix completion via convex optimization. *Found. Comput. Math.*, 9(6):717–772, 2009.

[5] S. Carmi, S. Havlin, S. Kirkpatrick, Y. Shavitt, and E. Shir. A model of internet topology using k-shell decomposition. In *PNAS*, 2007.

[6] D.-L. Couturier and M.-P. Victoria-Feser. Zero-inflated truncated generalized Pareto distribution for the analysis of radio audience data. *Annals of Applied Statistics*, 2010.

[7] T. Fawcett. An introduction to ROC analysis. *Pattern Recogn. Lett.*, 2006.

[8] Z. M. Mao, J. Rexford, J. Wang, and R. H. Katz. Towards an accurate AS-level traceroute tool. In *Proceedings of ACM SIGCOMM*, August 2003.

[9] T. G. Martin, B. A. Wintle, J. R. Rhodes, P. M. Kuhnert, S. A. Field, S. J. Low-Choy, A. J. Tyre, and H. P. Possingham. Zero tolerance ecology: improving ecological inference by modelling the source of zero observations. *Ecology Letters*, 2005.

[10] W. Mühlbauer, S. Uhlig, B. Fu, M. Meulle, and O. Maennel. In search for an appropriate granularity to model routing policies. In *SIGCOMM*, 2007.

[11] J. Rexford, J. Wang, Z. Xiao, and Y. Zhang. BGP routing stability of popular destinations. In *SIGCOMM Workshop on Internet measurment*, 2002.

[12] RIPE routing information service raw data project. `http://www.ripe.net/data-tools/stats/ris/ris-raw-data`.

[13] M. Roughan, W. Willinger, O. Maennel, D. Perouli, and R. Bush. 10 lessons from 10 years of measuring and modeling the internet's autonomous systems. *IEEE Journal on Selected Areas in Communications*, 2011.

[14] University of Oregon route views project. `http://www.routeviews.org/`.

[15] Z. Wen, W. Yin, and Y. Zhang. Solving a low-rank factorization model for matrix completion by a nonlinear successive over-relaxation algorithm. Technical report, Rice University, 2010.

[16] Y. Zhang, M. Roughan, W. Willinger, and L. Qiu. Spatio-temporal compressive sensing and internet traffic matrices. *SIGCOMM Comput. Commun. Rev.*, 2009.

Anatomy of a Large European IXP

Bernhard Ager*
ETH Zurich
bernhard.ager@tik.ee.ethz.ch

Nikolaos Chatzis
TU Berlin / T-Labs
nikos@net.t-labs.tu-berlin.de

Anja Feldmann
TU Berlin / T-Labs
anja@net.t-labs.tu-berlin.de

Nadi Sarrar
TU Berlin / T-Labs
nadi@net.t-labs.tu-berlin.de

Steve Uhlig*
Queen Mary, University of London
steve@eecs.qmul.ac.uk

Walter Willinger
AT&T Labs–Research
walter@research.att.com

ABSTRACT

The largest IXPs carry on a daily basis traffic volumes in the petabyte range, similar to what some of the largest global ISPs reportedly handle. This little-known fact is due to a few hundreds of member ASes exchanging traffic with one another over the IXP's infrastructure. This paper reports on a first-of-its-kind and in-depth analysis of one of the largest IXPs worldwide based on nine months' worth of sFlow records collected at that IXP in 2011.

A main finding of our study is that the number of actual peering links at this single IXP exceeds the number of total AS links of the peer-peer type in the entire Internet known as of 2010! To explain such a surprisingly rich peering fabric, we examine in detail this IXP's ecosystem and highlight the diversity of networks that are members at this IXP and connect there with other member ASes for reasons that are similarly diverse, but can be partially inferred from their business types and observed traffic patterns. In the process, we investigate this IXP's traffic matrix and illustrate what its temporal and structural properties can tell us about the member ASes that generated the traffic in the first place. While our results suggest that these large IXPs can be viewed as a microcosm of the Internet ecosystem itself, they also argue for a re-assessment of the mental picture that our community has about this ecosystem.

Categories and Subject Descriptors

C.2 [**Computer Systems Organization**]: Network Operations

Keywords

Internet Exchange Points, Internet topology, traffic characterization

1. INTRODUCTION

The basic role of Internet eXchange Points (IXPs) dates back to the establishment of Network Access Points (NAPs) as part of the decommissioning of the National Science Foundation Network (NSFNET) around 1994/95, a carefully orchestrated plan for transitioning the NSFNET backbone service to private industry. The vehicle that evolved in support of this transition was a set of four NAPs (i.e., MAE-East, Sprint NAP, PacBell NAP, and Ameritech NAP) that acted as connection points for the commercial carriers that were vying for offering backbone services (e.g., MCInet, Sprintlink, AGIS) and ensured that the network would remain connected at the top level once the NSFNET was retired.

Over the past 15 years, as the Internet grew by leaps and bounds by any imaginable metric, the original four NAPs were replaced by a steadily increasing number of modern IXPs. Originally providing largely just the bare necessities for supporting easy interconnection between their member ASes (e.g., physical space, caches, cabling, power, A/C, or secure access), IXPs themselves have evolved over time. Numbering now more than 300 worldwide [18], many of these IXPs are offering an array of different services that rely on advances in networking technology (e.g., VLANs or MPLS), exploit existing routing protocols in innovative ways (e.g., use of BGP for prefix-specific peering), or provide the economic incentives for an ever-increasing number of networks to join as paying members (e.g., remote peering offerings, support for IXP resellers).

In fact, large IXPs such as AMS-IX, situated in Amsterdam, and DE-CIX, in Frankfurt, offer high-end Service Level Agreements (SLAs) to their members that cover not only the initial provisioning and daily availability of a member's port(s) but also the level of performance of key service parameters. Such innovation on parts of the IXPs has enabled them to compete more directly with the traditional carriers and has led to today's environment where some of the largest IXPs worldwide (e.g., AMS-IX, DE-CIX, LINX, MSK-IX) reportedly carry on a daily basis similar amounts of traffic as some large ISPs (e.g., AT&T, Deutsche Telekom[1]). The traffic volumes at those IXPs are generated by some 300-500 networks that cover the whole spectrum of players in today's Internet marketplace. While there may be regional differences in how extensive in coverage or aggressive in the uptake of new members IXPs are, the critical role they have played in the Internet ecosystem has until recently gone largely unnoticed by the research community whose focus has traditionally been on large carriers and large content.

This paper reports on a first-of-its-kind measurement-based study of a large IXP and complements a body of existing literature that has focused squarely on large carriers or large content. To this end, we analyze a unique dataset consisting of nine months' worth of anonymized sFlow records that were collected at one of the largest IXPs in Europe, and worldwide, in 2011. We present our dataset, describe our data analysis and illustrate how our observations and findings contribute to an improved understanding of

*Part of this work was done while employed at TU Berlin / T-Labs.

[1] AT&T reports carrying 28.9 petabytes of data traffic on an average business day [2], Deutsche Telekom reports 422 petabytes per month corresponding to 14 petabytes per day on average [13].

SIGCOMM'12, August 13–17, 2012, Helsinki, Finland.

- the AS-level Internet; that is, the structure and dynamics of the Internet as a network of networks or ASes;
- the Internet peering ecosystem; that is, the practices and economic incentives that drive the market for Internet interconnection and peering between ASes; and
- the Internet inter-domain traffic; that is, the quantity and quality of the traffic exchanged among ASes.

Specifically, the main contributions of our work fall into three categories. First, we show that this large IXP exhibits a surprisingly rich peering fabric in support of the many business objectives of its members. In particular, in terms of AS links of the peer-peer type that are typically established among member AS pairs we show that this IXP has close to 400 members which have established some 67 % (or more than 50,000) of all possible such peerings and use them for exchanging some 10 PB of IP traffic daily. *To put this number in perspective, note that as of 2010, the number of inferred AS links of the peer-peer type in the Internet was reported to be around 40,000 – less than what we observe at this particular IXP alone!*

To explain this startling difference between the number of peerings observed at this IXP (i.e., ground truth) and the number of known peer-peer AS links Internet-wide, we show which portion of the IXP's actual peering matrix is and is not visible when relying on the publicly available BGP data that has formed the basis for much of the past and recent work on inferring the Internet's AS-level connectivity. To further highlight this issue, we illustrate why a large portion of this IXP's actual peering matrix remains invisible even to measurement efforts that go beyond the current state-of-the-art, either with respect to BGP-derived data or traceroute-based measurements, or a combination of the two. By combining our finding of an enormously rich peering fabric among the members of this IXP with an accurate picture of their upstream connectivity (i.e., customer-provider relationships) that is reportedly quite accurate [11, 38], we are able to reconcile the traditionally-assumed hierarchical structure of the Internet with recent claims about a flattening of that structure. Indeed, while the traditional tier-structure of the Internet is still recognizable and can be largely recovered, the observed rich peering fabric at this IXP enables connectivity among networks of all different types and is essentially agnostic of any tier structure. Thus, at least as far as connectivity for the part of the Internet that involves this IXP is concerned, the observed IXP-related peerings provide a myriad of shortcuts and essentially complement any perceived or real hierarchical structure. Note that this realization of a much more elaborate interconnect structure than previously assumed says nothing about how these existing peering links are used to carry traffic.

To address the issue of how much and what type of traffic is traversing this IXP's infrastructure via public peering links, our second main contribution consists of an in-depth analysis of the available sFlow records. By examining the members of this large IXP and with which other members they peer and exchange traffic with, we highlight that large IXPs are a microcosm of the Internet as a whole in terms of types of networks, business relationships, or traffic. We observe various types of networks, from tier-1 to regional and local ISPs, large/medium/small content, host and service providers, content distribution networks (CDN), and a spectrum of academic and enterprise networks. With which other networks these networks establish peering connections at this IXP and exchange what type of traffic has nothing to do with their standing within the traditionally-assumed tier structure, but is largely dictated by economic considerations and business objectives and reflects a wealth of reasons and incentives for why the different types of networks make use of the various service offerings at this IXP.

Our third contribution is motivated by the existing large body of literature on traffic engineering for ISPs that relies critically on understanding how routing policies internal and external to the ISP affect the traffic flow over the ISP's infrastructure and ultimately result in what is commonly referred to as the ISP's intra-domain traffic matrix. In contrast to large ISPs, an IXP's infrastructure as well as the logical connectivity and routing at an IXP are significantly less complex and make it relatively easy to compute an IXP's traffic matrix and use it as a key input to IXP-specific methods in support of an efficient and effective management and operation of its infrastructure in a dynamic IXP marketplace. Despite the significant amount of traffic that the largest IXPs carry, especially when compared to the largest global ISPs, we are not aware of any research paper that provides an even remotely realistic picture of the traffic traversing an IXP. To fill this void, this paper is the first to obtain and characterize the traffic matrix of one of the largest IXPs, with a particular focus on traffic variability and dependency over time and in space (i.e., across ASes), and application mix. Knowledge of an IXP's traffic matrix and its properties in conjunction with emerging new routing strategies at IXPs is critical for exploring what-if scenarios that an IXP may want to run before announcing new services, encouraging members to send more traffic through the IXP, or for deciding when to upgrade its infrastructure and how.

The remaining part of the paper is structured as follows. In Section 2, we provide information about the large European IXP that we study and discuss the available sFlow records. Using this data, we examine in Section 3 the IXP's peering fabric and contrast our findings with results that rely exclusively on data that have been collected without the active participation of the IXP. Studying the members at this IXP as well as how and why they connect to other member ASes, we provide in Section 4 a detailed account of this IXP's ecosystem. In Section 5, we focus on the IXP's traffic matrix and report on some of its key characteristics. We elaborate on some of the implications of our findings and the new challenges they pose in Section 6, discuss related work in Section 7, and conclude in Section 8 with a summary of our main findings.

2. A LARGE EUROPEAN IXP

In this section, we provide a high-level overview of the infrastructure and operations of a large European IXP. We discuss the data that we obtained from this IXP, and present some basic facts about the member ASes of this IXP and about its overall traffic.

2.1 IXP overview

The main business model of an IXP is to operate and manage a physical infrastructure in support of public and private Internet interconnection. In this paper, we focus on the public part of an IXP's infrastructure where the IXP's revenues derive mainly from selling network interfaces or ports to customer networks (i.e., ASes) and supporting different types of interconnection arrangements. Such customer networks are referred to as member ASes. A member AS has the advantage to gain network connectivity to all other members of the IXP. However, interconnection arrangements reflect bi-lateral agreements[2] between a pair of member ASes, and these networks may want to impose certain conditions to ensure that they connect only to certain other networks or connect with them in ways that reflect their business model and support their market strategies.

What makes the IXP substrate of the Internet (i.e., the IXPs, their member ASes, and the peerings among these member ASes at those IXPs) such a vibrant marketplace is that the incentives for

[2]For the purposes of this paper, we view multi-lateral peering agreements as a collection of bi-lateral peering agreements.

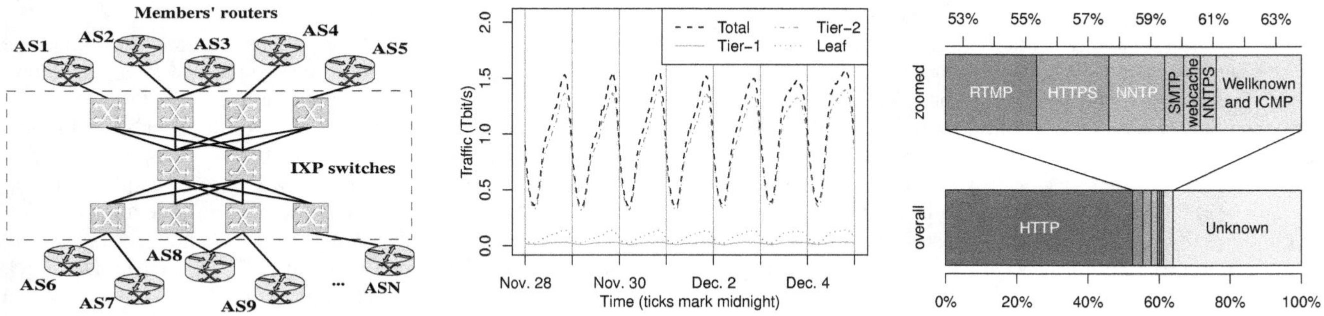

(a) A typical IXP architecture. (b) Traffic volume distribution across time. (c) Application mix.

Figure 1: IXP architecture and traffic statistics for the Nov/Dec week.

networks to become members at such public peering platforms are as diverse as the growing number of increasingly diverse ASes. For example, a CDN interested in optimizing its performance while keeping its cost low might want to choose an open peering policy to encourage direct and settlement-free traffic exchange at an IXP with as many networks as possible. On the other hand, large ISPs are likely to be interested in establishing peering relationships with other ISPs of about the same size. To achieve this objective, they may want to base their peering decision on a selective peering policy that allows them to deny peering with small ISPs, thus retaining them as paying customers in customer-provider type interconnection arrangements that are more lucrative. Transit networks have yet different objectives for using an IXP – they look at an IXP as a point of sale of their upstream connectivity offerings. In general, the larger the number of member ASes at an IXP, the more attractive that IXP is as a peering platform. This explains to a large degree the high level of innovation that the IXP marketplace has experienced in the process of becoming a vital component of the Internet ecosystem.

2.2 IXP infrastructure and data

Figure 1(a) illustrates a high-level overview of the architecture of our IXP. Although complex to maintain and scale, the infrastructure of this large IXP is typical of large IXPs in general, and the IXP's operation can be described in simple terms. The IXP provides a layer-2 switching fabric and each of the member ASes connects its access router to that switching fabric. When a pair of member ASes decides to peer at the IXP, they establish a BGP session between their access routers which, in turn, enables the exchange of IP traffic over this peering link across the IXP's infrastructure.

The volume and properties of the traffic exchanged at an IXP depend on the number of member ASes, the location and scope of the activities of the IXP, the IXP's service offerings, and if the IXP operates for profit or as a non-profit organization [18]. In this paper, we consider the traffic that is exchanged over the public peering fabric supported by the switching infrastructure of the IXP. In particular, for this study, we rely on nine months' worth of continuous sFlow [47] records that were collected in 2011 at the IXP's infrastructure using a random sampling of 1 out of 16k packets. Our sFlow records capture the first 128 bytes of each sampled packet, thus giving us access to the IP and TCP headers. The sFlow capturing process includes an anonymization step in which IP addresses are scrambled while maintaining prefix consistency [19].

The efforts we made to assess the quality of the available sFlow records included checking for sampling bias and identifying and filtering out less than 1 % of the total traffic that was immaterial for our study. For example, since sFlow sampling is performed simultaneously and independently by multiple switches within the IXP's infrastructure, there may exist a bias toward such flows that traverse multiple sampling points. When counting the number of different sFlow probes that capture packets exchanged between the same pair of member router interfaces (MAC addresses), we found that more than 99 % of these flows were only sampled by a single probe, providing hard evidence that our data is not corrupted by this sampling bias. As for immaterial traffic, we filtered out all traffic contributed by the IXP's management machines (e.g., route servers) as well as broadcast and multicast traffic, except for ARP packets. Finally, we also eliminated all IPv6 traffic as it constitutes less than 1 % of the overall traffic (in bytes or packets) at this IXP.

Table 1: Overview of IXPs sFlow dataset.

	Apr 25 May 1	Aug 22 Aug 28	Oct 10 Oct 16	Nov 28 Dec 4
Identified member ASes	358	375	383	396
Router IPs	426	445	455	474
MAC addresses	428	448	458	474
Tier-1	13	13	13	13
Tier-2	281	292	297	306
Leaf	64	70	73	77
Countries of member ASes	43	44	45	47
Continents of member ASes	3	3	3	3
Average packet rate (Mpps)	142	150	166	174
Average bandwidth (Gbps)	838	863	954	992
Daily avg volume (PB)	9.0	9.3	10.3	10.7

2.3 IXPs: A moving target

Studying one of the largest IXPs means chasing a moving target. Large IXPs present a changing environment, with a number of different dynamic factors acting on different time scales. Over large time scales (i.e., annual or monthly), there are changes due to new IXP policies. On more medium time scales (i.e., weekly), there is churn in IXP membership (e.g., new members join, but there are also potential departures from the IXP associated with mergers and acquisitions), number of switch ports, and peerings (e.g., new peerings are established, de-peerings, or peering changes such as switching from a public peering arrangement to a private peering). On small time scales (e.g., daily or hourly and below), traffic variations are the main cause for changing IXP conditions.

To address this aspect, instead of analyzing the entire nine months of essentially uninterrupted sFlow measurements from our IXP, we selected four one week-long periods during late April, late August, mid-October, and late November/early December of 2011. We selected weekly periods based on the fact that the AS membership at our IXP was by and large stable during the course of a week. At the same time, choosing four one week-long periods from the nine months long sFlow measurements results in four snapshots that – as seen from Table 1 – capture some of the churn that our IXP faces on the medium to large time scales. In particular, we note a steady increase in the number of members of our IXP and in the traffic volume they generated during the nine months long measurement period. How these and other changes manifest themselves in the IXP's peering fabric and its use is the theme of the next sections.

In the rest of this paper, we use the Nov/Dec data to illustrate our main findings. Where appropriate, we also include the results for the data of the other three weeks. Overall, we find that their analysis is consistent with the results we report here for the Nov/Dec data. In addition, we also spot-checked our results against a number of additional one week-long data and found no inconsistencies.

2.4 Membership and traffic statistics

To identify the active member ASes at our IXP during a given time period, we had to determine between which member ASes an observed IP packet is being forwarded. To this end, we relied on layer-2 information (i.e., MAC addresses) since the IP addresses in the header of the observed packets were those of the communication endpoints, not the routers on the path. We mapped MAC addresses to router IP addresses and their respective AS numbers by combining link-layer information from sampled ARP packets with routing data obtained from a publicly available looking glass at the IXP. This allowed us to identify 98 % of the members' routers. In the end, we succeeded in determining for more than 99 % of all observed sFlow packet samples their respective originating and receiving member ASes and as a result, we were able to identify for each week more than 350 AS members, each using between one and three logical router interfaces (see the first three rows in Table 1). The remaining less than 1 % of the exchanged traffic volume consists of IPv6 traffic and traffic that we could not associate with any member AS.

Being able to identify the IXP's members for a given time period, we examined next the member ASes of the IXP in each of the four weeks in more detail and report in rows 4 to 8 in Table 1 overall information about their tier level, country and continent. We considered the networks listed in Renesys "baker's dozen" [41] to be tier-1 ISPs and used the AS rank data provided by CAIDA [5] to classify the remaining members as tier-2 or leaf networks[3]. To this end, we classified a member AS as a tier-2 network iff it has both provider as well as customer ASes and as a leaf network iff it has only provider ASes. Based on this straightforward classification scheme that is largely agnostic to network specifics such as business, size, or traffic, we observed that irrespective of the considered time period, the vast majority of members of the IXP are tier-2 networks. At the same time, all the tier-1 ISPs in Renesys "baker's dozen" list are members of our IXP. To determine the country and continent of each member AS, we relied on the country code field that can be found in the AS's whois data. While the IXP members are from more than 40 countries in three continents, most of the member ASes are part of the European Internet scene. To highlight the geographic concentration even more, a majority of those European member ASes offer services in the same country in which our IXP is situated.

Figure 1(b) shows that for the Nov/Dec week, the total daily traffic volume generated by the member ASes and exchanged over the IXP's public switching infrastructure was in the petabyte range and followed a pronounced time-of-day pattern that is well-synchronized with the daily business or user activities in the country where our IXP is situated. In agreement with the observed tier-membership of the IXP's member ASes, the tier-2 networks were responsible for most of the total traffic volume. Rows 9, 10, and 11 in Table 1 provide information about the total traffic volume for each of the four weeks considered, namely the average packets per second, average bits per second, and average daily volume.

Not surprisingly, when breaking down the total traffic volume that traverses the IXP by member AS, we observe a skewed distribution, irrespective of whether we consider sent or received bytes. Indeed, less than 3 % of the member ASes were responsible for about 30 % of the total traffic and less than 30 % of the member ASes were responsible for close to 90 % of the traffic in the Nov/Dec week. Especially noteworthy is that while all tier-1 ISPs are members at this IXP and, with one exception, do exchange traffic over the IXP's public peering infrastructure, contrary to other parts of the Internet, they contribute relatively little to the total traffic volume, presumably because their high volume traffic travels over private peering links supported by the non-public part of this IXP's infrastructure. At the same time, the observed link load of public peerings involving tier-1 ISPs at this IXP was typically higher than average and causes some of those tier-1 ISPs to be included in the 30 % of members that generated close to 90 % of the total traffic volume.

Lastly, an added benefit of working with sFlow records is that it enabled us to examine the IXP traffic by applications. To this end, we separated the ICMP from the TCP and UDP packet samples by looking at their protocol field, and (when possible) associated the TCP and UDP packet samples with an application by looking at their source and destination port numbers and relying on the publicly available lists of port numbers used by the most popular applications. Figure 1(c) shows the application mix for the Nov/Dec week. While this straightforward approach cannot account for roughly 35 % of the bytes, we clearly see that HTTP is the most dominant application, accounting for more than 50 % of the bytes. This observation is consistent with recent reports from inter-AS traffic studies [28] and measurements of residential networks [30]. The next most popular applications are RTMP, HTTPS, and NNTP, the last of which has been reported to be used as a file sharing alternative [26].

3. AN IXP-CENTRIC VIEW OF AS-LEVEL CONNECTIVITY

The logical construct known as the AS-level Internet where nodes represent ASes and links denote AS relationships has no room for directly accounting for physical and geographically well-defined components of the Internet's infrastructure such as IXPs. As a result, an IXP's public peering fabric; i.e., the set of (bi-directional) AS relationships of the peer-peer type (P-P links[4], for short) that exist among pairs of member ASes of the IXP and express routing policies in support of settlement-free traffic exchange among pairs of members over the IXP's public infrastructure is not directly discernible and has received little attention in the past [3]. In this section, we rely on sFlow records from our IXP to obtain the ground truth of this IXP's public peering fabric. We call the compact description that summarizes which member AS is publicly peering with which other member ASes at this IXP the IXP's peering matrix. We then contrast this actual peering matrix with its counterparts derived from analyzing various BGP and traceroute datasets that have been used in the past for inferring AS-level connectivity and generating inferred AS-level maps of the Internet.

[3] In this paper, we decided against using the terms tier-3 and stub because of the different possible interpretations of their meaning.

[4] We are aware that the two parties of a bi-lateral peering agreement are free to (mis)use it as they see fit, e.g., as a customer-provider link. However, it is commonly assumed that most links at IXPs are of the peer-peer type [36, 50]. This has been confirmed by the IXP operator and is supported by our findings in Section 4.2 and 4.4.

3.1 Peering fabric seen from within the IXP

According to a commonly-used definition, two ASes are connected (at a particular time) in the logical AS graph if they can exchange routing information directly, i.e., without the help of an intermediary AS that provides transit, presumably for the purpose of exchanging IP traffic. In the case of our IXP where we know its topology (mapping of MAC and IP addresses to member ASes) and have access to its sFlow records, we use a more pragmatic definition and say that there exists a P-P link between a pair of member ASes if – during a given period of time – we see IP traffic being exchanged between these two member ASes over the IXP's public infrastructure. This pragmatic definition expresses our intention to focus on those P-P links of the IXP's peering fabric that matter; that is, carry actual IP traffic, e.g., BGP packets only in the case of backup links or IP packets generated by genuine application-level traffic. We call the thus-defined peering matrix the "ground truth" for our IXP as it provides the most useful and complete information about the actual status of the peerings between its member ASes.

After filtering the Nov/Dec sFlow records as described in Section 2 and analyzing the resulting traffic, we found that out of a total of $396 \times 395 / 2 = 78{,}210$ (bi-directional) P-P links that the 396 IXP member ASes could potentially establish at the IXP in that time period, more than 50,000 P-P links were actually established and were used to exchange IP traffic. This corresponds to a "peering rate" at our IXP or a "fill degree" of this IXP's (symmetric) peering matrix of about 67 %, meaning that on average, each member AS exchanges IP traffic over the IXP's public infrastructure with some 270 other member ASes. In total, the observed ground truth of this IXP's peering fabric with its more than 50,000 active P-P links is responsible for about 10 PB of traffic that traverses this IXP's public infrastructure daily. Next, we examine how well this IXP's actual peering matrix can be replicated when instead of relying on IXP-provided sFlow records, we are limited by measurements that do not involve the IXP and are obtained from outside the IXP.

3.2 Peering fabric seen from outside the IXP

In the past, BGP routing information (i.e., control-plane data) as well as traceroute measurements (i.e., data-plane information) have been widely used to analyze the structure and evolution of the AS-level Internet. Access to our IXP's actual peering fabric gives us a unique opportunity to evaluate how the various inferred peering matrices for this IXP that result from relying on these different IXP-external datasets compare to the IXP's ground truth.

In terms of BGP routing information, we relied on two well-known sources, i.e., Route-Views (RV) [45] and RIPE NCC (RIPE) [42], and on a non-public dataset (NP). For RV and RIPE, we relied on all their available route collectors, and used both BGP table dumps and updates from the same period when the Nov/Dec sFlow records were collected. NP consists of BGP dumps collected from about 70 routers worldwide which receive BGP information from 724 different ASes also covering the full week. Table 2 provides details about the total number of ASes from which the various datasets obtained BGP data and shows that despite varying significantly in magnitude, the three datasets are by and large complementary and contain routing information from almost 1,000 different ASes.

With respect to traceroute measurements, we used a dataset that resulted from a re-run of the targeted traceroute experiment described in [3]. This experiment was especially designed with the goal of discovering P-P links at IXPs and relied critically on the availability of publicly available traceroute-enabled looking glass (LG) servers throughout the Internet. The re-run was performed during Nov/Dec of 2011 using an updated list of available LG servers. The dataset we considered is derived from all traceroute probes launched as part

Table 2: Overview of routing and looking glass datasets for November. The numbers show P-P links.

Dataset	Unique LGs / ASN	Visible links	only in this dataset
RV	78	5,336	1,084
RIPE	319	10,913	5,460
NP	723	3,419	684
RV+RIPE+NP	997	13,051	10,472
LG	821 / 148	4,892	2,313
RV+RIPE+NP+LG	1,070	15,364	15,364

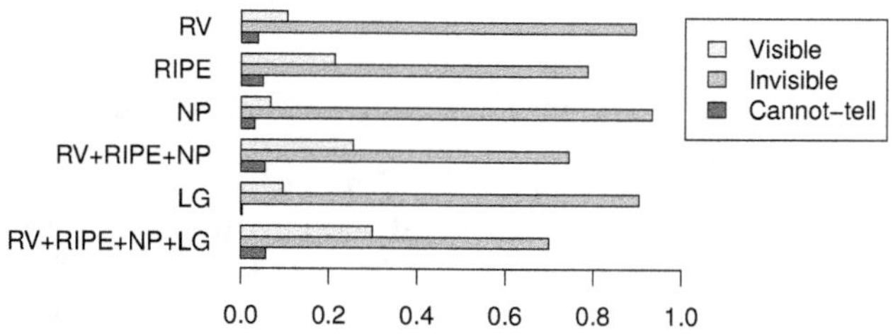

Figure 2: Peering links and visibility in control/data plane (normalized by number of detected P-P links).

of this recent campaign and consists of all inferred P-P links that involve our IXP and have an associated high confidence level of representing actual P-P links at our IXP (see [3] for details).

To systematically examine which P-P links at our IXP can and which cannot be discovered with the help of which IXP-external datasets, we classify these links into three categories. A **visible P-P link** is a P-P link that is observed both in the IXP-provided sFlow records and the IXP-external datasets (e.g., BGP or traceroute data). A P-P link is called an **invisible P-P link** if it is visible from the IXP-provided traffic data (i.e., IP packets traverse the link), but not visible from the IXP-external datasets. Lastly, a **cannot-tell P-P link** is a P-P link that is visible in BGP data but no traffic exchange is observed between the two member ASes in question from our IXP-provided data. This scenario is typical for private peering arrangements supported by the IXP's non-public infrastructure, but could also arise in those rare situations where a peering is not established at the IXP, or simply not visible in the traffic due to packet sampling. Note that the visible and invisible P-P links add up to the more than 50,000 P-P links that constitute the ground truth of our IXP's peering fabric. Furthermore, since the cannot-tell P-P links cannot be seen from the IXP-provided data, they are not a subset of either the visible or invisible peerings.

Using each of the IXP-external datasets, separately and in different combinations, Table 2 gives *(i)* the total number of visible P-P links that can be seen from the different IXP-external data and *(ii)* the number of unique visible P-P links; that is, those P-P links that can only be seen from exactly one of the IXP-external datasets. When compared to the ground truth, we see that each of the IXP-external datasets misses the vast majority of the observed links, and even when pooling all this available control- and data-plane information, we can still only account for a limited fraction of this IXP's actual peering fabric. A more detailed account of our findings is provided in Figure 2 and illustrates the breakdown of the P-P links into the three different categories of P-P links introduced above. We observe that even when relying on all the available datasets, about 70 % of the P-P links at this IXP remain invisible.

3.3 Some food for thought

A survey of the recent literature on measuring the AS-level Internet shows that as of late 2009, the total number of P-P links in

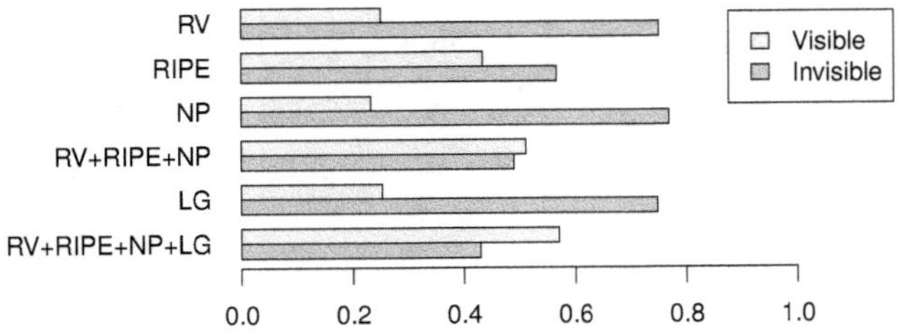

Figure 3: Peering traffic and visibility in control/data plane (normalized by total traffic volume).

the entire Internet was estimated to be in the 35,000-45,000 range. The low end of this range results from adding to the 15,000-20,000 P-P links reported in [16] the roughly 20,000 new P-P links that were discovered in [3] and passed very strict validation criteria. The high end of this range is reported by Chen et al. [11] who used less stringent criteria for validating newly detected P-P links. In stark contrast to these recent estimates, our above analysis shows that the more than 50,000 P-P links that we encountered in this single large European IXP exceed the total number of P-P links assumed to exist Internet-wide. In view of arguments that suggest that many of these P-P links at IXPs are not critical in topology inference [53] or for understanding the evolution of the Internet, for example due to their possible role as backup links [16], we use again our IXP as an example. We show in Figure 3 the fraction of the total traffic traversing the IXP infrastructure that would not be accounted for if we only knew about the visible links; that is, the P-P links whose existence at this IXP can be inferred from the various BGP or traceroute data. Figure 3 shows that when using these IXP-external datasets individually to infer the visible links, each of them misses between 56–78 % of the total traffic (in bytes or packets) handled by this IXP. Even when pooling all the IXP-external datasets, close to half of the total traffic would be missed, due to the large number of P-P links that are not seen. Therefore, trying to gain insight into the economic incentives and business reasons of the various member ASes of this IXP for establishing the encountered peering fabric would be very hard knowing the visible peerings only.

Table 2 and Figure 2 show why efforts to unveil the peering fabric at this IXP (or others) by using publicly available or even privately collected BGP data or relying on measurements obtained from carefully designed traceroute experiments are essentially doomed. The main reason is the well-known problem of vantage points [44]. On the one hand, as shown in [38], the locations within the AS-level Internet of the monitors traditionally used to collect the widely-used BGP data provide a relatively accurate picture of the Internet AS-level connectivity as far as AS links of the customer-provider type are concerned, reportedly missing less than 11,000 out of a total of about 94,000 of such links Internet-wide [11]. On the other hand, these monitors have hardly any visibility into the Internet's IXP substrate consisting of the various IXPs, their member ASes and the P-P links among them at those IXPs [37] and thus miss the majority of those links. At the same time, even when trying to use traceroute measurements and launch probes from LG servers close to an IXP to a target in an AS "on the other side" of the IXP, due to AS-specific routing policies, there is no guarantee that the probes traverse the IXP and improve the discovery of P-P links at the IXP.

4. DIVERSITY OF THE IXP ECOSYSTEM

In this section, we take a closer look at our IXP's ecosystem; that is, its member ASes, the rich peering fabric we described in Section 3, and various aspects of the traffic that is exchanged among the IXP's member ASes over this peering fabric.

4.1 Member ASes

We have already noted that the traditional classification of networks into tiers says little about the nature and nothing about the business types of the networks that are members at this IXP (see Section 2). Unfortunately, there is no readily available dataset which lets us determine the business type(s) of each member AS. Therefore, we manually examined the information available on each of the member ASes' web sites and present in Figure 4(a) their business type(s). Clearly, the business model of the member ASes differ significantly, and it is not uncommon to encounter member ASes that are in multiple business types. Focusing on their main business type, we further classified each of the member ASes as a *large ISP* (LISP), *small ISP* (SISP), *hosting/service and content distribution network* (HCDN), and an *academic and enterprise network* (AEN). A large ISP is providing transit, connectivity, eyeball access and additional services such as hosting or even content distribution. A small ISP is an access provider and may also provide transit services. Hosting and service providers are hosting content, either indirectly through providing web-space or rack-space to actual creators of content, or as content owners. Some of them also provide special services such as DNS. The AEN category comprises all networks that are solely used to connect enterprises and universities.

4.2 Peering

For each of our IXP's member ASes, Figure 4(b) shows the number of its P-P links; that is, the number of other members with which it peers at this IXP. The member ASes are ordered (x-axis) according to our classification as introduced in Section 2 (i.e., tier-1, tier-2, and leaf networks), with no particular order within the resulting groups. Figure 4(b) reveals an enormous diversity with respect to the number of peers – some member ASes peer only with a few other members, while others peer with almost all of them. In particular, we see that the tier-1 ISPs have a small number of P-P links, typically peering with less than 25 % of all member ASes. This observation is consistent with their stated intention of offering only restrictive peering (e.g., on peeringDB [39], on the IXPs web site, or on the companies' web sites). Tier-1 ISPs apparently use the IXP, among other reasons, to augment their existing peerings, but need to do this with care because most other member ASes are either transit customers or potential transit customer for them.

Non-tier-1 members typically peer with a large number of other members, with about 71 % of the tier-2 and leaf member ASes peering with at least 70 % of the other members. Among the tier-2 and leaf networks (more than 96 % of the members), there are less than 10 % which peer with less than 25 % of all members. Tier-2 and leaf member ASes usually have an open peering policy, meaning that when asked what they look for in a peering, their answers are mainly performance and reducing transit costs. Some prefer selective over open peering policies, especially if setting some standards for a potential peer's network in terms of criteria such as traffic exchange ratio, geographic scope, backbone capacity, or traffic volume is in their interest. We find that most member ASes at this IXP use open peering policies and peer massively with other members. However, we also encounter ASes with open peering policies that do not peer with that many other members. Possible reasons for their low peering rate are *when* they joined the IXP, *if* they offer desirable content or *if* they provide Internet access to a significant number of eyeballs.

Figure 4(b) also shows a classification of the member ASes in the four business categories defined above: LISP, SISP, HCDN, and

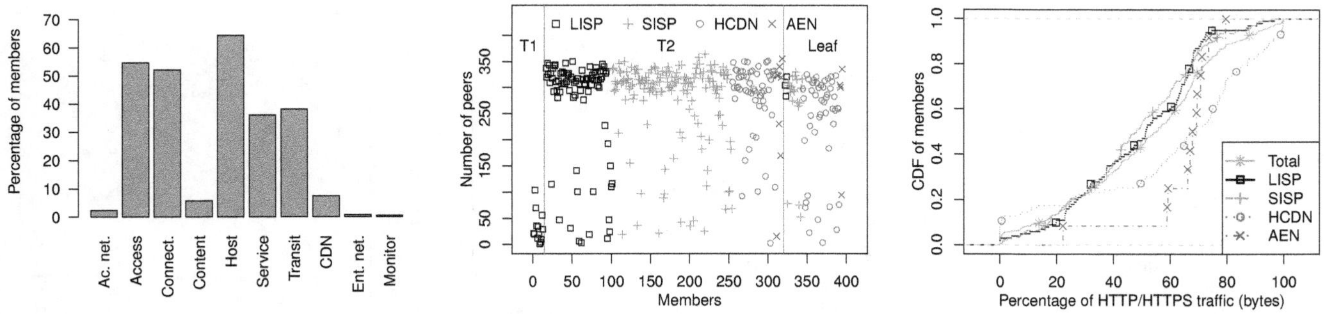

(a) AS classification by business types. (b) Scatter-plot of num. of peers per member. (c) Fractions of web-traffic across members.

Figure 4: Diversity in members: business type, number of peerings, and application mix exemplified by web-traffic.

AEN. Based on this classification, we find that in the LISP group, the member ASes with a small number of peerings are the tier-1 ISPs and those ISPs with a selective peering policy. In the HCDN group, the networks with a few peerings include some of the large players, but also small hosting providers (e.g., for banks or online games). The picture is less clear for the SISP group. In general, the observed large number of member ASes that have a large number of peers at this IXP is testimony for the ease with which member ASes can peer at this (and other) IXP. In fact, the findings of a recent survey [50] provide compelling reasons – some 99 % of the surveyed peerings were a result of "handshake" agreements (with symmetric terms) rather than formal contracts, and an apparent prevalence of multi-lateral peering agreements; that is, the exchange of customer routes within groups of more than two parties.

4.3 Traffic

The contributions to the IXP's overall traffic by the individual member ASes is highly skewed, with the top 30 % of member ASes contributing close to 90 % of the overall IXP traffic. Examining in more detail the traffic volume that each member AS contributes to the IXP's overall traffic, we first investigate what role the traffic exchange ratio plays in establishing P-P links. To this end, we consider the traffic asymmetry across all peerings between any two member ASes and show in Figure 5(a) the empirical cumulative probability distribution of this asymmetry. For improved readability we only show the part of the curve for ratios up to 100:1 (75 % of all peerings). The figure reveals a high variability in terms of exchanged traffic between the two member ASes of a peering. Indeed, only 27 % of the links have a traffic ratio of up to 3:1 (see support lines), where a 3:1 ratio is often stated as a typical requirement in common formal peering agreements [35]. Moreover, for 8 % of the peerings the ratio exceeds 100:1, and for another 17 % we observe traffic in only one direction. Figure 5(a) also depicts the empirical cumulative probability distribution for the P-P links at this IXP involving only tier-1 ISPs and shows that these peerings are less asymmetric, with more than 33 % of them having a ratio below 3:1.

Figure 5(b) shows the traffic asymmetry of the member ASes (i.e., the ratio of outgoing bytes vs. incoming bytes of a given member AS). The traffic of 52 % of the member ASes is more or less symmetric and within the range of 1:3 to 3:1. However, a significant number of member ASes fall in the 3:20 to 20:3 range[5]. In agreement with expectations, HCDNs have more outgoing than incoming traffic, while the opposite is true for LISPs and SISPs. However, there are various exceptions to this rule, and we find HCDNs with significantly more incoming than outgoing traffic and LISPs and SISPs where the opposite holds true. Note that despite the significant diversity in the ratio of incoming and outgoing traffic, more than half of the member ASes that send most of the traffic also receive most of the traffic. Indeed, there is a 50 % overlap among the top 50 member ASes according to bytes sent and the top 50 member ASes according to bytes received.

We can also examine how similar or dissimilar the overall application mix (see Section 2) is across all the IXP member ASes. For example, when computing for each member AS the fraction of HTTP/HTTPS traffic relative to the total number of bytes sent and received, we find in Figure 4(c) that this application mix differs significantly across the member ASes and follows almost a uniform distribution, indicating that without additional information, it would be difficult to predict which percentage of a member AS's traffic is HTTP. However, as soon as we include for example information about the member AS's business type, we observe that as expected, hosting providers and CDNs tend to send a larger fraction of HTTP traffic. However, rather unexpectedly, we also see more than 10 % of the hosting providers and CDNs with only marginal fractions of HTTP traffic. Closer inspection shows that these member ASes are primarily service providers that do not provide web content.

4.4 Prefixes

We next consider the prefix exchange ratio. For this purpose, we say that a prefix is *served* by a member AS if the member AS receives traffic for that prefix. Vice verse, we say that a prefix is *used* by a member AS if output traffic of its access router is destined toward that prefix. Figure 5(c) depicts a scatter-plot of the ratio of the number of prefixes used vs. the number of prefixes served by each member AS and provides clear evidence that the vast majority of the member ASes of our IXP use more than 10-times the number of prefixes they serve. Specifically, we see that hosting providers and CDNs have a tendency to serve a smaller number of prefixes but to use some two orders of magnitude more prefixes. Focusing on the ISPs, we can identify two groups. The first, larger group, serves a diverse but limited set of prefixes, from a few tens to a few thousands. The second, smaller group, serves and uses a large number of prefixes, some tens of thousands. Members that serve such large numbers of prefixes are likely acting as transit networks for other member ASes. However, we again observe exceptions to these general observations in almost all categories.

4.5 Geographical aspects

Conventional wisdom about IXPs states that ASes join regional

[5]To illustrate, if we had a member AS that would only deliver content using 1,500 byte-sized packets, the ratio could be as bad as 1:58, assuming on average one ACK of 52 bytes for every two data packets of 1,500 bytes and no overhead for the TCP connection establishment.

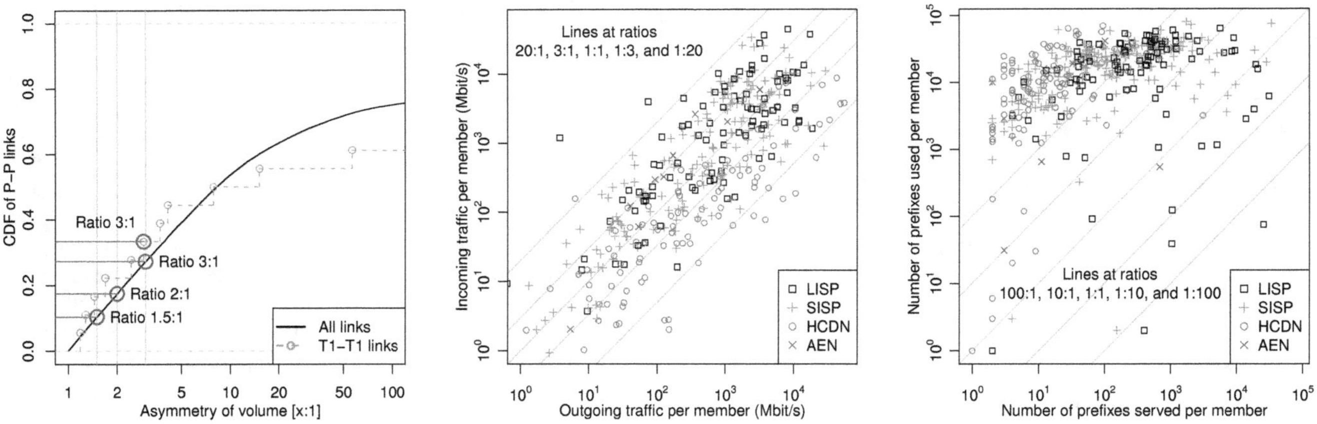

(a) Traffic asymmetry across P-P links. (b) Traffic asymmetry (out/in) per member. (c) Prefix ratio per member AS.

Figure 5: Diversity in traffic asymmetry and use of prefixes.

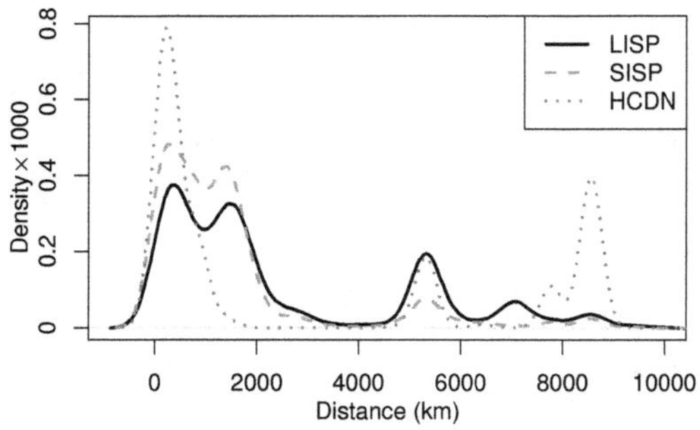

Figure 6: Geographic distances of IP endpoints to IXP.

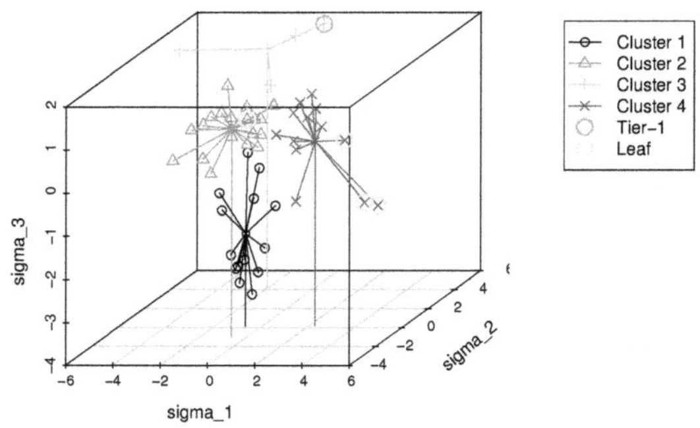

Figure 7: SVD-based 3D projection based on 12 features of the top 50 member ASes by bytes sent.

IXPs to exchange local traffic. To examine this general belief, we used the MaxMind GeoLite City database [31] to identify the geographic coordinates of both source and destination IP addresses for each sampled packet. Despite known inaccuracies of this geolocation database, using it for the needs of this study is appropriate, as we are only interested in approximate distances at the country level [40]. Contrary to our expectations, we found that only 10 % of the traffic is exchanged within the country in which the IXP is situated, while another 26 % originates from that country, and another 3 % is destined for that country. However, when relaxing the geographic constraint and considering a geographic region within a radius of 2,000 km from our IXP, we confirm the local nature of IXP traffic – almost 80 % and 72 % of the traffic terminates and originates, respectively, within this relatively close proximity.

To better understand the geographic reach of our IXP, Figure 6 depicts the density of the distances of traffic originated by member ASes in the LISP, SISP, and HCDN groups, weighted by byte volume. The distances shown in this figure are measured from the IP source address to the IXP, i.e., they represent the geographic range from which the IXP attracts traffic. We find that HCDNs have the largest fractions of very short distance traffic and at the same time the largest fraction of very long distance traffic–37 % of traffic volume with a distance larger than 5,000 km suggests the presence of significant intercontinental traffic. This indicates either mismatches in the IP address location [40] and/or that some of the traffic is indeed being served from remote locations. Likewise, member ASes in the LISP group show strong presence at around 5,000 km, which is mainly contributed by a small number of large international ISPs. SISPs and AENs (not shown) typically send traffic from closer to the IXP than members in the other business groups.

4.6 Tiers without tears

The above analysis highlights the diversity of the member ASes in terms of business types, the number of peerings, as well as their traffic characteristics. We have already seen indications that the traditional classification of networks by tiers cannot account for this observed diversity, mainly because it is agnostic to features of the member ASes such as their business type, traffic, peerings, prefixes, and geographic properties. Clearly, these and other features have the potential of painting a much more interesting and relevant picture of networks compared to what is possible knowing only the presence or absence of provider and customer networks.

In the rest of this section, we explore the possibility of combining some of these features and identify meaningful clusters. To this end, we consider 12 features in an attempt to characterize the member ASes' peerings and traffic characteristics: number of bytes sent, number of bytes received, number of peers, number of ASes and prefixes they send traffic to and receive traffic from, percentage of HTTP traffic that they send and receive, and 25-percentile of the distances from the traffic source to the destination, as well as from the IXP location to the destination for outbound traffic, and from the traffic source to the IXP location for inbound traffic.

We consider the Singular Value Decomposition [27] (SVD) of the 396 × 12 matrix and look at its projection into the 3D space defined by the three eigenvectors corresponding to the three largest singular values. Intuitively, the SVD produces a set of combined features (i.e., linear combinations of the original variables) so that the variability of the values of the first few combined features is maximized. Figure 7 shows the resulting 3D figure as a scatter-plot.

To keep the number of points reasonable, we sub-selected the top 50 member ASes according to sent bytes. Similar plots result if we sub-select according to number of bytes received or if we increase the number of member ASes to the top-100. In generating Figure 7, we repeatedly clustered the selected member ASes using the k-means clustering algorithm with random starting points into four clusters. Increasing or decreasing the number of clusters considered (i.e., value k) had no major impact on the nature of the results.

The output of this combined SVD/clustering method typically consists of one small cluster (~5 member ASes), two medium clusters (~10 member ASes) and one large cluster (~20 member ASes). When examining the clusters in more detail, we find that (i) the small cluster contains only large content and service providers, (ii) one of the medium cluster contains mainly small to medium ISPs that provide access to residential and enterprise customers and are all located in a country different from the IXP, (iii) the second medium cluster has mainly large ISPs and backbone networks that provide transit, and (iv) the big cluster has mainly data centers, big hosting providers, CDNs and some ISPs that provide web and server hosting. To highlight one such example clustering, we use in Figure 7 different symbols and colors and connect the clusters with a spider and mark their centers by support lines. We view Figure 7 as evidence that it is possible to classify an IXP's member ASes in ways that are practical and meaningful and respect the real-world diversity among networks that are critical elements of an IXP's ecosystem. Importantly, annotating the individual points in Figure 7 with tier-information shows why conventional tier classification is uninformative and of little help when trying to understand the Internet ecosystem locally (i.e., at this IXP) or globally.

5. IXP TRAFFIC MATRIX

Many IXPs report up-to-date traffic statistics on their web sites, and some of the largest European IXPs show daily traffic volumes for their public switching infrastructure that have been consistently in the petabyte range for some time. In Section 2, we confirm this publicly available but little-known fact for our IXP. A breakdown of this overall traffic by member ASes can be compactly described by an IXP's traffic matrix that specifies for example the hourly or daily traffic volumes exchanged between all member ASes that have a P-P link at this IXP. Despite the many similarities with an ISP's intra-AS traffic matrix, we are not aware of any published research paper that has considered IXP-specific traffic matrices and their properties. This is largely a reflection of the attention that large ISPs (and big content) have received from researchers and an indication that IXPs have been viewed as relatively uninteresting in terms of their topology, traffic, and routing. As is the case with ISP-provided measurements for computing or inferring an ISP's intra-domain traffic matrix, access to IXP-provided data for studying IXP-specific traffic matrices is similarly critical, and we rely in the following on our IXP-provided sFlow measurements. We report below on a first-of-its-kind analysis of the actual traffic matrix of one of the largest IXPs worldwide, examine in detail properties such as the diurnal pattern, sparsity, and (approximately) low rank, and discuss possible applications of our findings in support of managing and operating a large IXP's infrastructure.

5.1 Temporal properties

We have seen in Section 2 (see Table 1) that during our nine months-long measurement period, the overall traffic seen by our IXP steadily increased, mainly due to a similarly steady increase in the number of its member ASes. In addition, Figure 1(b) shows that the total traffic volume over time is dominated by a pronounced time-of-day characteristic, where the observed diurnal cycle coincides with the daily business cycle in the country where this IXP is located. Such diurnal behavior has long been a trademark of the temporal nature of measured intra-domain ISP traffic matrices.

To explain this diurnal behavior for our IXP, note that *(i)* the tier-2 member ASes are responsible for a majority of the overall traffic (see Figure 1(b)) *(ii)* the top-10 of these tier-2 member ASes generate much of the IXP's traffic (more than 33 %), and *(iii)* despite the tier-2 member ASes being a very heterogeneous group of networks, a majority of them cover the city, region, or country where our IXP is situated. Therefore, when plotting the incoming traffic volume for the top-10 receivers among the tier-2 member ASes in Figure 8(a), we see that the temporal behavior of their traffic is well-correlated with the overall traffic and well-aligned with the daily business cycle of the area where our IXP is located. An exception to this rule is the traffic of the two member ASes shown with solid lines in Figure 8(a) that are shifted by some 1–5 hours as a result of representing the traffic of ISPs serving geographic areas in different time zones, one in Europe and one outside of Europe. The traffic of the member AS plotted with dashed lines in Figure 8(a) visually reaches the bandwidth limits of its IXP network interfaces. We were able to confirm this observation by closer inspecting the sFlow records, which revealed that this particular member AS connects with five physical 10 Gbps links to the switching fabric of the IXP, thus it is indeed reaching its 50 Gbps capacity limit.

Knowing such temporal properties of an IXP's traffic matrix takes on a new meaning in an environment where innovation in the IXP marketplace in the form of new types of service offerings or refined routing policies is likely to increase the volatility and decrease the predictability of the traffic seen by the IXP. For example, in the presence of IXP port re-sellers, the IXP has limited visibility into the networks that use the re-seller as an intermediary and only indirect control over the traffic that these networks send or receive at the IXP. Similarly, members that can select prefix-specific peering to send certain traffic at certain times through the IXP instead of handing it off to their upstream provider(s) are likely to be a significant source for increased traffic volatility at IXPs. These are instances where a detailed understanding of the temporal dynamics of an IXP's traffic matrix will be critical for accurate prediction and successful root cause analysis of observed infrastructure performance problems.

5.2 Structural properties

A readily observable structural property of measured ISP-specific traffic matrices is their sparsity. The fact that generally only a small number of entries in these traffic matrices are populated or non-zero is largely due to a carefully planned network and complex routing decisions. In stark contrast, as reported in Section 3, with a "fill degree" of more than 65 %, our IXP-specific traffic matrix cannot be called sparse, and the main reason for this observed non-sparsity is economics. More precisely, an IXP's infrastructure and routing requirements are purposefully kept simple to facilitate easy network interconnection among member ASes, resulting in a rich peering fabric in support of traffic exchanges at the IXP between as many member ASes as dictated by "good" business practices.

Next, we examine how our IXP-specific traffic matrix compares to measured ISP-specific traffic matrices with respect to their widely-reported property of having approximately low rank. Recall that in SVD terminology, a matrix X of (algebraic) rank k has approximately or nearly low rank if only a small number $k^* \ll k$ of the largest singular values are needed to well-approximate X by a $k^* \times k^*$ matrix under a certain norm of the approximation error (see for example [52] for details). In practice, to check if a matrix has approximately low rank, we consider its singular values $\sigma_1 \geq \sigma_2 \geq \sigma_3 \geq \cdots \geq \sigma_n$, compute the energy function defined

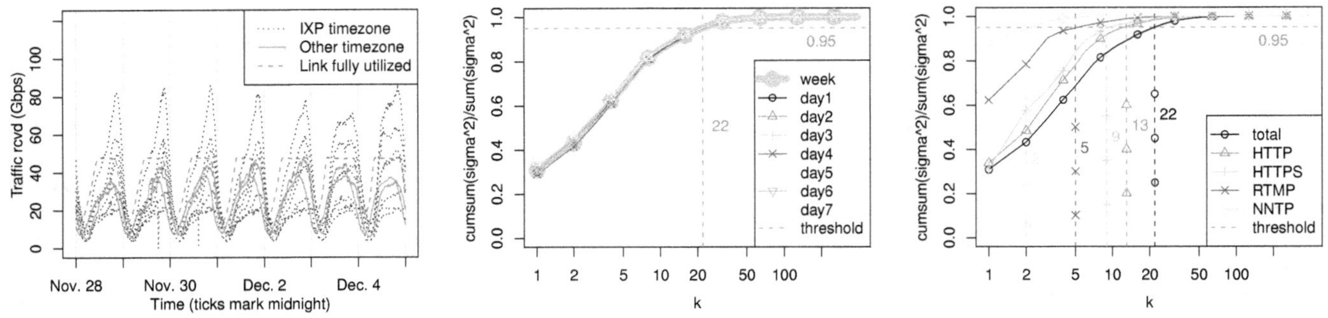

(a) Daily pattern of top-10 tier-2 members. (b) Traffic matrix energy. (c) Traffic matrix energy by application.

Figure 8: Traffic matrix properties.

by $f(k) = (\sum_{i=1}^{k} \sigma_i^2)/(\sum_{i=1}^{n} \sigma_i^2)$, and find the smallest k^* that captures, say, 95 % of the total energy.

Plotting the energy $f(k)$ as a function of k, Figure 8(b) shows the results of applying this method to a week-long traffic matrix (trace from Nov/Dec, thick line) and the corresponding seven daily traffic matrices (thin lines). In addition to confirming the approximately low rank of these traffic matrices (i.e., out of some 380 non-zero singular values, only the 22 largest are needed to capture 95 % of the energy), the plot also illustrates consistency among the week-long trace and the different daily traces. Similarly, recalling the application mix of the total traffic (see Figure 1(c)), Figure 8(c) shows the approximately low rank nature of the week-long application-specific traffic matrices. We observe that the low rank nature becomes more pronounced as we consider, in order, HTTP, HTTPS, RTMP, and NNTP, even though all these application-specific traffic matrices have almost full rank. In fact, in the case of NNTP, a simple 2×2 matrix can successfully capture most of the NNTP portion of the traffic in the entire original IXP-specific traffic matrix of size 396×396. Even for the case of RTMP, only a handful of singular values are needed to well approximate the portion of the overall traffic generated by this application. Another more specialized set of traffic matrices that also have approximately low rank can be constructed by considering only that portion of the overall IXP traffic that is produced by the top-50 member ASes (in terms of sent bytes) that were used as input to the clustering study described in Section 4 (see Figure 7) and ended up in one and the same cluster. A reason to examine such specialized traffic matrices is to look for any connection between the different business types that roughly specify these clusters and the low rank nature of the corresponding traffic matrices, and preliminary results (not shown here) suggest that the answer is affirmative.

While there are many other types of IXP-specific traffic matrices that can be considered, having approximately low rank is a common property among them and hints at the presence of enormous amounts of structure that is hidden in real-world IXP-specific traffic matrices and begs the question how this structure could possibly be exploited for practical purposes. One promising such application concerns efficient data acquisition. Specifically, as an IXP's infrastructure grows in terms of member ASes, peerings, and traffic, existing monitoring infrastructures that rely on increasingly lower sampling rate to keep up with the encountered growth in traffic may no longer be viable, especially in view of more stringent performance criteria that are promised by the IXP and typically require higher-resolution IXP measurements. The approximately low rank nature of IXP traffic matrices suggests a viable alternative whereby fewer but more intelligently collected (and hence more expensive) measurements can provide as much information as the current brute-force method of throwing more hardware at the problem to support the collection of maximal amounts of highly redundant data.

6. DISCUSSION

It is generally known and understood that BGP-based efforts for discovering P-P links in the AS-level Internet have somewhat limited success and provide at best a lower bound for the number of such links in the Internet [16]. However, that the number of P-P links at a single IXP exceeds even very recently reported such lower bounds [3, 11, 16] has come as a big surprise. This and the realization that even when laboriously combining the most up-to-date publicly available BGP data with hard-to-get non-public control-plane measurements and the latest available state-of-the-art data-plane measurements, our pooled data can only account for some 30 % of all known P-P links at our IXP begs the question how we can miss so many actual P-P links and why. Clearly, obscuring the existence of a "live" P-P link can occur with control-plane data (e.g., the route server from which BGP data is pulled not being close enough to the IXP) and with data-plane measurements (e.g., due to routing policies that prevent direct traffic exchange via an existing P-P link at the IXP and force traffic between two member ASes to take the upstream path). We plan to study the role of routing policies in hiding existing P-P links at IXPs as part of our future work and expect that an in-depth understanding of the root causes will suggest novel measurement experiments that have the potential of providing an accurate and near-complete peering matrix for each IXP and, in turn, an approximately valid snapshot of the AS-level Internet.

Irrespective of the reasons for the enormous number of encountered P-P links at our IXP, we have seen that the observed rich peering fabric by and large defies the well-known tiered structure of the Internet. As such, it is much more a reflection of the varied economic incentives and business benefits that drive the different member ASes to massively peer at this IXP and an indication of the ease with which such P-P links can be established. The resulting rich peering fabric supports massive direct interconnections or "shortcuts" as viable alternatives to sending traffic upstream and by and large agrees with recently reported findings of a "flattening" of the Internet. In fact, while some of our results are largely complementary to those of reported in [28], they do provide a different perspective. Relying exclusively on IXP data (as compared to the use of non-IXP-only data), we use an analysis of the IXP members' business types and a port-based application classification of their traffic to observe a similar consolidation of content and applications. However, our explanation for this consolidation centers more around the discovered massive peerings among all kinds of networks at this IXP and relies less on the presence and formation of "hyper-giants".

While this "flattening" of the Internet can fully co-exist with the Internet's traditionally assumed hierarchical structure, our IXP-specific findings and the following cautious extrapolations to the Internet as a whole suggest an even more radical change in perspective of the AS-level Internet. Indeed, considering only the European IXP scene (see [18] for details) and being conservative in using our large IXP as a baseline (i.e., assuming only a 50 % peering rate at IXPs), counting up the P-P links we expect to encounter at the four largest IXPs (with, say, 400 unique members each) and at the 10 next-largest IXPs (with, say, 100 unique members each), we obtain a realistic lower bound for the estimated number of P-P links for just the European portion of the Internet of some 200,000. This number is more than 100 % larger than the number of *all* AS links (i.e., customer-provider and peer-peer) in the entire Internet in 2010 as reported in [16] or the number of *all AS links of the customer-provider type* Internet-wide in 2008 as reported in [11][6].

Despite being extremely conservative, this estimate by itself should give reasons to pause. First, it indicates that there are easily an order of magnitude more P-P links in today's Internet than previously assumed. Second, contrary to conventional wisdom, there are many more P-P links in today's Internet than customer-provider type peerings, with twice as many being a conservative estimate. Third, judging from what we have seen at our IXP, most of these massive amounts of P-P links are of critical importance as they carry significant traffic. Given that past studies of the Internet peering ecosystem assumed exactly the opposite of what our findings support, there is a need for a major overhaul of the mental picture that our community has about the AS-level Internet, not only in terms of local and overall structure, but also with respect to its evolution in response to often rapidly changing business conditions locally at an IXP or within the larger Internet. In particular, we argue that assessing the "standing" of a network within the Internet's ecosystem has to account for network-specific features (e.g., business type) and some notion of traffic that this network is responsible for, either as a source, sink, or transit entity. This renewed focus on traffic requires novel approaches for the measurement, modeling, analysis, and inference of the Internet's inter-AS traffic matrix. Despite some initial efforts dealing with this matrix and a large body of existing work on intra-AS traffic matrices, the Internet inter-AS traffic matrix has remained a big enigma, but the availability of IXP-specific traffic matrices promises to invigorate research activities in this area.

7. RELATED WORK

Over the past years, the AS-level Internet has been a much-studied graph structure and continues to fascinate networking and non-networking researchers alike, though typically for different reasons. Instead of attempting to provide a necessarily incomplete overview of the existing literature on this topic, we refer the reader to a number of recent studies that serve as useful surveys [14, 16, 22, 44]. A majority of published research in this area has focused on measuring, inferring, modeling, and characterizing the AS-level Internet, often for the purpose of applying inferred AS graphs or carefully-tuned models to specific problems; e.g., [15, 17, 21, 32, 46].

Given that as logical constructs, AS topologies have no explicit space for physical infrastructure components, IXPs have long been neglected and generally viewed as an unimportant "detail". This view has slowly started to change with the gradual realization of the existence of "hidden" links in the Internet AS topology [8, 12, 23, 24, 37, 51]. [3] is the first traceroute-based study that purposefully targeted the existing IXPs worldwide. In conjunction with improvements on the measurement side, there has also been an increasing awareness of the important role that IXPs play in the Internet ecosystem, e.g., see [6, 7, 48, 49]. Ironically, among network operators, this realization has been there pretty much from the beginning of the commercial Internet [25, 33, 34], and there are signs that the research community is starting to give the Internet's IXP substrate the attention it deserves.

As far as traffic matrix research is concerned, despite some recent efforts [4, 9, 28, 44], little (if anything) is known about the Internet's inter-domain or inter-AS traffic matrix. The pieces of this puzzle that have received most attention by researchers have been the tier-1 ISPs and their intra-domain traffic matrices (e.g., see [1, 10, 53] and references therein). In fact, the latter have been a key ingredient for many ISP-critical tasks such as traffic engineering, capacity planning, and anomaly detection [20, 29, 43]. In stark contrast, when it comes to IXP-specific traffic matrices, only very recent work [6] has considered peering and traffic trends through a longitudinal study of a small European IXP across a period of 14 years. However, no study has gone into the same level of details as for intra-domain ISP traffic matrices. As is the case with ISP-provided measurements for computing or inferring an ISP's intra-domain traffic matrix, access to IXP-provided data for studying IXP-specific traffic matrices is similarly critical and requires close collaborations with IXPs. For the work reported in this paper, we have been fortunate to be able to work closely with one of the largest IXPs worldwide.

8. CONCLUSION

Examining readily available public information about a number of large European IXPs shows that they are very similar, not only with respect to the make-up of their constituents (i.e., member ASes) and overall traffic volumes, but also in terms of offered services, underlying technologies, business models, and overall purpose. As such, access to detailed internal measurements from even just one such IXP can highlight the important role that these largely ignored entities play for the Internet as a whole and for the particular geographic regions where they are located.

To this end, we analyze in this paper a unique data set of nine months' worth of continuous sFlow measurements from one of the largest IXPs in Europe, and worldwide, and clarify in the process some common misconceptions that exist regarding IXPs and the AS-level Internet. These include, among others, that tier-1 ISPs do not peer at IXPs (they do), IXPs are not used for transit (they are), the number of peer-peer links in the Internet is small (it is at least an order of magnitude larger than what has been assumed), the number of customer-provider links in the Internet is much larger than the number of peer-peer links (there are easily twice as many peer-peer links than customer-provider links), and IXP peerings are mostly used for back-up (they are not). In particular, we examine in detail the peering fabric and traffic matrix of our IXP and show the existence of a very diverse ecosystem in terms of the member ASes' business types, peering strategies, traffic exchanges, and geographic coverage that mimics the Internet's AS ecosystem as a whole. We argue that these findings are a proof that the mental picture our community has about IXPs and the AS-level Internet is in much need for a major overhaul.

Acknowledgments

We want to express our gratitude towards the IXP for their generous cooperation and support. We thank Mario Sanchez (Northwestern University) for re-designing and re-running the traceroute experiment originally described in [3] and for providing an up-to-date dataset (i.e., LG in Table 2). We also thank our shepherd Sergey

[6]Note that the reported numbers involving customer-provider links are from the published literature and cannot be derived from our IXP data.

Gorinsky and the anonymous reviewers who provided useful feedback. This work was supported by a G-Lab grant from the Federal Ministry of Education and Research of Germany (BMBF) and a Leibniz Prize grant from the German Research Foundation (DFG).

9. REFERENCES

[1] V. K. Adhikari, S. Jain, and Z.-L. Zhang. YouTube Traffic Dynamics and Its Interplay with a Tier-1 ISP: An ISP Perspective. In *Proc. of ACM IMC*, 2010.

[2] AT&T Company Information, 2012. http://www.att.com/gen/investor-relations?pid=5711.

[3] B. Augustin, B. Krishnamurthy, and W. Willinger. IXPs: Mapped? In *Proc. of ACM IMC*, 2009.

[4] V. Bharti, P. Kankar, L. Setia, G. Gürsun, A. Lakhina, and M. Crovella. Inferring Invisible Traffic. In *Proc. of ACM CoNEXT*, 2010.

[5] CAIDA AS Rank:AS Ranking. http://as-rank.caida.org/.

[6] J. C. Cardona Restrepo and R. Stanojevic. A history of an Internet eXchange Point. *ACM CCR*, Apr. 2012.

[7] I. Castro and S. Gorinsky. T4P: Hybrid Interconnection for Cost Reduction. In *Proc. of NetEcon*, 2012.

[8] H. Chang, R. Govindan, S. Jamin, S. Shenker, and W. Willinger. Towards Capturing Representative AS-Level Internet Topologies. *Computer Networks*, 2004.

[9] H. Chang, S. Jamin, Z. Mao, and W. Willinger. An Empirical Approach to Modeling Inter-AS Traffic Matrices. In *Proc. of ACM IMC*, 2005.

[10] H. Chang, M. Roughan, S. Uhlig, D. Alderson, and W. Willinger. The Many Facets of Internet Topology and Traffic. *Networks and Heterogeneous Media*, 2006.

[11] K. Chen, D. R. Choffnes, R. Potharaju, Y. Chen, F. E. Bustamante, D. Pei, and Y. Zhao. Where the Sidewalk Ends: Extending the Internet AS Graph Using Traceroutes From P2P Users. In *Proc. of ACM CoNEXT*, 2009.

[12] R. Cohen and D. Raz. The Internet Dark Matter—On the Missing Links in the AS Connectivity Map. In *Proc. of IEEE INFOCOM*, 2006.

[13] Deutsche Telekom ICSS, 2012. http://ghs-internet.telekom.de/dtag/cms/content/ICSS/en/1222498.

[14] A. Dhamdhere and C. Dovrolis. Ten Years in the Evolution of the Internet Ecosystem. In *Proc. of ACM IMC*, 2008.

[15] A. Dhamdhere and C. Dovrolis. The Internet is Flat: Modeling the Transition from a Transit Hierarchy to a Peering Mesh. In *Proc. of ACM CoNEXT*, 2010.

[16] A. Dhamdhere and C. Dovrolis. Twelve Years in the Evolution of the Internet Ecosystem. *IEEE/ACM Trans. Netw.*, 2011.

[17] D. Dolev, S. Jamin, O. Mokryn, and Y. Shavitt. Internet Resiliency to Attacks and Failures under BGP Policy Routing. *Computer Networks*, 2006.

[18] Euro-IX—European Internet Exchange Association. https://www.euro-ix.net/resources.

[19] J. Fan, J. Xu, M. H. Ammar, and S. Moon. Prefix-Preserving IP Address Anonymization: Measurement-Based Security Evaluation and a New Cryptography-Based Scheme. *Computer Networks*, 2004.

[20] A. Feldmann, A. Greenberg, C. Lund, N. Reingold, J. Rexford, and F. True. Deriving Traffic Demands for Operational IP networks: Methodology and Experience. *IEEE/ACM Trans. Networking*, 2001.

[21] P. Gill, M. Schapira, and S. Goldberg. Let the Market Drive Deployment: A Strategy for Transitioning to BGP Security. In *Proc. of ACM SIGCOMM*, 2011.

[22] H. Haddadi, G. Iannaccone, A. Moore, R. Mortier, and M. Rio. Network Topologies: Inference, Modelling and Generation. *IEEE Communications Surveys and Tutorials*, 2008.

[23] Y. He, G. Siganos, M. Faloutsos, and S. Krishnamurthy. A Systematic Framework for Unearthing the Missing Links: Measurements and Impact. In *Proc. of NSDI*, 2007.

[24] Y. He, G. Siganos, M. Faloutsos, and S. Krishnamurthy. Lord of the Links: A Framework for Discovering Missing Links in the Internet Topology. *IEEE/ACM Trans. Networking*, 2009.

[25] G. Huston. Interconnections, Peering and Financial Settlements. In *Proc. of INET*, 1999.

[26] J. Kim, F. Schneider, B. Ager, and A. Feldmann. Today's Usenet Usage: Characterizing NNTP Traffic. In *Proc. of IEEE Global Internet*, 2010.

[27] V. Klema and A. Laub. The singular value decomposition: Its computation and some applications. *IEEE Transactions on Automatic Control*, 1980.

[28] C. Labovitz, S. Lekel-Johnson, D. McPherson, J. Oberheide, and F. Jahanian. Internet Inter-Domain Traffic. In *Proc. of ACM SIGCOMM*, 2010.

[29] A. Lakhina, M. Crovella, and C. Diot. Diagnosing Network-Wide Traffic Anomalies. In *Proc. of ACM SIGCOMM*, 2004.

[30] G. Maier, A. Feldmann, V. Paxson, and M. Allman. On Dominant Characteristics of Residential Broadband Internet Traffic. In *Proc. of ACM IMC*, 2009.

[31] GeoLite City. http://www.maxmind.com/app/geolitecity/.

[32] W. Muhlbauer, A. Feldmann, O. Maennel, M. Roughan, and S. Uhlig. Building an AS-topology Model that Captures Route Diversity. In *Proc. of ACM SIGCOMM*, 2006.

[33] W. B. Norton. The Evolution of the U.S. Internet Peering Ecosystem. *Equinix White Papers*, 2004.

[34] W. B. Norton. A Business Case for Peering in 2010, 2010. http://drpeering.net/.

[35] W. B. Norton. A Study of 28 Peering Policies, 2010. http://drpeering.net/.

[36] W. B. Norton. Public vs Private Peering : The Great Debate, 2010. http://drpeering.net/.

[37] R. Oliveira, D. Pei, W. Willinger, B. Zhang, , and L. Zhang. In Search of the Elusive Ground Truth: The Internet's AS-level Connectivity Structure. In *Proc. of ACM SIGMETRICS*, 2008.

[38] R. Oliveira, D. Pei, W. Willinger, B. Zhang, and L. Zhang. The (In)completeness of the Observed Internet AS-Level Structure. *IEEE/ACM Trans. Networking*, 2010.

[39] PeeringDB. http://www.peeringdb.com/.

[40] I. Poese, S. Uhlig, M. A. Kaafar, B. Donnet, and B. Gueye. IP Geolocation Databases: Unreliable? *ACM CCR*, 2011.

[41] Renesys. http://renesys.com/.

[42] Routing Information Service (RIS)—RIPE Network Coordination Center. http://www.ris.ripe.net/.

[43] M. Roughan. Robust Network Planning. In *Guide to Reliable Internet Services and Applications*. Springer, 2009.

[44] M. Roughan, W. Willinger, O. Maennel, D. Perouli, and R. Bush. 10 Lessons from 10 Years of Measuring and Modeling the Internet's Autonomous Systems. *IEEE J. on Selected Areas in Communications*, 2012.

[45] University of Oregon Route Views Project. http://www.routeviews.org/.

[46] M. Schapira, Y. Zhu, and J. Rexford. Putting BGP on the Right Path: Better Performance via Next-Hop Routing. In *Proc. of SIGCOMM HotNets*, 2010.

[47] InMon—sFlow. http://sflow.org/.

[48] R. Stanojevic, I. Castro, and S. Gorinsky. CIPT: Using Tuangou to Reduce IP Transit Costs. In *Proc. CoNEXT*, 2011.

[49] V. Valancius, C. Lumezanu, N. Feamster, R. Johari, and V. Vazirani. How Many Tiers? Pricing in the Internet Transit Market. In *Proc. of ACM SIGCOMM*, 2011.

[50] B. Woodcock and V. Adhikari. Survey of Characteristics of Internet Carrier Interconnection Agreements, 2011. http://www.pch.net/docs/papers/peering-survey/.

[51] K. Xu, Z. Duan, Z.-L. Zhang, and J. Chandrashekar. On Properties of Internet Exchange Points and their Impact on AS Topology and Relationship. In *Networking*, 2004.

[52] Y. Zhang, M. Roughan, W. Willinger, and L. Qiu. Spatio-Temporal Compressive Sensing and Internet Traffic Matrices. In *Proc. of ACM SIGCOMM*, 2009.

[53] Y. Zhang, Z. Zhang, Z. Mao, C. Hu, and B. Maggs. On the Impact of Route Monitor Selection. In *Proc. of IMC*, 2007.

Measuring and Fingerprinting Click-Spam in Ad Networks

Vacha Dave *
MSR India and UT Austin
vacha@cs.utexas.edu

Saikat Guha
Microsoft Research India
saikat@microsoft.com

Yin Zhang
Univ. of Texas at Austin
yzhang@cs.utexas.edu

ABSTRACT

Advertising plays a vital role in supporting free websites and smartphone apps. Click-spam, i.e., fraudulent or invalid clicks on online ads where the user has no actual interest in the advertiser's site, results in advertising revenue being misappropriated by click-spammers. While ad networks take active measures to block click-spam today, the effectiveness of these measures is largely unknown. Moreover, advertisers and third parties have no way of independently estimating or defending against click-spam.

In this paper, we take the first systematic look at click-spam. We propose the first methodology for advertisers to independently measure click-spam rates on their ads. We also develop an automated methodology for ad networks to proactively detect different simultaneous click-spam attacks. We validate both methodologies using data from major ad networks. We then conduct a large-scale measurement study of click-spam across ten major ad networks and four types of ads. In the process, we identify and perform in-depth analysis on seven ongoing click-spam attacks not blocked by major ad networks at the time of this writing. Our findings highlight the severity of the click-spam problem, especially for mobile ads.

Categories and Subject Descriptors

K.6.5 [**Management of Computing and Information Systems**]: Security and Protection—*Online Advertising Fraud*

Keywords

Click-Spam, Click-Fraud, Invalid Clicks, Traffic Quality

1. INTRODUCTION

Background and motivation: Click-spam costs online advertisers on the order of hundreds of millions of dollars each year [4]. Instead of supporting free smartphone apps and websites, this money ends up in the pocket of click-spammers. Click-spam [1] subsumes a number of scenarios that all have two things in common: (1) the advertiser is charged for a click, and (2) the user delivered to the ad's target URL has no actual interest in being there. Click-spam can be generated using a variety of approaches, such as (i) botnets (where malware on the user's computer clicks on ads in the background), (ii) tricking or confusing users into clicking ads (e.g., on parked domains), and (iii) directly paying users to click on ads.

*Work done while at Microsoft Research India

[1] or equivalently, click-fraud. Since a fraudulent motive is often difficult to conclusively prove, we use the term click-spam instead.

SIGCOMM'12, August 13–17, 2012, Helsinki, Finland.

Incentives for click-spam are linked directly to the flow of money in online advertising — advertisers pay ad networks for each click on their ad, and ad networks pay publishers (websites or phone apps that show ads) a fraction (typically around 70% [2]) of the revenue for each ad clicked on their website or app. A publisher stands to profit by attracting click-spam to his site/app. An advertiser stands to inflict losses on his competitor(s) by attracting click-spam to his competitors' ads. An advertising network stands to increase revenues (but lose reputation) by not blocking click-spam.

Reputed online ad networks have in-house heuristics to detect click-spam and discount these clicks [39]. No heuristic is perfect. Advertisers pay for false negatives (click-spam missed by the heuristic). None of the ad networks we checked release any specifics about click-spam (e.g., which keywords attract click-spam, which clicks are click-spam, etc.) that would otherwise allow advertisers to optimize their campaigns, or compare ad networks.

Research goals and approach: We have two main research goals in this paper. Our primary goal is to design a methodology that enables advertisers to independently measure and compare click-spam across ad networks. The basic idea behind our approach is simple: since the user associated with click-spam is, by definition, not interested in the ad, he would be less likely to make any extra effort to reach the target website than a user legitimately interested in the ad. The advertiser can measure this difference and use it to estimate the click-spam fraction. Of course some legitimately interested users may not make the extra effort (false positives), or some uninterested users may still make the extra effort (false negatives). We correct for both types of errors by using a Bayesian framework, and by performing experiments relative to control experiments. Section 3 details our methodology.

Validating the correctness of our methodology is challenging because there is no ground truth to compare against. Ad networks do not know the false negative rate of their heuristics, and thus tend to underestimate click-spam on their network [12]. The accuracy of heuristics used by third-party analytics companies (e.g., Adometry) is unknown since their methodology and models are not open to public scrutiny; indeed, ad networks contend they overestimate click-spam [16]. We manually investigate tens of thousands of clicks we received (in Section 5). We present incontrovertible evidence of dubious behavior for around half of the search ad clicks and a third of the mobile ad clicks we suspect to be click-spam, and circumstantial evidence for the rest, thus establishing a tight margin-of-error in our methodology. In the process, we discover seven ongoing click-spam attacks not currently caught by major ad networks, which we reported to the parties concerned.

Our secondary goal is to measure the magnitude of the click-spam problem today. To this end, in Section 4, we apply our methodology to measure click-spam rates across ten major ad networks (including Google, Bing, AdMob, and Facebook) and four ad types. Our work represents the first measurement study of click-spam.

We also identify key research problems that can have a measurable impact in tackling click-spam.

Contributions: We make three main contributions in this paper. (i) We devise the first methodology that allows advertisers

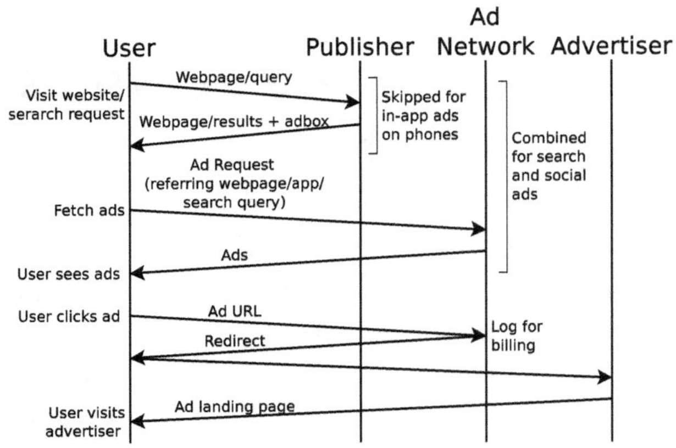

1: Time-line for serving ads.

to independently measure and compare click-spam rates across ad networks. We validate the correctness of the methodology using real-world data. (ii) We report on the sophistication of ongoing click-spam attacks and present strategies for ad networks to mitigate them. (iii) We conduct a large-scale in-depth measurement study of click-spam today.

2. ONLINE ADVERTISING PRIMER

Search vs. contextual: Keyword-based advertising is broadly classified into two categories: (i) *search advertising*, which are ads based on search keywords that show up on the side of search results, and (ii) *contextual advertising*, which are ads that show up on Web pages or in applications based on keywords extracted from the context. Search ads may be *syndicated — i.e.*, they are shown not only on the search engine operated by the ad network (*e.g.*, on www.google.com), but also on affiliate websites that offer customized search engines (*e.g.*, www.ask.com). The term *publisher* refers to the party that showed the ad (*e.g.*, website for contextual ads, smartphone application for mobile ads, affiliate for syndicated search ads). We do not consider other kinds of ads (*e.g.*, ads in videos, banner ads, etc.) in this paper.

Mobile vs. non-mobile: Both search and contextual advertising can be further classified as mobile or non-mobile based on what device the search or webpage request originated from. Mobile includes smart-phones and other mobile devices that have "full browser capabilities", as well as feature-phones with limited WAP browsers. The reason we draw a distinction between mobile and non-mobile is because we found ad networks internally seem to have very different systems for serving the same ads to mobile vs. to non-mobile users. We do not know the reason for this difference, but speculate it is because ad networks tend to expand through mergers and acquisitions, resulting in multiple technology stacks operating concurrently.

Ad delivery: Figure 1 illustrates the time-line for serving online ads. When the user visits a publisher website, the website returns an adbox (*e.g.*, embedded iframe), which causes the user's browser to contact the ad network. The request to the ad network identifies the referring website through the HTTP Referer (sic) header. The ad network then populates the adbox with contextual ads. In an alternate mechanism (not shown), premium publishers may directly query the ad network for relevant ads and seamlessly integrate them into the website content.

Charging model: While there are multiple charging models for online advertising (*e.g.*, impression-based, action-based), by far the most common is pay-per-click (PPC or CPC), where the advertiser is charged (and publisher paid) only if the user clicks on the ad. The publisher gets some fraction (typically 70%) of the revenue that the ad network collects from the advertiser. The accounting is performed as follows. The ad URL points to the ad network's domain with information about which ad was clicked (encoded in the GET parameters). When the user clicks an ad, the browser contacts the ad network, which logs the encoded information for billing purposes and redirects the user to the advertiser's site. This redirection is typically performed through an HTTP 302 response, which preserves the publisher's URL in the HTTP Referer seen by the advertiser's Web server. Black-hat techniques such as Referer-Cloaking [5] by the publisher, ad network policies [14], or bugs in browsers and proxies may result in empty or bogus Referer values being sent to the advertiser.

User engagement: The ad network can track limited user engagement (*i.e.*, ads viewed or clicked) for multiple ads shown across multiple publishers (*e.g.*, using cookies), but cannot, in general, track user engagement after the click. The advertiser, on the other hand, can track detailed user actions (only) on the advertiser's own website, but cannot track user engagement with other ads. Thus while the ad network has a broad-but-shallow view, the advertiser has a narrow-but-deep view into user engagement.

Click-spam discounts: Ad networks internally discount clicks based on in-house heuristics. The user is still redirected to the advertiser, but the advertiser is not charged for the click. Ad networks *do not* indicate which clicks were charged and which not in the advertiser's billing report.

3. ESTIMATING CLICK-SPAM

In this section we design a method that any party (*e.g.*, advertisers, ad agency, or researchers) can use to estimate click-spam rates for a given ad without explicit cooperation from the ad network.

3.1 Challenges

No ground truth: It is not possible to first identify (definitively) which clicks are click-spam, and then compute what fraction of the total traffic click-spam accounts for. A click is click-spam if the user did not *intend* to click the ad. There is no way to conclusively determine user intent without explicitly asking the user.

No global view: As mentioned, the ad network cannot track user engagement on the advertiser site, and the advertiser has no knowledge of the user's engagement with other advertisers. For example, the ad network does not know if the user never loads the advertiser's site after the click (we saw botnets exhibiting this behavior), and the advertiser does not know if the same user is implicated in click-spam attacks on another advertiser. The obvious solution is for the ad network and advertiser to cooperate. But financial disincentives and legal concerns prevent them from doing so: the advertiser could lie in his favor (claiming fraud where there is none) in order to gain deeper discounts from the ad network; the ad network could be held liable if sharing user history with advertisers has unforeseen privacy consequences.

Granularity: The granularity over which click-spam is estimated is important. An ad network may have low click-spam overall, say, but certain lucrative segments (*e.g.*, mortgage) may be experiencing orders of magnitude more click-spam. Advertisers require fine-grained measurements for their selected set of keywords.

Noise: As in any Internet-scale system, data is extremely noisy. We encountered users where Referer headers are inexplicably omitted, or browser User-Agents, IP addresses, cookies etc. change inexplicably within the same session (perhaps a buggy browser or proxy), and bad publishers that behave non-deterministically (perhaps to avoid detection). At the same time, because clicking an ad is a rare event, gathering good data is time-consuming. The combination results in very low signal-to-noise ratios.

3.2 Our Approach

We apply a Bayesian approach to work around the lack of ground truth. Instead of attempting to conclusively identify which clicks are click-spam, for a given ad, we create two scenarios detailed below where the (unknown) fractions of click-spam traffic is *different*. We link the two using a Bayesian formula to effectively cancel out quantities we cannot measure. The remaining quantities are those an advertiser can measure locally without requiring a global view. We control for noisy data (*i.e.*, adjust for false-positives and false-negatives) using a control experiment. Our approach *does not* (and indeed cannot) report whether a specific click is click-spam or not. The output is a single number representing our estimate of the fraction of clicks *for the given ad* that click-spam accounts for.

3.2.1 Data Collection

The detailed data collection procedure is as follows.

1. The advertiser (or a researcher signed-up as an advertiser with an ad network) creates his ads of interest in the usual way. That is, enters the text of the ad, selects ad targeting criteria (*i.e.*, search keywords, user demographics, mobile vs. non-mobile, etc.), sets his advertising budget (*i.e.*, auction bid amount, daily-budget), and sets the destination URL of the ad to his *landing-page* — a page on the advertiser's website the user should be sent to.

2. When a user clicks the ad, the advertiser website either loads the intended landing-page for the ad, or with some (small) probability first loads an interstitial page before eventually loading the landing-page. The point of the interstitial page is to *turn away* some fraction of users (legit clicks or not). To this end, the interstitial page may require some passive or active user engagement before continuing on to the landing-page. An example of passive engagement is to require the user to stay on the page for a few seconds before continuing on to the landing page (the interstitial page in this case may simply show a "loading..." message; it is known that a few seconds increase in page loading time can result in a measurable drop in traffic [6]). Some examples of active engagement is to require the user to click a link (*e.g.*, "This page has moved. Click here to continue.") or, in the extreme case, solve a CAPTCHA [15], both of which also result in significant drops in traffic.

Assumption 1. An implicit assumption is that the interstitial page will turn away a (unknown) larger portion of the click-spam traffic than the (unknown) portion of legitimate traffic it turns away. Specifically, we *do not assume* for example that all good users will patiently wait a few seconds; rather we assume that a smaller portion of click-spam traffic will wait the duration than the portion of good users doing so. Indeed user dwell time and willingness to click have long been considered an implicit measure of user interest [29, 35]. Interstitial pages leverage this knowledge to enhance the quality of the traffic reaching the landing-page by some (unknown) amount as compared to without the interstitial page.

3. Next the advertiser defines some advertiser-specific user engagement that he considers ultimate proof that the user intended to click the ad. For example, if the user makes a financial transaction on the advertiser's site, or signs up for some mailing list, etc. We call such users *gold-standard* users. Certainly not all users are gold-standard. But as long as a handful of users coming directly to the landing-page, and a handful coming to the landing-page via the interstitial page are gold-standard, we can apply our Bayesian approach below.

Assumption 2. We make an implicit assumption that for users that are *not turned away* by the interstitial page, their likelihood of becoming gold-standard users is not significantly changed. We *do not assume* that the interstitial page will not turn away users who may have become gold-standard users. Rather we assume for example that some users will not wait on the interstitial page, but if they do, they will not hold a grudge. We empirically find this assumption to hold in practice as we report later.

4. Lastly, extending the technique in [25], the advertiser creates a second ad that is identical to the ad created in step 1 above with the exception of the text of the ad. That is, the second ad has the same targeting criteria, advertising budget, and destination URL, but the text of the ad is junk (*e.g.*, a random nonsensical combination of words). We use this ad to adjust for false positives.

Assumption 3. We assume that very few users will intentionally click the second ad. While some users may be curious about the random set of words, we find this assumption is borne out in practice. We also assume for now that the click-spam *click-through-ratio i.e.*, the ratio of click-spam clicks to impressions of the ad) is independent of the text of the ad (we relax this assumption later).

3.2.2 Bayesian Estimation

Let G_d and G_i be the event that the user is a gold-standard user that arrived either directly, or via the interstitial page respectively. Let I_d and I_i be the event that the user intended to click the ad (*i.e.*, not click-spam) out of all users directly reaching the landing-page, or all users reaching via the interstitial page respectively.

Bayesian equation for $P(I_d)$: The advertiser is interested in learning $P(I_d)$. G and I are linked using Bayes theorem as follows: $P(G|I) = P(I|G) \times P(G)/P(I)$. Note that a gold-standard user implies that the click is not click-spam, *i.e.*, $P(I|G) = 1$. As discussed above in Assumption 2, $P(G_d|I_d) \simeq P(G_i|I_i)$; substituting and simplifying yields: $P(I_d) = \frac{P(G_d) \times P(I_i)}{P(G_i)}$.

$P(G)$ is computed as the ratio of the number of gold-standard clicks (g; known) to the number of clicks (n; known). $P(I_i)$ is the ratio of the number of non-click-spam clicks (i_i; unknown) to the number of clicks (n_i; known) arriving via the interstitial page. The above equation reduces to:

$$P(I_d) = \frac{\frac{g_d}{n_d} \times \frac{i_i}{\cancel{n_i}}}{\frac{g_i}{\cancel{n_i}}} = \frac{g_d \times i_i}{n_d \times g_i} \tag{1}$$

Only i_i on the right-hand-side is unknown.

Estimating i_i: As mentioned, the interstitial page enhances the traffic quality. If it did a perfect job — *i.e.*, all unintentional clicks would be turned away, and all intentional clicks would pass through — computing i_i would be trivial. In practice, the interstitial page has false-negatives (turns away users intentionally clicking the ad) and false-positives (not turning away click-spam). False-negatives do not affect i_i since it is the number of non-click-spam clicks *actually reaching* the landing-page, but false-positives result in an inflated value of i_i. We use a *control ad* to estimate the number of false-positives, and adjust for it.

Let F_i and T_i be the false-positive and true-positive click-through-ratios for the original ad and the interstitial page i. Similarly, let F_i' and T_i' be the false- and true-positive click-through-ratios for the control ad (identical ad except with junk ad text as mentioned above). We have four unknowns, and need four equations to solve. As discussed above in Assumption 3, we assume $T_i' \simeq 0$ and $F_i = F_i'$. The advertiser can measure $F_i + T_i = l_i/d$ where l_i is the number of clicks for original ad reaching the landing page through the interstitial page, and d is the number of impressions of the original ad as reported by the ad network, and the corresponding equation $F'_i + T'_i = l'_i/d'$ for the control ad.

The estimate for i_i is simply $T_i \times d$. Solving the four equations above for T_i we get the value of i_i adjusted downwards to account for false-positives as: $i_i = l_i - l'_i \times \frac{d}{d'}$

Final estimation formula: Combining the above with Eq. (1) we can estimate the click-spam rate for the original ad as:

$$P(I_d) = \frac{g_d \times (l_i - l'_i \times \frac{d}{d'})}{n_d \times g_i} \quad (2)$$

where:

- g_d, g_i : numbers of gold-standard users arriving directly and through the interstitial page respectively
- d, d' : number of impressions of the original ad and control ad respectively
- l_i, l'_i : number of clicks reaching (via the interstitial page) the landing-page for the original ad and control ad respectively
- n_d : number of clicks on the original ad directly reaching the landing-page

All these quantities can either be measured directly by the advertiser, or are present in billing reports ad networks generate today.

3.3 Limitations

A key limitation of our approach is that the advertiser must *actively* measure click-spam. The advertiser must interpose the interstitial page on live traffic (that he has paid-for), create a control ad (that he needs to pay for) to correct for false-positives, etc. Both the interstitial ad and control ad harm the user experience. It would be far more desirable to be able to passively look at logs and be able to estimate the click-spam rate from them.

One way to minimize the user experience impact is to apply our approach reactively when click-spam is suspected, but that runs into a second limitation — the rarity of data. Any estimation technique requires statistically significant data. The crucial factor in Eq. (2) is g_i and g_d — the number of gold-standard users. If these are small, the click-spam estimate can swing wildly. Suppose the advertiser manages to identify two gold-standard users, one arriving through the interstitial page and one directly, and computes a click-spam estimate based on it. If one new gold-standard user arrives through the interstitial site (or directly), the new click-spam estimate is half the previous (or will double). For a statistically significant estimate, as we report later, the advertiser must wait for roughly 25 gold-standard users via the two paths. This is especially an issue for small advertisers. Small advertisers may have to wait a long time to get gold-standard users — low advertising budgets means their ads don't get shown as much, even if they get shown users may not click on poorly ranked ads, even if they click they may not engage in a financial transaction with the advertiser, etc. The need to gather data over such an extended period is clearly at odds with minimizing the impact on user experience.

A group of small advertisers targeting similar keywords/users (or an ad-agency representing them) can apply our approach in the aggregate. Doing so has two benefits. First, due to the aggregation effect the group accretes statistically significant data more quickly. And second, the user experience impact is amortized across many advertisers. The downside, however, is that advertisers lose the ability to individually define what a gold-standard user means (which our approach otherwise allows) and have to depend on someone other than themselves to estimate click-spam rates.

Finally, our approach is naturally sensitive to the three choices the advertiser needs to make: 1) what his definition of a gold-standard user is, 2) what interstitial page approach he wishes to use, and 3) what the text of the control ad is. We discuss the implication of each design decision in turn. First, if the advertiser sets too high a bar for the gold-standard user he may not get statistically significant data; if he sets too low a bar that even click-spam users get classified as gold-standard he will underestimate click-spam rates. Second, if the advertiser picks too easy an interstitial page (everyone gets through), in Eq. (2) g_d/g_i will approach n_d/l_i and the estimate will approach 1 (*i.e.*, all clicks are legitimate) if the advertiser doesn't use a control ad; or 0 if he uses a control ad (*i.e.*, no clicks are valid). If the advertiser picks too hard an interstitial page (no one gets through), g_i and l_i will both approach 0, and the click-spam estimate will become undefined. Thus there is clearly some sweet-spot in designing the interstitial page, which we do not discuss. Third, if the control ad is not independent of the original ad (*e.g.*, the random choice of words happens to be related to the original ad), false-positives may be over- or under- corrected for. Making the right design choices is advertiser-specific.

To address the above issues to some degree, we report in the next section our experience with multiple types of interstitial pages, different definitions and numbers of gold-standard users. While our data shows much promise in our approach, we stress that a more thorough evaluation is needed.

4. MEASURING CLICK-SPAM TODAY

In this section we first validate the correctness of our approach from the previous section. We then conduct a large-scale measurement study of ten major ad networks and four types of ads.

Validation strategy: We assume that reputed search ad networks (specifically Google and Bing) are mature enough that their in-house algorithms are able to detect and discount for most of the click-spam on their search affiliate network. Validating our measurement approach then involves computing our click-spam estimate and comparing it to the *charged clicks* for Google and Bing search ads. Note that our algorithm does not have access to any data (including historical and aggregate data) that in-house algorithms at Google and Bing have access to, and Google and Bing do not have access to the detailed user-engagement data we collect as advertisers for user clicking our ads (specifically, we do not use any of the analytics products offered by Google or Bing). Given the datasets are completely different, if the click-spam rates we compute match that computed by leading ad networks (which they do as we report below), we have a strong reason to believe that our measurement approach is sound.

4.1 Methodology

We sign-up with ad networks as three different advertisers (each targeting different keywords) and follow the methodology from the previous section. The first advertiser targets a highly popular keyword (celebrity). The second, a medium-popularity keyword (yoga). And the third, a low-popularity keyword (lawnmower). We pick the keywords from a ranked list of popular keywords that the advertising tools of these ad networks provide.

For each keyword we create a realistic looking landing page, since the policies of the ad networks require us to use keywords that are relevant to the landing page. We instrument the landing-page to track mouse-movement, time spent on the page, switching browser tabs into or away from our page, whether any link on the page was clicked or not, page scroll, etc. Our instrumentation is through Javascript on the page; we detect browsers that don't support Javascript or have it disabled and exclude those data points. We direct on-path proxies (if any) to not cache any response so our server logs accurately reflect accesses. Unless otherwise mentioned, we pick a lax definition of gold-standard users based on

2: A control ad; content is a random set of English words

this telemetry, which is, that the user stays on the page for at least five seconds, and (for non-mobile browsers) produces at least one mouse/cursor move event. We are unable to define gold-standard users based on financial transactions, even though we expect that to be very strong signal of intent for real advertisers.

Next we create three interstitial pages: the first shows a loading message for five seconds before automatically redirecting to the landing-page. The second asks the user to click a link to continue to the landing-page. And the third asks the user to solve a CAPTCHA. We do not test the CAPTCHA interstitial for Google traffic since their advertiser policies restrict us from doing so.

We then create four ads for each target landing-page. The first ad directly takes the user to the landing page. The second, third and fourth ads first take the user to the three interstitial pages respectively, before continuing on to the landing page. All ads target the same keyword(s), user demographics, device and platform types, etc. The reason we create four separate ads (instead of a single one and interposing the interstitial page after the click) is so that Google/Bing produce fine-grained billing reports and statistics for each ad, which we can then validate our design choices against.

We create four additional (control) ads for each landing-page that correspond to the four original ads, but with junk ad text. The ad text was generated by picking five random words from an English dictionary (*e.g.*, Figure 2).

4.2 Scale

We repeat the above for 10 ad networks. For search ads we measure Google Search, Bing Search, and 7Search. For contextual ads we measure Google AdSense and Bing Contextual. For mobile ads we measure Google Mobile, Bing Mobile, AdMob (now owned by Google), and InMobi. And lastly for social ads, we measure Facebook. Altogether this adds up to 216 ads across all the networks.

We run the ads for a period of 50 days as needed to gather enough data. The majority of the ads were flighted in early January 2012. We continually adjust bids (mostly revising them higher) to help the lower popularity ads quickly attract enough data.

In all our ads were shown 26M times across all ad networks. They resulted in a total of 85K clicks (17K charged). Our ads were shown at at-least 1811 publisher websites and mobile apps (but the true number is likely much higher since we cannot determine the publisher for over 65% of our traffic). The landing pages were fetched by a total of 33K unique IP addresses located in 190 countries. We encountered over 7200 browser User-Agent strings (after sanitizing them to remove browser plugin version numbers).

4.3 Data

We log all web requests made to our server. The logs used in this study are standard Apache webserver logs that include the user's IP address, date and time of access, URL accessed (of a page on our webserver) along with any GET parameters, the HTTP Referer value and User-Agent value sent for that request, and a cookie value we set the first time we see a user to identify repeat visits from the same user. The raw logs including user engagement telemetry weighs in at over 3 GB.

A sanitized version of our raw logs is available online[2].

[2] http://www.cs.utexas.edu/~vacha/sigcomm12-clickspam.tar.gz

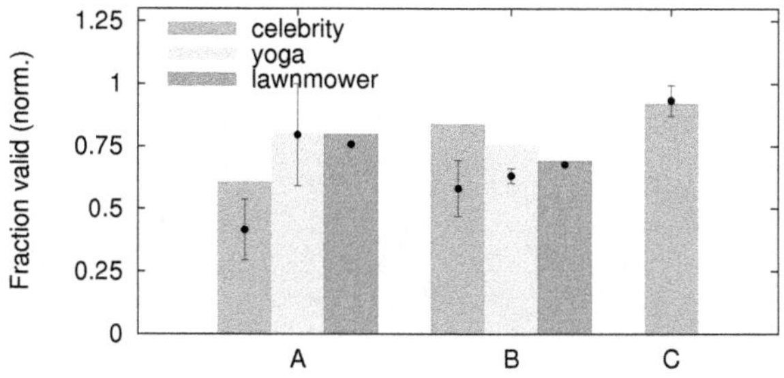

3: Normalized estimates for search ads

4.4 Ethics

Throughout our study we followed the advertiser terms-of-service (current as of when we did the measurement) for each of the ad networks we measured. Whenever our ads were rejected by the ad network (due to policy reasons) we fixed the issue so as to be compliant; if we couldn't fix it, we simply dropped that data-point.

High click-spam is an embarrassment for ad networks. Our goal in this paper is to systematically design a methodology, highlight the severity of the click-spam problem, and give researchers the tools and knowledge to further the state of the art. Our goal is not to embarrass ad networks. As a result, we prefer to report normalized or relative numbers whenever possible, and anonymize ad network names whenever it does not affect the core message of this paper.

Lastly, we expressly try to minimize adversely impacting user experience on these ad networks. For example, in order to get enough clicks on an ad, we have two options: run the campaign for longer, or increase the bid amount. We always choose the latter to minimize the time our ad is active on the network. For ads where despite increasing the bid we cannot gather traffic fast enough, we prefer to give up on that data-point and stop running that ad. Minimizing the time our ads are active also minimizes our contribution to the existing auction volatility for the keywords we bid on. As of this writing we have not received any complaints from ad networks, users, or advertisers regarding our study.

4.5 Validation

Figure 3 compares (normalized) complementary click-spam rates computed by our approach (plotted as error bars) and the (normalized) complementary click-spam rates as reported by Bing and Google for their search ad networks (plotted as bars). We ran another experiment where where we explicitly set our ad campaign to exclude syndicated search partners for one of the search ad networks (plotted as C in Figure 3). The Bing and Google estimates are the ratio between the number of clicks we were charged for (from the billing report), and the total number of landing-page fetches (from our logs). For our approach, we calculate two separate estimates based on the delay and click interstitial pages. The spread of the error bar plots the max and the min of the estimates we compute. The center tick plots the average. The figure plots the estimates for all three ads we flighted. In line with our goals, this figure (and all other figures in this section) are normalized so one of the data points is 1.

As is evident from Figure 3, our estimate for the yoga and lawnmower ads are in the same ball-park as that reported by Google and Bing. We manually investigated the difference between our estimates and that for the celebrity ad. We found over 50 clicks from sites associated with well-known search redirection viruses where browser toolbars hijack normal user searches and funnel them through affiliate search programs (Section 5.3.1 has more de-

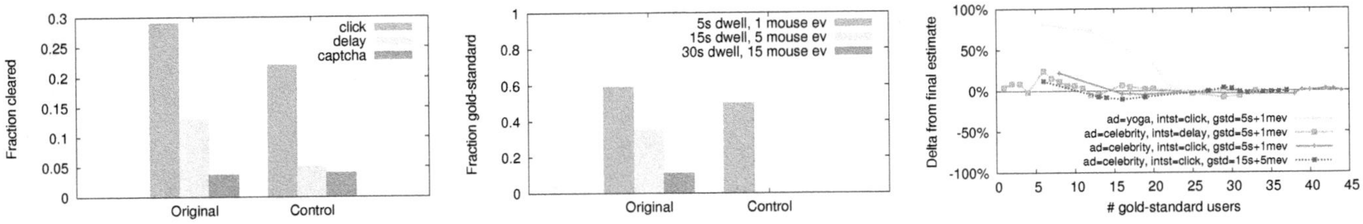

(a) Interstitial page performance (b) Gold-standard user definition (c) Convergence

4: Design space exploration

tails). We were charged for at least 48 of these clicks. Our estimates match the search ad network's estimates recomputed after discounting these clicks. Furthermore, as seen for network C where syndicated search partners are excluded, our estimates closely match that reported by the network. There were no clicks on the control ads on network C, which further supports our high estimate.

Our absolute numbers also agree with public estimates of average click-spam for these networks [3]. Next we drill deeper to validate our design decisions.

4.5.1 Interstitial Pages

Figure 4a plots the fraction of clicks for each interstitial page that reach the landing page for the celebrity ad, and for the corresponding control ad. Note that this fraction drops as the interstitial page changes from clicking a link (29%), to waiting 5 seconds (13%), to solving a CAPTCHA (4%), demonstrating the increasingly higher bar set by the interstitial pages. Interestingly, we find users are more likely to click through to the landing-page than wait 5 seconds. Except for the CAPTCHA interstitial, the fraction reaching the landing page is significantly lower for the control ad than for the original ad; this validates Assumption 1 from the previous section that the interstitial page concentrates non-click-spam traffic (by some unknown amount). Despite the varied interstitial page performance, the estimates computed from the delay and click interstitial converge (in Figure 3) for the experiments where we have a balanced number of converters through the interstitial and direct path, which supports Assumption 2. The CAPTCHA seems to reduce both normal and control traffic to the same low base level regardless of user intent; as a result, it is unsuitable for use in our framework.

4.5.2 Gold-Standard Users

Figure 4b plots the fraction of gold-standard users for the original celebrity ad and the control ad, for three different definitions of gold-standard users. The first definition is, as before, 5s of dwell time and 1 mouse event. The second definition is 15s of dwell time and 5 mouse events. The third definition is 30s of dwell time and 15 mouse events. Note that the fraction of gold-standard users for the control ad is zero for the second and third definition. This validates Assumption 3 that very few users are curious enough to click the control ad. In a real-world setting, we expect advertisers to define gold-standard users based on financial transactions (much tighter than any of our definitions).

We focus next on the sensitivity of the click-spam estimate to the number of gold-standard users. Figure 4c plots the convergence of our click-spam estimate as a function of the number of gold-standard users for various combinations of our ad, interstitial page, and definition of gold-standard user. X-values are driven by the periodicity of ad network reports. Y-values are deltas from our best estimate (last data-point for that series). In each case our estimate converges at or before 25 gold-standard users.

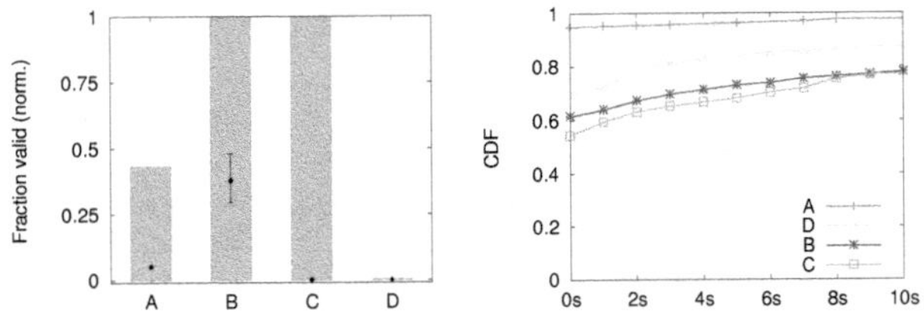

(a) Mobile Ad Network (b) Dwell Time

5: Normalized estimates and dwell times for mobile ads

4.6 Search Ad Networks

Figure 3 shows that while reputed search ad networks generally have a good handle on click-spam, a single average click-spam metric across the entire network is of little use due to different keywords experiencing different levels of click-spam. An advertiser cares only about click-spam rates for keywords he is interested in bidding on. The difference in click-spam rates between the celebrity ad and lawnmower ad is up to 20% (normalized).

We omit discussion of 7Search since we did not get any gold-standard users through that network to base our estimates on.

4.7 Mobile Ad Networks

Figure 5a plots our click-spam estimates and the networks' own estimates for the celebrity ad across the four mobile ad networks we measured. Despite running our ads for over a month, and weakening our definition of gold-standard users to only 5s of dwell time (*i.e.*, user spent 5s on our landing page; no tap event required), we failed to attract even five gold-standard users for the yoga and lawnmower ads, and attracted fewer than twenty for the celebrity ad. While, our estimates are below the convergence threshold, to investigate the huge difference in our interim estimates and ad network numbers, we plot the CDF of user dwell-time in Figure 5b.

Ad network A charged us for over a third of the clicks (non-normalized), yet as illustrated by the y-intercept in Figure 5b, over 95% of network A users spent *under a second* on our landing-page! We find evidence of an attack that would result in such a signature in Section 5.4.1. Network D appears to be quite well aware of the poor traffic quality on their network; they charged us for less than 1% of the clicks. Mobile ad network C is a curious case. There is practically no difference between the click-through-rate (CTR) of our original ad and the CTR of our control ad with junk text, suggesting that the content of the ad is irrelevant for users clicking ads on this network. Our Bayesian formula understandably estimates click-spam to be nearly 100% for this network despite the network charging us for most of these clicks.

Our data, although inconclusive, suggests that charges on mobile ad networks do not currently reflect actual user intent.

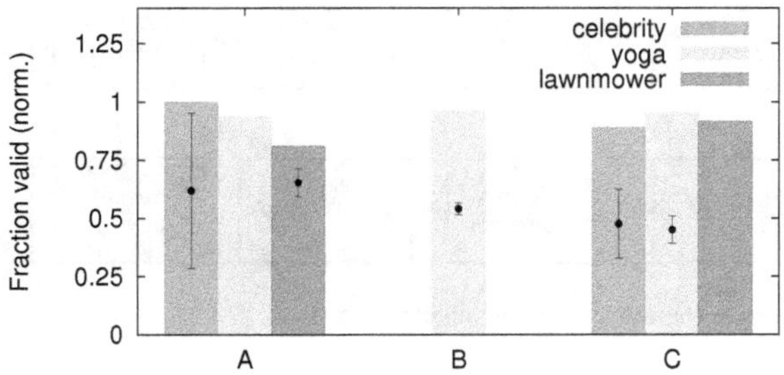

6: Normalized estimates for contextual and social ads

4.8 Contextual and Social Ad Networks

Figure 6 plots our click-spam estimates, and those reported by contextual and social ad networks we measured. Network B approved only our yoga ad. Click-spam is uniformly higher than that on reputed search ad networks (not apparent due to normalization). The networks do a better job than mobile ad networks in tracking this higher rate of click-spam. Nevertheless, our estimates are uniformly lower than the numbers reported suggesting that these networks do not yet discount all click-spam. Interestingly, network B consistently charged us for more unique clicks than we logged on our server. We speculate network B does not follow the standard practice of suppressing duplicate/double-clicks by a user [39].

5. FINGERPRINTING CLICK-SPAM

Recall in previous sections we assumed that very few people would intentionally click on our control ads (*e.g.*, Figure 2), and substantiated the assumption through the lack of gold-standard users for these ads. Nevertheless, convincing ad networks requires incontrovertible evidence of fraud. One needs to discover the full sequence of events that culminate in the fraudulent click. We manually investigate clicks we receive on control ads. The sophistication and diversity of attacks makes this non-trivial. In this section we describe seven ongoing click-spam attacks we discovered.

5.1 Scale

We were charged approximately $1000 for about 30,000 clicks on all the control ads we created across all ad networks. Our investigations cover 26% of the traffic our control ads attracted on reputed networks; as expected by design, all these clicks were found to be fraudulent in nature. The ad networks typically discounted substantially less than this fraction (between 6–20%). Thus we can confidently claim that some of this fraudulent traffic is currently not caught by ad networks. Note that 26% covers only the traffic we actually investigated; we expect the disparity in discounts vs. fraudulent traffic to grow as we investigate more clicks. That being said, our manual approach is too laborious and not scalable. More automated methods for investigating click-spam are needed.

5.2 Methodology

To prioritize manual investigation of the large number of clicks, we use simple graph-clustering over features in the HTTP request, and detecting heavy-hitting clusters. A naïve approach would be to use the HTTP Referer domain. We found groups of websites on unrelated domains but with nearly-identical layouts (Figure 7), all driving click-spam traffic to our site. This is done presumably to spread out the click-spam through multiple sources in order to operate below detection-thresholds of existing ad networks. Using additional features to cluster such publishers allows us to aggregate them back together and do proper heavy-hitter accounting.

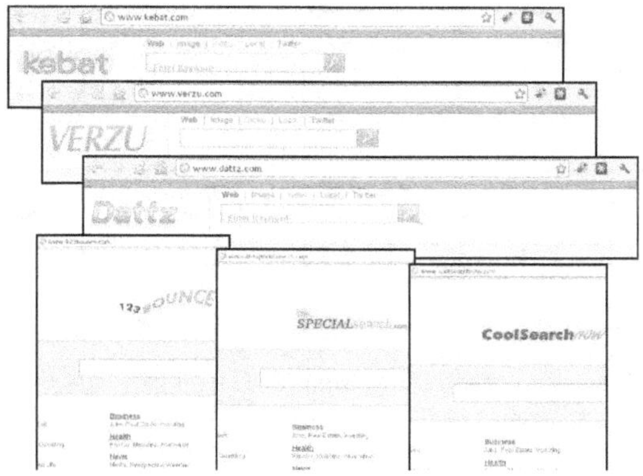

7: Some Sybil [21] websites driving click-spam to our ads

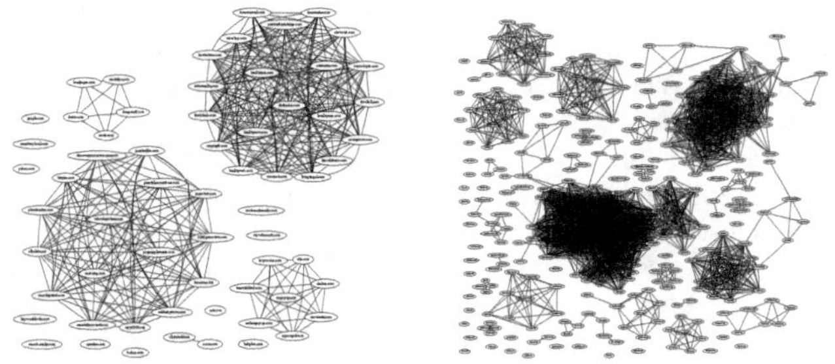

(a) Clusters (search) **(b)** Clusters (mobile)

Search Cluster	Signature Type	%
thespecialsearch.com + 2	Malware, Affiliates	5%
scour.com + 3	Badware, Affiliates	14%
Sedo-parked (58+)	Parked Domain	(cloaked)
NS-parked (51+)	Parked Domain	6%
dotellall.com + 20	Arbitrage	18%

(c) Top 5 heavy-hitter cluster for search ads

8: Graph-clustering and heavy-hitter detection output

Graph-clustering: We induce a graph that spans all publisher domains we see. We do this as follows. For any pair of publishers, we compute a similarity score. We construct a feature vector that consists of various network-level attributes (*e.g.*, Web host IP address, subnet, hosting provider, domain registrar, whois information) as well as HTTP-level attributes in our logs. We assign a weight to each attribute and compute a cosine similarity between the feature vectors of the pair of domains. The similarity score ranges between 1 (identical) and 0 (dissimilar). We add a graph edge between the domains if the similarity score is above some threshold. We find this simple technique is surprisingly robust to our selection of weights and thresholds. In our data distinct cliques emerge at thresholds as low as 0.2 and stay intact beyond 0.9, thus giving us much wiggle room in picking the initial weights and thresholds (which we manually refine iteratively).

Heavy-hitter detection: We use a conductance metric to detect heavy-hitters especially when the clusters do not neatly fall out as distinct cliques. Each node in the graph shares responsibility for clicks originating from another node up to 2-hops away. We found 2-hops to be quite effective since the 1-hop neighborhood was too sparse (due to sparsity in the underlying data), and 3-hop neighborhood resulted in clusters too large for them to represent real-world collusion between bad domains. We compute a badness score for each node as the number of clicks originating in their 2-hop neigh-

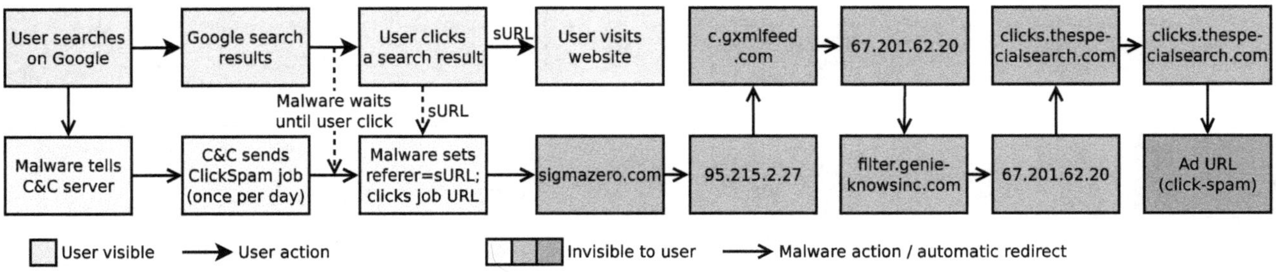

9: Click-spam through TDL-4 botnet. TDL-4 is stealthy (one click per day), mimics user behavior (click gated on user-click); money flows up affiliate chain (up to 10 redirects seen). Who to blame: sigmazero (chain start) or thespecialsearch (publisher for ad network)?

borhood. We then partition the graph into disjoint clusters by considering nodes in decreasing order of badness as cluster centers, and collapsing nodes within 2-hops from it into its cluster.

We believe better techniques based on learning and mining literature can be designed to find patterns in click-spam data (*e.g.*, [17, 18, 22]). We leave this for future investigations both by us and other researchers. To this end, as mentioned earlier, our raw logs are available online for other researchers to use.

That being said, even our simple technique was able to find meaningful clusters. Figure 8a plots our clustering and heavy-hitter output applied to control ad clicks on Google's and Bing's syndicated search ad networks; all clusters also happen to be cliques. Figure 8c lists the top 5 heavy-hitter clusters. Figure 8b plots the clusters from control ads on mobile ad networks. Next, we dig deeper and discuss some case studies chosen specifically to depict the wide variety and sophistication in current click-spam techniques.

5.3 Click-spam in Search Ads

Click-spam we observed in search ads can be attributed to three main attack vectors: (1) malware and badware, (2) parked domains, and (3) arbitrage.

5.3.1 Malware and BadWare

thespecialsearch.com affiliates: We noticed a large number of clicks in our logs that fit the pattern clicks.thespecialsearch .com/xtr_new?q=.... What followed the q= parameter changed from click to click, but almost always was a simple combination of English words (*e.g.*, Team Building or Saving more). 5% of the search clicks in our logs matched this pattern.

Searching online we found malware reports [11] for the Win32/-Olmarik (aka TDSS, TDL) botnet that had been observed fetching URLs fitting the above pattern. This particular malware family is incredibly sophisticated [23]. The malware is a generic task execution platform — it contacts its command-and-control server (C&C), downloads an arbitrary task meant specifically for that infection instance, executes it, and repeats the process. The malware hooks into all popular browsers (IE, Firefox, Safari, Chrome), through which it can inject clicks that appear indistinguishable from normal traffic generated by these browsers. The malware can also inject malicious code into iframes the user is browsing, or modify search results before they are shown to the user. It even attempts to cleanse the infected host of *other* malware so it has sole control over the host (and to disrupt other competing botnets).

We found a copy of the malware binary and installed it in a virtual machine. We routed all traffic from the virtual machine through a transparent proxy (running on the VM host) and logged all traffic. We configured the proxy to block SMTP traffic to block malware-generated spam campaigns. We also apply a strict network rate-limit to prevent DoS attacks, and configured our proxy to block requests to the *click URL* of Google, Bing, and other major ad networks to prevent advertisers being charged for clicks made by the malware instance.

How it works: Figure 9 illustrates the process.

1. For every search we performed through the browser, the malware contacted a specific IP address with the following URL: http://-63.223.106.16/bV03tDze8...JpdHk=08h. The string of random characters is encoded using base64. It decodes as follows: ver=4.2&-bid=noname&aid=50018&sid=0&rd=1307260520&eng=www.bing.co-m&q=celebrity. Thus, for each search, the malware reported back to its C&C server the version number of the bot, affiliate ID, the search engine where the query was submitted, and the query keywords.

2. In response the C&C server sent back an XML file that directed the bot to effectively click an ad. The XML file encoded the URL to click, the HTTP Referer value to use (*i.e.*, traffic will appear to come from this site), and some accounting information.

It is important to note that the bot did not immediately perform the click after receiving the XML file.

3. The browser showed the (unmodified) search results we had requested as usual.

4. Only when we clicked a search result, the bot kicked into action in the background (*i.e.*, the user doesn't notice anything unusual). The bot contacted the URL it was directed to contact with the appropriate Referer header. This led to a sequence of HTTP and JavaScript redirects (we observed upto 10) that culminated in the final redirection to the click URL for one of the major ad networks. Figure 9 plots the sequence of redirects.

Note: From the user's perspective the malware is completely transparent. The user's search results and subsequent click were not tampered with. The user wasn't shown any extra ads or popups. The user wasn't redirected to an advertiser etc. The malware performed all its activity stealthily in the background.

5. The bot then ceased this behavior for 24 hours (as long as the IP address stayed the same). The following day the bot would repeat this activity, perform one click, and become dormant again.

When we acquired a new external IP address (easy to do since the ISP uses DHCP), the bot came out of dormancy, performed one click, and resumed dormancy.

Discussion: It is clear that the C&C server is tracking which bots are active and from where, and ensuring that across the botnet each IP address is used only for one click in a 24 hour period — an extremely low threshold that would likely not raise any flags. Furthermore, when the bot does click, it is gated by a legitimate user click (on the search page), which would defeat click-spam detection mechanisms that look for deviations from normal user behavior (*e.g.*, [35]). Lastly, since it hooks into a regular browser, and forges the referer of a legitimate site, click-spam detection techniques that look for deviation at the HTTP or HTML layer would fail.

Who made money: The penultimate website (thespecialsearch.com) made money from the ad network. Note this website also shows up in our list of Sybils (Figure 7). The long chain of redirects to thespecialsearch.com presumably identify the chain of affiliates, each making some fraction of the money the next one in the chain did. We noted that the malware used different affiliate chains based on geography (*i.e.*, it used one when in the US, and a different one when in another country). This suggests that the same malware is performing click-fraud for different "customers". Thus the botmaster controlling the botnet likely made money as well.

scour.com affiliates: A large number of clicks were through clicks.scour.com. Scour is a meta-search engine that aggregates results from Google, Yahoo, Bing into a single search result page. It pays users to search through it, and to vote or comment on individual results. It has an affiliate program where registered affiliates are paid for users they refer to Scour.

How it works: We found a browser toolbar that hijacks the user's searches through Scour (but doesn't automatically click on ads.) The affiliate ID (6678) is hardcoded in the scour.com URL the toolbar points the browser to. The same affiliate ID shows up in our logs. The toolbar, which many anti-virus companies classify as the Scour redirect virus, is extremely hard to remove [9]. Additional search redirect viruses (unrelated to Scour) that we see clicks from include search-results.com, mywebsearch.com, search.babylon.com, search.alot.com and search.conduit.com. These sites explain the discrepancy between our estimate in Figure 3 and the ad networks'.

Who made money: As before, the publisher (scour.com) made money from the ad network, and the affiliate presumably made some fraction of that from Scour.

5.3.2 Parked Domains

sedo.com parkers: While investigating a set of about 35 clicks from a particular domain registered by Sedo, a domain registrar, we stumbled across 57 other domains also hosted by Sedo and in our various logs. All these domains are parked domains. A parked domain is a domain name that is registered, but not in use. The registrar typically points DNS for that domain name to a Web server that serves up a "This site is under construction" or similar message, followed by a set of ads that the user may or may not click.

For these specific parked domains, however, Sedo would automatically redirect the browser to the ad click URL.

How it works:

1. A domain name registered by Sedo was either never used, or its previous owner vacated it before the registration expired. Sedo serves the parked page to users reaching the domain name.

2. A user may reach the parked domain name through many ways. He may have mistyped a domain (*e.g.* nsdi.com is a Sedo parked domain that appeared in our logs; other examples include blogdpot.com a mistyping of blogspot.com, and icicbank.com a typo for icicibank.com)

In other cases where the owner was using the domain but is no longer, links to the domain when it was active may have been posted on forums, exchanged in emails, indexed by search engines etc., and users may click on these links in the present. There are also reports of adult link-exchange networks that launder traffic through parked domains [13].

3. The user is usually shown an ad-laden parking page.

4. If, however, the parked domain is one of these 58 domains (possibly more) and the user is geolocated to a certain set of countries, which include UK, Brazil, Italy, India, China, Spain and Argentina (but notably, *not* the US), Sedo serves up a piece of JavaScript that in effect automatically clicks the first ad link without ever serving up the parking page.

5. The automatic click initiates a chain of redirects (we observed upto 4) many of which (roughly one in three) culminate in a redirect to clicks.scour.com (but lacks an affiliate ID). Scour then shows ads from major ad network.

Based on the referrer we see in our logs (which appears to be a search query on Scour) we found that the domain of the Sedo parked page is linked to the search query on Scour. For example, the publisher URL for ad clicks originating at icicbank.com is ....scour.com?q=icic+bank. Indeed this is how we discovered the set of 58 domains that auto-redirect users. For each Scour query in our logs we attempted to guess the Sedo parked domain by appending common top-level domains (.com, .net, etc.), and checking which were parked, and then determined which countries they auto-redirect for using PlanetLab nodes located in 45 countries.

6. The remaining auto-redirects reach either thespecialsearch.com — encountered earlier in the context of malware, or searchmirror.com. Both these sites further auto-redirect to the ad click-URL.

The ultimate ad URL in these cases is encoded in each of the intermediate redirects starting from the very first redirect initiated at the Sedo parked domain. Thus the decision for which ad to click was made right at the onset.

7. In many cases, the first ad is for the correct version of the mistyped domain (*e.g.*, the ad posted by icicibank.com is the top ad on icicbank.com). Thus when the user is automatically redirected, he may never realize that he mistyped the domain, although under the covers icicibank.com (as an advertiser) has had to pay for a user that was about to reach his Web page anyway, and several parties made money in the process.

Discussion: Since Sedo parked domains redirect via a chain of affiliates, detecting Sedo (or the Sedo customer) as the root-cause requires reverse-engineering the chain (in this case through a query parameter on scour.com, which taken out of context appears to be a normal search query). More deviously, the user (*e.g.*, that typed icicbank.com) would engage normally with icicibank.com since he anyway meant to type the latter. Thus any advertiser driven engagement metrics would appear perfectly normal. Discovering such patterns automatically is likely to be highly challenging, but would illuminate a fraction of click-spam that is virtually undetectable. Detection is only part of the problem however.

Sedo is benefiting from a ad network policy that does not forbid its mode of operation. Parked domains are not only allowed to show ads, ad networks expose special APIs to help them in doing so [7]. Worse, even though ad networks have mechanisms to allow advertisers to block certain classes of traffic (*e.g.*, traffic through proxy-servers), ad networks *do not* allow advertisers to block traffic from parked domains.

Who made money: thespecialsearch.com and scour.com made money from major ad networks, some fraction of which, as before, traveled through the affiliate chain to the Sedo parked page.

networksolutions.com: NetworkSolutions, another domain registrar, has a similar model as Sedo, but does not automatically redirect. They account for 6% of the clicks we see for control ads.

In one scenario we found that even though the owner of www.noblenet.org (a library website) is actively using it, NetworkSolutions is showing a parked page for noblenet.org that, at first glance, appears to be a library page except all links are ads that direct the user away from their intended URL. Note that here the user did not make a typo; he simply omitted the www, which is often acceptable.

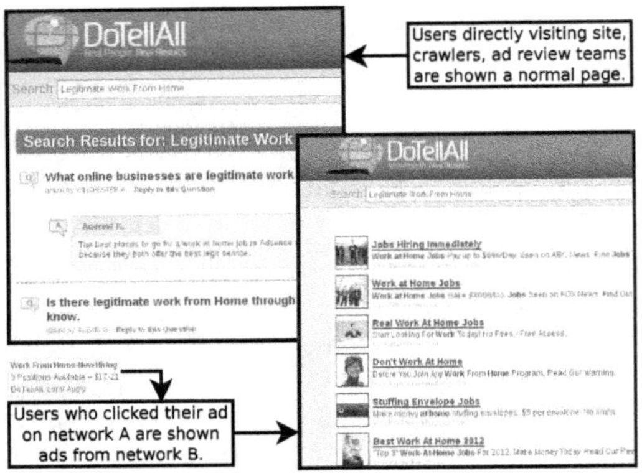

10: Arbitraging click-spam traffic through a fake site

Discussion: As before, this is largely a policy issue. Major ad network policies for parked domain affiliates states that they must not violate trademarks and copyrights [1]. NetworkSolutions does reserve for themselves the right to served parked pages for a domain (or sub-domain) in its terms-of-service (TOS) that customers must agree to. It is unclear whether benefiting from someone else's domain constitutes copyright or trademark infringement, and if it does, whether it can be overridden by the TOS. This is a loop-hole NetworkSolutions benefits from.

Who made money: NetworkSolutions made money from major ad networks if the user clicked a link on the parked domain.

5.3.3 Advertising Arbitrage

dotellall.com family: We next focus on the cluster of dotellall.com and 20 other related domains that account for 18% of the traffic for our search control ads. The entire cluster of websites on the surface appear to be lively social question answer forums (users ask questions, and post answers), but when we posted questions and answers on one of the sites, it disappeared after a few days, and the site was restored to its pristine condition. We noticed that over time the questions and answers do not change. No question has the date/time when it was asked or answered. For one of the sites, we found the content was blatantly copied from other locations on the web. As best as we can tell, the entire family of sites is an incredibly elaborate (and realistic) sham.

It was extremely puzzling as to how they attract traffic. Clearly users wouldn't frequent a fake social site. We couldn't find links to malware. The sites weren't typos of other popular sites (although one is named livingfrugal.com, which is similar to the popular livingsocial.com). Confusing us further, we (initially) couldn't find ads on their pages. It took us a long time (and a considerable amount of serendipity) to determine how this family of sites makes money.

How it works:

1. We serendipitously discovered that the family of sites *advertises* heavily on search and contextual ad networks. It advertises to the tune of thousands of ads, for a wide spread of (long-tail) keywords. As a result, they show up on low-popularity searches or low-quality publishers. On low-quality publishers ad links are almost indistinguishable from content. Being the only ad for many low-popularity searches, their ad is often placed above search results. Whenever there is competition, however, their ad is typically ranked much lower (*e.g.*, 7th or 8th position in the sidebar).

This suggests that they likely bid mere pennies for these thousands of ads, but nevertheless manage to acquire long-tail traffic.

2. When a user clicks one of these ads, he is taken to the site. The site *shows ads only when the user arrives through an ad-click* (identified through a URL parameter in the landing-page URL the site registered for its ads in the previous step). Figure 10 shows screenshots of the site for users arriving either directly or through an ad-click.

This second set of ads is from a different major search ad network. Based on the keywords highlighted, we believe this second set of ads are more expensive. The site filter-ins these higher-value ads by stuffing keywords into the ad request.

Thus the family of sites acts as an advertiser with one search/contextual ad network, and as a publisher with another search ad network.

3. Note that the users reaching the site have already displayed a propensity to click ads. Presented with low-quality content, and prominent ads on the top, it is likely a large fraction of these users click the ad on the site. When they do, the second ad network pays the site some fraction of the ad revenue from the lucrative ad, say a dollar, while the site likely pays the first ad network mere cents for getting the user (and pockets the difference).

Discussion: Arbitrage has been long known to be an issue in ad networks [24]. However, such elaborate fake sites can be incredibly hard for a human at an ad network to detect (given limited time to investigate publishers). Recall how the site does not even show ads if navigated to directly. Discovery is only half the problem.

The second half is that these sites are not violating ad network policy. An advertiser may show ads on the landing-page. A publisher may advertise his site. A publisher may provide useful content hints in the ad request. A poor quality page and a prominent ad box is bad user-experience, and an SEO optimized publisher ultimately costs the advertiser, but does not violate current policy.

Who made money: The dotellall.com family of sites likely made a lot of money from one search ad network, for inexpensive traffic it bought from the other ad network.

5.4 Click-spam in Mobile Ads

We next turn our attention to mobile ads, which as we found in Section 4, are challenging even for reputable ad networks to detect click-spam in. Figure 8b pictorially shows why. First, because mobile advertising is a relatively new market, large legitimate content providers have not yet replaced fly-by-night operators that exist to make a quick buck. Indeed many mobile sites on which our ads were shown serve primarily adult content; the abundance of these sites mirrors the state of the web two decades ago when banner ads first started appearing on similar sites. The two large cluster are different adult entertainment networks, one hosted in Turkey, and one in Denmark. We do not investigate these clusters.

5.4.1 Mobile Games

Ant-smasher and similar games: At least 2% of clicks on control ads came from smartphone games that all require the user to tap the screen close to where the ad is displayed. One such example is the Ant-smasher iPhone app where ants randomly walk around the screen up to (and under) where the ad is shown in the game, and the user must tap the ant before it disappears from the screen to progress in the game. We installed the games directing the most traffic and confirmed the following modus operandi.

How it works:

1. A mobile game developer accidentally (or intentionally) places the in-app advertising control close to where the user must tap, or drag things to, in order to succeed in the game.

2. Given the tiny screen real-estate, the user is prone to mistapping. When he does so, the browser navigates to the ad-click URL.

3. The user may realize his error and switch back to the game. The browser, which in the mean time has already begun fetching the ad landing-page, aborts the attempt. As a result, the user will appear to have spent very little time on the advertiser's page. We saw exactly this behavior on our mobile ads — 95% of users spent less than a second as mentioned earlier.

Discussion: The core issue here is the advertiser being charged despite the user not spending any time on the landing page. It is hard for an ad network to know how long the user spent on the advertiser's site. If it relied on the advertiser to get this information, the advertiser could easily lie to get a discount. Solving this without modifying the browser, and without hurting the user experience is a non-trivial problem.

One mitigating approach would be to audit games and apps that trick users into mistapping on the ad. Doing so would likely spark an arms race for apps intentionally exploiting this loop-hole, but would at least protect advertisers from apps accidentally triggering this. Unfortunately, ad networks are making it harder for advertisers and independent third-parties to identify bad apps. During the course of our study, one major mobile advertising network stopped sending the application ID in the HTTP Referer.

Who made money: The app made money from the ad network.

5.4.2 WAP Phones

waptrick.com and other sites: There is a sizable number of WAP phones (phones with a limited browser that access the web via a WAP proxy) that mobile ads are shown to. Nearly 42.1% of traffic on our mobile control ads are from these sources. We loaded a number of implicated sites with our browser's user-agent set to that of a WAP browser.

How it works:

1. Sites that cater to WAP users as well as WAP proxies optimize the page content for display on feature phones, potentially stripping away icons, colors, or sidebar content that visually differentiates an ad from a normal link.

2. Combined with the extremely small screen, and clunky (keypad-based) navigation on non-smart phones, the user clicks a link either unintentionally, or without knowing that it is an ad.

Discussion: Proxies are the biggest hurdle in tracking down bad WAP sites that confuse the user. As mentioned, less than 36% of our clicks had the HTTP Referer we need to track it back to the originating website and confirm that it intermixed ads with content. While one might wish for legacy phones to die out, it is unlikely to do so in developing countries in the near future. Advertisers wishing to reach a global market will have to contend with click-spam originating through these vectors.

Who made money: The WAP website made money from the ad network. For websites that have arrangements with proxies, the proxy operator potentially made some fraction of that money.

5.5 Epilogue and Future Work

Investigating 26% of clicks on our control ads, we find the five classes of invalid clicks discussed above. We believe there are more classes of dubious traffic lurking in our data, and are investigating more automated means of reconstructing the attacks. In any event, we find that click-spam is by no means a solved problem.

We also find that while there is a policy component to many of the case-studies we presented, there is also an associated technology (and research) component to proactively discover attacks.

Mobile is a particularly tricky case where much of the telemetry needed for detecting click-spam doesn't exist. Given the large role mobile advertising is expected to play in the coming future, research in this space is both important and timely.

6. RELATED WORK

Related work falls into three distinct categories.

Measuring Traffic Quality: There is surprisingly little past work in systematically measuring the quality of click traffic. [26] develops a learning algorithm for estimating the true CTR of an ad in the presence of click-spam. [41] measures traffic from bulk traffic providers and finds some providers to be qualitatively worse than ad networks. Startups including Adometry, Visual IQ and ClearSaleing that claim to be able to estimate click-spam rates provide no transparency into the specifics of their methods; furthermore, these approaches apply only at the granularity of entire ad networks, which we found is insufficient information for advertisers. Our click-spam estimation approach, which is grounded in our Bayesian framework and validated through extensive measurements, is the first principled approach an advertiser can independently apply at the granularity of his individual ads.

Documenting Click-Spam: The second category of related work is a snapshot-in-time of click-spam attacks, much like the case-studies presented in this paper. Daswani et. al. [19] give a good introduction to online advertising, pricing models, and online advertising fraud. Botnets like Clickbot.A [20], TDL-4 [36] and other botnets [34] have been used for click fraud. More recent work describes fraud in ad exchanges [38]. Individual advertisers, and security researchers have documented many more attacks in blog posts and white-papers [8, 10, 13]. Each of these has been an ad-hoc targeted investigation given a specific publisher or attack vector. Our generic clustering and heavy-hitter detection approach instead starts from raw click logs to automatically identify (and prioritize) potential publishers/attack vectors for targeted investigations.

Mitigating Click-Spam: The third category of related work aims to identify individual clicks as click-spam so they can be discounted. Bluff Ads [25], on which we base our control ad design, are ads with unrelated targeting information (*e.g.*, dog food ads for cat lovers). Clicks on Bluff ads are assumed to be click-spam, which the ad network should discount. While we subscribe to this assumption, we differ in how such ads should be used. [25] suggests blacklisting users that have above-threshold clicks on bluff ads. There are two problems. First, this only applies to click-spam driven by malware. In the non-malware scenarios we discovered, blacklisting the user serves little purpose since the bad publishers get a steady stream of unwitting users (false-negatives for Bluff ads); furthermore, the legitimate clicks of blacklisted users on good publishers would also get discounted (false-positives). The second problem is that even for click-spam driven by malware, it wouldn't work. The malware we analyzed performs one click per day. If Bluff Ads were to be shown 1% of the time, it would take on the order of a 100 days to blacklist a user. The cost to the ad network would be 1% of their revenue (hundreds of millions of dollars for reputed networks), which would be unacceptably high. We use control ads in a different way; we use it sparingly to collect data ($1000 represents a negligible fraction of ad network revenue), from which we then extract click-spam signatures that apply more broadly.

Other approaches to mitigating click-spam include SbotMiner [40], Sleuth [33] and Detectives [32]. SbotMiner tries to identify bot activity by using KL-divergence to detect change in query distributions, followed by pruning of false positives due to flash crowds, by leveraging heterogenity for genuine users. Sleuth uncovers single publisher fraud by finding correlation in multi-dimensional data; however, they claim that the technique is suitable only when the botnet uses tens of hundreds of IP addresses. Detectives detects coalition hit inflation attacks by their similarity seeker algorithm; it discovers coalitions made by pairs of fraudsters, which is then enhanced in [31] by finding groups of fraudsters. All these approaches apply only to botnet and malware driven click-spam, which is dwarfed by other sources of click-spam in our data.

Premium Clicks [27], access control gadgets (ACG) [37] and CDN fraud prevention [30] focus on mitigation strategies that go beyond botnets. Premium clicks employs economic disincentives that devalue clicks from non-gold-standard users. ACGs ensure authentic UI interactions by users clicking a link. CDN fraud prevention proposes a heavy-weight challenge-response protocol for publisher-payee CDN models. While the first assumes an alternate ad economy, the second and third (applied to ad networks) require re-architecting the browser, or the ad network infrastructure. None of these approaches apply to click-spam in existing ad networks.

Focusing squarely on existing ad networks, Camelot [28] is Google's click-fraud penetration system. It can test the susceptibility of the network to known click-spam signatures, but does not itself detect new signatures. [39] describes the invalid click detection system inside Google, without identifying the specific heuristics that are used to identify invalid clicks. No heuristic is perfect. Our data shows click-spam is still an open problem despite these deployed systems.

7. CONCLUSION

In this paper, we take a systematic look at click-spam. We propose the first methodology for advertisers to independently measure click-spam rates on their ads. We also develop an automated methodology for ad networks to proactively fingerprint different simultaneous click-spam attacks. We validate both methodologies using data from major ad networks. We then conduct a large-scale measurement study of click-spam across ten major ad networks and four types of ads. In the process, we identify and perform in-depth analysis on seven ongoing click-spam attacks not currently caught by major ad networks. We conclude that even for the *largest* ad networks, click-spam is a serious problem, and is especially rampant in the mobile advertising context. Given the evolving nature of click-spam, we believe that click-spam is an open problem that requires a concerted effort from the research community to tackle. To this end we have publicly released the data gathered for this paper to aid other researchers in the design of novel click-spam defense techniques.

Acknowledgments

We'd like to thank Jigar Mody, Matt Graham, our shepherd Kirill Levchenko, and our anonymous reviewers. This paper is much improved thanks to their valuable feedback and suggestions.

8. REFERENCES

[1] AdSense for domains program policies. http://support.google.com/adsense/bin/answer.py?answer=96332.
[2] The adsense revenue share. http://adsense.blogspot.com/2010/05/adsense-revenue-share.html.
[3] Click Fraud Falls in Q4 2010. http://searchenginewatch.com/article/2050117/Click-Fraud-Falls-in-Q4-2010.
[4] Click fraud rampant in online ads, says bing. http://www.theaustralian.com.au/media/click-fraud-rampant-in-online-ads-says-bing/story-e6frg996-1226056349034.
[5] Cloaking and Faking the Referrer. http://kbeezie.com/view/cloaking-and-faking-referrer/.
[6] For Impatient Web Users, an Eye Blink Is Just Too Long to Wait. http://www.nytimes.com/2012/03/01/technology/impatient-web-users-flee-slow-loading-sites.html.
[7] Google AdSense for Domains. http://www.google.com/domainpark/index.html.
[8] Google Click Fraud Inflates Conversion Rates and Tricks Advertisers into Overpaying. http://www.benedelman.org/news/011210-1.html.
[9] Google Redirect Virus: How to Remove. http://www.pcmag.com/article2/0,2817,2370676,00.asp.
[10] International Cyber Ring That Infected Millions of Computers Dismantled. http://www.fbi.gov/news/stories/2011/november/malware_110911.
[11] Malware connection report. http://www.malware-control.com/statics-pages/03aaf7c8e47ef32e8de23dfe9215d4a5.php.
[12] Stealing Clicks. http://www.forbes.com/2007/09/21/google-click-forensics-tech-secure-cx_ag_0924fraud.html.
[13] Uncovering an advertising fraud scheme. Or "the Internet is for porn". http://www.behind-the-enemy-lines.com/2011/03/uncovering-advertising-fraud-scheme.html.
[14] Upcoming changes in Google's HTTP Referrer. http://googlewebmastercentral.blogspot.com/2012/03/upcoming-changes-in-googles-http.html.
[15] L. V. Ahn, M. Blum, N. J. Hopper, and J. Langford. Captcha: using hard ai problems for security. EUROCRYPT'03, 2003.
[16] Click Quality Team, Google Inc. How Fictitious Clicks Occur in Third-Party Click Fraud Audit Reports. http://www.google.com/adwords/ReportonThird-PartyClickFraudAuditing.pdf.
[17] G. Cormode and S. Muthukrishnan. What's hot and what's not: Tracking most frequent items dynamically. In *Proceedings of ACM PODC*, July 2003.
[18] G. Cormode and S. Muthukrishnan. Improved data stream summaries: The count-min sketch and its applications. *Journal of Algorithms*, 2004.
[19] N. Daswani, C. Mysen, V. Rao, S. Weis, K. Gharachorloo, and S. Ghosemajumder. *Crimeware: Understanding New Attacks and Defenses*, chapter Online Advertising Fraud. 2008.
[20] N. Daswani and M. Stoppelman. The anatomy of clickbot.A. In *Proceedings of HotBots*, 2007.
[21] J. R. Douceur. The Sybil Attack. In *Proceedings of IPTPS '02*.
[22] C. Estan and G. Varghese. New Directions in Traffic Measurement and Accounting. In *Proceedings of ACM SIGCOMM*, Aug. 2002.
[23] A. M. Eugene Rodionov. The evolution of tdl: Conquering x64. http://go.eset.com/us/resources/white-papers/The_Evolution_of_TDL.pdf.
[24] B. Geddes. Search arbitrage: Web blight or brilliant marketing strategy? http://searchengineland.com/search-arbitrage-web-blight-or-brilliant-marketing-strategy-10768.
[25] H. Haddadi. Fighting online click-fraud using bluff ads. *SIGCOMM Comput. Commun. Rev.*, 2010.
[26] N. Immorlica, K. Jain, M. Mahdian, and K. Talwar. Click Fraud Resistant Methods for Learning Click-Through Rates. In *Proceedings of the Workshop on Internet and Network Economics (WINE '05)*.
[27] A. Juels, S. Stamm, and M. Jakobsson. Combating click fraud via premium clicks. In *Proceedings of 16th USENIX Security Symposium on USENIX Security Symposium*, 2007.
[28] C. Kintana, D. Turner, J.-Y. Pan, A. Metwally, N. Daswani, E. Chin, and A. Bortz. The goals and challenges of click fraud penetration testing systems. International Symposium on Software Reliability Engineering, 2009.
[29] H. Lieberman. Letizia: An agent that assists web browsing. In *International Joint Conference on Artificial Intelligence*, 1995.
[30] S. Majumdar, D. Kulkarni, and C. V. Ravishankar. Addressing click fraud in content delivery systems. In *In Proceedings of the 26th IEEE INFOCOM International Conference on Computer Communications*, 2007.
[31] A. Metwally, D. Agrawal, A. El Abbad, and Q. Zheng. On hit inflation techniques and detection in streams of web advertising networks. ICDCS '07, 2007.
[32] A. Metwally, D. Agrawal, and A. El Abbadi. Detectives: detecting coalition hit inflation attacks in advertising networks streams. WWW '07, 2007.
[33] A. Metwally, F. Emekçi, D. Agrawal, and A. El Abbadi. Sleuth: Single-publisher attack detection using correlation hunting. *VLDB*, 2008.
[34] B. Miller, P. Pearce, C. Grier, C. Kreibich, and V. Paxson. What's clicking what? techniques and innovations of today's clickbots. In *Proceedings of the 8th international conference on Detection of intrusions and malware, and vulnerability assessment*, DIMVA'11, 2011.
[35] B. Mordkovich and E. Mordkovich. Click fraud and how to counteract it in ad campaigns. In *Pay-Per-Click Search Engine Marketing Handbook*, 2005.
[36] E. Rodionov and A. Matrosov. The evolution of TDL: Conquering x64. Technical report, 2011.
[37] F. Roesner, T. Kohno, A. Moshchuk, B. Parno, H. Wang, and C. Cowan. User-driven access control Rethinking permission granting in modern operating systems. IEEE Symposium on Security and Privacy, 2012.
[38] B. Stone-Gross, R. Stevens, A. Zarras, R. Kemmerer, C. Kruegel, and G. Vigna. Understanding fraudulent activities in online ad exchanges. Internet Measurement Conference, 2011.
[39] A. Tuzhilin. The lane's gift v. google report. http://googleblog.blogspot.in/pdf/Tuzhilin_Report.pdf.
[40] F. Yu, Y. Xie, and Q. Ke. Sbotminer: large scale search bot detection. WSDM '10, 2010.
[41] Q. Zhang, T. Ristenpart, S. Savage, and G. M. Voelker. Got traffic?: an evaluation of click traffic providers. In *Proceedings of the 2011 Joint WICOW/AIRWeb Workshop on Web Quality*, WebQuality '11, 2011.

FairCloud: Sharing the Network in Cloud Computing

Lucian Popa
HP Labs

Gautam Kumar
UC Berkeley

Mosharaf Chowdhury
UC Berkeley

Arvind Krishnamurthy
U. Washington

Sylvia Ratnasamy
UC Berkeley

Ion Stoica
UC Berkeley

ABSTRACT

The network, similar to CPU and memory, is a critical and shared resource in the cloud. However, unlike other resources, it is neither shared proportionally to payment, nor do cloud providers offer minimum guarantees on network bandwidth. The reason networks are more difficult to share is because the network allocation of a virtual machine (VM) X depends not only on the VMs running on the same machine with X, but also on the other VMs that X communicates with and the cross-traffic on each link used by X. In this paper, we start from the above requirements–payment proportionality and minimum guarantees–and show that the network-specific challenges lead to fundamental tradeoffs when sharing cloud networks. We then propose a set of properties to explicitly express these tradeoffs. Finally, we present three allocation policies that allow us to navigate the tradeoff space. We evaluate their characteristics through simulation and testbed experiments to show that they can provide minimum guarantees and achieve better proportionality than existing solutions.

Categories and Subject Descriptors

C.2.0 [**Computer-Communication Networks**]: General

Keywords: cloud computing, network sharing

1. INTRODUCTION

Cloud computing is the platform of choice for deploying and running many of today's businesses. Central to cloud computing is its ability to share and multiplex resources across multiple tenants. Cloud networks, however, are shared in a best-effort manner making it hard for both tenants and cloud providers to reason about how network resources are allocated.

We argue that a desirable solution for sharing cloud networks should meet three requirements. The first is to provide tenants guarantees on the minimum network bandwidth they can expect for each VM they buy, irrespective of the network utilization of other tenants. Such guarantees are common for resources like CPU and memory, and having the same for the network is key to achieving lower bounds for the worst-case performance of an application. We refer to this requirement as *min-guarantee*.

The second desirable requirement, referred to as *high utilization*, aims to maximize network utilization in the presence of unsatisfied demands. For example, we would like an application to use the entire available bandwidth when no other application is active. This can significantly improve the performance for applications with bursty (on/off) traffic patterns, such as MapReduce. Improving application performance may, in turn, allow the provider to charge more.

The third and last requirement is that network resources should be divided among tenants in proportion to their payments, similar to CPU or memory. Under the current flat-rate per VM payment model, this means that two tenants with the *same* number of (identical) VMs should get the *same* aggregate bandwidth assuming they both have sufficient demands, since they paid the *same* amount of money. We refer to this allocation requirement as *network proportionality*. Note that the min-guarantee requirement does not achieve network proportionality, as it only refers to the *minimum* bandwidth guarantee of VMs; however, a VM can get a lower allocation than its guarantee when it has a lower demand, or a higher allocation when the other VMs have lower demands than their guarantees.

Unfortunately, none of the traditional network sharing policies (e.g., fairness among flows, source-destination pairs, or sources alone) can meet either of the min-guarantee or network proportionality requirements, while more recent proposals such as Oktopus [10] can only provide minimum guarantees. We argue that the difficulty of developing solutions to achieve these requirements stems from the following *fundamental tradeoffs*:

- There is a hard tradeoff between min-guarantee and network proportionality: if one aims to achieve min-guarantee, she cannot achieve network proportionality, and vice versa.
- Even without requiring min-guarantees, there is a tradeoff between network proportionality and high utilization.

To this end, we propose a set of properties to help us navigate the tradeoff space and present three allocation policies that obtain the maximal sets of non-conflicting desirable properties. There are two key concepts at the core of these policies. First, we allocate bandwidth along congested links in proportion to the number of VMs of each tenant, not to the number of flows, sources, or source-destination pairs of the tenant. This allows us to meet (restricted versions of) the network proportionality requirement. Second, we use the VM proximity to a link to compute a tenant's share on that link. Specifically, in tree-based network topologies, the share of a tenant on a link is computed as the number of VMs of that tenant in the sub-tree delimited by that link. This allows us to provide minimum guarantees by trading off network proportionality.

In summary, we make two contributions in this paper.

1. We expose the fundamental tradeoffs in network resource allocation in cloud and data center environments, and we provide a set of requirements and properties that allow us to explicitly express these tradeoffs (§2 and §3).
2. We develop a set of resource allocation policies to best navigate these tradeoffs (§4) and evaluate them using simulation and testbed experiments (§5).

SIGCOMM'12, August 13–17, 2012, Helsinki, Finland.

$\mathbf{D_A}$	Demand matrix of tenant A
$\mathbf{R_A}$	Allocation matrix for tenant A
$\vert\mathbf{R_A}\vert$	Aggregate bandwidth of tenant A

Table 1: Notation

2. REQUIREMENTS AND TRADEOFFS

In this section, we elaborate on the desirable requirements for bandwidth allocation across multiple tenants. Then, we show that, even though all these requirements are desirable at the same time, they cannot be simultaneously achieved. In this context, we discuss tradeoffs between these requirements and place traditional allocation policies in this space.

2.1 Assumptions and Notation

We assume an Infrastructure-as-a-Service (IaaS) cloud model such as Amazon EC2 [1], where tenants pay a fixed flat-rate per VM. Hence, our goals for network sharing are defined from a per VM viewpoint, akin to how other cloud resources are allocated today. For the simplicity of exposition, we assume in this section that all VMs are identical (in terms of hardware resources) and have the same price. We will consider heterogeneous VMs and expand on alternate pricing models in §4.

Our discussion is agnostic to VM placement and routing, which we assume are implemented independently. Also, our work is largely orthogonal to work on network topologies aimed at improving bisection bandwidth [7, 16, 18], because the possibility of congestion (and hence the need for sharing policies) remains even in full bisection bandwidth networks (e.g., many-to-many communication in MapReduce can congest any of the links in the network).

We abstract a cloud provider's network by a graph $\mathbf{G} = (\mathbf{V}, \mathbf{E})$, where $\mathbf{V}$ is the set of physical machines and $\mathbf{E}$ is the set of links that connect them. A machine is either a switch/router or a server. A server can host one or more VMs, possibly belonging to different tenants. We use the term congested to refer to a fully utilized link.

A tenant, K, consists of N_K VMs, and has an instantaneous $N_K \times N_K$ demand matrix $\mathbf{D_K} = [D_K^{i,j}]$, where $D_K^{i,j}$ represents the bandwidth demand from tenant K's VM i to VM j. An allocation policy P allocates a set of rates $\mathbf{R} = \{\mathbf{R_1}, \ldots, \mathbf{R_m}\}$ to the set of m tenants with demands $\mathbf{D} = \{\mathbf{D_1}, \ldots, \mathbf{D_m}\}$, i.e.,

$$P(\mathbf{G}, \mathbf{D}) = \{\mathbf{R_1}, \ldots \mathbf{R_m}\}, \qquad (1)$$

where the $N_K \times N_K$ matrix $\mathbf{R_K} = [R_K^{i,j}]$ is the instantaneous bandwidth allocation for tenant K and $R_K^{i,j} \leq D_K^{i,j}\ \forall i, j$. Finally, let $|\mathbf{R_K}| = \sum_{i,j} R_K^{i,j}$ denote the aggregate bandwidth allocation of tenant K (see Table 1).

2.2 Allocation Requirements

We desire a bandwidth allocation policy that meets the following three requirements.

• ***Min-Guarantee***: Provide a minimum absolute bandwidth guarantee for each VM. We consider the hose model [13] shown in Figure 1, where each VM is connected to a non-blocking switch by a dedicated connection whose capacity is equal to the minimum bandwidth guarantee. A similar model is assumed by other efforts such as Oktopus [10] or Gatekeeper [24]. This requirement is key for achieving predictable application performance and is usually enforced through admission control.

• ***High Utilization***: Do not leave network resources underutilized when there is unsatisfied demand. Consider a statically reserved, non work-conserving hose model in which tenants cannot exceed their guaranteed bandwidth; while it is a good fit for low-latency and predictable traffic, it is not well suited for bursty traffic. Note that high utilization is more general than work conservation; for instance, as we will discuss in §2.4, tenants could be disincentivized to use free network resources even if they have unsatisfied demands. This would lead to a lower utilization of the network even for a work conserving allocation.

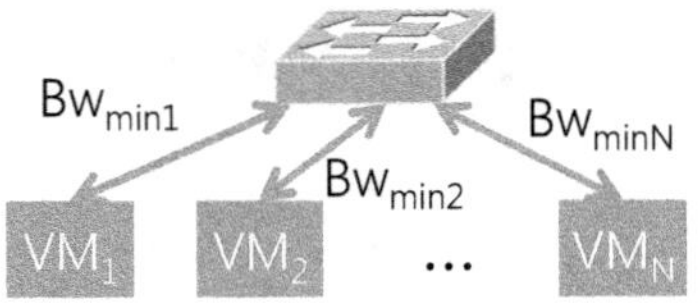

Figure 1: Guaranteed bandwidth of each VM in the hose model.

High utilization is particularly important for throughput-sensitive applications. For example, a tenant running MapReduce jobs can utilize excess bandwidth (i.e., bandwidth unused by other tenants) to improve the completion times for her jobs.[1]

• ***Network Proportionality***: Share bandwidth between tenants based on their payments, just as any other resource in the cloud. Given today's flat-rate per VM payment and assuming the VMs to be identical, this requires the network share of a tenant to be proportional to the total number of her VMs. Thus, given two tenants A and B, an ideal solution for network proportionality would allocate $|\mathbf{R_A}|/|\mathbf{R_B}| = N_A/N_B$. Unfortunately, this is not always possible due to different communication patterns, capacity constraints and demands of the tenants. For example, assume $N_A = N_B$, and both tenants have infinite demands, but all tenant A's traffic traverses a link of 1Gbps, while tenant B's traffic a 10Gbps link. If these are the only two tenants in the system, then the "natural" allocation would be $|\mathbf{R_A}| =$ 1Gbps and $|\mathbf{R_B}| =$ 10Gbps, respectively, which would violate the naive definition of network proportionality.

Formally, we define network proportionality as a generalization of max-min fairness. Let W_K be the weight associated to tenant K, e.g., $W_K = N_K$. Let F_K be the normalized allocation of tenant K, i.e., $F_K = |\mathbf{R_K}|/W_K$. Now, let $F^{ord}(\mathbf{R}) = \{F_{q_1}, \ldots, F_{q_m}\}$ be the sorted vector of normalized tenant allocations ($F_{q_i} \leq F_{q_{i+1}}$). The network proportional allocation $\mathbf{R}^*$ corresponds to the maximal normalized allocation $F^{ord,*}$ in increasing order, i.e., for any other feasible allocation F^{ord}, there exists q_i such that $F^{ord}_{q_i} < F^{ord,*}_{q_i}$ and $\forall q_j < q_i$, $F^{ord}_{q_j} = F^{ord,*}_{q_j}$. In essence, this allocation maximizes the minimum (normalized) tenant allocation.[2]

Next, we discuss the tradeoffs between these requirements.

2.3 Tradeoff Between Network Proportionality and Min-Guarantee

In this section we show that there is a tradeoff between achieving network proportionality and providing each VM a useful (i.e., large enough) bandwidth guarantee.

To illustrate this tradeoff, consider the example in Figure 2 showing two tenants A and B. A employs two VMs, while B employs eleven VMs. VMs A_1 and B_1 are hosted on the same machine; A_1 communicates with A_2, while B_1 communicates with the rest of the ten VMs of B. We assume that the access link of this machine is the only congested link in the system. According to the network proportionality requirement, A_1 should get 2/13 of the access link

[1] While several providers (e.g., Amazon EC2) do not allow statistical multiplexing for the CPU and memory resources, others, such as RackSpace do [6].

[2] Previous work [22] has shown that the maximal lexicographical allocation is equivalent to max-min for multi-path routing, a problem to which network proportionality can be reduced to.

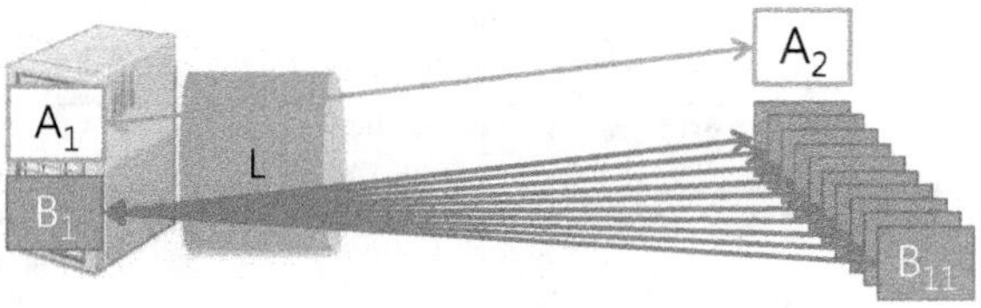

Figure 2: Network proportionality vs. min-guarantee. With proportionality, as B_1 communicates with more VMs, A_1's bandwidth can be decreased arbitrarily.

because A has two VMs and there are 13 VMs in total, while B_1 should get $11/13$. Unfortunately, one can arbitrarily reduce A_1's bandwidth by simply increasing the number of VMs that B_1 communicates with. While strictly speaking, proportionality still provides a min-guarantee, since the number of VMs in the data center is finite, the resulting guarantee is too low to be useful in practice.

If we consider the min-guarantee requirement alone, and assume no other VM can be admitted on the server, both A_1 and B_1 should each be guaranteed half of the capacity of the shared access link, just as each VM would be guaranteed half of the resources of the machine they are running on (recall that we are assuming identical VMs for now). This guarantee should not be affected by the traffic of other tenants in the network. However, as illustrated by our example, there is a hard tradeoff between network proportionality and min-guarantee: one can either achieve network proportionality or provide a useful min-guarantee, but not both!

2.4 Tradeoff Between Network Proportionality and High Utilization

We now show that even in the absence of the min-guarantee requirement, network proportionality is hard to achieve. In particular, we show that there is a tradeoff between network proportionality and high utilization.

To illustrate this tradeoff, consider the example in Figure 3 depicting two tenants A and B, each employing four VMs. Figure 3(a) shows a scenario in which their flows traverse the same congested link L of capacity C; each tenant gets $C/2$ of the aggregate bandwidth, and the network proportionality requirement is met. Now assume that VMs A_1 and A_3 start communicating along an uncongested path, P (Figure 3(b)). In order to preserve network proportionality, we need to *decrease* tenant A's allocation along link L. Unfortunately, if A deems its traffic along L more important than that between A_1 and A_3, A is disincentivized to use path P, which hurts network utilization.

We will refer to the ability of a tenant to use an uncongested path without being penalized on another (disjoint) congested path as the *utilization incentives* property. Thus, the tradeoff between network proportionality and high utilization can be reduced to a tradeoff between network proportionality and utilization incentives.

2.4.1 Congestion Proportionality

It might appear that one could get around this tradeoff by restricting the network proportionality requirement only to the traffic traversing congested links. We define *congestion proportionality* as network proportionality restricted to congested paths that involve more than one tenant. In other words, for each tenant K, congestion proportionality considers only $\sum_{i,j} R_K^{i,j}$, where there is at least a congested link along path $i \rightarrow j$ traversed by the flows of multiple tenants. For the example in Figure 3(b), since the path between A_1 and A_3 is used only by tenant A, it does not count towards A's use of congested resources when comparing the aggregate bandwidths of

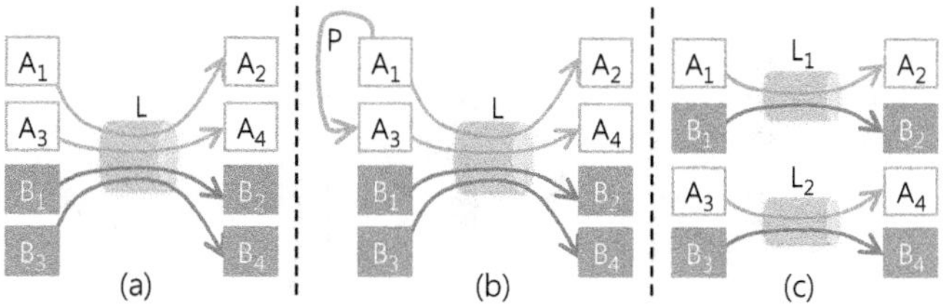

Figure 3: Network proportionality vs. high utilization. (a) VMs of tenants A and B have equal shares on link L. (b) If A_1 starts communicating with A_3, A's allocation on L decreases; thus, A may be incentivized to refrain from using the free path P. (c) Even with congestion proportionality, tenants can have incentives to keep links underutilized to increase their total allocation.

A and B. Thus, congestion proportionality does not disincentivize a tenant from using free resources.

Unfortunately, even congestion proportionality conflicts with high utilization, since it can incentivize tenants to artificially inflate or deflate their real demands. Consider Figure 3(c), where the traffic of tenants A and B is split across two congested links L_1 and L_2 with the same capacity C. Initially, assume that tenants have high demands and each tenant receives half of each congested link. This allocation trivially meets congestion proportionality.

Now assume the demand from B_3 to B_4 drops to a small value ϵ (all the other demands remain very high). As a result, $B_3 \rightarrow B_4$ will get ϵ on L_2, while $A_3 \rightarrow A_4$ will get $C-\epsilon$. In turn, congestion proportionality will change the allocation on L_1 so that $B_1 \rightarrow B_2$ gets $C - \epsilon$ and $A_1 \rightarrow A_2$ gets ϵ. By doing so, each tenant will still get same aggregate capacity C, and the system remains fully utilized.

However, congestion proportionality can still violate the utilization incentives property. For example, A may choose to send only $C - 2\epsilon$ on L_2. Since L_2 is no longer a congested link, congestion proportionality will only allocate L_1, by giving $C/2$ to both tenants. As a result, A ends up getting an aggregate allocation of $3C/2 - 2\epsilon$, while B will get only $C/2 + \epsilon$. By doing so, A increases her allocation, while the system utilization decreases from $2C$ to $2C - \epsilon$. This example also illustrates the fact that a tenant can artificially modify her demand to get a higher allocation. We refer to the ability to prevent such behavior as *strategy-proofness*.

2.4.2 Link Proportionality

We have shown that both network proportionality and its restricted version, congestion proportionality, end up compromising high utilization. To avoid this tradeoff, we further constrain the network proportionality requirement to *individual* links.

We define *link proportionality* as network proportionality restricted to a single link. More concretely, if the link is congested, link proportionality translates into max-min fairness between different tenants that communicate on that link.[3,4] A remaining question is what weight to associate with a tenant on a given link. The only constraint we impose for link proportionality is that the weight of any tenant K on a link L is the same for any communication pattern between K's VMs communicating over L and any distribution of the VMs as sources and destinations. For example, one can define the tenant K's weight on link L as the number of VMs of K that communicate on L. In another example, K's weight can be defined

[3]The max-min allocation is equivalent to the lexicographic maximum for both network proportionality and link proportionality.

[4]One can generalize link proportionality to the granularity of each VM instead of the tenant granularity, as we will discuss in §4.

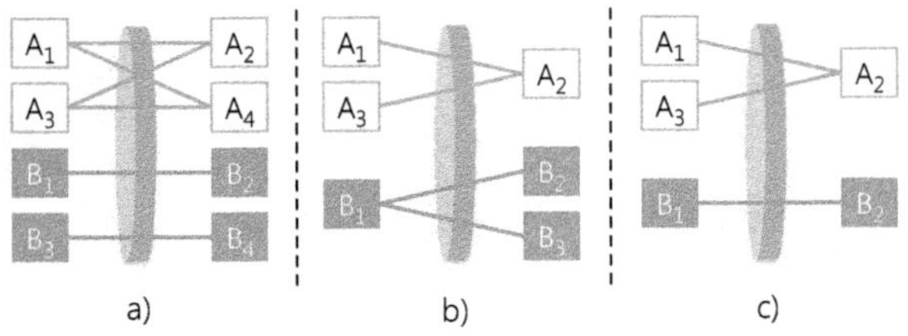

Figure 4: Two tenants sharing a single link using different communication patterns.

Property	Description
P1. Work Conservation	Full utilization of bottleneck links.
P2. Strategy-Proofness	By being dishonest a tenant cannot improve her utility.
P3. Utilization incentives	Tenants are not disincentivized to use uncongested resources.
P4. Communication-Pattern Independence	The allocation does not favor some communication patterns compared to others.
P5. Symmetry	Reversing demands of all flows in the network does not change their allocations.

Table 2: Network sharing properties desirable for any bandwidth allocation policy in clouds.

as N_K, i.e., total number of K's VMs in the network. Note that the latter allocation is similar to that of NetShare [19].

Since the allocation is independent across different links, link proportionality can achieve high utilization. However, link proportional allocations can be substantially different than network wide allocations such as those meeting congestion proportionality, since each VM can compete on a different number of congested links in the network.

More generally, we can classify all allocation policies to be applicable either at the link level or at the network level. We define link level allocations to have the *link-independence* property, meaning that the allocation at each congested link L is based only on the demands known at L and on static information about the VMs communicating on L (i.e., information that does not change during the lifetime of a VM). In the absence of link-independence, there can exist a congested link L whose bandwidth allocation can change due to a change in the allocation of another congested link L' or due to a change in the communication pattern of one of the VMs communicating on L. Network proportionality and congestion proportionality are not link-independent, while link proportionality is. All link-independent allocations provide incentives to use free resources.

Traditional allocation policies, such as per-flow fairness, are link-independent, and thus are capable of achieving high utilization. However, we next show that they cannot provide even link proportionality, nor do they provide min-guarantees.

2.5 Traditional Allocation Policies

The traditional approach to sharing the network is to apply *per-flow* fairness, where a flow is characterized by the standard five-tuple in packet headers. However, the Per-Flow mechanism can lead to unfair allocations at the VM granularity [11]. For instance, two VMs can increase the allocation between them at the expense of other VMs by simply instantiating more flows.

A natural "fix" is to use a *per source-destination pair* (Per-SD) allocation policy, where each source-destination pair is allocated an equal share of a link's bandwidth regardless of the number of flows between the pair of VMs. However, this policy does not provide link proportionality either, because a VM that communicates with many VMs gets more bandwidth than a VM that communicates with fewer VMs. For example, a tenant that employs an all-to-all communication pattern between N VMs will get a bandwidth share of $O(N^2)$, while a tenant that performs one-to-one communication between the same number of N VMs will get a share of only $O(N)$. Figure 4 (a) shows one such example, where the bandwidth allocated to tenant A is twice that of tenant B.

To address this problem, previous solutions (e.g., Seawall [25]) have proposed using a *per source* (Per-Source) allocation policy. Per-Source assigns equal weights to all the sources communicating over a given link, and the bandwidth is divided accordingly. While this is fair to sources, it does not meet link proportionality since it is not fair to destinations. For example, in Figure 4(b), if VMs A_1, A_3, and B_1 are the sources, then the bandwidth allocation of tenant A would be twice that of tenant B. However, if the direction of the communication is reversed, tenant A would receive only half the bandwidth allocated to B. Therefore, there is a mismatch between the amount of traffic sourced and the amount received by a VM.

We refer to this mismatch between allocations in opposite directions as *asymmetry* and consider Per-Source to be asymmetric. Asymmetry is undesirable because it can result in application-level inefficiencies. For example, in a MapReduce setting, a VM hosting both a mapper and a reducer can experience a significant discrepancy between the bandwidth available for the mapper to send and for the reducer to receive, slowing down the application to the slowest component. More generally, VMs can experience large variations between the incoming or outgoing bandwidths, without showing a preference for one of them. A *per destination* (Per-Destination) allocation policy is asymmetric as well for similar reasons.

We have shown that both Per-Source and Per-Destination allocation policies fail to provide link proportionality. In addition, neither satisfies min-guarantee. Referring back to Figure 2, we can easily see that A_1's incoming bandwidth can arbitrarily be reduced by tenant B for Per-Source, and same applies to A_1's outgoing bandwidth for Per-Destination.

To summarize, we have established that there are fundamental tradeoffs between our requirements and that the traditional allocation policies are not satisfactory. We will next express these tradeoffs using a set of lower-level desirable properties. Based on the requirements and properties, we will propose new allocation policies to be implemented in cloud data centers.

3. NETWORK SHARING PROPERTIES

In this section, we describe a set of desirable properties that enable us to examine the above tradeoffs more explicitly. Table 2 summarizes these properties. We do not claim this to be a complete set of desirable properties, but rather a set that enables us to better understand the tradeoffs. Figure 5 captures the relationship between these properties, the requirements and tradeoffs discussed in §2.

***P1. Work conservation*:** As long as there is at least a tenant that has packets to send along link L, L cannot be idle. More formally, consider m tenants with demands $\mathbf{D} = \{\mathbf{D_1}, \ldots, \mathbf{D_m}\}$, and let P be an allocation policy that provides allocations $\mathbf{R} = \{\mathbf{R_1}, \ldots, \mathbf{R_m}\}$ (see Eq. 1). We say that P is work-conserving, iff for any flow of K that traverses an *uncongested* path $i \rightarrow j$, $R_K^{i,j} = D_K^{i,j}$. In other words, a link is either fully allocated, or it satisfies all demands. Surprisingly, unlike the case of a single resource, in a distributed setting, work conservation does not guarantee high utilization. The next two properties illustrate this point.

***P2. Strategy-proofness*:** Tenants cannot improve their allocations by lying about their demands [14]. Consider allocation policy P that provides allocation $\mathbf{R_K}$ to tenant K with demand $\mathbf{D_K}$. With each tenant K, we associate utility $U_K(\mathbf{R_K})$, a scalar function defined on K's

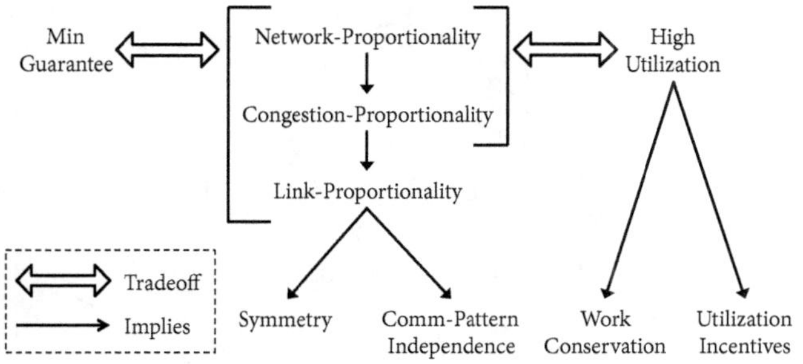

Figure 5: Requirements, properties, and tradeoffs between them.

allocation. We say that P is strategy-proof if tenant K cannot change its demands to get a better allocation with higher utility. That is, for any demand $\widehat{\mathbf{D_K}} \neq \mathbf{D_K}$, we have $U_K(\mathbf{R_K}) \geq U_K(\widehat{\mathbf{R_K}})$, where $\widehat{\mathbf{R_K}}$ is the allocation corresponding to $\widehat{\mathbf{D_K}}$. In other words, tenant K has no incentives to lie about her demands.

Unfortunately, any allocation policy that is unaware of the utility functions cannot satisfy the above definition. Moreover, even if we restrict the utility function of any tenant K to represent the total allocation of K, $U_K(\mathbf{R_K}) = |\mathbf{R_K}|$, i.e., each byte has the same utility for K, this property is still very challenging to achieve for any work conserving allocation.

For example, one can show that link proportionality fails to satisfy even this restricted version of strategy-proofness. Consider the example in Figure 6 consisting of two links with capacities C_1, and C_2, respectively. Assume a link proportionality policy, where each tenant has the same weight on each link, i.e., $A_1 \rightarrow A_2$, $A_3 \rightarrow A_4$ and $B_1 \rightarrow B_2$ have all equal weights. Now assume the demand from A_1 to A_2 is ϵ, while the other demands are infinite, and that $C_1 < C_2 < 2C_1$. In this case, only L_2 is congested. As a result, A gets an aggregate allocation of $\epsilon + \frac{C_2}{2}$ (i.e., ϵ on L_1 and $\frac{C_2}{2}$ on L_2), while B gets $\frac{C_2}{2}$. However, by artificially congesting L_1, A can reduce B's allocation to $\frac{C_1}{2}$, and *increase* her useful aggregate allocation to $\epsilon + C_2 - \frac{C_1}{2}$.

We believe that the above example can be extended to any work-conserving link-independent allocation. In fact, we are not aware of any work-conserving bandwidth allocation policy that is strategy-proof. We leave this challenge for future, and, instead, we next focus on a restricted version of this property, utilization incentives.

P3. Utilization incentives: Tenants are never incentivized to reduce their actual demands on uncongested paths or to artificially leave links underutilized. This property aims to preclude the scenarios described in §2.4, in which a tenant can improve her allocation, and, as a result, decrease the system utilization. More formally, let again $\mathbf{D_K}$ denote the true (real) demand of tenant K, and let $\widehat{\mathbf{D_K}}$ denote any demand, such that $D_K^{i,j} \geq \widehat{D_K^{i,j}}$, for any VM pair (i, j) for which the path $i \rightarrow j$ remains or becomes uncongested under demand $\widehat{D_K^{i,j}}$ ($i \rightarrow j$ can be either congested or not under $\mathbf{D_K}$), and $\widehat{D_K^{i,j}} = D_K^{i,j}$, otherwise. We say that an allocation satisfies the utilization incentives property, if it provides allocations $\mathbf{R_K}$ and $\widehat{\mathbf{R_K}}$, such that $U_K(\mathbf{R_K}) >= U_K(\widehat{\mathbf{R_K}})$, for any monotonic utility function of K.[5] In other words, tenant K will only reduce her utility by decreasing her real demand on an already uncongested link or on a link that becomes uncongested.

Note that the utilization incentives property is a particular case

[5] If a link L between $i \rightarrow j$ is congested for $D_K^{i,j}$ but not for $\widehat{D_K^{i,j}}$, we assume that the utility function is $U_K(\mathbf{R_K}) = |\mathbf{R_K}|$; for general utility functions it can be shown that this property is not achievable in this case by any work-conserving policy, since K can increase her most valued flows on link L.

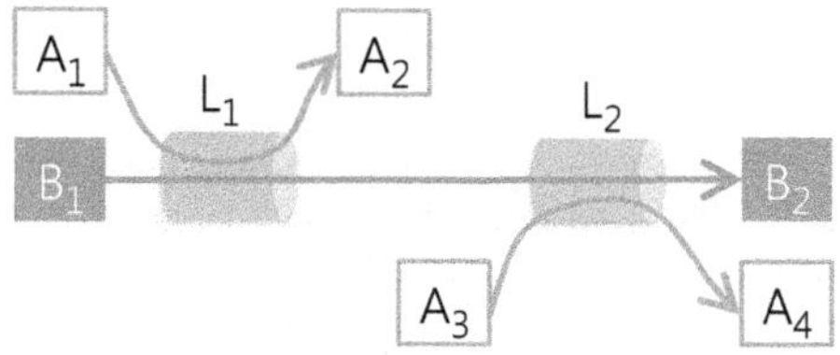

Figure 6: Link-independence vs. strategy-proofness: by increasing the $A_1 \rightarrow A_2$ traffic, A may increase her $A_3 \rightarrow A_4$ traffic.

of the strategy-proofness property, in which a tenant can only lie by reducing her demands on uncongested paths.

We have shown that network proportionality and congestion proportionality violate this property. Almost all link-independent allocations satisfy this property. Note that both work conservation and utilization incentives properties are necessary to satisfy the high utilization requirement.

P4. Communication-pattern independence: The allocation of a VM depends only on the VMs it communicates with and not on the communication pattern. Consider a set of source VMs, $\mathbb{Q}$ that communicates with a set of destination VMs, $\mathbb{P}$, where the VMs in $\mathbb{Q}$ and $\mathbb{P}$ do not communicate with any other VMs. Assuming sufficient demands, any communication pattern involving $\mathbb{Q}$ and $\mathbb{P}$ should result in the same aggregate allocation.

It is not always possible to achieve this property at the network level. For example, assume $\mathbb{Q} = \{A_1, A_3\}$, $\mathbb{P} = \{A_2, A_4\}$, and effective bandwidths along $A_1 \rightarrow A_2$ and $A_3 \rightarrow A_4$ to be much higher than that along $A_1 \rightarrow A_4$ and $A_3 \rightarrow A_2$ (e.g., due to different link capacities and/or background traffic). In this case, the first communication pattern ($A_1 \rightarrow A_2$, $A_3 \rightarrow A_4$) would receive a higher throughput than the second ($A_1 \rightarrow A_4$, $A_3 \rightarrow A_2$).

Consequently, we consider a simpler formulation of this property that is limited to a single congested link L. Some allocation policies still cannot support this property. In Figure 4(a), assume $\mathbb{Q} = \{A_1, A_3\}$, $\mathbb{P} = \{A_2, A_4\}$. Using Per-SD allocation, if A_1 and A_3 communicate with both A_2 and A_4, they get a larger share than a one-to-one communication such as $A_1 \rightarrow A_2$ and $A_3 \rightarrow A_4$. Per-Source allocation achieves this property since A_1 and A_3 will get $\frac{1}{2}$ of L's capacity regardless of the communication pattern (given sufficient demand).

We note that link proportionality implies communication-pattern independence; otherwise, allocations will not remain proportional when communication patterns change.

P5. Symmetry: If we switch the direction of all the flows in the network, then the reverse allocation of each flow should match its original (forward) allocation. More formally, assume G's routing is symmetric, and the capacities of every link are the same in both directions. If the demand of each tenant is transposed (i.e., the demand from i to j is equal to the original demand from j to i) and $\mathbf{R_K^T}$ is the resulting allocation for the transposed demands for tenant K, then $\mathbf{R_K^T} = \mathbf{R_K}\ \forall K$, where $\mathbf{R_K}$ is the original allocation.

Existing allocation policies make an implicit assumption as to whether the allocation is receiver- or sender-centric. However, it is difficult to anticipate application-level preferences. For example, server applications might value more the outgoing traffic while client applications might value more the incoming traffic. In the absence of application-specific information, we prefer allocations that provide equal weights to both incoming and outgoing traffic. As shown in §2, Per-Source and Per-Destination allocations do not provide the symmetry property. Proportionality requirements imply symmetry by definition, since the share of a tenant does not depend on the direction of its communication.

Figure 5 summarizes the desirable requirements, corresponding properties, and the tradeoffs discussed in §2 and §3.

4. PROPOSED ALLOCATION POLICIES

In the previous sections, we have described a set of requirements and properties desired of a network allocation policy and identified fundamental tradeoffs between them. In this section, we discuss how to navigate the tradeoffs shown in Figure 5 and describe three allocation policies that take different stands on these tradeoffs.

The first policy, PS-L, achieves link proportionality and can satisfy all the properties mentioned in §3 (except strategy-proofness). The second, PS-N, provides better proportionality at the network level (congestion proportionality in a restricted setting), but it does not fully provide utilization incentives. Finally, PS-P, provides minimum bandwidth guarantees in tree-based topologies (hence it does not provide proportionality). At the end of the section, we discuss how these policies can be implemented in practice as well as alternate pricing models.

Table 3 summarizes the properties achieved by these three policies as well as by the traditional network sharing policies.

Heterogeneous VMs: Before presenting the allocation policies, we first remove the assumption of all VMs being identical. We generalize to a model where tenants pay different flat-rate prices for individual VM's network share. Just as today's cloud providers offer VMs with different CPU and memory configurations at different prices, we consider a setting where each VM has a (positive) *network weight* associated with it based on the tenant's payment. Thus, each VM is also characterized by its network weight, in addition to its CPU and memory capacities. Intuitively, higher weights should result in higher bandwidth allocations. It is not difficult to extend link proportionality and min-guarantee to the heterogeneous model; for brevity we discuss these extensions as part of the presentation of our proposed allocation policies.

4.1 Proportional Sharing at Link-level

Proportional Sharing at Link-level (PS-L) is an allocation policy that provides link proportionality. The simplest way to understand PS-L is by considering a model in which each switch implements weighted fair queuing (WFQ) and has one queue for each tenant.

The weight of the queue for tenant A on link L is the sum of the weights of A's VMs that communicate through link L. For instance, let A's VMs have a communication pattern such that a set of VMs $\mathbb{Q}$ sends traffic to the set $\mathbb{P}$ over link L. Then, A's (unnormalized) weight will be $W_A = \sum_{X\in\mathbb{Q}} W_X + \sum_{Y\in\mathbb{P}} W_Y$. Consider Figure 4(b) and assume that all VMs have unit weights. Both tenants will be assigned a weight of 3, leading to an equal bandwidth distribution in both directions between them. In another example, on link L_1 in Figure 7, tenant A will have a weight of four since it has four VMs communicating and tenant B will have a weight of two.

One drawback of the above version of PS-L is that there is a simple strategy for tenants to increase their allocations. By sending an ϵ amount of data between all her VMs, a tenant can achieve a weight equal to her total number of VMs on any congested links. One fix is to simply use a weight for tenant A equal to the total weight of all of A's VMs, which can in fact be seen as an application of the NetShare model [19]. Another fix is to apply PS-L at a per VM granularity, rather than per tenant, as we describe next.

In this case, PS-L assigns to a communication between VMs X and Y on link L a weight of:

$$W_{X-Y} = \frac{W_X}{N_X} + \frac{W_Y}{N_Y}$$

where N_X is the number of other VMs X is communicating with on link L (similarly N_Y). For example in Figure 4(c), the per VM PS-L assigns weights of 1.5, 1.5, and 2 to $A_1 - A_2$, $A_3 - A_2$, and $B_1 - B_2$, respectively. So the flows between $A_1 - A_2$ and $A_3 - A_2$ would receive

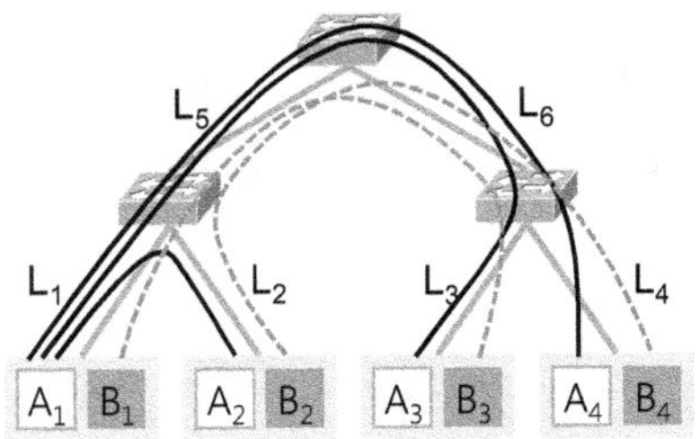

Figure 7: Example for illustrating the allocation policies.

$\frac{1.5}{5}$ of the link capacity, since A_2's weight is split across its two flows. This approach removes the incentives to send traffic between all of one VMs.[6]

As discussed in §2, link proportionality only offers a link level view of proportionality, which can be far from approximating proportionality at the network level. The next allocation policy aims to address this concern.

4.2 Proportional Sharing at Network-level

Proportional Sharing at Network-level (PS-N) is an allocation that aims to approximate congestion proportionality and can achieve it in a severely restricted setting. As a consequence, however, PS-N does not provide full utilization incentives (Table 3).

The intuition behind PS-N is that the communication between the VMs in a set $\mathbb{Q}$ has the *same total weight through the network* irrespective of the communication pattern between the VMs in $\mathbb{Q}$. This weight equals the sum of the weights of the VMs in $\mathbb{Q}$. For example, in Figure 7, the total weight of each tenant A or B is four regardless of the communication pattern between their VMs.

To achieve this, we extend PS-L to incorporate information regarding the global communication pattern of a tenant's VMs. PS-N uses a weight model that is similar to that of PS-L, as it sets the weight of the communication between VMs X and Y to be $W_{X-Y} = \frac{W_X}{N'_X} + \frac{W_Y}{N'_Y}$, with the difference that N'_X is the number of VMs that X communicates with across the *entire network* and not just over a particular link. For example, in Figure 7, PS-N would provide A_1-A_3 with a weight of $\frac{1}{4} + 1 = \frac{5}{4}$ on any link, while PS-L would provide it a weight of $\frac{1}{2} + 1 = \frac{3}{2}$ on link L_5, because A_1 communicates with only two VMs on L_5, but with four VMs in total.

PS-N strives to achieve proportionality in the absence of detailed knowledge about the load on each bottleneck link, assuming uniform load conditions throughout the network. Specifically, a set of VMs $\mathbb{Q}$ communicating using PS-N across a set $\mathbb{S}$ of bottleneck links achieves at least its proportional share of the total bottleneck capacity if the demand on each VM-to-VM communication is high enough and one of the following condition holds: (a) all bottleneck links have the same capacity and background weight (i.e., weight of traffic not involving VMs in $\mathbb{Q}$), or (b) all congested links are proportionally loaded, i.e., any two congested links of capacities C_1 and C_2 are loaded with total weights W_1 and W_2 such that $\frac{C_1}{C_2} = \frac{W_1}{W_2}$.[7]

For example, assume that the network shown in Figure 7 is fully provisioned, tenant A communicates all-to-all (i.e., each VM communicates with all other VMs of A), and tenant B communicates

[6] Note that the properties exhibited by the described per VM PS-L allocation are different for different demands. For example, in Figure 4(c), if the flow between A_1 and A_2 has a very small demand ϵ, the allocation between A_3-A_2 and B_1-B_2 will respect the ratio of 1.5 : 2 instead of a 1 : 1 ratio for a per VM proportionality. This can be addressed by taking into account the demands when assigning weights, but for brevity we do not detail that extension.

[7] There is a subtle difference between the two cases; for case (a), the weight does not include the VMs in $\mathbb{Q}$; for (b), it does.

	Per-Flow	Per-SD	Per-Source (Per-Destination)	Reservations (e.g., [10])	PS-L	PS-P	PS-N
Proportionality	×	×	×	×	link-proportionality	×	congestion-proportionality*
Min-Guarantee	×	×	×	√	×	√	×
Work Conservation	√	√	√	×	√	√	√
Utilization Incentives	√	√	√	√†	√	√	×
Comm-Pattern Indep.	×	×	√	√	√	√	√
Symmetry	√	√	×	√	√	√	√

Table 3: Properties achieved by different network sharing policies. (*in a restricted setting, † also strategy-proof)

one-to-one (e.g., as shown in the figure). In this case, the two tenants achieve the same total bandwidth allocation (assuming high enough demands), since the congestion is on the access links, which have the same capacity and the same weight load. Indeed, the two tenants will have equal weights on the access links: $W_A = 8 \times \frac{1}{4} = 2$, since each VM will contribute with a weight of $\frac{1}{4}$ to each VM to VM flow, while $W_B = 4 \cdot \frac{1}{2}$. Note that PS-L would not provide network wide proportionality in this case, since A's weight would be twice that of B on the access links.

A drawback of PS-N is that each VM's weight is statically divided across the flows with other VMs, irrespective of the traffic demands. This further constraints the situations when PS-N achieves proportionality (requiring high demands on all flows), as well as makes PS-N lack the utilization incentives property. For example, assume that all the links in Figure 7 have the same capacity and L_5 is the only congested link. If A deems A_1's traffic to A_3 and A_4 more important than that to A_2, A may not send traffic between A_1 and A_2 to get a larger share for its traffic to A_3 and A_4 on L_5. We believe this could potentially be addressed if we include the demands when assigning weights, e.g., by not providing any weight to flows traveling uncongested paths. However, such a policy will be significantly more difficult to deploy, and we leave its exploration to future work.

4.3 Proportional Sharing on Proximate Links

The previous policies strive to achieve forms of proportionality and thus do not provide minimum bandwidth guarantees. To offer (useful) minimum bandwidth guarantees one would want to prioritize the allocation of a link based on the "importance" of that link with respect to the VMs using it. For example, on the access link of one host, we might want to allocate the link bandwidth based more on the weights of the VMs residing on that host and less on the weights of the remote VMs using the link.

Based on the above observation we generalize PS-L to derive the following weighting scheme: $W_{X-Y} = W_{Y-X} = \alpha \frac{W_X}{N_X} + \beta \frac{W_Y}{N_Y}$. The coefficients α and β allow different weights for VMs located on the two sides of the link. By setting specific values for α and β at different links in the network, one can use the generalized PS-L to achieve bandwidth guarantees for different network topologies.

In this paper, we present *Proportional Sharing on Proximate Links (PS-P)*, which is suitable for tree-based topologies (e.g., traditional data center architectures, VL2 [15], and multi-tree structures such as fat trees [7]). PS-P prioritizes VMs that are close to a given link. More precisely, PS-P uses $\alpha = 1$ and $\beta = 0$ for all links in the tree that are closer to X than Y, and $\alpha = 0$ and $\beta = 1$ for all links closer to Y than X.

In practice, PS-P translates into applying per-source fair sharing for the traffic towards the root of the tree and per-destination fair sharing for the traffic from the root. For example, on link L_1 in Figure 7, the three flows communicating with A_1 will share A_1's weight irrespective of the weights of A_2, A_3, and A_4, while on link L_6, A_1-A_3 will have A_3's weight and A_1-A_4, A_4's weight. These weights apply equally to both directions of the traffic.

PS-P provides absolute bandwidth guarantees for any VM to/from the root of the physical tree, since that VM competes on a given link to the root only with the other VMs in the same subtree. Obviously, cloud providers must deploy admission control to ensure that the total bandwidth guarantees of the VMs hosted within the subtree of any link L does not exceed L's capacity, similar to how CPU guarantees are offered on a given host. Guarantees are computed assuming all hosts are fully loaded with VMs.

For example, in Figure 7, on link L_1, B_1 competes only with A_1's weight irrespective of A_1's communication pattern. Thus, assuming all VMs have have equal weights and that the maximum weight on each host is 2, B_1 is always allocated at least $\frac{1}{2}$ of its access link capacity (L_1). Similarly, B_1 will get at least $\frac{1}{4}$ of L_5.

These guarantees can be used to offer different service models to tenants. The basic model offered by PS-P is similar to Oktopus's Virtual Cluster [10] (i.e., hose model), with the difference that PS-P is work conserving. The cloud provider can associate a minimum guaranteed bandwidth for every unit of VM weight and advertise it to customers through different VM configurations (CPU, memory and bandwidth). The guarantee is computed as the minimum share across all the layers of the tree. Tenants entirely collocated within a higher capacity subtree can achieve higher guarantees. The value of the guarantees can vary from VM to VM and from tenant to tenant.

PS-P can also expose a service model similar to Oktopus's virtual oversubscribed cluster (VOC) [10]. Specifically, the model can expose higher guarantees between each group of VMs collocated within a high-capacity subtree; the guarantees when communicating with VMs from a different group are scaled down with the oversubscription factor. Unfortunately, all models exposed by PS-P would have the same oversubscription ratio, that of the physical network. To fix this, PS-P can be applied within *virtual* topologies with different oversubscription characteristics. (The virtual topologies could themselves be build with a PS-P-like mechanism.)

We assume that if VMs can communicate via multiple paths, the routing protocol performs load balancing of the traffic across the available paths. This assumption holds for many of the newly proposed multi-tree topologies that use multi-path routing to fully utilize the network bisection bandwidth, e.g., [7, 8, 15, 23].

Similar to PS-L, PS-P can be implemented *per tenant*, which reduces the hardware requirements when PS-P is implemented at switches (§4.5). In this case, the weight of tenant A's queue through link L equals the number of VMs of A located in the subtree communicating through L. For example, in Figure 7, on link L_5, the weight of A will be $W_{A_1} + W_{A_2}$ (W_{A_1} is A_1's weight). The disadvantage of the per-tenant implementation is that the provided guarantees are not for each VM, i.e., when buying a VM, the VM is not guaranteed a minimum amount of traffic to/from the network core; rather, the tenant in aggregate is guaranteed that the bandwidth from all its VMs to/from the core equals the sum of the minimum guarantees that would be offered to each of its VMs by the per-VM PS-P. We believe that by selecting suitable values for α and β, one can generalize

PS-P to provide guarantees for other topologies, such as BCube [16] or DCell [18]. We leave this to future work.

4.4 Summary

Table 3 summarizes the properties achieved by the allocation policies presented in this section. All the described policies are work conserving, symmetric, and offer communication-pattern independence, but they make different choices for the rest of the properties. PS-L achieves link proportionality, does not provide guarantees, but does provide utilization incentives; PS-N achieves better proportionality at the network level, but does not provide full incentives for high utilization; lastly, PS-P provides guarantees and incentives for high utilization but not proportionality.

4.5 Deploying PS-L, PS-P and PS-N

We identify three deployment paths for the presented policies:

1. *Full switch support.* For this deployment, each switch must provide a number of queues equal to or greater than the number of tenants communicating through it, and must support weighted fair queuing. All the described policies can be implemented with such switch support.

2. *Partial switch support.* PS-N and PS-P are amenable to implementation using CSFQ [26], which does not require support for per-tenant or per-VM queues at switches.

3. *No switch support.* There are two types of hypervisor-only implementations. First, a centralized controller can enforce rate-limiters at hypervisors based on the current network traffic and implement any of the proposed policies. Second, PS-N (and we believe PS-P as well) could be implemented through a distributed mechanism similar to Seawall [25].

In this paper, we focus on the full and partial switch support deployments and only sketch the hypervisor-only deployment, leaving it to future work. We present our evaluation in §5 and we present more details related to practical deployment issues in §6.

4.6 Other Models

In addition to the flat-rate per VM pricing model, PS-P can also accommodate a per-byte pricing model, such as paying for the bandwidth above the minimum guarantees (like DRP [9]), or simply paying for all bytes. We also note that if the per-byte price is constant across tenants and traffic volume, proportionality still is a desirable property. However, congestion pricing or other forms of price discrimination can offer alternative ways of sharing the network, which would not benefit from proportionality.

Finally, we note that it is possible to create a more complex allocation policy that provides guarantees and shares proportionally *only* the bandwidth unused by guarantees. In a nutshell, this allocation would work as follows. Assume the total weight of all the VMs in the subtree delimited by a link L (for which we provide guarantees on L) is W_G and the weight of the VMs currently active in the subtree is W_C. We note by $W_P = W_G - W_C$ as the weight to be divided proportionally (of the VMs not active). In this context, the weight of tenant A on link L is $W_A = W_{A_S} + WP \times \frac{W_{A_T}}{W_T}$, where W_{A_S} is the weight of A's VMs in the subtree, while W_{A_T} and W_T are the weights of all A's VMs, and all VMs, respectively, active on L.

5. EVALUATION

In this section we evaluate the allocation policies presented in §4. We divide the results in three parts.

- First, we consider a single congested link and explore the behavior of these policies under various scenarios (§5.1).

- Second, we extend our experiments the network level, exemplify the aforementioned tradeoffs using suitable small-scale examples, and observe how these policies behave (§5.2).

- Finally, we leverage traces obtained from a 3200-node production cluster at Facebook to validate the properties exhibited by these policies on a large scale (§5.3).

We generated the results using a flow-level simulator written in Java and validated the simulation results by experiments in the DETERlab testbed [4]. For this purpose, we implemented switch support for PS-L, PS-P and PS-N using per-flow queues in a software router implemented in Click [20]. The implementation is using kernel-mode Click and consists of ~500 lines of C++ code for the new elements and ~2000 lines of python scripts to generate configuration files. We also developed support for PS-N with CSFQ with additional ~400 lines of C++ for the Click elements. To achieve high performance, we deploy Click in kernel mode and approximate CSFQ using fixed point operations. Unless otherwise specified, the presented results are obtained in simulation.

Since Per-Flow's allocation can be arbitrary, depending on the number of flows, we only consider a maximum of one flow between any source-destination pair. In this case, Per-SD and Per-Flow are equivalent, and therefore, we omit Per-SD from the results. For ease of exposition, we assume all VMs to have a unit weight. The highlights of our findings are as follows:

- The tradeoff between proportionality and bandwidth guarantee is also evident at the network level. While PS-P is able to provide the highest guarantee, PS-N exhibits maximum proportionality in network allocation.

- The relative behaviors of these policies scale to large-scale clusters. PS-N is very close to achieving network proportionality for the small MapReduce jobs, and both PS-P and PS-N improve the minimum bandwidth allocations of the tasks of individual jobs and therefore result in an approximate speed-up of ~15× in the shuffle time of the small jobs.[8]

5.1 Link-Level Scenarios

For the first set of experiments, we focus on a single congested link and evaluate multiple allocation policies by varying workloads. Since PS-N would provide the same allocation as PS-L on a single link (assuming that the VMs in our experiments only communicate over that congested link), we omit it from the following discussion.

Figure 8 depicts a scenario similar to the one in Figure 2, where two VMs, A_1 and B_1, are collocated on the same physical host; A_1 communicates with VM A_2, while B_1 communicates with N other VMs. We assume VMs have high demands and the link capacity is 1Gbps. Figure 8 plots the bandwidth allocation of tenant A for increasing values of N. Figure 8(a) presents simulation results, while Figure 8(b) presents actual allocations achieved on the testbed. Remember that Per-Source is asymmetric, and so we present both incoming and outgoing allocations for it.

We make four observations from these results.

- First, PS-P maintains the same throughput for tenant A regardless of N. This is because A_1 is guaranteed a minimum bandwidth of half its access link capacity; the same is true for the outgoing bandwidth allocated by Per-Source.

[8]During the shuffle phase, mappers of a MapReduce job transfer intermediate data to the reducers over the network.

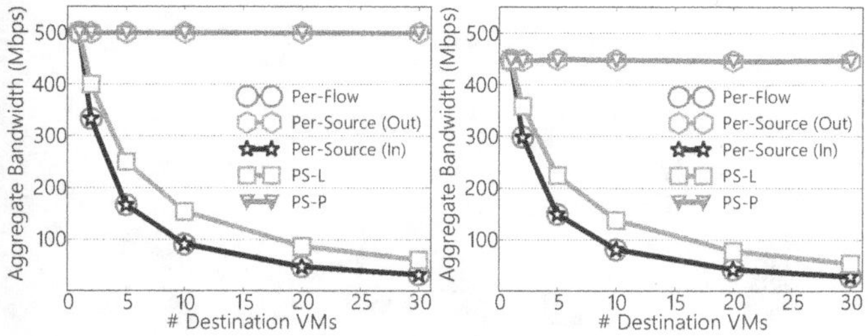

(a) Simulation results (b) Experimental results

Figure 8: Link level bandwidth allocation of tenant A, while tenant B has an increasingly larger number of VMs (similar to the scenario in Figure 2). PS-P maintains a constant throughput for tenant A.

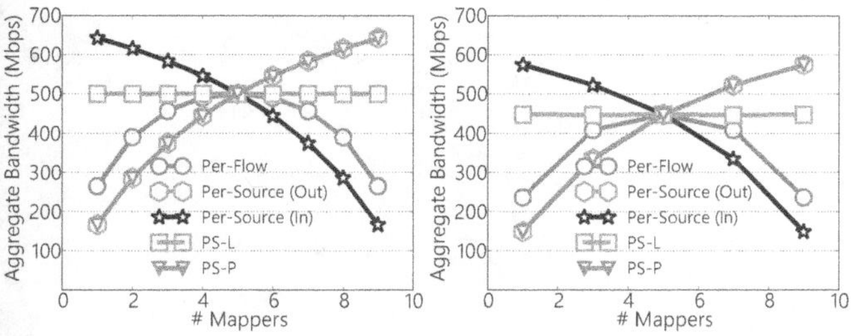

(a) Simulation results (b) Experimental results

Figure 9: Network allocation of tenant B on a congested link, while tenant A varies the number of mappers (senders). Both tenants use the many-to-many communication pattern commonly found in MapReduce shuffles.

- Second, PS-L provides tenant A its proportional share.
- Third, Per-Flow and Per-Source policies provide neither proportionality nor guarantees.
- Lastly, our implementation matches the simulation results, modulo the fact that the link capacities available on DETERlab are smaller (the available bandwidth is ~900Mbps instead of advertised 1Gbps).

In Figure 9, we consider two MapReduce jobs A and B (jobs play the role of tenants in this discussion), each with 10 VMs. While job B has 5 mappers and 5 reducers communicating over the congested link under consideration, we vary the ratio between the number of mappers (M) and reducers (R) of job A while keeping $(M+R) = 10$. As before, we show both simulation and testbed results.

We observe that PS-L achieves a proportional allocation that is not affected by the change in M, while PS-P's allocation increases as $\frac{M}{M+5}$ (5 is the number of mappers of tenant B). We also notice that Per-Flow achieves proportionality only when the distribution of mappers and receivers is the same between the two jobs (at $M = 5$), since only at this point both the jobs have equal number of flows; remember that the number of flows is $M \times R$.

5.2 Network-Level Scenarios

We now turn our attention to the network-wide allocation under different scenarios. For these experiments we include PS-N, because its network-wide allocations will be different than those of PS-L.

We start off with the simple scenario illustrated in Figure 10(a). In this particular example, we have a small tree with eight servers and two tenants A and B, each with one VM in each of the servers.

Strategy	Full Bisection BW			4× Oversubscribed		
	Sim	Exp		Sim	Exp	
Per-Flow	7.00	6.99		18.98	18.91	
Per-Source	1.00	1.00		6.99	6.75	
PS-L	4.00	4.03		7.00	7.20	
PS-P	1.00	0.99		7.00	7.13	
PS-N	1.00	WFQ 1.00	CSFQ 1.15	5.29	WFQ 5.30	CSFQ 5.55

Table 4: Simulation vs. Experimental Results (Ratio of Aggregate Bandwidth of the tenants, B/A)

We assume that tenant A communicates using a pairwise one-to-one communication pattern between its VMs (i.e., $A_i \leftrightarrow A_{i+4}$, where $i = \{1, \ldots 4\}$), while tenant B communicates all-to-all (i.e., B_i communicates with all B_j, where $j \neq i$). For simplicity, we assume all the VMs have equal weights and infinite bidirectional demands. From the network proportionality point of view, an ideal allocation would provide both tenants the same bandwidth.

Figure 10(b) presents the bandwidth allocation for the two tenants in Figure 10(a) when the core is fully provisioned (i.e., the network has full bisection bandwidth). In this case, the access links are the congestion points. We see from Figure 10(b) that PS-P, PS-N and Per-Source policies are able to match the desirable equal allocation. However, PS-L provides B twice as much bandwidth as A on each access link, since B competes with four VMs on each access link against only two VMs of A—we use the version of PS-L proportional the number of active tenant VMs on the link. Consequently, at the network level, Per-Flow and PS-L favor dense communication patterns such as all-to-all over sparse communication patterns.

Figure 10(c) presents the allocations for tenants A and B when the core is under-provisioned by a factor of 4× (each of the aggregation and core layers being oversubscribed by 2×). Given our setup, core links are the bottlenecks for tenant A. In this case, PS-P, PS-L and Per-Source policies allocate the core bandwidth equally between the two tenants, while allowing tenant B to fully utilize the aggregation-level bandwidth for free. However, PS-N penalizes tenant B for utilizing the aggregation-level bandwidth because half the weights of each of the tenant B's VMs is utilized in aggregation-level communication, and it provides twice as much core bandwidth to tenant A. Thus, PS-N provides the best proportionality at the network level (see Table 4 for quantitative results).

We validate the simulation results presented in the above two cases through experiments on a testbed with the same topology. Table 4 presents the comparison between the simulation and the experimental results. The experimental results closely match those from the simulation, with an error margin lower than than 4%.

In the next two experiments, we focus on highlighting the tradeoff between proportionality and bandwidth guarantees at the network level. First, consider a scenario, where two tenants A and B have the same number of VMs (64) deployed on a fully provisioned tree network —we use a 32 server cluster, each server containing one VM of each tenant. VMs of tenant A communicate in a (random) one-to-one fashion. B runs a MapReduce job, and his VMs are divided in two sets M and R, such that all VMs in M communicate to all in R. For ease of exposition, we consider the communications to be bidirectional.[9]

Figure 11 presents the ratio between the aggregate bandwidths of B and A for PS-P and PS-N in this setting.[10] The fully proportional

[9] Otherwise, congestion will not appear uniformly, and the results will become more difficult to understand.

[10] We did not include the other policies since they would offset the chart, making it significantly harder to visualize.

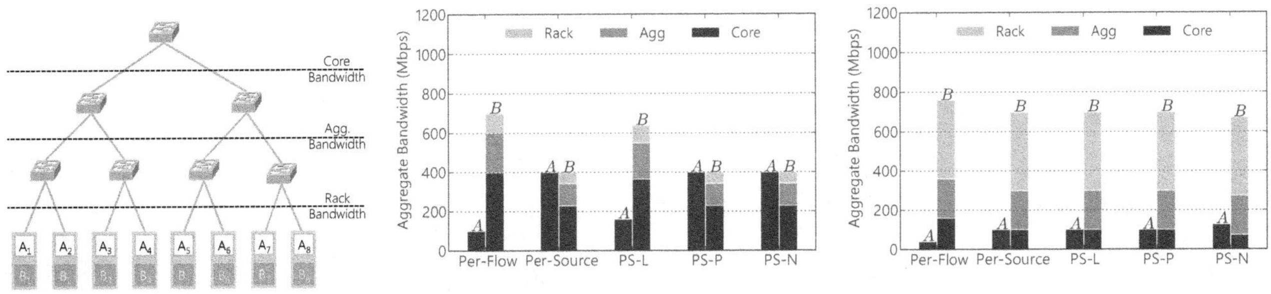

(a) Experiment (b) Full bisection bandwidth network (c) 4× Oversubscribed network

Figure 10: Simple scenario with two tenants having equal number of VMs and a uniform deployment. Tenant A communicates using a pairwise one-to-one communication pattern and B all-to-all. (b) PS-P, PS-L and Per-Source policies achieve perfect proportionality in presence of full bisection bandwidth; (c) PS-N provides the best allocation ratio among the compared policies when the network is oversubscribed.

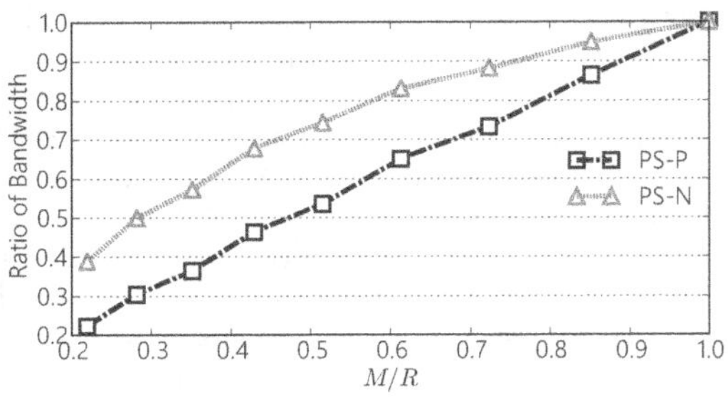

Figure 11: PS-N is more proportional than PS-P.

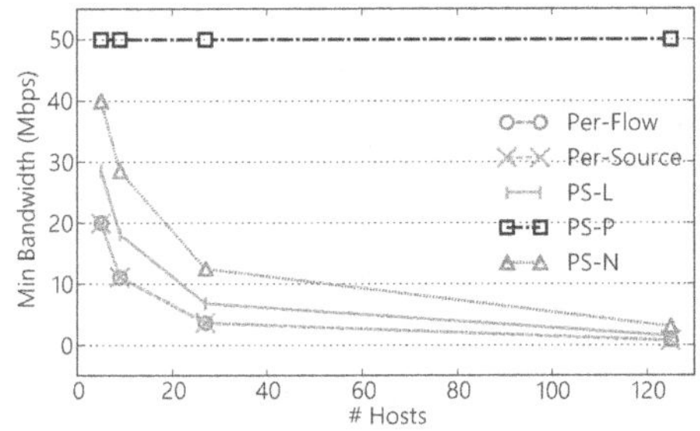

Figure 12: PS-P gives the highest bandwidth guarantee.

allocation would allocate equal network shares to both A and B, and would appear as a horizontal line at $y = 1$. However, this allocation is not always possible, since the MapReduce communication pattern is limited by $\min(M, R) \times C$, where C is the access link capacity. Figure 11 shows that PS-P provides direct proportionality between $\min(M, R)$ and tenant B's bandwidth: each VM is guaranteed its fair share of the network up to the core, and MapReduce will communicate with an aggregate throughput of $\min(M, R) \times Bw$, where Bw is the per-VM bandwidth guarantee ($Bw = \frac{C}{2}$ in this case). PS-N, on the other side, provides better proportionality than PS-P, i.e., the ratio of allocated bandwidths is closer to 1.0.

Proportionality, however, comes at a cost as expected from the tradeoffs presented in §2; the minimum bandwidth allocation of a VM provided by PS-N can become arbitrarily small in practice. Figure 12 presents the minimum guarantee provided by PS-N and PS-P for a scenario similar to the one shown in Figure 7 but with a slightly different communication pattern. In addition to the *N-to-one* communication pattern employed by A (i.e., all of A's VMs send to A_1), each of the N VMs sending traffic to A_1 also communicates with another of the N VMs, so that they can utilize the rest of the links. Without this additional flow from each sender, the throughput of A's VMs would be limited by the traffic going into A_1, and the minimum bandwidth would be affected by the hose model rather than the allocation scheme. We study the variation of minimum of the bandwidth allocations in such a scenario with increasing network size. Figure 12 shows that PS-P can maintain the same minimum guarantee regardless of the network size (for the same oversubscription ratio), while PS-N's minimum bandwidth decreases with the network size given this "bad" communication pattern.

5.3 Trace-driven Simulations

So far we have evaluated the proposed and existing network allocation policies in synthetic, small-scale scenarios. To understand their large-scale implications, we performed some experiments using MapReduce traces from a 3200-node Facebook production data center [12]. Our goal is to identify the network shares that different MapReduce jobs would achieve given different network allocation techniques.

We consider a one hour window in the trace and observe the number of jobs involved in active shuffle at a minute's interval in that hour. In our trace, an average number of 73 jobs were in active shuffle at any given time, with a maximum of 100 jobs and a minimum of 49. We then create a snapshot in time at the point when most jobs are in active shuffle, generate corresponding traffic matrices and observe the allocation of different policies. We infer the network position of each mapper and reducer based on the IP addresses and the names in the log files. The inference shows a heterogeneous cluster with an average of four active tasks per server, with some servers having up to 12 tasks. Due to the lack of knowledge about the oversubscription factor, we consider both full bisection bandwidth and 4× oversubscribed topologies with 20 aggregate switches and 160 racks. Note that the trace does not contain information about the processing times of each job; hence, we do not try to emulate entire running times of the jobs.

Figure 13(a) presents the network allocations for all the active jobs based on their size (measured by the number of mapper and reducer tasks) for the full bisection bandwidth case. The total network utilization was the same since all evaluated allocations are work conserving disciplines, but the distribution of bandwidth among jobs is different. During the shuffle phase, a MapReduce job with M mappers and R reducers can launch $M \times R$ flows, one from each of the mappers to each of the reducers. In such a scenario, a Per-Flow allocation can potentially give a network share proportional to $M \times R$ to a job whose payment was proportional to $M + R$ (the number of map and reduce slots it was allotted). Thus network allocations can be quadratic making allocations significantly unfair to small jobs.

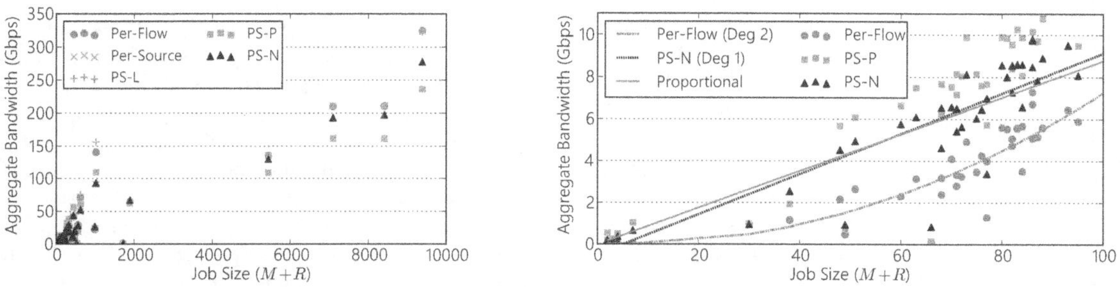

(a) All Jobs

(b) Small Jobs Only

Figure 13: Network allocation of MapReduce jobs on a 3200-node Facebook production cluster under different policies.

Strategy	Full Bisection BW		4× Oversubscribed	
	RMSD	Speed-up	RMSD	Speed-up
Per-Flow	1.76	1.00	2.60	1.00
Per-Source	1.22	9.64	1.31	11.91
PS-L	1.72	2.49	2.39	10.73
PS-P	1.61	16.61	1.19	14.37
PS-N	1.00	9.96	1.00	11.77

Table 5: Normalized deviation from proportional allocation measured using RMSD and speed-up in the shuffle completion time for small jobs.

To better understand this unfairness against small jobs, we consider jobs with $(M+R) < 100$, that form a substantial fraction of the total jobs. Figure 13(b) shows the manifestation of this quadratic allocation problem with Per-Flow, which gives considerably lower shares to the small jobs. We also note that PS-N almost closely matches the proportionality line$-(M+R)\frac{BW}{N}$, where BW is the total bandwidth and N is the total number of VMs of this data subset.

Table 5 presents quantitative results for small jobs for: (i) proportionality and (ii) speed-up in the shuffle completion time compared to Per-Flow, for both fully subscribed and oversubscribed topologies. We quantify proportionality by using the Root-Mean-Squared Deviation (RMSD) of the allocations from the proportional allocation and normalize them w.r.t. PS-N (which exhibits the least deviation for both topologies). The shuffle completion time of a job is bottlenecked by the last finishing flow [12], which in turn is dictated by the task having the minimum bandwidth allocation across all the tasks in a job. We thus approximate the shuffle time by dividing the bytes transferred per task by the bandwidth of the slowest transferring task. We report the median speed-up value relative to the Per-Flow (TCP) case which performs the worst.[11] PS-P performs the best (since it maximizes the minimum bandwidth guarantee for a task) with speed-ups of $\sim$15× for both topologies. Per-Source and PS-N also give significant improvements for the small jobs.

6. PRACTICAL CONSIDERATIONS

In this section, we present more details for the practical challenges towards deploying the allocation policies described in §4.

Full Switch Support: Today, there are switches that already have hardware capabilities matching our requirements (one queue per tenant, weighted fair queueing) such as routers with per-flow WFQ support [2], however most data center switches today have 8-64 hardware queues [5]. Determining the cost of this extra support compared to today's data center switches is a place for future work.

We also note that the weights for the tenant queues must be updated in time. For PS-P, weighs on a link L must be updated only when a new VM is started in the subtree delimited by L. For PS-L and PS-N the weight of tenant A through link L needs to be updated when there is a new pair of VMs belonging to A communicating on L. Updating weights can be done by (1) a centralized controller or (2) through the data-plane packets, which need to contain the tenant ID and weight information in packets. Providers can implicitly encode tenant IDs into different sets of IP or (virtual) MAC address ranges (e.g., using something like NetLord [21]). The weight of a VM can be encoded in packets through the use of the QoS bits in the IP header, filled out by hypervisors (for security reasons), which allows for 256 weights.

To support PS-L, switches require no additional information. For PS-P, switches need to be configured with one bit for each interface, identifying whether the interface is facing hosts or facing the network core.

Deploying PS-N is the most challenging, since it requires coordination between the source and the destination hypervisors for setting the weight of a source-destination pair. This weight must also be communicated to switches. Note that all the flows between a source and a destination contain the same weight. For the first packet between two VMs, the source's hypervisor inserts its allocated weight and the destination adds its weight to the returned packets. The subsequent packets contain the full weight of the source destination pair. When one of the endpoints starts communicating with another VM, its hypervisor updates the weights of the ongoing flows to the other endpoints. For practical purposes, the presence or absence of a communication between two VMs can be identified by using a threshold for the outgoing/incoming rate. Note that the 8 QoS bits in the IP packet header may offer too little weight granularity for PS-N. One approach to increase the number of available bits is to use encapsulation, e.g., something like NetLord [21]. In this case, only the MAC addresses are used for switching from the encapsulated header and many more other bits can be used for this purpose. VLAN tags can also be used to carry weights.

Partial Switch Support: CSFQ [26] was proposed to implement weighted fair queueing between flows without requiring per-flow state in the core switches but only in the edge switches. In data centers, we can use CSFQ and only maintain per-VM (or even per-tenant) state inside hypervisors at end points, and no such state in switches. PS-N can directly be implemented using vanilla CSFQ. PS-P can also be implemented using CSFQ but requires a slight change in the mechanism. In particular, each flow needs to contain two weights, the source and the destination weights, and switches

[11]The mean has a similar behavior albeit a few outliers with large speed-ups.

need to swap the two weights when the direction of packets changes from flowing towards the core to flowing away from the core (towards servers). However, a CSFQ-based deployment does require CSFQ support in switches. While we do believe hardware support for CSFQ can be inexpensive and fast, this claim remains to be proven by hardware manufacturers.

No Switch Support: TCP is known to provide per-flow fairness. If we use a single flow between a source and a destination or we impose an aggregate equivalent behavior in hypervisors, we achieve per source-destination fairness. If we are able to use weights between these flows (e.g., two TCP flows weight twice as much as a single flow and get twice as much bandwidth on a congested link) we can effectively implement PS-N. Seawall [25] aims to achieve this purpose. Thus, we expect one would be able to use Seawall, and instead of using per-source weights, use the weights given by PS-N. Weighted Flow Assignment (WFA) [12] can also approximate PS-N shares at the application layer using multiple TCP flows. We believe PS-P could be approximated with similar (but more complex) mechanisms.

7. RELATED WORK

Recently, there have been a few proposals for sharing cloud networks. Seawall [25] proposes a hypervisor-based mechanism for enforcing a generalized TCP-like behavior between VMs, where each TCP-like flow can have an arbitrary weight (rather than a single weight as in the case of TCP). Using this mechanism Seawall implements a per-source allocation policy. Therefore, Seawall is mostly orthogonal to our paper; in fact, Seawall's mechanism may be used to implement PS-N and PS-P. We leave this as future work.

Oktopus [10] and SecondNet [17] propose static reservations throughout the network to implement bandwidth guarantees for the hose model and pipe model, respectively. The main drawback of reservation systems is that they do not achieve the work conservation property, since the unused bandwidth is not shared between tenants. On the other hand, the advantage of reservation systems is that they can achieve more complex virtual topologies regardless of the physical location of the VMs. PS-P can support different bandwidth guarantees for different tenants by using carefully selected weights, but cannot support virtual topologies that are different than the physical topologies. For this purpose, reservation systems could be combined with our proposed allocation policies, which can be applied within each reserved virtual topology.

Gatekeeper [24] proposes a per-VM hose model with work conservation. Gatekeeper uses a hypervisor-based mechanism, which, however, works only for full bisection-bandwidth networks. In this paper we have described the PS-P allocation policy which supports a similar model for arbitrary tree networks, and described possible deployments using switch support; we are currently investigating how to implement PS-P using only hypervisors as well.

NetShare [19] advocates network sharing through the use of per-tenant weights that are constant throughout the network. This model can be used to implement a form of link proportionality.

Congestion Exposure (ConEx) [3] is a recent IETF effort that aims to equalize the number of dropped packets (congestion-volume) of different entities. By applying the ConEx mechanism between VMs one could achieve a Per-SD allocation. By applying ConEx between tenants it appears that the closest abstraction achieved is some form of congestion proportionality (but which also considers links congested by a single tenant). However, the precise set of properties of this approach remain to be determined.

8. CONCLUSIONS

In this paper, we have focused on understanding and exploring several key requirements and properties for network allocation in data centers. In summary, we have identified three main requirements: min-guarantee, proportionality (ranging from the network level to the link level) and high utilization, and a set of properties to guide the design of allocation policies in the tradeoff space.

In addition, we have introduced three allocation policies—PS-L, PS-P and PS-N—to navigate the tradeoff space. We have evaluated the proposed allocation policies using simulation and a software switch implementation. Through hand-crafted examples and traces of MapReduce jobs from a production cluster, we have shown that they achieve their intended properties. However, much more remains to be done. The allocation policies we have proposed in this paper should be seen as merely starting points in exploring the tradeoff space.

Acknowledgments: We thank the anonymous reviewers and our shepherd for their valuable feedback. Part of this work was supported by the NSF Award CNS-1038695.

9. REFERENCES

[1] Amazon web services. http://aws.amazon.com.
[2] Cisco 7500 series. http://goo.gl/m0Ve0.
[3] Congestion Exposure. http://datatracker.ietf.org/wg/conex/.
[4] DETERlab. http://www.isi.deterlab.net.
[5] HP 5900 ToR switch. http://goo.gl/kcycc.
[6] Rackspace Cloud Servers vs. VPS Platforms. http://goo.gl/LPxIJ.
[7] M. Al-Fares, A. Loukissas, and A. Vahdat. A scalable, commodity data center network architecture. In *SIGCOMM*. ACM, 2008.
[8] M. Al-Fares, S. Radhakrishnan, B. Raghavan, N. Huang, and A. Vahdat. Hedera: Dynamic Flow Scheduling for Data Center Networks. In *NSDI*, 2010.
[9] H. Ballani, P. Costa, T. Karagiannis, and A. Rowstron. The price is right: Towards location-independent costs in datacenters. In *Hotnets*, 2011.
[10] H. Ballani, P. Costa, T. Karagiannis, and A. Rowstron. Towards Predictable Datacenter Networks. In *ACM SIGCOMM*, 2011.
[11] B. Briscoe. Flow rate fairness: Dismantling a religion. *ACM SIGCOMM Computer Communication Review*, 2007.
[12] M. Chowdhury, M. Zaharia, J. Ma, M. I. Jordan, and I. Stoica. Managing data transfers in computer clusters with Orchestra. In *SIGCOMM*, 2011.
[13] N. G. Duffield, P. Goyal, A. G. Greenberg, P. P. Mishra, K. K. Ramakrishnan, and J. E. van der Merwe. A flexible model for resource management in virtual private networks. In *SIGCOMM*, 1999.
[14] A. Ghodsi, M. Zaharia, B. Hindman, A. Konwinski, S. Shenker, and I. Stoica. Dominant resource fairness: fair allocation of multiple resource types. In *USENIX NSDI*, 2011.
[15] A. Greenberg, J. R. Hamilton, N. Jain, S. Kandula, C. Kim, P. Lahiri, D. A. Maltz, P. Patel, and S. Sengupta. VL2: A Scalable and Flexible Data Center Network. *ACM SIGCOMM*, August 17 - 21 2009.
[16] C. Guo, G. Lu, D. Li, H. Wu, X. Zhang, Y. Shi, C. Tian, Y. Zhang, and S. Lu. BCube: A High Performance, Server-centric Network Architecture for Modular Data Centers. *ACM SIGCOMM*, 2009.
[17] C. Guo, G. Lu, H. J. Wang, S. Yang, C. Kong, P. Sun, W. Wu, and Y. Zhang. Secondnet: a data center network virtualization architecture with bandwidth guarantees. In *CoNEXT*. ACM, 2010.
[18] C. Guo, H. Wu, K. Tan, L. Shi, Y. Zhang, and S. Lu. Dcell: A Scalable and Fault-tolerant Network Structure for Data Centers. In *SIGCOMM*, 2008.
[19] T. Lam, S. Radhakrishnan, A. Vahdat, and G. Varghese. NetShare: Virtualizing Data Center Networks across Services. *Technical Report, UCSD*, 2010.
[20] R. Morris, E. Kohler, J. Jannotti, and M. F. Kaashoek. The click modular router. *SIGOPS Oper. Syst. Rev.*, 33(5):217–231, 1999.
[21] J. Mudigonda, P. Yalagandula, J. Mogul, B. Stiekes, and Y. Pouffary. Netlord: a scalable multi-tenant network architecture for virtualized datacenters. In *ACM SIGCOMM*, 2011.
[22] B. Radunović and J.-Y. L. Boudec. A unified framework for max-min and min-max fairness with applications. *IEEE/ACM Trans. Netw.*, Oct. 2007.
[23] C. Raiciu, S. Barre, C. Pluntke, A. Greenhalgh, D. Wischik, and M. Handley. Improving Datacenter Performance and Robustness with Multipath TCP. In *ACM SIGCOMM*, 2011.
[24] H. Rodrigues, J. R. Santos, Y. Turner, P. Soares, and D. Guedes. Gatekeeper: Supporting bandwidth guarantees for multi-tenant datacenter networks. In *USENIX WIOV*, 2011.
[25] A. Shieh, S. Kandula, A. Greenberg, C. Kim, and B. Saha. Sharing the Data Center Network. In *Usenix NSDI*, 2011.
[26] I. Stoica, S. Shenker, and H. Zhang. Core-stateless fair queueing: achieving approx. fair bandwidth allocations in high speed networks. In *SIGCOMM'98*.

The Only Constant is Change: Incorporating Time-Varying Network Reservations in Data Centers

Di Xie
Purdue University

Ning Ding
Purdue University

Y. Charlie Hu
Purdue University

Ramana Kompella
Purdue University

ABSTRACT

In multi-tenant datacenters, jobs of different tenants compete for the shared datacenter network and can suffer poor performance and high cost from varying, unpredictable network performance. Recently, several virtual network abstractions have been proposed to provide explicit APIs for tenant jobs to specify and reserve virtual clusters (VC) with both explicit VMs and required network bandwidth between the VMs. However, all of the existing proposals reserve a fixed bandwidth throughout the entire execution of a job.

In the paper, we first profile the traffic patterns of several popular cloud applications, and find that they generate substantial traffic during only 30%-60% of the entire execution, suggesting existing simple VC models waste precious networking resources. We then propose a fine-grained virtual network abstraction, Time-Interleaved Virtual Clusters (TIVC), that models the time-varying nature of the networking requirement of cloud applications. To demonstrate the effectiveness of TIVC, we develop PROTEUS, a system that implements the new abstraction. Using large-scale simulations of cloud application workloads and prototype implementation running actual cloud applications, we show the new abstraction significantly increases the utilization of the entire datacenter and reduces the cost to the tenants, compared to previous fixed-bandwidth abstractions.

Categories and Subject Descriptors: C.2.3 [Computer-Communication Networks]: Network Operations

General Terms: Algorithms, Design, Performance

Keywords: Datacenter, Network Reservation, Allocation, Bandwidth, Profiling

1. INTRODUCTION

Cloud computing has transformed the enterprise computing landscape significantly. By offering virtually unlimited resources without any upfront capital investment and a simple pay-as-you-go charging model, cloud computing provides a compelling alternative to enterprises constructing and maintaining their own cluster-computing infrastructure. The long-term viability of cloud computing depends, among others, on two major factors—cost and performance. Unless cloud platforms prove cheaper than enterprise ones, enterprises have no incentive to migrate their computational requirement to the cloud. Similarly, cloud platforms need to provide some guarantees about job performance, otherwise enterprises may be apprehensive to migrate to the cloud to begin with.

Unfortunately, today's public cloud platforms such as Amazon EC2 do not provide any performance guarantee, which in turn affects tenant cost. Specifically, the resource reservation model in today's clouds only provisions CPU and memory resources but ignores networking completely. Because of the largely oversubscribed nature of today's datacenter networks (*e.g.,* [19]), network bandwidth is a scarce resource shared across many tenants. When networking intensive phases of applications collide and compete for the scarce network resources, their running times become unpredictable. The uncertainty in execution time further translates into unpredictable cost as tenants need to pay for the reserved virtual machines (VMs) for the entire duration of their jobs.

Recent works such as SecondNet [20] and Oktopus [11] noted this lack of networking SLA in cloud environments and proposed new network abstractions for tenants to specify their networking needs along with their CPU and memory needs, so that their applications can obtain predictable performance. For example, SecondNet proposes bandwidth reservation between every pair of VMs. Oktopus proposes a simpler virtual cluster (VC) model where all VMs are connected to a virtual switch with links of bandwidth B. While both models provide a good start, they fail to capture the *temporal dimension* of network resource requirement. Specifically, we observe that the networking requirement of many applications experience significant changes throughout their execution. Hence provisioning a single bandwidth B for an entire cluster or for each pair of VMs throughout the job execution is clearly wasteful.

To illustrate this, we show the instantaneous network throughput of the MapReduce Sort application in Figure 1(a) (details are in §2.2). We observe from the figure that the Sort application utilizes the network only during the first half of the execution (during the shuffle phase), while the second half (merge sort and reduce phase) requires very little networking resource. Effectively the network utilization is in the form of a simple square wave. Figure 1 further shows the network utilization characteristics of several other applications such as Word Count, Hive Join and Hive Aggregation. While they exhibit more complicated network characteristics than Sort, a common takeaway is the relatively sparse, time-varying networking requirement.

Using a fixed bandwidth reservation can potentially waste precious resources of the datacenter. Consider several Sort jobs that have bandwidth requirement B during the first half of their execution. If we use Oktopus' virtual cluster (VC) model for specifying the network requirement of Sort, the fixed bandwidth B is provisioned during the entire duration of the job, preventing another Sort

SIGCOMM'12, August 13–17, 2012, Helsinki, Finland.

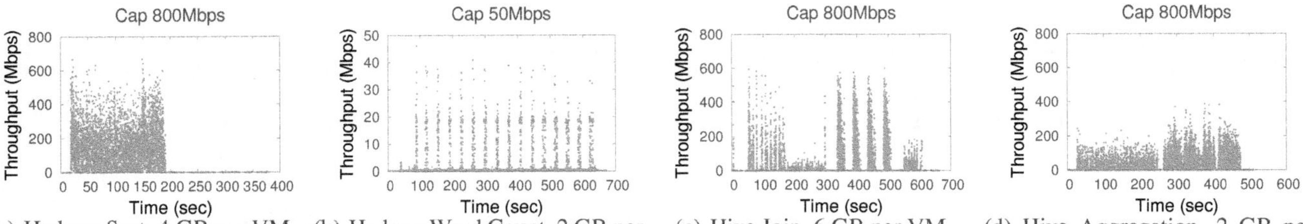

(a) Hadoop Sort, 4 GB per VM. (b) Hadoop Word Count, 2 GB per VM. (c) Hive Join, 6 GB per VM. (d) Hive Aggregation, 2 GB per VM.

Figure 1: Time-varying traffic demand of cloud applications.

job from using the unused bandwidth B during the second half of the job. Clearly, if there are enough VMs, we can schedule a second Sort job such that the first half of the second job overlaps with the second half of the already running first job.

Our main contribution in this paper is to exploit this key observation and explore a new, fine-grained network reservation abstraction called *temporally-interleaved virtual clusters* (TIVC), that allows specifying the time-varying networking requirement of cloud applications. By capturing the temporal networking usage, the TIVC abstraction reduces over-reservation of network bandwidth in fixed-bandwidth abstractions, allowing more jobs to be admitted into the datacenter. Thus the new abstraction not only improves the utilization of the network resource but also potentially increases the VM utilization and hence the overall datacenter utilization. The increased utilization of the entire datacenter in turn translates into either increased cloud provider revenue, *e.g.,* under today's VM-time only charging model, or both increased provider revenue and reduced tenant cost, when the cloud provider adopts a bandwidth charging model (*e.g.,* [30, 10]), compared to fixed-bandwidth abstractions.

For the TIVC abstraction to be viable in practice, there are two fundamental questions that we must answer: How can we obtain the TIVC model corresponding to a job? How do we allocate and provision jobs with TIVC specifications in a physical datacenter? While a similar question to the first confounded prior network abstractions [11, 20], they simply assumed that the customer will specify them somehow. In this paper, we propose a systematic, profiling-based methodology for automatically generating TIVC model parameters for a given application. There are several challenges that we need to tackle to make this profiling-based methodology practical.

The first challenge concerns the elastic nature of networking requirement for many jobs; as we provide more bandwidth, job will finish faster. So how should one determine the peak bandwidth requirement for an application in deriving its TIVC model, *e.g.,* in profiling the application? Using a detailed study using real cloud applications, we provide insights into this question, which previous works have ignored completely.

The second challenge concerns designing TIVC model functions so that they can be easily generated, they capture the time-varying networking requirement of an application well, and yet they can be practically enforced in the datacenter network. We propose to convert the networking requirement output of the profiling stage into coarse-grained pulse functions (square waves) with different widths and heights. Our conversion scheme ensures the derived bandwidth reservation model has little impact on the job completion time.

The final challenge concerns a practical allocation algorithm that not only needs to identify where a TIVC job can be admitted spatially (in the network topology), but also temporally (when the bandwidth will be freed in the future to accommodate this job throughout its execution). We propose an allocation algorithm based on dynamic programming that is highly scalable and finds valid allocations with the best spatial locality, *i.e.,* in the lowest subtree available.

We have developed PROTEUS, a system that implements the TIVC abstraction. PROTEUS derives TIVC models for applications, accepts jobs online and allocates them to the physical datacenter, and provisions network bandwidth according to the TIVC models. We deployed our prototype on an 18-node datacenter with a ternary tree topology using NetFPGA-based switches. Our simulations of a large-scale datacenter using real MapReduce application workload show our TIVC abstraction can increase the datacenter batched job throughput by 34.5%, and reduce the rejection rate of dynamically arrived jobs from 9.5% to 3.4%, which translates into 22% higher cloud provider revenue than the fixed-bandwidth abstraction under today's VM-time based charging model. When the cloud provider moves to a bandwidth charging model, TIVC can both improve cloud provider revenue by 11% and reduce tenant cost by 12%, compared to the fixed-bandwidth abstraction.

2. BACKGROUND AND MOTIVATION

In this section, we first review the state of the art, and then motivate the time-varying networking abstraction proposed in this paper.

2.1 State of the Art

To provide performance guarantees, several recent works [20, 11] have proposed *virtual network cluster* abstractions, which allow cloud users to specify not only the type and number of VMs requested, but also the associated networking requirement, *i.e.,* the bandwidth requirement between the VMs. Such a virtual cluster gives cloud users an assurance of their job performance based on the VM and networking SLAs.

SecondNet [20] proposes APIs that let users specify either end-to-end bandwidth for each pair of VMs (suitable for strong guarantees), or ingress/egress bandwidth for each VM (for better than best-effort sharing). Oktopus [11] proposes two simplified abstractions: virtual clusters and virtual oversubscribed clusters. A virtual cluster is a one-level logical tree topology, specified as $<N, B>$, which asks for N VMs and each VM is connected to a virtual switch by a bidirectional link of bandwidth B. A virtual oversubscribed cluster request, $<N, B, S, O>$, specifies a two-level logical tree topology with an oversubscription ratio of O.

All the models above, however, only allow *fixed* bandwidth guarantees, since they fundamentally assume the applications have the same bandwidth requirement throughout the entire execution, which is rarely true in practice. Specifically, our *key observation* is that typical cloud applications generate varying amount of traffic in different phases of their execution. Thus reserving their peak bandwidth requirement for the entire execution wastes scarce network resources in the datacenter, which reduces the number of jobs that can fit in the datacenter to run concurrently, and hence the overall utilization of the datacenter.

2.2 Profiling Networking Demand

To assess the extent of networking requirement variations of cloud applications, we conduct a measurement study. Since MapReduce [17] is a popular programming paradigm for large-scale data-intensive operations and many cloud applications have been ported to MapReduce, such as search indexing, database, and machine learning, we profile the networking patterns of several typical MapReduce jobs, including Sort, Word Count, Hive Join, and Hive Aggregation, which have been used as the primary benchmarks in recent datacenter studies (*e.g.,* [16, 36, 24, 28]). While this list of applications is not exhaustive, we believe they represent an important class of applications found in datacenters.

Methodology. We conducted the profiling experiments on our 33-machine cluster, using open-source Hadoop 0.20.0 as the MapReduce platform. The 33 machines are connected to a Gigabit switch. Each of the 33 machines is equipped with a 4-core Xeon 2.4GHz processor, 4GB memory, and running 2 Xen virtual machines (domU) and each VM is allocated 1.5GB memory and a dedicated hard drive. To obtain the network traffic, we deployed tcpdump at all the Hadoop slave VMs to log all communication (packet headers) between them. Since the bidirectional traffic are almost identical for the set of applications, for each application, we plot the access egress bandwidth of one VM in each time bin of 10ms, which is in the same order of RTTs under load and hence sufficiently small to capture the transient peak throughput. The throughput for different VMs are very similar, and we study the variation in Section 4.

Traffic Patterns of Cloud Applications. We observe that the traffic patterns of the set of applications fall into the following three categories, of increasing generality. In all categories, a base bandwidth needs to be reserved to facilitate job maintenance tasks such as communicating with the job scheduler, and to support low volume network traffic among the VMs.

Type 1: Single peak. The Hadoop *Sort* benchmark sorts randomly generated records with 10-byte keys and 90-byte values. Figure 1(a) plots the network throughput over the execution with 4 GB input data per VM. We observe that most of the network traffic were produced in the first half of the job execution. In other words, the traffic pattern exhibits a single peak, defined as a continuous period of throughput above some base amount.

Type 2: Repeated fixed-width peaks. The Hadoop *Word Count* benchmark counts the number of word occurrences in the input data. Figure 1(b) shows that there was only a small amount of data been shuffled over the network periodically throughout the execution. The small amount of data is due to reduced map output from enabling the combiner in map tasks, and the periodic network traffic is because the map tasks finished in batches, and hence their output were shuffled periodically. Thus the traffic demand exhibits repeated bursts with a similar amount of traffic volume per burst, and consequently similar duration.

Type 3: Varying-width peaks and *Type 4: Varying height and width peaks. Hive* [2] implements a data warehouse system built on top of Hadoop. It provides an SQL-like language for data queries. The Hive performance benchmark [5] is used to compare the performance of Hadoop, Hive, and PIG [3]. It consists of four queries: grepping from a table, selecting columns from a table, aggregating a table, and joining two tables. We generated a 4GB UserVisits table and a 2GB Rankings table per node. Figures 1(c)-(d) plot the throughput for the two queries Join and Aggregation. We observe that Hive Join exhibits a brief burst of network activity at the beginning, followed by a longer duration of less-intensive traffic from map tasks, followed by five bursts, corresponding to five reduce tasks. Hive Aggregation exhibits a long duration of moderate traffic demand, followed by a long duration of more intense traffic

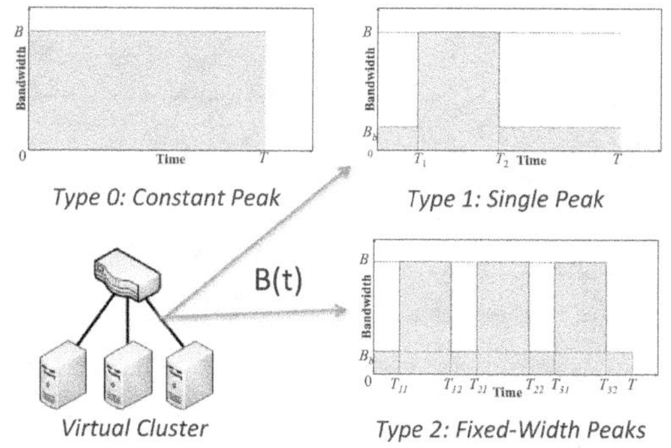

Figure 2: TIVC models contain time-varying bandwidth specifications $B(t)$.

demand. These two applications fall into the more general category of a sequence of traffic bursts of varying durations and of varying intensities.

3. THE TIVC ABSTRACTION

Our key observations in the previous section clearly indicate the need for a new networking abstraction that can express the time-varying nature of application networking requirement. To this end, we propose a novel network abstraction called *temporally-interleaved virtual cluster (TIVC)* that captures the temporal variations in the network behavior of cloud applications.

The TIVC abstraction (shown in Figure 2) consists of a virtual cluster of N nodes connected to a switch, via links of bandwidth B, similar to the VC abstraction proposed in [11]. However, the key difference is that the bandwidth for each link is a time-varying function $B(t)$ instead of a constant value in prior work. This allows capturing the actual networking requirement of applications much more precisely, which enables the cloud provider to achieve better utilization of datacenter resources, and ultimately improves the provider revenue and reduces the tenant cost without sacrificing the job performance, as we will show in our experiments.

3.1 Different TIVC Models

Given the networking requirement profile of an application, *e.g.,* shown in Figure 1, one can potentially derive some complicated function (*e.g.,* high-order polynomials) to precisely model the changing requirement over time. However, such smooth functions significantly complicate the process of allocating a TIVC in the physical datacenter network, as well as provisioning of the continuously changing bandwidth requirement in the physical network. To strike a balance between modeling precision and implementation difficulties[1], we choose to model the networking requirement as simple pulse functions in this paper. We leave exploring other tradeoffs in the spectrum as future work.

Since TIVC is a generalization of the VC abstraction, we call the model with a fixed bandwidth $B(t) = B$ as Type 0 as shown in Figure 2. To capture the several general time-varying patterns observed in our profiling study, we propose the following model functions, also shown in Figure 2.

Type 1: Single peak. A Type 1 model captures the networking demand of applications that only generate network traffic in a certain interval, and has a format of <N, T, B_b, P>, where P=(T_1, T_2, B). The bandwidth function is given as

$$B(t) = \begin{cases} B_b & : t \in [0, T_1] \; or \; [T_2, T] \\ B & : t \in [T_1, T_2] \end{cases}$$

[1]The fixed bandwidth specification in [20, 11] can be viewed as going to one extreme in this tradeoff, ease of implementation.

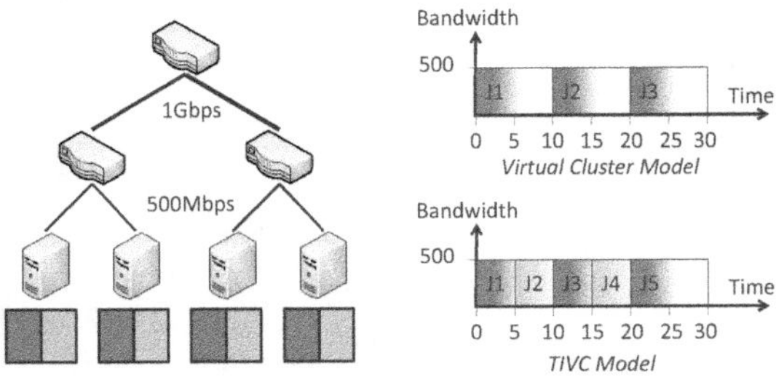

Figure 3: Simple tree with full-bisection bandwidth.

For example, the Sort application in Figure 1(a) would request for a Type 1 TIVC with <N, $382s$, $4Mbps$, ($18s$, $191s$, $800Mbps$)>.

Type 2: Fixed-width peaks. The Type 2 TIVC model captures the networking requirement of applications that have repeated traffic peaks. A request of this type has the format <N, T, B_b, P_1, ..., P_K>[2], where $P_i = (T_{i1}, T_{i2}, B)$. The bandwidth function is

$$B(t) = \begin{cases} B & : t \in [T_{i1}, T_{i2}], i \in [1, K] \\ B_b & : otherwise \end{cases}$$

The request specifies K repeated peaks of bandwidth B and the same width to be provisioned, *i.e.,* the widths of the peaks ($T_{i2} - T_{i1}$) are the same, and the base bandwidth B_b during the rest of time. For example, the Word Count application in Figure 1(b) would request for a Type 2 TIVC with <N, $672s$, $2Mbps$, ($80s$, $92s$, $50Mbps$), ..., ($620s$, $632s$, $50Mbps$)>.

Type 3: Varying-width peaks. The Type 3 TIVC model is more general than previous two types and captures the networking requirement of applications that have varying-width traffic peaks. The request format and the bandwidth function are the same as with Type 2, except the durations of the peaks ($T_{i2} - T_{i1}$) can differ. For example, Hive Join in Figure 1(c) would request for a Type 3 TIVC with <N, $672s$, $50Mbps$, ($52s$, $172s$, $800Mbps$), ($336s$, $361s$, $800Mbps$), ($387s$, $412s$, $800Mbps$), ($434s$, $459s$, $800Mbps$), ($485s$, $508s$, $800Mbps$), ($550s$, $607s$, $800Mbps$)>.

Type 4: Varying height and width peaks. The Type 4 TIVC model refines the Type 3 model to allow varying heights for different peaks. The format is the same as in Type 3, except $P_i = (T_{i1}, T_{i2}, B_i)$, which specifies a bandwidth cap of B_i is requested from time T_{i1} to T_{i2}.

The four models based on pulse functions are of increasing generality. While we do not claim they are universal, we find they capture well the traffic patterns of the applications we have studied (which are also widely used in previous datacenter networking studies, *e.g.,* [16, 36, 24, 28]).

3.2 Implication on Job Scheduling

By more precisely capturing the networking requirement of applications, TIVC enables the cloud provider to schedule more jobs to run concurrently not only in over-subscribed networks, but also in networks with full-bisection bandwidth, such as fat-trees [7, 19].

Consider the simple tree datacenter network with full-bisection bandwidth shown in Figure 3. Consider a sequence of Sort-like jobs, each requesting 4 VMs and access bandwidth of 500Mbps during the first half of their 10-second execution. Under the VC model which reserves the constant 500Mbps bandwidth throughout the job execution, only one job can be scheduled to run every 10 seconds. The four VMs have to be allocated on the four servers since each server has an access bandwidth of 500Mbps. In contrast, under TIVC, after the first job has run for 5 seconds and thus finished the networking phase, the second job can be scheduled to run on the other VM of each of the four servers. This results in doubling the resource utilization and hence the job throughput of the whole system.

[2]We enumerate the peaks in all types for consistency.

4. TIVC MODEL GENERATION

In this section, we study the key challenge in using TIVC abstractions in practice: How to automatically generate the TIVC model for a given cloud application? Since the quantitative program behavior, *e.g.,* networking requirement, is typically dependent on the MapReduce framework configurations at runtime (*e.g.,* [22]) and potentially on input parameters, in this paper, we propose a "black-box" approach to modeling the traffic requirement of a cloud application. The general idea is to collect the traffic trace of the application during profiling runs,[3] and use it in model generation. We discuss the generality of our approach in §4.5.

Realizing the above profiling-based approach faces two immediate challenges. First, there exists a tradeoff between the bandwidth cap and the execution time, since tightening the bandwidth constraint elongates the networking component of a job and affects the completion time. The question then is what bandwidth cap should be used during the profiling run? Second, given a traffic profile collected from the profiling run, how to automatically derive the most suitable TIVC model.

4.1 Impact of Bandwidth Capping

Under the TIVC abstraction, the cloud provider charges for both the VMs and the network usage[4]. This raises the question of how to balance the VM cost and the networking cost, *e.g.,* in trying to minimize the total cost of a tenant's job. Intuitively, a job can request for a lower bandwidth limit which lowers the networking cost per unit time and potentially the total networking cost, which however may lead to longer networking time and hence longer job execution time, increasing the total VM cost. To our knowledge, this important question has not been studied before, even in immediately relevant prior works such as Oktopus [11]. In the following, we conduct experiments to characterize the first-order impact that bandwidth capping has on the cloud application execution time from which we draw implications to TIVC model generation.

We repeat the profiling experiments in §2.2 while gradually reducing the access bandwidth limit per VM. Figure 4 shows the measured application execution times. We make two observations. First, for each application, until the bandwidth cap is reduced to a certain threshold (*e.g.,* 300 Mbps for Hive Join), there is virtually no impact on the application execution time and network throughput. Second, once the cap crosses below the threshold, denoted as the *no-elongation threshold bandwidth,* the execution time is elongated monotonically. We empirically confirmed that the elongation of the execution time is due to the slowdown of the application networking activities, as we measured the execution time slowdown to be equal to the elongation on the network active periods, *i.e.,* the width of the pulses captured in the TIVC model shown in §4.2.

The main reason for the above behavior is that cloud applications have mixed communication and computation even in network-intensive phases. For example, we calculated the average throughput per VM of Hive Join during each of the four high pulses in Figure 1(c) which had a bandwidth cap of 800 Mbps to be only 160 Mbps. This indicates that it may be unnecessary to set the bandwidth cap to be much higher than the average application traffic generation rate. However, we found that capping the bandwidth to

[3]We note profiling traffic demand is required even in Oktopus [11] to meaningful decide on the constant bandwidth parameter B.

[4]Designing bandwidth charging models is currently being studied [30, 10] and is beyond the scope of this paper.

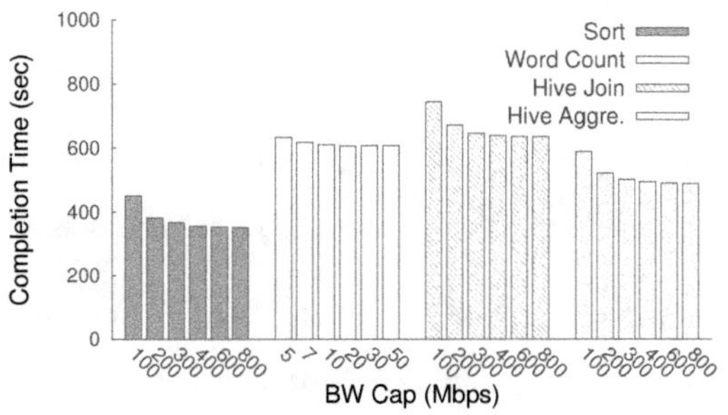

Figure 4: Application execution time under different bandwidth caps.

be exactly the same as the application data generation rate is too rigid, *i.e.,* it actually slows down the application network-intensive phases. This is shown in Figure 4 where Hive Join under 200 Mbps runs 25 seconds longer than under 300 Mbps. The reason is mainly due to applications' bursty networking behavior. Specifically, we observe that applications tend to interleave computation and data transfer phases, *i.e.,* it generates traffic in bursts during data transfer phases, at a rate higher than the average data generation rate. Further, the computation has certain dependence on the data transfer. For example, in MapReduce jobs, the data shuffling phase is interleaved with map tasks which generate processed data. If the bursty data transfer is slowed down due to bandwidth capping, it pushes back the subsequent computation.

The above threshold behavior suggests that the tenant should always pick the bandwidth cap to be the same or lower than the no-elongation threshold bandwidth, as higher capping wastes networking cost without making the application run faster. However, below the threshold, automatically deriving the bandwidth cap that optimizes the total cost requires a precise modeling of the relationship between the bandwidth cap and application execution time, which is beyond the scope of this paper. Instead, in §5.1, we implement a profiling-based approach to help the user to pick the bandwidth cap that optimizes the total cost while meeting the performance goal.

The above study further helps us to draw an important implication in terms of automatic TIVC model generation. If the user picks a bandwidth cap B_{cap} equal to or below the threshold value in the profiling run to be used for model generation, in the generated TIVC model, the ceiling of the pulses should be conservatively set to be the same as the bandwidth cap, as using a lower ceiling will elongate the application execution time relative to the profile run. However this rule can be loosened for certain pulses which will be become clear in §4.3.

4.2 Model Generation

The model generation algorithm takes as input the traffic profile of an application profiling run under bandwidth cap B_{cap}, and derives the TIVC parameters that achieves the highest efficiency in the following two steps. Here, we define *efficiency* of a TIVC model as the ratio of the total application traffic volume over the total traffic volume under the bandwidth reserved by the TIVC model, *i.e.,* $\int_0^T B(t)dt$.

Since TIVC models use pulse functions, *i.e.,* square curves, the main idea of automatic model generation is to derive square curves to cap the continuous bandwidth demand curve from the profiling run. There are potentially many ways of generating such bounding square curves, and different bounding curves may have different efficiencies. We note high efficiency fittings (and hence low total bandwidth volume) may not necessarily translate into fitting more jobs in the datacenter, as how well the TIVC models of competing jobs complement each other also plays an important role. However,

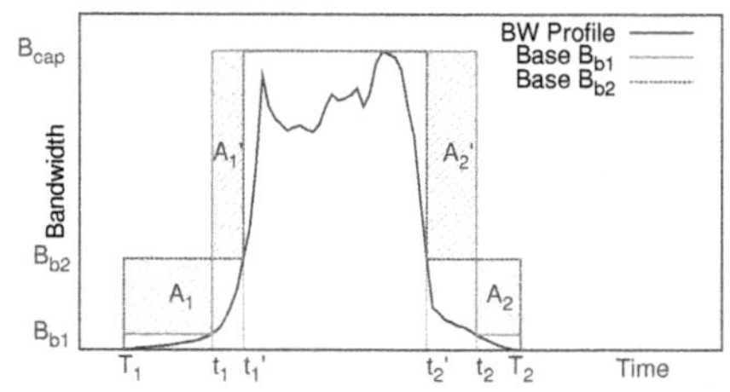

Figure 5: Model fitting by varying the base bandwidth.

we envision it is more practical that TIVC models are generated offline, *i.e.,* oblivious to the competing jobs at (future) scheduling time, and hence set maximizing the efficiency as the main objective in model generation.

We generate the most efficient Type 4 model in two steps. First we show how to generate the bounding square curves that maximize the efficiency under Type 3 TIVC (which generalizes Type 1 and 2). We then show the conditions when a Type 3 model is refined into Type 4 in §4.3. Recall a Type 3 TIVC model specifies a base bandwidth B_b and a list of fixed-height pulses $P_i = (T_{i1}, T_{i2}, B)$. For a fixed B_b value, to meet the application's bandwidth need, all periods in which the bandwidth usage is observed to be greater than B_b is conservatively rounded up to B_{cap}, for reasons discussed in §4.1. Thus B_b effectively controls the relative amount of bandwidth volume under the base bandwidth periods and under the pulse periods. In general, the higher the B_b value, the more area under the base bandwidth, and the less area under the (narrower) pulses, as shown in Figure 5, which shows two tentative square curve fittings with base bandwidth B_{b1} and B_{b2}, respectively. Now given an application bandwidth profile, the Type 3 TIVC parameters that maximize the model efficiency can be found by searching through different values of B_b.

Figure 6 shows the bandwidth profiles for the four applications shown in Figure 1 but under the no-elongation threshold bandwidth cap, and the corresponding TIVC models generated for these profiles. We make two observations. First, compared to Figure 1, using threshold bandwidth capping did not elongate the application execution, but smoothed the traffic peaks. Second, the TIVC models generated for the four applications are of Types 1, 2, 3 and 3, as expected, and achieve 22.9%, 8.9%, 13.1%, and 6.3% efficiencies, respectively.

4.3 Model Refinement

The Type 3 TIVC model generated for a given application traffic profile consists of square curves of two different bandwidth limits, base bandwidth B_b during valleys and capping bandwidth B_{Cap} during peaks. A close look at the individual pulses and valleys in Figure 6 reveals two findings. First. if we calculate the efficiencies of different pulses (shown in Figure 6 next to each pulse), the values can differ significantly, with the peak efficiency being around 30%. Second, the profile for the same application under a lower bandwidth cap (not shown due to page limit) shows the pulses with the highest efficiency get elongated first, and with low efficiencies can sustain a lower bandwidth cap without being elongated.

To improve the efficiency of the TIVC model, which allows the cloud provider to potentially fit more jobs in the datacenter, we refine the generated Type 3 model by lowering the bandwidth cap for pulses that have very low bandwidth efficiencies, using the following heuristic. If the efficiency is lower than a threshold γ, we lower the bandwidth cap so that the efficiency is around α. We empirically found for all the applications we studied, setting $\gamma = 8\%$ and $\alpha = 20\%$ is sufficiently conservative, *i.e.,* it will not elongate the pulses when running under the new bandwidth caps. Figure 7

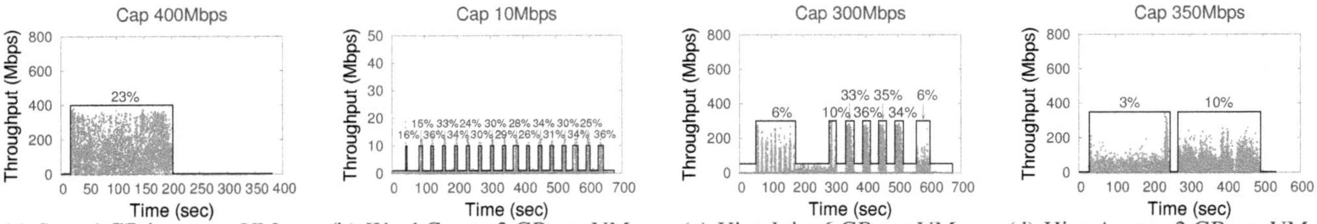

(a) Sort, 4 GB input per VM. (b) Word Count, 2 GB per VM. (c) Hive Join, 6 GB per VM. (d) Hive Aggre., 2 GB per VM.

Figure 6: Bandwidth profiles under no-elongation threshold bandwidth cap and their Type 3 models.

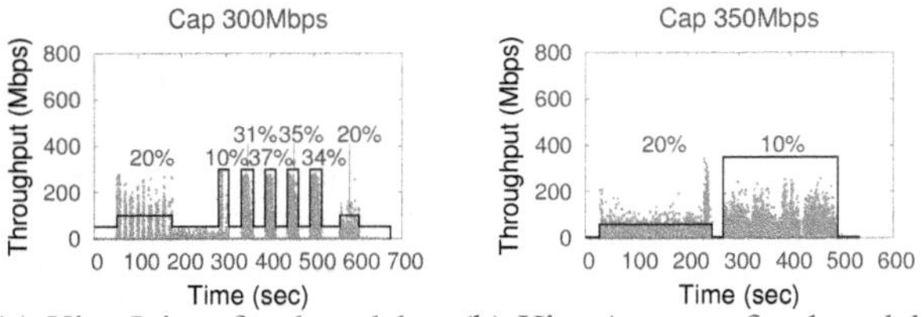

(a) Hive Join refined model. (b) Hive Aggre. refined model.

Figure 7: Hive Join and Hive Aggre. refined model.

shows the Type 4 models for Hive Join and Hive Aggre. which are refined from their corresponding models in Figure 6. We do not claim the two threshold values to be general and suitable threshold values are likely dependent on the networking and computation mix of the class of applications. Rather, the key point is that Type 4 models can be systematically derived by refining Type 3 models from lowering the bandwidth cap for pulses with low efficiencies.

4.4 Incorporating Model Offsets

A final challenge faced by TIVC model generation is concerned with the TIVC models for different VMs of the same job. There are two related questions: how much do the traffic demand and hence the TIVC models generated for different VMs differ, and if the TIVC models are the same, how well are their timings aligned. Our profiling study shows that for the set of MapReduce jobs, the traffic demand and hence the TIVC models generated across the VMs are of the same type and with the same number of pulses. However, the rising and falling edges of the pulses can be offset, potentially due to the delay in task dispatching from the task scheduler. Table 1 lists the standard deviation of the rise and fall timings of the pulses for the four applications. We see that the standard deviation of the pulses across 32 VMs is less than 9 seconds.

Table 1: TIVC offsets, measured as the standard deviation of the rise/fall timings of pulses across 32 VMs.

	Sort	WC	Hive Join	Hive Aggre.
Pulse 1 (rise, fall)	5.8, 7.9	3.5, 4.0	2.0, 3.2	0.1, 2.3
Pulse 2 (rise, fall)	N/A	5.3, 5.6	8.1, 3.6	0.1, 0.3
Pulse 3 (rise, fall)	N/A	5.4, 5.8	1.9, 2.7	N/A
Pulse 4 (rise, fall)	N/A	5.1, 5.5	1.9, 2.4	N/A
Pulse 5 (rise, fall)	N/A	5.7, 5.4	1.9, 2.4	N/A
Pulse 6 (rise, fall)	N/A	5.5, 5.5	1.9, 2.4	N/A
Pulse 7 (rise, fall)	N/A	5.3, 5.5	2.1, 2.6	N/A

The above small misalignment between TIVCs of different VMs suggests that instead of generating individual TIVCs for the VMs, we can generate a single TIVC which is easier to provision, as follows. We first process per-VM profiles to generate per-VM refined TIVC models. We then calculate the max of their base bandwidths, and regenerate per-VM TIVC models using this new base bandwidth. Next we merge the new per-VM TIVC models by merging the corresponding pulses, taking the max of their widths and heights. Figures 8(a)-(c) show the traffic profiles and TIVCs for three randomly sampled VMs and Figure 8(d) shows the merged TIVC, for Hive Join. Since the threshold bandwidth cap for Word Count, 10Mbps, is already very low, it should be provisioned throughout the application execution. The final TIVC models generated for the rest three applications are shown in Table 2.

4.5 Discussions

We discuss the generality and limitation of our profiling-based model generation. Our model generation uses the enveloping technique (§4.4) to tolerate small offsets among the traffic demand by different VMs of a job. This works well for the class of MapReduce-type applications which tend to be highly regular in nature – the worker VMs are performing similar tasks and hence are likely to generate traffic of similar volume and at similar times. For applications that generate non-uniform traffic, we can generate and enforce per-VM TIVC models.

Our approach assumes the input data size per VM stays the same during profiling runs and production runs. One potential source of variation between the traffic patterns during profiling runs and production runs is the input data, as the processing time of different data items and hence across the VMs could be uneven. In our experiments with input data generated using random seeds, we did not find much difference in the traffic characteristics across runs. Specifically, the standard deviation of the pulse edge timings across 5 runs of the four applications using input data generated using random seeds is less than 10 seconds (not shown due to page limit). In general, it is important to validate this assumption for any candidate application across multiple sample profile runs before using the TIVC models.

We envision the primary use scenarios of TIVC models are when customers repeatedly run the same type of jobs with the same input size (and hence same number of VMs), with potentially similar data sets from run to run. Such a scenario is common in iterative data processing (*e.g.,* [14]) such as PageRank [29], HITS (Hypertext-Induced Topic Search) [25], recursive relational queries [12], social network analysis, and network traffic analysis where much of the data stay unchanged from iteration to iteration, and is observed in many production environments (*e.g.,* in Bing's production clusters [6]), where the same job needs to be repeated day in and day out, and the data change slightly. In such scenarios, the jobs could be profiled on each run or periodically, with the TIVC models generated to help schedule the cluster during the next run.

Finally, as with all other network reservation approaches (*e.g.,* [20, 11]), TIVC faces a number of uncertainties during job execution: the data processed by each VM (*e.g.,* a map task) may not be on the local disk, a job execution may experience stragglers or task failures which require some tasks to be re-executed (*e.g.,* [9]), or some VMs or network elements may fail. In general, predictable performance in the presence of such uncertainties requires adding fault tolerance to applications and overprovisioning of not only network resources, but also extra VMs, for accommodating backup tasks. We leave dealing with such uncertainties as future work.

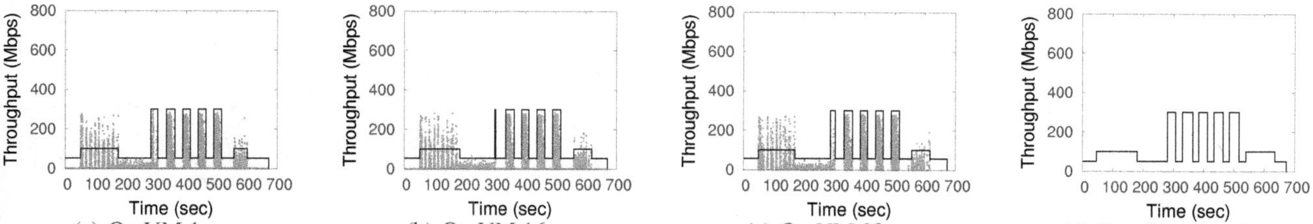

(a) On VM 1. (b) On VM 16. (c) On VM 32. (d) Envelop model.

Figure 8: Bandwidth profile and generated Type 4 TIVC models of Hive Join.

Table 2: TIVC models generated for the applications.

App.	Type	TIVC	Efficiency
Sort	1	<N, $382s$, $4Mbps$, ($17s$, $202s$, $400Mbps$)>	22.8%
Hive Join	4	<N, $672s$, $50Mbps$, ($46s$, $183s$, $100Mbps$), ($284s$, $301s$, $300Mbps$), ($329s$, $363s$, $300Mbps$), ($383s$, $413s$, $300Mbps$), ($434s$, $464s$, $300Mbps$), ($485s$, $517s$, $300Mbps$), ($539s$, $634s$, $100Mbps$)>	17.6%
Hive Aggre.	4	<N, $535s$, $4Mbps$, ($27s$, $253s$, $60Mbps$), ($268s$, $492s$, $350Mbps$)>	10.7%

5. THE PROTEUS SYSTEM

To demonstrate the effectiveness of TIVC models, we have developed a cloud sharing system called PROTEUS that implements the TIVC models.

5.1 Overview

The goal of PROTEUS is to allow cloud customers to obtain predictable performance and cost guarantees for their applications. This is achieved via three steps, as outlined in Figure 9. In the first step, the customer's application is profiled under different configurations, *i.e.,* of input data size per VM and bandwidth cap, and TIVC models are generated for the profiling runs using the techniques presented in §3. We note the profiling overhead can be drastically reduced when customers repeatedly run the same type of jobs with the same input size (see §4.5).

Each profiling run under a configuration in the first step results in a TIVC model and a service completion time. In the second step, the charging model published by the cloud provider is used to estimate the cost for the candidate TIVC models under different configurations. The customer can then pick whichever configuration that best suits her performance/cost objective.

In the final step, given a TIVC job configuration that the customer picks, the cloud provider runs a spatial-temporal TIVC allocation algorithm to place the job in the physical datacenter in a way that maximizes the utilization and hence the revenue of the cloud datacenter, and configures the datacenter network to enforce the requested time-varying bandwidth specified in the TIVC specification. We describe the details of these two components next.

5.2 Spatial-Temporal Allocation

The job manager implements the TIVC allocation algorithm to allocate VM slots on the physical machines in an online fashion. It achieves this by maintaining up-to-date information of (1) the datacenter network topology; (2) the empty VM slots in each physical machine; and (3) the residual bandwidth for each link, calculated from tallying the TIVC allocations of currently running jobs. We focus on tree-like topologies such as multi-rooted tree topologies which are typical of today's datacenters. In such a topology, machines are grouped into racks and the Top-of-Rack (ToR) switches are in turn connected to higher level switches.

We present a generic allocation algorithm for all TIVC models, each of which can be viewed as a sequence of pulses of different bandwidth and duration. We first show how to find a valid allocation for a TIVC request, then show how to find a good allocation out of all the valid allocations. Our allocation algorithm improves the Oktopus allocation algorithm [11] with a novel dynamic programming solution in searching valid allocations, which not only significantly improves the search efficiency, but also guarantees to find the most localized allocation, *i.e.,* in the lowest subtree.

Bandwidth requirement of a valid allocation. Before presenting the allocation algorithm, we first ignore the time dimension and explain on how a fixed access bandwidth B per VM in a TIVC request translates into the bandwidth requirement on the internal links in the physical network.

If the N VMs required by a request can be found at a level-0 subtree, *i.e.,* within a physical machine, they can be allocated right away, as there should be enough bandwidth between the VMs in the same machine. Otherwise, the N VMs will reside in multiple subtrees, and the traffic between them will travel up and down the tree. This poses a subtle challenge as to how much bandwidth needs to be reserved on the tree links. Consider a link L that connects a left subtree containing m allocated VMs and a right subtree with $(N-m)$ VMs. Since each VM cannot send or receive at a rate more than B, the maximum bandwidth needed on link L is $min(m, N-m)*B$. Thus a valid allocation needs to satisfy $min(m, N-m)*B < R_L$, where R_L is the residual bandwidth of link L.

Now, taking the time dimension into account, a valid allocation for a TIVC request is an embedding of the VMs into a subtree of the datacenter where each link L connecting parts of the subtree satisfies the bandwidth requirement of each pulse and valley in the TIVC request. Specifically, for each $P_i = (T_{i1}, T_{i2}, B_i)$ in the TIVC request, link L should have enough residual bandwidth during interval (T_{i1}, T_{i2}), *i.e.,*

$$min(m, N-m)*B_i < R_L(T_{i1}, T_{i2}) \quad (1)$$

Efficient search via dynamic programming. As discussed above, when the VMs of a TIVC request cannot be satisfied within a rack, they may paternally be divided into subgroups which are then allocated out of different racks under a depth-1 subtree, and if not, out of different depth-2 subtrees, and so on. The cost for searching for such possible valid allocations can quickly become combinatorial. We make a key observation that a valid suballocation of K_1 VMs in a depth-$(d-1)$ subtree can be reused in searching for a valid suballocation of K_2 VMs, $K_2 > K_1$, in the parent depth-d subtree, and hence we can formulate the searching algorithm as a dynamic programming problem which runs very efficiently.

Finding a good allocation. Given a TIVC request, there can be many possible valid allocations in the physical network. There are two dimensions in the physical network that quantify a good allocation. First, in the vertical dimension, a good allocation should exhibit good locality, *i.e.,* the VMs allocated to it should be as

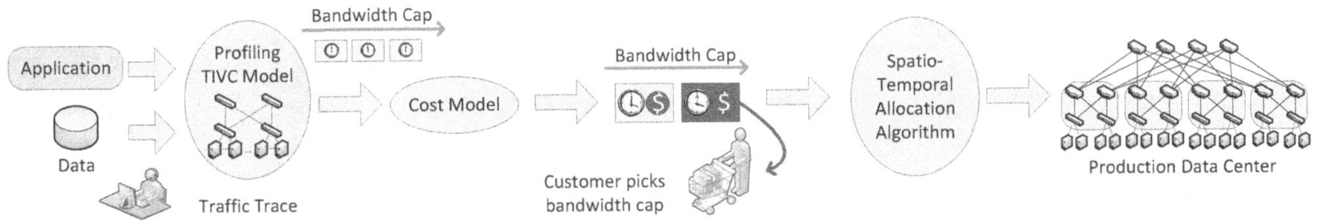

Figure 9: The Proteus system.

localized to a subtree as possible, as good locality conserves the bandwidth of the links in the upper levels of the tree. Second, it is possible that multiple equally localized allocations exist, *i.e.,* each within a depth-i subtree. A major consideration in choosing an allocation out of them is fragmentation, as allocating a TIVC to a subtree may result in the subtree having few VMs and little link bandwidth left to fit any future TIVCs. This fragmentation problem resembles the classic dynamic memory allocation problem in operating systems, for which a number of classic heuristic allocation strategies, including first-fit, best-fit, and worst-fit, exist. Our empirical experiments have shown there is no clear winner, and hence PROTEUS uses the local, first-fit strategy; it picks the first fitting lowest-level subtree out of all such lowest-level subtrees.

The algorithm. Figure 10 shows the TIVC allocation algorithm in pseudo code. The set of all possible numbers of VMs out of the N VMs needed by a job that can be allocated in the subtree rooted at v form the M set for v, M_v. Because of the bandwidth constraint (Equation 1), the numbers in M_v may not be continuous. Hence, we need to record the valid allocation out of each subtree during dynamic programming search, using a few data structures. Let $L_v[k]$ denote the set that contains the numbers of VMs that could be accommodated in the first k children of the subtree rooted at v, without considering the uplink bandwidth constraint of v. Then M_v contains all the values in $L_v[n]$ that can satisfy the uplink bandwidth constraint. To record the allocation in the traversed subtrees, *i.e.,* each possible value h in $L_v[k]$, we record in $D_v[k, h]$ the number of VMs assigned to the kth child of v, when it assigns h VMs in the first k children. The dynamic programming step is shown in lines 5–10, which calculates $L_v[k]$ and $D_v[k, h]$ recursively. Afterwards, all the candidate numbers of VMs are added to M_v if they pass the uplink bandwidth requirement of v (lines 11-14). If N can be allocated out of M_v, *Alloc*() is called which performs recursion according to $D_v[k, h]$ while recording the bandwidth reservation of the relevant links, and eventually outputs the number of VMs per machine (level 0), and the algorithm terminates. We can easily show the algorithm outputs the first allocation (from left to right) that fits in the lowest-level subtree.

```
Algorithm: Allocation for TIVC request r
Input: Datacenter topology tree T
1. For level l from 0 to height(T)
2.     For each subtree v at level l
3.         If (l==0) L_v[0] = {0, ..., # avail. VMs} // leaf machine
4.         Else L_v[0] = {0}
5.         For v's child k from 1 to n
6.             L_v[k] = {0}
7.             For each possible value e in v's kth child's M set (M_{v_k})
8.                 For each possible value h in L_v[k − 1]
9.                     L_v[k] = L_v[k] ∪ {e + h}
10.                    D_v[k, e + h] = e
11.        M_v = ∅
12.        For each value h in L_v[n]
13.            If (bandwidth check of v's uplink per Eq. (1) == true)
14.                M_v = M_v ∪ {h}
15.        If N ∈ M_v
16.            Alloc(r, v, N)
17.            Return true
18. Return false
function Alloc(r, v, m):
19. If v is a machine
20.     allocate m VMs in v
21. Else
22.     For v's child k from n to 1
23.         Alloc(r, v_k, D_v[k, m])
24.         record bw reservation on v's kth link
25.         m = m − D_v[k, m]
```

Figure 10: The TIVC Allocation algorithm.

5.3 Enforcing TIVC Reservations

After allocating the VMs for a job, PROTEUS needs to configure the network elements to enforce the reserved bandwidth on the access links and internal links that connect the VMs. Prior works [11, 34, 26] have opted for an end-host only approach, which reserves per-VM access bandwidth in the hypervisor. To enforce reservations in in-network links, such approaches add significant complexity in the hypervisor, which needs to perform online rate monitoring for all VM pairs, communicate the rates to a centralized optimizer which calculates the max-min fairness for each VM pair, *etc.*, making such approaches less scalable. Bloating the hypervisors is also typically disliked since it adds complexity in the critical path and compromise robustness and security [13].

We make a key observation that even in a large-scale datacenter, the number of jobs that share a link at the same time is generally low. Consider a typical rack in the production environment with 40 machines and hence 160 VMs assuming conservatively 4 VMs per machine. If most jobs are small enough to fit within a rack, few jobs need to straddle the rack boundaries. On the other hand, very few large jobs which need to cross the core of the network can be scheduled to run concurrently since each of them consumes a large number of VMs. In our simulation runs over a datacenter with 16,000 machines (§6), we found fewer than 26 concurrent jobs per link (see Figure 17). Rate limiting such a low number of jobs sharing a link can be easily implemented using network switches available today. For example, the Cisco Nexus 7000 Series 32-port 10Gb module already supports up to 16K policers (for rate limiting 16K aggregates) [4] and 64K ACL entries (for defining the traffic aggregates). Furthermore, the above low number of jobs sharing a link, as well as the fact the edges of bandwidth pulses of different jobs happen at different times, suggest that reconfiguring the policers can be done with low overhead.

Multi-path routing. The TIVC allocation algorithm in §5.2 assumes a simple tree topology where the traffic between a VM pair follows a single path up and down the tree. However, datacenters may have networks with richer connectivities such as multi-rooted trees (*e.g.,* [27]) and fat-trees [7, 19]. These networks typically use hash-based or randomized techniques such as ECMP and Valiant Load Balancing to spread traffic across multiple equal cost paths, and can use more involved techniques such as Hedera [8] and MPTCP [31] to ensure uniform traffic spread despite flow length variations. Therefore, such topologies can be incorporated into the TIVC allocation algorithm by treating the multiple links from each physical machine or switch that are used in multiple equal cost paths (to the same destination) as a single aggregation link, and enforcing bandwidth reservation of an aggregation link boils down to enforcing equal reservation split among the multiple physical links in the aggregate. We leave a detailed experimental study of such physical networks as future work.

6. EVALUATION

We use both simulations of large scale datacenter networks and our implementation on a testbed running real MapReduce applications to show PROTEUS exhibits significant advantage over a fixed-bandwidth reservation scheme such as Oktopus.

6.1 Simulation Setup

To show the effectiveness of PROTEUS in large scale datacenter settings, we developed a simulator that models VM and network bandwidth reservations in a shared datacenter. The simulator simulates a datacenter of three-level tree topology. There are 16,000 machines at level 0, each with 4 VM slots. 40 machines form a rack and are linked with a Top-of-Rack (ToR) switch with 1 Gbps links. Every 20 ToR switches are connected to a level-2 aggregation switch, and 20 aggregation switches are connected to the core switch of the datacenter. The default oversubscription of the physical network is 4, *i.e.,* ToR switches are connected to aggregation switches with 10 Gbps links, and aggregation switches to the core switch with 50 Gbps links.

Alternate abstractions. We compare PROTEUS with Oktopus, the state-of-the-art network abstraction [11]. Oktopus supports two network abstractions: virtual clusters (VC) and virtual oversubscribed clusters (VOC). However, VOC places a significant burden on the cloud users who not only have to specify the constant bandwidth constraint B, but also explicit subclustering of VMs that exhibit local communication and their oversubscription factors. Further, we do not observe any such communication locality in the applications we have studied. Thus we leave comparison with VOC as future work. Ideally, we should also compare TIVC with a baseline model that schedules jobs solely based on the number of available VMs. However, it is difficult to model in simulations the execution time elongation when jobs compete for networking freely.

Workload. We simulate tenant jobs based on the network workload extracted from the MapReduce applications studied in §2.2: Sort, Hive Join, and Hive Aggregation. We do not include Word Count as it has insignificantly low bandwidth requirement and will not benefit from bandwidth reservations provided by network abstractions like VC and TIVC. We use the no-elongation threshold bandwidth cap (§4.1) as the bandwidth requirement B under VC, and as the capping bandwidth in application profiling and model generation under TIVC. Using the same bandwidth cap B this way ensures that the job running times stay the same under the two abstractions during production runs. The generated TIVC parameters are shown in Table 2. To simulate a datacenter with diverse job mixes, we vary the number of VMs needed by each job; in our experiments by default the number of VMs per job request is exponentially distributed around a mean of 49 (following [11]).

6.2 Simulation Results

We compare PROTEUS with Oktopus under two scenarios: (1) A large number of tenant jobs are pooled at the job queue waiting to be scheduled to run. This workload captures production datacenters that host time-insensitive jobs, *e.g.,* data processing jobs to be run overnight. (2) Tenant jobs arrive dynamically and are accepted only if they could be scheduled at the moment of arrival. This workload is representative of shared clouds that host time-sensitive jobs. For each scenario, we first simulate 5,000 jobs running a single application, and then simulate 5,000 jobs of equally mixed applications. The four workloads are denoted as Sort, Hive Join, Hive Aggregation and Mixed.

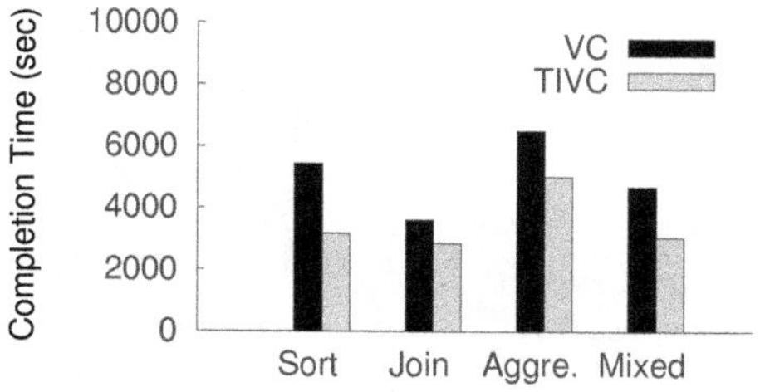

Figure 11: Total completion time.

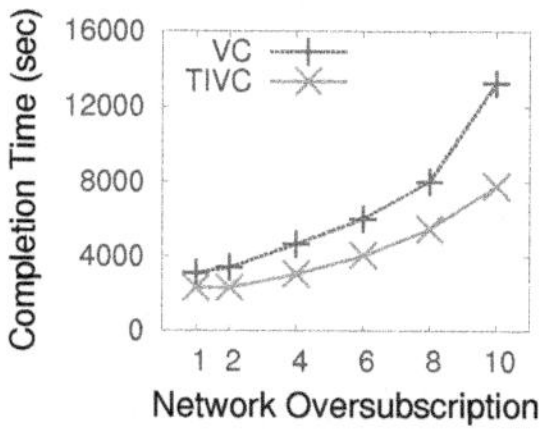

Figure 12: Completion time with varying oversub. for mixed workload.

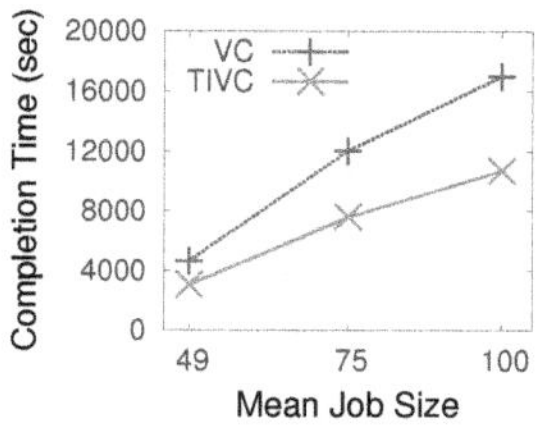

Figure 13: Completion time with varying mean job size for mixed workload.

6.2.1 Batched Jobs

For batched jobs, the job scheduling policy tries to maximize the job throughput. Under VC, this is achieved by going through the job queue and schedule the jobs that can be scheduled to run, whenever a job is finished. Under TIVC, however, this can be inefficient, as before any running job is finished, there can be enough residual bandwidth freed up so that a new job can be scheduled. Instead, the job scheduler rescans the job queue every 10 seconds, which is 20% of the average inter-job completion time.

Job completion time. Figure 11 plots the time to complete all 5,000 jobs under VC and TIVC. We see for all workloads, TIVC significantly improves the completion time, and hence job throughput of the datacenter, over VC. In particular, compared to VC, TIVC reduces the completion time by 41.5%, 20.8%, 23.1%, and 34.5% for Sort, Hive Join, Hive Aggre., and Mixed, respectively.

Varying the oversubscription rate and job size. We repeat the above experiments with varying oversubscription rates in the physical datacenter network. Figure 12 shows TIVC provides greater advantage over TC when the oversubscription rate is larger, reducing the total completion time by 35.2%, 36.0%, and 41.5% under oversubscription rates 6, 8, and 10, respectively. Similarly, Figure 13 shows increasing the mean job size N further increases the performance advantage of TIVC over VC since larger jobs are more likely to traverse the oversubscribed core network links.

6.2.2 Dynamically Arriving Jobs

We now consider the cloud scenario where job requests arrive over time. Assume the job arrival follows a Poisson process with rate λ, then the load on a datacenter with M VMs total is $\lambda \cdot N \cdot Tc/M$ where N is the mean job request size (*i.e.,* 49) and Tc is the mean job completion time. If a job cannot be allocated upon its arrival, it is rejected, as is the case with Amazon's EC2 job admission control [1] today. We again simulate 5,000 job requests under VC and TIVC while varying the load factor.

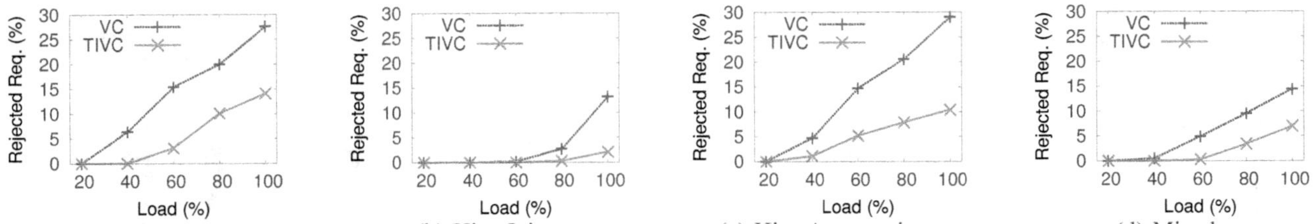

(a) Sort. (b) Hive Join. (c) Hive Aggregation. (d) Mixed.

Figure 14: Percentage of rejected requests with varying datacenter load.

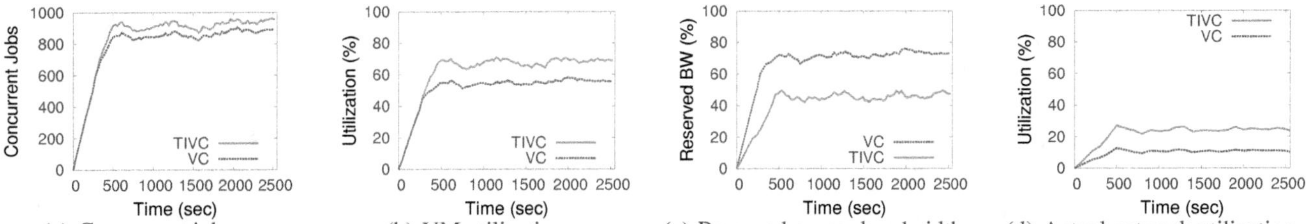

(a) Concurrent jobs. (b) VM utilization. (c) Reserved access bandwidth. (d) Actual network utilization.

Figure 15: Concurrent jobs, VM utilization, reserved and actual utilization of bandwidth, for Mixed and 80% load.

Job rejection rate. Figures 14(a)-14(d) plot the rejection rates for the three application workloads and the mixed workload. We observe that under low load, *e.g.,* 20%, the total networking reservation under both VC and TIVC can be met and hence both accept all jobs. As the load increases, VC rejects far more requests than TIVC. For example, at 80% load, 20.0%, 2.7%, 20.6%, and 9.5% of the requests are rejected under VC compared to 10.1%, 0.3%, 7.9%, and 3.4% under TIVC, for the four workloads, respectively.

Job concurrency and VM/network utilization. To understand how TIVC achieves much lower rejection rates than VC, we look at the number of concurrent jobs scheduled and the VM and network utilization under the two models. Due to page limit, we only show the results for the mixed workload under one load factor, 80%. Figure 15(a) shows that after the initial job arrival ramp-up phase, TIVC consistently achieves about 7% higher job concurrency than VC. Since the extra jobs accepted by TIVC tend to be larger than average, the 7% higher job concurrency under TIVC translates into on average close to 13% higher VM utilization (of the total 64,000 VMs in the datacenter) than under VC, as shown in Figure 15(b).

Finally, the reason TIVC is able to fit more jobs is by exploiting lower networking periods of a job to schedule other jobs. Figure 15(c) shows the average reserved access bandwidth over time under VC and TIVC. We see VC reserves on average 26.4% (of the link capacity) higher bandwidth than TIVC. However, Figure 15(d) shows TIVC achieves on average about 20.1% actual network utilization, calculated by adding the instantaneous traffic demand of individual jobs over each access link, and then averaged over all access links, much higher than the 8.9% under VC. This confirms that by capturing the time-varying nature of application traffic demand, TIVC is able to achieve much more efficient bandwidth reservation than VC. We note the overall low actual network utilization under the explicit network reservations may seem counter-intuitive. This is precisely the price to pay for predictable performance, *i.e.,* to reserve enough bandwidth so that the execution of real world applications, which can have diverse, bursty traffic phases, is not elongated (§4.1).

Job locality and link sharing. To assess the spatial locality of the TIVC jobs allocated by PROTEUS, we plot the average number of concurrent jobs allocated at different subtree levels in the datacenter, for the mixed workload runs. Figure 16 shows that after the ramp-up phase, under 80% load, on average around 795 out of the 920 total concurrent jobs are allocated within level-1 subtrees,

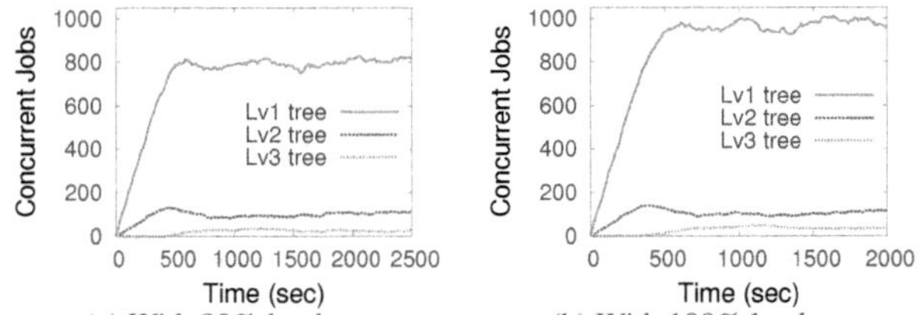

(a) With 80% load. (b) With 100% load.

Figure 16: Number of concurrent jobs allocated to diff. tree levels for Mixed.

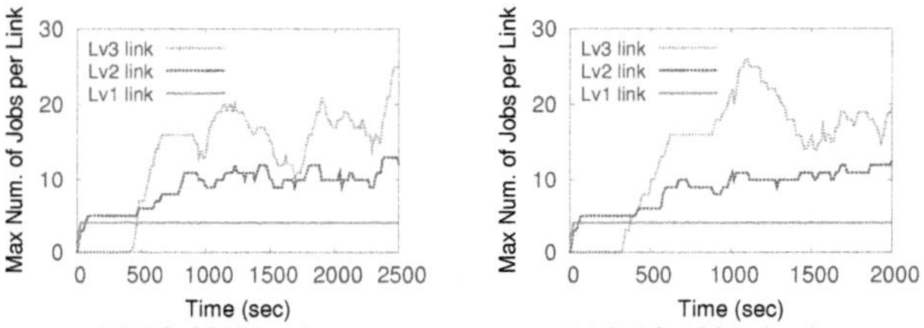

(a) With 80% load. (b) With 100% load.

Figure 17: Max number of jobs sharing a link at diff. tree levels for Mixed.

i.e., confined to ToR switches, and on average fewer than 99 and 26 jobs are within level-2 and level-3 subtrees, respectively. Under 100% load, there are fewer than 102 and 37 jobs within level-2 and level-3 trees. These numbers confirm that the vast majority of the jobs are localized to small subtrees. The immediate consequence of such locality is that there are few jobs sharing any link at the same time in the datacenter network. Figure 17 shows the maximum numbers of jobs sharing a link are less than 13 and 26 at level 2 and 3, and stay at 4 at level 1 (*i.e.,* the 4 VMs of a machine are allocated to 4 different jobs), under 80% and 100% load. We also measured the locality in separate experiments when the workload has only large jobs, *i.e.,* requiring thousands of VMs, and when the workload has mixed small and large jobs, and again found there are few jobs sharing any given link in the whole network. The reason is when jobs are large, few can be scheduled to run concurrently. In summary, the low number of jobs sharing any link in the network despite job size mixes suggests per-job bandwidth reservation in the internal links can be easily accomplished using policers in off-the-shelf switches (§5.3).

Tenant cost and provider revenue. Today's cloud providers such as Amazon EC2 charge tenants solely based on the consumed VM-

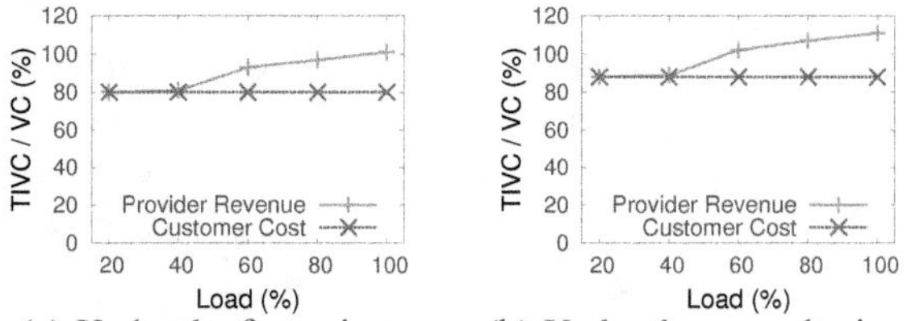

(a) Under the first price. (b) Under the second price.

Figure 18: Relative provider revenues and tenant costs for mixed workload.

time. In this case, the fraction of job requests that are accepted and the costs for them determine the cloud provider's revenue. Since the number of VMs allocated to each job and its execution time stay the same under VC and TIVC, the lower rejection rate under TIVC compared to VC directly translates into increased revenue for today's cloud provider, while the individual tenant's cost stays the same. For example, for the mixed workload in Figure 14(d), VC rejects 4.6%, 6.1%, and 7.4% more jobs under 60%, 80%, and 100% load, respectively. Since the rejected jobs tend to be larger than average, the extra rejected jobs translate into 16%, 22%, and 27% lower provider revenue under VC.

We envision that tomorrow's cloud providers will and should explicitly charge for networking bandwidth in providing tenants with explicit bandwidth reservations such as VC and TIVC. Since developing fair yet efficient charging model is still ongoing research [30, 10], we adopt the simple charging model in [11] which effectively charges networking based on the total reserved bandwidth volume over time in such a way that the cloud provider (*e.g.,* Amazon EC2) remains revenue neutral in transitioning from the VM-only charging model to the new model. Specifically, a tenant using N VMs for time T will be charged $N(T \cdot k_v + k_b \cdot V)$, where k_v is the unit-time VM cost, k_b is the unit-volume bandwidth cost, and V is the total bandwidth volume reserved over the time period T.

Under the above charging model, we compare the cloud provider revenue and the tenant cost under VC and TIVC for two sample estimated k_v and k_b prices: (0.04$/hr, 0.00016$/GB) and (0.04$/hr, 0.00008$/GB), for the mixed workload run in Figure 14(d). We calculate the ratio of the total cloud provider revenue and the ratio of the tenant job cost, under TIVC versus under VC. Figure 18(a) shows the two ratios under the first price. We see that TIVC allows tenants to pay on average about 20% less than VC, for accepted jobs, independent of the load, from reduced network usage. At low load, *e.g.,* 20–40%, the cloud provider revenue under TIVC is about 20% lower than under VC because the cloud provider accepts almost all jobs under both schemes while the tenants under TIVC on average pay 20% less than under VC. At close to 100% load, however, not only do tenants under TIVC pay 20% less than under VC, the cloud provider stays revenue neutral under TIVC compared to under VC. This is because the provider is able to accept about 7% more jobs under TIVC than under VC, which corresponds to about 13% higher VM utilization (again since these extra jobs tend to be larger than average). Finally, Figure 18(b) shows under the second price, TIVC not only allows tenants to pay on average about 12% less than under VC at all loads, but also allows the cloud provider to make more revenue when the load crosses 60%.

6.3 Testbed Experiment

We implemented PROTEUS following the description in §5 on a datacenter testbed consisting of 18 machines (specification in §2.2) forming a 3-tier tree topology as shown in Figure 19. Each machine runs 2 VMs, and the testbed switches are implemented using servers with NetFPGA cards with 4 1Gbps ports. We use the rate limiter module provided in the base package of NetFPGA reference router to limit the capacity of each internal tree link to emulate an oversubscribed datacenter. The level-1 (*i.e.,* between the machines and ToR switches), level-2, and level-3 link capacities are 230, 700, and 1000 Mbps, respectively.

The bandwidth provisioning for access links is implemented via the Linux traffic control API `tc`. For internal links, since it is not straight-forward to implement per-job rate limiters, we configured the combined rate of jobs allocated to the sub-tree. For example, assuming nodes $1-6$ are allocated to jobs 1 and 2, then the link between A and G, shared between the two jobs, would be configured with the sum of bandwidth requirements of the two jobs.

We use a mixed workload of 30 jobs, 10 each of the three applications, Sort, Hive Join, and Hive Aggre. PROTEUS profiles the 3 applications on the testbed, generates the TIVC and VC models using no-elongation threshold bandwidth cap, and allocates the jobs accordingly. We also run the same 30 jobs under a baseline model, which schedules the jobs solely based on the number of available VMs. Figure 20 shows the completion time for the workload are 2405, 3770, 5140 seconds, for Baseline, TIVC, and VC, respectively. TIVC reduces VC's completion time by 27% from more efficient bandwidth reservation and hence scheduling. However, Baseline has the shortest completion time since it aggressively schedules jobs to compete freely for the network which results in higher overall networking utilization, but however can lead to unpredictable application performance. Figure 21 shows the CDF of the per-job execution time under Baseline and under VC relative to that under TIVC. We see that in the median case, per-job execution time under Baseline is 10% longer than that under TIVC. Except for a few variations, TIVC results in similar per-job execution times as VC, because both models reserve the threshold bandwidth and avoid unpredictable competition for the network.

To evaluate the scalability of the TIVC allocation algorithm, we measure the time to allocate each of the 5,000 job requests in the large datacenter with 64,000 VMs used in §6.1. Our allocation algorithm is highly scalable; the single-threaded code running on an 8-core Intel Xeon E5410 2.33 Ghz processor and 16 GB RAM has a median time of 18.0ms and the 99th percentile time of 28.0ms.

7. RELATED WORK

Our work is closely related to the recently proposed virtual network abstractions [20, 11, 33]. We discussed [20, 11] in detail in §2. Like Oktopus, Gatekeeper [33] also proposes a per-VM hose model but for full bisection networks and focuses on managing servers' access bandwidth. The hose model was originally introduced in [18] for wide-area VPNs and did not consider allocating physical or virtual machines. Compared to these work, our work proposes TIVC, which extends the per-VM hose model to model the time-varying nature of networking requirement of cloud applications. More importantly, our work takes the first step towards automatically deriving the model parameters for a representative class of cloud applications.

Our work is also related to previous work on mechanisms for sharing datacenter networks. As discussed in §5.3, most of previous work (*e.g.,* [11, 33] use a hypervisor-based framework for enforcing bandwidth reservation in the network which can suffer poor scalability. Seawall [34] and Netshare [26] propose bandwidth slicing mechanisms that aim to provide fair sharing of networks with minimum bandwidth guarantee and statistical multiplexing, but do not provide deterministic bandwidth guarantees.

PROTEUS shares the same profiling methodology with Elastisizer [22], StarFish [23] and CBO [21] which focus on choosing the type and number of VMs for MapReduce jobs to balance cost/performance objectives. These work ignore networking requirement and hence complement our network profiling technique.

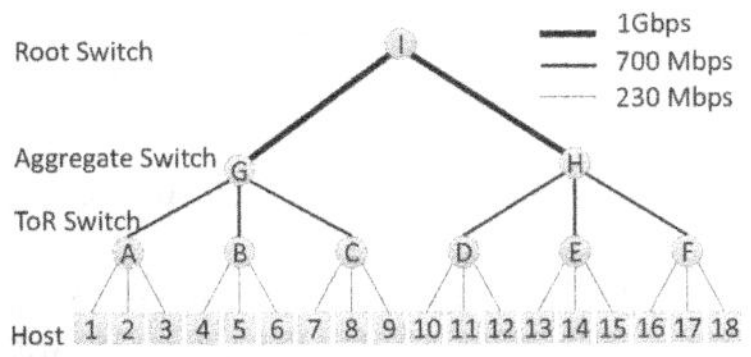

Figure 19: Testbed topology.

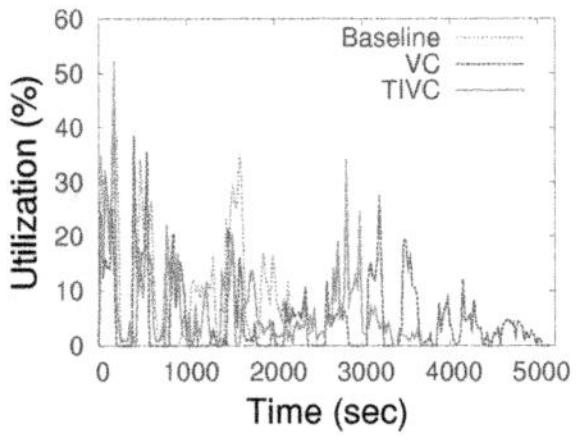

Figure 20: Network utilization.

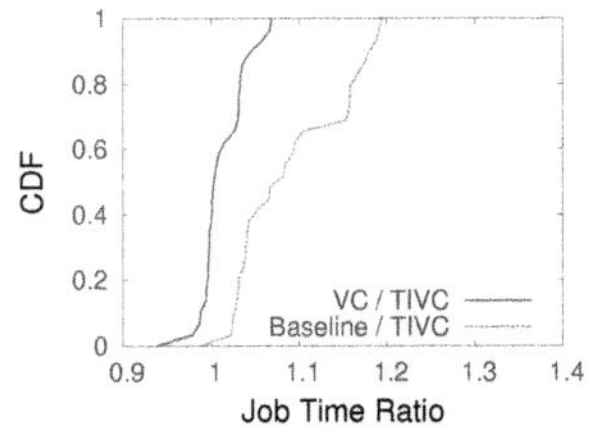

Figure 21: CDF of relative job time.

Our TIVC allocation algorithm is related to previous work on virtual network embedding (*e.g.*, [15, 35]) and testbed mapping [32]. These work resort to heavy-weight optimization solvers such as linear programming and can not scale to the larger number of VMs in modern datacenters.

Finally, several very recent work [30, 10] start to study fair and efficient charging model in sharing datacenter networks. These studies are complementary to PROTEUS and can be incorporated into PROTEUS in choosing cost-effective TIVC models.

8. CONCLUSIONS

In summary, the primary contributions of this paper are the design of the first network abstraction (to our best knowledge), TIVC, that captures the time-varying nature of cloud applications, and a systematic profiling-based methodology for making the abstraction practical and readily usable in today's datacenter networks. Our experimental evaluation using real MapReduce applications shows that TIVC significantly outperforms previous fixed-bandwidth network abstractions in improving job throughput and hence cloud provider revenue and reducing tenant cost. Our work takes a significant step forward towards efficient and cost-effective sharing of datacenter networks in providing cloud customers with predictable performance and cost.

The PROTEUS system which implements the TIVC abstraction can be readily used to extend today's dominant utility computing model offered by public clouds, which requires the customers to explicitly request for, and manage, virtual machines for their jobs, to support an *extended utility computing model* that directly meets the service time objectives of cloud customers. In this model, PROTEUS directly allocates an application slice of the datacenter, *i.e.*, a TIVC specification, that meets the target service time of a given application at the minimum cost to the customer.

Acknowledgements

We thank the anonymous reviewers, Chuangxiong Guo, and especially our shepherd, Ant Rowstron, for their helpful comments. We thank Pawan Prakash for his help with the testbed experiments. This work was supported in part by NSF grants CNS-1054788 and CRI-0751153.

9. REFERENCES

[1] Amazon ec2 api ec2-run-instances. http://docs.amazonwebservices.com/AWSEC2/latest/CommandLineReference/ApiReference-cmd-RunInstances.html.

[2] Apache hive. http://hive.apache.org/.

[3] Apache pig. http://pig.apache.org/.

[4] Cisco nexus 7000. http://www.cisco.com/en/US/prod/collateral/switches/ps9441/ps9402/ps9512/Data_Sheet_C78-437757.pdf.

[5] Hive performance benchmarks. https://issues.apache.org/jira/browse/HIVE-396.

[6] S. Agarwal, S. Kandula, N. Bruno, M.-C. Wu, I. Stoica, and J. Zhou. Re-Optimizing data-parallel computing. In *Proc. of USENIX NSDI*, 2012.

[7] M. Al-Fares, A. Loukissas, and A. Vahdat. A scalable, commodity data center network architecture. In *Proc. of ACM SIGCOMM*, 2008.

[8] M. Al-fares, S. Radhakrishnan, et al. Hedera: Dynamic flow scheduling for data center networks. In *Proc. of USENIX NSDI*, 2010.

[9] G. Ananthanarayanan, S. Kandula, A. Greenberg, I. Stoica, Y. Lu, , B. Saha, and E. Harris. Reining in the outliers in map-reduce clusters using mantri. In *Proc. of USENIX OSDI*, 2010.

[10] H. Ballani, P. Costa, T. Karagiannis, and A. Rowstron. The price is right: Towards location-independent costs in datacenters. In *Proc. of ACM HotNets*, 2011.

[11] H. Ballani, P. Costa, T. Karagiannis, and A. I. T. Rowstron. Towards predictable datacenter networks. In *Proc. of ACM SIGCOMM*, 2011.

[12] F. Bancilhon and R. Ramakrishnan. An amateur's introduction to recursiv query rocessing strategies. In *In Proc. of SIGMOD*, 1986.

[13] C. Brenton. Hypervisor vs host based security. https://cloudsecurityalliance.org/wp-content/uploads/2011/11/hypervisor-vs-hostbased-security.pdf.

[14] Y. Bu, B. Howe, M. Balazinska, and M. D. Ernst. HaLoop: Efficient Iterative Data Proecssing on Large Clusters. *Proc. of the VLDB Endowment*, 3(1), 2010.

[15] N. Chowdhury et al. Virtual Network Embedding with Coordinated Node and Link Mapping. In *IEEE INFOCOM*, 2009.

[16] T. Condie, N. Conway, P. Alvaro, J. M. Hellerstein, J. Gerth, J. Talbot, K. Elmeleegy, and R. Sears. Online aggregation and continuous query support in mapreduce. In *ACM SIGMOD*, 2010.

[17] J. Dean and S. Ghemawat. MapReduce: Simplified Data Processing on Large Clusters. In *Proc. of USENIX OSDI*, December 2004.

[18] N. G. Duffield, P. Goyal, A. Greenberg, et al. A flexible model for resource management in virtual private networks. In *Proc. of ACM SIGCOMM*, 1999.

[19] A. Greenberg, J. R. Hamilton, et al. VL2: a scalable and flexible data center network. In *Proc. of ACM SIGCOMM*, 2009.

[20] C. Guo, G. Lu, H. J. Wang, S. Yang, C. Kong, et al. Secondnet: a data center network virtualization architecture with bandwidth guarantees. In *Proc. of ACM CoNEXT*, 2010.

[21] H. Herodotou and S. Babu. Profiling, what-if analysis, and cost-based optimization of mapreduce programs. *PVLDB*, 4(11):1111–1122, 2011.

[22] H. Herodotou, F. Dong, and S. Babu. No one (cluster) size fits all: Automatic cluster sizing for data-intensive analytics. In *Proc. of ACM SOCC*, Oct. 2011.

[23] H. Herodotou, H. Lim, G. Luo, N. Borisov, L. Dong, F. B. Cetin, and S. Babu. Starfish: A self-tuning system for big data analytics. In *Proc. of CIDR*, 2011.

[24] B. Hindman, A. Konwinski, M. Zaharia, A. Ghodsi, A. D. Joseph, R. Katz, S. Shenker, and I. Stoica. Mesos: a platform for fine-grained resource sharing in the data center. In *Proc. of USENIX NSDI*, 2011.

[25] J. Kleinberg. Authoritative sources in a hyperlinked environment. *J. ACM*, 46(5), 1999.

[26] T. Lam and G. Varghese. Netshare: Virtualizing bandwidth within the cloud. *UCSD Technical Report*, 2009.

[27] J. Mudigonda, P. Yalagandula, M. Al-Fares, and H. L. Jeffrey Mogul. SPAIN: COTS data-center ethernet for multipathing over arbitrary topologies. In *Proc. of USENIX NSDI*, 2010.

[28] D. G. Murray, M. Schwarzkopf, C. Smowton, S. Smith, A. Madhavapeddy, and S. Hand. Ciel: a universal execution engine for distributed data-flow computing. In *Proc. of USENIX NSDI*, 2011.

[29] L. Page, S. Brin, R. Motwani, and T. Winograd. The PageRank citation ranking: Bringing order to the Web. *Technical Report 1999-66, Stanford Infolab*, 1999.

[30] L. Popa, A. Krishnamurthy, S. Ratnasamy, and I. Stoica. Faircloud: Sharing the network in cloud computing. In *ACM HotNets*, 2011.

[31] C. Raiciu, S. Barre, C. Pluntke, A. Greenhalgh, D. Wischik, and M. Handley. Improving datacenter performance and robustness with multipath tcp. In *Proc. of ACM SIGCOMM*, 2011.

[32] R. Ricci, C. Alfeld, and J. Lepreau. A Solver for the Network Testbed Mapping Problem. *SIGCOMM CCR*, 33(2), 2003.

[33] H. Rodrigues et al. Gatekeeper: Supporting bandwidth guarantees for multi-tenant datacenter networks. *HP Technical Report*, 2011.

[34] A. Shieh, S. Kandula, A. Greenberg, C. Kim, and B. Saha. Sharing the data center network. In *Proc. of USENIX NSDI*, 2011.

[35] M. Yu, Y. Yi, J. Rexford, and M. Chiang. Rethinking virtual network embedding: substrate support for path splitting and migration. *ACM SIGCOMM CCR*, 38(2), 2008.

[36] M. Zaharia, D. Borthakur, J. Sen Sarma, K. Elmeleegy, S. Shenker, and I. Stoica. Delay scheduling: a simple technique for achieving locality and fairness in cluster scheduling. In *Proc. of EuroSys*, 2010.

It's Not Easy Being Green

Peter Xiang Gao, Andrew R. Curtis, Bernard Wong, S. Keshav
Cheriton School of Computer Science
University of Waterloo

ABSTRACT

Large-scale Internet applications, such as content distribution networks, are deployed across multiple datacenters and consume massive amounts of electricity. To provide uniformly low access latencies, these datacenters are geographically distributed and the deployment size at each location reflects the regional demand for the application. Consequently, an application's environmental impact can vary significantly depending on the geographical distribution of end-users, as electricity cost and carbon footprint per watt is location specific. In this paper, we describe FORTE: Flow Optimization based framework for request-Routing and Traffic Engineering. FORTE dynamically controls the fraction of user traffic directed to each datacenter in response to changes in both request workload and carbon footprint. It allows an operator to navigate the three-way tradeoff between access latency, carbon footprint, and electricity costs and to determine an optimal datacenter upgrade plan in response to increases in traffic load. We use FORTE to show that carbon taxes or credits are impractical in incentivizing carbon output reduction by providers of large-scale Internet applications. However, they can reduce carbon emissions by 10% without increasing the mean latency nor the electricity bill.

Categories and Subject Descriptors.
C.2 [**Internetworking**]: Network Architecture and Design
General Terms. Design, Management, Performance
Keywords. Green computing, Energy

1. INTRODUCTION

Internet-scale applications, such as social networks, video distribution networks, and content distribution networks, provide service to hundreds of millions of end users. They achieve their enormous scale, while simultaneously reducing access latency, by routing service requests to a set of geographically distributed servers, typically located in datacenters. Such datacenters, which host up to 200,000 servers, are large-scale consumers of electricity, which is used not only for powering servers but also for keeping them cool. Currently, datacenters that power Internet-scale applications consume about 1.3% of the worldwide electricity supply and this fraction is expected to grow to 8% by 2020 [22]. Forrest, Kaplan, and Kindler in the McKinsey Quarterly, November 2008 showed data center carbon emissions were 0.6% of the global total, nearly those of the Netherlands. By 2020, the fraction is expected to reach 2.6%, exceeding the carbon emission of Germany. A single provider of Internet-scale services, Google, consumed 2.26×10^6MWh in 2010 [14]. In the United States, on average, generating a kWh of electricity emits about 500g of carbon [41], so Google's carbon emission in 2010 is equivalent to that emitted by 280,000 cars, assuming that each car runs 10,000 miles per year and emits 4 tons of carbon [1].

Given the rapid growth in the scale and usage of Internet-based applications, our primary goal is to help socially-aware companies reduce the carbon footprint of their infrastructure. We would also like to factor in the carbon cost of upgrades to this infrastructure. Importantly, we realize that carbon cost is only one factor in a complex decision process faced by application providers: what is needed is a way to navigate the three-way tradeoff between carbon footprint, electricity cost, and access latency.

This is a difficult problem. In addition to the inherent large scale, the carbon footprint of a datacenter varies both spatially and temporally, as does request load. Although requests can be easily routed to datacenters [33], optimal request-routing requires a complex joint optimization of both request-routing and data placement. Moreover, the continuing growth of request load implies that it is also necessary to periodically upgrade the infrastructure, keeping in mind both the available budget and the expected future workload. These inherent complexities have not been addressed in recent work, which focus essentially only on either the electricity cost [30,35,37] or the carbon footprint of datacenters [9,26,29].

In this paper, we introduce FORTE, a Flow Optimization based framework for Request-routing and Traffic Engineering and Fast-FORTE, a heuristic that closely approximates FORTE but runs about 20 times faster. FORTE takes a principled approach to the problem by using an objective function that balances the weighted sum of access latency, electricity costs, and carbon footprint. Our major contributions are:

- FORTE, a request-routing framework that provides a three-way tradeoff between access latency, electricity cost, and carbon footprint;
- Using FORTE to analyze the costs of carbon emission reduction for a large-scale Internet application; and,
- Using FORTE to determine a green datacenter upgrade and expansion plan for an Internet application.

Our approach is scalable and leads to three non-intuitive results. First, unless the price of carbon is set unrealistically high, application providers have no incentive to reduce their carbon footprint by more than a few percent. Second, by exploiting regional variations in carbon footprint and electricity costs, an application provider such as Akamai can reduce their carbon footprint by nearly 10% using FORTE, with no impact on their electricity cost or access latency. Third, FORTE can find green upgrade plans that can reduce the carbon footprint by over 25% over three years, compared to a carbon-oblivious upgrade plan, again, with no impact on electricity cost or a bounded increase in access latency.

SIGCOMM'12, August 13–17, 2012, Helsinki, Finland.

	Coal	Oil	Gas	Nuclear	Hydro	Other	Total
Elec.(TWh)	8,263	1,111	4,301	2,731	3,288	568	20,261
Proportion	41%	5%	21%	13%	16%	3%	100%

Table 1: Source of Electricity (World total, year 2008)

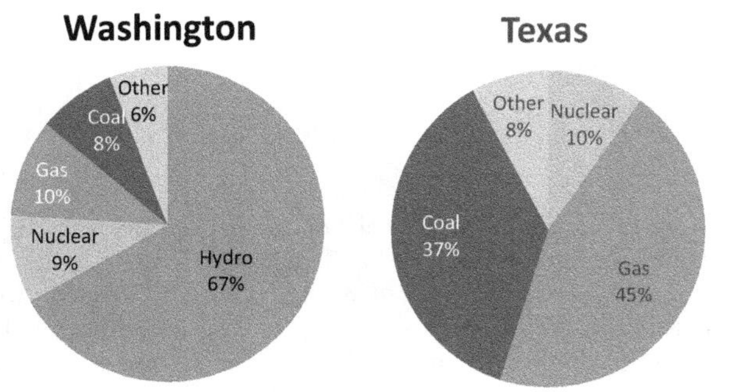

Figure 2: Generator fuel type in Washington and Texas

2. BACKGROUND

This section describes large-scale Internet services, the datacenters that host them, and the nature of electricity generation.

2.1 Datacenters and Request-Routing

We focus on large-scale Internet applications whose components run at multiple geographically distributed locations to provide scalability and reliability, and to reduce access latency (high access latency has been shown to have a negative economic impact [39]). For example, Google has more than 30 datacenters in at least 15 countries with an estimated 900K servers [31] and Akamai has more than 95,000 servers in nearly 1,900 networks in 71 countries [2].

Application providers usually place copies of data items such as video files or entire websites at one or more datacenters. Then, a request-routing system uses a set of metrics to direct end-hosts to datacenters that can best serve their requests. Current best practices for request-routing are *carbon oblivious*, that is, they ignore the amount of carbon emissions they generate. For example, a datacenter may be selected based on content availability, latency to the end-host, and load at that datacenter [33]. Our work modifies the request-routing algorithm to additionally take both the electricity cost and the carbon footprint of the datacenter into account.

2.2 Spatial and Temporal Variability in Electricity Carbon Footprint

Because of their scale, large-scale Internet services consume considerable amounts of electricity, which is delivered by the electricity grid. This grid inter-connects generator plants with consumers, including datacenters, in a geographical region. A region is usually served by several generators, which use different fuel types, such as gas, oil, coal, nuclear, and wind. Generation by burning fossil fuels emits much more carbon than generation with renewable energy, such as hydroelectricity, or nuclear plants. Two thirds of electricity is generated by fossil fuel today, as can be seen in Table 1, which shows fuel sources of electricity worldwide in 2008. However, there is significant difference among the fuel mix in different regions. A datacenter in Washington State uses cleaner electricity fuel sources than one in Texas. This is shown in Figure 2, which shows the fuel mix for these two states.

The carbon output of electricity generation varies temporally as well. The electricity grid has no storage, so the supply of electricity must match demand. Some generators are turned off in periods of

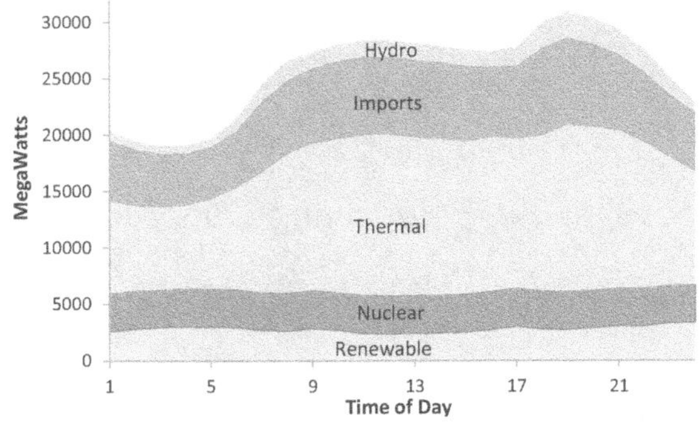

Figure 3: Hourly breakdown of total production by resource type, California, Jan 23rd 2012 [5]

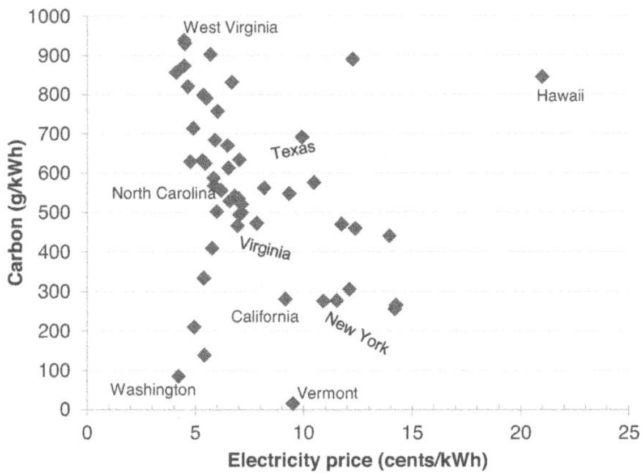

Figure 4: Grams of carbon per kWh vs. the cost of electricity in the 50 U.S. states.

low demand, and turned back on during peak hours. This is illustrated in Figure 3, which shows both the peak loads and the generation fuel mix in California on 23 January, 2012. The generators turned on during peak demand are usually thermal generators that emit more carbon. It is difficult to adjust the output of a nuclear plant, so they typically serve the base load.

There is no correlation between the "cleanness" of a region's electricity fuel mix and the price of electricity in the region. This is shown in Figure 4, which plots the cost and carbon output of electricity in each of the 50 U.S. states. The electricity in states like Washington is both clean and cheap, in contrast to states like West Virginia, where electricity is cheap but has a large carbon footprint. Consequently, schemes to reduce the electricity cost of powering an Internet service using request-routing (such as [35]) do not also reduce the carbon emissions of their operations.

2.3 Power Proportionality and PUE

Another way to reduce the carbon output of an Internet-scale application is to reduce the amount of energy its datacenters consume. There are two main ways to do this: power proportionality and lowering energy overheads. A *power-proportional* device uses power directly proportional to its utilization, e.g., if it is 10% utilized, then its power use will be 10% of its maximum power use [4]. A power-proportional datacenter can be built using power-proportional servers or by dynamically shutting off unnecessary equipment. Another approach to reducing datacenter power is to improve the effectiveness with which it uses electricity. This notion is captured by the *power usage effectiveness* (PUE) metric, which is defined as:

$$\text{PUE} = \frac{\text{Total IT equipment power}}{\text{Total facility power}} \quad (1)$$

The most efficient datacenters have a PUE of around 1.07–1.2 [15, 34], while the industry average is around 2.0 [10]. That means the average datacenter uses nearly 50% of its power on cooling and power transformation.

A power-proportional datacenter with a low PUE is clearly desirable. However, a datacenter's carbon emissions still depend on its regional electricity fuel mix. Therefore, a service operator that wants to minimize its carbon output should build and use datacenters in regions with green sources of electricity. In the remainder of this paper, we design algorithms to navigate the three-way tradeoff between carbon footprint, access latency, and electricity cost.

3. FORTE

The location and time-specific nature of carbon emission rates offers a currently untapped opportunity for Internet applications, deployed across multiple datacenters, to reduce their carbon footprint by directing traffic to cleaner locations. In this section, we describe FORTE, a Flow Optimization based framework for request Routing and Traffic Engineering, that offers a principled approach to assigning users and data objects to datacenters. FORTE performs user assignment by weighting each assignment's effect on three metrics: access latency, electricity cost, and carbon footprint. By making this three-way tradeoff explicit, FORTE enables Internet application providers to determine their optimal operating point that balances performance with cost and carbon footprint. Any application that uses dynamic request-routing can be easily modified to use FORTE to assign users to datacenters.

3.1 Model

Determining the impact of different user assignment decisions requires data on the location of users and datacenters, regional electricity costs, and the carbon emission per watt at each datacenter location. Additionally, one must also consider what data users are interested in and the location of the data. To encapsulate this information, we use a simple abstract model representing the relationship between the following three entities:

1. **User Group**: A user group is a collection of users who are geographically close or in the same autonomous system, for example: users from the same city or customers of a regional ISP.
2. **Datacenter**: A datacenter is a collection of servers that can serve user requests frequently. We use the term datacenter generically, and use it also to represent a cluster in a content delivery network (CDN).
3. **Data**: A data is an abstract object that users request, such as a video in an online video website. It can be a webpage, including all associated images and text files. Data can also be associated with a cloud service such as search or email for a cloud service provider. Note that we only model popular data, as the transfer of popular data accounts for most of Internet traffic.

The relationship between these three entities is illustrated in Figure 5. User groups indicate their desire for data through data requests, and these requests are served by any datacenter with a copy of the data. Note that content distribution systems are complicated, so our model necessarily makes some simplifying assumptions. However, we believe that this model largely captures the behavior of content distribution systems, and therefore our conclusions are robust to changes in the model's parameters.

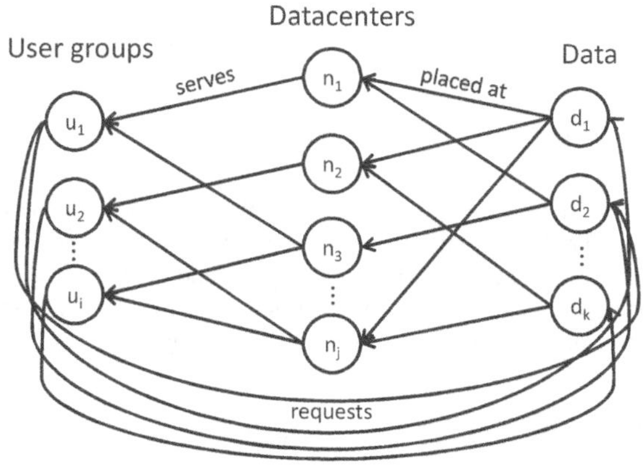

Figure 5: The relationship between user groups, datacenters, and data

We model the relationship between users and data as a minimum cost flow circulation problem. Here, the user group to datacenter assignment is re-evaluated periodically to adapt to changing request patterns. In addition to mapping user groups to datacenters, FORTE can also migrate or copy data to different datacenters with changes in the workload. The following is a complete list of the model parameters:

- $\mathbf{c(n_j)}$ is the average server carbon emission rate of the datacenter n_j.
- $\mathbf{e(n_j)}$ is the average server electricity cost of datacenter n_j.
- $\mathbf{f(u_i, n_j, d_k)}$ is the size of the flow from user group u_i, served by datacenter n_j, and requesting data d_k.
- $\mathbf{l(u_i, n_j, d_k)}$ is the average latency cost of the link between user group u_i and datacenter n_j for data d_k.
- $\mathbf{m(n_j)}$ is the number of servers running in datacenter n_j
- $\mathbf{p(n_j)}$ is the average processing power of servers in datacenter n_j. It is normalized to the number of requests it can process concurrently.
- $\mathbf{r(d_k, u_i)}$ is the number of users in user group u_i requesting data d_k.
- $\mathbf{rep(d_k, n_j)}$ equals $true$ if data d_k has a replica at data center n_j.
- $\mathbf{u(n_j)}$ is the number of servers in datacenter n_j.

Note that we model latency cost as a two part function $l(u_i, n_j, d_k)$. When latency is less than $l_{\max}$, the latency cost is linearly proportional to latency. Beyond $l_{\max}$, latency cost grows quadratically in order to model the user tendency to abandon Internet applications with perceptible interaction delays [39].

The FORTE model also differentiates between different data types. Specifically, it categorizes data as either throughput-sensitive or latency-sensitive, and enables different assignment decisions to be made based on the data category. For a data d_k that is throughput-sensitive, the latency cost will be 0 when access latency is less than $l_{\max}$. This significantly relaxes the constraint on feasible assignment decisions and enables throughput-sensitive traffic to be directed to any datacenter within the cut-off latency without paying latency costs.

3.2 Design

Our model captures the relationship between user groups, datacenters, and data, which enables FORTE to reason about the optimal mapping of users to datacenters *and* data to datacenters. The following sections describe FORTE's algorithms for optimally performing these two mappings. We also describe Fast-FORTE, an approximation algorithm that significantly improves upon the performance and scalability of FORTE while maintaining near-optimal assignments.

3.2.1 Assigning Users to Datacenters

The user to datacenter assignment problem, as we had previously mentioned in Section 3.1, can be modeled as a flow circulation problem. There are two main objectives when performing this assignment: all user requests must be satisfied and the weighted summation of access latency, carbon emission, and electricity cost must be minimized. We formulate this as a linear program:

$$\underset{f(u_i,n_j,d_k)}{\text{minimize:}} \quad (2)$$

$$\sum_{u_i,n_j,d_k} f(u_i,n_j,d_k)l(u_i,n_j,d_k)$$

$$+\lambda_1 \sum_{n_j} m(n_j)c(n_j) + \lambda_2 \sum_{n_j} m(n_j)e(n_j)$$

subject to:

$$\forall n_j, \sum_{u_i,d_k} f(u_i,n_j,d_k) \leq m(n_j)p(n_j) \quad (3)$$

$$\forall u_i, d_k, \sum_{n_j} f(u_i,n_j,d_k) = r(d_k,u_i) \quad (4)$$

$$\forall n_j, 0 \leq m(n_j) \leq u(n_j) \quad (5)$$

$$\forall u_i, n_j, d_k, f(u_i,n_j,d_k) \geq 0 \quad (6)$$

$$\forall u_i, n_j, d_k, \text{ s.t. } rep(n_j,d_k) = false, f(u_i,n_j,d_k) = 0 \quad (7)$$

This linear program minimizes its objective function, Equation 2, which represents the sum of access latency, electricity costs, and carbon emissions. The relative weight between the three metrics can be adjusted by changing the weight parameters λ_1 and λ_2. Equation 4 represents the constraint that all flows are served by some datacenters.

Additionally, we can easily incorporate several other constraints in this linear program to address pragmatic, deployment-related requirements. This includes: ensuring that the total load assigned to a datacenter is less than the datacenter's service capacity (Eq. 3), limiting the number of active servers in each datacenter (Eq. 5), and restricting users from requesting data d_k from datacenter n_j if a replica of d_k does not reside in n_j (Eq. 7). To ensure that the linear program return feasible results, we also restrict the flow and load balancing cost variables to be positive values (Eq. 6).

3.2.2 Assigning Data to Datacenters

In the previous section, we assumed that the data assignment is fixed and pre-determined, and use linear programming to solve the user assignment problem only. However, the data assignment is actually a parameter under our control, and in this section, we show how FORTE determines the optimal data assignment while assuming that each user has already been assigned to a datacenter.

As part of our data to datacenter problem formulation, we first define a temporary variable:

$$f(n_j,d_k) = \sum_{u_i} f(u_i,n_j,d_k) \quad (8)$$

which is the flow size of edge (n_j, d_k). Given this definition of edge flows size, we further define f_{th} as a percentile threshold across a set of flows.

From our analysis of Akamai traffic data (the dataset is described in Sec. 4.1), we found that flow sizes follow the Pareto Principle,

Flow size percentile	Percentage of total bytes
10%	90.998%
20%	97.941%
30%	99.520%

Table 6: Flow sizes follow the Pareto Principle for our Akamai workload.

Algorithm 1 Algorithm for determining the initial data placement.

Input: User Data Request Matrix $R(d_k, u_i)$
Input: Network Latency Cost Matrix $L(u_i, n_j)$
Input: Percentile Threshold f_{th}
Output: Need of Data Matrix $NEED(d_k, n_j)$
  Assume $rep(d_k, n_j) = true$ for all (d_k, n_j)
  Find out $F(u_i, n_j, d_k)$ by linear programming
  for each (n_j, d_k) **do**
    if $f(n_j, d_k)$ is a f_{th} percentile large flow **then**
      $need(n_j, d_k) = true$
    else
      $need(n_j, d_k) = false$
    end if
  end for

that is, a small percent of large flows account for a large percent of total bytes transferred. We see in Table 6 that 10% of the largest flows account for 90.998% of the total bytes transferred. Therefore, the size of a flow is the primary factor in determining data placement; we decide whether we need a replica of data d_k at datacenter n_j by the size of $f(n_j, d_k)$. More precisely, in our problem formulation, data d_k is needed by datacenter n_j if $f(n_j, d_k)$ is among the f_{th} percentile of large flows.

To solve the data assignment problem without regarding the user assignment as a priori, we use the following algorithm.

First, we perform user assignment using the linear programming approach we introduced in Section 3.2.1 with the added assumption that all data is available at every datacenter. Using this user assignment, we then determine whether data d_k is needed at datacenter n_j. Algorithm 1 shows the details of this approach[1].

However, this algorithm does not account for changes in user request patterns. Accommodating user assignment changes in response to request pattern changes is relatively simple and straightforward; users would be redirected to a different datacenter by changing request-routing. Unfortunately, changes in user assignment will lead to changes in data assignment, which potentially requires migrating a large volume of data. Such data migration is unreasonable if the data is only needed a small number of times before being migrated to another location. Therefore, we limit data migration frequency in our data assignment algorithm by first defining two additional matrices and parameters:

- *rep_ttl*(d_k, n_j) indicates the time to live of data replica d_k at data center n_j. *rep_ttl*$(d_k, n_j) = 0$ if the data d_k has no replica at data center n_j. *rep_ttl*$(d_k, n_j) > 0$ if there is a replica and the value of *rep_ttl*(d_k, n_j) indicates the time to live.
- *rep_counter*(d_k, n_j) is a counter that tracks how frequently $need(d_k, n_j) = true$ recently.
- *rep_initial_ttl* is the initial time to live.
- *rep_threshold* is a parameter deciding when to copy data. If *rep_counter*(d_k, n_j) > *rep_threshold*, data d_k will be actually copied to data center n_j.

[1]Please note that we use upper case letters to represent matrices and lower case letters to represent entries in a matrix.

Algorithm 2 Update Data Replication

```
Input: Need of Data Matrix NEED(d_k, n_j)
Input: Original TTL Matrix REP_TTL(d_k, n_j)
Input: Original Replication Counter Matrix REP_COUNTER(d_k, n_j)
Output: Updated Replication Status Matrix REP_TTL(d_k, n_j)
Output: Updated Replication Counter Matrix
  REP_COUNTER(d_k, n_j)
  for each (d_k, n_j) do
    if need(d_k, n_j) = true then
      if rep_ttl(d_k, n_j) = 0 then
        rep_counter(d_k, n_j) + +
        if rep_counter(d_k, n_j) = rep_threshold then
          Copy data d_k to data center n_j
          rep_ttl(d_k, n_j) = rep_initial_ttl
        end if
      else
        rep_ttl(d_k, n_j) = rep_initial_ttl
      end if
    else
      rep_ttl(d_k, n_j) − −
      if rep_counter(d_k, n_j) > 0 then
        rep_counter(d_k, n_j) − −
      end if
      if rep_ttl(d_k, n_j) = 0 then
        Remove data d_k from data center n_j
        rep_counter(d_k, n_j) = 0
      end if
    end if
  end for
```

At every time step, we obtain the updated user data request matrix $R(d_k, u_i)$ and determine the $NEED(n_j, d_k)$ matrix by Algorithm 1. We increment $rep\_counter(d_k, n_j)$ by 1 if data d_k has no replica at data center n_j and $need(n_j, d_k) = true$. If the counter reaches $rep\_threshold$, we replicate the data d_k to data center n_j. The $rep\_ttl(d_k, n_j)$ will decrease one if $need(n_j, d_k) = false$ at this time step. This process is described in Algorithm 2. By using Algorithm 2, we guarantee that only data replicas that tend to be used frequently will be replicated and only data that is infrequently used for a long time will be removed. This reduces unnecessary data migration across the datacenters.

3.2.3 *Fast-FORTE*

FORTE optimally assigns users to datacenters to minimize the cost function. However, its reliance on linear programming limits its scalability. In our experiments, FORTE's optimizer takes 2 minutes to run on a standard personal computer using a partial trace of Akamai traffic data as input. Because measurements are taken every 5 minutes, this allows us to essentially solve the assignment problem in real time. However, this is unlikely to be fast enough for a complete CDN network, given the rapid growth rates of Internet services. Hence, the number of data objects and user groups in the user-datacenter-data model (as shown in Figure 5) may be much larger than those in our simulations. For these scenarios, FORTE can be very slow as the complexity of the Simplex method grows quadratically. We therefore designed an alternative solution, called Fast-FORTE, that uses the same model as FORTE but replaces the linear programming flow solver with a heuristic algorithm based on entropy. Before describing the algorithm in detail, we first define the aggregate cost of a link between user group u_i and datacenter n_j as:

$$a\_cost(u_i, n_j) = l(u_i, n_j) + \lambda_1 \frac{c(n_j)}{p(n_j)} + \lambda_2 \frac{e(n_j)}{p(n_j)} \tag{9}$$

which is the cost to route user u_i to datacenter n_j. We also define

Algorithm 3 Fast-FORTE User Assignment

```
Input: User Data Request Matrix R(d_k, u_i)
Input: Network Latency Cost Matrix L(u_i, n_j)
Input: Data Replication Matrix REP(d_k, n_j)
Output: Flow Assignment Matrix F(u_i, n_j, d_k)
  Set all f(u_i, n_j, d_k) = 0
  Calculate A_COST(u_i, n_j, d_k)
  Calculate E(d_k)
  Sort A_COST(u_i, n_j, d_k) in ascending order and put them in queue
  Sort E(d_k) in descending order
  while Queue of A_COST(u_i, n_j, d_k) not empty do
    pop out link (u_i, n_j)
    put E(d_k) in queue
    while Queue of E(d_k) not empty do
      Pop out data d_k
      if rep(n_j, d_k) == true then
        if remaining(n_j) ≥ r(d_k, u_i) then
          f(u_i, n_j, d_k) = r(d_k, u_i)
        else
          f(u_i, n_j, d_k) = remaining(n_j)
        end if
      end if
    end while
  end while
  Output flow assignment
```

the weighted entropy of data d_k

$$\begin{aligned} &E(d_k) \\ =&P(d_k)H(U|D = d_k) \\ =&-P(d_k) \sum_{u_i \in U} P(u_i|d_k) \log P(u_i|d_k) \end{aligned} \tag{10}$$

where $P(u_i|d_k)$ is the probability that, given a request is asking for data d_k, it comes from user group u_i. The weight $P(d_k)$ is the probability that a request is asking for data d_k. The entropy value $H(U|D = d_k)$ measures the uncertainty of the requests. When the entropy reaches its maximum, we are uncertain about the location of requests as the location probability of each request is evenly distributed.

The Fast-FORTE algorithm is as follows. First, links between u_i and n_j are sorted by their aggregate cost $a\_cost(u_i, n_j)$. Data objects are also sorted by their weighted entropy $E(d_k)$. At each step, we select the link with the minimum aggregate cost and remove it from the graph. We then attempt to fill this link with data requests until we have exhausted all data requests. As we fill the link with data requests, we also attempt to fulfill requests asking for data with larger weighted entropy. The details of the algorithm is described in Algorithm 3. We use $remaining(n_j)$ to keep track the remaining capacity of datacenter n_j.

We evaluated Fast-FORTE by using the same experimental data we used for evaluating FORTE (described in Sec. 4.1). Fast-FORTE finds a solution in approximately 6 seconds compared to 2 minutes for FORTE. To estimate the quality of approximation, we calculate the approximation ratio:

$$Ratio = \frac{\text{Objective in Equation 2 by Approximation}}{\text{Objective in Equation 2 by LP}}$$

The average of approximation ratio is 1.00320 with a standard deviation 0.00558. Among the data we tested, the approximation ratio is less than 1.03 for 99.5% of the data points.

3.3 Using FORTE for Upgrading Datacenters

Determining a user and data assignment strategy that strikes a good balance between latency, electricity cost, and carbon footprint only addresses half of the assignment problem. To datacenter operators, determining how to upgrade their datacenters in response to growth is equally important, as Internet traffic has maintained an explosive growth rate of 40 percent per year in the past decade (see, e.g., [25]). Specifically, datacenter operators need to determine which datacenters should be upgraded, and how many servers within each datacenter should be upgraded. Simply upgrading the greenest datacenters first is not optimal, as it does not properly weigh the three-way tradeoff. For example, upgrading a green datacenter in a remote area will provide minimal benefits as, due to the high latency to end-users, most traffic will be directed away from it. In this section, we extend FORTE to help decision makers determine the optimal upgrade plan. The objective of upgrading is to ensure that an application service provider can meet user traffic demands while minimizing electricity cost and carbon emissions. This is made easier by assuming the continued adherence to Moore's law, which allows us to assume that servers in datacenter can be replaced with new models that consume the same amount of energy but provide twice the computing power. In this modified model, we must introduce three new variables:

- $\mathbf{a(n_j)}$ is the amortized purchasing cost of upgrading a server at datacenter n_j. The one time upgrading investment on the server is divided by its service period. e.g.: 5 years.
- $\mathbf{budget}$ is the maximum number of server that are allowed to upgrade by the budget.
- $\mathbf{up(n_j)}$ is the percentage of server that are upgraded.

The new model includes the cost and benefit of upgrading a datacenter. If the benefit of upgrading, such as reduction in latency, offsets the amortized purchasing cost $a(n_j)$, the datacenter will be upgraded. The following is the linear program formulation for solving the upgrade problem. The parts that differ from our previous user assignment linear program are marked in bold font.

$$\underset{f(u_i,n_j,d_k),up(n_j)}{\text{minimize:}} \tag{11}$$

$$\sum_{u_i,n_j,d_k} f(u_i,n_j,d_k)l(u_i,n_j,d_k)$$

$$+\lambda_1\sum_{n_j} m(n_j)c(n_j) + \lambda_2\sum_{n_j} m(n_j)e(n_j)$$

$$-\lambda_1\sum_{\mathbf{n_j}} \mathbf{u(n_j)up(n_j)c(n_j)} - \lambda_2\sum_{\mathbf{n_j}} \mathbf{u(n_j)up(n_j)e(n_j)}$$

$$+\lambda_\mathbf{3}\sum_{\mathbf{n_j}} \mathbf{u(n_j)up(n_j)a(n_j)}$$

subject to:

$$\forall n_j, \sum_{u_i,d_k} f(u_i,n_j,d_k) \le m(n_j)p(n_j) \tag{12}$$

$$\forall u_i, d_k, \sum_{n_j} f(u_i,n_j,d_k) = r(d_k,u_i) \tag{13}$$

$$\forall \mathbf{n_j}, \mathbf{0} \le \mathbf{m(n_j)} \le \mathbf{u(n_j)(1+up(n_j))} \tag{14}$$

$$\forall u_i, n_j, d_k, f(u_i,n_j,d_k) \ge 0 \tag{15}$$

$$\forall u_i, n_j, d_k, \text{ s.t. } rep(n_j,d_k) = false, f(u_i,n_j,d_k) = 0 \tag{16}$$

$$\forall \mathbf{n_j}, \mathbf{0} \le \mathbf{up(n_j)} \le \mathbf{1} \tag{17}$$

$$\sum_{\mathbf{n_j}} \mathbf{u(n_j)up(n_j)} \le \mathbf{budget} \tag{18}$$

We highlight four equations that are changed in this model:

Equation 11: The new expression adds the total amortization cost

$$\sum_{n_j} u(n_j)up(n_j)a(n_j)$$

into the objective. If the amortized purchasing cost of adding a server is smaller than the extra latency cost to direct further away traffic to this datacenter, it will then be worthwhile to add a new server. Since the new model consumes the same amount of energy as the original one, upgrading a server will not lead to extra electricity cost nor carbon emission. The term

$$-\lambda_1\sum_{n_j} u(n_j)up(n_j)c(n_j) - \lambda_2\sum_{n_j} u(n_j)up(n_j)e(n_j)$$

removes the extra carbon counted by

$$+\lambda_1\sum_{n_j} m(n_j)c(n_j) + \lambda_2\sum_{n_j} m(n_j)e(n_j)$$

when $up(n_j) > 0$.

Equation 14: Note that in the new model, $m(n_j)$ is the effective computing power rather than the number of running servers. Therefore, it is possible that $m(n_j) > u(n_j)$. When 20% of servers in datacenter n_j are upgraded ($up(n_j) = 0.2$), the effective computing power $m(n_j) = 1.2 \times u(n_j)$, as 20% of the servers double their computing power. $up(n_j)$ can be viewed as a "slack variable" to $m(n_j)$. The value of the slack variable $up(n_j)$ determines the amount of growth $m(n_j)$ that is needed to reduce the total cost (Equation 11). In other words, the value of $up(n_j)$ is the optimal upgrading percentage to minimize cost.

Equation 17,18: These two constraints limit the range of upgrading percentages and the total number of servers to upgrade.

Linear programming will find the trade-off between the cost and benefit of adding new servers at each location. The output of $up(n_j)$ indicates the percentage of server that needs to be upgraded in datacenter n_j. The input traffic matrix should be the average value of a relatively long period, say a month, to give stable results.

4. RESULTS

We now evaluate the carbon emission reductions that are possible using FORTE. Throughout our evaluation, we use the Akamai content distribution network (CDN) as a case study. We begin by describing our simulation methodology and the Akamai dataset; we then find the maximum amount of carbon emission reductions possible before evaluating the three-way tradeoff between carbon emissions, latency, and electricity costs. Finally, we give the results of using FORTE to upgrade datacenters.

4.1 Methodology

We implemented a custom discrete time simulator to evaluate FORTE, using a 24-day workload from Akamai. In the following sections, we describe Akamai's infrastructure and this dataset, how we estimate the carbon emissions of each datacenter, and our approximation of latency between a user and datacenters.

4.1.1 Akamai CDN

We use Akamai as a case study to evaluate the carbon emission reduction potential of an Internet-scale system. Akamai is the largest content delivery network in the world. It delivers 15–20 percent of worldwide Internet traffic [33] and it has more than 95,000 servers in 1,900 networks across 71 countries [2].

The Akamai workload we use as a case study was collected from 18 December 2008 through 11 January 2009, and was collected from a subset of their datacenters. The datacenters included are Akamai's public datacenters, which are generally located inside of co-location centers. As opposed to Akamai's private CDN nodes, these public datacenters can serve requests for any client. The dataset does not include logs for private Akamai datacenters. These are typically located within a university or enterprise network. The workload contains logs for users and CDN datacenters worldwide; however, the information given on users outside of North America is not detailed enough for our purposes, so we omit it from our simulations.

The dataset consists of measurements taken over a 5 minute interval. Each measurement contains the geographical origin of user requests, the datacenter that served the requests, and the aggregate size of the requests. We use FORTE to compute an updated user-to-datacenter assignment for each measurement. Then, this assignment is used for the duration of the 5 minute interval represented by the measurement. Because these logs contain Akamai's actual assignment of users to datacenters, we can compare FORTE to Akamai's proprietary request-routing scheme, even though we have no details about their scheme. In Section 4.2, we show that FORTE can closely approximate this scheme.

The logs include the load factor for each datacenter. This is estimated by the combination of CPU, memory, and network utilizations. We use the load factor together with the user request logs to estimate the capacity of each datacenter.

The user request logs contain only aggregate statistics, and do not provide us with the type or name of content requested by the user. Akamai handles interactive requests, which are latency-sensitive, as well as streaming requests, which are throughput-sensitive. To account for this, we assume that 90% of user requests are latency-sensitive, and that the remaining 10% of requests are not sensitive to latency. Recall that FORTE models user access latency costs for latency-sensitive traffic with a two-part function. For distances less than $l_{\max}$, the user access latency cost is linearly proportional to latency. Beyond $l_{\max}$, latency cost grows quadratically in order to discourage FORTE from routing users to far away datacenters. For our Akamai case study, we set $l_{\max}$ equal to 2,000 km which is half of the distance across the length of the U.S.. FORTE does not include access latency in its objective function for throughput-sensitive traffic, so these users are routed to datacenters with low carbon footprints, regardless of their distance.

4.1.2 Power proportionality and bandwidth

For all simulations, we assume that the datacenters are fully power-proportional . Most datacenters today are not power-proportional, though the industry is heading that direction [4]. One challenge here is that servers are not power-proportional. However, a datacenter can be made power-proportional by shutting off unused servers [28] and switches [18]. There are also proposals for power-proportional storage systems [3, 24]. By assuming that datacenters are fully power-proportional, we find an upper bound on the amount of carbon emission reductions possible with FORTE.

Further, we assume that bandwidth cost increases are insignificant. FORTE does not increase the number of user requests, and its optimization procedure minimizes the number of times a data item must be retrieved. However, because it routes users to minimize carbon emission, it is possible that it increases bandwidth costs. Currently, bandwidth costs are based on the 95/5 bandwidth pricing scheme, which divides traffic into 5 minutes intervals and uses the 95th percentile for pricing. Because bandwidth prices are not geographically differentiated for Akamai [35], re-routing traffic to another datacenter does not increase the overall bandwidth cost. However, affecting the 95th percentiles does. In our simulations,

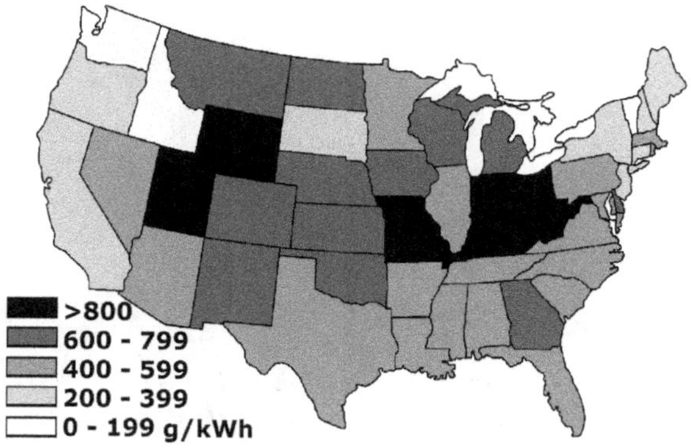

Figure 7: Carbon emission rate of each state in U.S. The average carbon emission rate of U.S. is 562g/kWh. Most of Akamai datacenters are in states with carbon emission rate between 200-599g/kWh.

	Nuclear	Coal	Gas	Oil	Hydro	Wind
$CO_2\ g/kWh$	15	968	440	890	13.5	22.5

Table 9: Carbon dioxide emission per kilowatt-hour for the most common fuel types [19].

we found that the sum of the 95/5 values for each datacenter using the request-routing assignments found by FORTE was 97% of the sum of 95/5 values found by Akamai's scheme. This indicates that FORTE does not significantly affect the 95th percentiles. Furthermore, bandwidth is relatively inexpensive today, with network bandwidth expenses only contributing a couple of percent to the total cost of operating a datacenter [16]. Finally, if bandwidth costs are a concern, then they can easily be incorporated into FORTE's optimization constraints.

4.1.3 Carbon footprint data

To estimate the carbon emission rate of each datacenter, we use electricity generation data from the U.S. Energy Information Administration's website [41]. They report the average electricity fuel mix of all the states in U.S. for the seven major types of fuel. Then, the average carbon emission rate of a state is found by summing the weighted contribution from each fuel type. This is defined as:

$$\text{state's avg. carbon emission rate} = \frac{\sum e_i \times r_i}{\sum e_i} \tag{19}$$

where e_i is the electricity generated from fuel type i and r_i is the carbon emission rate of fuel type i. The carbon emission rate of the most common fuel types is shown in Table 9. Of course, data at a finer grain than state-level would improve the fidelity of our results. However, as carbon emission rates are often governed by state-regulated legislation and power-generation, we believe that state-level carbon emission data is sufficiently fine grain for our model.

Figure 7 shows our estimate of the carbon emissions caused by electricity generation in each state. The carbon emissions of locations with an Akamai datacenters is summarized in Figure 8. We see that, despite not directly optimizing for carbon emissions, Akamai's datacenters are typically in low-polluting locations, and most of its datacenters are located in states with carbon emission rate less than U.S. average (562 g/kWh).

As described in Section 2.2, the carbon emission rate of electricity generation varies with both location and time, because the fuel mix changes with the turning on/off of peak plants. We do not have hourly fuel mix data of states in the U.S.. However, we know that most of peak plants are powered by oil and gas [8], so peak plants emit more carbon than base load plants. We estimated the daily

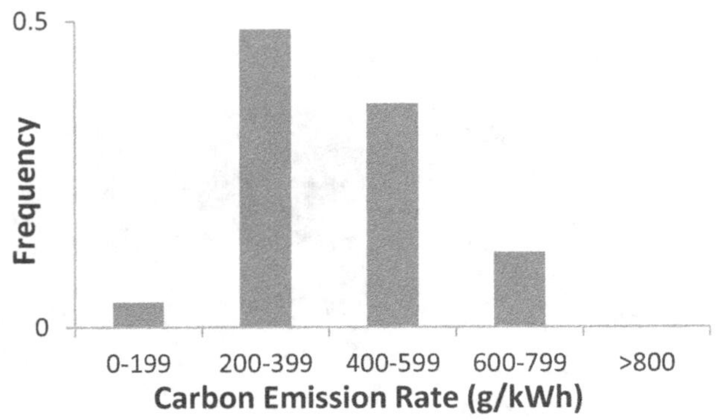

Figure 8: Distribution of the carbon emissions of Akamai datacenters.

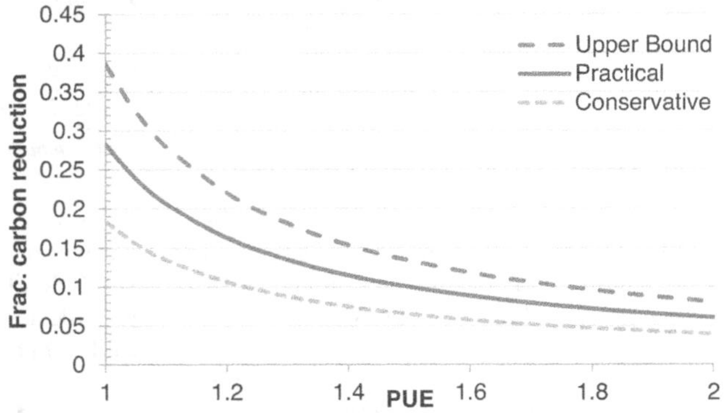

Figure 10: Percentage of carbon emission reduction with various PUEs and different schemes. The practical scheme increases the average user distance and the conservative scheme fixes both distance and electricity cost to their respective values under Akamai's routing scheme.

change on the carbon emission rate of each state by the change of total load and composition of fuel source. On average, the maximum carbon emission hour for a state emits 63% more carbon than its daily average.

4.1.4 Approximating latency with distance

The Akamai dataset does not include latency measurements, and latency estimation is a challenging problem. Therefore, we approximate access latency with geographical distance in our simulations. Although distance is not a perfect estimator of network latency, we believe that it is sufficiently accurate to determine the relative rank in latency from datacenters to each end-user. Therefore, for the remainder of this section, we use geographical distance as an approximation for latency. In practice, CDN providers have accurate measurements on network latency, so they can use their datasets to improve latency estimation.

4.2 Carbon Emission Reduction

We begin by determining the maximum amount of carbon reduction that is possible for Akamai. The reductions possible depends on the PUE of the datacenters in the distribution network, because the carbon output per unit of work is lower in datacenters with a lower PUE. We found the upper bound on carbon reduction possible as the PUE of the datacenters varies, assuming all datacenters have the same PUE. The results are shown in Figure 10. Here, carbon emission reduction is the main objective of FORTE's optimization—electricity costs and latency have a relatively smaller weight. Therefore, we can estimate the upper bound of carbon reduction for different PUEs.

As expected, as datacenter efficiency increases, FORTE achieves greater carbon emission reductions. For inefficient datacenters with a PUE of 2, we found that FORTE can reduce carbon emissions by 8%, while carbon emissions can be reduced over 38% if the datacenters have a PUE of 1. This graph demonstrates the incentive to further reducing PUE for a system whose current PUE is small. We do not have the PUE of the datacenters in the Akamai dataset; however, the industry average PUE was 2.0 in 2007 [10] and was 1.91 in 2009 [40]. The best datacenters have a PUE as low as 1.07 [13,34].

Figure 10 also shows the sensitivity of FORTE to datacenter PUE. We vary the PUE of the CDN datacenters and found the carbon reduction achieved by two schemes, a practical scheme and a conservative scheme, as well as the upper bound on carbon reduction. For the practical scheme, we fixed the electricity cost at the current Akamai level. We found that this scheme increased the average distance from 745 km to 1000 km; however, FORTE does not minimize distance for throughput-sensitive traffic, so this traffic disproportionally affects the average distance. The conservative scheme fixes both distance and electricity cost to the current Akamai levels. From the figure, we find that PUEs of less than 1.5 is the sensitive region, which motivates datacenter owners to further reduce their PUE. For the remainder of our simulations, we assume that all datacenters have a PUE of 1.2. In the 2007 E.P.A. report [10], they indicate that best-practices yield datacenters with PUEs of 1.3. Therefore, we believe that a 1.2 PUE is a reasonable estimate for a modern best-practice datacenter.

Next, we evaluated the effects of changing the weight of carbon emissions in FORTE's objective function. That is, we evaluate the effect of changing λ_1 on carbon emissions. We used four values for λ_1 and fixed all other parameters. The results are shown in Figure 11. The chart shows carbon emission levels, normalized to Akamai's carbon emissions, for the duration of the Akamai trace. We see that even if λ_1 is set to a very small value, we have around 4% carbon emission reduction. This is because users can be assigned to cleaner datacenters within a similar distance, so carbon emission reductions are found without increasing latency. When λ_1 is set to a large value, carbon emissions can be reduced up to 20% on average.

We also observed that:

- The major difference in carbon reduction between λ_1 = small and λ_1 = medium is at non-peak hours.
- The major difference in carbon reduction between λ_1 = medium and λ_1 = large is at peak hours.

This shows that when the load is low, FORTE has more freedom to assign users to cleaner datacenters. Therefore, as presented in Section 3.3, FORTE indicates where are the desirable places for datacenter upgrading. The result of datacenter upgrading is demonstrated in Section 4.4.

4.3 Costs of Carbon Reduction

We have shown that FORTE can reduce an Internet application's carbon emissions; however, these reductions may come at the price of increased access latency and electricity cost. To explore the tradeoffs an operator can make, we show the effects of varying distance and electricity costs on carbon emissions in Figure 12.

We find that we can reduce carbon emissions caused by Akamai's datacenters by 10% without increasing latency nor electricity cost from their existing levels. When carbon emissions are at 81% of the current Akamai level, the electricity cost reduces with the reduction of carbon emissions. This is because for some regions, carbon emission is positively correlated with electricity prices. In this situation, assigning users to datacenters powered by cleaner electricity will reduce both carbon emissions and electricity cost.

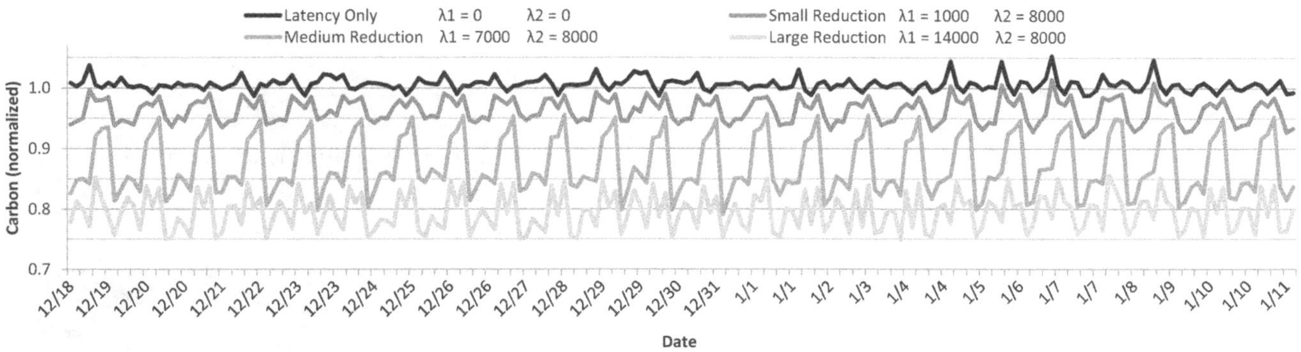

Figure 11: Carbon emissions with various weights of carbon in the FORTE objective function. The emissions shown are normalized against the carbon emissions by carbon-oblivious request-routing. The "latency only" one is our simulation on Akamai request-routing, which is very close to Akamai's emission level.

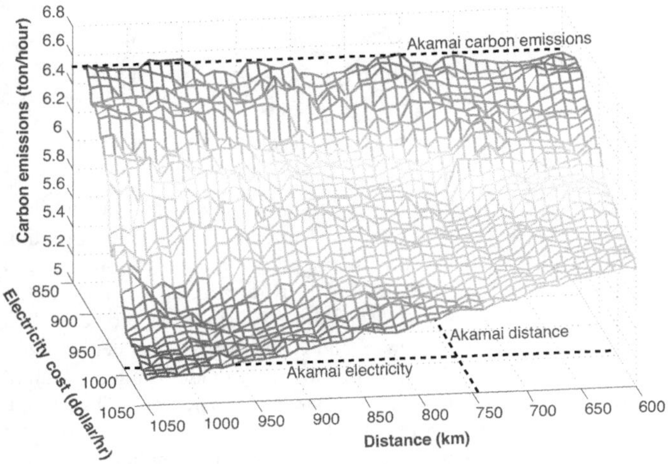

Figure 12: Tradeoff between carbon emissions, average distance, and electricity costs.

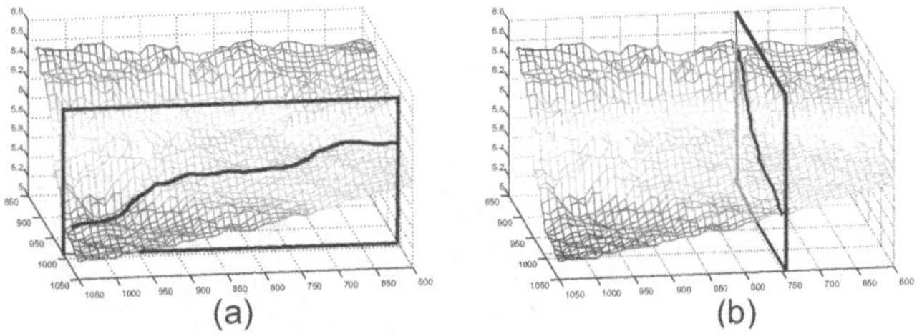

Figure 13: Locations of the slices plotted in Figures 14 and 15.

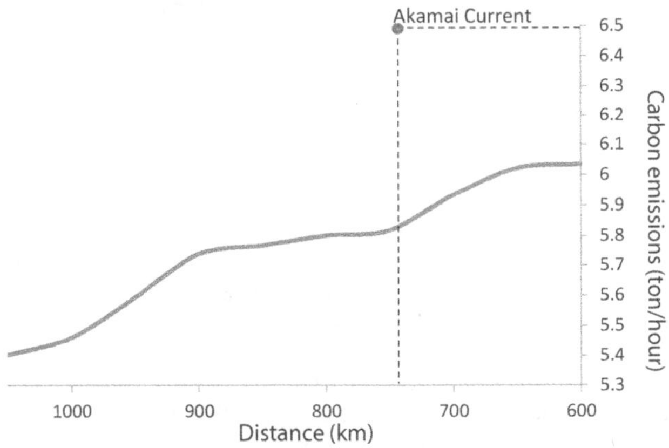

Figure 14: Carbon emissions vs. average user distance when electricity costs are fixed to Akamai's current costs.

To better understand the 3D surface plotting carbon emissions, electricity cost, and distance, shown in Figure 12, Figures 14 and 15 show the results of fixing electricity cost to the current Akamai cost and of fixing distance to the current Akamai distance respectively. Figure 13 shows the locations of these fixed points.

Figure 14 shows carbon emission vs. distance when we fix the electricity cost to Akamai's current electricity cost. We can reduce carbon emission by 10% without affecting distance. Electricity cost reductions beyond 20% increase distance linearly. After this point, further carbon emission reductions are cost prohibitive.

Figure 15 shows carbon emission vs. electricity cost when the average distance is fixed to Akamai's current distance. We observe that FORTE reduces carbon emissions by nearly 10% without increasing electricity cost. However, after this point, further carbon reduction is very expensive, as the reduction curve flattens out. To reduce the carbon emission by 0.1 ton per hour, the extra electricity cost is $23. The cost of unit carbon reduction is $230/ton, which is significantly higher than the current carbon tax ($5/ton to $39/ton [21]). Hence, paying the carbon tax is less expensive than reducing carbon emission.

4.4 Upgrading Datacenters

To evaluate FORTE's datacenter upgrade algorithm, we again use Akamai as a case study; however, we now assume that their traffic growth rate is 40% per year. (This growth rate is consistent with recent measurements [25].) We assume that each year, the datacenter operator has a budget large enough to increase the total number of servers by 15%. To simplify our analysis, we assume that upgrades are performed in June each year.

Each month's traffic estimation is used by FORTE to find the upgrading plan for each month. Let $up_{y,m}(n_j)$ denote the percentage of servers to be upgraded in datacenter n_j at month m of year y. We use the 12 month running average of $up_{y,m}(n_j)$ to calculate the upgrading plan for year y. Mathematically, we upgrade datacenter n_j in year y by:

$$up_y(n_j) = \alpha \sum_{m=1}^{12} (1-\alpha)^{12-m} up_{y,m}(n_j) \tag{20}$$

where α is a parameter controlling the weight of recent months.

We compared FORTE's upgrading algorithm with a carbon oblivious uniform upgrading scheme, which adds the same percentage of server each month. The total number of servers added for FORTE and the carbon oblivious scheme are the same. Figure 16 shows the carbon emissions of both schemes over a 36 month period. Up-

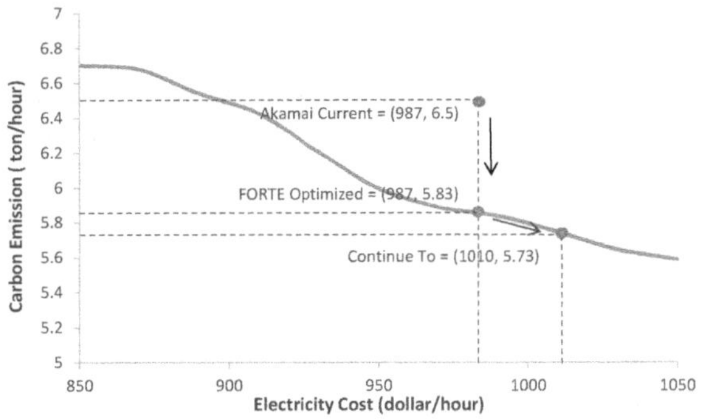

Figure 15: Carbon emissions vs. electricity costs when average distance is fixed at current Akamai levels.

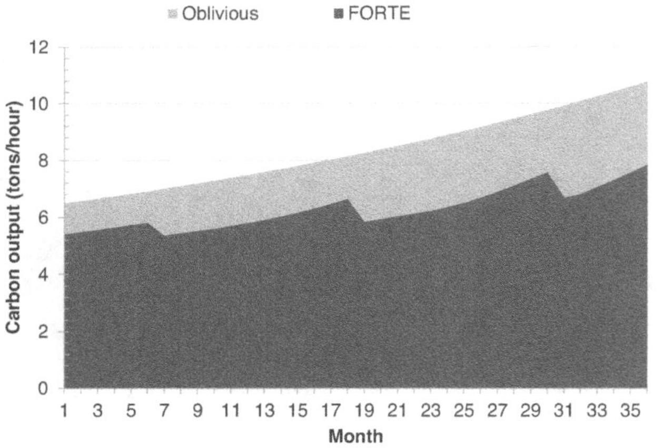

Figure 16: Carbon emissions over time in the datacenter upgrade scenario. FORTE performs upgrades every June, hence the drop in carbon emissions every 12 months.

grades take place at months 7, 19, and 31, which can be seen as dips in the carbon emission. The actual percentage of servers purchased each year is shown in Table 17. The carbon emission growth rate is also lower using FORTE's algorithm. At month 1, FORTE's solution has 83.22% of the carbon emission that the oblivious uniform upgrading scheme's solution has. At the end of month 36, FORTE's carbon emission is 72.80% of the carbon oblivious one. Therefore, FORTE lowers the carbon emission growth rate.

The datacenters upgraded are primarily near the most densely populated regions in United States such as the west coast, the northeast and the south. We selected some typical upgraded datacenters and plot the percentage of servers upgraded each year in Figure 18. When we compare Figure 18 with Figure 7, we see that datacenters selected for upgrades are primarily from the states with cleaner electricity generation. The state of Washington, California, New York, and New Jersey all have carbon emission rates less than 400 g/kWh. Although Texas has a emission rate of 562 g/kWh, it is a relatively clean state in the southern U.S., so upgrading is beneficial because it reduces latency for users in the southern part of the country.

Month	Upgrading Percentage
6	5.365%
18	9.872%
30	10.107%

Table 17: Percentage of servers upgraded among all servers

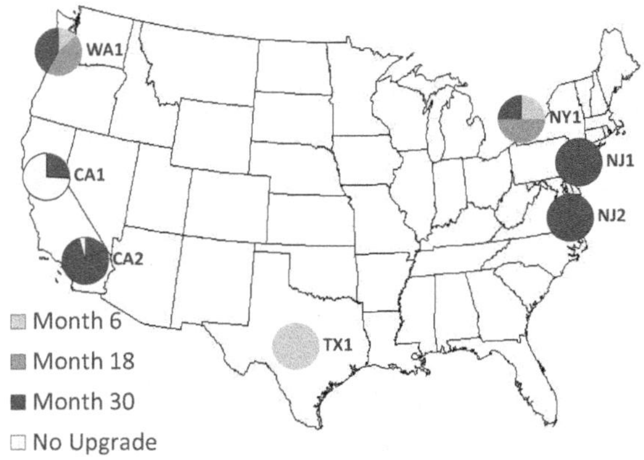

Figure 18: Three year upgrade plan for selected datacenters

5. DISCUSSION

Impact on application service providers: Application service providers fall into three broad classes. The first class, like Amazon, Google, Facebook, and Microsoft, are vertically integrated and own both the application as well as the underlying request-routing framework and datacenters. Such providers can use FORTE as-is to optimize their request-routing algorithms and datacenter upgrades. The second class of service providers, like Amazon Cloudfront, Microsoft Azure, AT&T, and Telefonica, own and operate datacenters that host applications and also provide a request-routing framework, but do not own the applications that they host. Such providers can also use FORTE as-is by monitoring the usage of the applications that they host. The third class of service providers, like Akamai, Limelight, Cotendo, and CDNetworks, own neither datacenters nor applications, and primarily provide a request-routing service and access to servers collocated at datacenters around the world. Although such providers can use FORTE to modify their request-routing algorithms, our approach is less applicable to them because they do not actually pay for energy but the capacity of power provisioned to the servers (a fact also noted by [35]). However, they can still use FORTE to decide which collocation sites to upgrade in an effort to minimize their overall carbon footprint.

Will carbon taxes or credits work? The price of electricity today does not reflect externalities such as its carbon footprint. It has been suggested[2] that this can be corrected by a carbon tax (or its converse, a carbon credit for green generators). Can these affect the behavior of Internet-scale application providers?

We believe that, at least for the near future, the answer is negative, because the carbon cost of a server is under 2% of the electricity cost. To see this, consider a typical server that uses approximately 250 W = 0.25 kW [23] in a region where the electricity cost is 11.2 c/kWh, the US average [41]. The hourly electricity cost of operating the server would be 0.25 * 0.112 = $0.28/hour. Recall that the mean carbon footprint of electricity in the US is 562g/kWh. The server's carbon emissions are therefore 0.25*562g = 140.5g/hour. The US currently does not impose a carbon tax. However, the European Union suggests a carbon tax of $5 to $39 per 1000 kg = $0.000039/g [21]. Even at the high end of this scale, the carbon cost of the server would be only $0.00548/hour, which is less than 2% of the electricity cost. Given this disparity, unless carbon taxes are at least an order of magnitude larger, which is unlikely for the next few years, they will have little effect on application service providers.

We believe, instead, that carbon footprint reductions can be achieved using a combination of intelligent request-routing and non-cash incentives. Section 4 demonstrates that, using FORTE, a service

[2]http://www.carbontax.org/

provider can reduce its carbon footprint by up to 10% with no impact on its electricity cost or mean access latency. The increased electricity cost of additional reductions could be balanced by the goodwill this can generate with the general public: a cost, like advertising, that corporations are often willing to bear.

A principled approach to datacenter location: FORTE provides a principled approach to decide where to build new datacenters and how large these should be. A service provider indicates candidate locations to FORTE by placing zero-size data centers in these locations and then running the upgrade algorithm presented in Section 3.3. An upgrade to a datacenter with zero servers to a datacenter with n servers is readily interpreted as a selection of this location for a new datacenter with n servers. In reality, location decisions are often heavily influenced by various incentives offered by authorities trying to attract datacenters to their location. It is straightforward to reflect these incentives in the objective function by adjusting the cost to build a datacenter in a given location.

Datacenter emergy: Electricity consumption is only part of an Internet service's carbon footprint. A more complete life-cycle assessment should incorporate additional environmental impacts, such as the energy and materials used to produce the datacenters it is deployed on. *Emergy* is the amount of energy used to manufacture a product, that is, it is the embodied energy of the product. The emergy of a product can be significant; for example, a study from 2004 estimated production to account for 81% of a computer and display's total life-cycle energy use [43]. Additionally, Raghavan and Ma estimate that embodied power is 53% of the Internet's total power use [36]. The relative contribution of emergy to a server's life-cycle energy use is less: it accounts for at most 21% to 10% of its total energy use, assuming an average power of 250 W to 500 W and a three-year lifespan and that a typical server has 5 GJ of emergy [6]. Because a datacenter is not just servers, the emergy of a datacenter may be considerably higher. Therefore, estimating the relative importance of datacenter emergy is an interesting direction for future work.

6. RELATED WORK

Request-routing and electricity: There has been considerable interest in reducing the electricity consumption of datacenters and networks in recent years. Qureshi et al. proposed a request-routing scheme to minimize the electricity bill of multi-datacenter systems [35]. They showed that this sort of optimization can reduce the electricity bill by 13% when assuming power-proportional datacenters. Later work, such as [30, 37], has improved the algorithms of Qureshi et al.. Other proposals consider how request-routing can be used to help with load balancing in the electric grid [32].

Recently, there has been work investigating modifying request-routing to reduce carbon emissions. Le et al. [26] considered the joint optimization problem of minimizing carbon emission and electricity cost. Their user assignment algorithm does not attempt to minimize the distance between users and the datacenters that serve them, so they are not able to capture the full space of tradeoffs. Liu et al. [29] also proposed an algorithm to geographically load balance users while taking carbon into account. Their algorithm assigns users to datacenters based on access latency and costs only; however, they put a dollar cost on carbon emissions for their analysis. As previously discussed, the per dollar cost of carbon is low compared to the cost of electricity, so carbon emission needs to be more explicitly considered in the optimization procedure. Liu et al. use their algorithm to investigate whether user load balancing can facilitate the introduction of stochastic energy sources (like wind and solar) to the electric grid. In [9], Doyle et al. proposed adjusting the number of servers running in datacenters to tradeoff between latency and carbon emissions. In this preliminary work, the authors use a simple model of the problem, which does not take into account load distribution, network latency, and the temporal dynamics of carbon emission rates.

None of the previously proposed request-routing schemes allow operators to tune all three parameters: electricity costs, access latency, and carbon emissions. FORTE solves this problem, and specifically we make the following contributions beyond previous works: (1) we model data at datacenters, whereas previous work ignore this and assume full replication of the data; (2) we propose a new algorithm for user assignment that accounts for the location of data, whereas previous works ignore data locality; (3) we propose an algorithm to plan datacenter upgrades such that carbon emissions are minimized; and, (4) we use this algorithm to thoroughly explore the tradeoffs between latency, electricity costs, and carbon emissions.

New architectures for service energy reduction: Others have considered new architectures to reduce the electricity costs of content distribution. Vytautas et al. proposed the use of nano datacenters [42], which act similar to peer-to-peer distribution networks, to replace the current centralized model. They argue that fixed costs such as cooling consume a large proportion of energy in datacenters, which is avoided by nano datacenters, because a nano datacenters are freely cooled by ambient air. Jourjon et al. [20] also proposed a peer-to-peer-based architecture; however, they have not yet evaluated their proposal. Uichin et al. [27] and Guan et al. [17] proposed architectures to reduce content distribution energy costs using content-centric networking (CCN). A CCN incorporates storage components into switches and routers in the network, enabling them to cache content. This may reduce energy use for serving static content distribution, but does not work for dynamic content.

These alternative approaches are promising, but it is unclear whether they will save energy. Others have argued that peer-to-peer content delivery architectures use more electricity overall than centralized models like CDNs [11]. However, these architecture do reduce the electricity use of datacenters, but because they increase overall electricity use, they will increase carbon emissions overall.

Datacenter upgrade and expansion: Site selection for new datacenters has traditionally been performed based on factors such as susceptibility to natural disasters, electricity prices, land prices, tax breaks, workforce, and availability of Internet connection points [38]. Recently, Goiri et al. [12] proposed a rigorous framework for datacenter site selection. Their solution automates site selection using optimization, and their algorithms simultaneously optimize many objectives, including minimizing costs and carbon emissions. Expanding the capacity of existing datacenters has not been as widely studied in the literature. However, there has been work on datacenter network upgrades and expansions [7].

7. CONCLUSIONS

The carbon footprint of Internet-scale applications is a relatively small but still rapidly growing fraction of total emissions. To address this issue, we propose FORTE, a principled approach based on flow optimization to route users to datacenters. FORTE allows application service providers, such as content distribution networks, to navigate the three-way tradeoff between carbon footprint, access latency and electricity cost. To deal with the scale of the problem, we also describe Fast-FORTE, a greedy algorithm that obtains nearly the same results as FORTE but runs faster by at least an order of magnitude. Due to their underlying principled approach, both algorithms can be easily enhanced to take other factors, such as tax incentives and upgrade budgets, into account.

Using FORTE and Fast-FORTE to analyze traces from the Akamai CDN, we find that our approach to request-routing can reduce Akamai's carbon footprint by 10% without increasing electricity cost, while simultaneously bounding the access latency. We also find that further reductions in carbon footprint come at the expense of either latency or electricity cost, and that these costs are unlikely to be offset by carbon credits or taxes. We modify FORTE to determine how best to upgrade datacenters in response to increases in request traffic. We find that, using our approach, under some simplifying assumptions, Akamai can reduce its carbon footprint by about 25% over three years, compared to a carbon-oblivious upgrade algorithm.

Acknowledgement

We thank Bruce Maggs for giving us access to the Akamai trace. We also thank our shepherd Fabián Bustamante and the anonymous reviewers for their insightful comments. This work is supported by HP Inc., National Science and Engineering Research Council, Canada, and the Canada Research Chairs Program.

8. REFERENCES

[1] Light-duty automotive technology, carbon dioxide emissions, and fuel economy trends: 1975 through 2011. *U.S. Environmental Protection Agency.*

[2] Akamai. Akamai reports fourth quarter 2010 and full-year 2010 financial results, 2011. http://tiny.cc/afyu9, Last visited Jan. 2012.

[3] H. Amur, J. Cipar, V. Gupta, G. R. Ganger, M. A. Kozuch, and K. Schwan. Robust and flexible power-proportional storage. In *SoCC*, 2010.

[4] L. Barroso and U. Holzle. The case for energy-proportional computing. *Computer*, 40(12):33 –37, dec. 2007.

[5] California ISO. Renewables watch, 2011. http://www.caiso.com/green/renewableswatch.html.

[6] J. Chang, J. Meza, P. Ranganathan, C. Bash, and A. Shah. Green server design: beyond operational energy to sustainability. In *HotPower*, 2010.

[7] A. R. Curtis, S. Keshav, and A. López-Ortiz. LEGUP: Using heterogeneity to reduce the cost of data center network upgrades. In *CoNEXT*, 2010.

[8] B. Dennis. Estimating a data center's electrical carbon footprint. *Schneider Electric White Paper 66*, 2011.

[9] J. Doyle, D. O'Mahony, and R. Shorten. Server selection for carbon emission control. In *GreenNet*, 2011.

[10] EPA. EPA report to congress on server and data center energy efficiency. Technical report, U.S. Environmental Protection Agency, 2007.

[11] A. Feldmann, A. Gladisch, M. Kind, C. Lange, G. Smaragdakis, and F.-J. Westphal. Energy trade-offs among content delivery architectures. In *CTTE*, 2010.

[12] I. Goiri, K. Le, J. Guitart, J. Torres, and R. Bianchini. Intelligent placement of datacenters for internet services. In *ICDCS*, 2011.

[13] Google. Efficient computing—step 2: efficient datacenters. http://www.google.com/corporate/green/datacenters/step2.html.

[14] Google green. http://www.google.com/green/, Last visited Jan. 2012.

[15] Google's Data Center Efficiency. http://www.google.com/about/datacenters/inside/efficiency/power-usage.html, Last visited Jan. 2012.

[16] A. G. Greenberg, J. R. Hamilton, D. A. Maltz, and P. Patel. The cost of a cloud: research problems in data center networks. *SIGCOMM CCR*, 39(1):68–73, 2009.

[17] K. Guan, G. Atkinson, D. Kilper, and E. Gulsen. On the energy efficiency of content delivery architectures. In *ICC*, 2011.

[18] B. Heller, S. Seetharaman, P. Mahadevan, Y. Yiakoumis, P. Sharma, S. Banerjee, and N. McKeown. Elastictree: saving energy in data center networks. In *NSDI*, 2010.

[19] B. H. Joseph V. Spadaro, Lucille Langlois. Greenhouse gas emissions of electricity generation chains: Assessing the difference. *IAEA Bulletin*, 2000.

[20] G. Jourjon, T. Rakotoarivelo, and M. Ott. Models for an energy-efficient p2p delivery service. In *PDP*, 2010.

[21] J. Kanter. Europe considers new taxes to promote 'clean' energy, 2010. http://nyti.ms/xSlobE, Last visited Jan. 2012.

[22] J. Koomey. Growth in data center electricity use 2005 to 2010. *Analytics Press*, Aug. 2011.

[23] J. G. Koomey. Worldwide electricity used in data centers. *Environmental Research Letters*, 3(3):034008, 2008.

[24] A. Krioukov, P. Mohan, S. Alspaugh, L. Keys, D. Culler, and R. H. Katz. Napsac: design and implementation of a power-proportional web cluster. In *GreenNet*, 2010.

[25] C. Labovitz, S. Iekel-Johnson, D. McPherson, J. Oberheide, and F. Jahanian. Internet inter-domain traffic. In *SIGCOMM*, 2010.

[26] K. Le, R. Bianchini, T. Nguyen, O. Bilgir, and M. Martonosi. Capping the brown energy consumption of internet services at low cost. In *IGCC*, 2010.

[27] U. Lee, I. Rimac, and V. Hilt. Greening the internet with content-centric networking. In *e-Energy*, 2010.

[28] M. Lin, A. Wierman, L. L. H. Andrew, and E. Thereska. Dynamic right-sizing for power-proportional data centers. In *INFOCOM*, 2011.

[29] Z. Liu, M. Lin, A. Wierman, S. H. Low, and L. L. Andrew. Greening geographical load balancing. In *SIGMETRICS*, 2011.

[30] V. Mathew, R. Sitaraman, and P. Shenoy. Energy-aware load balancing in content delivery networks. In *INFOCOM*, 2012.

[31] R. Miller. Report: Google Uses About 900,000 Servers. Data Center Knowledge, http://tiny.cc/gservers, 2011. Last visited Jan 2012.

[32] A. Mohsenian-Rad and A. Leon-Garcia. Coordination of cloud computing and smart power grids. In *SmartGridComm*, 2010.

[33] E. Nygren, R. K. Sitaraman, and J. Sun. The akamai network: a platform for high-performance internet applications. *SIGOPS OSR*, 44:2–19, August 2010.

[34] J. Park. Designing a Very Efficient Data Center. Facebook.com, http://tiny.cc/fbservers, 2011. Last visited Jan 2012.

[35] A. Qureshi, R. Weber, H. Balakrishnan, J. Guttag, and B. Maggs. Cutting the electric bill for Internet-scale systems. In *SIGCOMM*, 2009.

[36] B. Raghavan and J. Ma. The Energy and Emergy of the Internet. In *HotNets*, 2011.

[37] L. Rao, X. Liu, L. Xie, and W. Liu. Minimizing electricity cost: optimization of distributed internet data centers in a multi-electricity-market environment. In *INFOCOM*, 2010.

[38] J. Rath. Data center site selection. In *Rath Consulting White Paper*, 2007.

[39] E. Schurman and J. Brutlag. The user and business impact of server delays, additional bytes, and HTTP chunking in web search. Presentation at the O'Reilly Velocity Web Performance and Operations Conference, 2009.

[40] A. Sullivan. ENERGY STAR for Data Centers. Technical report, U.S. Environmental Protection Agency, 2009.

[41] U.S. Energy Information Administration. http://www.eia.gov, Last visited Jan. 2012.

[42] V. Valancius, N. Laoutaris, L. Massoulié, C. Diot, and P. Rodriguez. Greening the internet with nano data centers. In *CoNEXT*, 2009.

[43] E. Williams. Energy intensity of computer manufacturing: hybrid assessment combining process and economic input-output methods. *Environ. Sci. Technol.*, 38(22), Nov 2004.

Flashback: Decoupled Lightweight Wireless Control

Asaf Cidon, Kanthi Nagaraj, Sachin Katti
Stanford University
{cidon, kanthicn, skatti}@stanford.edu

Pramod Viswanath
University of Illinois at Urbana-Champaign
pramodv@illinois.edu

ABSTRACT

Unlike their cellular counterparts, Wi-Fi networks do not have the luxury of a dedicated control plane that is decoupled from the data plane. Consequently, Wi-Fi struggles to provide many of the capabilities that are taken for granted in cellular networks, including efficient and fair resource allocation, QoS and handoffs. The reason for the lack of a control plane with designated spectrum is that it would impose significant overhead. This is at odds with Wi-Fi's goal of providing a simple, plug-and-play network.

In this paper we present Flashback, a novel technique that provides a decoupled low overhead control plane for wireless networks that retains the simplicity of Wi-Fi's distributed asynchronous operation. Flashback allows nodes to reliably send short control messages *concurrently* with data transmissions, while ensuring that data packets are decoded correctly without harming throughput. We utilize Flashback's novel messaging capability to design, implement and experimentally evaluate a reliable control plane for Wi-Fi with rates from 175Kbps to 400Kbps depending on the environment. Moreover, to demonstrate its broad applicability, we design and implement a novel resource allocation mechanism that utilizes Flashback to provide efficient, QoS-aware and fair medium access, while eliminating control overheads including data plane contention, RTS/CTS and random back offs.

Categories and Subject Descriptors

C.2 [**Computer Systems Organization**]: Computer-Communication Networks

Keywords

Wireless Network, Wireless Control

1. INTRODUCTION

Due to their distributed and asynchronous nature, Wi-Fi networks, unlike cellular networks, function without a dedicated and separate control plane for node coordination. Wi-Fi uses implicit and explicit control mechanisms that are always coupled with the data plane. For example, CSMA's random back off is a form of implicit coordination between nodes, while RTS/CTS and association requests are explicit control packets that are multiplexed with data transmissions. The lack of a separate and decoupled control plane results in data plane performance problems, and makes it hard to provide desirable features such as QoS, handoffs, access control and power duty cycling. For example, random back offs render packet scheduling inefficient due to collisions and exposed terminals, while explicit control mechanisms such as RTS/CTS consume significant channel time [6, 11, 9, 15].

In contrast, cellular networks have none of these problems, since they use a separate and decoupled control plane. Cellular networks (e.g. LTE [13, 8]) designate dedicated spectrum for explicit control signaling between the base station and clients. Such a decoupled control plane has many benefits. For example, the base station can leverage the control plane to centrally schedule user's data flows, ensuring both channel efficiency (e.g. no collisions or exposed terminals) and the required QoS for traffic classes (voice, video and data). Several other useful functions, such as client power saving modes and seamless handoffs, are enabled by the separate control plane. However, the cellular control plane adds considerable overhead, because cellular networks have to reserve a significant share of their spectrum for the control channels [13, 8]. Furthermore, since the control channels are accessed using techniques such as OFDMA, which require microsecond level synchronization [1], cellular networks have to rely on accurate centralized time synchronization. Wi-Fi's goal is to be simple, cheap and asynchronous, and therefore the Wi-Fi standards cannot afford to adopt expensive and centrally coordinated mechanisms. This raises a natural question: Can we achieve similar control functionality in Wi-Fi without paying the high overhead of cellular networks, while retaining the simplicity and flexibility of a distributed asynchronous network?

In this paper, we present the design and implementation of Flashback, a novel technique that provides Wi-Fi networks with a *sepa-*

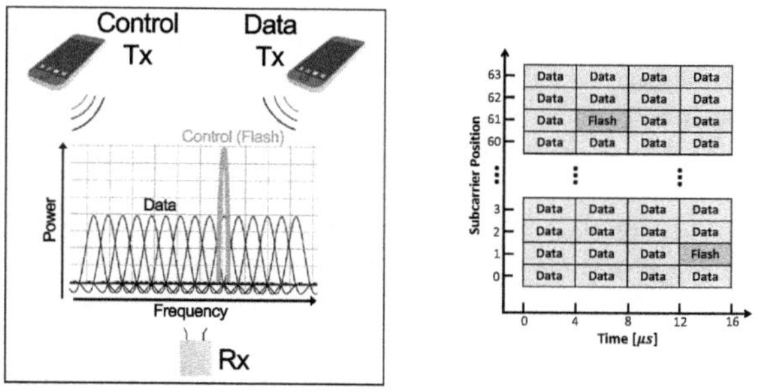

Figure 1: The first figure depicts a node concurrently sending control messages using flashes. The second figure visualizes OFDM packets as a time-frequency grid. Flashes are sent over specific slots on the grid.

rate, decoupled and low-overhead control plane, while preserving Wi-Fi's desirable distributed asynchronous nature. The key building block of Flashback is a novel mechanism that allows nodes to send short control messages on the same channel concurrently with data transmissions, without harming the data transmission. Moreover, the control messaging is asynchronous, and does not require precise clock synchronization among all the nodes. Flashback's control messaging adds very little resource overhead since no extra spectrum or time slots are used, yet as our experiments show, provides fast and reliable control plane messaging rates of up to $400Kbps$, with average rates of $175Kbps$. Flashback is general purpose and can be used for a wide gamut of applications, from centralized resource allocation to handoffs and network association.

How can nodes send short control messages concurrently with data transmissions without causing harm to the data? The key insight is to use short high-powered flashes that are localized in frequency and time, which only interfere with a very small number of symbols from the data transmission. When the data receiver detects these flashes, it simply erases the bits that were corrupted from the bits of the data packet. Flashback uses the location of the flashes as a way to modulate bits that represent the control message. To make this clearer, consider the scenario, where a client is transmitting a data packet to the AP using Wi-Fi OFDM. The packet can be visualized as a two-dimensional grid of time and frequency slots, with dimensions equal to the length of the packet and the number of OFDM subcarriers respectively, as depicted by the second picture in Fig. 1. Another client concurrently transmits short flashes that are localized to one of the slots in the time-frequency grid. As Fig. 1 shows, the flash is overlaid on top of one of the subcarriers in the received data packet. In practice, flashing is equivalent to sending a sinusoid that has a time length equal to that of a Wi-Fi OFDM symbol ($4\mu s$), at a frequency equal to the specific subcarrier on which the flash is sent. The client can transform a series of flashes into a small control message (32 bits) by varying the relative location of flashes. The AP can then read the client's message, by detecting the time-frequency slots in which it sees high-powered flashes and transforming them back to the control message.

Flashback exploits two unique properties of OFDM signaling to ensure that data transmissions do not fail due to interference from flashes. First, OFDM divides data into small chunks that are modulated separately on each time-frequency slot. By flashing a small number of specific time-frequency slots, Flashback ensures that any interference is localized and does not corrupt the entire packet. Second, successful packet transmissions invariably have some *link margin*, i.e. the SNR of the received packet is typically greater than the minimum SNR needed to decode the packet. Due to this margin, the loss of a few data bits does not affect decoding: the receiver can simply erase the bits from the slots where the flash was sent and recover the data from the other symbols.

We design and implement Flashback on a Virtex-5 based FPGA software radio test bed, including a full implementation of a $20MHz$ Wi-Fi OFDM receiver and transmitter with convolutional coding. Flashback is implemented by applying minimal firmware modifications to the OFDM PHY receive and transmit chains. Our experimental evaluations show that Flashback supports typical control message rate of about $175Kbps$ and a maximum rate of $400Kbps$ (depending on the environment), while causing minimal impact to concurrent data packets (less than 0.5% overhead). In addition, Flashback's flash detection can be configured so that it rarely misses a flash (less than 1% false negative rate) under almost all channel conditions, ensuring that control messages are received reliably.

The main contribution of Flashback is to exploit its flashing capability to create a novel control plane for asynchronous wireless networks that is decoupled from the data plane. Such a control plane can be used for a variety of network control functions. First, we show how Flashback can facilitate efficient, high throughput and QoS-aware medium access. We then demonstrate that Flashback can be used to design a virtual wireless enterprise network that provides seamless association, flow setup, teardown and scheduling. Finally, we show how Flashback can be used by Wi-Fi clients to efficiently duty cycle.

In order to evaluate these applications we ran trace-driven simulations under several network scenarios. Our simulation results show that Flashback can outperform the traditional Wi-Fi MAC protocols significantly, providing a throughput increase of more than 5× over CSMA/CA and 1.8× over RTS/CTS in congested networks, and 70% energy savings compared to Wi-Fi Power Saving Mode (PSM). Furthermore, Flashback can be used to easily enforce QoS policies. Using Flashback, latency-sensitive traffic (e.g. VoIP) would experience less than $1ms$ of delay even under highly congested networks.

2. OVERVIEW

This section provides an overview of Flashback. We first describe the general motivation for our control channel design. We then provide a short primer on OFDM wireless PHY and SNR link margins, which are fundamental for understanding Flashback's control messaging mechanism. We finally describe an overview of Flashback's design.

2.1 Motivation: Control Plane On Top of the Data Plane

Before we start describing Flashback's design, we would like to answer a simple question: Why does our control plane send flashes on top of concurrent data transmissions? Can't we implement a control plane similar to cellular networks, where spectrum is pre-allocated for a dedicated control channel?

The main reason is that spectrum is becoming scarce in the unlicensed band, especially due to the proliferation of dense wireless deployments. Specifically, if we were to pre-allocate a narrow $2Mhz$ chunk for a control channel, according to Wi-Fi specifications, we would need to reserve an additional $1.875MHz$ as guard band [12]). Consequently, in a $20MHz$ channel, almost 25% of the spectrum would be consumed by the control channel. In contrast, by sending control messages on top of the data plane spectrum, Flashback does not consume any extra spectrum.

Alternatively, we could use a protocol like OFDMA, which lets multiple nodes transmit data and control packets on different frequencies at the same time, without requiring extra spectrum [1]. However, OFDM requires tight synchronization (on the order of a microsecond) among all client nodes. Wi-Fi's current protocols do not support such tight time synchronization. In contrast to OFDMA, our scheme does not require any clock synchronization.

Finally, messages such as RTS and CTS packets, beacons and probes are all designed to implement control functions. However, these messages occupy the network time of the data packets. Furthermore, since these messages have to be delivered reliably, they are encoded using the lowest bitrates and consume significant portions of the channel time. As our own simulations and recent studies have shown [7], such messages can consume up to 40-50% of the channel's time. In addition, this overhead makes it hard for QoS standards like 802.11e [14] to enforce QoS requirements in congested networks, because they multiplex their control messages on the data plane. As a result of the extra overhead, many network managers prefer to disable most of these control functions.

Minimum Required SNR	Bitrate	Modulation	Coding
3.5 dB	6 Mbps	BPSK	1/2
4.5 dB	9 Mbps	BPSK	3/4
5 dB	12 Mbps	QPSK	1/2
9.5 dB	18 Mbps	QPSK	3/4
12 dB	24 Mbps	16-QAM	1/2
17.5 dB	36 Mbps	16-QAM	3/4
21 dB	48 Mbps	64-QAM	2/3
22 dB	54 Mbps	64-QAM	3/4

2.2 OFDM Primer

OFDM is a multi-carrier wideband modulation technique. OFDM divides the entire bandwidth available for transmission into evenly spaced orthogonal bands, which are called subcarriers. The data is split into parallel streams, one for each subcarrier. Specifically, in Wi-Fi, a 20MHz channel is split into 64 subcarriers, each spaced 312.5KHz apart.

To modulate data onto OFDM transmissions, a single constellation symbol is sent on each subcarrier of an OFDM symbol, depending on the bitrate being used (BPSK, QPSK, etc.). The constellation symbols are passed through an IFFT to produce time domain samples, which are then sent over the channel. The typical OFDM symbol length is $4\mu s$. OFDM symbols are sent one after the other. The symbols can therefore be pictured as a two-dimensional time-frequency grid, where each frequency index represents a single subcarrier, and each time index represents a period of $4\mu s$, as depicted by Fig. 1. Packets can be visualized as constellations being mapped to different points on this grid. This abstraction applies to any OFDM based system, including Wi-Fi, LTE and WiMAX.

The Wi-Fi OFDM transmitter sets the constellation modulation and channel coding rate according to the channel conditions, in order to protect against channel distortions such as noise and interference. For example, a dense constellation such as 64-QAM is less robust to errors than a sparser one like 16-QAM. Similarly, Wi-Fi transmitters can increase the data redundancy, by controlling the channel code rate. Wi-Fi has three different coding rates (1/2, 2/3, 3/4), where the rate is defined as the ratio of data bits to actual encoded bits that are sent. Thus, the lower the coding rate, the higher the redundancy.

To maximize throughput, Wi-Fi transmitters perform rate adaptation, i.e. they estimate the channel SNR to the receiver, and for that SNR, pick the densest constellation and highest coding rate that still ensure that the receiver can decode. Table 2.2 shows different combinations of Wi-Fi constellations and coding rates and their corresponding SNR thresholds, as we measured on our Wi-Fi receiver. SNR measurement in current networks is based on packet loss estimates, i.e. if a packet sent at a particular rate is lost, then the sender assumes that the channel SNR is lower than the threshold for that rate, and consequently the sender switches to a lower rate. There are more sophisticated techniques to measure SNR that tradeoff increased measurement overhead for better accuracy.

2.3 Link Margin

The previous discussion raises two observations. First, since SNR fluctuates over time on the order of milliseconds, SNR measurement is inherently error-prone. Consequently, even the most accurate SNR estimation techniques cannot adapt to instantaneous SNR variations, and therefore bitrate adaptation algorithms prefer to err on the conservative side. Algorithms minimize the risk of packet loss by using a lower bitrate than the threshold permits, since packet loss is expensive. Second, to simplify the implementation of Wi-Fi hardware, the bitrates available for transmission are discrete. Therefore, in order to protect against the uncertainty in measuring SNR, and due to the discreteness of bitrates, for every successful Wi-Fi transmission there is some *link margin*. Link margin is the difference between the instantaneous channel SNR and the minimal SNR required to decode the packet. The link margin's value is dependent on the accuracy of the SNR measurement, how conservative the rate adaptation protocol is, the available discrete bitrates and the amount of external interference.

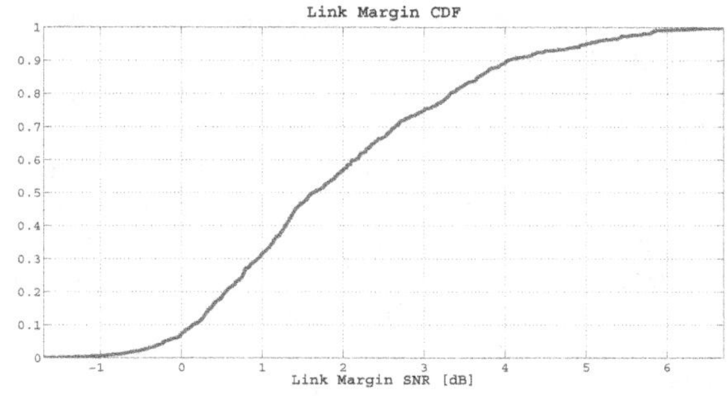

Figure 2: The CDF of the link margin between the actual SNR measured by the channel sounder, and the SNR threshold chosen by SoftRate's rate adaptation algorithm.

To measure the link margin in a realistic setting, with varying channel SNRs and a practical rate adaptation algorithm, we perform the following experiment. We use RUSK Stanford Channel Sounders [19] to measure a $20MHz$ channel on the $2.4GHz$ ISM band. One sounder is fixed, while the second one is moved at walking speed. Next, we simulate the rate adaptation decisions a state-of-the-art rate adaptation algorithm, SoftRate [20], would make if it were making decisions for the transmitter based on the channel trace. For each bitrate SoftRate picks for transmitting a packet, we measure the difference between the actual channel SNR (from the trace) and the SNR threshold required for the packet to be decoded. We collected a trace of about $70,000$ packets from our sounders. Fig. 2 plots the CDF of these margins, measured over the packets that were transmitted successfully.

The measured median link margin is $1.63dB$ and the average is $1.94dB$. In other words, typically there is a considerable amount of slack between the data rate used by the transmission and the data rate the channel could actually support. Furthermore, the estimate that we measured is most likely conservative. In commercial systems, Wi-Fi transmitters are configured to use higher link margins, due to interference from legacy radios (e.g. microwave ovens) or neighboring APs. Note that the link margin is negative for about 7% of the packets. For these packets, even though SoftRate chose a bitrate that was too high for the SNR of the transmitted packet, the algorithm 'got lucky', and the packet was decoded successfully.

2.4 Flashes: Flashback's Control Signals

The goal of Flashback is to provide a control plane for asynchronous wireless networks with negligible overhead. To this end, Flashback requires a messaging technique, which would allow nodes to concurrently send control messages without harming the ongoing data transmission, and without requiring any synchronization.

Flashback exploits the link margin to create the desired control messaging mechanism. The basic idea is that since typical transmissions have some link margin, any node could send a control packet even while a concurrent data transmission is taking place, as long as the interference from the packet is smaller than the tolerable margin.

However, simply sending regular Wi-Fi packets as control messages on top of concurrent data packets is doomed to fail for two reasons. First, the receiver needs to have a way to detect and decode the control packet. Since the power of the data packet would typically be much stronger than the control packet, it would be difficult to reliably determine whether a control packet has actually been sent, if it is limited to a signal that is weaker than $1.94dB$ on average. Second, the amount of interference that the control packet would cause is a function of both the power at which it is transmitted, as well as the channel between the flash transmitter and the receiver. To ensure that the interference stays within the required tight SNR margin, the control transmitter would have to precisely estimate the channel and set its transmission power to meet the link margin requirements. Given that the uncertainty involved in link SNR estimation is the raison d'être of link margins, it would be risky to rely on high precision SNR estimation to ensure that the margin is met.

Flashback's insight is to utilize the fundamental properties of OFDM modulation to concurrently send easily decodable control messages, while limiting the interference of the control messages to a level below the link margin. Flashback leverages the time-frequency abstraction of OFDM to localize the interference the control messages cause to the data packets. Instead of sending a regular OFDM packets over the entire band, the control message transmitter sends *flashes*. Flashes are simple sinusoids that have a frequency equal to the frequency of a specific subcarrier, and duration equal to the OFDM symbol time. The flash appears as a power spike, localized to a specific point in the time-frequency OFDM grid, corresponding to the frequency of the flash sinusoid and the time slot in which the flash was transmitted. Since the flash transmitter focuses all its power on a single subcarrier, in most cases the flash has much higher power relative to the data transmission symbols. Consequently, the flash's position on the time-frequency OFDM grid is easy to detect using a simple peak detection algorithm. Note that Flashback doesn't synchronize the flashes to the data symbols, so there might be time and frequency offsets between the grids of the transmitter and the receiver. We address these later in our design.

In order to decode the data transmission, the receiver first detects the flashes, and then instead of reading the erroneous bits from the symbols that were flashed, the receiver *erases them*. In other words, instead of causing bit errors due to concurrent control messaging, Flashback exploits the fact that flashes are easily detectable to convert them to bit erasures. Channel codes can correct twice as many bit erasures as bit errors [16, 18]. An additional advantage of flashing is that it causes bit erasures in short bursts, and wireless channel codes are designed to handle bursty interference. Due to all these factors, as long as the number of flashes is limited, a receiver can use simple algorithms to decode both the data and flash transmissions concurrently.

How can nodes leverage flashes to send control messages? The basic idea is to use the relative subcarrier position of the flashes as a way to signal a small set of bits. For example, a transmitter can send two consecutive flashes at subcarriers 2 and 27 and use the difference (25) as the means to encode information. Using the relative distance between subcarriers has the advantage of being robust to frequency offset errors. In Sec. 3.2 we describe a simple and robust modulation scheme that uses the relative subcarrier distance idea to design a control messaging algorithm. We also describe a protocol that mediates which node can send a control message at any point.

While this discussion provides the basic intuition behind Flashback, several practical questions remain.

- How many flashes can a Wi-Fi data transmission tolerate? How does the link margin relate to the number of allowed flashes?
- How does the receiver reliably detect flashes and decode control messages? How do we prevent the high powered flashes from saturating the ADC?
- How do we ensure that the number of flashes is under the tolerable limit when multiple nodes want to send control packets? What is the protocol for accessing the shared control plane?
- How can the MAC layer leverage this new decoupled control channel?

In the next section, we describe the design of Flashback that addresses these questions.

3. DESIGN

In this section we describe the control plane's three components: the basic flashing transmission and detection mechanisms, the technique that transforms flashes into short binary control messages, and the protocol for nodes to asynchronously access the control plane. We also describe the minor changes the data plane receiver has to implement to decode concurrently transmitted data packets.

Before we proceed, we define several notations for the rest of the paper. We assume that nodes use Wi-Fi OFDM for data packet transmission with 64 subcarriers spread over a $20MHz$ channel, where each OFDM symbol lasts $4\mu s$. As discussed before, OFDM can be abstracted as a time-frequency coordinate system (i, j), where $0 \leq i \leq 63$ identifies the subcarrier and j is the time slot. To send a data packet, a node fills these coordinates with different constellation samples. The magnitude of the constellations is a function of the transmit power P with the following constraint:

$$\frac{\sum_{i=0}^{63} \sum_{j=1}^{L} |x_{ij}|^2}{L} \leq P \qquad (1)$$

Where x_{ij} is the complex symbol transmitted on the (i, j) coordinate, and L is the number of symbols in the packet.

We define R, as the maximum number of flashes that all the nodes can send per second over the network, without harming the performance of the data transmission (i.e., without introducing more than 1% of errors). The value of R is a function of the link margin, which depends on the PHY's rate adaptation algorithm, as well as the accuracy of the flash detection algorithm. In our experimental evaluation we found that for Wi-Fi, even a conservative approach that minimizes flash interference with a state-of-the-art rate adaptation algorithm, allows Flashback to use a flash rate R of *50,000 flashes per second.*

3.1 Flashing

Flash transmission is simple to implement. Using the OFDM coordinate abstraction, a flash transmission amounts to transmitting a complex constellation f on a specific co-orindate (i, j), and zeroing all other coordinates on the same time slot (an entire column in the grid). In practice, this requires the node to send a single sinusoid that lasts for $4\mu s$ with frequency corresponding to the i'th subcarrier. The actual time domain message is generated using the standard OFDM modulator (i.e. it is passed through an FFT and the rest of the transmission chain). Flash transmission therefore requires no extra transmit hardware or signal processing logic.

The magnitude of the flash f is larger than the magnitudes used for sending data symbols. To see why, consider the power constraint in Eq. 1. Since for a given j, the node sends zeros on all of the i coordinates except one, the power constraint will be satisfied even if the magnitude of the flash is much higher than the one used for data symbols. The higher magnitude aids detection, as explained in the next subsection. In our current implementation, flashes transmit at $64\times$ greater power than the power used for individual data symbols. Prior work has made a similar observation on how using higher magnitude signals with narrower width Wi-Fi channels helps improve range [5].

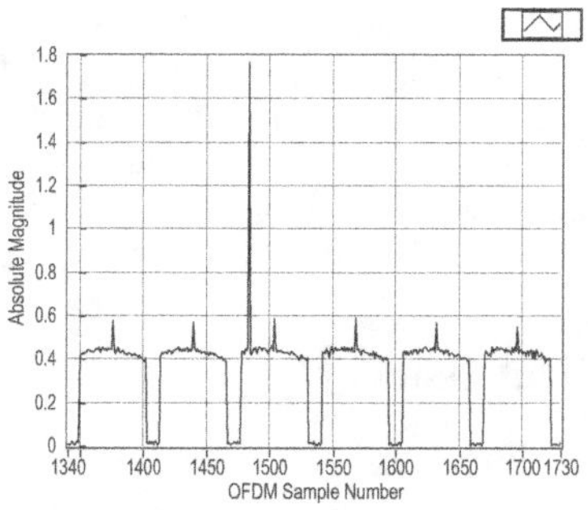

Figure 3: The graph contains a sequence of frequency-domain samples taken from a measurement conducted on our Flashback implementation. The flash is simple to discern from the data packets.

3.1.1 Flash Detection

The relatively high magnitude of flashes relative to data symbols makes them typically easy to detect. This is demonstrated by Fig. 3 where the magnitude of a single flash is clearly visible against a backdrop of regular OFDM data symbols. Flashes are of course even easier to detect when the channel is already idle. Using the same time-frequency OFDM abstraction, if the receiver sees a peak value in one of the time-frequency slots, it declares that a flash is detected in that slot. To reliably detect the flash and reduce false positives that can be caused by practical effects such as frequency-selective fading, Flashback uses a technique that uses the differential rather than the absolute value of the complex symbols.

Specifically, the receiver computes the following equation at every (i, j) coordinate:

$$D(i,j) = \frac{\sum_{m=-1}^{1} \sum_{n=-1}^{1} |(|y(i,j)| - |y(i-m, j-n)|)|}{8} \quad (2)$$

Where $y(i, j)$ is the constellation value received on coordinate (i, j).

The receiver detects a flash at coordinate (i, j) i.f.f. $D(i, j) > T$, where T is an empirically determined threshold. In effect, the equation above is computing the average differential value at (i, j). D will be very high when there is a flash, compared to a normal data symbol. The detection threshold T depends on practical system parameters such as receiver dynamic range and receiver gain.

There are a couple of practical issues with this detection algorithm. First, in order to simplify the implementation, we assumed that the flash transmitter does not synchronize to the packet of the data transmitter. Therefore, their respective time-frequency grids won't align perfectly. However, the receiver will synchronize to the data transmitter's grid since its primary goal is to decode the data packet. Practically, this means that a flash might span two time-slots at the receiver. Since the flash is now split, the flash's constellation value might be half the maximum value. However, since flashes are sent at $64\times$ the data symbol power, they are still likely to be received at a much higher power magnitude than the value of the data symbol and should be easy to detect. Performance can be further improved if the receiver runs two parallel flash detectors offset from each other in time by half a time slot, i.e. $2\mu s$. Intuitively, one of the detectors will usually have a better alignment with the flashing node, and the receiver can pick the grid with the higher flash value, and send its corresponding slots to the flash decoder. Second, in order to reliably detect flashes, we empirically measured that the receiver requires the flash power to be at least $6dB$ higher than the power of the concurrent data packet. In some cases, this might cause flashes by a node that is far away from the AP to be 'drowned' by nodes that are closer to the AP, and have significantly higher link SNRs. To mitigate this issue, the nodes with low channel SNRs might need to delay their flashes until the data channel is not occupied by data packets with high SNRs (e.g. packets with a high bitrate).

3.1.2 Data Packet Decoding

Flashback requires minimal changes to data packet decoding in the presence of concurrent flashes. As discussed before, the receiver erases the data symbols that are present at the same coordinates as the flashes. Practically, the data decoding PHY is instructed to ignore the demodulated bits for those symbols, and inform the decoder that these bits have been erased. It is known that erasures are easier to handle than bit errors, and require half the redundancy to correct [16, 18].

An important benefit of this scheme is that we do not have to change the existing OFDM PHY decoding hardware for decoding data in the presence of flashes, since soft Viterbi decoders support erasures.

3.2 Messaging with Flashes

The goal of the modulation algorithm described below is to communicate a series of bits that represent the control message, and communicate it to the receiver using a sequence of flashes. Note that Flashback offers a generic short messaging system that we can use for implementing any form of control functionality. In this section, we will describe a technique that provides control messaging rates of $175Kbps$ assuming that the maximum flashing rate is $50,000$ per second. However, as well see in Sec. 5 depending on interference conditions, nodes can flash at rates as high as $100,000$ per second, and the scheme below can be appropriately modified to provide a messaging rate of $400Kbps$.

We make three assumptions for the flash modulation algorithm:

- Flashback cannot assume any synchronization between nodes. Specifically, we cannot assume that the time-frequency grids of the data transmitter and the flash transmitter are tightly aligned.
- At any point, only one node can send a flash-based control message on the control plane. This assumption simplifies the messaging design as we show below.
- Nodes have the ability to sense whether the AP is currently transmitting a data packet. The reason for this requirement is that we assume the AP is half-duplex, and therefore any flashes that would be sent while the AP is transmitting cannot be decoded by the AP.

Flashback's modulation scheme transforms a 32 bit control message into a series of flashes, by varying the distance between subsequent flashed subcarriers. We also add an 8-bit CRC, which results in a total message size of 40 bits. Control messages are a series of 9 flashes sent one after the other in fixed time intervals of $20\mu s$. Using a fixed time interval between flashes simplifies the scheme, and helps other nodes sense that a flash message is being sent. The scheme also ensures that on average, nodes do not exceed R flashes per second.

For this entire section, assume that we do not flash on any pilot subcarriers, zero padded subcarriers, as well as the subcarriers adjacent to them. This leaves us with 36 subcarriers, which we logically number from 0 to 35. In addition, the first flash in the control message signals the start of the message, and is always sent on a pre-designated subcarrier (34). We also do not flash one subcarrier on either side of the start-of-message subcarrier (i.e. $33 - 35$).

The relative distance between two consecutive flashed subcarriers is used to encode the control message. We use the relative distance between consecutive flashes, rather than the absolute position of each one of the subcarriers, because in practice the carrier frequency of the flashing node could be slightly offset relative to the receiver's. Hence a messaging technique that uses the relative positions of the flashes will be robust to carrier frequency offset (CFO).

We take the bits that comprise the control message, and convert the binary number they represent into a number composed of 8 digits using base-32. A 40 bit message can always be represented using 8 digits in base-32. For example, assume the binary message is represented by $x_1, \ldots, x_8$ where x_i is the i'th digit in the base-32 number. Therefore, if the start-of-message flash was on subcarrier f_1, then the first flash (after the start-of-message flash) would be flashed on subcarrier $f_2 = (f_1 + x_1) \mod 32$, the second on subcarrier $f_3 = (f_1 + x_1 + x_2) \mod 32$ and so forth. To decode the message, the receiver detects flashed subcarriers $f'_1, f'_2, \ldots, f'_9$, and calculates the values of $x_1, \ldots, x_8$. For example, the formula for x_1 is given by $(f'_1 - f'_2) \mod 32 = x_1$. The same formula is applied to the other flashes. After it recovers the digits in base-32, the receiver can then read the message by converting the base-32 number into its binary representation. Finally, it checks the message's integrity using the 8 bit CRC.

Therefore, using our very simple coding scheme, Flashback can send on average 32 bits every $180\mu s$, which yields an overall bitrate of about $175Kbps$. 32 bit messages can be used for sending short control requests. For example, the first 10 bits in each message can be used to encode the sender's ID, and the last 22 bits can be used for specifying the contents of the control message (e.g. flow identifier, QoS request or a network association request). While $175Kbps$ is low by data transmission standards, our experiments demonstrate that this rate is sufficient support a practical Wi-Fi control channel.

We chose this simple scheme for a couple of reasons. First, it ensures that there is a fixed interval of $20\mu s$ between consecutive flashes, and as a result maintains a fixed flash rate, consistent with $R = 50,000$ flashes per second. Second, by using the relative distance between flashes, the scheme is robust to CFO errors. Finally, it provides receivers a simple mechanism to detect errors, since they know that a flash should be expected only once every $20\mu s$. In practice, we find that flash detection is fairly accurate, and has low false negative and negligible false positive rates. Hence the previous technique performs quite well.

3.3 Protocol for Sending a Flash

Flashback uses a carrier sense and random back off approach to regulate which node flashes at a particular time slot. Each node, before it flashes, performs a random back off, and counts down the back off counter. Meanwhile, it carrier senses for the *start-of-message flash subcarriers* for any other node flashing. If a flash is detected, it stops the countdown and restarts the counter after waiting for $180\mu s$, which is the time it takes to send a control message. If no flashes are detected and the counter goes to zero, then the node flashes the message. The average value of the random back off is set so that nodes do not exceed $R = 50,000$, the maximum number of flashes per second.

The protocol described above has similar problems to standard carrier sense based MAC protocols. However, unlike traditional Wi-Fi protocols, Flashback effectively exports the contention from the data plane to the control plane. Therefore, control plane access problems such as collisions, congestion and hidden nodes do not directly affect the overall performance of the data plane. This architecture is similar to most cellular control planes, where medium access in the data plane is centrally scheduled, while the control plane access is based on contention [13, 8].

Note that nodes can flash only when the AP is not transmitting, assuming the AP is half-duplex. Therefore, nodes have to decode the data packets of data transmitters, in order to verify whether that the AP is not transmitting a message. To acknowledge the control messages, the AP piggybacks on the regular ACK it sends after receiving a data plane transmission. This is an optimization; the AP could also have used Flashback to send the control message ACK.

3.4 Practical Issues

3.4.1 Interaction with Neighboring Networks

A practical concern is whether Flashback may interact destructively with neighboring networks, if they happen to be transmitting on the same Wi-Fi channel (this problem usually occurs in very dense networks). For example, an AP might detect flashes from a neighboring network and obtain incorrect control information. Further, flashing might hurt data plane transmissions on a neighboring network, since the overall flash rate might be higher than the limit picked per network.

Flashback's flash encoding technique naturally combats the problem of neighboring APs mistakenly decoding flash messages not intended for them. Since each control message provides 32 bits, several bits at the start of the message can be used to signal the identity of the sending node. As long as the identity spaces don't overlap, each AP should be able to tell whether the control message originated from its own network.

The second problem of multiple nodes concurrently flashing is largely avoided because of the "start-of-message" flash, and due to the fixed length of control messages. Since before sending the control message flashes, nodes always send a start-of-message flash on special subcarriers (40), any node sensing it will refrain from flashing for $180\mu s$. This applies to nodes across networks too. Of course there are corner cases where such sensing fails, as in any carrier sense mechanism.

Another concern is whether Flashback interacts poorly with narrowband protocols like Bluetooth or Zigbee. Since we designed Flashback to transmit flashes on single Wi-Fi subcarriers (i.e., $312.5KHz$ Bluetooth and Zigbee, which occupy a much wider bandwidth ($1MHz$ and $2MHz$ respectively), do not interfere with the correct decoding of Flashback control messages.

3.4.2 Size of Flash Messages

The messaging system described above assumes packets are long enough to accommodate the nine flashes that are needed to encode a 32 bit message. If we assume a flash rate of $50,000$ flashes per second, it translates to one control message every $180\mu s$. However, Flashback cannot send flashes while the AP is transmitting, since the AP won't be able to decode the control message. Therefore, the flashing node detects if the AP starts transmitting in the middle of sending a flash message. If so, it abandons the incomplete flash message and retries to send it at the next available opportunity.

3.4.3 Automatic Gain Control (AGC)

Another concern is that flashing may interact negatively with automatic gain control at the receiver. Specifically, the AGC tries to exploit the full dynamic range of the ADC, by automatically amplifying the incoming signal by the right amount, so that it fits the entire ADC range. However, when a flash arrives, since its value will be higher than the normal data transmission, the ADC can get saturated. In order to tackle this problem, we use static gain control. Specifically, our current system uses a 14 bit ADC that provides $86dB$ of dynamic range. Assuming that in the best case, we will see data transmissions with an SNR of $35dB$ ($\tilde{6}$ ADC bits), we can set the static gain so that the best case data message will consume

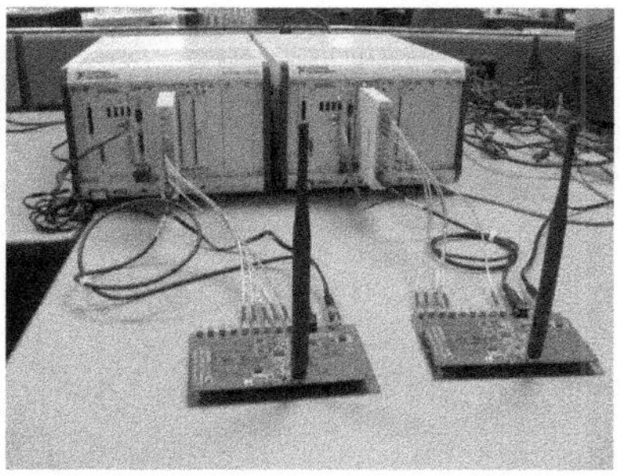

Figure 4: The National Instruments software defined radios we used to implement Flashback and carry our experiments.

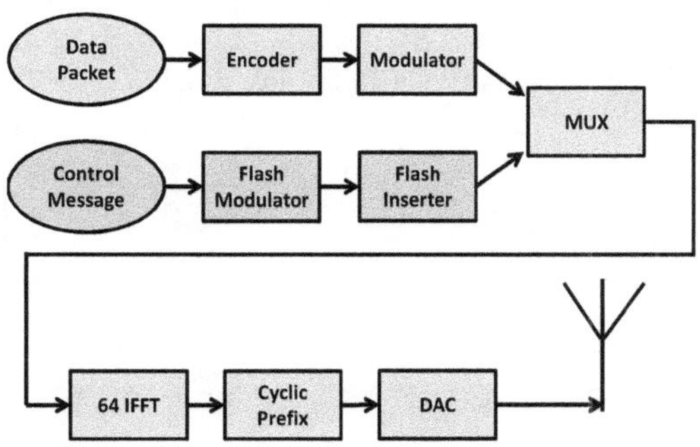

Figure 5: Flashback transmitter implementation.

8 bits of ADC resolution (leaving us with a margin of 2 extra bits). This leaves 6 bits or $36dB$ for decoding flashes, which is more than sufficient in practice.

4. IMPLEMENTATION

We implemented the Flashback receiver and transmitter using Virtex-5 LX30 FPGA based software radios from National Instruments [4], shown in Fig. 4. The FPGA has basic LUT-FF and 32 DSP48E arithmetic unit resources. The Flashback receiver was implemented on two units: the 7695 FPGA implements the FFT logic of the receiver chain, while the flash detection and decoding is performed on a real-time processor (NI PXIe-8133 RT Module). The FPGA is connected to an NI 5781 Baseband Transceiver, which serves as both the ADC and the DAC. The ADC has 14 bits of resolution, and the DAC has 16 bits. We used the National Instruments LabVIEW system design language [3] to implement and debug Flashback on the programmable radios.

4.1 Flash Transmitter Implementation

Fig. 5 presents the implementation of Flashback's transmitter. We added two blocks to the standard OFDM transmit chain: the flash modulator and the flash inserter. The flash modulator transforms a 32 bit binary number into a series of 9 flashes on the time-frequency grid. It does this, by adding an 8 bit CRC to the 32 bit binary sequence, and converting the entire sequence into eight 32-base numbers. The flash inserter then encodes the message using the technique described in Sec. 3.2 into a series of 9 flashes, adds the flashes generated by the encoder onto the transmission in the appropriate slots and zeros out the rest of the slots. Other than these two blocks, the Flashback transmitter is identical to a Wi-Fi transmitter. When a node wishes to flash, it switches on the flash transmitter logic, and when it has to transmit regular data messages, it can simply turn the Flashback logic off.

4.2 Flash Receiver Implementation

The implementation of the receiver is shown in Fig. 6. For the

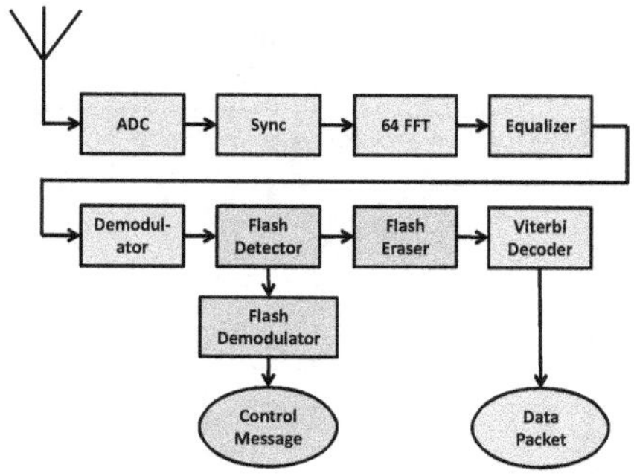

Figure 6: Flashback receiver implementation.

receiver, we added three blocks to the receiver chain: the flash detector, flash demodulator and flash eraser. The flash detector is always turned on by the receiver, because flashes can be received at any time.

The flash detector was implemented by utilizing the simple peak detector described in Sec. 3.1.1. The peak detector runs twice over the same samples, by reusing the same DSP hardware blocks of the OFDM receiver. The first peak detector is synchronized with the data packet, and the second one is offset from the first one by half a time slot, i.e. it is synchronized to $2\mu s$ after the first peak detector. If we detect a flash in the first flash detector, we only erase one subcarrier from the data packet. If we detect a flash in the offset flash detector, we erase the flashed subcarrier from the two data symbols it affected. We found that this 'aggressive' flash detection technique improves our flash detection error rates while reducing data errors.

The flash demodulator is very similar to the transmitter's flash modulator: it transforms the time-frequency flash grid into a binary sequence by converting a base-32 number into a binary number. Before sending the data into the Viterbi decoder, the flash eraser sets the confidence level of all the flashed bits (i.e. bits belonging to the flashed subcarriers) to 0. This signals to the Viterbi decoder that it should erase these bits from the data stream.

5. EXPERIMENTAL BENCHMARKS

In order to test Flashback, we benchmark Flashback's performance with two metrics.

Maximum Flash Rate (R): The maximum number of flashes per second that can be sent without harming data throughput.

Flashback Detection Accuracy: The false positive and false negative rates of detecting flashes at the receiver.

5.1 Maximum Flash Rate

As we discussed in Sec. 2, the link margin depends primarily on two factors: the discrete bitrates and the conservative SNR estimation of the bitrate adaptation algorithm. In practice, it also depends on a third factor, which is the level of interference. Bursty external interference will hurt the SNR estimation of the bitrate adaptation algorithm (i.e., the algorithm will interpret the interference as noise), which will cause the algorithm to behave more conservatively and increase the link margin.

We experimentally evaluate the impact these three factors on the maximum flash rate. All experiments were conducted during nighttime at our lab on the Wi-Fi band of $2.48Ghz$. In order to estimate the SNR for the data packets, we sent known packets. We then estimated the channel's noise by subtracting the known symbols from the received symbols. In order to estimate the effect of interference, we generated a trace of Zigbee and Bluetooth and added it as noise to our channel sounder's trace.

Discrete Bitrates: Our experimental setup consists of three nodes.

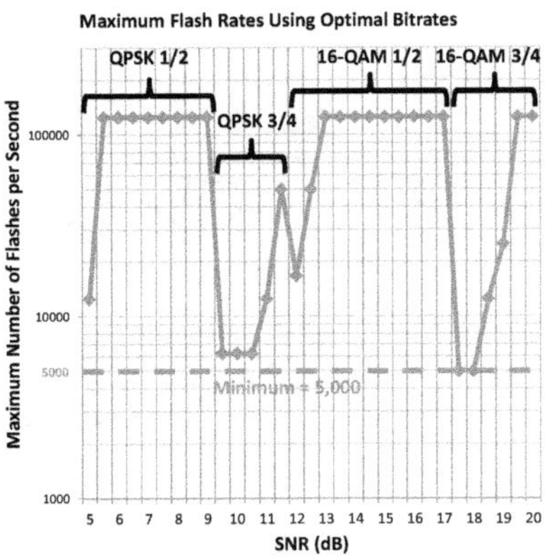

Figure 7: Number of flashes Flashback can support for different SNRs, with 'oracle' bitrate adaptation, i.e. when the transmitter is sending at the optimal bitrate for our receiver. In a real network setting, the transmitter will typically transmit data at a lower bitrate than optimal, and Flashback will be able to support a higher flash rate than the rates depicted here.

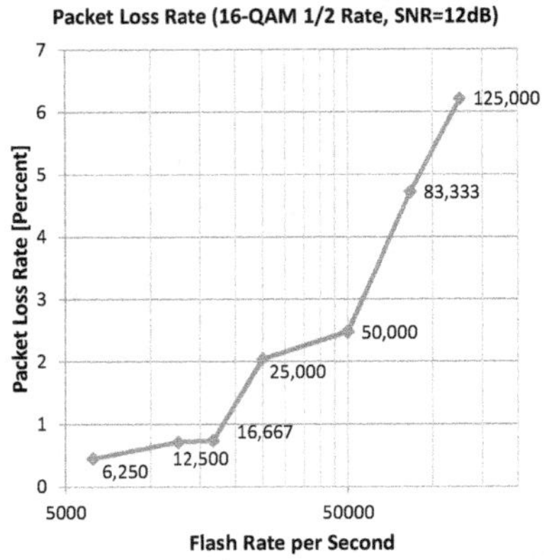

Figure 8: Packet loss rate of a $12dB$ link, as a function of the flash rate.

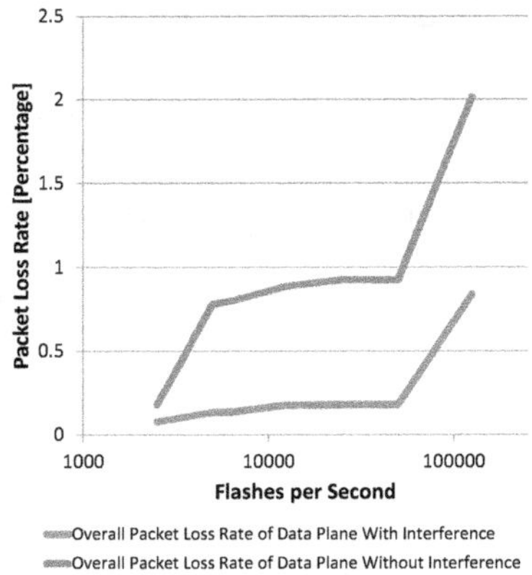

Figure 9: Overall packet loss rate as a function of the flash rate, measured by applying the SoftRate bit adaptation algorithm on a $75,000$ packet trace collected by our channel sounders.

The first node is a regular data transmitter. It keeps sending Wi-Fi packets with a fixed length of 40 symbols. The second node is the flash transmitter. The flash transmitter continuously transmits flashes at a specified rate at the same power. The third node is the receiver. The receiver decodes both the data packet and the flashes. In our experiments, the flashes are received at $6-10dB$ higher than the data transmission. The location of the data transmitter is varied to obtain different SNRs. For each SNR, we first tried transmitting hundreds of packets at all the bitrates, and picked the bitrate which maximizes the throughput for the data transmitter. Next, we varied the flash rate, and picked the highest rate, which doesn't affect the throughput of the data transmitter by more than 1%. Fig. 7 plots the maximum flash rate as a function of the SNR measured between the data transmitter and receiver, when the transmitter is transmitting at the most optimal discrete bitrate (i.e. this is an oracle transmitter; it can estimate the exact SNR for every packet ahead of time).

Analysis: As we can see from Fig. 7, since we are using an oracle transmitter that utilizes perfect SNR estimation, the channel can only support a maximum of $5,000$ flashes per second, because it is the minimum of all the flash rates across all SNRs. However, since the channel SNR is dynamic, and the bitrate adaptation algorithms cannot estimate SNR perfectly as we will show below, we can typically support much greater flash rates of $50,000$ flashes per second.

An interesting feature of the graph is that it looks like a 'chainsaw'; it starts low when Flashback switches to a new bitrate, and increases when the SNR is increased. The reason we see the chainsaw effect, is that every time Flashback switches to a higher bitrate, the data plane loses some redundancy in its channel code, since it is trying to send more data on a given channel. Since there is less redundancy, the flash eraser cannot erase as many bits as it could before, and Flashback can only support a lower flash rate. However, when the SNR increases, the message is received with fewer errors at the receiver, and the Viterbi decoder can tolerate more bit erasures due to flashes. As an aside, it is likely that the line in Fig. 7 could be increased beyond $125,000$ flashes a second at high SNRs. In our experiments, we did not attempt to flash at higher rates than $125,000$, since we only have one flash transmitter.

To further explain the previous results, we focus on one experiment where the SNR of the data transmission is $12dB$ at the receiver. For this SNR, the data transmitter uses a 16-QAM, 1/2 rate encoding for its packet to maximize throughput. Note that this SNR lies right at a transition point for data bitrates. For a slightly lower SNR of $11dB$, the data transmitter would have picked QAM, 3/4 rate encoding. We vary the flashing rate from the flash transmitter and measure the packet loss rate for the data transmission. The results are presented in Fig. 8. As expected, as we increase the flash rate, the packet loss rate of the data packets increases. The reason is that the increasing number of flashes causes the receiver to erase more bits from the bit stream. If the receiver erases too many bits, it will start seeing errors in the Viterbi decoder.

Note that even at the most minimal flash rate, there is a small probability of packet loss (around 0.45%), which most probably occurs because sometimes a flash can interfere with the packet header itself. In a practical system this would be avoided, since the flashing node could utilize the carrier sense mechanism to ensure it does not interfere with the header.

Conservative Bitrate Adaptation: Bitrate adaptation algorithms have to choose the transmitted bitrates conservatively, because of the difficulty in measuring channels under time-varying contention-prone environments. This increases the link margin in practical settings. To evaluate this property, we need to conduct benchmarks on a network where there are time-varying environmental conditions and multiple contending nodes. However, we cannot conduct such experiments with our software radios, since they are large static nodes (as we can see in Fig. 4) and too expensive for emulating an entire network. Hence, we resort to trace-driven experimentation.

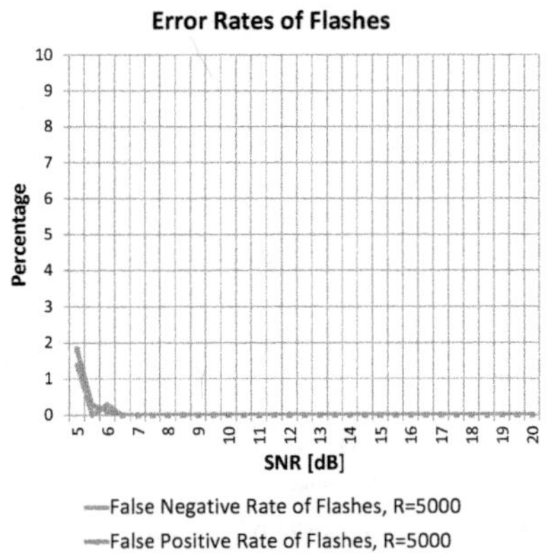

Figure 10: The false negative and positive rates of Flashback when received at roughly $6dB$ higher than the data packets.

We use the traces collected by the RUSK channel sounder, which we described in Sec. 2, to simulate time-varying and contention conditions, and a state-of-the-art bitrate adaptation algorithm, SoftRate [20]. We simulate a Wi-Fi transmitter sending packets over the channel trace, and collect the bitrate decisions SoftRate makes. We then vary the flash rate and measure the effect on the data throughput, by applying the packet loss rates we measured on our Flashback implementation. We also added interference by generating a trace of Bluetooth and Zigbee signals, received at SNRs between $3dB$ and $10dB$.

Fig. 9 plots the overall packet loss rate of the data plane as a function of the flash rate. Without interference, Flashback can send $50,000$ flashes per second. This number is 10 times higher than the number of flashes per second when Flashback was using the optimal discrete bitrate for each SNR. With external interference from sources such as Bluetooth and Zigbee radios, Flashback can send flashes at a rate as high as $100,000$ flashes per second. The reason is that rate adaptation algorithms cannot accurately estimate the channel SNR and interference, and are forced to make conservative bitrate decisions to avoid packet loss.

As we can see, the right flash rate depends on the environment Flashback operates in. In settings where contention, external interference and time-varying conditions are common, Flashback can transmit at a flash rate of $100,000$. However, if the environment is calmer (e.g., a home with several Wi-Fi clients), then Flashback supports a lower flash rate of $50,000$. These correspond to data rates of 400Kbps and 175Kbps respectively. By default, we assume Flashback has a data rate of $175Kbps$.

5.2 Flash Detection Accuracy

Next, we investigate the accuracy of flash detection. We sent flashes at $6dB$ on a known subcarrier over a varying data plane, while modifying the SNR of the received data packet. At each SNR, we calculated the false negative (number of flashes missed by the receiver) and false positive rate (flashes that the receiver detected which weren't actually sent). As depicted by Fig. 10, the error rates are very low for all SNRs. The false negative and positive rates are below 0.1% for all SNRs other than in the range of $5-6.5dB$. This is probably due to the fact that in low SNRs, there is more variation among the power of the different subcarriers and it is difficult to correctly detect the flash.

In the rare occasions that the flash detector produces false positives, it is always on a subcarrier adjacent to the real subcarrier that was sent by the flash transmitter. These false positives are caused by the carrier frequency offset between the flash transmitter and the receiver. This is why the flash messaging technique uses the relative distances between subcarriers to modulate the message, and is therefore resilient to such false positives, as described in Sec. 3.2.

6. APPLICATIONS

Flashback provides a generic decoupled control plane that can be used to improve data plane performance for a variety of applications. The decoupled control plane allows Flashback not only to significantly improve existing scheduling protocols like RTS/CTS and CSMA/CA, but also enable novel applications. This section presents how Flashback can be used to significantly improve existing Wi-Fi MAC protocols. We then show how Flashback enables a virtualized enterprise Wi-Fi network with better handoffs, duty-cycling, load and interference management.

6.1 Flashback-MAC

Flashback Šs control plane enables us to eliminate almost all of the control overhead in the current Wi-Fi MAC protocol, which uses CSMA/CA or RTS/CTS to mediate medium access. In contrast with these schemes, nodes use Flashback Šs control plane to notify the AP of their outstanding transmission requests, while the AP is solely responsible for determining the schedule for the entire network. We describe this simple and generic MAC protocol, which we call Flashback-MAC, and explain how it eliminates a number of chronic problems associated with the current Wi-Fi MAC protocols.

The **Demand Map** is the key novel abstraction that Flashback provides the AP to determine the schedule for the entire network. The demand map is a list of all outstanding transmission requests from all nodes in the network. The list is sorted in order of the arrival of the requests. Each request is associated with a flow identifier at that node, and is also annotated with the SNR of the node requesting a transmission opportunity. Each request can also include a few extra annotations, including bits that signify QoS requests, such as latency or throughput sensitive traffic. Note that the demand map also keeps track of traffic that the AP has to transmit to various nodes.

The demand map for uplink traffic is constructed using flashes. Whenever a node has a packet queued for transmission, it flashes a short control message to the AP. The message contains a flow identifier, and the corresponding amount of outstanding traffic and any associated QoS requests. Our current implementation allows two different classes: latency sensitive and regular data. Latency sensitive traffic is for interactive applications such as VoIP. The first 10 bits in the control message are used to signal the identity of the flashing node, the next 4 are reserved for the flow identifier, the next 8 for outstanding data in multiples of $100B$ for that flow, the next 8 for specifying deadlines, and the final 2 for specifying the QoS level. There are certain bits reserved in the ID and flow identifier fields for association requests for new nodes that wish to join the network.

The control message is sent using the control plane protocol just described. The AP decodes the control message, and also estimates the approximate SNR of the flashing node from the flash message itself. Whenever the AP has the opportunity to send a packet (either data or ACK), it piggybacks a short message at the end of those packets signaling the receipt of the control messages. Hence, if a node does not see an acknowledgement it can resend the control message using flashes.

The demand map decouples the medium access mechanism from the scheduling policy, i.e. it allows the AP to flexibly use any scheduling algorithm on top of the demand map. To demonstrate the demand map's flexibility, we simulated two simple and generic scheduling algorithms.

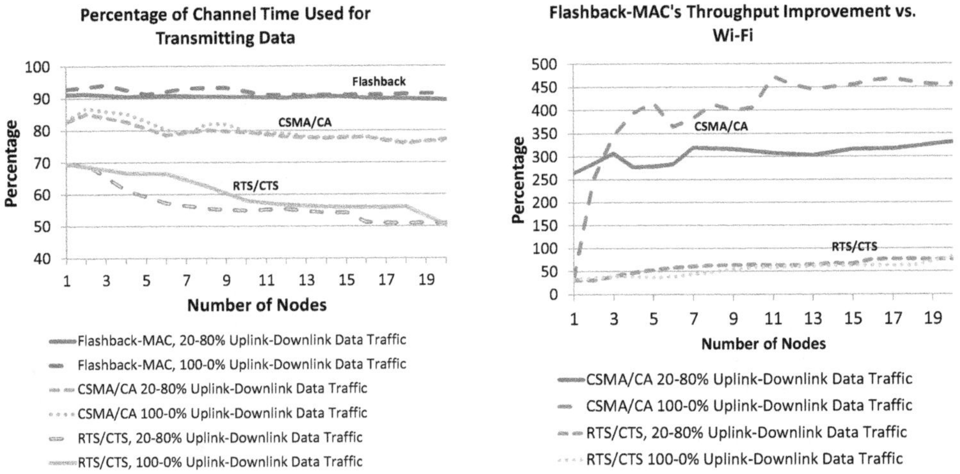

Figure 11: Flashback-MAC using FIFO scheduling compared to CSMA/CA and RTS/CTS with 1,000 byte packets.

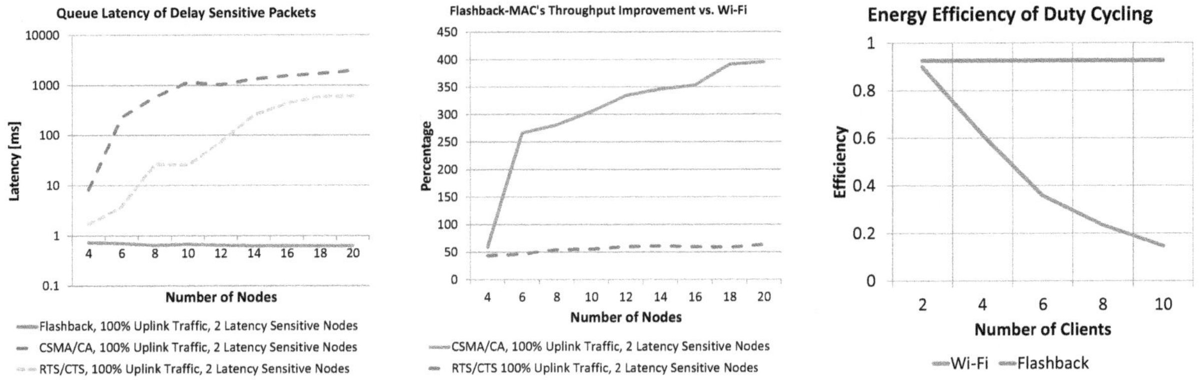

Figure 12: Flashback-MAC using QoS-aware scheduling and duty cycling, compared to RTS/CTS and Wi-Fi PSM. In the two graphs on the left, two nodes are sending latency-sensitive packets, while the rest are sending normal packets. Latency is measured as the delay between the time the node requests to send the packet and when it starts sending the packet. The third figure shows the energy efficiency gains of using Flashback compared to Wi-Fi PSM.

First, we designed a simple FIFO scheme where the AP schedules transmission requests in the order in which they arrive. Second, we simulated a QoS-aware scheduling algorithm that prioritizes latency sensitive traffic ahead of regular traffic.

Once the schedule is determined, the AP uses the next opportunity for a data plane transmission (either data or ACK) to signal which node gets to transmit next, and from which flow identifier.

Flashback-MAC provides significant benefits over the Wi-Fi MAC protocols. First, it eliminates overheads related to medium access. Nodes no longer have to perform carrier sense or exponential back off at the data plane, nor use any control messages such as RTS and CTS. These mechanisms add significant overhead, since they have to be sent at the lowest bitrate ($6Mbps$) compared to the data transmission, which can reach bitrates as high as $54Mbps$. Our simulation results show that such contention overhead consumes nearly 50% of the channel time in congested networks.

Second, it eliminates hidden terminals on the data plane, since nodes no longer have to use the inaccurate carrier sense mechanism to determine if a channel is idle. Instead, with Flashback-MAC the AP explicitly controls which node is transmitting and can actively prevent hidden terminals.

Third, it allows the AP to implement QoS on the uplink and downlink efficiently. Explicit coordination allows Flashback-MAC to guarantee QoS in a centralized fashion, while paying minimal overhead.

Finally, Flashback-MAC is backwards compatible. Even if only the AP and some of the nodes in a network support Flashback-MAC, the AP can still service the requests from the nodes that do not support Flashback, as long as the AP ensures there is enough idle time in the network. The only caveat is that the old nodes may still impact the desired QoS of the Flashback-MAC nodes.

6.1.1 *Performance and Latency*

We simulated Flashback-MAC using an NS-3 Wi-Fi simulation. Our simulated network includes one AP and a varying number of clients that are randomly uniformly distributed on a 80×80 grid. We applied to the simulation the packet loss and flash error rate values that we collected from our Flashback receiver implementation and from our channel sounder traces. The nodes collectively flash at a maximum rate of $50,000$ flashes per second. We implemented the demand map abstraction for the AP, and simulated the scheduling policies described above.

In order to analyze the amount of control overhead Flashback-MAC can eliminate, we simulated a congested network of nodes, each trying to schedule a large amount of traffic ($80Mbps$ of $1,000$ byte packets). We compare the amount of time the network spends on transmitting data with the amount of time the network spends on being idle or transmitting control messages. We measured this time for CSMA/CA, RTS/CTS and Flashback-MAC, under two scenarios: full uplink traffic, and a mix of 20% uplink traffic and 80% downlink. Fig. 11 shows the results of the simulation. As we can see, as the network grows larger, the overhead for RTS/CTS becomes greater, up to a level of 50%, while CSMA/CA falls to a level of about 77%. For CSMA/CA, as the number of nodes increases, more collisions occur due to hidden nodes, and nodes experience more packet loss and their bitrates deteriorate, which causes their packets to occupy the channel for a longer time. For RTS/CTS, the channel utilization decreases significantly, because the nodes spend more time contending on the channel in order to send their RTS or CTS packets. Flashback-MAC completely eliminates the hidden node problem of CSMA/CA, since it centrally schedules all the nodes in the network. In addition, Flashback-MAC does not suffer from RTS and CTS packet overhead, because it can send all the node requests over the control channel, and it piggybacks the request acknowledgements and scheduling decisions on the AP's ACKs. The reason Flashback-MAC has an overhead of about 8-10%, is that it still spends the Wi-Fi SIFS inter-packet idle time between sending a packet and receiving an ACK.

Flashback-MAC has significantly higher throughput than CSMA/CA (up to $5.5\times$) in congested networks, because CSMA/CA suffers from collisions due to hidden nodes, and hence nodes transmit at lower bitrates. All the nodes in our simulator use a bitrate adaptation algorithm that is based on packet loss. CSMA/CA's throughput can be improved, if we use a bitrate adaptation algorithm that discerns between SNR based packet loss and collision based packet loss. In comparison with RTS/CTS, the extra time that Flashback-MAC gains by not sending RTS and CTS packets, is translated into an increase of up to $1.8\times$ in throughput.

FIFO is of course a very basic scheduling algorithm. In order to demonstrate a slightly more sophisticated scheme that utilizes the demand map abstraction, we simulated a QoS-aware scheme, where incoming requests are inserted into two queues: a delay sensitive queue and a normal data queue. The delay sensitive queue is emptied at $4\times$ the rate of the normal data queue. Nodes can request to schedule a delay sensitive message by simply flashing a control message with the corresponding QoS bits.

The QoS simulation uses a similar network to the FIFO simulation. The only difference is that only two of the nodes send $1,000$ byte latency-sensitive packets at a relatively low rate ($800kbps$ per node), while the rest of the nodes are attempting to send regular $1,000B$ packets at the high rate of $80Mbps$. Fig. 12 depicts the results. As we can see, Flashback-MAC's QoS aware schedule can enforce that the latency of the delay-sensitive packets will be kept below $1ms$, while maintaining a significantly higher throughput than the Wi-Fi MAC protocols. Note that the low latency access is enforced, despite the fact that the control plane is congested. This is because control messages are small ($180\mu s$) relative to data packets. As a result, the QoS requests are promptly delivered to, and serviced by the AP. In conclusion, this simulation emphasizes the importance of decoupling the control and data planes.

6.1.2 Energy Proportional Data Plane

Flashback-MAC enables clients to operate the radios in an energy proportional manner. Energy proportionality means that a radio has to be awake only when it is sending or receiving useful traffic, and does not have to spend energy contending for channel access or waiting to be served by the AP. The basic idea is simple: the default mode for the client radio is to sleep. If a client wants to send packets to the AP, it wakes up and flashes a control message. The AP replies with the schedule, telling the client when it will have an opportunity to send its packet. The AP also informs the client if it has any downlink packets destined to it. If the client has no packets to send to the AP, it periodically (every $100ms$) wakes up and flashes a control message to the AP asking if there are any outstanding packets. The AP informs the client if it has any outstanding traffic and the schedule for that traffic. The client then wakes up at the appropriate time and waits for its packets to be delivered.

Fig. 12 plots the energy efficiency of a client using this protocol when compared to the state-of-the-art Wi-Fi Power Save Mode (PSM) used to conserve energy. Energy efficiency is defined as the fraction of time a client radio is sending and receiving useful traffic relative to the total time it is awake with both protocols. Wi-Fi PSM allows nodes to sleep with $100ms$ duty cycles, but if they have either uplink or downlink traffic, clients have to stay up until they get served. As we can see, clients obtain near optimal energy efficiency with Flashback, since clients do not have to stay up and contend to access the network like in normal Wi-Fi.

6.2 Centralized Enterprise Wi-Fi

We can use Flashback-MAC to enable an enterprise Wi-Fi network with the following properties.

- Seamless Mobility: Once a client authenticates with the network, it should be able to roam around the entire enterprise network and send and receive packets from the closest AP. It should not have to spend time and energy in re-associating.
- Better Load and Interference Management: The network should be able to assign clients dynamically to different APs to balance load. Within an AP, client performance should not suffer due to interference from other nodes.
- QoS: The network should be able to provide QoS to prioritize latency-sensitive traffic.

Current networks struggle to offer such capabilities. First, if clients are mobile, they are forced to re-associate with new APs as they move around. The association protocol is time consuming, since nodes need to wait for the beacons from the neighboring APs, find the one with the best signal strength, and then go through the association protocol even though they need to connect to the same enterprise network. Further, load balancing is hard, since clients make decisions on which AP to connect to and the network manager has little control over the allocation of clients to APs. Once connected, interference problems are common due to scenarios such as hidden terminals. Finally, the network has no capability of providing QoS. For example, there is no mechanism to provide low latency connectivity for VoIP traffic.

Flashback provides a general primitive to enable this network architecture, namely, a decoupled control plane. Nodes can utilize Flashback with the following protocol. When associating and authenticating with the network for the first time, nodes use Flashback to send control messages that include their address and authentication information. These messages can be sent at any time, without the node having to wait for the beacon and then contend for sending an association request packet to the AP. Further, nodes can just broadcast the Flashback association message, without specifying the AP. Any AP which hears the association message forwards it to a central controller. The controller assigns the client to the best available AP depending on the load on each AP and provides it with an authentication token. When the client wishes to initiate a flow, it flashes a control message with the request, which includes the authentication token. Since the token is recognized by all the APs in the network, any AP can immediately schedule the request.

This protocol can ensure that client association and mobility is seamless. First, clients do not have to associate with a particular AP to be able to send and receive traffic. Second, clients are presented with a simple flow abstraction: clients request a flow setup, and the controller just assigns the client the best possible AP to relay that traffic. This naturally allows for mobility, since there is no longer a notion of being connected to an AP. Further, since the controller can decide which APs to assign to which clients, it can dynamically adjust to varying load and ensure that no AP is overloaded.

Note that this functionality is very difficult or impossible to obtain with normal RTS/CTS style control signaling that is coupled with the data plane. Any type of control message requires a node to contend for access with other transmissions, whether it's for association or load-balancing.

7. RELATED WORK

Our flashing mechanism and its name were mainly inspired by FlashLinQ [21], a centrally synchronized peer-to-peer cellular protocol that also utilizes the positions of narrowband signals on different frequencies to encode scheduling information. Unlike Flashback, FlashLinQ requires tight and centralized symbol-level synchronization (it uses cellular GPS based time synchronization), which is not available in a distributed asynchronous network like Wi-Fi.

Flashback is related to prior work on pulse based messaging in Zigbee networks [10], which allows a node to concurrently send control messages with data transmissions. However, this technique sends a time domain pulse across the entire bandwidth and corrupts the whole packet. It requires the data transmission to be very resilient to interference, and therefore only works with protocols such as Zigbee, which do not perform rate adaptation and have very robust link margins. Flashback on the other hand, exploits the unique properties of OFDM to flash in order to localize interference. Moreover, Flashback works under Wi-Fi's rate adaptation mechanism, where link margins are much smaller than Zigbee.

Most synchronized or centralized wireless networks use some kind of independent control channel, which in many cases is physically separated from the data channel. For example, LTE has dedicated physical control channels both for downlink and uplink traffic [13, 8]. These control channels have pre-allocated frequencies, which creates a clear demarcation between the control and data channel bands. This type of frequency based separation is not possible for asynchronous distributed networks like Wi-Fi, because such networks are not centrally managed. Hence, in Wi-Fi, the receiver needs to re-synchronize each time to the transmitter's transmission (using the time-based preamble). This makes it very complicated for the receiver to decode transmissions of multiple senders on different bands in the same time slot, as base stations do in LTE. In particular, LTE and WiMAX [1, 2] both utilize OFDMA, a multi-user OFDM scheme that requires tight synchronization, where different nodes can transmit simultaneously on different subcarriers.

In addition, similar to Flashback, Sen et al. [17], utilize a back off arbitration mechanism based on randomly selecting subcarriers. The main difference between the two control techniques is that Sen et al. only use their technique when the data channel is idle, where Flashback allows nodes to flash concurrently with data transmissions and therefore creates a decoupled control plane.

8. CONCLUSION

Flashback follows the classic networking architecture of decoupling the control and data planes. Moreover, it provides this separate control plane at minimal overhead. We believe Flashback can be applied to a number of problems in wireless networks, ranging from fast network association, to mobility and power saving client modes. Further, Flashback can serve as a substrate for distributed coordination to tackle coexistence in dense network deployments. We plan to explore these applications in our future work.

9. ACKNOWLEDGEMENTS

We would like to thank Aditya Gudipati and Steven Hong for their help in setting up the experimental framework, and Manu Bansal, Dinesh Bharadia, Israel Cidon and Jeffrey Mehlman for their valuable feedback. We would also like to thank our shepherd, Krishna Chintalapudi, and the anonymous reviewers for their comments. Asaf Cidon is supported by the Leonard J. Shustek Stanford Graduate Fellowship.

10. REFERENCES

[1] Ieee standard for local and metropolitan area networks, part 16: Air interface for fixed wireless broadband systems.

[2] Ieee standard for local and metropolitan area networks, part 16: Air interface for fixed wireless broadband systems, amendment 2: Physical and medium access control layers for combined fixed and mobile operation in licensed bands and corrigendum 1.

[3] Labview system design software,http://www.ni.com/labview/.

[4] Ni pxie-8133 user manual, http://www.ni.com/pdf/manuals/372870b.pdf.

[5] CHANDRA, R., MAHAJAN, R., MOSCIBRODA, T., RAGHAVENDRA, R., AND BAHL, P. A case for adapting channel width in wireless networks. In *ACM SIGCOMM* (2008).

[6] CHENG, Y.-C., BELLARDO, J., BENKÖ, P., SNOEREN, A. C., VOELKER, G. M., AND SAVAGE, S. Jigsaw: solving the puzzle of enterprise 802.11 analysis. In *Proceedings of the 2006 conference on Applications, technologies, architectures, and protocols for computer communications* (New York, NY, USA, 2006), SIGCOMM '06, ACM, pp. 39–50.

[7] EUGENIO MAGISTRETTI, KRISHNA CHINTALAPUDI, B. R., AND RAMJEE, R. Wifi-nano: Reclaiming wifi efficiency through 800ns slots. In *Proc. of MOBICOM 2011*.

[8] GHOSH, A., RATASUK, R., XIAO, W., CLASSON, B., NANGIA, V., LOVE, R., SCHWENT, D., AND WILSON, D. Uplink control channel design for 3gpp lte. In *Personal, Indoor and Mobile Radio Communications, 2007. PIMRC 2007. IEEE 18th International Symposium on* (sept. 2007), pp. 1 –5.

[9] JUDD, G., AND STEENKISTE, P. Using emulation to understand and improve wireless networks and applications. In *Proceedings of the 2nd conference on Symposium on Networked Systems Design & Implementation - Volume 2* (Berkeley, CA, USA, 2005), NSDI'05, USENIX Association, pp. 203–216.

[10] KAISHUN WU, HAOYU TAN, Y. L. J. Z. Q. Z., AND M.NI, L. Free side channel: Bits over interference. In *ACM MOBICOM* (2010).

[11] KHURANA, S., KAHOL, A., AND JAYASUMANA, A. Effect of hidden terminals on the performance of ieee 802.11 mac protocol. In *Local Computer Networks, 1998. LCN '98. Proceedings., 23rd Annual Conference on* (oct 1998), pp. 12 –20.

[12] KUN TAN, JI FANG, Y. Z. S. C. L. S. J. Z., AND ZHANG, Y. Fine grained channel access in wireless lan. In *Proc. of SIGCOMM 2010*.

[13] LOVE, R., KUCHIBHOTLA, R., GHOSH, A., RATASUK, R., CLASSON, B., AND BLANKENSHIP, Y. Downlink control channel design for 3gpp lte. In *Wireless Communications and Networking Conference, 2008. WCNC 2008. IEEE* (31 2008-april 3 2008), pp. 813 –818.

[14] MANGOLD, S., CHOI, S., MAY, P., KLEIN, O., HIERTZ, G., STIBOR, L., POLL CONTENTION, C., AND POLL, F. Ieee 802.11e wireless lan for quality of service.

[15] NG, P. C., LIEW, S. C., SHA, K. C., AND TO, W. T. Experimental study of hidden-node problem in ieee802.11 wireless networks *, 2005.

[16] PLESS, V. *The Theory of Error Correcting Codes*. Wiley Interscience Series in Discrete Mathematics and Optimization. New York: John Wiley and Sons, 1989. Second Edition.

[17] SEN, S., ROY CHOUDHURY, R., AND NELAKUDITI, S. No time to countdown: migrating backoff to the frequency domain. In *ACM MobiCom* (2011).

[18] SINGLETON, R. Maximum distance -nary codes. *Information Theory, IEEE Transactions on 10*, 2 (apr 1964), 116 – 118.

[19] TELEKOMMUNIKATION, F., CZINK, N., B, B., VAZQUEZ-VILAR, G., JALLOUL, L., AND PAULRAJ, A. Stanford july 2008 radio channel measurement campaign, 2008.

[20] VUTUKURU, M., BALAKRISHNAN, H., AND JAMIESON, K. Cross-layer wireless bit rate adaptation. In *Proceedings of the ACM SIGCOMM 2009 conference on Data communication* (New York, NY, USA, 2009), SIGCOMM '09, ACM, pp. 3–14.

[21] WU, X., TAVILDAR, S., SHAKKOTTAI, S., RICHARDSON, T., LI, J., LAROIA, R., AND JOVICIC, A. Flashlinq: A synchronous distributed scheduler for peer-to-peer ad hoc networks. In *Communication, Control, and Computing (Allerton), 2010 48th Annual Allerton Conference on* (29 2010-oct. 1 2010), pp. 514 –521.

JMB: Scaling Wireless Capacity with User Demands

Hariharan Rahul Swarun Kumar Dina Katabi
Massachusetts Institute of Technology

ABSTRACT

We present *joint* multi-user beamforming (JMB), a system that enables independent access points (APs) to beamform their signals, and communicate with their clients on the same channel as if they were one large MIMO transmitter. The key enabling technology behind JMB is a new low-overhead technique for synchronizing the phase of multiple transmitters in a distributed manner. The design allows a wireless LAN to scale its throughput by continually adding more APs on the same channel. JMB is implemented and tested with both software radio clients and off-the-shelf 802.11n cards, and evaluated in a dense congested deployment resembling a conference room. Results from a 10-AP software-radio testbed show a linear increase in network throughput with a median gain of 8.1 to 9.4×. Our results also demonstrate that JMB's joint multi-user beamforming can provide throughput gains with unmodified 802.11n cards.

Categories and Subject Descriptors

C.2.2 [**Computer-Communication Networks**]: Network Protocols

Keywords

Wireless Networks, Multi-user MIMO, Distributed MIMO

1. INTRODUCTION

Wireless spectrum is limited; wireless demands can, however, grow unlimited. Busy Wi-Fi networks, for instance, in conference rooms, hotels, and enterprises are unable to keep up with user demands [40, 16], even causing high profile failures like the wireless network collapse during the Steve Jobs iPhone 4 keynote. Cellular networks are in a similar predicament, with their demands forecast to exceed available capacity within the next few years [32]. This is not for lack of improvement in the performance of wireless devices. Indeed, individual wireless devices have improved dramatically in recent years through innovations like the introduction of multi-antenna systems, better hardware, and lower receiver noise. The problem however is that there is a mismatch between the way user demands scale and network throughput scales; user demands scale with the number of devices in the network but network throughput does not. Unless network throughput also scales with the number of devices, wireless networks will always find it hard to keep up with their demands, and the projected demands will keep exceeding the projected capacity.

SIGCOMM'12, August 13–17, 2012, Helsinki, Finland.

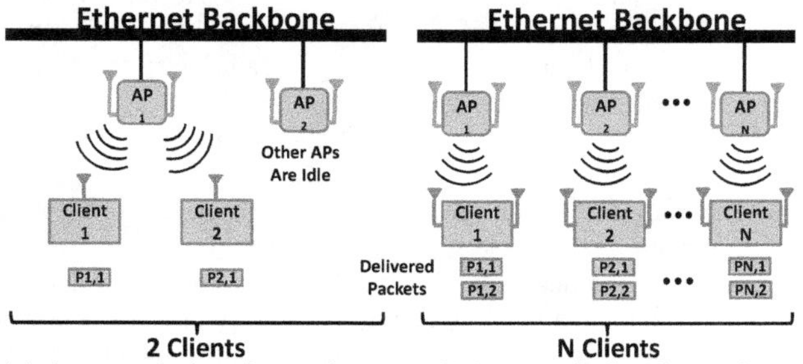

(a) 1 transmission for each antenna on a *single* AP

(b)1 transmission for each antenna on *all APs*

Figure 1: **Traditional vs. Joint Multi-User Beamforming.** In a traditional multi-user beamforming system with multiple 2 antenna APs, only 1 AP can transmit on a given channel at any given time. This leads to a maximum of 2 simultaneous packet transmissions regardless of the total number of APs. In contrast, JMB enables all APs to transmit on the same channel, allowing up to $2N$ simultaneous packet transmissions if there are N 2-antenna APs.

In this paper, we present a system that enables a network to scale its throughput with the number of transmitting devices. We focus on the scenario of typical busy wireless environments such as multiple users in a conference room, enterprise, hotel *etc.* We enable a wireless LAN to keep increasing its total throughput by continuously adding more access points (APs) on the same channel.

The key technical idea behind our system is joint multi-user beamforming (JMB). Multi-user beamforming is a known technique that enables a MIMO transmitter to deliver multiple independent streams (i.e., packets) to receivers that have fewer antennas, as shown in Fig. 1(a), where a 2-antenna access point delivers two packets concurrently to two single antenna receivers. In contrast, as shown in Fig. 1(b), JMB enables multiple access points on the same channel to deliver their packets concurrently to multiple receivers, without interfering with each other. This system scales network throughput with the number of devices, and delivers as many concurrent streams/packets as the total number of antennas on all APs. Furthermore, it can leverage the continuing performance and reliability improvements of individual devices (e.g., more antennas per device).

The main challenge in implementing JMB stems from the need to synchronize the phases of distributed transmitters. Specifically, the goal of beamforming is to ensure that each client can decode its intended signal without interference. Thus, at each client, the signals intended for the other clients have to cancel each other out. This requires the transmitters to control the relative phases of their transmitted signals so that the desired cancellation can be achieved. Such a requirement is naturally satisfied in the case of a single device performing multi-user beamforming. However, in the case of JMB, the transmitters have independent oscillators, which are bound to have differences in their carrier frequencies. If one simply tries to jointly beamform these independent signals from different transmitters, the drift between their oscillators will make the signals rotate at different speeds relative to each other, causing the phases to diverge and hence preventing beamforming.

At first blush, it might seem that it would be sufficient to estimate the frequency offset (i.e., the drift) $\Delta\omega$ between the transmitters, and compensate for the beamforming phase errors as $\Delta\phi = \Delta\omega t$, where t is the elapsed time. However, such an approach is not practical. It is well known [1, 36, 15] that frequency offset estimates have errors due to noise, and using such estimates to compute phases causes rapidly accumulating errors over time. Even a small error of, say, 10 Hz (4×10^{-3} ppm, which is several orders of magnitude smaller than the mandated 802.11 tolerance of 20 ppm, or cellular tolerance of 1-2 ppm), can lead to a large error of 20 degrees (0.35 radians) within a short time interval of 5.5 ms. Such a large error in the phase of the beamformed signals will cause significant interference at the receivers, preventing them from decoding.

JMB presents a simple and practical approach for synchronizing the phases of multiple distributed transmitters. The key idea underlying JMB is to elect one of the APs as a lead and use its phase as a reference for the whole system. Other APs (i.e., the slaves) directly measure the phase of the lead AP and change the phase of their signals to maintain a desired alignment with respect to the lead. In particular, JMB precedes every data packet with a couple of symbols transmitted by the lead AP. The slave APs use these symbols to directly measure the required phase correction for proper beamforming. Since this is a direct phase measurement as opposed to a prediction based on frequency offsets, it has no accumulated errors. After correcting for this phase error, the slave APs use the estimate for their frequency offset to predict any phase changes throughout the packet and correct for it. This bounds the maximum phase error accumulation to the duration of a packet. One can use a simple long term average for the frequency offset to ensure that the phase error accumulated for the duration of a packet is within the desired performance bounds.

In the rest of the paper, we expand on this basic idea and demonstrate that it can deliver accurate joint beamforming across distributed transmitters. Further, we also extend this idea to work with off-the-shelf 802.11 cards. This would allow organizations to directly leverage JMB by simply upgrading their AP infrastructure, without requiring any modification to the clients.

We implemented JMB in two environments:

- The first environment consists of USRP2 APs and receivers, where both APs and clients can be modified. We use this environment to verify the scaling properties of JMB, and also to perform finer grained analysis of the individual components of JMB.
- The second environment consists of USRP2 APs and receivers with Intel Wi-Fi Link 5300 adapters. Each AP in this second testbed consists of two USRP2s connected via an external clock and configured to act as a 2-antenna MIMO AP. Correspondingly, each receiver Wi-Fi card has 2 antennas enabled. We use this testbed to verify that JMB can provide throughput gains with off-the-shelf 802.11n cards, and further, that JMB can provide these gains with multi-antenna devices.

We evaluated JMB in an indoor testbed using APs and receivers deployed densely in a room to simulate a conference room scenario. Our results reveal the following findings:

- **USRP testbed:** JMB's throughput increases linearly with the number of APs. In particular, in our testbed, which has 10 APs, JMB can achieve a median throughput gain of $8.1 - 9.4\times$ over traditional 802.11 unicast, across the range of 802.11 SNRs.
- **802.11 testbed:** JMB's ability to linearly scale the network throughput with the number of transmitters applies to off-the-shelf 802.11 clients. Specifically, JMB can transmit simultaneously from two 2-antenna APs to two 2-antenna 802.11n clients to deliver a median throughput gain of $1.8\times$ compared to traditional 802.11n.
- **Phase Synchronization:** JMB's distributed phase synchronization algorithm is accurate. The 95th percentile misalignment between APs observed at the receiver is less than 0.05 radians. Further, for the whole range of operational SNRs of 802.11 (5-25 dB), the reduction in SNR at each client due to misalignment, (*i.e.*, the total power of interference from all signals not intended for this client to the noise floor) increases on average by 0.13 dB for every additional AP-client pair.

Contributions: This work presents the first system that scales wireless throughput by enabling joint beamforming from distributed independent transmitters. To achieve this, we design a simple and practical approach for performing phase synchronization across multiple distributed transmitters. Finally, we also show that our system can deliver throughput gains from joint beamforming with off-the-shelf 802.11n cards.

2. RELATED WORK

(a) Empirical systems: Recent years have seen a few systems that tried to capture the gains promised by distributed multi-user beamforming [8, 31, 7, 27]. These systems, however, do not address phase synchronization, which is a basic problem in achieving such a system. In particular, they either require the base stations to be tightly synchronized with a Global Positioning System (GPS) clock[1], or assume that all the transmit antennas are driven by a single oscillator [7], or even assume that the receivers can jointly decode the data by exchanging all the received signals [27]. The closest to our work is [17], which addresses phase synchronization, but does not perform distributed joint transmission and achieves large errors (around 20 degrees) that cannot support distributed MIMO. In contrast, JMB provides the first system that achieves phase synchronization using independent oscillators at the devices in the network. As a result, JMB can enable devices to operate independently without having to share a common clock or use external clocks such as GPS. Finally, since JMB does not require any modifications to existing hardware, it can work with off-the-shelf 802.11n cards.

JMB is related to work on enabling concurrent transmissions across different nodes in the network like MU-MIMO in LTE and WiMAX [24, 21], SAM [37], IAC [10], multi-user beamforming [3] , and n+ [19]. However, these systems do not scale with the number of devices in the network. In particular, the throughput of these systems is limited either by the number of antennas on a single AP [24, 21, 37, 3], or the maximum number of antennas on any device in the network [19], or twice the number of antennas on any device in the network [10]. In contrast, JMB is the first system that enables the number of concurrent transmissions to scale with the number of APs, independent of the number of antennas on a single device. This allows JMB to support multiple independent APs communicating simultaneously with multiple independent clients.

JMB is also related to work on harnessing channel diversity gains such as distributed antenna systems [22, 5], and SourceSync [30], as well as work on phased arrays [11], which provide directional gain by sending the same signal on different antennas with different, carefully calibrated delays. However, these systems can not provide multiplexing benefits and hence, unlike JMB, cannot scale network throughput with the number of APs. Finally, recent work has shown how to synchronize concurrent transmissions in time and frequency [36, 30]. JMB builds on these results to deliver a distributed

[1]While promising, GPS typically does not work indoors, rendering such a GPS-based system hard to use in practice.

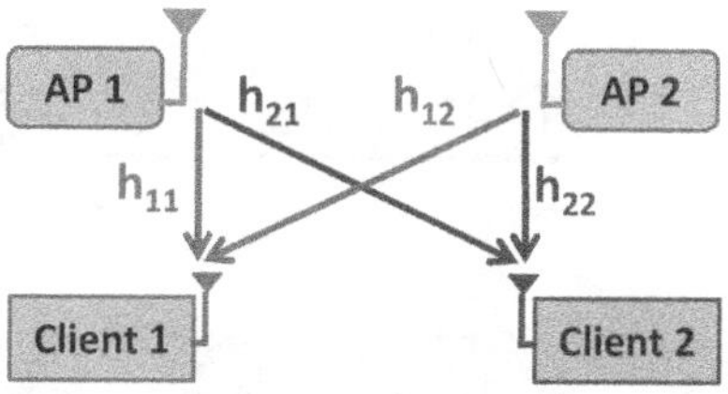

Figure 2: Channel matrix with 2 APs transmitting to 2 clients.

MIMO system. However, time and frequency synchronization alone are not sufficient, since joint multi-user beamforming intrinsically depends on the ability of the distributed APs to achieve phase synchronization, without which it is impossible to allow independent clients to decode simultaneously.

(b) Theoretical results: There is some theoretical work [38, 4] that addresses distributed phase synchronization, but assumes frequency synchronous oscillators and only provides one-time phase offset calibration. Further, the promise of distributed MIMO to improve the scalability of wireless networks has been explored in the theoretical community [2, 35, 41]. Work by Ozgur *et al.* [28] theoretically proved that such a setup can scale wireless capacity with the number of nodes. While JMB builds on this foundational work, JMB is the first empirical system that shows that linear scaling of throughput with the number of transmitters is possible in practical systems with unsynchronized oscillators and resulting time-varying phase differences.

3. JMB OVERVIEW

JMB is designed for the wireless downlink channel. It is applicable to wireless LANs, especially in dense deployments like enterprises, hotels, and conference rooms. JMB APs can operate with off-the-shelf WiFi client hardware. The techniques in JMB are also applicable to cellular networks, but the potential of integrating them with off-the-shelf cellular clients and evaluating them in the cellular context are beyond the scope of this paper.

JMB APs are connected by a high throughput backend, say, Gigabit Ethernet, like they are today. Packets intended for receivers are distributed to all APs through the shared backend. JMB enables these APs to transmit concurrently to multiple clients, as if they were one large MIMO node, potentially delivering as many streams (i.e., packets) as the total number of antennas on all APs.

In the next few sections, we describe how JMB works. We start with the basic idea that enables distributed phase synchronization. We then describe our protocol implementing this basic idea for emulating a large MIMO node. We then extend our system to integrate our design with off-the-shelf WiFi cards.

4. DISTRIBUTED PHASE SYNCHRONIZATION

The chief goal of distributed phase synchronization is to enable different transmitters powered by different oscillators to emulate a single multi-antenna transmitter where all antennas are driven by the same oscillator. Intuitively our solution is simple: We declare one transmitter the lead, and make all other transmitters synchronize to the oscillator of the lead transmitter, *i.e.*, each transmitter measures the offset between its oscillator and the lead oscillator and compensates for the offset by appropriately correcting the phase of its transmitted signal. This behavior makes all transmitters act as if they were antennas on the same chip controlled by the same oscillator.

We now demonstrate how this intuitive design can deliver the proper MIMO behavior and hence enable each receiver to correctly decode its intended signal without interference. For simplicity, we consider a scenario of 2 single-antenna APs transmitting to 2 single-antenna clients, as shown in Fig. 2. Let h_{ij}, where, $i,j \in \{1,2\}$ be the channel to client i from AP j, $x_j(t)$ the symbol that needs to be delivered to client j at time t, and $y_j(t)$ the symbol that is received by client j at time t. Correspondingly, let $\mathbf{H} = [h_{ij}], i,j \in \{1,2\}$ be the 2x2 channel matrix, $\vec{x(t)} = [x_1(t)\ x_2(t)]^T$ be the desired symbol vector, and $\vec{y(t)} = [y_1(t)\ y_2(t)]^T$ be the received symbol vector.

No Oscillator Offset: Assume first that there are no oscillator offsets between any of the APs and clients. If each AP i simply transmits the signal $x_i(t)$, each client will receive a linear combination of the transmitted signals. Since each client has only one antenna, client 1 receives $y_1(t) = h_{11}x_1(t) + h_{12}x_2(t)$ and client 2 receives $y_2(t) = h_{21}x_1(t) + h_{22}x_2(t)$. Each of these equations has two unknowns, and hence, neither client can decode its intended data.

In order to deliver two concurrent packets to the two clients, the APs need to ensure that each client receives only the signal intended for it (*i.e.*, it experiences no interference from the signal intended for the other client). Specifically, we need the effective channel experienced by the transmitted signal to be diagonal, *i.e.*,, it should satisfy:

$$\begin{pmatrix} y_1(t) \\ y_2(t) \end{pmatrix} = \begin{pmatrix} g_{11} & 0 \\ 0 & g_{22} \end{pmatrix} \begin{pmatrix} x_1(t) \\ x_2(t) \end{pmatrix}, \tag{1}$$

where g_{11} and g_{22} are any non-zero complex numbers. In this case, the received signal will simply appear at each receiver as if it has experienced the channel g_{ii}, which each receiver can estimate using standard techniques.

The APs can achieve this result by using *beamforming*. In beamforming, the APs measure all the channel coefficients from the transmitters to the receivers at time 0. Then, instead of transmitting $x_1(t)$ and $x_2(t)$ directly, the APs transmit:[2]

$$\begin{pmatrix} s_1(t) \\ s_2(t) \end{pmatrix} = \mathbf{H}^{-1} \begin{pmatrix} x_1(t) \\ x_2(t) \end{pmatrix} \tag{2}$$

In this case, the two clients receive:

$$\begin{aligned} \begin{pmatrix} y_1(t) \\ y_2(t) \end{pmatrix} &= \mathbf{H} \begin{pmatrix} s_1(t) \\ s_2(t) \end{pmatrix} \\ &= \mathbf{H}\mathbf{H}^{-1} \begin{pmatrix} x_1(t) \\ x_2(t) \end{pmatrix} \end{aligned}$$

Since $\mathbf{H}\mathbf{H}^{-1} = \mathbf{I}$, the effective channel experienced by the clients in this case is a diagonal matrix, *i.e.*, Eq. 1 is satisfied. Hence, each client can now decode its intended data without interference from the signal intended for the other client.

With Oscillator Offset: What happens when the oscillators of the APs and clients have different frequencies? Let ω_{Ti} be the oscillator frequency of AP i, and ω_{Rj} be the oscillator frequency of client j, $i,j \in \{1,2\}$. In this case, the channel at time t, $\mathbf{H(t)}$, can be written as:

$$\mathbf{H(t)} = \begin{pmatrix} h_{11}e^{j(\omega_{T1}-\omega_{R1})t} & h_{12}e^{j(\omega_{T2}-\omega_{R1})t} \\ h_{21}e^{j(\omega_{T1}-\omega_{R2})t} & h_{22}e^{j(\omega_{T2}-\omega_{R2})t} \end{pmatrix},$$

where $j = sqrt(-1)$. Because the oscillators rotate with respect to each other, the channel no longer has a fixed phase.

Now, if the APs try to perform beamforming as before, using the channel value they computed at time $t = 0$ and transmitting $\mathbf{H}^{-1}\vec{x}$, the clients receive:

$$\begin{pmatrix} y_1 \\ y_2 \end{pmatrix} = \mathbf{H(t)}\mathbf{H}^{-1} \begin{pmatrix} x_1 \\ x_2 \end{pmatrix},$$

[2]The APs also need to normalize $\mathbf{H}^{-1}$ to respect power constraints, but we omit that detail for simplicity.

The product $\mathbf{H(t)H^{-1}}$ is no longer diagonal, and hence the receivers cannot decode their intended signal. Thus, standard MIMO beamforming does not work in this case.

So how can one do beamforming with such a time varying channel? A naive approach would try to make each transmitter compute $\mathbf{H(t)}$ at every t and then multiply its time signal by $\mathbf{H(t)}^{-1}$. Say that the network has N APs and N clients. Then such an approach would require each transmitter to maintain accurate estimates of N^2 frequency offsets of the form $\Delta\omega_{ij} = \omega_{Tj} - \omega_{Ri}$. (Further since nodes can only measure frequency offsets relative to other nodes, but not the absolute frequencies of their oscillators, the number of estimates cannot be reduced to N.) Measurement errors from all of these estimates will accumulate, prevent accuracy of beamforming, and create interference at the receivers. However, according to our intuition at the beginning of this section, we can make multiple transmitters act as if they were one big MIMO node, and hence do accurate beamforming, by having each transmitter estimate only its frequency offset to the lead transmitter. Said differently, our intuition tells us that it should be possible to reduce the number of frequency offset estimates that each transmitter maintains from N^2 to one. Let us see how we can achieve this goal.

Observe that we can decompose the channel matrix at time t as $\mathbf{H(t)} = \mathbf{R(t)HT(t)}$, where $\mathbf{H}$ is time invariant, and $\mathbf{R(t)}$ and $\mathbf{T(t)}$ are diagonal matrices defined as:

$$\mathbf{R(t)} = \begin{pmatrix} e^{-j\omega_{R1}t} & 0 \\ 0 & e^{-j\omega_{R2}t} \end{pmatrix}$$

and

$$\mathbf{T(t)} = \begin{pmatrix} e^{j\omega_{T1}t} & 0 \\ 0 & e^{j\omega_{T2}t} \end{pmatrix}$$

Since $\mathbf{R(t)}$ is diagonal, it can function analogous to the $\mathbf{G}$ matrix in Eq. 1. Thus, if the transmitters transmit the modified signal $\mathbf{T(t)^{-1}H^{-1}}\vec{x}$ at time t, then the received signal can be written as:

$$\begin{pmatrix} y_1 \\ y_2 \end{pmatrix} = \mathbf{R(t)HT(t)T(t)^{-1}H^{-1}} \begin{pmatrix} x_1 \\ x_2 \end{pmatrix} \quad (3)$$

which reduces to the desired form of Eq. 1

$$\begin{pmatrix} y_1 \\ y_2 \end{pmatrix} = \mathbf{R(t)} \begin{pmatrix} x_1 \\ x_2 \end{pmatrix} \quad (4)$$

Note that $\mathbf{T(t)}$ is also diagonal, and as a result the transmitter phase correction matrix

$$\mathbf{T(t)}^{-1} = \begin{pmatrix} e^{-j\omega_{T1}t} & 0 \\ 0 & e^{-j\omega_{T2}t} \end{pmatrix} \quad (5)$$

is also diagonal. Further, the phase correction entry for each AP depends only on the oscillator phase of that AP. This means that if each AP, i, knows its phase, $e^{j\omega_{Ti}t}$, at time t, it can simply compensate for that phase and the AP will not need any additional frequency or phase measurements. Unfortunately, this is not practical. An AP has no way to measure the exact phase change of its oscillator locally.

We address this difficulty by observing that the channel equation is unchanged when we multiply by $1 = e^{j\omega_{T1}t}e^{-j\omega_{T1}t}$, *i.e*,

$$\begin{aligned} \mathbf{H(t)} &= e^{j\omega_{T1}t}\mathbf{R(t)HT(t)}e^{-j\omega_{T1}t} \\ &= \begin{pmatrix} e^{j(\omega_{T1}-\omega_{R1})t} & 0 \\ 0 & e^{j(\omega_{T1}-\omega_{R2})t} \end{pmatrix} \mathbf{H} \begin{pmatrix} 1 & 0 \\ 0 & e^{(j(\omega_{T2}-\omega_{T1})t} \end{pmatrix} \end{aligned}$$

Since the new observed channel matrix is still diagonal, the clients can still continue to decode the received signal as before.

The resulting system implements our intuition at the beginning of this section.

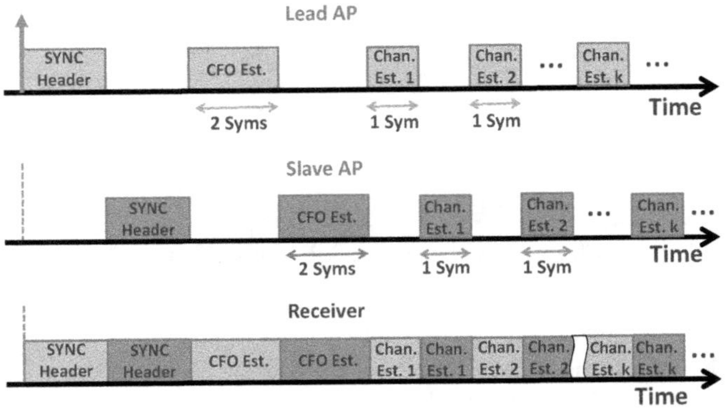

Figure 3: Packet Structure from the perspective of APs and the receiver. Symbols in blue are transmitted by the lead AP, symbols in red by the slave AP, and symbols in white reflect silence periods.

5. JMB PROTOCOL

We start by describing the protocol at a high level, and follow by the detailed explanation. JMB's distributed transmission protocol works in two phases:

- JMB starts with a **channel measurement phase**, in which the APs measure two types of channels: 1) the channels from themselves to the receivers (i.e., the matrix $\mathbf{H}$), which is the beamforming channel whose inverse the APs use to transmit data concurrently to their clients; and 2) the channels from the lead AP to the slave APs (the h_i^{lead}'s), which enables each slave AP to determine its relative oscillator offset from the lead AP.
- The channel measurement phase is followed by the **data transmission phase**. In this phase, the APs transmit jointly to deliver concurrent packets to multiple receivers. Data transmission uses beamforming after having each slave AP corrects for its frequency offset with respect to the lead AP.

Note that a single channel measurement phase can be followed by multiple data transmissions. Channels only need to be recomputed on the order of the coherence time of the channel, which is several hundreds of milliseconds in typical indoor scenarios [9]. §7 describes how JMB reduces channel measurement overhead in greater detail.

We now describe the channel measurement and data transmission phases in greater detail. (The description below assumes symbol level time synchronization, for which we use the scheme in [30], which provides tight synchronization up to a few nanoseconds. Our experimental results also incorporate an implementation of that scheme).

5.1 Channel Measurement

The goal of channel measurement is to obtain a snapshot of the channels from all APs to all clients, i.e., $\mathbf{H}$ and the reference channels from the lead AP to the slave APs, i.e., the h_i^{lead}, $\forall i$.

The key point is that *all these channels have to be measured at the same time*, which is the reference time $t = 0$. Otherwise the channels would rotate with respect to each other due to frequency offsets and hence be inconsistent. Below, we divide channel measurement into a few sub-procedures.

(a) Collecting Measurements. The lead AP starts the channel measurement phase with a synchronization header, followed by channel measurement symbols, *i.e.*, known OFDM symbols that the clients can use to estimate the channel. The channel measurement symbols are separated by a constant gap, whose value is chosen to permit the slave APs to send their channel measurement symbols interleaved with the symbols from the lead AP. When the slave APs hear the synchronization header, they know to transmit their channel measurement symbols in the gap, one after another, as shown in Fig. 3.

Thus, channel measurement symbols are repeated and interleaved. They are repeated to enable the clients to obtain accurate channel measurements by averaging multiple estimates to reduce the impact of noise. They are interleaved because we want the channels to be measured as if they were measured at the same time. Since exactly simultaneous transmissions will lead the APs to interfere with each other, JMB performs a close approximation to simultaneous transmission by interleaving symbols from different APs.

(b) Estimating H at the clients. Upon reception of the packet in Fig. 3, each client performs three tasks: it computes its carrier frequency offset (CFO) to each AP; it then uses its knowledge of the transmitted symbols and the CFO to compute the channel from each AP to itself; and finally it uses its knowledge of the CFOs to rotate the phase of the channels so that they look as if they were measured exactly at the same time. We detail these tasks below.

Different transmitters (i.e., APs) have different oscillator offsets to the receivers, and the receiver needs to measure the frequency offset from each transmitter to correct the corresponding symbols from that transmitter appropriately. To enable this, the channel measurement transmission uses CFO symbols from each receiver followed by channel estimation symbols similar to traditional OFDM [15]. The only departure is that the receiver computes and uses different CFO and channel estimates for symbols corresponding to different APs.

Note that these channel estimates are still not completely simultaneous, in particular, the channel estimation symbols of slave AP i is separated from the symbol of the lead AP by $i-1$ symbol widths, as shown in Fig. 3. The receiver compensates for this by rotating the estimated channel for AP i by $e^{-j\Delta\omega_i(i-1)kT+D}$ (in each OFDM subcarrier), where T is the duration of one OFDM symbol, k is the index of the interleaved symbol, and D is the duration of the lead AP synchronization header. This ensures that all channels are measured at one reference time, which is the start of the synchronization header. The receiver averages the channel estimates (in each OFDM subcarrier) from each AP to cancel out the noise and obtain an accurate estimate. The receivers then communicate these estimated channels back to the transmitters over the wireless channel.

(c) Estimating the h_i^{lead}'s at the Slave APs. Each slave AP uses the synchronization header to compute the value of the channel from the lead AP to itself at the reference time $h_i^{lead}(0)$.

Note that at the end of the channel measurement phase, each slave AP i has the entire channel matrix to be used for beamforming, as well as a reference channel, $h_i^{lead}(0)$ from the lead AP which it will use during data transmissions, with all channels measured with respect to one reference time.

5.2 Data Transmission

Now that the channels are measured, the APs can use beamforming to transmit data concurrently without interference.

(a) AP Coordination: The APs need to agree on which packets are sent concurrently in one beamforming frame. To do this we leverage the bandwidth of the backend Gigabit Ethernet to send all client packets to all APs. The lead AP makes all control decisions and communicate them to the slave APs over the Ethernet. In particular, it determines which packets will be combined in a data transmission and communicates it to the slave APs over the wired backend.

(b) Beamforming: Client packets are transmitted by joint beamforming from the JMB APs participating in the system. Note that slave APs need to correct the phase of their signal prior to transmission. One way to do this would be for each slave to estimate the frequency offset $\omega_{lead}-\omega_{slave}$ from the lead to itself (using the synchronization header from the previous phase) and then compute the net elapsed phase by calculating $(\omega_{lead}-\omega_{slave})t$, where t is the time elapsed since the channel measurement was taken. However, this would lead to large accumulated errors over time because of any inaccuracies in the measurement of the initial frequency offset. For example, even a small error of 100 Hz in the measurement of the initial frequency offset can lead to a large phase error of π radians in as short a timespan as 20 ms, and hence significantly affect the phase alignment required for correct beamforming. Unless addressed, this error would prevent JMB from amortizing the cost of a single channel measurement over the coherence time of the channel, e.g., 250 ms, and would force the system to repeat the process of measuring **H** every few milliseconds, which means incurring the overhead of communicating the channels from all clients to the APs almost every packet.

JMB avoids this issue of accumulating error over large timescales by directly measuring the phase difference between the lead AP and the slave AP. Said differently instead of multiplying the frequency offset $\Delta\omega(=\omega_{lead}-\omega_{slave})$ by the elapsed time (which leads to errors that accumulate over time), JMB directly measures the phase difference $\Delta\phi(t)(=(\omega_{lead}-\omega_{slave})t)$.

In JMB the lead AP initiates data transmission using a synchronization header, as in channel estimation. Each slave AP use this synchronization header to measure the current channel, $h_i^{lead}(t)$ from the lead AP to itself. Note that the current channel will be rotated relative to the reference channel because of the oscillator offset between the lead AP and slave AP. In particular, $h_i^{lead}(t)=h_i^{lead}(0)e^{j(\omega_{T1}-\omega_{T2})t}$. Each slave can therefore compute $e^{j(\omega_{T1}-\omega_{T2})t}$ directly, from its two measurements of the lead AP channel. Such an estimate does not have errors that accumulate over time because it is purely a division of two direct measurements. The slave then multiplies its transmitted signal by this quantity, as described in §4.

Now that all the AP oscillators are synchronized at the beginning of the data transmission, the slave AP also needs to keep its oscillator synchronized with the lead transmitter throughout the actual data packet itself. It does this by multiplying its transmitted signal by $e^{j(\omega_{T1}-\omega_{T2})t}$ where t is the time since the initial phase synchronization at the beginning of the joint transmission. Note that this offset estimate only needs to be accurate within the packet, *i.e.*, for a few hundred microseconds or about 2 ms at most. JMB APs maintain a continuously averaged estimate of their offset with the lead transmitter across multiple transmissions to obtain a robust estimate that can maintain accurate phase synchronization within a packet.

Two additional points about JMB's synchronization are worth noting.

First, for ease of exposition, we have discussed the entire system so far in the context of correcting carrier frequency offsets. However, any practical wireless system has to also account for the sampling frequency offsets. Note that any offset in the sampling frequency just adds to the phase error in each OFDM subcarrier. Since our phase offset estimation using the synchronization header, described in §5, estimates the overall phase, it automatically accounts for the initial phase error accumulated from sampling frequency offset. Within each packet, the JMB slave APs correct for the effect of sampling frequency offset during the packet by using a long-term averaged estimate, similar to the carrier frequency offset.

Second, as mentioned earlier, in §5, JMB APs are synchronized in time using [30]. As described in [30], due to differences in propagation delays between different transmitters and different receivers, one cannot synchronize all transmitted signals to arrive exactly at the same time at all receivers. It is important to note that JMB works correctly even in the presence of different propagation delays between different transmitters and receivers. This is because the signals from different JMB APs will arrive within a cyclic prefix of each other at all receivers.[3] The delay differences between the signals from different APs at a receiver translate to a relative phase difference between the channels from these APs to that receiver. JMB's channel measurement phase captures these relative phase differences in the channel matrix, and JMB's beamforming then applies the effect of these phase differences while computing the inverse of the channel matrix.

5.3 Overarching Principles

In summary, the core challenge met by JMB's design is to accurately estimate and track the phase differences between each of the N clients and N APs. This challenge is particularly arduous for two reasons: (1) each receiver must simultaneously track the phase of N independent transmitters, and (2) errors in the estimates in the CFO result in phase offsets that accumulate over time, quickly leading to very large errors. Our general approach to tackling these challenges is to have all transmitters and receivers synchronize their phase to that of a single lead transmitter. Our implementation of this approach has been guided by following three overarching principles:

- **Between APs and within a packet we can use estimated frequency and sampling offsets to track phase:** We can measure the frequency and sampling offsets *between APs* accurately enough that the accumulated phase differences within a single packet (10s to a few 100s of *microseconds*) are not significant enough to harm performance. Specifically, since APs are a part of the infrastructure, and CFOs do not change significantly over time, we can get very accurate estimates of the CFO between APs by averaging over samples taken across many packets.
- **Between APs and *across* packets we *cannot* use estimated frequency and sampling offsets to track phase:** The across packet time scales (10s to 100s of *milliseconds*) are large enough that even with extremely accurate estimates of the frequency and sampling offsets, the accumulated phase differences from residual errors will lead to significant performance degradation. To handle this, JMB uses a single header symbol to directly estimate the total *phase offset* and re-sync the phases of all nodes at the beginning of each packet.
- **Between a client and an AP, we cannot use estimated frequency and sampling offsets to track phase even through a packet:** Since clients are a transient part of the network, we cannot get accurate enough estimates of frequency and sampling offsets to use for phase tracking even within a single packet. Thus each client uses standard OFDM techniques to track the phase of the lead AP symbol by symbol, Additionally when performing channel estimation, the APs interleave their packets so that the correction of the channels to a common reference time has minimal error.

6. COMPATIBILITY WITH 802.11

In order for JMB to work with clients using off-the-shelf 802.11 cards, JMB needs to address two challenges:

1. **Sync header:** The sync header transmitted by the lead AP to allow the slave APs to compute their oscillator offset, and trigger their transmission, is not supported by 802.11.
2. **Channel measurement:** Recall that JMB requires a snapshot of the channel from all transmitters to all receivers measured at the same time. In §4, we described how to do this with a custom channel measurement packet format with interleaved symbols that allows a receiver to measure channels from all transmitters. However, such a packet format is not supported by 802.11, and hence 802.11 cards cannot simultaneously measure channels from all APs at the same time.

JMB solves these issues in the context of 802.11n by leveraging 802.11n channel state information (CSI) feedback for beamforming. We now describe JMB's solutions to each of the challenges listed above.

6.1 Sync Header

The lead AP in JMB needs to prefix each transmission with a sync header that allows the slave transmitters to measure their relative oscillator offset from the lead, and also triggers their joint transmission. A mixed mode 802.11n packet essentially consists of an 802.11n packet prefixed with 5 legacy symbols. These legacy symbols are only intended to trigger carrier sense in 802.11a/g nodes, and are not used by 802.11n receivers. Thus, the lead JMB can use these legacy symbols as a sync header. JMB slave APs use the legacy symbols to measure their oscillator phase offset from the lead, correct their transmission signal, and join the lead AP's transmission after the legacy symbols when the actual 802.11n symbols are transmitted.

6.2 Channel Measurement

802.11n does not support the interleaved packet format that allows JMB to measure a snapshot of the channels from all the transmitters to a receiver simultaneously. Even more fundamentally, an 802.11n receiver with K (at most 4) antennas can measure at most K channels at a time. In a JMB system, the total number of transmit antennas across all APs is larger than the number of antennas on any single receiver. Thus, a receiver with off-the-shelf 802.11n cards will be unable to simultaneously measure channels from all transmit antennas to itself.

Naively, one could measure the channels from all transmit antennas by transmitting a separate packet from each AP, and then correcting these channel measurements using the estimated frequency offsets to the receiver like in §5.1. Unlike the scenario in §5.1 where the transmissions from different APs are separated from each other by only a few symbols (using interleaving), the transmissions from different APs here are separated by at least one packet width. As discussed in §5.3, it is not practical to compute receiver frequency offsets accurately enough to ensure that the accumulated phase error across packets will be tolerable.

JMB instead performs robust channel measurement by "tricking" the receiver into measuring channels from different AP antennas simultaneously. This trick allows JMB to measure the channel from each AP antenna to the receiver in conjunction with a common reference channel to the receiver. Using such a common reference across all measurements allows JMB to avoid measuring the *receiver frequency offset*, and instead directly estimate the oscillator *phase offset* between different channel measurements, and therefore compensate for it, as we describe below.

For simplicity, we focus on the scenario in Fig. 4 with 2 APs (a lead and a slave) and 1 client, where all nodes have 2 antennas each. In the rest of this discussion, we will focus only on the channel measurements to R_1 since the channels to R_2 are naturally measured simultaneously with R_1 in exactly the same manner.

At time t_0, L_1 and L_2 transmit a 2-stream packet jointly to R_1. This measurement gives us the channels $L_1 \rightarrow R_1$ and $L_2 \rightarrow R_1$

[3] In fact, since the common design scenario for JMB is confined locations like conference rooms and auditoriums, the propagation delay differences between different APs to a receiver are in the tens of nanoseconds, which is smaller than the 802.11 cyclic prefix of 400 or 800 ns, which is designed for worst case multipaths.

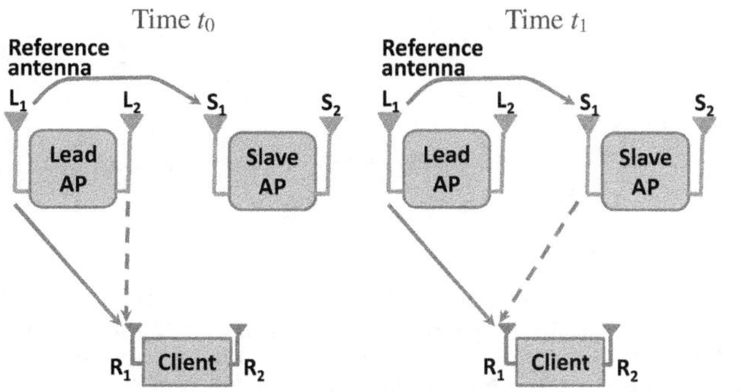

Figure 4: 802.11 Channel Measurement: JMB measures channels with 802.11 clients by sending a series of two-stream transmissions. Every transmission includes the reference antenna, L_1, as well as one other antenna (either L_2 or S_1 in our example). Note that for clarity, the figure does not show the transmissions to/from R_2 and S_2, but JMB naturally measures the channels to R_2 simultaneously.

at time t_0. In addition, S_1 measures the channel $L_1 \rightarrow S_1$ using the synchronization header.

At time t_1, L_1 and S_1 trick the receiver by jointly transmitting a 2-stream packet from 2 different APs. This measurement gives us the channels $L_1 \rightarrow R_1$ and $S_1 \rightarrow R_1$ at time t_1. Again, S_1 measures the channel $L_1 \rightarrow S_1$ using the synchronization header.

The challenge is that we would like to obtain the channel $S_1 \rightarrow R_1$ at time t_0 but we have only the channel $S_1 \rightarrow R_1$ measured at t_1.

We therefore need to correct our measured channel by the accumulated phase offset between S_1 and R_1 in the time interval t_0 to t_1. To do this, we take advantage of the fact that we can compute the accumulated phase offset between both L_1 and R_1, and between L_1 and S_1 in the time interval t_0 to t_1.

- L_1 **and** R_1**:** We can compute this accumulated phase offset using the measurements of the channel $L_1 \rightarrow R_1$ at time t_0 and time t_1.
- L_1 **and** S_1**:** We can compute this accumulated phase offset using the measurements of the channel $L_1 \rightarrow S_1$ at time t_0 and time t_1.

The difference between these two accumulated phase offsets gives us the desired accumulated phase offset between S_1 and R_1 in the time interval t_0 to t_1.

We can similarly measure the channel $S_2 \rightarrow R_1$ in the next time slot, say t_2, and rotate it back to time t_0. We can repeat this process for all AP antennas.

7. DECOUPLING MEASUREMENTS TO DIFFERENT RECEIVERS

The scheme described in §4 assumed that the channels from all APs to all receivers are all measured at the same time. In §6.2, we showed how JMB could measure channels from different APs to a single receiver at different times and compensate for differences in oscillator offset by using a shared reference measurement across all APs for that receiver. But what about the channels to a different receiver? If this receiver joins the wireless network after the channels to the first receiver are measured, there is no opportunity for a shared reference measurement between the two receivers. It might therefore seem that JMB's requirement for all channels to be measured at the same time would necessitate measurement of channels to all receivers whenever a receiver joins the network, or when a single receiver's channels change.

In fact, we can show that such full measurement is not necessary, and that JMB can decouple channel measurements to different receivers. The key idea is that JMB can use the channels from the lead AP to slave APs as a shared reference in this case, instead of the channel from the lead AP to a receiver as was the case in §6.2. We prove in the appendix that using such a shared reference allows JMB to measure channels to different receivers at different times, and still correctly perform multi-user beamforming using distributed phase synchronization.

8. DIVERSITY

MIMO systems can provide both multiplexing and diversity gains. So far, we have described the use of JMB for multiplexing. The same discussion applies to diversity except that in this case, we have all the APs transmitting jointly to a single client, say client 1. Each AP then computes its beamformed signal as $\frac{h_{1i}^*}{\|h_{1i}\|}x_1$ and slaves continue to perform distributed phase synchronization as before.

9. JMB'S LINK LAYER

So far, we have described JMB's physical layer, which enables multiple APs to transmit simultaneously to multiple receivers. We now describe how JMB's link layer is designed to use this capability.

MAC and Carrier Sense: In JMB, all downlink packets are sent on the Ethernet to all JMB APs. Thus, all APs in the network have the same downlink queue. Each packet in the queue has a *designated AP*, which is the AP with the strongest SNR to the client to which that packet is destined. JMB always uses the packet at the head of the queue for transmission, and nominates the designated AP of this packet as the lead AP for this transmission. The lead AP then chooses additional packets for joint transmission with this packet in order to maximize the network throughput. There are a variety of heuristics [43, 33, 42] that can be adopted for selecting the packets for joint transmission, and we leave the exact algorithm for making this choice for future work.

The lead AP contends on behalf of all slave APs, with its contention window weighted by the number of packets in the joint transmission as described in [29]. Clients contend for the medium as they do today using 802.11 CSMA. When the lead AP wins a contention window, it starts transmitting its synchronization header, which causes the slave APs to join the transmission. We note that contention for joint transmission by the lead AP is robust to hidden terminals for two reasons. First, JMB is intended for dense deployments like conference rooms where access points can hear each other, and the overall wireless capacity is limited by interference between access points. Further, even in the unlikely event of hidden terminals, situations causing persistent packet loss due to repeated collisions can be detected using mechanisms like in [34], and the lead AP can ensure that JMB access points that trigger hidden terminal packet loss above a threshold are not part of the joint transmission. This ensures that both JMB joint transmissions, as well as other transmissions, do not encounter persistent hidden terminals.

Rate Selection using Effective SNRs: In systems like JMB where different sets of APs transmit concurrently for different packets, the rate to a client can change from packet to packet as the effective channel at each client changes as a result of beamforming. Such systems therefore need to use a rate-selection algorithm [19]. JMB uses the **effective SNR** algorithm, which is designed for rate selection for 802.11-like frequency selective wideband channels [13]. Since APs in JMB know the full channel matrix, $\mathbf{H}$, prior to transmission. APs multiply the signals by $k\mathbf{H}^{-1}$ (k accounts for the maximum power constraint at APs). Thus, the effective channel is $k\mathbf{H}^{-1}\mathbf{H} = k\mathbf{I}$, giving a signal strength of k^2 at each client. Client cards report noise N as in [11]. Clients send the noise N to APs along with the measured channels. Since APs know the signal strength, k^2, in each subcarrier, and the associated noise N, they can compute the SNR in each subcarrier as $\frac{k^2}{N}$. They can then map this set of SNRs to rate by performing a table lookup [13]. Thus, each client in a JMB joint transmission gets the same rate, which is similar to traditional 802.11 fairness.

Acknowledgments: JMB disables synchronous ACKs at clients and uses higher layer asynchronous acknowledgments like in prior work such as MRD and ZipTx [25, 20]. Further, similar to systems like Maranello [14], JMB can modify the firmware on clients to implement an optimized joint synchronous acknowledgment protocol consisting of a single SIFS, followed by back-to-back acknowledgments from all the clients.

Packet losses and retransmissions: It is important to note that, even though APs transmit packets jointly to different receivers, packet losses at different clients are decoupled. Specifically, if APs have stale channel information to a client, only the packet to that client is affected, and packets at other clients will still be received correctly. As in regular 802.11, APs in JMB keep packets in the queue until they are ACKed. If a packet is not ACKed, they can be combined with other packets in the queue for future concurrent transmissions.

10. TESTBED AND IMPLEMENTATION

We implement JMB's AP design in software radios and evaluate it with both off-the-shelf 802.11 clients and software-radio clients.

(a) Implementation for the software radio testbed: In this testbed, each node is equipped with a USRP2 board [6], and an RFX2400 daughterboard, and communicates on a 10 MHz channel in the 2.4 GHz range. We implement OFDM in GNURadio, using various 802.11 modulations (BPSK, 4QAM, 16QAM, and 64QAM), coding rates, and choose between them using the effective-SNR bitrate selection algorithm [13].

Our JMB implementation includes the following modules: distributed phase alignment, beamforming for multiplexing and diversity, and bitrate selection. We do not implement ACKs, CSMA, or retransmissions. To perform correct phase alignment, concurrent transmitters must be synchronized tightly at the sample level. We do this by using USRP2 timestamps to synchronize transmitters despite delays introduced by software. Before every data packet, the lead AP sends a trigger signal on the medium at $t_{trigger}$. All other APs log the timestamp of this signal, add a fixed delay t_Δ to it, and then transmit concurrently at this new time. We select t_Δ as $150\mu s$ based on the maximum delay of our software implementation. Finally, to optimize the software turnaround, we did not use GNURadio, but wrote our own C code that directly interacts with the USRP hardware.

(b) Implementation for the 802.11 testbed: There are two main differences between this testbed and the one above. First, each client in this testbed uses an off-the-shelf 802.11 card. Second, each node in this testbed has two antennas and can act as a MIMO node. Our objective is to show that JMB extends beyond single antenna systems; For example, it can combine two 2x2 MIMO systems to create a 4x4 MIMO system.

Each AP is built by connecting two USRP2 nodes via an external clock and making them act as a 2-antenna node. Each client is a PC equipped with a Intel Wi-Fi Link 5300 a/b/g/n wireless network adapter on which 2 antennas are enabled. The Intel Wi-Fi Link 5300 adapters are updated with a custom firmware and associated `iwlwifi` driver in order to obtain the channel state information in user space [12].

The AP software implementation is similar to the other testbed except that we make the channel width 20 MHz to communicate with actual 802.11 cards. The packet format is also changed to match 802.11 packets. The client software collects the channel measurements from the firmware and logs correctly decoded packets.

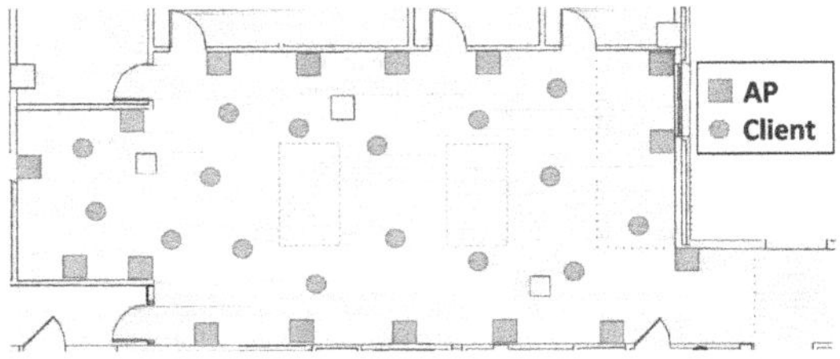

Figure 5: Testbed Topology. Client locations are marked by red circles, and AP locations by blue squares. *Note that the figure shows the set of possible locations for clients and APs, and different subsets of locations are picked for different experiments.*

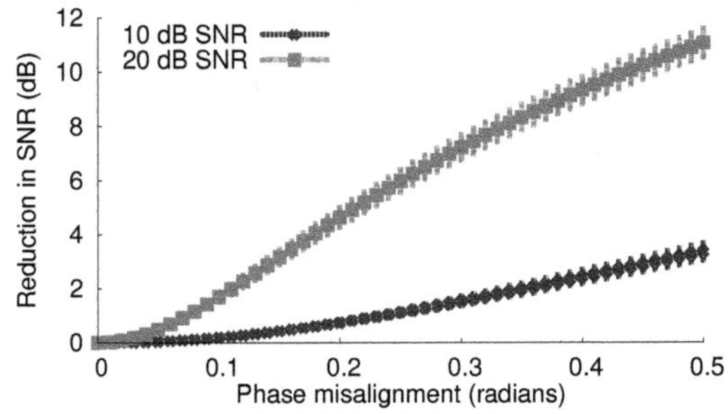

Figure 6: Degradation of SNR due to phase misalignment. Even with only 2 transmitters, a misalignment of just 0.35 radians can reduce the SNR by almost 8 dB at the receivers due to interference.

(c) Testbed Topology: We evaluate JMB in an indoor wireless testbed that simulates a conference room or classroom, with APs deployed on ledges near the ceiling, and clients scattered through the room. Fig. 5 shows node locations in the experimental environment. In every run, the APs and clients are assigned randomly to these locations. Note that the testbed exhibits significantly diverse SNRs as well as both line-of-sight and non line-of-sight paths due to obstacles such as pillars, furniture, ledges etc. The APs transmit 1500 byte packets to the clients in all experiments.

11. RESULTS

We evaluate JMB both through microbenchmarks of its individual components, as well as an integrated system on both USRP and 802.11n testbed.

11.1 Microbenchmarks

(a) Necessity of Phase Alignment: A key challenge for a distributed MIMO system is that it must compensate for oscillator offsets between the transmitters. In this section, we demonstrate the impact of misalignment between transmitters on the received SNR.

Method. We simulate a simple 2-transmitter, 2-receiver system where different data is intended for each receiver. The transmitters measure the initial channel matrix to the receivers, and use this matrix to compute their beamforming vectors. We then introduce a phase misalignment at the slave transmitter, and compute the reduction in SNR at each receiver as a result of this misalignment. We repeat this process for 100 different random channel matrices, phase misalignments from 0 to 0.5 radians, and for two systems - one in which the average SNR is 10 dB, and other in which the average SNR is 20 dB.

Results. Fig. 6 shows the average reduction in SNR in dB, as a function of phase misalignment. As one would expect, an increase in phase misalignment increases the interference at each receiver. As the graph shows, even a phase misalignment as small as 0.35

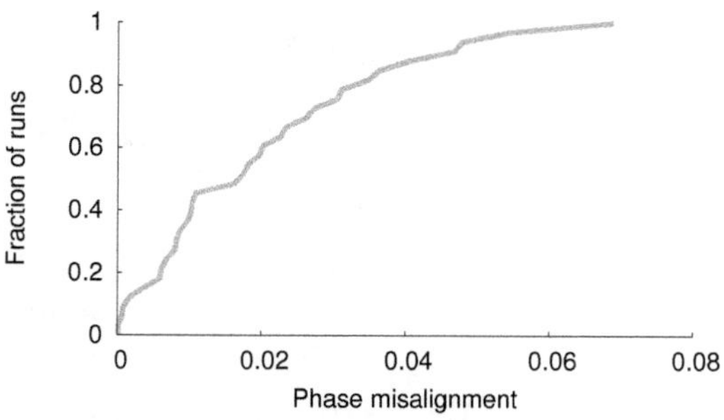

Figure 7: **CDF of observed phase misalignment.** The median misalignment is 0.017 radians, and the 95th percentile misalignment is 0.05 radians.

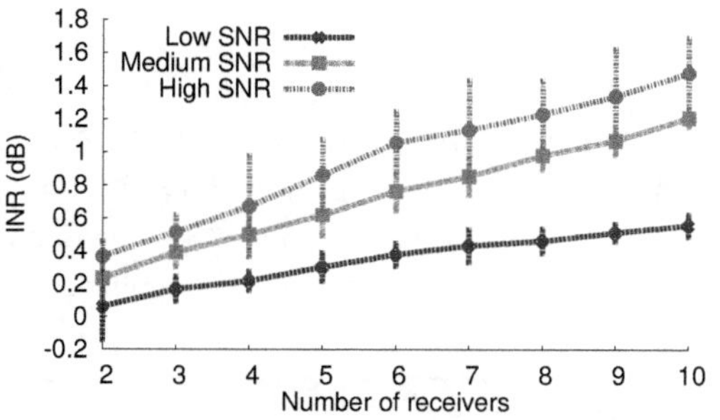

Figure 8: **Accuracy of Phase Alignment.** Average INR at receivers for various numbers of receivers across different SNR ranges. INR stays below 1.5dB across SNRs even with 10 receivers.

radians[4] , can cause an SNR reduction of almost 8 dB at an SNR of 20 dB. This SNR reduction will be greater as we add more and more transmitters to the system. Further, phase misalignment causes a greater reduction in SNR when the system is at higher SNR. This is because the impact of additional noise added by interference is higher when the original noise itself is low, *i.e.*, at high SNR.

(b) Accuracy of JMB's Phase Alignment: We now examine the accuracy of JMB's lightweight distributed phase alignment algorithm.

Method. In this experiment, we place two JMB nodes at random AP locations and a third JMB node at a receiver location. We randomly pick one of the two APs to be the lead and the other to be the slave. The slave transmitter implements JMB's distributed phase synchronization algorithm, and performs phase correction on its transmission before joining the lead transmitter's data transmission. In order to measure the accuracy of JMB's phase synchronization algorithm, we make the lead and the slave APs alternate between transmitting OFDM symbols. In particular, each transmitter's transmission consists of pairs of an OFDM symbol followed by an OFDM symbol length of silence. The transmissions of the lead and slave transmitter are offset by 1 symbol so that the receiver sees alternating symbols from lead and the slave transmitter. The receiver estimates the lead and slave transmitter's channels separately, and computes the relative phase between them. We then perform several rounds of this measurement. The receiver uses the first measurement of relative phase as a reference, and computes the deviation of relative phase from this reference in subsequent transmissions.

Result. Fig. 7 plots the CDF of the absolute value of the deviation in relative phase across all the experiments. If the lead and slave transmitter are always perfectly aligned, the deviation should be zero. However, estimation errors due to noise and oscillator drift due to the delay from when the slave measures the lead's channel and turns around to jointly transmit data will induce misalignment. As can be seen, however, the median misalignment is less than 0.017 radians, and the 95th percentile misalignment is less than 0.05 radians. Based on Fig. 6, with two transmitters, JMB's phase alignment algorithm can ensure that the SNR of joint transmission is not reduced by 0.4 dB at the 95th percentile.

(c) How does SNR reduction scale? The previous experiments examined in depth the impact of misalignment and JMB's precise alignment performance in the case of a 2x2 distributed MIMO system. In this experiment, we observe how the SNR reduction grows as we increase the number of transmitters in the system.

Method. We evaluate the SNR reduction in JMB in three effective SNR [13] ranges: low (6-12 dB), medium (12-18 dB) and high (> 18 dB). For each range, we place several JMB nodes in random AP locations in the testbed. We then place the same number of JMB nodes in random client locations, such that all clients obtain an effective SNR in the desired SNR range. For each placement, we then choose a client at which all APs null their interference, *i.e.* the expected signal at that client is zero and measure the received signal power at that client. If phase alignment is perfect, the received signal power should be comparable to noise, *i.e.* the ratio of the received signal power to noise should be 0 dB. Any inaccuracy in phase misalignment will lead to interference, manifest as a higher ratio of received power to noise, and produce a corresponding reduction in SNR if data were actually transmitted to that client. For each topology, we null at each client, and compute the average interference to noise ratio (INR) across clients. We repeat this experiment for different topologies, and different numbers of JMB APs at low, medium and high SNRs.

Result. Fig. 8 shows the INR as a function of the number of JMB APs at low, medium and high SNRs. Note that, as before, the reduction in SNR increases with SNR, but is below 1.5 dB even at high SNR. The INR also increases with the number of APs, as the number of interferers increases, but increases gradually: only ~ 0.13 dB for every additional AP-client pair even at high SNR.

11.2 Increase of Network Throughput with the Number of APs

JMB's key goal is to increase the network throughput as the number of APs increases. This experiment verifies if JMB delivers on that promise.

Method. We evaluate JMB's performance in three effective SNR ranges: low (6-12 dB), medium (12-18 dB) and high (> 18 dB). For each range, we place a certain number of JMB nodes in random AP locations in the testbed. We then place the same number of nodes in random client locations such that all clients obtain an effective SNR in the desired range. For each such topology, we measure the throughput obtained both with regular 802.11 and JMB. Since USRP2 cannot perform carrier sense due to software latency, we measure the throughput of 802.11 by scheduling each client so that it gets an equal share of the medium. We repeat the experiment for 20 different topologies and also vary the number of JMB APs for each SNR range.

Results. Figs. 9(a), (b), and (c) show the total throughput obtained by 802.11 and by JMB for different numbers of APs, and different SNR ranges. Note that, as one would expect, the obtained throughput increases with SNR (802.11 throughput at low SNR is 7.75 Mbps, at medium SNR is around 14.9 Mbps, and at high SNR is 23.6 Mbps). There are two main points worth noting:

- 802.11 cannot benefit from additional APs operating in the same channel, and allows only one AP to be active at any given time. As

[4]0.35 radians is much smaller than the misalignment expected with the mandated 802.11 tolerance of 20 ppm.

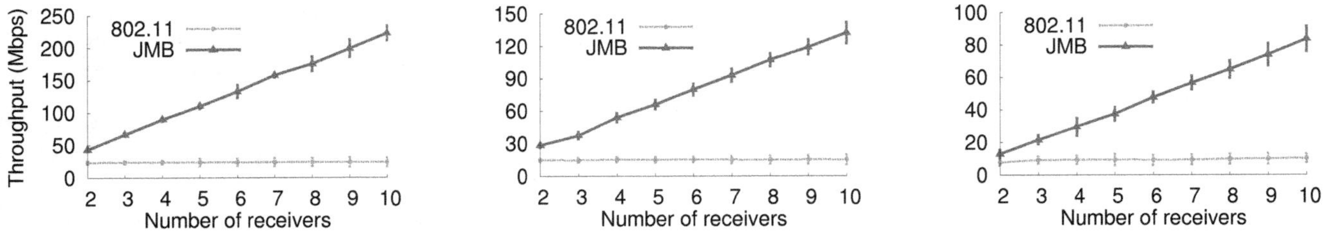

(a) High SNR (> 18 dB) (b) Medium SNR (12-18 dB) (c) Low SNR (6-12 dB)

Figure 9: **Scaling of throughput with the number of APs.** In this experiment, the number of APs equals the number of receivers. At all SNRs, JMB's network throughput increases linearly with the number of APs while total 802.11 throughput remains constant.

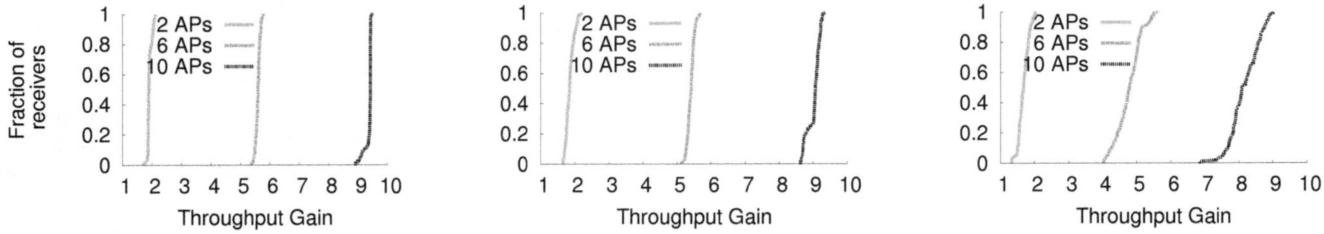

(a) High SNR (> 18 dB) (b) Medium SNR (12-18 dB) (c) Low SNR (6-12 dB)

Figure 10: **Fairness.** CDFs of per-client throughput gain. Across all SNRs, JMB provides all clients with very similar gains.

a result, its throughput stays constant even as the number of APs increases. This throughput might vary with the number of APs in a real 802.11 network due to increased contention; however, since USRPs don't have carrier sense, we compute 802.11 throughput by providing each client with an equal share of the medium. In contrast, with JMB, as we add more APs, JMB can use these APs to transmit concurrent packets to more receivers. As a result, we see that the throughput of JMB increases linearly with the number of APs.

- The absolute gains provided by JMB are higher at high (~9.4× for 10 APs) and medium (~9.1×) SNRs, than at low SNRs (~8.1×). This is a consequence of the theoretically predicted throughput of beamforming. In particular, the beamforming throughput with N APs scales as $N\log(\frac{\text{SNR}}{K}) = N\log(\text{SNR}) - N\log(K)$, where K depends on the channel matrix $\mathbf{H}$ and is related to how well conditioned it is [39]. Natural channel matrices can be considered random and well conditioned, and hence K can essentially be treated as constant for our purposes. The 802.11 throughput scales roughly as $\log(\text{SNR})$ [39]. The expected gain of JMB over 802.11 can therefore be written as $N(1 - \frac{\log(K)}{\log(\text{SNR})})$ and hence becomes closer to N as SNR increases. This is why, JMB's gains at lower SNR grow at a lower rate than the gains at high SNR.

11.3 Fairness

In this experiment, we verify if JMB is fair, *i.e.*, it delivers the above throughput gains to all nodes.

Method. We perform the same experiment as in §11.2. We then compute the throughput gain of each node as the ratio of its throughput with JMB to its throughput with 802.11. As before, we perform this experiment varying the number of APs from 2 to 10, and across the full range of SNRs.

Results. Figs. 10(a), (b), and (c) plot the CDF of the throughput gain for 2, 6 and 10 APs at high, medium, and low SNRs. The results show that JMB is fair *i.e.* all nodes see roughly the same throughput gains, and these match the gains in total throughput shown in §11.2. Note that the CDF is wider at lower SNR. This is a consequence of greater measurement noise at low SNR causing larger throughput differences between clients.

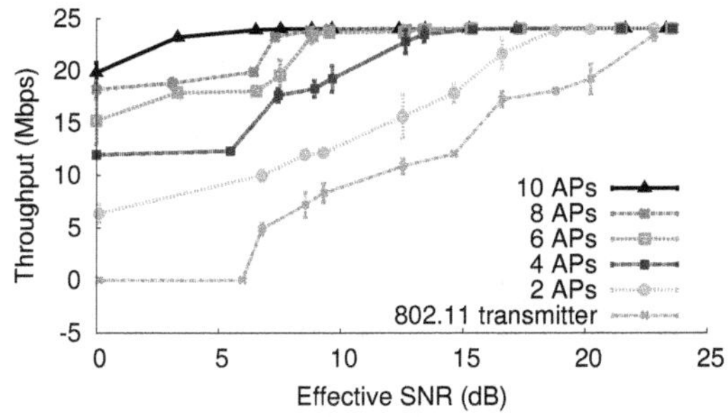

Figure 11: **Diversity Throughput.** Throughput of a JMB client when using diversity with 2, 4, 6, 8 and 10 APs. JMB can achieve close to maximum rate even to a client unable to receive any packets with 802.11.

11.4 Diversity

As described in §8, in addition to providing multiplexing gains, JMB can also provide throughput gains through diversity. In this section, we investigate JMB's diversity gains.

Method. We place several APs in random AP locations in the testbed, and one node at a client location, ensuring it has roughly similar SNRs to all APs. We then compute the throughput with regular 802.11 and JMB. We repeat the experiment with the number of APs varying from 2 to 10, and plot the results for the range of operational SNRs of 802.11.

Results. Fig. 11 shows throughput of 802.11 and JMB as a function of SNR for 2, 4, 6, 8 and 10 APs. Note that JMB provides significant gains over 802.11, especially at low SNRs. For instance, a client that has 0 dB channels to all APs (*i.e.* its received power from each AP is about the same as the noise) cannot get any throughput with 802.11. However the figure shows that, with 10 APs, such a client can achieve a throughput of 21 Mbps with JMB. Thus, using JMB for diversity can significantly expand the coverage range of an 802.11 deployment, and alleviate dead spots. This is expected because with JMB's coherent diversity, using APs to coherent combine the signal can provide a multiplicative increase in the SNR of N^2 [39]. This results in significant throughput improvements in the low SNR regime.

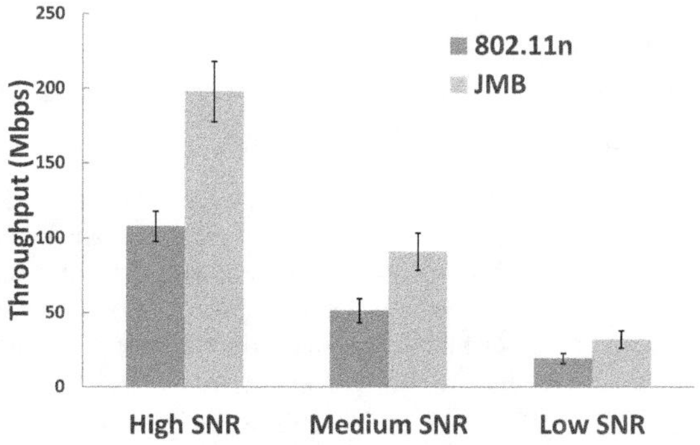

Figure 12: Throughput achieved using JMB on off-the-shelf 802.11n cards. JMB significantly improves the performance of off-the-shelf 802.11n cards at high (>18 dB), medium (12-18 dB) and low (6-12 dB) SNRs.

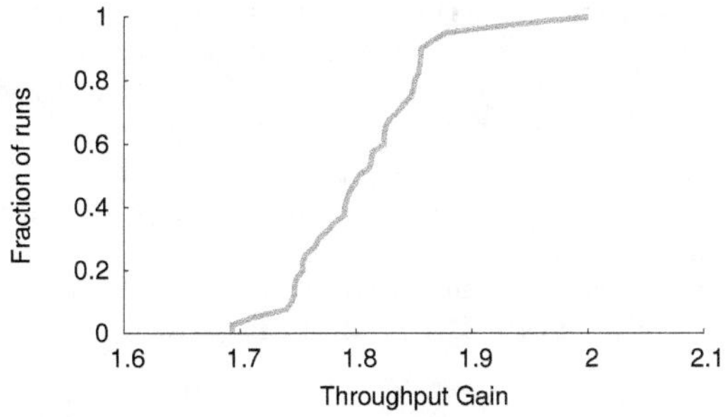

Figure 13: Fairness Results. For all nodes in our testbed, JMB delivers a throughput gain between 1.65-2×, with a median gain of 1.8× across SNRs. This shows that JMB provides similar throughput gains for every node in the network.

11.5 Compatibility with 802.11

Finally, as described in §6, JMB is compatible with existing 802.11n cards. In this section, we investigate whether JMB can deliver significant throughput gains when used with commodity 802.11n cards. Further, since each AP and each 802.11n card in this system has 2 antennas, this experiment also verifies that JMB can provide its expected gains with multi-antenna transmitters and receivers.

Method. We place 2 JMB nodes at random AP locations in the testbed, and 2 802.11n receivers at random client locations in the testbed. For each topology, we compute the total throughput with 802.11n and with JMB. As before, we compute the 802.11n throughput by giving each transmitter an equal share of the medium. We repeat the experiment across multiple topologies and the entire range of SNRs.

Results. Fig. 12 shows the total throughput with and without JMBat high, medium and low SNRs. Since we have two receivers in this experiment, the theoretically throughput gain compared to 802.11n is 2×. The chart shows that JMB delivers an average gain of 1.67-1.83× across all SNR ranges. Similar to the case with USRP receivers, the gains in the high SNR regime are larger than gains in the low SNR regime.

We now investigate JMB's fairness, *i.e.* whether JMB can deliver its throughput gains for every receiver in the network across all locations and SNRs. Fig. 13 shows the CDF of the throughput gain achieved by JMB as compared to 802.11n across all the runs. The results show that JMB delivers throughput gains between 1.65-2× for all the receivers and hence is fair to the receivers in the network.

12. CONCLUSION

This paper enables joint beamforming from distributed independent transmitters. The key challenge in delivering this system is to perform accurate phase synchronization across multiple distributed transmitters. The lessons learnt from building the system and testing it with real hardware are : 1) Estimates of frequency offset can be made accurate enough to predict (and hence correct) phase misalignment within an 802.11 packet; however, these estimates cannot be used across multiple packets due to large build-ups in phase errors over time; and 2) Joint multi-user beamforming can be achieved by synchronizing the phases of all senders to one lead sender, and does not impose any phase synchronization constraints on the receivers.

We believe that the design of JMB has wider implications than explored in this paper. In particular, several areas of information theory like lattice coding, noisy network coding, and transmitter cooperation for cognitive networks [26, 18, 23] assume tight phase synchronization across transmitters. We are optimistic that the algorithms presented in this paper can bring these ideas closer to practice.

Acknowledgments: We thank Dan Halperin for answering various questions about CSI feedback from the Intel cards, and Shyam Gollakota and Nate Kushman for their helpful comments on the paper. This work is funded by NSF.

13. REFERENCES

[1] IEEE standard for information technology–telecommunications and information exchange between systems–local and metropolitan area networks–specific requirements part 11: Wireless LAN medium access control (MAC) and physical layer (PHY) specifications amendment 5: Enhancements for higher throughput. *IEEE Std 802.11n-2009 (Amendment to IEEE Std 802.11-2007 as amended by IEEE Std 802.11k-2008, IEEE Std 802.11r-2008, IEEE Std 802.11y-2008, and IEEE Std 802.11w-2009)*, pages c1 –502, 29 2009.

[2] S. Aeron and V. Saligrama. Wireless ad hoc networks: Strategies and scaling laws for the fixed SNR regime. *IEEE Transactions on Inf. Theor.*, 53(6), 2007.

[3] E. Aryafar, N. Anand, T. Salonidis, and E. Knightly. Design and experimental evaluation of multi-user beamforming in wireless LANs. In *Mobicom 2010*.

[4] S. Berger and A. Wittneben. Carrier phase synchronization of multiple distributed nodes in a wireless network. In *8th IEEE Workshop on Signal Processing Advances for Wireless Communications (SPAWC), Helsinki, Finland*, June 2007.

[5] Distributed Antenna Systems. http://medicalconnectivity.com/2008/02/05/distributed-antenna-systems-no-replacement-for-wireless-strategy.

[6] USRP. http://www.ettus.com. Ettus Inc.

[7] A. Forenza, R. W. H. Jr., and S. G. Perlman. *System and Method For Distributed Input-Distributed Output Wireless Communications*. U.S. Patent Application number 20090067402.

[8] System for increasing capacity in future mobile communications networks. http://www.hhi.fraunhofer.de/fileadmin/hhi/downloads/BM/PR_Demonstration_Network_MIMO.pdf. Fraunhofer Heinrich Hertz Institute.

[9] A. Goldsmith. *Wireless Communications*. Cambridge University Press, 2005.

[10] S. Gollakota, S. Perli, and D. Katabi. Interference alignment and cancellation. *ACM SIGCOMM*, 2009.

[11] Greentouch consortium. https://www.youtube.com/watch?v=U3euDDr0uvo. GreenTouch Demonstrates Large-Scale Antenna.

[12] D. Halperin, W. Hu, A. Sheth, and D. Wetherall. Tool release: gathering 802.11n traces with channel state information. *SIGCOMM Comput. Commun. Rev.*, 41:53–53.

[13] D. Halperin, W. Hu, A. Sheth, and D. Wetherall. Predictable 802.11 packet delivery from wireless channel measurements. In *ACM SIGCOMM*, 2010.

[14] B. Han, A. Schulman, F. Gringoli, N. Spring, B. Bhattacharjee, L. Nava, L. Ji, S. Lee, and R. Miller. Maranello: practical partial packet recovery for 802.11. In *Proceedings of the 7th USENIX conference on Networked systems design and implementation*, NSDI'10, pages 14–14, Berkeley, CA, USA, 2010. USENIX Association.

[15] J. Heiskala and J. Terry. *OFDM Wireless LANs: A Theoretical & Practical Guide*. Sams Publishing, 2001.

[16] The iPad and its impact on hotel owners and operators. http://www.ibahn.com/en-us/public/docs/The_Impact_of_iPad.pdf. iBAHN.

[17] D. C. Jenn, J. H. Ryu, T. Yen-Chang, and R. Broadston. Adaptive phase synchronization in distributed digital arrays. In *Adaptive Hardware and Systems (AHS), 2010 NASA/ESA Conference on*, pages 199 –204, june 2010.

[18] S. Lim, Y. Kim, A. El Gamal, and S. Chung. Noisy network coding. In *IEEE Info. Theor. Workshop*, 2010.

[19] K. C.-J. Lin, S. Gollakota, and D. Katabi. Random access heterogeneous mimo networks. In *Proceedings of the ACM SIGCOMM 2011 conference*, SIGCOMM '11, pages 146–157, New York, NY, USA, 2011. ACM.

[20] K. C.-J. Lin, N. Kushman, and D. Katabi. ZipTx: Harnessing Partial Packets in 802.11 Networks.

[21] LTE: MIMO techniques in 3GPP-LTE. http://lteportal.com/Files/MarketSpace/Download/130_LTEMIMOTechniquesFreescaleNov52008.pdf.

[22] Z. Ma, M. Zierdt, J. Pastalan, A. Siegel, T. Sizer, A. J. de Lind van Wijngaarden, P. R. Kasireddy, and D. M. Samardzija. Radiostar: Providing wireless coverage over gigabit ethernet. *Bell Lab. Tech. J.*, 14:7–24, May 2009.

[23] I. Maric, N. Liu, and A. Goldsmith. Encoding against an interferer's codebook. In *Allerton*, 2008.

[24] MIMO schemes in 16m (WiMAX 2.0). http://www.wimax360.com/profiles/blogs/mimo-schemes-in-16m-wimax-20. WiMax360.

[25] A. Miu, H. Balakrishnan, and C. E. Koksal. Improving loss resilience with multi-radio diversity in wireless networks. In *MobiCom '05: Proceedings of the 11th annual international conference on Mobile computing and networking*, 2005.

[26] B. Nazer and M. Gastpar. The case for structured random codes in network capacity theorems. *European Transactions on Telecommunications*, 19(4), 2008.

[27] Network MIMO. http://www.alcatel-lucent.com/wps/DocumentStreamerServlet?LMSG_CABINET=Docs_and_Resource_Ctr&LMSG_CONTENT_FILE=Data_Sheets/Network_MIMO.pdf. Alcatel-Lucent.

[28] A. Ozgur, O. Leveque, and D. Tse. Hierarchical cooperation achieves optimal capacity scaling in ad hoc networks. *IEEE Trans. on Info. Theor.*, 2007.

[29] H. Rahul, F. Edalat, D. Katabi, and C. Sodini. Frequency-Aware Rate Adaptation and MAC Protocols. In *ACM MOBICOM 2009*, Beijing, China, September 2009.

[30] H. Rahul, H. Hassanieh, and D. Katabi. Sourcesync: a distributed wireless architecture for exploiting sender diversity. *SIGCOMM*, 2010.

[31] Distributed-input distributed-output wireless technology. http://www.rearden.com/DIDO/DIDO_White_Paper_110727.pdf. Rearden Companies.

[32] Mobile broadband capacity constraints and the need for optimization. http://rysavy.com/Articles/2010_02_Rysavy_Mobile_Broadband_Capacity_Constraints.pdf. Rysavy Research.

[33] L. Shen-fa and W. Wei-ling. Fast antenna selection algorithms for distributed mimo systems. In *Journal of Beijing University of Posts and Telecommunication*, volume 30, pages 50–53, Jun. 2007.

[34] V. Shrivastava, N. Ahmed, S. Rayanchu, S. Banerjee, S. Keshav, K. Papagiannaki, and A. Mishra. Centaur: realizing the full potential of centralized wlans through a hybrid data path. In *Proceedings of the 15th annual international conference on Mobile computing and networking*, MobiCom '09, pages 297–308, New York, NY, USA, 2009. ACM.

[35] O. Simeone, O. Somekh, H. Poor, and S. Shamai. Distributed MIMO in multi-cell wireless systems via finite-capacity links. In *ISCCSP*, 2008.

[36] K. Tan, J. Fang, Y. Zhang, S. Chen, L. Shi, J. Zhang, and Y. Zhang. Fine-grained channel access in wireless lan. In *ACM SIGCOMM 2010*.

[37] K. Tan, H. Liu, J. Fang, W. Wang, J. Zhang, M. Chen, and G. M. Voelker. SAM: enabling practical spatial multiple access in wireless LAN. In *MobiCom 2009*.

[38] I. Thibault, G. Corazza, and L. Deambrogio. Phase synchronization algorithms for distributed beamforming with time varying channels in wireless sensor networks. In *Wireless Communications and Mobile Computing Conference (IWCMC), 2011 7th International*, pages 77 –82, july 2011.

[39] D. Tse and P. Vishwanath. *Fundamentals of Wireless Communications*. Cambridge University Press, 2005.

[40] Turn off your wi-fi network! http://www.youtube.com/watch?v=fFiJ5rnIPVw. Steve Jobs iPhone4 Keynote.

[41] S. Venkatesan et al. A WiMAX-based implementation of network MIMO for indoor wireless. *EURASIP*, '09.

[42] Z. Xin-sheng, Y. Xiao-hu, and Z. Ding-qian. An investigation on dynamic rau selection method for distributed radio mobile communications system. In *Journal of Electronics ïijĘ Information Technology*, volume 28, pages 2334–2338, Dec. 2006.

[43] L. Ying-Dong and Z. Guang-Xi. Transmit-receive antenna selection for distributed multi-user mimo downlink channels. In *Information Science and Engineering (ICISE), 2009 1st International Conference on*, pages 2767 –2770, dec. 2009.

APPENDIX

A. DECOUPLING MEASUREMENTS TO DIFFERENT RECEIVERS.

For simplicity, we focus on the example of two APs and two clients in Fig. 2. Let us consider a system where the channels, h_{11} and h_{12}, to receiver 1 are measured at time t_1 and the channels, h_{21} and h_{22}, to receiver 2 are measured at time t_2. For a subsequent transmission at time t, the channels experienced to receiver 1 experience a rotation corresponding to the time $t - t_1$, while the channels experienced to receiver 2 experience a rotation corresponding to time $t - t_2$. In particular, the channel matrix experienced at time t can be written as:

$$\mathbf{H(t)} = \begin{pmatrix} h_{11}e^{j(\omega_{R1}-\omega_{T1})(t-t_1)} & h_{12}e^{j(\omega_{R1}-\omega_{T2})(t-t_1)} \\ h_{21}e^{j(\omega_{R2}-\omega_{T1})(t-t_2)} & h_{22}e^{j(\omega_{R2}-\omega_{T2})(t-t_2)} \end{pmatrix} \tag{6}$$

Recall that for JMB to perform distributed phase synchronization, we need to decompose $\mathbf{H(t)}$ into the form $\mathbf{R(t)HT(t)}$ where $\mathbf{H}$ is time-invariant, and the time-dependent matrices $\mathbf{R(t)}$ and $\mathbf{T(t)}$ are diagonal, and the ith diagonal entry of $\mathbf{T(t)}$ (similarly) depends only on parameters that the ith AP (similarly i^{th} receiver) can estimate locally. The APs can then all use the time invariant matrix $\mathbf{H}$ to calculate their beamforming signal, and perform correction using the relevant entry of $\mathbf{T(t)}$.

We observe that $\mathbf{H(t)}$ can indeed be written in this desired form. Specifically, we can write $\mathbf{H(t)}$ as $\mathbf{R(t)HT(t)}$, where

$$\mathbf{R(t)} = \begin{pmatrix} e^{j(\omega_{R1}-\omega_{T1})(t-t_1)} & 0 \\ 0 & e^{j(\omega_{R2}-\omega_{T1})(t-t_2)} \end{pmatrix} \tag{7}$$

$$\mathbf{H} = \begin{pmatrix} h_{11} & h_{12} \\ h_{21} & h_{22}e^{-j(\omega_{T1}-\omega_{T2})(t_2-t_1)} \end{pmatrix} \tag{8}$$

$$\mathbf{T(t)} = \begin{pmatrix} 1 & 0 \\ 0 & e^{j(\omega_{T1}-\omega_{T2})(t-t_1)} \end{pmatrix} \tag{9}$$

Note that $\mathbf{H}$ is now time invariant as desired. The entries only depend on the oscillator offset between times t_1 and t_2. Slave AP i can easily compute this offset by using the reference channel, h_i^{lead} from the lead measured at time t_1 and t_2.

Further, note that the diagonal entries of the matrix $\mathbf{T(t)}$ only depend on the frequency offset of the corresponding slave AP from the lead AP, and hence each slave AP can, as before, observe the channel of the sync header, compute the oscillator offset using the channel measured at time t_1 as reference, and correct its transmission appropriately.

Similarly the diagonal entries of the matrix $\mathbf{R(t)}$ only depend on the frequency offset of the corresponding receiver from the lead AP, and hence each receiver can independently decode its packet as if it were sent from a single transmitter.

Intuitively, this scheme can be understood as the slave AP rotating its measured channel to receiver 2 back to the time t_1 by multiplying h_{22} by $e^{-j(\omega_{T1}-\omega_{T2})(t_2-t_1)}$, and then performing all future channel corrections relative to the time t_1. This is why it corrects by the time dependent quantity $e^{-j(\omega_{T1}-\omega_{T2})(t-t_1)}$ shown in $\mathbf{T(t)}$.

TUBE: Time-Dependent Pricing for Mobile Data

Sangtae Ha[⋆] Soumya Sen[⋆] Carlee Joe-Wong[⋆] Youngbin Im[⋆,†] Mung Chiang[⋆]

[⋆]Princeton University [†]Seoul National University

{sangtaeh, soumyas, cjoe, chiangm}@princeton.edu ybim@mmlab.snu.ac.kr

ABSTRACT

The two largest U.S. wireless ISPs have recently moved towards usage-based pricing to better manage the growing demand on their networks. Yet usage-based pricing still requires ISPs to over-provision capacity for demand at peak times of the day. Time-dependent pricing (TDP) addresses this problem by considering *when* a user consumes data, in addition to how much is used. We present the architecture, implementation, and a user trial of an end-to-end TDP system called TUBE. TUBE creates a price-based feedback control loop between an ISP and its end users. On the ISP side, it computes TDP prices so as to balance the cost of congestion during peak periods with that of offering lower prices in less congested periods. On mobile devices, it provides a graphical user interface that allows users to respond to the offered prices either by themselves or using an "autopilot" mode. We conducted a pilot TUBE trial with 50 iPhone or iPad 3G data users, who were charged according to our TDP algorithms. Our results show that TDP benefits both operators and customers, flattening the temporal fluctuation of demand while allowing users to save money by choosing the time and volume of their usage.

Categories and Subject Descriptors: C.2.3[**Computer-Communication Networks**]: Network Operations—*Network Management*

General Terms: Economics, Human Factors, Management

Keywords: Time-dependent pricing, User trial, Wireless

1. INTRODUCTION

While researchers have proposed different plans for pricing data usage for many years, wireless ISPs have traditionally used only flat-rate, usage-based, or simple day/night charging. However, the recent rapid growth in demand for data [2] is forcing them to explore new ways to match revenues to costs. Dynamic *time-dependent pricing* (TDP) is one way to do so. With TDP, an ISP can offer lower prices in less-congested periods, incentivizing users to shift their usage from peak to off-peak periods. In fact, many applications today, e.g., movie or software downloads and cloud data synchronization, have significant delay tolerances and can be deferred to low usage periods if proper incentives are provided. Other applications can tolerate shorter, but still useful, delays. These deferrals can reduce the peak traffic: our partner ISP data shows that the demand in peak hours can be ten times that in off-peak hours. Even within ten minutes, demand can vary by a factor of two. TDP leverages this traffic pattern to help ISPs reduce the cost of peak-load provisioning for their networks, while allowing users to save money by choosing the time of their usage.

Implementing such a TDP plan requires *architecting* and *prototyping* a fully functional system that enables ISPs to offer prices acceptable to both themselves and end users. Additionally, it requires developing simple and intuitive GUIs that let users view and respond to the offered prices. In this paper, we present the design, implementation, and pilot trial evaluation of the TUBE (Time-dependent Usage-based Broadband price Engineering) system for mobile data [18].

1.1 Matching Price to Cost

The proliferation of high-speed LTE, smartphones, tablets, bandwidth-hungry apps, and cloud-based services has brought about an explosive growth in wireless Internet usage. The heavy tail of this usage distribution, which largely drives ISPs' operational costs, has led to the demise of 'flat rate' unlimited data plans in countries like the U.S. [5, 24, 28]. ISPs are now pursuing measures such as tiered usage pricing, overage charges, aggressive throttling, and service discontinuation to alleviate congestion on their networks [1]. By the spring of 2012, both AT&T and Verizon Wireless had announced updates to their mobile data usage policies, effectively imposing usage-based pricing of about \$10/GB [6]. But as Clark [3] observed:

> *The fundamental problem with simple usage fees is that they impose usage costs on users regardless of whether the network is congested or not.*

Moreover, usage-based fees fail to address ISPs' real problem. Heavy users typically congest the network at the same time, resulting in large demand peaks that force ISPs to over-provision network capacity and incur costs accordingly. Solving this problem requires a viable economic model and system capability for charging users by not only how much data they consume but also when they do so. This idea of 'responsive pricing' was advocated by Mackie-Mason et al. [14] as early as 1995 when the commercial Internet was still evolving:

SIGCOMM'12, August 13–17, 2012, Helsinki, Finland.

We argue that a feedback signal in the form of a variable price for network service is a workable tool to aid network operators in controlling Internet traffic. We suggest that these prices should vary dynamically based on the current utilization of network resources.

The general concept of congestion pricing has been studied in the energy, transportation, telephone, ATM, and IP network research communities for several decades. Only recently, however, has congestion-dependent pricing become realistic for mobile data. For example, Uninor in India and MTN in Uganda already offer congestion- and location-dependent pricing for voice calls [25].

1.2 Opportunities and Challenges

Time-dependent pricing for mobile data is a natural step in the transition from simple usage pricing to congestion-dependent pricing. Unlike voice calls, many mobile apps have different degrees of time sensitivity and do not require real-time data transfers or user interactivity. Time-dependent pricing can exploit these features to create effective price incentives for users to flatten their temporal demand profile by adjusting prices to user response. But TDP for mobile data also presents several new technological and social challenges:

- How can ISPs compute price incentives that they are willing to offer and users are ready to accept?
- Can we develop economic models that can be easily estimated from real data?
- What are the key system design challenges?
- How can we assess the benefits of TDP?
- Will real users respond to TDP favorably?
- How can we minimize user interaction from the client side?

Answering each of these questions requires significant effort in conducting consumer surveys, developing analytical models, building a system prototype, and finally running trials with real users. The aim and scope of such an undertaking echoes those of the Berkeley INDEX project for usage-based wireline pricing [27] and the work in [21] for voice calls. Yet the present context of wireless TDP introduces new requirements, challenges, and opportunities, including the following:

Dynamic TDP: To compute time-dependent prices, ISPs must estimate the amount of usage that will be shifted to lower-price periods as a function of the prices offered. However, this estimation should change over time: as an ISP offers time-dependent prices and observes the resulting network demand, it must adjust and improve its estimates of user reaction to the prices offered to better reflect the usage observed. These changes in user behavior estimates then prompt changes in the prices offered, forming a *feedback loop* interaction shown in Fig. 1.

User behavior estimation: Given the prices offered over a day, users will shift different amounts of different types of traffic to lower-price periods. For instance, some users may wait for five minutes but not an hour to stream movies. Similarly, a user may wait to refresh a personalized news magazine, but not to download urgent email attachments. The ISP's model of user behavior must therefore account for this heterogeneity in users' reactions to the offered prices. It must also be readily adjustable to observed changes in *aggregate* demand across users.

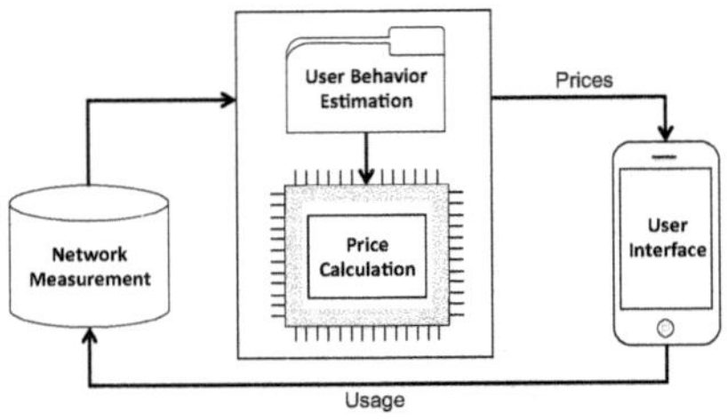

Figure 1: Feedback-loop schematic of TUBE.

User experience: When prices are published to users' mobile devices, each user optimizes her behavior to satisfy her goals, e.g., spending less than a certain monthly budget. Most users, however, will not manually optimize their usage to do so. Thus, TDP requires an "autopilot" mode that can automatically schedule usage for users. To make the system scalable to multiple users and prevent privacy violations, this autopilot should operate separately from the ISP.

1.3 TUBE Contributions

While prototyping and deploying TUBE in a trial with 3G data users, we focus on five main features:

A fully functional system for offering TDP for mobile data. Deploying TDP as a new mobile data pricing plan requires taking this idea from economic theory to a fully integrated system. We build a model for dynamic TDP that incorporates evolving user behavior, populate the model parameters from user surveys, design a supporting architecture, implement a prototype, and, finally, run a pilot trial with real users.

An architecture using feedback control. TUBE creates a feedback loop between the ISP server, which computes the prices to offer users, and the users who respond to the offered prices. The ISP offers prices on a *day-ahead* basis: at any given time, users know the prices for the next twenty-four hours. Day-ahead prices provide some certainty for users to plan ahead, while allowing ISPs to adjust prices each day according to revised user behavior estimates.[1]

User behavior models and optimized price computation. We propose an economic model of user behavior, as well as an algorithm for dynamically estimating relevant parameters from aggregate demand. The model helps predict usage for subsequent TDP periods, allowing the ISP to optimize its prices. These prices balance the cost of handling high demand relative to capacity with that of offering price discounts to users.

An user interface design. We study psychological aspects of user interaction and offer an autopilot mode for scheduling apps that minimizes "human-in-the-loop" issues. Users are thus able to optimize their usage with respect to the time-dependent prices independent of the ISP.

A realistic evaluation with real users. We recruited 50 iPhone/iPad 3G users on our university campus as trial participants. We charged participants according to TDP by deploying the TUBE prototype on our server and participant devices. Our results indicate that users indeed respond to prices: when offered monetary discounts, they will shift their demand from peak- to off-peak periods and even consume more in off-peak periods.

The overall TUBE architecture and the details of our user

[1]Of course, the prices need not be offered a full day ahead of time; TUBE allows ISPs to choose any time-window size. Pricing for sudden increases in demand due to special events can also be accommodated.

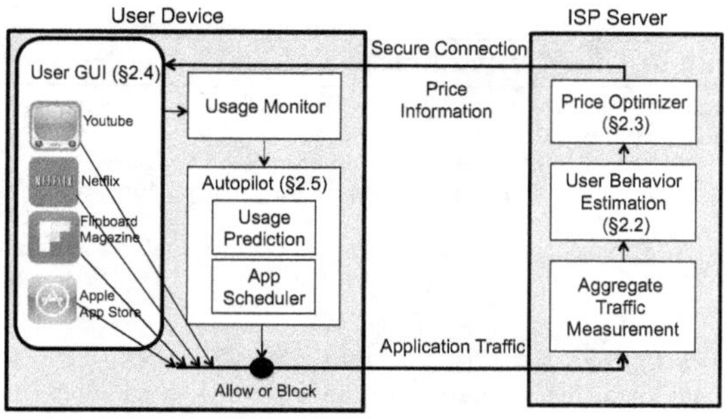

Figure 2: TUBE architecture showing the user-side and ISP-side components and interaction.

behavior modeling and price computation are described in Section 2. Section 3 next discusses our prototype implementation on the server and user devices, and Section 4 discusses the setup, logistics, and results of our pilot trial. Section 5 points out limitations of our work and future directions, while Section 6 discusses related work in this field. Finally, Section 7 concludes the paper.

2. ARCHITECTURE AND DESIGN

The TUBE architecture consists of two main components, as shown in Fig. 2. One component resides on the ISP server and computes the prices offered to users, while the other resides on user devices.

The ISP-side component of Fig. 2 measures individual usage and determines the prices offered to the end users.[2] These prices optimize the ISP's profit given its evolving estimates of user behavior.

At regular intervals, the user device pulls the price information over a secure connection and displays the prices computed on the ISP server. The usage monitor in the device measures the timing and volume of each application's usage. If activated by the users, the autopilot mode uses these measurements to schedule applications on their behalf.

In this section, we first discuss our guiding design principles (Section 2.1). Next, we consider two important modules on the ISP server: how TUBE estimates user behavior (Section 2.2) and how it optimizes the TDP prices (Section 2.3). Finally, we discuss our GUI designs (Section 2.4) and an autopilot algorithm (Section 2.5) that runs on user devices.

2.1 Design Guidelines

Our design choices stem from the following principles:

Separating functionality: The ISP server estimates users' reactions to prices and solves a large-scale nonlinear optimization to compute the prices offered to users. Since user devices have limited computing power, we run this price calculation on a central server. To allow users to view the prices and automatically respond by scheduling their usage, we implement a user interface on their individual devices.

Scaling up the system: To maintain scalability, our behavior estimation algorithm requires only aggregate, and not individual, usage data. We formulate the price calculation as a convex optimization problem, which can be rapidly solved for many TDP periods.

Protecting user privacy: TUBE requires neither monitoring of users' application source/destination addresses, nor any Deep Packet Inspection (DPI). The only data exchanged between a user device and ISP servers are the prices and usage in each period, which are secured with TLS/SSL connections.

Empowering user control: The user interface displays the prices offered and corresponding device usage, allowing users to educate themselves about their data consumption and schedule their usage according to the prices offered. To facilitate this educational component, we use simple and intuitive GUI designs.

2.2 Estimating User Behavior

In order to set time-dependent prices, ISPs monitor users' traffic patterns, i.e., the volume of traffic at different times, with and without TDP. This data is then used to estimate users' willingness to shift their traffic in exchange for a monetary discount. The estimates are used to calculate the time-dependent prices for the next day, linking the prices and user behavior in the system's feedback loop (Figs. 1 and 2).

In this section, we first discuss a parameterized model of user behavior and use consumer survey results to initialize the parameter values. We then provide an algorithm to update these user behavior estimates and evaluate its effectiveness on simulated data.

2.2.1 *Modeling Delay Tolerances*

Users' willingness to wait for data usage depends on the type of session under consideration: for instance, iTunes downloads can often be more readily delayed than streaming YouTube videos. Thus, we view each user as a set of *application sessions*, e.g., streaming, browsing, and file transfer sessions. Sessions are assumed to have a fixed minimum bandwidth requirement, which is particularly appropriate for streaming sessions.

In general, users' willingness to defer any given application session depends on two factors: the monetary reward for deferring the session and the time for which the session is deferred. For instance, users may wait for 1 hour to watch a video in exchange for \$2, but may not wait if offered only \$1. At the same time, users may not wait more than one hour, even if offered the \$2 reward.

In order to quantify users' willingness to defer application sessions, we introduce the concept of *waiting functions*. These functions give the probability that users will defer an application session for a given amount of time τ in exchange for a discount d, e.g., d \$/GB from some baseline metered price. Since different users and applications may have distinct waiting functions, we choose a parameterized form of the waiting functions, with different parameters corresponding to different levels of user patience. These parameters thus quantify the various price-delay tradeoffs corresponding to different users and application sessions.

The chosen form of the waiting functions should be decreasing in the time deferred, τ, and concave and increasing in the discount offered d: users will be less likely to defer as the time deferred increases but will be more likely to defer if offered a larger monetary reward. While many functional forms are reasonable, we choose the simple form

$$w_\rho(d, \tau) = \frac{d}{\lambda_\rho(\tau + 1)^\rho}, \quad (1)$$

where ρ is a parameter measuring patience, the *patience index*, and λ_ρ is an appropriate normalization constant. A

[2]Although the behavior estimation requires only aggregate data, ISPs can of course keep track of individual usage data in order to calculate users' monthly bills.

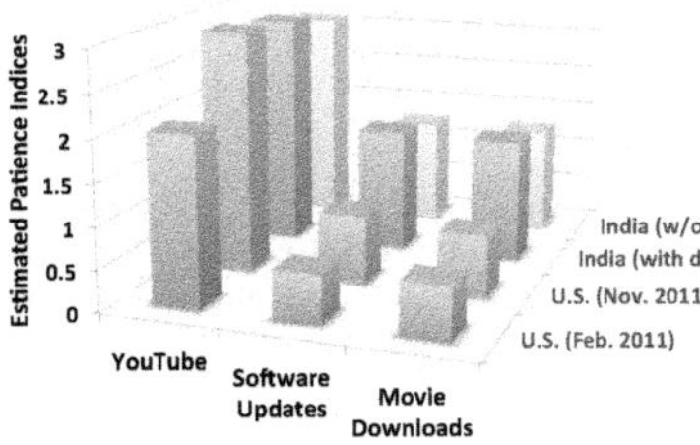

Figure 3: Patience indices from survey results.

larger ρ indicates less patience, while $\rho = 0$ indicates indifference to the time deferred τ: as τ increases, willingness to wait w_ρ drops off faster for larger ρ. The normalization constant is chosen so that the $w_\rho(d, \tau)$ could represent the probability that users will defer for τ amount of time, given the discount d. Thus, we generally choose $\lambda_\rho = \sum_{\tau=1}^{n-1} w_\rho(D, \tau)$, where D denotes the maximum possible discount offered in each period and $w_\rho(D, \tau)$ is the unnormalized value of (1).[3] The summation of τ from 1 to $n-1$ represents the possible times deferred (1 period, 2 periods, etc.) up to one day, or n periods. We expect that users will not have significant incentive to delay their traffic for more than one day.

2.2.2 Initializing Patience Indices

In this section, we give reasonable estimates of the patience indices ρ used in (1) to model users' willingness to delay their data consumption. In practice, such estimates may be achieved with a calibration period of pilot trials in which the ISP offers a wide variety of prices to users and observes their responses. In the present work, we take a different approach with initial market surveys in the U.S. and India. We conducted two U.S. surveys, one in February 2011 and one in November 2011, each with 130 participants from 25 states. The India survey included 546 participants from 5 large cities.

In each survey, we asked the respondents whether they would delay a given application type, e.g., YouTube videos or software updates, for specified time intervals. Participants were told that waiting would reduce their bill by two-thirds in the first U.S. and India surveys and by one-third in the second U.S. survey. The time interval choices ranged from 5 minutes (e.g., for YouTube videos) to 48 hours (e.g., for software updates). The survey questions are given in [8].

We first use the survey responses to find the fraction of users willing to wait for each of these time intervals. Given these fractions, we then compute their discrete derivative with respect to time (i.e., the differences between the fractions divided by the interval duration) to find the waiting function values for each traffic class at the survey-specified discount. Figure 3 shows the resulting patience indices for three different traffic classes. As expected, in all three surveys, participants are much more likely to wait for software updates or movie downloads than YouTube streaming, reflecting streaming's more "immediate" utility to users.

2.2.3 Updating Patience Indices

We now give an algorithm for ISPs to update their estimates of (1)'s user patience indices for different applications. We assume that the ISP has several days of aggregate (cross-user and cross-application) usage data and the corresponding prices offered in each period.

The main idea of our algorithm is to compare the observed usage data with a baseline usage trace, taken for time-independent prices (TIP). We assume that changes between the baseline usage and observed usage with TDP are due to the time-dependent prices offered. These changes can then be expressed in terms of the waiting function parameters and volume of different application types; we can use standard curve-fitting methods to estimate the optimal waiting function parameters. The following discussion details this procedure, which is summarized in Algorithm 1.

Since the baseline traces have only aggregate usage data, the ISP does not know the usage for each application. Yet for our purposes, grouping sessions by application type has an important disadvantage: depending on the user, a given application may correspond to different patience indices. We therefore group applications by *traffic class*, defined as a group of sessions with the same patience index, rather than application type. After choosing a number of traffic classes, we must estimate both the patience index and the fraction of traffic corresponding to each traffic class.

Our estimation algorithm relies on calculating the expected amount of traffic deferred from a given period i to another period k in terms of the waiting function parameters.[4] Thus, the amount of traffic in period i without any traffic shifting is simply the average amount of traffic in period i under TIP. Given a set of discounts d_i, $i = 1, 2, \ldots, n$, offered over one day, the expected amount of traffic deferred from period i to period $k \neq i$ is then

$$A_{ik} = Y_i \sum_{j=1}^{m} \mu_j w_{\rho_j}\left(d_k, |k-i|_n\right), \tag{2}$$

where Y_i is the TIP usage in period i, and $|k-i|_n$ is understood to be modulo n, representing the time difference between period i and the nearest period k after period i. If $k < i$, period k will occur on the day after period i. There are m traffic classes, with the jth traffic class having patience index ρ_j and taking up a proportion μ_j of the traffic. Denoting by S_i the difference between the TIP and TDP traffic in period i, we see that each $S_i = \sum_{k \neq i} A_{ik} - A_{ki}$.

Each of the n S_i values is a linear combination of the $n(n-1)/2$ A_{ik} variables. One equation may be eliminated, since we assume the sum of the S_i is zero (no traffic is lost with TDP). We can thus reduce this system of $n-1$ linear equations for the S_i to one equation, by solving for $n-2$ of the A_{ik} in terms of other A_{ik} variables. The ISP can then estimate the parameters μ_j and ρ_j using Algorithm 1.

Algorithm 1: Waiting function estimation

Data: Prices offered and TDP traffic pattern.
Result: Estimated waiting functions.
Compute the S_i, to obtain n linear equations for the A_{ik};
Solve for $n-2$ of the A_{ik}, such that for each period j, at least one A_{jk} is not solved for;
Substitute these expressions back into the original equations for S_i, so that only one equation, linear in the A_{ik}, remains;
// This equation is a function of parameters μ_j and ρ_j and the offered discounts.
Use the TIP and TDP data for this function to estimate (e.g., with nonlinear least-squares) all the μ_j and ρ_j parameters it contains;
// These μ_j and ρ_j values give the waiting functions.

[3]For instance, D may be the baseline metered price.

[4]For simplicity, we assume that the average traffic volume over each day remains the same when TDP is introduced, i.e., that no traffic is lost due to TDP.

Period	Actual			Estimated			Error
	ρ_1	ρ_2	μ_1	ρ_1	ρ_2	μ_1	(%)
1	1	2	0.17	1.03	2.48	0.46	11.8
2	1	2.33	0.5	1.02	2.49	0.45	9.0
3	1	2.67	0.83	0.90	2.15	0.71	0.5

Table 1: Actual and estimated parameter values in a simulation of waiting function estimation.

Since baseline TIP demand may change over time, we use the following procedure to adjust the baseline. The ISP estimates the waiting functions using TDP data from several days, e.g. one month, and uses these estimates to solve for the demand under TIP, Y_i, in each period i. The n equations (2) are linear in the Y_i, and all other variables are known upon choosing a set of discounts offered and traffic observed.[5] We then estimate the Y_i values.

To illustrate Algorithm 1, we consider a simple example with two traffic classes and three periods. Actual parameter values are given in Table 1. Our simulation takes a set of given discounts and computes the traffic if waiting functions are perfectly followed, adding noise to these results. Table 1 shows the μ_j and ρ_j values estimated by nonlinear least squares. The maximum percent difference between actual and estimated waiting function values $\mu_1 w_{\rho_1} + (1-\mu_1)w_{\rho_2}$, measured on a 1000-point grid of discounts and times deferred, remains small at under 12%.

2.3 Optimizing Prices

We now describe how to use the estimated waiting functions to calculate time-dependent prices over the next day. An ISP wishes to set prices that balance two types of costs: that of exceeding the maximum capacity, and that of offering discounts to users in less-congested time periods. We take the "maximum capacity" to mean the maximum amount of traffic that can be handled by the network with an acceptable amount of congestion. Thus, demand may "exceed capacity" in the sense that over a certain capacity, user response time due to congestion becomes unacceptably high.

Suppose that there are n periods in a day, each lasting one unit (e.g., hour or half-hour) of time. We assume that the ISP's network has a bottleneck link of capacity C_i in period i, defined as the link capacity less any background traffic or excess capacity "cushion." The cost of exceeding capacity is assumed to be piecewise-linear and convex; we denote the cost in period i as $g(y_i - C_i)$, where y_i is the predicted usage with TDP discounts d_i in period i. For ease of notation, $j \in i$ indexes all traffic classes j in period i. Moreover, the time between periods i and k is given by $|i-k|_n$ as in (2).

We can now calculate the costs of exceeding capacity Γ_1 and offering discounts Γ_2 as

$$\Gamma_1 = \sum_{i=1}^{n} g\Bigg(Y_i \Bigg(1 - \sum_{j \in i} \mu_j \sum_{k \neq i}^{n} w_j\left(d_k, |k-i|_n\right) \Bigg) + \sum_{k=1, k\neq i}^{n} Y_k \sum_{j \in k} \mu_j w_j\left(d_i, |i-k|_n\right) - C_i \Bigg) \quad (3)$$

$$\Gamma_2 = \sum_{i=1}^{n} d_i \Bigg(\sum_{k \neq i}^{n} Y_k \sum_{j \in k} \mu_j w_j\left(d_i, |i-k|_n\right) \Bigg). \quad (4)$$

[5]Noise in the data results in different Y_i for different sets of discounts; the ISP can use an average of these Y_i.

Algorithm 2: Price determination.

Data: Estimated waiting functions.
Result: Optimized TDP prices.
Start with a set of discounts for the next n periods, determined with initial waiting function estimates;
while *TDP is offered* **do**
  for $k = 1 \to n$ **do**
    Choose the discount for the nth period after period k so as to minimize $\Gamma_1 + \Gamma_2$ in (3-4).
    if $k == n$ **then**
      Run the waiting function estimation (Algorithm 1) to find updated waiting function parameters.

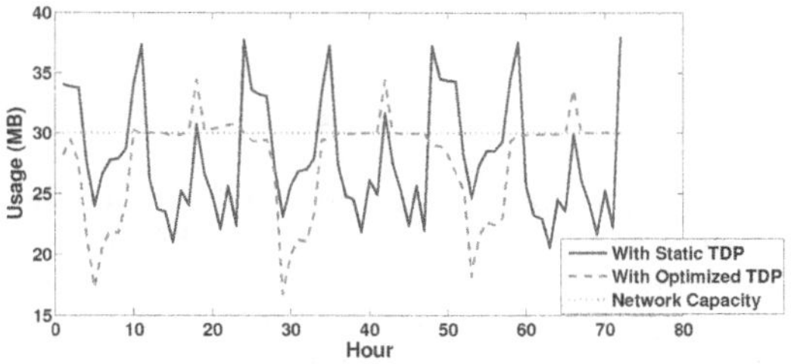

Figure 4: Static TDP vs. dynamic TDP.

The ISP attempts to minimize $\Gamma_1 + \Gamma_2$ with respect to the discount variables d_i, $i = 1, 2, \ldots, n$ [10]. For our trial, TUBE offers *day-ahead* pricing: when TDP is first introduced, the ISP publishes a full day of time-dependent prices to users. After each subsequent period, a new price for the period one day ahead of the current one is published. We thus have the online price determination in Algorithm 2.

To test the efficacy of TUBE's feedback loop, we compare simulated results of our feedback loop with results with static TDP, i.e., time-dependent prices that do not change from day to day. Figure 4 shows the results over three days. We see that the peak periods under static TDP are about 8 MB above the maximum capacity, while those under dynamic TDP are only 4 MB above the maximum capacity. The ISP underestimates users' reaction to TDP in setting the static prices; thus, many users shift to lower-price periods, resulting in a new peak period. Dynamic TDP, however, adjusts to this underestimation by correcting the prices offered, thus flattening out the peak usage.

2.4 Graphical User Interface

Users react to the prices offered through our GUIs, which are designed to be simple and intuitive. We include the following components in our design:

Price display: Users check the prices for the next 24 hours on the GUI home screen. Each price is color-coded by its discount rate, e.g., red (< 10%), orange (10 – 19%), yellow (20 – 29%), green ($\geq$ 30%). Since users likely will not open our application to check each period's price, we also embed a colored price indicator on the status bar at the top right corner of the device screen.

Self-education: Users can view their top 5 bandwidth consuming applications to better understand their mobile data usage. The device stores usage and price history for the last three months so that the users can look up their past usage and prices by day, week, and month. The push notifier running on the ISP server informs users when they consume significant amounts of data in high-price periods.

User control: The device learns each application's usage pattern over time, enabling it to schedule applications in the autopilot mode so as to keep users from exceeding their specified budget. Users can also override the scheduling decisions computed by the autopilot algorithm (e.g., parental control on certain applications).

2.5 Autopilot Mode

The autopilot algorithm schedules the top five applications by usage, identified from the usage monitor in Fig. 2, to keep users below their specified monthly budget. The user ranks these applications by delay tolerance, and we provide default rankings if the user does not provide them. We denote the delay tolerances by ν_j, where $j = 1, \ldots, 5$ indexes the applications and a larger ν_j indicates a higher delay tolerance. For privacy reasons, these delay tolerances, as well as usage statistics for each application, are *not* shared with the ISP server. The behavior estimation (Algorithm 1) neither requires nor utilizes this data.

Usage Prediction: To schedule application usage, we first predict usage in each period using triple exponential smoothing (TES), which incorporates *trend, periodic,* and *smoothing* factors into the next day's usage prediction. We use a variation of the traditional TES equations [15, 23] to incorporate periodicity over one day and over one week.

Application Scheduling: The full scheduling algorithm is presented in Algorithm 3 and outlined below. At the beginning of each day, we calculate a daily budget from the amount of money the user has left to spend, divided by the number of days left in the month. For flexibility, we scale up the budget by a factor that increases with the number of days remaining in the month: even if the user consumes a large amount of data on the current day, she can remain within the monthly budget by reducing usage in the remaining days. This scaling can be, for example, exponential. We then compare the daily budget to the expected amount spent over one day, as calculated from the predicted usage. If the expected amount spent is larger than the daily budget, we use Algorithm 3 to schedule usage so that users are predicted to remain within their daily budget. The algorithm begins with the most delay tolerant app, and defers it from some period k to period $k+1$. Here k is chosen so as to save the most money. If the user is still predicted to exceed her budget, the same app is deferred from another period k' to period $k'+1$.

If the user can no longer save money by delaying by one period, the algorithm checks whether the user would prefer to delay sessions for this application by two periods, or would prefer deferring the application of next-highest delay tolerance by one period. It continues in this fashion until the expected amount spent is lower than the daily budget. On any given day, the user is prohibited from exceeding the weekly budget, which is calculated analogously to the daily budget but over one week's duration.

3. SYSTEM IMPLEMENTATION

In this section, we discuss the implementation details of TUBEOpt and TUBEApp, the two main components of the TUBE system. Figure 5 shows the modules in both TUBEOpt, which resides on the ISP server, and TUBEApp, which resides on user devices.

3.1 TUBEOpt

We implemented TUBEOpt on a Linux system with an Intel Xeon 2.0 GHz CPU and 8GB of RAM. It provides a web-based API so that any device supporting web can exchange data with the server. Figure 5a shows TUBEOpt's component diagram, including the price optimization components discussed in Sections 2.2 and 2.3; the shaded blocks are part of our current implementation. All of the blocks represent dynamic modules and can be reloaded on the fly.

Algorithm 3: Autopilot app. scheduling algorithm

Data: Prices offered and predicted usage.
Result: Usage schedule.
Initialize $h = 1$;
// h is the number of periods deferred.
Calculate predicted amount spent using the prices p_i and predicted usage u_i in each period i;
while *projected spending exceeds the daily budget* **do**
    // Choose the app k with highest delay tolerance.
    $k \leftarrow \text{argmax}_i \left\{\nu_i : \exists \sum_{q=0}^{h-1} u_{i,j+q} > 0 \text{ s.t. } p_{j+h} < p_j\right\}$;
    // Choose the period l which will save the most money.
    $l = \text{argmax}_j \left\{\sum_{q=0}^{h-1} u_{k,j+q} : p_{j+h} < p_j\right\}$;
    Block app k from period l to $l+h-1$, inclusive;
    // Users can choose whether or not a notification is sent at the beginning of period $l+h$ to say that the app is no longer blocked.
    Update the projected usage values;
    $h \leftarrow h+1$;
    Recalculate the predicted amount spent;
    if *app k does not have the lowest delay tolerance* **then**
        if *the user prefers to defer the app with next-highest delay tolerance by 1 period rather than defer app k by $h+1$ periods* **then**
            $h \leftarrow 1$;
            $\nu_k \leftarrow 0$;
    The user will exceed the budget under all allowable scheduling constraints;

In the following discussion, we detail TUBEOpt's scalability of usage monitoring and the computational overhead for its user behavior estimation and price computation.

3.1.1 Improving Scalability of Usage Monitoring

To measure individual usage, we assigned a unique IP address to each user and created a `Netfilter` rule. When TUBEOpt records the usage, it retrieves the byte and packet counts from each rule. Unfortunately, this approach scales linearly with the number of users; each user requires one rule, and the computational cost increases linearly with the number of rules. While `ipset` can improve performance by combining multiple rules into one hash table, its use here is limited, as usage with `ipset` is tracked for the hash table, not for the individual rules. To improve scalability, we therefore implemented a separate kernel module that hooks the `LOCAL_IN` of the IPSEC/VPN interface (`ipsec0`). It creates a hash table and records the usage for each IP address, requiring only O(1) running time.

3.1.2 Computational Overhead

We next examine the computational overhead of using Matlab and Python to estimate user behavior and compute the optimized TDP prices. TUBE requires the runtime of these codes to be relatively short, since each period's price computation should finish before the next TDP period, at which time the server advertises the newly computed price. Therefore, we evaluate the overhead of these two computations as we increase the optimization complexity.

We measure the computational overhead (the total runtime in Matlab) as we increase the number of periods from 12 to 144 (2 hour to 10 minute periods). Table 2 shows

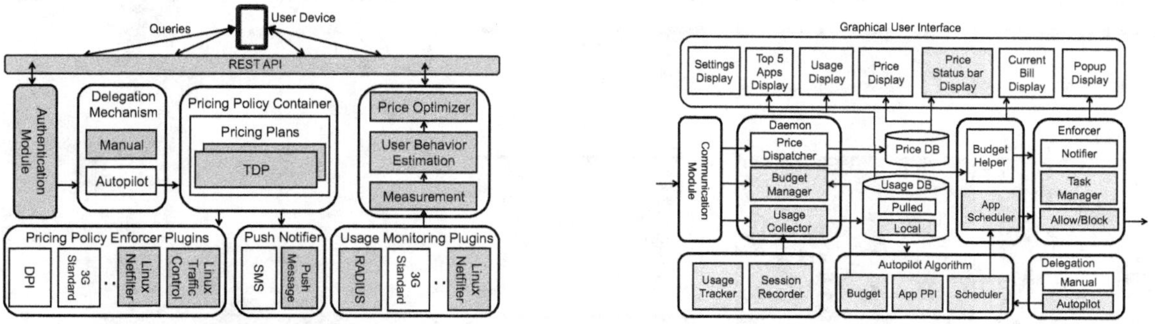

(a) TUBEOpt design (b) TUBEApp design

Figure 5: Building blocks of TUBE: (a) TUBEOpt on the server side and (b) TUBEApp on the mobile device. The shaded blocks in (a) are TDP-specific modules, while those in (b) require system-level modification.

Number of Periods	12	24	48	96	144
Behavior Estimation	12.76	200.0	959.6	1967	15040
Price Calculation	1.67	1.69	1.70	1.81	1.84

Table 2: Runtime of the behavior estimation and price calculation in seconds.

	Number of Application Types		
Number of Periods	2	4	8
12	0.21	12.99	21.52
24	3.33	47.08	75.47
48	15.99	197.22	215.42

Table 3: Runtime of the behavior estimation (mins).

the measured running time of the behavior estimation and price calculation. Even with 144 periods, the price calculation is quite fast (1.84 seconds); the estimation algorithm performs adequately (4.2 hours), as it runs only once a day for day-ahead TDP.

We also measure the effect of adding multiple traffic classes to the behavior estimation algorithm. Table 3 shows the algorithm running time on our Intel Xeon server. The computation with 48 periods and 8 application types still takes less than 4 hours (215.42 minutes), which is more than fast enough, as the estimation runs once a day. Our estimation uses one month of simulated data, which was generated by perturbing the usage predicted from given waiting functions by up to 50%. The running times were averaged across five computations with random data and starting points. With more powerful hardware and optimized code, significant further acceleration can be achieved.

3.2 TUBEApp

We implemented TUBEApp on the iOS, Android, and Windows platforms, although all trial participants were iOS users due to these devices' popularity on our campus. We therefore focus our discussion on the iOS implementation. We consider both the manual and autopilot modes, as shown in Fig. 5b's design and Fig. 6's GUI screenshots. Due to the iOS platform's closed nature, implementing the shaded blocks in Fig. 5b, e.g., monitoring each application's usage, requires jailbreaking the devices. We demo our app in [26].

3.2.1 OS Limitations

The Windows, Android, and iOS platforms each have various limitations; the iOS platform offers the most restrictions. While all platforms support showing the prices offered and the device's aggregate data usage, TUBEApp's autopilot mode additionally requires 1) measuring the volume of each application's usage, 2) displaying the price for the current period on a status bar, and 3) allowing and blocking bandwidth for individual applications. These features heavily depend on the openness of the platform.

Type	Status bar	App usage	Daemon support	Code size (# lines)
iPhone	No	No	Partial	25K
Android	Yes	Yes	Yes	5.4K
Windows	Yes	Yes	Yes	5.3K

Table 4: TUBEApp on different platforms.

Table 4 shows the TUBEApp features supported on each platform before hacking the device, as well as the code size for implementing the full TUBEApp. In particular, for the iOS platform, we hook several internal functions to track the usage per application, run a daemon process to dispatch and show TDP prices, and block applications. The iOS implementation thus requires 25K lines of code, while the Android and Windows implementations need only 5.4K and 5.3K lines, respectively.

3.2.2 Enhancing the User Experience

The autopilot mode minimizes user interactions by estimating device usage patterns and scheduling applications for the user. To inform users of the autopilot actions, we send pop-up notifications when usage is blocked, as shown in Fig. 7's iPhone screenshots. The warning and blocking pop-ups are displayed when the user's usage reaches the expected daily and weekly budgets (Fig. 7a and b, respectively).

To ensure that the autopilot's implementation is practical, we measure its energy usage. TUBEApp with autopilot running consumes only 4% more battery power than the device without our TUBEApp installed, indicating that autopilot does not drain too much power.

4. TRIAL DESIGN AND RESULTS

We conducted a small scale pilot trial of the TUBE system at Princeton University from May 2011 to January 2012. This section provides an overview of the trial goals, setup, and limitations, followed by a discussion of some key conclusions drawn from the trial data.

4.1 Goals and Structure

The goals of our trial are to demonstrate the TUBE system's functionality and benefits, and to provide an initial verification of TUBE's deployment feasibility with real users. Throughout the trial, we effectively acted as a resale ISP, paying the participants' regular 3G data bills to AT&T while

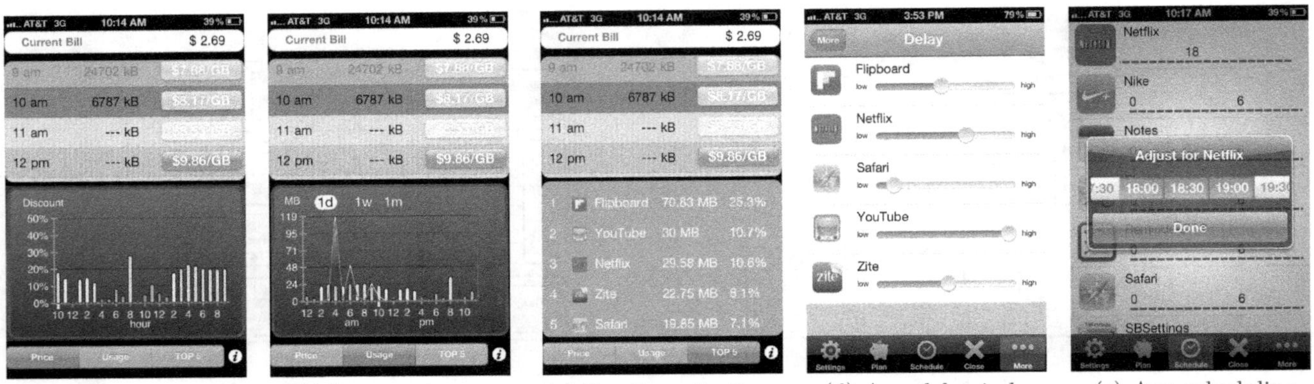

(a) Price display (b) Price and usage (c) Top 5 applications (d) App delay index (e) App scheduling

Figure 6: Screenshots of TUBEApp on iPhone. iPhone users can (a) check the prices for next 24 hours, (b) learn from price and usage history, (c) identify top 5 apps by bandwidth usage, (d) modify each app's delay tolerance, and (e) check when their apps are scheduled during the day and manually override the results.

charging them according to our TUBE algorithms. To assess the benefits of TDP, we divide the trial into two phases: first, we simply monitor the usage patterns without TDP (i.e., collect TIP data). We then offer TDP and study its impact in the second phase of the trial. The following aspects of the trial are addressed in this section:

Baseline Traffic Statistics: Section 4.4.1 reports on three months of TIP usage statistics from our trial participants. We ask if the sample population has a representative mix of heavy and light users and bandwidth-hungry applications, so as to realistically assess the benefits of TDP.

Price Sensitivity: In Section 4.4.2, we examine users' price sensitivity to static TDP patterns: if we offer low and high price periods alternately, will users defer their traffic to use more in the lower-price period? This question tests TDP's basic premise that users will delay their traffic in exchange for a monetary discount.

Effectiveness of GUI Design: Section 4.4.3 analyzes the effectiveness of displaying numerical values versus color codes (red: high, orange: medium, green: low) to indicate TDP prices on the user device.

Benefits of Optimized TDP: Finally, Section 4.4.4 studies whether TUBE's optimized prices benefit ISPs in reducing peak-to-average ratios of network usage.

4.2 Trial Setup

We recruited 50 users (27 iPhones and 23 iPads) of AT&T's 3G Corporate Data Plan as our trial participants. They were faculty and staff from 14 academic and administrative divisions. During the trial, we acted as a resale ISP, charging participants after every billing cycle according to TUBE's TDP. We excluded measurements from development devices to avoid bias.

To record participants' usage, we separated their 3G traffic from that of other AT&T customers using an Access Point Name (APN) setup, which tunneled the participant's 3G traffic from the AT&T core to the TUBE servers in our lab (Fig. 8). Participants installed and used the TUBEApp on their iOS devices. WiFi usage, voice calls, and SMS were not included in the trial traffic as they are not 3G data.

4.3 Trial Limitations

We were limited by logistics to recruiting only AT&T data plan users with iOS devices, out of which 16 were jailbroken (JB) and 34 were non-jailbroken (non-JB) devices. The

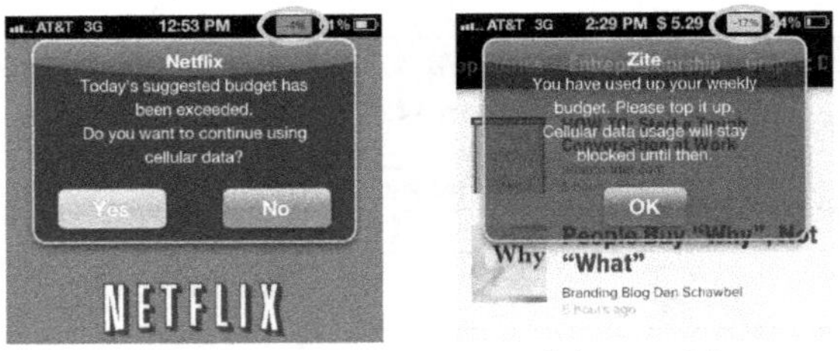

(a) **Warning** (b) **Blocking**

Figure 7: Screenshots of auto-pilot in action on the iPhone. The warning (a) is displayed when the daily budget is reached, and usage is blocked in (b) when the user reaches the weekly budget.

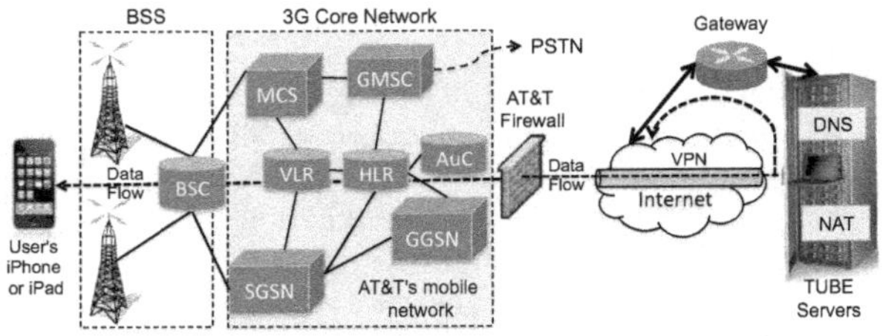

Figure 8: TUBE APN Setup. User traffic is routed through the SGSN and GGSN.

non-JB devices gave us less flexibility in experimentation, and hence ran a TUBEApp with limited features. In particular, only users with JB devices could see the current price/discount directly from the home screen's status bar (circled in Fig. 7's screenshots) without manually launching the TUBEApp, and the autopilot algorithm only runs on JB devices. On non-JB devices, we used push notifications to alert participants during high-price periods.

Since our trial included only 50 participants, even peak traffic from trial users did not congest our lab's access link. To demonstrate TDP's benefits, we simulated congestion conditions by logically scaling up the traffic volume in TUBE-Opt's price computation.[6]

4.4 Results and Evaluation

Following Section 4.1's outline, we now present the trial results. In many cases, we use *Wilcoxon's signed rank test* [12] against the null hypothesis that a set of values is sym-

[6]If users had experienced real congestion due to this scaled traffic, we expect they would have been even more willing to delay their traffic to off-peak periods.

metrically distributed with mean zero. In our case, we apply the test to the difference between the changes in usage in high- and low-price periods. A higher probability of a symmetric distribution then indicates a lack of response to the price signals, as the expected change in usage is likely the same for both high and low prices.

4.4.1 Baseline Traffic Statistics

Question: *Do our participants include both heavy and light bandwidth users? Which applications use the most data?*

Method: We measured usage for both iPad and iPhone users from July to September and used `tcpdump` to record application-specific traffic.

Results: Our participants are a mix of light- and heavy-bandwidth users. Video streaming applications accounted for most of the traffic, corroborating the reported trends of growing demand for mobile video.

Figure 9a shows the CDF of total traffic per user for uploads and downloads from July to September 2011. While 90% of the users uploaded less than 0.5GB, some users had large download volumes: 20% of users consumed 2.1 – 5.3 GB over three months.

Figure 9b shows the distribution of total traffic by application type for the three month period, normalized with respect to the number of iPhone, JB, and non-JB iPad users. Not surprisingly, iPads show a higher usage than iPhones for most application types, and a large part of the mobile traffic for all device types comes from movie streaming.

4.4.2 Price Sensitivity

Question: *Do users wait to use mobile data in return for a monetary discount?*

Method: We conducted a three week experiment on iPad and iPhone trial participants in October 2011, in which we offered a basic TDP pattern of consecutive high, high, and low price periods, repeated throughout the day. The high-price periods offered approximately a 10% discount, while the low-price periods offered a 40% discount on the baseline price of \$10/GB. If monetary incentives do induce usage deferrals, we expect that average usage should decrease in high-price and increase in low-price periods.

To measure users' response to prices, we sent messages at ten minute intervals during *high-price* periods if the user exceeded 2 MB of usage in the previous ten minutes. We first analyzed the data for each user by calculating the percentage change in usage for each one-hour time period when compared to the mean usage in that same period before TDP (i.e., under TIP pricing). We then weighted these percent changes by the proportion of TDP usage in that period to account for diurnal variations. This gives the weighted average percent change in usage under TDP for high- and low-price periods.

Results: We found that *users did shift their traffic* from high- to low-price periods under TDP. For most users, the average usage decreased in high-price periods relative to the changes in low-price periods.

Figure 10 shows the weighted average percent change in usage for iPad users for high- and low-price periods. The reference line indicates an equal change in both types of periods. Each dot on the scatter plot represents values for an individual user, and its size is proportional to the user's TDP usage volume. With the given static TDP pattern, usage increased more in low-price periods relative to high-price periods for almost all users. Interestingly, about half of the users decreased their overall usage in both high- and low-price periods, while the other half increased their usage in both periods.

We further verify these results by using Wilcoxon's test on the differences between each user's percent change in high- and low-price periods. We find only a 0.56% probability that the null hypothesis is valid, indicating that the users' observed responses are statistically significant. A similar plot may be observed for the iPhone users.

The overall iPhone usage changed by −11.3% in high-price and −5.7% in low price periods, while overall iPad usage changed by −10.1% in high-price and 15.7% in low-price periods. Thus, iPad users generally decreased their usage in high-price periods and increased it in low-price periods. The overall decrease in iPhone usage is likely due to limited user notification and display options on non-JB iPhones. However, the greater usage decrease in high- relative to low-price periods indicates that iPhone users attempted to use less in high-price periods.

Next, we examine the effect of the number of notification messages sent to users on their usage in high-price periods. Multiple consecutive notifications were sent to a user only if usage in each preceding 10 minute interval exceeded 2 MB. We examine the percent change in usage in the ten minute span before and after each notification. Figure 11 shows the CDF of the percent change in usage due to a first, second, etc. notification. About 80–90% of iPad and iPhone users did not increase their usage after the first notification, indicating that notifications can effectively reduce peak usage. For all subsequent notifications, about 60–80% of the active users responded by decreasing their usage.

4.4.3 Effectiveness of User Interface Design

Question: *Do users respond more to the numerical values of TDP prices or to the color of the price indicator bar on the home screen?*

Method: In December 2011, we installed a price indicator bar on the home screen of all JB devices. The indicator displays the numerical value of the price discounts available in the current period and changes its color according to these discounts. It is green for discounts over 30%, orange for 10–29% discounts, and red for discounts below 10%.

Our experiment had two stages: in the first stage, we offered discounts of approximately 40% every third period of the day, starting with a 40% discount at midnight. The other periods offered discounts of about 10%. After two weeks of following this pattern, we began the second stage, repeating the pattern of a 9% discount at midnight, followed by 28%, 30%, 28%, 9%, and 30% discounts.

We compare three types of periods to assess the effect of the indicator color and numerical discount: hours deemed as Type 1 periods offered a 10% discount in the first stage of the experiment and 28% discount in the second stage; the indicator remained orange despite this increase in the discount. Type 2 periods offered a 10% (orange) discount in the first stage and 30% (green) discount in the second stage, while Type 3 periods offered a 10% discount in the first and 9% discount in the second stage of the experiment (the indicator is orange in both periods). Table 5 summarizes the combinations of discounts and colors used in the two stages that characterize each type of period.

We calculated the percent changes in usage for each period

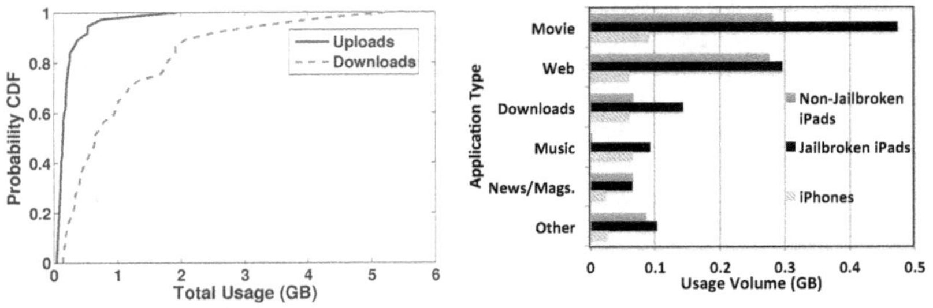

(a) Total traffic for each user. (b) Total usage by app type.

Figure 9: Usage statistics from July - Sept. 2011.

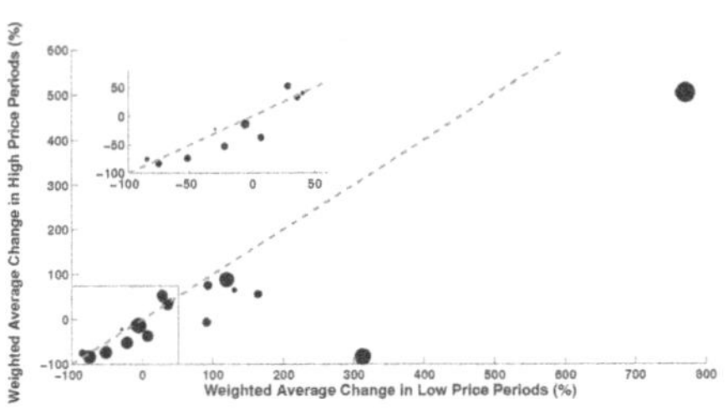

Figure 10: iPad response to static TDP.

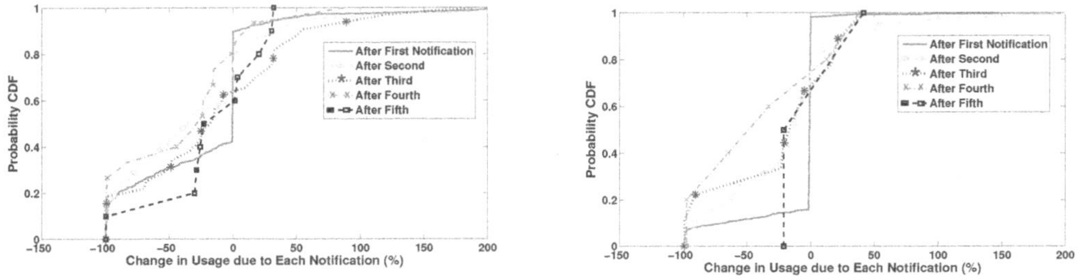

(a) Usage changes (iPad users). (b) Usage changes (iPhone users).

Figure 11: User response to notifications sent.

Type	Periods	First Stage		Second Stage	
		Color	Disc.	Color	Disc.
1	2, 8, 14, 20	Orange	10%	Orange	28%
2	3, 6, ..., 24	Orange	10%	Green	30%
3	5, 11, 17, 23	Orange	10%	Orange	9%

Table 5: Period types in the color experiment.

type between the first and second stages of the experiment. To do so, we first found the average usage in each period of the day (i.e., each hour) for the first stage of the experiment. We then calculated the percent change in usage of each period in the second stage of the experiment from the average usage in the same period during the first stage. The *average change* in each type of period is then defined as the weighted average of these percent changes for each period of the given type. The weights were proportional to the usage in that period.

Results: We found that *users paid more attention to indicator color than to the numerical discount value.* When discounts increased significantly with no change in indicator color, only half of the users increased their usage relative to other periods. However, when the indicator color also changed, almost all users increased their usage in those periods relative to others. In Fig. 12, each data point represents one user's average change in each period type, with the size of the data point indicating the volume of usage in the second stage of the experiment. The reference line represents equal changes in both period types considered.

Figure 12a shows the average change in usage for each user in Type 1 periods versus Type 3 periods. For both period types, the color did not change, but the discount in Type 1 periods increased significantly. Thus, if users had reacted to the numerical prices, we would expect usage to increase in Type 1 and decrease in Type 3 periods: users' data points should lie above the reference line. Figure 12a shows that this is the case with only half of the users. Moreover, some users increased their usage dramatically in both types of periods, while most decreased their usage in both types of periods. Wilcoxon's test reveals an 82% probability that the null hypothesis is valid on the differences in usage changes in both types of periods: since the indicator color did not change, users were mostly agnostic to the numerical values of the discounts.

We next compare these results to those obtained when the indicator color and discount offered both change. Figure 12b plots the average change in usage in Type 2 versus Type 1 periods. The discounts in both periods increased by comparable amounts, but the indicator color changed from orange to green only in Type 2 periods. We see that most users' data points lie above the reference line, so that usage increased more (or decreased less) in Type 2 as compared to Type 1 periods. Wilcoxon's test yields only a 9.8% probability that the null hypothesis is valid on the differences in usage changes in both types of periods. Thus, users responded to the indicator color despite the comparable numerical discounts.

4.4.4 Benefits of Optimized TDP

Question: *Does peak usage decrease with time-dependent pricing? And does this decrease come at the expense of an* overall *decrease in usage?*

Method: We offered optimized time-dependent prices to all users with JB devices over a two week period in January 2012. The waiting functions used to calculate these prices were estimated from usage data for the static time-dependent prices. To measure the peak reduction, we calculated the *peak-to-average ratio* (PAR), i.e., the ratio of usage in the peak period to average per-period usage, for each day. We then compared the PARs from our TIP data to those observed with TUBE's optimized TDP.

Results: *Optimized time-dependent prices reduce the peak-to-average ratio* from TIP usage. Moreover, *overall usage significantly increased* after TDP was introduced, partially because people used more in the discounted valley periods.

Figure 13a shows the distribution of daily PARs both before and after TDP was introduced. The maximum PAR decreases by 30% with TDP, and approximately 20% of the

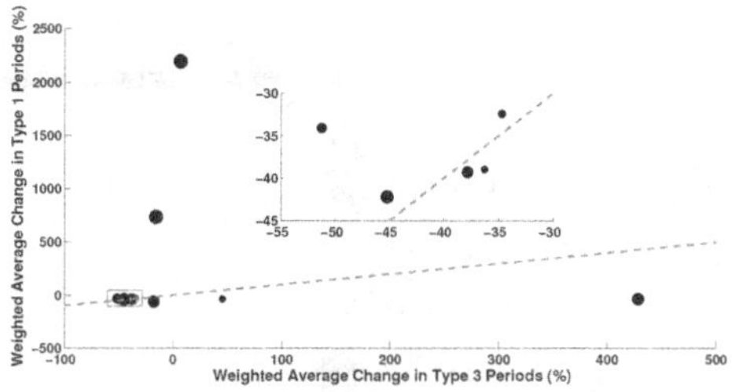

(a) Period types 1 and 3.

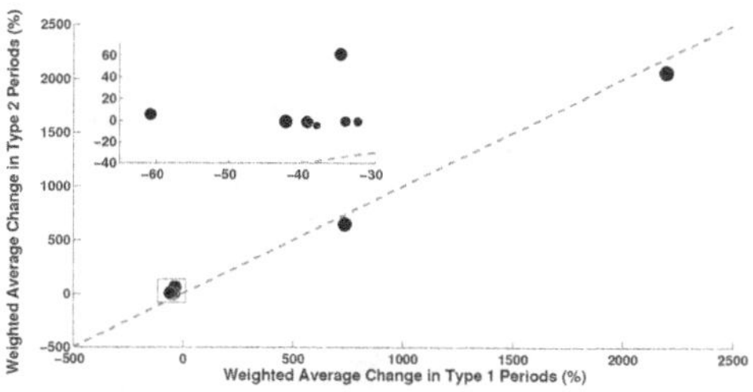

(b) Period types 1 and 2.

Figure 12: Changes in usage for the period types in Table 5.

PARs before TDP are larger than the maximum PAR with TDP. Thus, TDP significantly reduced the peak-to-average ratio, flattening demand over the day.

We next show that this decrease in PAR is not due to a net loss of traffic. Figure 13b shows the peak daily usage observed before and after TDP. Though the maximum peak usage is about the same, peak usage is generally larger with TDP. Since the PARs also decreased, usage in non-peak periods increased. In fact, we observe a 130% increase in usage from TIP to TDP. Part of this increase may be due to the time of year–we measured the TIP usage from July to September, and the TDP usage in January. TDP, however, is likely a major factor: the discounts during off-peak periods allowed users to consume more data while still spending less money and decreasing the PAR.

Finally, we examine the changes in application distribution due to the introduction of TDP. Figure 13c shows the average daily usage by application before TDP (i.e., during our TIP calibration period) and when TDP is introduced. We see that movie streaming nearly quadrupled, while usage of other applications stayed about the same. Since streaming is generally used for entertainment, the discounts may have motivated people to consume more data during low-price periods. Thus, the "valleys" of TIP usage were further filled up by an increase in demand, creating a win-win-win across end users, ISPs, and content providers.

5. DISCUSSION AND FUTURE WORK

We now discuss several limitations of TUBE's model and our trial, and then identify future extensions of our work.

Mobility: Since mobile users may switch their base stations frequently, the proposed TUBE system requires user devices to keep track of different prices across distinct base stations. However, user mobility can often be predicted from location history [22]; thus, the user device can predict its location over the next day and pull prices from the appropriate base stations. The device itself can then keep track of the user's bill, an approach scalable to many users.

Single bottleneck: Our model assumes a single bottleneck in the network. This assumption is consistent with a wireless base station in an urban area or a middle-mile bottleneck in rural areas.

Time granularity: Our initial trial uses hour-long periods, since users are more familiar with hourly prices, but we can also shrink the periods to, say, 10 minutes. Shorter time periods allow users to wait less, possibly enhancing TDP's effectiveness. If the timescale is further reduced to several seconds, the autopilot mode on user devices effectively responds to real-time congestion conditions, turning time-dependent pricing into congestion-dependent pricing.

Trial scale and additional functionalities: Different demographics likely have different price and time sensitivities for mobile data, making our trial only the first step towards understanding TDP's effectiveness. We are currently conducting larger-scale trials with U.S. and Indian ISPs that will further illuminate users' price-delay tradeoffs and directly test the effect of the autopilot mode.

Control group: While a trial control group is highly desirable, it is difficult to compare the TDP usage of one group of users with another control group of TIP users unless they are matched properly. Such matching is especially challenging in a small population of users, as in our trial. Our planned large-scale trials will address this issue.

6. RELATED WORK

Internet pricing models have been debated since the 1990s. Several pricing schemes, both *static* and *dynamic*, have been suggested by networking researchers to alleviate congestion. Static pricing plans charge users according to predetermined rates without adapting to customers' usage behavior, e.g., metered, flat price, cap then metered, and two-period time of day pricing plans [17]. Other proposals include Clark's Expected Capacity Pricing [3], Cocchi's Edge Pricing [4], and Odlyzko's Paris Metro Pricing [16], with the former two admitting dynamic versions as well.

Dynamic pricing has the advantage of adapting prices to the network condition, as shown in Gupta et al.'s Priority Pricing [7], Hayer's Transport Auction [9], Kelly et al.'s Proportional Fair Pricing [11], Varian's Smart Market Pricing [13], MacKie-Mason et al.'s Responsive Pricing [14], Semret et al.'s Market Pricing [19], etc. Sen et al. [20] provides a detailed overview of these various pricing proposals and their realization in current data plans. The social and ethical dimensions of dynamic pricing have also been widely studied, and its consumer adoption and benefits in electricity and transportation networks are well documented [20]. However, there have been no documented trials of dynamic pricing for mobile data.

7. CONCLUSION

Though time-dependent pricing for mobile data has been discussed for several decades, no experimental study has been conducted to investigate a functional prototype. To this end, we developed and implemented TUBE, an architecture that takes TDP from economic theory to a system implementation. TUBE creates a feedback loop between the ISP's price computation and users' ever-changing response to these prices. To link these components, we estimate and predict users' future behavior each day with aggregate usage data. Users respond to the prices via a GUI that resides on their devices, either manually or using an autopilot mode.

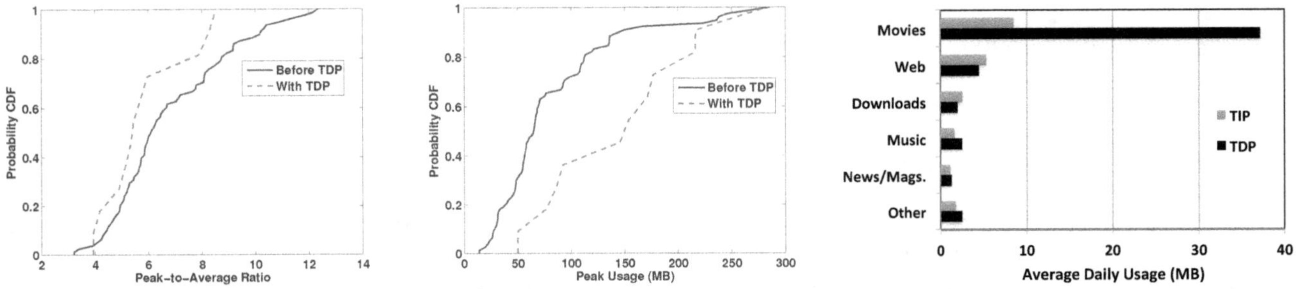

(a) PARs for TIP and TDP. (b) Peak traffic for TIP and TDP. (c) App distribution with TIP and TDP.

Figure 13: Usage statistics for TIP and TDP.

To confirm TDP's feasibility, we conduct a trial with 50 iPhone/iPad 3G data plan users, acting as a resale ISP and charging them according to our TDP algorithms. Our trial results indicate that people are sensitive to time-dependent prices and indeed shift their Internet usage to off-peak periods, while increasing the monthly total of data usage. TDP flattens the temporal distribution of user demand for data, thus reducing ISP cost while allowing users to save money. This implementation and pilot trial motivates future study on time-dependent pricing for different markets and demographics.

Acknowledgements

We thank Rudiger Rill and Dana Butnariu for their iOS development efforts, Jennifer Rexford for her comments on an earlier version of this work, and many colleagues for discussions on smart mobile pricing. We also thank AT&T for providing the APN and our trial customers for their participation. Part of the work reported here was supported by NSF CNS-1117126 and AFOSR FA9550-09-1-0643. C. Joe-Wong was supported by the NDSEG fellowship.

8. REFERENCES

[1] Cheng, R. Verizon to curb highest data users, Feb. 4 2011. Wall Street Journal.

[2] Cisco Systems. Cisco visual networking index: Forecast and methodology, 2011-2016, May 30 2012. http://tinyurl.com/VNI2012.

[3] Clark, D. D. Internet cost allocation and pricing. In *Internet Economics*, L. W. McKnight and J. P. Bailey, Eds. The MIT Press, 1997, pp. 215–252.

[4] Cocchi, R., Shenker, S., Estrin, D., and Zhang, L. Pricing in computer networks: Motivation, formulation, and example. *IEEE/ACM Transactions on Networking 1* (1993), 614–627.

[5] Falaki, H., Mahajan, R., Kandula, S., Lymberopoulos, D., Govindan, R., and Estrin, D. Diversity in smartphone usage. In *Proc. of ACM MobiSys* (2010), ACM, pp. 179–194.

[6] Goldman, D. AT&T hikes rates on smartphone plans, Jan. 19 2012. CNN Money.

[7] Gupta, A., Stahl, D., and Whinston, A. Priority pricing of integrated services networks. In *Internet Economics*, L. W. McKnight and J. P. Bailey, Eds. The MIT Press, 1997, pp. 323–352.

[8] Ha, S., Sen, S., Joe-Wong, C., Im, Y., and Chiang, M. TUBE survey questions and demographics, Jan. 2012. http://scenic.princeton.edu/tube/TUBE_Survey.pdf.

[9] Hayer, J. Transportation auction: A new service concept. *M.Sc./M.B.A. Thesis, TR-93-05* (1993).

[10] Joe-Wong, C., Ha, S., and Chiang, M. Time-dependent broadband pricing: Feasibility and benefits. *Proc. of IEEE ICDCS* (June 2011).

[11] Kelly, F., Maulloo, A. K., and Tan, D. H. K. Rate control for communication networks: Shadow prices, proportional fairness, and stability. *Journal of Operational Research Society 49* (1998), 237–252.

[12] Langley, R. *Practical Statistics Simply Explained.* Dover Books Explaining Science Series. Dover Publications, 1971.

[13] MacKie-Mason, J., and Varian, H. Pricing the Internet. In *Public Access to the Internet*, B. Kahin and J. Keller, Eds. Prentice-Hall, 1995, pp. 269–314.

[14] MacKie-Mason, J. K., Murphy, L., and Murphy, J. Responsive pricing in the Internet. In *Internet Economics*, L. W. McKnight and J. P. Bailey, Eds. The MIT Press, 1997, pp. 279–303.

[15] NIST. *Engineering Statistics Handbook.* http://itl.nist.gov/div898/handbook/pmc/section4/pmc435.htm.

[16] Odlyzko, A. Paris metro pricing for the Internet. *Proc. of the 1st ACM Conf. on Electronic Commerce* (Nov. 1999).

[17] Parris, C., Keshav, S., and Ferrari, D. A framework for the study of pricing in integrated networks. Tech. rep., Tenet Group, ICSI, UC Berkeley, 1992. TR-92-016.

[18] Princeton EDGE Lab. TUBE website, 2012. http://scenic.princeton.edu/tube/.

[19] Semret, N., Liao, R. R.-F., Campbell, A. T., and Lazar, A. Pricing, provisioning and peering: Dynamic markets for differentiated internet services and implications for network interconnections. *IEEE Journal on Selected Areas in Communications 18* (2000), 2499–2513.

[20] Sen, S., Joe-Wong, C., Ha, S., and Chiang, M. Pricing data: A look at past proposals, current plans, and future trends. *arXiv* (Feb. 2012). http://arxiv.org/abs/1201.4197.

[21] Shih, J., Katz, R., and Joseph, A. Pricing experiments for a computer-telephony-service usage allocation. In *Proc. of IEEE Globecom* (2001), vol. 4, IEEE, pp. 2450–2454.

[22] Song, C., Qu, Z., Blumm, N., and Barabási, A. Limits of predictability in human mobility. *Science 327*, 5968 (2010).

[23] Taylor, J. Short-term electricity demand forecasting using double seasonal exponential smoothing. *Journal of the Operational Research Society* (2003), 799–805.

[24] Taylor, P. AT&T imposes usage caps on fixed-line broadband, March 14 2011. Financial Times.

[25] The Economist. The mother of invention: Network operators in the poor world are cutting costs and increasing access in innovative ways, Sept. 24 2009. Special Report.

[26] TUBE Project. Enabling mobile time-dependent pricing, 2012. http://www.youtu.be/lXuJw4tWH40.

[27] Varaiya, P. P., Edell, R. J., and Chand, H. INDEX project report, Aug. 1996. http://people.ischool.berkeley.edu/~hal/index-project/R98_005P.PDF.

[28] Welch, C. Verizon to kill grandfathered unlimited data plans for customers seeking upgrades, May 16 2012. The Verge.

CarSpeak: A Content-Centric Network for Autonomous Driving

Swarun Kumar, Lixin Shi, Nabeel Ahmed, Stephanie Gil, Dina Katabi and Daniela Rus
Massachusetts Institute of Technology
{swarun, lixshi, n3ahmed, sgil, dk, rus}@mit.edu

ABSTRACT

This paper introduces CarSpeak, a communication system for autonomous driving. CarSpeak enables a car to query and access sensory information captured by other cars in a manner similar to how it accesses information from its local sensors. CarSpeak adopts a content-centric approach where information objects – i.e., regions along the road – are first class citizens. It names and accesses road regions using a multi-resolution system, which allows it to scale the amount of transmitted data with the available bandwidth. CarSpeak also changes the MAC protocol so that, instead of having nodes contend for the medium, contention is between road regions, and the medium share assigned to any region depends on the number of cars interested in that region.

CarSpeak is implemented in a state-of-the-art autonomous driving system and tested on indoor and outdoor hardware testbeds including an autonomous golf car and 10 iRobot Create robots. In comparison with a baseline that directly uses 802.11, CarSpeak reduces the time for navigating around obstacles by 2.4×, and reduces the probability of a collision due to limited visibility by 14×.

Categories and Subject Descriptors C.2.2 [**Computer Systems Organization**]: Computer-Communications Networks

Keywords Autonomous Vehicles, Content-Centric, Wireless

1. INTRODUCTION

Autonomous vehicles have been the topic of much recent research [5, 31, 18]. The goal of these systems is to drive from point A to point B in an efficient and safe manner, while dealing with continuous changes in the environment due to pedestrian and object movements, and the potential of unexpected events, such as roadwork and accidents. To achieve their goal, autonomous vehicles need detailed realtime information about their surroundings [22]. They typically use laser rangefinder sensors to discover the surfaces of nearby objects and represent this information as a *3D-point cloud* similar to that shown in Fig. 1. Using only the car's on-board sensors, however, prevents autonomous vehicles from uncovering hidden objects that are not directly in their line-of-sight, e.g., a kid running around the corner, or a car pulling out of an occluded driveway. These sensors also cannot deliver long-range data with sufficient accuracy, which limits the car's ability to plan ahead [31]. Further, they are costly (e.g., the sensors alone on an autonomous vehicle can cost several hundred thousand US dollars [17, 18]). For these reasons, the report from the recent DARPA Urban Challenge identifies the need for information sharing between autonomous vehicles as a key lesson learned from the contest [3]. However, 802.11 is ill-suited for this application. Navigation sensors can generate realtime streams at Gb/s from each car, leading to a scenario where there is always more data than bandwidth to send it. Furthermore, a communication protocol that cannot capture the importance of different pieces of information for the application will end up inundating the medium with irrelevant or stale data, and potentially denying access to important and urgent information.

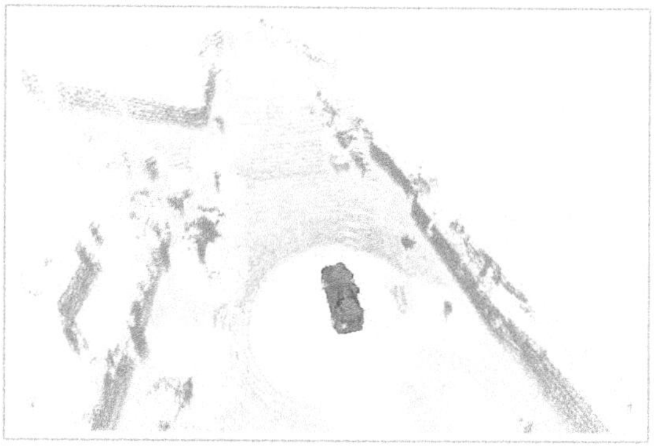

Figure 1—Example of sensory information used in autonomous driving. The figure shows a 3D-point cloud of a road obtained by a Velodyne laser sensor, where colors refer to elevation from the ground. Note that a 3D-point cloud provides the (x, y, z) coordinates of points lying on the surface of obstacles.

This paper introduces CarSpeak, a communication system that addresses the needs of autonomous vehicles. CarSpeak enables cars to request and access sensory information from other cars, as well as static infrastructure sensors, in a manner similar to how they access their own sensory information. To achieve its goal, CarSpeak adopts a content-centric design, where information objects are first class citizens. CarSpeak's information objects are regions in the car's environment (e.g., a cube of 1 m^3). In CarSpeak, a car can request a realtime stream of a 3D-point cloud data from a particular region along the road. It can also zoom in to get a more detailed description, or zoom out for a wider view.

CarSpeak delivers its design via three components that address the main challenges in sharing navigation sensor data:

- *How does a car describe the information it wants, at a particular resolution, if that information describes a region along the road?* In order to name and find road regions, CarSpeak divides the world recursively into cubes; smaller cubes provide a finer description of the encompassing cube. Each cube refers to a region. To efficiently represent this data, CarSpeak uses an Octree, a data structure commonly used in graphics to represent 3D objects [28, 11]. Each node in the Octree refers to a cube, and the sub-tree rooted at that node refers to finer details inside that cube, as shown in Fig. 2. The Octree representation allows CarSpeak to name regions at different resolution and according to their lo-

SIGCOMM'12, August 13–17, 2012, Helsinki, Finland.

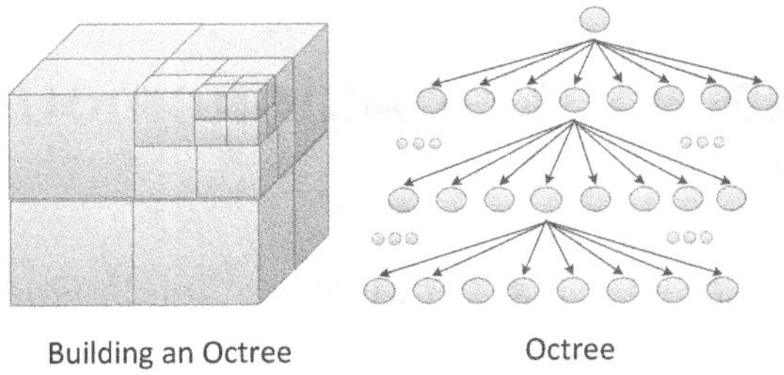

Figure 2—Representation of regions using Octree.

cation in the world. Specifically CarSpeak names a region by referring to the root of the region's sub-tree; it expresses the resolution of the region using the depth from the root of the region's sub-tree. The Octree also enables a car to store its data efficiently because, though the world is huge, each car needs to only expand the part of the Octree in its neighborhood.

- *How does the system allocate the wireless bandwidth to the most recent data from the region, given that multiple cars may sense the same region and each car does not know what information other cars know?*
  CarSpeak adopts a content-centric MAC where information objects, as opposed to senders, contend for medium access. Further, each information object (i.e., 3D-point cloud stream) obtains a share of the medium proportional to the number of requests it receives.
  CarSpeak implements this abstraction using a distributed protocol, where nodes that sense a region contend on its behalf. Requests for region data are broadcast on the medium. Nodes compute a summary value of the quality of the information they have of each region (which is a measure of the timeliness and completeness of this information). CarSpeak uses a low overhead protocol to share this information among the nodes as annotations on their transmitted data packets. Each car uses these annotations to compute how much sensory data it should transmit so that its contribution to each stream is proportional to the completeness and freshness of the data it has from the corresponding region. CarSpeak then enforces this allocation by controlling the 802.11 contention window appropriately.
- *How does the system compress the redundancy in the transmitted sensor data while being resilient to packet loss?*
  CarSpeak makes each packet self-contained by assigning it an independent set of branches in the Octree that are derived from the root. As a result, each received packet can be correctly inserted into the tree independent of other packets. CarSpeak also reduces the overlap between data transmitted by cars that sense the same region. Recall that each region is a cube that encompasses many smaller cubes, whose values keep changing in realtime due to the arrival of new sensor data. In CarSpeak even if multiple cars receive a request for the same region (i.e., the same encompassing cube), each of them will pick a different permutation according to which they transmit the sub-cubes in the region. Thus, if only one car has sensor data about the region, it will eventually transmit all the sub-cubes from the region. However, if multiple cars have data about the same region, then they are likely to cover all sub-cubes in the region, while limiting the overlap in their transmissions.

We built a prototype of CarSpeak in ROS, the Robot OS [26] and integrated it with a state of the art path planner, whose earlier version was used in the DARPA Urban Challenge. We evaluated CarSpeak on two testbeds: 1) an indoor testbed of iRobot Create programmable robots connected to netbooks with Atheros AR9285 cards and gathering sensor data from Xbox 360 Kinects, and 2) an outdoor testbed composed of an autonomous Yamaha G22E golf car mounted with Hokuyo laser range sensors, and exchanging sensory information with the Create robots. We compared CarSpeak with a baseline inter-vehicle communication protocol that directly uses the existing 802.11 protocol.

Experiments from the indoor testbed show that compared to the 802.11 baseline, CarSpeak reduces the time taken to navigate an environment with obstacles by 2.4×, and the probability of a collision due to limited visibility by 14×.

Outdoor experiments with the a Yamaha golf car tests the role of communication in enabling cars to react safely to pedestrians who suddenly exit a blind spot and cross the car's path. Empirical results show that use of CarSpeak allows for the receiver on the golf car to issue a stop command with a maximum average delay of 0.45 seconds which is 4.75× smaller than the *minimum* delay of 2.14 seconds using 802.11. These relatively small delays using CarSpeak allow the vehicle to safely stop before the crosswalk if the pedestrian appears at distances as small as 1.4 meters on average, even when the vehicle is traveling at its maximum velocity of 2 meters per second. In contrast, using 802.11 the vehicle is unable to stop before reaching the crosswalk if the pedestrian appears when the vehicle is closer than four meters from the crosswalk on average.

Contributions: To our knowledge, CarSpeak is the first communication system for multiple autonomous vehicles that focuses on maximizing the utility of information for this application, and that is fully integrated with autonomous vehicle systems. It is evaluated on a testbed of autonomous vehicles, and demonstrated to reduce path length and the probability of collisions. Its content-centric design that operates on realtime rich sensory data sets it apart from past work on VANET. This design is delivered via three components including a multi-resolution naming and addressing scheme, a content-centric MAC, and a new approach to compressing rich sensory data that is suitable for lossy and dynamic environments.

2. RELATED WORK

Recent years have witnessed major advances in designing and building autonomous vehicles to realize safer and more fuel efficient future cars [5, 31, 18]. Past work in this domain [14, 8, 17], including the DARPA Urban Challenge and the Google autonomous car, focuses on issues related to perception, efficient path planning, obstacle detection, etc. In contrast, this paper focuses on designing a communication protocol that is most suitable for sharing sensory data between autonomous vehicles.

Our work is related to a broad area in robotics that studies networks of robots. Past work in this area can be divided into two categories: The first category uses communication as a black-box, and focuses on algorithms that enable robots to collaborate on a desired task, for instance, cooperative exploration [23] or pursuit evasion [15]. The second category considers the application as a black-box and focuses on harvesting robot mobility to improve network connectivity or throughput [24, 9]. In contrast, our work is based on designing the communication protocols around the needs of the application, and takes neither as a black box.

A large number of research papers have focused on the problem of Vehicular ad-hoc networks (VANETs). Work in this area focuses on efficient routing [19, 27, 30], delay tolerant networks [20], reliable delivery of emergency messages [2, 6], or specific applications such as detecting accidents [13]. None of these papers, however, present a content-centric architecture or design a MAC protocol where information objects contend for the medium. Also, none of them present a solution that is particularly suitable for autonomous driving.

Our work builds on past work on content-centric networking.

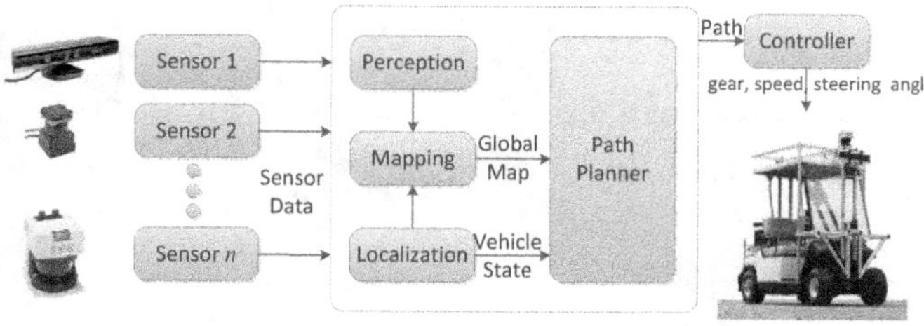

Figure 3—High-Level Architecture of Autonomous Vehicular Systems. The path planner module uses information from various sensors to compute a safe path for the vehicle.

Past work in this domain is mostly focused on the Internet [12, 16]. The few papers that apply this concept in the wireless domain are focused on storage or routing information content [25, 29, 4]. Our work differs from all these papers in that it is focused on resource sharing at the MAC layer. Also, it uses a multi-resolution naming system and is fully integrated with an autonomous driving in terms of design, implementation and evaluation.

3. PRIMER ON AUTONOMOUS VEHICLES

In this section, we provide a quick background of autonomous driving software so that it is clearer how CarSpeak interfaces with these systems. Successful performance of autonomous vehicles relies on their ability to sense and process information about the environment around them. To obtain this information, autonomous vehicles and robots are typically equipped with ranging sensors, which deliver realtime measurements of the distance of the vehicle to the surrounding 3D objects. The vehicle may use laser scanners, ultrasonic range finders for outdoor settings and Kinect for indoor settings [31, 5, 18, 10]. Other sensors like cameras and light detectors are also used for additional information.

Most autonomous vehicles use the Robot Operating System (ROS) framework [26]. ROS provides a publish/subscribe architecture, where a module (e.g., sensor) publishes a topic (e.g., /sensor_data) that can be subscribed to by multiple modules. We discuss the commonly defined high-level modules below (Figure 3):

- **Sensor Infrastructure:** Each sensor attached to the autonomous vehicle has an associated module which converts raw sensor information obtained from the driver into a generic sensor format. The most widely used format is a *3D-point cloud* which provides the 3-D (x, y, z) coordinates of points lying on the surface of obstacles. The point cloud, along with a timestamp t denoting the time of retrieval of sensor data, is published by each sensor module.
- **Planner:** The planner's goal is to use sensory information to plan an obstacle-free path for the vehicle to navigate along. The planner typically has access to a detailed global map of the environment. The planner is sub-divided into four modules:
  - *Perception module* subscribes to point cloud information from the sensors and applies complex obstacle detection algorithms to recognize obstacles in the frame of reference of the vehicle. It publishes a map of these obstacles.
  - *Localization module* publishes the vehicle's position within the global map based on GPS, odometry or more advanced sensory infrastructure, some of which can be as accurate as a few centimeters [18].
  - *Mapper* subscribes to information from the localization and perception modules and publishes a global map incorporated with locations of obstacles.
  - *Path planner* subscribes to the vehicle's location and the global obstacle map and publishes a path for the vehicle to travel along.

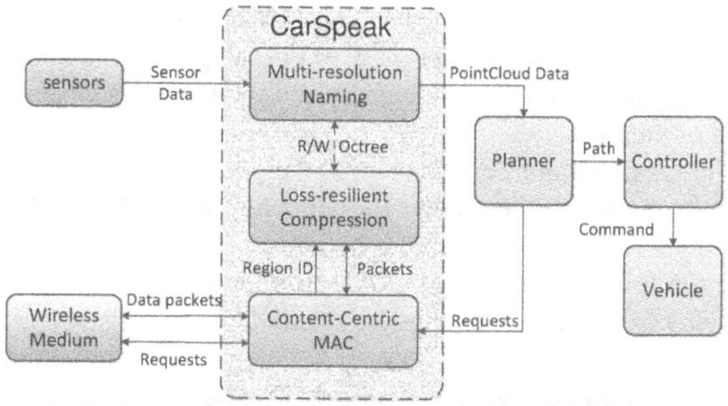

Figure 4—Information Flow in CarSpeak. CarSpeak has three components: A Multi-Resolution Naming and Addressing system, A Loss-Resilient Compression system and a Content-Centric MAC

- **Controller:** The controller subscribes to the vehicle's path and issues steering and velocity-control commands to the vehicle, so that it navigates along the computed path. In many cases, the controller may execute emergency maneuvers if there is substantial change in the obstacle map due to moving obstacles.

4. CARSPEAK'S ARCHITECTURE

CarSpeak's design aims to interface effectively with the ROS (Robotic Operating System) architecture for autonomous vehicles. From the perspective of the ROS planner, CarSpeak looks like a local sensor that streams sensory information obtained from other vehicles and static infrastructure sensors.[1] CarSpeak receives requests from the car's planner by subscribing to /query_region topic. It propagates these requests over the network to CarSpeak modules on other vehicles to direct them to transmit information from the requested region. When it receives sensory data in response to requests it sent, like other sensors, CarSpeak publishes this data as a stream of 3D point-cloud data (under the topic /car_speak). The planner may now subscribe to the information from the requested regions. Unless refreshed, a subscription (as well as the corresponding requests broadcast on the medium) expires after one minute. Timing out subscriptions is done for efficient use of bandwidth as cars are expected to lose interest in some regions and gain interest in others, as they move around.

CarSpeak's guarantees are best effort, i.e. CarSpeak aims to make the best use of the available bandwidth to send as much relevant information as possible, in a loss-resilient manner. CarSpeak has three components: A Content-Centric MAC, a Multi-Resolution Naming and Addressing system, and a Loss-Resilient Compression system. Fig. 4 illustrates how these components interact with each other, the planner, and the wireless channel.

The MAC receives region requests from the planner and broadcasts these requests on the medium. It also keeps track of requests received from other cars over the wireless medium. It evaluates the importance of different regions based on how many requests they have recently received and tries to satisfy these requests by working with the other CarSpeak components.

The multi-resolution naming and addressing system subscribes to 3D-point cloud information published by local sensors and builds an Octree-representation of this data. The Octree is read by the compression module whenever CarSpeak sends data packets in response to outside requests, and is written by the compression module whenever CarSpeak receives data packets in response to requests generated by the car. The multi-resolution naming and ad-

[1] Sharing information requires a notion of trust. One option is to use the IEEE 1609.2 security standard for inter-vehicular networks to digitally sign and verify all messages. However, the details are beyond the scope of this paper.

dressing system also publishes updates to the Octree caused by the arrival of external data as a 3D-point cloud under the topic /car_speak, which is subscribed to by the planning system of the autonomous vehicle.

The loss-resilient compression is triggered by the MAC to generate compressed data packets for transmission on the medium or to decode received data packets and insert them in the Octree.

In the following sections, we discuss the design and functioning of each of these modules in more detail.

5. MULTI-RESOLUTION NAMING

In autonomous vehicles, sensory information is typically represented as a 3D-point cloud. The point-cloud representation however is unstructured, and hence does not facilitate requesting information about a specific region. It is also inefficient because it does not compress the information by leveraging the fact that points close to each in space other tend to have similar properties: empty or occupied. Sharing information across vehicles requires a naming scheme in which a car can name a specific region of interest. It also requires an efficient representation that compresses the exchanged data and reduces bandwidth consumption.

5.1 Information Naming and Representation

CarSpeak uses the *Octree naming system* to identify and represent sensor information from the environment. Specifically, CarSpeak divides the world recursively to cubes. It starts with a known bounding cube that encompasses all points observed by vehicles in the environment. Thus this bounding cube, or root cube, is known and agreed upon by all vehicles . Each cube is then recursively divided into 8 smaller cubes as shown in Figure 5. A cube is set to be either: (1) *occupied*, if the point cloud representation has points within it (i.e., the cube has some object and the car should not drive through it); (2) *unoccupied*, if there are no points within it (i.e., the cube is vacant and the car may drive through it); and (3) *unknown* if there is no sufficient sensory information about it (i.e., the cube may have some object but the car does not yet have sensor data to identify it). We note that a parent is occupied if any of its descendants are occupied. A parent is unoccupied only if all of its descendants are unoccupied. Otherwise the parent is unknown.

CarSpeak maintains this recursive structure in an Octree, where each vertex in the Octree represents a cube, and the sub-tree rooted at that vertex refers to the recursive divisions of the cube. The Octree representation allows CarSpeak to name road regions at different resolutions and according to their locations in the world. Specifically, in CarSpeak a region is an encompassing cube, which is nothing but a sub-tree in the Octree, truncated to L levels, where L is the resolution at which the region is described. In principle, one can allow regions of any size to have any resolution. This would allow a car to request the whole world at the finest resolution. Such a design is both inefficient and unnecessary.

Thus, CarSpeak expresses large regions at coarse resolutions and smaller regions at finer resolutions. Specifically, CarSpeak partitions the Octree into mutually disjoint sub-trees, where each subtree is truncated to L levels. Each of these truncated sub-trees denotes a region and forms a hierarchy, as shown in Fig. 5. All regions are described completely by their corresponding truncated sub-tree, which contain up to $8^L - 1$ vertices, labeled either "occupied", "unoccupied", or "unknown". Regions at a higher-level in the hierarchy provide a zoom-out view and are represented at a coarser spatial granularity, whereas regions at a lower level in the hierarchy provide a zoom-in view and are represented at a finer spatial granularity. A key point to note is that truncated sub-trees corresponding to any pair of regions, regardless of their hierarchy, do not overlap.

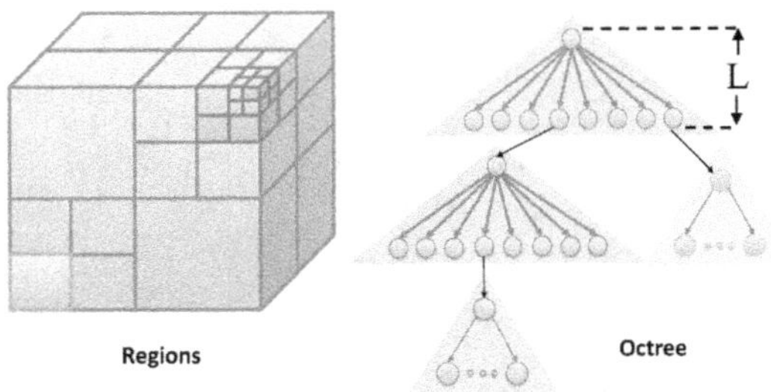

Figure 5—Naming Regions in CarSpeak. The figure on the left shows four different regions at different hierarchies with $L = 2$. The figure on the right depicts the truncated sub-trees corresponding to these regions in the Octree.

To assign regions globally unique names, we number them in ascending order in a top-down, breadth-first manner starting from the region containing the root of the Octree (that provides a coarse view of the entire environment). The number of regions and L are design parameters that can be chosen to fit the specific application.

Benefits: CarSpeak Octree-based naming system has two advantages. First, it facilitates requesting and accessing sensory data at different resolutions. Second, it compresses the data both for storage and transmission. In particular, while the world may be huge, each vertex need not expand regions that are far from its location. Further, when exchanging information, a large subtree can be sent as one vertex if all vertices in the subtree have the same value (e.g., if a whole subtree has occupied vertices, CarSpeak sends a single value expressing the fact that the whole subtree is occupied).

5.2 Information Retrieval and Maintenance

The multi-resolution naming and addressing module subscribes to information from local sensors and incorporates this data into its Octree. It also populates its Octree using sensor data received from other vehicles over the wireless medium. This octree provides information to other nodes that may failed to obtain it from the original sender due to packet loss or disconnectivity. Further, it publishes the data it receives from other vehicles through the topic /car_speak after converting it to the 3D point cloud representation. (Note that data from local sensors is already published by the sensors themselves.)

Garbage Collection. Each vertex of the the Octree datastructure is annotated with a time_stamp field that is set to the timestamp of the most recent sensor information stored in this cube. CarSpeak runs a thread to periodically traverse the Octree and removes vertices which are outdated. If time_stamp is older than the current time by over a fixed threshold, then the entire sub-tree is dropped. We note that in CarSpeak, all cars use NTP to synchronize their time to within tens of milliseconds, which is sufficient for the purpose of the application.

5.3 Information Quality

The 3D stream of a region contains data from multiple nodes that sense that region. On average, for each region, CarSpeak aims to provide each node a share of the medium proportional to the *quality* of information it possesses about that region. Let $Q(i,r)$ denote the quality of information that node i has regarding region r. CarSpeak evaluates this function based on the following metrics:

- *Time*: Sensor information gathered more recently is of greater value, than sensor information gathered in the past. CarSpeak decreases the quality of sensor information $Q(i,r)$ based on delay, by a factor of $\mu^{(t_{curr}-t_{sense})}$; where t_{curr} is the current timestamp,

t_{sense} denotes the timestamp at which sensor information was last obtained from the region and $0 < \mu < 1$ is a constant.

- *Completeness*: The more complete the information a node has about a region, the higher the value of its information. Hence, CarSpeak linearly increases the quality of sensor information based on its completeness, i.e., $Q(i,r) \propto C(i,r)$, where $C(i,r)$ is a measure of completeness of information that node i possesses about region r. The Octree presents an effective metric for measuring $C(i,r)$. Let $C(i,r)$ denote the number of vertices not labeled unknown in the section of the Octree, possessed by node i, representing region r divided by the number of possible vertices in region r. Hence, $C(i,r)$ represents the fraction of region r that node i has sensed.

Thus, CarSpeak's multi-resolution naming and addressing system on node i keeps track of the quality of sensor information, $Q(i,r)$, possessed by its node for each region r, as:

$$Q(i,r) = C(i,r)\mu^{(t_{curr}-t_{sense})}, \qquad (1)$$

where t_{sense}, t_{curr}, and $C(i,r)$ are as defined above and $0 < \mu < 1$ is a constant ($\mu = 0.5$ in our implementation).

6. LOSS-RESILIENT COMPRESSION

3D streams of region data are highly redundant. The information at each node is correlated because occupied cubes tend to be co-located. The same applies to unoccupied and unknown cubes. Further, different nodes that sense the same region may have overlapping information (though the overlap is typically not complete due to differences in perspective and viewing distances). Thus, efficient transmission requires compressing the data. A good compression scheme however should deal with the following issues: 1) The compression algorithm should have a low computational overhead to maintain the realtime nature of the data; 2) It should not require the cars to track how their information relate to each other; and 3) It should be resilient to packet loss. We describe how CarSpeak meets these requirements below.

State-of-the-art compression schemes in the graphics community for 3D-point clouds leverage the fact that the corresponding Octrees can be efficiently encoded to provide a compressed high-entropy representation [28, 11]. Similarly, CarSpeak exploits this compression capability of the Octree.

The standard Octree encoding technique is as follows: Let each vertex in the Octree be represented by a tuple of length 8 representing the occupancy of each of its children. We traverse the tree in a top-down and breadth-first fashion and read off the corresponding tuples to obtain the compressed data. The root vertex can always be assumed to be occupied. Then, for each vertex, its occupied and unknown children are recursively encoded. In contrast, no further information needs to be encoded for unoccupied vertices (as they are assumed to be entirely unoccupied). Additionally, vertices whose descendants are entirely occupied or unknown are terminated by a special 8-tuple (with all entries marked occupied or unknown respectively) and are not further encoded, as shown in Fig. 6. The decoder can then faithfully recover the Octree by following the same traversal rules as the encoder.

This algorithm is very efficient since it simply walks the Octree representation. The problem however with using traditional Octree compression is that the loss of even a single byte in the Octree representation affects the correctness of all data following that byte. Consequently, Octree based compression frameworks are highly sensitive to packet loss. Furthermore, the above algorithm cannot deal with data overlap across nodes that sense the same region.

We introduce a new algorithm that leverages the above standard Octree compression, but deals with packet loss and overlap. Specifically: Let r be the region from which a packet is requested by the Content-Centric MAC. Let $\mathcal{O}_r$ denote the truncated sub-tree in the Octree corresponding to this region. Let $v_0, v_1, \ldots, v_{K-1}$ denote the leaf vertices of this sub-tree. The algorithm to generate a packet proceeds as follows: Let variable j be set to a random integer between 0 and $K-1$, and variable $\mathcal{T}$ be set to an empty tree.

(1) Add vertex v_j to the tree $\mathcal{T}$ as well as all its ancestors tracing all the way up to the root of $\mathcal{O}_r$.
(2) Set $j \leftarrow j+1$ and repeat step (1), while the encoding of $\mathcal{T}$ fits within the maximum size of a packet.

The algorithm writes the Octree encoding of $\mathcal{T}$ to a packet and transmits it. Note that all packets are not generated in one shot, i.e. even if two packets are describing the same region, the one with the more recent time stamp will have more current information. As nodes pick different random seeds, it is highly probable that different nodes that sense the same region do not transmit overlapping information, at any point in time.

The compression algorithm can be further improved by using predictive coding techniques [28, 11] as the sub-graph $\mathcal{T}$ preserves edges between parents and children in the original Octree $\mathcal{O}_r$ (Our current implementation however does not use predictive coding).

The resulting compression algorithm has the following desirable properties: 1) it is computationally efficient because it is linear in the size of the region's Octree; 2) it is robust to losses because each packet is self-contained; 3) packets received from different cars that sense the same region have minimal overlap because at any point in time, they transmit different parts of the region's Octree; and 4) it supports a form of unequal error protection because the transmitted subtrees contain paths of all vertices to the root of the region. Hence, vertices at higher levels in the tree are less likely to be lost than vertices at lower levels. Thus, the loss of a packet typically results in the loss of resolution as opposed to complete loss of information.

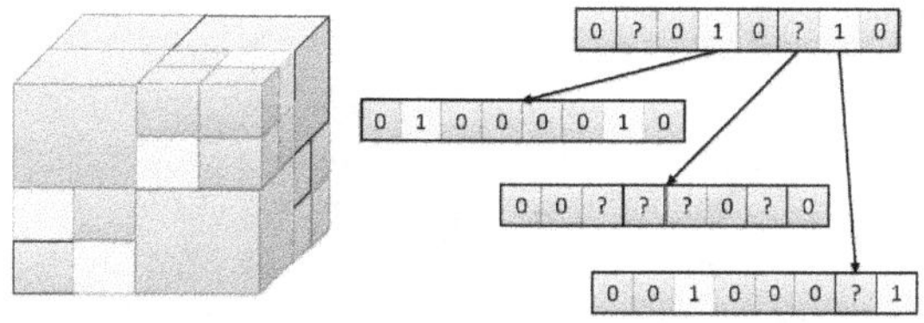

Figure 6—Octree Representation Representation of 3D-point cloud using a 2-level Octree. Vertices that are unoccupied are not expanded; vertices that are completely occupied or unknown are terminated by special leaf vertices; these have been omitted for clarity.

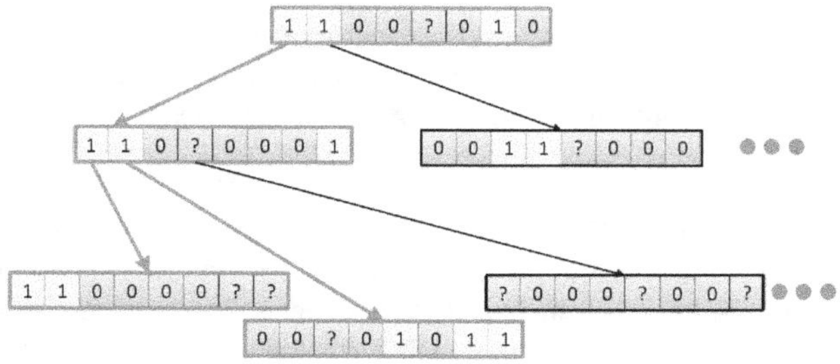

Figure 7—Picking tree $\mathcal{T}$ for compression Tree $\mathcal{T}$ picked with j initialized to 0 is highlighted by red contours.

7. CONTENT CENTRIC MAC

CarSpeak adopts a content-centric MAC that focuses on the application's goals and requirements. In CarSpeak, regions, as opposed to senders, contend for the medium. The MAC tracks region requests and allocates to each region a medium share proportional

Symbol	Meaning
S_i	Node i's desirable share of the medium
$CW_{min,i}$	Contention window of node i
R_r	Region r's desirable share of the medium
$P_{i,r}$	Probability of transmitting packets from region r at node i
$Q(i,r)$	The quality-function at node i for region r

Table 1— Table of Notations. We use i to iterate over nodes, and r to iterate over regions

to how often it is requested. Further, it ensures that the number of transmissions each node makes on behalf of a region is proportional to the quality of information that the node has about the region as measured by the function $Q(i,r)$. Below we describe how the MAC performs these functions.

7.1 Tracking Region Requests

The content-centric MAC handles requests for different regions both from its own vehicle and other vehicles. The module records REQ_s, a measure of requests made for each region s by various nodes. In our implementation, REQ_s is set to one plus the number of requests made for region s. This is to ensure that in the absence of requests all regions get equal share of the medium.

Internal Requests: When the MAC receives requests from its own car, it broadcasts them to other vehicles over the wireless medium. It also keeps track of past requests and times them out after a minute. When packets with region data arrive, the module checks whether they answer a request that has not timed out, in which case it passes them to the compression module for decoding.

External Requests: The MAC actively listens on the medium to track the requests made for various regions, and to identify which regions are observed by vehicles in the network. When a vehicle receives a request for region s, it updates REQ_s accordingly, which biases it to transmit more information about region s, if available.

7.2 Region Contention

CarSpeak aims to share the medium among regions proportionally to the number of requests they receive.

(a) Sharing the Medium Among Regions. Let R_r be region's r share of the wireless medium, i.e., the percentage of transmissions that should describe region r. We can write:

$$R_r = \frac{\text{pos}(\sum_i Q(i,r))REQ_r}{\sum_s \text{pos}(\sum_i Q(i,s))REQ_s} \quad (2)$$

where REQ_s is a measure of requests made for region s, and $\text{pos}(x) = 1$ if $x > 0$ and 0 otherwise. The function $\text{pos}(\sum_i Q(i,r))$ ensures that only regions for which some node has information acquire a share of the medium. Regions that no node has sensed (i.e., $Q(i,r) = 0, \forall i$) do not get a share of the medium.

But how does a node obtain the information it needs to substitute in the above equation in order to compute R_r? CarSpeak disseminates this information as annotation on the data packets transmitted by each node. Specifically, every CarSpeak packet sent by node j includes a list of region ids for which node j has information and their corresponding $Q(j,r)$'s. By default this list has 5 entries for a total of 40 bytes (6 bytes for region ids and 2 bytes for $Q(j,r)$).

CarSpeak nodes listen on the medium and collects information about the different regions and quality of information that other nodes have for these regions. They use this information to populate a table of region ids, and the quality of information the various nodes have for each region. A garbage collection thread that runs every 10 seconds multiplies $Q(i,r)$ values by a factor μ, $(0 < \mu < 1)$ in order to age-out quality information that is outdated it also timeout requests that have not been refreshed in the past minute.

(b) Controlling Medium Access. Using its estimate of the share of the various regions of the medium, R_r's, a node can estimate how often it should transmit, i.e., its own share of the medium. Let S_i be the medium share of node i. Node i's share of the medium is the sum of its contribution to the transmissions related to all regions for which i has data. This contribution is also proportional to the quality of information the node has about each of these regions. Thus:

$$S_i = \sum_r R_r \frac{Q(i,r)}{\sum_j Q(j,r)}. \quad (3)$$

Conceptually, once a node knows its share of the wireless medium, it should be able to transmit according to that share. At first, it seems that the node can achieve this goal by simply waiting for a transmission opportunity – i.e., the medium being idle – and using such opportunities as often as its share permits. For example, if its share is 20% of the medium time, it then transmits once every five times it senses the medium to be idle. Unfortunately, this approach does not work in practice. In practice, the decision to transmit upon the detection of an idle medium is performed in the card itself and cannot be controlled in software.

Thus, we will enforce the node share indirectly by controlling its contention window CW_{min}. The relation between the contention window and the resulting share of the wireless medium is given in [1] as:

$$CW_{\text{min},i} = \frac{2 - S_i}{S_i}. \quad (4)$$

The above relation is derived from a detailed Markov chain model of the progression of the contention window in 802.11 [1]. Intuitively, however, one can understand it as follows: In 802.11 a node picks a random value between 0 and CW_{min}. Thus, the average contention value is $\frac{CW_{min}+1}{2}$. Thus, on average the node accesses the medium once every $\frac{CW_{min}+1}{2}$, and hence its share of the medium $S_i = \frac{2}{1+CW_{min}}$.

(c) Partitioning a Node's Transmissions Among Regions. While the above ensures that the node gets the proper share of the medium, the node still has to divide this share between various regions depending on: 1) each region's share of the medium, and 2) the quality of information the node has about the region. To achieve this goal, whenever the node has an opportunity to transmit a packet, it picks the packet from region r with the following probability:

$$P_{i,r} = \frac{R_r \times \frac{Q(i,r)}{\sum_j Q(j,r)}}{\sum_s R_s \times \frac{Q(i,s)}{\sum_j Q(j,s)}} = \frac{R_r}{S_i} \times \frac{Q(i,r)}{\sum_j Q(j,r)} \quad (5)$$

Clearly $\sum_r P_{i,r} = 1$, for every wireless node i.

The above is implemented using a non-blocking UDP socket. Whenever the socket has space for new packets, the node picks those packets from the regions according to the probabilities $P_{i,r}$'s.

7.3 Scaling

The above design has an important side benefit: it provides congestion control for 802.11 broadcast mode. Specifically, the presence of many 802.11 senders can lead to excessive collisions and a congestion collapse. This effect is countered in 802.11 unicast mode by the fact that a node that does not receive an ACK for its packet, backs off and doubles its contention window. Hence, during congestion, nodes tend to back off and reduce the number of collisions. In contrast, 802.11 broadcast mode does not have ACKs and hence it cannot use the lack of ACK as a signal of congestion to which it reacts by backing off. This leaves the broadcast mode with no protection against medium congestion. The resulting problem is typically referred to as a broadcast storm [19, 30]. In contrast, CarSpeak scales with a large number of senders because senders do not contend for the medium. It also scales with a large number of regions because as the number of regions increases the share of each region decreases because R_r depends on a region's share of the total number of requests.

8. DISCUSSION

In this section, we discuss some design considerations in implementing CarSpeak:

Communicating Processed Information. An important design decision is whether CarSpeak nodes should send processed sensor information, such as locations of pedestrians or whether a road is congested, instead of raw sensor information. While this approach may be sufficient for specialized scenarios, they are not suitable for general-purpose communication between autonomous vehicles. In the most general applications, transmitting nodes in networks of autonomous vehicles need not know how receivers plan to process this information. Furthermore, different receivers may process the same sensor information to achieve different objectives. Native sensor information, available at different resolutions, is the only representation generic enough to cater to varied objectives, such as evaluating road congestion, detecting pedestrians, avoiding vehicles, enabling better localization, route planning, and curb-detection amongst others.

One Hop vs. Multi-Hops. One design decision is whether CarSpeak nodes should relay requests, in an attempt to find the relevant information at vehicles that are multiple hops away from the originator. We chose not to do so, i.e., we do not make vehicles forward region requests. Our reasoning is based on the tradeoff between bandwidth consumption and the value of information about relatively distant locations. CarSpeak targets urban environments and speeds lower than 20 miles per hour. For autonomous driving applications, and even with a conservative estimate, a car should not need information from locations that are farther than half to one minute away. At the above speeds, this translates into locations that are 100 to 200 meters away, which are typically within radio range.[2] Hence, we believe that limiting access to only information that is within the radio range of the requester is a reasonable design choice that enables each region to expend its wireless bandwidth on serving its local, and hence most urgent, requests.

Regular Traffic. CarSpeak can support 802.11 traffic unrelated to autonomous driving as well. Such traffic can be represented simply as a virtual region in space. The designer can decide how to weigh this region in comparison to autonomous driving regions. For example, one may want to divide the medium equally between autonomous driving and other applications by setting $R_{virtual} = 0.5$, in which case the autonomous driving application can use half the medium share (as well as any resource unused by the virtual region).

9. IMPLEMENTATION

We implement CarSpeak's multi-resolution naming, addressing, and information sharing system as a module ("ROS node") in the Robot Operating System. We operate ROS on the Ubuntu 11.04 distribution (with linux kernel version 2.6.38-8-generic), that runs on the ASUS netbooks attached to the iCreate robots. Our implementation of CarSpeak's multi-resolution naming system maintains the Octree datastructure with $L = 8$ and three levels of region sub-trees. We also implement CarSpeak's garbage collection as a ROS timer thread with a threshold of 10 seconds for the freshness of sensor information.

[2]For example, the Dedicated Short Range Communication (DSRC) technology, which was adopted by the intelligent transportation system (ITS) has a radio range of up to 1000 meters [21].

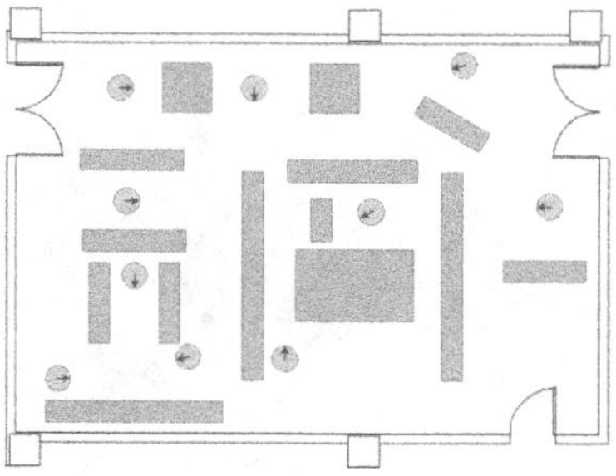

Figure 8—CarSpeak's Indoor Testbed. The blue circles denote candidate locations for robots and gray boxes denote obstacle locations.

Our implementation of CarSpeak's multi-resolution naming system subscribes to multiple topics containing sensor information in ROS's PointCloud format. It publishes the /car_speak topic, in ROS's PointCloud format, based on UDP packets received from the MAC layer. In this sense, CarSpeak behaves as any other sensor module in ROS. We implement CarSpeak's Octree-based compression framework to sub-sample the Octree and generate UDP packets to be forwarded to the MAC module.

CarSpeak's content centric MAC implementation has two key requirements: 1) The ability to modify channel access parameters such as the contention window size and, 2) Accurate timing to ensure packets are transmitted by the driver with minimum queuing delay. We chose the open-source ath9k driver+firmware for Atheros 802.11n based chipsets because it met our requirements. In our implementation, whenever the driver receives a packet (over-the-air or from userspace), it searches for a CarSpeak header within the payload of the packet to identify it as a CarSpeak packet. If the packet is from userspace, the driver places it in a queue corresponding to the region for which the packet contains information. The driver does not directly transmit the packet because the next packet to transmit (based on region sampling probabilities) may not correspond to the region for which the packet contains information. For actual transmission, we create a separate high priority thread within the driver to schedule packets based on the region sampling probabilities discussed in Section §3. Once a region is chosen for transmission, the thread dequeues the packet from the region's queue, sets the CWMin for the hardware's queue, and writes the packet into the hardware's queue. To minimize waisted airtime, we schedule this thread as fast as possible with the help of *High-Resolution Timers* available in the 2.6.x version of the Linux kernel. HR Timers are very accurate, with scheduling errors as low as 10us.

10. EVALUATION ENVIRONMENT

Below we describe the testing environment and the evaluated schemes.

Testbeds: We evaluate CarSpeak in both indoor and outdoor settings. Our indoor testbed uses a Vicon motion capture system for robot localization, and contains 10 iRobot Create robots equipped with Xbox 360 Kinect sensors. Asus EEPC 1015PX netbooks equipped with Atheros AR9285 wireless network adapters are mounted on each robot. Our testbed includes several large and small

Figure 9—CarSpeak's Outdoor Setup. Image of the actual golf car route demonstrating the lobby area that poses a hazardous blind spot for the golf car and makes visual confirmation of a pedestrian difficult before he enters the road.

obstacles as shown in Fig. 8. The testbed is divided to 40 high resolution regions. Low resolution regions are specified per experiment.

Our outdoor testbed contains an autonomous Yamaha G22E golf car mounted with various sensors, such as cameras, SICK and Hokuyo range finders. The autonomous car, navigating in a campus-like environment, needs to detect pedestrians and other vehicles. We implement CarSpeak on the golf car and several iRobot Create robots equipped with Kinect sensors situated in multiple locations. The setup was deployed over an area of 20×40 m. The robots assist the golf car's navigation system by providing sensor information useful in detecting pedestrians in the environment. Figure 9 shows the actual pedestrian crosswalk and depicts that the lobby adjacent to the crosswalk is a blind spot for the vehicle.

Compared Schemes: We evaluate three schemes including CarSpeak and two baseline implementations:

- **802.11**: An 802.11 based inter-vehicle communication system, which allows vehicles to make requests for regions, Responses are in the form of UDP/broadcast packets and are provided by all wireless nodes which possess any information about the given set of regions. The system uses the standard 802.11 MAC protocol to transmit information. The protocol keeps track of requests and causes requests older than one minute to expire. It also discards sensor data older than 10 seconds. The system however does not implement Octree-based naming or compression and instead transmits raw 3D-point cloud information. It also does not implement the functionalities of the content centric MAC.
- **802.11+Naming**: This baseline includes CarSpeak's Octree based naming and compression modules. It tracks requests and transmits packets from each region proportionally to the number of requests it received for that region, i.e., REQ_r. It also times out requests after one minute and discards sensor data older than 10 seconds. However it does not implement region-based contention or other CarSpeak MAC functions.
- **CarSpeak**: CarSpeak with all of its components including the content-centric MAC.

Metric: We compare CarSpeak against these baseline implementations based on a utility function, computing the *rate of useful sensor information*, received per second. A 3D point cloud is considered *useful*, if it contains sensor information only from the requested region(s), at the right resolution. For e.g., if a region is requested at a coarse resolution, fine grained high resolution information from that region are aggregated into the requested resolution and then their contribution to the useful information is computed. If all the fine grained information covers only 1 point in the requested coarse resolution, their contribution to the utility metric will be 1 point.

11. RESULTS

We evaluate CarSpeak in both indoor and outdoor environments. Our indoor testbed contains several obstacles that create blind-spots for the robots. Figure 8 depicts candidate robot locations in the test-bed. Experiments are repeated with robots assigned to different randomly chosen locations and moving towards different randomly chosen destinations. Our experiments allow robots to obtain sensor information from a diverse set of regions at various points in time.

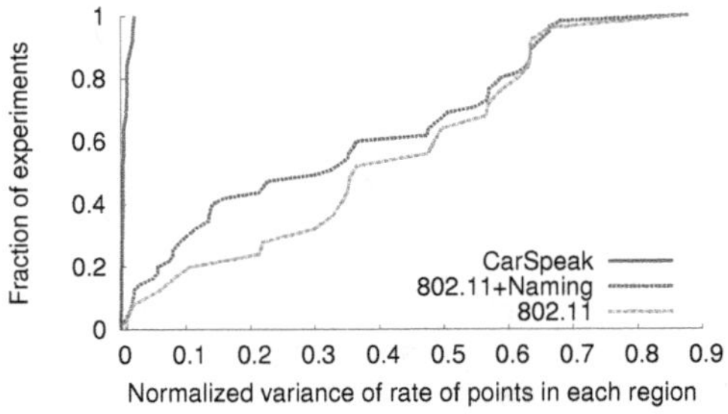

Figure 10—Region Contention. CDFs of variance in rate of points received from two regions. The variance obtained using CarSpeak is significantly smaller, across a variety of topologies and in the presence of mobile robots.

11.1 Region Contention

CarSpeak's key goal is to enable regions to share the medium efficiently, regardless of the location or number of nodes. We verify if CarSpeak delivers on that promise.

Method. We place robots in randomly chosen locations in the indoor testbed. We issue an equal number of requests for two different regions in the environment at regular intervals from two wireless nodes in the testbed. We measure the variance of the rate of 3D-points received from the two regions by both robots, by CarSpeak the standard 802.11 MAC protocol, and a hybrid approach 802.11+Naming. We repeat the experiment for 20 different topologies, with requests generated from different pairs of robots.

Results. Figure 10 shows the cumulative distribution function (CDF) of the variance (normalized by the average square) of the rate of points received from the two regions by the robots. The mean variance obtained using CarSpeak is 0.0015, while that of the standard 802.11 protocol and 802.11+Naming are 0.101 and 0.081 respectively. The higher 802.11 variance is due to the fact that 802.11 allocates bandwidth to senders not regions. Hence, the region that was observed by more robots received a greater share of the medium compared to the other region. The exact difference in the shares of the two regions varied from one experiment to another depending on the topology and mobility pattern. 802.11+Naming had a slightly lower variance. This is because the protocol enforces the desired region rates locally – i.e., if one robot has information from both regions the amount of data it transmits is balanced between the two regions – but cannot guarantee the desired medium allocation across different nodes. In contrast, CarSpeak's region based contention mechanism ensures that the medium is shared equally between the two requested regions, across a variety of topologies and mobility patterns.

11.2 Region Requests

In this experiment, we test CarSpeak's region request module and verify an increased number of requests for a given region leads to a proportional increase in the number of 3D points received from that region.

Method. We place robots in randomly chosen locations in the indoor testbed. We issue queries for two regions in the environment. We fix the query rate for the first region (5 requests/sec) and vary the query rate for the second region across experiments. We measure the ratio of the number of points received from the two regions at the requesting robots, when the experiments are carried using

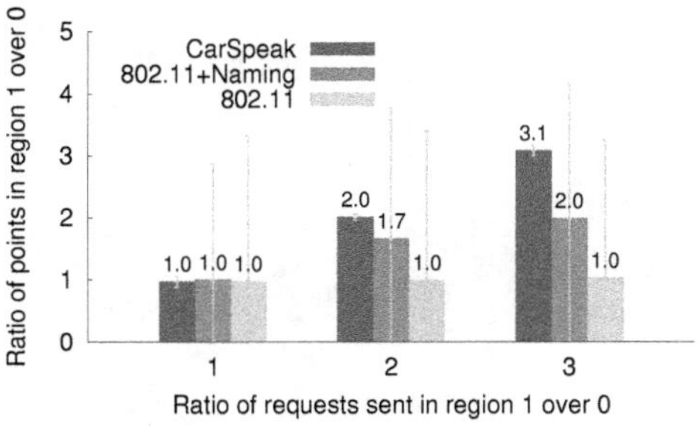

Figure 11—Region Requests. Ratio of the number of points received from two requested regions plotted against ratio of the number of requests made for the two regions. CarSpeak ensures the wireless bandwidth is allocated to region proportionally to the number of requests they receive.

CarSpeak, 802.11 and 802.11+Naming. We repeat the experiment for 20 different topologies, with requests generated from different pairs of robots.

Results. Figure 11 plots the ratio of the number of 3D points received from the two requested regions as a function of the ratio of the number of requests made for the two regions. The figure shows that, for CarSpeak, the ratio of received points is roughly equal to the ratio of requests. This holds across a variety of topologies and mobility patterns. In contrast, for 802.11, the ratio of points is totally independent of the ratio of requests. 802.11+Naming performs slightly better showing some correlation between the ratio of points from the two regions and the ratio of their requests.

11.3 Scaling

In this experiment, we demonstrate that CarSpeak scales to environments with a large number of vehicles.

Method. We conduct the experiment with two regions that have equal request rates. However, we increase the number of transmitters and explore the impact on the protocols. We measure the number of points received by the requesting receivers for CarSpeak and the two baselines. We repeat the experiment for different topologies and pairs of regions.

Results. Figure 12 plots the number of received 3D points, with CarSpeak, 802.11, and 802.11+Naming as a function of the number of contending nodes. While CarSpeak's performance scales gracefully, the performance of both the 802.11 baselines deteriorates when there are over 6 nodes. This is due to the large number of collisions that occur when multiple nodes transmit using the 802.11 broadcast mode, causing a broadcast storm. CarSpeak's content centric MAC protocol solves this problem by adapting the nodes' contention window so that it stays independent of the number of transmitters.

11.4 Compression

In this experiment, we evaluate the performance of CarSpeak's compression module. We verify if our compression scheme is robust to packet loss while providing significant compression over sending uncompressed point cloud data.

Method. Since the level of possible compression depends on the scene, we place the robots in a typical outdoor setting containing several buildings and obstacles, with Kinect sensors receiving depth information.[3] We vary the distance between the robots to achieve a wide range of loss rates. We evaluate CarSpeak's compression module against the following two compression schemes:

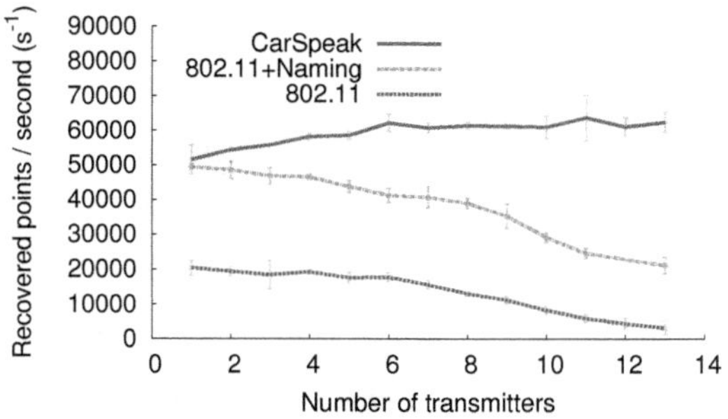

Figure 12—Scaling. Number of 3D points received at a receiver by CarSpeak, 802.11 and 802.11+Naming as a function of the number of contending nodes. We observe that while CarSpeak scales gracefully, 802.11's performance deteriorates when there are over 6 nodes, due to an excessive number of collisions and the lack of a backoff mechanism in the broadcast mode.

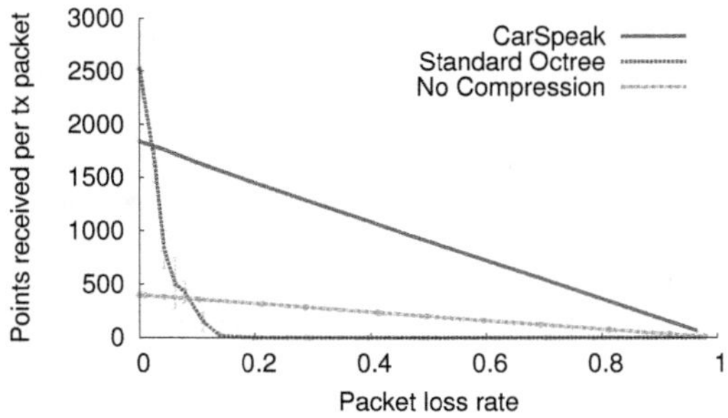

Figure 13—Compression module. Ratio of the number of points received over the number of transmitted points measured across packet loss rates when packetizing compressed point cloud data, using CarSpeak's compression module and without using compression. CarSpeak's compression module provides a consistent gain of 4.5x over sending uncompressed data, while packetizing compressed point cloud data performs poorer than CarSpeak for packet loss rates as low as 2%.

- *No Compression*: 3D-point cloud information is transmitted directly without any compression but with random sub-sampling.
- *Standard Octree Compression*: 3D-point cloud data from the environment obtained from the sensor at regular intervals is compressed using the standard Octree compression algorithm described in §6. The resultant data is packetized and broadcast on the medium.

We repeat the experiment for different locations of the robots in an outdoor setting.

Results. Figure 13 plots the number of received 3D points divided by the number of transmitted packets, as a function of the packet loss rate. CarSpeak's compression module provides a consistent gain of 4× over sending uncompressed data. While packetizing compressed point cloud data achieves a greater compression at very low loss rates, the scheme deteriorates to poorer than sending uncompressed data at a packet loss rate of 10% (which we found to be typical in our mobile outdoor scenarios). Since point cloud data is sought by several receivers whose channel to the transmitter varies with time due to mobility, a practical compression scheme must be robust to a wide range of packet loss. CarSpeak delivers on this promise.

11.5 Resolution

In this experiment, we evaluate the performance of CarSpeak when observing regions at different resolutions. We verify if Car-

[3]Kinect does not work in sunny outdoor settings. Hence, we pick afternoon hours and locations a lot of shades.

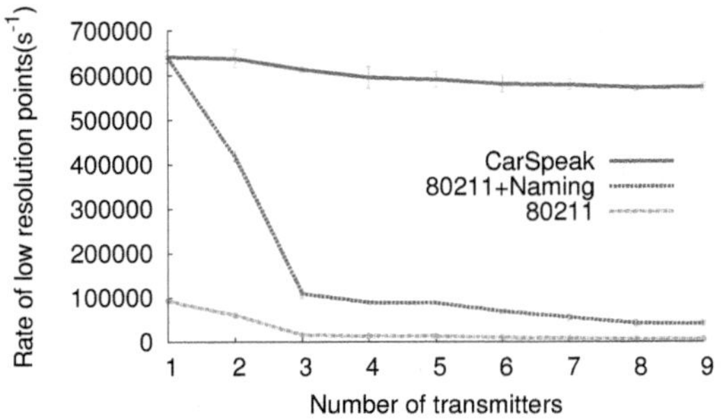

Figure 14—Resolution. The figure shows the rate at which information is received from a requested low-resolution region, for different schemes. Increasing the number of robots that do not have the requested resolution can dramatically impact the performance of 802.11 and 802.11+Naming. In contrast, CarSpeak maintains high rate of information from the desired region.

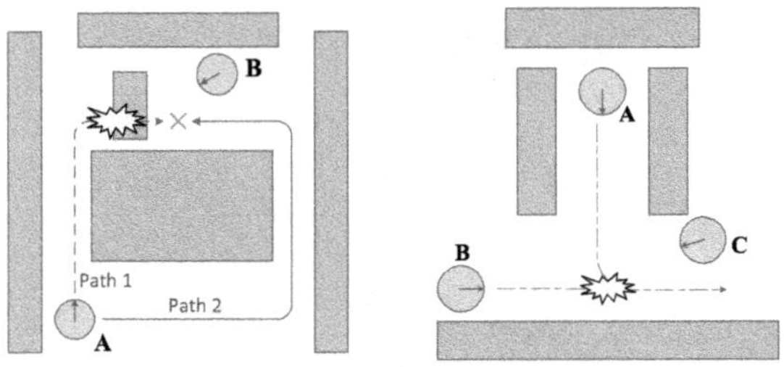

(a) Planning efficiency setup (b) Safety setup

Figure 15—Experiment setup for (a), The shortest path 1 from A to destination is blocked, efficient communication from sensor B should enable A to take path 2; for (b), There is a potential collision of A with B when A tries to merge into the traffic in another road at the T intersection. The collision can be avoided if A can hear from C about the other side of the road.

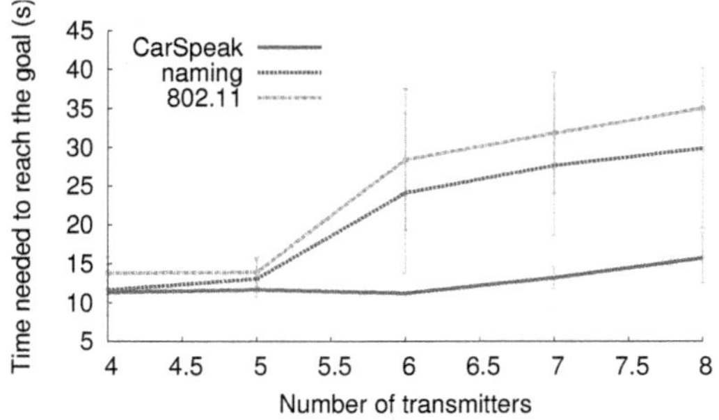

Figure 16—Planning Efficiency. Time taken by the robot to navigate an obstacle ridden environment across number of contending wireless nodes. CarSpeak performs 2.4× better, on average, than the baseline 802.11 implementation in a network of over 6 robots.

Speak responds with high quality information at the appropriate resolution, when a region is requested.

Method. We experiment with a scenario in which a robot requests a region at a low resolution. The environment has one robot who has the region at the proper resolution and many other robots that have incomplete and higher resolution information of the region. We measure the number of the 3D points received from the large region at the requester, in each of the three compared schemes. Note that fine grained high resolution information from within the requested region are aggregated into the requested resolution and then their contribution to the useful information is computed. For example, if all the fine grained information ends up covering only one point in the requested coarse resolution, their contribution to the utility metric will be one point. We repeat the experiment 20 times under different topologies.

Results. Figure 14 plots the rate at which the requester receives points from the desired resolution. The figure shows that adding robots observing smaller regions does not reduce CarSpeak's performance, as it recognizes that the robot observing the entire region has a greater quality of sensor information and deserves greater access to the medium. However, 802.11's performance is reduced as the medium is increasingly shared by wireless nodes observing only a small fragment of the requested region. Note that as the 802.11 baseline does not implement Octree-based compression, its rate of received sensor information is lower, compared to 802.11+Naming or CarSpeak. Overall, across experiments, CarSpeak delivery rate of the desired data is 4.5× higher than 802.11+Naming and over 29× higher than 802.11.

11.6 Planning Efficiency

In this experiment, we demonstrate CarSpeak's capability to provide the path planner with more efficient routes in an environment with obstacles.

Method. Consider a topology of the robots as shown in figure 15(a). Robot A seeks to navigate to location X, via the shortest possible path. However, the road ahead of X is blocked, and this information is available only with Robot B. Robot A does not have a line-of-sight view of the road block. The environment also has several other robots positioned at various other locations with sensor information of lower importance, also contending for the medium. Robot A makes several requests for regions close to X, for which its own sensors have no information. In the presence of timely sensor information from Robot B, Robot A can make a detour at the intersection to reach its destination via a marginally longer route. However, without this information, Robot A reaches the road-block and must U-turn to take the detour. We repeat the experiment with a different number of vehicles contending for the medium in the environment.

Results. Figure 16 plots the time taken by robot A to navigate to location X vs. the number of contending nodes when running CarSpeak as well as the two 802.11 benchmarks. We observe that in a network of over 6 contending wireless transmitters, CarSpeak performs, on average, 2.4× better than the 802.11 baseline and 2.1× better than 802.11's MAC with CarSpeak's multi-resolution naming system. In this network, 802.11 predominantly picks the incorrect path to the destination, while CarSpeak correctly picks the detour at the intersection, with high probability. While, 802.11+Naming performs marginally better than 802.11 due to a more effective compression scheme, its performance remains poor as much of the available wireless bandwidth is used by other nodes, with sensor information of much lower importance.

11.7 Safety

In this experiment, we evaluate CarSpeak's effectiveness in improving the safety of autonomous driving by detecting obstacles outside the field of view of the vehicle.

Method. Consider a topology of the robots as shown in figure 15(b) emulating the common scenario of vehicles at an intersection. Robot A is navigating towards a T-intersection and seeks to merge with other traffic on the main roadway. Ideally, Robot A must yield to Robot B (emulating a human-driven car without sensors), which is currently traveling on the main road. However, Robot A's sensors have a limited field of view and cannot detect Robot B. Negotiating such intersections is one of the most challenging problems in designing autonomous vehicles, often requiring human intervention or additional information regarding obstacles on the road [7]. In this topology, Robot C has access to sensor information capturing

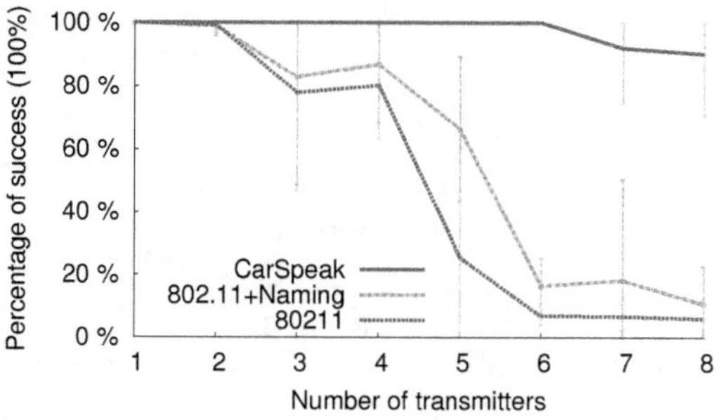

Figure 17—Safety. Percentage of successful detection of Robot B across number of wireless nodes contending for the medium. For a network of over 6 contending transmitters, CarSpeak's probability of successfully detecting Robot B is 14× that of 802.11 and 6.5× that of 802.11+Naming.

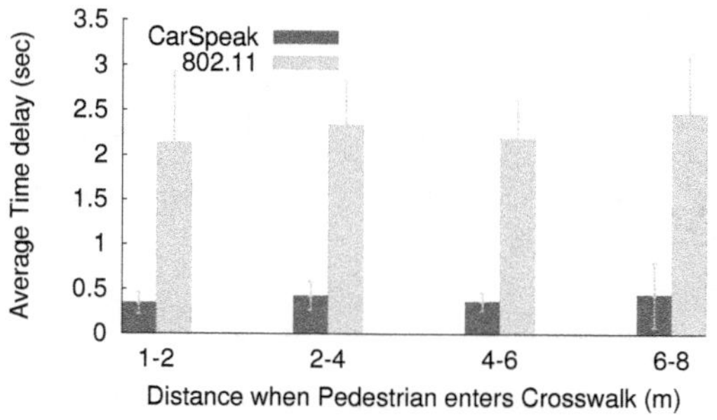

Figure 18—Time Delay Averages for CarSpeakand 802.11. Distances on the x-axis are grouped into bins of two meters representing distances of the golf car from the crosswalk at the time the pedestrian enters the crosswalk.

Robot B. The network has several other robots contending for the wireless medium, placed in randomly chosen locations. We evaluate the ability of Robot A to detect Robot B, while implementing CarSpeak against the 802.11 baseline implementations. We repeat the experiment with different numbers of vehicles contending for the medium in the environment.

Results. Figure 17 plots the percentage of successful detection of Robot B vs. the number of wireless nodes contending for the medium. While the performance of 802.11 and 802.11+Naming deteriorate to as low as 6.1% and 11.9% as the number of robots increases, CarSpeak successfully detects Robot B with 91% probability. In a network of over 6 transmitters, CarSpeak's probability of detecting Robot B is 14× that of 802.11 and 6.5× that of 802.11+Naming.

11.8 Outdoor Experiments on an Autonomous Vehicle

CarSpeak was implemented in an outdoor setting at a pedestrian crosswalk in a campus-like environment. This pedestrian crosswalk presents a hazardous setting where the two buildings on either side of the crosswalk completely block the view such that vehicles on the road are not aware of pedestrians before they emerge onto the street. See Figure 9. We present empirical results demonstrating CarSpeak's capability of improving the stopping time of an autonomous Yamaha G22E golf car over 802.11 when point cloud sensor data for pedestrians in the vehicle's blind spot is transmitted to the vehicle. In particular, our results show that CarSpeak enables the vehicle to make a stop decision before the crosswalk even at full speeds, if a pedestrian appears when the vehicle is one to two meters away from from the crosswalk.

Method. Our setup consists of a total of six Kinect sensors placed adjacent to the pedestrian crosswalk, i.e., the vehicle blind spot. The experiments were conducted in the presence of multiple collision domains, and hidden terminals.[4] Five out of six of these Kinects are monitoring a different section of the environment and thus are inconsequential for detecting pedestrians entering the crosswalk. Only one of the Kinect sensors is strategically placed to monitor the pedestrian crosswalk blind spot and thus obtains information relevant for the vehicle. Each Kinect broadcasts its point cloud sensor information using the Asus netbook described in §10. A receiver node on the autonomous golf car, a Vaio VPCF23BFX laptop with an Intel Core I7-2670QM processor, processes the sensor data that it receives from the Kinects to infer the presence of a pedestrian in the critical region (i.e., vehicle blind spot) viewed by the high priority Kinect. The pedestrian detection module issues a positive reading if the number of point cloud data points within the critical region is above a threshold of 1000 points. Upon detecting the presence of the pedestrian, the receiver node immediately publishes a stop command to the golf car through a ROS publish/subscribe interface. We compare CarSpeak against the benchmark of the traditional 802.11 protocol for data transmission.

For the purposes of obtaining our performance metric, we make the node attached to the Kinect log the sensor data to detect the exact time when the pedestrian appears in the lobby in front of the transmitting Kinect. This time is then compared against the timestamp of when the receiver issues a stop command to the vehicle. Using the vehicle's on-board localization paired with the two timestamps recorded, we also compare the distance of the vehicle from the crosswalk when the pedestrian enters the crosswalk and when the golf car is issued a stop command by the receiver. We note however, that processing is not necessary at the transmitter and is only done for computing our performance metrics.

The golf car drives from 15 meters away towards the crosswalk. We perform the experiment by allowing the pedestrian to enter the crosswalk's blind spot when the golf car is traveling at a full speed of two meters per second at distances of roughly ten meters, eight meters, six meters, four meters, and two meters from the crosswalk. For all of our results we assume the pedestrian takes an additional 0.5 seconds to enter the crosswalk from the time he is detected at the Kinect in the lobby and this is the time value we use on the x-axis of our plots. The results of these experiments are averaged over five runs for each of these distances using both CarSpeak and the traditional 802.11 protocols and are compared in the next section.

Results. Our results, in Figure 18, show a clear improvement in the vehicle's ability to safely stop before the crosswalk using CarSpeak as compared to 802.11. In particular CarSpeak allows for the receiver to issue a stop command with a minimum average delay of as little as 0.3 seconds from when the pedestrian appears in the field of view of the Kinect and a maximum average delay of 0.45 seconds. The maximum average delay of a positive pedestrian detection using CarSpeak is 4.75 times smaller than the *minimum* delay of 2.14s using 802.11.

These relatively small delays using CarSpeak allow the vehicle to safely stop before the crosswalk even when it is one to two meters away and traveling at a speed of two meters per second when the pedestrian appears. Use of the traditional 802.11 protocol, however, fails to stop the car before the crosswalk if a pedestrian appears when the vehicle is closer than four meters from the crosswalk, on average. See Figure 19. Using CarSpeak allows for a larger portion of critical information requested by the golf car from the priority Kinect sensor to reach the receiver, whereas an 802.11 protocol floods the receiver with proportionally more data from the irrelevant

[4]Pairwise pings show that only a subset of the pairs can directly hear each other, and in some pairs, the two nodes do not receive each other's pings though a third node can receive pings from both nodes, which indicates a hidden terminal scenario.

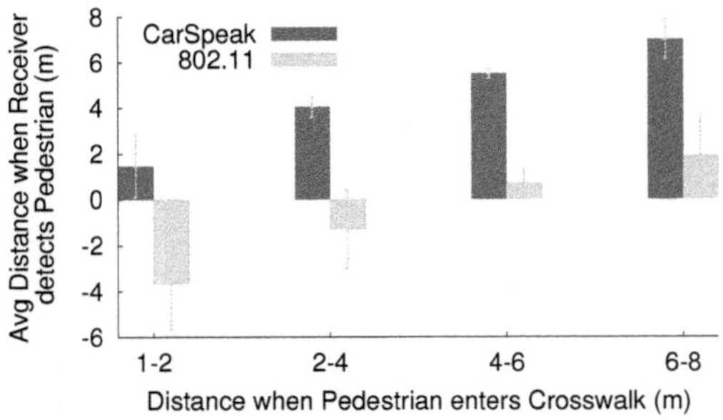

Figure 19—Comparison of golf car distance from the crosswalk when the pedestrian enters the crosswalk to the distance from the crosswalk at which the receiver issues a stop command to the car. Distances on the x-axis are grouped into bins of 2 m.

Kinect sensors, inhibiting the receiver's ability to process a positive pedestrian detection. In particular, using CarSpeak the receiver obtains 7.5× as many pedestrian critical 3D points as 802.11, averaged over twenty runs. Thus, using CarSpeak allows the receiver to gain several folds more information about regions of the environment that it considers important, even with several contending non-relevant transmitters, allowing more timely usage of important data to make critical decisions on actual autonomous vehicles.

12. CONCLUSION

This paper introduces CarSpeak, a content-centric communication system for autonomous driving, enabling cars to query and access sensory information captured by other cars in a manner similar to how they access information from their local sensors. Field tests using a combination of iRobot robots and a Yamaha instrumented car show that, in comparison with a baseline that directly uses 802.11, CarSpeak improves safety, increases information throughput, and provides several folds reduction in the time to navigate an obstacle-ridden environment.

Acknowledgments: We thank Brice Rebsamen and Tirthankar Bandyopadhyay for assisting us with the golf car in Singapore. We thank Sam Powers, Edward Mugica, Sachithra Hemachandra for helping us with our indoor testbed. We also thank Haitham Al-Hassanieh, Fadel Adib, and the NETMIT group for their support. This work is funded by NSF and SMART-FM. We thank the members of the MIT Center for Wireless Networks and Mobile Computing, including Amazon.com, Cisco, Intel, Mediatek, Microsoft, and ST Microelectronics, for their interest and support.

13. REFERENCES

[1] G. Bianchi. Performance analysis of the ieee 802.11 distributed coordination function. *Selected Areas in Communications, IEEE Journal on*, pages 535 –547, 2000.

[2] L. Bononi and M. Di Felice. A cross layered mac and clustering scheme for efficient broadcast in vanets. In *MASS*, pages 1–8, 2007.

[3] M. Campbell, M. Egerstedt, J. P. How, and R. M. Murray. Autonomous driving in urban environments: approaches, lessons and challenges. *Philosophical Transactions of the Royal Society Series A*, 368:4649–4672, 2010.

[4] J. Chesterfield and P. Rodriguez. Deltacast: efficient file reconciliation in wireless broadcast systems. MobiSys '05, 2005.

[5] Z. Chong, B. Qin, T. Bandyopadhyay, T. Wongpiromsarn, E. Rankin, M. Ang, E. Frazzoli, D. Rus, D. Hsu, and K. Low. Autonomous personal vehicle for the first- and last-mile transportation services. In *CIS 2011*, pages 253 –260, 2011.

[6] C. Cuyu, X. Yong, S. Meilin, and L. Liang. Performance observations on mac protocols of vanets in intelligent transportation system. In *CMC '09.*, pages 373 –379, 2009.

[7] M. Darms, P. Rybski, and C. Urmson. Classification and tracking of dynamic objects with multiple sensors for autonomous driving in urban environments. In *Proceedings of the 2008 IEEE Intelligent Vehicles Symposium*, pages 1197–1202, 2008.

[8] D. Dolgov, S. Thrun, M. Montemerlo, and J. Diebel. Path planning for autonomous vehicles in unknown semi-structured environments. *Int. J. Rob. Res.*, pages 485–501, 2010.

[9] S. Gil, M. Schwager, B. Julian, and D. Rus. Optimizing communication in air-ground robot networks using decentralized control. In *ICRA 2010*, pages 1964 –1971, 2010.

[10] A. S. Huang, A. Bachrach, P. Henry, M. Krainin, D. Maturana, D. Fox, and N. Roy. Visual odometry and mapping for autonomous flight using an rgb-d camera. In *ISRR*, 2011.

[11] Y. Huang, J. Peng, C.-C. Kuo, and M. Gopi. A generic scheme for progressive point cloud coding. *Visualization and Computer Graphics, IEEE Transactions on*, 2008.

[12] V. Jacobson, D. K. Smetters, J. D. Thornton, M. F. Plass, N. H. Briggs, and R. L. Braynard. Networking named content. In *CoNEXT '09*, pages 1–12, 2009.

[13] S. Jiang and H.-L. Chao. Linear cooperative detection for alarm messages in cluster-based vehicular ad hoc networks. In *WCNC, 2010 IEEE*, pages 1 –6, 2010.

[14] S. Karaman and E. Frazzoli. Sampling-based algorithms for optimal motion planning. *CoRR*, 2011.

[15] A. Kolling and S. Carpin. Multi-robot pursuit-evasion without maps. In *ICRA 2010*, pages 3045 –3051, 2010.

[16] T. Koponen, M. Chawla, B.-G. Chun, A. Ermolinskiy, K. H. Kim, S. Shenker, and I. Stoica. A data-oriented (and beyond) network architecture. In *SIGCOMM '07*, pages 181–192, 2007.

[17] J. Leonard, J. How, S. Teller, M. Berger, S. Campbell, G. Fiore, L. Fletcher, E. Frazzoli, A. Huang, S. Karaman, O. Koch, Y. Kuwata, D. Moore, E. Olson, S. Peters, J. Teo, R. Truax, M. Walter, D. Barrett, A. Epstein, K. Maheloni, K. Moyer, T. Jones, R. Buckley, M. Antone, R. Galejs, S. Krishnamurthy, and J. Williams. A perception-driven autonomous urban vehicle. *J. Field Robot.*, pages 727–774, 2008.

[18] J. Levinson, J. Askeland, J. Becker, J. Dolson, D. Held, S. Kammel, J. Kolter, D. Langer, O. Pink, V. Pratt, M. Sokolsky, G. Stanek, D. Stavens, A. Teichman, M. Werling, and S. Thrun. Towards fully autonomous driving: Systems and algorithms. In *IEEE IV 2011*, pages 163 –168, 2011.

[19] F. Li and Y. Wang. Routing in vehicular ad hoc networks: A survey. *Vehicular Technology Magazine, IEEE*, pages 12 –22, 2007.

[20] P. Luo, H. Huang, W. Shu, M. Li, and M.-Y. Wu. Performance evaluation of vehicular dtn routing under realistic mobility models. In *WCNC 2008. IEEE*, pages 2206 –2211, 2008.

[21] B. Maitipe and M. I. Hayee. Development and field demonstration of dsrc-based v2i traffic information system for the work zone. Technical report, 2010.

[22] F. Maurelli, D. Droeschel, T. Wisspeintner, S. May, and H. Surmann. A 3d laser scanner system for autonomous vehicle navigation. In *ICAR 2009.*, pages 1 –6, june 2009.

[23] Mendez-Polanco and Munoz-Melendez. Collaborative robots for indoor environment exploration. In *ICARCV 2008*, 2008.

[24] J. L. Ny, A.Ribiero, and G. Pappas. Robot deployment with end-to-end communication constraints. In *Proceedings of the Conference on Decision and Control*, 2011.

[25] S. Oh, D. Lau, and M. Gerla. Content centric networking in tactical and emergency manets. In *Wireless Days IFIP*, pages 1 –5, 2010.

[26] M. Quigley, K. Conley, B. P. Gerkey, J. Faust, T. Foote, J. Leibs, R. Wheeler, and A. Y. Ng. Ros: an open-source robot operating system. In *ICRA Workshop on Open Source Software*, 2009.

[27] D. Reichardt, M. Miglietta, L. Moretti, P. Morsink, and W. Schulz. Cartalk 2000: safe and comfortable driving based upon inter-vehicle-communication. In *Intelligent Vehicle Symposium, 2002*, pages 545 – 550 vol.2, 2002.

[28] R. Schnabel and R. Klein. Octree-based point-cloud compression. In *Symposium on Point-Based Graphics*, 2006.

[29] N. Thompson, R. Crepaldi, and R. H. Kravets. Locus: A location-based data overlay for disruption-tolerant networks. In *Fifth Workshop on Challenged Networks (CHANTS 2010)*, 2010.

[30] O. Tonguz, N. Wisitpongphan, F. Bai, P. Mudalige, and V. Sadekar. Broadcasting in vanet. In *2007 Mobile Networking for Vehicular Environments*, pages 7 –12, 2007.

[31] C. Urmson, C. Baker , J. Dolan, P. Rybski, B. Salesky, W. Whittaker, D. Ferguson, and M. Darms. Autonomous driving in traffic: Boss and the urban challenge. *AI Magazine*, pages 17–29, 2009.

Vitamin C for your Smartphone: The SKIMS Approach for Cooperative and Lightweight Security at Mobiles

Matthias Wählisch*, Sebastian Trapp*, Jochen Schiller*, Benjamin Jochheim†, Theodor Nolte†, Thomas C. Schmidt†, Osman Ugus†, Dirk Westhoff†, Martin Kutscher‡, Matthias Küster‡, Christian Keil◇, Jochen Schönfelder◇

*Institut für Informatik, Freie Universität Berlin, Berlin, Germany

†Dept. Informatik, Hamburg University of Applied Sciences, Hamburg, Germany

‡escrypt GmbH, Bochum, Germany

◇DFN-CERT Services GmbH, Hamburg, Germany

{first.last}@fu-berlin.de, {first.last}@haw-hamburg.de, {first.last}@escrypt.com, {last}@dfn-cert.de

ABSTRACT

Smartphones are popular attack targets, but usually too weak in applying common protection concepts. SKIMS designs and implements a cooperative, cross-layer security system for mobile devices. Detection mechanisms as well as a proactive and reactive defense of attacks are core components of this project. In this demo, we show a comprehensive proof-of-concept of our approaches, which include entropy-based malware detection, a mobile honeypot, and spontaneous, socio-inspired trust establishment.

Categories and Subject Descriptors

C.2.0 [**Computer-Comm. Networks**]: General—*Security and protection (e.g., firewalls)*

General Terms

Security

Keywords

Mobile security, malware detection, mobile honeypot, ad hoc trust

1. INTRODUCTION

Mobile, wireless-based end devices represent a significant part of our current networks, both with respect to the number of deployed equipment and their economic impact. Limited hardware resources and the per se unprotected transmission medium air turn such devices into attractive targets for attacks. These are additionally motivated because handhelds commonly store or exchange confidential data (e.g., calendar, contact lists).

In contrast to wired end systems, mobile devices exhibit three characteristic differences: (a) they normally are equipped with several different network interfaces; (b) their capabilities are significantly limited, so that they are not able to permanently maintain strong protection mechanisms; (c) physical vicinity with the help of appropriate access technologies allow for the establishment of separate, cooperative delivery structures. Mobile end devices, thus, need a lightweight, environment-adaptive protection mechanism that exploits the heterogeneous technologies available on-board.

In this demo, we present SKIMS, a digital immune system for smartphones. Similar to the biological immune system, our framework tries to protect the mobile device on its own. In cases where this is insufficient, cooperation between neighbors is established. For the detection and defense, SKIMS is guided by the principles of infrastructureless and lightweight approaches.

From a general perspective, we demonstrate a user-friendly security system that independently estimates apparent and existing threats. From a detailed perspective, we show (1) entropy-based malware detection for zero-day shellcode, (2) a mobile honeypot, (3) socio-inspired trust-establishment for reliable ad hoc communication to protect the mobile, and (4) secure data distribution using fountain coding.

2. SKIMS IN A NUTSHELL

The SKIMS protection system consists of multiple components that allow for detection as well as proactive and reactive defense of attacks. These are controlled by the mobile sandbox. Mobiles are consumer devices. We therefore hide most of the inherent complexity from the end user.

Detection—Malware Identification We employ a statistical analysis of the information content obtained from network streams to detect potential anomalies in real-time. Unlike previous work, our time-frequency analysis extracts the non-stationary properties of entropy signals. From this context-adaptive technique, we obtain a clear signature of binary instructions and can also detect embedded shellcode. Preliminary details of our approach are sketched in [2].

Detection—Mobile Honeypot A low-interaction honeypot collects suspicious connections. It emulates multiple network services (FTP, POP3, etc.) and collects valuable information from the observed attacks. It allows for universal deployment on mobiles as well as on common Linux systems, which enables a comprehensive analysis of attacks on multiple platforms.

Defense—Ad hoc Trust Establishment This component evaluates spontaneous trust between peers to establish reliable ad hoc communication. Compared to other approaches, our scheme [3] applies only data inherent at mobiles and does not require any central component. Using a commutative encryption protocol two peers exchange their address books in a privacy-friendly way and find mutual contacts without

SIGCOMM'12, August 13–17, 2012, Helsinki, Finland.
ACM 978-1-4503-1419-0/12/08.

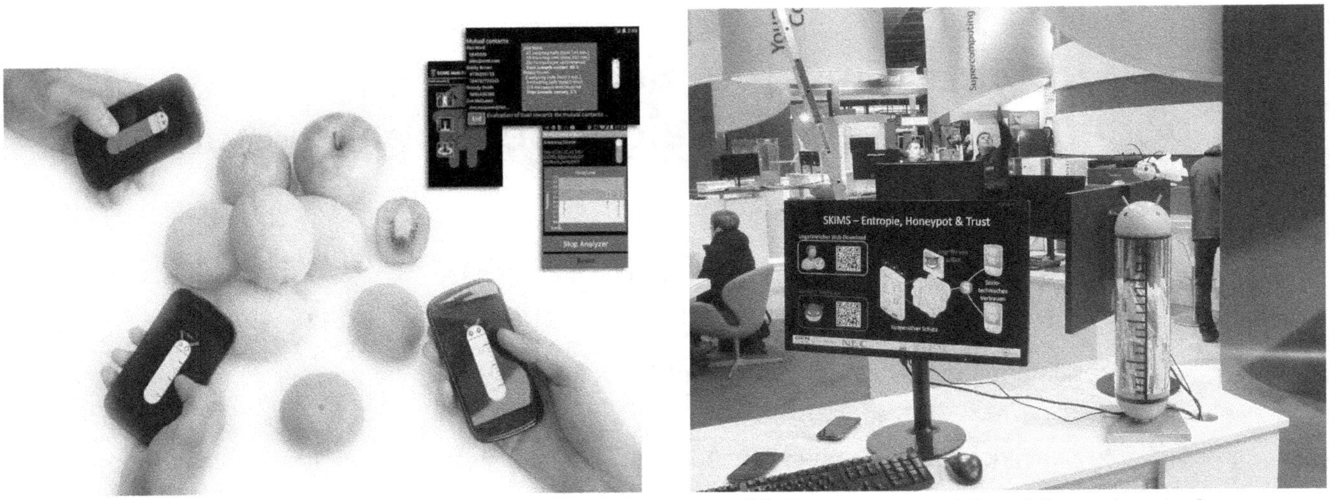

(a) SKIMS GUI reduces complexity and presents different threat levels and details in a user-friendly way

(b) Demonstrator with mobiles and control screen

Figure 1: Deployment and demonstration of the SKIMS application

revealing different entries. Mutual entries will be weighted locally using data from communication logs [4]. An overall, transitive trust value is derived from sociological principles (tie strengths).

Defense—Secure Data Transmission This component allows for a secure and reliable data broadcast between the mobiles. It is achieved by adapting the techniques which are originally proposed in [1] to mobile devices. The communication is based on fountain coding to increase the efficiency in places where the connectivity is unreliable due to high amount of wireless interferences. It enables the encrypted distribution of any kind of data such as security logs and programs to adjacent mobile devices running the SKIMS app.

Controlling—SKIMS App Each component for detection reports its current threat level to the SKIMS app, which calculates an overall state. Depending on the severity, the SKIMS app activates autonomously defense strategies or interacts with the end user.

A core design objective while developing the SKIMS app was the lightweight integration and interaction of the different building blocks. We implemented the internal signaling based on a content provider.

3. DEMONSTRATION

Setup An attendee can test all SKIMS components, either by itself or in combination. A typical procedure is the following: The user starts with the threat level "no attacks" (green) and may choose to download an uninfected or malicious file via a QR code. The user observes the live entropy analysis. In case of embedded malware, the SKIMS app switches to yellow. From an external PC we emulate an attack by connecting to the mobile and poke the local file system. This is handled by the mobile honeypot, which changes the current state to red. The user is asked to cancel its current Internet connection and to establish an ad hoc link. The SKIMS app visualizes the spontaneous trust calculation with an adjacent mobile.

On-Site Requirements For our demonstration, we will need a connection for power supply and a table. An Internet connection is not required. We will bring an access point, a notebook, multiple Android smartphones, and a tablet, which will communicate among each other via 802.11.

Acknowledgements

We would like to thank the students who helped in the implementation and testing of the SKIMS application. This work is supported by the German BMBF within the project SKIMS (http://skims.realmv6.org).

4. REFERENCES

[1] Bohli, J. M., Westhoff, D., Hessler, A., and Ugus, O. Security Enhanced Multi-Hop over the Air Re-programming with Fountain Codes. In *Proc. of the 34th IEEE LCN. 4th IEEE International Workshop on Practical Issues in Building Sensor Network Applications (SenseApp)* (Piscataway, NJ, USA, October 2009), IEEE Press.

[2] Schmidt, T. C., Wählisch, M., Jochheim, B., and Gröning, M. WiSec 2011 Poster: Context-adaptive Entropy Analysis as a Lightweight Detector of Zero-day Shellcode Intrusion for Mobiles. *ACM SIGMOBILE Mobile Computing and Communications Review (MC2R) 15*, 3 (July 2011), 47–48.

[3] Trapp, S., Wählisch, M., and Schiller, J. Short Paper: Can Your Phone Trust Your Friend Selection? In *Proc. of the 1st ACM CCS Workshop on Security and Privacy in Mobile Devices (SPSM)* (New York, 2011), ACM, pp. 69–74.

[4] Trapp, S., Wählisch, M., and Schiller, J. Bridge the Gap: Measuring and Analyzing Technical Data for Social Trust between Smartphones. Technical Report arXiv:1205.3068, Open Archive: arXiv.org, May 2012.

Energino: Energy Saving Tips for Your Wireless Network

Roberto Riggio, Karina Mabell Gomez, Tinku Rasheed
CREATE-NET
Trento, Italy
first.lastname@create-net.org

Cigdem Sengul
Telekom Innovation Labs,
TU-Berlin Berlin, Germany
cigdem@net.t-labs.tu-berlin

ABSTRACT

The energy wasted in wireless networks is a serious concern and the main challenge lies in determining when and where the energy is wasted. In this demo, we present Energino, an energy measurement and control system designed to deliver high performance while remaining a cheap solution.

Categories and Subject Descriptors

C.2.3 [Computer Communication Networks]: Network Operations

General Terms

Experimentation, Measurement, Performance

Keywords

Arduino, Energy consumption monitoring, Open hardware, Wireless

1. INTRODUCTION

Current wireless networks are not designed with energy conservation in mind. They generally cater to the worst case scenario, which mandates for an overprovisioning of resources. For instance, the enterprise WLAN in Intel in Portland, OR, has 125 APs over a four floors building with an inter–node distance of 5 m to provide sufficient capacity for four very close users using voice, data, and multimedia applications simultaneously. However, measurements showed that these APs spend ≈ 20 to 80% of their time idling and consume 8.76MWh of energy per year in total [1]. Similar stories repeat in other deployments. The core of the problem is that current networks do not gracefully adjust their energy consumption accordingly to the "amount of work".

In this demo, we show a system, called Energino that can be used to both monitor and control the energy consumption of WLANs. Energino is an Arduino-based plug–load meter designed to monitor, in real-time, the energy consumption of DC devices. Compared to its commercial counterparts, such as Watts Up? [1], Energino was shown to provide a cheaper solution with better performance in terms of resolution and sampling rate [2]. This demo presents the prototype of a more complex system with mechanical relays and a controller to adapt the energy consumption of devices under control.

[1] https://www.wattsupmeters.com

SIGCOMM'12, August 13–17, 2012, Helsinki, Finland.
ACM 978-1-4503-1419-0/12/08.

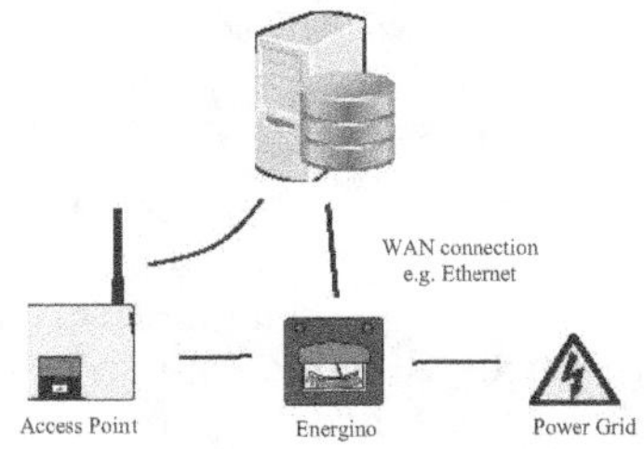

Figure 1: Energino network architecture.

We release both the hardware schematics and the software with a permissive BSD license[2] to encourage the research community to use and extend it.

2. THE ENERGINO SYSTEM

In this section, we detail Energino main components.

2.1 Energino Devices

To measure and control energy, we extended the Arduino with a custom module integrating a voltage sensor (based on a voltage divider), a current sensor (based on the Hall effect), and a mechanical relay. Each input can be sampled with a period of 100µs, which results in a maximum sampling rate of 10kHz, using the Arduino 10–bit Analog–to–Digital Converter (ADC). Energino can monitor DC loads up to 60V and absorbing up to 5A. The resolution for the voltage is 54mV while the resolution for the current is configurable and depends on the maximum current. More specifically, for currents up to 5A the resolution is 26mA, while for currents up to 280mA the resolution is 0.5mA.

During each sampling period the Arduino continuously polls the voltage and the current sensors and accumulates the values into two separate registers. At the end of each period, the average power consumption is computed and the result is sent over the USB interface. This is done in order to filter out fluctuations in the values read from the analog inputs. For example, if the sampling period is set to 1s, both the voltage and the current readings will be the average of ≈ 5000 samples.

2.2 Measurement Framework

The measurement framework conceptually consists of three layers: (1) The network layer contains the devices to be monitored. (2) The energy monitoring layer contains the Energino devices that measure the energy consumption and implement the actions scheduled by the controller. Notice

[2] http://www.wing-project.org/energino/

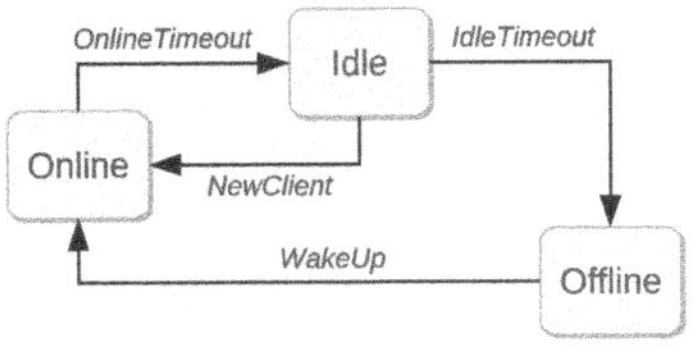

Figure 2: *Energino* finite state machine.

that *Energino* and the APs share both the connection to the Internet and the actual power source. Finally, (3) *the control layer* contains the network controller responsible for gathering network and energy usage statistics from both the network and the energy monitoring layers. The controller also schedules the necessary actions to switch the operating mode of the different network devices. The network architecture is sketched in Fig. 1.

We currently support three operating modes. In the *online* mode, an AP and its wireless interfaces are always on. This is the default mode for a standard WiFi AP. We introduce an *idle* mode, in which the AP is powered on but its wireless interfaces are periodically switched on/off according to a programmable duty-cycle. This mode is similar to the Power-Save Mode typically reserved for client devices in IEEE 802.11. In the *offline* mode, the entire AP is turned off and only the *Energino* is powered.

The controller uses a Finite State Machine to transition the network devices to different modes (see Fig. 2). If the AP has been inactive for at least T_{online} seconds, an *OnlineTimeout* event is triggered. Here, inactive means either the AP has no associated clients or, its clients are not sending or receiving. In this case, the AP switches from *Online* to *Idle*. If the AP has been in the *Idle* mode for at least T_{idle} seconds, this generates an *IdleTimeout* event and the AP is switched to *Offline*. In the *Idle* mode, if a new client has just made an association request to the AP or an existing client started traffic, the AP returns to the *Online* mode. Finally, the controller can trigger a *WakeUp* event to power–up an AP, and switch the AP from *Offline* to *Online*.

While both the operating modes and the events we support currently are quite general, the framework can be easily extended to using different modes and state machines to realize different energy-saving solutions.

3. DEMO SCENARIO

In this demo, we show the operation of *Energino* in a high-density deployment scenario (as the Intel deployment). In the network layer, we will have 3 APs and $3-5$ clients. The energy monitoring layer will be composed of 3 *Energino* devices, and one laptop will act as the controller. We assume the APs form a cluster, and one AP is enough to provide sufficient network coverage at least at the basic transmission rates (i.e., 1/6 Mbps for IEEE 802.11abg and MCS 0 for IEEE 802.11n). This node can be considered as the cluster-head. Within this scenario, we plan to show how the controller monitors and controls the energy consumption of the network. A management backend written in Python is used to both configure *Energino*'s operating parameters, e.g., sampling rate and resolution, and to gather the network and the energy consumption statistics.

In our scenario, the network transitions through different traffic load stages.

- *Low load:* At this stage, all APs, except the cluster-head, are off.

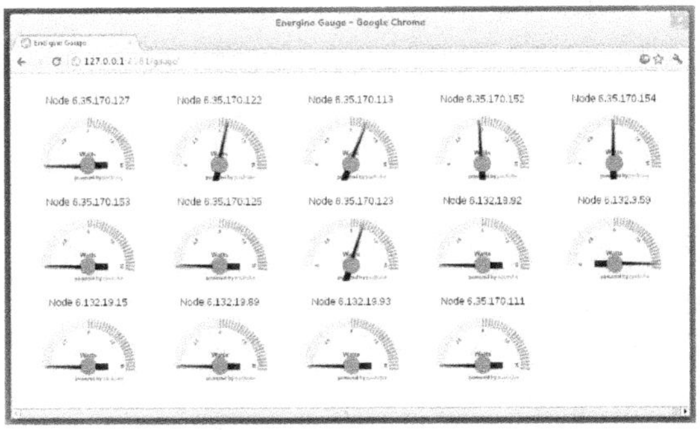

(a) Energy consumption.

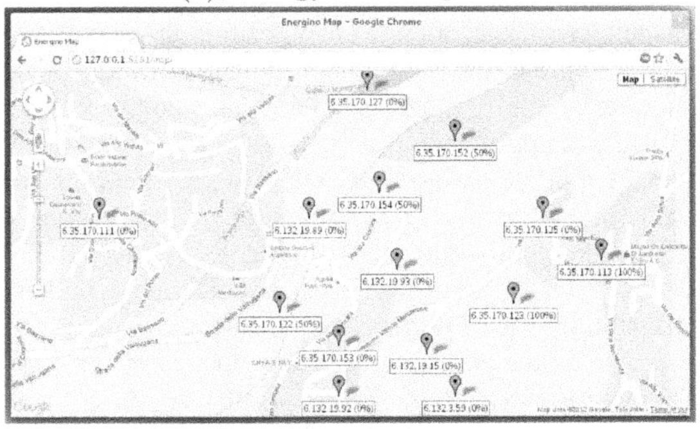

(b) Network topology and duty cycle.

Figure 3: The *Energino* monitoring dashboard.

- *Increasing load:* As more clients join the network the controller starts turning on APs. At the beginning, the clients will be able to associate only to the cluster-head, however, as more APs turn on, the clients may re-associate to these APs.
- *Decreasing load:* All APs, except the cluster-head, gradually switch first to *Idle*, then *Offline*, after *OnlineTimeout* and *IdleTimeout*, respectively.

Both the energy consumption and the current duty cycle of each APs will be visualized using a web-based dashboard. Fig. 3 show example screenshots of the dashboard, detailing the energy status of the corporate 14-AP WLAN at CREATE-NET. The figures show that the APs with 0% duty cycle (e.g., 6.132.19.89) are powered down, and the APs with the highest duty cycle (e.g. 6.35.170.113) have the highest energy consumption.

4. DISCUSSION

In this demo, we presented an affordable solution for monitoring and controlling the energy consumption of wireless deployments. Our goal is to eventually use *Energino* as a benchmarking tool to evaluate performance of energy saving solutions under different reference traffic conditions seen commonly in wireless networks. This will allow understanding the energy vs. performance trade-offs, and consequently design more performant and low-power solutions.

5. REFERENCES

[1] A. Jardosh et al., "Green WLANs: On-demand WLAN infrastructures," *Mobile Networks and Applications*, vol. 14, no. 6, pp. 798–814, Dec. 2009.

[2] K. M. Gomez et al., "Energino: An hardware and software solution for energy consumption monitoring," in *WinMee*, 2012.

MultiNet: Usable and Secure WiFi Device Association

Anthony Brown Richard Mortier Tom Rodden
School of Computer Science, University of Nottingham, UK
firstname.lastname@nottingham.ac.uk

ABSTRACT

This demo presents *MultiNet*, a novel method for joining devices to a domestic Wi-Fi network. MultiNet dynamically reconfigures the network to accept each device, rather than configuring each device to fit the network as is the norm. It does so by assuming that each device is pre-configured with a cryptographically generated WPA2 network SSID/passphrase pair, and then providing a lightweight interaction through which the user creates a new network for each device. This approach makes securely adding devices to a wireless network straightforward without compromising security or burdening the user, and maintaining backward compatibility with existing deployed standards and protocols.

The demo deploys a MultiNet Access Point (AP) and a number of Wi-Fi enabled consumer devices to allow viewers to dynamically construct and deconstruct the network via the MultiNet controller currently implemented as an app on an Android phone (Figure 1). The code for MultiNet is publicly available under open-source licenses.[1]

Categories and Subject Descriptors

C.2.1 [**Network Architecture and Design**]: Wireless communication

Keywords

Usable security, domestic environments, 802.11, infrastructure intervention

1. MULTINET OVERVIEW

Usability of domestic network security mechanisms remains a challenge, with the HCI community suggesting *interface veneers* alone cannot solve all the apparent issues [1]. MultiNet is an infrastructure intervention which changes the way in which devices are joined to the network, *configuring of the infrastructure to the device* rather than configuring each device to fit the infrastructure as is the norm. This approach has allowed us to redesign the joining interaction to be consistent across all devices while retaining backwards compatibility with legacy equipment. In our first prototype the interaction relies on a visual out-of-band channel created

[1]Available at `https://github.com/Toshbrown/`

SIGCOMM'12, August 13–17, 2012, Helsinki, Finland.
ACM 978-1-4503-1419-0/12/08.

Figure 1: Joining a printer using MultiNet.

Step One	Use the camera on the MultiNet controller to read the QR Code on the joining device.
Step Two	Request that the joining device is added to the network by sending the SSID/passkey pair for the virtual network to the AP.
Step Three	Establish the virtual network allowing the pre-configured joining device to connect.

Figure 2: MultiNet dataflow for adding a device.

via the use of an intermediary device possessing a camera[2]. We assume devices are pre-configured, and we then modified *hostapd* to dynamically create and destroy WPA2 secured networks on demand. The result is that the router offers multiple networks, approximately one per device.

Credential exchange takes place via an intuitive "capture image and connect" interaction. The MultiNet controller connects to the router over a dedicated WPA2 secured *control network*, which forms a secure signalling channel over which the credentials can be passed. Devices are assumed to be preconfigured by the manufacture with an SSID and

[2]intuitively named the "MultiNet Controller"

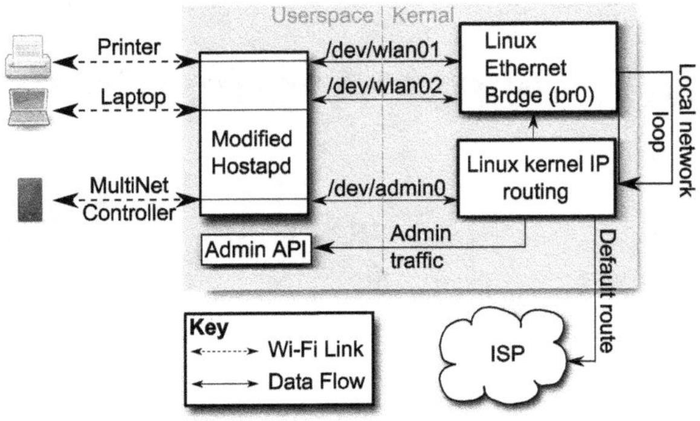

Figure 3: Overview of MultiNet implementation.

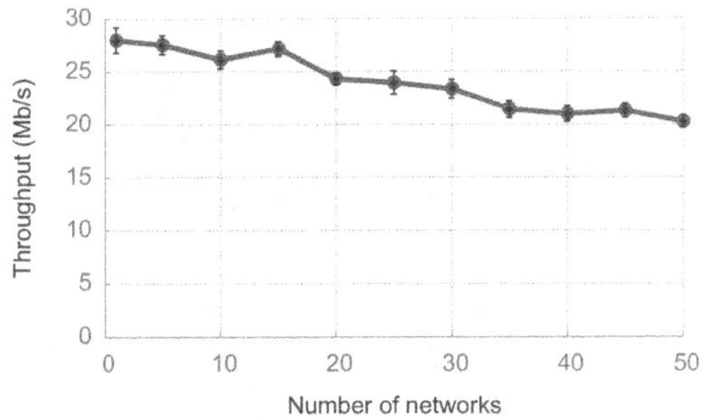

Figure 4: Device to AP throughput (Mb/s) as the number of configured networks on the AP increases.

a passphrase; in our prototype these are encoded visually as a QR Code, which the MultiNet controller acquires through its built-in camera. These credentials are then securely conveyed to the router. On receipt the router creates a virtual AP with the required SSID/passphrase pair and the joining device joins this newly created network. The overall protocol flow is depicted in Figure 2. This approach effectively configures the AP to accept the device and creates a "one network per device" infrastructure.

The AP uses the *hostapd* user-space daemon to provide wireless access point with authentication via IEEE 802.11i WPA2. We modified *hostapd* to enable dynamic creation and destruction of WPA2 secured networks, and to change the configuration re-load code to only de-authenticate stations when the SSID or passphrase of a network is changed or removed from the configuration file. After creation, these virtual network interfaces are connected to the standard Linux layer 2 bridge (*br0*), enabling communication between them. An overview of the design is depicted in Figure 3. The AP is controlled through a set of RESTful web services allowing the MultiNet controller to add and remove networks by configuring *hostapd*. These services are only accessible over HTTPS using the trusted controller network. The admin network is completely isolated from the device specific networks, protecting the admin functions from all non-trusted devices connected to the network. In our prototype the MultiNet controller is implemented as a simple Android application; other implementations are possible.

2. PROTOTYPE EVALUATION

Performance. We configured the MultiNet AP to offer the required number of networks and to act as an Iperf traffic sink. Once configured a device was added to act as a Iperf traffic source and measurements of throughput, latency and jitter were made. The AP to device throughput shows approximately linear reduction as the number of networks and associated overheads increases, as expected. At 20 networks there is a 13% reduction in maximum throughput, and by 50 networks this figure has risen to 27%, this is shown in Figure 4. Results for per-packet latency show a generally linear upward trend from from 8 ms to 15 ms for up to 20 networks. Jitter also rises slowly from 3 ms to 5 ms in the 1 to 20 network region. While MultiNet has some impact on throughput, latency and jitter it is acceptably small for up to 20 devices (a recent survey [2] put the UK household average at 4.3).

Usability. We performed an initial usability evaluation comparing MultiNet to Wi-Fi Protected Setup (WPS) with 16 participants. The trial consisted of one task building a small home network with three consumer devices under two conditions: (*C1*) using WPS, (*C2*) using MultiNet. All participants experienced both conditions in a lab environment. Several metrics were used to asses the overall usability of both systems. The effectiveness and efficiency were measured using task completion time, while user satisfaction was measured using SUS scales and post trial interviews. In all cases MultiNet showed a statistically significant improvement over WPS.

3. SECURITY CONSIDERATIONS

Recently discovered brute force attacks on the In-band WPS configuration highlight the issues of trading usability for security [3] and the danger of hiding security related features from the user: MultiNet exposes security related features making them more easily perceivable. Use of a visual out-of-band channel reduces the opportunity for undetected attacks, as reading the credentials from the QR codes requires reasonable physical proximity and line of sight to acquire a decodable image. By removing the choice of SSID and passphrase from the users we are also able to use securely generated values which are much longer and use a larger alphabet than could be entered by a human. Finally, MultiNet effectively maps network security to the physical security of devices and MultiNet controller which should help users better manage the risks associated with their actions as they manage the physical security of objects everyday.

4. ACKNOWLEDGEMENTS

This work was supported by the Doctoral Training Centre in Ubiquitous Computing, RCUK grant EP/G037574/1 and by Horizon Digital Economy Research, RCUK grant EP/G065802/1.

5. REFERENCES

[1] K. Edwards, R. Grinter, R. Mahajan, and D. Wetherall. Advancing the state of home networking. *Communications of the ACM*, 54(6):62–71, June 2010.

[2] TP-Link. 2011 connectivity report. http://thenextweb.com/uk/2011/12/09/uk-households-have-an-average-of-4-6-devices-connected-to-wifi, 2011.

[3] S. Viehböck. Us-cert vulnerability note vu#723755 - wifi protected setup (wps) pin brute force vulnerability, 2011.

Demo: Runtime MAC Reconfiguration Using a Meta-compiler Assisted Toolchain

Xi Zhang, Junaid Ansari and Petri Mähönen

RWTH Aachen University, Institute for Networked Systems,
Kackertstrasse 9, D-52072 Aachen, Germany
{xzh, jan, pma}@inets.rwth-aachen.de

ABSTRACT

A rapid reconfiguration of medium access scheme is required in order to achieve runtime performance optimization for dynamic spectrum access and fulfilling varying Quality of Service (QoS) demands. We have developed TRUMP, a toolchain which allows composing MAC solutions at runtime [1] [2]. In this demonstration, we will show how MAC reconfiguration can be achieved efficiently using TRUMP. Inspired by the optimum route calculation method used in car navigation systems, the compiler toolchain in TRUMP realizes an appropriate MAC solution at runtime. TRUMP allows expressing various types of constraints and options such as speed, energy consumption and packet delivery rate which leads to different MAC compositions. The live demonstration of MAC reconfiguration will be carried out on WARP SDR platform [3].

Categories and Subject Descriptors

C.2.1 [**Network Architecture and Design**]: Wireless communication

General Terms

Design, Experimentation, Performance

Keywords

MAC, Reconfiguration, Compiler assisted, SDR platform

1. INTRODUCTION

With a growing number and complexity of wireless communication applications, new challenges have arisen in terms of providing the required stable and high QoS due to the ever crowding spectrum, changing application requirements and the need to coexist with other wireless applications. Since the spectral environment and the performance expectation can be highly dynamic, a runtime reconfigurable medium access scheme is needed in order to optimize the performance characteristics. For instance, a resource constrained device typically gives energy efficiency high priority while for video streaming applications, maximizing the data rate would become more important. Therefore, in order to achieve the desired performance characteristics, the application should be allowed to specify the preferences. This in turn will enable the medium access control protocol to reconfigure at runtime.

Classically, protocols are implemented in a monolithic fashion which restricts reconfigurability as required in a dynamic spectrum access paradigm. A common approach for reconfiguration is through appropriately setting the parameters of a particular protocol such as the selection of the modulation and coding schemes as in LTE [4]. MultiMAC [5] switches among a few pre-defined standalone protocols depending on the QoS requirements. Although these approaches aim at providing flexible and adaptable configurations at runtime, the design choice is limited to the subset of the tunable parameter values and the pre-selected options. Tinnirello *et al.* [6] have realized runtime composition of various IEEE 802.11 WLAN MAC behaviours on commodity hardware. However, this approach is limited to 802.11 group of MAC protocols implemented on WLAN cards. Our demonstration uses TRUMP [1] which allows composing a wide range of MAC protocols for various types of wireless networks at runtime by simply binding reusable MAC functional components [7].

In order to enable runtime protocol realization tailored to a wide range of user-specified application preferences, with the support of a MAC meta-compiler [2] we have adopted an approach similar to the car navigation system. The car navigation systems plan optimal route based on the position of a car in real time. The driver can specify a wide range of preferences and constraints such as minimum cost, minimum time, minimum distance, etc. The optimum solution is concluded based on the prior knowledge of the length of all the route, the toll costs and the traffic condition. It allows runtime update, route recalculation as the car position and driver preference changes. In our demonstration, a user can specify multiple criteria for MAC realization such as maximum energy efficiency, minimum latency and maximum data reliability through the meta-compiler. The meta-compiler requires pre-fed knowledge such as the execution time of each individual functional component, energy consumption at different protocol and hardware states, etc. Using this information, the meta-compiler selects the optimum combination of components and parameters.

2. SYSTEM DESIGN AND IMPLEMENTATION

Using a meta-compiler, TRUMP supports protocol realization based on user preferences both at design stage and during runtime as shown in Figure 1. Only two criteria are shown in the figure for simplicity. At the design stage, the meta-compiler can provide the MAC protocol designer with the predicted protocol performance in terms of execution speed and power consumption based on the prior knowledge of the MAC components without having to deploy the MAC code on the target platform. The compiler also selects MAC parameters based on designer specified preferences at the design stage. If there are multiple implementations for the same functionality available and the user does not specify a choice, the compiler selects the implementation which is optimized for user preferences.

SIGCOMM'12, August 13–17, 2012, Helsinki, Finland.
ACM 978-1-4503-1419-0/12/08.

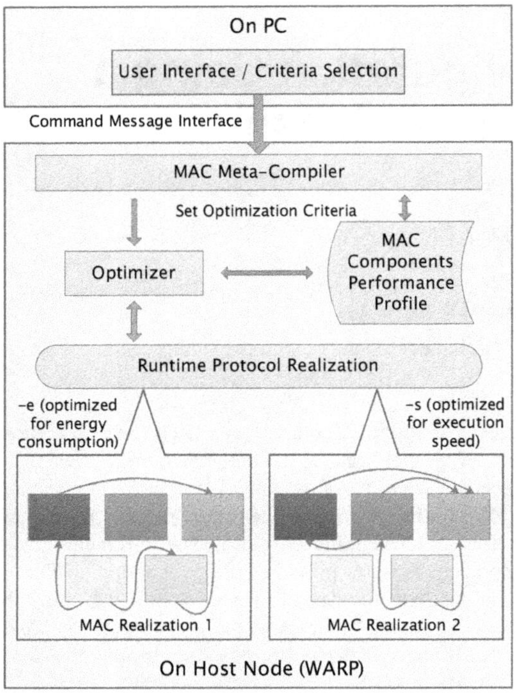

Figure 1: System architecture.

As a simple example, for sending a packet the user may specify the modulation scheme used on the packet payload. If the designer wishes to be able to send packets as fast as possible without having an insight on how each and every modulation option affects the MAC performance, one can set the optimization criteria for minimum execution speed and leave the modulation parameter for the `SendPacket()` interface to be `ANY`. The compiler queries the performance database for the `SendPacket()` component and replaces `ANY` with `QAM16` on our platform or any other modulation scheme which results in the highest data rate. If the user specifies maximum data reliability as the top priority, the compiler will return `BPSK` for the modulation parameter instead.

At runtime, an optimizer monitors the performance of the currently executing MAC. With feedbacks from the meta-compiler, it adapts the protocol according to the preference set by changing protocol functionalities and parameters. At the same time, it also monitors the spectral conditions by sampling received signal strength indicator periodically. For example, when a particular wireless channel is used by other networks, chances for the medium to be free are low. The optimizer will be aware of the increasing latency of delivering a packet due to the lack of opportunity to send packet in a free channel. If the application gives minimizing latency high priority, the protocol can deploy functionalities for multiple-channel support. It essentially provides a higher possibility within a time frame to locate a non-occupied medium for the packet to be delivered [8]. However, using multiple channels leads to additional control overhead and message exchanges which results in more energy consumption. Therefore, if the application prefers to be highly energy saving, the protocol can choose not to adopt the multichannel scheme. In conclusion, based on the selected criteria and the spectral condition, the MAC protocol adapts itself to the best suited behaviour by modifying composition of MAC components at runtime.

We have implemented our solution on WARP, a SDR platform with Virtex Pro II FPGA [3] (cf. Figure 2) based on the v16 OFDM reference design. The host compiler is developed using Lex&Yacc [9] and modified for our target platform.

Figure 2: Demonstration setup.

3. DEMONSTRATION DESCRIPTION

This demonstration will provide opportunity to the audience to interactively modify a protocol realization criteria on-the-fly and correspondingly observe the triggered reconfiguration and the resulting MAC performances. The demonstration will also show an efficient runtime adaptation of the MAC protocol realization depending on to the channel condition. A controlled external interferer will be used to influence the spectral environment and thereby initiate auto MAC reconfiguration. In this demonstration, two WARP boards will be connected to a PC. An interactive Graphical User Interface (GUI) is used to display live MAC performance statistics. The demonstration will use two channels at 2.4 GHz ISM band.

Acknowledgment

We thank partial financial support from RWTH Aachen University, DFG through UMIC research centre (Nucleus project) and EU through 2PARMA project. We also thank Mr. Guangwei Yang for his technical support.

4. REFERENCES

[1] X. Zhang, J. Ansari, G. Yang and P. Mähönen. "TRUMP: Efficient Realization of Protocols for Cognitive Radio Networks" In *Proc. of IEEE DySPAN*, 2011.

[2] J. Ansari, X. Zhang and P. Mähönen. "A compiler assisted approach for component based reconfigurable MAC design" In *Proc. of Med-Hoc-Net (invited paper)*, 2011.

[3] A. Khattab, J. Camp, C. Hunter, P. Murphy, A. Sabharwal, and E. Knightly. "WARP: a flexible platform for clean-slate wireless medium access protocol design" In SIGMOBILE Mob. Comput. Commun. Rev, 2008.

[4] 3rd Generation Partnership Project (3GPP). The LTE Protocol Specification, 3GPP Rel8 In http://www.3gpp.org/Release-8.[Last visited: 10.06.2012]

[5] C. Coerr, M. Neufeld, J. Fifield, T. Weingart, D. C. Sicker, and D. Grunwald. "MultiMAC - an adaptive MAC framework for dynamic radio networking" In *Proc. of IEEE DySPAN*,2005.

[6] I. Tinnirello, G. Bianchi, P. Gallo, D. Garlisi, F. Giuliano, F. Gringoli. "Wireless MAC Processors: Programming MAC Protocols on Commodity Hardware" In *Proc. of IEEE INFOCOM*, 2012.

[7] J. Ansari, X. Zhang and P. Mähönen. "A Flexible MAC Development Framework for Cognitive Radio Systems" In *Proc. of IEEE WCNC*, 2011

[8] J. Ansari, X. Zhang and P. Mähönen. "A Decentralized MAC for Opportunistic Spectrum Access in Cognitive Wireless Networks," In Proceedings of the ACM SIGMOBILE Workshop on Cognitive Wireless Networking, 2010

[9] The Lex & Yacc Page In http://dinosaur.compilertools.net. [Last visited: 10.06.2012]

Demo: Programming Enterprise WLANs with Odin

Lalith Suresh
INESC-ID/Instituto Superior Tecnico
Lisbon, Portugal
lalith.puthalath@ist.utl.pt

Julius Schulz-Zander
Telekom Innovation Laboratories/TU Berlin
Berlin, Germany
julius@net.t-labs.tu-berlin.de

Ruben Merz
Telekom Innovation Laboratories/TU Berlin
Berlin, Germany
ruben.merz@telekom.de

Anja Feldmann
Telekom Innovation Laboratories/TU Berlin
Berlin, Germany
anja@net.t-labs.tu-berlin.de

ABSTRACT

We present a demo of Odin, an SDN framework to program enterprise wireless local area networks (WLANs). Enterprise WLANs need to support a wide range of services and functionalities. This includes authentication, authorization and accounting, policy, mobility and interference management, and load balancing. WLANs also exhibit unique challenges. In particular, access point (AP) association decisions are not made by the infrastructure, but by clients. In addition, the association state machine combined with the broadcast nature of the wireless medium requires keeping track of a large amount of state changes. To this end, Odin builds on a light virtual AP abstraction that greatly simplifies client management. Odin does not require any client side modifications and its design supports WPA2 Enterprise. With Odin, a network operator can implement enterprise WLAN services as network applications.

Categories and Subject Descriptors

C.2.1 [**Network Architecture and Design**]: Network communications; C.2.1 [**Network Architecture and Design**]: Wireless communication; C.2.3 [**Network Operations**]: Network management

Keywords

Odin, SDN, Enterprise WLANs

1. INTRODUCTION

Deployments of modern IEEE 802.11 [1] enterprise wireless local area networks (WLANs) serve today a multitude of client devices such as smart-phones, laptops or tablets. Large deployments provide resilience, fault-tolerance and fail-over capabilities. Scalability in this context is paramount. Independent of the size, what most enterprise WLAN deployments have in common is the support for features such as authentication, authorization and accounting (AAA), policy management, mobility management, interference management with dynamic channel reconfigurations, load balancing and providing QoS guarantees. Their management is usually centralized. These systems are also proprietary, with each vendor offering its own closed-source platform.

In this paper, we demonstrate Odin[1], a software defined networking (SDN) framework for enterprise WLANs. The objective of Odin is to empower network operators to program and deploy typical enterprise WLAN services and features as network applications. In this context, WLANs exhibit specific challenges. The 802.11 standard lets clients make AP association decisions on the basis of locally made decisions. The infrastructure has no control over these decisions made by the client. In addition, the association state machine at the AP, combined with the dynamic, broadcast and time-varying nature of the wireless medium can require to keep track of state information changes continuously. Furthermore, not only associated, but also all interfering 802.11 devices need to be taken into account.

To effectively express applications that implement high-level services in enterprise WLANs, the programmer needs simple and powerful abstractions. This is essential if the programmer operates on a central view of the entire network. Hence, a design goal of Odin is to prevent the programmer from keeping track of changes to the authorizer, authenticator and client state machine. Indeed, the programmer cannot make assumptions about the endpoints of the link between the client and the infrastructure (MAC and IP address). To shield the programmer from this burden and to simplify the programming model, a central component of Odin, is the light virtual access point (LVAP) abstraction. LVAPs virtualize association state and separate them from the physical AP. Multiple clients connected to a single physical AP are treated as a set of logically isolated clients connected to different ports of a switch. The LVAP abstraction enables Odin to offer a straighforward programming model. Typically, an application running on Odin does not need to directly handle the association state. But LVAPs also greatly facilitate mobility management, allowing for the infrastructure to handoff clients without triggering the client's re-association mechanism.

[1]Odin is a major god in Norse mythology.

SIGCOMM'12, August 13–17, 2012, Helsinki, Finland.
ACM 978-1-4503-1419-0/12/08.

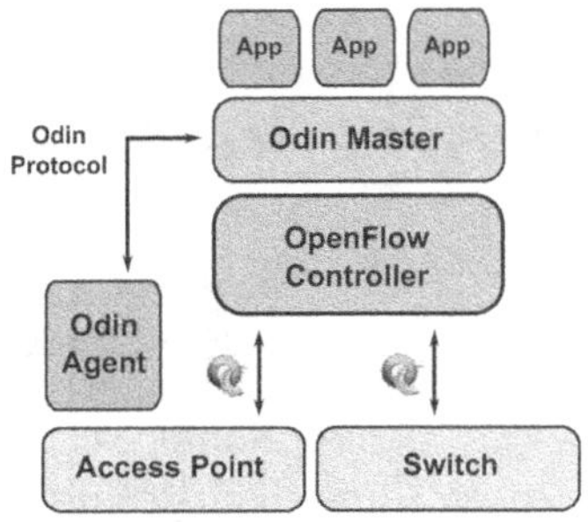

Figure 1: Architecture of Odin. The Odin Master (an OpenFlow application), speaks OpenFlow to the switches and the APs, and uses a custom protocol to talk to each Odin Agent running on APs. Odin applications use Odin's primitives to implement enterprise specific services.

Odin is work in progress, under heavy development and operates under the following assumptions: It targets fully centralized and reasonably sized deployments with one controller. It does not expect any client side modifications and supports WPA2 Enterprise by design. It is fully transparent to clients, which connect to it as regular IEEE 802.11 stations in infrastructure mode.

2. ODIN: AN SDN FRAMEWORK FOR ENTERPRISE WLANS

2.1 Architecture of Odin

Odin's architecture consists of a single master, multiple agents and a set of applications (Figure 1). The master itself is an application on top of an OpenFlow [2] controller. It has a global view of flows in the network, clients connected to the network, and the infrastructure that comprises the network. Odin agents run on the APs. Together, the master and agents implement a Wi-Fi split-MAC.

2.2 Light Virtual AP (LVAP)

LVAPs are a central primitive within the Odin framework. When Wi-Fi clients in managed mode scan APs, probe request messages are generated. APs responding with probe response messages become potential association candidates. A client then proceeds to perform a connection handshake with a locally selected AP. The LVAP abstraction enables to take control on association decisions from client and leads to a logical isolation of clients with respect to the 802.11 MAC. With LVAPs, every client receives a unique BSSID to connect to, i.e, every client is given the illusion of owning its own AP. In Odin, the LVAP is the client specific AP. Each *physical* AP hosts an LVAP for each connected client. Removing a client LVAP from a physical AP and spawning it on another achieves the effect of handing off a client without the client performing a re-association, generating additional layer 2 or 3 messages, and most importantly, without requiring any special software or hardware at the client. Thus, Odin always provides clients a consistent link to the network, and the programmer of an Odin application needs not be concerned with *how* the client's link to the network changes. The end-point of a link always corresponds to the client's IP and MAC addresses and the unique BSSID assigned by Odin.

3. DEMO PLAN

The demo setup will comprise three APs connected via a switch to a laptop. The APs are x86 based Alix 3D2 boards running OpenWrt with Wistron DNMA92 802.11n interfaces (Atheros AR9220 chipset). Each AP runs an Odin agent and the laptop runs the Odin master with a set of Odin applications. In addition, the laptop will run necessary services for the demo. Namely, a visualization interface and, depending on the local availability, an upstream connectivity through NAT.

Our demo of Odin will display various aspects of a typical enterprise system.

- A complete Odin deployment with WPA support where demo attendees can use their own devices as clients.
- Using a visualization interface, we will demonstrate the inner working of the LVAP concept. It will show at which physical AP each client is associated by means of LVAPs. In addition, the visualization interface will display the various statistics that are collected by the measurement collection component of Odin. For instance, the received signal strength of different clients as observed at the APs, the noise, and data rates.
- We will demonstrate various Odin applications. In particular, the mobility manager and the load balancer.
- Demonstration of network slicing based on policies, such as separation of a guest WiFi network, isolation of malicious nodes or traffic restrictions.
- Demonstrate connection restorations in light of AP failures using LVAP migrations.
- In order to show the ease with which Odin applications can be written, we will provide another laptop to let demo attendees program their own Odin applications. We will support them and help them to run their application on our Odin deployment.

4. SUMMARY

There is a large body of work on enterprise WLAN architectures and system, see for instance [3, 4] and the references therein. For Odin, these works contain typical applications that we expect to eventually support.

5. REFERENCES

[1] *IEEE Std 802.11-2007 (Revision of IEEE Std 802.11-1999)*, 2007.

[2] N. McKeown, T. Anderson, H. Balakrishnan, G. Parulkar, L. Peterson, J. Rexford, S. Shenker, and J. Turner. Openflow: enabling innovation in campus networks. *SIGCOMM Comput. Commun. Rev.*, 38:69–74, March 2008.

[3] V. Shrivastava, N. Ahmed, S. Rayanchu, S. Banerjee, S. Keshav, K. Papagiannaki, and A. Mishra. CENTAUR: realizing the full potential of centralized wlans through a hybrid data path. In *ACM Mobicom 09*, 2009.

[4] R. Murty, J. Padhye, A. Wolman, and M. Welsh. Dyson: an architecture for extensible wireless LANs. In *USENIX ATC 10*, 2010.

Supporting Network Evolution and Incremental Deployment with XIA

Robert Grandl*, Dongsu Han, Suk-Bok Lee, Hyeontaek Lim, Michel Machado†, Matthew Mukerjee, David Naylor

Carnegie Mellon University * University of Wisconsin-Madison † Boston University

ABSTRACT

eXpressive Internet Architecture (XIA) [1] is an architecture that natively supports multiple communication types and allows networks to evolve their abstractions and functionality to accommodate new styles of communication over time. XIA embeds an elegant mechanism for handling unforeseen communication types for legacy routers.

In this demonstration, we show that XIA overcomes three key barriers in network evolution (outlined below) by (1) allowing end-hosts and applications to start using new communication types (e.g., service and content) before the network supports them, (2) ensuring that upgrading a subset of routers to support new functionalities immediately benefits applications, and (3) using the same mechanisms we employ for 1 and 2 to incrementally deploy XIA in IP networks.

Categories and Subject Descriptors

C.2.1 [**Computer-Communication Networks**]: Network Architecture and Design

General Terms

Design

Keywords

Internet architecture, evolution, multiple communication styles

1 Introduction

What makes network evolution so hard? The "narrow waist" design of the Internet has been tremendously successful. However, the Internet does not facilitate a clean, incremental path for the adoption of new capabilities at the narrow waist. Two main barriers against evolution are:

B1 It is not possible for an end-host or an application to use new functionality until the network supports it.

B2 There is no built-in mechanism to support incremental deployment of new functionality. For example, enabling a single router in a IP network to support content-centric networking is fruitless if that traffic will simply be tunneled through the network using IP.

XIA [1] is a clean-state architecture that solves these issues. However, a clean-state design has its own problems:

SIGCOMM'12, August 13–17, 2012, Helsinki, Finland.
ACM 978-1-4503-1419-0/12/08.

B3 Incrementally deploying a new architecture over IP is non-trivial and often requires careful configuration.

In this demo, we show that XIA effectively overcomes these three barriers. The rest of this section provides a brief summary of XIA's core mechanism for evolution.

How does XIA solve this? The key architectural element that XIA adds to provide evolvability is the notion of an extensible set of *principal* communication types (e.g., host-to-host, service-oriented, and content-oriented). By selecting the appropriate principal for a given communication, applications can directly express the *intent* of that communication.

XIA's addresses can simultaneously express both identifiers for a "new" principal type (e.g., a content or service identifier) and one or more backwards compatible identifiers (e.g., host identifiers) called *fallbacks*. XIA allows for this flexibility by specifying addresses as *directed acyclic graphs* (DAGs). For example, the following DAG specifies that the intent is to reach a service ("Srv"), but routers may instead forward the packet to host "Hst" in destination network "Dom" if they do not support services.

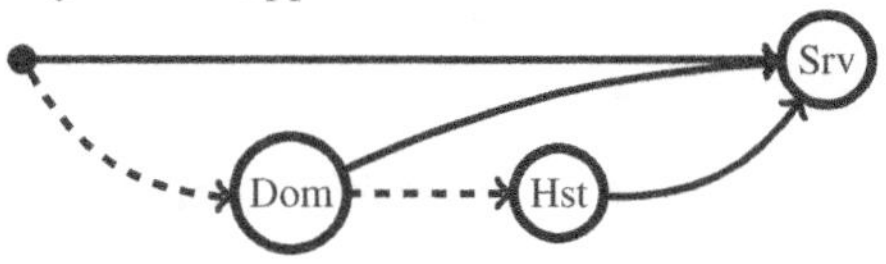

Thus, by using fallbacks, two hosts that have been upgraded to understand the service principal can begin to use it even if no routers support the principal type, overcoming **B1**. As some routers are upgraded to support services, applications (without any modification) will begin to see benefits (**B2**).

How can XIA be deployed? To address **B3**, we introduce a new principal type: IPv4 addresses, or "4IDs." 4IDs leverage the power of the XIA addressing scheme (namely DAGs and fallbacks) to support old communication styles (IPv4). We do this by encoding IPv4 addresses within these 4ID identifiers, which are included in DAGs as fallbacks.

Using 4IDs, we can effectively connect an XIA island to the XIA backbone. A 4ID specifies an IPv4 entry point (a dual-stack router) of an XIA network, and is obtained through name resolution. Thus, we eliminate the burden of manually configuring tunnels between XIA islands. When a packet with a 4ID fallback can no longer progress though the XIA network, we use the IPv4 address and dynamically tunnel the packet to the XIA network whose IPv4 entry address is specified by the 4ID.

This begs the question: where do 4IDs come from? The sender must know of an IPv4 address that serves as an ingress point to the destination XIA network. We handle this by having each domain register a set of IPv4 addresses that it is willing to use as ingress points. These addresses are returned as 4IDs during the name resolution process.

2 System Overview

We have implemented support for host-, service- and content-oriented communication over XIA. This includes an XIA router with principal-specific processing for each of these principals as well as each type's end-host stack and socket API.

Router: We have implemented a prototype router in Click that processes XIA packets. All routers support host-based communication. Additionally, routers may support the content, service, and/or 4ID principal types. If content support is enabled, our routers can act as in-network caches. The router forwards packets that contain content to the content module, which stores content and inserts a routing entry for the cached content to direct future requests to itself. If services are supported, anycast and failover support are implemented.

End-host: We have implemented end-host support for service- and content-oriented communication and exposed this functionality to application developers through a sockets API resembling standard Berkeley sockets. To invoke a particular transport type (and thereby choose either service- or content-oriented communication), developers pass an extra parameter when creating a socket.

Applications: See below.

3 Demo Scenarios

We will demonstrate our solutions to the three barriers discussed in §1, using both the service and content principal types as examples of new functionality we wish to add to the network incrementally. To do so, we will use two small XIA networks we have deployed on GENI and at Carnegie Mellon University (CMU).

Applications We have developed some simple applications that run over XIA and are in the process of developing more. Some subset of these will be used to demo XIA:

- Web server (and proxy)
- Video streaming server (uses content support)
- "Stock Ticker" server (uses service support)
- "VoIP" (VoXIA, really) — in development
- A fun, interactive, application that you too can try (see "Attendee Participation" below) — in development

Demonstrating Solutions to Evolution Barriers We will first demonstrate that XIA allows applications to begin using new principal types without network support (**B1**) by running applications that use the service and content principals over a network of XIA routers in GENI that support only the host principal. The applications run without problems because the routers use fallbacks to satisfy the application's intent. The audience will be able to observe the amount of traffic (by principal type) flowing on each link through our real-time visualization tool that monitors activity on GENI. Figure 1(a) illustrates this example; none of the packets are being processed using CID- or SID-specific processing, meaning the "legacy" routers are using the host ID fallbacks.

We will then demonstrate that applications can benefit by upgrading even just one router in the network to support a new principal type (**B2**). By enabling support for the content principal type at one of the routers in our demo network, our video streaming application can immediately benefit from in-network content caching. The audience will be able to observe that the video is being served

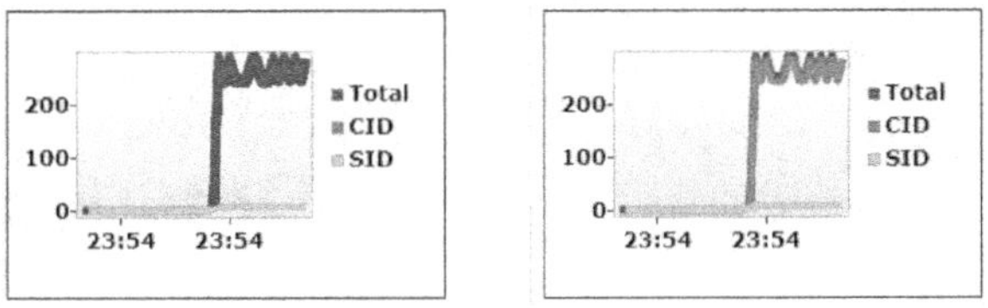

(a) Routers forward based on fallback host IDs (b) Upgraded routers now forward packets based on CIDs

Figure 1: Video streaming over XIA networks with and without support for the content principal.

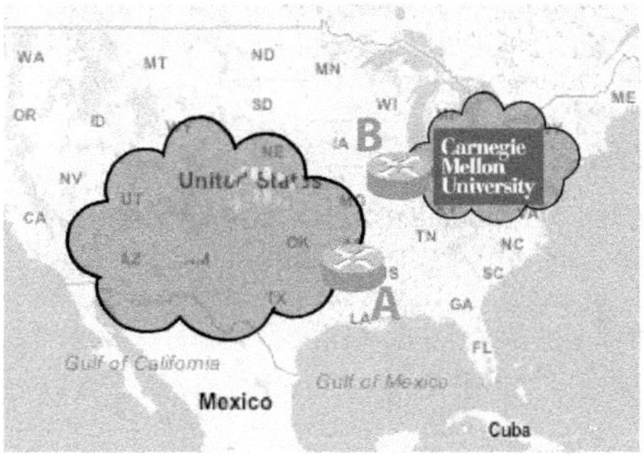

Figure 2: We use the IPv4 principal type to communicate between XIA networks without explicit tunnels.

from a nearby cache instead of the server we fell back to previously. On the upgraded router the traffic will look like Figure 1(b): the same packets are passing through the link, but our router now understands that we are using content IDs (CIDs).

Finally, we can show how XIA might be incrementally deployed in today's IP Internet (**B3**). By adding 4IDs as fallback nodes in our DAG addresses (see below), we will allow XIA-enabled dual-stack machines at CMU to communicate with our XIA network on GENI (Figure 2). The GENI XIA backbone and the CMU XIA island will be connected using 4IDs instead of a manually configured tunnel.

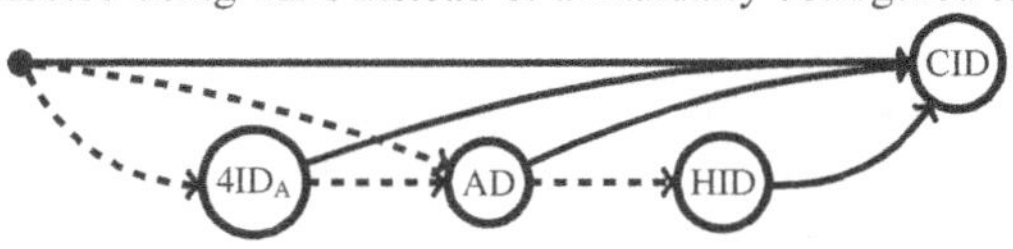

Attendee Participation We are developing an interactive application to run over XIA that will allow us to demonstrate 4IDs on a larger scale. The application will be added to the XIA source code (already available[1]) and, environment permitting, we hope to allow conference attendees to download our VM and join the fun. Each XIA-enabled VM would function as its own singleton XIA network; client applications connecting to our servers would use 4IDs as fallbacks to traverse the IPv4 Internet. The demo will provide entertainment as well as first-hand experience with how two disjoint XIA networks can be connected using 4IDs.

4 Acknowledgments

This research was supported in part by the National Science Foundation under awards CNS-1040757, CNS-1040800, and CNS-1040801.

5 References

[1] D. Han, A. Anand, F. Dogar, B. Li, H. Lim, M. Machado, A. Mukundan, W. Wu, A. Akella, D. G. Andersen, J. W. Byers, S. Seshan, and P. Steenkiste. XIA: Efficient support for evolvable internetworking. In *Proc. 9th USENIX NSDI*, San Jose, CA, Apr. 2012.

[1] `http://www.cs.cmu.edu/~xia/`

Picasso: Flexible RF and Spectrum Slicing

Steven Hong
Stanford University
hsiying@stanford.edu

Jeffrey Mehlman
Stanford University
jmehlman@stanford.edu

Sachin Katti
Stanford University
skatti@stanford.edu

Categories and Subject Descriptors: C2.1 [**Computer Systems Organization**]: Computer Communication Networks

Keywords: Radio Virtualization, Interference Cancellation

1. INTRODUCTION

Many applications can benefit from the capability to simultaneously and independently use arbitrarily sized but separate spectrum fragments with a single radio and antenna. By this capability we mean that the radio can simultaneously transmit, simultaneously receive, or simultaneously transmit and receive on arbitrary but separate spectrum fragments. For example, we can use it for *spectrum aggregation* in fragmented ISM bands as shown in as shown in Fig. 2(A). A WiFi AP can run independent OFDM PHY and CSMA MAC protocols on two WiFi channels to simultaneously serve two legacy WiFi clients assigned to different channels and achieve significantly higher throughput than a legacy AP that is restricted to use only one channel at a time. Similarly, a WiFi client radio with such a capability can use it to simultaneously connect to multiple WiFi APs on different channels and obtain a much higher aggregate throughput than current radios that can transmit or receive on only one channel at a time.

Such a capability could also be used for *radio sharing* and *coexistence*. Portable consumer devices such as smartphones must accommodate a growing list of separate ISM band protocols, such as WiFi, WiFi-Direct, Zigbee, NFC, Bluetooth, etc. Current practice is to use a separate radio and antenna for each protocol, shown in Fig. 2(B), but as the number of radios increase, it becomes increasingly difficult to find enough space to separately place all the antennas these radios would need (e.g. IPhone 4 "antenna-gate" was caused by antennas placed too closely [1]). Instead, a radio with a single antenna that allows simultaneous TX/RX on arbitrary spectrum fragments could be shared among all the above protocols, WiFi would use one fragment, Zigbee another and so on, saving valuable real estate on space constrained devices. They can operate their own independent PHY/MAC protocols on the shared radio without interfering with each other.

Picasso provides a clean architecture for multiple such networked applications to co-exist by allowing one to decouple their operation from each other using slicing. Specifically, Picasso enables us to slice each device's radio into three slices. The first slice would be assigned to the WiFi network, which

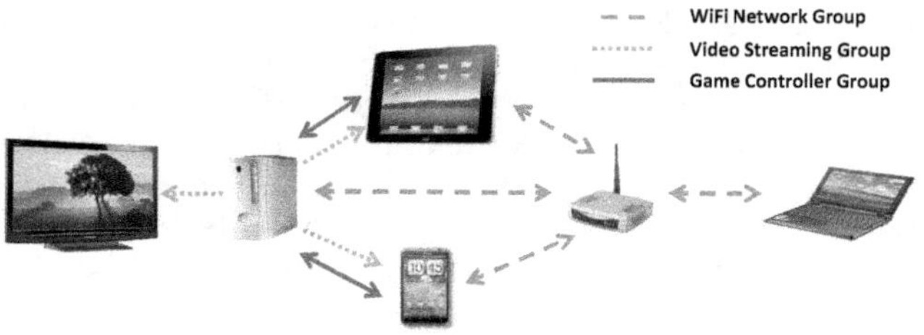

Figure 1: Coexistence in Dense Environments with Picasso: Picasso enables devices operating in multiple networks to flexibly share fragmented spectrum and RF frontends, yet operate in a decoupled manner.

is now free to operate its own optimized PHY and MAC protocols to maximize throughput. The second slice would be for the gameplay video streaming application (Xbox, smartphone, tablet and TV), and if it so chooses, would operate independent PHY and MAC protocols optimized to maximize video quality. Finally, a third slice is for the gameplay controller application (Xbox, smartphone, tablet), with an independent PHY/MAC optimized for low latency. Thus Picasso *decouples the operation of the networked applications from each other*, and enables them to gracefully coexist yet independently design their protocols to optimize their respective objectives.

However, it is non-trivial to design such a radio. The key obstacle is that current radios cannot simultaneously transmit and receive on different arbitrary spectrum fragments with a single shared RF frontend and antenna. The reason for this is that the transmitted signal causes high powered self-interference, which ends up *saturating the RX antenna and ADC*, consequently nulling the received signal. While the standard solution is to utilize static analog RF filters to eliminate the interference, such an option is infeasible because spectrum fragmentation is dynamic - available spectrum in the ISM band varies in space and time, depending on the presence of other wireless networks. Consequently, if a radio wants to leverage all the available spectrum and be able to simultaneously transmit and receive on different fragments, the shared analog frontend would need programmable analog filters that can be dynamically configured to let only the received signals through and filter out the self interference. Analog filters however are typically statically configured and programmable analog filters that can be changed dynamically are expensive and impractical to deploy in current radios [2].

2. OUR APPROACH AND CONTRIBUTIONS:

In this demonstration, we will present the design and implementation of Picasso, a novel full duplex circuit design that

SIGCOMM'12, August 13–17, 2012, Helsinki, Finland.
ACM 978-1-4503-1419-0/12/08.

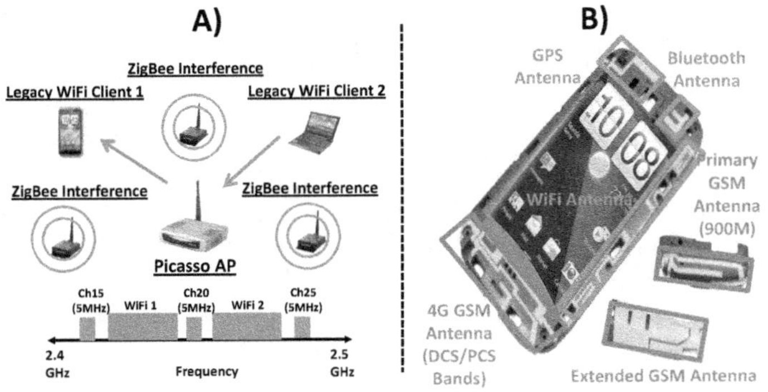

Figure 2: A) Fragmented Spectrum is common in ISM band:Picasso AP can aggregate multiple fragments and simultaneously service multiple clients on different channels. B) Samsung Galaxy Printed Antenna Layout: Instead of having an antenna per protocol, Picasso enables multiple protocols to use the same radio and antenna, saving space on the device.

sufficiently *cancels (instead of filters)* the self-interference in analog and prevents antenna and ADC saturation, enabling the radio to cleanly recover the received signal. Our key contribution here is a novel circuit design that leverages a circulator [3] to isolate TX and RX signals from a single antenna, and exploits the fact that self interference to the RX signal travels through a fixed known circuit to design a statically configured self interference cancellation circuit. This allows a *radio to simultaneously transmit and receive on arbitrary spectrum fragments even while using a single RF-frontend and antenna.*

Picasso leverages this full duplex capability to build a radio slicing abstraction that allows one to flexibly slice a single radio and antenna into separate independent slices operating on different spectrum fragments. Each slice is associated with a specific spectrum fragment in the ISM band (whose width/position can be programmatically specified). The key property is that the *operation of each slice is decoupled from the other slices*, i.e. the slice is free to run whatever narrowband PHY and MAC protocols it chooses, and the protocol behavior is not impacted by any other slice that may be present on the shared radio and antenna. Thus in the above scenarios, the AP would have two slices corresponding to the two spectrum fragments and run two independent WiFi OFDM/CSMA protocols on the two slices in parallel. Similarly, a radio could be shared amongst multiple protocols (e.g. WiFi, Bluetooth, and NFC) by assigning independent slices to the appropriate spectrum fragment and running the corresponding protocol.

Further, to ensure that each slice can use existing well-engineered narrowband PHY and MAC protocols on each slice, Picasso designs a novel reconfigurable filter engine that transparently takes signals spread over different spectrum fragments, and efficiently filters and resamples them so that the higher layers just see a simple sample stream consisting of narrowband digital samples. The higher layers are then free to process these samples with any narrowband PHY technique they choose, and schedule access to the slice with a MAC protocol of their choice. The slice for all intents and purposes appears as their own piece of spectrum centered at zero operating on their own radio, Picasso thus completely abstracts out the complexity of spectrum fragmentation.

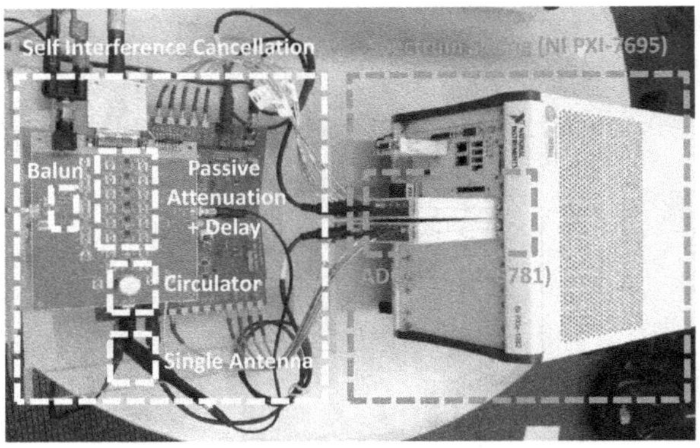

Figure 3: Picasso Radio Physical Setup: Built with off the shelf components and software radios from National Instruments.

3. DEMONSTRATION SETUP

We design and implement a prototype of Picasso on Xilinx Virtex-5 FPGA based software radios [4]. Our implementation consists of both the radio design that provides the slicing abstraction, as well as a WiFi style contiguous OFDM PHY and CSMA MAC to operate on top of the slices. We show that Picasso's implementation of the slicing abstraction provides strong decoupled operation, i.e. there's no SNR loss because of either programmable filtering or simultaneous TX/RX. In other words, a Picasso radio achieves the same throughput as one would have achieved by using several independent radios statically configured to operate on individual slices.

We have also implemented a WiFi style OFDM PHY and a CSMA MAC split across the FPGA and the realtime OS. The OFDM PHY can be configured to operate over different bandwidths (from 20MHz to 5MHz), and runs in realtime on the SDR platform. It supports all the WiFi constellations (from BPSK to 64-QAM) as well as channel coding rates (1/2, 2/3, 3/4 convolutional coding). We have not implemented an automatic rate adaptation algorithm, but in the experiments we empirically pick the combination of constellation size and channel coding rate that maximizes the link throughput. Thus our prototype implements all the relevant features from a WiFi PHY/MAC. Note that we can run multiple instances of the WiFi PHY and MAC one for each slice on the device.

We will be demonstrating this physical setup, first demonstrating that a typical node without Picasso cannot transmit and receive on different arbitrary bands. Without sharp analog filters (which are fixed in nature), the transmit signal overwhelms the received signal even when the two are on different bands. We then test the performance of a Picasso node in the same scenario - demonstrating that it can simultaneously operate independent connections over multiple spectrum slices.

4. REFERENCES

[1] Gizmodo, "IPhone 4 Antenna-Gate", *http://gizmodo.com/5846638/giz-explains-whats-so-smart-about-the-iphone-4ss-antenna*, 2011.

[2] Matthaei, George L., Jones, E. L. and Young, Leo, "Microwave filters, impedance-matching networks, and coupling structures", *Dedham, Mass: Artech House Books. ISBN 0-89006-099-1.*.

[3] D. Pozar, "Microwave Engineering: 3rd Edition", *Wiley*, 2004.

[4] NI PXIe-8133 User Manual, *http://www.ni.com/pdf/manuals/372870b.pdf*, 2011.

Dismantling Intrusion Prevention Systems

Olli-Pekka Niemi
Stonesoft Corporation
Helsinki, Finland
olli-pekka.niemi@stonesoft.com

Antti Levomäki
Stonesoft Corporation
Helsinki, Finland
antti.levomaki@stonesoft.com

Jukka Manner
Aalto University
Helsinki, Finland
jukka.manner@aalto.fi

ABSTRACT

This paper introduces a serious security problem that people believe has been fixed, but which is still very much existing and evolving, namely evasions. We describe how protocols can still be misused to fool network security devices, such as intrusion prevention systems.

Categories and Subject Descriptors

C.2.0 [**Computer Communication Networks**]: Security and protection.

General Terms

Security.

Keywords

Network, intrusion prevention, evasion, IDS, IPS.

1. INTRODUCTION

The Internet has evolved from the closed environment of academics to a worldwide network. The resources and services available to anyone at any time are huge. Protecting our digital assets from misuse is a major business and area of active research.

Hosts and services on the Internet can be protected by e.g., firewalls, virus scanners, intrusion detection systems (IDS) and intrusion prevention systems (IPS). Our interest are IPS devices (sometimes called network intrusion detection) that do real-time analysis of the traffic and seek to close and raise alarms about flows that include harmful content, e.g., attacks exploiting remote code execution vulnerabilities on server services such as HTTP and MSRPC.

To make attacks unnoticed by IPS devices, the attacker can use so-called evasion techniques, ways to obfuscate the harmful content. Typical tricks include fragmenting the payload into small packets and sending these packs in disorder [2], desynchronizing protocol streams [3], using insertion attacks [2][5] and using various encodings [4][6][8]. The reason that evasions work is the old robustness principle stated by Jon Postel in RFC793 [1] "be conservative in what you do, be liberal in what you accept from others". When the sender deliberately breaks this principle, he can send data that is eventually accepted by the receiver but not necessarily fully understood by an IPS.

SIGCOMM'12, August 13–17, 2012, Helsinki, Finland.
ACM 978-1-4503-1419-0/12/06.

Most work on evasions is ten years old. The standard way to reveal evasions is to use traffic normalization to rebuild an obfuscated data stream. The challenge is that obfuscated data streams can be very complex and resource wise expensive to open up. This has led vendors to do anomaly-based detection with minimum amount of normalization and TCP/IP reassembly. The device vendors seem to trust that all known techniques have been disclosed to date.

In our research we have gone further than the current state of the art. We have devised hundreds of atomic tricks on protocols such as IPv4/v6, TCP, NetBIOS, SMB, MSRPC, HTTP and also HTML encoding. These evasions can be combined to form complex evasions [7]. Furthermore, by doing combinations of these atomic evasions on multiple protocol layers and targeting the evasions into detection wise critical stages, we can build advanced evasions techniques (AET) that can by-pass any IPS on the market today. In this paper and demo, we show the Evader tool built to experiment on various evasions. The tool hides harmful content from IPS systems and injects it to the target host. As a result all commercial and open source IPS devices are penetrated, most of them very easily.

2. EVASION RESEARCH THUS FAR

Research on evasions is being done by the scientific and security communities. One of the first papers about evasions was published by Ptacek and Newsham [2] 14 years ago. Many IPS devices still do not handle properly the mentioned methods. In 1998 Horizon wrote an article to Phrack Magazine [3] about desynchronizing TCP stream. Rain Forest Puppy introduced multiple HTTP evasions in [4] and this work was further refined by Daniel Roelker [6]. An insertion evasion called SealMa was disclosed in Phrack [5] and combinations of evasions were introduced by Gorton and Champion [7]. Caswell and Moore [8] summarized at the time known evasion techniques and introduced some new evasions regarding SMB and MSRPC protocols. Examples of more scientific works to solve the evasion problem include a Sigcomm 2006 paper by Varghese et al. [9], and the work by Watson et al. [10]. ACM CCS has had many papers in the past on the topic, e.g., Shunting [11].

Despite all this work, there is no thorough and systematic discussion of ways to evade network intrusion detection. Our work focuses on understanding in detail the width and size of the problem. Once this is better understood, we can start devising ways to counter the attacks.

3. THE EVADER TOOL

We have implemented a tool we call Evader. It applies network level evasions to send a payload into a remote host through the IPS device under test (DUT). Evader first sends non-malicious payloads that should not be prevented. This is called the false positive test. If this is successful, the malicious payload will be sent. Depending on the selected malicious payload, the remote system is either crashed or compromised via remote code execution. If this happens we know that the evasion was functional against the DUT.

3.1 Features

Evader contains a multilayer network protocol stack. When sending the payload, Evader can apply multiple evasions on various protocols. If the payload e.g. exploits the Microsoft Server Service Vulnerability MS08-067 [12], the evasions applied can be from IP, TCP, NetBIOS, SMB and MSRPC layers. Evader can divide the connection into several stages and every stage can have its own evasions applied. In theory the Evader can produce every possible obfuscated data stream transmitting the payload, but in practice this cannot be tested since there are hundreds of atomic evasions implemented, and virtually endless amount of combinations and stage permutations.

3.2 Test Results

We have used the tool on several commercial IPS systems. The DUTs have been updated to latest software, firmware and signature level. Almost every DUT had to be tuned as well since they either failed to detect the exploit completely or missed it if any evasion were applied. Almost every DUT can be bypassed even with a tuned policy by previously disclosed evasions and AET is not even actually needed. With carefully devised AET, all tested devices fail.

We run evasion tests using different vulnerabilities. Usually a DUT detects the attack and is able to name it with a CVE reference. Yet, with evasions DUTs do not identify the attack anymore even though it may generate protocol anomaly events. The problem with protocol anomalies is that they commonly generate false positives and cannot be used to drop/terminate connections. An example is the prevention of small TCP segments that some vendors use to block evasions. Table 1 shows simple tests where the recent Remote Desktop Vulnerability [13] was exploited through eight commercial IPS appliances. The exploit was taken from Metasploit and the evasions applied were: no evasion (Base), small TCP segmentation (Seg), reverse order sent TCP segments within the congestion window (Reverse), time-wait evasion where multiple decoy connections are opened and closed using same connection tuple before time-wait counter expires (Time-Wait) and overlapping TCP segments with timestamp manipulation (PAWS).

Table 1 Example of experiments (FAIL: successful evasion)

Vendor	Base	Seg	Reverse	Time-Wait	Paws
A	FAIL	FAIL	FAIL	FAIL	FAIL
B	OK	FAIL	FAIL	FAIL	FAIL
C	OK	FAIL	FAIL	OK	OK
D	OK	OK	FAIL	OK	FAIL
E	OK	OK	OK	OK	FAIL
F	OK	FAIL	FAIL	OK	OK
G	OK	OK	OK	FAIL	OK
H	OK	OK	FAIL	OK	FAIL

4. CONCLUSIONS

We have shown how network intrusion prevention is very much behind the means of the attacker and we expect that evasions are being used all the time. We hope to trigger a serious discussion on how to perform network intrusion prevention and protect the society's digital assets. We plan to write a series of articles on these findings in the future.

5. REFERENCES

[1] J. Postel, Transmission Control Protocol. RFC793, IETF, 1981.

[2] T. Ptacek, T. Newsham, Insertion, Evasion, and Denial of Service: Eluding Network Intrusion Detection, Technical report, Secure Networks Inc., 1998.

[3] Horizon jmcdonal@unf.edu, Defeating Sniffers and Intrusion Detection Systems, Phrack Magazine Volume 8, Issue 54 Dec 25th, 1998, article 10 of 12.

[4] Rain Forest Puppy, A look at whisker's anti-IDS tactics

[5] NIDS Evasion Method named "SeolMa", Phrack Magazine Volume 11, issue 57, Phile 0x03, 2001

[6] Daniel J. Roelker , HTTP IDS Evasions Revisited, 2003, http://docs.idsresearch.org/http_ids_evasions.pdf

[7] A. Gorton and T. Champion, Combining Evasion Techniques to Avoid Network Intrusion Detection Systems, Skaion, 2004.

[8] B. Caswell, H D Moore, Thermoptic Camouflage: Total IDS Evasion, Blackhat, 2006

[9] Varghese, et al., Detecting Evasion Attacks at High Speeds without Reassembly, Sigcomm, 2006.

[10] Watson, et al. Protocol Scrubbing: Network Security Through Transparent Flow Modification, IEEE/ACM TON, vol. 12, no. 2, April, 2004.

[11] Gonzalez, et al., Shunting: A Hardware/Software Architecture for Flexible, High-Performance Network Intrusion Prevention. ACM CCS, 2007.

[12] Microsoft, Vulnerability in Server Service Could Allow Remote Code Execution, 2008, MS08-067

[13] Microsoft Remote Desktop Protocol CVE-2012-0002

namehelp: Intelligent Client-Side DNS Resolution

John S. Otto, Mario A. Sánchez, John P. Rula, Ted Stein, and Fabián E. Bustamante
Northwestern University
{jotto, msanchez, john.rula, ted.stein, fabianb}@eecs.northwestern.edu

ABSTRACT

The Domain Name System (DNS) is a fundamental component of today's Internet. Recent years have seen radical changes to DNS with increases in usage of remote DNS and public DNS services such as OpenDNS. Given the close relationship between DNS and Content Delivery Networks (CDNs) and the pervasive use of CDNs by many popular applications including web browsing and real-time entertainment services, it is important to understand the impact of remote and public DNS services on users' overall experience on the Web. This work presents a tool, *namehelp*, which comparatively evaluates DNS services in terms of the web performance they provide, and implements an end-host solution to address the performance impact of remote DNS on CDNs. The demonstration will show the functionality of *namehelp* with online results for its performance improvements.

Categories and Subject Descriptors

C.2.4 [**Computer Communication Networks**]: Distributed Systems

Keywords

Domain Name System, Content Delivery Networks

1. MOTIVATION

The Domain Name System (DNS) is a critical component of today's Internet. DNS provides a distributed lookup service primarily used to resolve human-readable machine names to Internet Protocol addresses. The pervasiveness of the Web and the Web's dependence on DNS to resolve the hostname portion of URLs has further accentuated its importance [7].

Recent years have brought radical changes to DNS with the increasing use of remote DNS and the growth of public DNS services such as OpenDNS, UltraDNS and Google DNS. Public DNS services promise higher reliability and faster lookup times as well as a number of security features (e.g. phishing protection).

Several studies have shown the potential advantages of a "personalized" selection of DNS servers based on responsiveness [1,3,4]. The close relationship between many web applications and DNS demands a careful analysis of the impact of remote and public DNS usage on users' web experience. An increasing fraction of the Web relies on Content Delivery Networks (CDNs) for scalability and performance.

In a survey of CDN usage by popular websites, we found that 75% of the top 1000 sites (as ranked by Alexa) use a CDN. CDNs cache content at several replica servers throughout the network and redirect clients to a nearby replica for low latency, high bandwidth access to content. The selection of a replica to serve a given client's request is commonly done based on different features including the client's approximate location, and network and server loads. To approximate the client's location, CDNs adopt the location of the client's DNS resolver as a proxy on the assumption that clients are near their DNS resolvers (DNS-based redirection). The recent growth of public DNS services and the use of remote DNS by ISPs break this assumption potentially resulting in worse end-to-end web performance.

SIGCOMM'12, August 13–17, 2012, Helsinki, Finland.
ACM 978-1-4503-1419-0/12/08.

2. END-HOST SOLUTION: namehelp

The potential negative impact of remote DNS usage on web performance motivates the recently proposed EDNS0 extension [2]. This extension, put forward by a collaboration of several CDNs and DNS service providers [6], allows DNS recursive resolvers to pass along clients' subnet information that could be used by CDNs to improve replica server selection. A key challenge to this approach is adoption, as it requires the commitment of both CDN and DNS services to be effective.

We present *namehelp*,[1] an end-host based system that comparatively evaluates the CDN and web performance obtained when using different DNS services, helping users to choose the best DNS service. In addition, it implements a client-side technique for obtaining good CDN and web performance even when using remote DNS services.

namehelp has two major components: the DNS proxy for improving web performance with remote DNS and a DNS server recommendation engine; we discuss each in turn.

2.1 DNS proxy functionality

namehelp functions as a transparent DNS proxy running on the user's local machine. After installing *namehelp*, users have only to configure "localhost" as their primary DNS server to start using it.

As the machine's primary DNS server, *namehelp* receives the full stream of DNS queries coming from the operating

[1] http://aqualab.cs.northwestern.edu/projects/namehelp

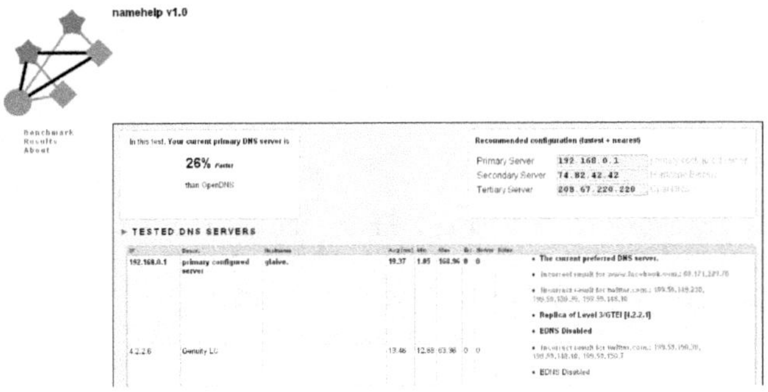

Figure 1: The *namehelp* DNS configuration recommendation interface, showing a ranked list of DNS servers.

system. By default, it simply passes on these queries to the configured recursive DNS servers such as the DHCP-provided DNS configuration or a specific IP address, such as Google's DNS server (8.8.8.8).

When the proxy receives the DNS answer from the recursive resolver, it examines the response to determine whether there was a redirection to a CDN by looking for a canonical name (`CNAME`) record in the reply. In the simple case – when no CDN redirection occurred – *namehelp* returns the answer directly to the operating system. However, if a CDN redirection took place, *namehelp* discards the response and directly queries the CDN's authoritative DNS server for the same `CNAME` record, returning the answer from the second query. Since the client queries the CDN directly, the CDN knows the actual location of the client resulting in potentially improved server selection and better performance.

namehelp acts also as a local DNS cache and prefetches frequently-requested queries for improved performance. This caching also allows us to skip the queries to other resolvers for names that are known to be CDN redirections, reducing the total overhead in the vast majority of cases to a fraction of a millisecond of local processing.

2.2 Configuration recommendations

To enable users to make an informed decision regarding which DNS server to configure, *namehelp* periodically runs a battery of tests that compare DNS servers. The result is a ranked list of DNS servers, as shown in Fig. 1, which tells users which DNS service is best in terms of several factors, including responsiveness and the quality of CDN redirections. We incorporate code from the popular *namebench* DNS benchmarking utility [5] for this task. *namehelp* schedules tests periodically to produce suggested configurations based on both DNS performance and HTTP performance.

Since each user has a unique set of websites she visits, *namehelp*'s benchmarking tests personalize the list of sites to test based on the user's habits, incorporating information from the user's browser history. As a result, the DNS configuration recommended by *namehelp* is personalized for best web performance.

After running the tests, *namehelp* generates a report that includes the ranked list of DNS servers and justifications for its results. A pane of the user interface is dedicated to presenting the current and historical reports to the user.

Once the report has been generated, *namehelp* has two options for acting on the results: manual or automatic configuration. With manual configuration, the user simply examines the results and decides which server to use. However, the best DNS service for a user may vary over time for several factors, including changes in the sites a user visits to variations in the performance of the selected DNS service (e.g. due to fluctuations in load). To handle this scenario, *namehelp* can be configured to automatically update the user's DNS configuration to match the recommendation from the latest test.

3. DEMONSTRATION

Our demonstration of *namehelp* will include the following elements:

- Online analysis of the DNS proxy's performance improvement
- Fine-grained tool to examine the DNS proxy's behavior
- Examples of varying web browsing patterns and their impact on recommended DNS
- Results of running benchmarking tests in several different network environments

Acknowledgements

We would like to thank the reviewers for their comments. This work is supported by the National Science Foundation through Awards CNS 0917233 and CNS 0855253, and by a generous Google Research Award.

4. REFERENCES

[1] B. Ager, W. Mühlbauer, G. Smaragdakis, and S. Uhlig. Comparing DNS resolvers in the wild. In *Proc. of IMC*, 2010.

[2] C. Contavalli, W. van der Gaast, S. Leach, and E. Lewis. Internet-draft: Client subnet in DNS requests, 2012. `http://tools.ietf.org/html/draft-vandergaast-edns-client-subnet-01`.

[3] C. Huang, D. A. Maltz, A. Greenberg, and J. Li. Public DNS system and global traffic management. In *Proc. of IEEE INFOCOM*, 2011.

[4] R. Khosla, S. Fahmy, and Y. C. Hu. Content retrieval using cloud-based DNS. In *Proc. of IEEE Global Internet Symposium*, 2012.

[5] namebench. http://code.google.com/p/namebench/.

[6] The Global Internet Speedup. A Faster Internet. `http://www.afasterinternet.com`.

[7] M. Walfish, H. Balakrishnan, and S. Shenker. Untangling the Web from DNS. In *Proc. of USENIX NSDI*, 2004.

Scalable Software Defined Network Controllers

Andreas Voellmy
Yale University
New Haven, CT, U.S.A.
andreas.voellmy@yale.edu

Junchang Wang
University of Science and Technology of China
Hefei, China
wangjc@mail.ustc.edu.cn

ABSTRACT

Software defined networking (SDN) introduces centralized controllers to dramatically increase network programmability. The simplicity of a logical centralized controller, however, can come at the cost of control-plane scalability. In this demo, we present McNettle, an extensible SDN control system whose control event processing throughput scales with the number of system CPU cores and which supports control algorithms requiring globally visible state changes occurring at flow arrival rates. Programmers extend McNettle by writing event handlers and background programs in a high-level functional programming language extended with shared state and memory transactions. We implement our framework in Haskell and leverage the multicore facilities of the Glasgow Haskell Compiler (GHC) and runtime system. Our implementation schedules event handlers, allocates memory, optimizes message parsing and serialization, and reduces system calls in order to optimize cache usage, OS processing, and runtime system overhead. Our experiments show that McNettle can serve up to 5000 switches using a single controller with 46 cores, achieving throughput of over 14 million flows per second, near-linear scaling up to 46 cores, and latency under 200 μs for light loads and 10 ms with loads consisting of up to 5000 switches.

Categories and Subject Descriptors

C.2.1 [**Computer-Communication Networks**]: Network Architecture and Design—*Centralized networks* ; D.3.2 [**Programming Languages**]: Language Classifications—*Applicative (functional) languages, Haskell*; D.1.3 [**Programming Techniques**]: Concurrent Programming; C.5.5 [**Computer System Implementation**]: Servers

Keywords

Software-defined Networking, Haskell, OpenFlow, Multicore

1. INTRODUCTION

Network systems are becoming more feature-rich and complex, and system designers often need to modify network software in order to achieve their requirements. Software-defined networking attempts to move as much network functionality as possible into user-definable software, making more of the network system components programmable. In particular, SDN architectures introduce a centralized control server (controller) to allow potentially dramatically simplified, flexible network programming.

Unfortunately, as the network scales up—both in the number of switches and the number of end hosts—the SDN controller can become a key bottleneck. Specifically, Tavakoli et al. [7] estimate that a large data center consisting of 2 million virtual machines may generate 20 million flows per second. On the other hand, current controllers, such as NOX [2] or Nettle [8], about 10^5 flows per second [7].

Previous work [7, 4] has made use of distributed controllers to achieve scalability, preventing programmers from sharing state at high transaction rates. With McNettle, we demonstrate a highly scalable SDN control framework that executes on shared-memory multicore servers, preserving a simple and natural programming model for controller developers.

2. PROGRAMMING MODEL

McNettle programs consist of a collection of message handlers, one for each switch in the network. The message handlers include a function which is applied whenever a packet-miss message is sent by a switch in the network. The packet-miss function updates *switch-local* and *network* state variables, and may decide on actions to provision flows in the network. Messages from each switch are handled sequentially, while messages from different switches are executed concurrently. As a result, switch-local state can be accessed without synchronization, while access to network state must be typically be synchronized in order to preserve correctness. To simplify programming concurrent global state access, we emphasize the use of *memory transactions*, which allows programmers to delineate sections of code that should execute atomically. Memory transactions are especially appropriate in this domain, since shared state is often accessed in unpredictable ways, depending on dynamically maintained graph data structures. Consider a controller which accepts requests to reserve bandwidth between pairs of hosts. Such a controller must keep track of available bandwidth on each link in the network, and must update these amounts when reservations are fulfilled. The particular links updated for a reservation depend on the network routing, which typically varies dynamically, and is therefore difficult to predict.

Memory transactions allow us to easily implement a concurrent reservation program that uses fine-grained, optimistic concurrency. Figure 1 shows the code to reserve bandwidth along a path, and consists of two functions. *reservePath* ex-

SIGCOMM'12, August 13–17, 2012, Helsinki, Finland.
ACM 978-1-4503-1419-0/12/08.

reservePath capTable amt path
  $= forM\ path\ (\lambda link \rightarrow reserveLink\ capTable\ amt\ link)$
reserveLink capTable amt link =
  do *current* ← *linkCapacity capTable link*
    let *remaining* = *current* − *amt*
    when (*remaining* > 0) *retry*
    updateCapacity capTable link remaining

Figure 1: Bandwidth reservation using STM.

ecutes the *reserveLink* function on each link in a path and *reserveLink* checks if the link has enough available bandwidth and if so updates the link's available bandwidth. These functions execute as part of a transaction, which can be delineated and executed using an *atomically* statement:

atomically (*reservePath capTable amt path*)

McNettle provides a library of concurrent data structures that can be used to track network state and execute efficiently on multicore architectures. These libraries include several different concurrent hash tables that provide different atomicity guarantees and performance. We rely on GHC's STM system [3] to implement memory transactions.

3. IMPLEMENTATION

While McNettle draws on our earlier work on Nettle [8], achieving scaling on ccNUMA systems with tens of cores required redesigning many system components, from high-level programming APIs, to Haskell standard libraries, and Haskell runtime system components. In particular, McNettle avoids contention on sockets and ensures that each message is processed on a single CPU core, reducing inter-core synchronizations to only those required by user-specified controller logic. McNettle's API is redesigned to allow the McNettle implementation to safely reuse various message buffers, allowing the implementation to improve cache behavior and reduce load on the garbage collector, and more than double throughput of McNettle controllers. Furthermore, we eliminated a bottleneck in GHC's runtime system by fully parallelizing the waiting and dispatching of threads on IO devices, and stabilizing the multi-core load balancing algorithm of GHC to prevent excessive cache-thrashing due to repeated work migration. These improvements lead to a 50-fold reduction in latency in our controllers.

4. RESULTS

We evaluated throughput and latency of McNettle using a lightly modified version of the cbench [6] program, which simulates a collection of OpenFlow switches. We ran our controllers on a DELL Poweredge R815 server with 48 cores of AMD Opteron 1.7GHz 6164 processors and 64 GB memory, using 8 1Gbps and two 10Gbps NICs.

Figure 2 shows maximum throughput scaling results for McNettle, Beacon [1], a multithreaded OpenFlow controller platform in Java, and multi-threaded NOX [5], a multi-threaded OpenFlow platform written in C++, on a workload of 1000 switches with a very simple "learning switch" controller. We see that the McNettle controller outperforms the NOX controller for all loads with performance up to over six times of NOX. We see that the McNettle controllers scale up through 46 cores, while NOX controllers scale to 10 cores. Beacon obtains better throughput than McNettle for fewer than 30 cores, but stops scaling at 20 cores and obtains lower peak throughput than McNettle.

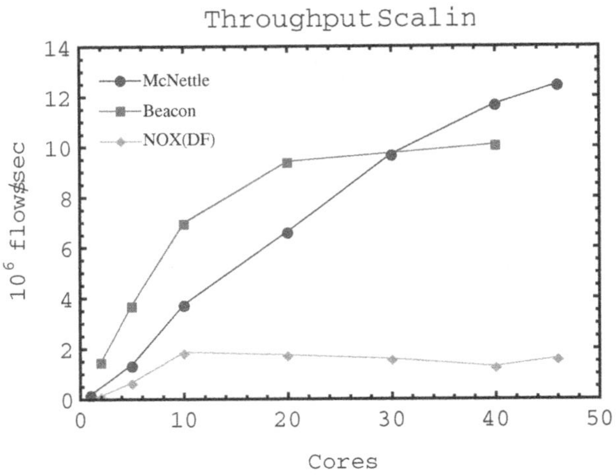

Figure 2: Throughput for McNettle, NOX destiny-fast branch, and Beacon learning controllers as a function of the number of cores.

5. DEMO

We will demonstrate the features of McNettle using example controllers including learning switch controllers, a topology discovery module, a bandwidth-on-demand controller using STM, and a parallelized shortest path routing computation. These controllers demonstrate (1) the basic structure of controllers, (2) switch-local and network state, (3) non-blocking synchronization, (4) transactional memory, and (5) deterministic parallel programming. We will run controllers on a local laptop and on a multicore server running at Yale University.

Acknowledgements This work was supported in part by a gift from Microsoft Research.

6. REFERENCES

[1] D. Erickson. Beacon. `https://openflow.stanford.edu/display/Beacon/Home`, 2012. [Online].

[2] N. Gude, T. Koponen, J. Pettit, B. Pfaff, M. Casado, N. McKeown, and S. Shenker. Nox: towards an operating system for networks. *SIGCOMM Comput. Commun. Rev.*, 38(3):105–110, July 2008.

[3] T. Harris, S. Marlow, S. Peyton-Jones, and M. Herlihy. Composable memory transactions. In *Proceedings of the tenth ACM SIGPLAN symposium on Principles and practice of parallel programming*, PPoPP '05, pages 48–60, New York, NY, USA, 2005. ACM.

[4] T. Koponen, M. Casado, N. Gude, J. Stribling, L. Poutievski, M. Zhu, R. Ramanathan, Y. Iwata, H. Inoue, T. Hama, and S. Shenker. Onix: a distributed control platform for large-scale production networks. In *Proceedings of the 9th USENIX conference on Operating systems design and implementation*, OSDI'10, 2010.

[5] noxrepo.org. Nox destiny-fast. `http://www.noxrepo.org/`, 2012. [Online; git commit e9c3da6bb12ad3fa0d2b609e697a50ce44ca19f4].

[6] R. Sherwood and K.-K. Yap. Cbench. `http://www.openflow.org/wk/index.php/Oflops/`, 2012. [Online; accessed 10-April-2012].

[7] A. Tavakoli, M. Casado, T. Koponen, and S. Shenker. Applying NOX to the datacenter.

[8] A. Voellmy and P. Hudak. Nettle: taking the sting out of programming network routers. In *Proceedings of the 13th international conference on Practical aspects of declarative languages*, PADL'11, pages 235–249, Berlin, Heidelberg, 2011. Springer-Verlag.

RaptorStream: Boosting Mobile Peer-to-Peer Streaming with Raptor Codes

Philipp M. Eittenberger
University of Bamberg, Germany
philipp.eittenberger@ieee.org

ABSTRACT

As mobile devices and cellular networks become ubiquitous, first apps for popular P2P video streaming networks emerge. We have observed that when these applications operate in cellular networks, they don't upload video traffic back to other peers. This paper presents a reason for this behavior and proposes a viable solution to exploit the uplink capacity of mobile devices more efficiently. To the best of our knowledge, this paper is the first to propose the usage of Raptor codes to increase the upload throughput of mobile P2P applications.

Categories and Subject Descriptors

C.2.4 [**Computer-Communication Networks**]: Distributed Systems

General Terms

Design, Measurement, Performance

Keywords

Android, Mobile P2P Streaming, Raptor Codes

1. INTRODUCTION

Recently, client applications of popular P2P video streaming services have been published for mobile devices (e.g. TVU-Player and SopCast). To assess the performance of this new breed of application, we conducted a measurement campaign with the SopCast client that is operating in a High Speed Packet Access (HSPA) network. One of our key findings is that the client doesn't upload any video data back to other peers. To investigate the reasons of this behavior, we have implemented a measurement framework for Android and conducted several measurement runs in the HSPA network of T-Mobile in Bamberg (Germany). Our measurement framework comprises one Android device that either sends packets to one server or receives packets from it. Packet loss is one metric that has been gathered, in particular, we wanted to investigate the influence of the packet inter-arrival time and packet size on the resulting packet loss rate. The packet inter-arrival time is generated by the sender by delaying each packet transmission for a specified time span. Figure 1 illustrates an excerpt of the outcome of one measurement run, for each combination of packet size and inter-arrival time 1,000 UDP packets have been send from a stationary mobile device to the server. Despite HSPA's inherent hybrid ARQ (HARQ) mechanism, the packet loss rate is excessively high (around 60 - 70 % without a clear effect pattern). More importantly, for smaller batch sizes (about 100 packets) the packet loss drops towards zero, which means that for this number of sent packets the underlying HARQ is working. Yet, when batch sizes increase, the Android device is not able to process the retransmission request messages in a timely manner and thereby, a overflow of the UDP buffer happens, which is responsible for the high packet loss. As shown in Figure 1, even relatively large packet inter-arrival times of 20 ms can't redeem this failure. Not surprisingly, the high packet loss negatively affects the achievable uplink throughput. For instance, Figure 2 shows the outcome of one measurement run conducted with our framework to estimate the available instantaneous uplink throughput in relation to different packet sizes. It presents the 25 % quantile of instantaneous TCP uplink throughput measurements for different packet sizes. The instantaneous throughput T is computed by

$$T = \frac{l\,[Byte]}{\Delta t\,[s]}$$

with frame size l and the inter-arrival time Δt between two consecutive packets, such that $\Delta t = t_{i+1} - t_i$. On the receiving side lost packets are ignored when calculating T_r such that $\Delta t^r = t^r_{i+1} - t^r_i$ is the difference of two successively received packets. Due to buffering effects on the path between the stationary measurement server and the Android client, it can be observed that sometimes the 25 % quantile of the received instantaneous throughput values is higher compared to the sending throughput.

Due to retransmissions and large latencies commonly encountered in cellular networks, the uplink TCP throughput suffers and doesn't even reach 128 Kbit/s in this measurement run. We were able to reach under identical conditions a maximum uplink throughput of 480 Kbit/s in our experiments by using UDP as transport protocol. However, it is hardly possible to transmit useful video data in time by UDP due to the high packet loss, because UDP doesn't ensure in-order delivery or retransmissions. It is thus no surprise that the investigated mobile P2P streaming clients didn't upload any video data.

SIGCOMM'12, August 13–17, 2012, Helsinki, Finland.
ACM 978-1-4503-1419-0/12/08.

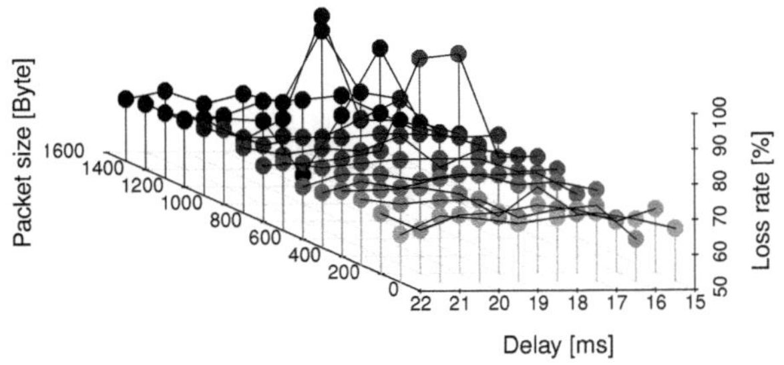

Figure 1: Packet loss in a HSUPA network for different packet sizes and inter-arrival times.

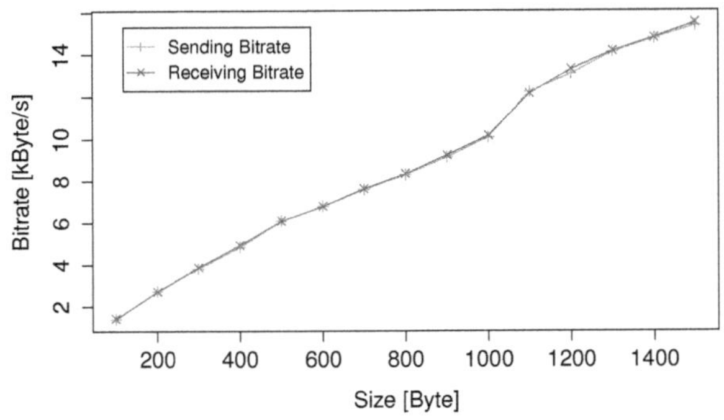

Figure 2: Uplink TCP throughput in a HSUPA network for different packet sizes.

2. RAPTORSTREAM

One possible solution to exploit the uplink capacity of mobile devices in cellular networks is given by the usage of one particular class of forward error correction (FEC) codes. Due to their desirable property of ratelessness, Fountain codes have the ability to generate theoretically an unlimited amount of uniquely encoded data on-the-fly. This property eliminates completely the need for content reconciliation, as no redundant content exists in the network. Fine grained chunk scheduling would not be needed any more, since a receiving peer could restore the original data needing just a slightly larger amount of encoded data from any set of peers.

Raptor codes have been developed by Amin Shokrollahi [4] as an advancement of Tornado codes and represent the first practical class of a Fountain code with near optimal error correction functionality. It is possible to reduce the scheduling complexity of the P2P dissemination and to cope with the packet loss in cellular networks by using Raptor codes as application layer forward error correction (AL-FEC). For further practical details regarding the usage of Raptor codes in the context of P2P streaming, we would like to refer the reader to [2].

As every data unit is uniquely encoded by a Raptor code, any arbitrary subset of encoded data is sufficient to restore the original data, given that a small price, i.e. overhead, has to be paid. Therefore, the usage of Raptor codes in combination with UDP as transport protocol would be a viable option to increase the uplink throughput. We implemented a first base line version of the R10 Raptor code [3] in plain Java. The implementation was not specifically optimized for encoding/decoding speed, e.g. by using a native language like C or GPU instructions. However, it is able to run on every Android certified device without further adjustments due to the pure Java implementation. First performance measurements conducted on a commodity Android device (Galaxy Tab 7 - GT-P1000 with a 1.0 GHz ARM processor) yielded average encoding throughputs of approx. 700 KBit/s and mean decoding throughputs of approx. 800 KBit/s (by using the fastest parameter combinations for encoding and decoding). This performance is already enough to fill the measured available uplink capacity of recent 3.5G networks without further optimizations.

We have already presented RapidStream [1], which is a first prototypical implementation of a mobile P2P video streaming application for Android devices. Right now, we are in the process to incorporate our Raptor code implementation into the streaming engine of RapidStream, which will be then called RaptorStream.

3. CONCLUSION

Our measurements have raised some questions that need further investigations with regard to the future deployment of P2P applications in cellular networks: For instance, at which number of sent packets does the high packet loss emerge? How can that behavior be avoided? In addition, we have proposed the usage of Raptor codes as AL-FEC to exploit the upload capacity of mobile devices and to reduce the complexity of P2P chunk scheduling. In general, Raptor codes require only a small overhead rate and are able to support the necessary streaming rates. For future work, we plan to conduct extensive field trials to evaluate the performance of our proposal after the completion of RaptorStream.

Acknowledgments

The author would like to thank Marcel Großmann for his assistance with the measurements and Todor Mladenov for his guidance through Raptor coding. Furthermore, the author is grateful to Dr. Ingo Dahm of the Deutsche Telekom AG for the provided access to the T-Mobile 3G network.

4. REFERENCES

[1] P. M. Eittenberger, M. Herbst, and U. R. Krieger. RapidStream: P2P Streaming on Android. In *19th International Packet Video Workshop.*, 2012.

[2] P. M. Eittenberger, T. Mladenov, and U. R. Krieger. Raptor Codes for P2P Streaming. In *20th Euromicro International Conference on Parallel, Distributed, and Network-Based Processing.*, 2012.

[3] M. Luby, A. Shokrollahi, M. Watson, and T. Stockhammer. Raptor Forward Error Correction Scheme for Object Delivery. RFC 5053, 2007.

[4] A. Shokrollahi. Raptor codes. *IEEE Transactions on Information Theory*, 52:2551–2567, June 2006.

Enabling Dynamic Network Processing with ClickOS

Mohamed Ahmed, Felipe Huici, Armin Jahanpanah
NEC Europe Ltd

Categories and Subject Descriptors

C.2 [**Computer-Communication Networks**]: General

Keywords

networks, virtualization, xen, minimalistic OS, click

1. INTRODUCTION

Software Defined Networking (SDN) is improving how flexible networks are by allowing their basic components (e.g., switches and routers) to be easily programmable. Most of the work so far has focused on network routing and on managing SDNs; less attention has been paid to the packet processing that is actually done in the network.

The original Internet architecture was based on the end-to-end principle, with packets being forwarded through the network largely unmodified. In the current Internet this model is largely a thing of the past: NATs and firewalls were the first commonly deployed devices that placed layer 4 (or higher) functionality in the middle of the network, not just at the endpoints, and devices such as DPI boxes, rate-limiters, transparent web proxies, application accelerators and many others have followed. In fact, a recent study shows that as many as 33% of paths tested keep state and perform some level of L4+ functionality [2].

What's clear is that network processing, as embodied by such middleboxes, forms a crucial part of today's networks. As a result, any sensible SDN should have mechanisms to not only program the network, but also the packet processing that takes place in it. Such a system should be able to easily cover a wide range of middlebox functionality and be extensible. In addition, it should have the ability to be quickly and dynamically instantiated so that when the network changes, the middlebox processing that takes place in it can change along with it.

Finally, a critical requirement is that the system provide isolation. In a SDN world where slices of networks are given to different entities and users, it becomes increasingly important that network processing from such users that happens to run on common hardware do not affect each other, both from a security and performance point of view.

Towards these requirements we introduce ClickOS, a minimalistic network operating system that runs on top of the Xen hypervisor [1] and that is based on the Click modular router software [3].

SIGCOMM'12, August 13–17, 2012, Helsinki, Finland.
ACM 978-1-4503-1419-0/12/08.

2. SYSTEM DESCRIPTION

Ideally we would like to have an OS that can do middlebox processing without all the overheads of a full-fledged OS like Linux. While several minimalistic OSes exist, their limited driver support would present a difficult obstacle for a network-focused system. ClickOS takes advantage of the best from two worlds: the small overheads of a minimalistic OS (mini-os, which comes with the Xen source code) and the driver support and isolation mechanisms provided by Xen. In the rest of this section we discuss ClickOS's design, including its control and data planes.

2.1 Control Plane

The system consists of a set of ClickOS virtual machines (vms), each composed of Click version 2.0.1 running on top of mini-os (figure 1). In addition, Xen's privileged domain, called dom0, contains the ClickOS CLI and the Xen store. The Xen store keeps control information about running vms, such as their ids and what virtual interfaces they have.

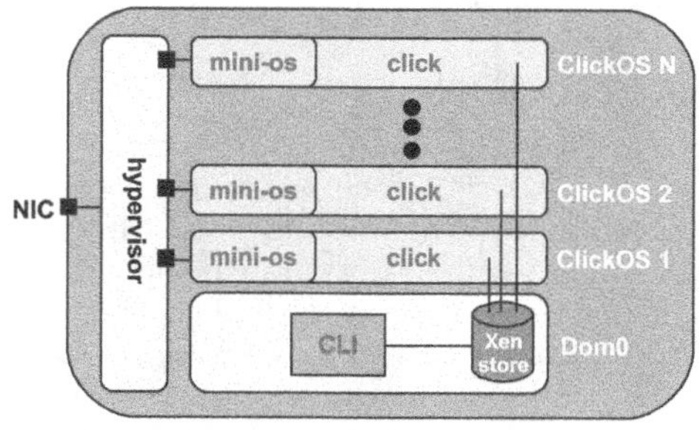

Figure 1: ClickOS Architecture Overview

In addition to these basic parts, ClickOS's control plane consists of two mechanisms. The first is the control thread, a mini-os thread that gets created at system start-up. This thread adds an entry to the Xen store and then watches for changes to it. When a Click configuration string is written to it, it takes care of creating and running a Click instance, where by instance we mean a running Click configuration such as, for example, a firewall or NAT. The second mechanism is a new Click element called *ClickOSControl* which takes care of reading and writing to element handlers from the CLI (an element handler is a Click element internal variable, e.g., a packet count for the element *AverageCounter*).

2.2 Data Plane

Xen has a split driver model, where a netback driver running in a driver domain (usually dom0) talks to hardware devices; and a netfront driver running in a guest domain (e.g., ClickOS) talks to the netback driver via shared memory, more specifically a ring.

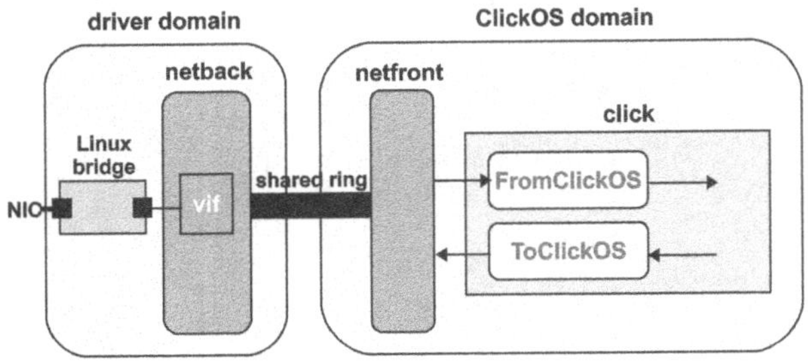

Figure 2: ClickOS Data Plane.

Under a typical Xen set-up, a network card (in our case an Intel X520-T2 dual-port 10Gb card using the ixgbe driver) is linked to a virtual network device called a *vif* via a regular Linux bridge (figure 2). When a packet is received, the Linux kernel hands it to the vif, which queues it at the netback driver. At a later point in time, one of the netback driver threads picks up the packet and puts it on the shared ring, notifying the netfront driver in the process.

In order to receive packets, our modified netfront driver (see section 3) has a dedicated mini-os thread per vif. This thread takes a packet from the shared ring and puts it in the *FromClickOS* element's queue (this is a new element we created that knows how to talk to the netfront driver).

In addition to these per-vif threads, ClickOS creates one mini-os thread for each Click instance. Such a Click thread schedules *FromClickOS* to run, taking the packet from its queue and sending it to the next element down the line. The transmit process is similar but simpler: when scheduled, the *ToClickOS* element takes a packet from an upstream element and sends it directly to the shared ring via the netfront's transmit function.

3. EVALUATION

Image Size: ClickOS is compiled with most of the available Click elements (221/267), the remaining ones requiring a file system to work.[1] The uncompressed ClickOS image is 12MB, the compressed one 2.9MB, and the virtual machine needs a minimum of 5MB to run. This shows ClickOS's small footprint: in a quick test we were able to have as many as 1,010 dummy virtual machines without network devices before errors (not related to memory exhaustion) occurred.
Start-up Times: ClickOS boots quickly. Creating a ClickOS virtual machine and starting a Click instance within in takes roughly 7.9 seconds. Instantiating a Click instance in an existing ClickOS vm takes only 1 second.
Migration Time: ClickOS's small image size and fast boot-time mean that migrating a virtual machine without much state (e.g., only a few forwarding rules for an IP router or a few firewall rules) takes about 7.5 seconds (0.5 to stop, 0.9 to copy, 6.1 to restart the vm).
Netfront Performance: Out of the box, the mini-os netfront driver performs rather poorly. To improve it, we introduce three mechanisms: (1) *poll*: we change the driver's receive function to poll for packets from a dedicated mini-os thread, rather than be interrupt driven; (2) *GRU*: we re-use the grants that receive buffers are given and keep them for the lifetime of the network device (a grant is Xen's way of allowing two domains to share memory); and (3) *FR*: we do fast re-fill of request entries in the shared ring, since the netback driver cannot send packets to the netfront unless requests are available on the shared ring.

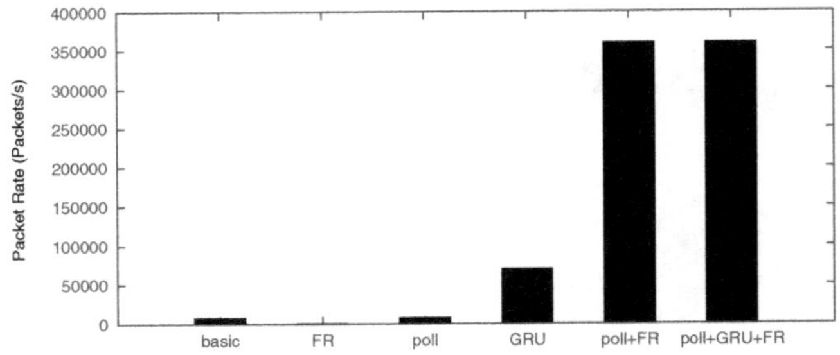

Figure 3: Netfront Performance.

Figure 3 shows netfront receive performance results for maximum-sized packets under different set-ups. As mentioned, without improvements ("basic") the netfront achieves only 8,000 pkt/s, which equates to 96Mb/s, or less than 1% of the 10Gb/s that our system's NIC can achieve. The next three bars are for rates when only one of the three mechanisms is used, and show that an important gain can be had by including only GRU (70,000 pkt/s). Using poll and FR yields the best performance, equivalent to when all three mechanisms are in place. We hypothesize that this is because at this point the bottleneck is somewhere else (possibly the netback driver), and so the gains that adding GRU gives go unnoticed. The final rate of about 360,000 packets per second, equivalent to 4.3Gb/s, represent a 45x increase with respect to the basic netfront driver; this is better than the 2.9Gb/s reported in [4] for a Linux vm.

4. CONCLUSION

We presented ClickOS, a tiny network operating system based on Xen and the Click modular router system. ClickOS's small size and quick start-up and migration times allow for fast and dynamic instantiation of network processing in programmable networks such as SDNs. Further, we have shown that ClickOS provides good performance, receiving packets at rates of 4.3Gb/s on a single ClickOS vm. As future work, we are partly through extending the netfront driver to support multiple vifs per ClickOS vm. We're further working to scale this performance with increasing number of ClickOS vms, as well as seeing what the performance is like when large number of vms are running in the system. We are also in the process of implementing middlebox functionality in ClickOS such as firewalls and carrier-grade NATs.

5. ACKNOWLEDGMENTS

We would like to thank Adam Greenhalgh for the original idea behind ClickOS.

6. REFERENCES

[1] Barham, P., Dragovic, B., Fraser, K., Hand, S., Harris, T., Ho, A., Neugebauer, R., Pratt, I., and Warfield, A. Xen and the Art of Virtualization. In *Proc. ACM SOSP, 2003*.

[2] Honda, M., Nishida, Y., Raiciu, C., Greenhalgh, A., Handley, M., and Tokuda, H. Is it still possible to extend tcp? In *Proc. ACM IMC* (2011).

[3] Kohler, E., Morris, R., Chen, B., Jannotti, J., and Kaashoek, M. F. The click modular router. *ACM Transactions on Computer Systems, August 2000*.

[4] Ram, K. K., Santos, J. R., Turner, Y., Cox, A. L., and Rixner, S. Achieving 10 gb/s using safe and transparent network interface virtualization. In *Proc. ACM VEE, 2009* (2009), VEE '09.

[1] We are in the process of porting a simple file system to increase the number of compiled elements.

Predicting Location Using Mobile Phone Calls

Daqiang Zhang
Nanjing Normal University
Nanjing, 210046, China
dqzhang@njnu.edu.cn

Athanasios V. Vasilakos
National Technical University
Athens, 15780, Greece
vasilako@ath.forthnet.gr

Haoyi Xiong
Institute Mines Telecom
Paris, 91011, France
hyxiong@ieee.org

ABSTRACT

Location prediction using mobile phone traces has attracted increasing attention. Owing to the irregular user mobility patterns, it still remains challenging to predict user location. Our empirical study in this paper shows that the call patterns are strongly correlated with co-locate patterns (*i.e., visiting the same cell tower at the same period*), and the call patterns mainly affect user short-time mobility. On top of these findings, we propose *NextMe* — a novel scheme to enhance the location prediction accuracy by leveraging the social interplay revealed in the cellular calls. To identify when the social interplay will affect user mobility, we introduce the concepts of the Critical Call Pattern (CCP), and the Critical Call (CC). We validate *NextMe* with the MIT Reality Mining dataset, involving 350,000-hour activity logs of 106 persons, and 112,508 cellular calls. Experimental results show that the social interplay significantly improves the accuracy.

Categories and Subject Descriptors

J.4 [**Computer Application**]: Social and Behaviorial Sciences; D.4.8 [**Performance**]: Prediction

General Terms

Algorithms, Experimentation, Measurement

Keywords

Social Networks, Mobile Phone Calls, Social Interplay

1. INTRODUCTION

Mobile devices have become a ubiquitous part of user daily lives such that they capture a great many of mobile traces including user cell IDs, cellular calls, short messages and activity logs. These traces provide insights into user behavior, location and proximity and thus open a new avenue for location prediction.

A variety of location prediction schemes have been proposed in recent years. Some schemes, e.g., PMM [2], foretell user location based on an intuition that the majority of human mobility is periodic among a small set of location. Some schemes make location prediction by the social relationships, e.g., HCMM [1]. However, existing schemes do not clearly tell when the social relationships affect user mobility. Moreover, they rely on other data source than mobile phone traces to identify user relationship, e.g., *extracting friendships from Facebook connections.*

SIGCOMM'12, August 13–17, 2012, Helsinki, Finland.
ACM 978-1-4503-1419-0/12/08.

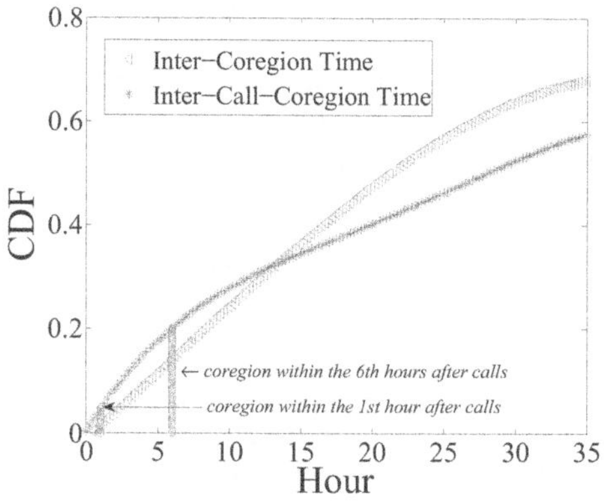

Figure 1: The CDF diagrams

To this end, we investigate the location prediction using the MIT Reality Mining dataset that is one of most popular mobile traces, involving 1,000,000 GSM traces and 112,508 mobile calls [1]. We find that the cellular call pattern between two users is highly correlated with their co-location (*i.e., simultaneously co-locate at the same cell tower*), and it mainly influences user mobility in a short period. User pairs who follow a certain call pattern will co-locate soon after calls with a great probability. Consequently, we introduce the **social interplay** concept to reveal the social relationship embedded in mobile calls, and propose a predictor named ***NextMe*** on top of it. To invoke the *NextMe* predictor, we further introduce the concepts of the Critical Call Pattern (CCP) and the Critical Call (CC).

2. PRELIMINARY FINDINGS

We report our preliminary findings by two questions: 1) does the social interplay exist in mobile calls, and 2) when will the social interplay affect the user mobility. We extract two kinds of events from the MIT Reality Mining dataset — *the mobile call* that denotes a direct call from one use to another, and *the co-locate event (i.e., coregion event)* that stands for two users visiting the same cell tower at the same period of time.

In order to answer the first question, we present two con-

[1]N. Eagle, PNAS'2009, http://reality.media.mit.edu/

cepts — the inter-coregion time and the inter-call-coregion time. The former concept is the interval between two consecutive co-locate events, whereas the later one denotes the interval between the cellular call time and the next co-locate time. Their CDFs are illustrated as Fig. 1. When the prediction time varies from 1 to 12, the CDF of inter-call-coregion time is higher than that of the inter-coregion time. This figure has two important implications. Firstly, the social interplay as one kind of social relationship does exist in mobile calls. Secondly, it has significant impact on user short-time mobility. This indicates that *the social interplay will drive user to move for co-location within next few hours after critical mobile calls.*

Regarding the second question, we plan to investigate what kinds of cellular call patterns would coincide with the co-locate event. We find a certain Cellular Call Pattern (CCP), which makes a call pair have an immediate co-locate event with a very high likelihood. Let n be the number of cellular calls occurred during the last co-locate event and the current call time and λ be the interval between the last two mobile calls. Apparently, all n follows the Gaussian distribution. Thus, we leverage the Gaussian Parameter Estimation to get its confidence interval $[n_1, n_2]$ with the confidence level as 0.95. For the same reason, we obtain the confidence interval $[\lambda_1, \lambda_2]$ for the parameter λ.

DEFINITION 1. *A Critical Call Pattern (CCP) between two successive co-locate events is a call sequence that satisfies: 1)* $n_1 \preceq n \preceq n_2$*, and 2)* $\lambda_1 \preceq \lambda \preceq \lambda_2$*.*

DEFINITION 2. *Given a CCP, the Critical Call (CC) refers to the last cellular call.*

With the CCP and the CC, we can easily detect the moment social interplay might affect user mobility. Note that there may be other call patterns such that the call pair will co-locate soon after calls.

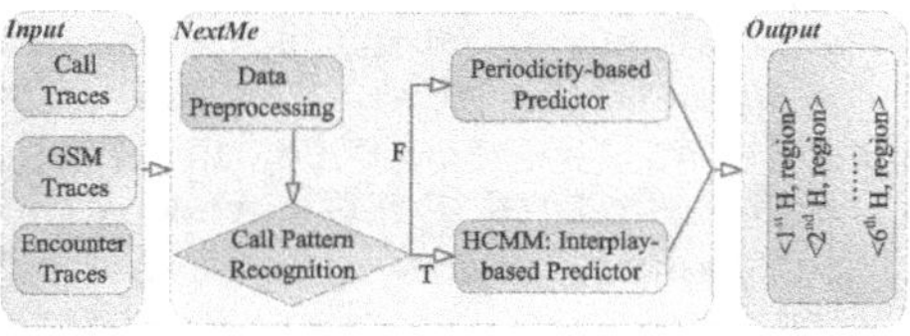

Figure 2: The designed architecture

3. NEXTME SCHEME

In this section, we propose *NextMe* scheme that leverages both user behavior regularity and social interplay to predict user location at cell tower level.

Fig. 2 illustrates the system architecture of the proposed scheme, comprising *data preprocessing, call pattern recognition, the periodicity-based predictor PMM* [2] and *the social interplay-based predictor HCMM* [1]. The call pattern recognition module identifies when the social interplay-based predictor works. By Definition 1, it will recognize the CCP. By Definition 2, it easily identifies the CC that triggers the social interplay-based predictor. Once a CCP is detected, *NextMe* will predict user location using the social interplay, otherwise using the behavior regularity.

Specifically, *NextMe* collects the inter-call-coregion time of a given call pair from the Reality dataset. Then, it obtains

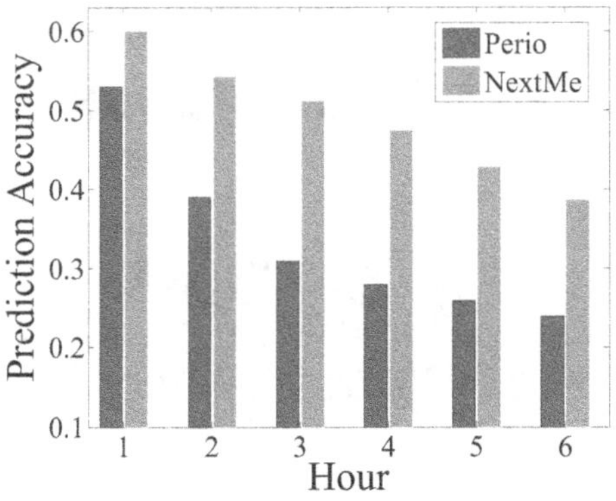

Figure 3: The overall performance of the *PMM* and *NextMe*

the Gaussian distribution of the inter-call-coregion time by parameter estimation. It computes the PDF of the inter-call-coregion time. According to the time that CCs occurred and the PDF, *NextMe* infers the moment that the call pair will co-locate.

Given a call pair (u_x, u_y), let τ be the prediction duration (*i.e., its value as one of 1 to 6 hours*), T_i be the ith call in the nth successive calls in a CCCP, $K^{u_x,u_y}(\tau)$ is a likelihood function for the τth prediction hour, and G_{u_x,u_y} be the Gaussian function. *NextMe* obtains the coregion probability $P_{u_x,u_y}(\tau)$ for user pair u_x and u_y at the τth hour as Formula 1.

$$\begin{gathered}\arg\max_{\tau} P_{u_x,u_y}(\tau) \text{ and } \tau \in \{1, 2, \ldots, 6\} \\ P_{u_x,u_y}(\tau) = P^{1-6}_{u_x,u_y} \bullet \frac{K_{u_x,u_y}(\tau)}{\sum_{\tau \in \{1,2,\ldots,6\}} K_{u_x,u_y}(\tau)}\end{gathered} \quad (1)$$

, where $K_{u_x,u_y}(\tau)$ is computed as Formula 2:

$$K_{u_x,u_y}(\tau) = \prod_{1 \le i \le n} G_{u_x,u_y}(T_n - T_i + \tau), \tau \in (1, 6) \quad (2)$$

We select the PMM as the baseline, and the aggregated accuracy as the metric that is computed as the average prediction accuracy of all participants. Fig. 3 illustrates the overall performance of *NextMe*, where the x-axis is the prediction hour after calls. It shows that *NextMe* achieves higher accuracy than *PMM* in all predictions. This implies that the social interplay dramatically affects user short-time mobility and contributes 13% on average to accuracy.

4. CONCLUSIONS

In this paper, we have uncovered the social interplay revealed in mobile calls. We further propose *NextMe* scheme on top of user behavior regularity and social interplay. Initial results show that the social interplay has a significant affect on user short-time mobility.

5. REFERENCES

[1] C. Boldrini and A. Pass. Modelling spatial and temporal properties of human mobility driven by users' social relationships. *Comp. Comm.*, 33:1056–1074, 2010.

[2] E. Cho et al. Friendship and mobility: user movement in location-based social networks. In *Proc. of the SIGKDD*, pages 1082–1090, 2011.

SmartDiet: Offloading Popular Apps to Save Energy

Aki Saarinen, Matti Siekkinen, Yu Xiao, Jukka K. Nurminen, Matti Kemppainen
Aalto University, School of Science, Finland
aki@akisaarinen.fi, {matti.siekkinen, yu.xiao, jukka.k.nurminen}@aalto.fi,
matti.kemppainen@iki.fi

Pan Hui
Deutsche Telekom Labs, Berlin, Germany
pan.hui@telekom.de

ABSTRACT

Offloading computation to cloud has been widely used for extending battery life of mobile devices. However, little effort has been invested in applying the offloading techniques to communication-related tasks. We propose SmartDiet, a toolkit to identify the constraints that reduce offloading opportunities and to calculate the energy-saving potential of offloading communication-related tasks. SmartDiet traces the method-level application execution and estimates the allocation of communication energy cost from traffic traces. We discuss key features of SmartDiet and show some preliminary results using a prototype implementation.

Categories and Subject Descriptors

C.2.4 [**Computer-Communication Networks**]: Distributed Systems—*Distributed Applications*; D.2.8 [**Software Engineering**]: Metrics—*performance measures*

Keywords

Offloading, energy consumption, constraint analysis

1. INTRODUCTION

A seemingly straightforward way to conserve battery life of a mobile device is to reduce workload by migrating the whole or part of the application execution to a more powerful machine called *surrogate*. This method is called *offloading* or sometimes *cyber foraging*. Several frameworks including MAUI [3], Cuckoo [4], CloneCloud [2], ThinkAir [5], and Scavenger [6] have been used for implementing computation offloading that aims at reducing energy cost by CPU and memory. Although most popular apps today involve intensive communications that consume a significant part of the overall energy consumption on mobile devices, little effort has been put on offloading communication-related tasks and it is unclear whether and in which way such offloading can save energy. To address this issue, we investigate the feasibility of such offloading in this work, taking a set of open source applications as examples. We refer to such offloading as *communication offloading*.

There are two means by which energy savings could be achieved from communication offloading. First is to reduce the amount of the network traffic that needs to be handled by the mobile device. For example, signaling traffic such as keep-alive messages can be partly suppressed. Second is to optimize the pattern shown in the traffic and to improve the overall latency and/or throughput. For instance, packet interval pattern and throughput have a significant impact on communication energy cost[9].

In this study we take the first look into the feasibility of communication offloading through case studies. We use a specific method-level application partitioning framework called ThinkAir [5]. In general, existing frameworks for method-level application partitioning provide APIs for specifying which methods can be possibly offloaded, while requiring programmers with expert knowledge to manually annotate these methods. To better utilize these frameworks, tools that can automatic the partitioning are desperately needed.

The methods that can be marked as offloadable should first be suitable for remote execution. In practice, there are many constraints that limit the remote execution of certain methods. These constraints are non-trivial and laborous to identify manually. Based on case studies of open source apps, we summarize the constraints into three types and develop a toolkit called SmartDiet for identifying such constraints automatically. Our toolkit also provides suggestion of code modifications for releasing certain constraints such as serialization issue in method-level offloading. Furthermore, as communication cost is heavily depedent on traffic pattern, traffic pattern can be shaped through code reconstruction in order to reduce communication cost. To assist developers in making such improvement, our toolkit provides energy estimation at method level. It can be used for evaluating the energy-efficiency and performance of code reconstruction at development stage, and can therefore guide programmers to improve application implementation for better energy-efficiency.

2. SMARTDIET

Based on our experiences in offloading open-sourced popular apps, described in [8], we conclude that for offloading to be feasible in practice, further assistance for developers in profiling and modifying existing programs is needed. We share many concerns that Balan et al. have presented in [1]. Their goal was to enable rapid modification of applications for cyber foraging so that the developer first creates a so-called tactics file corresponding to the program being modified, after which the actual program code is modified. This is still a pretty laboursome process. Our approach is to provide the SmartDiet toolkit to help developers who are using

SIGCOMM'12, August 13–17, 2012, Helsinki, Finland.
ACM 978-1-4503-1419-0/12/08.

Table 1: Constraint statistics for methods within 16 open source apps.

Statistic	Median	Min	Max
Number of methods	431	121	4411
Directly migratable	0.17%	0.00%	3.70%
Migratable with minor changes	15.7%	0.00%	46.8%
Hardware access constraints	14.2%	2.28%	41.3%
Potential unexpected behavior because of access to file system	10.7%	0.00%	30.3%

highly automated offloading frameworks such as ThinkAir and MAUI. SmartDiet comprises two tools: energy profiling and constraint identification. We have designed and implemented prototypes of both. Detailed descriptions are in [8].

Energy profiling tool: This tool finds and visualizes parts of application code that could yield energy savings if offloaded. The tool collects two kinds of information while the program is running: the traffic trace, and a trace of the program execution flow to later produce class and method level statistics for the developer. We use statistical methods to match each collected network packet into one single method call in the program. We apply power models [9] to get detailed time series of energy consumption of the wireless network interface in use. Similar to Eprof [7], the models we use take tail power state into account. Moreover, our power models account for the impact of power consumption differences in each active state during the data transmission. Program execution in each thread can be viewed as a hierarchical call tree, where a method calls another method which calls another and so on. Our tool reconstructs this tree, carrying along the information of the detected network usage. It then aggregates the traffic of the nodes up in the tree, so that the root method, where the execution starts, gets associated with all packets that have been sent or received within each thread.

Constraint identification tool: This tool identifies offloading constraints through static analysis of the source code, determines which methods can be offloaded as such and points out trouble spots in the code. The developer would apply this tool after first profiling the energy consumption of the app to identify the candidate methods for offloading. For each method of the application, it points out problems that can prevent offloading unless the code is modified. The tool currently uses heuristics that identify problems associated with Android platform and Java Serialization API which is used to implement the remote execution of methods. Similar heuristics can be crafted to other remote execution mechanisms, e.g., the Android Parcelable mechanism or .NET serialization.

Hardware constraints: The first set of constrained methods are those that require access to the hardware of the local device. We currently identify method as having this constraint if it tries to show, for instance, notifications to the user, update anything on the screen, vibrate the phone, access the Bluetooth, wifi or usb subsystem, and so on. We identified 20 constrained subsystems while going through Android system APIs.

We ran our constraint analysis tool with a set of programs that are non-trivial in size and include either communication or non-trivial computation. Results are shown in Table 1. Maximum of 3% of methods are directly migratable. SmartDiet can improve the situation by pointing trouble spots in the source code. If all of these issues were fixed, it would enable the migration of 15% to 47% of methods. SmartDiet can also guide the developer into fixing the issues regarding hardware or file system access. The results show that typical existing apps are heavily constrained in terms of what can be offloaded but with a little assistance, developers could leverage offloading much more in practice.

3. CONCLUSIONS AND FUTURE WORK

SmartDiet has shown promising results in helping the development of offloadable code, although some challenges still remain to complete all the features we believe would be useful. We will integrate energy estimator of CPU usage to the same toolkit using existing models. We also study ways to statistically analyze the dependencies between execution of multiple threads and to track packet flow inside a program. Programming styles and application structures are also interesting topics for future work.

4. ACKNOWLEDGMENTS

This work was supported by the Academy of Finland, grant number 253860.

5. REFERENCES

[1] R. K. Balan, D. Gergle, M. Satyanarayanan, and J. Herbsleb. Simplifying cyber foraging for mobile devices. In *MobiSys 2007*, pages 272–285.

[2] B.-G. Chun, S. Ihm, P. Maniatis, M. Naik, and A. Patti. Clonecloud: elastic execution between mobile device and cloud. In *EuroSys 2011*, pages 301–314.

[3] E. Cuervo, A. Balasubramanian, D.-k. Cho, A. Wolman, S. Saroiu, R. Chandra, and P. Bahl. Maui: making smartphones last longer with code offload. In *MobiSys 2010*, pages 49–62.

[4] R. Kemp, N. Palmer, T. Kielmann, and H. Bal. Cuckoo: A computation offloading framework for smartphones. In *Mobile Computing, Applications, and Services*, volume 76, pages 59–79. Springer Berlin Heidelberg, 2012.

[5] S. Kosta, A. Aucinas, P. Hui, R. Mortier, and X. Zhang. Thinkair: Dynamic resource allocation and parallel execution in the cloud for mobile code offloading. In *INFOCOM 2012*, pages 945–953.

[6] M. Kristensen. Scavenger: Transparent development of efficient cyber foraging applications. In *PerCom 2010*, pages 217–226.

[7] A. Pathak, Y. C. Hu, and M. Zhang. Where is the energy spent inside my app?: fine grained energy accounting on smartphones with eprof. In *EuroSys 2012*, pages 29–42.

[8] A. Saarinen, M. Siekkinen, Y. Xiao, J. K. Nurminen, M. Kemppainen, and P. Hui. Offloadable apps using smartdiet: Towards an analysis toolkit for mobile application developers. *CoRR*, abs/1111.3806, 2011.

[9] Y. Xiao, P. Savolainen, A. Karppanen, M. Siekkinen, and A. Ylä-Jääski. Practical power modeling of data transmission over 802.11g for wireless applications. In *e-Energy 2010*, pages 75–84.

Revealing Contact Interval Patterns in Large Scale Urban Vehicular Ad Hoc Networks

Yong Li
Electronic Engineering Dept.
Tsinghua University
Beijing 100084, China
liyong07@tsinghua.edu.cn

Depeng Jin
Electronic Engineering Dept.
Tsinghua University
Beijing 100084, China
jindp@tsinghua.edu.cn

Pan Hui
Deutsche Telekom Lab.
TU-Berlin
Berlin 10587
Pan.Hui@cl.cam.ac.uk

Li Su
Electronic Engineering Dept.
Tsinghua University
Beijing 100084, China
lisu@tsinghua.edu.cn

Lieguang Zeng
Electronic Engineering Dept.
Tsinghua University
Beijing 100084, China
zenglg@tsinghua.edu.cn

ABSTRACT

Contact interval between moving vehicles is one of the key metrics in vehicular ad hoc networks (VANETs), which is important to routing schemes and network capacity. In this work, by carrying out an extensive experiment involving tens of thousands of operational taxis in Beijing city, we find an invariant character that the contact interval can be modeled by a three-segmented distribution, and there exists a characteristic time point, up to which the contact interval obeys a power law distribution, while beyond which it decays as an exponential one. This property is in sharp contrast to the recent empirical data studies based on *Shanghai* vehicular mobility, where the contact interval exhibits only exponential distribution.

Categories and Subject Descriptors

C.2.0 [**Computer-Communication Networks**]: General; C.4 [**Performance of Systems**]: [Modeling techniques, Performance attributes]

Keywords

Vehicular networks, contact interval patterns, mobility trace

1. INTRODUCTION

Urban vehicular ad hoc networks (VANETs) are recognized as an important element of the future intelligent transportation system to alleviate serious problems, such as traffic jams and accidents, as well as to enable new mobile applications to the public [1]. Since urban VANETs are highly mobile and sometimes sparse, it is hard to maintain a connected network to communicate. In order to transfer data from the source to the destination, vehicles need to first wait until they meet some other vehicles, i.e. until they are within the communication ranges of other vehicles. The performance of these networks, which are based on such type of data transmission, depend strongly on the vehicular mobility characteristics, especially on how often such communication opportunities occurs and on how long they last. In this work, we focus on the metric of *contact interval*, which is the time interval between two contacts. It is one of the most important factors directly affecting the amount of data that can be transferred between vehicles when they opportunisticly encounter each other. Therefore, revealing its fundamental properties, which hold in large scale vehicular environment, can provide important hints on the characteristics of mobile nodes behaviors in VANETs, and would greatly benefit the applications design like data dissemination schemes in such network environment.

Until now, there exists no credible model reported on the urban vehicular environment based on large scale realistic vehicular mobility trace. Consequently, the current proposed forwarding protocols can only be investigated based on the results obtained from human mobility [3], which may lead to either over-pessimistic or over-optimistic results in the vehicular environment due to the significant differences between human and vehicular mobilities. Therefore, it is in pressing demand to have a profound understanding of practical constraints imposed by the contact interval on opportunistic vehicular data transmission.

However, contact intervals between highly mobile vehicles are usually casual, and existing most largest vehicular trace of *Shanghai* [2] includes 2,019 operational taxis, based on which Ref. [2] reveals the exponential distribution of contact interval. Since this trace only involves about two thousands of vehicles, it is not enough for studying larger scale vehicular system in modern urban environment which may include tens of millions vehicles, and therefore the obtained result may be not very credible and should vary from different cities. Consequently, a large data record is required to obtain an objective characterization. Due to the costly investment for such experiments, there is no further work studying the contact interval distribution in a realistic large scale urban environment to the best of our knowledge.

2. DATA COLLECTION

In collecting *Beijing* trace, we used the mobility track logs obtained from 27,000 participating *Beijing* taxis carrying GPS receivers and GPRS module shown in Fig. 1 (a) during May of 2010. The reason for us to choose taxis as vehicular devices is that they are more sensitive to urban environments in terms of underlying road topology, traffic control and urban planning, and they have broader cover-

SIGCOMM'12, August 13–17, 2012, Helsinki, Finland.
ACM 978-1-4503-1419-0/12/08.

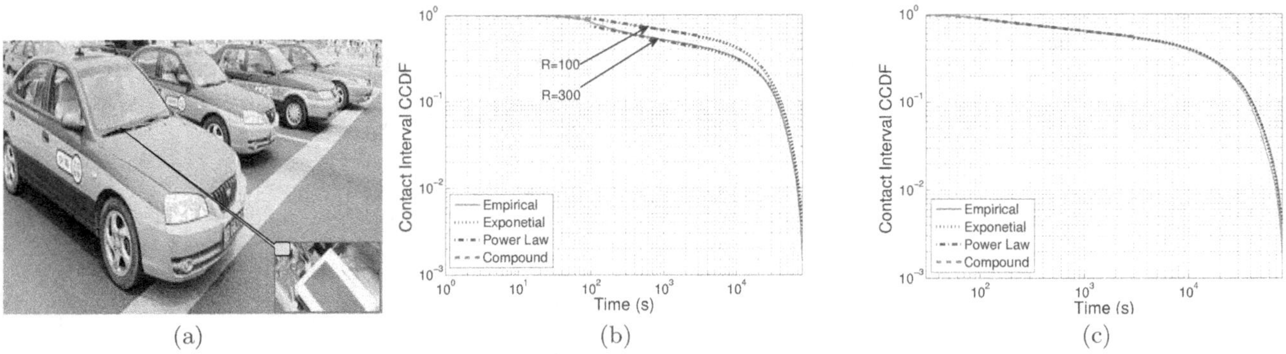

Figure 1: (a) The illustration of deployed vehicular mobility trace collection system. (b)CCDF of contact interval for the whole month; and (c) CCDF of contact interval for each day in a week ($R = 200$).

age in terms of space and operation time than buses and private cars. Specifically, we utilized the GPS devices to collect the taxis locations and timestamps and GPRS modules to report the records every 15 seconds for moving taxis. The specific information contained in such a report includes: the taxi's ID, the longitude and latitude coordinates of the taxi's location, timestamps, instant speed and heading. Our collected *Beijing* trace is the largest vehicular data trace available.

3. MAIN RESULT

To obtain an accurate contact information, we use the method of linear interpolation to extract the exact beginning and ending times of each contact. The aggregated empirical distributions of contact interval between vehicles were examined, and the obtained CCDFs of contact interval, i.e., $P(X > t)$, are plotted using log-log scale in Fig. 1, where Fig. 1(b) present the aggregate contact intervals of all vehicles under different communication ranges of $R = 100$ and $R = 300$ during the whole trace collection time while Fig. 1(c) presents the distribution of aggregate contact interval in the time scale of one day under the communications range of $R = 200$. By observing the distribution property, we divide the curves by three segments as follows, $[0-100]s$, $[100-3000]s$, and $[3000-60000]s$. We use the compound function of $y = b * e^{x^a}$ to fit the first part, and use power law and exponential distribution to fit the middle and tail parts respectively. From the empirical distributions of contact intervals that are plotted by the red curves, it can be observed that between the middle and tail parts, there is a "knee" in the distribution plot, which is referred to as the *characteristic time*. We observe that, the contact interval exhibits a power low distribution before the characteristic time and an exponential tail after the characteristic time, which is in sharp contrast to the results in Ref. [2].

In order to verify the accuracy of the three-segmented distribution model for the contact interval of urban VANETs, we use the respective distributions to fit the first, middle and tail parts, partitioned by the characteristic time, of all the empirical curves in the two figures, respectively. The goodness of fit is measured quantitatively by the R-square statistics, shown in Table 1. It can be seen from Table 1 that the average adjusted R-square statistics is all over 99% for the three parts. This confirms the accuracy of the three-segmented distribution model for contact interval.

Table 1: The adjusted R-square statistics of the exponential, power law and compound fittings.

Transmission range	$R = 100$	$R = 200$	$R = 300$
Exponential Part	99.61%	99.99%	99.99%
Power law Part	99.32%	99.69%	99.46%
Compound Part	99.29%	99.41%	99.68%

4. DISCUSSION AND FUTURE WORK

Different from the existing results found in *Shanghai* vehicular mobility, in our studies, the first and the second part of contact intervals do not obeys exponential distribution, and only the last tail part obeys the exponential distribution. Our future work will address the fundamental questions of what are the essential factors that generates the three-segmented distribution behavior of the contact interval of taxis as well as how to utilize the three-segmented distribution based contact interval model for helping the routing and forwarding protocol design in urban VANETs.

5. ACKNOWLEDGMENT

This work is supported by National Nature Science Foundation of China (No. 61171065, No. 61021001) and Chinese National Major Scientific and Technological Specialized Project (No. 2010ZX03004-002-02), as well as Changjiang Scholars and Innovative Research Team in University (PCSIRT) and Tsinghua National Laboratory for Information Science and Technology (TNLIST).

6. REFERENCES

[1] M. Khabazian, S. Aissa, and M. Mehmet-Ali, "Performance modeling of message dissemination in vehicular ad hoc networks with priority," *IEEE J. Selected Areas in Communications*, vol. 29, no. 1, pp. 61–71, Jan. 2011.

[2] H. Zhu, M. Li, L. Fu, G. Xue, Y. Zhu, and L. Ni, "Impact of traffic influxes: Revealing exponential inter-contact time in urban VANETs," *IEEE Trans. Parallel and Distributed Systems*, vol. 22, no. 8, pp. 1258–1266, Aug. 2011.

[3] Y. Li, Z. Wang, D. Jin, L. Zeng, and S. Chen, "Collaborative Vehicular Content Dissemination with Directional Antennas," *IEEE Transactions on Wireless Communications*, vol. 11, no. 4, pp. 1301–1306, April 2012.

fs-PGBR: A Scalable and Delay Sensitive Cloud Routing Protocol

Julien Mineraud, William Donnelly
Sasitharan Balasubramaniam
Telecommunications Software & Systems Group
Waterford Institute of Technology, Ireland
{jmineraud, wdonnelly, sasib}@tssg.org

Jussi Kangasharju
Department of Computer Science
University of Helsinki, Finland
Jussi.Kangasharju@cs.helsinki.fi

ABSTRACT

This paper proposes an improved version of a fully distributed routing protocol, that is applicable for cloud computing infrastructure. Simulation results shows the protocol is ideal for discovering cloud services in a scalable manner with minimum latency.

Categories and Subject Descriptors

C.2.2 [**Network Protocols**]: Routing Protocols

General Terms

Performance

Keywords

Scalable route discovery, Cloud computing infrastructure

1. MOTIVATION

Cloud computing is recognized as the next generation infrastructure that would revolutionize the way users access services [3]. The driving force behind the concept of cloud computing has been fueled by the growth of computing devices (in particular mobile devices) as well as rich services which execute on power and memory constrained devices. The vision of cloud computing will enable users to offload these services into high powered storage facilities, where users will be able to access infrastructures, platforms, as well as software within the cloud. In essence, the cloud will represent a network of high powered data centers [2] that provides a multitude of services to end users based on a pay as you go model (Figure 1). Although the cloud will enable users to elastically consume resources, this will bring along new challenges for the underlying infrastructure. One of these challenges include a dynamic traffic pattern in the underlying network, which will require more scalable, efficient, and adaptive routing algorithms. In this paper, we propose a scalable routing algorithm called fast search - Parameterised Gradient Based Routing (fs-PGBR) that is fully distributed, scalable and reduces considerable latency during route discovery. The algorithm is an extension to the Parameterised Gradient Based Routing (PGBR) [1], which suffers from scalability performance once the network topology size increases.

SIGCOMM'12, August 13–17, 2012, Helsinki, Finland.
ACM 978-1-4503-1419-0/12/08.

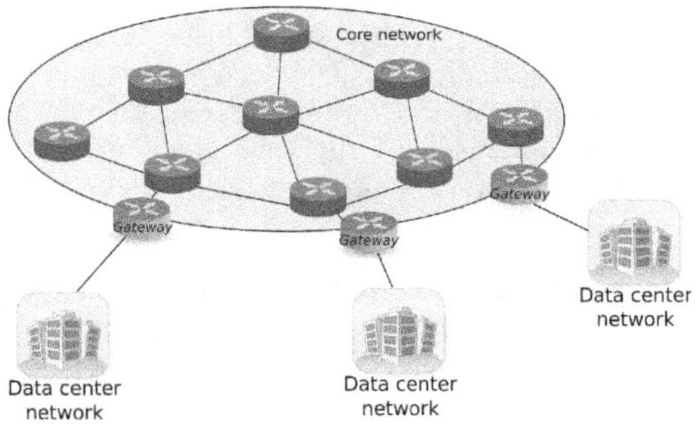

Figure 1: Cloud Computing Infrastructure

2. PROPOSED SOLUTION

2.1 PGBR

The original PGBR algorithm is a bio-inspired algorithm that is inspired from the bacteria motility process known as *chemotaxis*. The process of *chemotaxis*, enables bacteria to mobilize itself towards a destination by sniffing through an attractant gradient field formed in the environment. Mimicking this process, the PGBR algorithm generates an increasing chemical gradient field that attracts packets to a destination node. The equation for generating the gradient field can be represented as:

$$G_i^d(j) = \alpha nl(j) + \beta ll_i(j) + \gamma h_i^d(j) \qquad (1)$$

where $G_i^d(j)$ is the gradient value of link $i \rightarrow j$ for a packet to destination d, $nl(j)$ represents the node load value of neighbour node j, $ll_i(j)$ is the load value of the link $i \rightarrow j$, $h_i^d(j)$ is the normalised hop count neighbour node j of i for destination d, and α, β and γ are weighting parameters. The weighting parameters provides an extra feature of service oriented routing, where different weight sets can correspond to different service types.

The value for $h_i^d(j)$ is calculated during the topology formation, and will remain static. The equation for $h_i^d(j)$ is represented as:

$$h_i^d(j) = 1 - \frac{w^d(j)}{W^d}, \text{ where } W^d = max(w^d(k)), \forall k \qquad (2)$$

where $w^d(j)$ is the minimum distance of node j to destination d.

While PGBR provides numerous benefits that are suitable for the underlying network supporting the cloud infras-

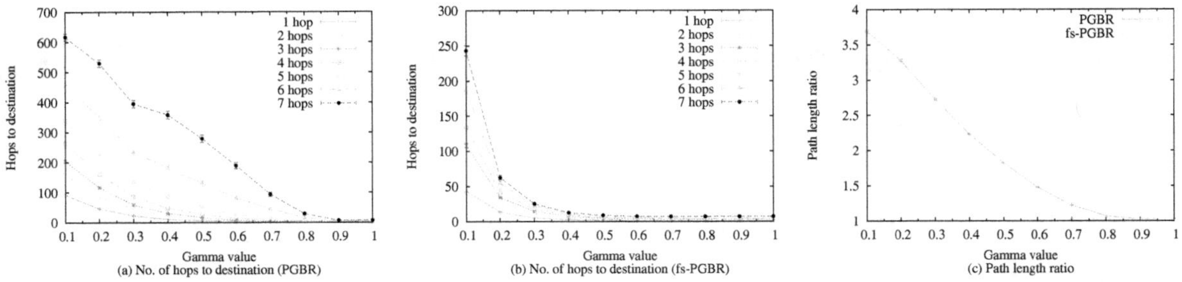

Figure 2: Performance evaluation of the new metric

tructure, there is a slight shortfall. This shortfall is in the approach used to calculate $h_i^d(j)$, which is not scalable once the topology reaches a large number of nodes. In particular, this could be an issue for large-scale cloud infrastructure supported by a network topology with a high number of nodes.

2.2 Hop count value for fs-PGBR

In order to counter the scalability problem, we have modified the $h_i^d(j)$ calculation, and this is represented as:

$$h_i^d(j) = \frac{max(w_i^d(k)) - w_i^d(j)}{max(w_i^d(k)) - min(w_i^d(k))}, \forall k \text{ neighbour of } i \quad (3)$$

This modification enables each node i to compare the shortest path weight of its neighbour k and evaluate it against the maximum and minimum weights of its neighbours. In the case of $max(w_i^d(k)) = min(w_i^d(k))$, the hop count $h_i^d(j)$ is set to 1. Therefore, the new calculation evaluates its hop count not from the relational distance of nodes i and j to the destination d, but rather from the immediate neighboring nodes, and its relation to shortest path. This leads to a perturbation based route discovery process that diverts depending on local conditions of neighboring node (e.g. node or link load).

3. PRELIMINARY EVALUATION

We have evaluated and compared fs-PGBR to the original PGBR discovery mechanism, where simulations were conducted in our custom developed Java simulator. The simulation evaluated two performance metrics based on a real-world large scale topology (i.e. Sprintlink network). The weights that were used to calculate the shortest path were provided by the Rocketfuel project and the load of each link were uniformly selected between 0 and 100% (loads are not symmetric). The simulation scenario generated 100 source-destination pairs for each hop count distance, ranging from 1 to 7 hops. The simulations were reproduced 100 times with different loads, and source-destination pairs selection. For each simulation, we varied the value of γ from 0.1 to 1.0, which is the weighting for $h_i^d(j)$ and reflects the importance of this weight with respect to the shortest path approach. In addition, α and β have similar weights, and set to make the sum of the three weights equal to 1.

As shown in Figure 2, we evaluated the number of hops required to successfully reach the destination as well as the discovered path length ratio (Figure 2 (a) and (b)).

We observed, in Figure 2 (a) and (b) that using the new $h_i^d(j)$ calculation approach leads to minimal discovery time compared to the original approach, in particular for long distance sources. This is due to a number of factors, including minimal discovery rejection that may result from Time-To-Live (TTL) expiration. Consequently, the new extension also enables more diversity in the choice of parameters for fs-PGBR discoveries. Secondly, as shown in Figure 2 (c), the resulting path is maintained under smaller and more reasonable path length ratio compared to the original approach. This demonstrates the effectiveness of fs-PGBR and its suitability towards timely route discovery for cloud computing infrastructures.

4. CLOUD COMPUTING INTEGRATION

Cloud computing requires fast discovery of services that are populated in large scale data centre networks. Once discovered, the paths for the services' must also ensure that Quality-of-Service (QoS) and Quality-of-Experience (QoE) requirements are met. fs-PGBR provides individual routing paths that can adapt to each service's requirements in a fully distributed manner, while minimizing the cost of discovery. In particular, fs-PGBR is ideal for cloud computing infrastructure that are built over large network topologies, and is able to support high service churn rates or varying user demand.

5. ACKNOWLEDGEMENTS

This work has received support from Science Foundation Ireland via "A Biologically inspired framework supporting network management for the Future Internet" starting investigator award (grant no. 09/SIRG/I1643).

6. REFERENCES

[1] S. Balasubramaniam, J. Mineraud, P. Mcdonagh, P. Perry, L. Murphy, W. Donnelly, and D. Botvich. An Evaluation of Parameterized Gradient Based Routing With QoE Monitoring for Multiple IPTV Providers. *IEEE Transactions on Broadcasting*, 57(2):183–194, 2011.

[2] H. Ballani, P. Costa, T. Karagiannis, and A. Rowstron. Towards predictable datacenter networks. In *Proceedings of the ACM SIGCOMM 2011 conference*, SIGCOMM '11, pages 242–253, New York, NY, USA, 2011.

[3] H. T. Dinh, C. Lee, D. Niyato, and P. Wang. A survey of mobile cloud computing: architecture, applications, and approaches. *Wireless Communications and Mobile Computing*, accepted, 2011.

Accelerating Last-Mile Web Performance with Popularity-Based Prefetching

Srikanth Sundaresan, Nazanin Magharei, Nick Feamster, Renata Teixeira
{srikanth.sundaresan,nazanin,feamster}@cc.gatech.edu, renata.teixeira@lip6.fr

Categories and Subject Descriptors

C.2.3 [**Computer-Communication Networks**]: Network Operations—*Network Management*; C.2.3 [**Computer-Communication Networks**]: Network Operations—*Network Operations*

General Terms

Measurement, Performance

Keywords

Broadband Networks, Web performance, Pre-fetching

1. INTRODUCTION

As broadband speeds increase, latency is becoming a bottleneck for many applications—especially for Web downloads. Latency affects many aspects of Web page load time, from DNS lookups to the time to complete a three-way TCP handshake; it also contributes to the time it takes to transfer the Web objects for a page. Previous work has shown that much of this latency can occur in the last mile [2]. Although some performance bottlenecks can be mitigated by increasing downstream throughput (*e.g.*, by purchasing a higher service plan), in many cases, latency introduces performance bottlenecks, particularly for connections with higher throughput.

To mitigate latency bottlenecks in the last mile, we have implemented a system that performs DNS prefetching and TCP connection caching to the Web sites that devices inside a home visit most frequently, a technique we call *popularity-based prefetching*. Many devices and applications already perform DNS prefetching and maintain persistent TCP connections, but most prefetching is predictive based on the content of the page, rather than on past site popularity. We evaluate the optimizations using a simulator that we drive from traffic traces that we collected from five homes in the BISmark testbed [1]. We find that performing DNS prefetching and TCP connection caching for the twenty most popular sites inside the home can double DNS and connection cache hit rates.

2. POPULARITY-BASED PREFETCHING

Figure 1 shows the design of our system, which we have implemented and deployed on BISmark. We augment `dnsmasq`, a caching DNS resolver, to maintain a *popular domain list* that users in the home network resolve most frequently, and `polipo`, a HTTP proxy, to maintain a *popular connection list* of domains users visit. A helper script refreshes the DNS entries for expired domains, and maintains open connections to servers in the connection list by periodically sending dummy `GET` requests to each site. (The prototype ensures that neither of these actions induce a positive feedback loop that keeps unpopular domains in the popular domains list.) The following parameters affect the tradeoff between page load time and overhead:

- *Sizes of the popular domains and popular connections list.* The number of domains for which the system actively prefetches DNS records and the number of sites to which it maintains active connections. The router maintains this list using an (LRU) cache.
- *Domain and connection timeout thresholds.* If no device in the home network looks up a domain in the popular domain list within a *domain timeout threshold*, the system removes the domain from the list and no longer actively prefetches the DNS records for this domain. Similarly, if no HTTP request occurs within a *connection timeout threshold*, the system no longer keeps TCP connections alive for that domain. The system maintains only one active connection per domain.

Deploying these optimizations on the router itself, rather than relying solely on browser-based optimizations, offers several benefits, since: (1) not all devices and browsers may implement these optimizations; (2) users and devices across a household may have common browsing activity.

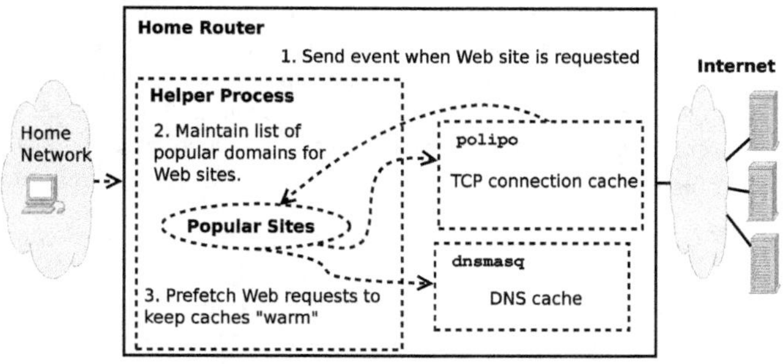

Figure 1: *Augmenting the home router to automatically prefetch DNS records and maintain TCP connections to popular Web sites.*

3. EVALUATION

We analyze the effects of popularity based DNS prefetching and connection caching using a trace-driven discrete event simulator that models the optimizations as described in the previous section.

3.1 Simulator Setup

We evaluate the benefits of popularity-based prefetching using a simulator driven by traces from five homes in the BISmark testbed [1]. These traces contain all DNS lookups and HTTP connection requests (anonymized) across all devices. Table 1 shows

SIGCOMM'12, August 13–17, 2012, Helsinki, Finland.
ACM 978-1-4503-1419-0/12/08.

	Home 1	Home 2	Home 3	Home 4	Home 5
			Statistics		
People	1	3	1	2	1
Days	81	75	88	103	24
DNS Lookups	159K	185K	82K	219K	60K
New HTTP transactions	450K	254K	146K	347K	99K
			Traffic Properties		
Unique Lookups	23	5	38	9	20
Unique HTTPs	32	8	23	10	14
DNS Int-req.	840s	450s	850s	180s	502s
Non-simultaneous HTTP intervals	210s	63s	135s	72s	340s
Non-simultaneous HTTP requests	59%	58%	37%	60%	63%

Table 1: *Properties of traces we have analyzed in our simulations.*

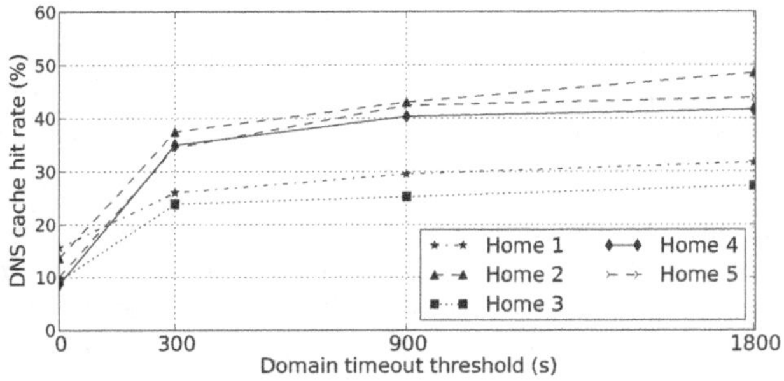

Figure 2: *Daily DNS cache hit ratio with DNS prefetching when the popular domain list size is 20.*

the characteristics of the homes we study. Unique lookups and HTTPs are the median number of unique domains and new HTTP connections every hour. DNS inter-request time is the median inter-arrival time between DNS requests. Non-simultaneous HTTP interval is the median time between the end of one HTTP flow and the start of the next to the same server, for non-overlapping flows. Non-simultaneous HTTP requests is the percentage of HTTP flows that do not overlap with the previous request to the same server.

DNS caching and prefetching The simulator maintains the DNS cache using the timestamp, domain, and TTL of resolved DNS responses from traffic logs. When prefetching a domain, the simulator introduces a random delay between 50 and 200 ms; the trace determines the TTL that we use for each DNS record. We consider DNS requests for domains that wait for a response as cache misses.

TCP connection caching The simulator maintains the TCP connection cache based on the timestamp, five-tuple, and duration of both the entire TCP connection and its connection establishment of all port 80 requests from the passive traces. For each new request, we update the popularity of the corresponding IP address and compare it with the entries in connection cache. In the event of a cache miss, or when a connection to the same IP address is still in use by another request, we establish a new connection, with the flow and TCP connection establishment durations that correspond to the TCP connection in the trace. Upon a cache hit, the simulator denotes the cache entry as in use for the duration of TCP connection, minus the connection establishment time.

3.2 DNS prefetching

The size of the popular domains list and the value of the domain timeout threshold determine the benefits and the overhead of DNS prefetching. The overhead is the ratio of the number of prefetched DNS requests to the total number of DNS cache hits. First, we study how the domain timeout threshold affects the DNS cache hit ratio and overhead when the size of popular domain list is 20.

Figure 2 shows the results of our experiment. The median daily DNS cache hit ratio without prefetching varies from 7% to 15% depending on the home, but setting a prefetching threshold of 30 minutes results in a cache hit ratio of about 30–50%. By default, the prefetching overhead varies from 5 to 150, but applying the backoff algorithm described in Section 2 reduced the prefetching overhead to less than 10 for all homes, with less than a 2% reduction in the hit ratio. The DNS cache hit ratio and prefetching overhead depend on traffic characteristics in each home, such as number of concurrent Internet users, diversity of the looked up domains, inter-arrival times of of the requests, browser, and distribution of TTLs across domain names. Larger DNS request interarrival times decrease the DNS cache hit ratio. Homes with higher prefetching overhead have domains with TTL values as low as one second. To test the sensitivity of our results to the domain list size, we repeated the experiment for sizes ranging from 10 to 320. Sizes beyond 20 results in only a marginal improvement in cache hit ratio in most cases.

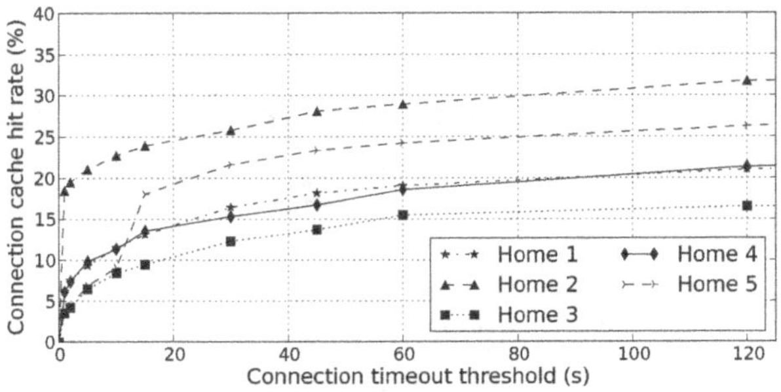

Figure 3: *Daily TCP connection cache hit ratio when the popular connection list size is 20.*

3.3 TCP connection caching

We explore whether popularity-based TCP connection caching can improve the cache hit ratio. We use a popular connection list size of 20. We define overhead as the number of keepalive packets required to keep the TCP connections to the server open.

We find that maintaining an active connection cache of popular connections can improve connection cache hit ratio of as much as 30%, even by maintaining no more than one active connection to each server. Figure 3 shows that hit ratio of the connection cache can improve by up to 30%, even for connection timeout values of as small as two minutes. For a timeout of two minutes, the connection hit ratios are 16–32%, compared with a baseline of 4–19% across homes. The overhead varies from two to nine keep-alives per cached connection per day with a timeout value of two minutes. Increasing the timeout to larger than one minute yields only marginally better connection hit ratio. Even small threshold values result in significant improvements because HTTP requests tend to be quite bursty. Due to the low diversity in HTTP requests, we observe only a 2% improvement in daily connection cache hit ratio for popular connection list size of larger than 20.

Acknowledgments This project is supported by NSF CNS-1059350 and a Google Focused Research grant.

REFERENCES

[1] BISMark: Broadband Internet Service Benchmark. `http://projectbismark.net/`.

[2] S. Sundaresan, W. de Donato, N. Feamster, R. Teixeira, S. Crawford, and A. Pescapè. Broadband internet performance: A view from the gateway. In *Proc. ACM SIGCOMM*, Toronto, Canada, Aug. 2011.

First Insights from a Mobile Honeypot

Matthias Wählisch*, Sebastian Trapp*, Christian Keil†, Jochen Schönfelder†,
Thomas C. Schmidt‡, Jochen Schiller*,
*Institut für Informatik, Freie Universität Berlin, Germany
†DFN-CERT Services GmbH, Hamburg, Germany
‡Dept. Informatik, Hamburg University of Applied Sciences, Hamburg Germany
{first.last}@fu-berlin.de, {last}@dfn-cert.de, t.schmidt@ieee.org

ABSTRACT

Computer systems are commonly attacked by malicious transport contacts. We present a comparative study that analyzes to what extent those attacks depend on the network access, in particular if an adversary targets specifically on mobile or non-mobile devices. Based on a mobile honeypot that extracts first statistical results, our findings indicate that a few topological domains of the Internet have started to place particular focus on attacking mobile networks.

Categories and Subject Descriptors

C.2.0 [**Computer-Comm. Networks**]: General—*Security and protection (e.g., firewalls)*

General Terms

Security

Keywords

Mobile vs. non-mobile attacks, mobile honeypot

1. INTRODUCTION

Common attacks on Internet devices start with a connection to a random or particular TCP/UDP port. An adversary attempts to perform a denial of service (DoS) attack or to overcome system barriers with the intend to gain unauthorized access. The attack gets more effective when the attacker exploits a specific vulnerability of the target.

Mobile phones are particularly threatened by attacks. Their limited hardware resources allow for easy DoS disruptions, and the local system protection is less mature as compared to notebooks or PCs. However, it still remains an open question whether typical adversaries conduct context-specific port scans for intrusions.

In this paper, we report on a mobile honeypot, which collects external requests to an end device connected by a mobile operator. We compare the measurements of more than one month with the logs of four probes that are attached to different (non-mobile) ISPs. Our analysis reveals that a mobile device on average suffers from the same amount of attacks as a home network device. However, the distribution of attacks across autonomous systems, as well as the number of attackers from the same ASes is significantly more pronounced and could indicate Internet regions that specialize on attacking mobiles.

Current studies focused on the identification of attacker-friendly ASes [3] or the network-level behaviour of spammers [2] but did not differentiate between mobile and stationary access. Work in the field of mobile honeypots [1] deals with the secure *implementation* of honeypots on smartphones. We are not aware of any attack analysis based on a mobile honeypot.

SIGCOMM'12, August 13–17, 2012, Helsinki, Finland.
ACM 978-1-4503-1419-0/12/08.

2. MOBILE HONEYPOT

2.1 Background & Design

A honeypot is a trap for collecting data from unauthorized system access—in this analysis via IP—initiated by remote parties. Its intention is to learn about the nature and the characteristics of attacks. The novel term *mobile* honeypot is not yet well-defined. It is used for a probe that either (a) resides on a mobile device, (b) is running on a mobile operating system, or (c) is operated in the network of mobile devices. We argue that remote attacks are bound to the network layer and moreover do not focus on specifics of mobile hardware, but solely target at the system level.

For the mobile honeypot, we build our subsequent analysis of monitoring attacks on a Linux-based system that is connected to a mobile operator network for two reasons: First, a major part of currently deployed smartphones runs Android, which makes this platform appropriate for observations. However, current OS fingerprinting tools such as Nmap or Xprobe do not reveal specifics of the Android OS, but only report about a Linux system. Second, this approach bears the advantage of fully compatible results across platforms, as we use the identical honeypot tools for both, the mobile and the fixed Internet domain.

2.2 Measurement Setup

We deploy a low-interaction server honeypot based on the standard tools Honeytrap and Dionaea, because we are interested in the statistical analysis of attacks and not in dedicated threats. This mainly concentrates on attacks from the 'background noise', but even more intricated threats require the establishment of a connection to the target.

The honeypot runs at four Linux hosts, each connected via a different network access type: One UMTS network, an open university network, a DSL home network, and a darknet. At each site, the honeypot listens at a single *public*

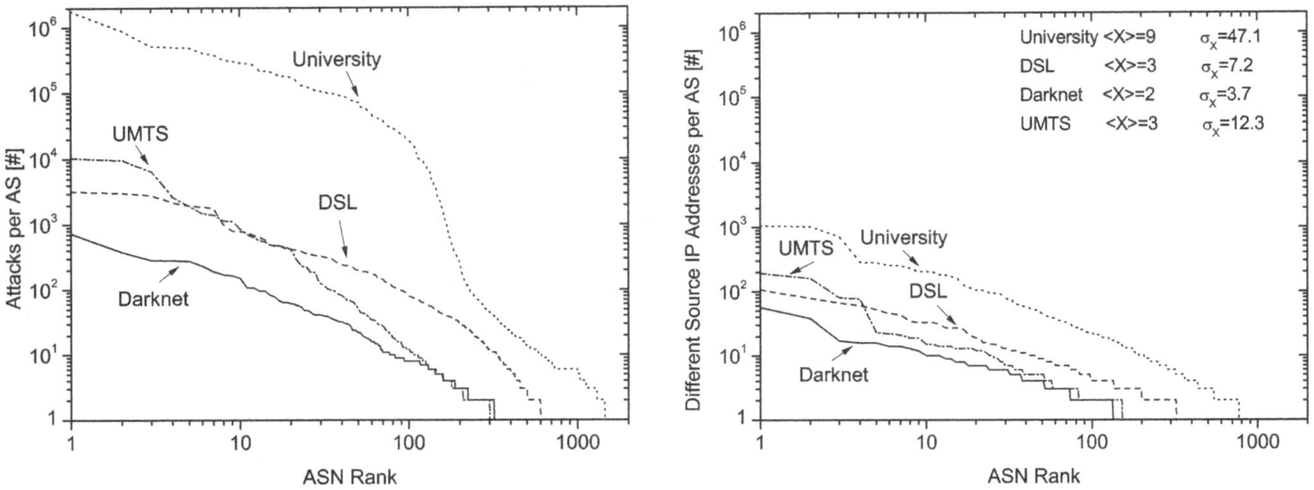

(a) Attacks per AS

(b) Attackers per AS

Figure 1: Comparing the amount of requests per autonomous system separately ranked per network access

IP address. Every external IP connect is considered as an attack, its source IP address is called the attacker.

3. PRELIMINARY RESULTS

The subsequent results represent measurements from mid December 2011 to the end of January 2012.

General observations Most of the attacks target at the (open and stable) university host and the minority at the darknet node. The DSL and UMTS probes measure on average 46 bogus requests per hour. Between 93% and 99% of the attacks use TCP with ports 22, 80, and 222 among the top-ten. Only 35 IP source addresses have been seen on all four monitors.

Attacks per AS We map the source IP addresses of the attackers to their origin autonomous systems (AS) and rank the ASes by the number of attacks. It is worth noting that the ranking is conducted separately for each type of network access. For example, AS 23650 and AS 8402 are ranked first in case of the UMTS and university network, respectively.

In general, most of the attacks have been initiated from the same small set of ASes (cf., Fig. 1(a)). The top-5 ASes mainly originate from China and Russia over all providers. It is clearly visible that these few ASes have a more pronounced impact on the mobile regime. In general, the distributions of attacks among ASes is of similar shape for the darknet and the home network, while the university encounters an enhanced and widened distribution of attacks. In contrast, attack statistics from the mobile network are significantly narrowed.

Attackers per AS We further quantify the number of different source IP addresses per AS to evaluate the number of distinct adversaries (cf., Fig. 1(b)). The AS ranking is conducted, again, separately. This measure balances individual intensities of attackers and consequently reduces the maximum values by some orders of magnitude. Nevertheless, the characteristic shape of the curves of Fig. 1(a) becomes even more significant. Attacks on the UMTS network remain significantly more concentrated to specific ASes than those of the fixed networks.

4. DISCUSSION

This paper presented first ideas towards a better understanding of the nature of mobile-specific attacks. Our preliminary results for an UMTS-connected device show that on the overall offenders show an intensity similar to home networks, whereas regions and originators of attacks are better pronounced and operate at higher intensity. This could be an indication of specific topological regimes that start to focus on mobile attacks.

We admit that any IP-level analysis is biased due to the problem of spoofed source addresses. However, identifying the spoofing of active networks at the end system is fuzzy, as well. Estimating the error is part of our future work. On the other hand, most of the addresses belong to the same IP range/origin AS (cf., also [2]) and thus do not affect our observations. In the future, we plan to perform more subtle correlation analysis of how specific groups of attackers behave with the aim to identify individual patterns of mobility-related aggressions. We will also analyse attacks per port in more detail. Due to limited statistics, though, these considerations will require a much longer range of observation.

Acknowledgements

This work is supported by the German BMBF within the project SKIMS (http://skims.realmv6.org).

5. REFERENCES

[1] MULLINER, C., LIEBERGELD, S., AND LANGE, M. Poster: HoneyDroid - Creating a Smartphone Honeypot, 2011. Poster at IEEE Security & Privacy.

[2] RAMACHANDRAN, A., AND FEAMSTER, N. Understanding the Network-level Behavior of Spammers. In *Proc. of ACM SIGCOMM'06* (New York, NY, USA, 2006), ACM, pp. 291–302.

[3] SHUE, C. A., KALAFUT, A. J., AND GUPTA, M. Abnormally Malicious Autonomous Systems and their Internet Connectivity. *IEEE/ACM Trans. Netw. 20*, 1 (2012), 220–230.

uvNIC: Rapid Prototyping Network Interface Controller Device Drivers

Matthew P. Grosvenor
University of Cambridge Computer Laboratory
matthew.grosvenor@cl.cam.ac.uk

Categories and Subject Descriptors

D.4.4 [**Operating Systems**]: Communications Management – *Network communication*

General Terms

Design, Experimentation, Verification.

Keywords

Hardware, Device Driver, Emulation, Userspace, Virtualisation

1. INTRODUCTION

Traditional approaches to NIC driver design focus on commodity network hardware, which exhibit slow moving feature sets and long product life cycles. The introduction of FPGA based network adapters such as [1][2] alter the status-quo considerably. Whereas traditional ASIC based NICs may undergo minor driver interface revisions over a timespan of years, FPGA based NIC interfaces can be totally reimplemented in months or even weeks. To the driver developer this presents a considerable challenge: Driver development cannot seriously begin without hardware support, but is now expected to take place simultaneously with hardware development.

To solve this problem, I present the userspace, virtual NIC framework (uvNIC). uvNIC implements a custom virtual NIC as a standard userspace application. To the driver developer, it presents a functional equivalent to a physical device. Only minor modifications are required to switch a uvNIC enabled driver over to operating on real hardware. To the hardware designer, uvNIC presents a rapid prototyping environment for initial specifications and a fully functional model against which HDL code can later be verified.

2. Design and Implementation

Typical NIC device drivers implement two interfaces; a device facing PCI interface and kernel facing network stack interface. Ordinarily, a device driver would send/receive packets by interacting with real hardware over the PCI interface. Instead of (or addition to) regular PCI operations, uvNIC forwards interactions with hardware to the uvNIC virtual NIC application. This application implements a software emulation of the hardware NIC and responds appropriately by sending and receiving packets over a commodity device operated in raw socket mode.

Implementing the uvNIC PCI virtualisation layer is not trivial. OS kernels are designed with strict one way dependencies. That is, userspace applications are dependent on the kernel, the kernel is dependent on the hardware. Importantly, the kernel is not designed for, nor does it easily facilitate dependence on userspace applications. For the uvNIC framework, this is problematic. The virtual NIC should appear to the driver as a hardware device, but to the kernel it appears as a userspace application.

SIGCOMM'12, August 13–17, 2012, Helsinki, Finland.
ACM 978-1-4503-1419-0/12/08.

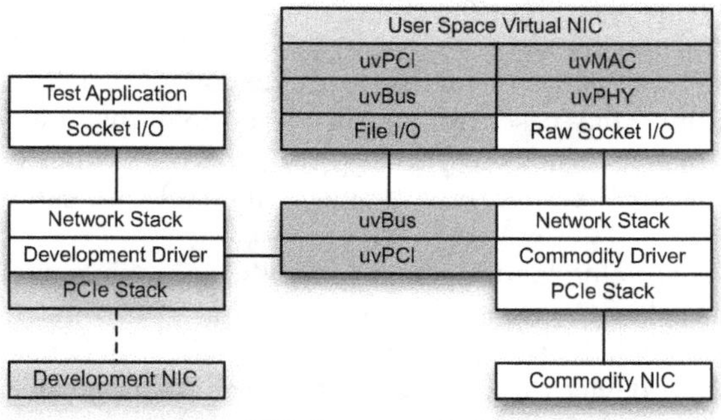

Figure 1: The uvNIC framework design.

Figure 1. illustrates the uvNIC implementation in detail. At the core is a message transport layer (uvBus) that connects the kernel and the virtual device. uvBus uses file I/O operations (`open()`, `ioctl()`, `mmap()`) to establish shared memory regions between the kernel and userspace. Messages are exchanged by enqueuing and dequeueing fixed size packets into lockless circular buffers. Message delivery order is strictly maintained. uvBus also includes an out of band, bi-directional signalling mechanism for alerting message consumers about incoming data. Userspace applications signal the kernel by calling `write()` with a 64 bit signal value, likewise, the kernel signals userspace by providing a 64 bit response to `poll()`/ `read()` system calls.

A lightweight PCIe like protocol (uvPCI) is implemented on top of uvBus. uvPCI implements posted (non-blocking) write and non-posted (blocking) read operations in both kernel and userspace. In kernel space, non-posted reads are implemented by spinning and kept safe with timeouts and appropriate calls to `yield()`. An important aspect of uvPCI is that it maintains read and write message ordering in a manner that is consistent with hardware PCIe implementations.

In addition to basic PCI read and write operations, uvPCI implements x86 specific PCIe restrictions such as 64 bit register reads/writes, message signalled interrupt generation and 128B, 32bit aligned DMA operations. DMA operations appear to the driver as they would in reality. That is, data appears in DMA mapped buffers asynchronously without the driver's direct involvement.

In practice, uvPCI implements a functionally equivalent, parallel implementation of the PCI stack. Driver writers perform little more than a search/replace and a recompile to switch over to using the real PCI stack.

3. Related Work

Userspace networking has a long history [3][4] and continues to be employed widely, especially in high performance situations [5][6]. Whilst uvNIC shares the basic concept of implementing a part of the network stack in userspace, it is distinct from previous attempts because it implements the network hardware in userspace software rather than parts of network driver or IP stack as is commonly the case. It is crucial to note that network performance is not the primary goal of uvNIC, rather, the primary goal is rapid prototyping at the software/hardware interface.

The structure and function of uvNIC is highly similar to the virtual NICs found in hypervisors and virtual machines (VMs). Both VMware [6] and Xen [7] expose virtualised hardware devices to their guest OSes and hence the guest OS drivers. It is possible that custom hardware could be designed and written in a VM context instead of uvNIC. However, the approach has the distinct disadvantage that development and integration of a new virtual device into a VM is a complex and time consuming task. This is in direct opposition to the stated goal, which is which is to aid rapid prototyping of the NIC and device driver.

uvNIC is also similar to File System in User Space (FUSE) [9] systems. Like uvNIC, FUSE requires that kernel become dependent on userspace applications. In contrast, however, FUSE systems are simpler than uvNIC because no direct emulation of hardware timing, ordering and consistency parameters are required.

4. Preliminary Results

A uvNIC virtual network device has been implemented on a Linux 2.6 host. The virtual device was tested with simple linux tools such as `ping` and `traceroute` and found to be functional. This test confirmed that the uvNIC framework is suitable for writing simple, but functional network devices and device drivers. Since performance was not a consideration, `iperf` tests were not performed.

Additionally, a uvNIC driver has been written for and tested against a NetFGPA 10G card [2] which was running a simple register and MSI interface firmware module. This test confirmed that the uvNIC framework is capable of writing simple device drivers that are portable to real hardware platforms.

Efforts are currently underway to port the feature rich Intel IXGBE 10G driver to a uvNIC virtual device using an Intel e1000 1G as a physical device. This comprehensive test will demonstrate uvNIC's ability to augment simple physical hardware with additional virtual functionality.

5. Conclusions and Next Steps

uvNIC is a simple but unique approach to a real problem. Work so far has indicated that the approach is valid and functional. Future steps may include migrating the virtualised hardware model to a cycle accurate simulation of hardware RTL code or using the framework as a measurement and debugging tool for real device drivers. uvNIC affords designers a unique opportunity to rapidly explore the NIC software/hardware interface at low cost. It is hoped that this ability will lead to faster and more stable NIC designs.

More information on the uvNIC project can be found at http://www.cl.cam.ac.uk/research/srg/netos/mrc/projects/uvNIC

6. Acknowledgements

This work was jointly supported by the EPSRC INTERNET Project EP/H040536/1 and the Defense Advanced Research Projects Agency (DARPA) and the Air Force Research Laboratory (AFRL), under contract FA8750-11-C-0249. The views, opinions, and/or findings contained in this article/presentation are those of the author/presenter and should not be interpreted as representing the official views or policies, either expressed or implied, of the Defense Advanced Research Projects Agency or the Department of Defense.

7. REFERENCES

[1] John W. Lockwood, Nick McKeown, Greg Watson, Glen Gibb, Paul Hartke, Jad Naous, Ramanan Raghuraman, and Jianying Luo. 2007. NetFPGA--An Open Platform for Gigabit-Rate Network Switching and Routing. In *Proceedings of the 2007 IEEE International Conference on Microelectronic Systems Education* (MSE '07). IEEE Computer Society, Washington, DC, USA, 160-161.

[2] NetFPGA 10G Project, *NetFPGA website*, http://www.netfpga.org

[3] T. von Eicken, A. Basu, V. Buch, and W. Vogels. 1995. U-Net: a user-level network interface for parallel and distributed computing. *SIGOPS Oper. Syst.* Rev. 29, 5 (December 1995), 40-53.

[4] Nanette J. Boden, Danny Cohen, Robert E. Felderman, Alan E. Kulawik, Charles L. Seitz, Jakov N. Seizovic, and Wen-King Su. 1995. Myrinet: A Gigabit-per-Second Local Area Network. *IEEE Micro15*, 1 (February 1995), 29-36.

[5] Ian Pratt, Keir Fraser. 2001. Arsenic: A user-accessible gigabit ethernet interface. *Proceedings of the Twentieth Annual Joint Conference of the IEEE Computer and Communications Societies,* INFOCOM01.

[6] David. Riddoch, Steven. Pope. 2008. OpenOnload, A user-level network stack. *Google Tech Talk,* http://www.openonload.org/openonload-google-talk.pdf

[7] Jeremy Sugerman, Ganesh Venkitachalam, and Beng-Hong Lim. 2001. Virtualizing I/O Devices on VMware Workstation's Hosted Virtual Machine Monitor. In *Proceedings of the General Track: 2002 USENIX Annual Technical Conference*, Yoonho Park (Ed.). USENIX Association, Berkeley, CA, USA, 1-14.

[8] Aravind Menon, Alan L. Cox, and Willy Zwaenepoel. 2006. Optimizing network virtualization in Xen. In Proceedings of the annual conference on USENIX '06 Annual Technical Conference (ATEC '06). USENIX Association, Berkeley, CA, USA, 2-2.

[9] Miklos Szeredi. 2012, *File System in User Space.* http://fuse.sourceforge.net

Policy Transformation in Software Defined Networks

Nanxi Kang, Joshua Reich, Jennifer Rexford, David Walker
Princeton University
{nkang, jreich, jrex, dpw}@cs.princeton.edu

ABSTRACT

A Software Defined Network (SDN) enforces network-wide policies by installing packet-handling rules across a distributed collection of switches. Today's SDN platforms force programmers to decide how to decompose a high-level policy into the low-level rules in each switch. We argue that future SDN platforms should support automatic transformation of policies by moving, merging, or splitting rules across multiple switches. This would simplify programming by allowing programs written on one abstract switch to run over a more complex network topology, and simplify analysis by consolidating a policy spread over multiple switches into a single list of rules. This poster presents our ongoing work on a sound and complete set of axioms for policy transformation, to enable rewriting of rules across multiple switches while preserving the forwarding policy. These axioms are invaluable for creating and analyzing algorithms for optimizing the rewriting of rules.

Categories and Subject Descriptors

C.2.m [**Computer-Communication Networks**]: Miscellaneous

Keywords

software defined networks, network virtualization, OpenFlow

1. INTRODUCTION

The Software Defined Network (SDN) paradigm enables flexible programming of packet-switched networks. Openflow [2] provides a key SDN building-block by defining a standardized protocol for a controller to interact with switches: installing *rules* for forwarding and modifying packets, querying traffic counters, and directing packets to a *controller* for further processing. Today's controller applications manage the network at the level of individual switches. Yet, a truly powerful SDN system should enable reasoning about the rules installed across *multiple* switches, and their cumulative effect on packets traversing the network.

The goal of our work is to address this challenge by building machinery for *global policy transformation*. Global policy transformation operates by moving, merging, and/or splitting rules across multiple switches, while preserving the overall forwarding behavior of a network. Using this machinery, policies written for a single switch can be distributed across a set of switches, as shown in Figure 1, and policies already distributed over multiple switches can be consolidated. This is important for several reasons:

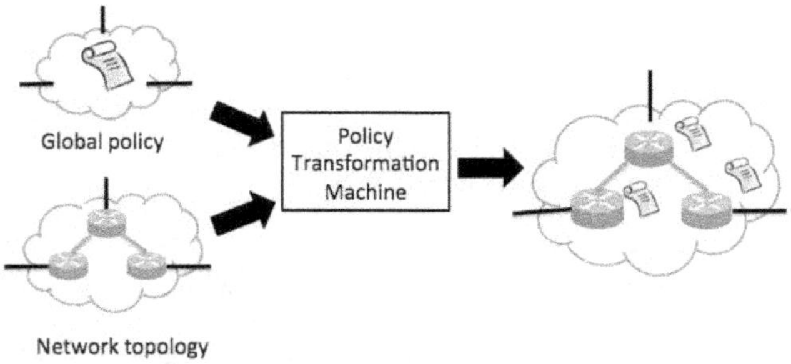

Figure 1: Policy transformation.

- Resource usage: Policies too large to fit on any one switch can be spread over multiple switches.
- Ease of programming: Policies for a single switch are easier for programmers to write.
- Portability: Policies written for one network topology can be transformed to run on another.
- Analysis: A policy spread over multiple switches can be transformed into a single list of rules, to more easily check that the policy satisfies invariants.

Our goal is to enable SDN controllers to transform policies automatically, shielding the programmer from the details.

For many years, researchers have studied firewalls represented as a prioritized list of access-control rules that permit or deny traffic, with the goal of checking whether two policies are equivalent, or to minimize the number of rules. For instance, Appelgate *et al.* [1] show that minimizing a single access control list is NP hard, and present an optimal algorithm for one particular case of access-control lists, while Yuan *et al.* [4] analyze the correctness of distributed firewalls. Other related work [3] develop general algorithms to optimize a single list. Our work is the first to examine the general problem of network-wide *policy transformation*.

We identify two critical challenges with respect to *policy transformation*. The first is *correctness*: each packet `p` must be treated under a transformed policy `A'`, exactly as it would under the original policy `A`. The second is that of *efficiency*: transformed policies should minimize the total number of rules, minimize the maximum number of rules on any switch, or, in the case of a network of switches with different capacities, find a way to pack rules in to differ-

SIGCOMM'12, August 13–17, 2012, Helsinki, Finland.
ACM 978-1-4503-1419-0/12/08.

ent switches so that no switch overflows its capacity. We plan to address these problems of correctness and efficiency by developing a general-purpose policy rewriting framework that characterizes all possible, sound policy transformations. This framework will provide a formal foundation that can be used to analyze the correctness of specific policy rewriting algorithms.

2. POLICY TRANSFORMATION

We begin our exploration of global network policy optimization by analyzing two categories of network policy: the single switch policy and the chain-of-switches policy. Other topologies are certainly interesting, but as the topology grows in complexity, so too grows the complexity of the transformation system. We continue by discussing the rewriting axioms that are necessary and sufficient to convert from one policy to another (in the same category). We conclude with a brief mention of our ongoing work to use these rewriting systems to develop efficient and correct deterministic optimization algorithms.

2.1 Policy Categories

A policy is a collection of packet-forwarding rules where each rule r is a $(pattern, action)$ pair. Typical patterns include exact-match patterns, wildcard patterns and prefix-match patterns. Since any pattern p will match some set of packets (perhaps a singleton set), we write $p \subseteq p'$ when p matches a smaller set of packets than p'. Typical actions include forward, drop, and modify actions.

A *single-switch policy* is a list $l = r_1, r_2, \ldots, r_n$ of rules. The rules appear in the list with decreasing priority. Therefore, to compute the action for a packet h in a rule list l, one can linearly scan l from left to right until hitting a rule (p, a) where the packet header h matches p. In such a case, the action a is applied to the packet.

A *chain-of-switches policy* c is organized as a list of single-switch policies (*i.e.*, $c = l_1; l_2; \ldots; l_m$). Switch chains can be used to model both cascading flowtables within a switch as well as a set of switches placed in series.

In general, when describing a generic policy we use an upper-case letter P. The semantics of a policy P relative to a packet h, written $P[h]$, is the set of observable actions applied by the policy to that packet as it traverses the network.

2.2 Policy Rewriting

Policy rewriting is specified via a collection of simple, local rewriting axioms. These simple local axioms may be composed with one another to produce arbitrary semantics-preserving policy transformations. As an example, consider the Shadow axiom below.

$$\frac{p_2 \subseteq p_1}{(p_1, a_1), (p_2, a_2) \leftrightarrow (p_1, a_1)} \quad (Shadow)$$

One should read the statements above the horizontal line as assumptions and statements below the line as conclusions. Hence, the axiom states that if pattern p_2 is a subset of pattern p_1 then the pair of adjacent rules $(p_1, a_1), (p_2, a_2)$ can be rewritten to the single rule (p_1, a_1), and vice-versa.[1] Such local rewrites may be applied correctly in a large context using the Congruence axiom as follows.

$$\frac{l_2 \leftrightarrow l_2'}{l_1, l_2, l_3 \leftrightarrow l_1, l_2', l_3} \quad (Congruence)$$

The Congruence axiom states that if a local rewrite applies then that local rewrite may be used within the context of a longer list of rules. In addition to Shadow and Congruence, there are a few other axioms for single switch policies including Reordering and Join.

Rewriting rules for switch chains relies on the rewriting axioms for the single-switch case to optimize individual switches in the chain. In addition, there are further axioms for shuffling functionality back and forth along the chain.

2.3 Soundness and Completeness Properties

Any good rewriting system should satisfy two key properties: (1) A rewriting system is *Sound* if any combination of rewriting axioms is guaranteed to produce a semantically equivalent list of rules, and (2) a rewriting system is *complete* if any semantically equivalent policy can be produced by applying the axioms. More precisely:

Definition 1. Soundness: If $P_1 \leftrightarrow P_2$ then for all packet headers h, $P_1[h] = P_2[h]$.

Definition 2. Completeness: If for all packet header h, $P_1[h] = P_2[h]$, then $P_1 \leftrightarrow P_2$.

To date, we have proven our rewrite system for single-switch policies is sound and complete and our rewrite system for switch-chains is sound. In the future, we hope to develop a completeness argument for switch-chain rewriting. We are also looking at richer topologies including trees and DAGs.

2.4 Ongoing Work: Optimization Algorithms

Our rewrite system does not itself constitute a deterministic algorithm for generating optimized policies. Rather, it describes the search space any such algorithm must navigate, and provides a framework within which one may analyze the correctness of such algorithms.

The next phase of the research involves developing multiphase algorithms for generating optimized, global network policies. Our current plan involves three phases: (1) a heuristic algorithm to distribute global policy to particular switches, (2) balancing across neighbors via switch-to-switch rule movement and (3) optimization on individual switches.

3. REFERENCES

[1] D. L. Applegate, G. Calinescu, D. S. Johnson, H. Karloff, K. Ligett, and J. Wang. Compressing rectilinear pictures and minimizing access control lists. In *ACM-SIAM SODA*, pages 1066–1075, 2007.

[2] N. McKeown, T. Anderson, H. Balakrishnan, G. Parulkar, L. Peterson, J. Rexford, S. Shenker, and J. Turner. Openflow: Enabling innovation in campus networks. *SIGCOMM Computer Communications Review*, 38(2):69–74, Mar. 2008.

[3] C. R. Meiners, A. X. Liu, and E. Torng. TCAM Razor: A systematic approach towards minimizing packet classifiers in TCAMs. *IEEE/ACM Trans. Netw.*, 18(2):490–500, Apr. 2010.

[4] L. Yuan, J. Mai, Z. Su, H. Chen, C.-N. Chuah, and P. Mohapatra. FIREMAN: A toolkit for firewall modeling and analysis. In *IEEE Symposium on Security and Privacy*, pages 199–213, 2006.

[1]Note that sometimes an optimization is enabled by first expanding a rule in to multiple rules that are easier to manage or move around.

Perspectives on Network Calculus – No Free Lunch, but Still Good Value

Florin Ciucu
T-Labs / TU Berlin

Jens Schmitt
University of Kaiserslautern

ABSTRACT

ACM Sigcomm 2006 published a paper [26] which was perceived to unify the deterministic and stochastic branches of the network calculus (abbreviated throughout as DNC and SNC) [39]. Unfortunately, this seemingly fundamental unification—which has raised the hope of a straightforward transfer of all results from DNC to SNC—is invalid. To substantiate this claim, we demonstrate that for the class of stationary and ergodic processes, which is prevalent in traffic modelling, the probabilistic arrival model from [26] is quasi-deterministic, i.e., the underlying probabilities are either zero or one. Thus, the probabilistic framework from [26] is unable to account for statistical multiplexing gain, which is in fact the raison d'être of packet-switched networks. Other previous formulations of SNC can capture statistical multiplexing gain, yet require additional assumptions [12, 22] or are more involved [14, 9, 28], and do not allow for a straightforward transfer of results from DNC. So, in essence, there is no free lunch in this endeavor.

Our intention in this paper is to go beyond presenting a negative result by providing a comprehensive perspective on network calculus. To that end, we attempt to illustrate the fundamental concepts and features of network calculus in a systematic way, and also to rigorously clarify some key facts as well as misconceptions. We touch in particular on the relationship between linear systems, classical queueing theory, and network calculus, and on the lingering issue of tightness of network calculus bounds. We give a rigorous result illustrating that the statistical multiplexing gain scales as $\Omega(\sqrt{N})$, as long as some small violations of system performance constraints are tolerable. This demonstrates that the network calculus can capture actual system behavior tightly when applied carefully. Thus, we positively conclude that it still holds promise as a valuable systematic methodology for the performance analysis of computer and communication systems, though the unification of DNC and SNC remains an open, yet quite elusive task.

Categories and Subject Descriptors

C.4 [**Computer Systems Organization**]: Performance of Systems—*Modeling techniques*

General Terms

Theory, Performance

Keywords

Network Calculus, Statistical Multiplexing Gain

1. INTRODUCTION

Queueing theory is an important theory for the performance analysis of resource sharing systems such as communication networks. One of the success stories of queueing theory is Erlang's formula for the computation of the blocking probability that some shared resource is occupied [20]; this formula has been used for nearly a century to dimension telephone networks. Concomitantly, queueing theory has been generalized from Erlang's primordial single queue model with Poisson arrivals and exponential service times to the class of *product-form queueing networks* which can account for multiple service time distributions, scheduling, or routing (e.g., [3, 29]).

Notwithstanding the advances made in the classical branch of queueing theory [25], which is primarily concerned with exact models and solutions, the class of tractable queueing networks is largely constrained by the technical assumption of Poisson arrivals. This apparent limitation has motivated the development of alternative theories to queueing, especially over the past three decades witnessing a rapid growth of high-speed data networks. The relevance of the emerging theories, especially for the Internet community, has become evident with the discovery that Internet traffic is fundamentally different from Poisson [32, 40]. Moreover, as it became clear that improper traffic models can lead to bogus results, the necessity to overcome the Poisson assumption limitation has reached a wide consensus.

One of the alternatives to the classical queueing theory is the network calculus. This was conceived by Cruz [16] in the early 1990s in a deterministic framework, and extended soon after by Chang [11], Kurose [30], and Yaron and Sidi [49] in a probabilistic or stochastic framework. Subsequently, many researchers have contributed to both formulations of the network calculus (see the books of Chang [12], Le Boudec and Thiran [5], and Jiang [28]). While DNC was motivated by the need for a theory to compute deterministic (worst-case) bounds on performance metrics, the raison d'être of SNC was to additionally capture statistical multiplexing gain when some violations of the deterministic bounds are tolerable. This feature enables a much more efficient dimensioning of resource sharing systems, such as packet-switched networks, and continues to play a pivotal role in the evolution of SNC.

The promise of the combined branches of network calculus is to jointly overcome the technical barriers of queueing networks on all fronts. Achieving this rather daunting task is enabled by two key features:

- *Scheduling abstraction.* At a single queue with multiplexed arrival flows, the specific properties of many scheduling algorithms, and also of many arrival classes, can be abstracted away by suitably constructing the so-called *service processes* (technical details are deferred to Section 3).
- *Convolution-form networks.* The service processes from single queues can be *convolved* across a network of queues, and thus a multi-node network analysis can be drastically simplified by reducing it to a single-node analysis.

Equipped with these two features, the network calculus can analyze many scheduling algorithms and arrival classes, over a multi-node network, in a *uniform manner.* That means that the Poisson model, in particular, plays no special role anymore in facilitating the analytical tractability of a whole network. Compared to classical queueing theory, which separately analyzes various combinations of arrivals and scheduling, network calculus conceivably offers a much more simplified and uniform framework. For this reason, the network calculus has been applied in many recent areas such as IntServ [6], switched Ethernets [44], systems-on-chip [10], avionic networks [41], the smart grid [48], etc.

This versatile applicability, however, is only possible at the expense of providing *bounds* on performance metrics. The bounds are a manifestation of resorting to inequalities, whenever exact derivations become intractable. The tightness of the bounds is certainly a major concern, since loose bounds may be more misleading than wrongly fitted Poisson models. The tightness issue has several dimensions depending on the nature of the bounds (deterministic or probabilistic) or the number of flows/queues. Deterministic bounds are generally tight for single queues ([5], p. 27), but can be very loose in some queueing networks with arbitrary multiplexing [43]. Moreover, the deterministic bounds can be very inefficient for network dimensioning when some violation probabilities are tolerable (e.g., running IntServ for many flows could result in very low network utilization). Probabilistic bounds are generally asymptotically tight (in terms of scaling laws in the number of queues) [8, 34], whereas numerical tightness ranges from reasonable [13] to quite loose [34], depending on the arrival model.

In this paper we touch on the lingering issue of tightness, as part of a broader perspective on network calculus. Concretely, we address the asymptotic tightness of probabilistic bounds, in the number of flows N, and demonstrate that such bounds improve upon corresponding deterministic bounds by a factor of $\Omega\left(\sqrt{N}\right)$. That means that, e.g., implementing a probabilistic extension of IntServ could significantly increase the network utilization. Our result not only rigorously reveals the magnitude of the statistical multiplexing gain achieved with SNC, but clearly highlights the fundamental advantage of SNC over DNC.

Our broader goal is to deliver an intuitive and yet comprehensive perspective of the two core concepts in network calculus, i.e., service and envelope processes, by focusing on subtleties and raising awareness of inherent pitfalls. Along the discussion we attempt to make a suggestive statement that there is no free lunch in the framework of the network calculus, yet it brings good value as a companion/alternative to the classical queueing theory. This perspective is motivated by a large effort in the literature to develop SNC formulations which reproduce in particular the 'convolution-form networks' feature from DNC. Arguably the simplest of such formulations has appeared in a Sigcomm 2006 paper [26], and has since raised the hope that DNC results can be transferred into SNC in a straightforward manner. Unfortunately, the formulation from [26] is based on a quasi-deterministic arrival model[1], which roughly means that the proposed SNC cannot capture statistical multiplexing gain. We believe that exposing this pitfall, through a rigorous analysis, is essential to the comprehensive understanding of SNC arrival models.

After introducing notations, the rest of the paper is structured as follows. In Section 3 we provide a comprehensive perspective on service processes by making a multilateral analogy of network calculus with linear systems and classical queueing theory. In Section 4 we present representative envelope processes and elaborate on the quasi-deterministic aspects of the one from [26]. In Section 5 we lead together envelope and service processes in order to shed some light on the often raised concern about the tightness of network calculus bounds. We conclude the paper in Section 6.

2. NOTATIONS

The time model is discrete starting from zero. The time indices are denoted by the symbols i, k, n. The *cumulative* arrivals and departures at/from a (queueing) node up to time n are denoted by non-decreasing processes $A(n)$ and $D(n)$. The doubly-indexed extensions are $A(k,n) = A(n) - A(k)$ and $D(k,n) = D(n) - D(k)$. The associated *instantaneous* arrival and departure processes are $a_n = A(n-1,n)$ and $d_n = D(n-1,n)$, respectively; by convention, $a_0 = d_0 = A(0) = D(0) = 0$. The vector representations are $\mathbf{A} = (A(0), A(1), \dots)$ and $\mathbf{a} = (a_0, a_1, \dots)$ for the arrivals, and $\mathbf{D} = (D(0), D(1), \dots)$ and $\mathbf{d} = (d_0, d_1, \dots)$ for the departures. These processes have primarily a spatial interpretation, e.g., a_n quantifies the number of data units (referred to as bits throughout) arrived at time n; with abuse of notation, $\mathbf{a}$ and $\mathbf{d}$ will also have a temporal meaning to be made locally clear.

The sets of natural, integer, real, and positive real numbers are denoted by $\mathbb{N}$, $\mathbb{Z}$, $\mathbb{R}$, and $\mathbb{R}_+$, respectively; their restriction to non-zero numbers are denoted by $\mathbb{N}^*$, $\mathbb{Z}^*$, $\mathbb{R}^*$, and $\mathbb{R}_+^*$. The integer part of a number $x \in \mathbb{R}$ is denoted by $\lfloor x \rfloor$. For $x \in \mathbb{R}$, the positive part is denoted by $[x]_+ = \max\{x, 0\}$. For some boolean expression E, the indicator function is denoted by 1_E and takes the values 1 or 0 depending whether E is true or false, respectively.

[1]The authors of [9] also point out the quasi-determinism issue in [26], but for a service model and without proof; as a further side remark, we have ourselves experienced a similar quasi-determinism pitfall in SNC [42].

For two functions $f, g : \mathbb{N} \to \mathbb{R}$, the (min, +) convolution operator '$*$' is defined as

$$f * g(n) := \min_{0 \le k \le n} \{f(k) + g(n-k)\} \ \forall n \ge 0 .$$

If the function g is bivariate, i.e., $g : \mathbb{N} \times \mathbb{N} \to \mathbb{R}$, then $f * g(n) := \min_{0 \le k \le n} \{f(k) + g(k, n)\} \ \forall n \ge 0$.

3. SERVICE PROCESSES

Network calculus operates by reducing a 'complex' non-linear (queueing) system into a *'somewhat looking'* linear system. Because the reduced system is often analytically tractable—linearity conceivably implies simplicity—network calculus is regarded as an attractive approach to analyze complex queueing systems. In this section we elaborate on this key reduction operation by exploring conceptual similarities with the more traditional linear system and queueing theories, as well as on its main diverging point from the two. The final goal is to highlight the emergence of the concept of a *service process*, which is instrumental for abstracting away some of the technical challenges characteristic of non-linear systems.

The 'complex' system is a node, or a network of nodes, in which bits arrive and depart according to various factors such as probability distributions for arrival processes, scheduling, routing, etc. A fundamental networking and queueing problem which is at the core of the philosophy of network calculus is the following:

System Identification (SI) Problem: Is it possible to characterize a random process (the departures) based on another random process (the arrivals) while accounting for yet another random process determined by other arrivals, scheduling, routing, etc. (the noise)?

To answer, let us formalize the system by an operator (a.k.a. filter)

$$T : \mathcal{F} \to \mathcal{F}, \ T(\mathbf{a}) = \mathbf{d} ,$$

where $\mathcal{F}$ is the set of discrete-time sequences, i.e., $\mathcal{F} = \{\mathbf{a} = (a_0, a_1, \dots) : a_i \in \mathbb{N}\}$. The physical interpretation of T is that it takes $\mathbf{a} = (a_0, a_1, \dots)$ as input, it accounts for the noise, and outputs $\mathbf{d} = (d_0, d_1, \dots)$. The sequences $\mathbf{a}$ and $\mathbf{d}$ have two networking interpretations, depending on the type of information they quantify:

1. Spatial quantification (SQ): a_n and d_n quantify the number of bits which arrive and depart from the network system at time n.
2. Temporal quantification (TQ): a_n and d_n quantify the arrival and departure times of the nth bit.

The **SI** problem requires thus the construction of T such that for *any* input $\mathbf{a}$, the output $\mathbf{d}$ can be completely determined as $\mathbf{d} = T(\mathbf{a})$. The problem is difficult not only because *all* inputs must be accounted for by a *single* expression of T, but also because T should account for noise and its correlations with output and possibly input as well.

The next two sections, 3.1 and 3.2, present two partial solutions for the **SI** problem by exploiting key properties from linear system and queueing theories, respectively. Then Section 3.3 combines the ideas from these partial solutions into a more general, though approximative solution.

3.1 Linear System Theory

The **SI** problem has a direct correspondent in linear system theory [31]. Assume that T is *linear* and *time-invariant* (LTI), i.e.,

$$\begin{cases} T(c_1\mathbf{a}_1 + c_2\mathbf{a}_2) = c_1 T(\mathbf{a}_1) + c_2 T(\mathbf{a}_2) \\ T(\mathbf{a}_{(-k)}) = T(\mathbf{a})_{(-k)} \end{cases} \tag{1}$$

for all signals $\mathbf{a}, \mathbf{a}_1, \mathbf{a}_2$, scalars c_1, c_2, and integers k. Here we tacitly assume 'signal' interpretations of the input and output sequences, and also their extension to doubly infinite sequences such that the shifted version $\mathbf{a}_{(-k)}$ of $\mathbf{a}$, i.e., $a_{(-k)n} := a_{n-k} \ \forall n, k \in \mathbb{Z}$, is well defined. Define the Kronecker input signal $\mathbf{u}$ (also called impulse signal) and its corresponding output signal $\mathbf{v}$ (also called impulse-response)

$$u_n = \begin{cases} 0 & , \ n \neq 0 \\ 1 & , \ n = 0 \end{cases} , \ \mathbf{v} = T(\mathbf{u}) . \tag{2}$$

The impulse signal $\mathbf{u}$ is (technically) motivated by the convolution property $a_n = \sum_k a_k u_{n-k} \ \forall \mathbf{a}, n$.

With these assumptions, it can be shown that for *any* input signal $\mathbf{a}$, the corresponding output signal $\mathbf{d}$ can be completely determined by the following convolution

$$d_n = \sum_{k=-\infty}^{\infty} a_k v_{n-k} \ \forall n \in \mathbb{Z} . \tag{3}$$

This result is central in linear system theory, as it simply solves for T. This solution for the **SI** problem, however, relies on the strong LTI assumption from Eq. (1).

3.2 Queueing Theory and (min,+) vs. (max,+) Algebras

Here we present another solution for the **SI** problem by inspecting basic queueing properties in a simplified scenario. First, let us readopt one of the networking interpretations of the sequences $\mathbf{a}$ and $\mathbf{d}$.

Consider a single work-conserving server (node) with capacity C and unlimited queueing (buffering) space; all arrivals have the same size of one bit. The construction of the operator T, that completely determines $\mathbf{d} = T(\mathbf{a}) \ \forall \mathbf{a} \in \mathcal{F}$, follows directly from basic queueing dynamics, and is almost analogous to the one from Eq. (3). Under SQ, the output sequence d is determined by

$$d_1 + d_2 + \cdots + d_n = \min_{0 \le i \le n} \{a_1 + \cdots + a_i + C(n-i)\} ,$$

for all $n \ge 1$. It is convenient to represent the partial sums by considering the input and output (*cumulative*) sequences $\mathbf{A} = (A(0), A(1), \dots)$ and $\mathbf{D} = (D(0), D(1), \dots)$, where $A(n) = a_1 + \cdots + a_n$, $D(n) = d_1 + \cdots + d_n \ \forall n \ge 1$, and $A(0) = D(0) = 0$. With these notations, the operator T satisfies $\mathbf{D} = T(\mathbf{A})$, where

$$\begin{aligned} D(n) &= \min\{nC, A(1) + (n-1)C, \dots, A(n)\} \\ &= \min_{0 \le k \le n} \{A(k) + (n-k)C\} \ \forall n \in \mathbb{N} . \end{aligned} \tag{4}$$

In turn, under TQ, $\mathbf{d}$ is determined by

$$\begin{aligned} d_n &= \max\left\{a_1 + \frac{n}{C}, a_2 + \frac{n-1}{C}, \dots, a_n + \frac{1}{C}\right\} \\ &= \max_{1 \le i \le n} \left\{a_i + \frac{n-i+1}{C}\right\} \ \forall n \in \mathbb{N} . \end{aligned} \tag{5}$$

These equations are fundamental elementary identities in queueing theory.

The operations from Eqs. (4) and (5) on input sequences resemble much with the convolution operation from Eq. (3), except for the underlying algebra. Concretely, while Eq. (3) is formed according to the traditional convolution involving the addition of products, Eqs. (4) and (5) are formed by minimizing and maximizing, respectively, sums. For this reason, it is said that the operator T operates in a $(\min,+)$ algebra in Eq. (4), and in a (max,+) algebra in Eq. (5).

We have thus presented another complete characterization of T. What makes this second solution partial as well is that the considered queueing system is noiseless, i.e., it assumes a constant-rate server, no scheduling, etc.

3.3 Emergence of the Service Process Concept

We now combine the ideas from the previous two subsections in order to present a much more general solution for the **SI** problem, and concomitantly to highlight the emergence of one of the key modelling concepts in network calculus: the *service process*.

We tailor the **SI** problem, for some general queueing system, in terms of an *unknown* operator

$$T : \mathcal{F} \to \mathcal{F},\ T(\mathbf{A}) = \mathbf{D}\ , \tag{6}$$

where $\mathbf{A}$ and $\mathbf{D}$ have the interpretations from the previous subsection, i.e., cumulative sequences counting bits. Recall that T has to be constructed in such a way that it completely determines the output $\mathbf{D}$ for *any* input $\mathbf{A}$.

Inspired from the previous two subsections, it is intuitive to reproduce the steps for the construction of an impulse-response from Section 3.1, but in the modified $(\min,+)$ algebra which was shown to be appropriate, in Section 3.2, to represent input-output relationships in queueing systems. This approach could be viewed as a merge between linear system theory and queueing theory.

As in Section 3.1, we first enforce an LTI assumption on T by reproducing Eq. (1) in the $(\min,+)$ algebra:

$$\begin{cases} T\left(\min\left\{c_1+\mathbf{A}_1, c_2+\mathbf{A}_2\right\}\right) \\ \qquad = \min\left\{c_1+T\left(\mathbf{A}_1\right), c_2+T\left(\mathbf{A}_2\right)\right\} \\ T\left(\mathbf{A}_{(-k)}\right) = T(\mathbf{A})_{(-k)} \end{cases} , \tag{7}$$

for all sequences $\mathbf{A}, \mathbf{A}_1, \mathbf{A}_2$, scalars c_1, c_2, and shifts $k \in \mathbb{Z}$ (whether such an apparently strong assumption holds for typical queueing systems will be clarified in two follow-up examples). The second step is to define the analogue of the Kronecker impulse signal and its shifted version in the $(\min,+)$ algebra, i.e.,

$$\delta(n) = \begin{cases} 0 & , \ n=0 \\ \infty & , \ n\neq 0 \end{cases} ,\ \delta_{-k}(n) = \delta(n-k),\ \forall\, n,k \in \mathbb{Z}\ . \tag{8}$$

Similarly to the Kronecker impulse signal, the newly defined impulse sequence $\boldsymbol{\delta}$ is motivated by the fact that any input sequence $\mathbf{A}$ can be expressed as the $(\min,+)$ convolution of itself with the impulse function, i.e.,

$$\begin{aligned} &\mathbf{A} = \mathbf{A} * \boldsymbol{\delta},\ \text{or, equivalently,} \\ &A(n) = \min_{0\leq k\leq n} \left\{A(k) + \delta(n-k)\right\}\ \forall n \in \mathbb{N}\ . \end{aligned}$$

When the input to the system is the impulse $\boldsymbol{\delta}$, define the corresponding output as the impulse-response

$$\mathbf{S} = T(\boldsymbol{\delta})\ . \tag{9}$$

Under the assumption that T is $(\min,+)$ LTI (in the sense of Eq. (7)), it follows from the $(\min,+)$ linear system theory (see [5], p. 136, or [2], p. 276) that for *any* input sequence $\mathbf{A}$ the corresponding output sequence $\mathbf{D}$ satisfies

$$D(n) = \min_{0\leq k\leq n} \left\{A(k) + S(n-k)\right\}\ , \tag{10}$$

where $\mathbf{S} = T(\boldsymbol{\delta})$ is the impulse-response. Therefore, the unknown operator T is now fully characterized: for *any* input sequence $\mathbf{A}$, the corresponding output is $T(\mathbf{A}) = \mathbf{A} * \mathbf{S}$.

Note that despite the cyclic dependence between T and S, i.e., $T(\mathbf{A}) = \mathbf{A} * \mathbf{S}$ and $\mathbf{S} = T(\boldsymbol{\delta})$, T is well-defined. The reason is that the impulse-response $\mathbf{S}$, which induces the cyclic dependence, has a well defined physical meaning as the system's output for the input $\boldsymbol{\delta}$.

The key observation to make, however, is that T is yet another partial solution for the **SI** problem, due to the underlying LTI assumption from Eq. (7). In the following we present two queueing examples which show that, in spite of the fact that many queueing systems are generally *not* $(\min,+)$ linear, the unknown operator T, and thus a solution for the **SI** problem, can be constructed in great generality. Since there is no free lunch, this promising increase in generality is only possible at an inevitable price: sacrificing exactness, or, more concretely, replacing the equality from Eq. (6) by an inequality, i.e., $\mathbf{D} \geq T(\mathbf{A})$.

3.3.1 Example 1

Consider the (noiseless) queueing system from Figure 1.(a) (see next page) and recall the relationship from Eq. (4) between departures and arrivals. Fitting this relationship with Eq. (10) yields $S(n) = nC\ \forall n \in \mathbb{N}$. As mentioned, $\mathbf{S}$ has also the physical interpretation of the cumulative output from the queue when the input is the impulse $\boldsymbol{\delta}$ from Eq. (8).

This example, although repetitive, is meant to illustrate the (almost perfect) analogy in the arguments used in Sections 3.1 and 3.2. Concretely, it is apparent that there is a match between the construction of $\mathbf{S}$ from 1) the $(\min,+)$ linear system theory (as in Eqs. (9) and (10)), and 2) elementary queueing properties (as in Eq. (4)). The missing element for a perfect analogy is that the queueing system is $(\min,+)$ LTI under an artificial interpretation of the required 'plus' property $T(c+\mathbf{A}) = c + T(\mathbf{A})$ from Eq. (7): the addition of scalars c occurs, both in the input and output, before the queueing system actually starts. This is clearly an inconvenient physical system interpretation, but it enables the view of a constant-rate queueing server as a $(\min,+)$ LTI system. As a side remark, the physically more meaningful interpretation of the 'plus' property as a burst of c bits at time zero has the negative consequence that any system of practical interest, in particular the constant-rate work-conserving server, would be $(\min,+)$ non-linear. This can be seen by a result in ([5], Proposition 8.3.1) on a system with non-empty initial buffer clearly exhibiting a $(\min,+)$ non-linear behavior.

3.3.2 Example 2

Making an analogy between $(\min,+)$ LTI systems and elementary queueing properties is even more compounded for the FIFO queueing system from Figure 1.(b), which includes noise in the form of cross-traffic. The reason is that this queueing system is *not* anymore $(\min,+)$ linear, even under the artificial interpretation of the 'plus' operation. First, the 'min' property $T(\min\{\mathbf{A}_1, \mathbf{A}_2\}) = \min\{T(\mathbf{A}_1), T(\mathbf{A}_2)\}$

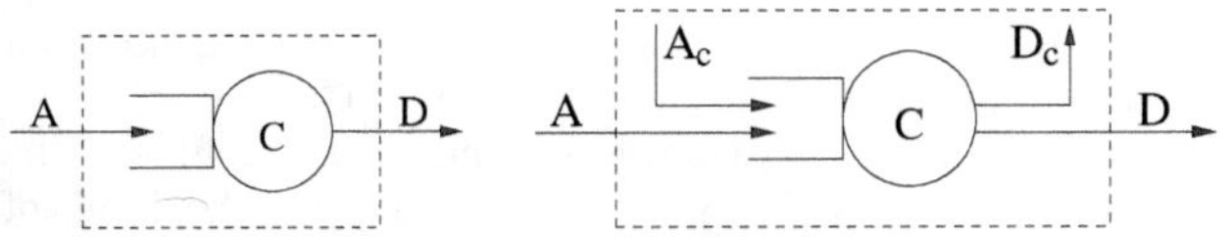

(a) No multiplexing (b) Multiplexing

Figure 1: Two queueing systems from the perspective of flow A: in (a) the flow is isolated, in (b) the flow shares the queue with another (cross) flow $\mathbf{A}_c$

fails; a quick example is $C = 3$, $A_c(t) = (0, 1, 3, 4)$, $A_1(t) = (0, 3, 4, 6)$, and $A_2(t) = (0, 1, 7, 10)$. Second, the time invariance property fails as well; a quick example is $C = 1$, $\mathbf{A} = (0, 1, \ldots)$, $\mathbf{A}_c = (0, 1, 3, \ldots)$, and the right shift $k = 2$. Because the queueing system is not $(\min, +)$ LTI, one cannot follow the construction of the impulse-response sequence $\mathbf{S}$ in order to exactly characterize the queueing system as in Eq. (10).

At this apparent impasse, the network calculus slightly diverges from LTI systems and queueing theory. The key idea is to transform the non-linear queueing system into a 'somewhat looking' linear system. The actual transformation occurs by directly constructing a 'somewhat analogous' impulse-response $\mathbf{S}$ (a.k.a. *service process*) [1] satisfying

$$\mathbf{D} \geq \mathbf{A} * \mathbf{S} \ \forall \mathbf{A} . \tag{11}$$

Therefore, instead of *exactly* characterizing the system as in Eq. (6), the network calculus makes the crucial concession of *inexactly* characterizing the system by resorting to an *inequality*, as in Eq. (11). For the FIFO multiplexing example, one choice of a service process $\mathbf{S}$ is the bivariate random process $S_i(k, n) = [C(n-k) - A_c(k, n-i)]_+ 1_{\{n-k>i\}}$, for some $i \geq 0$ [18], which satisfies

$$D(n) \geq \min_{0 \leq k \leq n} \{A(k) + S_i(k, n)\} \ \forall \mathbf{A}, n \in \mathbb{N} . \tag{12}$$

Except for the inequality, this characterization resembles much with both Eq. (10) (from $(\min, +)$ LTI systems theory) and Eq. (4) (from queueing theory). Note, however, the non-trivial expression of $S_i(k, n)$ stemming from non-trivial characteristics of FIFO multiplexing. In particular, the bivariate form is due to the lack of time invariance.

The FIFO multiplexing example reflects a fundamental tradeoff in network calculus. On one hand, as queueing systems are generally neither linear nor time invariant, network calculus resorts to inequalities for their characterization, as in Eqs. (11) or (12). On the other hand, the service processes $\mathbf{S}$ ought to be reasonably concise and also provide tight bounds in Eq. (11); otherwise, they may render a cumbersome analysis (see, e.g., [33] for the above choice of $\mathbf{S}$) or simply arbitrarily loose bounds. Constructing such 'nice' service processes for existing scheduling algorithms is a key challenge in network calculus; see [36] for some state-of-the-art examples of service processes concerning Δ-scheduling, which generalizes FIFO, static priority, and earliest deadline first (EDF).

Nevertheless, the underlying methodology of constructing service processes to abstract away the details of scheduling algorithms, in queueing scenarios with many flows, renders two central features of network calculus: *scheduling abstraction* and *convolution-form networks*. With the former, many classes of scheduling policies and arrival processes are amenable to a *uniform analysis*. In other words, once service processes are suitably constructed, the network calculus analysis does not conceptually differentiate between, e.g., FIFO and EDF policies, or Poisson and Markov arrival processes, for the purpose of computing *per-flow (or per-class) performance metrics*. With the latter feature, the multi-node queueing analysis is drastically simplified. Concretely, once service processes $\mathbf{S}^i$ are constructed at each node along a network path, the entire network analysis can be reduced to the analysis of a single-node, which is characterized by the following service process

$$\mathbf{S} = \mathbf{S}^1 * \mathbf{S}^2 * \ldots * \mathbf{S}^n , \tag{13}$$

i.e., the convolution of the service processes along the network path. What makes this reduction particularly appealing is that the multi-node performance bounds obtained in this manner are asymptotically tight in the number of nodes (see, e.g., [14, 22]).

In conclusion, network calculus provides a methodology to solve the **SI** problem by transforming a non-linear system (subject to various arrivals, scheduling, or multi-node) into a 'somewhat looking' linear system which is amenable to a quite straightforward analysis. The key challenge is the transformation itself, i.e., the construction of 'nice' service processes.

Most of the interpretations on network calculus illustrated in this section appear in the literature in isolation: for the analogy with linear systems see [19, 12, 5], for the analogy with queueing theory see [27], for a discussion on the non-linearity of FIFO systems see [35]. For a comprehensive survey of service processes we refer to [23]. Our contribution herein was to present a comprehensive perspective on the emergence and central role of service processes in network calculus, by weaving together linear systems, queueing theory, and network calculus.

4. ENVELOPE PROCESSES

We now shift the discussion to the other fundamental concept in network calculus: *envelope processes*. While their role is to model a very broad class of arrival processes, achieving this generality comes at the price of sacrificing exactness in the arrivals' representation. The goal of this section is to highlight the key aspects of envelope processes; it is not meant to provide a review of the types of envelopes, for which we refer to [37].

A (cumulative) arrival process $A(n)$ is typically described by either a complementary cumulative distribution function (CCDF) or a moment generation function (MGF), i.e.,

$$F_{A(n)}(\sigma) := \mathbb{P}\Big(A(n) > \sigma\Big) , \quad M_{A(n)}(\theta) := E\left[e^{\theta A(n)}\right] ,$$

respectively, for all $n \in \mathbb{N}$, $\sigma \in \mathbb{R}$, and $\theta \in \Theta$, where Θ is some space over $\mathbb{R}$. The two descriptions silently assume that $A(n)$ is a stationary random process, which means that the CCDF is invariant under time shift.

The existence of the MGF is equivalent to an exponentially bounded CCDF, in which case the CCDF uniquely determines the MGF according to the identity relating expectations and tails, i.e., $E[X] = \int_0^\infty \mathbb{P}(X > x)dx$ for positive random variable (r.v.) X. Conversely, in the case when Θ is an open interval including zero, the MGF uniquely determines the CCDF according to analytic function theory ([21], p. 274). Throughout we consider arrival processes which have an MGF.

Let us consider the following example of a compound Bernoulli arrival process

$$A(n) = \sum_{k=1}^{n} X_k \ \forall n \in \mathbb{N} \ , \quad (14)$$

where X_k's are i.i.d. Bernoulli(p) r.v.'s taking the values 1 and 0 with probabilities p and $1-p$, respectively. The simplicity of $A(n)$ will enable the illustration of some key insights in an intuitive and yet rigorous manner. The corresponding expressions for the CCDF and MGF are

$$\begin{aligned} F_{A(n)}(\sigma) &= \begin{cases} 1, \ \sigma < 0 \\ \sum_{k=\lfloor\sigma\rfloor+1}^{n} \binom{n}{k} p^k (1-p)^{n-k}, \ \sigma \geq 0 \end{cases} , \\ M_{A(n)}(\theta) &= e^{\theta r n} \end{aligned} \quad (15)$$

where $r = \frac{\log(pe^\theta + 1 - p)}{\theta}$ is a rate and $\theta \in \mathbb{R}_+^*$.

Although the process $A(n)$ is fully described, the network calculus provides more 'flexibility' by offering further modelling alternatives depending on two factors:

1. type of analysis: deterministic (a.k.a. worst-case) or probabilistic. The former seeks to yield statements like "The (queueing) delay is smaller than some number." The latter seeks statements like "The delay is smaller than some number *with some probability*."
2. tradeoff between accuracy and elegance of the analysis itself.

For instance, in the case of a deterministic analysis, one must resort to deterministic models to (partially) suppress the uncertainty of arrivals. The suppression process occurs by replacing random processes with deterministic functions, which are referred to as deterministic envelopes (see Section 4.2). In turn, in the case of a probabilistic analysis, one can carry on with probabilistic arrival models, like CCDF or MGF, throughout the analysis.

A word of caution is in place regarding carrying on with *exact* probabilistic arrival models, like CCDF or MGF. Although they can lend themselves to tight (or possibly exact) bounds on performance metrics, the obtained results and the analysis itself may lack elegance and insight. The CCDF from Eq. (15) conceivably lends itself to quite a messy analysis. As another example, the very broad class of Markov arrival processes can be modelled by an exact MGF expression, as a weighted hyperexponential (see, e.g., [15])

$$M_{A(n)}(\theta) = \sum_{l=1}^{L} w_l e^{\theta r_l n} \quad (16)$$

with L terms, where $\sum w_l = 1$ and r_l's are rates. Carrying on with the entire sum of exponentials can be very cumbersome, and also prone to numerical problems, especially when L is large. Often, a much simplified arrival model consisting of an MGF bound with very few (one or two) dominant exponentials is sufficient with only a negligible loss in tightness. MGF or CCDF bounds are even more appropriate when exact expressions are difficult to derive.

This bounding approach is instrumental to the philosophy of network calculus concerning the modelling of arrival processes. On one hand, it significantly widens the modelling scope of arrivals. On the other hand, it lends itself to an elegant analysis, not only in the sense of carrying out concise formulas, but also in the sense that the final formulas may be amenable to, e.g., convex optimizations encountered in dimensioning problems. For the rest of this section we present some key arrival models in network calculus. The goal of the presentation is to give insight on the issue of 'What are suitable bounding models for arrival processes?'.

4.1 A Pitfall: The Simplest Arrival Bound

Perhaps the most tempting bound for an arrival process $A(n)$ is the following

$$A(n) \leq \mathcal{G}(n) \ \forall n \in \mathbb{N} \ , \quad (17)$$

where $\mathcal{G}(n)$ is a non-random function. It is important to remark that for every time n, the model bounds a r.v., i.e., $A(n)$, by a non-random number, i.e., $\mathcal{G}(n)$. As an example, if $A(n)$ is the compound Bernoulli arrival process from Eq. (14), then any function $\mathcal{G}(n)$ satisfying $\mathcal{G}(n) \geq n$ would fit Eq. (17); note that $\mathcal{G}(n) = n$ would be the tightest achievable deterministic bound. As another example, if $A(n)$ is a (discretized) Poisson process, or any random process taking arbitrarily large values with non-zero probabilities, then *no* bounded function $\mathcal{G}(n)$ would fit Eq. (17).

The actual drawback of the bounding model from Eq. (17) does not stand in its apparent weak modelling power, but rather in its incompleteness for computing performance measures, e.g., the queueing backlog. Let us reconsider the queueing scenario from Figure 1.(a). The backlog process B_n is defined as

$$B_n = A(n) - D(n) \ ,$$

i.e., the amount of bits in the system at time n. Recalling the expression for $D(n)$ from Eq. (4), one may immediately derive Reich's equation:

$$B_n = \max_{0 \leq k \leq n} \{A(k,n) - C(n-k)\} \ . \quad (18)$$

This equation clearly indicates that the bounding model from Eq. (17) is insufficient to get a non-degenerate bound on the backlog process. That is because a bound on $A(n)$ does not necessarily induce a bound on $A(k,n)$, which is what is actually needed. For a quick example, take the arrival process $\mathbf{A} = (0,0,20)$ and $\mathcal{G}(n) = 10n$; clearly, $A(n) \leq \mathcal{G}(n)$ but $A(1,2) = 20 > 10 = \mathcal{G}(1)$. A more intuitive argument is that the backlog process depends on the entire history of the process, and relative to *any* time point, whereas Eq. (17) only captures the process relative to the origin.

Let us next make an additional stationarity assumption on $A(n)$. Then, if a deterministic computation of queueing measures was the goal, the model from Eq. (17) would remain insufficient; in fact, the model would only be sufficient for constant-rate arrivals. However, if a probabilistic computation was the goal, then the model from Eq. (17) could become useful. Indeed, a computation of the backlog tail would be for instance

$$\begin{aligned} \mathbb{P}(B_n > \sigma) &= \mathbb{P}\left(\max_{0 \leq k \leq n} \{A(k,n) - C(n-k)\} > \sigma\right) \\ &\leq \sum_{k=0}^{n} \mathbb{P}(A(k,n) - C(n-k) > \sigma) \ . \end{aligned} \quad (19)$$

The last line follows from Boole's inequality[2]. As a side remark, we point out that this apparently loose inequality is

[2]For some probability events E and F, Boole's inequality is $\mathbb{P}(E \cup F) \leq \mathbb{P}(E) + \mathbb{P}(F)$.

not so bad if the r.v.'s $X_k := A(k,n)$ are rather uncorrelated (e.g., when $A(n)$ is a (discretized) Poisson process), but it is quite loose if they are highly correlated [46]. In such a case one can make use of more sophisticated techniques with refined martingale inequalities (see [12], pp. 339-343). For the purpose of our presentation it is sufficient to adopt the simplified technique with Boole's inequality.

Due to the stationarity assumption, the last line can be continued by replacing $A(k,n)$ with $A(n-k)$, and further by $\mathcal{G}(n-k)$ according to the bounding arrival model from Eq. (17). Note however that $\mathcal{G}(n)$ should be defined as a random process, and the inequality from Eq. (17) should hold a.s. (almost surely, i.e., $\mathbb{P}(A(n) < \mathcal{G}(n)) = 0$). Otherwise, if $\mathcal{G}(n)$ was a non-random function, then the previous derivation would be quasi-deterministic since all the probabilities would evaluate either to 0 or 1.

4.2 Classic Deterministic Arrival Model

The previous pitfall indicates that, in general, it is not sufficient to bound arrivals on all intervals $(0,n)$ but rather on all intervals (k,n). This observation suggests the following bounding arrival model

$$A(k,n) \le \mathcal{G}(n-k) \ \forall 0 \le k \le n \ , \tag{20}$$

where $\mathcal{G}(n)$ is a non-random function. Network calculus was essentially founded on this arrival model [16].

With this arrival model, the deterministic continuation of Eq. (18) is straightforward:

$$B_n \le \max_{0\le k\le n} \{\mathcal{G}(k) - Ck\} \ . \tag{21}$$

The RHS term can be computed explicitly in $\mathcal{O}(1)$ time if $\mathcal{G}(n)$ is a sufficiently 'nice' expression, e.g., $\mathcal{G}(n) = rn + b$ where r and b have the meanings of rate, and burst, respectively. Otherwise, if $\mathcal{G}(n)$ is given pointwise, then the RHS term can be computed in $\mathcal{O}(n)$ time.

We emphasize that there is no requirement of stationarity on the arrival process $A(n)$. In fact, the regularity constraint from Eq. (20) is satisfied by infinitely many (and possibly unknown) arrival processes, thus illustrating the high modelling potential of Eq. (20). Moreover, despite the apparent tradeoff between modelling potential and accuracy of representation, the derivation from Eq. (21) is actually *tight* ([5], p. 27). Tightness means that there exists an arrival process which 1) satisfies the arrival bound from Eq. (20), and 2) induces a backlog process which matches with the predicted bound from Eq. (21). Even more remarkably, the tightness of the backlog bound holds even when multiplexing many possibly 'conspiring' flows, e.g., producing large bursts at the same time.

To more concretely elaborate on the *tightness* of the deterministic modelling and analysis from Eqs. (20)-(21), let us consider an aggregate $A(n) = \sum_{j=1}^{N} A_j(n)$, where $A_j(n)$'s are compound Bernoulli processes as in Eq. (14). The flows are multiplexed at a server with capacity $C < N$, and there is no statistical independence assumption amongst them. The performance metric of interest is again a bound on the backlog process B_n. According to Eq. (21), to get the tightest bound on B_n, one must first construct the smallest non-random function satisfying Eq. (20), which is $\mathcal{G}(n) = Nn$. When plugged into Eq. (21), this yields the bound

$$B_n \le \max_{0\le k\le n} \{N(n-k) - C(n-k)\} = (N-C)n \ .$$

Because $N > C$ the bound diverges and clearly becomes useless when $n \to \infty$. We point out, however, that the bound is *tight* even under a statistical independence assumption on $A_j(n)$'s. Indeed, $\forall p, n > 0$ there exists a positive probability such that $A_j(n) = n \ \forall j$, i.e., there exists a sample-path which attains the apparently very loose bound on B_n.

Another illustrative example concerns possible degenerate results obtained from deterministic modelling and analysis. This is the case when the arrival process $A(n)$ can take infinitely large values, i.e., $\forall\, K > 0 \ \exists \varepsilon_K > 0$ such that $\mathbb{P}(A(n) > K) > \varepsilon_K$ (the Poisson process is an example). For such arrival processes, the regularity constraint from Eq. (20) is only satisfied by the degenerate function $\mathcal{G}(n) = \infty$, which clearly yields the degenerate, and useless, backlog bound $B_n \le \infty$. However, reiterating the previous argument, this degenerate bound is also tight.

We conclude here by pointing out that if one seeks deterministic bounds from deterministic or even stochastic arrival models, then DNC is an attractive theory (illustrated here by Eqs. (20)-(21)): it has a high modelling potential and it (mostly) yields tight bounds. Otherwise, if one seeks *probabilistic* bounds, e.g.,

$$\mathbb{P}(B_n > \sigma) \le \varepsilon(\sigma), \text{ where } \varepsilon(\sigma) \text{ is to be determined,}$$

then DNC is an inopportune theory. The main reason is that a purely deterministic analysis can yield extremely loose bounds due to not leveraging from statistical multiplexing gain. This discussion will be continued in Section 5.

4.3 Stochastic Arrival Models

A probabilistic analysis generally requires a probabilistic arrival model. Here we consider three of the main stochastic arrival models proposed in the SNC literature. For the sake of presentation we use the names SBB, S^2BB, and S^3BB, and omit original names. The goal of this subsection is to explain their motivation and benefits.

$$\text{SBB}: \ \mathbb{P}\Big(A(k,n) - \mathcal{G}(n-k) > \sigma\Big) \le \varepsilon(\sigma) \ \forall\, k, n, \sigma$$

$$\text{S}^2\text{BB}: \ \mathbb{P}\Big(\max_{0\le k\le n} \{A(k,n) - \mathcal{G}(n-k)\} > \sigma\Big) \le \varepsilon(\sigma) \ \forall\, n, \sigma$$

$$\text{S}^3\text{BB}: \ \mathbb{P}\Big(\max_{0\le k\le n\le\infty} \{A(k,n) - \mathcal{G}(n-k)\} > \sigma\Big) \le \varepsilon(\sigma) \ \forall\, \sigma$$

$\mathcal{G}(n)$'s are non-random and are called *envelope functions*. $\varepsilon(\sigma)$'s are called *error functions*. Some technical and quite intuitive conditions are that the envelope functions are non-decreasing, whereas the error functions are non-increasing. Note also that a degree of freedom of the bounding approach is that for all three models the arrival process $A(n)$ does not need to be stationary, although the bounds themselves (the envelopes $\mathcal{G}(n)$) are so.

Before we explain the three models, it is important to observe the formation of the bottom two: S^2BB [17] is formed by inserting the free variable k from SBB (short-hand for stochastically bounded burstiness) [45] into the probability, whereas S^3BB [26] is further formed by inserting the free variable n as well. Informally, S^3BB measures events consisting of all past histories of the process $A(n)$, i.e., relative to *all* times. In turn, S^2BB measures events consisting of a single past history, i.e., relative to a fixed time, whereas SBB measures events as single fragments of past histories.

Although the SBB model seems the simplest amongst the three, it is actually the S^2BB model which is the natural extension of the classic deterministic model from Eq. (20). To see the reason, rewrite Eq. (20) as

$$\max_{0\leq k\leq n} \{A(k,n) - \mathcal{G}(n-k)\} \leq 0 \; \forall \, n \; . \tag{22}$$

Note that the S^2BB model enforces a bound on the CCDF of the LHS term above. In other words, the S^2BB model quantifies, with an upper bound, the probability that the deterministic model is violated by more than σ. An attractive property of S^2BB is that it immediately lends itself to the calculation of performance bounds. Indeed, a straightforward manipulation of Eq. (18) and S^2BB yields the bound

$$\mathbb{P}\left(B_n > \sigma_0 + \sigma\right) \leq \varepsilon(\sigma) \; , \tag{23}$$

where $\sigma_0 = \max_{k\geq 0} \{\mathcal{G}(k) - Ck\}$ is exactly the deterministic backlog bound from Eq. (21). To recapitulate, S^2BB quantifies the violation probabilities of the deterministic model (see S^2BB and Eq. (22)), whereas the probabilistic backlog bound quantifies the violation probabilities of the deterministic backlog bound (see Eqs. (23) and (21)). As these violation probabilities are identical, one can argue that S^2BB is the 'natural' probabilistic extension of the deterministic arrival model from Eq. (20) (or, equivalently, from Eq. (22)).

In practice, the choice of SBB vs. S^2BB depends on the arrivals' input. If the input is a measurement trace, then S^2BB should be chosen since it immediately lends itself to performance bounds, as shown earlier. Given a trace $A(n)$ with n elements, an S^2BB fitting algorithm would follow the steps: 1) make a guess on $\mathcal{G}(n)$ (e.g., $\mathcal{G}(n) = (r+\delta)n$, where r is the average rate of the trace and $\delta > 0$ is a tuning parameter), 2) compute the partial sums $A(k,n)$ and the LHS terms in Eq. (22), and 3) fit a distribution function. Ignoring the accuracy of the fitting, i.e., the range of values σ, the algorithm runs in $O(n^2)$ time. There is no specific rule for the tuning parameter δ, which is to be optimized numerically.

If the arrivals' input is some random process $A(n)$, then it is generally easier to first fit the SBB model. A typical way is to derive an MGF bound, e.g., $M_{A(n)}(\theta) \leq e^{\theta r n}$, for some $\theta > 0$ (see Eq. (16) for Markov arrival processes). If $A(n)$ is also stationary, then an SBB model can be fitted using the Chernoff bound[3], i.e.,

$$\mathbb{P}\left(A(k,n) > r(n-k) + \sigma\right) \leq e^{-\theta\sigma} \; \forall k,n,\sigma \; . \tag{24}$$

This SBB model further lends itself to the S^2BB model. Indeed, one can write for some $\delta > 0$:

$$\begin{aligned} &\mathbb{P}\left(\max_{0\leq k\leq n} \{A(k,n) - (r+\delta)(n-k)\} > \sigma\right) \\ &\quad \leq \sum_{0\leq k<n} \mathbb{P}\left(A(k,n) > r(k-n) + \delta(k-n) + \sigma\right) \\ &\quad \leq \sum_{k\geq 1} e^{-\theta\delta k} e^{-\theta\sigma} \leq \frac{1}{\theta\delta} e^{-\theta\sigma} \; . \end{aligned}$$

The second line follows from Boole's inequality, whereas the exponential bounds in the last line follow from Eq. (24). What is important to remark is that the transition from the SBB to S^2BB involves a rate increase from r to $r+\delta$. This penalty is due to the need to obtain a bounded error function for S^2BB; note that, if δ was zero, then the obtained error function would be unbounded.

[3]For some r.v. X and $x, \theta \in \mathbb{R}_+$, the Chernoff bound is $\mathbb{P}(X > x) \leq M_X(\theta) e^{-\theta x}$.

4.4 Quasi-Determinism in the S^3BB Model

So far we have only commented on the applicability of the SBB and S^2BB models. The reason is that, as we demonstrate in this section, the S^3BB model is quasi-deterministic for the class of stationary and ergodic arrival processes. Formally, the quasi-determinism means that the corresponding violation probabilities, set through the error function $\varepsilon(\sigma)$, can only take the extreme values, i.e.,

$$\varepsilon(\sigma) \in \{0,1\} \; \forall \sigma \; . \tag{25}$$

The immediate consequence is that the resulting SNC formulation from [26] is essentially quasi-deterministic, and does not capture statistical multiplexing gain. In fact, multiplexing quasi-deterministic S^3BB flows yields quasi-deterministic aggregates, by using the Superposition Property from [26]; for the precise meaning of 'statistical multiplexing gain' we refer to Section 5.

We next prove the quasi-determinism claim for stationary and ergodic processes, and then construct two rather contrived arrival processes for which the S^3BB model is not necessarily quasi-deterministic.

4.4.1 Stationary and Ergodic Processes

First we give some definitions (see Breiman [7], pp. 104-120). Consider a random process $\mathbf{X} = (X_1, X_2, \dots)$ defined on some joint probability space $(\Omega, \mathcal{F}, \mathbb{P})$; the Borel σ-field of the subsets of $\mathbb{R}$ is denoted by $\mathcal{B}$. We denote $I = \{1, 2, \dots\}$, and the product spaces $\mathbb{R}^I = \{\mathbf{x} = (x_1, x_2, \dots) : x_i \in \mathbb{R}\}$ and $\mathcal{B}^I = \{\mathbf{B} = (B_1, B_2, \dots) : B_i \in \mathcal{B}\}$.

By definition, the process $\mathbf{X}$ is (strongly) stationary if

$$\begin{aligned} &\mathbb{P}\Big(X_{i_1} \leq x_1, X_{i_2} \leq x_2, \dots, X_{i_n} \leq x_n\Big) \\ &\quad = \mathbb{P}\Big(X_{i_1+k} \leq x_1, X_{i_2+k} \leq x_2, \dots, X_{i_n+k} \leq x_n\Big) \; , \end{aligned}$$

for all $n, k \in \mathbb{N}^*$, $0 \leq i_1 \leq i_2 \leq \dots \leq i_n$, and $x_1, x_2, \dots, x_n$. In other words, stationarity means that the distribution of any sequence $(X_{i_1}, X_{i_2}, \dots, X_{i_n})$ is invariant under shift.

Further, the notion of ergodicity requires the introduction of an explicit *shift operator* $T : \mathbb{R}^I \to \mathbb{R}^I$, defined as

$$T(x_1, x_2, \dots) = (x_2, x_3, \dots) \; ,$$

for all sequences $\mathbf{x} = (x_1, x_2, \dots)$. The stationarity of $\mathbf{X}$ implies that T is *measure preserving*, i.e., by definition

$$\mathbb{P}\left(\mathbf{X} \in B\right) = \mathbb{P}\left(T\mathbf{X} \in B\right) \; \forall B \in \mathcal{B}^I \; .$$

For $B \in \mathcal{B}$, the event $\{\mathbf{X} \in B\}$ is said to be *invariant* if

$$\{\mathbf{X} \in B\} = \{T\mathbf{X} \in B\} \; \mathbb{P}\text{-a.s.} \; ,$$

i.e., the events $\{\mathbf{X} \in B\}$ and $\{T\mathbf{X} \in B\}$ differ by a set of probability zero. In other words, the event $\{\mathbf{X} \in B\}$ is invariant if its incidence does not depend (a.s.) on any finite prefix of $\mathbf{X}$. Finally, the process $\mathbf{X}$ is *ergodic* if any invariant event has probability 0 or 1.

The following lemma will be used to prove the claim of quasi-determinism.

LEMMA 1. *Consider a stationary and ergodic process* $\boldsymbol{X} = (X_1, X_2, \dots)$. *Then*

$$\mathbb{P}\Big(\max\{X_1, X_2, \dots\} > \sigma\Big) \in \{0,1\} \; \forall \sigma \; .$$

The lemma implies that $\max\{X_1, X_2, \dots\} = K$ a.s., where K is a constant or $K = \infty$.

PROOF. Fix σ and let $B = (-\infty, \sigma]^I$. We shall prove that

$$\{\mathbf{X} \in B\} = \left\{\max_{i \geq 1} X_i \leq \sigma\right\}$$

is an invariant event, which is equivalent to showing that

$$\mathbb{P}(\{\mathbf{X} \in B\} \,\Delta\, \{T\mathbf{X} \in B\}) = 0 \;, \qquad (26)$$

where 'Δ' denotes the symmetric difference and $\{T\mathbf{X} \in B\} = \{\max_{i \geq 2} X_i \leq \sigma\}$.

Let us first note that

$$\begin{aligned}
&\mathbb{P}(\max\{X_1, X_2, \ldots\} > \sigma) \\
&= \lim_{n \to \infty} \mathbb{P}(\max\{X_1, X_2, \ldots X_n\} > \sigma) \\
&= \lim_{n \to \infty} \mathbb{P}(\max\{X_2, X_3, \ldots, X_{n+1}\} > \sigma) \\
&= \mathbb{P}(\max\{X_2, X_3, \ldots\} > \sigma) \;. \qquad (27)
\end{aligned}$$

The second and last lines follow from the monotone convergence theorem (if B_n is a non-decreasing sequence of events, then $\mathbb{P}(\lim_n B_n) = \lim_n \mathbb{P}(B_n)$). The third line follows from the stationarity of X_n.

Expanding the symmetric difference from Eq. (26) into the union of two events, we have for the first one

$$\mathbb{P}\left(\left\{\max_{i \geq 1} X_i \leq \sigma\right\} \cap \left\{\max_{i \geq 2} X_i > \sigma\right\}\right) = \mathbb{P}(\emptyset) = 0 \;.$$

For the second one we use the inclusion-exclusion formula:

$$\begin{aligned}
&\mathbb{P}\left(\left\{\max_{i \geq 1} X_i > \sigma\right\} \cap \left\{\max_{i \geq 2} X_i \leq \sigma\right\}\right) \\
&= \mathbb{P}\left(\max_{i \geq 1} X_i > \sigma\right) + \mathbb{P}\left(\max_{i \geq 2} X_i \leq \sigma\right) - 1 \\
&= 0 \;.
\end{aligned}$$

In the last line we applied Eq. (27). Collecting terms implies that Eq. (26) holds and thus the event $\{\mathbf{X} \in B\}$ is invariant. Because $\mathbf{X}$ is ergodic, it follows that $\mathbb{P}(\mathbf{X} \in B) \in \{0, 1\}$, which completes the proof. □

We are now ready to demonstrate the quasi-determinism claim. Let the S^{3}BB model from Section 4 for an arrival process $A(n)$, some envelope $\mathcal{G}(n)$, and error function $\varepsilon(\sigma)$. Assume that $a_n := A(n-1, n)$ is stationary and ergodic. It then follows that for any $m \in \mathbb{N}^*$, the *block* process $\left(X_n^{(m)}\right)_{n \geq 1}$ comprising blocks of m consecutive instances of a_n and defined as

$$X_n^{(m)} := A(n-1, n+m-1) - \mathcal{G}(m) \;\forall n \geq 1$$

is also stationary and ergodic (cf. [7], Propositions 6.6 and 6.31). According to Lemma 1, there exists $K^{(m)}$'s such that

$$\max\left\{X_1^{(m)}, X_2^{(m)}, \ldots\right\} = K^{(m)} \text{ a.s. } ,$$

for all $m \geq 1$. Taking $K = \max_m K^{(m)}$ we obtain that

$$\max_{0 \leq k \leq n \leq \infty} \{A(k, n) - \mathcal{G}(n-k)\} = K \text{ a.s. } ,$$

thus concluding that

$$\varepsilon(\sigma) \in \{0, 1\} \;\forall \sigma$$

in the definition of the S^{3}BB model.[4]

[4]We remark that S^{3}BB is not necessarily quasi-deterministic under the additional assumptions of restricting the 'max' operator to a finite interval $0 \leq k \leq n$ and letting $\varepsilon(\sigma)$ depend on the right margin n (see Definition 3.2.1 in [28]).

4.4.2 *Non-Stationary or Non-Ergodic Processes*

Here we give two examples of arrival processes for which the S^{3}BB model is not necessarily quasi-deterministic. Such processes have to be non-stationary or non-ergodic.

For an example of a non-stationary process consider

$$a_1 = r + X, \; a_n = r \;\forall n \geq 2 \;,$$

where X is some r.v. satisfying $E[X] > 0$. Note that the cumulative process $A(n) = rn + X \;\forall n$ does not have stationary increments because $E[a_1] \neq E[a_2]$. Moreover, for $\sigma > 0$, the probability

$$\mathbb{P}\Big(\max_{0 \leq k \leq n \leq \infty} \{A(k, n) - r(n-k)\} > \sigma\Big) = \mathbb{P}\Big([X]_+ > \sigma\Big)$$

can be different from zero and one.

For an example of a non-ergodic process consider

$$a_n = X \;\forall n \geq 1 \;,$$

for some r.v. X. The cumulative process $A(n) = nX$ is stationary but non-ergodic, as there are many realizations of the process for which the time averages are different. To construct a non-quasi-deterministic S^{3}BB model, one can take $\mathcal{G}(n) = rn$ and X be any Bernoulli r.v. with $E[X] = r$ but $\mathbb{P}(X \neq r) > 0$.

What the two examples have in common is that the sample-paths are completely determined from some time scale on. In particular, in the second example, the sample-paths are completely determined once time starts. We speculate that more compounded examples would also account for randomness but for a finite time scale *only*, in order to avoid the limiting argument in the preceding quasi-determinism proof. Due to this rather unnatural restricted capability in capturing randomness, the relevance of such models is unclear.

5. STATISTICAL MULTIPLEXING GAIN

In this section we justify the raison d'être of SNC; concretely, we present a result which rigorously reveals the magnitude of the statistical multiplexing gain, as a scaling law, achieved by SNC in the single-node case. Then we discuss on the existence of multiplexing gain in the multi-node case, and present numerical results.

5.1 Single-Node Case

Statistical multiplexing is an essential property of packet-switched networks, which are based on the principle of resource sharing. It basically says that the number of resources needed to support service for (say N) flows is *much smaller* than N times the number of resources needed to support service for a single flow. The raison d'être of SNC is to capture the gap between these two quantities, i.e., the *statistical multiplexing gain*, while closely reproducing the elegant methodology of DNC.

To illustrate the magnitude of the statistical multiplexing gain achieved with SNC we consider a node of capacity C serving N flows $A_j(n)$, each modelled with the envelope

$$A_j(k, n) \leq r(n-k) + b \;\forall\; k, n \;, \qquad (28)$$

where $r > 0$ is a rate and $b \geq r$ is a burst size. Consider now the design question **Q1:** *"How large should C be such that the delay is smaller than some value, normalized to 1?"* To answer, it is convenient to derive a backlog bound. Assuming the stability condition $\rho := \frac{Nr}{C} \leq 1$, the bound follows directly from Eq. (21), where $\mathcal{G}(k) = Nrk$:

$$B_n \leq Nb \; \forall n \; .$$

Because of the delay normalization to 1, which implies that C and the backlog scale identically, we conclude that the required capacity scales as $C = \mathcal{O}(N)$ in the burst b. Although this conclusion is based on a *tight* bound (recall the discussion from Section 4.2), the intuition is that a much smaller capacity would be sufficient under broad statistical assumptions on the flows $A_j(n)$, and as long as some violations of the delay constraint are tolerable.

Let us additionally assume that $A_j(n)$'s are stationary and statistically independent, and enforce the (tolerable) constraint $\mathbb{P}(\text{delay} > 1) \leq \varepsilon$, where ε is some small value, e.g., $\varepsilon = 10^{-3}$. With these assumptions, we can use a stochastic bound on the aggregate $A(n) := \sum_{j=1}^{N} A_j(n)$, i.e., [38]

$$\mathbb{P}(A(n) > Nrn + \sigma) \leq e^{-\frac{\sigma^2}{2Nb^2}} \; \forall \sigma \geq 0 \; . \qquad (29)$$

A backlog bound can then be computed as in Eq. (19):

$$\begin{aligned} \mathbb{P}(B_n > \sigma) &\leq \mathbb{P}\left(\max_{0 \leq k < n} \{A(k,n) - C(n-k)\} > \sigma\right) \\ &\leq \sum_{k \geq 1} \mathbb{P}(A(k) > Nrk + (C - Nr)k + \sigma) \\ &\leq \int_0^{\infty} e^{-\frac{1}{2}\left(\frac{(C-Nr)s+\sigma}{\sqrt{N}b}\right)^2} ds \; . \end{aligned}$$

In the last line we used the bound from Eq. (29) and bounded a sum of non-increasing terms by an integral. With the change of variable $u = \frac{(C-Nr)s+\sigma}{\sqrt{N}b}$ the last term becomes

$$\frac{\sqrt{N}b}{C - Nr} \int_{\frac{\sigma}{\sqrt{N}b}}^{\infty} e^{-\frac{u^2}{2}} du \leq \frac{b^2}{r(\rho^{-1} - 1)\sigma} e^{-\frac{\sigma^2}{2Nb^2}} \; .$$

Here we used Gordon's inequality for the standard normal density function, i.e., $\int_x^{\infty} e^{-\frac{u^2}{2}} du \leq \frac{1}{x} e^{-\frac{x^2}{2}}$ [24]. Setting the last term to ε yields

$$\sigma^2 = 2Nb^2 \left(\log \frac{b^2}{\varepsilon r(\rho^{-1} - 1)} - \log \sigma\right) \; .$$

From here it follows that $C = \mathcal{O}\left(\sqrt{N}\right)$ in the burst b (recall that C scales identically with the backlog σ); note, however, that $C = \mathcal{O}(N)$ in the rate r in order to satisfy the stability condition. The $\mathcal{O}\left(\sqrt{N}\right)$ law can also be deduced from a result from [4], which is however obtained using an approximative application of the Central Limit Theorem, and hence not rigorous. Several other probabilistic bounds on multiplexed deterministically regulated arrivals (as in Eq. (28)) exist in the literature, e.g., [47, 50]; however, they do not appear to easily lend themselves to the $\mathcal{O}\left(\sqrt{N}\right)$ law.

The difference in the scaling laws $C = \mathcal{O}(N)$ (obtained with DNC) vs. $C = \mathcal{O}\left(\sqrt{N}\right)$ (obtained with SNC) reveals thus the magnitude of the statistical multiplexing gain achieved with SNC as $\Omega\left(\sqrt{N}\right)$.

5.2 Multi-Node Case

Lastly, we discuss on an unconventional type of statistical multiplexing which, to the best of our knowledge, has not been raised previously. Consider the tandem network from

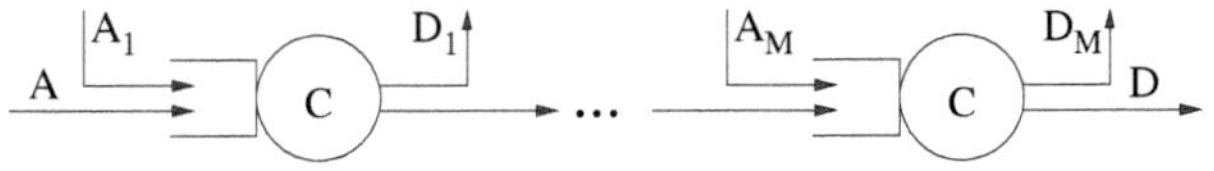

Figure 2: A tandem network with cross traffic

Figure 2 in which a flow A crosses M nodes in series; at each node $j = 1, \ldots, M$ along the end-to-end (e2e) path, A shares the local resource (the capacity C) with a local cross flow A_j. All flows are stationary and statistically independent. This type of resource sharing looks similar to the conventional one, except that the 'resource' is now a distributed one (i.e., all the capacities) and the cross flows do not share the same resource with each other. The arising question concerns the existence of a *distributed* multiplexing gain.

To answer, we apply and compare DNC and SNC for the following scenario: A is bounded by the envelope from Eq. (28) with rate r and burst b, and A_j's are bounded by the same envelope but with rate Nr and burst Nb (N will be used for tuning conventional (per-node) multiplexing gain). We enforce the stability condition $C \geq (N+1)r$ and assume that flow A gets lower priority at each node. We ask the design question **Q2**: *"How large should C be such that the e2e delay of A is smaller than 1?"*.

To deal with the additional complexities due to scheduling and multi-node (i.e., the system's 'noise'), we run the network calculus engine, i.e., transform the network system into a 'somewhat looking' linear system. The first step is to derive the service processes at each node, i.e., $S^j(n) = [(C - Nr)n - Nb]_+$, and then apply the convolution formula from Eq. (13) yielding $S(n) = \left[\left(n - \frac{NMb}{C-Nr}\right)\right]_+ (C - Nr)$. From the transformed system, consisting of the input $\mathbf{A}$ and the service process $\mathbf{S}$, the deterministic e2e delay bound is $W \leq \frac{(NM+1)b}{C-Nr}$. A probabilistic e2e delay bound can be also derived (not shown here) using the SNC formulation from Fidler [22] and the representation from Eq. (29).

Figure 3 illustrates the required capacities C, computed with DNC and SNC, for the questions **Q1** in (a) and **Q2** in (b). Note that (a) clearly shows the (conventional) multiplexing gain when $b = 3r$. Note also that the multiplexing gain kicks-in around ten flows. In turn, (b) illustrates that there is no distributed multiplexing gain when there is only one cross flow $N = 1$; however, for sufficiently large N, e.g., 50, the conventional multiplexing gain kicks-in and compensates for the lack of distributed multiplexing gain.

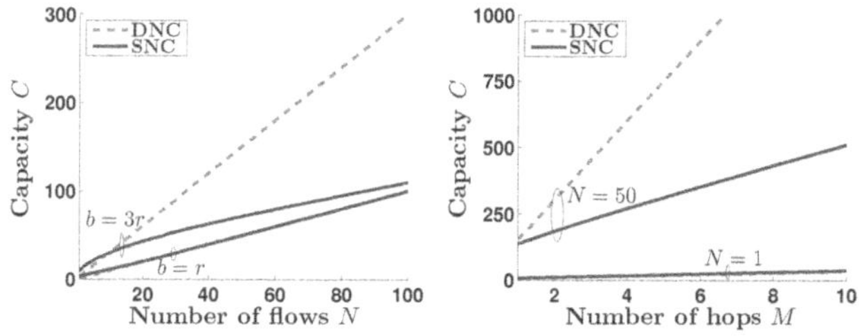

(a) Single-node (b) Multi-node ($b = 3r$)

Figure 3: Capacity dimensioning, with DNC and SNC, for the delay constraints (delay $\leq$ 1) and $\mathbb{P}$(delay $> 1) \leq \varepsilon$, respectively ($r = 1$, $\varepsilon = 10^{-3}$)

6. CONCLUSIONS

Network calculus (NC) has seen a lot of research over the last 20 years, e.g., Google Scholar yields $\approx$ 3000 hits when searching for "network calculus", comparing relatively well to its 'older brother' queueing theory, which yields $\approx$ 60000 hits. The broad perception on NC is that it holds good promises as an alternative/complementary methodology to classical queueing theory. Yet, in the camp of NC researchers as well as the larger audience, the mood somewhat oscillates between "Hooray, we found the Holy Grail!" and "Oh no, it's not gonna work." In this paper, we have tried to better locate where the truth stands and aimed at reducing the level of confusion—and hopefully not creating new one—by clarifying some important issues related to the core modelling abstractions in network calculus: arrival envelopes and service processes. On this mission, we also provided some new insights into network calculus in the larger frame of things. Specifically, we collected the following facts:

- NC is about approximating a complex (typically non-linear) queueing system by a min-plus linear one. The approximation closely follows the traces of both traditional LTI system theory and of elementary properties from classical queueing theory; the analogy is good, but not perfect.
- With respect to arrival envelopes we provided a number of pitfalls as well as advice on how to avoid them. The most salient and profound observation is that the S^3BB envelope model presented at ACM Sigcomm 2006 delivers quasi-deterministic performance bounds for a large class of arrivals (stationary and ergodic), which means that it cannot generally capture statistical multiplexing gain.
- Statistical multiplexing gain can be captured well by SNC, under carefully defined envelope and service processes. In particular, for the application scenario of multiplexing independent regulated arrival processes, we rigorously showed the gain to be on the order of $\Omega(\sqrt{N})$, which before had only been shown approximately. So there is light at the end of the tunnel and hopefully it is not by a train railing towards us.
- We have also discussed on the tightness of bounds, a lingering issue surrounding almost any discussion on NC. We clarified issues about the tightness of the DNC and provided insights into that of SNC. In short, DNC generally delivers tight bounds, i.e., the bounds can be attained; SNC is as tight as the underlying probability inequalities being used (Boole, Chernoff, martingale inequalities). As these inequalities can be plugged into SNC in a modular fashion, one may argue that the SNC analysis can be made as tight as the state-of-the-art in probability theory allows.

Summing up, while using NC does not provide us with a free lunch, it still seems to be of good value in analyzing traditionally hard fundamental queueing problems due to its scalable tradeoff between accuracy and ease of analysis.

7. ACKNOWLEDGMENTS

We thank the following people for their constructive influence on this manuscript: Jörg Liebeherr, Almut Burchard, Felix Poloczek, Yuming Jiang, Yashar Ghiassi-Farrokhfal, Markus Fidler, Hao Wang, Karthik Rajkumar, and all the students from the Stochastic Network Calculus class given at TU Berlin (WS/11-12). We also thank our shepherd Mark Crovella and the anonymous Sigcomm reviewers for their comments.

8. REFERENCES

[1] R. Agrawal, R. L. Cruz, C. Okino, and R. Rajan. Performance bounds for flow control protocols. *IEEE/ACM Transactions on Networking*, 7(3):310–323, June 1999.

[2] F. L. Baccelli, G. Cohen, G. J. Olsder, and J.-P. Quadrat. *Synchronization and Linearity: An Algebra for Discrete Event Systems*. John Wiley and Sons, 1992.

[3] F. Baskett, K. M. Chandy, R. R. Muntz, and F. G. Palacios. Open, closed and mixed networks of queues with different classes of customers. *Journal of the ACM*, 22(2):248–260, Apr. 1975.

[4] R. Boorstyn, A. Burchard, J. Liebeherr, and C. Oottamakorn. Statistical service assurances for traffic scheduling algorithms. *IEEE Journal on Selected Areas in Communications. Special Issue on Internet QoS*, 18(12):2651–2664, Dec. 2000.

[5] J.-Y. Le Boudec and P. Thiran. *Network Calculus*. Springer Verlag, Lecture Notes in Computer Science, LNCS 2050, 2001.

[6] R. Braden, D. Clark, and S. Shenker. Integrated services in the Internet architecture: an overview. IETF RFC 1633, July 1994.

[7] L. Breiman. *Probability*. SIAM, 1992.

[8] A. Burchard, J. Liebeherr, and F. Ciucu. On superlinear scaling of network delays. *IEEE/ACM Transactions on Networking*, 19(4):1043–1056, Aug. 2011.

[9] A. Burchard, J. Liebeherr, and S. D. Patek. A min-plus calculus for end-to-end statistical service guarantees. *IEEE Transactions on Information Theory*, 52(9):4105–4114, Sept. 2006.

[10] S. Chakraborty, S. Kuenzli, L. Thiele, A. Herkersdorf, and P. Sagmeister. Performance evaluation of network processor architectures: Combining simulation with analytical estimation. *Computer Networks*, 41(5):641–665, Apr. 2003.

[11] C.-S. Chang. Stability, queue length and delay, Part II: Stochastic queueing networks. In *31st IEEE Conference on Decision and Control*, pages 1005–1010, 1992.

[12] C.-S. Chang. *Performance Guarantees in Communication Networks*. Springer Verlag, 2000.

[13] F. Ciucu. Network calculus delay bounds in queueing networks with exact solutions. In *International Teletraffic Congress (ITC)*, pages 495–506, 2007.

[14] F. Ciucu, A. Burchard, and J. Liebeherr. Scaling properties of statistical end-to-end bounds in the network calculus. *IEEE Transactions on Information Theory*, 52(6):2300–2312, June 2006.

[15] C. Courcoubetis and R. Weber. Effective bandwidths for stationary sources. *Probability in Engineering and Informational Sciences*, 9(2):285–294, Apr. 1995.

[16] R. Cruz. A calculus for network delay, parts I and II. *IEEE Transactions on Information Theory*, 37(1):114–141, Jan. 1991.

[17] R. L. Cruz. Quality of service management in integrated services networks. In *1st Semi-Annual Research Review, CWC, University of California at San Diego*, June 1996.

[18] R. L. Cruz. SCED+: Efficient management of quality of service guarantees. In *IEEE Infocom*, pages 625–634, 1998.

[19] R. L. Cruz and C. Okino. Service guarantees for window flow control. In *34th Allerton Conference on Communications, Control and Computating*, 1996.

[20] A. Erlang. Solution of some problems in the theory of probabilities of significance in automatic telephone exchanges. *Elektrotkeknikeren*, 13, 1917 (in Danish).

[21] W. Feller. *An Introduction to Probability Theory and Its Applications, Vol. 2.* Wiley, 1971.

[22] M. Fidler. An end-to-end probabilistic network calculus with moment generating functions. In *IEEE International Workshop on Quality of Service (IWQoS)*, pages 261–270, 2006.

[23] M. Fidler. A survey of deterministic and stochastic service curve models in the network calculus. *IEEE Communications Surveys & Tutorials*, 12(1):59–86, Feb. 2010.

[24] R. D. Gordon. Values of Mills' ratio of area to bounding ordinate and of the normal probability integral for large values of the argument. *Annals of Mathematical Statistics*, 12(3):364–366, Sept. 1941.

[25] M. Harchol-Balter. *Performance Modeling and Design of Computer Systems. Queueing Theory in Action.* Cambridge University Press, 2012.

[26] Y. Jiang. A basic stochastic network calculus. In *ACM Sigcomm*, pages 123–134, Sept. 2006.

[27] Y. Jiang. Network calulus and queueing theory: Two sides of one coin. In *ICST ValueTools*, 2009.

[28] Y. Jiang and Y. Liu. *Stochastic Network Calculus.* Springer, 2008.

[29] F. P. Kelly. Networks of queues with customers of different types. *Journal of Applied Probability*, 3(12):542–554, Sept. 1975.

[30] J. Kurose. On computing per-session performance bounds in high-speed multi-hop computer networks. In *ACM Sigmetrics*, pages 128–139, 1992.

[31] E. A. Lee and P. Varaiya. *Structure and interpretation of signals and systems.* Addison-Wesley, 2003.

[32] W. E. Leland, M. S. Taqqu, W. Willinger, and D. V. Wilson. On the self-similar nature of Ethernet traffic. *IEEE/ACM Transactions on Networking*, 2(1):1–15, Feb. 1994.

[33] L. Lenzini, E. Mingozzi, and G. Stea. A methodology for computing end-to-end delay bounds in FIFO-multiplexing tandems. *Performance Evaluation*, 65(11-12):922–943, Nov. 2008.

[34] J. Liebeherr, A. Burchard, and F. Ciucu. Delay bounds in communication networks with heavy-tailed and self-similar traffic. *IEEE Transactions on Information Theory*, 58(2):1010–1024, Feb. 2012.

[35] J. Liebeherr, M. Fidler, and S. Valaee. A system-theoretic approach to bandwidth estimation. *IEEE/ACM Transactions on Networking*, 18(4):1040–1053, Aug. 2010.

[36] J. Liebeherr, Y. Ghiassi-Farrokhfal, and A. Burchard. On the impact of link scheduling on end-to-end delays in large networks. *IEEE Journal on Selected Areas in Communications*, 29(5):1009–1020, May 2011.

[37] S. Mao and S. S. Panwar. A survey of envelope processes and their applications in quality of service provisioning. *IEEE Communications Surveys & Tutorials*, 8(3):2–20, Dec. 2006.

[38] L. Massoulié and A. Busson. Stochastic majorization of aggregates of leaky bucket-constrained traffic streams. *Preprint, Microsoft Research*, 2000.

[39] V. Misra. Public review on 'A Basic Stochastic Network Calculus' by Y. Jiang, ACM Sigcomm 2006. `http://conferences.sigcomm.org/sigcomm/2006/discussion/showpaper.php?paper_id=14`.

[40] V. Paxson and S. Floyd. Wide-area traffic: The failure of Poisson modelling. *IEEE/ACM Transactions on Networking*, 3(3):226–244, June 1995.

[41] J.-L. Scharbarg, F. Ridouard, and C. Fraboul. A probabilistic analysis of end-to-end delays on an AFDX avionic network. *IEEE Transactions on Industrial Informatics*, 5(1):38–49, Feb. 2009.

[42] J. Schmitt and I. Martinovic. Demultiplexing in network calculus - a stochastic scaling approach. In *International Conference on the Quantitative Evaluation of Systems (QEST)*, pages 217–226, 2009.

[43] J. Schmitt, F. A. Zdarsky, and M. Fidler. Delay bounds under arbitrary multiplexing: When network calculus leaves you in the lurch... In *IEEE Infocom*, pages 1669–1677, 2008.

[44] T. Skeie, S. Johannessen, and O. Holmeide. Timeliness of real-time IP communication in switched industrial ethernet networks. *IEEE Transactions on Industrial Informatics*, 2(1):25–39, Feb. 2006.

[45] D. Starobinski and M. Sidi. Stochastically bounded burstiness for communication networks. *IEEE Transactions on Information Theory*, 46(1):206–212, Jan. 2000.

[46] M. Talagrand. Majorizing measures: The generic chaining. *Annals of Probability*, 24(3):1049–1103, July 1996.

[47] M. Vojnovic and J.-Y. Le Boudec. Bounds for independent regulated inputs multiplexed in a service curve network element. *IEEE Transactions on Communications*, 51(5):735–740, May 2003.

[48] K. Wang, F. Ciucu, S. Low, and C. Lin. A stochastic power network calculus for integrating renewable energy sources into the power grid. *IEEE Journal on Selected Areas in Communications; Smart Grid Communications Series*, 30(6), July 2012.

[49] O. Yaron and M. Sidi. Performance and stability of communication networks via robust exponential bounds. *IEEE/ACM Transactions on Networking*, 1(3):372–385, June 1993.

[50] Y. Ying, F. Guillemin, R. Mazumdar, and C. Rosenberg. Buffer overflow asymptotics for multiplexed regulated traffic. *Performance Evaluation*, 65(8):555–572, July 2008.

Abstractions for Network Update

Mark Reitblatt
Cornell

Nate Foster
Cornell

Jennifer Rexford
Princeton

Cole Schlesinger
Princeton

David Walker
Princeton

ABSTRACT

Configuration changes are a common source of instability in networks, leading to outages, performance disruptions, and security vulnerabilities. Even when the initial and final configurations are correct, the update process itself often steps through intermediate configurations that exhibit incorrect behaviors. This paper introduces the notion of consistent network updates—updates that are guaranteed to preserve well-defined behaviors when transitioning between configurations. We identify two distinct consistency levels, per-packet and per-flow, and we present general mechanisms for implementing them in Software-Defined Networks using switch APIs like OpenFlow. We develop a formal model of OpenFlow networks, and prove that consistent updates preserve a large class of properties. We describe our prototype implementation, including several optimizations that reduce the overhead required to perform consistent updates. We present a verification tool that leverages consistent updates to significantly reduce the complexity of checking the correctness of network control software. Finally, we describe the results of some simple experiments demonstrating the effectiveness of these optimizations on example applications.

Categories and Subject Descriptors

C.2.4 [**Computer-Communication Networks**]: Distributed Systems—*Network Operating Systems*

General Terms

Design, Languages, Theory

Keywords

Consistency, planned change, software-defined networking, OpenFlow, network programming languages, Frenetic.

1. INTRODUCTION

"Nothing endures but change."

—Heraclitus

SIGCOMM'12, August 13–17, 2012, Helsinki, Finland.

Networks exist in a constant state of flux. Operators frequently modify routing tables, adjust link weights, and change access control lists to perform tasks from planned maintenance, to traffic engineering, to patching security vulnerabilities, to migrating virtual machines in a datacenter. But even when updates are planned well in advance, they are difficult to implement correctly, and can result in disruptions such as transient outages, lost server connections, unexpected security vulnerabilities, hiccups in VoIP calls, or the death of a player's favorite character in an online game.

To address these problems, researchers have proposed a number of extensions to protocols and operational practices that aim to prevent transient anomalies [8, 2, 9, 3, 5]. However, each of these solutions is limited to a specific protocol (*e.g.*, OSPF and BGP) and a specific set of properties (*e.g.*, freedom from loops and blackholes) and increases the complexity of the system considerably. Hence, in practice, network operators have little help when designing a new protocol or trying to ensure an additional property not covered by existing techniques. A list of example applications and their properties is summarized in Table 1.

We believe that, instead of relying on point solutions for network updates, the networking community needs foundational principles for designing solutions that are applicable to a wide range of protocols and properties. These solutions should come with two parts: (1) an abstract interface that offers strong, precise, and intuitive semantic guarantees, and (2) concrete mechanisms that faithfully implement the semantics specified in the abstract interface. Programmers can use the interface to build robust applications on top of a reliable foundation. The mechanisms, while possibly complex, would be implemented once by experts, tuned and optimized, and used over and over, much like register allocation or garbage collection in a high-level programming language.

Software-defined networks. The emergence of Software Defined Networks (SDN) presents a tremendous opportunity for developing general abstractions for managing network updates. In an SDN, a program running on a logically-centralized controller manages the network directly by configuring the packet-handling mechanisms in the underlying switches. For example, the OpenFlow API allows a controller to install rules that each specify a *pattern* that matches on bits in the packet header, *actions* performed on matching packets (such as drop, forward, or divert to the controller), a *priority* (to disambiguate between overlapping patterns), and *timeouts* (to allow the switch to remove stale rules) [10]. Hence, whereas today network operators have (at best) indirect control over the distributed implementations of routing, access control, and load balancing, SDN platforms like OpenFlow provide programmers with direct control over the processing of packets in the network.

However, despite the conceptual appeal of centralized control,

Example Application	Policy Change	Desired Property	Practical Implications
Stateless firewall	Changing access control list	No security holes	Admitting malicious traffic
Planned maintenance [1, 2, 3]	Shut down a node/link	No loops/blackholes	Packet/bandwidth loss
Traffic engineering [1, 3]	Changing a link weight	No loops/blackholes	Packet/bandwidth loss
VM migration [4]	Move server to new location	No loops/blackholes	Packet/bandwidth loss
IGP migration [5]	Adding route summarization	No loops/blackholes	Packet/bandwidth loss
Traffic monitoring	Changing traffic partitions	Consistent counts	Inaccurate measurements
Server load balancing [6, 7]	Changing load distribution	Connection affinity	Broken connections
NAT or stateful firewall	Adding/replacing equipment	Connection affinity	Outages, broken connections

Table 1: Example changes to network configuration, and the desired update properties.

an OpenFlow network is still a distributed system, with inevitable delays between the switches and the controller. To implement a transition from one configuration to another, programmers must issue a painstaking sequence of low-level install and uninstall commands that work rule by rule and switch by switch. Moreover, to ensure that the network behaves correctly during the transition, they must worry about the properties of every possible intermediate state during the update, and the effects on any packets already in flight through the network. This often results in a combinatorial explosion of possible behaviors—too many for a programmer to manage by hand, even in a small network. A recent study on testing OpenFlow applications shows that programmers often introduce subtle bugs when handling network updates [11].

Our approach. This paper describes a different alternative. Instead of requiring SDN programmers to implement configuration changes using today's low-level interfaces, our high-level, abstract operations allow the programmer to update the configuration of the entire network in one fell swoop. The libraries implementing these abstractions provide strong semantic guarantees about the observable effects of the global updates, and handle all of the details of transitioning between old and new configurations efficiently.

Our central abstraction is *per-packet consistency*, the guarantee that every packet traversing the network is processed by exactly one consistent global network configuration. When a network update occurs, this guarantee persists: each packet is processed either using the configuration in place prior to the update, or the configuration in place after the update, but never a mixture of the two. Note that this consistency abstraction is *more* powerful than an "atomic" update mechanism that simultaneously updates all switches in the network. Such a simultaneous update could easily catch many packets in flight in the middle of the network, and such packets may wind up traversing a mixture of configurations, causing them to be dropped or sent to the wrong destination. We also introduce *per-flow consistency*, a generalization of per-packet consistency that guarantees all packets in the same flow are processed with the same configuration. This stronger guarantee is needed in applications such as HTTP load balancers, which need to ensure that all packets in the same TCP connection reach the same server replica to avoid breaking connections.

To support these abstractions, we develop several *update mechanisms* that use features commonly available on OpenFlow switches. Our most general mechanism, which enables transition between any two configurations, performs a two-phase update of the rules in the new configuration onto the switches. The other mechanisms are optimizations that achieve better performance under circumstances that arise often in practice. These optimizations transition to new configurations in less time, update fewer switches, or fewer rules.

To analyze our abstractions and mechanisms, we develop a simple, formal model that captures the essential features of OpenFlow networks. This model allows us to define a class of network properties, called *trace properties*, that characterize the paths individual packets take through the network. The model also allows us to prove a remarkable result: if *any* trace property P holds of a network configuration prior to a per-packet consistent update as well as after the update, then P also holds continuously throughout the update process. This illustrates the true power of our abstractions: programmers do not need to specify *which* trace properties our system must maintain during an update, because a per-packet consistent update preserves *all* of them! For example, if the old and new configurations are free from forwarding loops, then the network will be loop-free before, during, and after the update. In addition to the proof sketch included in this paper, this result has been formally verified in the Coq proof assistant [12].

An important and useful corollary of these observations is that it is possible to take any verification tool that checks trace properties of *static* network configurations and transform it into a tool that checks invariance of trace properties as the network configurations evolve *dynamically*—it suffices to check the static policies before and after the update. We illustrate the utility of this idea concretely by deploying the NuSMV model checker [13] to verify invariance of a variety of important trace properties that arise in practice. However, other tools, such as the recently-proposed header space analysis tool [14], also benefit from our approach.

Contributions. This paper makes the following contributions:

- **Update abstractions:** We propose per-packet and per-flow consistency as canonical, general abstractions for specifying network updates (§2,§7).
- **Update mechanisms:** We describe OpenFlow-compatible implementation mechanisms and several optimizations tailored to common scenarios (§5,§8).
- **Theoretical model:** We develop a mathematical model that captures the essential behavior of SDNs, and we prove that the mechanisms correctly implement the abstractions (§3). We have formalized the model and proved the main theorems in the Coq proof assistant.
- **Verification tools:** We show how to exploit the power of our abstractions by building a tool for verifying properties of network control programs (§6).
- **Implementation:** We describe a prototype implementation on top of the OpenFlow/NOX platform (§8).
- **Experiments:** We present results from experiments run on small, but canonical applications that compare the total number of control messages and rule overhead needed to implement updates in each of these applications (§8).

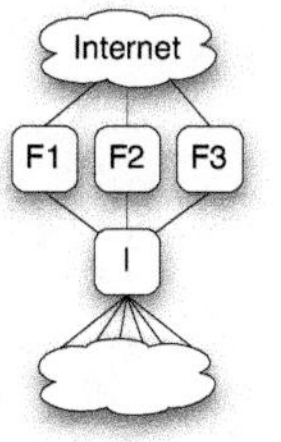

Configuration I

	Type	Action
I	U, G	Forward F_1
	S	Forward F_2
	F	Forward F_3
F_1	SSH	Monitor
	$*$	Allow
F_2	$*$	Allow
F_3	$*$	Allow

→

Configuration II

	Type	Action
I	U	Forward F_1
	G	Forward F_2
	S, F	Forward F_3
F_1	SSH	Monitor
	$*$	Allow
F_2	SSH	Monitor
	$*$	Allow
F_3	$*$	Allow

Figure 1: Access control example.

2. EXAMPLE

To illustrate the challenges surrounding network updates, consider an example network with one ingress switch I and three "filtering" switches F_1, F_2, and F_3, each sitting between I and the rest of the Internet, as shown on the left side of Figure 1. Several classes of traffic are connected to I: untrustworthy packets from Unknown and Guest hosts, and trustworthy packets from Student and Faculty hosts. At all times, the network should enforce a security policy that denies SSH traffic from untrustworthy hosts, but allows all other traffic to pass through the network unmodified. We assume that any of the filtering switches have the capability to perform the requisite monitoring, blocking, and forwarding.

There are several ways to implement this policy, and depending on the traffic load, one may be better than another. Suppose that initially we configure the switches as shown in the leftmost table in Figure 1: switch I sends traffic from U and G hosts to F_1, from S hosts to F_2, and from F hosts to F_3. Switch F_1 monitors (and denies) SSH packets and allows all other packets to pass through, while F_2 and F_3 simply let all packets pass through.

Now, suppose the load shifts, and we need more resources to monitor the untrustworthy traffic. We might reconfigure the network as shown in the table on the right of Figure 1, where the task of monitoring traffic from untrustworthy hosts is divided between F_1 and F_2, and all traffic from trustworthy hosts is forwarded to F_3. Because we cannot update the network all at once, the individual switches need to be reconfigured one-by-one. However, if we are not careful, making incremental updates to the individual switches can lead to intermediate configurations that violate the intended security policy. For instance, if we start by updating F_2 to deny SSH traffic, we interfere with traffic sent by trustworthy hosts. If, on the other hand, we start by updating switch I to forward traffic according to the new configuration (sending U traffic to F_1, G traffic to F_2, and S and F traffic to F_3), then SSH packets from untrustworthy hosts will incorrectly be allowed to pass through the network. There is one valid transition plan:

1. Update I to forward S traffic to F_3, while continuing to forward U and G traffic to F_1 and F traffic to F_3.
2. Wait until in-flight packets have been processed by F_2.
3. Update F_2 to deny SSH packets.
4. Update I to forward G traffic to F_2, while continuing to forward U traffic to F_1 and S and F traffic to F_3.

But finding this ordering and verifying that it behaves correctly requires performing intricate reasoning about a sequence of intermediate configurations—something that is tedious and error-prone, even for this simple example. Even worse, in some examples it is impossible to find an ordering that implements the transition simply by adding one part of the new configuration at a time (*e.g.*, if we swap the roles of F_1 and F_3 while enforcing the intended security policy). In general, more powerful update mechanisms are needed.

We believe that any energy the programmer devotes to navigating this space would be better spent in other ways. The tedious job of finding a safe sequence of commands that implement an update should be factored out, optimized, and reused across many applications. This is the main achievement of this paper. To implement the update using our abstractions, the programmer would simply write:

```
per_packet_update(config2)
```

Here `config2` represents the new global network configuration. The per-packet update library analyzes the configuration and topology and selects a suitable mechanism to implement the update. Note that the programmer does not write any tricky code, does not need to consider how to synchronize switch update commands, and does not need to consider the packets in flight across the network. The `per_packet_update` library handles all of the low-level details, and even attempts to select a mechanism that minimizes the cost of implementing the update.

Further, suppose the programmer knows that the security policy holds initially but wants to be sure it is enforced continuously through the update process and also afterwards when `config2` is in force. In this case, the programmer can execute another command:

```
ok = verify(config2, topo, pol)
```

If the boolean `ok` is true, then the security policy represented by `pol` holds continuously. If not, the programmer has made a mistake and can work on debugging it. The policy `pol` itself is expressed in a common specification language called CTL and is verified with the help of a model checker. We supply a library of common network properties such as loop-freeness for use with our system and programmers can write their own custom properties.

To implement the update, the library could use the safe, switch-update ordering described above. However, in general, it is not always possible to find such an ordering. Nevertheless, one can always achieve a per-packet consistent update using a two-phase update supported by configuration versioning. Intuitively, this universal update mechanism works by stamping every incoming packet with a version number (*e.g.*, stored in a VLAN tag) and modifying every configuration so that it only processes packets with a set version number. To change from one configuration to the next, it first populates the switches in the middle of the network with new configurations guarded by the next version number. Once that is complete, it enables the new configurations by installing rules at the perimeter of the network that stamp packets with that next version number. Though this general mechanism is somewhat heavyweight, our libraries identify and apply lightweight optimizations.

This short example illustrates some of the challenges that arise when implementing a network update with strong semantic guarantees. However, it also shows that all of these complexities can be hidden from the programmer, leaving only the simplest of interfaces for global network update. We believe this simplicity will lead to a more reliable and secure network infrastructure. The following sections describe our approach in more detail.

3. THE NETWORK MODEL

This section presents a simple mathematical model of the essential features of SDNs. This model is defined by a relation that describes the fine-grained, step-by-step execution of a network. We write the relation using the notation $N \xrightarrow{us}{}^{\star} N'$, where N is the network at the beginning of an execution, N' is the network after some number of steps of execution, and us is a list of "observations" that are made during the execution.[1] Intuitively, an observation should be thought of as a message between the controller

[1]When a network takes a series of steps and there are no observa-

Bit	$b ::= 0 \mid 1$
Packet	$pk ::= [b_1, ..., b_k]$
Port	$p ::= 1 \mid \cdots \mid k \mid \mathit{Drop} \mid \mathit{World}$
Located Pkt	$lp ::= (p, pk)$
Trace	$t ::= [lp_1, ..., lp_n]$
Update	$u \in \mathit{LocatedPkt} \rightharpoonup \mathit{LocatedPkt\ list}$
Switch Func.	$S \in \mathit{LocatedPkt} \to \mathit{LocatedPkt\ list}$
Topology Func.	$T \in \mathit{Port} \to \mathit{Port}$
Port Queue	$Q \in \mathit{Port} \to (\mathit{Packet} \times \mathit{Trace})\ \mathit{list}$
Configuration	$C ::= (S, T)$
Network State	$N ::= (Q, C)$

(a)

T-PROCESS

if p is any port (1)
and $Q(p) = [(pk_1, t_1), (pk_2, t_2), ..., (pk_j, t_j)]$ (2)
and $C = (S, T)$ (3)
and $S(p, pk_1) = [(p'_1, pk'_1), ..., (p'_k, pk'_k)]$ (4)
and $T(p'_i) = p''_i$, for i from 1 to k (5)
and $t'_1 = t_1 +\!\!+ [(p, pk_1)]$ (6)
and $Q'_0 = \mathit{override}(Q, p \mapsto [(pk_2, t_2), ..., (pk_j, t_j)])$ (7)
and $Q'_1 = \mathit{override}(Q'_0, p''_1 \mapsto Q(p''_1) +\!\!+ [(pk'_1, t'_1)])$

$\vdots$

and $Q'_k = \mathit{override}(Q'_{k-1}, p''_k \mapsto Q(p''_k) +\!\!+ [(pk'_k, t'_1)])$
then $(Q, C) \longrightarrow (Q'_k, C)$ (8)

T-UPDATE

if $S' = \mathit{override}(S, u)$ (9)
then $(Q, (S, T)) \xrightarrow{u} (Q, (S', T))$ (10)

(b)

Figure 2: The network model: (a) syntax and (b) semantics.

and the network. In this paper, we are interested in a single kind of message—a message u that directs a particular switch in the network to update its forwarding table with some new rules. The formal system could easily be augmented with other kinds of observations, such as topology changes or failures. For the sake of brevity, we elide these features in this paper.

The main purpose of the model is to compute the *traces*, or paths, that a packet takes through a network that is configured in a particular way. These traces in turn define the properties, be they access control or connectivity or others, that a network configuration satisfies. Our end goal is to use this model and the traces it generates to prove that, when we update a network, the properties satisfied by the initial and final configurations are preserved. The rest of this section will make these ideas precise.

Notation. We use standard notation for types. In particular, the type $T_1 \to T_2$ denotes the set of total functions that take arguments of type T_1 and produce results of type T_2, while $T_1 \rightharpoonup T_2$ denotes the set of partial functions from T_1 to T_2; the type $T_1 \times T_2$ denotes the set of pairs with components of type T_1 and T_2; and T *list* denotes the set of lists with elements of type T.

We also use standard notation to construct tuples: (x_1, x_2) is a pair of items x_1 and x_2. For lists, we use the notation $[x_1, ..., x_n]$ for the list of n elements x_1 through x_n, $[\,]$ for the empty list, and $xs_1 +\!\!+ xs_2$ for the concatenation of the two lists xs_1 and xs_2. Notice that if x is some sort of object, we will typically use xs as the variable for a list of such objects. For example, we use u to represent a single update and us to represent a list of updates.

Basic Structures. Figure 2(a) defines the syntax of the elements of the network model. A *packet pk* is a sequence of bits, where a *bit b* is either 0 or 1. A *port p* represents a location in the network where packets may be waiting to be processed. We distinguish two kinds of ports: ordinary ports numbered uniquely from 1 to k, which correspond to the physical input and output ports on switches, and two special ports, *Drop* and *World*. Intuitively, packets queued at the *Drop* port represent packets that have been dropped, while packets queued at the *World* port represent packets that have been forwarded beyond the confines of the network. Each ordinary port will be located on some switch in the network. However, we will leave the mapping from ports to switches unspecified, as it is not needed for our primary analyses.

Switch and Topology Functions. A network is a packet processor that forwards packets and optionally modifies the contents of those packets on each hop. Following Kazemian *et al.* [14], we model packet processing as the composition of two simpler behaviors: (1) forwarding a packet across a switch and (2) moving packets from one end of a link to the other end. The *switch function* S takes a located packet lp (a pair of a packet and a port) as input and returns a list of located packets as a result. In many applications, a switch function only produces a single located packet, but in applications such as multicast, it may produce several. To drop a packet, a switch function maps the packet to the special *Drop* port. The *topology function* T maps one port to another if the two ports are connected by a link in the network. Given a topology function T, we define an ordinary port p to be an *ingress port* if for all other ordinary ports p' we have $T(p') \neq p$. Similarly, we define an ordinary port p to be an *internal port* if it is not an ingress port.

To ensure that switch and topology functions are reasonable, we impose the following conditions:

(1) For all packets pk, $S(\mathit{Drop}, pk) = [(\mathit{Drop}, pk)]$ and $S(\mathit{World}, pk) = [(\mathit{World}, pk)]$;

(2) $T(\mathit{Drop}) = \mathit{Drop}$ and $T(\mathit{World}) = \mathit{World}$; and

(3) For all ports p and packets pk if $S(p, pk) = [(p_1, pk_1), ..., (p_k, pk_k)]$ we have $k \geq 1$.

Taken together, the first and second conditions state that once a packet is dropped or forwarded beyond the perimeter of the network, it must stay dropped or beyond the perimeter of the network and never return. If our network forwards a packet out to another network and that other network forwards the packet back to us, we treat the return packet as a "fresh" packet—*i.e.*, we do not explicitly model inter-domain forwarding. The third condition states that applying the forwarding function to a port and a packet must produce at least one packet. This third condition means that the network

tions (*i.e.*, no updates happen), we omit the list above the arrow, writing $N \longrightarrow^{\star} N'$ instead.

cannot drop a packet simply by not forwarding it anywhere. Dropping packets occurs by explicitly forwarding a single packet to the *Drop* port. This feature makes it possible to state network properties that require packets either be dropped or not.

Configurations and Network States. A *trace* t is a list of located packets that keeps track of the hops that a packet takes as it traverses the network. A *port queue* Q is a total function from ports to lists of packet-trace pairs. These port queues record the packets waiting to be processed at each port in the network, along with the full processing history of that packet. Several of our definitions require modifying the state of a port queue. We do this by building a new function that overrides the old queue with a new mapping for one of its ports: $override(Q, p \mapsto l)$ produces a new port queue Q' that maps p to l and like Q otherwise.

$$override(Q, p \mapsto l) = Q'$$
$$\text{where } Q'(p') = \begin{cases} l & \text{if } p = p' \\ Q(p') & \text{otherwise} \end{cases}$$

A *configuration* C comprises a switch function S and a topology function T. A *network state* N is a pair (Q, C) containing a port queue Q and configuration C.

Transitions. The formal definition of the network semantics is given by the relations defined in Figure 2(b), which describe how the network transitions from one state to the next one. The system has two kinds of transitions: packet-processing transitions and update transitions. In a packet-processing transition, a packet is retrieved from the queue for some port, processed using the switch function S and topology function T, and the newly generated packets are enqueued onto the appropriate port queues. More formally, packet-processing transitions are defined by the T-PROCESS case in Figure 2(b). Lines 1-8 may be read roughly as follows:

(1) If p is any port,
(2) a list of packets is waiting on p,
(3) the configuration C is a pair of a switch function S and topology function T,
(4) the switch function S forwards the chosen packet to a single output port, or several ports in the case of multicast, and possibly modifies the packet
(5) the topology function T connects each output port to an input port,
(6) a new trace t'_1, which extends the old trace and records the current hop, is generated,
(7) a new queue Q'_k is generated by moving packets across links as specified in steps (4), (5) and (6),
(8) then (Q, C) can step to (Q'_k, C).

In an update transition, the switch forwarding function is updated with new behavior. We represent an *update* u as a partial function from located packets to lists of located packets (*i.e.*, an update is just a "part" of a global (distributed) switch function). To apply an update to a switch function, we overwrite the function using all of the mappings contained in the update. More formally, $override(S, u)$ produces a new function S' that behaves like u on located packets in the domain[2] of u, and like S otherwise.

$$override(S, u) = S'$$
$$\text{where } S'(p, pk) = \begin{cases} u(p, pk) & \text{if } (p, pk) \in \mathsf{dom}(u) \\ S(p, pk) & \text{otherwise} \end{cases}$$

[2]Domain of an update is the set of located packets it's defined upon.

Update transitions are defined formally by the T-UPDATE case in Figure 2(b). Lines 9-10 may be read as follows: if S' is obtained by applying update u to a switch in the network then network state $(Q, (S, T))$ can step to new network state $(Q, (S', T))$.

Network Semantics. The overall semantics of a network in our model is defined by allowing the system to take an arbitrary number of steps starting from an *initial state* in which the queues of all internal ports as well as *World* and *Drop* are empty, and the queues of external ports are filled with pairs of packets and the empty trace. The reflexive and transitive closure of the single-step transition relation $N \xrightarrow{us}{}^{\star} N'$ is defined in the usual way, where the sequence of updates recorded in the label above the arrow is obtained by concatenating all of the updates in the underlying transitions in order.[3] A network *generates* a trace t if and only if there exists an initial state Q such that $(Q, C) \longrightarrow^{\star} (Q', C)$ and t appears in Q'. Note that no updates may occur when generating a trace.

Properties. In general, there are myriad properties a network might satisfy—*e.g.*, access control, connectivity, in-order delivery, quality of service, fault tolerance, to name a few. In this paper, we will primarily be interested in *trace properties*, which are prefix-closed sets of traces. Trace properties characterize the paths (and the state of the packet at each hop) that an individual packet is allowed to take through the network. Many network properties, including access control, connectivity, routing correctness, loop-freedom, correct VLAN tagging, and waypointing can be expressed using trace properties. For example, loop-freedom can be specified using a set that contain all traces except those in which some ordinary port p appears twice. In contrast, timing properties and relations between multiple packets including quality of service, congestion control, in-order delivery, or flow affinity are not trace properties.

We say that a port queue Q satisfies a trace property P if all of the traces that appear in Q also appear in the set P. Similarly, we say that a network configuration C satisfies a trace property P if for all *initial* port queues Q and all (update-free) executions $(Q, C) \longrightarrow^{\star} (Q', C)$, it is the case that Q' satisfies P.

4. PER-PACKET ABSTRACTION

One reason that network updates are difficult to get right is that they are a form of concurrent programming. Concurrent programming is hard because programmers must consider the interleaving of every operation in every thread and this leads to a combinatorial explosion of possible outcomes—too many outcomes for most programmers to manage. Likewise, when performing a network update, a programmer must consider the interleaving of switch update operations with every kind of packet that might be traversing their network. Again, the number of possibilities explodes.

Per-packet consistent updates reduce the number of scenarios a programmer must consider to just two: for every packet, it is as if the packet flows through the network *completely before* the update occurs, or *completely after* the update occurs.

One might be tempted to think of per-packet consistent updates as "atomic updates", but they are actually better than that. An atomic update would cause packets in flight to be processed partly according to the configuration in place prior to the update, and partly according to the configuration in place after the update. To understand what happens to those packets (*e.g.*, whether they get dropped), a programmer would have to reason about every possible

[3]The semantics of the network is defined from the perspective of an omniscient observer, so there is an order in which the steps occur.

trace formed by concatenating a prefix generated by the original configuration with a suffix generated by the new configuration.

Intuitively, per-packet consistency states that for a given packet, the traces generated during an update come from the old configurations, or the new configuration, but not a mixture of the two. In the formal definition of per-packet consistency, we introduce an equivalence relation $\sim$ on packets. We extend this equivalence relation to traces by considering two traces to be equivalent if the packets they contain are equivalent according to the $\sim$ relation (similarly, we extend $\sim$ to properties in the obvious way). We then require that all traces generated *during* the update be equivalent to a trace generated by either the initial or final configuration. For the rest of the paper, when we say that $\sim$ is an equivalence relation on traces, we assume that it has been constructed like this. This specification gives implementations of updates flexibility by allowing some minor, irrelevant differences to appear in traces (where $\sim$ defines the notion of irrelevance precisely). For example, we can define a "version" equivalence relation that relates packets pk and pk' which differ only in the value of their version tags. This relation will allow us to state that changes to version tags performed by the implementation mechanism for per-packet update are irrelevant. In other words, a per-packet mechanism may perform internal bookkeeping by stamping version tags without violating our technical requirements on the correctness of the mechanism. The precise definition of per-packet consistency is as follows.

Definition 1 (Per-packet $\sim$-consistent update) *Let $\sim$ be a trace-equivalence relation. An update sequence us is a per-packet $\sim$-consistent update from C_1 to C_2 if and only if, for all*

- *initial states Q,*
- *executions $(Q, C_1) \xrightarrow{us}{}^{\star} (Q', C_2)$,*
- *and traces t in Q',*

there exists

- *an initial state Q_i,*
- *and either an execution $(Q_i, C_1) \longrightarrow^{\star} (Q'', C_1)$ or an execution $(Q_i, C_2) \longrightarrow^{\star} (Q'', C_2)$,*

such that Q'' contains t', for some trace t' with $t' \sim t$.

From an implementer's perspective, the operational definition of per-packet consistency given above provides a specification that he or she must meet. However, from a programmer's perspective, there is another, more useful side to per-packet consistency: *per-packet consistent updates preserve every trace property.*

Definition 2 ($\sim$-property preservation) *Let C_1 and C_2 be configurations and $\sim$ be a trace-equivalence relation. A sequence us is a $\sim$-property-preserving update from C_1 and C_2 if and only if, for all*

- *initial states Q,*
- *executions $(Q, C_1) \xrightarrow{us}{}^{\star} (Q', C_2)$,*
- *and properties P that are satisfied by C_1 and C_2 and do not distinguish traces related by $\sim$,*

we have that Q' satisfies P.

Universal $\sim$-property preservation gives programmers a strong principle they can use to reason about their programs. If programmers check that a trace property such as loop-freedom or access control holds of the network configurations before and after an update, they are guaranteed it holds of every trace generated throughout the update process, even though the series of observations us may contain many discrete update steps. Our main theorem states that per-packet consistent updates preserve all properties:

Theorem 1 *For all trace-equivalence relations $\sim$, if us is a per-packet $\sim$-consistent update of C_1 to C_2 then us is a $\sim$-property-preserving update of C_1 to C_2.*

The proof of the theorem is a relatively straightforward application of our definitions. From a practical perspective, this theorem allows a programmer to get great mileage out of per-packet consistent updates. In particular, since per-packet consistent updates preserve *all* trace properties, the programmers do not have to tell the system *which* specific properties must be invariant in their applications.

From a theoretical perspective, it is also interesting that the converse of the above theorem holds. This gives us a sort of completeness result: if programmers want an update that preserves all properties, they need not search for it outside of the space of per-packet consistent updates—any universal trace-property preserving update is a per-packet consistent update.

Theorem 2 *For all trace-equivalence relations $\sim$, if us is a $\sim$-property-preserving update of C_1 to C_2 then us is a per-packet $\sim$-consistent update of C_1 to C_2.*

The proof of this theorem proceeds by observing that since us preserves all $\sim$-properties, it certainly preserves the following $\sim$-property P_{or}:

$$\begin{array}{l}\{t \mid \text{there exists an initial } Q \text{ and a trace } t' \\ \quad \text{and } ((Q, C_1) \longrightarrow^{\star} (Q', C_1) \text{ or } (Q, C_2) \longrightarrow^{\star} (Q', C_2)), \\ \quad \text{and } t \sim t', \\ \quad \text{and } t' \in Q'\}\end{array}$$

By the definition of P_{or}, the update us generates no traces that cannot be generated either by the initial configuration C_1 or by the final configuration C_2. Hence, by definition, us is per-packet consistent.

5. PER-PACKET MECHANISMS

Depending on the network topology and the specifics of the configurations involved, there may be several ways to implement a per-packet consistent update. However, all of the techniques we have discovered so far, no matter how complicated, can be reduced to two fundamental building blocks: the one-touch update and the unobservable update. For example, our two-phase update mechanism uses unobservable updates to install the new configuration before it is used, and then "unlocks" the new policy by performing a one-touch update on the ingress ports.

One-touch updates. A one-touch update is an update with the property that no packet can follow a path through the network that reaches an updated (or to-be-updated) part of the switch rule space more than once.

Definition 3 (One-touch Update) *Let $C_1 = (S, T)$ be the original network configuration, $us = [u_1, ..., u_k]$ an update sequence, and $C_2 = (S[u_1, ..., u_k], T)$ the new configuration, such that the domains of each update u_1 to u_k are mutually disjoint. If, for all*

- *initial states Q,*
- *and executions $(Q, C_1) \xrightarrow{us}{}^{\star} (Q', C_2)$,*

there does not exist a trace t in Q' such that

- *t contains distinct trace elements (p_1, pk_1) and (p_2, pk_2),*
- *and (p_1, pk_1) and (p_2, pk_2) both appear in the domain of update functions $[u_1, ..., u_k]$,*

then us is a one-touch update from C_1 to C_2.

Theorem 3 *If us is a one-touch update then us is a $\sim$-per-packet consistent update for any $\sim$.*

The proof proceeds by considering the possible traces t generated by an execution $(Q, C_1) \xrightarrow{us}{}^\star (Q', C_2)$. There are two cases: (1) There is no element of t that appears in the domain of an update function in us, or (2) some element lp of t appears in the domain of an update function in us. In case (1), t can also be generated by an execution with no update observations: $(Q, C_1) \longrightarrow^\star (Q'', C_1)$, and the definition of per-packet consistency vacuously holds. In case (2), there are two subcases:

(i) lp appears in the trace prior to the update taking place and so t is also generated by $(Q, C_1) \longrightarrow^\star (Q'', C_1)$.

(ii) lp appears in the trace after the update has taken place and so t is also generated by $(Q, C_2) \longrightarrow^\star (Q'', C_2)$.

The one-touch update mechanism has a few immediate, more specific applications:

- *Loop-free switch updates:* If a switch is not part of a topological loop (either before or after the update), then updating all the ports on that switch is an instance of a one-touch update and is per-packet consistent.
- *Ingress port updates:* An ingress port interfaces exclusively with the external world, so it can not be a part of an internal topological loop and is never on the same trace as any other ingress port. Consequently, any update to ingress ports is a one-touch update and is per-packet consistent. Such updates can be used to change the admission control policy for the network, either by adding or excluding flows.

When one-touch updates are combined with unobservable updates, there are many more possibilities.

Unobservable updates. An unobservable update is an update that does not change the set of traces generated by a network.

Definition 4 (Unobservable Update) *Let $C_1 = (S, T)$ be the original network configuration, $us = [u_1, ..., u_k]$ an update sequence, and $C_2 = (S[u_1, ..., u_k], T)$ the new configuration. If, for all*

- *initial states Q,*
- *executions $(Q, C_1) \xrightarrow{us}{}^\star (Q', C_2)$,*
- *and traces t in Q',*

there exists

- *an initial state Q_i,*
- *and an execution $(Q_i, C_1) \longrightarrow^\star (Q'', C_1)$,*

such that the trace t is in Q'', then us is an unobservable update from C_1 to C_2.

Theorem 4 *If us is an unobservable update then us is a per-packet consistent update.*

The proof proceeds by observing that every trace generated during the unobservable update $(Q, C_1) \xrightarrow{us}{}^\star (Q', C_2)$ also appears in the traces generated by C_1.

On their own, unobservable updates are useless as they do not change the semantics of packet forwarding. However, they may be combined with other per-packet consistent updates to great effect using the following theorem.

Theorem 5 (Composition) *If us_1 is an unobservable update from C_1 to C_2 and us_2 is a per-packet consistent update from C_2 to C_3 then $us_1 \mathbin{+\!\!+} us_2$ is a per-packet consistent update from C_1 to C_3.*

A simple use of composition arises when one wants to achieve a per-packet consistent update that extends a policy with a completely new path.

- *Path extension:* Consider an initial configuration C_1. Suppose $[u_1, u_2, ..., u_k]$ updates ports $p_1, p_2, ..., p_k$ respectively to lay down a new path through the network with u_1 updating the ingress port. Suppose also that the ports updated by $us = [u_2, ..., u_k]$ are unreachable in network configuration C_1. Hence, us is an unobservable update. Since $[u_1]$ updates an ingress port, it is a one-touch update and also per-packet consistent. By the composition principle, $us \mathbin{+\!\!+} [u_1]$ is a per-packet consistent update.

Notice that the path update is achieved by first laying down rules on switches 2 to k and then, when that is complete, laying down the rules on switch 1. A well-known (but still common!) bug occurs when programmers attempt to install new forwarding paths but lay down the elements of the path in wrong order [11]. Typically, there is a race condition in which packets traverse the first link and reach the switch 2 before the program has had time to lay down the rules on links 2 to k. Then when packets reach switch 2, it does not yet know how to handle them, and a default rule sends the packets to the controller. The controller often becomes confused as it begins to see additional packets that should have already been dealt with by laying down the new rules. The underlying cause of this bug is explained with our model—the programmer intended a per-packet consistent update of the policy with a new path, but failed to implement per-packet consistency correctly. All such bugs are eradicated from network programs if programmers use per-packet consistent updates and never use their own ad hoc update mechanisms.

Two-phase update. So far, all of our update mechanisms have applied to special cases in which the topology, existing configuration, and/or updates have specific properties. Fortunately, provided there are a few bits in packets that are irrelevant to the network properties a programmer wishes to enforce, and can be used for bookkeeping purposes, we can define a mechanism that handles arbitrary updates using a two-phase update protocol.

Intuitively, the two-phase update works by first installing the new configuration on internal ports, but only enabling the new configuration for packets containing the correct version number. It then updates the ingress ports one-by-one to stamp packets with the new version number. Notice that the updates in the first phase are all unobservable, since before the update, the ingress ports do not stamp packets with the new version number. Hence, since updating ingress ports is per-packet consistent, by the composition principle, the two-phase update is also per-packet consistent.

To define the two-phase update formally, we need a few additional definitions. Let a version-property be a trace property that does not distinguish traces based on the value of version tags. A configuration C is a version-n configuration if $C = (S, T)$ and S modifies packets processed by any ingress port p_{in} so that after passing through p_{in}, the packet's version bit is n. We assume that the S function does not otherwise modify the version bit of the packet. Two configurations C and C' coincide internally on version-n packets whenever $C = (S, T)$ and $C' = (S', T')$ and for all internal ports p, and for all packets pk with version bit set to n, we have that $S(p, pk) = S'(p, pk)$. Finally, an update u is a refinement of S, if for all located packets lp in the domain of u, we have that $u(lp) = S(lp)$.

Definition 5 (Two-phase Update) *Let $C_1 = (S, T)$ be a version-1 configuration and $C_2 = (S', T)$ be a version-2 configuration. Assume that C_1 and C_2 coincide internally on version-1 packets. Let $us = [u_1^i, ..., u_m^i, u_1^e, ..., u_n^e]$ be an update sequence such that*

- *$S' = override(S, us)$,*
- *each u_j^i and u_k^e is a refinement of S',*
- *p is internal, for each (p, pk) in the domain of u_j^i,*
- *and p is an ingress, for each (p, pk) in the domain of u_k^e.*

Then us is a two-phase update from C_1 to C_2.

Theorem 6 *If us is a two-phase update then us is per-packet consistent.*

The proof simply observes that $us_1 = [u_1^i, ..., u_m^i]$ is an unobservable update, and $us_2 = [u_1^e, ..., u_n^e]$ is a one-touch update (and therefore per-packet consistent). Hence, by composition, the two-phase update $us_1 +\!\!+\ us_2$ is per-packet consistent.

Optimized mechanisms. Ideally, update mechanisms should satisfy *update proportionality*, where the cost of installing a new configuration should be proportional to the size of the configuration change. A perfectly proportional update would (un)install just the "delta" between the two configurations. The full two-phase update mechanism that installs the full new policy and then uninstalls the old policy lacks update proportionality. In this section, we describe optimizations that substantially reduce overhead.

Pure extensions and retractions are one important case of updates where a per-packet mechanism achieves perfect proportionality. A pure extension is an update that adds new paths to the current configuration that cannot be reached in the old configuration—*e.g.*, adding a forwarding path for a new host that comes online. Such updates do not require a complete two-phase update, as only the new forwarding rules need to be installed—first at the internal ports and then at the ingresses. The rules are installed using the current version number. A *pure retraction* is the dual of a pure extension in which some paths are removed from the configuration. Again, the paths being removed must be unreachable in the new configuration. Pure retractions can be implemented by updating the ingresses, pausing to wait until packets in flight drain out of the network, and then updating the internal ports.

If paths are not only added or removed but are modified then more powerful optimizations are available. Per-packet consistency requires that the active paths in the network come from either of the configurations. The subset mechanism works by identifying the paths that have been added, removed or changed and then updating the rules along the entire path to use a new version. This optimization is always applicable, but in the degenerate case it devolves into a network-wide two-phase update.

6. CHECKING PROPERTY INVARIANCE

As per-packet consistent updates preserve all trace properties, programmers can turn any trace property checker that verifies individual, static network configurations into a verification engine that verifies the *invariance* of trace properties as configurations evolve over time. In this section, we demonstrate this idea concretely using a model checker, but it applies quite broadly; any kind of static network analysis can also benefit from our abstractions.

Model Checking Trace Properties. Intuitively, trace properties describe the paths that an individual packet is allowed to travel through the network, as well as the packet state at each point in the path. Temporal logic is a natural fit for the specification of such properties.

We use Computation Tree Logic (CTL) [15], a branching time temporal logic, to specify the legal paths that a packet may take through the network. Simple formulae describe attributes of packets in the network. For example, the formula

```
port = 3 & src_ip = 10.0.0.1
```

describes a packet with a source IP 10.0.0.1 at port 3. As a second example, recall that our network model contains two special ports, *Drop* and *World*. A dropped packet is one that arrives at the *Drop* port; packets forwarded through our network correctly arrive at the *World* port.[4] Hence the formula `port = DROP` simply states that the packet in question has been dropped.

While simple formulae describe a packet at one place in the network, the following quantified formulae describe a packet's path through the network.

- `AX` ϕ or `EX` ϕ. The formula ϕ holds at the next position on all paths (`AX`) from the current position, or at the next position along at least one path (`EX`).
- `AF` ϕ or `EF` ϕ. On all paths (`AF`) or on some path (`EF`) from the current position, ϕ holds on *some* future position.
- `AG` ϕ or `EG` ϕ. On all paths (`AG`) or on some path (`EG`) from the current position, ϕ holds on *all* future positions.

With these operators, we can specify many interesting trace properties. For example, the 'no black holes' property (incoming traffic must not be dropped) can be specified as `AG (port != DROP)`. Read aloud, this formula says, "On all paths, and at all future positions on those paths, the current port is never *Drop*."

Below, we state other useful trace properties as CTL formulae.

- **No loops:** every packet will eventually be dropped or leave the network. We represent this with the formula

  `AF (port = DROP | port = WORLD).`

  That is, on all paths through the network, a packet either reaches the outside world or is dropped.

- **Egress:** all packets with a header field `h = 1` reach the external world:

  `h = 1 -> AF (port = WORLD).`

  That is, all paths for packets with `h = 1` eventually lead to the *World* port.

- **Waypointing:**[5] all packets with a header field `h = 1` reach switch `s4`.

  `h = 1 -> AF (switch = s4).`

  That is, all paths for packets with `h = 1` eventually pass through switch `s4`.

[4]Recall that in our model, no packet is ever silently dropped—packets are only explicitly dropped by virtue of being sent to the *Drop* port.

[5]Though the model from the previous section did not mention switches, our implementation contains a notion of a set of ports belonging to a switch.

- **Blacklisting:** no packets with header `h = 0` traverse link (`s2`, `s3`).

```
(h = 0) ->
  AG (switch = s2 -> ~(EX switch = s3))
```

That is, if a packet header matches `h = 0`, then on all paths where the packet reaches switch `s2`, it will not reach switch `s3` on the next hop—i.e. it cannot traverse the edge between `s2` and `s3`, if such an edge exists.

Blacklisting, which stipulates that a packet must not reach an edge or switch, is roughly the negation of waypointing, which stipulates that a packet must pass through an edge or switch. Thus, both switches and edges can be waypointed or blacklisted. Moreover, both properties generalize to sets of edges and switches by adding a clause for each element to the conclusion of the implication. For example, the waypointing property where packets with headers `h = 0` must pass through switches `s1` and `s2` is:

```
h = 0 -> (AF switch = s1 & AF switch = s2).
```

Deploying Trace Verification. Static analyses play a significant role in the traditional software development cycle, because they offer an opportunity to verify program properties before the final software deliverable ships. Software-defined networking provides a similar opportunity for network programmers: network updates are programs and can be analyzed before being deployed to update the network. As in the traditional software development cycle, the judicious use of static analyses in network programming can pinpoint bugs before deployment, reducing time spent diagnosing performance problems and patching security vulnerabilities. We envision verification procedures, like the trace verification described above, being deployed as part of the following scenarios:

- **Planned change.** Before deploying a planned network update, programmers verify that the new configuration maintains expected security and reliability properties.
- **Debugging.** Trace verification can serve as an "assert" statement, allowing the programmer to quickly check properties of the configuration under development.
- **Background verification.** The programmer may also verify a network configuration after deployment, reducing the delay in making updates while still catching bad configurations in a reasonably timely manner.

We implemented a library of canonical formulae based on the properties presented above, which we used to verify the example benchmarks we report on in this paper. On small examples—with tens of switches and hundreds of rules—the NuSMV tool reports results in under a second. Larger examples—such as guaranteeing no loops for shortest-path routing over Waxman graphs [16] with hundreds of switches and hundreds of thousands of rules—take over an hour.We leave it as an open question for the model checking community to determine how far verification of network trace properties can scale.

7. PER-FLOW CONSISTENCY

Per-packet consistency, while simple and powerful, is not always enough. Some applications require a stream of related packets to be handled consistently. For example, a server load-balancer needs all packets from the same TCP connection to reach the same server replica. In this section, we introduce the per-flow consistency abstraction, and discuss mechanisms for per-flow consistent updates.

Per-flow abstraction. To see the need for per-flow consistency, consider a network where a single switch S load-balances between two back-end servers A and B. Initially, S directs traffic from IP addresses starting with 0 (*i.e.*, source addresses in 0.0.0.0/1) to A and 1 (*i.e.*, source addresses in 128.0.0.0/1) to B. At some time later, we bring two additional servers C and D online, and re-balance the load using a two-bit prefix, directing traffic from addresses starting with 00 to A, 01 to B, 10 to C, and 11 to D.

Intuitively, we want to process packets from new TCP connections according to the new configuration. However, all packets in existing flows must go to the same server, where a *flow* is a sequence of packets with related header fields, entering the network at the same port, and not separated by more than n seconds. The particular value of n depends upon the protocol and application. For example, the switch should send packets from a host whose address starts with "11" to B, and not to D as the new configuration would dictate, if the packets belong to an ongoing TCP connection. Simply processing individual packets with a single configuration does not guarantee the desired behavior.

Per-flow consistency guarantees that all packets in the same flow are handled by the same version of the configuration. Formally, the per-flow abstraction preserves all path properties, as well as all properties that can be expressed in terms of the paths traversed by *sets* of packets belonging to the same flow.

Per-flow mechanisms. Implementing per-flow consistent updates is much more complicated than per-packet consistency because the system must identify packets that belong to active flows. Below, we discuss three different mechanisms. Our system implements the first of the three; the latter two, while promising, depend upon technology that is not yet available in OpenFlow.

Switch rules with timeouts: A simple mechanism can be obtained by combining versioning with rule timeouts, similar to the approach in [7]. The idea is to pre-install the new configuration on the internal switches, leaving the old version in place, as in per-packet consistency. Then, on ingress switches, the controller sets soft timeouts on the rules for the old configuration and installs the new configuration at lower priority. When all flows matching a given rule finish, the rule automatically expires and the rules for the new configuration take effect. When multiple flows match the same rule, the rule may be artificially kept alive even though the "old" flows have all completed. If the rules are too coarse, then they may never die! To ensure rules expire in a timely fashion, the controller can refine the old rules to cover a progressively smaller portion of the flow space. However, "finer" rules require more rules, a potentially scarce commodity. Managing the rules and dynamically refining them over time can be a complex bookkeeping task, especially if the network undergoes a *subsequent* configuration change before the previous one completes. However, this task can be implemented and optimized once in a run-time system, and leveraged over and over again in different applications.

Wildcard cloning: An alternative mechanism exploits the wildcard *clone* feature of the DevoFlow extension of OpenFlow [17]. When processing a packet with a *clone* rule, a DevoFlow switch creates a new "microflow" rule that matches the packet header fields exactly. In effect, clone rules cause the switch to maintain a concrete representation of each active flow. This enables a simple update mechanism: first, use clone rules whenever installing configurations; second, to update from old to new, simply replace all old clone rules with the new configuration. Existing flows will continue to be handled by the exact-match rules previously generated by the old clone rules, and new flows will be handled by the new clone rules, which themselves immediately spawn new microflow rules.

While this mechanism does not require complicated bookkeeping on the controller, it does require a more complex switch.

End-host feedback: The third alternative exploits information readily available on the end hosts, such as servers in a data center. With a small extension, these servers could provide a list of active sockets (identified by the "five tuple" of IP addresses, TCP/UDP ports, and protocol) to the controller. As part of performing an update, the controller would query the local hosts and install high-priority microflow rules that direct each active flow to the assigned server replica. These rules could "timeout" after a period of inactivity, allowing future traffic to "fall through" to the new configuration. Alternatively, the controller could install "permanent" microflow rules, and explicitly remove them when the socket no long exists on the host, obviating the need for any assumptions about the minimum interval time between packets of the same connection.

8. IMPLEMENTATION AND EVALUATION

We have built a system called Kinetic that implements the update abstractions introduced in this paper, and evaluated its performance on small but canonical example applications. This section summarizes the key features of Kinetic and presents experimental results that quantify the cost of implementing network updates in terms of the number of rules added and deleted on each switch.

Implementation overview. Kinetic is a run-time system that sits on top of the NOX OpenFlow controller [18]. The system comprises several Python classes for representing network configurations and topologies, and a library of update mechanisms. The interface to these mechanisms is through the `per_packet_update` and `per_flow_update` functions. These functions take a new configuration and a network topology, and implement a transition to the new configuration while providing the desired consistency level. Both functions are currently based on the two-phase update mechanism, with the `per_flow_update` function using timeouts to track active flows. In addition to this basic mechanism, we have implemented a number of optimized mechanisms that can be applied under certain conditions—*e.g.*, when the update only affects a fraction of the network or network traffic. The runtime automatically analyzes the new configuration and topology and applies these optimizations when possible to reduce the cost of the update.

As described in Section 5, the two-phase update mechanism uses versioning to isolate the old configuration and traffic from the updated configuration. Because Kinetic runs on top of OpenFlow 1.0, we currently use the VLAN field to carry version tags (other options, like MPLS labels, are available in newer versions of OpenFlow). Our algorithms analyze the network topology to determine the ingress and internal ports and perform a two-phase update.

Experiments. To evaluate the performance of Kinetic, we developed a suite of experiments using the Mininet [19] environment. Because Mininet does not offer performance fidelity or resource isolation between the simulated switches and the controller, we did not measure the time needed to implement an update. However, as a proxy for elapsed time, we counted the total number of install OpenFlow messages needed to implement each update, as well as the number of extra rules (beyond the size of either the old or new configurations) installed on a switch.

To evaluate per-packet consistency, we have implemented two canonical network applications: routing and multicast. The routing application computes the shortest paths between each host in the topology and updates routes as hosts come online or go offline and switches are brought up and taken down for maintenance. The multicast application divides the hosts evenly into two multicast groups and implements IP multicast along a spanning tree that connects all of the hosts in a group. To evaluate the effects of our optimizations, we ran both applications on three different topologies each containing 192 hosts and 48 switches in each of three different scenarios. The topologies were chosen to represent realistic and proposed network topologies found in datacenters (fattree, small-world), enterprises (fattree) and a random topology (waxman). The three scenarios can be divided up into:

1. Dynamic hosts and static routes
2. Static hosts and dynamic routes
3. Dynamic hosts and dynamic routes

In each scenario, we moved between 3 different configurations, changing the network in a well-prescribed manner. In the dynamic host scenario, we randomly selected between $10\% - 20\%$ of the hosts and added or removed them from the network. In the dynamic routes scenario, we randomly selected 20% of the routes in the network, and forced them to re-route as if one of the switches in the route had been removed. For the multicast example, we changed one of the multicast groups each time. Static means that we did not change the host or routes.

To evaluate per-flow updates, we developed a load-balancing application that divides traffic between two server replicas, using a hash computed from the client's IP address. The update for this experiment involved bringing several new server replicas online and re-balancing the load among all of the servers.

Results and analysis. Table 2 compares the performance of the subset optimization to a full two-phase update. Extension updates are not shown: whenever an extension update is applicable, our subset mechanism performs the same update with the same overhead. The two-phase update has high overhead in all scenarios.

We subject each application to a series of topology changes—adding and dropping hosts and links—reflecting common network events that force the deployment of new network configurations. We measure the number of OpenFlow operations required for the deployment of the new configuration, as well as the overhead of installing extra rules to ensure per-packet consistency. The overhead is the ratio of the number of extra rules installed during the per-packet update of a switch divided by the (maximum) number of rules in the old or new configuration. For example, if the old and new configurations both had 100 rules and during the update the switch had 120 rules installed, that would be a 20% overhead. The *Overhead* column in Table 2 presents the maximum overhead of all switches in the network. Two-phase update requires approximately 100% overhead, because it leaves the old configuration on the switch as it installs the new one. Because both configurations may not be precisely the same size, it is not always exactly 100%. In some cases, the new configuration may be much smaller than the old (for example, when routes are diverted away from a switch) and the overhead is much lower than 100%.

The first routing scenario, where hosts are added or removed, demonstrates the potential of our optimizations. When a new host comes online, the application computes routes between it and every other online host. Because the rules for the new routes do not affect traffic between existing hosts, they can be installed without modifying or reinstalling the existing rules. Similarly, when a host goes offline, only the installed rules routing traffic to or from that host need to be uninstalled. This leads to update costs proportional to the number of rules that changed between configurations, as opposed to a full two-phase update, where the cost is proportional to the size of the entire new configuration.

Application	Toplogy	Update	2PC		Subset		
			Ops	*Max Overhead*	*Ops*	*Ops %*	*Max Overhead*
Routing	Fat Tree	Hosts	239830	92%	119003	50%	20%
		Routes	266234	100%	123929	47%	10%
		Both	239830	92%	142379	59%	20%
	Waxman	Hosts	273514	88%	136230	49%	66%
		Routes	299300	90%	116038	39%	9%
		Both	267434	91%	143503	54%	66%
	Small World	Hosts	320758	80%	158792	50%	30%
		Routes	326884	85%	134734	41%	23%
		Both	314670	90%	180121	57%	41%
Multicast	Fat Tree	Hosts	1043	100%	885	85%	100%
		Routes	1170	100%	634	54%	57%
		Both	1043	100%	949	91%	100%
	Waxman	Hosts	1037	100%	813	78%	100%
		Routes	1132	85%	421	37%	50%
		Both	1005	100%	821	82%	100%
	Small World	Hosts	1133	100%	1133	100%	100%
		Routes	1114	90%	537	48%	66%
		Both	1008	100%	1008	100%	100%

Experimental results comparing two-phase update (2PC) with our subset optimization (Subset). We add or remove hosts and change routes to trigger configuration updates. The *Ops* column measures the number of OpenFlow install operations used in each situation. The Subset portion of the table also has an additional column (Ops %) that tabulates (Subset Ops / 2PC Ops). *Overhead* measures the extra rules concurrently installed on a switch by our update mechanisms. We pessimistically present the maximum of the overheads for all switches in the network – there may be many switches in the network that never suffer that maximum overhead.

Table 2: Experimental results.

Our optimizations yield fewer improvements for the multicast example, due to the nature of the example: when the spanning tree changes, almost all paths change, triggering an expensive update.

We have not applied our optimizations to the per-flow mechanism, therefore we do not include an optimization evaluation of the load balancing application.

9. RELATED WORK

This paper builds on our earlier workshop paper [20], which did not include a formal model or proofs, the optimized per-packet update mechanisms, formal verification of network configurations, or an implementation and evaluation.

The problem of avoiding undesired transient behavior during planned change has been well studied in the domain of distributed routing protocols. Most prior work focused on protocol-specific approaches of adjusting link metrics to minimize disruptions [8, 2, 3]. The recent work by Vanbever *et. al* [5] also handles more significant intradomain routing changes, such as switching to a different routing protocol. Unlike our approach, these methods can only preserve basic properties such as loop-freedom and connectivity. In addition, these approaches are tied to distributed routing protocols, rather than the logically-centralized world of SDN.

Consensus Routing [9] seeks to eliminate transient errors, such as disconnectivity, that arise during BGP updates. In particular, Consensus Routing's "stable mode" is similar to our per-packet consistency, though computed in a distributed manner for BGP routes. On the other hand, Consensus Routing only applies to a single protocol (BGP), whereas our work may benefit any protocol or application developed in our framework. The BGP-LP mechanism from [21] is essentially per-packet consistency for BGP.

The dynamic software update problem is related to network update consistency. The problem of upgrading software in a general distributed system is addressed in [22]. The scope of that work differs in that the nodes being updated are general purpose computers, not switches, running general software.

The related problem of maintaining safety while updating firewall configurations has been addressed by Zhang *et. al* [23]. That work formalized a definition of safety similar in spirit to per-packet consistency, but limited to a single device.

In the past, a number of tools have been developed to analyze static network configurations [24]. Our main contribution is not to show that it is possible to analyze static network configurations (or that we can do so using a model checker). Rather, it is the fact that *any* such static analysis technique (such as header space analysis [14]) can be combined with our per-packet consistent update mechanisms to guarantee that trace properties are preserved as configurations evolve over time.

The model developed by Kazemian *et. al* [14] was the starting point for our own model. Since their model only spoke of a single, static configuration, we extended the network semantics to include updates so we could model a network changing dynamically over time. In addition, while their model was used to help *describe* their algorithms, ours was used to help us state and prove various correctness properties of our system.

10. CONCLUSIONS AND FUTURE WORK

Reasoning about concurrency is notoriously difficult, and network software is no exception. To make fundamental progress, the networking field needs simple, general, and reusable abstractions for changing the configuration of the network. Our per-packet and per-flow consistency abstractions allow programmers to focus their attention on the state of the network before and after a configuration change, without worrying about the transition in between. The update abstractions are powerful, in that the programmer does not need to identify the properties that should hold during the transition, since *any* property common to both configurations holds for *any* packet traversing the network during the update. This enables lightweight verification techniques that simply verify the properties of the old and new configurations. In addition, our abstractions are practical, in that efficient and correct update mechanisms exist and are implementable using today's OpenFlow switches. Our implementation and Coq proofs are available at our website www.frenetic-lang.org.

In our ongoing work, we are exploring new mechanisms that make network updates faster and cheaper, by limiting the number of rules or the number of switches affected. In this investigation, our theoretical model is a great asset, enabling us to prove that our proposed optimizations are correct. We also plan to extend our formal model to capture the per-flow consistent update abstraction, and prove the correctness of the per-flow update mechanisms. In addition, we will make our update library available to the community, to enable future OpenFlow applications to leverage these update abstractions. Finally, while per-packet consistency and per-flow consistency are core abstractions with excellent semantic properties, we want to explore other notions of consistency that either perform better (but remain sufficiently strong to provide benefits beyond eventual consistency) or provide even richer guarantees.

Acknowledgments. The authors wish to thank Hussam Abu-Libdeh, Robert Escriva, Mike Freedman, Tim Griffin, Mike Hicks, Eric Keller, Srinivas Narayana, Alan Shieh, the anonymous reviewers, and our shepherd Ramana Kompella for many helpful suggestions. Our work is supported in part by ONR grant N00014-09-1-0770 and NSF grants 1111698, 1111520, 1016937 and 0964409.

11. REFERENCES

[1] P. Francois and O. Bonaventure, "Avoiding transient loops during the convergence of link-state routing protocols," *IEEE/ACM Trans. on Networking*, Dec 2007.

[2] P. Francois, P.-A. Coste, B. Decraene, and O. Bonaventure, "Avoiding disruptions during maintenance operations on BGP sessions," *IEEE Trans. on Network and Service Management*, Dec 2007.

[3] S. Raza, Y. Zhu, and C.-N. Chuah, "Graceful network state migrations," *IEEE/ACM Trans. on Networking*, vol. 19, Aug 2011.

[4] D. Erickson *et al.*, "A demonstration of virtual machine mobility in an OpenFlow network," Aug 2008. Demo at *ACM SIGCOMM*.

[5] L. Vanbever, S. Vissicchio, C. Pelsser, P. Francois, and O. Bonaventure, "Seamless network-wide IGP migration," in *ACM SIGCOMM*, Aug 2011.

[6] N. Handigol, S. Seetharaman, M. Flajslik, N. McKeown, and R. Johari, "Plug-n-Serve: Load-balancing web traffic using OpenFlow," Aug 2009. Demo at *ACM SIGCOMM*.

[7] R. Wang, D. Butnariu, and J. Rexford, "OpenFlow-based server load balancing gone wild," in *Hot-ICE*, Mar 2011.

[8] P. Francois, M. Shand, and O. Bonaventure, "Disruption-free topology reconfiguration in OSPF networks," in *IEEE INFOCOM*, May 2007.

[9] J. P. John, E. Katz-Bassett, A. Krishnamurthy, T. Anderson, and A. Venkataramani, "Consensus routing: The Internet as a distributed system," in *NSDI*, Apr 2008.

[10] N. McKeown, T. Anderson, H. Balakrishnan, G. Parulkar, L. Peterson, J. Rexford, S. Shenker, and J. Turner, "Openflow: Enabling innovation in campus networks," *SIGCOMM CCR*, vol. 38, no. 2, pp. 69–74, 2008.

[11] M. Canini, D. Venzano, P. Peresini, D. Kostic, and J. Rexford, "A NICE way to test OpenFlow applications," in *NSDI*, Apr 2012.

[12] Y. Bertot and P. Casteran, "Interactive theorem proving and program development: Coq'Art the calculus of inductive constructions," in *EATCS Texts in Theoretical Computer Science*, Springer-Verlag, 2004.

[13] A. Cimatti, E. Clarke, E. Giunchiglia, F. Giunchiglia, M. Pistore, M. Roveri, R. Sebastiani, and A. Tacchella, "NuSMV 2: An opensource tool for symbolic model checking," pp. 359–364, Springer, 2002.

[14] P. Kazemian, G. Varghese, and N. McKeown, "Header space analysis: Static checking for networks," in *NSDI*, Apr 2012.

[15] E. M. Clarke, E. A. Emerson, and A. P. Sistla, "Automatic verification of finite-state concurrent systems using temporal logic specifications," *ACM Transactions on Programming Languages and Systems*, vol. 8, pp. 244–263, 1986.

[16] B. M. Waxman, "Broadband switching," ch. Routing of Multipoint Connections, pp. 347–352, IEEE Computer Society Press, 1991.

[17] J. Mogul, J. Tourrilhes, P. Yalagandula, P. Sharma, A. Curtis, and S. Banerjee, "DevoFlow: Scaling flow management for high-performance networks," in *ACM SIGCOMM*, Aug 2011.

[18] N. Gude, T. Koponen, J. Pettit, B. Pfaff, M. Casado, N. McKeown, and S. Shenker, "NOX: Towards an operating system for networks," *SIGCOMM CCR*, vol. 38, no. 3, 2008.

[19] B. Lantz, B. Heller, and N. McKeown, "A network in a laptop: Rapid prototyping for software-defined networks," in *HotNets*, Oct 2010.

[20] M. Reitblatt, N. Foster, J. Rexford, and D. Walker, "Consistent updates for software-defined networks: Change you can believe in!," in *HotNets*, Nov 2011.

[21] D. Katabi, N. Kushman, and J. Wrocklawski, "A Consistency Management Layer for Inter-Domain Routing," Tech. Rep. MIT-CSAIL-TR-2006-006, Cambridge, MA, Jan 2006.

[22] S. Ajmani, B. Liskov, and L. Shrira, "Modular software upgrades for distributed systems," in *Proceedings of the 20th European conference on Object-Oriented Programming*, ECOOP'06, (Berlin, Heidelberg), pp. 452–476, Springer-Verlag, 2006.

[23] C. C. Zhang, M. Winslett, and C. A. Gunter, "On the safety and efficiency of firewall policy deployment," in *IEEE Symp. on Security and Privacy*, 2007.

[24] N. Feamster and H. Balakrishnan, "Detecting BGP configuration faults with static analysis," in *NSDI*, May 2005.

A Smart Pre-Classifier to Reduce Power Consumption of TCAMs for Multi-dimensional Packet Classification

Yadi Ma
University of Wisconsin
Madison, WI, USA
yadi@cs.wisc.edu

Suman Banerjee
University of Wisconsin
Madison, WI, USA
suman@cs.wisc.edu

ABSTRACT

Ternary Content-Addressable Memories (TCAMs) has become the industrial standard for high-throughput packet classification. However, one major drawback of TCAMs is their high power consumption, which is becoming critical with the boom of data centers, the growing classifiers and the deployment of IPv6. In this paper, we propose a practical and efficient solution which introduces a smart pre-classifier to reduce power consumption of TCAMs for multi-dimensional packet classification. We reduce the dimension of the problem through the pre-classifier which pre-classifies a packet on two header fields, source and destination IP addresses. We then return to the high dimension problem where only a small portion of a TCAM is activated and searched for a given packet. The smart pre-classifier is built in a way such that a given packet matches at most one entry in the pre-classifier, which make commodity TCAMs sufficient to implement the pre-classifier. Furthermore, each rule is stored only once in one of the TCAM blocks, which avoids rule replication. The presented solution uses commodity TCAMs, and the proposed algorithms are easy to implement. Our scheme achieves a median power reduction of 91% and an average power reduction of 88% on real and synthetic classifiers respectively.

Categories and Subject Descriptors

C.2.6 [**Computer-Communication Networks**]: Internetworking—*Routers*

Keywords

Packet classification, SmartPC, power consumption

1. INTRODUCTION

Given a set of strictly ordered rules in a classifier, the packet classification problem is to find out the first (highest priority) rule that matches each incoming packet at a router.

Many common packet classification techniques define rules on multiple header fields, e.g., a 5-tuple on source and destination IP addresses, source and destination port numbers and protocol. Packet classification plays a significant role in many networking services and functions that require network traffic (i.e. packets) to be distinguished and isolated into different flows for suitable processing. Examples of such services include: QoS, security, packet filtering (e.g., deny all packets from a known source), policy routing (e.g., route all VoIP traffic over a separate network), traffic shaping (e.g., ensure that no one source overloads the network), and so on.

There are two major trends of previous work on packet classification: RAM-based algorithmic solutions and hardware-based TCAM solutions. The most prevailing RAM-based solutions include decision-tree based schemes such as HiCuts [8], HyperCuts [19], EffiCuts [23], et al. Algorithmic methods do not scale effectively to high performance systems that must process tens of millions of packets per second.

TCAMs allow simultaneous match of an incoming packet against all the stored rules of a classifier within a single memory access and a fixed number of clock cycles. TCAM has become the *de facto* industrial standard for packet classification in high performance routers. However, one major disadvantage of using TCAMs is the high power consumption. The main component of power consumptions of TCAMs is proportional to the number of searched entries [26]. A typical 18Mbit TCAM device can consume up to 15 Watts of power when all the entries are searched. This can be a significant power overhead for the router, switch, or other networking hardware that embeds them. Furthermore, as enterprises and operators continue to seek larger and bigger classifiers, the energy footprint of these devices would continue to grow even further. In this paper, we present a new solution for reducing the energy footprint of TCAMs when applied to the problem of multi-dimensional packet classification by carrying out the classification in two pipelined stages.

1.1 Our pre-classifier approach

TCAM vendors have been providing mechanisms to address and activate smaller portions of the TCAMs, called blocks. A TCAM block is a contiguous, fixed-sized chunk of TCAM entries, much smaller than the size of the entire TCAM, as shown in Figure 1. If the width of a TCAM block is 80 bits and there are 2K entries in each block, the size of a block is 160K bits. Further suppose the 104 bit header fields of an IPv4 packet (32 bit source or destination IP address, 16 bit source or destination port, plus 8 bit protocol) would occupy two entries in a block. Therefore, each block can hold up to 1K rules. A large classifier of size 50K requires 50 such blocks to store the rules. To reduce power consumption of TCAMs, we take advantage of this mechanism to select and activate a small number of blocks, instead of all the 50 blocks, for each packet classification.

Activating a small number of blocks is only feasible if we can somehow ensure that the incoming packet will be correctly matched by searching the entries stored in these active blocks. We achieve

SIGCOMM'12, August 13–17, 2012, Helsinki, Finland.

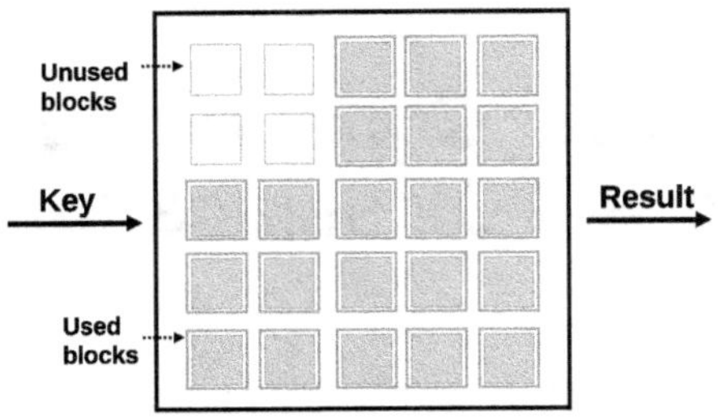

(a) TCAM is composed of a number of blocks. For each incoming packet, all the blocks that are used for packet classification are activated.

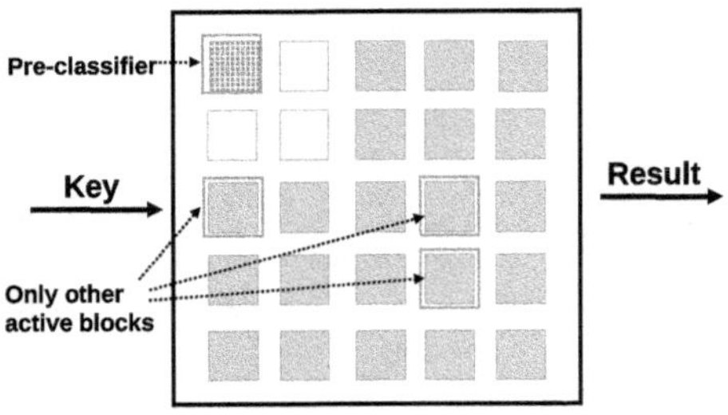

(b) A TCAM block is used for pre-classifier. For each incoming packet, the pre-classifier is consulted first. Based on its result, only 3 blocks are activated.

Figure 1: An overview of TCAM power saving.

this goal by using a *pre-classifier.* In particular, we construct a specific pre-classifier and shuffle the rules in the original classifier into different blocks so that only the pre-classifier and a small number of TCAM blocks need to be activated to classify a packet. The novelty of our approach lies in finding a good way to form pre-classifier entries so as to minimize the number of blocks activated for each packet classification.

In Figure 1(a), we show an example TCAM composed of 25 blocks. In this example, 21 blocks are used to store the rules in a classifier while 4 remain unused (could be used for other purposes, e.g., IP lookups for packet forwarding). All the 21 blocks that are used to store the original classifier need to be activated for every packet classification. In our modified system, shown in Figure 1(b), one block is used to store pre-classifier entries [1]. The pre-classifier, in this example, requires only three other blocks be activated for correctly matching the incoming packet. This achieves a power reduction of $(21 - 4)/21 * 100\% = 81\%$. For a 18Mbit TCAM, this approach could bring down its power consumption from 15 Watts to less than 3 Watts.

Our proposed pre-classifier is motivated by previous work [8, 7, 19, 9] which suggest that although large classifiers could contain hundreds of thousands of rules, for *any individual packet* there are usually a small number of "intercepting" rules — rules that match the given packet. (Note that different packets can have completely different intercepting rules and hence it is possible a large fraction of the rules are actually compared to classify a large number of packets.) The power consumption could be reduced if we can somehow shuffle the rules in a classifier into TCAM blocks, such that for each incoming packet, only a subset of blocks that contain all the rules intercepting with the packet are searched.

In this paper we create this pre-classifier by reducing the dimensionality of the problem. Instead of the typical 5-dimensional rules, we construct the pre-classifier by creating rules in two dimensions only: the source and destination IP address ranges. This choice provides an adequate reduction in storage requirements, yet provides adequate ability to classify packets. In fact, the results of the pre-classifier ensure that the packet can then be correctly classified by comparing against only a few other TCAM blocks (which contain suitably shuffled rules). We present a detailed analysis to explain the efficacy of this choice of two dimensions in Section 3.

Finally, this design and use of the pre-classifier is made possible by new features of TCAMs, which support turning on and off a specific subset of blocks. The time and overheads of turning on and off such blocks is negligible. Furthermore, such hardware supports multiple searches in parallel, e.g., Cisco's TCAM4 which allows four parallel searches [3].

The notion of using multiple stages to implement packet classification techniques is not entirely new. Two of the most related examples of this approach for TCAMs are CoolCAMs [26], and packet classification using extended TCAMs [20]. Our work differs from such prior approaches in the specific algorithmic techniques in creating pre-classifiers, and our approach is very deployable with current TCAMs for the problem of packet classification. For instance, CoolCAMs[26] was designed for the problem of IP lookups, in which classification is performed in a single dimension — destination IP address. Their proposed hash-based and trie-based approaches for solving this one dimensional problem do not extend to the multi-dimensional packet classification problem that we are addressing, and required new algorithmic considerations. Similarly, extended TCAMs assumes a new type of ternary match hardware that are not available in current commodity TCAMs. The specifics of our multi-dimensional packet classification problem and the specific restrictions of commodity TCAMs make our solution uniquely applicable to this important problem domain.

1.2 Summary of our contributions

In this paper, we propose a **Smart Pre-C**lassifier to reduce power consumption of TCAMs for packet classification (SmartPC). With SmartPC, the actual packet classification occurs in two stages: first a given packet is classified by a pre-classifier which provides information on which TCAM block needs to be activated in the next stage; in the second stage, the block from first stage plus a few *general* blocks are activated and searched in parallel for a match for the packet. SmartPC can typically bring down the TCAM's power consumption from 15 Watts to less than 3 Watts. The two-dimensional pre-classifier contains non-overlapping entries, which are formed by combining and expanding rules in the original classifier. The rules in the original classifier are thus reorganized into TCAM blocks so that the rules covered by an entry in the pre-classifier are stored in the same TCAM block. Those rules that do not fit in a block are marked as general rules and stored in general blocks.

We present a detailed design of SmartPC and evaluate it on a large amount of real classifiers from ISPs and synthetic classifiers generated by ClassBench [21]. The experimental results suggest huge power reductions on both real and synthetic classifiers. With block size 128, SmartPC achieves a median of 91% and an average of 88% power reductions on these classifiers with the highest reduction reaching 98%.

2. BACKGROUND

2.1 Problem statement of packet classification

Packet classification is performed using a packet classifier, also called a policy database, flow classifier, or simply a classifier. A

[1]This figure shows one way of implementing a pre-classifier. In practice, a pre-classifier could also be implemented in software, as discussed later in Section 4.

Table 1: A simple example with 8 rules on 5 fields

Rule	F_1	F_2	F_3	F_4	F_5	Action
$R0$	000*	111*	10	*	UDP	$action_0$
$R1$	000*	10*	01	10	TCP	$action_1$
$R2$	000*	01*	*	11	TCP	$action_0$
$R3$	0*	1*	*	01	UDP	$action_2$
$R4$	0*	0*	10	*	UDP	$action_1$
$R5$	000*	0*	*	01	UDP	$action_1$
$R6$	*	*	*	*	UDP	$action_3$
$R7$	*	*	*	*	TCP	$action_4$

classifier is a collection of rules or policies. Given a set of strictly ordered rules in a classifier, the packet classification problem is to find out the first rule that matches each incoming packet at a router. Each rule is associated with an action. After classification, the corresponding action will be performed on each packet.

Suppose a classifier at a router contains a set of N rules, and each rule contains K fields. We consider the case where $K = 5$. The five fields are source IP address, destination IP address, source port, destination port and protocol type, respectively (shown as F_1 through F_5 in Table 1). There are three types of matches that a field can have: prefix match (source or destination IP address), exact match (protocol type), or range match (source or destination port).

Table 1 shows a simple example with eight rules on five fields. In a prefix match, the rule field should be a prefix of the header field. Suppose header field 2 of a packet is 1010. In Table 1, it matches the second field, F_2, of rule $R1$. In exact match, the header field of a packet should match the rule field exactly. For example, protocol type could be TCP or UDP. In a range match, the header value should lie in the range specified by the rule. If each of the header fields of a packet P matches each of the corresponding fields in a rule R, the packet P is said to match rule R. If P matches multiple rules, the first rule (with the minimum index) is returned.

2.2 Geometric view of packet classification

We can view a 32-bit prefix like 001* as a range of addresses from $0010 \cdots 00$ to $0011 \cdots 11$ on the number line from 0 to 2^{32}. If prefixes can be viewed as line segments geometrically, two-dimensional rules correspond to rectangles, three-dimensional rules to cubes, and so on. A given packet header is a point. The problem of packet classification reduces to finding the highest-priority multi-dimensional region that contains a given point. This is a classic problem in computational geometry, named point location problem. The point location problem is defined as follows: given a partition of K-dimensional space into N disjoint regions, determine the region where a query point lies. Numerous results have been reported in the literature [10, 6], most of which deal with the case of non-overlapping regions or specific arrangement of hyperplanes or hypersurfaces of bounded degree.

2.3 Basics of TCAMs

Ternary Content Addressable Memories (TCAMs) are fully associative memories that allow three matching states, "0", "1" or "X"(wildcard). A "X" state matches both 0s and 1s in the corresponding input bit. This feature makes TCAMs particularly attractive for packet classification and route lookup applications which require longest prefix matches.

The rules are stored in the TCAM array in the order of decreasing priority. Given a packet header to classify, the TCAM performs a comparison against all of its entries in parallel, and a priority encoder selects the first matching rule. TCAMs allow simultaneous match of a packet against a large number of rules within a single memory access, while conventional trie-based designs may require multiple memory accesses for a single packet. Therefore, TCAMs are very popular for designing high-throughput packet classification solutions. However, TCAMs do suffer from some deficiencies such as high power consumption and so on.

3. PROPERTIES OF REAL CLASSIFIERS

A number of previous work have mentioned important properties of real classifier [8, 7, 19, 9]. However, these studies are based on a very small number of classifiers (e.g., in [9], the authors analyzed four ACLs), which may not represent the wide range of real classifiers.

We performed an analysis of more than 200 real classifiers ranging in size from 3 to 15,181. These classifiers were provided to us by a large networking vendor and these are sample classifiers they use for testing their own classification systems. However, the specifics of the classifier are under an NDA and unfortunately cannot be released.

Rule overlapping.

A classifier could contain a large amount of rules, but the number of rules intercepting with a given packet is usually significantly smaller than the theoretical upper bound [8, 7, 19, 9]. Even though only a small number of rules are intercepting with a packet, all the multi-dimensional rules stored in a TCAM need to be searched, which leads to high power consumption. It is reasonable to reduce the dimension for the purpose of power efficiency. Since source and destination addresses are more specific than the other three header fields, we consider these two dimensions as candidates. Therefore we are more interested in what is the maximum number of source and destination IP prefix pairs that need to be searched for any incoming packet.

This question is equivalent to the following geometry problem: given a set of axis-aligned rectangles, where each rectangle represents a two-dimensional rule with a range of values in each dimension, what is the maximum number of rectangles that overlap at any point? We employ a line sweep algorithm [22] in which a vertical sweep line moves from left to right. When it crosses the left edge of a rectangle, the rectangle is added to an active set. When it crosses the right edge, the rectangle is removed from the active set. In the inner loop, a horizontal sweep line moves top-down. When it crosses the upper or lower horizontal edge of a rectangle in the active set, we increment or decrement a counter that says how many rectangles overlap at the current point. The maximum value of the counter is the number we are looking for.

We run this line sweep algorithm on the classifiers and the results are shown in Figure 2(a), where x-axis represents classifier size, and y-axis shows the maximum number of overlapping rules in the two-dimensional space (i.e., source and destination IP addresses). Similarly, we plot the number of rules that contain wildcards in the two dimensions as shown in Figure 2(b). The wildcard rules match any packet and intercept with any other rules in the two-dimensional space, therefore, they form a subset of the set of maximum overlapping rules.

From the plots, we observed that:

- In the two-dimensional space, the maximum number of overlapping rules and the number of wildcard rules of a classifier is an order of magnitude smaller than the classifier size.

Based on these properties, in Section 4, we build a two-dimensional pre-classifier and design an efficient classification algorithm to reduce the power consumption of TCAMs.

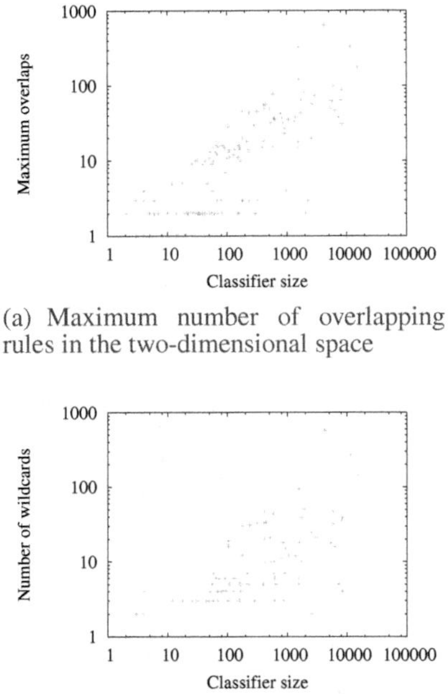

(a) Maximum number of overlapping rules in the two-dimensional space

(b) Number of wildcard rules in the two-dimensional space

Figure 2: Properties of real classifiers (both axes are in log scale)

4. DESIGN

To reduce the power consumption of TCAMs, we propose SmartPC, in which we pre-process a classifier and shuffle its rules into blocks in a TCAM, so that only a few TCAM blocks, instead of all the blocks that are used to store the classifier, will be searched to find the desired matching rule for any given packet.

SmartPC reduces the complex multi-dimensional packet classification problem to a much simpler problem by introducing a *pre-classifier* which classifies each incoming packet on two header fields, source and destination IP addresses. Hereafter, we use "rules" to refer to five-dimensional rules in the original classifier, and "entries" to refer to two-dimensional entries in the pre-classifier. The rules are shuffled into TCAM blocks such that each pre-classifier entry covers a number of rules in one of the blocks. Note that when rules are shuffled to TCAM blocks, the relative order of the rules is preserved in each TCAM block, since commodity TCAMs select the match from the first (highest priority) rule.

4.1 Design overview

SmartPC is motivated by our large-scale analysis of real classifiers as shown in Section 3, from which we observe that in a classifier, the maximum number of source-destination pairs matching a given packet is much smaller than the size of the classifier.

The main idea behind our work is that we group source-destination pairs in a classifier into clusters based on their locations in the two-dimensional space. The two-dimensional space is thus divided by non-overlapping rectangles of different sizes. Each rectangle *covers* a cluster of rules and represents an entry in the pre-classifier for the classifier, as discussed in Section 4.2.

Given the ranges of the source and destination IP addresses of a pre-classifier entry p,

$$p = \{[sl_p, sh_p], [dl_p, dh_p]\}$$

and the ranges of the two header fields of a rule r,

$$r = \{[sl_r, sh_r], [dl_r, dh_r]\}$$

we say p covers r if $sl_p \leq sl_r \leq sh_r \leq sh_p$ and $dl_p \leq dl_r \leq dh_r \leq dh_p$, where sl, sh represent lower and upper bound of source address, and dl and dh stand for lower and upper bound of destination address.

We then shuffle the rules in the classifier such that each pre-classifier entry is associated with a TCAM block, named *specific block*, which contains up to a block size of rules that intercept with the entry in the two-dimensional space. If the number of rules that intercept with the pre-classifier entry is larger than one block size, the extra rules are stored in TCAM blocks named *general blocks*. When classifying a packet, the pre-classifier is first consulted. If a match is found in the pre-classifier, then its associated specific block plus the general blocks are activated and searched in parallel for the final result; otherwise, if no match is found in the pre-classifier, only the general blocks are searched.

We propose a classification system for SmartPC as shown in Figure 3. The classification process of a packet occurs in the following steps:

- Each incoming packet is first matched against Index TCAM, which contains pre-classifier entries. The Index TCAM indexes into an associated Index SRAM, which contains the ID of the specific TCAM block. This block will be searched in the second stage lookup.
- In the second stage, only the specific block found in the first stage, if any, together with general blocks are activated and involved in two parallel searches.
- Finally the matches from the specific block and general blocks are resolved by priority, and the action of the higher priority match is returned as the final result.

The above steps can be pipelined to maintain the same operating frequency as commodity TCAMs. It is worth noting that although latency grows slightly due to pipelining, throughput remains unchanged. Throughput is the main metric of interest.

4.2 Construction of a pre-classifier

In SmartPC, we propose the concept of *pre-classifier*, which is motivated by Storm [14], where sampled incoming traffic (i.e., packets) are expanded and combined to form evolving rules which are stored in a rule cache. Those evolving rules are updated over time and are used as popular rules to classify incoming packets. In SmartPC, we adapt the idea of expanding rule, while we expand and combine original rules in a classifier to construct entries in the proposed pre-classifier. Our approach is not dependent on traffic pattern and the process only occurs once for each classifier [2]. Although this insight about expanding and combining rules is derived out of Storm, it is worth noting that SmartPC employs an intelligent pre-classifier to cluster overlapping rules while Storm is focusing on a multi-core approach.

How to build an effective and efficient pre-classifier is of crucial importance. In SmartPC, each entry in a pre-classifier is constructed by expanding and combining the rectangles formed by source and destination fields of the rules in the original classifier. The rules in the classifier are reorganized, with each pre-classifier entry pointing to a TCAM block that contains all the rules covered by the corresponding pre-classifier entry. The number of original

[2]The pre-classifier does not change over time unless the rules in the classifier change.

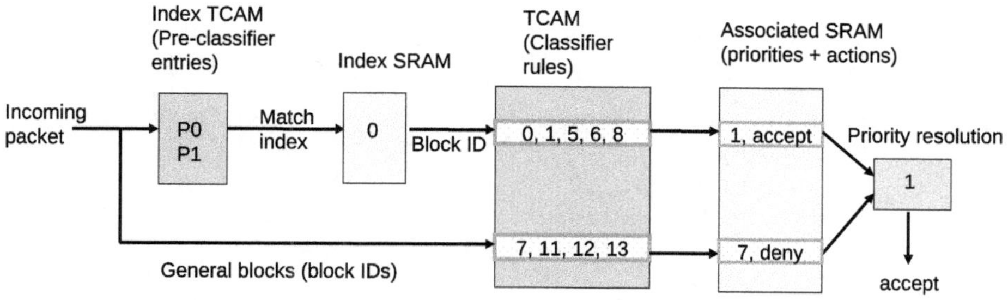

Figure 3: Packet classification system for SmartPC.

rules covered by a pre-classifier entry is restricted to be less than a pre-defined parameter, named *block size* in SmartPC. If the inclusion of a rule causes the number of rules covered by a pre-classifier entry to exceed the block size, the rule will be marked as *general* [3]. All the rules marked as general are stored in general blocks, which will be searched in parallel with the specific block associated with the matching pre-classifier entry, if there is any, for each incoming packet.

Our experimental results show that usually only a small portion of rules in a classifier are marked as general (e.g., the fraction of general rules ranges from 0 to around 20% for all the classifiers we evaluated), therefore the number of general blocks will be much smaller than the total number of TCAM blocks used to store the whole classifier.

Example on how to build a pre-classifier.

Table 2 shows a five-dimensional classifier containing 14 rules. The source and destination addresses are in the format of address/prefix-length. Figure 4(a) is a graphical representation of the source and destination address fields of this classifier, where x-axis and y-axis represent destination address and source address, respectively.

Figure 4 demonstrates the process of building a pre-classifier for the given classifier, assuming a block size of 5 (note that real block size could be much larger, and this small number is for demonstration purposes only). Initially there is no existing pre-classifier entry, thus rule 0 becomes the first pre-classifier entry $P0$ as shown in Figure 4(a). Then we try to expand $P0$ to cover other rules. In Figure 4(b), $P0$ is expanded to cover rule 1. Since rule 5 intercepts with the expanded pre-classifier entry $P0$, we keep expanding $P0$ to cover rule 5 as shown in Figure 4(c). Rule 6 lies inside $P0$. Expansion of P0 to cover rule 7 fails, since the expanded $P0$ would cover more than 5 rules. Therefore 7 is marked as general and will be stored in general blocks. Rule 8 lies inside $P0$, thus so far $P0 = \{0, 1, 5, 6, 8\}$. The size of $P0$ reaches block size. Other intercepting rules 11, 12 and 13 are also marked as general.

We continue to form new pre-classifier entries. Rule 2 forms a new pre-classifier entry $P1$. Similarly, $P1$ is expanded to cover rules 3, 4, 9 and 10. Thus $P1 = \{2, 3, 4, 9, 10\}$ as shown in Figure 4(d).

Heuristic algorithm.

We present the pseudo code on how to build a pre-classifier in Algorithm 1.

The inputs to the algorithm include a classifier C, and an empty pre-classifier P. The properties *isDone* and *isGeneral* of each rule in the classifier are initialized to false, where isDone indicates whether the rule is already covered by a pre-classifier entry, and isGeneral indicates whether it is marked as a general rule. The *size* property of each pre-classifier entry is updated to show the current number of rules covered by the pre-classifier entry. When this number reaches BLOCK-SIZE, which is the pre-defined TCAM block size, the pre-classifier entry can not be expanded any further.

In the outer loop of the algorithm, each rule in the classifier is processed to form or expand a pre-classifier entry. In the inner loop, the existing pre-classifier entries are checked one by one, until a pre-classifier entry is found which can be expanded to include the current rule without exceeding BLOCK-SIZE. If none is found at the end of the inner loop, and the rule is not marked as general, this rule forms a new pre-classifier entry by itself, which may be expanded later.

Algorithm 1 BuildPreClassifier()

```
for each rule i in C do
  success = false
  if (i.isDone) || (i.isGeneral) then
    continue
  end if
  for each pre-classifier entry j in P do
    if (j.size = BLOCK-SIZE) then
      continue
    end if
    expandedEntry = j
    if ExpandPreEntry(i, j, expandedEntry) then
      success = true
      j = expandedEntry
      for each rule k in C do
        if k is covered by j then
          k.isDone = true
        end if
      end for
      break {j is successfully expanded to cover i}
    end if
    if i.isGeneral then
      break
    end if
  end for
  if ((!success) && (!i.isGeneral)) then
    a new pre entry k = i{no existing pre entry can be ex-
    panded and i does not intercept with any existing pre-
    entry}
    k.size = 1
  end if
end for
```

Algorithm 1 calls ExpandPreEntry, shown as Algorithm 2, in which a rule is marked as general if it intercepts with the current pre-classifier entry but the inclusion of the rule causes the size of

[3]Though a rule with more wildcards (or shorter prefix lengths on source and destination addresses) is more likely to be marked as general, rules that are marked as general are not necessarily the rules with more wildcards.

Table 2: A classifier which contains 14 rules

Rule#	Src_addr	Dst_addr	Src_port	Dst_port	protocol
0	228.128.0.0/9	0.0.0.0/0	0 : 65535	0 : 65535	0x01
1	223.0.0.0/9	0.0.0.0/0	0 : 65535	0 : 65535	0x06
2	0.0.0.0/1	175.0.0.0/8	0 : 65535	0 : 65535	0x06
3	0.0.0.0/1	225.0.0.0/8	0 : 65535	0 : 65535	0x01
4	0.0.0.0/1	225.0.0.0/8	0 : 65535	0 : 65535	0x06
5	128.0.0.0/1	123.0.0.0/8	0 : 65535	0 : 65535	0x06
6	128.0.0.0/1	37.0.0.0/8	0 : 65535	0 : 65535	0x06
7	0.0.0.0/0	123.0.0.0/8	0 : 65535	0 : 65535	0x06
8	178.0.0.0/7	0.0.0.0/1	0 : 65535	0 : 65535	0x06
9	0.0.0.0/1	172.0.0.0/7	0 : 65535	0 : 65535	0x06
10	0.0.0.0/1	226.0.0.0/7	0 : 65535	0 : 65535	0x06
11	128.0.0.0/1	120.0.0.0/7	0 : 65535	0 : 65535	0x01
12	128.0.0.0/1	120.0.0.0/7	0 : 65535	0 : 65535	0x06
13	128.0.0.0/1	38.0.0.0/7	0 : 65535	0 : 65535	0x06

the entry to exceed BLOCK-SIZE, or it causes entries in the pre-classifier to be overlapping, or it results in conflicts with other rules. This algorithm keeps expanding a pre-classifier entry to cover non-general rules intercepting with it until its size reaches BLOCK-SIZE, or all the intercepting rules in this round are included. This function returns true if the expansion succeeds, and false otherwise.

Algorithm 2 ExpandPreEntry(i, j, expandedEntry)

```
tmp = expandedEntry
expandedEntry = new entry formed by expanding j to cover i
expandedEntry.size += 1
if expandedEntry.size > BLOCK-SIZE then
   i.isGeneral = true
   expandedEntry = tmp
   return false {fail to expand}
end if
if expandedEntry overlaps with other pre entries || RuleConflict(expandedEntry) then
   if i intercepts with j then
      i.isGeneral = true
   end if
   expandedEntry = tmp
   return false {fail to expand}
end if
return true {succeed to expand}
```

Algorithm 2 and Algorithm 3 calls each other recursively. Algorithm 3 checks all the non-general rules in classifier C that are intercepting with expandedEntry for conflicts. It returns true if a conflict is found or BLOCK-SIZE is exceeded; otherwise, it returns false.

When Algorithm 1 terminates, a list of pre-classifier entries are formed, and those rules that are not covered by the pre-classifier are marked as general. These general rules are stored in general TCAM blocks. All the rules covered by a pre-classifier entry are stored in the same TCAM block, with their orders in the original classifier preserved. In spite of this, the rules covered by other entries can also be stored in the same TCAM block if they can fit in the block. This is because the rules covered by different entries will not overlap. We only need to separate the blocks that store general rules from other blocks. It is a bin-packing problem to minimize the number of TCAM blocks when organizing the rules covered by pre-classifier entries into TCAM blocks. Simple and efficient heuristics such as best-fit decreasing and first-fit decreasing could be used to solve the problem.

Algorithm 3 RuleConflict(expandedEntry)

```
for each rule k in C do
   if k.isGeneral || k is already covered || k does not intercept with expandedEntry then
      continue
   end if
   if k lies inside expandedEntry then
      expandedEntry.size += 1
      if expandedEntry.size > BLOCK-SIZE then
         return true
      end if
   else if k intercepts with expandedEntry then
      tmp = expandedEntry
      if ExpandPreEntry(k, j, expandedEntry) then
         break {succeed}
      else
         expandedEntry = tmp
      end if
   end if
end for
return false {no conflict}
```

Algorithm analysis.

Our goal is to build a pre-classifier which minimizes the number of general rules. However, an optimal solution is hard to find. Though our proposed heuristic algorithm may not be optimal, it is proved to be effective and efficient, achieving a median power reduction of 91% on real classifiers. Smarter heuristics might exist (and the slack in our experiments is obviously upper bounded by around 9%).

In the worst case, all the rules are overlapping in the 2-dimensional space and all are marked general, resulting in no power reduction. As expected, this is not common in practice since the maximum number of overlapping rules is an order of magnitude smaller than classifier size (see Section 3).

4.3 Packet classification with SmartPC

When a packet as shown in Figure 4(a) arrives, the pre-classifier entries $P0$ and $P1$ are consulted first, and a matching entry $P0$ is found. Figure 3 shows the whole process of the classification. The TCAM block pointed to by $P0$, which contains rules (0, 1, 5, 6, 8), together with the general TCAM block are activated and searched in parallel. Finally, the priorities of two matches, rule 1 and 7, are compared, and the action of the higher priority rule 1, "accept", is returned as the final classification decision for the packet.

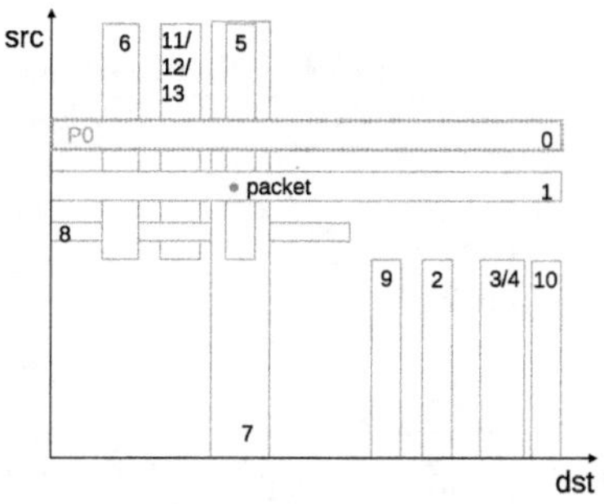

(a) Rule 0 in the classifier forms the first pre-classifier entry P0.

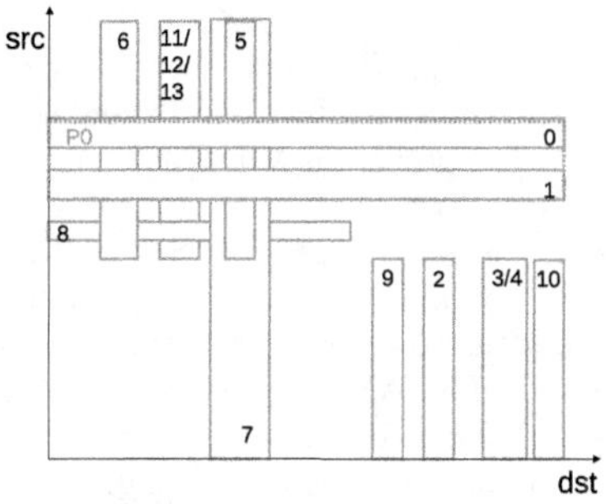

(b) Expand P0 to cover rule 1.

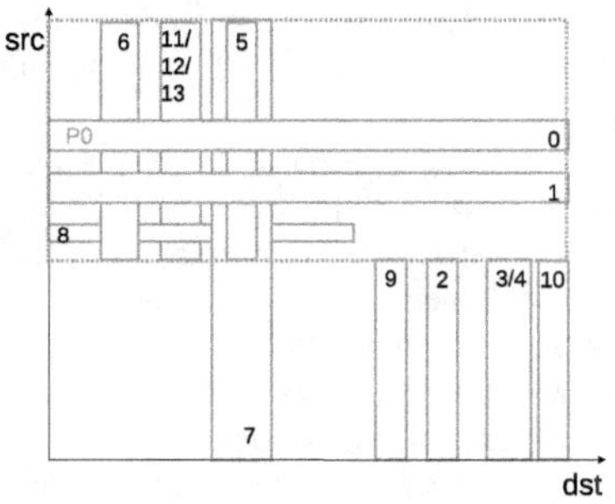

(c) Rule 5 intercepts with P0, expand P0 to cover 5. P0={0,1,5,6,8}.

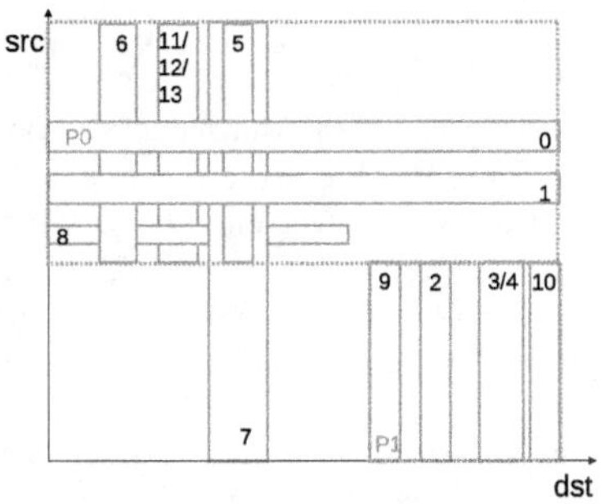

(d) The second pre-classifier entry P1={2,3,4,9,10}.

Figure 4: An example on how to build pre-classifier (assume block size = 5). Two pre-classifier entries are formed and four rules (7, 11, 12, 13) are marked as general.

4.4 Properties of pre-classifiers

We build an intelligent pre-classifier that satisfies the following properties for the purpose of simplicity and power-efficiency. Note that pre-classifier entries do not necessarily cover the whole two-dimensional space.

Entries in a pre-classifier are non-overlapping. This property guarantees that any incoming packet matches at most one entry in the pre-classifier. If the pre-classifier is implemented in TCAM, there is no need to consider the priorities of pre-classifier entries. While if pre-classifier entries are allowed to overlap, the situation becomes complicated when a given packet matches multiple entries in the pre-classifier, since we need to decide which entry has higher priority. In addition, this property provides more flexibility to the implementation of a pre-classifier as discussed below.

Each rule in a classifier is either covered by only one pre-classifier entry, or marked as general. In other words, each rule is associated with no more than one pre-classifier entry. This property ensures that rules in a classifier are not replicated in TCAM blocks, i.e., each rule is stored only once in one of the blocks. Therefore, our approach is storage-efficient. The only extra storage in SmartPC is the storage used for storing the entries in a pre-classifier.

4.5 Implementation of a pre-classifier

Considering the properties of pre-classifiers, there are a number of approaches to implement a pre-classifier.

For the special case of two-dimensional non-overlapping rectangles (which is the case in our proposed pre-classifier), a number of geometric solutions have been reported with logarithmic time complexity and near-linear space complexity [10, 6]. However, geometric approaches are not efficient enough for pre-classifiers. For the special case of pre-classifiers on just two header fields, Grid of Tries [24] is an efficient algorithmic solution.

TCAM is another alternative where a pre-classifier is stored in TCAM block(s). Our experimental results suggest that a pre-classifier usually contains a small number of entries which can fit in one or a few TCAM blocks. Plus TCAMs are much faster than SRAMs (used in geometric and algorithmic solutions). Though geometric and algorithmic solutions may incur lower power consumption than TCAM-based methods, we decide to store pre-classifiers in TCAMs.

4.6 Rule update

Rule update overhead of SmartPC is generally less than that of regular TCAMs. In SmartPC, the ordering of TCAM entries is kept within one specific block or within a small number of general blocks when rules are added or deleted, while with regular TCAMs, the ordering of TCAM entries has to be kept throughout all the blocks.

To add a rule, the simplest approach is to insert the rule into the general blocks [4]. We can rerun our algorithms and construct from scratch if the number of general blocks exceeds some predefined number. A slightly sophisticated approach could be that if the rule is covered by a pre-classifier entry (i.e., the pre-classifier entry matches the rule), and the specific block associated with this pre-classifier entry is not full, the rule is stored into the specific block; otherwise, the rule can be simply inserted into the general blocks. Depending on rule add/delete frequencies, we may leave some holes in each block (e.g., by under loading the blocks when constructing pre-classifiers) to reduce the overhead of moving rules.

To delete a rule, the block that stores the rule is first found and the rule is deleted. Only when the deletion of the rule changes the ranges of the corresponding pre-classifier entry, the pre-classifier entry is updated.

[4]Please note that the ordering of rules (i.e. rule priorities) must be kept when inserting rules into blocks.

Table 3: Summary of real classifiers

Name	Size	MaxOverlaps	Wildcard
R1	5233	49	18
R2	5626	63	32
R3	5874	98	48
R4	6339	47	16
R5	7356	38	5
R6	8063	64	35
R7	8475	31	4
R8	10054	1	0
R9	11574	334	271
R10	15181	177	143

An update operation can be considered as a delete operation followed by an add operation, as described above.

In summary, only one specific block or a few general blocks plus at most one pre-classifier entry need to be modified for each rule add/delete, while regular TCAMs may have to modify all the blocks.

5. EXPERIMENTAL RESULTS

We start this section with an overview of our experimental methodology. We then evaluate the power reduction of our scheme on real classifiers and synthetic classifiers as well. We also look into the performance of pre-classifiers.

5.1 Experimental methodology

We evaluate the performance of SmartPC using a number of real classifiers and synthetic classifiers generated by ClassBench [21]. As we discussed in Section 3, the real classifiers were provided to us by a large networking vendor and the specifics of the classifier are under an NDA and unfortunately cannot be released. So we are presenting some statistics in Table 3.

Since the power consumption problem is more critical for large classifiers, we only show results from 10 real classifiers with more than 5,000 rules. The largest real classifier we have access to has a size of 15,181. Larger real classifiers are hard to find since ISPs or CAM venders do not release their classifiers. Therefore, we also show results on synthetic classifiers of sizes 50,000 and 100,000. Even larger synthetic classifiers show similar trend and the results are not shown in the paper. Though our focus is on large classifiers, we also evaluated the performance of small real classifiers. SmartPC could achieve significant power reductions on these classifiers as well.

An important parameter is TCAM block size, which is usually dependent on hardware design. In this paper, we show results with different block sizes, to demonstrate how the performance of our scheme is affected by block sizes. This work might aid TCAM designers in choosing the right block size.

Table 3 summarizes the properties of the real classifiers used in the evaluations. In this table, we show the size of each classifier, the maximum number of overlapping rules in two dimensions: source and destination IP addresses, and the number of rules that are wildcard(covers the whole space) on the two dimensions (as discussed in Section 3). These wildcard rules need to be searched for every packet classification and will be marked as general. MaxOverlaps indicates the maximum number of rules that need to be searched for a given packet. This number gives us a rough estimation of block size. As seen here, MaxOverlaps of most classifiers are smaller than 128, and all of them are smaller than 512. We evaluated real classifiers with six different block sizes, 32, 64, 128, 256, 512 and 1024, respectively.

Table 4: Summary of synthetic classifiers

Name	Size	MaxOverlaps	Wildcard
S1	9802	22	4
S2	9416	126	57
S3	9497	76	18
S4	9624	82	12
S5	7255	28	0
S6	99823	27	5
S7	87039	249	79
S8	99836	89	47
S9	99866	81	38
S10	99220	10	0

Similarly, we summarize the properties of ten synthetic classifiers in Table 4. We generate these classifiers using ClassBench by setting rule size parameters to 10k and 100k and using five different classifier seeds. The properties MaxOverlaps and Wildcard are similar to those of real classifiers.

To evaluate our scheme, we use power reduction as the main metric, since the goal of this work is to reduce power consumptions of TCAMs. As mentioned in Section 1, the main component of power consumptions of TCAMs is proportional to the number of searched entries [26]. Therefore, we employ a simple linear power model to estimate the power reductions, though the real reductions may be slightly different. Suppose a classifier contains N rules, and TCAM block size is B. In the default scheme without using SmartPC, all the TCAM blocks that are used to store the classifier, defined as X, must be searched for any packet classification, where $X = \lceil N/B \rceil$.

To classify a packet using SmartPC, the pre-classifier, and at most one specific TCAM block plus general TCAM blocks are activated. The power consumption of pre-classifiers depends on how pre-classifiers are implemented. Here we assume pre-classifier entries are stored in TCAM blocks, and we count its power consumption as well. Suppose G out of N rules are marked as general, and P pre-classifier entries are formed. With SmartPC, at most $Y = \lceil P/B \rceil + 1 + \lceil G/B \rceil$ blocks must be activated for each packet classification. Therefore, the percentage of power reduction with SmartPC is $\frac{X-Y}{X} \times 100\%$. We use this definition to evaluate power reductions of SmartPC.

5.2 Power reductions

We evaluate the percentage of power reductions on real and synthetic classifiers by SmartPC. We first evaluate how block sizes affect the power reductions on all the classifiers. We evaluate these classifiers using five different block sizes, 32, 64, 128, 256, 512, and 1024, respectively. We plot the power reductions in Figure 5, where x-axis represents block size and y-axis shows the percentage of power reductions.

As shown in Figure 5(a), regardless of classifier sizes, our scheme achieves huge power reductions on real classifiers, ranging from 78% to 97%. We further observe that the power reduction of each classifier fluctuates with block sizes. For most classifiers, the maximum power reduction occurs at block sizes 64, 128 and 256. With block size 128, the average and median power reductions are 92% and 87%, respectively.

Similarly, we evaluate the percentage of power reductions on synthetic classifiers. As shown in Figure 5(b), SmartPC also achieves huge power reductions on synthetic classifiers. For every classifier and block size combination, the power reduction is more than 60%, with the highest reduction reaching 98%. With block size 128, the average and median power reductions are 91% and 88%, respectively. It is worth noting that power reductions on the five larger

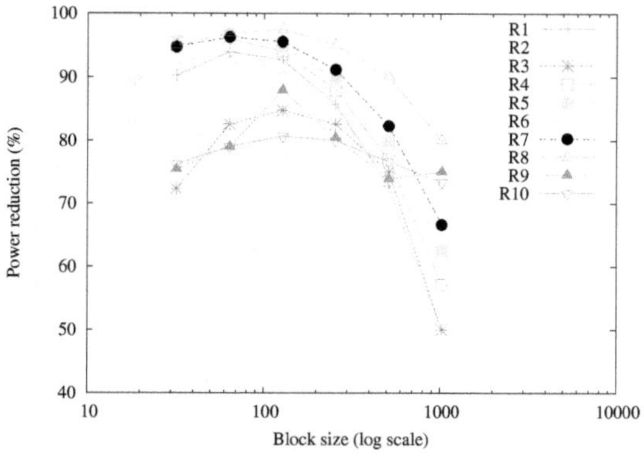

(a) Real classifiers

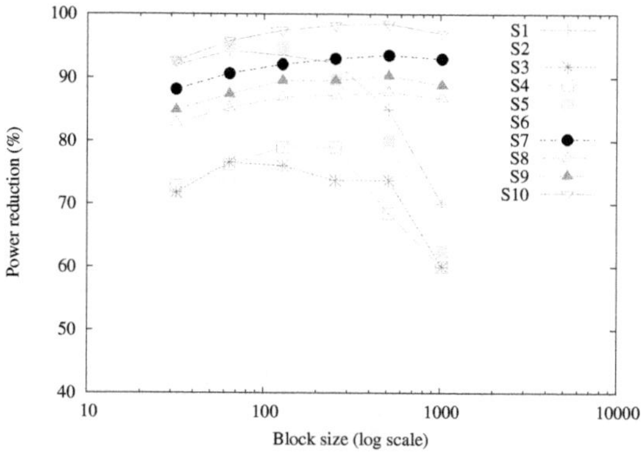

(b) Synthetic classifiers

Figure 5: The percentage of power reductions (note that y-axis starts at 40).

Table 5: Number of general rules and pre-classifier entries for real classifiers with block sizes 64, 128 and 256.

Name	# general rules			# pre entries		
	64	128	256	64	128	256
R1	85	101	159	101	50	25
R2	146	146	146	103	52	28
R3	824	520	302	103	55	29
R4	62	62	62	103	52	27
R5	11	11	11	123	62	31
R6	1513	1462	516	108	53	35
R7	11	11	11	179	88	48
R8	0	0	0	234	120	58
R9	2055	1004	1792	224	131	62
R10	2799	2479	2380	304	169	86

classifiers (S6 through S10) are bigger than those on the smaller ones.

By comparing Figure 5(a) and Figure 5(b), we find that the five 10K synthetic classifiers (S1 through S5) present similar trend of power reductions as the real classifiers (R1 through R10). However, the five 100k classifiers (S6 through S10) show bigger power reductions, as they appear in the upper part in Figure 5(b). This suggests that SmartPC achieves more power reductions for larger classifiers. The intuition behind this is that the number of intercepting rules with a given packet and the number of wildcard rules in the two-dimensional space do not increase proportionally with classifier size (as discussed in Section 3). Consequently, the number of pre-classifier entries and general blocks do not scale as well. According to the definition of the percentage of power reduction ($\frac{X-Y}{X}\times 100\%$), in the case of larger classifiers, Y does not increase proportionally with X. Therefore, SmartPC usually achieves more power reductions on larger classifiers.

To further understand the behaviors of our proposed algorithms for building pre-classifiers, we summarize the number of general rules, and the number of pre-classifier entries of real classifiers and synthetic classifiers, as shown in Table 5 and Table 6, for block sizes 64, 128, and 256 respectively.

We make the following observations from these tables:

- With the increase of block size, there is a general trend of non-increasing number of general rules. This is because that with smaller block sizes, rules have bigger chances to be marked as general, while with larger block sizes pre-classifier entries are more likely to be expanded successfully to cover more rules.

- With the increase of block size, the pre-classifier sizes decrease, since the number of entries in a pre-classifier is roughly proportional to N/B.

Table 6: Number of general rules and pre-classifier entries for synthetic classifiers with block sizes 64, 128 and 256.

Name	# general rules			# pre entries		
	64	128	256	64	128	256
S1	203	130	91	252	129	64
S2	2075	2030	1903	223	110	54
S3	1971	1931	1877	178	90	42
S4	2028	1776	1530	179	88	45
S5	119	73	67	167	97	48
S6	344	316	290	2855	1561	824
S7	5810	5490	4883	2302	1153	593
S8	11509	11191	11318	3147	1604	861
S9	9401	8699	9198	3091	1573	819
S10	787	558	379	3286	1748	866

- The number of general rules and pre-classifier entries are much smaller than the size of a classifier. Therefore, SmartPC can achieve huge savings in power consumptions since only the pre-classifier, general rules and a block of specific rules will be activated.

5.3 Storage overhead

In SmartPC, we need extra storage to store pre-classifier entries. Pre-classifier entries are two-dimensional on source and destination addresses, so each pre-classifier entry occupies 64 bits, while each five-dimensional rule occupies 104 bits. We show the percentage of pre-classifier size compared to the size of a whole classifier on real and synthetic classifiers in Figure 6. For example, R1 contains 5233 rules. With block size 128, there are 50 pre-classifier entries for R1. Therefore, we calculated the percentage of storage overhead as $\frac{50\times 64}{5233\times 104}\times 100 = 0.59\%$. Though the actual extra storage depends on TCAM specifications (e.g., width of TCAM, and etc), the numbers shown in Figure 6 provide an estimation of the storage overhead of pre-classifiers. As we can see, the percentage of extra storage decreases as block size increases, since the number of pre-classifier entries is proportional to N/B. We observe that the overhead is pretty small, less than 4% for every classifier. More than 80% of power reductions on the classifiers justify these storage overhead.

SmartPC may introduce holes in blocks, since some blocks may not be full. As discussed in Section 4.2, classifier rules associated with multiple pre-classifier entries can be stored into the same TCAM block as long as the block can fit them. We implemented a

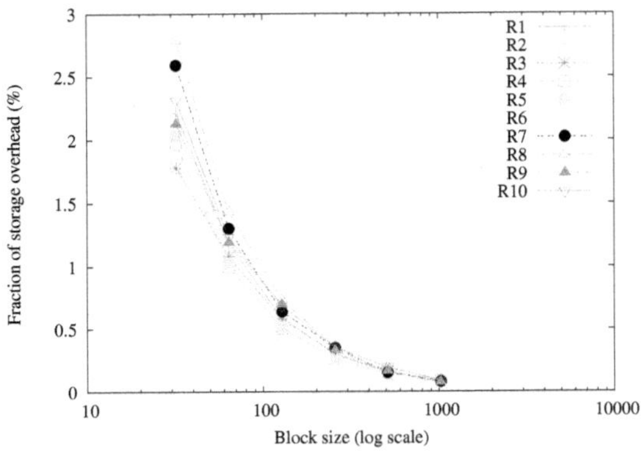

(a) Real classifiers

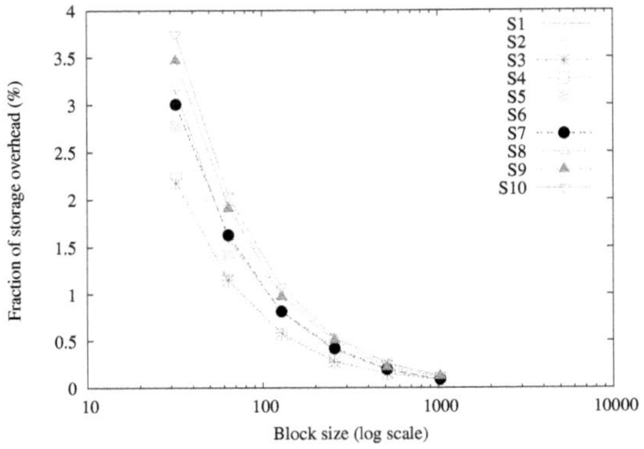

(b) Synthetic classifiers

Figure 6: The storage overhead of SmartPC.

variant of first-fit algorithm to minimize the number of blocks used and found no significant increase of blocks. Considering regular TCAMs also leave holes in blocks to reduce overhead of rule insertions, we only focus on the extra storage incurred by pre-classifiers.

5.4 Comparison of SmartPC with naive-divide

To prove that the huge power reductions come from the intelligence of the proposed pre-classifiers, we compare SmartPC with a naive approach, named naive-divide, which recursively divides the multi-dimensional space into smaller non-overlapping regions. As in the extended TCAMs paper [20], up to a block size number of rules that lie entirely in each region are assigned to the region and these rules are stored in the same TCAM block. Different from the extended TCAMs paper, naive-divide executes one phase, rather than multiple phases, so that there will be only one match in the index TCAM, instead of multiple matches. Naive-divide eliminates the need for multi-match TCAMs. In naive-divide, those rules that do not fit in the blocks are treated as general rules as in SmartPC.

In Figure 7, we compare the power reductions of SmartPC and naive-divide on real and synthetic classifiers. Here we show results with block size 128, while other block sizes show similar performance. SmartPC outperforms naive-divide on every classifier we evaluated. On average, SmartPC achieves 87% power reductions on real classifiers, while naive-divide only achieves 67%. SmartPC outperforms naive-divide by 14% to 34% on real classifiers, with an average of 20%. With synthetic classifiers, SmartPC achieves 88% power reductions on average, while naive-divide achieves 65%. SmartPC outperforms naive-divide by 7% to 32% on synthetic classifiers, with an average of 23%.

The reductions of SmartPC come from the intelligence of the pre-classifiers. In SmartPC, the structures of the rules are taken into account when building the pre-classifier entries. While in naive-divide, the multi-dimensional space can be divided arbitrarily, resulting in a larger amount of general rules. Compared to naive-divide, SmartPC results in smaller number of active TCAM blocks by reducing the number of general rules. Therefore, SmartPC achieves higher power reductions.

5.5 Discussion

5.5.1 The effect of block size on power reductions

To understand how block sizes affect power reductions, we plot the number of blocks that need to be activated with default scheme (without using power reduction optimizations) and SmartPC in Figure 8, where x-axis is block size, y-axis represents the number of TCAM blocks that need to be activated, and both axes are in log

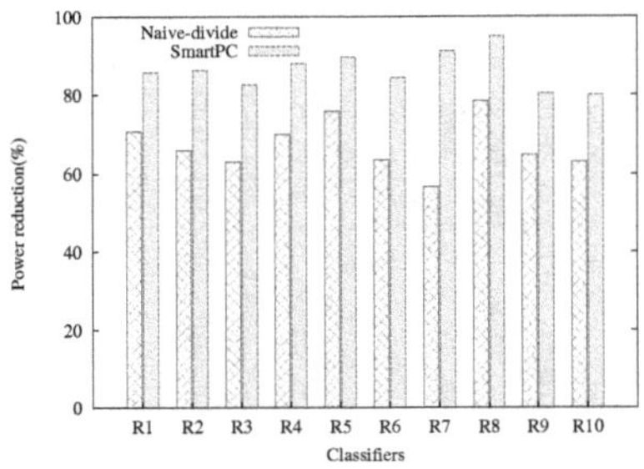

(a) Real classifiers

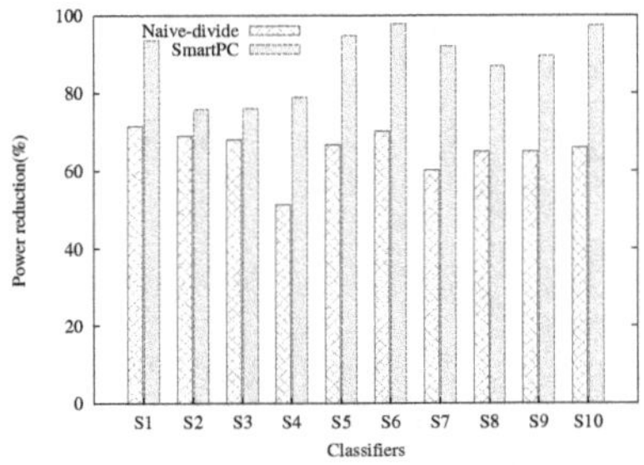

(b) Synthetic classifiers

Figure 7: Compare power reductions of naive-divide and SmartPC on real and synthetic classifiers with block sizes 128.

scale. For brevity, we use classifiers R1, R9, S4 and S10 as representatives, while other classifiers show similar trends.

In the default scheme, the number of active blocks decrease linearly with block size because the number of active blocks equals to $\lceil N/B \rceil$, resulting in a straight line for each classifier in Figure 8. While in SmartPC, there is a non-linear relationship between the number of active blocks and block size. Initially, the number of active blocks decreases quickly with block size, till it reaches some point when the decreasing speed reduces. We can see this point occurs at block size 64 for R1 (as shown in 8(a)), 128 for R9 (8(b)), 256 for S4 (8(c)) and 512 for S10 (8(d)). Accordingly, the maximum power reductions are achieved at these points as shown in Figure 5.

Moreover, we discover that the number of active blocks for R1 reaches a constant value of 3 when the block size exceeds 128. 3 blocks is the minimum number of blocks that have to be searched in our scheme [5], which include a block storing the pre-classifier, a general block and a specific block. Similarly, we also observe this

[5]This is generally true, while it is possible that the number of gen-

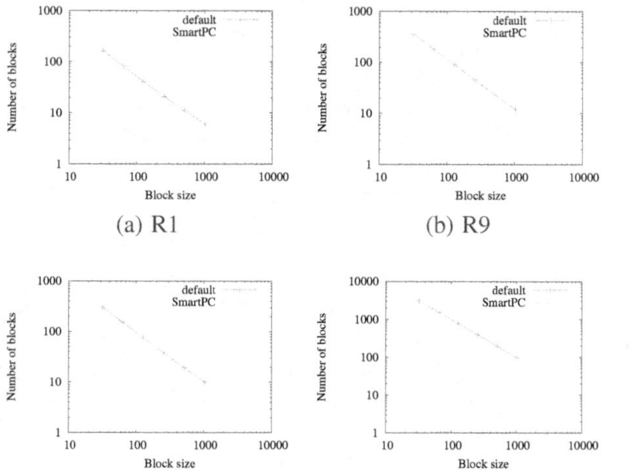

(a) R1 (b) R9

(c) S4 (d) S10

Figure 8: Number of blocks activated with default scheme and SmartPC (both axes are in log scale)

behavior with other classifiers. For example, the number of active blocks for S10 also reaches 3 when the block size is greater than 512.

5.5.2 *The effect of prefix distribution and prefix length*

Using ClassBench [21] tool, we generate classifiers with different prefix distributions by adjusting the *smoothness* parameter which may take a value from 0 to 64. A value 0 maintains the distributions specified in a parameter file to the rule generator, while a value of 64 models a uniform distribution. We observe that with the increase of smoothness from 0 to 64, both the number of general rules and the size of the pre-classifier decrease, resulting in larger power reductions. We set smoothness to 2 when generating the ten synthetic classifiers to evaluate the performance of SmartPC in a relatively rigorous environment.

We also generate classifiers with more or less specific address prefixes by varying *address scope*, a parameter in range of -1.0 to 1.0 which adjusts the average scope of the address prefixes. With less specific address prefixes (shorter prefix lengths), the number of rules marked as general increases, while the size of the associated pre-classifier decreases. Though the number of general rules and the size of pre-classifier fluctuate with the parameter, there is no obvious change in the percentage of power reductions in SmartPC.

5.5.3 *Power reductions on small classifiers and IPV6 classifiers*

Though the focus of our work is on large classifiers, small classifiers can also benefit from SmartPC. From above analysis, our scheme could provide power reductions on a classifier which occupies more than three TCAM blocks, since generally three blocks is the minimum number of blocks that are activated in our scheme.

Although we do not have access to IPV6 classifiers, the algorithms presented in Section 4 apply to IPV6 classifiers without any modifications. We expect similar performance on IPV6 classifiers.

6. RELATED WORK

There are two main threads of research on packet classification: RAM-based algorithmic approaches and TCAM-based approaches. A lot of intelligent algorithmic solutions are proposed and many of them are based on decision trees, e.g., HiCuts [8], HyperCuts [19], HyperSplit [18], Modular packet classification [25], Common Branches [4], and EffiCuts [23]. The core issue of algorithmic approaches centers on the tradeoff between memory usage and speed.

Wire speed packet classification motivated the development of hardware-based solutions. Ternary Content Addressable Memories (TCAMs) has become the *de facto* industrial standard for packet classification in high performance routers. However, TCAMs suffer from several primary deficiencies: high power consumption, high cost per bit relative to other memory technologies, and storage inefficiency. Accordingly, prior work in optimizing TCAM-based systems fall into four broad categories: power efficiency [26, 20], circuit modification [20, 11], classifier compression [5, 15] and range reencoding [17]. In this paper, we focus on power-efficient TCAM solutions, while reencoding or compression or reducing the number of entries is an orthogonal problem, and our solution can be combined with these types of work.

CoolCAMs [26] divides a TCAM device into multiple partitions. An IP lookup becomes a two stage process where only one partition is selected in the first stage and the partition is queried in the second stage. However, CoolCAMs is limited to the problem of IP forwarding where the destination address of an incoming packet is matched against the longest matching prefix in a routing table.

To apply TCAMs to the more difficult packet classification problem where incoming packets are matched against multi-dimensional rules in TCAMs, extended TCAMs [20] extends the partitioned TCAM concept in [26]. A filter set is partitioned into multiple blocks, each of which is associated with an appropriate index filter. Upon a query, all matching index filters are first identified and then the blocks associated with those matching index filters are queried. However, packets follow up on multiple matches in the index TCAM, which makes this approach infeasible in two ways. First, if legacy chips are used, the bitmap with the blocks can not be transferred to activate on each search. Second, it requires a new type of ternary match hardware that returns all matches because commodity TCAMs give only the first result.

Except [26] and [20], there is significant recent work in the space of power efficiency that we are aware of, but they are within companies and are of proprietary nature. We are under NDA with one company and can not refer to them.

Classifier compression optimizations convert a given classifier into another semantically equivalent classifier that requires fewer TCAM entries. While these techniques would also reduce power consumptions, SmartPC is complementary to compression techniques, and if combined, power consumptions could be further reduced. In [5], by expanding, trimming, adding and merging rules, the authors identify semantically equivalent classifiers that lead to fewer TCAM entries. The redundancy removal algorithm in [12] can reduce TCAM usage by eliminating all the redundant rules in a packet classifier. To address the prefix expansion problem of TCAMs, TCAM Razor [15] proposed a greedy algorithm that finds locally minimal solutions along each dimension and combines these solutions into a smaller equivalent packet classifier. In Bit Weaving [16], the authors proposed the first algorithm that can compress a given classifier into a non-prefix ternary classifier.

Range reencoding schemes cope with range expansion by developing a new representation for important packets and intervals. Previous range reencoding schemes fall into two categories: database independent encoding schemes [1, 11], where each rule is encoded according to standard encoding scheme, and database dependent encoding schemes [2, 17, 13], where the encoding of each rule depends on the intervals present within the classifier. While range

eral rules is zero for some classifiers, therefore bring down this number to 2.

reencoding schemes mitigate the effects of prefix expansion, they require either extra hardware or more per packet processing time.

In [9], the authors explored the structure and properties of four ACLs, and provided a guideline for designing classification algorithm which states that a multi-dimensional classification problem should be split into two stages. Though we also employ a two-stage process, there are major differences. The focus of [9] is on properties of ACLs and the authors did not design any specific packet classification algorithm. Furthermore, we employ a pre-classifier to pre-classify on source and destination addresses, and then a full classifier on all the five fields. While in [9], the first stage is classification on source-destination pairs and the second is on other fields. The results from the two stages are merged to get the final result.

7. CONCLUSION

In this paper, we performed a large scale analysis of important properties of more than 200 real classifiers. Based on our analysis, we propose SmartPC, a smart pre-classifier for power-efficient packet classification using TCAMs. In SmartPC, the rules in a classifier are shuffled into TCAM blocks such that each pre-classifier entry is associated with a TCAM block. To classify a packet in SmartPC, the two-dimensional pre-classifier is first consulted which directs the search to at most one specific TCAM block that contains rules intercepting with the packet. Then the specific block plus a few general blocks are searched in parallel and two matches are generated. Finally the action from the higher priority match is returned as the final result. SmartPC uses commodity TCAMs, and the algorithms for building pre-classifiers are easy to implement. We evaluated SmartPC with real and synthetic classifiers. SmartPC achieves more than 80% power reductions on most classifiers with less than 4% storage overhead. With block size 128, SmartPC achieves a median power reduction of 91% and an average power reduction of 88% on these classifiers. To demonstrate the effectiveness of pre-classifiers, we compared SmartPC with naive-divide, a naive approach that recursively divides the multi-dimensional space into smaller regions. SmartPC outperforms naive-divide for every classifier, with 20% more power reductions on average. SmartPC is a practical and promising solution to address the high power consumption of TCAMs and can find its applications in data centers.

8. ACKNOWLEDGEMENTS

All authors are supported in part by the following grants of the US NSF: CNS-1040648, CNS-0916955, CNS-0855201, CNS-0747177, CNS-1064944, and CNS-1059306.

9. REFERENCES

[1] A. Bremler-barr and D. Hendler. Space-efficient TCAM-based classification using gray coding. In *IEEE INFOCOM*, 2007.

[2] H. Che, Z. Wang, K. Zheng, and B. Liu. DRES: Dynamic range encoding scheme for tcam coprocessors. *IEEE Transactions on Computers*, 57:902–915, 2008.

[3] Cisco. Cisco catalyst 4500 series supervisor engine 6-e centerflex technology. http://www.cisco.com/en/US/prod/collateral/switches ps5718/ps4324/prod_white_paper0900aecd806dc821.html.

[4] E. Cohen and C. Lund. Packet classification in large ISPs: Design and evaluation of decision tree classifiers. In *ACM SIGMETRICS*, 2005.

[5] Q. Dong, S. Banerjee, J. Wang, D. Agrawal, and A. Shukla. Packet classifiers in ternary CAMs can be smaller. In *ACM SIGMETRICS*, 2006.

[6] H. Edelsbrunner, L. J. Guibas, and J. Stolfi. Optimal point location in a monotone subdivision. *SIAM J. Comput.*, 15:317–340, May 1986.

[7] P. Gupta and N. Mckeown. Packet classification on multiple fields. In *ACM SIGCOMM*, 1999.

[8] P. Gupta and N. Mckeown. Packet classification using hierarchical intelligent cuttings. In *Hot Interconnects VII*, 1999.

[9] M. E. Kounavis, A. Kumar, H. Vin, R. Yavatkar, and A. T. Campbell. Directions in packet classification for network processors. In *HPCA-9*, 2003.

[10] T. V. Lakshman and D. Stiliadis. High-speed policy-based packet forwarding using efficient multi-dimensional range matching. In *ACM SIGCOMM*, 1998.

[11] K. Lakshminarayanan, A. Rangarajan, and S. Venkatachary. Algorithms for advanced packet classification with ternary CAMs. In *ACM SIGCOMM*, 2005.

[12] A. X. Liu, C. R. Meiners, and Y. Zhou. All-match based complete redundancy removal for packet classifiers in TCAMs. In *IEEE INFOCOM*, 2008.

[13] H. Liu. Efficient mapping of range classifier into Ternary-CAM. In *HOT Interconnects*, 2002.

[14] Y. Ma, S. Banerjee, S. Lu, and C. Estan. Leveraging parallelism for multi-dimensional packet classification on software routers. In *ACM SIGMETRICS*, 2010.

[15] C. Meiners, A. Liu, and E. Torng. TCAM Razor: A systematic approach towards minimizing packet classifiers in TCAMs. In *ICNP*, 2007.

[16] C. R. Meiners, A. X. Liu, and E. Torng. Bit weaving: A non-prefix approach to compressing packet classifiers in TCAMs. In *ICNP*, 2009.

[17] C. R. Meiners, A. X. Liu, and E. Torng. Topological transformation approaches to optimizing TCAM-based packet classification systems. In *ACM SIGMETRICS*, 2009.

[18] Y. Qi, L. Xu, B. Yang, Y. Xue, and J. Li. Packet classification algorithms: From theory to practice. In *IEEE INFOCOM*, 2009.

[19] S. Singh, F. Baboescu, G. Varghese, and J. Wang. Packet classification using multidimensional cutting. In *ACM SIGCOMM*, 2003.

[20] E. Spitznagel, D. Taylor, and J. Turner. Packet classification using extended TCAMs. In *ICNP*, 2003.

[21] D. Taylor and J. Turner. ClassBench: A packet classification benchmark. http://www.arl.wustl.edu/~ det3/ClassBench/index.htm.

[22] TopCoder. Line sweep algorithms. http://community.topcoder.com/tc?module=Static&d1=tutorials&d2=lineSweep.

[23] B. Vamanan, G. Voskuilen, and T. N. Vijaykumar. EffiCuts: optimizing packet classification for memory and throughput. In *ACM SIGCOMM*, 2010.

[24] G. Varghese. *Network Algorithmics: An Interdisciplinary Approach to Designing Fast Networked Devices*. Morgan Kaufmann, 2005.

[25] T. Y. Woo. A modular approach to packet classification: algorithms and results. In *IEEE INFOCOM*, 2000.

[26] F. Zane, G. Narlikar, and A. Basu. CoolCAMs: Power-efficient TCAMs for forwarding engines. In *IEEE INFOCOM*, 2003.

ShadowStream: Performance Evaluation as a Capability in Production Internet Live Streaming Networks

Chen Tian* Richard Alimi† Yang Richard Yang*× David Zhang§
*Yale University †Google §PPLive ×SplendorStream
{tian.chen, yang.r.yang}@yale.edu ralimi@google.com davidzhang@pplive.com

ABSTRACT

As live streaming networks grow in scale and complexity, they are becoming increasingly difficult to evaluate. Existing evaluation methods including lab/testbed testing, simulation, and theoretical modeling, lack either scale or realism. The industrial practice of gradually-rolling-out in a testing channel is lacking in controllability and protection when experimental algorithms fail, due to its passive approach. In this paper, we design a novel system called ShadowStream that introduces evaluation as a built-in capability in production Internet live streaming networks. ShadowStream introduces a simple, novel, transparent embedding of experimental live streaming algorithms to achieve safe evaluations of the algorithms during large-scale, real production live streaming, despite the possibility of large performance failures of the tested algorithms. ShadowStream also introduces transparent, scalable, distributed experiment orchestration to resolve the mismatch between desired viewer behaviors and actual production viewer behaviors, achieving experimental scenario controllability. We implement ShadowStream based on a major Internet live streaming network, build additional evaluation tools such as deterministic replay, and demonstrate the benefits of ShadowStream through extensive evaluations.

Categories and Subject Descriptors: C.4 [**Performance of Systems**]: Measurement techniques.
General Terms: Measurement, Experimentation, Design.
Keywords: Performance Evaluation, Live Testing, Streaming.

1. INTRODUCTION

Motivation: Live streaming (*e.g.*, [11, 27, 45]) is a major Internet application that has been used to carry both daily and major events such as the Obama inauguration address, the 2010 Winter Olympics, and the 2010 World Cup. As live streaming continues to permeate into our daily lives [1], it is becoming increasingly important that Internet live streaming networks provide reliable performance, despite such networks becoming more complex, to operate in the increasingly more complex Internet.

A key capability to ensure that live streaming networks provide reliable performance is to subject them to large-scale, realistic performance evaluations. In our recent survey of the developers of a number of Internet-scale application sites, the ability to conduct large-scale, realistic performance evaluations is among the most desired, but the most difficult to achieve abilities to improve site reliability. In the context of live streaming, without such an ability, developers have to resort to theoretical modeling (*e.g.*, [8, 25, 46]), simulation (*e.g.*, [29, 39]), or lab/testbed testing (*e.g.*, [6, 10, 22, 31, 32]). But all of these existing methods are seriously limited either in scale or in realism. For example, no existing testbed can scale easily to tens of thousands of clients but many live streaming events have reached or even surpassed this scale. Further, it is becoming increasingly difficult for modeling, simulation, or testbed to capture the increasingly complex Internet, which includes diverse ISP network management practices and heterogeneous network features such as PowerBoost [2, 36] in cable networks, large hidden buffers in access networks [15, 23, 36], and shared bottlenecks at ISP peering or enterprise ingress/egress links [24].

A practice used by the live streaming industry is to use a testing channel to roll out new versions. However, the passive approach of this practice leads to serious limitations on scale and experimental feasibility. First, due to concerns on the failures of experimental algorithms, existing usage of testing channels is conservative, limiting tests to a small number of real users. Second, using real viewers as they naturally arrive and depart, may not provide the target operating conditions that stress the experimental system. A developer may always want to evaluate whether a modification will change the effectiveness of the live streaming system to handle a flash-crowd with 100,000 viewers. Although the testing channel may currently have more than 100,000 real viewers, the natural arrivals of the viewers do not conform to a flash-crowd arrival pattern.

Live streaming systems updated without going through realistic, large-scale evaluations may operate at sub-optimal states, and often do not achieve performance expectations formed at small-scale lab settings. For example, PPLive, a major live streaming distribution network, encountered surprising performance failures when it moved from an initial deployment in the university-based CERNET (China Education and Research Network) to the general Internet with many viewers in ADSL networks. A major category of viewer complaints in the viewer forum of PPLive [33] is poor performance after network updates.

The motivation of designing ShadowStream starts with a relatively extreme recognition that a production live streaming network can be a rarely available foundation for designing a testing system with both *scale* and *realism*, by taking advantage of its large number of real viewers distributed across the real Internet. Building on this foundation to introduce protection and controllability, we develop a novel Internet live streaming system that integrates performance evaluation as an intrinsic capability, a new capability that complements analysis, simulation, and lab testing. We refer to testing during production streaming as *live testing*.

Challenges: A live testing platform poses two key challenges that are not present in traditional testing platforms such as VINI [7], PlanetLab [12], or Emulab [41]. The objective of ShadowStream is to address both challenges.

The first is *protection*. Since real viewers are used, live testing needs to protect the real viewers' quality of experience from the performance failures of experimental systems. Hence, live testing needs to manage the challenge of letting performance failures of an experimental system happen so that poor performing algorithms can be identified, and at the same time masking these failures from real viewers. As a contrast, a traditional testing platform uses ded-

SIGCOMM'12, August 13–17, 2012, Helsinki, Finland.

icated clients in a non-production environment, and hence performance failures of an experimental system will not do harm.

The second is *orchestration*. In particular, live testing needs to orchestrate desired experimental scenarios (*e.g.*, flash-crowd) from production viewers, without disturbing their quality of experiences. In a traditional testing platform, clients are dedicated and fully controllable artificial viewers. In live testing, a real viewer expects that the live streaming service starts promptly, as soon as the viewer arrives; real viewers do not expect disruptions in the middle of the streaming to ask them to join experiments; as soon as a real viewer closes the live streaming program, the service should stop, not continuing as a hidden background process to finish the experiment.

Our approach: ShadowStream applies a general *Experiment→Validation→Repair* scheme to achieve protection and transparent orchestration; see Section 8 on discussions for generalization. Specifically, ShadowStream introduces, at each client, a simple *streaming hypervisor*, which manages multiple concurrent *virtual streaming machines*, one of which is an experimental streaming machine containing the experimental algorithms. The ShadowStream hypervisor first assigns each task (*i.e.*, downloading a video piece) to the experiment, as if the experimental system were running alone, and records any failures (*i.e.*, if a piece does not arrive before the target deadline of the experimental system), to obtain accurate experimental results. However, repair steps in to fix experimental failures before they become visible to real viewers. The hypervisor also implements *pre-reservation* of task space so that an experimental system can always be started without the interference of the production system (*i.e.*, inserting an experiment while the real viewer has already started with the production system). To make repair and pre-reservation feasible, in the context of hybrid live streaming systems, ShadowStream introduces a small additional lag on the playpoints of real viewers.

ShadowStream extends the basic scheme with novel algorithms to achieve scalability, for both protection and orchestration. In particular, it introduces a novel *production-CDN-experiment (PCE)* streaming machine layout, to achieve highly scalable protection; it also introduces novel, local orchestration algorithms to achieve highly scalable orchestration.

We implement ShadowStream based on one of the largest live streaming networks. Our implementation shows that ShadowStream is easy to implement, with code base around 8,000 lines. Based on real system traces, we demonstrate that ShadowStream can allow scalable testing reaching 100,000 clients while no existing testbed can easily reach even thousands.

2. MOTIVATION

2.1 Live Streaming Grows in Complexity

Many modern live streaming networks have become complex hybrid systems that use both peer-to-peer (P2P) networks and content delivery networks (CDN) [30]. Adobe Flash is one platform that has implemented support for a hybrid architecture in recent versions. In this paper, we target this general, widely used streaming architecture.

The basic design of a hybrid live streaming network is typically BitTorrent-like. Guided by a tracker, peers viewing the same live event connect with some neighbors to form an overlay network to relay video data through the overlay topology. A source encodes small clips of video data called *pieces*, which are made available to CDNs. Some peers in the overlay network download pieces from CDN networks. A peer may directly push new pieces to some neighbors and/or exchange piece bitmaps with its neighbors so that the neighbors can request pieces from it.

To handle practical challenges and improve performance, the preceding basic design is typically substantially enhanced. Hence, a production live streaming system often includes a large number of modules: (i) P2P topology management, (ii) CDN management, (iii) buffer and playpoint management, (iv) rate allocation, (v) download/upload scheduling, (vi) viewer-interfaces, (vii) shared bottleneck management (*e.g.*, [20]), (viii) flash-crowd admission control, and (ix) network-friendliness (*e.g.*, [5, 44]).

The performance of a network with a large number of peers each running a software system containing many modules can become difficult to evaluate. A main metric on the performance of live streaming is *piece missing ratio*, which is the fraction of pieces that are not received by their playback deadlines. A major factor affecting piece missing ratio is channel *supply ratio*, which is the ratio between total bandwidth capacity (CDN and P2P together) and total streaming bandwidth demand of all participating peers. Next, we use a few concrete examples to demonstrate the importance of large-scale realistic experiments.

2.2 Misleading Results at Small-Scale

Testbed experiments have difficulty replicating the scale of actual deployment, which can reach hundreds of thousands of viewers. But results obtained from small-scale networks can be quite different from those obtained from large scale networks. As a concrete example, we run the same production version (default version) of a major live streaming network at two scales in Emulab: 60 clients (smaller scale) and 600 clients (larger scale). In both cases, clients join sequentially with a 5-second interval; the streaming rate is 400 Kbps; each client's upload capacity is 500 Kbps, and a CDN server provides an upload capacity of 10,000 Kbps. Thus, the supply ratio of the smaller network is around 1.67 (= (60 * 500 + 10,000)/(60 * 400)), while for the larger network it is around 1.29. Since the smaller network has a larger supply ratio, one might think that it would perform better than the larger network.

	Small-Scale	Large-Scale
Production Default	3.5%	0.7%

Table 1: Piece missing ratio at different scales (default).

Table 1 shows the performance of the two networks measured by piece missing ratio. We observe a large performance difference: the piece missing ratio of the small network is 5 times that of the large network. A detailed trace analysis shows that content bottleneck (*i.e.*, diversity of data among neighbors) is a key reason for the poor performance of the small network. The average buffer map difference among peers is not sufficient to fully utilize available bandwidth. On the other hand, in the larger network, content bottleneck may not be an issue; the peer upload bandwidth together with the bandwidth scheduling algorithm dominate the performance. This observation is consistent with a recent measurement [42].

	Small-Scale	Large-Scale
With Connection Limit	3.7%	64.8%

Table 2: Piece missing ratio at different scales (variant).

We run a variant of the default version with a connection limit (*i.e.*, a client declines all subsequent connection requests when the connection limit is reached). Table 2 shows a surprising and interesting result: after imposing the limit, we again observe a large performance difference, but opposite from the preceding result. Specifically, the performance degrades significantly in the larger scale experiment, with piece missing ratio of the larger scale experiment being 17.5 times that of the smaller scale! Our detailed evaluation shows that with a connection limit, the CDN server's neighbor connections are exhausted by those clients that join earlier and the clients that join later cannot connect to the CDN server and experience starvation. Both experiments illustrate that observations from a smaller scale may not directly translate to those at a larger scale.

2.3 Misleading Results due to Missing Realistic Features

It is extremely challenging to capture all aspects of real networks in a simulator, emulator, or testbed. For example, end hosts may differ in type of their network connectivity, amount of system resources, and network protocol stack implementations; routers may differ in their queue management policies, scheduling algorithms, and buffer sizes; background Internet traffic may change dynamically and is hard to replicate. Failures to capture any of these effects may result in misleading results. As a motivating example, below we show that simply changing the access link types can lead to different performance results.

Specifically, typical academic testbeds (*e.g.*, PlanetLab, Emulab) are Ethernet based. On the other hand, most live streaming viewers are connected to the Internet using cable or DSL networks. We run the same live streaming network in two settings: LAN-like university networks versus ADSL-like networks. We set up 20 clients, where each client is a PPLive client, with 1-second piece request time out. Clients join sequentially with a 5-second inter-arrival time; the testing lasts 100 seconds after all peers have joined. Each client's upload capacity is equal to the streaming rate of 400 Kbps, and the CDN server provides an upload capacity of 2,400 Kbps.

	LAN	ADSL
% Piece missing ratio	1.5%	7.3%
# Timed-out requests	1404.25	2548.25
# Received duplicate pkts	0	633
# Received outdated pkts	5.65	154.20

Table 3: Different results at LAN vs ADSL.

Table 3 shows the results. We observe a large performance difference between the LAN-based and ADSL-based experiments: the piece missing ratio of ADSL-based is almost 5 times that of LAN-based. Our investigation shows that the ADSL-based experiment performs worse because of large hidden buffers that are not present in LAN-like networks [23]. As an ADSL modem has a large buffer but limited upload bandwidth, a large queue can be built up at the modem and the queueing delay can reach several seconds. On the other hand, the default piece request time out value is 1 second, due to testing in the LAN-based network. This shows that testing in an environment that is different from the real setting can give misleading results and yield inappropriate system configurations.

Note that once it is determined that this is an important feature, hidden buffers can be approximated by a FIFO queue using the Linux tc tool. other network features, such as PowerBoost [2], may involve internal and dynamic ISP-policies, and hence can be much harder to emulate [14].

3. SYSTEM COMPONENTS

ShadowStream makes two extensions to a traditional live streaming network: (1) it introduces a lightweight experiment orchestrator, which extends the functions of a typical tracker to coordinate experiments; (2) it introduces new capabilities in traditional streaming clients to support live testing. ShadowStream complements analysis, simulation, and testbed to provide a more complete experimentation framework.

Figure 1 shows the major components of a ShadowStream network, their interactions, and tools such as deterministic replay that we build on top of basic ShadowStream. For scalability, the main intelligence of ShadowStream is in individual ShadowStream clients.

4. STREAMING CLIENT DESIGN

We start with the design of ShadowStream clients. We assume that experimental algorithms have been first tested in lab, going through unit and regression tests to fix correctness errors and pro-

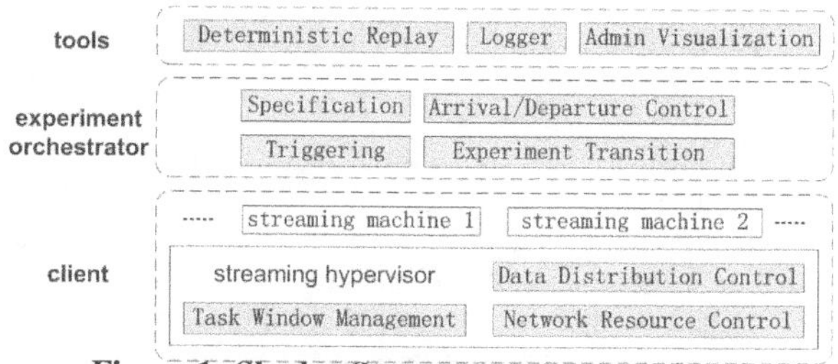

Figure 1: ShadowStream system components.

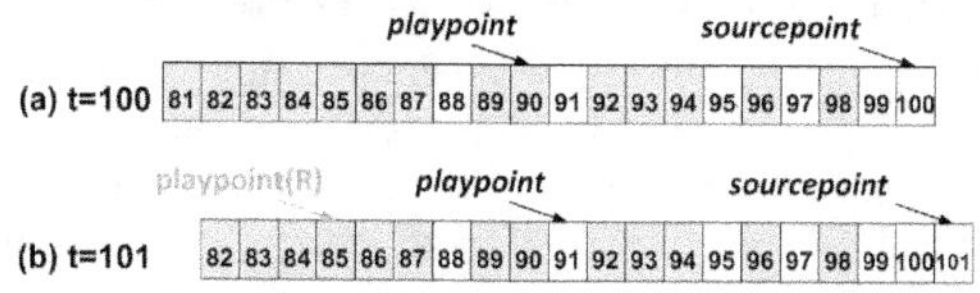

Figure 2: Streaming machine buffer: (a) at t=100; (b) at t=101.

gram crashes. Our focus is on performance. We focus on the main ideas in this section and choose to present our design in steps to make understanding easier. Implementation is in the next section.

4.1 Streaming Machine

We adopt the general, widely used piece based streaming structure, where a live streaming client downloads and uploads streaming data in units of pieces. For simplicity of explanation, we assume that each piece contains 1-second of streaming data.

We refer to a self-complete set of algorithms to download and upload pieces as a *streaming machine* or a machine for short. In a traditional setting, a streaming client of a viewer has one streaming machine. In ShadowStream, as we will see in this section, there can be multiple streaming machines executing concurrently. We refer to the streaming machine containing the experimental algorithms as `experiment`.

A key data structure of a streaming machine is its play buffer, which keeps track of pieces that are already downloaded as well as the pieces that the machine needs to download from its neighbors or CDN networks in order to feed a media player, which displays to a viewer. Figure 2 (a) is an example illustrating the play buffer status at time $t = 100$ of a client i. A shaded piece is one that has been downloaded. We follow a convention that the index of a piece is the time that the piece was produced at the source.

In a traditional setting without ShadowStream, the right most piece of play buffer is the most recent piece produced by the source. We refer to this piece as the *sourcepoint*. We also use sourcepoint to refer to the index of the sourcepoint piece. For the example in Figure 2 (a), the sourcepoint is 100. A client may not always communicate with the source to directly know about the sourcepoint. But during initialization, the client learns the time at the source and hence can always compute the time at the source through its local clock and hence the current sourcepoint. Another important piece at an instance of time is the next piece to be delivered to the media player at the client. We refer to it as the *playpoint*. For the example in Figure 2 (a), the playpoint is 90.

The time difference between the sourcepoint and the playpoint represents the *lag*, which is the delay from the time that a piece becomes available to the time that it should become visible to a viewer. A streaming machine may choose a lag when it starts. But as soon as it starts playback to a real viewer, it can no longer change the lag without disturbing the real viewer. The lag is essentially the total budget of time to finish downloading each piece. Note that the play buffer keeps some pieces older than the playpoint so that the client can serve other peers in a P2P mode.

The playpoint and sourcepoint advance in time. Figure 2 (b) shows the advancement of the playbuffer from Figure 2 (a). We see that at the next time $t = 101$, the playpoint becomes 91, and the sourcepoint becomes 101. If the client is not able to download piece 91 from time $t = 100$ to $t = 101$, then we say that piece 91 is missing.

We comment that one may consider the piece-based streaming machine model as a quite general program model, where each piece is a computational task, and a sequence of tasks arrive at a streaming machine to be computed and the results are revealed to real viewers. Hence, we also refer to each piece to be downloaded as a task; a piece missing its deadline as a task failure.

4.2 Starting Point: R+E to Mask Failures

The starting design of ShadowStream is simple and intuitive. To reveal the true performance of `experiment`, ShadowStream assigns a task first to `experiment` (E) as if it were running alone. To protect real viewers' quality-of-experience (QoE), ShadowStream assigns a failed task to another streaming machine `repair` (R) for repair; that is, if a piece is not received by its target playback deadline, ShadowStream records a performance failure event and the responsibility for downloading the piece is shifted to R. We refer to such a design as a R+E design pattern which is shown in Figure 3.

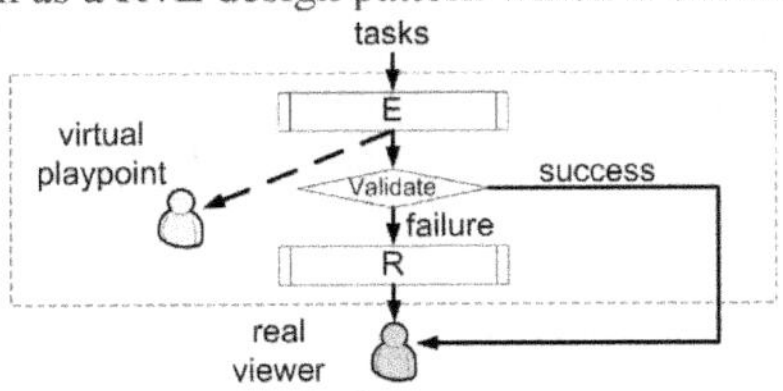

Figure 3: The general R+E design pattern.

Assume that Figure 2 (b) is the playbuffer state of a streaming machine `experiment`. Since `experiment` is not able to download piece 91, ShadowStream records that `experiment` has missed piece 91, and then assigns it to R. Assume that R is able to download piece 91 within one second. Then the piece becomes available at time $t = 102$.

To ensure that the failure of `experiment` on finishing piece 91 is not visible to the real viewer during our evaluation, we need that the playpoint of `experiment` is not the actual viewer visible playpoint. Hence, we say that the playpoint of an experimental streaming machine is a *virtual playpoint*. Specifically when $t = 101$, if the viewer visible playpoint actually is 85, then the real viewer will not see a missing piece 91.

In other words, in the R+E design, `experiment` is evaluated by its ability to finish each piece within its designed target deadline, which is 10 = (100-90) in this example, when `experiment` is running alone. But the R+E design delays the delivery of results to real viewers with an additional 5-second lag, to give R the time to repair errors, if any. We comment that the technique of introducing a slight delay in revealing results has been used in other contexts. In particular, radio and television broadcast introduces a 7-second profanity delay to handle profanity, bloopers, violence, or other undesirable material [9]. We are the first to introduce this technique to achieve our new goal of live testing.

4.3 R=`production` for Staging/Protection

A question that we have not answered is how to design R. One simple approach is that R is a machine which can check the status of each piece and download missing pieces from dedicated CDN resources provided for testing; we call this design R=`rCDN`. However, there can be limitations.

First, using R=`rCDN` requires dedicated CDN resources. Note that the dedicated CDN resources are independent of the CDN resources in other streaming machines. Suppose that a testing channel consists of 100,000 clients at a 1 Mbps streaming rate. To guarantee performance during the worst case when `experiment` cannot finish the majority of the tasks assigned to it, R=`rCDN` needs to reserve a total of 100 Gbps capacity from the CDN network. For a streaming technology company with multiple groups developing different algorithms in parallel, multiple testing channels might be running concurrently, hence the amount of reserved capacity on CDN could be too high.

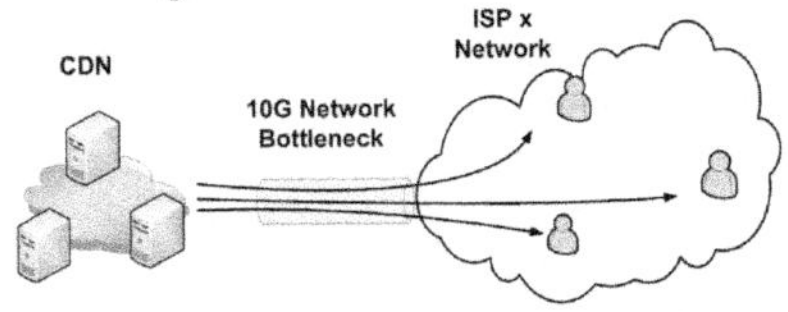

Figure 4: Network bottleneck between CDN and clients.

Second, R=`rCDN` may not work well in some network settings. Figure 4 shows a network bottleneck (*e.g.*, enterprise ingress) from the locations of the CDN edge servers to a group of streaming clients [24]. In this setting, even if the CDN has a large reserved capacity (*e.g.*, 100 Gbps), due to the bottleneck (*e.g.*, 10 Gbps), the repair capability of the CDN edge servers cannot be fully utilized.

To better understand the limitations of R=`rCDN`, we compare `rCDN` with the production streaming engine named `production`. We observe that `production` typically has more sophisticated, fine-tuned algorithms (*e.g.*, hybrid architecture) compared with a repair-only `rCDN`; `production` can use a much larger resource pool (the whole production infrastructure) than the dedicated CDN.

Hence, using R=`production` can lead to much more scalable protection. Further, it leads us to a unified approach to handle both protection and orchestration (*i.e.*, serving a client before `experiment` starts). Specifically, a ShadowStream client always uses `production` as the streaming machine to start. We can consider this state as a testing state with an `experiment` streaming machine that misses every task (*i.e.*, a no-op). When a client joins a test, a corresponding `experiment` starts. Hence, we can (conceptually) consider that a client is always in a testing phase. We refer to this as *pre-reservation* for testing. The virtual "arrival" and "departure" of an experimental streaming machine `experiment` are simple state operations. Figure 5 illustrates the transitions of states, and these virtual transitions are invisible to real viewers.

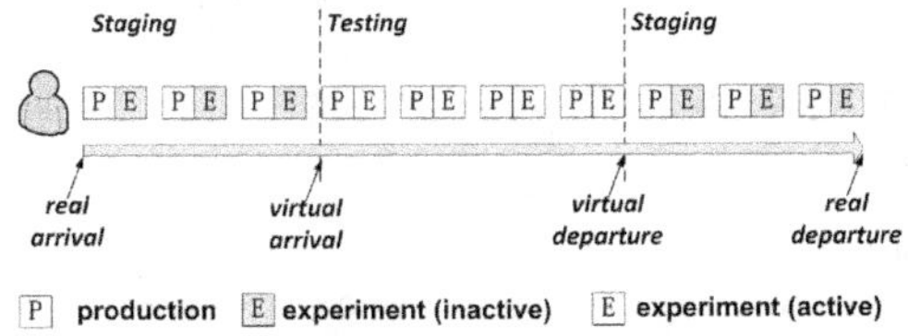

Figure 5: ShadowStream orchestration timeline: real arrival, virtual arrival, virtual departure, real departure.

4.4 Problems of R=`production`

One issue of using R=`production`, however, is that it has a systematic bias in underestimating the performance of `experiment`, due to competition between `experiment` and `production` on the shared resources: peer upload bandwidth. To protect viewers' QoE, it is desirable to give `production` a higher priority over `experiment` on using the shared upload bandwidth. However, the reduction of resources to `experiment` may lead to a systematic underestimation on the performance of `experiment`, leading a developer to unnecessarily reject `experiment` as poor performing. Below we provide a more detailed analysis on the bias. Readers who seek a solution directly can skip to the next subsection.

For a reasonably designed streaming machine, piece missing ratio $m(\theta)$ is a non-increasing function of the supply ratio θ. The curve labeled as *performance curve* in Figure 6 is an example curve of the piece missing ratio function of a typical live streaming machine. Then our goal of evaluating `experiment` is to get an accurate estimation of $m(\theta_0)$ for `experiment`, during a typical live testing with supply ratio θ_0.

However, competition between `production` and `experiment` may not allow us to observe $m(\theta_0)$. Instead, the dynamic competition between these two streaming machines will lead to a *stationary point* labeled as "misleading result" in Figure 6. The point is the intersection between the performance curve and the curve representing the points whose x and y-axis values sum to θ_0. We label the x-axis (effective supply ratio to `experiment`) of the "misleading result" point as θ^*.

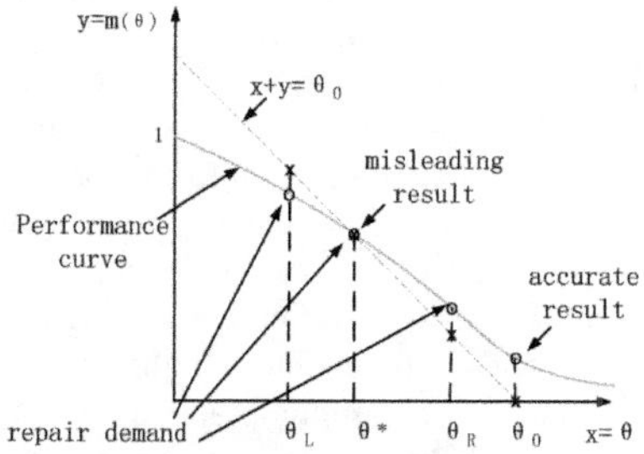

Figure 6: Systematic bias when using `production`.

To understand why θ^* is a stationary point that will be observed, first consider a supply ratio θ_R on the right of θ^*. Assume that at an instance of time the supply ratio of `experiment` is θ_R, the piece missing ratio of `experiment` will be $m(\theta_R)$. Hence, `production` takes away $m(\theta_R)$ resources to repair the losses of `experiment`. For simplicity, we assume that `production` has a repair efficiency η of 1; that is, it uses one unit of bandwidth resource to repair one unit of task. The available resources to `experiment` after the deduction become $\theta_0 - m(\theta_R)$. Since $m(\theta_R)$ is higher than $\theta_0 - \theta_R$, as one can verify by noticing from Figure 6 that the $x + y = \theta_0$ curve is a 45 degree line, we know that the remaining resources to `experiment` will become lower than θ_R $(= \theta_0 - (\theta_0 - \theta_R))$. In other words, the supply ratio of `experiment` moves from θ_R toward θ^*. One can make a similar observation when considering a supply ratio θ_L on the left of θ^*. Specifically, for a fixed η, we can conclude that θ^* satisfies:

$$\theta^* + \eta * m(\theta^*) = \theta_0.$$

In more complex settings, the efficiency of η could be a function of the specific missing pieces of `experiment` as well as the network supply ratio. Hence, one may solve for θ^* through the equation:

$$\theta^* + \eta[(m(\theta^*)] * m(\theta^*) = \theta_0.$$

To summarize, we know that the experimental results reveal $m(\theta^*)$, instead of the true $m(\theta_0)$.

4.5 Putting It Together: PCE

To remove the systematic bias of R=`production` and still provide scalable protection, ShadowStream introduces a simple, novel streaming system structure where R consists of two streaming machines: a repair CDN (referred to as `rCDN`) with bounded, dedicated repair resources and `production`. We refer to this scheme as the PCE scheme (*i.e.*, R=P+C).

Specifically, in this scheme, if `experiment` cannot download a piece by its deadline, responsibility for downloading the piece is shifted to `rCDN`. However, the dedicated CDN repair resources are only up to δ percentage of total demand, where δ is a control parameter depending on available CDN resources and the desire to obtain accurate results for a test. If a piece cannot be repaired by `rCDN`, due to either repair CDN capacity exhaustion or network bottlenecks, responsibility for downloading the piece is shifted to `production` as a final protection.

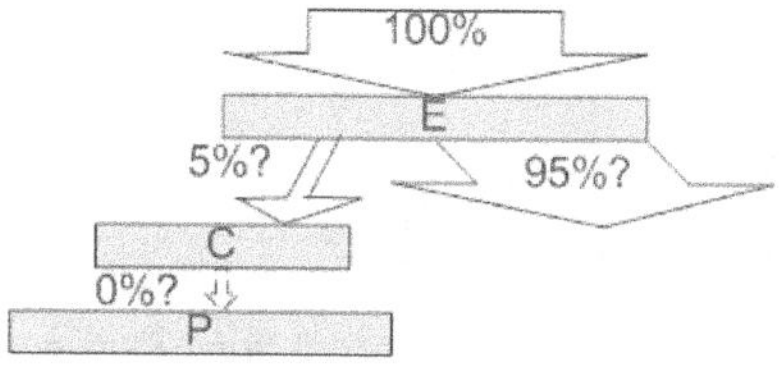

Figure 7: ShadowStream tasks processing percentages to PCE.

One way to understand the essence of the R=P+C design is to think of it as inserting a small `rCDN` "filter" between `experiment` and `production` to handle pieces missed by `experiment`. Figure 7 shows the flow. If the amount of missing pieces is lower than the capacity of `rCDN` (*e.g.*, lower than 5%), `rCDN` absorbs all leaked load of `experiment`, so that `production` is idle to avoid interference with `experiment`. On the other hand, if `experiment` has a major failure, repair load is handled by both `rCDN` and `production`, providing a guarantee to viewers' QoE.

Another way to understand the R=P+C design is that it "lowers" the piece missing ratio curve of `experiment` visible by `production` down by δ. The solid line in Figure 8 shows the actual missing ratio curve and the dashed line the lowered one. If $m(\theta_0) < \delta$, then all missing data can be retrieved by `rCDN`. Hence, `production` is idle and there is no competition on peers' upload bandwidth, resulting in an accurate experimental result. On the other hand, if $m(\theta_0) > \delta$, then `rCDN` cannot repair some missing pieces. The remaining missing pieces "overflow" to `production`.

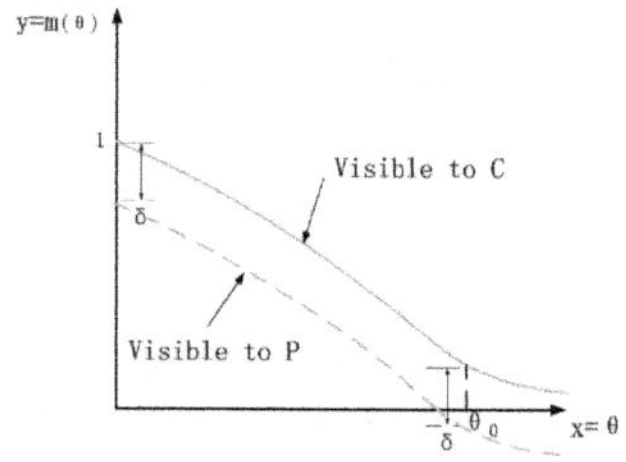

Figure 8: ShadowStream PCE design intuition.

We make a precise statement on the R=P+C design:

PROPOSITION 1. *Assume that the piece missing ratio of* `experiment` *is a non-increasing function of supply ratio. With an unknown θ value, PCE provides a bi-modal evaluation result: if $m(\theta) \leq \delta$, precise experiment accuracy can be achieved; otherwise, the system converges to a stationary point where the upper bound of $m(\theta)$ can be obtained, and we get an exact indication that $m(\theta) > \delta$.*

4.6 Extension: Dynamic Streaming

There are multiple ways to extend PCE. In particular, we present an extension of ShadowStream to dynamic streaming (*i.e.*, dynamic bitrate switching), which presents a larger design space for ShadowStream. We consider two perspectives. First, `experiment` may use dynamic streaming, and dynamic streaming experiments may be more sensitive to interference caused by protection (*i.e.*, `experiment` switching streaming rates). Second, ShadowStream protection can take advantage of dynamic streaming, which offers a new dimension for designing the protection scheme. Consider the case that both `experiment` and protection use dynamic streaming. A straightforward protection strategy is *Follow*: when ShadowStream detects that the pieces at time t is incomplete, it repairs the video for t, according to the encoding rate decided by `experiment`. However, the general idea of ShadowStream allows us to repair a different encoding rate, as long as it protects the viewer's QoE. One strategy is *Base*, which is different from *Follow* in that it always repairs the base (lowest) encoding rate. Furthermore, ShadowStream can use *Adaptive*, in which it considers both available protection bandwidth and the amount of remaining tasks at each rate left by `experiment`, and then picks the best repair.

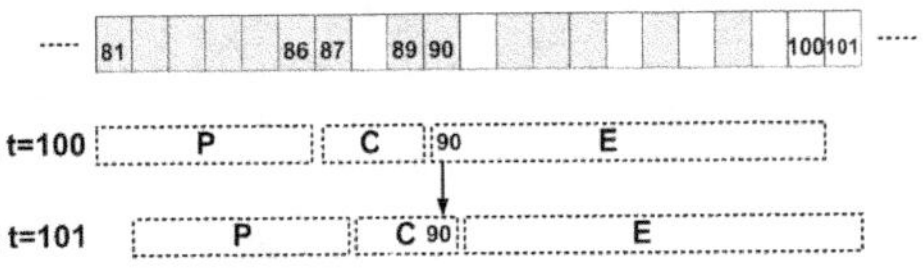

Figure 9: Streaming machine sliding download window: (a) at t=100; (b) at t=101.

5. CLIENT IMPLEMENTATION

There are challenges in implementing the novel PCE scheme designed in the preceding section. In particular, the PCE scheme consists of multiple streaming machines that may interact with each other. It is highly desirable that each streaming machine be a modular process, isolated from other streaming machines, without knowing whether it is in a test as `experiment` or `production`, or it is just running alone as `production`. The benefit of such isolation is that the production streaming machine performs the same when it has been evaluated.

A key observation of implementing ShadowStream is that we can implement the PCE scheme naturally, seamlessly, using a simple sliding window to partition the downloading tasks at a client. Based on the observation, ShadowStream imposes a minimal sliding window abstraction on streaming machines and introduces a simple *streaming hypervisor* to implement PCE.

5.1 Basic Idea

The objective of PCE is to assign a task first to `experiment`; if `experiment` fails, reassign the task to `rCDN`; if `rCDN` fails, reassign the task to `production`. We make an observation on this PCE behavior.

Specifically, at any instance of time, looking at the pieces which each streaming machine is responsible for, we observe that the streaming machines essentially partition the total downloading range spanning from the viewer playpoint to sourcepoint, with `experiment` responsible for a range of pieces that are adjacent to sourcepoint, and `production` responsible for a range that is adjacent to the viewer visible playpoint. Figure 9 (a) illustrates the partition at time $t = 100$. We call each partition the *task window* of the corresponding streaming machine.

Consider a simple scheme that we fix the size of each partition and slide each partition synchronously to the right at the playback speed. After each window movement, the oldest piece in the task window of `experiment` moves as the newest of the task window of `rCDN`. Figure 9 (b) shows that piece 90 moves from `experiment` to `rCDN`. If a piece that moves from `experiment` to `rCDN` has not been downloaded by `experiment`, it becomes a task to be finished by `rCDN`. In other words, using the simple sliding task window scheme, we achieve automatic shifting of results and failures from one streaming machine to the next in line!

The preceding observation leads to a simple, natural, modular abstraction of the key state of a streaming machine implementing the PCE scheme. Specifically, at an instance of time t, each streaming machine x is responsible for a subrange $[x_{left}(t), x_{right}(t)]$ of the total downloading range.

We introduce a streaming hypervisor to manage the PCE machines. Specifically, the streaming hypervisor provides four key sets of functions: (1) task window management (TWM) sets up sliding window; (2) data distribution control (DDC) copies data among streaming machines; (3) network resource control (NRC) conducts bandwidth scheduling among flows of streaming machines; and (4) experiment transition (ET) starts or stops experiments.

Table 4 lists key API functions between a streaming machine and the streaming hypervisor. We divide the APIs into four categories corresponding to the four sets of functions.

5.2 Task Window Management

Task window management informs a streaming machine about the pieces that it should download. We set the length of the task window of a streaming machine x as the max lag when the streaming machine is running alone:

$$x_{lag} = x.getMaxLag().$$

Hence, we focus on the right border of a task window. First consider `experiment`. We observe that $x_{right}(t)$ is actually equal to the sourcepoint t known to x. Hence, instead of notifying `experiment` $x_{right}(t)$, we notify it of the sourcepoint t. In other words, in our implementation, we provide a `getSourceTime()` function which each streaming machine can call to get the sourcepoint. For `experiment`, it is the real sourcepoint, but for `production`, it is actually the real sourcepoint minus $e_{lag} + c_{lag}$, where e_{lag} is the max lag of `experiment`, and c_{lag} is the max lag of `rCDN`. In other words, it is a "virtual sourcepoint". Notifying the right border of a task window through sourcepoint is highly desirable, because we have observed that it is a common practice for a streaming machine to define program behaviors in reference to the sourcepoint. For example, a streaming machine may move its playpoint forward to be closer to the sourcepoint after falling behind due to freezing.

Using the functions to manage the task windows is straightforward. When the streaming hypervisor is notified of a test, it gets the lag values p_{lag}, c_{lag} and e_{lag} for `production`, `rCDN` and `experiment` respectively. The start time and returned values of `getSourceTime()` are listed in Table 5.

machine	start time	`getSourceTime()`
`production`	Viewer arrival	sourcepoint - $(c_{lag} + e_{lag})$
`rCDN`	Enter testing	sourcepoint - e_{lag}
`experiment`	Enter testing	sourcepoint

Table 5: Results of calling `getSourceTime()`.

5.3 Data Distribution Control

Figure 10: Directional data propagation.

Data distribution control copies pieces among streaming machines. Figure 10 shows the direction of data flows among the three streaming machines. For example, we see that a piece downloaded by C or P should not be made available to E because E may redistribute it by uploading to others, leading to misleading experimental results. In other words, the propagation of a downloaded piece is directional.

A straightforward implementation of directional data propagation is that each streaming machine has its own data store. When an earlier machine finishes downloading a piece to its data store, the piece is also copied to the data stores of those streaming machines after it. In ShadowStream, we implement a shared data store, where a piece is stored only once, for efficiency. The data store of an individual streaming machine stores pointers to the actual data pieces in the shared data store. Figure 11 shows the states of the data stores of `experiment` and `rCDN`. For example, we observe that `rCDN` has a pointer to piece 88 downloaded by itself, but `experiment` does not due to directional data distribution control.

Two functions implementing directional data control and data sharing are `writePiece()` and `checkPiece()`. A streaming machine calls `deliverPiece()` to notify the hypervisor that a piece is ready to be delivered to player; the hypervisor decides when to actually deliver the piece.

5.4 Network Resource Control

Network resource control handles the network bandwidth allocation among streaming machines, to achieve the design objectives

Call	Category	Direction	Description
`getMaxLag()`	TWM	$H \mapsto M$	Streaming machine notifies ShadowStream its maximum lag, which defines the size of its subrange.
`getSourceTime()`	''	$M \mapsto H$	Streaming machine obtains its virtual sourcepoint, which defines the right border of its subrange.
`writePiece()`	DDC	$M \mapsto H$	Streaming machine notifies ShadowStream the successful download of a piece. It updates the data stores of streaming machines that should see the piece.
`checkPiece()`	''	$M \mapsto H$	Streaming machine obtains the success/failure status of a piece.
`deliverPiece()`	''	$M \mapsto H$	Streaming machine notifies that a piece is ready to be delivered to player.
`sendMessage()`	NRC	$M \mapsto H$	Streaming machine sends a protocol/data message. ShadowStream internally manages the priority queues and may drop the message.
`recvMessage()`	''	$M \mapsto H$	Streaming machine obtains the next received message.
`start()`	ET	$H \mapsto M$	Streaming hypervisor notifies the streaming machines to activate operations.
`stop()`	''	$H \mapsto M$	Streaming hypervisor notifies the streaming machines to deactivate operations.
`rejoin()`	''	$H \mapsto M$	Streaming hypervisor notifies the streaming machines to disconnect all neighbors and rejoin.
`fetchCDN()`	''	$M \mapsto H$	Streaming hypervisor and `rCDN` directly fetch pieces from dedicated CDN resources.

Table 4: APIs (H hypervisor, M machine): $H \mapsto M$ APIs are implemented by M and called by H; $M \mapsto H$ vice visa.

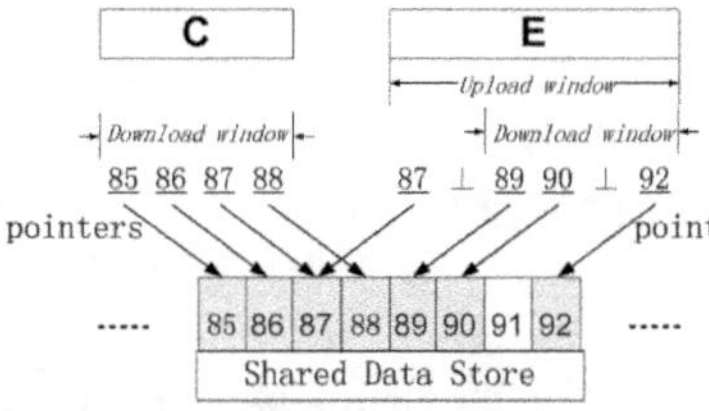

Figure 11: Using data store for data distribution control.

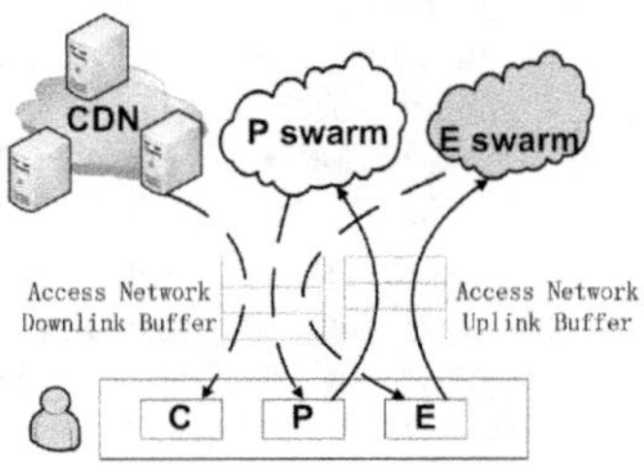

Figure 12: Network resource control

of ShadowStream. At any instance of time during a test when `experiment`, `rCDN`, and `production` are running concurrently, there are multiple sets of data flows originating from or terminating at a streaming client. Figure 12 shows 3 sets of downlink flows and 2 sets of uplink flows. In particular, consider the two sets of uplink flows due to `experiment` and `production` respectively. These two sets may compete on the shared network uplink when `experiment` performs poorly, as we have analyzed in Section 3.

Network resource control in ShadowStream assigns flows from `production` with a higher priority over those from `experiment`. However, simple priority is not sufficient as a misbehaving `experiment` may send at a high rate to create external congestions (*e.g.*, due to hidden buffers). Hence, ShadowStream resource control imposes pacing, based on an algorithm similar to LEDBAT [34] to perform dynamic bandwidth estimation.

No streaming machine can exceed estimated network bandwidth to create hidden network congestions. This feature has been implemented for both the UDP mode and the TCP mode. The provided APIs are `sendMessage()` and `recvMessage()`.

6. EXPERIMENT ORCHESTRATION

We have covered how a streaming machine can be started inside a streaming client. A capability still missing in an experimental platform is to compute the start and stop times of `experiment` in each client to create desired testing client behavior scenarios. Furthermore, when a client joins a test and changes from normal state to testing state, a smooth transition is needed to guarantee that the accuracy of the testing is not impaired.

6.1 Specification and Triggering

ShadowStream allows intuitive specification of testing client behavior patterns. Specifically, a testing behavior pattern defines one or multiple *classes* of clients, where the class of a client is defined by its properties (*e.g.*, cable or DSL, estimated upload capacity class, production software version or network location [44]).

For each class j, a behavior pattern defines:

- A class-wide arrival rate function $\lambda_j(t)$ during the interval $[0, t_{\text{exp}}]$, where t_{exp} is the duration of the testing. For example, to generate a flash-crowd arrival in class j, $\lambda_j(t)$ will have high values initially, and then decreases sharply.
- A client life duration function L is specified by the probability that a client stays less than a duration in the testing. For live streaming, it is important that client lifetime is dependent on the arrival time [38] and video quality [28, 38]. For example, if the streaming quality at a client is below a threshold (*e.g.*, piece missing ratio $> 3\%$), the client departs. Specifically, for a client arriving at time x after experiment starts, the client's lifetime is determined by both L_x and the streaming quality.

Given a specified client behavior pattern, experiment orchestration provides two functions:

- Triggering: The experiment orchestrator monitors the testing channel to wait for the network to evolve to a state where the testing client behavior pattern can be triggered. The triggered test starting time is referred to as t_{start}. Thus, test will run from t_{start} to $t_{start} + t_{exp}$.
- Arrival/Departure Control: The orchestrator selects clients to join and depart from the testing channel to create the target testing behavior pattern.

Triggering Condition: To simplify presentation, we assume a single global client class with arrival rate function $\lambda(t)$. If an experiment is triggered at time 0 and should continue for a duration of t_{exp}, then for t between 0 and t_{exp}, an upper bound of the expected number of clients active in the experiment, denoted as $\exp(t)$, is as follows:

$$\exp(\mathrm{t}) = \Lambda(t) - \int_0^t \lambda(x)\Pr\{L_x < t - x\}\,dx, \qquad (1)$$

where $\Lambda(t) = \int_0^t \lambda(x)\,dx$. The expression is an upper bound because clients may depart due to insufficient streaming quality, which is hard to direct model. Such departures only decrease the number of active clients in the experiment.

At any time t_0, the orchestrator predicts the number of active clients in the testing channel in the interval $[t_0, t_0 + t_{exp}]$. Let $\text{predict}(t)$ be the predicted value at time $t \in [t_0, t_0 + t_{exp}]$. To compute $\text{predict}(t)$, the orchestrator uses a simple extension of the autoregressive integrated moving average (ARIMA) method [43] that uses both recent testing channel states and the past history of the same program. To obtain current testing channel states, the orchestrator gathers channel state (arrivals and departures) from clients' low cost UDP reports. Specifically, at current time t_0, the orchestrator checks the condition:

$$\exp(t) \leq \text{predict}(t_0 + \Delta + t')\,\forall t' \in [0, t_{\exp}], \qquad (2)$$

where Δ is triggering delay. If the condition is satisfied, $t_0 + \Delta$ can be triggered as t_{start}.

6.2 Independent Arrivals Achieving Global Arrival Pattern

After the triggering condition, the orchestrator needs a mechanism to notify a large number of clients in real-time about their time to join and leave a testing scenario to create a specified testing behavior pattern.

A direct approach is that one or more orchestrators issue commands at appropriate times to each client to let it join or leave the test. This approach is taken by some experimental platforms (*e.g.*, [26, 37, 41]).

However, this approach is not scalable. Furthermore, since the connections between the orchestrator and the clients can be asymmetric (*e.g.*, many real viewers in the Internet are behind NAT devices [19]), it is easier for clients to send to the orchestrator than from the orchestrator to the clients.

Hence, ShadowStream introduces distributed orchestration, which decouples the scenario parameters from their execution, and thus relaxes the requirement that control messages are delivered to each client within a small delay, in contrast with direct control.

Specifically, in the distributed control mode, after computing t_{start}, the orchestrator embeds network-wide common parameters, including t_{start}, t_{exp}, and $\lambda(t)$, into keep-alive response messages and distributes to all clients. The orchestrator does not compute or control the arrival times of individual clients. It may first appear that achieving a defined testing client behavior such as flash-crowd requires global computation of client arrival times. However, we will see that there exists an effective, distributed algorithm where each client can *locally* decide and control its own arrival times. The key is the following theorem from Cox and Lewis [13]:

THEOREM 1. *Let $T_1, T_2, \ldots$ be random variables representing the event times of a non-homogeneous Poisson process with expectation function $\Lambda(t) = \int_0^t \lambda(x)\,dx$, and let N_t represent the total number of events occurring before time t in the process. Then, conditioned on the number of events $N_{t_{\exp}} = n$, the event times $T_1, T_2, \ldots, T_n$ are distributed as order statistics from a sample with distribution function $F(t) = \frac{\Lambda(t)}{\Lambda(t_{\exp})}$ for $t \in [0, t_{\exp}]$.*

An implication of Theorem 1 is that we can generate arrival times by drawing random numbers *independently* according to the same distribution function $F(t)$. Sorting these independent arrival times, we obtain the arrival times of clients following the desired arrival rate function $\lambda(t)$.

One remaining issue in a large-scale experiment is that the preceding theorem requires selecting a fixed n. Enforcing that exactly n clients is chosen can force the orchestrator to keep hard state about specific clients. To reduce orchestrator overhead, the orchestrator draws the total number of clients from a distribution $\hat{N}_{t_{\exp}}$ with the same mean ($\mathrm{E}[\hat{N}_{t_{\exp}}] = \mathrm{E}[N_{t_{\exp}}]$) but without increasing the variance ($\mathrm{Var}[\hat{N}_{t_{\exp}}] \leq \mathrm{Var}[N_{t_{\exp}}]$). This mode trades slight variation in arrival rate for higher scalability. It permits the variance on number of total clients to be reduced in the interest of tighter control over the number of clients.

Specifically, the orchestrator computes p as a ratio of the expected value of clients in the scenario ($\Lambda(t_{\exp}) = \mathrm{E}[N_{t_{\exp}}]$) to the total number of available clients M. Each client i independently participates in the scenario with probability p, and computes an arrival time $a_{e,i}$ at which it will become active in the test. This leads to a simple distributed algorithm shown in Figure 13.

Orchestrator:
01. Let M be the total number of available clients
02. Let $p = \frac{\Lambda(t_{\exp})}{M}$
03. Send t_{start}, $t_{\exp}$, $\lambda(t)$, and p to each client

Client i, upon receiving t_{start}, $t_{\exp}$, $\lambda(t)$, and p:
04. **if** $random() > p$ **then return**
05. Draw waiting time w_i according to $F(t) = \frac{\Lambda(t)}{\Lambda(t_{\exp})}$
06. Compute arrival time: $a_{e,i} = t_{\text{start}} + w_i$

Figure 13: Algorithm with decentralized control for each client i to choose arrival time $a_{e,i}$.

6.3 Experiment Transition

Suppose that the current time is t_0 and client i has computed according to the preceding section that it will (virtually) join the test at $a_{e,i}$. The preparation during $[t_0, a_{e,i}]$ to start `experiment` at client i consists of the following two simple steps.

Connectivity Transition: At $a_{e,i}$ when `experiment` starts at client i, one possibility that may cause unnecessary interference to `experiment` is that `production` uploads to a neighbor's `production` while the neighbor is not in test. A simple, clean solution is that during $[t_0, a_{e,i}]$, client i's `production` rejoins the testing channel, and the orchestrator returns only peers who are also in testing as neighbors to client i's `production`.

Playbuffer State Transition: Look at the target playbuffer state at time $a_{e,i}$ when `experiment` starts at client i. Figure 14 (a) gives an illustration. We have two observations: (1) pieces at the range for `experiment` should be empty; and (2) pieces for `production` and `rCDN` should be full. If these buffers are not full, `production` will download those pieces, causing interference to `experiment`. We refer to filling up `production` and `rCDN` windows directly from CDN before time $a_{e,i}$ as "legacy" removals.

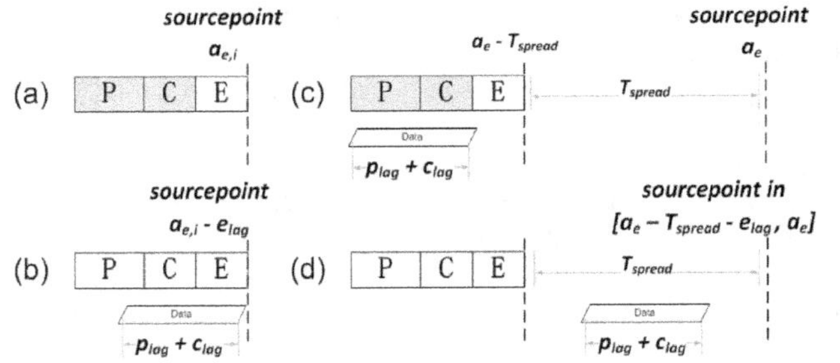

Figure 14: Playbuffer state transition.

Specifically, at time $a_{e,i}$, the piece range for `production` and `rCDN` together is $[a_{e,i} - p_{lag} - c_{lag} - e_{lag}, a_{e,i} - e_{lag}]$; note that these "legacy" pieces become all available to download at time $a_{e,i} - e_{lag}$ as shown in Figure 14 (b). For simplicity, the goal of playbuffer state transition is to pre-fetch pieces in this range during time $[a_{e,i} - e_{lag}, a_{e,i}]$. If the testing has a gradual arrival pattern, the fetching of the range is spread out evenly and there is no issue.

On the other hand, for a flash-crowd arrival, the demand of "legacy" removal would in-turn cause a flash-crowd to dedicated CDN resources. Consider the extreme case of a simultaneous flash-crowd arrival of n clients at time a_e. Then the total load due to the prefetching is $n(p_{lag} + c_{lag})$. This load should be finished from time $a_e - e_{lag}$ to a_e. The bandwidth demand rate then is $\frac{n(p_{lag}+c_{lag})}{a_e-(a_e-e_{lag})} =$

$\frac{n(p_{lag}+c_{lag})}{e_{lag}}$. If this rate is higher than the available CDN capacity, the transition cannot happen smoothly.

To avoid the issue, we can shift the PCE windows further behind in the time domain. As shown in Figure 14 (c), this extra window space is defined as T_{spread}. Note that all "legacy" pieces are available for download at time $a_e - T_{spread} - e_{lag}$. The CDN load can be finished from time $a_e - T_{spread} - e_{lag}$ to a_e, as shown in in Figure 14 (d); hence the requirement is that

$$\frac{n(p_{lag} + c_{lag})}{T_{spread} + e_{lag}} \leq R_{pre},$$

where R_{pre} is available dedicated CDN capacity. In ShadowStream, the orchestrator computes T_{spread} and announces it along with other arrival parameters. Provided APIs `start()`, `stop()`, `rejoin()` and `fetchCDN()` are shown in Table 4. In this mode, the returned value of `getSourceTime()` should be revised correspondingly.

6.4 Replace Early Departed Clients

A main challenge is that real viewers have their real departure behaviors, which cannot be controlled by ShadowStream. Clients participating in a testing scenario may switch to different channels or be terminated due to viewer-initiated operations. For example, a client could have chosen to join the test at a particular calculated time, but the real viewer has quit the testing channel before that time; such clients are called *early-departed clients*.

Capturing client state: When a client departs early, it piggybacks a small state snapshot in the disconnection message which includes the scheduled arrival time of the client. When the replacement client takes over, it "reconstructs" the state of the replaced client.

Substitution: In ShadowStream, an early-departed client is replaced by another client in the current network, which is not selected for joining testing yet. The orchestrator maintains a FIFO queue D_j for early-departed clients in class j. Upon detecting an early-departed client, the orchestrator appends the client's state to D_j. The orchestrator monitors the queue size of D_j, computes a probability that each client in class j locally chooses to become a replacement candidate, and piggybacks the probabilities in control messages. If a client in class j becomes a candidate, it connects to the orchestrator. The orchestrator checks if D_j is non-empty. If so, it dequeues the first state snapshot from the queue and includes it in the reply.

6.5 Independent Departure Control

The ShadowStream departure control has limitations due to the use of real viewers. ShadowStream cannot support arbitrary client departure patterns since the departure pattern of real viewers cannot be controlled. Instead, ShadowStream requires that the "departure" behavior pattern in testing can only be "equal to" or "more frequent" than the real viewer departure pattern. We argue that this limitation is acceptable since engineers mostly care about stressful scenarios.

The design of departure control is to fill the gap between the desired departure pattern and the real viewer departure pattern. The mechanism is similar to distributed arrival control mentioned above; we omit its details due to space limitation.

7. EVALUATIONS

In this section, we evaluate multiple aspects of ShadowStream. First, we evaluate the software architecture in terms of code size and implementation experience. Then, we evaluate the unique evaluation opportunities only available in production networks. The protection and the evaluation accuracy ability of ShadowStream is evaluated in the third part. The fourth part illustrates the performance of distributed control. In the fifth part we illustrate how ShadowStream works with dynamic streaming. Finally, we evaluate an additional feature, deterministic replay.

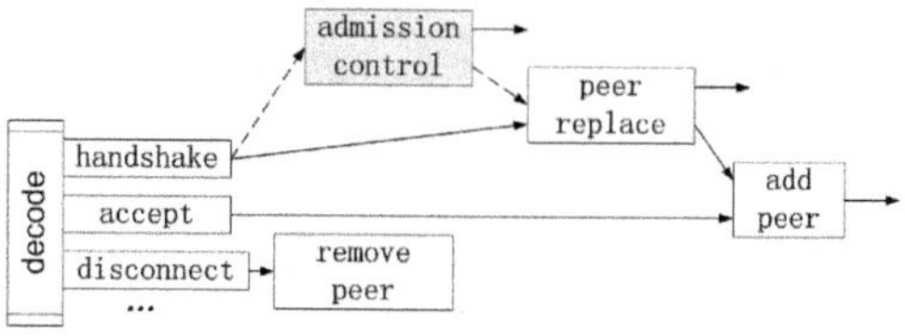

Figure 15: Adding an admission control component.

7.1 Software Framework

We implement a complete live streaming network with clients and an orchestrator using an architecture called Compositional Runtime. This block-based architecture not only supports evaluation with streaming machines and easier distribution of code for a testing scenario, but also matches well with the large distributed peer-to-peer live streaming networks, which typically consist of a set of key algorithmic components, such as connection management, upload scheduling, admission control, and enterprise coordination.

Overview: The key objective of the Compositional Runtime is that the software structure should allow modular design of algorithmic modules as well as easy composition of a system consisting of all algorithmic modules for a test. Our software architecture is inspired by prior architectures, such as Click [21], GNU Radio [17] and SEDA [40]. Algorithmic components are implemented as independent blocks that define a set of input and output ports over which messages are received and transmitted. A runtime scheduler is responsible for delivering packets between blocks.

System base: Since we have implemented the full system, we present statistics on the size of the code to illustrate that the framework is simple yet powerful. The framework and live streaming machine are implemented in C++. The full system is divided into multiple components:

- Compositional Runtime (including scheduler, dynamic loading of blocks, etc): 3400 lines of code;
- Pre-packaged blocks (HTTP integration, UDP sockets and debugging): 500 lines of code;
- Live streaming machine: 4200 lines of code.

Flexibility: Members of our research group have implemented application-layer rate limiting, modified download schedulers, and even push-based live streaming by simply writing new blocks and updating a configuration file.

Figure 15 illustrates a portion of a live streaming machine that is responsible for managing neighbor connections. A block *decode* is responsible for de-multiplexing received packets and sending them to connected blocks responsible for handling each type of message.

Next, the designer wishes to add an *admission control* algorithm to avoid reduced performance for existing clients during flash-crowd. Admission control may be implemented as an independent block, which reads handshake messages for newly-connected clients. The block emits either the handshake message if the new connection should be accepted or a disconnect message to the client if the connection should be rejected. The designer then composes the block as shown by the dotted line.

7.2 Experimental Opportunities

We demonstrate that live testing provides experiment scales that are not possible in any existing testbed. We use real traces from two live streaming testing channels, SH Sports channel and HN Satellite channel, from one of the largest live streaming networks for 1 week beginning on September 6, 2010.

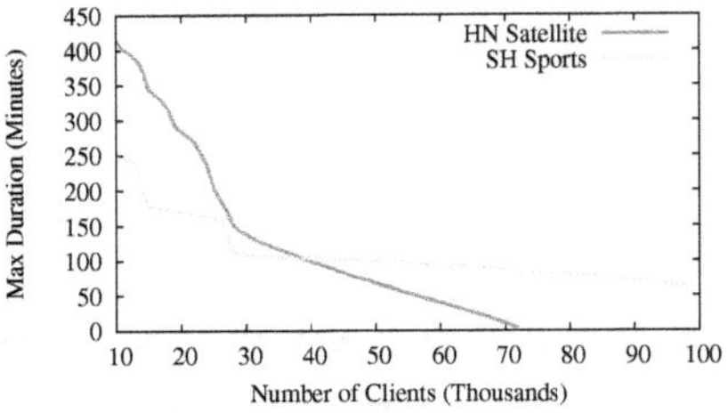

Figure 16: Experiment opportunities.

When evaluating experiment opportunities, we use a demanding testing behavior pattern: the arrival curve is a flash-crowd that all clients arrive within $\Delta = 1sec$, and no clients depart from the test. Figure 16 demonstrates the test size and test duration that can be triggered. In this figure, the x-axis is the triggered size and the y-axis is the possible duration at the size.

We make the following observations. First, the two real testing channels can allow us to test 60,000 clients for at least 40 minutes, a reasonable duration. In addition, the SH Sports channel can accommodate a test of 100,000 clients for at least 60 minutes. This scale is not possible with any testbed that we are aware of. Second, we observe that different channels can provide different tradeoffs. While the SH sports channel can provide a large channel size, the HN channel allows us to run experiments at longer durations.

7.3 Protection and Accuracy

We evaluate protection and accuracy by running fully implemented ShadowStream clients on Emulab. This is a major lacking of our evaluations, as results from production deployment may reveal unexpected issues, but production deployment of ShadowStream is currently in a work-in-progress state. On the other hand, Emulab allows us to conduct an experiment multiple times, with and without ShadowStream, all in the exact same setting. This is not possible using a production deployment. Hence, Emulab results can reveal better on ShadowStream accuracy. We plan to report production results as they become available.

Specifically, we customize a Fedora Linux image with Modelnet support for Emulab, in order to increase the number of clients. Each result is the mean value of 10 repeated evaluations.

In the test, there are 300 clients joining the testing channel with an inter-arrival time of 5 seconds. After all clients have joined, they play for 100 seconds. The channel rate is 400 Kbps and each peer's upload capacity is 500 Kbps. We use a stable P2P streaming machine as `production`. We use a standard HTTP server as the CDN server and fix the length of the `rCDN` window as 10 seconds. The capacity limitation of the CDN is set to 2.4 Mbps using the Linux tc tool. Note that for 300 clients, the average lifetime is 850 seconds; the total demand of the experiment is roughly $850*300 = 255,000$ seconds of data. The total supply of the CDN is bounded by $2400/400 * 1700 = 10,200$ seconds of data; hence the δ value is bounded by $10,200/255,000 = 4\%$.

A dedicated `experiment` version is customized for the purpose of evaluation. We inject a bug to the `experiment`, where pieces with the lowest digit equal to 1 are excluded from the download window until they enter the urgent window (*i.e.*, 3 seconds prior to the playpoint). We choose this simple setting to highlight the fact that even simple parameter changes can have a serious performance impact and therefore continuous testing is essential.

In our evaluation, we focus on two piece missing ratios for clients: (1) virtual playpoint e_{play} of the `experiment`, and (2) viewer playpoint c_{play}/p_{play} at the `rCDN`/`production`.

R=`rCDN` design: We start with getting the performance of the buggy version by running the `experiment` alone. As shown in the second column of Table 6, the buggy version yields poor performance, confirmed by the increased piece missing ratio of 8.73% when running alone. In the next test (third column in Table 6), we use R=`rCDN` and remove the CDN capacity limitation. The measured piece missing ratio is accurate with a negligible error (which we ascribe to the non-determinism the real network); viewer-observable miss at c_{play} is 0. These results are consistent with our proposition that when CDN capacity is not the bottleneck, R=`rCDN` can achieve very high experimental accuracy and viewers' QoE is guaranteed.

However, when we add a CDN limitation in the third test (forth column in Table 6), the missing ratio at c_{play} increases to 5.42%. Only around 3.4% pieces missed by the `experiment` are repaired by `rCDN`, which is less than the 4% capacity. This is because at the initial stage of testing, the number of clients in testing is small; at this stage CDN capacity can not be fully utilized. As we have stated before, when network bottlenecks exist, the viewer perceived quality could be seriously impacted.

	Buggy	R=`rCDN`	R=`rCDN` with bottleneck
e_{play} Missed	8.73%	8.72%	8.81%
c_{play} Missed	N/A	0%	5.42%

Table 6: Experiment accuracy v.s. CDN capacity.

PCE design: Next, we use our PCE design. The results are shown in the second column of Table 7. Due to the presence of a CDN bottleneck, `production` repair is activated; thus we see that the `experiment` evaluation is inaccurate. As we have expected, the network evolves to a stationary point; we get an upper bound value of missing ratio as 9.13%. A key observation is that, even when the `experiment` has poor performance, the piece missing ratio at p_{play} is only 0.15%. This test illustrates the graceful and scalable protection provided by the PCE design.

As the final test, we increase the CDN bottleneck to 9.6 Mbps. As shown in the third column of Table 7, the resulting piece missing ratio is a good approximation of the `experiment`. And again, the viewer perceived quality is guaranteed; there is no piece missing observed at p_{play}.

	PCE	PCE with higher CDN bottleneck
e_{play} Missed	9.13%	8.85%
p_{play} Missed	0.15%	0%

Table 7: PCE is more scalable and guarantee viewers' QoE.

7.4 Experiment Control

Next we evaluate experiment control. We use real client behaviors from one live streaming network captured during September 2010 to perform a trace-driven simulation. However, due to the lack of traces of viewers' departure behaviors, this part focuses on studying the distributed arrival performances. Since distributed departures perform similarly to arrivals, we believe this part is already sufficient for illustrating the effectiveness of distributed control.

Accuracy of Distributed Arrivals: We evaluate the accuracy of the distributed arrival algorithm in Section 6. Specifically, we use the arrival rate function shown in Figure 1 of [3]. When instantiated in ShadowStream, it has 349,080 arrivals. To test the impact of client clock synchronization, each client perturbs its generated arrival time by a uniform random value selected from the interval $[-\frac{s}{2}, \frac{s}{2}]$, where s is a parameter that the orchestrator announces.

First, we observe that distributed arrivals result in a close match to the target arrival rate function. The results are shown in Figure 17, which compares the target arrival rate with the actual arrival rate resulting from clients joining the channel with orchestrator. In this experiment, the orchestrator sets the clock skew parameter $s = 3$, so clients' clocks are desynchronized by up to 3 seconds.

Next, we further evaluate the effects of clock desynchronization on distributed arrivals. In particular, for varying values of s, we test

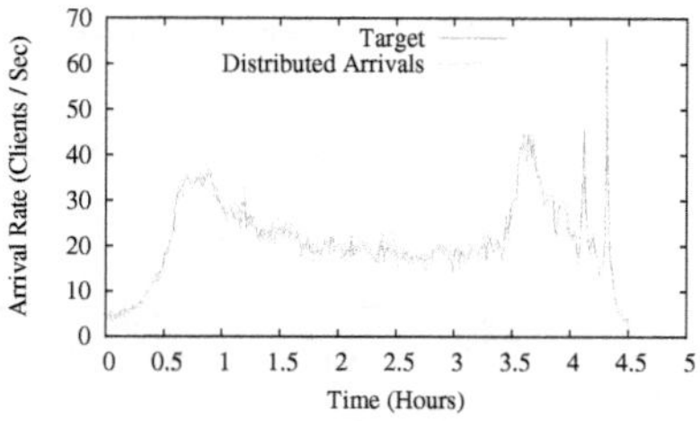

Figure 17: Experiment Control generates accurate arrival behaviors. Clients have clocks differing by up to 3 seconds.

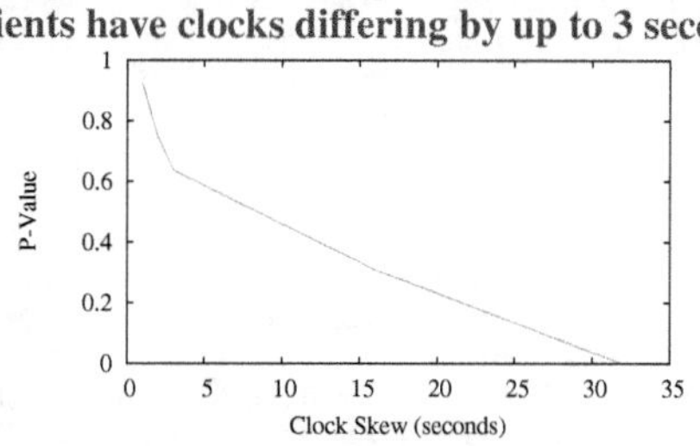

Figure 18: Experiment Control robust to desynchronization.

the hypothesis that an observed arrival behavior could have been generated by rate parameter $\lambda(t)$ using a Chi-Square test. The null hypothesis is that the arrivals in interval $[t_1, t_2)$ are distributed according to $\Lambda(t_2)-\Lambda(t_1)$. We place arrival times in an intervals of 3 seconds, and then merge adjacent intervals until both the expected number of arrivals and the number of arrivals in each are at least 5.

Figure 18 shows the results. The x-axis is the level of clock desynchronization and the y-axis is the corresponding p-value (*i.e.*, the probability that the deviation of the observed from that expected). We observe that if client clocks are synchronized within 3 seconds with the orchestrator, the distributed arrivals are accepted with high confidence (the p-value is at least 0.6); we reject the null hypothesis for clock skews $s > 20$ seconds. Synchronization to a precision of 3 seconds is readily doable: in ShadowStream, peers would use sourcepoint value from source as time reference.

7.5 Dynamic Streaming

We evaluate ShadowStream when `experiment` implements dynamic rate switching among 4 rates: 1x (400 Kbps), 2x (800 Kbps), 3x (1.2 Mbps) and 4x (1.6 Mbps). Specifically, `experiment` switches its streaming rate by measuring buffer-filling-ratio (*i.e.*, fraction of downloaded pieces) and adapts according to the following Adobe OSMF-like rule: (1) every peer starts from rate 1x (400 Kbps); every 120 seconds, it re-calculates the buffer-filling-ratio; (2) when the buffer-filling-ratio is above 80%, switch to a higher level rate (overlay) if possible; (3) when the buffer-filling-ratio is below 50%, switch to a lower level rate (overlay) if possible. The protection decision is made every 10 seconds, considering missing video pieces to be played in the next 10 seconds.

	Follow	*Base*	*Adaptive*
Accuracy	1.26x	1.26x	1.26x
Protected QoE	1.59x	1.42x	1.58x
Protection Overhead	1.49	3.69	1.39

Table 8: Evaluation of Repair schemes.

We evaluate ShadowStream performance from three perspectives: (1) Accuracy (to reflect interference by protection to `experiment`), measured by average streaming rate observed by `experiment`; (2) Protected QoE, measured by average streaming rate observed by `production`; and (3) QoE protection overhead, measured by per-client protection downloading rate (in Kbps).

We report the results of a setting of 300 clients with a supply ratio of 2.5 over the 1x rate. We first run `experiment` alone. The average downloading rate is 1.59x. However, if a segment, which contains 10-second video data, is incomplete, we do not count the segment when calculating the viewer QoE streaming rate. The average viewers' QoE streaming rate when running `experiment` alone is 1.26x.

Table 8 shows the results of using the three protection strategies discussed in Section 4.6. We make two observations. First, for the dynamic streaming `experiment`, the protection overhead is low. None of three protection strategies need more than 3.5 Kbps (<1% 1x rate) average protection downloading rate. Dynamic streaming experiments may need lower protection overhead because that they already try to do self-repair: when the buffer-filling-ratio of a client decreases, it switches to a lower rate.

Second, *Base* is not a good protection strategy. The protected QoE of *Base* is only 1.42x, while the other two strategies achieve 1.59x/1.58x. At the same time, the protection overhead of *Base* is higher than twice that of the other two. The reason is that *Base* often discards partially-downloaded pieces by `experiment` while the other two reuse them. In particular, *Adaptive* has the lowest protection overhead.

7.6 Deterministic Replay

We build a replay capability on top of ShadowStream so that real tests in production streaming can be played back step-by-step offline for detailed analysis. A main challenge for introducing this capability is how to minimize the logged data. Operating in a large-scale production environment, ShadowStream clients cannot log and then upload a large amount of data.

The hypervisor structure of ShadowStream allows us to implement efficient deterministic replay by controlling non-deterministic inputs. Specifically, a streaming machine is typically an event-driven system. Since external events are easier to log and replay, ShadowStream explicitly transforms a timer event input to an external input provided as a streaming hypervisor service. The event queue and message queue of each streaming machine are maintained by the streaming hypervisor. Random number generations always use the sourcepoint from the *getSourceTime()* call as seeds. Note that the logging implementation can be optimized: for protocol packets, we can save the whole payload; for data packets, only the packet headers are needed. Since our streaming machines are not computationally-intensive, we implement each streaming machine in a single thread. Using per-client input logs, we can replay any client's behavior.

Table 9 shows the per-client log size in two evaluations: one with 100 clients running for 650 seconds and the second with 300 clients running for 1800 seconds. The channel rate is 480 Kbps. We observe that the sizes of logged data to achieve replay are practical: only 223 KB for the first case, and 714 KB for the second.

	Log Size
100 clients; 650 sec	223 KB
300 clients; 1800 sec	714 KB

Table 9: Logged data size for deterministic replay.

8. RELATED WORK AND DISCUSSIONS

Q: [Related Work] How is ShadowStream related to previous work?
A: There are significant previous studies on the debugging and evaluation of distributed systems (*e.g.*, [4, 16, 18]). Compared with the preceding work, ShadowStream is the first system based on the key observation, instantiated in the context of live streaming, that both the `production` system and the `experiment` system can contribute to the objective of the same function: fetching streaming data. Another advantage of ShadowStream is that it allows specific scenarios during live testing, while previous studies focus on the current deployment state.

ShadowStream is also different from staging/provisioning/rollback, as used in industry. Such techniques cannot protect real viewers from visible disruptions, and have limited capability to orchestrate evaluation conditions.

A particularly related project is FlowVisor [35], which proposes to allocate a fixed portion of tasks and resources to `experiment` to evaluate performance. The scope of their approach, however, is limited to only linear systems. On the other hand, complex systems such as live streaming networks can be highly non-linear due to factors such as network bottlenecks.

Q: [General Approach] Is the ShadowStream approach general?
A: ShadowStream's *Experiment→Validation→Repair* scheme is general and can be extended to other applications. Specifically, it requires only that (1) failures of `experiment` can be efficiently identified; (2) one can introduce effective Repair to mask `experiment` failures. For (1), many computer science problems/systems have the asymmetric property that efficient validation of a solution is feasible, but better algorithms take time to develop (*e.g.*, NP problems). For (2), ShadowStream offers flexibility on designing Repair as long as it masks obvious user visible failures. Note that the additional lag is NOT a necessity in the general scheme.

Generalizing ShadowStream to other applications should consider the specific problem domain. Consider video-on-demand (VoD). One difference between live and VoD is data availability: all pieces are available in VoD at any time and a streaming machine hence can download pieces forward to the end of the video. Also, the playpoints of VoD viewers can be heterogenous due to arrival time differences and viewer seek/pause. To apply ShadowStream for P2P VoD, a developer can impose a downloading range to avoid overlapping tasks between `experiment` and `production`.

Q: [Limitations] Are there limitations on the experiments that can be conducted by ShadowStream?
A: Yes. First, ShadowStream introduces a small additional lag. Users accept such small lags in many production channels. If the required lag is unacceptable, ShadowStream cannot embed experiments. Second, the ability of ShadowStream protection depends on the Repair mechanism to mask user-visible failures of `experiment`. An experiment cannot be conducted if Repair and `experiment` use different codecs and media player cannot use both. Third, ShadowStream assumes that production is always running and that management traffic (monitoring and logging) is small. This is typically true (*e.g.*, 1 report packet per 30 seconds), but if not, the accuracy of ShadowStream may be reduced.

Acknowledgments: We are grateful to Ion Stoica (paper shepherd), Nicole Shibley, Lili Qiu, Ye Wang, Yin Zhang, Hongqiang Harry Liu, David Wolinsky, Peng Zhang, Haiwei Xue, Junchang Wang and anonymous reviewers for valuable suggestions.

9. REFERENCES

[1] Cisco Visual Networking Index: Forecast and Methodology, 2011-2016. www.cisco.com/en/US/solutions/collateral/ns341/ns525/ns537/ns705/ns827/white_paper_c11-481360.pdf.

[2] PowerBoost. broadbandreports.com/shownews/75298.

[3] S. Agarwal, J. P. Singh, A. Mavlankar, P. Baccichet, and B. Girod. Performance and Quality-of-Service Analysis of a Live P2P Video Multicast Session on the Internet. In *IWQoS 2008*.

[4] G. Altekar and I. Stoica. ODR: Output-Deterministic Replay for Multicore Debugging. In *ACM SOSP 2009*.

[5] IETF ALTO. datatracker.ietf.org/wg/alto/charter/.

[6] S. Banerjee, B. Bhattacharjee, and C. Kommareddy. Scalable Application Layer Multicast. In *SIGCOMM 2002*.

[7] A. Bavier, N. Feamster, M. Huang, L. Peterson, and J. Rexford. In VINI Veritas: Realistic and Controlled Network Experimentation. In *SIGCOMM 2006*.

[8] T. Bonald, L. Massoulie, F. Mathieu, D. Perino, and A. Twigg. Epidemic Live Streaming: Optimal Performance Trade-offs. In *SIGMETRICES 2008*.

[9] Broadcast Delay. en.wikipedia.org/wiki/Broadcast_delay.

[10] M. Castro, P. Druschel, A. marie Kermarrec, A. Nandi, A. Rowstron, and A. Singh. SplitStream: High-bandwidth Multicast in Cooperative Environments. In *ACM SOSP 2003*.

[11] H. Chang, S. Jamin, and W. Wang. Live Streaming Performance of the Zattoo Network. In *IMC 2009*.

[12] B. Chun, D. Culler, T. Roscoe, A. Bavier, L. Peterson, M. Wawrzoniak, and M. Bowman. PlanetLab: an Overlay Testbed for Broad-coverage Services. *ACM SIGCOMM CCR*, 33(3):3–12, 2003.

[13] D. R. Cox and P. A. W. Lewis. *The Statistical Analysis of Series of Events*. Methuen, 1966.

[14] M. Dischinger, K. P. Gummadi, A. Haeberlen, S. Saroiu, and I. Beschastnikh. SatelliteLab: Adding Heterogeneity to Planetary-Scale Network Testbeds. In *SIGCOMM 2008*.

[15] M. Dischinger, A. Haeberlen, K. P. Gummadi, and S. Saroiu. Characterizing Residential Broadband Networks. In *IMC 2007*.

[16] D. Geels, G. Altekar, P. Maniatis, T. Roscoe, and I. Stoica. Friday: Global Comprehension for Distributed Replay. In *NSDI 2007*.

[17] GNU Radio. www.gnu.org/software/gnuradio/.

[18] D. Gupta, K. V. Vishwanath, and A. Vahdat. DieCast: Testing Distributed Systems with an Accurate Scale Model. In *NSDI 2008*.

[19] Y. Huang, T. Z. J. Fu, D.-M. Chiu, J. C. S. Lui, and C. Huang. Challenges, Design and Analysis of a Large-scale P2P-VoD System. In *SIGCOMM 2008*.

[20] M. S. Kim, T. Kim, Y.-J. Shin, S. S. Lam, and E. J. Powers. A Wavelet-Based Approach to Detect Shared Congestion. In *SIGCOMM 2004*.

[21] E. Kohler, R. Morris, B. Chen, J. Jannotti, and M. F. Kaashoek. The Click Modular Router. *ACM Trans. on Computer Systems*, 18(3):263–297, Aug. 2000.

[22] D. Kostic, A. Rodriguez, J. Albrecht, and A. Vahdat. Bullet: High Bandwidth Data Dissemination Using an Overlay Mesh. In *ACM SOSP 2003*.

[23] C. Kreibich, N. Weaver, B. Nechaev, and V. Paxson. Netalyzr: Illuminating the Edge Network. In *IMC 2010*.

[24] R. Krishnan, H. V. Madhyastha, S. Jain, S. Srinivasan, A. Krishnamurthy, T. Anderson, and J. Gao. Moving Beyond End-to-End Path Information to Optimize CDN Performance. In *IMC 2009*.

[25] R. Kumar, Y. Liu, and K. Ross. Stochastic Fluid Theory for P2P Streaming Systems. In *INFOCOM 2007*.

[26] L. Leonini, E. Riviere, and P. Felber. SPLAY: Distributed Systems Evaluation Made Simple. In *NSDI 2009*.

[27] B. Li, G. Y. Keung, C. Lin, J. Liu, and X. Zhang. Inside the New Coolstreaming: Principles, Measurements and Performance Implications. In *INFOCOM 2008*.

[28] Z. Liu, C. Wu, B. Li, and S. Zhao. Why Are Peers Less Stable in Unpopular P2P Streaming Channels? In *Networking 2009*.

[29] N. Magharei and R. Rejaie. PRIME: Peer-to-Peer Receiver-drIven MEsh-based Streaming. In *INFOCOM 2007*.

[30] A. Mansy and M. Ammar. Analysis of Adaptive Streaming for Hybrid CDN/P2P Live Video Systems. In *ICNP 2011*.

[31] V. Pai, K. Kumar, K. Tamilmani, V. Sambamurthy, A. E. Mohr, and E. E. Mohr. Chainsaw: Eliminating Trees from Overlay Multicast. In *IPTPS 2005*.

[32] F. Picconi and L. Massoulie. Is there a future for mesh-based live video streaming? In *P2P 2008*.

[33] PPLive User Comments. tieba.baidu.com/f?kz=700224794.

[34] S. Shalunov, G. Hazel, J. Iyengar, and M. Kuehlewind. Low Extra Delay Background Transport (LEDBAT). IETF Draft.

[35] R. Sherwood, G. Gibb, K. kiong Yap, M. Casado, N. Mckeown, and G. Parulkar. Can the Production Network be the Testbed. In *OSDI 2010*.

[36] S. Sundaresan, W. de Donato, N. Feamster, R. Teixeira, S. Crawford, and A. Pescape. Broadband Internet Performance: A View From the Gateway. In *SIGCOMM 2011*.

[37] C. Tuttle, A. C. Snoeren, and A. Vahdat. PlanetLab Application Management Using Plush. *Operating Systems Review*, 40(1), Nov. 2006.

[38] I. Ullah, G. Bonnet, G. Doyen, and D. GaÃfti. Modeling User Behavior in P2P Live Video Streaming Systems through a Bayesian Network. In *MAMNS 2010*.

[39] V. Venkataraman, P. Francis, and J. Calandrino. Chunkyspread: Multi-tree Unstructured Peer-to-Peer Multicast. In *IPTPS 2006*.

[40] M. Welsh, D. Culler, and E. Brewer. SEDA: An Architecture for Well-Conditioned, Scalable Internet Services. In *ACM SOSP 2001*.

[41] B. White, J. Lepreau, L. Stoller, R. Ricci, S. Guruprasad, M. Newbold, M. Hibler, C. Barb, and A. Joglekar. An Integrated Experimental Environment for Distributed Systems and Networks. In *OSDI 2002*.

[42] C. Wu, B. Li, and S. Zhao. Diagnosing Network-wide P2P Live Streaming Inefficiencies. In *INFOCOM 2009*.

[43] C. Wu, B. Li, and S. Zhao. Multi-channel Live P2P Streaming: Refocusing on Servers. In *INFOCOM 2008*.

[44] H. Xie, Y. R. Yang, A. Krishnamurthy, Y. Liu, and A. Silberschatz. P4P: Provider Portal for Applications. In *SIGCOMM 2008*.

[45] H. Yin, X. Liu, T. Zhan, V. Sekar, F. Qiu, C. Lin, H. Zhang, and B. Li. Design and Deployment of a Hybrid CDN-P2P System for Live Video Streaming: Experiences with LiveSky. In *ACM MultiMedia 2009*.

[46] Y. Zhou, D.-M. Chiu, and J. C. Lui. A Simple Model for Analyzing P2P Streaming Protocols. In *ICNP 2007*.

A Case for a Coordinated Internet Video Control Plane

Xi Liu, Florin Dobrian, Henry Milner
Conviva

Junchen Jiang
CMU

Vyas Sekar
Intel Labs

Ion Stoica
Conviva
UC Berkeley

Hui Zhang
Conviva
CMU

ABSTRACT

Video traffic already represents a significant fraction of today's traffic and is projected to exceed 90% in the next five years. In parallel, user expectations for a high quality viewing experience (e.g., low startup delays, low buffering, and high bitrates) are continuously increasing. Unlike traditional workloads that either require low latency (e.g., short web transfers) or high average throughput (e.g., large file transfers), a high quality video viewing experience requires *sustained* performance over *extended* periods of time (e.g., tens of minutes). This imposes fundamentally different demands on content delivery infrastructures than those envisioned for traditional traffic patterns. Our large-scale measurements over 200 million video sessions show that today's delivery infrastructure fails to meet these requirements: more than 20% of sessions have a rebuffering ratio $\geq$ 10% and more than 14% of sessions have a video startup delay $\geq$ 10s. Using measurement-driven insights, we make a case for a video control plane that can use a global view of client and network conditions to dynamically optimize the video delivery in order to provide a high quality viewing experience despite an unreliable delivery infrastructure. Our analysis shows that such a control plane can potentially improve the rebuffering ratio by up to 2$\times$ in the average case and by more than one order of magnitude under stress.

Categories and Subject Descriptors

C.2.4 [**Computer-Communication Networks**]: Distributed systems—*Distributed applications*; C.4 [**Performance of Systems**]: [measurement techniques]

General Terms

Design, Performance, Measurement

Keywords

Video, CDNs, Control plane

1. INTRODUCTION

Over the last few years, video traffic has quickly become the dominant fraction of Internet data traffic. Studies show that Netflix alone accounts for more than 20% of the US Internet traffic [42] and projections show that by 2014, video traffic will constitute more than 90% of the total traffic on Internet [2]. These estimates are for *streaming* traffic (including both live and video-on-demand services) and does not include offline video downloads (e.g., via shared upload or P2P services).

User expectations in streaming video workloads impose fundamentally different service requirements on the network infrastructure compared to traditional data traffic. Traditional workloads focus either on latency (e.g., interactive sessions or short web transfers) or on transfer completion time (e.g., long file transfers). In contrast, latency is less critical in streaming video because application data units are large enough to amortize latency effects. Similarly, overall completion time does not really reflect the actual user experience as it does not capture rebuffering-induced interruptions; we know that users are sensitive to buffering as a 1% increase in buffering can lead to more than a 3 minutes reduction in expected viewing time [21]. Streaming video not only introduces new quality metrics at the network- and user-level, but also requires that this quality (e.g., high bitrates with low rebuffering) be *sustained* over extended periods of time as typical videos span multiple minutes.

In addition to the improvements last-mile connectivity, a key driver for the rapid explosion of streaming video traffic has been the shift from specialized streaming protocols and infrastructure (e.g., Windows Media Services, RealNetworks, RTMP) to HTTP chunk-based streaming protocols (e.g., [5,42]). This use of a commodity service dramatically decreases the cost of dissemination and the barrier of entry by allowing content providers to leverage existing HTTP CDN infrastructures to deliver content to a wide audience. Furthermore, the reliance on HTTP also implies the ability to support multiple viewing platforms as HTTP support is ubiquitous.

Unfortunately, there is a mismatch between the requirements of video streaming and the architecture of today's HTTP-based video delivery infrastructures, both at the ISP and CDN level. Using fine-grained client-side measurements from over 200 million client viewing sessions, we find that 20% of these sessions experience a rebuffering ratio of $\geq$ 10%, 14% of users have to wait more than 10 seconds for video to start up, more than 28% of sessions have an average bitrate less than 500Kbps, and 10% of users fail to see any video at all.

Analyzing the causes of these performance problems reveals:

- significant spatial diversity in CDN performance and availability across different geographical regions and ISPs,
- substantial temporal variability in the CDN performance and client-side network performance, and
- poor system response to overload scenarios when there are "hotspots" of client arrivals in particular regions or ISPs.

Our overarching goal is to meet the demands for a high-quality viewing experience despite an unreliable video delivery infrastructure. In this context, the design space for optimizing video delivery quality consists of three high-level dimensions:

1. *What* parameters can we adapt; e.g., bitrate, CDN?
2. *When* are these parameters optimized; e.g., at video startup or midstream?

SIGCOMM'12, August 13–17, 2012, Helsinki, Finland.

3. *Who* chooses these parameters; e.g., client or server?

The above observations regarding CDN variability across space and time suggest that purely server- or client-driven selection and adaptation are unlikely to be sufficient. To this end, we envision a video control plane that can use a *global* view of network and CDN performance to dynamically assign clients a suitable choice of CDN and bitrate that optimizes the video delivery. Beyond the performance benefits, such a control plane also offers content providers more flexibility in instrumenting fine-grained policies; e.g., providing higher quality service to premium customers under load, ensuring that certain types of content is only accessible with specific geographical regions, or taking into account the cost-performance tradeoffs that different CDNs have to offer [29].

Realizing such a control plane is challenging, and thus a natural first question is whether this exercise is worthwhile. To this end, we use a measurement-driven framework to extrapolate the potential for improvement in video quality. We observe that there is significant potential and that even just choosing a CDN more optimally can reduce the average rebuffering ratio by 2× in the common case and more than 10× under extreme scenarios.

We would also like to confirm that these gains are not merely hypothetical. To do so, however, we need to concretely specify aspects of the control plane such as the allocation algorithms, performance estimators, and policy functions. To this end, we present one specific realization of such a control plane to illustrate the benefits. Our choices in this respect are far from ideal and have to necessarily embed several simplifying assumptions. We believe this exercise is still valuable because our goal is to make a *case for* a control plane, rather than present a reference design and implementation. Our simulations confirm that such an approach can outperform other options in the design space for optimizing video delivery in both common and extreme load scenarios.

Contributions and Roadmap: To summarize, our key contributions are:

- Measurements to expose the shortcomings of today's video delivery infrastructure (Section 2) that motivate the need for a video control plane (Section 3).
- Using an extrapolation approach to establish the potential room for improvement (Section 4).
- Corroborating these potential gains under a concrete (but simplified) operation model (Section 5 and Section 6).

We discuss outstanding issues in Section 7 and place our work in the context of related work in Section 8 before concluding in Section 9.

2. MOTIVATION

Previous research has confirmed the impact of quality on user experience to show that users are quite sensitive to buffering and high startup latency, and prefer higher bitrate content [3,21]. Given this need for high-quality video delivery, we analyze how today's infrastructure performs.

In this section, we examine the performance of today's delivery infrastructure and highlight potential sources of inefficiencies. We begin by focusing on the end-user streaming video performance. Then, we identify three potential sources of performance problems: variability in client-side, variability within a single ISP or Autonomous System (AS), and variability in CDN performance.

2.1 Dataset

The dataset used in this paper is based on one week of client-side measurements from over 200 million viewing sessions or views (both successful and failed) from over 50 million viewers across 91 popular video content providers around the world. The chosen week includes a single large live event lasting two hours, but otherwise has normal traffic. Table 1 summarizes the data set. The content served by these providers includes both *live* (e.g., sports broadcasts) and *video-on-demand* (e.g., TV episodes and movies). Since we observe similar results from both live and video-on-demand traffic, we show results only on aggregate data from both types of traffic. The data was generated via client-side player instrumentation that collects statistics regarding the current network conditions (e.g., estimated bandwidth to the chosen CDN server) and the observed video quality (e.g., rebuffering ratio, chosen bitrate). Many of the content providers have the option to deliver content to their customers from multiple CDNs; the specifics of how they choose CDNs is proprietary. Due to business and anonymity considerations, we anonymize the providers, CDNs, ISPs, and cities in the following results. Our goal here is to highlight the overall problems in video delivery in general rather than pinpoint inefficiencies of specific ISPs or CDNs.

Time	2011.12.10 - 2011.12.17
Views	281M
Viewers	54M
View hours	30M
Content providers	91
Videos	2M
Countries / Regions	231

Table 1: Dataset Summary

Metrics: We focus on the following industry-standard video quality metrics [6,42]:

- *Average bitrate:* The average bitrate experienced by a view over its lifetime.
- *Rebuffering ratio:* This is computed the buffering time divided by buffering plus playing time, excluding paused or stopped time and buffering time before video start. (We use the term rebuffering ratio and buffering ratio interchangeably.)
- *Startup time:* This is the wait or buffering time before a video starts to play.
- *Failure rate:* This is the percentage of views that failed to start playing and experienced a fatal error during the process. In our experience, these fatal errors usually indicate CDN issues. One example is missing content that the CDN failed to populate to edge servers, and thus users cannot access the video. Another possibility is the CDN server rejecting new connections (e.g., due to overload).
- *Exits before video start:* This is the percentage of views that failed to play the video without experiencing a fatal error. There are generally two causes: (1) users are not interested in the content, and (2) users waited too long for the video to load and lose patience.

We choose these metrics because earlier work showed that they have a significant impact on user engagement [21]. Our goal is not to design an aggregate quality metric that can combine these factors (e.g., [8]). Rather, we want to show the inefficiencies of today's infrastructure and provide directions to improve the video quality. Thus, we consider each metric in isolation in this study.

2.2 Video quality today

We begin by analyzing the video quality that today's delivery infrastructure provides before looking at more in-depth analysis to understand reasons for this inefficiency.

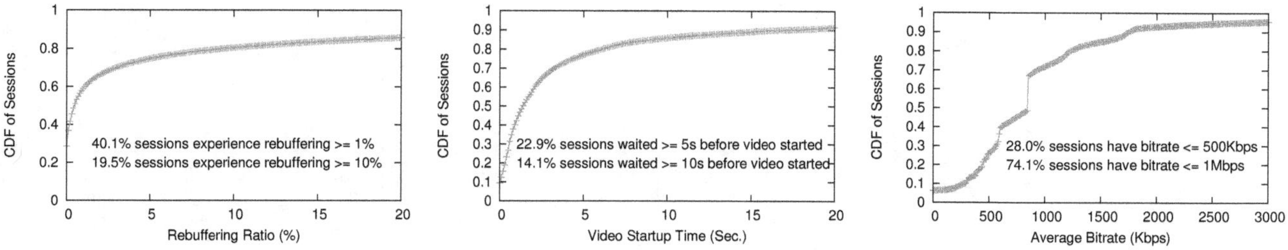

(a) Rebuffering Ratio (b) Video Startup Time (c) Average Bitrate

Figure 1: Distribution of three standard video quality metrics computed over > 200 million user views across 91 providers. The result shows that a non-trivial fraction of views suffer quality issues.

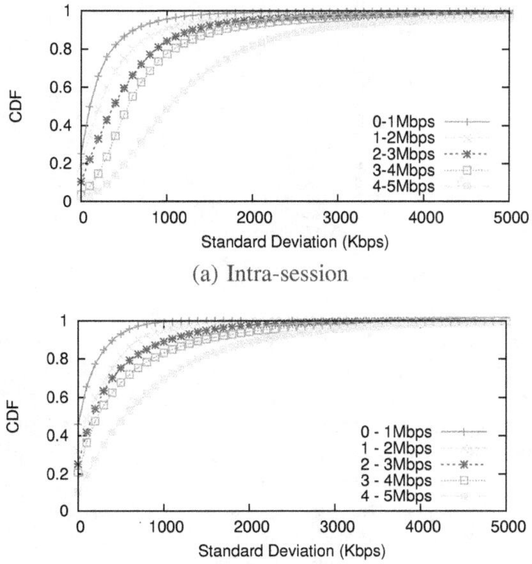

(a) Intra-session

(b) Inter-session

Figure 2: There is significant variability in client-side bandwidth both within and across sessions confirming the need for bitrate adaptation.

Figure 1 shows the distribution of rebuffering ratio, video start up time, and average bitrate from views that have started the video playing. Note that these are the observed performances in the wild with the default video players that the providers use. For rebuffering ratio and average bitrate, we remove sessions less than one minute, because they usually come from users that are not interested in the video.

The result shows that

- 40% of the views experience at least 1% rebuffering ratio, and 20% experience at least 10% rebuffering ratio.
- 23% of the views wait more than 5 seconds before video starts, and 14% wait more than 10 seconds.[1]
- 28% of the views have average bitrate less than 500Kbps, and 74.1% have average bitrate less than 1Mbps.

We also observe that 2.84% views failed to start due to fatal errors, and 14.43% without errors (not shown). Furthermore, we see that more than 9% of the views have actually waited at least 20 seconds before they lose patience in waiting for the video to start.

Implications: To put these results in perspective, previous work shows that a 1% increase in rebuffering ratio can reduce the total play time by more than 3 minutes, viewers who have low join times are more likely to return to the content providers, and viewers who receive higher bitrate videos are likely to watch the video longer [21]. Our analysis indicates that today's end user experience is far from perfect, and highlights the need for performance optimization.

2.3 Sources of quality issues

Next we identify and analyze three potential issues that could result in poor video quality.

Client-side variability: Figure 2 shows the distribution of the standard deviation of the client-side intra- and inter-session estimated bandwidth, which shows significant variability in client-side conditions. In this result, we rely on the client player's bandwidth estimation logic which effectively measures the observed bandwidth for the data transferred from the selected CDN server, and the data is collected every 10 seconds. For intra-session bandwidth, we compute the standard deviation of all the bandwidth samples across the entire lifetime of a view. Then we plot the CDF for all views, excluding views that have only one sample. For inter-session bandwidth, for each viewer, we compute the average bandwidth of each session and then compute the standard deviation across the different sessions initiated by that viewer. In both cases, we bin the different views (for intra-session) or viewers (for inter-session) based on their average bandwidth and show the distribution for the five bins from 0-1Mbps to 4-5Mbps. For views with bandwidth less than 1Mbps, more than 20% have a intra-session deviation of 400Kbps. The deviation is 2Mbps for views with bandwidth between 4-5Mbps. Furthermore, there is a fair amount of variability in the inter-session case as well. For example, more than 20% of the viewers with bandwidth less than 1Mbps have a deviation of 250Kbps. We also confirmed that such variability is a general phenomenon that occurs across all ISPs (not shown).

Implications: Given today's bitrate levels (e.g., 400, 800, 1000, 3000 Kbps), this naturally implies the need for intelligent bitrate selection and switching to ensure a smooth viewing experience. Specifically, we see that it is necessary to choose a suitable bitrate at the *start* of each session to account for inter-session variability and also dynamically adapt the bitrate *midstream* to account for intra-session variability.

CDN variability across space and time: The performance of CDN infrastructure for delivering video can vary significantly both spatially (e.g., across ISPs or across geographical regions) and temporally. Such variation can be caused by load, misconfiguration (e.g., content not reaching a CDN's edge servers), or other network conditions. Our goal is not to diagnose the root causes of these problems (e.g., [32]), but to show that they occur in the wild.

Figure 3 shows the average rebuffering ratio, video startup time, and video start failure rate experienced by clients with three major CDNs across different geographical regions during the busiest

[1]These are the views that have in fact started playing the video.

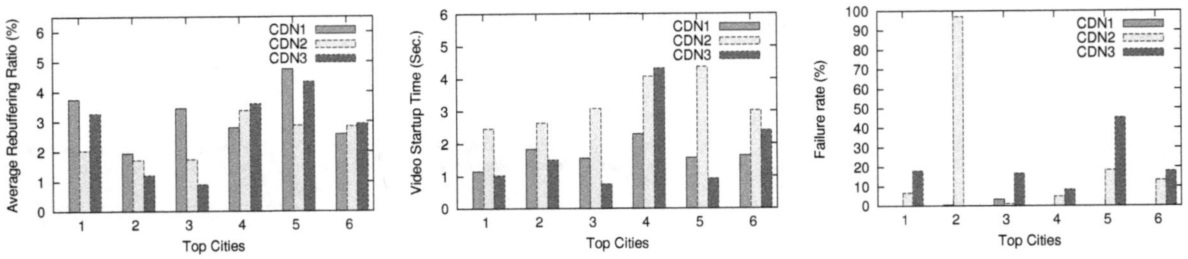

(a) Rebuffering Ratio (b) Video Startup Time (c) Video Start Failure Rate

Figure 3: CDN performance can vary substantially across different geographical regions

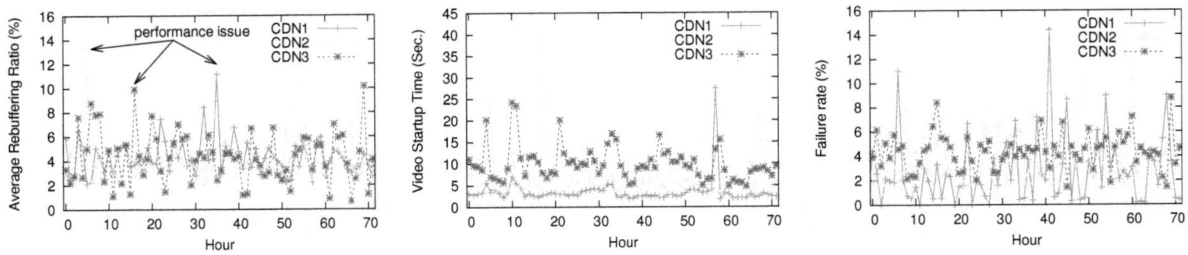

(a) Rebuffering Ratio (b) Video Startup Time (c) Video Start Failure Rate

Figure 4: CDN performance within a given geographical region can vary significantly over time as well

CDN	Rebuffering Ratio	Startup Time	Failure Rate
1	34.25%	79.08%	53.85%
2	25.22%	12.55%	37.50%
3	40.53%	8.37%	8.65%

Table 2: Percentage of scenarios where one of the CDNs performs the best in terms of each of the quality metrics.

hour on a weekday. Here, we choose the geographical regions corresponding to the top six cities by user population. Since there is a potential tradeoff between a session's bitrate and its performance under these quality metrics (higher bitrates will typically result in higher rebuffering ratios), we focus only on sessions having the same bitrate by choosing the most commonly used bitrate within that geographical region. We also remove sessions that cannot sustain the lowest bitrate (300Kbps) to rule out client-side effects in this analysis.

In summary, the results in Figure 3 show that:

- The performance of different CDNs can vary within a given city. For example, in City1, the rebuffering ratio of CDN1 is almost 2× that of users with CDN2.
- For each metric, no single CDN is optimal across all cities. For example, in the case of rebuffering ratio, CDN1 is optimal for City4 and City6, CDN2 for City1 and City5, and CDN3 for City2 and City3.
- CDNs may differ in their performance across metrics. For example, when we consider video startup time, CDN3 performs the best in all cases except City4. In contrast, when it comes to failure rate, CDN3 performs the worst.

Figure 4 shows the same metrics for one of these top cities over three days. (Each point is the average over several thousand sessions.) Here, we see that:

- For all three metrics, no CDN has the best performance all the time. Every CDN experiences some performance issues during the 3-day period. Table 2 shows how often each CDN is the best choice in a city-hour pair over the course of one weekday.[2]
- The rebuffering ratio and failure rate of a CDN may experience high fluctuations over time. For example, for roughly half of the time CDN3 has the lowest rebuffering ratio, and for the other half it has the highest rebuffering ratio.
- Most of the performance degradation is not correlated across CDNs, suggesting that these variations are not merely due to time-of-day effects but other factors.

One possible reason for such variability in the quality observed with CDNs is the load on the CDN. Figure 5(a) shows the rebuffering ratio vs. normalized CDN load for one CDN in one city over a week. Here, we measure the load as the number of unique sessions that we observe over each 5-minute interval. Since our clients represent only a fraction of the total load on the CDN, we normalize the observed load for each CDN by the maximum observed over the entire week for that CDN. Figure 5(a) shows that the rebuffering ratio generally increases with the normalized load.

Implications: This result highlights the need for providers to have multiple CDNs to optimize delivery across different geographical regions and over time. It also suggests that dynamically choosing a CDN can potentially improve the overall video quality.

AS under stress: Finally, ISPs and ASes can also experience quality issues under heavy load. Figure 5(b) shows the rebuffering ratio of one AS from all three CDNs during a 4-hour flash crowd period.[3] Each point shows the average buffering ratio across clients at a given time. We report the normalized load on the x-axis by dividing the current number of users by the maximum number of clients observed over time. During this flash crowd, the rebuffering ratio becomes quite high when the number of views increase.

Implications: These results suggest that heavy load can lead to ISP congestion. Ideally, we want the video delivery infrastructure to be

[2]Here we consider only city-hours where all three CDNs serve a reasonable number of views.

[3]This data comes from a known event which is not included in the data set presented before.

aware of these network hotspots to optimize video quality. In this case, the load increases on *all* CDNs and thus switching to a different CDN would not help. As a content provider, one reasonable policy is to reduce the bitrate for all views during these overload scenarios or provide higher quality only to "premium" customers.

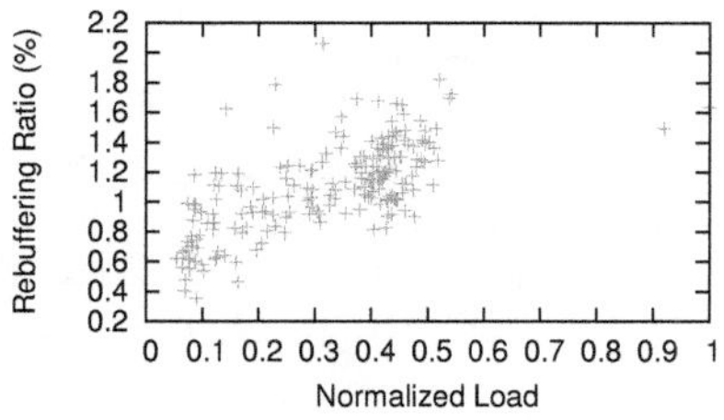

(a) One CDN under Load

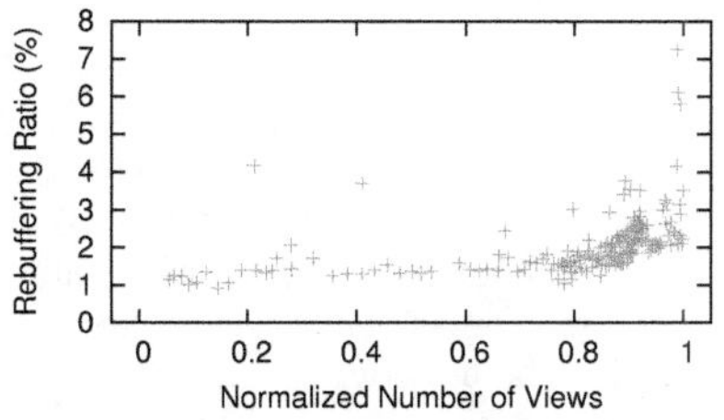

(b) Whole AS under Load

Figure 5: Rebuffering ratio under stress

2.4 Summary of key observations

The goal of this section was to analyze the current state of video delivery quality and analyze potential sources of performance problems. We see that:

- A significant fraction of sessions suffer quality issues, with more than 20% sessions having 10% rebuffering, and more than 14% sessions with 10 seconds of startup delay.
- There is significant variability in client-side bandwidth both within and across sessions, suggesting the need for intelligent bitrate adaptation.
- CDN quality varies considerably both across time and across space, which indicates the need for providers to dynamically choose different CDNs for different clients.
- When the streaming demand exceeds the capacity of CDNs or/and ISPs, content providers may need to enforce a global policy across clients to ensure a good viewing experience.

3. FRAMEWORK FOR OPTIMIZING VIDEO DELIVERY

The previous section highlighted that many video sessions today observe serious quality issues that arise as a consequence of client-side variability, spatio-temporal variability in CDN performance, and occasionally due to overload. The natural question then is how can we design an optimized video delivery mechanism that is robust to such conditions. In this section, we begin with an overview of the design space of optimizing video delivery and then sketch a high-level vision for a video control plane.

3.1 Design Space

The design space for optimizing video delivery quality has three natural dimensions:

1. *What* parameters can we control?
   There are two main parameters here: choice of bitrate and choice of CDN/server to serve the content. Because the specific video server is controlled by the CDN (e.g., based on load and latency), we only consider server selection at a CDN granularity.
2. *When* can we choose these parameters?
   There are two natural options here. We can select the parameters (i.e., CDN, bitrate) at *startup time* when the video player is launched or dynamically adapt these *midstream* in response to changing network conditions.
3. *Who* decides the values for these parameters?
   There are three high-level options we envision here: purely client-side mechanisms (the de-facto approach today), server-driven mechanisms (e.g., [28]), and an alternative *control plane* that selects these parameters based on global state.[4]

What parameter?	Who chooses?	When to choose?
CDN, Bitrate	Client	Startup
CDN Bitrate	Client Client	Startup Midstream
CDN, Bitrate	Control Plane	Startup
CDN Bitrate	Control Plane Client	Startup Midstream
CDN Bitrate	Control Plane Client	Midstream Midstream
CDN, Bitrate	Control Plane	Midstream

Table 3: Some examples from the overall design space for optimizing video delivery quality. We do not consider the cases where the client chooses the CDN and the control plane chooses the bitrate.

Note that this assumes the viability of two mechanisms—bitrate adaptation and CDN switching—which are already widely used. (The specific algorithms to implement these mechanisms are orthogonal to the focus of this paper.) Bitrate adaptation is already widely adopted in industry (e.g., [1, 4]). Similarly, CDN switching is already adopted in many industry players and with HTTP chunking, chunks can be requested from different CDNs without affecting user experience.

Table 3 looks at some example points in this design space by combining different options of these three variables. At the simplest end of the spectrum (row 1), we can think of a static selection of both CDN and bitrate by the client when the player is launched. This approach is not robust as both changes in CDN performance and client access bandwidth can impact the user experience. The de-facto approach today, shown in the second row in the table, is *client-side* bitrate adaptation but with the CDN/server fixed at start time [1, 4, 7]. There are two advantages of client-side adaptation: (1) clients are in the best position to observe local network effects and (2) the response time to react to network dynamics will be low. As we saw earlier, there is significant temporal and spatial variability in CDN performance that is difficult to detect and alleviate with purely client-side strategies.

To this end, we believe it will be helpful to have a *control plane* deployed either by a content provider or a third party on the behalf of the content providers that is aware of such temporal and spatial variations. (We discuss what such a control plane may look like in the next subsection.) Beyond these performance insights, a control plane also offers content providers more flexibility in instrumenting fine-grained policies; e.g., providing higher quality service to premium customers under load. In the ideal case (last row), we envision this control plane can dynamically adapt both the CDN and bitrate midstream based on global knowledge of network, distribution of active clients, and CDN performance.

[4]We do not consider server-driven mechanisms because these can be equivalently realized by via client- or control-plane mechanisms.

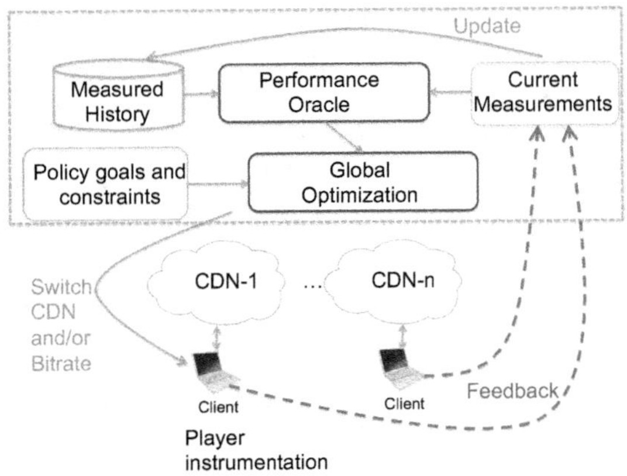

Figure 6: Overview of a video control plane

Of course, it may be unnecessary and impractical for this control plane to be continuously involved in adapting the CDN and bitrate. Thus, we can consider intermediate points in this design space as well. For example, CDN selection can be driven by the control plane because it has a global view of CDN performance, but bitrate adaptation may run purely at the client (rows 4 and 5).

3.2 Vision for a Video Control Plane

The notion of a centralized control plane to optimize content delivery is not new and has been used within CDNs and ISPs for server selection and content placement [12, 13, 36]. There are two key differences in the context of video optimization. First, we introduce a new dimension of *cross CDN* optimization and combining this with bitrate selection/adaptation. Second, we focus on the possibility of *midstream switching* of both parameters, whereas most CDN optimizations focus only on start-time selection. As a simple starting point, our current work assumes that this control plane operates *per content provider*. That is, a video content provider such as YouTube or Hulu runs such a control plane to monitor and improve the video experience for its customers. We discuss issues involving the interaction between multiple such providers and controllers in Section 7.

Figure 6 shows a high-level overview of the three key components in the video control plane: (1) a *measurement* component responsible for actively monitoring the video quality of clients, (2) a *performance oracle* that uses historical and current measurements to predict the potential performance a user will receive for a particular combination of CDN and bitrate at the current time, and (3) the *global optimization* engine that uses the measurement and performance oracle to assign the CDN and bitrate for each user. Next, we briefly highlight the main factors and challenges involved in the design of each component.

Measurement Engine: The measurement engine periodically collects quality statistics for currently active users. Because the client-side player is in the best position to measure the observed video quality, we envision the client player *periodically* (every few seconds) reporting such statistics. In addition to reporting the video quality metrics (e.g., buffering, join time, average bitrate), the measurement engine also collects user and session attributes such as the ISP, location, current CDN being used, and player version that will aid in the performance prediction. The challenge here is to choose a suitable granularity of attributes and quality metrics to measure, and to decide an appropriate frequency at which these reports are sent to the control plane.

Performance Oracle: The performance oracle plays a key role in answering *what-if* style questions at the control plane to predict the performance (e.g., rebuffering ratio, startup delay, failure rate) that a given user may observe at the current time if it chose a different combination of CDN and bitrate. By design, the oracle will have to *extrapolate* the performance based on past and current measurements. For example, it may cluster users based on a set of attributes (e.g., ISP, location) and use the empirical mean within this cluster as its prediction. The challenge here is that the extrapolation must be robust to noise and missing data; e.g., are there enough points within this cluster for this extrapolation to be statistically sound?

Global Optimization: At a high-level, we are solving a *resource allocation* problem, where the resources are the CDNs. Each CDN is characterized by a given network capacity (i.e., how many clients can it serve) and distribution costs. We want to assign each user a suitable CDN and bitrate that maximizes some notion of *global utility* for the content providers and consumers, while operating within the provider's cost constraints and the CDN capacities. There are three main challenges here. First, we want to choose a suitable utility and policy objective. For example, this utility can be a function of the bitrate, quality metrics such as buffering, and the providers' policy goals (e.g., premium customers get higher priority over non-paying users). Designing a good video utility metric that can combine different notions of quality (e.g., bitrate, rebuffering, startup delay) is an open challenge that is outside the scope of this paper [21, 35, 40]. The provider can also specify other policy constraints; e.g., should it admit new clients when all CDNs are overloaded. Our focus is to make a case for such a framework and present initial steps toward a practical realization rather than prescribe specific utility or policy functions. Second, this optimization must fast enough in order to periodically re-optimize the assignments in response to network dynamics. Third, we need to ensure that the optimization is stable and does not itself introduce biases or instability (see Section 5.1).

4. POTENTIAL FOR IMPROVEMENT

Before attempting to design a specific control plane, we want to first establish the improvement in video quality that we can achieve. To this end, in this section we analyze the potential improvement that clients could achieve by choosing a better CDN. As we will see later, the techniques described here can be extended to realize the performance oracle described in the previous section.

4.1 Approach

Our goal is to determine the potential performance improvement assuming each session makes the best possible choice. Ideally, each client will try all possible choices and pick the one with the best performance (e.g., rebuffering rate). Moreover, a client will constantly re-evaluate the performance and switch, if needed, to improve its performance. For example, a client can start with the configuration $(CDN_1, bitrate_1)$, and later switch to $(CDN_2, bitrate_2)$, if the new choice provides better performance. Of course, in practice we cannot have each client continuously probe all possible combinations. To get around this limitation, we *extrapolate* the performance a client could have achieved based on our observed performance of other clients that share similar *attributes*, such as ISP, location, device, and time-of-day. We follow previous work on non-parametric prediction [24, 33] with some simplifying modifications.

We make two simplifying assumptions. First, we do not consider bitrate selection in this section. Second, we assume that session outcomes are independent and that CDN performance does not degrade with load. We relax these assumptions in Section 5.2.

Our approach has two logical stages: *estimation* and *extrapolation* that we describe next.

Estimation: In the estimation step, we compute the empirical performance of each combination of attribute and parameter values.

Let a denote a set of values of a client's attributes, e.g., ISP = *AT&T*, City=*Chicago*, Device=*XBox*. Further, let S_a denote the set of clients sharing same attribute values a, and let $S_{a,p}$ denote the set of clients with attribute values a that have made the same choice or parameter p (i.e., CDN). An example of such set would be, XBox devices of Comcast's subscribers located in Chicago that stream content from Akamai.

For each set $S_{a,p}$, we compute the empirical distribution for the metric of interest, e.g., rebuffering ratio. Let $\mathit{PerfDist}_{a,p}$ denote this empirical distribution. Given two such performance distributions, we say that $\mathit{PerfDist}_{a,p_1}$ is better than $\mathit{PerfDist}_{a,p_2}$, if $\text{MEAN}(\mathit{PerfDist}_{a,p_1}) < \text{MEAN}(\mathit{PerfDist}_{a,p_2})$.[5]

Extrapolation: We use $p_a^* = \text{argmin}_p\{\text{MEAN}(\mathit{PerfDist}_{a,p})\}$ to denote the parameter with the best performance distribution for this specific value of the attribute a. Using this definition, we can extrapolate the best possible performance that can be achieved by a session with attribute values a by selecting parameter p_a^*, and assuming that the performance experienced by the session is randomly drawn from the distribution $\mathit{PerfDist}_{a,p_a^*}$ as shown in Figure 7(a).

Now, for such extrapolations to be statistically meaningful, the number of observations for each a, p setting $|S_{a,p}|$ should be reasonably large. Unfortunately, we run into the classic curse of dimensionality; that is, as the attribute space becomes more fine-grained, the available data becomes sparse [14]. This is particularly problematic because we will be picking the CDN with the highest extrapolated value using this methodology. If we pick the highest among several noisy predictions, we may show improvements even where there are none.

Hierarchical estimation and extrapolation: To address the problem of data sparsity at finer granularity, we use a hierarchical approach [25, 33]. We begin with an exhaustive enumeration of all possible combinations of attributes. That is, if $A = \{a_1 \ldots a_n\}$ is the set of client attributes, we construct the *powerset* 2^A of all attribute combinations. Let $\mathit{attrset} \in 2^A$ denote one such combination of attribute elements such as {isp, location, timestamp}, {isp, location} or {isp}; let a_s denote the values of the attributeset *attrset* for session s.

Note that a given video session's performance measurement will contribute to multiple attribute partitions corresponding to different granularities. That is, a session with ISP = *AT&T*, City=*Chicago* gets added to the partitions (ISP = *AT&T*, City=*Chicago*), (ISP = *AT&T*), and (City=*Chicago*). Then, for the partitions $S_{a,p}$ that have a sufficient number of data points (we use a threshold of 1000 sessions), we estimate the expected performance using the empirical mean.

The extrapolation step becomes slightly more involved. As discussed earlier, we want to use the most fine-grained attribute information available, but we may not have sufficient statistical confidence in predictions from very fine-grained groups. In such cases, we want to identify a *coarser* attribute combination (see Figure 7(b)) at which we have sufficient data. To get a suitable coarsened granularity, we consider a logical ordering on the powerset of all attribute combinations 2^A such that finer-granularity combinations come earlier. That is, if we had three attributes ISP, city, and timestamp, then {ISP,city,timestamp} < {ISP, city} < {ISP}.[6] Given

[5]Other possible metrics to compare two distributions can be median, or, more generally, the q-quantile of the distribution.

[6]Strictly speaking this is a partial order. In practice, we break the ties arbitrarily noting that it does not affect the results significantly.

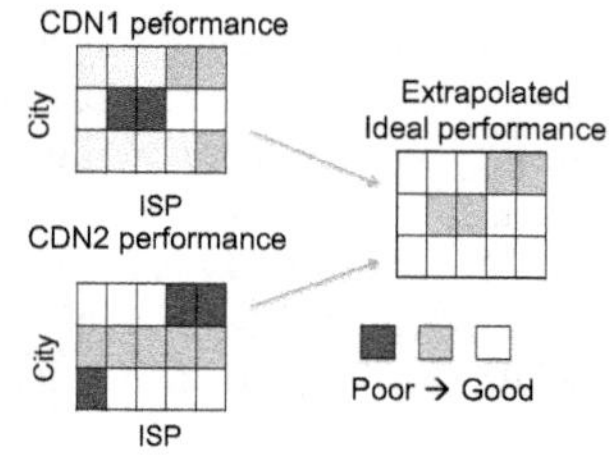

(a) Basic extrapolation

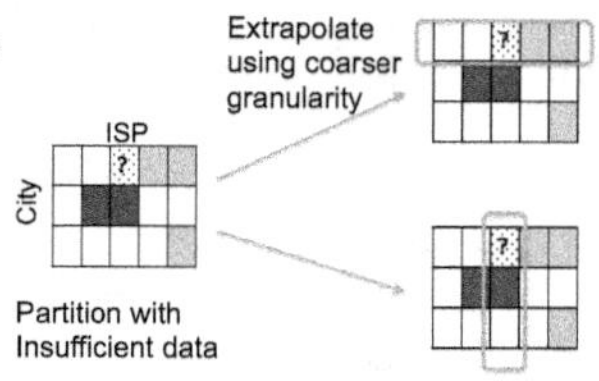

(b) Insufficient data

Figure 7: There are two user attributes: ISP and city. For each combination of attribute values, we want to analyze the potential improvement by choosing a better CDN (a). If a combination does not have sufficient data, we identify a suitable coarser level (b). In the simplest case, the extrapolation uses the mean of the distribution within each partition.

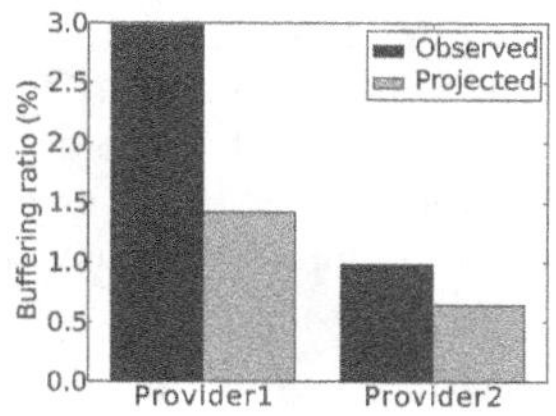

(a) Buffering ratio

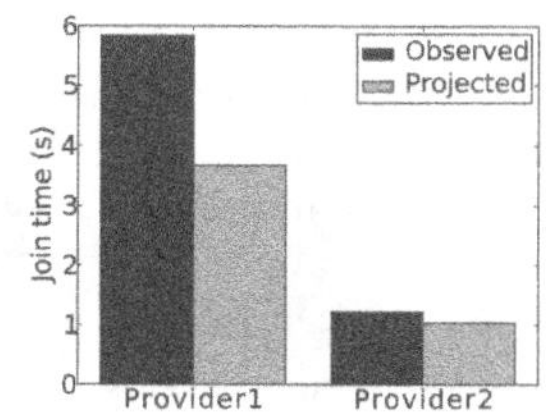

(b) Join time

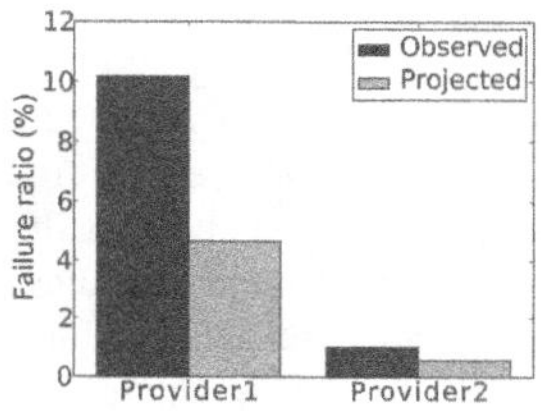

(c) Failure rate

Figure 8: Potential improvement in three quality metrics (rebuffering ratio, failure rate, and join time) for two providers.

this ordering, we proceed up the hierarchy from the finer to coarser partitions, until we have sufficient data to make a prediction.

Let a_s^* denote the value of session attributes at this point in the hierarchy. Based on the chosen level in the hierarchy and the parameter setting $(a_s^*, p_{a_s^*}^*)$, we extrapolate the performance by drawing a value from the empirical distribution $\mathit{PerfDist}_{a_s^*, p_{a_s^*}^*}$. In other words, by replacing one performance distribution with another we are simulating the effect of choosing a better parameter setting. As a further refinement, we can also imagine mapping each session into an appropriate percentile bin in the new distribution. That is, if the current session was within the 90-95th percentile of clients in $S_{a,p}$, then we draw from the same bin of 90-95th percentile of the performance of S_{a,p_a^*}. Intuitively, this emulates a policy of not switching unless the quality is likely to improve. For brevity, we do not show the percentile-based extrapolation, noting that this refinement will magnify the potential for improvement.

4.2 Improvement Analysis

For this analysis, we choose two popular providers that use multiple CDNs for video delivery but do not explicitly optimize the CDN based on observed quality or assign clients preferentially to CDNs based on their quality. This ensures that the potential improvement we extrapolate by choosing a better CDN is unbiased.

Validation: We acknowledge that any such extrapolation analysis is challenging and necessarily involves simplifying assumptions. That said, we tried to do a careful job in leveraging our large dataset. One specific type of validation we perform is to ensure that we do not spuriously predict improvements when there are none. To this end, we create an artificial dataset by replacing the actual CDN in each session with a different CDN chosen *uniformly at random*. The idea here is that this synthetic dataset should in theory have no scope for improvement (modulo small stochastic effects). We run our algorithm over this synthetic dataset and confirm that our extrapolation predicts negligible (0.05%) improvement.

Average improvement: We begin by computing the average improvement in video quality over a one week period for the two providers using the above extrapolation approach. Figure 8 shows the average improvement for three video quality metrics: buffering ratio, join time, and failure rate. The result shows that for Provider1, we see a significant (more than 2×) decrease in the buffering ratio from 3.0 to 1.4. Provider1 also shows significant potential for improvement in the failure rate (10% to 4.5%) and the join time (5.8s to 3.6s). In contrast, the delivery quality for Provider2 is already very good, and thus the scope for improvement there is comparatively lower in the average case. However, as we see next, even Provider2 shows significant improvement under more extreme scenarios.

Improvement under stress: We expect the room for improvement to be significantly higher under more extreme scenarios; e.g., a particular CDN performs poorly in a particular region or has a lot of failures. To this end, we pick specific time segments where we observe large incidents where Provider1 and Provider2 see marked degradation in performance. Then, we analyze the potential for improvement in the video quality under these more extreme scenarios in Table 4 and Table 5 respectively. The results show a dramatic improvement in the potential performance over fairly long durations: 10× for buffering ratio and 32× reduction in failure rate for Provider1 and up to 100× improvement in the failure rate for Provider2.

Main observations: To summarize, our analysis shows that better CDN selection can show marked improvement in video delivery quality. Specifically, we find

- More than 2× improvement in mean buffering ratio and startup time, and 1.6× reduction in failure rate in the average case.
- 10-32× improvement in the buffering ratio and failure rate over extended time periods under stress.

5. TOWARD A PRACTICAL DESIGN

The previous section establishes that there is a non-trivial potential for improvement. In practice, a control plane has to also take into account the impact of bitrate on performance, effect of CDN load, and also rely on past estimates to predict what the future performance will be. Furthermore, we had also ignored the tractability of global optimization and the specific utility functions or policy objectives.

In this section, we present a preliminary effort at addressing these issues. Our goal is not to realize an "ideal" control plane; rather, we want to establish a *feasible but concrete* control plane design that we can use to analyze if the benefits promised from the previous section can be translated into reality.

5.1 Optimization

To make our discussion concrete, we focus on a specific policy objective and utility function. Our policy goal is to achieve both *fairness* (i.e., do not deny clients if there is sufficient capacity) and *efficiency* (i.e., maximize aggregate utility). There is a rich literature on balancing efficiency-fairness tradeoffs that we can build on

Metric	Duration (hrs)	Current	Projected
Buffering ratio (%)	3	10.41	0.795
Start time (s)	2	6.41	1.997
Failure ratio (%)	1	16.57	0.213

Table 4: Potential improvement in the mean performance under extreme scenario for Provider1

Metric	Duration (hrs)	Current	Projected
Buffering ratio (%)	1	2.24	0.29
Start time (s)	7	1.56	0.39
Failure ratio (%)	3	35.6	0.3

Table 5: Potential improvement in the mean performance under extreme scenario for Provider2

here; as a starting point we choose a simple goal of first ensuring fairness and then optimizing for efficiency.

We fix the utility function to be a simple linear combination of the different performance metrics that captures the expected viewing time for a given session. We choose the function

$$Utility = -3.7 \times BuffRatio + \frac{Bitrate}{20}$$

where *BuffRatio* is in percentage and *Bitrate* is in Kbps. This utility function is based on prior observations on linear relationship between the expected play time and the different quality metrics reported in a previous measurement study; e.g., a 1% increase in buffering ratio caused a 3.7 minute drop in viewing time [21].

Having fixed the utility function and policy objective, we use a simple two-phase algorithm. First, we assign the clients a *fair share* of the CDN resources by computing the average sustainable bitrate. This allocation ensures that each client has been assigned to some *feasible* combination of bitrate and CDN and there is no unfairness in the allocation in terms of bitrates. Also, each client is assigned a CDN at random so that each CDN gets assigned a share of clients proportional to its capacity. Next, we use this allocation as a starting point and try to incrementally improve the total utility. This proceeds as a greedy algorithm where at each iteration, it picks the combination of client and CDN/bitrate setting that provides the largest incremental contribution to the global utility function.

The intuition behind this two-phase approach is as follows. A pure greedy approach optimizes efficiency (i.e., total utility) but may leave some clients unassigned; e.g., it might initially assign clients with a higher bitrate but may end up saturating the capacity and drop some clients. A pure fair-sharing approach on the other hand guarantees that all clients will be assigned if there is sufficient capacity, but it is agnostic to the performance variability across CDN-client-bitrate combinations. Our approach strikes a balance between these two extremes by ensuring all clients are assigned, and improves the efficiency from this starting point. We do not claim that this optimal in a theoretical sense but we do observe that it works well across a range of realistic workloads.

One subtle issue is that the optimization may itself introduce undesirable temporal biases. For example, if we discover that CDN1 is performing poorly and shift all clients to CDN2, then it impacts our ability to predict the performance of both CDN1 (we do not have any samples) and CDN2 (we have increased its load) in the future. This is a classical "exploration-exploitation" tradeoff (e.g., [20]). We face a particularly difficult form of this problem; our rewards are not independent (due to CDN load), and the characteristics of CDNs change over time even if we do not use them. However, we do have access to a large amount of data. A simple solution in this case is to use some form of randomization. Here, we choose a random subset of sessions that will not receive any explicit optimization so that we can observe their performance in the wild. With a large enough population even a small fraction (say 2-5%) of unoptimized clients would suffice to build a robust prediction model. There are other more sophisticated approaches to solve this problem, such as knowledge gradient algorithms [38], that will further reduce such biases. We currently use the simple randomization approach due to its ease of implementation.

5.2 Performance estimation

A key issue with performance extrapolation we already observed in Section 4 is the need for a prediction mechanism that is robust to data sparsity and noise at such finer granularities. We address this by building on the hierarchical extrapolation techniques described in Section 4. There are three additional practical challenges that need to be addressed here.

First, in Section 4, our goal was to estimate the potential for improvement. Hence, we assumed access to a *performance oracle* that has a current view of the performance. In practice, a real control plane will not have such an oracle and will have to rely on historical measurements. We extend the hierarchical approach from the previous section to make predictions based on recent historical measurements of the performance for specific CDN-client-bitrate combinations. Since CDN performance also shows significant temporal variability, we simply use the most recent information (e.g., last hour). In Section 6.3, we show that even this simple use of historical measurements works very well in practice.

Second, the extrapolation in the previous section ignores the effect of CDN load and how the performance degrades as a function of load. To this end, we augment the performance estimation step to also model the CDN load. Specifically, we observe that the CDN performance shows a roughly thresholded behavior where the performance is largely independent of the load up to some threshold T_1, after which the performance degrades roughly linearly, and at a higher load threshold T_2, the performance would drop significantly. We select these thresholds based on observed measurements (not shown for brevity).

Third, we did not consider bitrates in the previous section. Here, we simply treat bitrate as an additional attribute to our decision process. That is, in addition to characteristics such as ISP and city, we also partition clients based on their current bitrate when building the performance estimators and prediction models. We extend the measurements from the previous section to reflect bitrates in our prediction model.

6. TRACE-DRIVEN SIMULATIONS

In this section, we use trace-driven simulations to evaluate the qualitative benefits of the global control plane approach we instantiated in Section 5 over other choices in the design space from Table 3 under different scenarios.

6.1 Setup

We built a custom simulation framework that takes the following configurations as input: (1) client arrival and viewing time patterns obtained from measurements from the same dataset described in Section 2, and (2) empirically observed CDN performance distribution in different geographical regions at different (normalized) loads. In each epoch, a number of clients join the system and choose a viewing time drawn from empirical distribution. The client either stays until the end of viewing time, or leaves when its utility becomes 0, i.e., the performance becomes unbearable. In this simulation, we implement three strategies:

1. *Baseline:* Each client chooses a CDN and bitrate at random. This can be viewed as a strawman point of comparison.
2. *Global coordination:* The control plane algorithm proposed in Section 5.
3. *Hybrid:* Each client is assigned to a CDN with lowest load and a bitrate by the global optimization when it first arrives, but the subsequent adaptation is only limited to client-driven bitrate adaptation, i.e., no midstream CDN switching. This can be viewed as a simplified version of what many providers deploy today: start time CDN selection and client-side mid-stream bitrate adaptation.

The primary goal of these simulations is to analyze the qualitative benefits of a global control plane. These alternative points do not necessarily reflect exact strategies in operation today; we pick these as sample points from our design space in Table 3.

In each epoch, after the decisions are made, the simulator computes the resulting load on each CDN (by summing up bitrates from all clients), and then computes the expected performance of each client based on empirical measurements and the resulting load, as described in the previous section. In this simulation, we use three CDNs. The performance metrics of each CDN under normal load are obtained by taking its mean performance over a week, so that any overload effect is averaged out. Then the load thresholds for modeling the impact of CDN load on performance (see Section 5.2) are extrapolated from visually correlating the relationship between the load and quality metrics for each CDN. In order to scale our simulation framework, we normalize the CDN capacity and the number of clients proportionally compared to the actual capacity and number of clients. The normalization ensures that 1) the required capacity to serve the highest bitrate for all clients will exceed the combined capacity of all three CDNs, and 2) the required capacity to serve the lowest bitrate for all clients does not exceed that of any two CDN combined.

We simulate three scenarios: average case, CDN performance degradation, and flash crowd. In the average case scenario, arrival patterns and playback duration mimic the typical client arrival patterns observed in the dataset. In the CDN performance degradation scenario, we retain the same client arrival pattern, but emulate a sudden degradation in one of the CDNs. Finally, in the flash crowd scenario, a large number of clients arrive simultaneously.

6.2 Results

We focus on two metrics in our evaluation: *average utility* and *failure ratio*. Average utility is computed by the total utility divided by all clients in the system (including clients that failed to play video and thus has zero utility). Since the arrival pattern is the same for all three strategies, they all have the same denominator in all epochs. Failure ratio is the ratio of clients that could not receive any video due to either CDN exceeding capacity or unbearably high rebuffering ratio. Figure 9 shows the simulated time series of these two metrics for the three scenarios across different adaptation strategies.

Average case: In Figure 9(a), we observe that global coordination significantly outperforms the baseline strategy in terms of the average utility. One interesting observation is that in the common case, a hybrid strategy provides similar performance to global coordination. This is because in this strategy, a client chooses an ideal CDN-bitrate combination when it arrives, and under regular conditions, this strategy can maintain good performance. Unsurprisingly, the baseline approach performs poorly even in the common case and has both a high failure rate and low overall utility. The primary reason is that this approach is agnostic to CDN performance.

With regard to failure rates, we see that the baseline approach is consistently high. The capacity-aware global coordinator is able to assign clients to suitable CDNs and bitrates and keep a zero failure

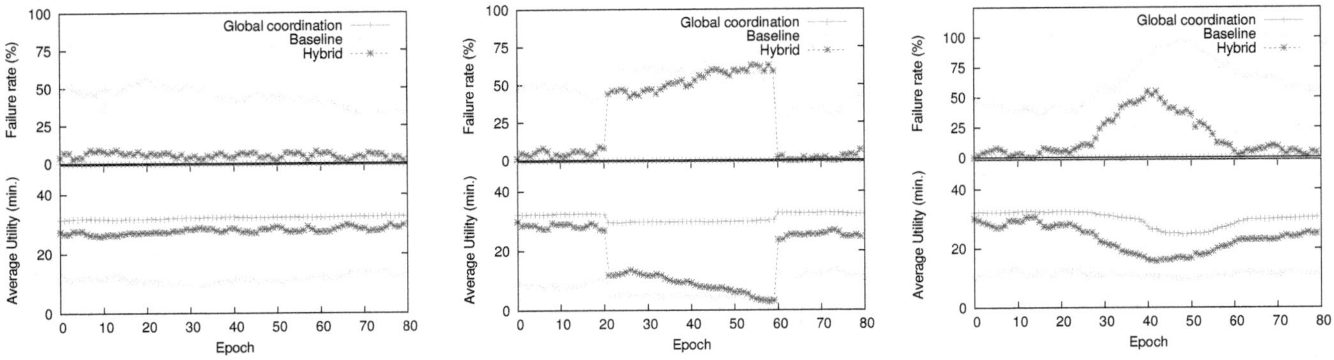

(a) Common case (b) CDN degradation (c) Flash crowd

Figure 9: Simulation results of three scenarios. For each scenario, we show the performance of baseline, hybrid and global coordination in terms of failure ratio and average utility value. (The utility metric is in units of expected minutes of playing time.)

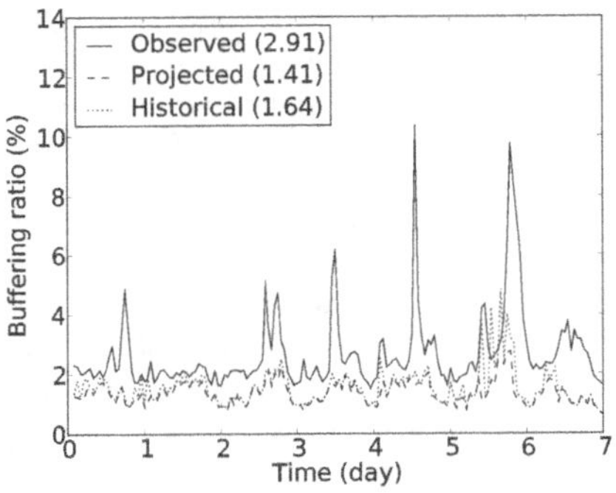

(a) Prediction vs. Optimal

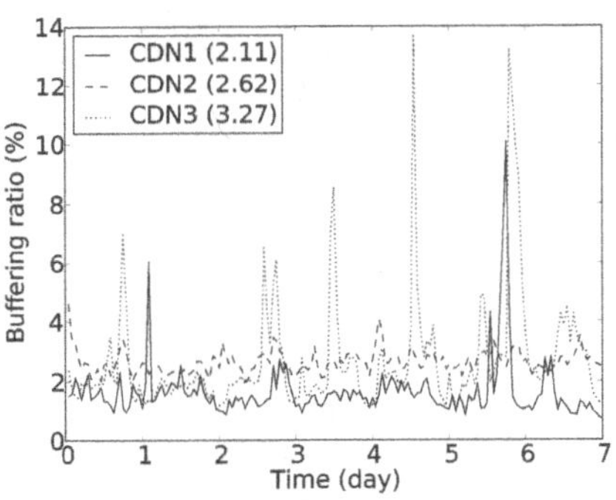

(b) Performance of different CDNs

Figure 10: Performance gap between a history-based and an oracle performance estimator. We also show the performance of the individual CDNs to visually confirm why history-based estimates work in most cases and highlight the specific epochs where it does not.

rate. The hybrid strategy has a small but non-zero failure rate; the reason is that some clients may use high bitrates so that a small number of clients are rejected.

CDN variability: In this scenario, in epochs 20–60, the previously best CDN experiences a huge degradation with the average rebuffering ratio going to 13%, before eventually recovering at epoch 60. Figure 9(b) shows how different strategies respond to CDN variability. In this case, global coordination still maintains zero failure rate (which is possible because of our normalization) and has a significantly higher average utility than both other approaches. However, it too suffers a mild drop in utility compared to the normal scenario during the degradation. We also see that the hybrid approach does almost as poorly as the baseline in both metrics. This is because in this strategy, the server is not aware of CDN performance degradation, and even when the clients can identify this degradation, it is unable to switch CDN midstream.

Flash crowd: Last, we emulate a sudden flash crowd in Figure 9(c), where a large number of clients try to join the system between epochs 20–60. In this case, the global control algorithm lowers the bitrate for many users in order to accommodate more new users. Recall that this is one of the policy decisions we imposed in our design to balance fairness in allocating users vs. efficiency. During the flash crowd, we see that the failure rate for global coordination remains at zero, whereas the baseline and hybrid have exceedingly high failure rates. The hybrid approach does not degrade as much as in previous case because the it is aware of the load on each CDN, but without midstream CDN switching, the performance is worse than global coordination.

6.3 History vs. Oracle prediction

As discussed earlier, a practical control plane will need to rely on historical estimates of the performance to choose the best CDN. A natural question is how far away from the optimal performance is such history-based prediction. To analyze this gap, we use the same dataset for Provider1 from Section 4 and compare a history-based vs. oracle prediction in Figure 10(a) over one week. We consider two options for using the historical estimates: using either the *previous hour* or using the *previous day*. That is, for each partition, we identify the best performing CDN using the previous hour or day measurements and use that as the prediction for the current epoch. The oracle uses predictions based on the *current hour*. The result shows that using the previous hour's predictions, the gap between the history and oracle predictor is small for many instances. However, we also see some cases where there is a non-trivial difference between the historical and optimal. Figure 10(b) visualizes how the individual CDNs' performance varies during this period to provide some intuition on why such large gaps occur. We do see that the CDNs' performance is largely predictable using the recent history but does occasionally spike. These performance spikes cause causes our estimate using the previous hour to become inaccurate. Our preliminary results (not shown) suggest that some of these gaps can be alleviated this using more fine-grained historical information (e.g., previous five minute epochs).

6.4 Summary of results

- Global control plane works well in all scenarios, including CDN performance variation and flash crowd.
- A hybrid approach of using the coordinator only at startup time and relying on pure client-side adaptation may work quite well in common scenarios.
- However, such a hybrid approach could suffer performance degradation under CDN variability and flash crowds. In such cases, a control plane can implement more flexible policies. For example, under flash crowd it maintains a zero failure rate by reducing all client bitrates.
- The benefits of a control plane can be realized with simple extrapolation by using predictions from previous epochs.

7. DISCUSSION

Next, we present preliminary insights on how we can address issues such as scalability, the interactions between such a control plane and CDNs, and interactions across multiple such controllers.

Scalability: A concern with global optimization is scalability vs. number of clients and the time to respond to network events. Our unoptimized implementation in Java takes $\approx$ 30s to run the global optimization for 10,000 clients, 4 CDNs, and 5 bitrates. We speculate that typical video utility functions will possess a natural diminishing property [23]. Intuitively, this means that the incremental utility in going from a 10% buffering ratio to a 5% buffering ratio will be higher than the increment going from 6% to 1%. In this case, there are known techniques to speed up the greedy step via "lazy evaluation" [30]. Beyond such algorithm optimizations, we also envision scaling the control plane by logically partitioning different geographical regions and running one instance per region.

Switching tolerance: A natural question is how much bitrate switching can users tolerate? Controlled studies suggest users are sensitive both to frequent switches (e.g., [18]) and also to sudden changes in bitrate (e.g., [35]). We do not, however, have a good quantitative understanding on the tradeoff between switching vs. the desire to maintain high bitrate and low buffering. As this tradeoff becomes clearer with future measurement studies, this can be incorporated into the control plane optimization function as well.

Interaction with CDNs: One question is whether CDNs can do (are doing) such optimizations themselves today. While we cannot speculate about their internal policies, measurement studies suggest that CDNs are largely optimizing for latency [27]. Furthermore, content providers increasingly use multiple CDNs and thus no single CDN can provide the required cross-CDN optimization. We do note, however, that the techniques we develop apply equally well in the context of individual CDNs.

Another concern is whether there can be undesirable interactions between such higher-level optimization and CDNs' optimizations. Of particular concern are possible oscillations caused by such interactions. Unfortunately, it is hard to answer this question due to the limited visibility we have into CDN policies. Nonetheless, we hope that this potential problem will be alleviated in the future, as we envision new generation architectures where CDNs expose APIs to content providers and such controllers. For example, more fine-grained information on the available capacity of the CDNs or current load for different geographical regions can inform better control plane optimization strategies.

Finally, an emerging direction is the concept of federated CDNs [19, 37]. A federated CDN incorporates a technology integration and business partnership between carriers and CDNs to provide a unified CDN offering that has a global presence and benefits to both CDNs and content providers. For example, a federated CDN eliminates the need for the content provider to publish to each CDN. The global coordinator proposed in this paper is complementary to a federated CDN and can be used to enable high quality video distribution across a federated CDN. In fact, we believe a coordinator is essential to delivering high quality in a federated CDN.

Multiple controllers: So far, we implicitly assumed a simple model, in which the different controllers are independent, and that one controller's decisions will have limited impact on others. In the future, we expect such controllers to expose APIs to exchange performance data and policy constraints/preferences, similar to the way ISPs use BGP to coordinate.

8. RELATED WORK

Client side measurements: The variability in client-side throughput–across ISPs, within a given viewing session and across multiple sessions–have been well documented in past measurement studies (e.g., [26,42]). The natural solution in a video streaming context is to adapt the video bitrate in response to changing bandwidth conditions to ensure an uninterrupted viewing experience.

Client-side adaptation: Several commercial products today perform some form of client-side adaptation to adapt to changing bandwidth conditions (e.g., [1,4]) and there are ongoing standardization efforts in this respect [7]. The key difference here is that these focus purely on bitrate adaptation. Recent analysis of commercial players suggest that there is room for improvement in client-adaptation strategies [17, 39]. As we saw in earlier sections, there is significant variability in network and CDN performance [31]. Furthermore, there is an inherent need for coordination under overload which means that even near-ideal client-side mechanisms will not be sufficient. A global control plane can alleviate these concerns by coordinating actions across multiple viewers.

Video Coding: Layered coding and multiple description coding offer alternatives for graceful degradation of video quality (e.g., [15, 34]). While these are attractive in theory, they impose significantly higher complexity on the provider, delivery, and player infrastructure. We do note that if these solutions do get deployed, a video control plane is well-positioned to leverage the additional flexibility that these offer as it can more smoothly degrade performance instead of having to choose from a discrete set of bitrates.

CDN and server selection: Server selection strategies within a CDN are based on proprietary algorithms, but measurement studies suggest that these are largely based on proximity and latency (e.g., [27]). Similarly, in the context of video delivery, the details of how particular providers choose CDNs or direct clients to different CDNs are proprietary. Preliminary measurements, however, suggest that the strategies are largely statically configured (e.g., [41]) and that there appears to be no concerted effort to choose CDNs either at startup (e.g., [10]) or midstream (e.g., [11]). In making a case for a global video control plane, our goal is not to pinpoint the inefficiency of particular providers' selection strategies. Rather, we want to design a general framework for high-quality video delivery.

Other video measurements: Given the growing dominance of video traffic, there have been many measurement studies of deployed systems that focus on: content popularity and access patterns (e.g., [16]), the user's desire for high quality and how it impacts play time (e.g., [21]), user viewing patterns (e.g., [9,22]), and extreme scenarios such as flash crowds (e.g., [43]). These works have been instrumental in exposing performance bottlenecks and implications of user behavior on system design. However, these do not directly focus on optimizing video quality by intelligent choice of CDNs and bitrates, which is the focus of our work.

9. CONCLUSIONS

User expectations of high quality video delivery—low buffering, low startup delays, and high bitrates— are continuously rising. While HTTP-based adaptive streaming technologies have dramatically decreased the barrier for content providers to reach a wide audience, the network and delivery infrastructure these rely on is fundamentally unreliable. Our measurements from over 200 million sessions confirm that this is indeed the case: more than 20% of sessions suffer quality issues such as more than 10% buffering or more than 5 seconds startup delay.

Our motivating question was whether it is possible to deliver high-quality video in such a dynamic environment. Given the significant variability in ISP and CDN performance, we argued the case for a video control plane that uses measurement-driven performance feedback to dynamically adapt video parameters such as the CDN and bitrate to improve the video quality. We established the potential for improvement using measurement-driven extrapolation and find that optimal CDN selection can improve the buffering ratio by up to 2× in normal scenarios and more than 10× under more extreme scenarios. We further augmented these results using trace-driven simulations and confirm the potential benefits of such a control plane.

In making a *case for* a video control plane, our work follows in the spirit of approaches for CDN and ISP management that show the benefit of network-wide views. There are several challenges that need to be addressed before these benefits can be realized in practice: scalability, interaction with CDNs, issues surrounding multiple providers and controllers among others.

Acknowledgments

We thank our shepherd Sujata Banerjee and the anonymous reviewers for their feedback that helped improve the final version of the paper. We also thank Ganesh Ananthanarayanan and Justin Ma for providing comments on early drafts.

10. REFERENCES

[1] Akamai HD Adaptive Streaming. http://wwwns.akamai.com/hdnetwork/demo/index.html.

[2] Cisco forecast. http://blogs.cisco.com/sp/comments/cisco_visual_networking_index_forecast_annual_update/.

[3] Driving Engagement for Online Video. http://events.digitallyspeaking.com/akamai/mddec10/post.html?hash=ZDlBSGhsMXBidnJ3RXNWSW5mSE1HZz09.

[4] Microsoft Smooth Streaming. http://www.microsoft.com/silverlight/smoothstreaming.

[5] Move networks. http://www.movenetworks.com/.

[6] Video quality metrics. http://www.akamai.com/html/solutions/stream_analyzer.html.

[7] I. Sodagar. The MPEG-DASH Standard for Multimedia Streaming Over the Internet. *IEEE Multimedia*, 2011.

[8] K. Chen, C. Huang, P. Huang, C. Lei. Quantifying Skype User Satisfaction. In *Proc. SIGCOMM*, 2006.

[9] L. Plissonneau and E. Biersack. A Longitudinal View of HTTP Video Streaming Performance. In *Proc. MMSys*, 2012.

[10] V. K. Adhikari, Y. Chen, S. Jain, and Z.-L. Zhang. Where Do You Tube? Uncovering YouTube Server Selection Strategy. In *Proc. ICCCN*, 2011.

[11] V. K. Adhikari, Y. Guo, F. Hao, V. Hilt, , and Z.-L. Zhang. A Tale of Three CDNs: An Active Measurement Study of Hulu and Its CDNs. In *Proc. IEEE Global Internet Symposium*, 2012.

[12] K. Andreev, B. M. Maggs, A. Meyerson, and R. Sitaraman. Designing Overlay Multicast Networks for Streaming. In *Proc. SPAA*, 2003.

[13] D. Applegate, A. Archer, V. Gopalakrishnan, S. Lee, and K. K. Ramakrishnan. Optimal Content Placement for a Large-Scale VoD System. In *Proc. CoNext*, 2010.

[14] R. E. Bellman. Adaptive control processes: A guided tour. Princeton University Press.

[15] J. Byers, M. Luby, and M. Mitzenmacher. A digital fountain approach to asynchronous reliable multicast. *IEEE JSAC*, Oct. 2002.

[16] M. Cha, H. Kwak, P. Rodriguez, Y.-Y. Ahn, and S. Moon. I Tube, You Tube, Everybody Tubes: Analyzing the World's Largest User Generated Content Video System. In *Proc. IMC*, 2007.

[17] L. D. Cicco and S. Mascolo. An Experimental Investigation of the Akamai Adaptive Video Streaming. In *Proc. USAB*, 2010.

[18] N. Cranley, P. Perry, and L. Murphy. User perception of adapting video quality. *International Journal of Human-Computer Studies*, 2006.

[19] D. Rayburn. Telcos and Carriers Forming new Federated CDN Group called OCX (Operator Carrier Exchange). June 2011. StreamingMediaBlog.com.

[20] D. P. de Farias and N. Megiddo. Exploration-Exploitation Tradeoffs for Experts Algorithms in Reactive Environments. In *Proc. NIPS*, 2004.

[21] F. Dobrian, V. Sekar, A. Awan, I. Stoica, D. A. Joseph, A. Ganjam, J. Zhan, and H. Zhang. Understanding the impact of video quality on user engagement. In *Proc. SIGCOMM*, 2011.

[22] A. Finamore, M. Mellia, M. Munafo, R. Torres, and S. G. Rao. Youtube everywhere: Impact of device and infrastructure synergies on user experience. In *Proc. IMC*, 2011.

[23] G. Nemhauser, L. Wosley, and M. Fisher. An analysis of the approximations for maximizing submodular set functions. *Mathematical Programming*, 14:265–294, 1978.

[24] A. George, W. B. Powell, S. R. Kulkarni, and S. Mahadevan. Value function approximation using multiple aggregation for multiattribute resource management. http://www.scientificcommons.org/53756787, 2009.

[25] I. Ryzhov and W. B. Powell. Bayesian Active Learning with Basis Functions. In *Proc. IEEE Workshop on Adaptive Dynamic Programming and Reinforcement Learning*, 2011.

[26] C. Kreibich, B. N. V. Paxson, and N. Weaver. Netalyzr: Illuminating The Edge Network. In *Proc. IMC*, 2010.

[27] R. Krishnan, H. V. Madhyastha, S. Jain, S. Srinivasan, A. Krishnamurthy, T. Anderson, and J. Gao. Moving Beyond End-to-End Path Information to Optimize CDN Performance. In *Proc. IMC*, 2009.

[28] L. De Cicco, S. Mascolo, and V. Palmisano. Feedback Control for Adaptive Live Video Streaming. In *Proc. of MMSys*, 2011.

[29] H. Liu, Y. Wang, Y. R. Yang, A. Tian, and H. Wang. Optimizing Cost and Performance for Content Multihoming. In *Proc. SIGCOMM*, 2012.

[30] M. Minoux. Accelerated Greedy Algorithms for Maximizing Submodular Set Functions. In *Proc. of 8th IFIP Conference, Springer-Verlag*, 1977.

[31] M. Venkataraman and M. Chatterjee. Effects of Internet Path selection on Video QoE. In *Proc. MMSys*, 2011.

[32] A. Mahimkar, Z. Ge, A. Shaikh, J. Wang, J. Yates, Y. Zhang, and Q. Zhao. Towards Automated Performance Diagnosis in a Large IPTV Network. In *Proc. SIGCOMM*, 2009.

[33] A. K. Mccallum. Learning to use selective attention and short-term memory in sequential tasks. In *Proc. Conference on Simulation of Adaptive Behavior*, 1996.

[34] S. McCanne, M. Vetterli, and V. Jacobson. Low-complexity video coding for receiver-driven layered multicast. *IEEE JSAC*, Aug. 1997.

[35] R. K. P. Mok, E. W. W. Chan, X. Luo, and R. K. C. Chang. Inferring the QoE of HTTP Video Streaming from User-Viewing Activities . In *Proc. SIGCOMM W-MUST*, 2011.

[36] R. S. Peterson and E. G. Sirer. Antfarm: Efficient Content Distribution with Managed Swarms. In *Proc. NSDI*, 2009.

[37] R. Powell. The Federated CDN Cometh. May 2011. TelecomRamblings.com.

[38] I. Ryzhov, P. Frazier, and W. Powell. The knowledge gradient algorithm for a general class of online learning problems. http://www.princeton.edu/~iryzhov/journal/online7.pdf, 2011.

[39] S. Akhshabi, A. Begen, C. Dovrolis. An Experimental Evaluation of Rate Adaptation Algorithms in Adaptive Streaming over HTTP. In *Proc. MMSys*, 2011.

[40] H. H. Song, Z. Ge, A. Mahimkar, J. Wang, J. Yates, Y. Zhang, A. Basso, and M. Chen. Q-score: Proactive Service Quality Assessment in a Large IPTV System. In *Proc. IMC*, 2011.

[41] R. Torres, A. Finamore, J. R. Kim, M. Mellia, M. M. Munafo, and S. Rao. Dissecting Video Server Selection Strategies in the YouTube CDN. In *Proc. ICDCS*, 2011.

[42] M. Watson. HTTP Adaptive Streaming in Practice. http://web.cs.wpi.edu/~claypool/mmsys-2011/Keynote02.pdf.

[43] H. Yin, X. Liu, F. Qiu, N. Xia, C. Lin, H. Zhang, V. Sekar, and G. Min. Inside the Bird's Nest: Measurements of Large-Scale Live VoD from the 2008 Olympics. In *Proc. IMC*, 2009.

Optimizing Cost and Performance for Content Multihoming

Hongqiang Harry Liu* Ye Wang* Yang Richard Yang*× Hao Wang† Chen Tian*
*Yale University †Google ×SplendorStream
{hongqiang.liu, ye.wang, yang.r.yang, tian.chen}@yale.edu wanghao@google.com

ABSTRACT

Many large content publishers use multiple content distribution networks to deliver their content, and many commercial systems have become available to help a broader set of content publishers to benefit from using multiple distribution networks, which we refer to as *content multihoming*. In this paper, we conduct the first systematic study on optimizing content multihoming, by introducing novel algorithms to optimize both performance and cost for content multihoming. In particular, we design a novel, efficient algorithm to compute assignments of content objects to content distribution networks for content publishers, considering both cost and performance. We also design a novel, lightweight client adaptation algorithm executing at individual content viewers to achieve scalable, fine-grained, fast online adaptation to optimize the quality of experience (QoE) for individual viewers. We prove the optimality of our optimization algorithms and conduct systematic, extensive evaluations, using real charging data, content viewer demands, and performance data, to demonstrate the effectiveness of our algorithms. We show that our content multihoming algorithms reduce publishing cost by up to 40%. Our client algorithm executing in browsers reduces viewer QoE degradation by 51%.

Categories and Subject Descriptors: C.2.3 [**Computer Communication Networks**]: Network Operations.

Keywords: Content Delivery, Multiple CDNs, Optimization.

1. INTRODUCTION

Many content publishers on the Internet use multiple content distribution networks (CDNs) to distribute and cache their digital content. We refer to content publishing using multiple content distribution networks as *content multihoming*. In our recent survey, we found that all major content publishers such as Netflix, Hulu, Microsoft, Apple, Facebook, and MSNBC use content multihoming.

Content publishers adopt content multihoming to aggregate the diversity of individual CDN providers on features, performance and commitment [7]. For example, one CDN may provide good coverage for locations 1 and 2, whereas another CDN provides good coverage for locations 2 and 3. To deliver content to viewers from all three locations, a content publisher may need to use both CDNs.

Given the wide usage and potential benefits of content multihoming, many commercial systems supporting content multihoming have recently been deployed (*e.g.*, [9, 11, 16, 17, 21, 27]), so that more content publishers can benefit from content multihoming. However, these commercial products either use ad hoc approaches or do not provide details on their designs. No previous studies on how to effectively utilize content multihoming are known.

In this paper, we attempt to provide a framework and a set of novel algorithms to optimize the benefits of content multihoming.

SIGCOMM'12, August 13–17, 2012, Helsinki, Finland.

We ask a simple question: Given that content multihoming allows a content object to be delivered from multiple CDNs, which CDN(s) should a content publisher use to deliver each object to each content viewer requesting this object, so that the publisher optimizes its benefits from content multihoming? This question is the key to efficiently utilizing content multihoming, since its solutions can be implemented directly with the flexible request routing mechanisms (*e.g.*, DNS CNAME, HTTP Redirect from servers, and client scripts in end hosts) in modern content delivery infrastructures.

An answer to this simple question, however, is not immediately obvious. Consider the current common approach of choosing, for each content viewer, the best performing CDN among all candidate CDNs. This approach, despite its simplicity, has multiple issues. First, although the chosen CDN may provide the highest level of performance, for example, satisfying that 99% viewers do not see quality of experience (QoE) degradation, the cost of the chosen CDN can be much higher than another CDN with a slightly lower, but still high enough level of 95%. Second, there are often multiple CDNs with comparable and sufficient levels of performance at a given region, *e.g.*, in US. One common approach to break ties in such cases is to pick the CDN with the lowest cost. However, the costs of CDNs, in particular, pay-as-you-go CDNs such as Amazon CloudFront, are volume based and non-linear. The cost of one object assignment depends on the other assignments. Third, there are locations where even the best performing CDN falls short. For example, a content publisher may have a QoE target of 95%, but the best performing CDN at some location achieves only 90%.

In this paper, we answer the preceding question by designing two algorithms: (1) an efficient optimization algorithm executing at content publisher to compute content distribution guidance, and (2) a simple algorithm executing at individual content viewers to follow the guidance with local adaptation. Either algorithm can be deployed alone, but together they benefit the most.

Specifically, the publisher optimization algorithm, named CMO, computes CDN assignments considering many real factors: nonlinear, multi-region CDN traffic charging, per-request charging, content licensing restrictions, CDN feature availability, and CDN performance variations. The CMO algorithm is novel and highly efficient. For example, when considering traffic cost only, the complexity of CMO is polynomial and independent of the number of content objects, whereas the complexity of simple enumeration is exponential in the number of content objects.

The local viewer algorithm provides a capability for a content viewer to make efficient usage of multiple servers from multiple CDNs, with a preference ordering on the usage of CDN edge servers provided by the content publisher. Inspired by TCP AIMD and using a simple prioritized assignment mechanism, the algorithm adapts the usage of multiple CDNs, achieving a performance level that no single CDN/server can achieve alone.

We implement both of the algorithms and conduct systematic, extensive evaluations using real charging data, content viewer demands, and CDN performance to demonstrate the effectiveness of our algorithms. We show that our content multihoming algorithms reduce publishing cost by up to 40%. Our implementation of the client algorithm running in Adobe Flash enabled browsers reduces viewer QoE degradation by 51%.

2. RELATED WORK

The importance of content multihoming has led to substantial industrial related work. We divide the industrial efforts into three categories [7]. The first category is software systems which we name CDN switchers (*e.g.*, [1, 21]) and integrators (*e.g.*, [19, 27]. For example, the One Pica Image CDN extension [21] of the Magento Commerce platform provides an API to support the integration with multiple CDNs, including Amazon S3, Coral CDN, Mosso/Rackspace Cloud Files, and any CDN server or service that supports FTP, FTPS, or SFTP. Their objective, however, is not on the algorithms to effectively use multiple CDNs, but rather on usability issues such as seamless switching from one CDN provider to another. Commercial systems such as [19, 27] provide CDN services based on aggregation of multiple CDNs. They can benefit from using our algorithms.

Going beyond the CDN switchers and integrators is a category of systems named CDN Load balancers. There are many CDN load balancers commercially available, including Cotendo CDN balancer, LimeLight traffic load balancer, Level 3 intelligent traffic management, and Dyn CDN manager. Some of these systems offer quite flexible rules to split CDN traffic among multiple CDNs. For example, Cotendo CDN balancer supports rules with weighted allocation, geographic location, geographic distance, time of day, and any combination. In particular, the weighted allocation rule allows a publisher to specify: x percent to CDN one, y percent to CDN two, and so on. A key missing component of these existing systems, however, is the key algorithms to compute the allocation (*e.g.*, the percentages). Hence, the output of our optimizer can be used to configure these systems.

There are also client based CDN load balancers. One interesting example is Conviva [9], whose video player plug-in performs continuous video quality monitoring, and could perform automatic CDN and/or source server switching during video playback. The exact details of their algorithm, however, are unknown.

A third category of related industrial efforts is CDN interconnect. In [20], a CDN interconnect (CDNi) architecture has been proposed so that a content publisher contracts with a few upstream CDNs, who may delegate some requests to downstream CDNs. The delegation relationship can be recursive to form a directional delegation graph, and all of the involved CDNs are said to form a CDN federation [5]. Our algorithms can be extended to the CDNi setting by considering a set of connected CDNs as a single logical CDN.

So far content multihoming has not been a focus of academic research. A related recent academic work is a measurement study of Netflix [2]. The paper shows that similar to many content publishers, Netflix uses content multihoming. The paper conducts a measurement study and shows that there are indeed potential performance benefits of using content multihoming.

There has been some recent studies on intra-CDN optimizations which can be used in conjunction with content multihoming. The authors of [3, 22, 23, 24] study how CDNs can utilize information on load of edge servers, network conditions, and locations/bandwidth of clients to improve CDN request routing. Our work complements these studies by optimizing the assignment of contents into CDNs.

We refer to our system as content multihoming to draw an analogy with traditional Internet multihoming (*e.g.*, [13]). However, content multihoming is quite different from traditional ISP multihoming. For example, while ISPs typically have a uniform price based on traffic, CDNs charge customers by regions.

3. BACKGROUND AND NOTATIONS

We start by introducing the background and notations. Table 1 provides a reference for the major notations used in this paper.

There are three key types of entities being managed in content multihoming: (1) content objects; (2) viewers of contents; (3) distribution networks that cache contents from origin networks to serve content viewers.

Content object: A content publisher can have a large number of content objects such as videos and images. Let N denote the total number of content objects. An object has many properties. In the context of content distribution, the performance requirement, the size, and the popularity of an object are its key properties [4]. Let s_i be the size of object i. We introduce object popularity when we next introduce content viewers.

Notation	Description
N	Number of content objects.
K	Number of CDNs.
A	Set of fine-grained location areas.
s_i	Size of object i.
n_i^a	Number of requests for object i from area a.
i^a	Location object: object i requested from location area a.
α_k^r	Set of location areas served by charging region r of CDN k.
t_k^r	Charging volume of CDN k at its charging region r.
$C_k^r()$	Charging function of CDN k for its charging region r.
F_k	Set of location objects that CDN k can serve.
$p_{i,k}^a$	CDN k's performance for object i at location area a: fraction of times CDN k can deliver i^a with sufficient QoE.
$x_{i,k}^a$	CDN guidance: fraction of n_i^a requests directed to CDN k.
ψ_k	CDN assignment: set of location objects assigned to CDN k.

Table 1: Summary of key notations.

Content viewer (client): There can be a large number of content viewers requesting content objects. These content viewers can be distributed across multiple geographical areas. The specific geographical areas depend on the particular requirements of a content publisher. For generality and conceptual clarity, let A be the set of all geographical areas, say all cities or countries. Note that in this challenging general case the size of A can be large, on the order of thousands. Let $a \in A$ denote a location area.

The popularity of an object among content viewers is location dependent [12]. Let n_i^a denote the number of times that object i will be requested, during a time interval (say a month), from content viewers located at location area a.

We also use n_i^a to encode *licensing restrictions* that a content publisher often needs to enforce in practice. Specifically, if content viewers from a location area a should not receive a content object i, n_i^a should be 0.

Content distribution network (CDN): A key reason of content multihoming is to aggregate the *capability-geography expertise* of different CDNs, as different CDNs can have quite different performance and cost characteristics, at different geographic regions. On the other hand, as we will see, such differences are a major source of intrinsic complexity when optimizing content multihoming. In this paper, we assume that the set of CDNs is given. Let K be the number of CDNs, and we use k and j to index individual CDNs.

First consider *performance*. Figure 1 shows the edge server footprints of three real CDNs: Amazon CloudFront, MaxCDN and ChinaCache. When a content viewer from a location area a requests a content object through CDN k, a well designed CDN will choose an edge server (or several servers) that is close to a to serve the request, since a short latency from edge servers to end users is typically needed to achieve good content delivery performance [15]. Comparing the geographical footprints of the three CDNs shown in Figure 1, one can anticipate that CloudFront and MaxCDN are more likely to provide better performance in US and Europe, while ChinaCache may perform well in China. None of the three covers regions such as Russia and Africa.

To quantify the performance, we conduct performance measurements using 600 PlanetLab nodes at different locations to request content objects from three CDNs (CloudFront, MaxCDN, and Liquid Web). Since performance metrics are dependent on the content

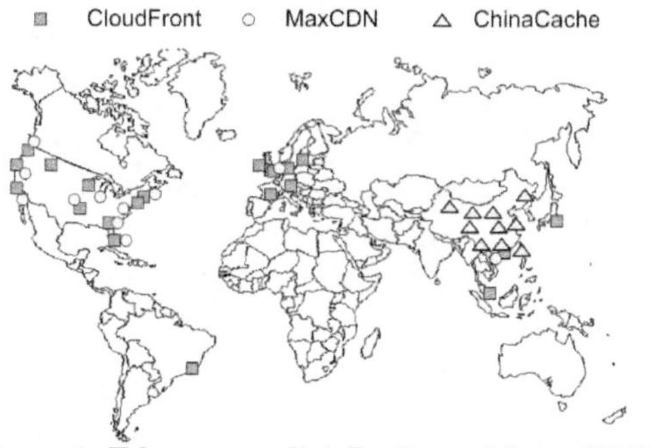

Figure 1: Edge server distributions of three CDNs.

type, and streaming media is a major content type [8], we evaluate the delivery of streaming media. Table 2 shows the measured success rates (fraction of clients with no video freeze) of the CDNs to deliver streaming objects encoded at three different streaming rates (1 Mbps, 2 Mbps, and 3 Mbps) to some representative locations. For example, the entry for Liquid Web/Spain shows the success rates when PlanetLab viewers from Spain request from Liquid Web: if the object is encoded at 1 Mbps, 99.4% of the viewers can receive at the encoding rate; for a 2 Mbps object, only 47.3% of the viewers can receive at the encoding rate; for a 3 Mbps object, almost no viewers can receive at the encoding rate. The measurement results show clearly that the usability of a CDN depends on both the object (*e.g.*, a video encoded at 1 Mbps or higher) and the location of the viewer. We refer to object i being requested by viewers at a specific location area a as a *location object*, denoted as location object i^a.

	CloudFront			MaxCDN			Liquid Web		
US	99.9	99.9	99.9	99.2	98.4	97.8	99.3	96.1	92.1
Brazil	100	100	99.9	98.6	70.5	24.4	99.6	0	0
Austria	99.9	99.9	99.8	97.6	96.7	95.3	97.0	42.2	0
Spain	99.9	99.9	99.9	98.7	96.6	95.1	99.4	47.3	0.2
Japan	99.9	99.9	99.9	97.5	95.8	77.1	99.7	0	0
China	99.9	99.9	99.8	91.1	24.7	0	1.6	0	0
Australia	100	100	99.9	94.7	89.5	0	99.7	0	0

Table 2: Measured CDN performance $p^a_{i,k}$ (3 content objects at streaming rates 1/2/3 Mbps).

To precisely characterize the performance of CDN k, in this paper, we define $p^a_{i,k}$ as the fraction of times (*e.g.*, 90%) that CDN k can deliver i^a at the encoding rate of the object i. One may define $p^a_{i,k}$ for other contexts (*e.g.*, for images) and use other metrics (*e.g.*, 95-percentile latency). Practically $p^a_{i,k}$ can be obtained from measurements or service level agreements (SLA) of CDNs.

Next consider *costs*. Different regions may have different resource (*e.g.*, bandwidth, electricity) costs. Different CDNs may operate at different scales at different regions to negotiate different volume discounts. Hence, different CDNs' prices might be various, and one CDN may charge differently at each region.

Figure 2 shows the real charging structures of two CDNs: Amazon CloudFront and MaxCDN. We show these two structures because they are public and represent typical CDN charging structures. We make three observations. First, each CDN groups the locations of its edge servers into multiple regions and each region may have a different pricing model. We refer to each such region as a *charging region*. For example, CloudFront divides into 5 charging regions: US, EU, South America (SA), Japan (JP), and Singapore/Hong Kong (SHK). MaxCDN divides into 2 charging regions: US/EU/SA, and Asia Pacific (AP). The total charge of a CDN to a content publisher is the sum of the charges at all of the charging regions. Second, denote the total traffic originated from the edge servers of a CDN located at a charging region during a billing period as the *charging volume* of the charging region; then the charging function of each charging region is a nonlinear concave function of the charging volume. Third, there can be large price diversity within a CDN as well as across CDNs. For example, CloudFront's charge for South America for "next 100 TB" is \$0.18/GB, which is 3 times that for US at the same traffic volume.

To precisely express the charging of CDNs, we let α^r_k be the set of location areas served by charging region r of CDN k. For viewers from a location area a, each CDN has its own strategy to select servers in some r to serve them. This strategy is controlled by CDN k but can be observed by content publishers [25]. For instance, in our measurements, requests from Beijing are redirected to CloudFront's JP region but to MaxCDN's US/EU/SA region. Let t^r_k denote the charging volume of CDN k at its charging region r, during a billing period. Specifically, the value of t^r_k is computed as the total traffic delivered during the billing period to content viewers at location areas who are assigned to α^r_k by CDN k. Then the total charge of CDN k to the content publisher is a sum of the charges at individual charging regions. Let $C^r_k()$ denote the charging function of CDN k for its charging region r.

4. CONTROL FRAMEWORK

We adopt a general, practical content publishing control framework shown in Figure 3. A central *Optimizer* computes configurations to direct viewer requests to CDNs. The configurations are sent to a DNS system, HTTP redirector, or a manifest-file server system to implement direction for specific viewer requests.

In particular, we distinguish two types of clients according to their capabilities. This distinction affects our problem formulation. The first type is passive clients. A key characteristic of passive clients is that they use one CDN edge server at a time. Although multiple CDNs and/or multiple servers from one CDN are available in content multihoming, such traditional clients at content viewers use only a single CDN server to serve a particular content object request [2]. For such clients, we assume that the results of content multihoming optimization are implemented only by server side mechanisms such as DNS redirection.

Specifically, suppose that content publisher `cp.com` uses DNS-based CDN redirection. For simplicity, say the publisher assigns content object i URL `http://obj-i.cp.com`. When a client requests this link, a DNS query from the client or the local DNS server of the client to resolve `obj-i.cp.com` can be sent to the DNS server of the content publisher, labeled `CPDNS` in Figure 3. By looking up the IP address of the client or the local DNS resolver of the client, `CPDNS` obtains the location area a. Using the output from the Optimizer, `CPDNS` returns to the client the CNAME of a chosen CDN k. As an example shown in Figure 3, CloudFront is chosen. Note in real implementations, as we will see in Section 6, URL assignment is not necessarily per-object. Alternative implementations (*e.g.*, using HTTP-based CDN redirection) are similar.

The second type is active clients. Such clients include an adaptation algorithm (*e.g.*, in Adobe Flash ActionScript) to utilize multiple CDN servers when retrieving a single content object. In particular, when the service rate of one CDN server is insufficient, an active client can use additional servers (from the same CDN or backup CDNs) to make up the deficit. The additional CDNs/servers are provided to the adaptation algorithm through a manifest file from the content publisher. Such manifest files are already used by some clients such as the Netflix clients. As shown in Figure 3, the CNAMEs of two CDNs are returned to an active client.

5. PROBLEM STATEMENTS

With the preceding background and control framework, our problems are easy to state. Note that there is much flexibility in the deployment of our control framework. There can be settings with only passive clients, or only active clients, or a hybrid. We first state the

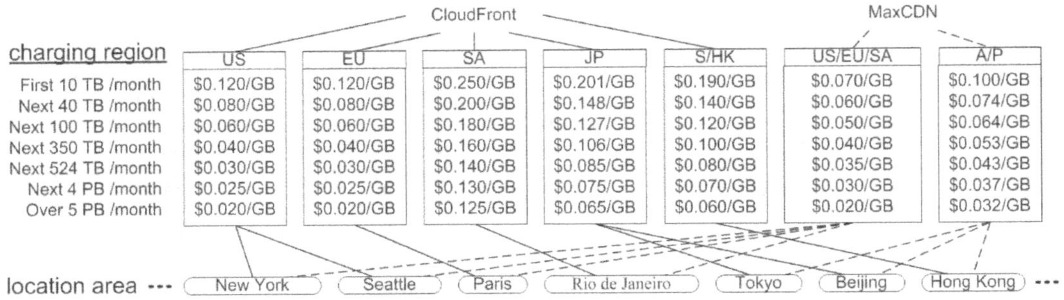

Figure 2: Charging structures of CloudFront and MaxCDN.

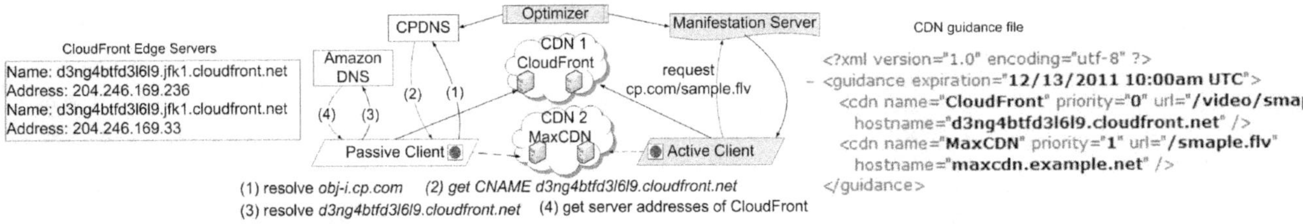

Figure 3: Content multihoming control framework (shaded components include our contributions).

problem in the passive client only setting. An extension to active client only setting is straightforward and follows. Combining them together is also straightforward and hence omitted.

5.1 Passive Client

Since a generic problem formulation has benefits, we define our problem using generic constrained programming. Specifically, we formulate the content multihoming optimization problem as consisting of two objectives:

QoE guarantee: First and foremost, for every object i and location area a, if $n_i^a > 0$, content multihoming should assign one or more CDNs for viewers from location area a requesting object i to achieve a QoE target. Each assigned CDN k should satisfy two requirements: (1) CDN k is providing the required features (*e.g.*, streaming vs download) to deliver object i; (2) $p_{i,k}^a$ exceeds the performance target ($\bar{p}$). Define F_k={i^a: CDN k can provide the features to deliver object i and $p_{i,k}^a \geq \bar{p}$.}. In other words, F_k is the set of location objects that CDN k can serve. Note that it is F_k that allows us to distinguish between streaming vs downloading content, as it depends on object QoE metrics and features (*e.g.*, HTTP-streaming).

Cost optimization: Under the QoE guarantee constraint, content multihoming may balance the load to multiple CDNs, in particular to minimize the total cost. Let $x_{i,k}^a$ denote the fraction of the n_i^a requests that is directed to CDN k. Hence, each $x_{i,k}^a$ is an optimization variable with a valid value range between 0 and 1. We state the problem **Q** as:

$$\begin{aligned}\underset{\{x_{i,k}^a\}}{\text{minimize}} \quad & C(\{x_{i,k}^a\}) = \sum_k \sum_{\alpha_k^r} C_k^r\Big(\sum_{a\in\alpha_k^r} \sum_i x_{i,k}^a n_i^a s_i\Big) \\ \text{subject to} \quad & \forall i, a, n_i^a > 0 : \sum_k x_{i,k}^a = 1, x_{i,k}^a \geq 0; \\ & \forall k, i, a, i^a \notin F_k : x_{i,k}^a = 0.\end{aligned}$$

The first constraint states that each demand for an object i at a location a should be served. The second constraint states the QoE constraint. In other words, if CDN k cannot provide sufficient performance or feature for a location object i^a, no content viewers from location area a for object i should be directed to CDN k. Note that the QoE constraint may lead to infeasibility. Since feasibility can be checked efficiently, we assume feasibility.

One can add additional constraints (*e.g.*, CDN capacity constraints) and consider additional cost factors (*e.g.*, storage, per request). We will first focus on the basic constraints and cost factors. In Section 6.2, we will discuss some extensions. Note that we start with a basic formulation based on longer-term statistics/prediction to select CDNs. In Section 7, we will add active clients for online adaptation to form a more complete 2-level optimization framework. Better traffic predictions (or publisher-knowledge) on n_i^a can be input plugins to the problem formulation.

After computing a solution (*i.e.*, $\{x_{i,k}^a\}$), the Optimizer sends the solution to `CPDNS` in the control framework of Section 4 to implement it. Typically, the computation should be performed each billing cycle (*e.g.*, a month) but can be more frequently.

5.2 Active Client

An active client allows the usage of multiple CDNs to serve the same request. One might think that this will add substantial complexity to the preceding formulation. But as we will see, it is a simple extension of the preceding problem definition.

Without loss of generality, consider that each active client uses two CDNs: one primary and one backup. Consider the following set of CDNs: each CDN is a "virtual CDN" that consists of a pair of CDNs, say $k^{'} = (k, j)$ where k is the primary CDN and j is the backup. Then we will have a similar problem formulation as the preceding formulation in problem **Q**.

First consider the QoE Guarantee. Define $F_{k'}$={i^a: both CDN k and CDN j can provide the features to deliver object i and $p_{i,k}^a \cup p_{i,j}^a \geq \bar{p}$.}, where $\cup$ denotes the joint reliability of the two CDNs.

Next consider the objective function. Assume that each primary CDN k still delivers the same amount of traffic. The backup CDN j incurs an additional traffic $1 - p_{i,k}^a$ fraction of the time. One may verify that $C(\{x_{i,k'}^a\})$ and $C(\{x_{i,k}^a\})$ have similar structure and can be solved with the same method.

After computing a solution(*i.e.*, $\{x_{i,k'}^a\}$), where each $k^{'}$ is a pair of CDNs, the Optimizer sends the solution to manifest file servers to return two CDNs for each active client request.

6. COMPUTING OPTIMIZATION

We now develop techniques to solve the problem defined in the preceding section. Since the problems for the passive clients and active clients have the same format, we use the passive client formulation: Problem **Q**. One might consider solving **Q** using standard LP, but this can be intractable: 500K objects, 200 locations, and 3 CDNs will imply 300M variables, 100M constraints, and to minimize a *concave* function. No generic solver we know can handle this case, so we develop our specific algorithm.

Our strategy is to first transform the problem to a combinatorial assignment problem in Section 6.1. Then, in Section 6.2 we develop a novel, efficient algorithm that computes an optimal assignment without enumerating all of the exponentially many possibilities. We discuss extensions in Section 6.3.

6.1 Location Object Assignment

At a first glance of the problem **Q**, one might think about using convex programming (*e.g.*, [6]) to solve the problem. Unfortunately, the objective function $C(\{x^a_{i,k}\})$, which we target to minimize, is a *concave* function. Hence, traditional, efficient convex programming does not apply.

On the other hand, the concavity of the objective does lead to one observation: there is a minimizer of problem **Q** such that each location object is put into only one CDN. Precisely, we have:

LEMMA 1. *There exists a minimizer of $C(\{x^a_{i,k}\})$ for problem* **Q***, in which for each location object i^a, there exists a CDN k^* such that $i^a \in F_{k^*}$, and $x^a_{i,k^*} = 1$.*

Consider one such solution, and let ψ denote the assignment (or mapping), according to the solution, from each location object i^a to its assigned single CDN k: $\psi(i^a) = k$. Let ψ_k denote the set of location objects that are assigned to CDN k. If $\forall k, \forall i^a \in \psi_k$ we have $i^a \in F_k$, we call ψ a *feasible assignment*.

The above interpretation of the solution allows us to change problem **Q** into an assignment-based formulation, as shown in Figure 4. For each location object, the figure shows the candidate CDNs that the location object can be assigned to. CDN k is a candidate for location object i^a only if $i^a \in F_k$. The problem then is to assign each location object to a single CDN to minimize the cost.

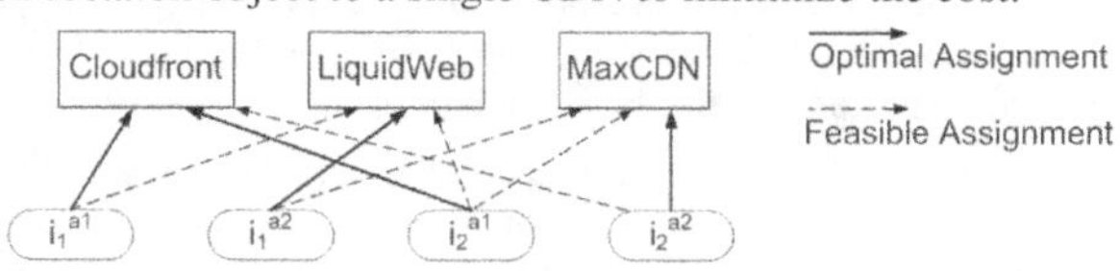

Figure 4: Q can be formulated as an assignment problem.

An advantage of the discrete assignment formulation is that it allows enumeration. A straightforward approach to finding an optimal assignment is to enumerate all possibilities, and select the best one among the feasible assignments. We know that each location object i^a can be assigned to one of the K CDNs. Hence, the total number of assignments is $K^{|A|N}$. For $K = 2$ or 3, $|A|$ on the order of thousands and N hundreds of thousands, direct enumeration is practically infeasible. In other words, the assignment formulation allows enumeration, but naive enumeration does not work.

6.2 Efficient and Optimal Assignment

Our key insight to substantially reduce the complexity is that the naive enumeration of all of the exponential number of assignments is unnecessary. Instead, we need to consider only a polynomial number of assignments.

Basic idea: Specifically, consider the space of all possible assignments illustrated by the space on the left of Figure 5(a), where each assignment is shown as a point. A feasible assignment is shown as an white box while an infeasible one is shown as a black box. Naive enumeration evaluates every assignment, ignores infeasible assignments, and picks a feasible one that gives the best outcome.

Now instead of looking at the space of assignments, we look at the space of the *outcomes* of the assignments, illustrated by the right space in Figure 5(a). Each assignment point in the left space has a corresponding outcome point in the right outcome space. Specifically, the outcome of an assignment ψ is a vector, with each element of the vector representing the charging volume at charging region of CDN. Let V_ψ denote the multi-dimensional charging volume vector representing the outcome of an assignment ψ. We will develop the exact representation shortly.

Since the objective function of our problem **Q** is a concave function of the charging-volume vector, we know from concave optimization theory that we need to evaluate the objective function only over the extremal points of the convex hull of the charging volume vectors produced by feasible assignments. In other words, if the charging volume vector V_ψ resulted from a feasible assignment ψ is not an extremal point, the vector can be expressed as a convex combination of those resulted from some other feasible assignments, and hence there is no need to evaluate ψ. Figure 5(a) illustrates that we need to consider those V_ψ marked with an "x". As we will see, the number of extremal points is polynomial and below we develop our efficient algorithm to identify them.

Representing each location object as a vector: The foundation of our basic idea is based on considering the resulting charging volumes of an assignment ψ as a vector V_ψ. We now introduce a representation of each location object $v = i^a$ as a vector to allow easy aggregation on the outcome of an assignment. This representation is quite simple but involves some notation complexity at the beginning. The benefit of the representation is that it provides essential insight and simplification during our development.

We first introduce charging region intersections. Recall that each CDN k defines a mapping from a location area a to one of its serving charging regions $\alpha^1_k, \cdots, \alpha^{R_k}_k$, where R_k is the number of charging regions of CDN k. Note that $\alpha^1_k, \cdots, \alpha^{R_k}_k$ provides a partition of all location regions A. An intrinsic complexity of multiple CDNs is the heterogeneity of their charging regions. Define the "intersections" of the charging regions of the K CDNs. Let $\alpha^{r_1 r_2 \cdots r_K}$ denote the intersection of the charging regions r_1 of CDN 1, r_2 of CDN 2, and r_K of CDN K. Then a total of $R = R_1 * R_2 \cdots R_K$ intersections are defined. Figure 6(a) illustrates a setting of two CDNs with 3 and 2 charging regions respectively. At most $R = 6$ non-empty intersections may be defined.

With charging region intersections, we can represent each location object as a vector. Specifically, given the set of charging region intersections, one can observe that each location area a belongs to one and only one of the charging region intersections. Fix one ordering of the intersections. Then we can convert the traffic of each location object $v = i^a$ to an R-dimension vector with all elements except one being 0. The position of the non-zero element is the intersection that the location area a belongs to, and the value at the position is $n^a_i s_i$. When it is clear from the context, we use v to either represent the name of a location object i^a or the vector. Figure 6(b) shows the vector representations of two location objects.

By representing each location object $v = i^a$ as a R dimensional vector, we introduce a simple, linear outer-production operator to reflect the effect of assigning v to CDN k. Let e_k be a unit K-dimensional vector whose only non-zero element is at the k-th position and the value at the k-th position is 1. Define $v \otimes e_k$, which reflects the effect of assigning v to CDN k, as producing a $R \times K$ matrix such that v is at the k-th column and the other columns are zero. Figure 6(c) shows four examples, when we assign two location objects to two CDNs. For example, the first example shows $v_1 \otimes e_1$; that is, assigning v_1 to CDN 1.

Given this definition of the outer-product and an assignment ψ, we can calculate the outcome of ψ. Recalling that ψ_k is the set of location objects assigned to CDN k, we have:

$$V_\psi = [\sum_{v \in \psi_1} v, \ldots, \sum_{v \in \psi_K} v] = \sum_v v \otimes e_{\psi(v)} \in \mathbb{R}^{R \times K}. \quad (1)$$

We now go back to identifying extremal assignments.

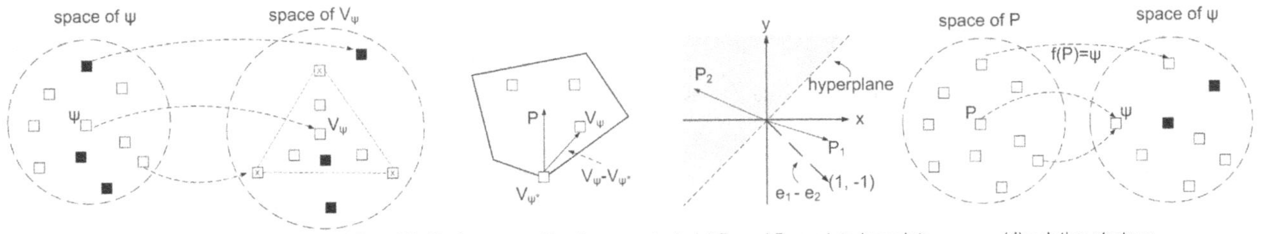

(a) mapping from ψ to V_ψ (b) P and V_ψ-V_{ψ^*} have a positive inner product (c) P_1 and P_2 are interior points (d) solution strategy

Figure 5: An example illustrating the basic idea to solve problem Q.

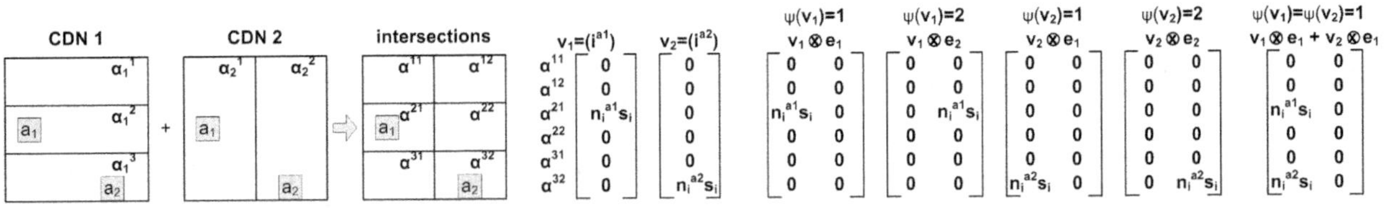

(a) charging region intersections (b) location-object as a vector (c) location-object can choose either CDN 1 or CDN 2 (d) both select CDN 1

Figure 6: An example illustrating the charging-intersections.

Extremal assignments v.s. separation vector: The basis of our technique to identify the extremal points of a set of points is through the fundamental Separation Lemma [14]. Specifically, define $V_\Psi = \{V_\psi\}$, where ψ is a feasible assignment. Then a sufficient and necessary condition for a specific $V_{\psi^*} \in V_\Psi$ to be an extremal point of V_Ψ is that there exists a vector P in the same dimension space whose inner product $\langle P, V_\psi - V_{\psi^*}\rangle > 0$ for any other $V_\psi \in V_\Psi$. In other words, the inner product of P and V_{ψ^*} is smaller than the inner product between P and any other point $V_\psi \in V_\Psi$. For instance, in Figure 5(b) , $\langle P, V_\psi - V_{\psi^*}\rangle > 0, \forall \psi \neq \psi^*$.

Hence, a strategy to identify the set of all extremal points is to compute a set of Ps and from each P we compute an extremal feasible assignment. Denote the computation from P to an assignment as $f(P)$. Figure 5(d) illustrates a set of Ps and a mapping function $f()$. We want the following conditions on the set of Ps and $f()$: (1) *computationally efficient*: both the set of Ps and $f(P)$ are easy to compute; (2) *valid*: each $f(P)$ is an extremal feasible assignment; (3) *exhaustive*: for each extremal feasible assignment ψ^*, there exists one P in the set of Ps, such that $f(P) = \psi^*$. Below, we show a set of Ps and a function $f(P)$ satisfying the preceding conditions.

Function $f(P)$: We start by developing $f(P)$. We have the following CDN Identification Lemma.

LEMMA 2 (CDN IDENTIFICATION LEMMA). *A feasible assignment ψ^* is extremal among the set of all feasible assignments if and only if $\exists P \in \mathbb{R}^{R\times K}$ such that $\forall v, k$, $k \neq \psi^*(v) \wedge v \in F_k$: $\langle P, v \otimes e_{\psi^*(v)}\rangle < \langle P, v \otimes e_k\rangle$.*

PROOF. The right to the left is straightforward, as the right condition is stronger than the Separation Lemma condition, when one applies Expression (1). We prove the left to right. Given the Separation Lemma, we know that $\exists P \in \mathbb{R}^{R\times K}$ such that $\langle P, V_{\psi^*}\rangle < \langle P, V_\psi\rangle$, for any feasible $\psi \neq \psi^*$. Consider any v and k such that $v \in F_k$ and $k \neq \psi^*(v)$, let ψ' be the assignment obtained from ψ^* by moving v to CDN k; that is, $\psi'(v) = k$ and $\psi'(u) = \psi^*(u)$, $\forall u \neq v$. Since ψ' is also a feasible assignment, it follows that $\langle P, V_{\psi^*}\rangle < \langle P, V_{\psi'}\rangle$, hence $\langle P, v \otimes e_{\psi^*(v)}\rangle - \langle P, v \otimes e_k\rangle = \langle P, V_{\psi^*}\rangle - \langle P, V_{\psi'}\rangle < 0$. □

We name the lemma CDN Identification Lemma because it is the foundation to develop $f(P)$. Given a P in the lemma, we can compute a corresponding extremal assignment ψ efficiently: for each location object $v = i^a$, iterate among all feasible CDNs k for the object (*i.e.*, $i^a \in F_k$), we compute $\langle P, v \otimes e_k\rangle$. We assign v to the (unique) CDN k attaining the minimal value: $\psi(v) = k$.

Set of P**s**: We consider the following set of Ps: a P satisfies that all of the elements in $\{\langle P, v \otimes e_k\rangle | \forall k : v \in F_k\}$ are distinct:

$$\{P : \forall k, j, v, k \neq j \wedge v \in F_k \cap F_j : \langle P, v \otimes (e_k - e_j)\rangle \neq 0\}. \quad (2)$$

Since the conditions are stronger than those from the CDN Identification Lemma, we know that each such P can compute an extremal assignment.

Geometrically, the condition that P satisfies $\langle P, v \otimes (e_k - e_j)\rangle \neq 0$ is equivalent to that P is not on the hyperplane that is orthogonal to $v \otimes (e_k - e_j)$. Denote this hyperplane as $[v \otimes (e_k - e_j)]^\perp$ and let $\mathbb{H}$ be the set of all these hyperplanes. Hence, a P satisfying all conditions in (2) is not on any of the hyperplanes in $\mathbb{H}$. In other words, P should be an interior point in a cell created with hyperplanes in $\mathbb{H}$ as boundaries. Efficient algorithms (*e.g.*, [26]) exist to enumerate one interior point from each cell.

Exhaustiveness: From the preceding development, it should be clear that we have developed a set of Ps and the function f which is computationally efficient, and $f(P)$ is valid. The only remaining issue is whether we satisfy the exhaustive requirement by enumerating an arbitrary interior P from each cell. First, we have

PROPOSITION 1. *If ψ^* is an extremal feasible assignment, $\exists P$ making $f(P) = \psi^*$ and P is an interior point of a cell in $\mathbb{H}$.*

PROOF. Suppose we find an extremal assignment ψ^* from a point Q that is not an interior point of any cell in $\mathbb{H}$. Our method is to construct another point Q', from Q, that is indeed an interior point of some cell in $\mathbb{H}$ and that $f(Q') = \psi^*$.

Since Q is not interior, $\exists v_0, k_0 \neq j_0$ such that $\langle Q, v_0 \otimes (e_{k_0} - e_{j_0})\rangle = 0$. Consider $P(\epsilon) = Q + \epsilon \cdot v_0 \otimes (e_{k_0} - e_{j_0})$ where $\epsilon \in \mathbb{R}^1$. Therefore $\forall v, k \neq j$, $\langle P(\epsilon), v \otimes (e_k - e_j)\rangle = \langle Q, v \otimes (e_k - e_j)\rangle + \epsilon \cdot \langle v_0 \otimes (e_{k_0} - e_{j_0}), v \otimes (e_k - e_j)\rangle$ is a continuous function of ϵ. When $|\epsilon'| > 0$ is small enough: (1) For $\forall v \neq v_0, k \neq k_0$ or $j \neq j_0$, $\langle P(\epsilon'), v \otimes (e_k - e_j)\rangle \neq 0$ and has the same sign ('+' or '-') with $\langle Q, v \otimes (e_k - e_j)\rangle$; (2)$\langle P(\epsilon'), v_0 \otimes (e_{k_0} - e_{j_0})\rangle = \epsilon' \cdot \|v_0 \otimes (e_{k_0} - e_{j_0})\|^2 \neq 0$.

This means that $P(\epsilon')$ is on one less hyperplanes in $\mathbb{H}$ than Q. Moreover, since ψ^* is extremal, we have $\langle Q, v \otimes (e_{\psi^*(v)} - e_k)\rangle < 0$ $\forall v, k, k \neq \psi^*(v) \wedge v \in F_k$, so it follows from (1) that $\langle P(\epsilon'), v \otimes (e_{\psi^*(v)} - e_k)\rangle < 0$ as well, hence $f(P(\epsilon')) = \psi^*$ by Lemma 2. This process can be repeated to yield a Q' that is not on any hyperplanes in $\mathbb{H}$ and that $f(Q') = \psi^*$. □

A potential issue is that the interior P from the preceding lemma may not be the one that our algorithm uses. However, we have the following result, and establish the exhaustiveness of our approach.

LEMMA 3. *Interior points from the same cell find the same extremal feasible assignment.*

PROOF. Let P_1 and P_2 are two interior points from the same cell. Suppose their corresponding extremal assignments $\psi_1^* \neq \psi_2^*$, then $\exists v_0$ which has $\psi_1^*(v_0) \neq \psi_2^*(v_0)$. According to Lemma 2:

$$\langle P_1, v_0 \otimes (e_{\psi_1^*(v_0)} - e_{\psi_2^*(v_0)}) \rangle < 0$$
$$\langle P_2, v_0 \otimes (e_{\psi_1^*(v_0)} - e_{\psi_2^*(v_0)}) \rangle > 0$$

which contradicts with that P_1 and P_2 are from the same cell. □

Redundancy elimination: On the surface, we need to enumerate all cells created by a total of $|A|NK(K-1)$ hyperplanes, where $|A|N$ is due to the number of possibilities for v, which is the number of location objects, and $K(K-1)$ is due to the number of possibilities for $(e_k - e_j)$. However, some of the $|A|NK(K-1)$ hyperplanes are redundant. Specifically, if one vector is just the scaling of another vector, they define the same hyperplane. For example, $v \otimes (e_k - e_j)$ and $v \otimes (e_j - e_k)$ give the same hyperplane. Hence, we need to consider only distinct pairs of k and j. Also, consider two location objects $v_1 = i_1^{a_1}$ and $v_2 = i_2^{a_2}$. If their vector representations satisfy $v_1 = \lambda v_2$, where λ is a scalar, then they define the same hyperplane, for each pair of k and j. In other words, all location objects mapped to the same charging region intersection define the same hyperplane for each pair of k and j. Hence, the number of unique hyperplanes is at most $R\binom{K}{2}$, which is independent of the number of content objects.

Algorithm 1: CMO(V, $\{F_k\}$)

```
Input: V: location objects to be assigned.
Input: {F_k}: K CDNs and their feasibility sets.
Output: optAs: optimal assignment
1  /* Step 1: Identify hyperplanes */
2  HPs ← ∅ ;
3  foreach v ∈ V do
4      vVec = vAsVector(v) ;
5      foreach distinct (k, j) pairs do
6          if (v ∈ F_k ∧ v ∈ F_j) then
7              hpCandidate = normalize([vVec ⊗ (e_k − e_j)]^⊥) ;
8              if (hpCandidate ∉ HPs) then
9                  HPs += hpCandidate
10 /* Step 2: Compute interior points from hyperplanes */
11 Ps ← computePs(HPs);
12 /* Step 3: Evaluate extremal assignments identified by Ps */
13 optAs ← null;
14 foreach P ∈ Ps do
15     /* compute extremal assignments ψ identified by P */
16     ψ ← null ;
17     foreach v ∈ V do
18         optOuter ← +∞;
19         foreach CDN k do
20             if v ∈ F_k ∧ ⟨P, v ⊗ e_k⟩ < optOuter then
21                 ψ(v) ← k ;
22                 optOuter ← ⟨P, v ⊗ e_k⟩
23     /* compare new extremal ψ with current optAs */
24     if (ψ > optAs) then
25         optAs ← ψ
26 return optAs;
```

Algorithm: For completeness, we specify the content multihoming optimization (CMO) algorithm in Algorithm 1. There are many ways implementing `computePs()`, and we choose [26] as it can be easily parallelized for multiple CPU cores and computers.

Example: To help readers understand the CMO algorithm, we apply it to the simplest setting of two CDNs and both use one global charging region. This is a setting where one can solve problem **Q** using intuition. Specifically, in this setting, we can divide the location objects into 3 categories: V_1: the location objects that can be assigned to only CDN 1; V_2: the location objects that can be assigned to only CDN 2; and V_3: the location objects that can be assigned to either CDN 1 and CDN 2. Then the only remaining issue is to determine the assignments of objects in V_3. One can verify that the correct strategy is that we compare the objective function values of two alternatives: (1) assigning all objects in V_3 to CDN 1, with objects in V_1 and V_2 preassigned to their respective CDNs; and (2) assigning all objects in V_3 to CDN 2, with objects in V_1 and V_2 preassigned to their respective CDNs.

Now, we see how CMO works. In Step 1 (Lines 2 to 9), the algorithm computes that there is only one hyperplane defined by $[1, -1]^{\perp}$. In Step 2, `computePs` computes two interior Ps P_1 and P_2, where P_1 is a vector in the lower right half-space (x-coordinate is larger than the y-coordinate) of Figure 5(c), and P_2 is one in the upper left half-space (x-coordinate is smaller than the y-coordinate). In Step 3, the algorithm first evaluates P=P_1 to compute an extremal assignment. For each location object v=i^a, if it is in V_1 or V_2, the algorithm assigns it to the only feasible CDN. If $v \in V_3$, the value of $\langle P_1, v \otimes e_1 \rangle$ is the x-coordinate of P_1 times the traffic volume of object v, and the value of $\langle P_1, v \otimes e_2 \rangle$ is the y-coordinate of P_1 times the traffic volume of v. Since P_1 is chosen in Step 2 such that the x-coordinate is larger than the y-coordinate, P_1 produces the extremal assignment of assigning all objects in V_3 to CDN 2. The algorithm next evaluates P=P_2 and produces the extremal assignment of assigning all objects in V_3 to CDN 1. At Line 24, the algorithm compares the two cases and picks the better one. Hence, it produces the intuitive result. For general settings that we can no longer appeal to intuition, the algorithm computes the optimal assignment efficiently.

6.3 Extensions

The CMO algorithm developed in the preceding section applies to concave charging functions or charging functions that can be converted to or approximated by concave charging functions. In this section, we discuss extensions to handle practical issues on CDN subscription levels, per-request costs and dynamic streaming.

CDN subscription levels: For some CDNs, a content publisher must subscribe to a usage level (*e.g.*, maximum traffic volume) and pay a fix fee to the subscription level. To handle such a charging model, we treat each subscription level as an individual CDN with a capacity constraint. Note that the basic CMO algorithm needs to be slightly extended to handle capacity constraints [18].

Per-request cost: Besides charging for traffic, some CDNs also include charges for the number of requests. For instance, CloudFront charges \$0.0075 per 10,000 HTTP requests in US. Consider that CloudFront charges \$0.12/GB for the first 10 TB traffic as shown in Figure 2. One can calculate that if the object sizes are less than 6.25 KB, then the per-request charge can be higher than the traffic charge. Hence, the per-request charge can be the major cost for content publishers providing small objects (*e.g.*, small images).

Extending Algorithm 1 to consider both traffic and per-request charge is relatively straightforward. Specifically, in the preceding section, each location object is represented as a R-dimension vector with one non-zero element at the charging region of the object. An extension to include per-request charge is to represent each location object as a $R+1$ dimension vector, with the one added dimension representing the number of requests for the object.

CDN	Traffic Charge	Per-request Charge
CDN 1	\$1.0 per GB	\$0.0 per request
CDN 2	\$0.1 per GB	\$0.1 per request

Object	Size	#Request	Traffic	Vector Representation
v_1	0.01 GB	30	0.3 GB	[0.30 GB, 30]
v_2	0.01 GB	49	0.49 GB	[0.49 GB, 49]
v_3	0.025 GB	20	0.5 GB	[0.50 GB, 20]
v_4	1.0 GB	1	1.0 GB	[1.00 GB, 1]

Table 3: Example of CMO with per-request cost extension.

We demonstrate the extension using an example setting (Table 3): 4 objects, 2 CDNs with one charging region (*i.e.*, $R = 1$):

- First the 4 objects are represented as 4 2D vectors (see Table 3).
- Lines 3 to 9 construct four corresponding hyperplanes: h_1 = $[0.3,-0.3,30,-30]^\perp$, h_2 = $[0.49,-0.49,49,-49]^\perp$, h_3 = $[0.5,-0.5,20,-20]^\perp$, h_4 = $[1,-1,1,-1]^\perp$. After normalization and de-duplication (Lines 7, 8), only three are left: $[1,-1,100,-100]^\perp$, $[1,-1,40,-40]^\perp$, $[1,-1,1,-1]^\perp$; *i.e.*, first two are redundant and only 1 is needed.
- Line 11 finds 6 interior points: P_1 = [1,0,1,0], P_2 = [-1,0,-1,0], P_3 = [70,0,-1,0], P_4 = [-70,0,1,0] , P_5 = [20,0,-1,0], P_6 = [-20,0,1,0].
- Lines 14 to 22 find 6 extremal object assignments (ψ :={CDN1 objects}{CDN2 objects}): $\psi_1 = \{\}\{v_1, v_2, v_3, v_4\}$, $\psi_2 = \{v_1, v_2, v_3, v_4\}\{\}$, $\psi_3 = \{v_1, v_2\}\ \{v_3, v_4\}$, $\psi_4 = \{v_3, v_4\}\ \{v_1, v_2\}$, $\psi_5 = \{v_1, v_2, v_3\}\ \{v_4\}$, $\psi_6 = \{v_4\}\ \{v_1, v_2, v_3\}$.
- Lines 23 to 25 enumerate the costs of the 6 extremal object assignments, and identify optimal assignment ψ_5. As a comparison, simple enumerations need to consider 16 assignment possibilities (4 objects each with 2 possible assignments).

Multiple streaming rates: A content publisher can encode the same video object at multiple rates, and a client can switch among the encoding rates dynamically online. To extend CMO for this setting, one can consider each video at each encoding rate as an independent object and then derive the number of requests to each such object. Specifically, suppose a total of n_i^a requests for video i from location area a when there is no multi-rate. Consider two encoding rates, say 1x and 2x. Then according to client access bandwidth distribution at location area a and the publisher dynamic streaming algorithm, one can derive $n_{i_1}^a$ and $n_{i_2}^a$ for the two encoding rates from n_i^a [18].

7. ACTIVE CLIENTS

An active client is provided with a list of CDNs to use when requesting a single content object. The list may come from the result of our optimization algorithm in the preceding section or another source of guidance. Even though our optimization algorithm considers performance constraints, the filtering is based on long-term statistics. Hence, the objective of the active client is to adapt to specific real-time CDN performance dynamics, in particular, to improve QoE during CDN server failures.

7.1 Adaptation Problem Statement

An active client receives guidance in the format of a list of CDNs where each CDN has a priority value. Without loss of generality, we consider two CDNs: the first *primary* CDN, and the second *backup* CDN. For example, the example in Figure 3 gives an active client two CDNs with priority 0 (primary) and priority 1 (backup) respectively. We assume that each individual CDN on the list provides a small number (say, 1 or 2) preferred edge servers through its request routing mechanism (*e.g.*, DNS resolution or HTTP redirect). Let H denote the set of all candidate edge servers. As an example, if the primary CDN provides two servers (h_{11} and h_{12}) and the backup CDN provides one server (h_{21}), then $H = \{h_{11}, h_{12}, h_{21}\}$. Each server in H has a priority value, which is the priority value of its CDN. At time t, an active client maintains a subset H_{active} of H under the following guidelines:

- *QoE protection (feasibility)*: the combined available bandwidth of the servers in H_{active} can provide the target QoE;
- *Prioritized guidance*: the available bandwidth of a higher priority server should be used before that of a lower priority server;
- *Low session overhead*: the overhead of redistributing load among same-priority servers is unnecessary unless it reduces the number of servers used concurrently to reduce the number of connections overhead.

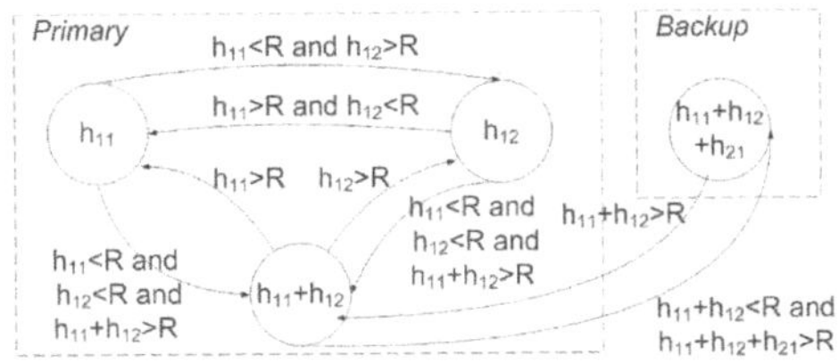

Figure 7: An active-client control diagram.

Based on the preceding guidelines, we derive a simple control state transition diagram as the control objective of an active client. Figure 7 shows an example control diagram where each node represents the current subset of edge servers that the active client uses. The client starts in the state $\{h_{11}\}$, indicating that the client starts from downloading from h_{11}.

Note that one can introduce other transitions using our control state diagram based approach.

7.2 Adaptation Algorithm

We implement the control diagram using the classic window-based AIMD (Additive Increase Multiplicative Decrease) scheme, based on a key observation that there is an analogy between the traditional congestion control and active client adaptation. In particular, if we consider the flow from each server to the client as a link, then we are solving a rate allocation problem among the links, where two essential mechanisms are needed: (1) the rate on a link should be reduced if the link is overloaded; and (2) a probing scheme is needed to utilize newly available capacity.

Specifically, for each sever h, the client maintains a *request window* size w_h (KB), which is the current upper bound on the outstanding request load from the client to h per T seconds, where T is a configuration parameter. For simplicity, we refer to each KByte in the request window as a *piece*. We use the classic AIMD algorithm as a base to adjust the window sizes: if h and the network from h to the client has enough capacity to finish the requested pieces, w_h is linearly increased; otherwise, w_h is multiplicatively decreased. Our client uses HTTP range request to query a set of pieces in one HTTP request message. After the range request is finished or a time out happens after T seconds, an `onResponse` event handler is triggered. We implement the AIMD algorithm in `onResponse` listed as Algorithm 2.

Algorithm 2: onResponse(h, reqPieceSet$_h$, rcvdPieceSet$_h$)

```
if rcvdPieceSet_h = reqPieceSet_h and w_h = reqPieceSet_h.size() then
    w_h ← w_h + 1;
if rcvdPieceSet_h ⊂ reqPieceSet_h then
    w_h ← max(1, w_h/2);
/* Update piece status */
foreach piece i ∈ reqPieceSet_h do
    if piece i ∈ rcvdPieceSet_h then
        // extract piece i's data, add to video buffer;
        pieceStatusMap[i] ← DOWNLOADED;
    else
        pieceStatusMap[i] ← NOT_ASSIGNED;
avail_h ← true; /* mark h avaiable for new assignment */
updateAssignment();
```

However, our problem is also different from the traditional congestion control problem, and hence we need to introduce two novel and interesting techniques.

Total workload control: Naive usage of traditional AIMD will imply that all servers can be fully utilized to achieve a download rate that is as high as possible. However, this can be unnecessary for the content viewer. Specifically, to achieve QoE of streaming content, the client only needs to sustain a sufficient downloading rate (*e.g.*, the video encoding rate ω KB/s).

Based on this observation, we apply *total workload control* to appropriately limit the usage of the CDN servers. In our design, every T seconds, a client calls `releaseLoad()`, shown in Algorithm 3, to release $\omega \cdot T$ new pieces to be downloaded. Note the client may download more than $\omega \cdot T$ pieces in T seconds if there are incomplete pieces in previous downloading periods.

Algorithm 3: releaseLoad()

```
/* reqLimit: the "newest" piece to be assigned for downloading */
reqLimit ← reqLimit + ω · T;
// set pieceStatusMap of new pieces NOT_ASSIGNED
updateAssignment();
```

Prioritized assignment: Total workload control does not yet achieve our adaptation goal. In particular, the load may still spread unnecessarily to too many servers. For example, a single primary server h_{11} has enough capacity to serve the client, but a backup server h_{21} may be also used unnecessarily for downloading, if h_{21} has an equal opportunity to fetch assigned pieces.

Our solution to this issue is *prioritized assignment*. Specifically, when assigning pieces to be downloaded, a client starts with servers with higher ranks. Algorithm 4 gives more details on how pieces to be downloaded are assigned to servers.

Algorithm 4: updateAssignment()

```
// sort serverList by rank_h (retrieved from manifest xml)
// and then w_h if same rank;
for h ← serverList[0] ... serverList[serverList.size() - 1] do
    /* avail_h is h's availability for assignment, see Algorithm 2 */
    if avail_h then
        /* request from h for up to w_h pieces */
        reqPieceSet_h ← ∅;
        /* playPoint: the "oldest" piece being consumed by player */
        i ← playPoint + 1;
        while i < reqLimit and reqPieceSet_h.size() < w_h do
            if pieceStatusMap[i] = NOT_ASSIGNED then
                reqPieceSet_h ← reqPieceSet_h ∪ {piece i};
                pieceStatusMap[i] = ASSIGNED;
            i++;
        if reqPieceSet_h != ∅ then
            avail_h ← false;
            asyncHTTPRequest(h, reqPieceSet_h, timeout(T));
```

Comparison with TCP: Our window-based downloading adaptation algorithm is different from traditional TCP congestion control. (1) To maintain client QoE, it requires (at application-level, during T seconds) that the total downloading rate across all servers to be at least video encoding rate. In particular, the sum of window sizes should satisfy $\sum_h w_h \geq \omega \cdot T$. The adaptation algorithm enforces this by setting initial window size for a primary server (*e.g.*, w_{h11}) to be $\omega \cdot T$, and for each backup server to be 1. (2) Different from TCP's per-segment window update, we apply AI on the window size after all requests of the entire window are completed. In steady state (client streaming smoothly), this allows the client to slowly probe higher-ranked servers. Upon primary server failure or congestion, the AI strategy increases the backup server's window size; due to self clocking and before reaching the streaming rate, the increase behavior is similar to TCP slow start, which is fast to allow request queue cleanup. (3) Although the algorithm maintains a window size for each server, it does not open a (TCP) connection to a lower-ranked server until necessary. Also, when the higher-ranked servers have sufficient capacity, the adaptation algorithm disconnects the lower-ranked servers.

Active multi-rate streaming: Just as CMO can be extended to handle multi-rate streaming, the preceding active clients can also be extended to take advantage of multi-rate. Specifically, consider a video encoded in a set of rates, say 1x, 2x, and 4x. As we derive Algorithm 3 and 4 from Figure 7, after a publisher defines its preference on video rates and the extended control diagram, we can similarly derive a control algorithm. As a simple example, a content publisher with a control goal of providing the highest video rate whenever possible will lead to a transition diagram that more servers are used until combined capacity is above 4x, if possible.

8. EVALUATIONS

In this section, we evaluate the cost and performance of our system design for content multihoming. In specific, we implement and test our optimization algorithm (CMO), the client adaptation algorithm, and the interactions between the two system components. We use real data to drive the run of our optimizer, and instrument clients on PlanetLab to conduct experiments.

8.1 Evaluation Methodology

Content publishers: We evaluate our algorithms using real traces of content requests to two production Video-on-Demand (VoD) publishers. We name the content publishers CP1 and CP2 respectively. We consider each video as a content object, and collect the following information about each video: its size, and the number of times that it is requested from each city per month for a 6-month duration. Table 4 shows the summary statistics of the content objects. Figure 8 plots detailed statistical distributions of object sizes (s_i), number of requests (n_i) to each object, and traffic volume ($s_i n_i$) of each object. These distributions are long-tail distributions.

We use the MaxMind GeoIP database to map a client IP address in the trace to a location area. Our evaluations define location areas as following: we start with each country as a location area; for a country with a large geographical span, we refine it to a next level; for example, we define each state in USA as a location area.

	# Objects	Sum of Obj Size	Total Traffic	#Request
CP1	667,856	71 TB	27,307 TB	390,235,440
CP2	529,411	40 TB	12,114 TB	153,129,348

Table 4: Summary statistics of content objects.

Content distribution networks: Our evaluation is based on three commercial CDNs: Amazon CloudFront, MaxCDN, and an anonymous private CDN which we refer to as CDN3. The geographic footprints of CloudFront and MaxCDN are shown in Figure 1, and the real charging structures and parameters of CloudFront and MaxCDN are shown in Figure 2. The server distribution and detailed price information for CDN3 are not shown due to privacy. Moreover, in our evaluations, we require that $p^a_{i,k} \geq 90\%$ for each CDN k and each location object i^a.

To obtain how each CDN maps a location area to its charging region, we deploy a measurement client on each one of 536 available PlanetLab machines to request objects from each CDN. We use `traceroute` to determine the locations of the CDN servers, as the GeoIP database can be inaccurate, *e.g.*, all CloudFront servers are always classified as in Seattle, WA, US. After computing the charging region intersections of the three CDNs, we pick the top 5 intersections that contain the most traffic. Table 5 shows the percentages of traffic to major geographical regions.

	US	EU	SA	Asia & Pacific	Japan
CP1	19 %	7 %	1 %	71 %	2 %
CP2	77 %	11 %	6 %	5 %	1 %

Table 5: Traffic distribution across major geo regions.

Optimizer: Our publishing optimizer is simple to implement ($\sim$2000 lines of C++ code) and runs on a commodity PC with 2 quad-core Intel Xeon 2.33 GHz CPUs and 3 GB of memory. It takes about 12 minutes for CMO to compute the optimal assignments for each charging period (one month) for our data set.

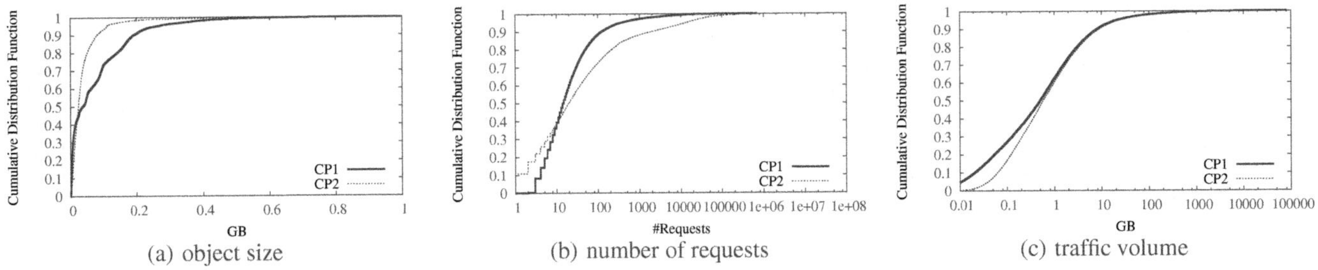

Figure 8: Statistics of object size, number of requests to each object, and traffic volume for each object.

Active clients and controller: We integrate our client adaptation algorithm into an Adobe Flash video player. Our player is deployed on CDN3 and can be accessed by any web browser on the Internet, including PlanetLab nodes. We leverage HTTP `range request` and Adobe Flash's `NetStream.appendBytes` (supported by version 10.1 and above) to integrate multiple CDN servers for one video streaming session. Our player periodically reports back through HTTP to a publisher controller. The reporting process allows the controller to obtain player IP, player local DNS server address, and the IP addresses of each CDN serving the player's content request. An implementation issue of the player is that the Flash browser sandbox does not allow UDP DNS queries. Hence, although a CDN typically resolves a DNS name to multiple server IPs, the player can have access to only one. We use two approaches to obtain backup server IPs for a CDN. The first is that the controller returns, in the manifiest file, additional server IPs reported recently by other closeby (same location area) players. The second is that the player makes a TCP connection to its LDNS or CDN DNS server (if allowed by Flash Socket Policy) to query DNS directly.

We install Mozilla Firefox with Adobe Flash Player (using Xvnc as XServer) on 412 PlanetLab nodes and some personal laptops with ADSL and WiFi, and instrument these clients to conduct video streaming experiments. Note that although most of our clients are PlanetLab nodes which typically have good network connectivity rather than real clients, we control the settings to reflect two key factors that contribute to real client QoE degradation: server overload and network bottlenecks. Both factors are included in our stress tests to impose challenging, oscillating limits on bandwidth.

Specifically, based on our traces, we deploy active clients on PlanetLab according to their availability and geographical locations. We select random videos from CP1 and generate active client request load according to video request patterns of CP1.

Content deployment: We deploy video objects testing active clients on both CloudFront and CDN3. CloudFront has the best performance according to our PlanetLab measurement (see Table 2). We are able to run Adobe Flash Socket Policy service on CDN3, and hence active clients can use customized and optimized TCP sockets for requests to this private CDN. We also run multiple pre-tests (before the evaluation) to warm up both CDNs, *i.e.*, the edge servers prefetched the video content before the experiments start.

CDN server capacity failure models: We evaluate the effectiveness of our client adaptation algorithm under both controlled stress tests and real server congestions. In the real experiments using two CDNs and PlanetLab clients, we do *not* inject any failures, but the CDN servers can get temporarily congested due to the bursty nature of client request load.

Performance metrics: We evaluate both the *CDN cost* and *client QoE* of our content multihoming optimization. CDN cost is simply the total charge (in USD) by all CDNs given a CDN assignment, *i.e.*, the value of function $C(\{x^a_{i,k}\})$ defined in problem **Q** in Section 5.1. For client QoE, we use three performance metrics: (1) *freezes*, the frequency (number of times per view) a viewer encounters rebuffering during a video view, excluding cases due to initial start or user drag (seek). As observed in [10], freezes are a major factor reducing viewer's QoE. (2) *smoothness ratio*, the percentage of the clients that never encounter freeze. This is a statistical performance metric. (3) *buffering time*, viewer visible buffering time (in seconds) per video view, including startup delay and seek delay.

CDN assignment algorithms: We evaluate 5 CDN assignment algorithms: (1) *CMO*: This is the CMO algorithm defined in Section 6.2 considering both traffic and per-request cost defined in Section 6.3; (2) *greedy*: This algorithm assigns location objects sequentially in a uniformly random order. When assigning the next object, the algorithm computes the cost to be reached when the object is put in each feasible CDN. The object is assigned to the CDN resulting the lowest cost among the alternatives; (3) *round-robin*: This algorithm also assigns location objects sequentially in a uniformly random order. A CDN index is maintained. When assigning the next object, the algorithm uses round-robin, starting from the current CDN index, to assign the object to a feasible CDN; (4) *cost-only*: This algorithm minimizes cost without any performance considerations. (5) *perf-only*: This algorithm selects the best performing CDN(s) at each location area regardless the cost.

8.2 Publishing Cost Optimization

We start by evaluating the CDN cost savings of our CMO algorithm. At the beginning of a month, for each location area a, we use the content traffic to a in last month as the traffic prediction in this month; we leverage the algorithms listed in Section 8.1 to decide how to redirect requests from various locations to the three CDNs; we then use the real traffic in the month to calculate the total monthly cost of CP1 and CP2.

We first compare the monthly costs to the two content publishers using CMO *vs* those using the other 4 algorithms. Figures 9(a) and 9(d) show the results. We observe that CMO saves around 30% $\sim$ 40% each month for both CP1 and CP2, compared with all algorithms that should satisfy real performance constraints (perf-only, round-robin, and greedy). For CP2, CMO's cost is close to cost-only because (1) the traffic of CP2 is mostly in US/EU, and (2) all 3 CDNs have good performance in US/EU and hence each can be used. Note that CMO achieves savings despite traffic fluctuations across billing periods. Figures 9(b) and 9(e) show the traffic demands from different regions in different months. From Figure 9(b), we see that the demand from US changes from 4000 TB to 3000 TB in the first 2 months. Despite such fluctuations and hence prediction variations, CMO achieves substantial cost savings.

Next we look at some details on how CMO assigns traffic among CDNs. Figure 9(c) shows the traffic assignments for CP1, during the 2nd month. As comparisons, we also show results for two simple extreme strategies: perf-only and cost-only. We observe that cost-only assigns all traffic to MaxCDN, but as we have seen in Table 2, this can lead to performance violations at locations such as China and Brazil. The perf-only strategy assigns a large fraction of traffic to CDN3, whose performance is the best in AP and JP. However, the cost of perf-only can be much higher. Specifically, we can see from Figure 9(f) that the cost of perf-only for CP1 during month 2 is 1.84 times that of CMO. CMO assigns most traffic to

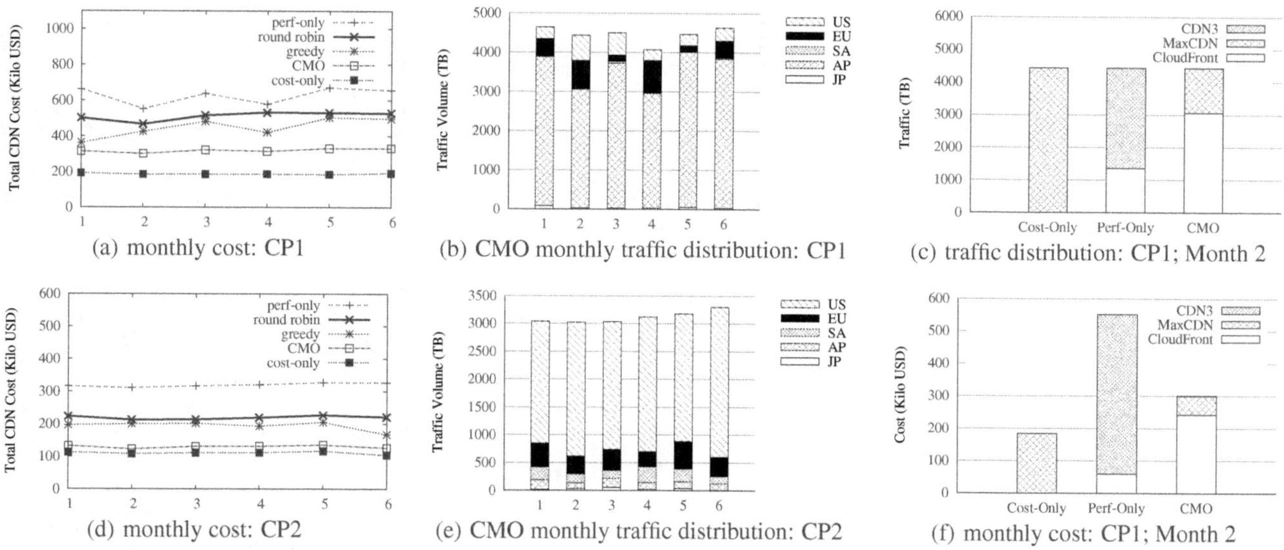

(a) monthly cost: CP1 (b) CMO monthly traffic distribution: CP1 (c) traffic distribution: CP1; Month 2

(d) monthly cost: CP2 (e) CMO monthly traffic distribution: CP2 (f) monthly cost: CP1; Month 2

Figure 9: Cost and traffic distributions in the 6 months with different CDN assignment algorithms.

CloudFront and MaxCDN, as they satisfy performance constraints at lower costs.

8.3 Active Clients

Passive clients cannot handle the failures or congestions of the primary CDN, and hence may encounter QoE degradation at times. We demonstrate that active clients can protect per-view QoE despite CDN server failures/congestions.

Stress tests: We start with stress tests when delivering a 1080p HD video object encoded at 480 KB/s. We run two sets of experiments: (1) only one CDN (primary), which has two servers named `primary1` and `primary2`; (2) two CDNs (one primary+one backup), each with one server, named `primary1` and `backup1` respectively. In each set, we vary the capacity of `primary1` in the following three cases: (1) step-down, in which the capacity of `primary1` is reduced down to only 10% of video encoding rate and then recovers after 2 min; (2) ramp-down: in which the capacity of `primary1` is linearly decreased (to 10% of the video encoding rate) in one minute and then linearly increased back; (3) oscillation, in which the capacity of `primary1` periodically falls down (to 10% of the encoding rate) and then recovers after 20 seconds. We plot detailed downloading rates to observe more behaviors. Figure 10 plots the results.

We make multiple observations on client behaviors. First, in all 6 cases, the client downloads at full speed at the beginning to build the required 4-second video buffer before playback can start. After playback starts, the active client continues to maintain a 16-second video buffer (total workload control).

Second, despite of `primary1`'s fluctuations, our active client achieves QoE protection by downloading from alternative servers. In (b) and (e), despite gradual `primary1` capacity changes, the aggregated downloading rate (labeled `total`) never falls below the streaming rate at any instance of time. In the other 4 cases, there are instantaneous dips when the total rate drops below the streaming rate. However, instantaneous dips may not lead to viewer visible freezes if the streaming buffer has enough data. In all 6 cases, the rebuffering wheel never appears during the entire video playback.

Third, the active client prefers `primary1` over `backup1`, following prioritized guidance. For example, in (d), `primary1` recovers at 250s and its utilization starts to increase, and at time 280s all requests have shifted from `backup1` back to `primary1`. One can also observe this "shifting-back" in (e) at time 160s-210s.

Fourth, the active client achieves "stickiness" for low session overhead. For example, comparing (a)(b) with (d)(e), we observe that in (a) and (b), there is no shifting-back to `primary1`, as `primary2` can handle the load alone and belongs to the same CDN as `primary1`.

Fifth, the active client performs relatively the worst in (f), when there is a *single* primary CDN server and the capacity of the server fluctuates widely. We observe multiple downloading spikes, as the client recovers from low rates resulted from HTTP request timeouts. The fluctuations of the client downloading rate reflects the primary server's capacity fluctuations. In practical deployment, it is recommended that a content publisher uses a primary CDN which offers *multiple* edge servers.

PlanetLab experiments: We next evaluate the statistical performance for both passive and active clients with real PlanetLab experiments. First, we measure smoothness. Figure 11(a) shows that with passive clients, 8% clients experience video freeze. Thus, the smoothness ratio is 92%, which is considered as high performance in industry. Active clients improve viewer QoE and reduce the percentage of clients observing video freeze to only 3.8%. Thus, active clients reduce QoE degradation by 51%. We also calculate that active clients reduce the average number of video freezes from 4.78 (for passive clients) to 2.19. Second, Figure 11(b) shows the buffering time performance. Active clients reduce the average buffering time from 9.6 seconds to less than half at 4.5 seconds.

Cost impact of active clients: Active clients might increase publisher cost for two reasons. First, the optimization algorithm considers the impact of active clients by predicting the amount of traffic shifted to backup. Reality may be different from the prediction. Second, for simplicity, during evaluations, our optimization algorithm does not consider the additional number of requests due to backup protection.

We evaluate the cost impact of client adaptation by comparing the computed optimal cost, the real cost during our PlanetLab experiments (after scaling up the traffic), and the cost of using round-robin CDN assignment. To better understand this impact, we further break down the cost into "traffic cost" and "per-request cost". Figure 11(c) shows the result. We observe that the cost impact in total is less than 5.6% ($\sim$8,000 USD added to $\sim$142,000). Traffic deviation is less than 2.1% from prediction and contributes to 1.4% of the total 5.6% difference. The additional number of requests accounts for 4.2% of the total cost.

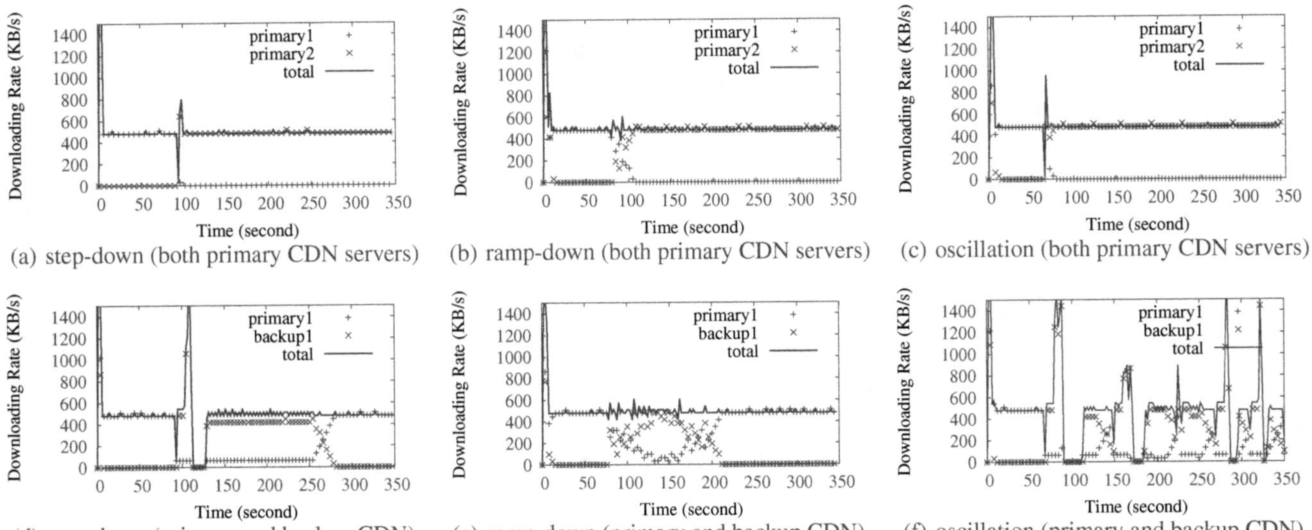

(a) step-down (both primary CDN servers) (b) ramp-down (both primary CDN servers) (c) oscillation (both primary CDN servers)

(d) step-down (primary and backup CDN) (e) ramp-down (primary and backup CDN) (f) oscillation (primary and backup CDN)

Figure 10: Stress tests of client adaptation in CDN server failure cases.

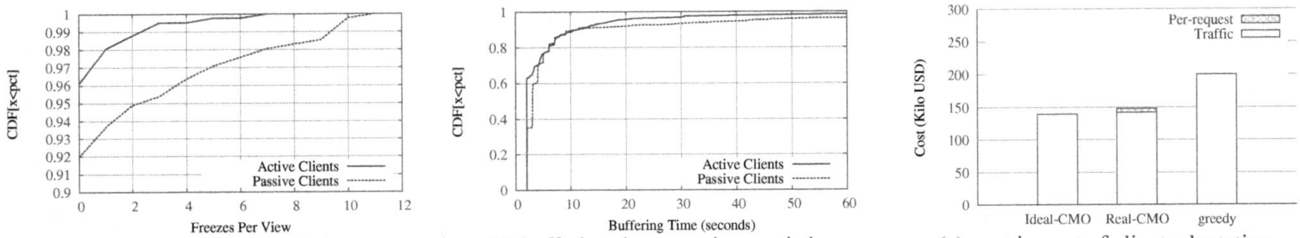

(a) freeze per view statistics (b) buffering time per view statistics (c) cost impact of client adaptation

Figure 11: Per-view QoE in PlanetLab experiments.

9. CONCLUSIONS

In this paper, we have conducted the first systematic study on content multihoming, by introducing the CMO algorithm and the client adaptation algorithm to optimize both the cost and the performance for content multihoming. Our realistic evaluations show that our content multihoming algorithms reduce publishing cost by up to 40%, and reduce viewer QoE degradation by 51%.

Acknowledgments: The research of Y. Richard Yang is supported in part by grants from NSF, and by a gift from Huawei. We are grateful to Aditya Akella (paper shepherd), Nicole Shibley, Peng Zhang, Haiwei Xue, Andreas Voellmy, Yin Zhang and SIGCOMM anonymous reviewers for valuable suggestions.

10. REFERENCES

[1] 01box. http://cdn.01box.net.

[2] V. K. Adhikari, Y. Guo, F. Hao, M. Varvello, V. Hilt, M. Steiner, and Z.-L. Zhang. Unreeling netflix: Understanding and improving multi-CDN movie delivery. In *IEEE INFOCOM'12*.

[3] H. A. Alzoubi, S. Lee, M. Rabinovich, O. Spatscheck, and J. Van Der Merwe. A practical architecture for an anycast CDN. *ACM Trans. Web*, 5(4):17:1–17:29, Oct. 2011.

[4] D. Beaver, S. Kumar, H. C. Li, J. Sobel, and P. Vajgel. Finding a needle in haystack: Facebook's photo storage. In *USENIX OSDI'10*.

[5] G. Bertrand, E. Stephan, G. Watson, T. Burbridge, P. Eardley, and K. Ma. Use cases for CDNi. IETF Draft, Jan. 2012.

[6] D. Bertsekas. *Convex Analysis and Optimization*. 2003.

[7] CDN expert. http://cdnexpertonline.com/node/45.

[8] Cisco Systems. Cisco Visual Networking Index: Forecast and Methodology, 2011-2016.

[9] Conviva. http://www.conviva.com.

[10] F. Dobrian, V. Sekar, A. Awan, I. Stoica, D. Joseph, A. Ganjam, J. Zhan, and H. Zhang. Understanding the impact of video quality on user engagement. In *ACM SIGCOMM'11*.

[11] Dyn CDN manager. http://dyn.com/.

[12] Geo best-of YouTube. http://geobestofyoutube.gmapify.fr/.

[13] D. Goldenberg, L. Qiu, H. Xie, Y. R. Yang, and Y. Zhang. Optimizing cost and performance for multihoming. In *ACM SIGCOMM'04*.

[14] A. Ioffe and V. Tihomirov. *Theory of Extremal Problems*. Elsevier Science Ltd, 1979.

[15] R. Krishnan, H. V. Madhyastha, and etc.. Moving beyond end-to-end path information to optimize CDN performance. In *ACM IMC'09*.

[16] Level 3 Intelligent Traffic Management. http://www.level3.com/~/media/Assets/brochures/brochure_intelligent_traffic_management.pdf.

[17] Limelight Traffic Load Balancer. http://www.limelight.com/traffic-load-balancer/.

[18] H. H. Liu, Y. Wang, Y. R. Yang, H. Wang, and C. Tian. Optimizing cost and performance for content multihoming. Technical Report YaleCS-TR1456, May 2012.

[19] MetaCDN. http://www.metacdn.com/.

[20] B. Niven-Jenkins, F. L. Faucheur, and N. Bitar. Content distribution network interconnection problem statement. IETF Draft, Jan. 2012.

[21] OnePica. http://www.magentocommerce.com.

[22] R. S. Peterson and E. G. Sirer. Antfarm: efficient content distribution with managed swarms. In *NSDI'09*.

[23] R. S. Peterson, B. Wong, and E. G. Sirer. A content propagation metric for efficient content distribution. In *ACM SIGCOMM'11*.

[24] I. Poese, B. Frank, B. Ager, G. Smaragdakis, and A. Feldmann. Improving content delivery using provider-aided distance information. In *IMC'10*.

[25] T. Ristenpart, E. Tromer, H. Shacham, and S. Savage. Hey, you, get off of my cloud: Exploring information leakage in third-party compute clouds. In *ACM CCS'09*.

[26] N. H. Sleumer. Output-sensitive cell enumeration in hyperplane arrangements. *Nordic J. of Computing*, 6:137–147, June 1999.

[27] XDN. http://www.xdn.com.

Private and Verifiable Interdomain Routing Decisions

Mingchen Zhao
University of Pennsylvania

Wenchao Zhou
University of Pennsylvania

Alexander J. T. Gurney
University of Pennsylvania

Andreas Haeberlen
University of Pennsylvania

Micah Sherr
Georgetown University

Boon Thau Loo
University of Pennsylvania

ABSTRACT

Existing secure interdomain routing protocols can verify validity properties about individual routes, such as whether they correspond to a real network path. It is often useful to verify more complex properties relating to the route decision procedure – for example, whether the chosen route was the best one available, or whether it was consistent with the network's peering agreements. However, this is difficult to do without knowing a network's routing policy and full routing state, which are not normally disclosed.

In this paper, we show how a network can allow its peers to verify a number of nontrivial properties of its interdomain routing decisions *without* revealing any additional information. If all the properties hold, the peers learn nothing beyond what the interdomain routing protocol already reveals; if a property does not hold, at least one peer can detect this and prove the violation. We present SPIDeR, a practical system that applies this approach to the Border Gateway Protocol, and we report results from an experimental evaluation to demonstrate that SPIDeR has a reasonable overhead.

Categories and Subject Descriptors

C.2 [**Computer-Communication Networks**]: Security and protection; C.2.2 [**Network protocols**]: Routing Protocols

Keywords

Routing, Privacy, Security, Accountability, Fault detection

1 Introduction

In interdomain routing, there is an inherent tension between verifiability and privacy: both properties are desirable, but they seem contradictory. Communicating networks have expectations about one another's routing decisions, but they are stymied from verifying these expectations because routing configurations are usually kept confidential.

Routing promises. Interdomain routing policies are routinely governed by formal agreements, such as peering and transit contracts, and the correct implementation of these policies is vital for allowing networks to achieve other contractual goals, such as maintaining traffic ratios [6]. In some cases, such as 'partial transit' relationships, the desired policy can be complex, placing additional cost on the implementors [33]. Less formally, networks often publish information on how customers and others can tweak the route selection process, using *BGP communities* [2, 3, 36]. Such capabilities represent an understanding between the networks about how certain routes should be treated. Negotiation about the provision of routing services, including undertakings about routing policy, is an important part of the Internet interconnection marketplace [6].

The value of verifiability. Unfortunately, these promises are not always kept, and violations are hard to detect. Promise-breaking may be deliberate, since networks may have economic incentives to lie about their routes [11]. Other examples of malicious behavior abound [16, 30]; one study found that 18 of 28 peering agreements contained clauses against abuse of the peering relationship through BGP configuration [28]. More innocently, misconfigurations [21], compromised routers [27], or equipment failures [42] can cause routing decisions to deviate from expectations.

Secure variants of BGP (e.g., S-BGP [17]) have been proposed as mechanisms for ISPs to check whether routing announcements correspond to the claimed path and destination, but these mechanisms do not address the important question of whether the *route decision process* matches expectations. Complete verification could be enabled by revealing all routing tables [14], but verifiability is not the only concern.

The value of privacy. For operational security or commercial reasons, ISPs have traditionally been reluctant to disclose details of their routing policy. Some aspects may be revealed to neighbors, included in a route registry, or exposed indirectly via looking glass services, but we cannot expect network operators to agree to use any system that reveals even more of their private information. Existing work has shown that it is possible to make deductions about which autonomous systems are connected, and even about some aspects of policy [4, 8, 19, 34, 39], but these inferences have limited accuracy [33] and require considerable effort to carry out, which makes them unsuitable for verifying routing decisions.

Can we have both? Intuitively, it seems that verifiability and privacy are conflicting goals—by revealing more information, we can improve verifiability, but we reduce privacy. In this paper, we show that this intuition is incorrect. We present a first step towards an interdomain routing system in which networks can verify each other's promises *without revealing any additional information*. Our approach is to enable the networks to verify promises *collaboratively*: Each promise is broken into small pieces such that a) each piece can be verified by some network using only information it already knows, and b) a successful verification of all pieces implies that the promise has been kept.

We show that collaborative verification is possible for an entire category of nontrivial promises, covering the relative preference assigned to different classes of route, and including the possibility of route filtering. We present a practical verification algorithm called

SIGCOMM'12, August 13–17, 2012, Helsinki, Finland.

VPref, as well as a formal proof that VPref both guarantees detection of broken promises and preserves privacy. We also present a data structure, called a *modified ternary tree (MTT)*, that can be used to run our algorithm efficiently for large numbers of prefixes.

To show that our approach is practical, we present *Secure and Private Inter-Domain Routing (SPIDeR)*, a collaborative verification system that can be deployed as a companion protocol to BGP, possibly on separate hardware, and that makes its decisions based on observing the BGP message flow. We report experimental results to show that SPIDeR's overhead is reasonable. SPIDeR is meant to be a proof of concept: it can verify a nontrivial set of control-plane actions, but it does not cover data-plane performance, and it cannot verify promises about certain aspects of BGP functionality, such as proxy aggregation.

Although this paper motivates and evaluates collaborative verification in the context of BGP, the VPref algorithm is not specific to BGP, or even to interdomain routing: it could be applied to any scenario where one entity privately chooses between multiple options presented by other entities. Thus, we speculate that VPref could be useful even beyond network routing.

In summary, this paper makes the following five contributions:

- a novel algorithm, VPref, for collaborative verification of promises about private choices (Section 4);
- a formal proof that VPref guarantees privacy and verifiability (Section 4.6; details in [43]);
- a data structure (MTT) that can be used to efficiently scale VPref (Section 5);
- SPIDeR, a companion protocol to BGP that applies VPref to interdomain routing (Section 6); and
- an experimental evaluation of SPIDeR (Section 7).

To provide some context, we begin with a general overview and roadmap (Section 2), followed by a definition of 'promises' in the context of BGP (Section 3).

2 Overview and Roadmap

Figure 1 illustrates the problem we are concerned about in this paper. Alice and Bob are operators of two autonomous systems (ASes) A and B, which are connected by a direct link. Alice receives interdomain routes from Bob via BGP [32], and these routes typically do not terminate in Bob's AS but rather traverse one of Bob's neighbors: Charlie, Doris, or Eliot. At any given time, Bob may have several routes to a prefix p; he chooses one of these routes according to his local policy and then decides whether or not to make that route available to Alice. However, Alice is unable to observe which routes (if any) Bob had available at that time.

We are specifically interested in situations where Bob has made some kind of *promise* to Alice regarding his routing policy. A promise is a statement that Bob will prefer certain routes over certain other routes; for instance, Bob could promise that he will always choose the shortest route to p that is available to him, or that he will prefer routes through Charlie's AS over routes through Eliot's. Since Bob's incoming routes are not visible to Alice, Bob can easily break his promise without Alice noticing—for instance, a configuration error might result in Eliot's route being chosen, instead of the preferred route that Charlie had offered. Our goal is to enable Alice to verify whether Bob is keeping his promise, *without* forcing Bob to reveal his incoming routes to Alice. We will refer to the first part as our *verifiability* goal and to the second part as our *privacy* goal.

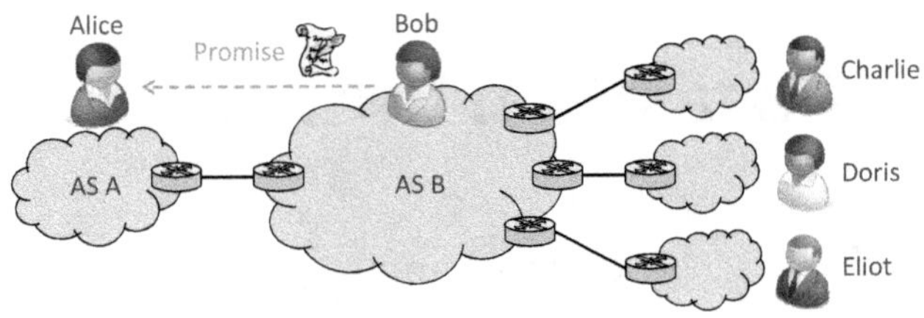

Figure 1: Motivating scenario. Bob has made a promise to Alice about his routing decisions, but Alice has no way to verify whether Bob is keeping his promise.

2.1 How much privacy do we need?

Before we can formalize these goals, we need to provide a more specific definition of privacy. A very strong definition could demand that the downstream neighbors of Bob's AS learn *nothing at all* about the routes available to Bob. However, this property seems too strong—certainly much stronger than what BGP offers today. For instance, if Bob announces a route to Alice via BGP, Alice is able to see (from the AS_PATH attribute) which upstream neighbor the route traverses, and she can infer that this neighbor had previously exported it to Bob. Since this seems acceptable to the ASes that exist today, we adopt a slightly weaker definition: assuming that BGP is already running and that all participants are correct, operation of the additional protocol should not enable any AS participant to deduce *additional* information about the routing state or policy of any other AS, *beyond what it can already learn via BGP*.

In particular, it should not be possible for Alice to determine which routes were available to Bob at any given time, except for those routes she has already learned from him through BGP; and she should not be able to deduce the relative preference of any two routes, other than as already specified by Bob in his promise. However, if Alice colluded with Charlie, then she could find out which routes he sent to Bob – but she could do this even if our system were not deployed. When some participants break their promises, it is acceptable to reveal new information, e.g., evidence that a promise has been broken.

2.2 What is a promise?

We also need to define what it means for an AS to make a promise about its routing policy. Section 3 contains more detail on this topic, and how our formalism can model existing policy for local preference groups, selective export, etc. The promises of interest to us relate to the routing decision procedure, in which several routes enter and at most one leaves; routing policy determines which route is the winner. A promise does not specify every possible nuance of route selection, but may give partial information about which routes will be considered 'better' than others.

Assume that an AS A, for a prefix p, has a total order over the set $R(A,p)$ of all possible routes from A to p, yielding its preference. During protocol operation, A will choose the best route, according to this order, from all those that are currently available. We envision a promise as dividing $R(A,p)$ into several *indifference classes*, where preferences exist between them, but not within each class. In this way, A could divide $R(A,p)$ into 'routes through customer networks' and 'all other routes', and promise that the customer routes constitute the more preferred class. This is an undertaking that if A ever has both a customer and a non-customer route, then it will choose the customer route; but it promises nothing about what will happen when two customer routes are both possible, or when only non-customer routes are available. In these last two cases, the candidate routes are all within the same class, and so no public preference is specified among them.

In the presentation, we assume that, for each pair of neighboring ASes, only a single promise is in effect for a given prefix. However, real routing policy may legitimately be different at each interconnection point, and the visible routes will differ as well: ASes are not atomic [24]. In Section 8, we discuss the proper handling of this fact in our system.

2.3 Goals

We are now ready to define (based on [13]) the four properties we want to provide in SPIDeR:

1. **Verifiability:** If an AS A breaks a promise and all of A's neighbors are correct, then at least one neighbor detects this.
2. **Privacy:** No AS can learn information by running SPIDeR that it could not already learn from running BGP.
3. **Evidence:** If an AS A detects that another AS B has broken a promise, A can obtain evidence against B that will convince a third party.
4. **Accuracy:** If an AS A is keeping all of its promises and is running SPIDeR correctly, no other AS can obtain valid evidence against A.

The first two properties correspond to the informal goals from the beginning of this section. Properties 3 and 4 are useful when holding ASes accountable for broken promises: evidence ensures that ASes cannot deny genuine violations, and accuracy protects correct ASes from spurious accusations.

Non-goals: To keep the problem manageable, we set two explicit non-goals for this paper. First, we focus strictly on verifying control-plane behavior; checking data-plane forwarding against control-plane announcements is a separate problem that is discussed, e.g., in [41]. We also do not attempt to handle *all* aspects of BGP operation (such as aggregation [35]), nor of peering agreements (such as cold potato manipulation). The latter can be handled by systems like BorderGuard [7], which is complementary to SPIDeR.

3 Policy and Promises

BGP best-route selection is carried out on the basis of routes' *attributes*, according to the local configuration of each router. The decision procedure is lexicographic, beginning with the *local preference* attribute and proceeding down a chain of tie-breakers as necessary. Pattern-match rules allow attribute values to be modified as routes are propagated. In the case of local preference, the numeric value might be chosen based on the AS neighbor from whom the route was received. Choosing shorter routes (those going through fewer ASes) is the next step, if local preference values were tied.

Route preference can also be influenced by BGP *communities* [2, 3, 36], which are numeric tags attached to route advertisements. For example, routes might get local preference 100 by default, but those with a special community tag might get a preference of 80, making them seem less attractive to the recipient. Other communities can be used to control the export of routes, or to cause AS-number prepending. They can also be attached on export to give additional information (e.g., a route's geographic origin [5]). These capabilities are determined by the pattern matching rules in the router configurations of the recipient AS.

Many lists of supported communities are publicly available, indicating that ASes are often willing to disclose certain aspects of their routing policy, at least at a high level. If an AS reveals its local preference tiers in this way (for example, the fact that transit routes are the least preferred) then it has committed to only part of the information in its actual route preference order. That is, although the AS will make its routing decisions based on many different route attributes, it has only made a public promise about the first attribute. Outside observers cannot say for sure, concerning two routes in the same tier, which one would actually be preferred. This observation motivates our order-theoretic model of an AS's promises.

3.1 Promises formalized

DEFINITION 1. *Suppose that $R(A,p)$ is the given total order on all routes that might ever be received by AS A for prefix p. A* promise *on $R(A,p)$ consists of a partial order $(C,\leq)$ and a partition $\biguplus_{c\in C} R_c$ of $R(A,p)$.*

The set C is used to label various classes of routes. It is endowed with a partial order, representing the definite preference of one class over another. Every route falls into exactly one of these classes. The semantics of this promise are that each R_c is an *indifference class*, meaning that no preference is stated among the routes within that class. However, if $c < d$ for some c and d in C, then a preference exists between routes in R_c and R_d: if r is in R_c and s is in R_d, then r is stated to be less preferred than s. Although this model is very general, we expect that typical use would not involve very many classes, and that the mapping from routes to classes would match contemporary BGP usage. One consequence is that the representation of a promise can be expected to be compact.

We include the *null route* $\perp$ among $R(A,p)$, and so it is also represented somewhere in the promise, perhaps but not necessarily in a class of its own. Because $\perp$ is always available, for a route to be ranked worse than $\perp$ means that the route should *never* be propagated by A. This capability allows the handling of promises about routes that should not be exported.

Promises to different neighbors. An AS may make different promises to different neighbors, each consistent with what it is actually doing. For example, suppose that the AS prefers customer routes over peer routes, and in addition prefers one customer over all others. The favored customer could be told that its routes will get the highest preference; other customers might only be told that customer routes are better than peer routes. Both promises can be kept at the same time, although one is more specific than the other.

Promise privacy. Not all neighbors need have access to the full definition of the promise. This is useful for cases where it is important to conceal which promises have been made, on the grounds of policy privacy, A producer of routes will only need to know the definitions of classes into which their routes might fall. It does not need to know the partial ordering of those classes, or the definitions of other classes. Therefore, the most they can tell is that the promise makes *some* preference distinction about the known classes. Further obfuscation is possible by splitting classes into mutually indifferent subclasses, to hide the true basis on which decisions are being made.

3.2 Examples

Next, we present a few examples of AS promises based on current interdomain routing practices. Our examples involve neither many route classes nor complex rules about how routes are assigned to their classes. However, all of the examples represent behavior that neighbors would expect to rely on, but where the AS involved might do something different. Figure 2 summarizes some supporting evidence that policies and promises of this kind are currently being used in practice, based on documentation provided by AS networks themselves. Note that this does not cover every possible use of communities (for example, to control prepending or blackholing). We briefly describe some example promises, ordered by how frequently they appear in [29]:

Set local preference: Many ASes offer neighboring networks a choice of community values to adjust the local preference of their

Method	ASes
Set local preference	57
Selective export by neighbor group	48
Selective export by specific AS	45
Information about route origin	45

Figure 2: Summary of BGP community actions supported by 88 autonomous systems [29].

routes. Of the ASes from Figure 2, 64% (57 ASes) supported this option by setting their local preferences, with a mode of three tiers and a maximum of twelve.

Selective export: Communities may be used to exclude certain routes from being given to specific neighbors or categories of neighbor [5]. In our model, this can be accomplished by placing the exportable routes in a separate, preferred class to the other routes. If the semantics are that these excluded routes should never be exported (as opposed to merely being de-preferenced) then the null route should be placed, in a class of its own, between the two main classes. In verifying this promise, the original sender of a route can confirm that it was indeed not exported, and the ultimate recipient can be sure that none of the routes that it was entitled to see were falsely excluded. As shown in Figure 2, selective export is popular; it may be based on naming the specific ASes to be excluded (51%), or may deal with them in groups, such as 'peers in Poland' or 'all transit providers' (54%).

Partial customer or transit relationship: A dual version of the above policy has the recipient of the routes, rather than the sender, asking for a particular export policy. In this 'partial customer' model, certain neighbors are only interested in a subset of the routing table; for example, they might want to receive only routes to destinations in Japan [33]. Correct implementation of these requirements is critical for enabling ASes to balance traffic across a diverse collection of peers [6]. The use of exported communities for geographic information [5] is additional evidence that connected networks are interested in making decisions on a geographic basis. As shown in Figure 2, such opportunities were provided by 51% (45) of the ASes.

Prefer customer: Following the Gao-Rexford guideline [9], and in accordance with commonly-declared AS practice, a network could promise that all customer routes would be preferred over all non-customer routes. This would yield just two indifference classes. An AS could also differentiate between peer and provider routes, the latter being least preferred of all, for three classes in total.

Path length: For any of these schemes, an AS might additionally want to make a promise about preferring shorter AS-level paths. In this case, each original class would be split: what was the 'peer route' class now becomes 'peer routes of length 2', 'peer routes of length 3', and so on, up to some small maximum. Note that making a promise about path length requires more complete disclosure of the true local preference values being used. In the 'favored customer' scenario, a longer route through that customer will be preferred over a shorter route through a different customer: if the AS has previously promised to deliver the shortest customer route regardless of that customer's identity, then this is a violation of the promise.

4 The VPref algorithm

Next, we introduce the VPref algorithm for collaborative verification of promises, and we sketch our proof of correctness. The basic algorithm works for routes to a single prefix at a single time point; in Section 5, we show how to extend VPref to multiple prefixes and handle route announcements and withdrawals.

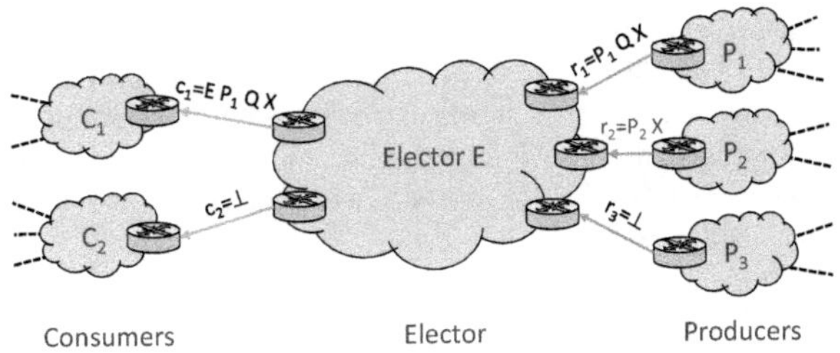

Figure 3: Simplified system model.

4.1 System model and definitions

In the following, we focus on the routing decision of a specific AS, which we will call the *elector* E (Figure 3). The elector is connected to several upstream ASes P_i, called *producers*, as well as to several downstream ASes C_i, called *consumers*. (These terms, unlike 'provider' or 'customer', are not intended to suggest any particular business relationship among the ASes, but only indicate the flow of routing data.) As in BGP, each producer P_i advertises to the elector a single route r_i, which can be the null route $\perp$. The elector then picks a single route e, which must either be $\perp$ or one of the advertised routes r_i. Finally, the elector advertises to each consumer C_j a route c_j, which must be either $c_j = e$ or $c_j = \perp$. Note that this model can represent import filtering (the elector is free to choose $\perp$ instead of an advertised route) as well as export filtering (the elector can offer $\perp$ to certain consumers instead of e).

As discussed in Section 3.1, we assume that the set of possible routes is divided into k *indifference classes* $R_1, \ldots, R_k$, which are known to all ASes. We further assume that the elector has made a promise to each consumer C_j, which we model as a partial order $\leq_j$ over the classes $R_1, \ldots, R_k$. We say that the elector has *broken his promise* to a consumer C_j iff one of the inputs r_i is in a class R_i that is strictly more preferred than the class R of c_j (that is, $R <_j R_i$): so the elector had a route available that was 'better' than the one he exported to C_j. If the elector has not broken any promise, we say that the elector is *correct*.

4.2 Assumptions

We make the following seven assumptions:

1. Each AS has access to the same collision-resistant cryptographic hash function H;
2. Each AS i has a private key, written σ_i, and a public key, written π_i;
3. No AS can invert the hash function H or forge a digital signature of a correct AS;
4. Signed messages contain timestamps and logical counters to prevent replay attacks;
5. The topology and the public keys are known to all ASes;
6. Each C_i has some representation of the promise $\leq_i$ that is signed by E; and
7. Correct ASes can eventually communicate with each other.

The first four assumptions can be satisfied, e.g., by using SHA-512 as the hash function and RSA as the signature scheme, and by applying standard security practices. The fifth assumption seems acceptable because the AS-level topology of the Internet can already be inferred [4]; for key distribution, the deployment of RPKI [15] would certainly suffice. The sixth assumption could be satisfied by exchanging a representation of $\leq_i$ out of band, for example as part of a peering agreement. The seventh assumption simply states that link failures and other disruptions are eventually repaired.

4.3 Intuition

Intuitively, VPref works by breaking the complex property that is to be verified into a number of simpler lemmata, such that 1) each lemma can be verified by *some* network, using information it already knows, and 2) in combination, the lemmata imply the original property. This is combined with a commitment mechanism to ensure that all lemmata are verified using the same state. Finally, we add cryptographic signatures to enable networks to prove violations to each other, or to a third party.

VPref has one lemma for each indifference class, which says that the elector has at least one route from that class *or* a more preferred route. When a producer gives the elector a route from some class R_i, it can ask the elector to prove that the corresponding lemma holds for its routing state, but the producer learns nothing from such a proof because it already follows from its own input. When a consumer receives a route from class R_i, it can ask for a proof that the lemmata for all more preferred classes do *not* hold; again, such a proof reveals no additional information because the elector is supposed to output a route from the most preferred class.

4.4 Commitment phase

VPref proceeds in two rounds, a mandatory *commitment phase* and an optional *verification phase*. The commitment phase involves the following six steps:

1. Each producer P_i picks his route r_i, signs it with his key σ_{P_i}, and then sends $\sigma_{P_i}(r_i)$ to the elector.
2. The elector responds to each producer with an acknowledgment $\sigma_E(\sigma_{P_i}(r_i))$.
3. The elector chooses e to be either $\perp$ or some r_i. It also chooses k *input bits* $b_1, \ldots, b_k$, one for each R_j, and sets $b_j = 1$ if either $r_i \in R_j$ for some i (i.e., at least one input is from class R_j) or $R_j \leq_i e$ for some i (i.e., R_j is worse than e according to at least one promise). Otherwise it sets $b_j = 0$.
4. The elector chooses k random bitstrings $x_1, \ldots, x_k$ and computes $h := H(H(b_1||x_1) \mid\mid \ldots \mid\mid H(b_k||x_k))$.
5. The elector sends a *commitment* $\sigma_E(h)$ to each producer and each consumer.
6. To each consumer C_j, the elector sends either
   - $\sigma_E(C_j, \perp)$, if $c_j = \perp$, or
   - $\sigma_E(C_j, \sigma_{P_i}(r_i), \sigma_E(r_i))$, if $c_j = r_i$.

If an expected message is not received or the message is not properly signed, the corresponding AS raises an alarm. If no alarm is raised at the end of the protocol, each producer holds a signed proof that the elector has acknowledged his input, each consumer holds a route announced by the elector, and each producer and elector holds an (opaque) commitment $\sigma_E(h)$ to the elector's bits. The acknowledgments and commitments are needed for the optional verification phase.

4.5 Verification phase

Any producer or consumer may trigger a verification of the elector's choice by broadcasting a VERIFY($\sigma_E(h)$) message to the other producers and consumers. Upon receiving this message, each AS compares $\sigma_E(h)$ to the commitment it has received from the elector earlier. If an AS has received a different commitment $\sigma_E(h')$, the elector has clearly misbehaved, and the AS broadcasts a INVALIDCOMMIT($\sigma_E(h), \sigma_E(h')$) message as proof of the elector's misbehavior. Otherwise the ASes are satisfied that they all have received the same commitment.

Next, each producer or consumer forwards the VERIFY message to the elector, which responds with a number of signed *bit proofs*. A bit proof for the ith bit consists of b_i, x_i, and, for all $j \neq i$, $H(b_j||x_j)$. The recipient of such a proof can (by recomputing h) verify that the ith bit really did have the value b_i when the commitment was produced. Specifically,

- A consumer C_j that was offered a route c_j earlier will now receive a bit proof for any b_x such that $c_j \leq_j R_x$ and $c_j \notin R_x$ (i.e., the indifference class R_x is preferred over the class containing c_j);
- A producer P_j that sent a route $r_j \neq \perp$ earlier will now receive a bit proof for b_x such that $r_j \in R_x$;
- A producer P_j that sent the empty route $\perp$ earlier will not receive any bit proofs.

When a producer P_i previously provided an input r_i from indifference class R_x but does not receive a valid bit proof for $b_x = 1$, it broadcasts a PROOFCHALLENGE message to the other ASes that contains the acknowledgment $\sigma_E(\sigma_{P_i}(r_i))$ and, if it exists, the invalid bit proof it received from the elector. The other ASes can then use the signed acknowledgment from the PROOFCHALLENGE message to challenge the elector to produce a valid bit proof for $b_x = 1$. A correct elector would always be able to answer such a challenge for any acknowledgment it has signed, so, if the elector refuses, it effectively admits its own guilt.

Finally, each consumer C_j checks whether it has (a) received all the bit proofs it is due, and whether (b) all the bits are being proven to be 0. If a proof is missing or the bit is being proven to be 1, C_j broadcasts a PROOFCHALLENGE message that contains (i) the message it previously received from the elector in step six, (ii) a signed representation of $\leq_j$, and (iii) any bit proofs received from the elector. Again, any other AS can use this message to challenge the elector to produce the corresponding proofs. Note that a signed representation of $\leq_j$ must exist (Assumption 6) and is included so that the other ASes know which bit proofs to request.

4.6 Correctness

Next, we briefly sketch our proofs that VPref satisfies all the goals we described in Section 2.3. The full proofs are available in [43].

THEOREM 1. **Verifiability:** *If (1) an AS is faulty during the commitment phase or breaks a promise, (2) both phases of the protocol are executed, and (3) all the neighbors of the faulty AS are correct, then at least one correct neighbor will detect the fault.*

Proof sketch: The proof is a case distinction that identifies, for each possible misbehavior, the step in VPref that detects it.

A producer is faulty if it provides an invalid signature in its route announcement; such behavior is trivially detected. A correct consumer does not send any messages during the commitment phase, and hence any message received from a consumer immediately reveals that it is faulty. The elector must send identical commitments to all correct ASes, or else an AS will discover the discrepancy when a VERIFY message is received.

The elector breaks a promise to C_j if there is a route r_q from producer P_i such that $c_j \leq_j r_q$ (where c_j is the elector's choice of route) and c_j and r_q are in different indifference classes. This case is always detected by a producer or consumer: if R_q is the indifference class that contains r_q, then either $b_q = 0$ (in which case P_i discovers that the elector is faulty) or $b_q = 1$ (in which case C_j discovers that the elector broke a promise). This follows from the collision-resistant property of the hash function; i.e., the elector is unable to provide a valid bit proof that inverts any bit used to compute in its previously sent commitment. □

THEOREM 2. **Evidence:** *If a correct AS X detects that another AS Y was faulty during the commitment phase, then either (1) Y did not send an expected message or did not properly sign a message with its key σ_Y in which case X raises alarm, or (2) X can convince any correct AS that Y is faulty.*

Proof sketch: The proof is a case distinction that lists the possible faults that may be detected in AS Y, and shows that the AS that detects the fault can convince another correct AS that Y is faulty.

Evidence against producers and consumers consist of malformed messages that are signed. These messages serve as *proof of misbehaviors* (PoMs) against the sender; i.e., they can convince any correct AS that the sender is faulty. A correct AS that detects a faulty elector who sends inconsistent commitments can construct a PoM by transmitting an INVALIDCOMMIT message.

A correct producer that detects an invalid commitment (i.e., that $b_z = 0$ where R_z contains its advertised route) can cause other correct ASes to also detect the fault by broadcasting a PROOFCHALLENGE message that contains the elector's signed acknowledgment of the omitted route. Similarly, a correct consumer C_j that detects a broken promise can cause other correct ASes to also detect the violation by sending a PROOFCHALLENGE message that contains the elector's choice from step six along with the encoding of $\leq_j$. In either case, the elector cannot offer a valid bit proof, indicating to the correct AS that it is faulty. □

THEOREM 3. **Accuracy:** *If an AS X is correct during the commitment phase, then no correct AS will detect a fault in X, and no valid evidence can exist against X.*

Proof sketch: The proof is a case distinction of all possible detectable faults and evidence, showing in each case that the fault cannot be detected, and the evidence cannot exist, if X is correct.

A correct AS detects faults (resp. collects evidence) against X if X sends an invalid (resp. and properly signed) message. However, since X is correct, it will send only valid and signed messages. Additionally, a correct AS detects faults against an elector if the elector does not send proper acknowledgments, commitments, or routing choices, or does not provide a valid bit proof. However, a correct elector will always properly acknowledge messages and send a single well-formed commitment to all ASes. From the definition of the protocol, a correct elector maintains sufficient information to construct a bit proof, and therefore can always provide valid bit proofs. Finally, an AS cannot have valid evidence against a correct elector because the elector will not send inconsistent commitments (hence no valid INVALIDCOMMIT messages can exist) or be unable to provide a valid bit proof (hence no PROOFCHALLENGE message will uncover faulty behavior). □

THEOREM 4. **Privacy:** *If all ASes are correct, then no AS can learn anything from running VPref that it could not already have learned from running BGP.*

Proof sketch: When a consumer receives c_j, it cannot use the information revealed in VPref to tell whether E had any other routes at all. Therefore, it cannot deduce unrevealed preferences of E, since it cannot tell whether E even had a choice to make. A producer P can only tell that its route was received: since this happens even when E has a trivial policy to discard every route, P cannot deduce anything about E's policy or state. Finally, E does not receive any data from VPref that it does not already know. □

THEOREM 5. **Inconsistent promises:** *If an elector makes inconsistent promises, i.e., there exist indifference classes R_i and R_j and consumers C_a and C_b such that $R_i \leq_a R_j$ and $R_j \leq_b R_i$, then there exists a set of inputs such that the elector either must choose $e = \perp$ or break at least one promise.*

Proof sketch: The proof considers inputs r_i, r_j that are respectively in classes R_i and R_j. The elector must choose $e \in \{\perp, r_i, r_j\}$. If $e = \perp$, the theorem trivially holds. If $e = r_i$, then the elector breaks its promise to C_a. Otherwise, if $e = r_j$, then the elector breaks its promise to C_b. □

In an extended technical report [43], we prove minor variations of the above theorems for cases in which faulty parties collude (i.e., coordinate their efforts). Briefly, if the elector colludes with some of the producers, detection is only guaranteed for violations that would exist for *any* combination of inputs from the colluding producers—if there is any combination that would make the elector's output conform to the promise, the elector can simply ask his confederates to pretend that this is what they provided. The proofs for the evidence and accuracy theorems remain unchanged.

Our privacy result says that VPref participants learn nothing that they could not have learned from BGP, so long as the elector's promise is not broken, i.e., the algorithm does not provide them with new information. This holds even if the participants share data: they cannot share any more than they have already obtained from BGP. Colluding parties could manipulate route advertisements in an attempt to discover the elector's hidden preferences, but they do not need VPref to help them.

5 Multi-prefix VPref

The basic VPref algorithm has two important limitations: it only considers routes to a single prefix, and it only works in a static setting. In this section, we extend VPref to support multiple prefixes as well as route announcements and withdrawals at runtime.

5.1 Additional challenges

Adding support for routing changes and multiple prefixes results in two additional challenges that are both related to confidentiality. One way to handle multiple prefixes would be to run a separate instance of VPref for each prefix the elector can reach. However, this would leak some information: If an AS A is invited by its neighbor B to participate in a VPref instance for some prefix p, A can conclude that B has a route to p, even if B never exports such a route to A via BGP. An alternative would be to always run a VPref instance for *every possible* prefix, regardless of whether B ever had a corresponding route, but this would be very expensive: there are $2^{33} - 1$ possible prefixes in IPv4! Thus, our first additional challenge is to run multiple parallel VPref instances efficiently, without leaking information. To this end, we introduce a special data structure called a *modified ternary tree (MTT)* that is based on a ternary Merkle hash tree (Section 5.2).

Another challenge results from the need to handle route announcements and withdrawals. To guarantee detection of all faults, we must re-verify the elector's routing decision whenever the underlying set of routes changes. However, if the elector simply re-runs VPref whenever a route is announced or withdrawn, it again leaks information: If a neighbor observes an execution of VPref, it can conclude that the elector has received an update from *some* producer, even if the elector has not changed the route it exports via BGP to that particular neighbor. In SPIDeR, we avoid this problem by running VPref *periodically* at short, fixed intervals (Section 5.3). This means that we cannot reliably detect violations whose duration is less than one interval, but in return, the fact that VPref runs at some particular time does not leak any information about the underlying routes.

5.2 The modified ternary tree

We now describe the modified ternary tree (MTT), which SPIDeR uses for efficient commitments. An MTT is a tree with four types

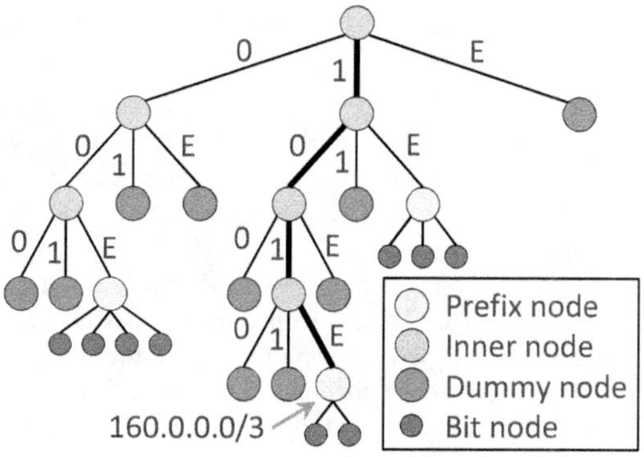

Figure 4: MTT with three prefixes: `0/2`, `160/3`, and `128/1`.

of nodes: *inner nodes*, *prefix nodes*, *bit nodes*, and *dummy nodes*. Each inner node has three children, and we will imagine the edges leading to these children as being labeled 0, 1, and E (for 'end of prefix'). The children at the end of the edges labeled 0 or 1 can be inner nodes, prefix nodes, or dummy nodes; the child at the end of the edge labeled E can be a prefix node or a dummy node. Each prefix node has at least one bit node as its child (but can have more). Bit nodes and dummy nodes do not have any children of their own.

Given an MTT, we can map each prefix node to a specific IP prefix simply by concatenating the bits on the path from the root to the node (but omitting the E label). Figure 4 shows an example MTT in which the highlighted node would be mapped to the prefix `160.0.0.0/3` (`101` in base 2). A full MTT of depth 32 (without dummy nodes) would contain $2^{33} - 1$ prefix nodes, one for each possible IPv4 prefix, but, as we will show in Section 7.3, MTTs for realistic routing tables are much smaller. Also, for any set P of prefixes and any function ϵ that maps each prefix to a set of indifference classes, there is a unique minimal MTT $M(P, \epsilon)$ that contains (a) a prefix node $\pi(p)$ for each prefix $p \in P$, and (b) for each element of $\epsilon(p)$, a bit node that is a child of $\pi(p)$.

The MTT can be used to efficiently run multiple instances of VPref in parallel. Recall that VPref requires two key primitives: 1) a commitment to a set of input bits for each prefix, and 2) a bit proof to show that a specific input bit had a certain value. If an elector has routes to a set of prefixes P, it can use $M(P, \epsilon)$ to maintain the corresponding input bits (in the bit nodes below $\pi(p)$). It can then produce a *single* commitment for all the bits, and it can construct a bit proof for each input bit *without* revealing the presence or absence of any particular prefix in P. Next, we describe these primitives in more detail.

5.3 Commitments and bit proofs with MTTs

To produce a commitment to a given MTT $M(P, \epsilon)$, the elector essentially uses the MTT as a Merkle Hash tree [22]. First, it assigns a label to each leaf node: each dummy node is labeled with a random bitstring, and each bit node is labeled with $H(b_i \,||\, x_i)$, where b_i is the value of the bit and x_i is a random bitstring. (All random bitstrings must be the same length as a hash value). Then it recursively labels each interior node with the hash $H(l_1 \,||\, \ldots \,||\, l_k)$ of the labels l_i of its children. Finally, the commitment is $\sigma_E(h)$, where h is the label of the root node.

A bit proof for a bit b_i then consists of a) the values of b_i and x_i, as well as b) the labels of all direct children of each node on the path from b_i's bit node and the root. As with basic VPref, the recipient of such a proof can verify it by recomputing h from the values in the proof. Crucially, the recipient cannot distinguish between hash values and random bitstrings, and thus has no way of knowing whether a given value in the proof is the label of an inner node or a dummy node. Therefore, a bit proof does not leak any information about the presence or absence of any prefixes in the MTT, other than the specific prefix to which the bit belongs.

An important detail is that the random bitstrings must be replaced for each new commitment. If the bitstrings were reused, neighbors could compare the labels of subtrees (which they can extract from the bit proofs) in consecutive commitments. If a subtree had the same label in two MTTs, this would mean that all the bits in that subtree were almost certainly identical in both commitments, i.e., all of the corresponding routes would have remained the same.

6 Application to BGP

The approach we have described so far is generic and could be applied to different routing protocols. Next, we present SPIDeR, a system that applies our approach to one specific protocol: BGP.

6.1 System overview

SPIDeR is designed as a companion protocol to BGP and could run alongside existing BGP equipment. It consists of three components: a *recorder*, a *proof generator*, and a *checker*. The recorder implements the commitment part of VPref: it opens BGP connections to the border routers in its local AS, mirrors their routing state, and then announces the same routes to the recorders in adjacent ASes, but with the appropriate signatures and acknowledgments added. The recorder also maintains a record of all recent messages it has sent or received, and it periodically computes a commitment, which it distributes to recorders in neighboring ASes.

The two other components implement the verification part of VPref. When a verification is triggered for some AS, the proof generator in that AS reconstructs the corresponding MTT from the recorded message trace and then generates a set of bit proofs, which can be distributed to the neighboring ASes. There, they are verified by the checker. If verification fails, the checker in the detecting AS produces evidence that can be distributed to other neighbors, which can validate the evidence with their own checker.

The recorder, proof generator, and checker implement all the SPIDeR-specific functionality, and they need not be run on the routers—they could run on a separate workstation in each AS. The only change that needs to be made to existing routers is a new pair of iBGP/eBGP sessions to the recorder.

6.2 Signatures and acknowledgments

When verification is triggered for a past commitment at time T, the proof generator needs to reconstruct its AS's precise routing state at time T; moreover, if some faulty AS reconstructs its state incorrectly, the other ASes need to be able to prove that they (or the faulty AS) were in fact im- or exporting a particular route at time T. For this purpose, the recorders timestamp, sign, and acknowledge each BGP UPDATE, and they keep a record of past updates and acknowledgments they have sent or received.

Specifically, a route announcement sent from an elector E to a consumer C has the form $m := \sigma_E(\text{ANNOUNCE}, t, C, p, \sigma_P(r'), \sigma_E(r))$, where t is a timestamp, C is the AS number of the recipient, p is the prefix, r' the underlying route that E itself has imported (if any), and r is the route itself, including all the attributes that are currently in a BGP UPDATE (AS_PATH, communities, etc.). σ_P denotes a signature by the producer who has exported the underlying route r' to E, and σ_E denotes a signature by E. Note that t serves as a nonce, and that both inner signatures are necessary: σ_P proves to C that the route actually exists (and was not fabricated by E); σ_E is necessary so that C can use it as the fifth element when it propagates the route further to its own consumers. A withdrawal has the form $\sigma_E(\text{WITHDRAW}, t, C, p)$.

When a recorder receives a message m that contains an announcement or a withdrawal, it checks whether m's timestamp is reasonably close to its own clock (within, say, a few seconds); if so, it returns a message $\sigma_r(\text{ACK}, t, C, H(m))$, where $H(m)$ is the hash of message m. If a router fails to acknowledge m after some time $T_{\max}$, even after several retransmissions, the sender raises an alarm, which must be handled outside of the system, e.g., by calling the faulty router's system administrator on the phone. The same process is used when a recorder detects that the signed messages it is receiving from some neighbor's recorder contain different routes than the BGP messages the local routers are receiving via BGP from that neighbor. Thus, when there have been no recent alarms, each non-faulty router must have a valid ACK for any message it has sent more than $T_{\max}$ ago. To limit the number of signatures that are required during message bursts, routers can sign messages in batches, e.g., using a variant of Nagle's algorithm [26].

6.3 Evidence

Recall from Section 4.5 that there are two situations in which an AS must provide something about the elector's past routing state during verification:

- if a producer did not receive a bit proof for a route r it was exporting to the elector, it must prove that the elector had acknowledged r (evidence of import); and
- if a consumer did not receive bit proofs for a route r it had received from the elector, it must prove that the elector had exported r to it (evidence of export).

With periodic commitments, the situation is more complicated than in Section 4.5: announcements and acknowledgments are no longer sufficient proof because they can be withdrawn at a later time. Therefore, we adopt an iterative approach: announcements and acknowledgments are considered initial evidence but can be disproven by other evidence, such as a WITHDRAW message with a higher timestamp. Since clocks are only loosely synchronized, we always use the timestamp of the elector that is being verified, i.e., outgoing announcements and withdrawals are considered effective when they are sent, and incoming ones are effective when they are acknowledged. (A malicious elector cannot change its timestamps by re-signing messages because it cannot forge matching ACKs from other ASes, and because signing multiple ACKs for the same message would constitute a proof of misbehavior.)

Evidence of import: If Alice wants to prove that she was exporting a route r to Bob at time t, Alice presents her ANNOUNCE for r, timestamped $t' < t$, and the matching ACK from Bob. If Alice has made a mistake and is presenting an announcement she has withdrawn again before t, Bob can refute her evidence by presenting Alice's WITHDRAW for r with a timestamp t'' such that $t' < t'' < t$.

Evidence of export: If Alice wants to prove that Bob was exporting a route r *to her* at time t, she presents Bob's ANNOUNCE for r, timestamped $t' < t$. If Alice has made a mistake and Bob has withdrawn r at some time $t'' < t$, Bob can refute her evidence by presenting his own WITHDRAW for r, timestamped $t' < t'' < t$, with Alice's matching ACK.

6.4 Handling loose synchronization

In an ideal world, an AS would update its routing decisions instantly whenever it receives an announcement or a withdrawal. In practice, it can take some time for changes to take effect, e.g., due to propagation delays or due to BGP features such as MRAI or route flap damping [38]. If a commitment is requested shortly after an incoming message, some of the outgoing routes may not yet be consistent, and this could appear to neighbors as if the AS had broken a promise.

To avoid this problem, we allow electors, for the purposes of a commitment at time T, to choose their inputs from a time window $[T-\delta, T]$. For instance, suppose Alice had received route r_1 from Bob at time t_1, but r_1 had been withdrawn at time t_2 and then replaced with r_2 at time t_3 ($T-\delta < t_1 < t_2 < t_3 < T$), then Alice would be free to choose whether she wants her input from Bob to be r_1, $\perp$, or r_2. Alice informs Bob of her choice during verification, so that Bob can know how to check the corresponding bit proofs. Alice may choose different times for different neighbors and different prefixes, as long as they all fall into $[T-\delta, T]$.

Electors could use this freedom to hide promise breaches shorter than δ, as long as *some* combination of inputs from $[T-\delta, T]$ would have satisfied the promise. However, this seems acceptable because it is restricted to periods of instability: when the routes for a given prefix are stable, the elector has no freedom at all!

If the elector is correct, a satisfactory combination of inputs must exist, as long as all delays are shorter than δ. This is because any output must be the result of the BGP decision process, and all inputs to the process must have been valid at some point within the last δ seconds. To find such a combination during verification, the proof generator can simply choose the actual output and, for each other producer, the first input from $[T-\delta, T]$ that would not have been preferred over the actual output. Such an input must exist because otherwise that producer would have offered the elector a better route during the entire time window, so the elector's output could not be correct.

6.5 Logging, checkpointing, and replay

The recorder maintains a log of all announcements, withdrawals, and acknowledgments that its AS has sent or received. This log is used to produce evidence in case a verification is triggered. To prevent the log from growing without bounds, verification is limited to commitments that are at most R days old; thus, ASes can safely discard log entries older than R days, since they are no longer needed for verification. We refer to R as the *retention time*, and we expect $R = 365$ to be a typical value.

SPIDeR also uses the log to save space for MTTs. Recall from Section 4.5 that electors must provide bit proofs on demand, which requires access to any MTT within the retention time. However, the MTTs are generated from the AS's routing tables at commitment time, and the log contains all the information needed to reconstruct the MTT. This can be done as follows: the recorder maintains a full checkpoint of its routing state at the beginning of the log, and optionally some additional checkpoints at various commitment times. Then, when verification is triggered for a commitment at some time t, the proof generator loads the most recent checkpoint prior to t, and replays all messages with timestamps up to t. This reproduces the same routing state that was used to create the MTT originally. Thus, the recorder can discard the MTTs after each commitment, and the proof generator can reconstruct them on demand.

However, the MTT contains not only the routing state, but also some random bitstrings. In principle, the recorder could store these bitstrings explicitly, but there is an easier way: it can generate the bitstrings using a cryptographically secure pseudo-random number generator (CSPRNG) with a secret seed. The proof generator can thus later reconstruct the bitstrings from the stored seed. As long as a new seed is chosen truly at random for each commitment, other ASes (who do not have knowledge of the seed) cannot predict the generated bitstrings or distinguish them from hash values, which is sufficient to maintain privacy.

6.6 Verification of withdrawals

Similar to S-BGP [17], SPIDeR uses signed route announcements to enable the consumers to verify that a received route actually exists in the underlying topology. However, as in S-BGP, the consumer cannot easily verify whether the route *continues* to be valid once it is announced. If the route is withdrawn by the producer and the elector fails to propagate the withdrawal to the consumer, the consumer has no way to detect this. Therefore, SPIDeR allows ASes to trigger an *extended verification* for a given commitment, which works exactly like a normal verification, except that the AS must also re-announce all routes it had announced originally, with timestamps equal to the commitment time. To prevent the re-announcements from being used in place of the original announcements, they contain a special RE-ANNOUNCE message type.

Since the elector cannot sign messages on behalf of the producers, the producers must send a set of RE-ANNOUNCE messages to the elector when extended verification is triggered, one for each route they were exporting at the time of the original commitment. (If a producer fails to provide an acceptable RE-ANNOUNCE message, the elector can use evidence of import to demonstrate that the producer is misbehaving.) The elector then selects the RE-ANNOUNCEs that correspond to the routes it had originally chosen, and distributes those to any consumers that had originally received the routes. Note that the elector cannot avoid asking for all messages, even though it must discard most of them; if it asked only for RE-ANNOUNCEs that corresponded to chosen routes, it would reveal to the producers which routes had been chosen, which would compromise privacy.

6.7 Incremental deployment

SPIDeR is a local protocol: all interactions occur between the AS that is being verified and its direct neighbors. Thus, SPIDeR does not need global adoption, or a global PKI, to be effective—it merely requires some local collaboration between ISPs, similar to what is done for route debugging today. Moreover, SPIDeR's guarantees strengthen gradually with the size of the deployment: if only a subset of the neighbors of an AS deploy SPIDeR, they can still detect and prove violations of promises that involve inputs and outputs from that subset. Thus, an initial deployment could be as small as three adjacent ASes: one AS that has made some of the promises from Section 3.2, and two customers or peers of that AS who would like to verify that the promises are being kept. Once such 'islands' of deployment exist, they enable the ASes on their perimeter to easily offer verification to their own neighbors, so the 'islands' can grow incrementally.

7 Evaluation

Next, we report results from an experimental evaluation of SPIDeR. Our goal is to answer two high-level questions: 1) is SPIDeR practical?, and 2) how expensive is SPIDeR?

To provide a baseline for comparisons, we aligned our experiments with those from the NetReview paper [14]. NetReview is a good baseline for SPIDeR because it can also verify promises about interdomain routing policies, and can also be deployed as a companion protocol to BGP. However, NetReview requires ASes to disclose a lot of sensitive information, whereas SPIDeR is designed to provide strong privacy guarantees.

7.1 Prototype implementation

For our experiments, we built a proof-of-concept implementation of SPIDeR, including a recorder, a proof generator, and a checker. For the recorder, we reused some code from NetReview [14], specifically the component for mirroring BGP routing state from existing

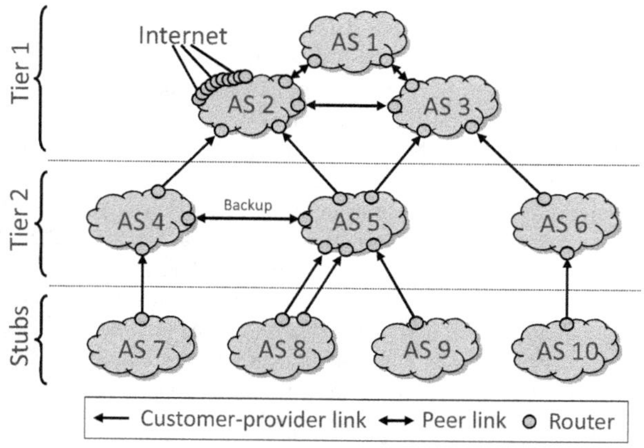

Figure 5: AS topology for our experiments (from [14]). A RouteViews trace is injected at AS 2.

routers and the component for maintaining a tamper-evident message log with signatures and acknowledgments (but not the code for auditing, which is different in SPIDeR). We added code for the MTT and for generating commitments; the proof generator and checker are written from scratch. Overall, we added or changed 8,012 lines of C++ code.

We chose RSA-1024 signatures and the SHA-512 hash function, but we use only the first 20 bytes of each digest to save space. The CSPRNG is implemented by encrypting sequences of zeroes with RC4, discarding the first 3,072 bytes to mitigate known weaknesses in RC4. Our recorder implementation uses separate threads for handling messages and for generating commitments; this prevents the message handler from blocking while MTTs are being labeled. The number c of commitment threads can be varied to take advantage of multiple cores; when $c > 1$, we break the MTT into subtrees that are each labeled completely by one of the threads.

7.2 Methodology and experimental setup

Our goal was to estimate the cost a typical Internet AS would incur by running SPIDeR. Since it was not feasible to replicate the Internet's entire AS topology in our lab, we decided to set up a small, synthetic topology (shown in Figure 5) using 36 Quagga BGP daemons in 10 ASes. However, as in [14], we injected BGP messages from a RouteViews trace into one of the ASes. Thus, the conditions in our synthetic topology were approximately as if the ASes had been a part of the global Internet: the routing tables contained routes to every reachable IP prefix, and the number and the arrival pattern of the BGP UPDATEs were similar to the conditions at the RouteViews collection point.

Specifically, we used a 15-minute RouteViews trace that was collected by a Zebra router at Equinix in Ashburn, VA, on January 18, 2012 at 10am. This trace contains 38,696 BGP messages, and the corresponding RIB snapshot contains 391,028 distinct IP prefixes. In our experiment, we first populated the routing tables by slowly announcing the prefixes from the snapshot over a period of 30 minutes; then we replayed the 15-minute message trace. We refer to the first phase as the *setup period* and to the second phase as the *replay period*. Unless otherwise specified, we report data that was collected during the replay period, and we focus on the AS in the middle (AS 5).

Each AS was configured with a simple routing policy based on Gao-Rexford [9], defined 50 indifference classes based on the number of hops, and promised to choose the shortest route to *all* prefixes in the BGP routing table. These simple choices are sufficient for measuring overhead because the cost depends mostly on the size of the MTT and thus on the *number* of prefixes and indifference classes. Recall from Section 3 that only very few ASes support more than five local-pref classes, and consider that promises could be made for a specific set of prefixes ("I will give you my shortest route to Google"); hence, 50 classes and all prefixes are both conservative choices.

We ran our experiments on a cluster of 11 machines that were connected by a 1 Gb Ethernet network. Each machine had a 2.4 GHz Intel X3220 CPU with four cores and 4 GB RAM, and ran Fedora Core 10 (Linux 2.6.27.41). For the BGP daemon, we used Quagga 0.99.20 with a 100-line patch that enables the daemon to bind to a specific IP. Commitments were generated every 60 seconds. Unless otherwise specified, we used $c = 3$ cores for commitments and the fourth core for message handling.

7.3 Microbenchmarks

We first ran a number of microbenchmarks to measure the size of a typical MTT, and the time needed to generate and verify proofs.

MTT size. The MTT from AS 5's last commitment in the experiment contains 22,333,767 nodes, including 389,653 prefix nodes, 950,372 inner nodes, 1,511,092 dummy nodes, and 19,482,650 bit nodes. This is expected because there is one prefix node for each IP prefix that is reachable at that point in the trace, and each prefix node has 50 bit nodes. In total, these nodes required about 137.5 MB of memory.

Labeling time. With $c = 3$ cores, computing the label of this MTT's root node took 13.4 seconds. This seems unproblematic because the computation is done by the recorder (and not by a border router!) and because it is asynchronous, i.e., does not block BGP. Thus, an AS could use our implementation to make a commitment every 15 seconds and catch any promise violations that last at least that long. For comparison, the same computation took 38.8 seconds with only $c = 1$ core, so the speed-up for $c = 3$ is 2.9. This is expected because MTT labeling is highly scalable. Because of this, shorter intervals could be achieved by adding more cores, or even additional machines.

Proof generation and proof size. When verification is triggered, the proof generator must reconstruct the MTT at the time of the commitment and then generate a set of bit proofs for each neighbor. For AS 5's last commitment, it took 13.4 s to reconstruct the MTT and 70.2 s to generate the proofs for the five neighbors. The average size of a proof was 449 MB. As a rough approximation, each bit proof with k indifference classes contributes k hashes, or $20 \cdot k$ bytes, plus potentially some hashes of dummy nodes.

For comparison, we also generated the proofs for an alternative promise about just one prefix: "I will give you the shortest route to Google." This took only 0.431 s to generate (after MTT reconstruction), and the corresponding proofs were 2.1 KB for the producers and 2.1 KB for the consumers. If an AS wants to make promises about very many prefixes and proof size is a concern, its neighbors could trigger verification for smaller subtrees, e.g., all prefixes in `32.0.0.0/8`.

Proof checking. With $c = 1$ core, verification of a single proof takes 27 s on average; we observed times from 8.6 s to 40 s. As a first step, the checker needs to rebuild the part of the MTT that is included in the proof and re-label it to verify the commitment; this step took 26 s on average. Then, the checker must verify that all the required bits are present and have the appropriate value; this step accounted for the remaining 1 s.

7.4 Functionality check

Next, we performed a number of sanity checks on our implementation. First, we ran the experiment to completion and then triggered verification; as expected, no broken promises were reported. Next, we re-ran the experiment, injecting different faults into AS 5 that caused its promise to be violated:

- **Overaggressive filter:** The AS incorrectly filters out a good route from an upstream AS.
- **Wrongly exporting:** A received route is marked as 'not for export' (with a promise where some routes are worse than the null route) but the AS passes it on.
- **Tampered bit proof:** The AS attempts to hide a good route by changing a bit in the bit proof sent to a downstream AS.

After each run, we triggered verification, and in each case the fault was detected by one of the ASes. In the first run, the upstream AS raised an alarm because it did not receive a bit proof for the route it had supplied. In the second run, the downstream AS noticed that it had a bit proof for the null route, which was better than the route it had actually received. In the third run, the downstream AS detected that the proof did not match the hash value from the commitment. These examples complement the proofs in Section 4.6 to give us confidence in our implementation.

7.5 Overhead: Computation

The SPIDeR recorder performs two kinds of operations that are computationally expensive: It signs messages and ACKs, and it generates and labels a MTT for each commitment. To quantify this overhead, we used the `getrusage` system call to measure the computation time the recorder's threads spent overall, and we separately instrumented the code to measure the time spent on generating and verifying signatures and on labeling MTTs. We excluded the first and the last minute to avoid startup/shutdown effects.

We found that, during the replay period, the recorder spent 634.5 s of computation time overall. 9.75 s were spent on generating and verifying 3,913 RSA-1024 signatures; note that this is lower than the total number of BGP updates in the trace because, when updates arrive in bursts, the recorder can batch several of them together and sign the entire batch. Generating 13 MTTs required 519 s. All other operations, e.g., BGP RIB maintenance, account for the remaining 105.75 s. Averaged over the entire 13 minutes, a single X3220 core would have been about 81.3% utilized. Since we reuse its messaging code, NetReview would have incurred exactly the same costs, except for the MTT generation; thus, NetReview's CPU utilization would have been about five times lower.

SPIDeR's computational cost increases with the commitment generation rate, and with the number of routing updates that need to be sent (which in turn depends on the number of neighbors), since there are more messages that need signing. For a small AS with five neighbors, like AS 5, the SPIDeR recorder could easily be run on a single commodity workstation. According to CAIDA's topology data [1], 89% of the current Internet ASes have five or fewer neighbors.

7.6 Overhead: Bandwidth

SPIDeR increases the amount of interdomain routing traffic because all BGP updates need to be re-announced via SPIDeR with additional signatures and acknowledgments. To quantify this overhead, we used `tcpdump` to capture all BGP packets and all SPIDeR packets that were sent from AS 5 during the replay period. We found that, on average, BGP sent traffic at a rate of 11.8 kbps and SPIDeR at a rate of 32.6 kbps. The relative increase (176%) may seem high, but compared to the amount of traffic ASes commonly handle, 20.8 kbps is not very much—it is about 2% of a single typical DSL upstream.

SPIDeR additionally sends bit proofs to neighboring ASes for verification; the amount depends on the frequency of verifications and the number of commitments checked, which in turn depend on perceived AS needs. Verifying 1% of commitments every minute would result in about 3.0 Mbps of traffic for AS 5.

7.7 Overhead: Storage

Each SPIDeR recorder requires some local storage for the message log, the information needed to reconstruct past MTTs, and a number of snapshots of its routing state. To quantify the amount of storage needed, we examined the storage of AS 5's recorder after the replay period. We excluded information that was stored during the setup period.

The log contained 2.95 MB of message data, excluding snapshots; a substantial fraction (24.4%) consisted of cryptographic signatures. Thus, it grew at an average rate of about 232.3 kB per minute. Complete snapshots of the routing state were about 94.1 MB. All of this information would be stored by NetReview as well. The only addition in SPIDeR is the MTT-related data, which was comparatively small: each commitment added only 32 bytes to the log. This is because the MTT can be regenerated from the message trace; only the CSPRNG's seed needs to be stored explicitly.

Based on these results, we estimate that an AS could keep a year's worth of logs, including one snapshot per day, in 145.7 GB of storage. This data would easily fit onto a commodity hard drive.

8 Discussion

AS atomicity. In our presentation of SPIDeR we have assumed that an elector has uniform policy towards its consumer, for a given prefix. In reality, policy (and therefore promises) may have geographic differences, and the sets of routes visible at each border router may legitimately vary: ASes are not atomic [24, 33]. Our system, operating at the AS level, treats such equivocation as evidence of misbehavior. In order to recover the ability to make and verify such promises, the protocol must be run not only for each consumer but for each consumer adjacency, so that the promise made to Alice in Europe can be differentiated from the promise made to her in Asia. This capability would also be necessary to support communities relating to specific adjacencies, such as 'do not advertise this route at LINX' [5]. If this were done, then additional information would be revealed to producers: we have assumed that the AS-level topology is public knowledge, but now the producers can find out how many interconnections there are between the elector and consumer. Adding extra dummy instances would conceal the true number of connections, at additional cost.

Aggregation. Aggregation of routing announcements is an important concern for network scalability [35]. Most aggregation is performed by the network originating the prefixes, and poses no problem for us. BGP also allows *proxy aggregation*, where some other AS can combine routes to adjacent prefixes in some circumstances. It would be possible to support proxy aggregation in the case of identical AS paths, with additional computational cost (but we do not propose to deal with the rare and deprecated `AS_SET` attribute [18]). The MTT could include aggregates as well as ordinary prefixes. For example, if p and q were two aggregatable $/24$ prefixes, then their immediate parent node in the tree, a $/23$ prefix, would contain a subtree for verifying promises about the aggregate. To achieve privacy, the elector must construct the $/23$ tree (including a 1 bit for the routes in question) and the producer must check it, whether or not aggregation actually occurred. This is in addition to the requirement to validate the individual $/24$ trees. Without this step, the producer could use the fact that aggregation happened to deduce that both of its routes had been adopted.

The addition of aggregate prefixes into SPIDeR would greatly increase the computational overhead, in terms of storage space, computation time, and volume of traffic. However, since correct SPIDeR operation requires the producer to reason about the possible aggregation, the producer would be better off simply doing the aggregation directly. Ultimately, it should be done by the originator in all cases, removing the need for the larger MTT.

9 Related Work

BGP security: There has been a great deal of work on BGP route and origin attestation, aimed at preventing prefix hijacking and related problems; example protocols include S-BGP [17], SO-BGP [40] and psBGP [37]. These proposals provide mechanisms for ensuring that an individual route is genuine, but do not aim to validate any aspect of the BGP decision process or of AS policy, beyond this minimum level. The IETF Secure Inter-domain Routing Working Group is engaged in an attempt to strengthen and standardize such proposals, including the provision of a Resource Public Key Infrastructure [15]. Our own system aims to check a complementary set of properties about the outcome of routing decisions, and in a local rather than end-to-end fashion. It could also make use of the RPKI, and be run alongside one of these other systems.

NetReview [14] does allow routing decisions to be checked, but it reveals the entire stream of BGP updates an AS has received from its neighbors, so it is considerably less private than SPIDeR. An earlier variant of VPref, previously published in [13], provided similar guarantees but was only able to support two simple operators, and only in a static setting. This paper substantially generalizes our earlier work, and it also presents a complete system design and an evaluation. BorderGuard [7] verifies a different type of promise, namely whether an ISP is advertising consistent routes at all peering points it shares with a given neighbor. In contrast to the promises we consider here, this can be done using information that is already available to the ISP, so privacy is not an issue.

Verifiable aggregation: In the context of distributed database queries, verifiable aggregation in the presence of adversaries has been considered by Garofalakis et al. [10]. Their 'proof sketches' idea allows parties to check if the outcome of a query (such as count, sum, or average) is close to the true value. The technique combines an authentication manifest—which uses cryptographic signatures to prevent input values from being forged—with a one-pass streaming algorithm for approximate aggregation; this is analogous to our combination of S-BGP and the MTT structure. They do not consider privacy of the aggregation function, or the idea of verifying the actual function against a promise.

Collaborative detection: SPIDeR is not the only system in which nodes work together to detect misbehavior; for instance, in Catch [20], nodes collaboratively detect free-riders in a mobile ad-hoc network. However, in Catch, there is no private information to be protected, whereas SPIDeR performs collaborative verification with privacy guarantees. Privacy-preserving detection of *traffic* anomalies is possible in the P3CA system [25], through a version of principal component analysis that employs secure multiparty computation. This is a data-plane counterpart to our work.

Zero-knowledge proofs: Following the seminal paper by Goldwasser, Micali, and Rackoff [12], a variety of cryptographic techniques have been developed to allow one node, the *prover*, to convince another node, the *verifier*, that a certain statement is true, without revealing any additional information. ZKPs are very general but also somewhat expensive; however, specialized but more efficient variants have been developed, e.g., for straight-line computations [31]. The bit proofs in VPref can be seen as special-purpose ZKPs, and our tree constructions resemble those used for zero-knowledge sets [23]. VPref's bit proofs are not as general as zero-knowledge sets, but they can be constructed efficiently using only hashing, i.e., without modular exponentiation.

10 Conclusion

This paper has shown that interdomain routing systems do not need to make a choice between verifiability and privacy: it is possible to have both. Using our VPref algorithm for collaborative verification, networks can verify a number of nontrivial promises about each others' BGP routing decisions without revealing anything that BGP would not already reveal. The results from our evaluation of SPIDeR show that the costs for the participating networks would be reasonable. VPref is not BGP-specific and could be applied to other routing protocols, or perhaps even to private verification tasks in other domains.

Acknowledgments

We thank our shepherd, Hovav Shacham, and the anonymous reviewers for their comments and suggestions. We also thank Jennifer Rexford for helpful comments on earlier drafts of this paper. This work was supported by NSF grants IIS-0812270, CCF-0820208, CNS-0845552, CNS-1040672, CNS-1054229, and CNS-1065130, and DARPA contracts N66001-11-C-4020 and FA8650-11-C-7189. Any opinions, findings, and conclusions or recommendations expressed herein are those of the authors and do not necessarily reflect the views of the funding agencies.

References

[1] AS relationships dataset from CAIDA. `http://www.caida.org/data/active/as-relationships/`.

[2] O. Bonaventure and B. Quoitin. Common utilizations of the BGP community attribute. Internet draft, 2003.

[3] E. Chen and T. Bates. An application of the BGP community attribute in multi-home routing. RFC 1998, Aug 1996.

[4] X. Dimitropoulos, D. Krioukov, M. Fomenkov, B. Huffaker, Y. Hyun, kc claffy, and G. Riley. AS Relationships: Inference and Validation. *ACM CCR*, (1):29–40, Jan 2007.

[5] B. Donnet and O. Bonaventure. On BGP communities. *ACM CCR*, 38(2):55–59, April 2008.

[6] P. Faratin, D. Clark, P. Gilmore, S. Bauer, A. Berger, and W. Lehr. Complexity of Internet interconnections: Technology, incentives and implications for policy. In *Proc. 35th Annual Telecomm. Policy Research Conf. (TPRC)*, 2007.

[7] N. Feamster, Z. M. Mao, and J. Rexford. BorderGuard: Detecting cold potatoes from peers. In *Proc. IMC*, Oct. 2004.

[8] L. Gao. On inferring autonomous system relationships in the Internet. *IEEE/ACM ToN*, 9:733–745, Dec. 2001.

[9] L. Gao and J. Rexford. Stable Internet routing without global coordination. *IEEE/ACM ToN*, 9(6):681–692, Dec. 2001.

[10] M. Garofalakis, J. Hellerstein, and P. Maniatis. Proof sketches: Verifiable in-network aggregation. In *Proc. ICDE*, Apr. 2007.

[11] S. Goldberg, S. Halevi, A. Jaggard, V. Ramachandran, and R. Wright. Rationality and traffic attraction: Incentives for honestly announcing paths in BGP. In *Proc. ACM SIGCOMM*, Aug. 2008.

[12] S. Goldwasser, S. Micali, and C. Rackoff. The knowledge complexity of interactive proof systems. *SIAM J. Comput.*, 18(1):186–208, 1989.

[13] A. J. T. Gurney, A. Haeberlen, W. Zhou, M. Sherr, and B. T. Loo. Having your cake and eating it too: Routing security with privacy protections. In *Proc. HotNets*, Nov. 2011.

[14] A. Haeberlen, I. Avramopoulos, J. Rexford, and P. Druschel. NetReview: Detecting when interdomain routing goes wrong. In *Proc. NSDI*, Apr 2009.

[15] IETF Working Group on Secure Inter-domain Routing. `http://tools.ietf.org/wg/sidr`.

[16] A. J. Kalafut, C. A. Shue, and M. Gupta. Malicious hubs: detecting abnormally malicious autonomous systems. In *Proc. INFOCOM*, Mar. 2010.

[17] S. Kent, C. Lynn, and K. Seo. Secure border gateway protocol (S-BGP). *IEEE JSAC*, 18(4):582–592, 2000.

[18] W. Kumari and S. Kotikalapudi. Recommendation for not using AS_SET and AS_CONFED_SET in BGP. RFC 6472.

[19] H. V. Madhyastha, E. Katz-Bassett, T. Anderson, A. Krishnamurthy, and A. Venkataramani. iPlane Nano: path prediction for peer-to-peer applications. In *Proc. NSDI*, Apr. 2009.

[20] R. Mahajan, M. Rodrig, D. Wetherall, and J. Zahorjan. Sustaining cooperation in multi-hop wireless networks. In *Proc. NSDI*, May 2005.

[21] R. Mahajan, D. Wetherall, and T. Anderson. Understanding BGP misconfiguration. In *Proc. ACM SIGCOMM*, Sep 2002.

[22] R. Merkle. Protocols for public key cryptosystems. In *Proc. Symposium on Security and Privacy*, Apr. 1980.

[23] S. Micali, M. Rabin, and J. Kilian. Zero-knowledge sets. In *Proc. FOCS*, Oct. 2003.

[24] W. Mühlbauer, A. Feldmann, O. Maennel, M. Roughan, and S. Uhlig. Building an AS-topology model that captures route diversity. In *Proc. ACM SIGCOMM*, Sept. 2006.

[25] S. Nagaraja, V. Jalaparti, M. Caesar, and N. Borisov. P3CA: Private anomaly detection across ISP networks. In *Proc. Privacy Enhancing Technologies Symp. (PETS)*, July 2011.

[26] J. Nagle. Congestion control in IP/TCP internetworks. RFC 896, Jan 1984.

[27] O. Nordstroem and C. Dovrolis. Beware of BGP attacks. *ACM CCR*, 34(2):1–8, Apr. 2004.

[28] W. B. Norton. A study of 28 peering policies. Technical report, DrPeering International.

[29] One Step Consulting, Inc. BGP community guides. `http://onesc.net/communities`, 2012.

[30] N. Patrick, T. Scholl, A. Shaikh, and R. Steenbergen. Peering Dragnet: anti-social behavior amongst peers, and what you can do about it. NANOG 38, 2006.

[31] M. O. Rabin, R. A. Servedio, and C. Thorpe. Highly efficient secrecy-preserving proofs of correctness of computations and applications. In *Proc. LICS*, July 2007.

[32] Y. Rekhter, T. Li, and S. Hares. A border gateway protocol 4 (BGP-4). RFC 4271, Jan 2006.

[33] M. Roughan, W. Willinger, O. Maennel, D. Perouli, and R. Bush. 10 lessons from 10 years of measuring and modelling the Internet's Autonomous Systems. *IEEE JSAC*, 29(9):1810–1821, 2011.

[34] R. Sherwood, A. Bender, and N. Spring. DisCarte: a disjunctive Internet cartographer. In *Proc. SIGCOMM*, 2008.

[35] P. Smith, R. Evans, and M. Hughes. Recommendations on route aggregation. Technical Report RIPE-399, RIPE Routing Working Group, Dec. 2006.

[36] P. Traina and R. Chandrasekeran. BGP communities attribute. RFC 1997, Aug 1996.

[37] P. C. van Oorschot, T. Wan, and E. Kranakis. On interdomain routing security and pretty secure BGP (psBGP). *ACM TISSEC*, 10(3), 2007.

[38] C. Villamizar, R. Chandra, and R. Govindan. BGP route flap damping. RFC 2439, Nov 1998.

[39] F. Wang and L. Gao. On inferring and characterizing Internet routing policies. In *Proc. IMC*, Oct. 2003.

[40] R. White. Securing BGP through Secure Origin BGP. *The Internet Protocol Journal*, 6(3):15–22, 2006.

[41] E. L. Wong, P. Balasubramanian, L. Alvisi, M. G. Gouda, and V. Shmatikov. Truth in advertising: lightweight verification of route integrity. In *Proc. PODC*, Aug. 2007.

[42] J. Wu, Z. M. Mao, J. Rexford, and J. Wang. Finding a needle in a haystack: Pinpointing significant BGP routing changes in an IP network. In *Proc. NSDI*, May 2005.

[43] M. Zhao, W. Zhou, A. J. T. Gurney, A. Haeberlen, M. Sherr, and B. T. Loo. Private and verifiable interdomain routing decisions. Technical Report MS-CIS-12-10, U. Penn, 2012.

LIFEGUARD: Practical Repair of Persistent Route Failures

Ethan Katz-Bassett
Univ. of Southern California
Univ. of Washington

Colin Scott
UC Berkeley

David R. Choffnes
Univ. of Washington

Ítalo Cunha
UFMG, Brazil

Vytautas Valancius
Georgia Tech

Nick Feamster
Georgia Tech

Harsha V. Madhyastha
UC Riverside

Thomas Anderson
Univ. of Washington

Arvind Krishnamurthy
Univ. of Washington

ABSTRACT

The Internet was designed to always find a route if there is a policy-compliant path. However, in many cases, connectivity is disrupted despite the existence of an underlying valid path. The research community has focused on short-term outages that occur during route convergence. There has been less progress on addressing avoidable long-lasting outages. Our measurements show that long-lasting events contribute significantly to overall unavailability.

To address these problems, we develop *LIFEGUARD*, a system for automatic failure localization and remediation. *LIFEGUARD* uses active measurements and a historical path atlas to locate faults, even in the presence of asymmetric paths and failures. Given the ability to locate faults, we argue that the Internet protocols should allow edge ISPs to steer traffic to them around failures, without requiring the involvement of the network causing the failure. Although the Internet does not explicitly support this functionality today, we show how to approximate it using carefully crafted BGP messages. *LIFEGUARD* employs a set of techniques to reroute around failures with low impact on working routes. Deploying *LIFEGUARD* on the Internet, we find that it can effectively route traffic around an AS without causing widespread disruption.

Categories and Subject Descriptors

C.2.2 [**Communication Networks**]: Network protocols

Keywords

Availability, BGP, Measurement, Outages, Repair

1. INTRODUCTION

With the proliferation of interactive Web apps, always-connected mobile devices and data storage in the cloud, we expect the Internet to be available anytime, from anywhere.

However, even well-provisioned cloud data centers experience frequent problems routing to destinations around the Internet. Existing research provides promising approaches to dealing with the transient unavailability that occurs during routing protocol convergence [18, 22–24], so we focus on events that persist over longer timescales that are less likely to be convergence-related. Monitoring paths from Amazon's EC2 cloud service, we found that, for outages lasting at least 90 seconds, 84% of the unavailability came from those that lasted over ten minutes.

We focus on disruptions to connectivity in which a working policy-compliant path exists, but networks instead route along a different path that fails to deliver packets. In theory this should never happen – if working paths exist, the Internet protocols are designed to find them, even in the face of failures. In practice, routers can fail to detect or reroute around a failed link, causing silent failures [35].

When an outage occurs, each affected network would like to restore connectivity. However, the failure may be caused by a problem outside the network, and available protocols and tools give operators little visibility into or control over routing outside their local networks. Operators struggle to obtain the topology and routing information necessary to locate the source of an outage, since measurement tools like traceroute and reverse traceroute [19] require connectivity to complete their measurements.

Even knowing the failure location, operators have limited means to address the problem. Traditional techniques for route control give the operators' network direct influence only over routes between it and its immediate neighbors, which may not be enough to avoid a problem in a transit network farther away. Having multiple providers still may not suffice, as the operators have little control over the routes other ASes select to it.

To substantially improve Internet availability, we need a way to combat long-lived failures. We believe that Internet availability would improve if data centers and other well-provisioned edge networks were given the ability to repair persistent routing problems, regardless of which network along the path is responsible for the outage. If some alternate working policy-compliant path can deliver traffic during an outage, the data center or edge network should be able to cause the Internet to use it.

We propose achieving this goal by enabling an edge network to disable routes that traverse a misbehaving network, triggering route exploration to find new paths. While accomplishing this goal might seem to require a redesign of the Internet's protocols, our objective is a system that works today with existing protocols, even if it cannot address all outages. We present the design and implementation of a system that enables rerouting around many long-lasting failures while being deployable on today's Internet. We call our system *LIFEGUARD*, for *Locating Internet Failures Effectively and Generating Usable Alternate Routes Dynamically*. *LIFEGUARD* aims to automatically repair partial outages in minutes, replacing the manual process that can take hours. Existing approaches often enable an edge AS to avoid problems on its forward paths to destinations but provide little control over the paths back to the AS. *LIFEGUARD* provides reverse path control by having the edge AS O insert the problem network A into path advertisements for O's ad-

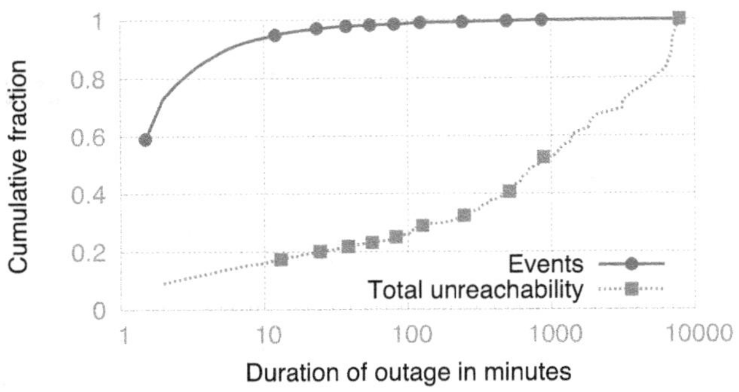

Figure 1: For partial outages observed from EC2, the fraction of outages of at most a given duration (solid) and their corresponding fraction of total unreachability (dotted). The x-axis is on a log-scale. More than 90% of the outages lasted at most 10 minutes, but 84% of the total unavailability was due to outages longer than 10 minutes.

dresses, so that it appears that A has already been visited. When the announcements reach A, BGP's loop-prevention mechanisms will drop the announcement. Networks that would have routed through A will only learn of other paths, and will avoid A. Using the BGP-Mux testbed [5] to announce paths to the Internet, we show *LIFEGUARD*'s rerouting technique finds alternate paths 76% of the time.

While this BGP *poisoning* provides a means to trigger rerouting, we must address a number of challenges to provide a practical solution. *LIFEGUARD* combines this basic poisoning mechanism with a number of techniques. *LIFEGUARD* has a subsystem to locate failures, even in the presence of asymmetric routing and unidirectional failures. We validate our failure isolation approach and present experiments suggesting that the commonly used traceroute technique for failure location gave incorrect information 40% of the time. We address how to decide whether to poison; will routing protocols automatically resolve the problem, or do we need to trigger route exploration? We show empirically that triggering route exploration could eliminate up to 80% of the observed unavailability. When it reroutes failing paths to remediate partial outages, *LIFEGUARD* carefully crafts BGP announcements to speed route convergence and minimize the disruption to working routes. Our experimental results show that 94% of working routes reconverge instantly and experience minimal ($\leq 2\%$) packet loss. After rerouting, *LIFEGUARD* maintains a *sentinel prefix* on the original path to detect when the failure has resolved, even though live traffic will be routing over an alternate path. When *LIFEGUARD*'s test traffic reaches the sentinel, *LIFEGUARD* removes the poisoned announcement.

2. BACKGROUND AND MOTIVATION

2.1 Quantifying Unreachability from EC2

To test the prevalence of outages, we conducted a measurement study using Amazon EC2, a major cloud provider. EC2 presumably has the resources, business incentive, and best practices available for combating Internet outages. We show that even EC2 data centers experience many long-term network connectivity problems.

We rented EC2 instances in the four available AWS regions from July 20, 2010 to August 29, 2010. Each vantage point issued a pair of pings every 30 seconds to 250 targets – five routers each from the 50 highest-degree ASes [33]. We selected the routers randomly from the iPlane topology [17], such that each router was from a distinct BGP prefix. We focus on paths to routers in major networks, which should be more reliable than paths to end-hosts. We define an outage as four or more consecutive dropped pairs of pings from a single vantage point to a destination. This methodology means that the minimum outage duration we consider is 90 seconds.

In 79% of the outages in our study, some vantage points had connectivity with the target (one of the routers), while others did not. Fig. 1 shows the durations of these 10,308 partial outages. By comparison, an earlier study found that 90% of outages lasted less than 15 minutes [13]. We also find that most outages are relatively short; more than 90% lasted less than 10 minutes (solid line). However, these short outages account for only 16% of the total unavailability (dotted line). The relatively small number of long-lasting problems account for much of the overall unavailability. Delayed protocol convergence does not explain long outages [23].

In fact, many long-lasting outages occur with few or no accompanying routing updates [13, 20]. With routing protocols failing to react, networks continue to send traffic along a path that fails to deliver packets. Such problems can occur, for example, when a router fails to detect an internal fault (e.g., corrupted memory on a line card causing traffic to be black-holed [35]) or when cross-layer interactions cause an MPLS tunnel to fail to deliver packets even though the underlying IP network is operational [21].

During partial outages, some hosts are unable to find the working routes to the destination, either due to a physical partition, due to a routing policy that restricts the export of the working path, or due to a router that is announcing a route that fails to deliver traffic. The techniques we present later in this paper rely on the underlying network being physically connected and on the existence of policy-compliant routes around the failure, and so we need to establish that there are long-lasting outages that are not just physical failures or the result of routing export policies.

We can rule out physical partitions as the cause in our EC2 study. All EC2 instances maintained connectivity with a controller at our institution throughout the study. So, physical connectivity existed from the destination to some EC2 instance, from that instance to the controller, then from the controller to the instance that could not reach the destination.

Since the problems are not due to physical partitions, either routing policies are eliminating all working paths, or routers are advertising paths that do not work. By detouring through our institution, the paths we demonstrated around EC2 failures violate the valley-free routing policy [15] – in not making those paths available, routers are properly enacting routing policy. However, if working policy-compliant paths also exist, it might be possible to switch traffic onto them.

2.2 Assessing Policy-Compliant Alternate Paths

Earlier systems demonstrated that overlays can route around many failures [2, 6, 16]. However, overlay paths tend to violate BGP export policies. We build on this previous work by showing that alternate policy-compliant paths appear to exist during many failures. Generally, the longer a problem lasted, the more likely it was that alternative routes existed.

Previous work found that many long-lasting failures occur outside of the edge networks [13, 20], and we focus on these types of problems. Every ten minutes for a week starting September 5, 2011, we issued traceroutes between all PlanetLab sites. This setting allowed us to issue traceroutes from both ends of every path, and the probes to other PlanetLab sites give a rich view of other paths that might combine to form alternate routes. We considered as outages all instances in which a pair of hosts were up and had previously been able to send traceroutes between each other, but all traceroutes in both directions failed to reach the destination AS for at least three consecutive rounds, before working again. This yielded nearly 15,000 outages.

We checked if the traceroutes included working policy-compliant routes around the failures. For each round of a failure, we tried to find a path from the source that intersected (at the IP-level) a path to the destination, such that the spliced path did not traverse the AS in which the failed traceroute terminated. We only considered a spliced path as valid if it would be available under observable export policies. To check export policies, when splicing together a potential path, we only accepted it if the AS subpath of length three centered at the splice point appeared in at least one traceroute during the week [17, 25]. This check suffices to encode the common valley-free export policy [15].

Our methodology may fail to identify some valid paths that exist. PlanetLab has somewhat homogeneous routing. We also required that spliced paths intersect at a shared IP address. Two traceroutes might intersect at a router or a PoP without sharing an IP address. We would not consider this intersection when trying to splice paths.

We found that, for 49% of outages, alternate paths existed for the duration of the failure. Considering only long outages that lasted at least an hour, we found alternate routes in 83% of failures. For 98% of the outages in which an alternate path existed during the first round of the failure, the path persisted for the duration.

2.3 Current Approaches to Address Failures

Lacking better options, operators rely on insufficient techniques to try to locate and resolve long-lasting outages, especially if the failure is in a network outside the operators' control. Asymmetric paths leave operators with a limited view even when paths work. Tools like traceroute require bidirectional connectivity to function properly, and so failures restrict their view further. Public traceroute servers and route collectors [26,31] extend the view somewhat, but only a small percentage of networks make them available. In fact, these challenges mean that operators frequently resort to asking others to issue traceroutes on their behalf to help confirm and isolate a problem [28].

If operators successfully identify a failure outside their own networks, they have little ability to effect repair:

Forward path failures: The source network's operators can select an alternative egress in an attempt to avoid the problem. When choosing, they can see the full BGP paths provided by their neighbors. Each of the source's providers announces its preferred path, and the source is free to choose among them. If the network's providers offer sufficiently diverse paths, the failure may be avoided. For example, we inspected BGP routes from five universities (University of Washington, University of Wisconsin, Georgia Tech, Princeton, and Clemson) [5] to prefixes in 114 ASes. If these universities were our providers, the routes are sufficiently diverse that, if the last AS link before the destination on one of the routes failed silently, we could route around it to reach the destination in 90% of cases by routing via a different provider. In §5.2, we present an equivalent experiment demonstrating that our techniques would allow us to avoid 73% of these links on reverse paths back from the 114 ASes, without disturbing routes that did not use that link. [1]

Reverse path failures: Using traditional techniques, however, having multiple providers may not offer much reverse path diversity. Under BGP, the operators can only change how they announce a prefix to neighbors, perhaps announcing it differently to different neighbors. They have no other direct influence over the paths other networks select. A major limitation of existing techniques for announcement-based route control is that they generally act on the next hop AS, rather than allowing a network to target whichever AS is causing a problem. We discuss the techniques below:

[1]The 114 ASes were all those that both announce prefixes visible at the universities, needed for the forward path study, and peer with a route collector [1, 26, 29, 31], needed for the reverse study.

Multi-Exit Discriminator (MEDs): An AS that connects to another AS at multiple points can use MEDs to express to the neighbor on which peering point it prefers to receive traffic. However, MEDs have meaning only within the context of that single neighbor, so they generally are effective only if the problem is in the immediate upstream neighbor.

Selective Advertising: An origin AS with multiple providers can advertise a prefix through only some providers. In variations on this approach, the origin can advertise more-specific prefixes through some providers and only less-specifics through others. Or, since many ASes use path length as a tiebreaker when making routing decisions, networks sometimes prepend routes they announce with multiple copies of their AS, to make that path longer and hence less preferred than shorter ones. With all these approaches, the origin can shift traffic away from providers it wants to avoid. If the problem is not in the immediate provider, these techniques may be deficient because (1) all working routes that had previously gone through that provider will change; and (2), even if all sources with failing paths had routed through a particular provider before selective advertising, forcing them to route via a different provider may not change the portion of the path containing the failure.

BGP communities: Communities are a promising direction for future experiments in failure avoidance but do not currently provide a complete solution. An AS can define communities that other networks can tag onto routes they announce to the AS. Communities instruct the AS on how to handle the routes. For example, SAVVIS offers communities to specify that a route should not be exported to a peer. However, communities are not standardized, and some ASes give limited control over how they disseminate routes. Further, many ASes do not propagate community values they receive [30], and so communities are not a feasible way to notify arbitrary ASes of routing problems. We announced experimental prefixes with communities attached and found that, for example, any AS that used a Tier-1 to reach our prefixes did not have the communities on our announcements.

Changes to BGP announcements and to local configuration may be unable to repair outages. In such cases, operators often must resort to phone calls or e-mails asking operators at other networks for support. These slow interactions contribute to the duration of outages. We now show how our approach enables an operator to avoid reverse path failures.

3. ENABLING FAILURE AVOIDANCE

Suppose an AS O wants to communicate with another AS Q but cannot because of some problem on the path between them. If the problem is within either O or Q, operators at that network have complete visibility into and control over their local networks, and so they can take appropriate steps to remedy the problem. Instead, consider a case in which the problem occurs somewhere outside of the edge ASes, either on the forward path to Q or on the reverse path back to O. Further suppose that O is able to locate the failure and to determine that an alternate route likely exists.[2]

O would like to restore connectivity regardless of where the problem is, but its ability to do so currently depends largely on where the problem is located. If the problem is on the forward path and O's providers offer suitable path diversity, O can choose a path that avoids the problem. By carefully selecting where to locate its PoPs and which providers to contract with, O should be able to achieve decent resiliency to forward path failures. However, having a diversity of providers may not help for reverse path failures, as O has

[2]We discuss how *LIFEGUARD* does this in §4.

little control over the routes other ASes select to reach it. As explained in §2.3, route control mechanisms like MEDs and selective advertising only let *O* control the PoP or provider through which traffic enters *O*. However, these BGP mechanisms give *O* essentially no control over how other ASes reach the provider it selects.

O needs a way to notify ASes using the path that the path is not successfully forwarding traffic, thereby encouraging them to choose alternate routes that restore connectivity. As a hint as to which paths they should avoid, *O* would like to inform them of the failure location. AS-level failure locations are the proper granularity for these hypothetical notifications, because BGP uses AS-level topology abstractions. In particular, when one of the notified ASes chooses an alternate path, it will be selecting from AS paths announced to it by its neighbors. Therefore, *O* needs to inform other ASes of which AS or AS link to avoid, depending on whether the failure is within a single AS or at an AS boundary.

Ideally, we would like a mechanism to let the origin AS *O* of a prefix *P* specify this information explicitly with a signed announcement we will call Avoid_Problem(X,P). Depending on the nature of the problem, *X* could either be a single AS (Avoid_Problem(A,P)) or an AS link $A - B$ (Avoid_Problem(A-B,P)). Note that AS *O* is only able to directly observe the problem with prefix *P*; it cannot determine if the issue is more widespread. Announcing this hypothetical primitive would have three effects:

- *Avoidance Property*: Any AS that knew of a route to *P* that avoided *X* would select such a route.
- *Backup Property*: Any AS that only knew of a route through *X* would be free to attempt to use it. Similarly, *A* would be able to attempt to route to *O* via its preferred path (through *B* in the case when *X* is the link *A-B*).
- *Notification Property*: *A* (and *B*, for link problems) would be notified of the problem, alerting its operators to fix it.

3.1 LIFEGUARD's Failure Remediation

Deploying Avoid_Problem(X,P) might seem to require changes to every BGP router in the Internet. Instead, we use mechanisms already available in BGP to perform the notifications, in order to arrive at a solution that is usable today, even if the solution is not complete. A usable approach can improve availability today while simultaneously helping the community understand how we might improve availability further with future BGP changes. We call our approach *LIFEGUARD*, for *Locating Internet Failures Effectively and Generating Usable Alternate Routes Dynamically*.

To approximate Avoid_Problem(A,P) on today's Internet, *LIFEGUARD* uses BGP's built-in loop prevention to "poison" a problem AS that is announcing routes but not forwarding packets. To poison an AS *A*, the origin announces the prefix with *A* as part of the path, causing *A* to reject the path (to avoid a loop) and withdraw its path from its neighbors [8, 10]. This causes ASes that previously routed via *A* to explore alternatives. Importantly, the poison affects only traffic to *O*'s prefix experiencing the problem. By allowing an AS to poison only prefixes it originates, our approach is consistent with the goals of work toward authenticating the origin of BGP announcements [27]. Proposals to verify the entire path [3] are consistent with the future goal for our approach, in which Avoid_Problem(X,P) would be a validated hint from the origin AS to the rest of the network that a particular AS is not correctly routing its traffic. By the time such proposals are deployed, it may be feasible to develop new routing primitives or standardized communities to accomplish what we currently do with poisoning.

Although BGP loop prevention was not intended to give *O* control over routes in other ASes, it lets us experiment with failure

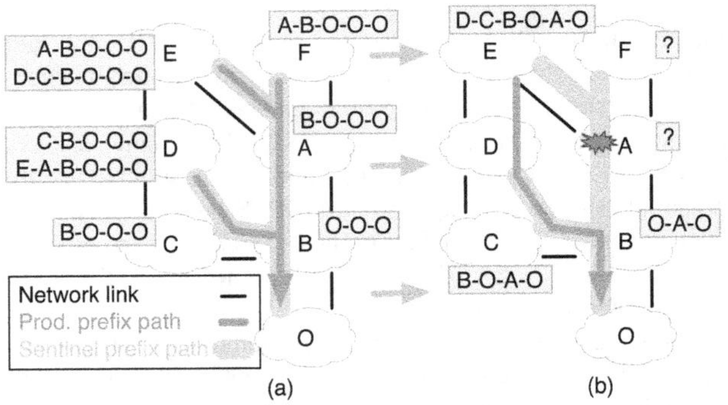

Figure 2: Routes and routing tables (a) before and (b) after *O* poisons *A* to avoid a problem. Each table shows *only paths to the production prefix*, with the in-use, most-preferred route at the top. Poisoning *A* for the production prefix causes it to withdraw its route from *E* and *F*, forcing *E* to use its less-preferred route through *D* and leaving *F* with only the sentinel. Routes to the sentinel prefix do not change, allowing *O* to check when the problem has resolved.

avoidance. In effect, poisoning *A* implements the *Avoidance Property* of Avoid_Problem(A,P), giving *O* the means to control routes to it. *A*'s border routers will receive the poisoned announcement and detect the poison, a form of the *Notification Property*.

On its own, poisoning is a blunt, disruptive instrument, a limitation that *LIFEGUARD* must overcome. Poisoning inserts *A* into all routes, so even ASes that were not routing through *A* may undergo route exploration before reconverging to their original route, leading to unnecessary packet loss [23]. Instead of providing the *Backup Property*, poisoning cuts off ASes that lack a route around *A*. Poisoning disables all paths through *A*, even if some work.

In the following sections, we show how *LIFEGUARD* overcomes what initially seem like limitations of poisoning in order to better approximate Avoid_Problem(X,P).

3.1.1 Minimizing Disruption of Working Routes

Inserting an AS to poison an announcement increases AS path length. Suppose that an origin AS *O* decides to poison *A* for *O*'s prefix *P*. The poisoned path cannot be *A-O*, because *O*'s neighbors need to route to *O* as their next hop, not to *A*. So, the path must start with *O*. It cannot be *O-A*, because routing registries list *O* as the origin for *P*, and so a path that shows *A* as the origin looks suspicious. Therefore, *O* announces *O-A-O*. Experiments found that BGP normally takes multiple minutes to converge when switching to longer paths, with accompanying packet loss to the prefix during this period [23]. This loss would even affect networks with working paths to the prefix.

To poison in a way that shortens and smooths this convergence period, *LIFEGUARD* crafts steady-state unpoisoned announcements in a way that "prepares" all ASes for the possibility that some AS may later be poisoned. Fig. 2 provides an example of an origin AS *O* with a *production prefix P* which carries real traffic. Fig. 2(a) depicts the state before the problem, and Fig. 2(b) depicts the state following a failure, after *O* has reacted to repair routing.

LIFEGUARD speeds up convergence and reduces path exploration by prepending to the production prefix *P*'s announcement, announcing *O-O-O* as the baseline path. If *O* detects that some networks (ASes *E* and *F* in Fig. 2) cannot reach *P* due to a problem in *A*, *O* updates the announcement to *O-A-O*. These two announcements are the same length and have the same next hop, and so, under default BGP, they are equally preferred. If an AS is using a route that starts with *O-O-O* and does not go through *A*, then receives an update that changes that route to start with *O-A-O* instead, it will likely switch to using the new route without exploring other

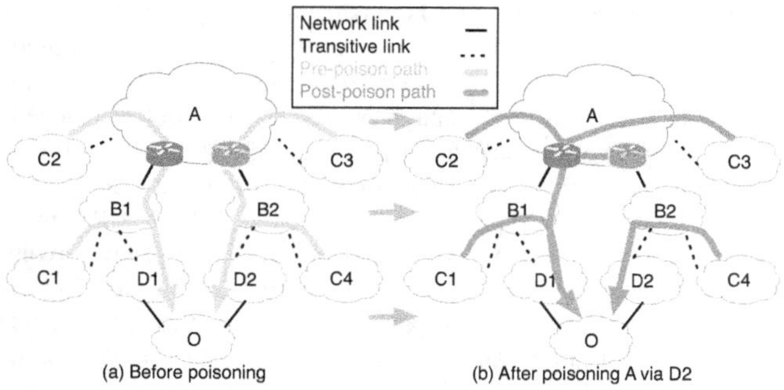

Figure 3: A case in which *LIFEGUARD* can use *selective poisoning*. By selectively poisoning *A* on announcements to *D2* and not on announcements to *D1*, *O* can cause traffic to avoid the link from *A* to *B2*, without disrupting how *C3* routes to *A* or how *C[1,2,4]* route to *O*.

options, converging instantly. We will show in a measurement study in §5.2 that this prepending smooths convergence, helping ease concerns that an automated response to outages might introduce needless routing instability. This approach is orthogonal to efforts to reduce convergence effects [18, 22, 24], which *LIFEGUARD* would benefit from.

3.1.2 Partially Poisoning ASes

LIFEGUARD tries to avoid cutting off an entire AS *A* and all ASes that lack routes that avoid *A*. We have three goals: (1) ASes cut off by poisoning should be able to use routes through *A* to reach *O* as soon as they work again; (2) if some paths through *A* work while others have failed, ASes using the working routes should be able to continue to if they lack alternatives; and (3) when possible, we should steer traffic from failed to working paths within *A*.

Advertising a less-specific sentinel prefix. While *O* is poisoning *A*, ASes like *F* that are "captive" behind *A* will lack a route [7]. To ensure that *F* and *A* have a route that covers *P*, *LIFEGUARD* announces a less-specific *sentinel prefix* that contains *P* (and can also contain other production prefixes). When *P* experiences problems, the system continues to advertise the sentinel with the baseline (unpoisoned) path. As seen in Fig. 2(b), ASes that do not learn of the poisoned path, because they are "captive" behind *A*, will receive the less specific prefix and can continue to try routing to the production prefix on it, through *A*, instead of being cut off. This effect is the *Backup Property* desired from AVOID_PROBLEM(A,P) and helps achieve goals (1) and (2).

Selectively poisoning to avoid AS links. Although most failures in a previous study were confined to a single AS, 38% occurred on an inter-AS link [13]. We use a technique we call *selective poisoning* to allow *LIFEGUARD*, under certain circumstances, to implement AVOID_PROBLEM(A-B,P). Poisoning does not provide a general solution to AS link avoidance, but, given certain topologies, selective poisoning can shift traffic within *A* onto working routes.

Specifically, under certain circumstances, *O* may be able to steer traffic away from a particular AS link without forcing it completely away from the ASes that form the link. Suppose *O* has multiple providers that connect to *A* via disjoint AS paths. Then *O* can poison *A* in advertisements to one provider, but announce an unpoisoned path through the other provider. Because the paths are disjoint, *A* will receive the poisoned path from one of its neighbors and the unpoisoned path from another, and it will only accept the unpoisoned path. So, *A* will route all traffic to *O*'s prefix to egress via the neighbor with the unpoisoned path. This *selective poisoning* shifts routes away from *A*'s link to the other neighbor, as well as possibly affecting which links and PoPs are used inside *A*.

Fig. 3 illustrates the idea. Assume *O* discovers a problem on the link between *A* and *B2*. This failure affects *C3*, but *C2* still has a working route through *A*, and *C4* still has a working route through *B2*. *O* would like to shift traffic away from the failing link, without forcing any networks except *A* to change which neighbor they select to route through. In other words, *O* would like to announce AVOID_PROBLEM(A-B2,P). If *O* only uses selective advertising without poisoning, announcing its prefix via *D1* and not *D2*, *C4*'s route will have to change. If *O* poisons *A* via both *D1* and *D2*, *C2* and *C3* will have to find routes that avoid *A*, and *A* will lack a route entirely (except via a less-specific prefix). However, by selectively poisoning *A* via *D2* and not via *D1*, *O* can shift *A* and *C3*'s routes away from the failing link, while allowing *C3* to still route along working paths in *A* and without disturbing any other routes. Selective poisoning functions like targeted prepending – prepending requires that *A* use path length to make routing decisions and potentially causes other ASes to shift from using routes through *D2*, whereas selective poisoning forces only *A* to change. In §5.2 we find that selective poisoning lets *LIFEGUARD* avoid 73% of the links we test.

4. APPLYING FAILURE AVOIDANCE

In the previous section, we described how *LIFEGUARD* uses BGP poisoning to approximate AVOID_PROBLEM(X,P). This allows us to experiment with failure avoidance today, and it gives ASes a way to deploy failure avoidance unilaterally. In this section, we describe how *LIFEGUARD* decides when to poison and which AS to poison, as well as how it decides when to stop poisoning an AS.

4.1 Locating a Failure

An important step towards fixing a reachability problem is to identify the network or router responsible for it. To be widely applicable and effective, we require our fault isolation technique to:

- be effective even if the system has control over only one of the endpoints experiencing a reachability problem.
- be accurate even if measurement probes to reachable targets appear to fail due to rate-limiting, chronic unresponsiveness, and/or being dropped on the reverse direction.
- integrate information from multiple measurement nodes, each with only partial visibility into routing behavior.

We assume that a routing failure between a pair of endpoints can be explained by a single problem. While addressing multiple failures is an interesting direction for future work, this paper focuses on single failures.

4.1.1 Overview of Failure Isolation

The failure isolation component of *LIFEGUARD* is a distributed system, using geographically distributed PlanetLab hosts to make data plane measurements to a set of monitored destinations. Because many outages are partial, *LIFEGUARD* uses vantage points with working routes to send and receive probes on behalf of those with failing paths. Because many failures are unidirectional [20], it adapts techniques from reverse traceroute [19] to provide reverse path visibility. In the current deployment, vantage points send pings to monitor destinations, and a vantage point triggers failure isolation when it experiences repeated failed pings to a destination.

LIFEGUARD uses historical measurements to identify candidates that could potentially be causing a failure, then systematically prunes the candidate set with additional measurements. We outline these steps first before describing them in greater detail.

1. **Maintain background atlas:** *LIFEGUARD* maintains an atlas of the round-trip paths between its sources and the monitored

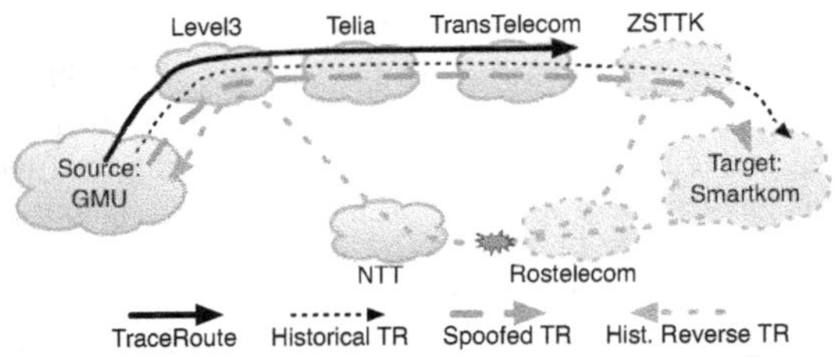

Figure 4: Isolation measurements conducted for an actual outage. With traceroute alone, the problem appears to be between TransTelecom and ZSTTK. Using spoofed traceroute, reverse traceroute, and historical path information, *LIFEGUARD* determines that the forward path is fine, but that Rostelecom no longer has a working path back to GMU.

targets to discern changes during failures and generate candidates for failure locations.

2. **Isolate direction of failure and measure working direction:** After detecting a failure, *LIFEGUARD* isolates the direction of failure to identify what measurements to use for isolation. Further, if the failure is unidirectional, it measures the path in the working direction using one of two measurement techniques.
3. **Test atlas paths in failing direction:** *LIFEGUARD* tests which subpaths still work in the failing direction by probing routers on historical atlas paths between the source and destination. It then remeasures the paths for responsive hops, thereby identifying other working routers.
4. **Prune candidate failure locations:** Routers with working paths in the previous step are eliminated. Then, *LIFEGUARD* blames routers that border the "horizon of reachability." This horizon divides routers that have connectivity to the source from those that lack that connectivity.

4.1.2 Description of Fault Isolation

We illustrate the steps that *LIFEGUARD* uses to isolate failures using an actual example of a failure diagnosed on February 24, 2011. At the left of Fig. 4 is one of *LIFEGUARD*'s vantage points, a PlanetLab host at George Mason University (labeled *GMU*). The destination, belonging to Smartkom in Russia, at the right, became unreachable from the source. For simplicity, the figure depicts hops at the AS granularity.

Maintain background atlas: In the steady state, *LIFEGUARD* uses traceroute and reverse traceroute to regularly map the forward and reverse paths between its vantage points and the destinations it is monitoring. During failures, this path atlas yields both a view of what recent paths looked like before the failure, as well as a historical view of path changes over time. These paths provide likely candidates for failure locations and serve as the basis for some of the isolation measurements we discuss below. Because some routers are configured to ignore ICMP pings, *LIFEGUARD* also maintains a database of historical ping responsiveness, allowing it to later distinguish between connectivity problems and routers configured to not respond to ICMP probes.

The figure depicts historical forward and reverse traceroutes with the dotted black and red lines, respectively. The thick, solid black line in Fig. 4 depicts the traceroute from GMU during the failure. Traceroute can provide misleading information in the presence of failures. In this case, the last hop is a TransTelecom router, suggesting that the failure may be adjacent to this hop, between TransTelecom and ZSTTK. However, without further information, operators cannot be sure, since the probes may have been dropped on the reverse paths back from hops beyond TransTelecom.

Isolate direction of failure and measure working direction: *LIFEGUARD* tries to isolate the direction of the failure using spoofed pings [20]. In the example, spoofed probes sent from GMU to Smartkom reached other vantage points, but no probes reached GMU, implying a reverse path failure. When the failure is unidirectional, *LIFEGUARD* measures the complete path in the working direction. Extending the spoofed ping technique, *LIFEGUARD* sends spoofed probes to identify hops in the working direction while avoiding the failing direction. For a reverse failure, *LIFEGUARD* finds a vantage point with a working path back from *D*, then has *S* send a spoofed traceroute to *D*, spoofing as the working vantage point. In the example, GMU issued a spoofed traceroute, and a vantage point received responses from ZSTTK and the destination. The blue dashed edges in Fig. 4 show the spoofed traceroute.

If the failure had been on the forward path, the system instead would have measured the working reverse path with a spoofed reverse traceroute from *D* back to *S*.

It is useful to measure the working direction of the path for two reasons. First, since the path is likely a valid policy-compliant path in the failing direction, it may provide a working alternative for avoiding the failure. Second, knowledge of the path in the working direction can guide isolation of the problem in the failing direction, as we discuss below.

Test atlas paths in failing direction: Once it has measured the working path, *LIFEGUARD* measures the responsive portion of the path. For forward and bidirectional failures, the source can simply issue a traceroute towards the destination.

For reverse failures, *LIFEGUARD* cannot measure a reverse traceroute from *D*, as such a measurement requires a response from *D* to determine the initial hops. Instead, *LIFEGUARD* has its vantage points, including *S*, ping: (1) all hops on the forward path from *S* to *D* and (2) all hops on historical forward and reverse paths between *S* and *D* in its path atlas. These probes test which locations can reach *S*, which cannot reach *S* but respond to other vantage points, and which are completely unreachable. *LIFEGUARD* uses its atlas to exclude hops configured never to respond. For all hops still pingable from *S*, *LIFEGUARD* measures a reverse traceroute to *S*.

In the example, *LIFEGUARD* found that NTT still used the same path towards GMU that it had before the failure and that Rostelecom no longer had a working path. We omit the reverse paths from most forward hops to simplify the figure. In this case, *LIFEGUARD* found that all hops before Rostelecom were reachable (denoted with blue clouds with solid boundaries), while all in Rostelecom or beyond were not (denoted with light-gray clouds with dashed boundaries), although they had responded to pings in the past.

Prune candidate failure locations: Finally, *LIFEGUARD* removes any reachable hops from the suspect set and applies heuristics to identify the responsible hop within the remaining suspects. For forward outages, the failure is likely between the last responsive hop in a traceroute and the next hop along the path towards the destination. *LIFEGUARD*'s historical atlas often contains paths through the last hop, providing hints about where it is trying to route.

For a reverse failure, *LIFEGUARD* considers reverse paths from *D* back to *S* that are in its atlas prior to the failure. For the most recent path, it determines the farthest hop *H* along that path that can still reach *S*, as well as the first hop *H'* past *H* that cannot. Given that *H'* no longer has a working path to *S*, contacting the AS containing *H'* or rerouting around it may resolve the problem.

If the failure is not in *H'*, one explanation is that, because *H'* lacked a route, *D* switched to another path which also did not work. In these cases, *LIFEGUARD* performs similar analysis on older historical paths from D, expanding the initial suspect set and repeating

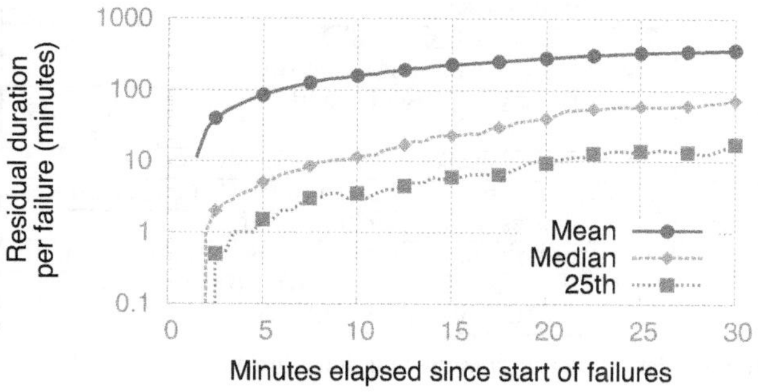

Figure 5: For our EC2 dataset, residual duration after outages have persisted for X minutes. The graph shows that, once a problem has persisted for a few minutes, it will most likely persist for at least a few more minutes unless we take corrective action.

the pruning. Since Internet paths are generally stable [37], the failing path will often have been used historically, and there will often be few historical paths between the source and destination.

Because both historical reverse paths from unresponsive forward hops traversed Rostelecom, it seems highly likely this is the point of failure. This conclusion is further supported by the fact that all unreachable hops except Rostelecom responded to pings from other vantage points, indicating that their other outgoing paths still worked. We provide details on this and other examples at `http://lifeguard.cs.washington.edu`, and we provide details on *LIFEGUARD*'s failure isolation in a tech report [32].

4.2 Deciding to Start and Stop Poisoning

Deciding whether to poison: As seen in Fig. 1, most outages resolve quickly. For the system to work effectively, it would be helpful to differentiate between outages that will clear up quickly and those that will persist. If routing protocols will quickly resolve a problem, then it would be better to wait, to avoid causing further routing changes. If the protocols will not restore connectivity quickly on their own, then poisoning may be a better approach.

The following analysis of our EC2 study (§2.1) shows that it is possible to differentiate between these cases with high likelihood. Fig. 5 shows the residual duration of these outages, given that they have already lasted for X minutes. The median duration of an outage in the study was only 90 seconds (the minimum possible given the methodology). However, of the 12% of problems that persisted for at least 5 minutes, 51% lasted at least another 5 minutes. Further, of the problems that lasted 10 minutes, 68% persisted for at least 5 minutes past that. *LIFEGUARD* triggers isolation after multiple rounds of failed pings, and it takes an average of 140 seconds to isolate a reverse path outage. If a problem persists through both those stages, then the results suggest that the problem is likely to persist long enough to justify using poisoning to fix it. We will show in §5.2 that poisoned routes converge within a few minutes in almost all cases, with little loss during the convergence period. So, if there are alternative paths that avoid the problem *LIFEGUARD* locates, our system should quickly restore connectivity.

Long-lasting problems account for much of the total unavailability. As a result, even if *LIFEGUARD* takes five minutes to identify and locate a failure before poisoning, and it then takes two minutes for routes to converge, we can still potentially avoid 80% of the total unavailability in our EC2 study. In §5.1, we will show that it is possible to determine (with high likelihood) whether alternate policy compliant paths will exist before deciding to poison an AS. If no paths exist, *LIFEGUARD* does not attempt to poison the AS.

Deciding when to unpoison: Once *LIFEGUARD* accurately identifies the AS *A* responsible for a problem, BGP poisoning can target it and cause other ASes to route around *A*. However, *A* will eventually resolve the underlying issue, at which point we would like to be able to revert to the unpoisoned path, allowing ASes to use paths through *A*, if preferred. When the poisoned announcement is in place, however, *A* will not have a path to the prefix in question.

LIFEGUARD uses a sentinel prefix to test reachability. Concerns such as aggregation and address availability influence the choice of sentinel. In our current deployment, the sentinel is a less specific prefix containing both the production prefix and a prefix that is not otherwise used. Responses to pings from the unused portion of the sentinel will route via the sentinel prefix, regardless of whether the hops also have the poisoned more-specific prefix. By sending active ping measurements from this prefix to destinations that had been unable to reach the production prefix prior to poisoning (e.g., *E* in Fig. 2), the system can detect when to unpoison the production prefix. If the sentinel is a less-specific without any unused prefixes, *LIFEGUARD* can instead ping the destinations within the poisoned AS (e.g., *A*) or within captives of the poisoned AS (e.g., *F*).

5. LIFEGUARD EVALUATION

To preview the results of our evaluation, we find that *LIFEGUARD*'s poisoning finds routes around the vast majority of potential problems, its approach is minimally disruptive to paths that are not experiencing problems, and its failure isolation can correctly identify the AS needing to be poisoned. Table 1 summarizes our key results; the following sections provide more details.

We deployed *LIFEGUARD*'s path poisoning using the BGP-Mux testbed [5], using its AS number and prefixes. *LIFEGUARD* connected to a BGP-Mux instance at Georgia Tech, which served as the Internet provider for the BGP-Mux AS (and hence for *LIFEGUARD*). For the poisoning experiments in this section, *LIFEGUARD* announced prefixes via Georgia Tech into the commercial Internet.

We assessed the effects of poisoning in the absence of failures. To obtain ASes to poison, we announced a prefix and harvested all ASes on BGP paths towards the prefix from route collectors [26, 31]. We excluded all Tier-1 networks, as well as Cogent, as it is Georgia Tech's main provider. In §5.2 and §7.1, we evaluate techniques to poison even these large ASes. For each of the remaining harvested ASes,[3] we first went from a baseline of *O* to a poisoned announcement *O-A-O*, then repeated the experiment starting from a baseline of *O-O-O*. We kept each announcement in place for 90 minutes to allow convergence and to avoid flap dampening effects. For the duration of the experiments, we also announced an unpoisoned prefix to use for comparisons.

5.1 Efficacy

Do ASes find new routes that avoid a poisoned AS? We monitored BGP updates from public BGP route collectors to determine how many ASes found an alternate path after we poisoned an AS on their preferred path. There were 132 cases in which an AS peering with a route collector was using a path through one of the ASes we poisoned. In 102 of the 132 cases (77%), the route collector peer found a new path that avoided the poisoned AS. Two-thirds of the cases in which the peer could not find a path were instances in which we had poisoned the only provider of a stub AS.

Do alternate policy-compliant routes exist in general? We analyzed a large AS topology to show that, in the common case, alternate paths exist around poisoned ASes. The topology, along

[3]We announced our experiments on the NANOG mailing list and allowed operators to opt out their ASes. None did. A handful opted out of an earlier Georgia Tech study, and we honored their list.

Criteria	Summary	Experimental Result
Effectiveness (§5.1)	Most edge networks have routes that avoid poisoned ASes, and we can calculate which do *a priori*	77% of poisons from BGP-Mux 90% of poisons in large-scale simulation
Disruptiveness (§5.2)	Working routes that already avoid the problem AS reconverge quickly after poisoning	95% of paths converge instantly
	Minimal loss occurs during convergence	Less than 2% packet loss in 98% of cases
Accuracy (§5.3)	Locates failures as if it had traceroutes from both ends	Consistent results for 93% of inter-PlanetLab failures
	Isolates problems that traceroute alone misdiagnoses	40% of cases differ from traceroute
Scalability (§5.4)	Quickly isolates problems with reasonable overhead	140 s for poisoning candidates, 280 probes per failure
	Reasonably small additional update load for addressing most of observed unavailability	< 1% if 1% of ISPs use *LIFEGUARD* < 10% if 50% of ISPs use *LIFEGUARD*

Table 1: Key results of our *LIFEGUARD* evaluation, demonstrating its viability for addressing long-duration outages.

with AS relationships, is from a dataset combining public BGP feeds with more than 5 million AS paths between BitTorrent (BT) peers [9]. To simulate poisoning an AS A on a path from a source S to an origin O, we remove all of A's links from the topology. We then check if S can restore connectivity while avoiding A (i.e., a path exists between S and O that obeys export policies).

We check policies using the three-tuple test [25], as in §2.2. This approach may miss alternate paths if the valley-free assumption is too strict,[4] and rarely used backup paths may not be in our topology. Conversely, it may identify valley-free paths not used in practice.

To establish the validity of this methodology, we simulated the Georgia Tech poisonings. In 92.5% of cases, the simulation found an alternate path if and only if the AS found such a path following our actual poisoning. In the remaining 7.5% of cases, the simulation found an alternate path, but in practice the AS failed to find one. In these cases, the source was an academic network connected to both a commercial provider and an academic provider (which only provides routes to academic destinations). These connections made the AS multi-homed in our simulation. In practice, however, the AS seems to reject routes from its commercial provider if the route contained an academic AS, and we had poisoned one.

Having established that our simulation closely predicts the results of actual poisonings, we simulated poisoning ASes on the BT and BGP feed paths. For each AS path with length greater than 3 (i.e., traversing at least one transit AS in addition to the destination's provider), we iterated over all the transit ASes in the path except the destination's immediate provider and simulated poisoning them one at a time.[5] An alternate path existed in 90% of the 10M cases. We then simulated poisoning the failures isolated by *LIFEGUARD* in June 2011. Alternate paths existed in 94% of them.

5.2 Disruptiveness

How quickly do paths converge after poisoning? We used updates from the BGP collectors to measure convergence delay after poisoning. We will show that, in most cases, if an AS was not routing through the AS we poisoned, it re-converges essentially instantly to its original path, requiring only a single update to pass on the poison. Global convergence usually takes at most a few minutes.

First, we assess how quickly ASes that were using the poisoned AS settle on a new route that avoids it. We also assess how quickly routes converge for ASes that were not using the poisoned AS. As explained above, we poisoned each harvested AS twice each using different pre-poisoning baseline paths. After each announcement, for each AS that peers with a route collector, we measured the delay from when the AS first announced an update to the route collector to when it announced its stable post-poisoning route. We leave out (route collector peer, poisoned AS) pairs if the peer AS did not have a path to the *LIFEGUARD* prefix following the poisoning.

[4]It is well known that not all paths are valley-free in practice, and we observed violations in the BitTorrent traceroutes.

[5]A multi-homed destination can use selective advertising to avoid a particular provider. A single-homed destination can never avoid having paths traverse its provider.

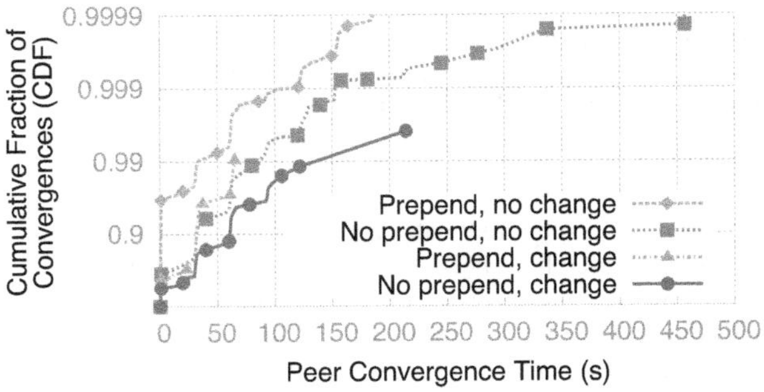

Figure 6: Convergence time for route collector peer ASes after poisoned announcements. A data point captures the convergence for one peer AS after one poisoned announcement. *Change* vs. *no change* indicates if the peer had to change its path because it had been routing through the poisoned AS. For *Prepend*, the baseline announcement before poisoning was *O-O-O*, whereas for *No prepend* it was *O*. The poisoned announcement was always *O-A-O*. Prepending reduces path exploration by keeping announcement length consistent.

As seen in Fig. 6, a baseline announcement of *O-O-O* greatly speeds convergence. More than 95% of the time, ASes that were not using the poisoned AS converged essentially instantly upon receiving the poisoned update, as they did not need to change their path, and 99% of them converged within 50 seconds (*Prepend, No Change* line). In comparison, if we simply announce O as the baseline before poisoning, less than 70% of the unaffected ASes converged instantly, and 94% converged within 50 seconds (*No Prepend, No Change* line). Similarly, using *O-O-O* as the baseline helps ASes that had been routing via A settle on a new route faster: 96% converged within 50 seconds, compared to only 86% if we use O as the baseline. Prepending keeps the announcements a consistent length, which reduces path exploration for ASes not routing through A. In fact, with prepending, 97% of unaffected ASes made only a single update, informing neighbors only of the insertion of the poisoned AS A into the route. Without prepending, only 64% of these ASes made only one update. The other 36% explored alternatives before reverting back to their original path.

These graphs capture per-AS convergence. Because announcements need to propagate across the Internet and are subject to protocol timers as they do so, different ASes receive a poisoned announcement at different times.

We also assessed how long global convergence took, from the time the first router collector receives an update for our prefix until all route collector peer ASes had converged to stable routes. With prepending, global convergence took at most 91 seconds in the median case, at most two minutes for 75% of poisonings, and at most 200s in 90% of poisonings. In contrast, without prepending, the

50th, 75th, and 90th percentiles are 133s, 189s, and 226s. Compared to the delay following a poisoning, it generally takes slightly less time for routes to converge globally when we remove the poison and revert to the baseline announcement. Because most ASes that were not using the poisoned AS reconverge to their original path without exploring other options, we would not expect them to experience transient loss during the global convergence period.

How much loss accompanies convergence? Our results indicate the packet loss during convergence is minimal. We calculated loss rate during convergence following our poisonings that used a baseline of *O-O-O*. Every ten seconds for the duration of the experiment, we issued pings from the poisoned *LIFEGUARD* prefixes to all 308 working PlanetLab sites. In general, many of the sites were not routing via any particular poisoned AS, and so this experiment lets us evaluate how much we disrupt working paths. We filtered out cases in which loss was clearly due to problems unrelated to poisoning, and we excluded a PlanetLab site if it was completely cut off by a particular poisoning. Following 60% of poisonings, the overall loss rate during the convergence period was less than 1%, and 98% of poisonings had loss rates under 2%. Some convergence periods experienced brief spikes in loss, but only 2% of poisonings had any 10 second round with a loss rate above 10%.

Can poisoning shift routes off an AS link without completely disabling either AS? We demonstrate selective poisoning using two BGP-Mux sites, University of Washington (UWash) and University of Wisconsin (UWisc). Paths from most PlanetLab nodes to UWash pass through the Internet2 (I2) Seattle PoP, then to Pacific Northwest Gigapop, and to UWash. Most PlanetLab paths to UWisc pass through I2's Chicago PoP, then through WiscNet, before reaching UWisc. So, the BGP-Mux AS has UWash and UWisc as providers, and they connect via disjoint paths to different I2 PoPs.

We tested if we could shift traffic away from the I2 Chicago PoP, supposing the link to WiscNet experienced a silent failure. We advertised the same two prefixes from UWash and UWisc. We announced the first prefix unpoisoned from both. We poisoned I2 in UWisc's announcements of the second prefix, but had UWash announce it unpoisoned.

First, we show that paths that were not using I2 would not be disrupted. We looked at paths to the two prefixes from 36 RIPE RIS BGP peers. For 33 of them, the paths to the two prefixes were identical and did not use I2. The other three ASes routed to the unpoisoned prefix via I2 and WiscNet. For the selectively poisoned prefix, they instead routed via I2 and PNW Gigapop, as expected.

We then compare traceroutes from PlanetLab hosts to the two prefixes, to show that paths through I2 avoid the "problem." For the unpoisoned prefix, over 100 PlanetLab sites routed through I2 to WiscNet. Focusing on just these PlanetLab sites, we assessed how they routed towards the other prefix, which we selectively poisoned for I2 from UWisc. For that prefix, all the sites avoided the link from I2 to WiscNet. All but three of the sites routed via PNW Gigapop and UWash, as we intended. The remaining three – two sites in Poland and one in Brazil – used Hurricane Electric to reach UWisc. Excepting these three sites, selective poisoning allowed us to shift paths within a targeted network without changing how networks other than the target routed.

Are Internet paths diverse enough for selective poisoning to be effective? This technique may provide a means to partially poison large networks, and services with distributed data centers may have the type of connectivity required to enable it. In our second selective poisoning experiment, we approximated this type of deployment, and we assess how many ASes we could selectively poison. To make this assessment, we need to identify whether ASes select from disjoint paths, which would allow us to selectively poison them. It is hard to perform this assessment in general, because we need to know both which path an AS is using and which paths it might use if that path became unavailable.

Route collectors and BGP-Mux let us experiment in a setting in which we have access to these paths. We announced a prefix simultaneously via BGP-Muxes at UWash, UWisc, Georgia Tech, Princeton, and Clemson. This setup functions as an AS with five PoPs and one provider per PoP. We iterated through 114 ASes that provide BGP feeds. For each pair of AS A and BGP-Mux M in turn, we poisoned A from all BGP-Muxes except M and observed how A's route to our prefix varied with M. We found that selective poisoning allowed us to avoid 73% of the first hop AS links used by these peers, while still leaving the peers with a route to our prefix. In §2.3, we found that these five university providers allowed us to avoid 90% of these links on forward paths to these same ASes.

5.3 Accuracy

Having demonstrated that *LIFEGUARD* can often use BGP poisoning to route traffic around an AS or AS link without causing widespread disruption, we assess *LIFEGUARD*'s accuracy in locating failures. We show that *LIFEGUARD* seems to correctly identify the failed AS in most cases, including many that would be not be correctly found using only traceroutes. Our tech report provides further analysis of *LIFEGUARD*'s failure isolation [32].

Are LIFEGUARD*'s results consistent with what it would find with control of both ends of a path?* In general, obtaining ground truth for wide-area network faults is difficult: emulations of failures are not realistic, and few networks post outage reports publicly. Due to these challenges, we evaluate the accuracy of *LIFEGUARD*'s failure isolation on paths between a set of PlanetLab hosts used as *LIFEGUARD* vantage points and a disjoint set of PlanetLab hosts used as targets.[6] Every five minutes, we issued traceroutes from all vantage points to all targets and vice versa. In isolating the location of a failure between a vantage point and a target, we only gave *LIFEGUARD* access to measurements from its vantage points. We checked if *LIFEGUARD*'s conclusion was consistent with traceroutes from the target, "behind" the failure. *LIFEGUARD*'s location was consistent if and only if (1) the traceroute in the failing direction terminated in the AS A blamed by *LIFEGUARD*, and (2) the traceroute in the opposite direction did not contradict *LIFEGUARD*'s conclusion. The traceroute contradicted the conclusion if it included responses from A but terminated in a different AS. The responses from A indicate that some paths back from A worked. We lack sufficient information to explain these cases, and so we consider *LIFEGUARD*'s result to be inconsistent.

We examined 182 unidirectional isolated failures from August and September, 2011. For 169 of the failures, *LIFEGUARD*'s results were consistent with traceroutes from the targets. The remaining 13 cases were all forward failures. In each of these cases, *LIFEGUARD* blamed the last network on the forward traceroute (just as an operator with traceroute alone might). However, the destination's traceroute went through that network, demonstrating a working path from the network back to the destination.

Does LIFEGUARD *locate failures of interest to operators?* We searched the Outages.org mailing list [28] for outages that intersected *LIFEGUARD*'s monitored paths [32]) and found two interesting examples. On May 25th, 2011, three of *LIFEGUARD*'s vantage points detected a forward path outage to three distinct locations. The system isolated all of these outages to a router in Level3's Chicago PoP. Later that night, a network operator posted the following message to the mailing list: "Saw issues routing through

[6] We cannot issue spoofed packets or make BGP announcements from EC2, and so we cannot use it to evaluate our system.

Chicago via Level3 starting around 11:05 pm, cleared up about 11:43 pm." Several network operators corroborated this report.

In the second example, a vantage point in Albany and one in Delaware observed simultaneous outages to three destinations: a router at an edge AS in Ohio and routers in XO's Texas and Chicago PoPs. *LIFEGUARD* identified all Albany outages and some of the Delaware ones as reverse path failures, with the remaining Delaware one flagged as bidirectional. All reverse path failures were isolated to an XO router in Dallas, and the bidirectional failure was isolated to an XO router in Virginia. Several operators subsequently posted to the Outages.org mailing list reporting problems with XO in multiple locations, which is likely what *LIFEGUARD* observed.

Does LIFEGUARD *provide benefit beyond traceroute?* We now quantify how often *LIFEGUARD* finds that failures would be incorrectly isolated using only traceroute, thus motivating the need for the system's more advanced techniques. In the example shown in Fig. 4, traceroutes from GMU seem to implicate a problem forwarding from TransTelecom, whereas our system located the failure as being along the reverse path, in Rostelecom. For the purposes of this study, we consider outages that meet criteria that make them candidates for rerouting and repair: (1) multiple sources must be unable to reach the destination, and these sources must be able to reach at least 10% of all destinations at the time, reducing the chance that it is a source-specific issue; (2) the failing traceroutes must not reach the destination AS, and the outage must be partial, together suggesting that alternate AS paths exist; (3) and the problem must not resolve during the isolation process, thereby excluding transient problems. During June 2011, *LIFEGUARD* identified 320 outages that met these criteria [32]. In 40% of cases, the system identified a different suspected failure location than what one would assume using traceroute alone. Further, even in the other 60% of cases, an operator would not currently know whether or not the traceroute was identifying the proper location.

5.4 Scalability

How efficiently does LIFEGUARD *refresh its path atlas? LIFEGUARD* regularly refreshes the forward and reverse paths it monitors. Existing approaches efficiently maintain forward path atlases based on the observations that paths converge as they approach the source/destination [12] and that most paths are stable [11]. Based on these observations, we implemented a reverse path atlas that caches probes for short periods, reuses measurements across converging paths, and usually refreshes a stale path using fewer probes than would be required to measure from scratch. In combination, these optimizations enable us to refresh paths at an average (peak) rate of 225 (502) reverse paths per minute. We use an amortized average per path of 10 IP option probes (vs. the 35 reported in existing work [19]) and slightly more than 2 forward traceroutes. It may be possible to improve scalability in the future by focusing on measuring paths between "core" PoPs whose routing health and route changes likely influence many paths,[7] and by scheduling measurements to minimize the impact of router-specific rate limits.

What is the probing load for locating problems? Fault isolation requires approximately 280 probe packets per outage. We found that *LIFEGUARD* isolates failures on average much faster than it takes long-lasting outages to be repaired. For bidirectional and reverse path outages, potential candidates for poisoning, *LIFEGUARD* completed isolation measurements within 140 seconds on average.

[7]For example, 63% of iPlane traceroutes traverse at least one of the most common 500 PoPs (0.3% of PoPs) [17].

		$d = 5$ minutes		$d = 15$ min.		$d = 60$ min.	
		$T = 0.5$	1.0	0.5	1.0	0.5	1.0
I	0.01	393	783	137	275	58	115
	0.1	3931	7866	1370	2748	576	1154
	0.5	19625	39200	6874	13714	2889	5771

Table 2: Number of additional daily path changes due to poisoning for fraction of ISPs using *LIFEGUARD* (I), fraction of networks monitored for reachability (T), and duration of outage before poisoning (d). For comparison, routers currently make 110K–315K updates per day.

What load will poisoning induce at scale? An important question is whether *LIFEGUARD* would induce excessive load on routers if deployed at scale. Our earlier study showed that, by prepending, we can reduce the number of updates made by each router after a path poisoning. In this section, we estimate the Internet-wide load our approach would generate if a large number of ISPs used it. While our results serve as a rough estimate, we find that the number of path changes made at each router is low.

The number of daily path changes per router our system would produce at scale is $I \times T \times P(d) \times U$, where I is the fraction of ISPs using our approach, T is the fraction of ASes each ISP is monitoring with *LIFEGUARD*, $P(d)$ is the aggregate number of outages per day that have lasted at least d minutes and are candidates for poisoning, and U is the average number of path changes per router generated by each poison. Based on our experiments, $U = 2.03$ for routers that had been routing via the poisoned AS, and $U = 1.07$ for routers that had not been routing via the poisoned AS. For routers using the poisoned AS, BGP should have detected the outage and generated at least one path update in response. For routers not routing through it, the updates are pure overhead. Thus, poisoning causes affected routers to issue an additional 1.03 updates and and unaffected routers an additional 1.07 updates. For simplicity, we set $U = 1$ in this analysis.

We base $P(d)$ on the Hubble dataset of outages on paths between PlanetLab sites and 92% of the Internet's edge ASNs [20]. We filter this data to exclude cases where poisoning is not effective (e.g., complete outages and outages with failures in the destination AS). We assume that the Hubble targets were actually monitoring the PlanetLab sites and define $P(d) = H(d)/(I_h T_h)$, where $H(d)$ is the total number of poisonable Hubble outages per day lasting at least d minutes, $I_h = 0.92$ is the fraction of all edge ISPs that Hubble monitored, and $T_h = 0.01$ is our estimate for the fraction of total poisonable (transit) ASes on paths from Hubble VPs to their targets. Because the smallest d that Hubble provides is 15 minutes, we extrapolate the Hubble outage distribution based on the EC2 data to estimate the poisoning load for $d = 5$.

Scaling the parameters estimates the load from poisoning under different scenarios. In Table 2, we vary the fraction of participating ISPs (I), the fraction of poisonable ASes being monitored (T), and the minimum outage duration before poisoning is used (d). We scale the number of outages linearly with I and T. We scale with d based on the failure duration distribution from our EC2 study.

LIFEGUARD could cause a router to generate from tens to tens of thousands of additional updates. For reference, a single-homed edge router peering with AS131072 sees an average of approximately 110K updates per day [4], and the Tier-1 routers we checked made 255K-315K path updates per day [26]. For cases where only a small fraction of ASes use poisoning ($I \leq 0.1$), the additional load is less than 1%. For large deployments ($I = 0.5, T = 1$) and short delays before poisoning ($d = 5$), the overhead can become significant (35% for the edge router, 12-15% for Tier-1 routers). We note that reducing the number of monitored paths or waiting longer to poison can easily reduce the overhead to less than 10%.

6. LIFEGUARD CASE STUDY

To demonstrate *LIFEGUARD*'s ability to repair a data plane outage by locating a failure and then poisoning, we describe a failure between the University of Wisconsin and a PlanetLab node at National Tsing Hua University in Taiwan. *LIFEGUARD* announced production and sentinel prefixes via the University of Wisconsin BGP-Mux. *LIFEGUARD* had monitored the PlanetLab node for a month, gathering historical data. On October 3-4, 2011, nodes in the two prefixes exchanged test traffic with the PlanetLab node, and we sent traceroutes every 10 minutes from the PlanetLab node towards the test nodes to track its view of the paths. We use the measurements from the Taiwanese node to evaluate *LIFEGUARD*.

After experiencing only transient problems during the day, at 8:15pm on October 3, the test traffic began to experience a persistent outage. When *LIFEGUARD* isolated the direction of the outage, spoofed pings from Wisconsin reached Taiwan, but spoofed pings towards Wisconsin failed, indicating a reverse path problem. *LIFEGUARD*'s atlas revealed two historical paths from Taiwan back to Wisconsin. Prior to the outage, the PlanetLab node had been successfully routing to Wisconsin via academic networks. According to the atlas, this route had been in use since 3pm, but, prior to that, the path had routed through UUNET (a commercial network) for an extended period. The UUNET and academic paths diverged one AS before UUNET (one AS closer to the Taiwanese site). *LIFEGUARD* issued measurements to see which of the hops on these reverse paths still could reach Wisconsin. These measurements establish a reachability horizon with UUNET and the ASes before it behind the failure. During the failure, all routers along the academic path from the divergence point to Wisconsin were still responsive to pings, meaning they had a route to the University of Wisconsin. However, hops in UUNET no longer responded to probes. In fact, hand inspection of traceroutes from the PlanetLab node to Wisconsin showed that, at 8:15pm, the path had switched to go via UUNET, and the traceroutes had been terminating in UUNET.

Because the hops on the academic route had paths to Wisconsin, it was a likely viable alternative to the broken path, and we used *LIFEGUARD* to poison UUNET. For a brief period after *LIFEGUARD* announced the poison, test traffic was caught in a convergence loop. After convergence, the test traffic and a traceroute from Taiwan successfully reached the production prefix via academic networks. Traceroutes to the sentinel prefix continued to fail in UUNET until just after 4am on October 4, when the path through UUNET began to work again. In summary, *LIFEGUARD* isolated an outage and used poisoning to re-establish connectivity until the outage was resolved. Once repaired, the poison was no longer necessary, and *LIFEGUARD* reverted to the baseline unpoisoned announcement.

7. DISCUSSION

7.1 Poisoning Anomalies

Certain poisonings cause anomalous behavior. Some networks disable loop detection, accepting paths even if they contain their AS. Other networks do not accept an update from a customer if the path contains a peer of the network. We discuss these two issues.

Some networks with multiple remote sites communicate between sites across the public Internet, using the same AS number across sites. One approach is to disable BGP's loop prevention to exchange prefixes from the remote sites, even though they share an origin AS. The paths for these prefixes allow the remote sites to route to each other using BGP. Fortunately, best practices mandate that, instead of disabling loop detection altogether, networks should set the maximum occurrences of their AS number in the path. For instance, AS286 accepts updates if it is already in the AS path. Inserting AS286 twice into the AS path, however, causes it to drop the update, thus enabling the use of poisoning. We expect ASes that use the public Internet to communicate between remote sites to be stubs (meaning they have no customers). We use poisoning to bypass faulty transit networks, so have no need to poison stubs.

Some networks do not accept an update from a customer if the path contains one of the network's peers. For example, Cogent will not accept an update that includes one of its Tier-1 peers in the path, meaning that announcements poisoning one of these ASes via Georgia Tech did not propagate widely. However, we could poison them via BGP-Mux instances at universities that were not Cogent customers, and 76% of route collector peers were able to find a path that avoided a poisoned AS through which they had previously been routing. While the filtering reduced the coverage of our poisoning studies and will prevent poisoning from rerouting around some failures, it is likely not a big limitation. First, most failures occur outside of large networks (such as Cogent and Tier-1s) [32, 36]. Second, we used poisoning to experiment with AS avoidance, but most of the techniques and principles still apply to other implementations, such as a modification of BGP to include a new signed AVOID_PROBLEM(X,P) notification.

7.2 Address Use by the Sentinel Prefix

We proposed using a less-specific prefix with an unused sub-prefix as a sentinel. We discuss the trade-offs when using three alternative approaches. First, absent any sentinel, it is not clear how to check for failure recovery or to provide a backup unpoisoned route. Second, if an AS has an unused prefix that is not adjacent to the production prefix, it can use the unused prefix as a sentinel, even though it may lack a super-prefix that covers both prefixes. This approach allows the AS to check when the failed route is repaired. However, this scheme does not provide a "backup" route to the production prefix for networks captive behind the poisoned AS. Third, if an AS lacks unused address space to serve as a sentinel, a less-specific prefix will ensure that the poisoned AS and ASes captive behind it still have a route, even though the less-specific prefix does not include an unused sub-prefix. Pings to the poisoned AS and its captives will return via routes to the unpoisoned less-specific prefix, and so they can be used to check for repairs.

Alternatively, a provider with multiple prefixes hosting the same service can use DNS redirection to test when a problem has resolved, without using additional addresses. Such providers often use DNS to direct a client to a nearby data prefix. Our scheme relies on clients using the same route to reach all prefixes in the absence of poison. To establish that this property holds, at least for Google, we resolved a Google hostname at 20 PlanetLab sites around the world, yielding a set of Google IP addresses from various data centers. We then issued traceroutes from the PlanetLab sites to the set of IP addresses. Each PlanetLab site used a consistent path to reach Google's network for all IP addresses. Google then routed the traffic to a particular data center. With this consistent routing, if a provider discovers a routing problem affecting a set of clients *C*, it could poison the prefix *P1* serving those clients. It need not poison its prefix *P2* that serves other clients (possibly from a different data center). Periodically, the provider's DNS resolvers could give a client from *C* an address from *P2* and an address from *P1*, with *P1* serving as a failover. By checking server logs, the provider could discern if the client was able to reach *P2*. When clients can reach *P2*, the provider can remove the poison on *P1*.

8. RELATED WORK

Failure isolation. Hubble [20] informs our failure isolation approach. *LIFEGUARD* makes heavy use of reverse traceroute [19].

Feamster et al. measured failures between vantage points using pings and traceroutes [13]. Upon detected a problem, PlanetSeer triggered measurements from multiple vantage points [36]. However, after identifying partial outages, all further probes and analysis focusing only on the local measurements.

Feldmann et al. located the cause of BGP updates by assuming that it was on the old path or the new path. They could pinpoint the likely cause of a BGP update to one or two ASes [14]. iSpy [39] identified prefix hijacking using a technique similar to *LIFEGUARD*'s reachability horizon, but assumed that paths were symmetric. *LIFEGUARD* uses reverse traceroute and spoofed probes to more accurately determine the reachability horizon.

Failure Avoidance. Previous research used poisoning as a measurement tool to uncover network topology [10] and default routes [8]. While inspired by that work, we propose using poisoning operationally as a means to improve availability.

In the absence of a solution to long-lasting, partial outages, researchers and companies have proposed systems to detour traffic around outages using overlay paths [2, 6, 16]. These approaches provide a great alternative when no other solutions exist. In contrast to detouring, our poisoning approach does not require the expense of an overlay and can carry traffic at core Internet data rates along policy-compliant BGP paths that avoid the identified problem.

Entact used overlapping BGP prefixes to simultaneously measure alternative paths [38]. Bush et al. used an anchor prefix to determine whether a test prefix should be reachable [8]. Similarly, *LIFEGUARD* uses a sentinel prefix to test for repairs along the failing path while rerouting traffic to a path avoiding the outage.

MIRO proposed to enable AS avoidance and other functionality by allowing ASes to advertise and use multiple inter-domain paths [34]. The proposal retains much of the quintessence of BGP and routing policies, but it requires modifications to protocols and routers, meaning it faces a slow path to adoption. Our deployable solutions and results on failure avoidance could perhaps present some of the argument for the adoption of MIRO-like modifications.

9. CONCLUSION

Increasingly, Internet and cloud-based services expect the Internet to deliver high availability. Nevertheless, partial outages that last for hours occur frequently, accounting for a large portion of the Internet's overall end-to-end downtime. Such an outage can occur when an ISP advertises a BGP route but fails to deliver packets on the route. To address these problems, we introduce *LIFEGUARD*, a system that locates the AS at fault and routes around it. Using multiple vantage points and spoofed probes, we show that *LIFEGUARD* can identify the failing AS in the wild. We show that we can use BGP poisoning to cause routes to avoid an AS without disrupting working routes. An ISP can deploy our approach unilaterally today. *LIFEGUARD* enables experiments that we hope will motivate the need for and design of new route control mechanisms.

Acknowledgments

We gratefully acknowledge our shepherd Craig Labovitz and the SIGCOMM reviewers. The BGP-Mux system would not be available without the help of researchers and network administrators at the sites: Scott Friedrich, Michael Blodgett, Jeff Fitzwater, Jennifer Rexford, Larry Billado, Kit Patterson, and Schyler Batey. This work was partially funded by Google, Cisco, FAPEMIG, NSF (CNS-0905568 and CNS-1040663), and the NSF/CRA Computing Innovation Fellowship program. We are thankful for their support.

10. REFERENCES

[1] Abilene Internet2 network. http://www.internet2.edu/network/.

[2] D. Andersen, H. Balakrishnan, F. Kaashoek, and R. Morris. Resilient overlay networks. In *SOSP*, 2001.

[3] R. Austein, S. Bellovin, R. Bush, R. Housley, M. Lepinski, S. Kent, W. Kumari, D. Montgomery, K. Sriram, and S. Weiler. BGPSEC protocol. http://tools.ietf.org/html/draft-ietf-sidr-bgpsec-protocol.

[4] The BGP Instability Report. http://bgpupdates.potaroo.net/instability/bgpupd.html.

[5] BGPMux Transit Portal. http://tp.gtnoise.net/.

[6] C. Bornstein, T. Canfield, and G. Miller. Akarouting: A better way to go. In *MIT OpenCourseWare 18.996*, 2002.

[7] M. A. Brown, C. Hepner, and A. C. Popescu. Internet captivity and the de-peering menace. In *NANOG*, 2009.

[8] R. Bush, O. Maennel, M. Roughan, and S. Uhlig. Internet optometry: assessing the broken glasses in Internet reachability. In *IMC*, 2009.

[9] K. Chen, D. R. Choffnes, R. Potharaju, Y. Chen, F. E. Bustamante, D. Pei, and Y. Zhao. Where the sidewalk ends: Extending the Internet AS graph using traceroutes from P2P users. In *CoNEXT*, 2009.

[10] L. Colitti. *Internet Topology Discovery Using Active Probing*. PhD thesis, University di "Roma Tre", 2006.

[11] I. Cunha, R. Teixeira, and C. Diot. Predicting and tracking Internet path changes. In *SIGCOMM*, 2011.

[12] B. Donnet, P. Raoult, T. Friedman, and M. Crovella. Efficient algorithms for large-scale topology discovery. In *SIGMETRICS*, 2005.

[13] N. Feamster, D. G. Andersen, H. Balakrishnan, and M. F. Kaashoek. Measuring the effects of internet path faults on reactive routing. In *SIGMETRICS*, 2003.

[14] A. Feldmann, O. Maennel, Z. M. Mao, A. Berger, and B. Maggs. Locating Internet routing instabilities. In *SIGCOMM*, 2004.

[15] L. Gao. On inferring autonomous system relationships in the Internet. *IEEE/ACM TON*, 2001.

[16] K. P. Gummadi, H. V. Madhyastha, S. D. Gribble, H. M. Levy, and D. Wetherall. Improving the reliability of Internet paths with one-hop source routing. In *OSDI*, 2004.

[17] iPlane. `http://iplane.cs.washington.edu`.

[18] J. P. John, E. Katz-Bassett, A. Krishnamurthy, T. Anderson, and A. Venkataramani. Consensus routing: The Internet as a distributed system. In *NSDI*, 2008.

[19] E. Katz-Bassett, H. V. Madhyastha, V. K. Adhikari, C. Scott, J. Sherry, P. van Wesep, A. Krishnamurthy, and T. Anderson. Reverse traceroute. In *NSDI*, 2010.

[20] E. Katz-Bassett, H. V. Madhyastha, J. P. John, A. Krishnamurthy, D. Wetherall, and T. Anderson. Studying black holes in the Internet with Hubble. In *NSDI*, 2008.

[21] R. R. Kompella, J. Yates, A. Greenberg, and A. C. Snoeren. Detection and localization of network black holes. In *INFOCOM*, 2007.

[22] N. Kushman, S. Kandula, and D. Katabi. R-BGP: Staying connected in a connected world. In *NSDI*, 2007.

[23] C. Labovitz, A. Ahuja, A. Bose, and F. Jahanian. Delayed Internet routing convergence. In *SIGCOMM*, 2000.

[24] K. K. Lakshminarayanan, M. C. Caesar, M. Rangan, T. Anderson, S. Shenker, and I. Stoica. Achieving convergence-free routing using failure-carrying packets. In *SIGCOMM*, 2007.

[25] H. Madhyastha, E. Katz-Bassett, T. Anderson, A. Krishnamurthy, and A. Venkataramani. iPlane Nano: Path Prediction for Peer-to-Peer Applications. In *NSDI*, 2009.

[26] D. Meyer. RouteViews. `http://www.routeviews.org`.

[27] P. Mohapatra, J. Scudder, D. Ward, R. Bush, and R. Austein. BGP prefix origin validation. http://tools.ietf.org/html/draft-ietf-sidr-pfx-validate.

[28] Outages mailing list. `http://isotf.org/mailman/listinfo/outages`.

[29] Packet clearing house. http://www.pch.net/home/index.php.

[30] B. Quoitin and O. Bonaventure. A survey of the utilization of the BGP community attribute. Internet draft, draft-quoitin-bgp-comm-survey-00, 2002.

[31] RIPE RIS. `http://www.ripe.net/ris/`.

[32] C. Scott. LIFEGUARD: Locating Internet Failures Effectively and Generating Usable Alternate Routes Dynamically. Technical report, Univ. of Washington, 2012.

[33] UCLA Internet topology. `http://irl.cs.ucla.edu/topology/`.

[34] W. Xu and J. Rexford. MIRO: Multi-path Interdomain ROuting. In *SIGCOMM*, 2006.

[35] J. Yates and Z. Ge. Network Management: Fault Management, Performance Management and Planned Maintenance. Technical report, AT&T Labs, 2009.

[36] M. Zhang, C. Zhang, V. Pai, L. Peterson, and R. Wang. PlanetSeer: Internet path failure monitoring and characterization in wide-area services. In *OSDI*, 2004.

[37] Y. Zhang, V. Paxson, and S. Shenker. The stationarity of Internet path properties: Routing, loss, and throughput. *ACIRI Technical Report*, 2000.

[38] Z. Zhang, M. Zhang, A. Greenberg, Y. C. Hu, R. Mahajan, and B. Christian. Optimizing cost and performance in online service provider networks. In *NSDI*, 2010.

[39] Z. Zhang, Y. Zhang, Y. C. Hu, Z. M. Mao, and R. Bush. iSpy: detecting IP prefix hijacking on my own. In *SIGCOMM*, 2008.

On-Chip Networks from a Networking Perspective: Congestion and Scalability in Many-Core Interconnects

George Nychis†, Chris Fallin†, Thomas Moscibroda§, Onur Mutlu†, Srinivasan Seshan†

† Carnegie Mellon University
{gnychis,cfallin,onur,srini}@cmu.edu

§ Microsoft Research Asia
moscitho@microsoft.com

ABSTRACT

In this paper, we present network-on-chip (NoC) design and contrast it to traditional network design, highlighting similarities and differences between the two. As an initial case study, we examine *network congestion* in bufferless NoCs. We show that congestion manifests itself differently in a NoC than in traditional networks. Network congestion reduces system throughput in congested workloads for smaller NoCs (16 and 64 nodes), and limits the scalability of larger bufferless NoCs (256 to 4096 nodes) even when traffic has locality (e.g., when an application's required data is mapped nearby to its core in the network). We propose a new source throttling-based congestion control mechanism with application-level awareness that reduces network congestion to improve system performance. Our mechanism improves system performance by up to 28% (15% on average in congested workloads) in smaller NoCs, achieves linear throughput scaling in NoCs up to 4096 cores (attaining similar performance scalability to a NoC with large buffers), and reduces power consumption by up to 20%. Thus, we show an effective application of a network-level concept, congestion control, to a class of networks – bufferless on-chip networks – that has not been studied before by the networking community.

Categories and Subject Descriptors

C.1.2 [**Computer Systems Organization**]: Multiprocessors – Interconnection architectures; C.2.1 [**Network Architecture and Design**]: Packet-switching networks

Keywords On-chip networks, multi-core, congestion control

1. INTRODUCTION

One of the most important trends in computer architecture in the past decade is the move towards multiple CPU cores on a single chip. Common chip multiprocessor (CMP) sizes today range from 2 to 8 cores, and chips with hundreds or thousands of cores are likely to be commonplace in the future [9, 55]. Real chips exist already with 48 cores [34], 100 cores [66], and even a research prototype with 1000 cores [68]. While increased core count has allowed processor chips to scale without experiencing complexity and power dissipation problems inherent in larger individual cores, new challenges also exist. One such challenge is to design an efficient and scalable interconnect between cores. Since the interconnect carries all inter-cache and memory traffic (i.e., all data accessed by the programs running on chip), it plays a critical role in system performance and energy efficiency.

Unfortunately, the traditional bus-based, crossbar-based, and other non-distributed designs used in small CMPs do not scale to the medium- and large-scale CMPs in development. As a result, the architecture research community is moving away from traditional centralized interconnect structures, instead using interconnects with distributed scheduling and routing. The resulting Networks on Chip (NoCs) connect cores, caches and memory controllers using packet switching routers [15], and have been arranged both in regular 2D meshes and a variety of denser topologies [29, 41]. The resulting designs are more network-like than conventional small-scale multicore designs. These NoCs must deal with many problems, such as scalability [28], routing [31,50], congestion [10,27,53,65], and prioritization [16, 17, 30], that have traditionally been studied by the networking community rather than the architecture community.

While different from traditional processor interconnects, these NoCs also differ from existing large-scale computer networks and even from the traditional multi-chip interconnects used in large-scale parallel computing machines [12, 45]. On-chip hardware implementation constraints lead to a different tradeoff space for NoCs compared to most traditional off-chip networks: chip area/space, power consumption, and implementation complexity are first-class considerations. These constraints make it hard to build energy-efficient network buffers [50], make the cost of conventional routing and arbitration [14] a more significant concern, and reduce the ability to over-provision the network for performance. These and other characteristics give NoCs unique properties, and have important ramifications on solutions to traditional networking problems in a new context.

In this paper, we explore the adaptation of conventional networking solutions to address two particular issues in next-generation bufferless NoC design: *congestion management* and *scalability*. Recent work in the architecture community considers bufferless NoCs as a serious alternative to conventional buffered NoC designs due to chip area and power constraints[1] [10,20,21,25,31,49,50,67]. While bufferless NoCs have been shown to operate efficiently under moderate workloads and limited network sizes (up to 64 cores), we find that with higher-intensity workloads and larger network sizes (e.g., 256 to 4096 cores), the network operates inefficiently

[1] Existing prototypes show that NoCs can consume a substantial portion of system power (28% in the Intel 80-core Terascale chip [33], 36% in the MIT RAW chip [63], and 10% in the Intel Single-Chip Cloud Computer [34]).

SIGCOMM'12, August 13–17, 2012, Helsinki, Finland.

and does not scale effectively. As a consequence, application-level system performance can suffer heavily.

Through evaluation, we find that congestion limits the efficiency and scalability of bufferless NoCs, even when traffic has locality, e.g., as a result of intelligent compiler, system software, and hardware data mapping techniques. Unlike traditional large-scale computer networks, NoCs experience congestion in a fundamentally different way due to unique properties of both NoCs and bufferless NoCs. While traditional networks suffer from congestion collapse at high utilization, a NoC's cores have a *self-throttling* property which avoids this congestion collapse: slower responses to memory requests cause pipeline stalls, and so the cores send requests less quickly in a congested system, hence loading the network less. However, congestion does cause the system to operate at less than its peak throughput, as we will show. In addition, congestion in the network can lead to increasing inefficiency as the network is scaled to more nodes. We will show that addressing congestion yields better performance scalability with size, comparable to a more expensive NoC with buffers that reduce congestion.

We develop a new congestion-control mechanism suited to the unique properties of NoCs and of bufferless routing. First, we demonstrate how to detect impending congestion in the NoC by monitoring *injection starvation*, or the inability to inject new packets. Second, we show that simply throttling all applications when congestion occurs is not enough: since different applications respond differently to congestion and increases/decreases in network throughput, the network must be application-aware. We thus define an application-level metric called *Instructions-per-Flit* which distinguishes between applications that should be throttled and those that should be given network access to maximize system performance. By dynamically throttling according to periodic measurements of these metrics, we reduce congestion, improve system performance, and allow the network to scale more effectively. In summary, we make the following **contributions**:

- We discuss **key differences** between NoCs (and bufferless NoCs particularly) and traditional networks, to frame NoC design challenges and research goals from a networking perspective.
- From a **study of scalability and congestion**, we find that the bufferless NoC's scalability and efficiency are limited by congestion. In small networks, congestion due to network-intensive workloads limits throughput. In large networks, even with *locality* (placing application data nearby to its core in the network), congestion still causes application throughput reductions.
- We propose a new low-complexity and high performance congestion control mechanism in a bufferless NoC, motivated by ideas from both networking and computer architecture. To our knowledge, this is the first work that comprehensively examines congestion and scalability in *bufferless* NoCs and provides an effective solution based on the properties of such a design.
- Using a large set of real-application workloads, we demonstrate improved performance for small (4x4 and 8x8) bufferless NoCs. Our mechanism improves system performance by up to 28% (19%) in a 16-core (64-core) system with a 4x4 (8x8) mesh NoC, and 15% on average in congested workloads.
- In large (256 – 4096 core) networks, we show that congestion limits scalability, and hence that congestion control is *required* to achieve linear performance scalability with core count, even when most network traversals are local. At 4096 cores, congestion control yields a 50% throughput improvement, and up to a 20% reduction in power consumption.

2. NOC BACKGROUND AND UNIQUE CHARACTERISTICS

We first provide a brief background on on-chip NoC architectures, bufferless NoCs, and their unique characteristics in comparison to traditional and historical networks. We refer the reader to [8, 14] for an in-depth discussion.

2.1 General NoC Design and Characteristics

In a chip multiprocessor (CMP) architecture that is built on a NoC substrate, the NoC typically connects the processor nodes and their private caches with the shared cache banks and memory controllers (Figure 1). A NoC might also carry other control traffic, such as interrupt and I/O requests, but it primarily exists to service cache miss requests. On a cache miss, a core will inject a memory request packet into the NoC addressed to the core whose cache contains the needed data, or the memory controller connected to the memory bank with the needed data (for example). This begins a data exchange over the NoC according to a cache coherence protocol (which is specific to the implementation). Eventually, the requested data is transmitted over the NoC to the original requester.

Design Considerations: A NoC must service such cache miss requests quickly, as these requests are typically on the user program's critical path. There are several first-order considerations in NoC design to achieve the necessary throughput and latency for this task: chip area/space, implementation complexity, and power. As we provide background, we will describe how these considerations drive the NoC's design and endow it with unique properties.

Network Architecture / Topology: A high-speed router at each node connects the core to its neighbors by links. These links may form a variety of topologies (e.g., [29, 40, 41, 43]). Unlike traditional off-chip networks, an on-chip network's topology is statically known and usually very regular (e.g., a mesh). The most typical topology is the two-dimensional (2D) Mesh [14], shown in Figure 1. The 2D Mesh is implemented in several commercial processors [66, 70] and research prototypes [33, 34, 63]. In this topology, each router has 5 input and 5 output channels/ports: one from each of its four neighbors and one from the network interface (NI). Depending on the router architecture and the arbitration and routing policies (which impact the number of pipelined arbitration stages), each packet spends between 1 cycle (in a highly optimized best case [50]) and 4 cycles at each router before being forwarded.

Because the network size is relatively small and the topology is statically known, global coordination and coarse-grain network-wide optimizations are possible and often less expensive than distributed mechanisms. For example, our proposed congestion control mechanism demonstrates the effectiveness of such global coordination and is very cheap (§6.5). Note that fine-grained control (e.g., packet routing) must be based on local decisions, because a router processes a packet in only a few cycles. At a scale of thousands of processor clock cycles or more, however, a central controller can feasibly observe the network state and adjust the system.

Data Unit and Provisioning: A NoC conveys packets, which are typically either request/control messages or data messages. These packets are partitioned into *flits*: a unit of data conveyed by one link in one cycle; the smallest independently-routed unit of traffic.[2] The width of a link and flit varies, but 128 bits is typical. For NoC performance reasons, links typically have a latency of only one or two cycles, and are pipelined to accept a new flit every cycle.

[2]In many *virtual-channel buffered* routers [13], the smallest independent routing unit is the packet, and flits serve only as the unit for link and buffer allocation (i.e., flow control). We follow the design used by other bufferless NoCs [20, 22, 36, 50] in which flits carry routing information due to possible deflections.

Unlike conventional networks, NoCs cannot as easily overprovision bandwidth (either through wide links or multiple links), because they are limited by power and on-chip area constraints. The tradeoff between bandwidth and latency is different in NoCs. Low latency is critical for efficient operation (because delays in packets cause core pipeline stalls), and the allowable window of in-flight data is much smaller than in a large-scale network because buffering structures are smaller. NoCs also lack a direct correlation between network throughput and overall system throughput. As we will show (§4), for the same network throughput, choosing differently which L1 cache misses are serviced in the network can affect system throughput (instructions per cycle per node) by up to 18%.

Routing: Because router complexity is a critical design consideration in on-chip networks, current implementations tend to use much more simplistic routing mechanisms than traditional networks. The most common routing paradigm is x-y routing. A flit is first routed along the x-direction until the destination's y-coordinate is reached; then routed to the destination in the y-direction.

Packet Loss: Because links are on-chip and the entire system is considered part of one failure domain, NoCs are typically designed as *lossless* networks, with negligible bit-error rates and no provision for retransmissions. In a network without losses or explicit drops, ACKs or NACKs are not necessary, and would only waste on-chip bandwidth. However, particular NoC router designs can choose to explicitly *drop* packets when no resources are available (although the bufferless NoC architecture upon which we build does not drop packets, others do drop packets when router outputs [31] or receiver buffers [20] are contended).

Network Flows & Traffic Patterns: Many architectures split the shared cache across several or all nodes in the system. In these systems, a program will typically send traffic to many nodes, often in parallel (when multiple memory requests are parallelized). Multithreaded programs also exhibit complex communication patterns where the concept of a "network flow" is removed or greatly diminished. Traffic patterns are driven by several factors: private cache miss behavior of applications, the data's locality-of-reference, phase behavior with local and temporal bursts, and importantly, self-throttling [14]. Fig. 6 (where the Y-axis can be viewed as traffic intensity and the X-axis is time) shows temporal variation in injected traffic intensity due to application phase behavior.

2.2 Bufferless NoCs and Characteristics

The question of **buffer size** is central to networking, and there has recently been great effort in the community to determine the right amount of buffering in new types of networks, e.g. in data center networks [2, 3]. The same discussions are also ongoing in on-chip networks. Larger buffers provide additional bandwidth in the network; however, they also can significantly increase power consumption and the required chip area.

Recent work has shown that it is possible to completely eliminate buffers from NoC routers. In such *bufferless NoCs*, power consumption is reduced by 20-40%, router area on die is reduced by 40-75%, and implementation complexity also decreases [20, 50].[3] Despite these reductions in power, area and complexity, application performance degrades minimally for low-to-moderate network intensity workloads. The general system architecture does not differ from traditional buffered NoCs. However, the lack of buffers requires different injection and routing algorithms.

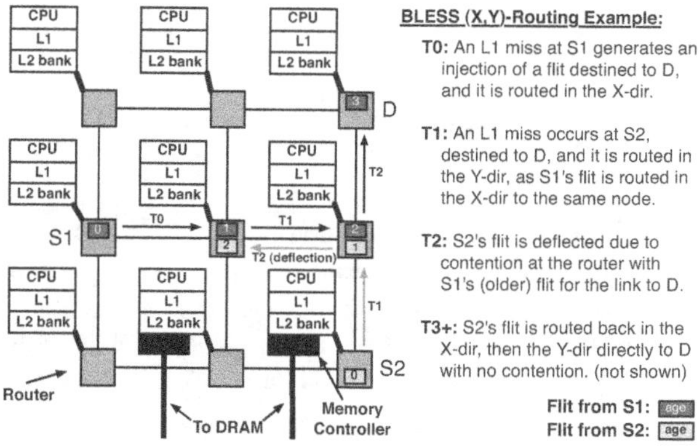

Figure 1: A 9-core CMP with BLESS routing example.

Bufferless Routing & Arbitration: Figure 1 gives an example of *injection*, *routing* and *arbitration*. As in a buffered NoC, injection and routing in a bufferless NoC (e.g., *BLESS* [50]) happen synchronously across all cores in a clock cycle. When a core must send a packet to another core, (e.g., *S1* to *D* at T0 in Figure 1), the core is able to inject each flit of the packet into the network as long as one of its output links is free. Injection requires a free output link since there is no buffer to hold the packet in the router. If no output link is free, the flit remains queued at the processor level.

An age field is initialized to 0 in the header and incremented at each hop as the flit is routed through the network. The routing algorithm (e.g, XY-Routing) and arbitration policy determine to which neighbor the flit is routed. Because there are no buffers, flits must pass through the router pipeline without waiting. When multiple flits request the same output port, *deflection* is used to resolve contention. Deflection arbitration can be performed in many ways. In the *Oldest-First* policy, which our baseline network implements [50], if flits contend for the same output port (in our example, the two contending for the link to D at time T2), ages are compared, and the oldest flit obtains the port. The other contending flit(s) are *deflected* (*misrouted* [14]) – e.g., the flit from *S2* in our example. Ties in age are broken by other header fields to form a total order among all flits in the network. Because a node in a 2D mesh network has as many output ports as input ports, routers never block. Though some designs drop packets under contention [31], the bufferless design that we consider does not drop packets, and therefore ACKs are not needed. Despite the simplicity of the network's operation, it operates efficiently, and is livelock-free [50].

Many past systems have used this type of deflection routing (also known as *hot-potato routing* [6]) due to its simplicity and energy/area efficiency. However, it is particularly well-suited for NoCs, and presents a set of challenges distinct from traditional networks.

Bufferless Network Latency: Unlike traditional networks, the injection latency (time from head-of-queue to entering the network) can be significant (§3.1). In the worst case, this can lead to starvation, which is a fairness issue (addressed by our mechanism - §6). In-network latency in a bufferless NoC is relatively low, even under high congestion (§3.1). Flits are quickly routed once in the network without incurring buffer delays, but may incur more deflections.

3. LIMITATIONS OF BUFFERLESS NOCS

In this section, we will show how the distinctive traits of NoCs place traditional networking problems in new contexts, resulting in new challenges. While prior work [20,50] has shown significant reductions in power and chip-area from eliminating buffers in the network, that work has focused primarily on low-to-medium network

[3]Another evaluation [49] showed a slight energy advantage for buffered routing, because control logic in bufferless routing can be complex and because buffers can have less power and area cost if custom-designed and heavily optimized. A later work on bufferless design [20] addresses control logic complexity. Chip designers may or may not be able to use custom optimized circuitry for router buffers, and bufferless routing is appealing whenever buffers have high cost.

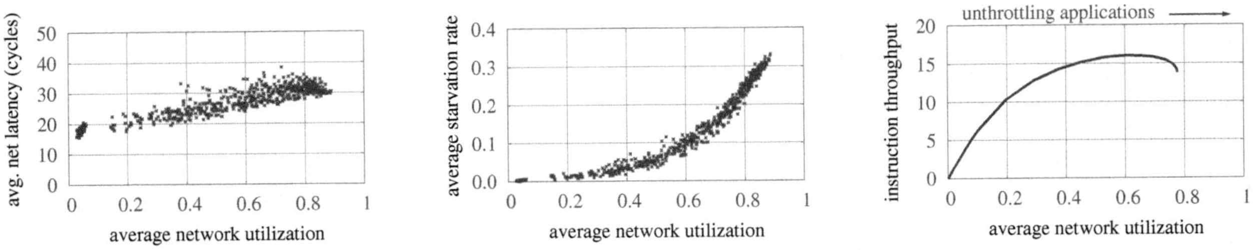

(a) Average network latency in cycles (Each point represents one of the 700 workloads).

(b) As the network becomes more utilized, the overall starvation rate rises significantly.

(c) We unthrottle applications in a 4x4 network to show suboptimal performance when run freely.

Figure 2: The effect of congestion at the network and application level.

load in conventionally sized (4x4 and 8x8) NoCs. Higher levels of network load remain a challenge in purely bufferless networks; achieving efficiency under the additional load without buffers. As the size of the CMP increases (e.g., to 64x64), these efficiency gains from bufferless NoCs will become increasingly important, but as we will show, new scalability challenges arise in these larger networks. Congestion in such a network must be managed in an intelligent way in order to ensure scalability, even in workloads that have high traffic locality (e.g., due to intelligent data mapping).

We explore limitations of bufferless NoCs in terms of network load and network size with the goal of understanding *efficiency* and *scalability*. In §3.1, we show that as network load increases, application-level throughput reduces due to congestion in the network. This congestion manifests differently than in traditional buffered networks (where in-network latency increases). In a bufferless NoC, network admission becomes the bottleneck with congestion and cores become "*starved*," unable to access the network. In §3.2, we monitor the effect of the congestion on application-level throughput as we scale the size of the network from 16 to 4096 cores. Even when traffic has locality (due to intelligent data mapping to cache slices), we find that congestion significantly reduces the scalability of the bufferless NoC to larger sizes. These two fundamental issues motivate *congestion control for bufferless NoCs*.

3.1 Network Congestion at High Load

First, we study the effects of high workload intensity in the bufferless NoC. We simulate 700 real-application workloads in a 4x4 NoC (methodology in §6.1). Our workloads span a range of network utilizations exhibited by real applications.

Effect of Congestion at the Network Level: Starting at the network layer, we evaluate the effects of workload intensity on network-level metrics in the small-scale (4x4 mesh) NoC. Figure 2(a) shows average network latency for each of the 700 workloads. Notice how per-flit network latency generally remains stable (within 2x from baseline to maximum load), even when the network is under heavy load. This is in stark contrast to traditional buffered networks, in which the per-packet network latency increases significantly as the load in the network increases. However, as we will show in §3.2, network latency increases more with load in larger NoCs as other scalability bottlenecks come into consideration.

Deflection routing shifts many effects of congestion from within the network to network admission. In a highly-congested network, it may no longer be possible to efficiently inject packets into the network, because the router encounters free slots less often. Such a situation is known as **starvation**. We define **starvation rate** (σ) as the fraction of cycles in a window of W, in which a node tries to inject a flit but cannot: $\sigma = \frac{1}{W}\sum_i^W starved(i) \in [0,1]$. Figure 2(b) shows that starvation rate grows superlinearly with network utilization. Starvation rates at higher network utilizations are significant. Near 80% utilization, the average core is blocked from injecting into the network 30% of the time.

These two trends – relatively stable in-network latency, and high queueing latency at network admission – lead to the conclusion that network congestion is better measured in terms of *starvation* than in terms of latency. When we introduce our congestion-control mechanism in §5, we will use this metric to drive decisions.

Effect of Congestion on Application-level Throughput: As a NoC is part of a complete multicore system, it is important to evaluate the effect of congestion at the application layer. In other words, network-layer effects only matter when they affect the performance of CPU cores. We define *system throughput* as the application-level instruction throughput: for N cores, **System Throughput** = $\sum_i^N IPC_i$, where IPC_i gives *instructions per cycle* at core i.

To show the effect of congestion on application-level throughput, we take a network-heavy sample workload and *throttle* all applications at a throttling rate swept from 0. This throttling rate controls how often all routers that desire to inject a flit are blocked from doing so (e.g., a throttling rate of 50% indicates that half of all injections are blocked). If an injection is blocked, the router must try again in the next cycle. By controlling the injection rate of new traffic, we are able to vary the network utilization over a continuum and observe a full range of congestion. Figure 2(c) plots the resulting system throughput as a function of average net utilization.

This static-throttling experiment yields two key insights. First, network utilization does not reach 1, i.e., the network is never fully saturated even when unthrottled. The reason is that applications are naturally *self-throttling* due to the nature of out-of-order execution: a thread running on a core can only inject a relatively small number of requests into the network before stalling to wait for replies. This limit on outstanding requests occurs because the core's instruction window (which manages in-progress instructions) cannot retire (complete) an instruction until a network reply containing its requested data arrives. Once this window is stalled, a thread cannot start to execute any more instructions, hence cannot inject further requests. This self-throttling nature of applications helps to prevent *congestion collapse*, even at the highest possible network load.

Second, this experiment shows that injection throttling (i.e., a form of congestion control) can yield increased application-level throughput, even though it explicitly blocks injection some fraction of the time, because it reduces network congestion significantly. In Figure 2(c), a gain of 14% is achieved with simple static throttling.

However, static and homogeneous throttling across all cores does not yield the best possible improvement. In fact, as we will show in §4, throttling the wrong applications can significantly reduce system performance. This will motivate the need for *application-awareness*. Dynamically throttling the proper applications based on their relative benefit from injection yields significant system throughput improvements (e.g., up to 28% as seen in §6.2).

Key Findings: *Congestion contributes to high starvation rates and increased network latency. Starvation rate is a more accurate indicator of the level of congestion than network latency in a*

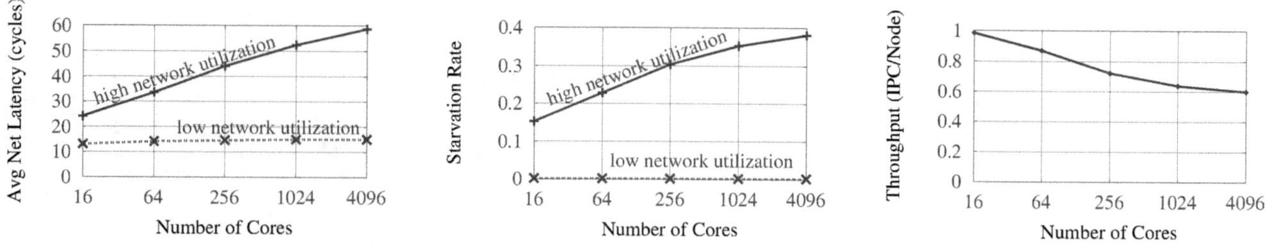

(a) Average network latency with CMP size. (b) Starvation rate with CMP size. (c) Per-node throughput with CMP size.

Figure 3: Scaling behavior: even with data locality, as network size increases, effects of congestion become more severe and scalability is limited.

bufferless network. Although collapse does not occur at high load, injection throttling can yield a more efficient operating point.

3.2 Scalability to Large Network Size

As we motivated in the prior section, scalability of on-chip networks will become critical as core counts continue to rise. In this section, we evaluate the network at sizes much larger than common 4x4 and 8x8 design points [20,50] to understand the scalability bottlenecks. However, the simple assumption of uniform data striping across all nodes no longer makes sense at large scales. With simple uniform striping, we find per-node throughput degrades by 73% from a 4x4 to 64x64 network. Therefore, we model increased data locality (i.e., intelligent data mapping) in the shared cache slices.

In order to model locality reasonably, independent of particular cache or memory system implementation details, we assume an exponential distribution of data-request destinations around each node. The private-cache misses from a given CPU core access shared-cache slices to service their data requests with an exponential distribution in distance, so most cache misses are serviced by nodes within a few hops, and some small fraction of requests go further. This approximation also effectively models a small amount of global or long-distance traffic, which can be expected due to global coordination in a CMP (e.g., OS functionality, application synchronization) or access to memory controllers or other global resources (e.g., accelerators). For this initial exploration, we set the distribution's parameter $\lambda = 1.0$, i.e., the average hop distance is $1/\lambda = 1.0$. This places 95% of requests within 3 hops and 99% within 5 hops. (We also performed experiments with a power-law distribution of traffic distance, which behaved similarly. For the remainder of this paper, we assume an exponential locality model.)

Effect of Scaling on Network Performance: By increasing the size of the CMP and bufferless NoC, we find that the impact of congestion on network performance increases with size. In the previous section, we showed that despite increased network utilization, the network latency remained relatively stable in a 4x4 network. However, as shown in Figure 3(a), as the size of the CMP increases, average latency increases significantly. While the 16-core CMP shows an average latency delta of 10 cycles between congested and non-congested workloads, congestion in a 4096-core CMP yields nearly 60 cycles of additional latency per flit on average. This trend occurs despite a fixed data distribution (λ parameter) – in other words, despite the same average destination distance. Likewise, shown in Figure 3(b), starvation in the network increases with CMP size due to congestion. Starvation rate increases to nearly 40% in a 4096-core system, more than twice as much as in a 16-core system, for the same per-node demand. This indicates that the network becomes increasingly inefficient under congestion, despite locality in network traffic destinations, as CMP size increases.

Effect of Scaling on System Performance: Figure 3(c) shows that the decreased efficiency at the network layer due to congestion degrades the entire system's performance, measured as IPC/node, as the size of the network increases. This shows that congestion is limiting the effective scaling of the bufferless NoC and system under higher intensity workloads. As shown in §3.1 and Figure 2(c), reducing congestion in the network improves system performance. As we will show in §6.3, reducing the congestion in the network significantly improves scalability with high-intensity workloads.

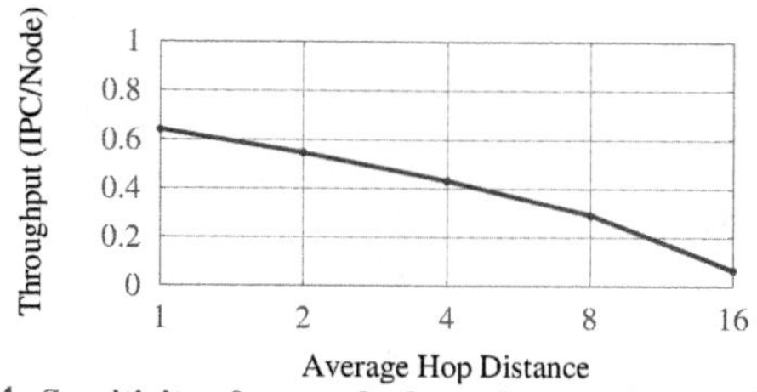

Figure 4: Sensitivity of per-node throughput to degree of locality.

Sensitivity to degree of locality: Finally, Figure 4 shows the sensitivity of system throughput, as measured by IPC per node, to the degree of locality in a 64x64 network. This evaluation varies the λ parameter of the simple exponential distribution for each node's destinations such that $1/\lambda$, or the average hop distance, varies from 1 to 16 hops. As expected, performance is highly sensitive to the degree of locality. For the remainder of this paper, we assume that $\lambda = 1$ (i.e., average hop distance of 1) in locality-based evaluations.

Key Findings: *We find that even with data locality (e.g., introduced by compiler and hardware techniques), as NoCs scale into hundreds and thousands of nodes, congestion becomes an increasingly significant concern for system performance. We show that per-node throughput drops considerably as network size increases, even when per-node demand (workload intensity) is held constant, motivating the need for congestion control for efficient scaling.*

4. THE NEED FOR APPLICATION-LEVEL AWARENESS IN THE NOC

Application-level throughput decreases as network congestion increases. The approach taken in traditional networks – to throttle applications in order to reduce congestion – will enhance performance, as we already showed in §3.1. However, we will show in this section that *which applications* are throttled can significantly impact per-application and overall system performance. To illustrate this, we have constructed a 4×4-mesh NoC that consists of 8 instances each of `mcf` and `gromacs`, which are memory-intensive and non-intensive applications, respectively [61]. We run the applications with no throttling, and then statically throttle each application in turn by 90% (injection blocked 90% of time), examining application and system throughput.

The results provide key insights (Fig. 5). First, *which application is throttled has a significant impact on overall system throughput.* When `gromacs` is throttled, overall system throughput drops by 9%. However, when `mcf` is throttled by the same rate, the overall system throughput increases by 18%. Second, *instruction throughput is not an accurate indicator for whom to throttle.* Although `mcf` has lower instruction throughput than `gromacs`, system through-

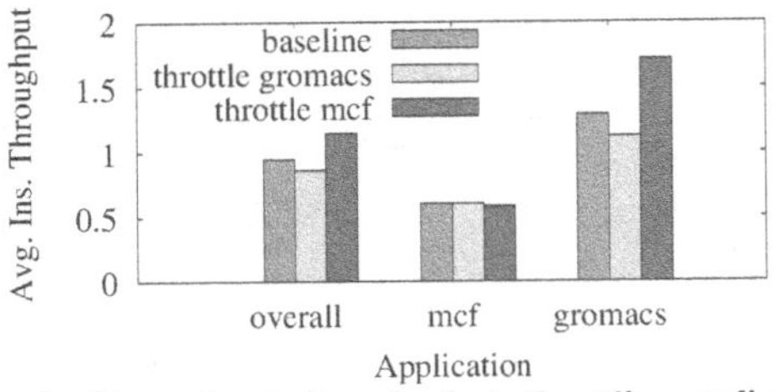

Figure 5: Throughput after selectively throttling applications.

put increases when `mcf` is throttled, with little effect on `mcf` (-3%). Third, *applications respond differently to network throughput variations.* When `mcf` is throttled, its instruction throughput decreases by 3%; however, when `gromacs` is throttled by the same rate, its throughput decreases by 14%. Likewise, `mcf` benefits little from increased network throughput when `gromacs` is throttled, but `gromacs` benefits greatly (25%) when `mcf` is throttled.

The reason for this behavior is that each application has a varying L1 cache miss rate, requiring a certain volume of traffic to complete a given instruction sequence; this measure depends wholly on the behavior of the program's memory accesses. Extra latency for a single flit from an application with a high L1 cache miss rate will not impede as much forward progress as the same delay of a flit in an application with a small number of L1 misses, since that flit represents a greater fraction of forward progress in the latter.

Key Finding: *Bufferless NoC congestion control mechanisms need application awareness to choose whom to throttle.*

Instructions-per-Flit: The above discussion implies that not all flits are created equal – i.e., that flits injected by some applications lead to greater forward progress when serviced. We define *Instructions-per-Flit* (IPF) as the ratio of *instructions retired* in a given period by an application *I* to *flits of traffic F* associated with the application during that period: **IPF** $= I/F$. IPF is only dependent on the L1 cache miss rate, and is thus independent of the congestion in the network and the rate of execution of the application. Thus, it is a stable measure of an application's current network intensity. Table 1 shows the *average* IPF values for a set of real applications. As shown, IPF can vary considerably: `mcf`, a memory-intensive application produces approximately 1 flit on average for every instruction retired (IPF=1.00), whereas `povray` yields an IPF four orders of magnitude greater: 20708.

Application	*Mean*	*Var.*	*Application*	*Mean*	*Var.*
matlab	0.4	0.4	cactus	14.6	4.0
health	0.9	0.1	gromacs	19.4	12.2
mcf	1.0	0.3	bzip2	65.5	238.1
art.ref.train	1.3	1.3	xml_trace	108.9	339.1
lbm	1.6	0.3	gobmk	140.8	1092.8
soplex	1.7	0.9	sjeng	141.8	51.5
libquantum	2.1	0.6	wrf	151.6	357.1
GemsFDTD	2.2	1.4	crafty	157.2	119.0
leslie3d	3.1	1.3	gcc	285.8	81.5
milc	3.8	1.1	h264ref	310.0	1937.4
mcf2	5.5	17.4	namd	684.3	942.2
tpcc	6.0	7.1	omnetpp	804.4	3702.0
xalancbmk	6.2	6.1	dealII	2804.8	4267.8
vpr	6.4	0.3	calculix	3106.5	4100.6
astar	8.0	0.8	tonto	3823.5	4863.9
hmmer	9.6	1.1	perlbench	9803.8	8856.1
sphinx3	11.8	95.2	povray	20708.5	1501.8

Table 1: Average IPF values and variance for evaluated applications.

Fig. 5 illustrates this difference: `mcf`'s low IPF value (1.0) indicates that it can be heavily throttled with little impact on its throughput (-3% @ 90% throttling). It also gains relatively less when other applications are throttled (e.g., <+1% when `gromacs` is throttled). However, `gromacs`' higher IPF value implies that its performance will suffer if it is throttled (-10%), but can gain from additional network throughput (+25%).

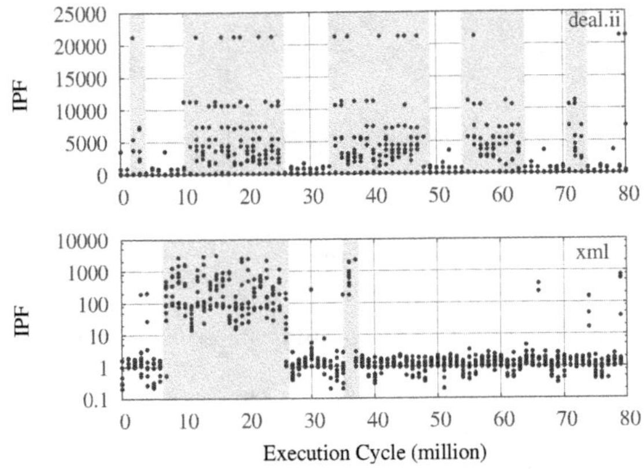

Figure 6: IPF is dynamic and varies with application phases.

Need for Dynamic Throttling: Until now, we have illustrated throttling as a static mechanism. In Figure 5, we throttled one application or the other throughout its entire execution. While applications can exhibit "average" behavior (e.g., an average IPF) by which they could be throttled, performing *only* static throttling based on average behavior can degrade performance and reduce efficiency relative to what could be possible because *application behavior is dynamic over time* (thus, IPF is also dynamic – also shown by the variance in Table 1). Throttling one application under congestion may help improve system throughput at one point in time, but hurt system performance at another point in time.

To illustrate this, we plot two applications' IPF values over time in Figure 6: **deal.II** (a finite element analysis application) and **xml** (an XML parser). As shown, application behavior changes over time, and in fact, applications can have "phases" through which they cycle (e.g., in the case of *deal.II*). Throttling either application during congestion in the periods where IPF is low (non-shaded) will improve system performance without significantly degrading the application's performance. However, throttling either in high-IPF phases will significantly degrade the application's performance (and thus the overall system's performance). Therefore, a successful throttling mechanism must continually monitor application intensity and the level of congestion, dynamically adjusting over time. IPF is a dynamic metric that enables this.

Key Findings: *IPF (Instructions-per-Flit) quanitifies application intensity and service requirements, enabling application-aware throttling mechanisms. Applications have phases that vary widely in network intensity. A dynamic throttling mechanism is thus needed, and this mechanism must monitor IPF dynamically.*

5. CONTROLLING CONGESTION

Section 4 defined a metric that determines an application's network intensity and its response to throttling. As shown in §4, *when the network is congested*, we must consider application-layer information to throttle effectively. We improve instruction throughput by throttling applications when they have low IPF. This works for three reasons: 1) applications with low IPF are relatively insensitive to throttling compared to applications with high IPF, 2) applications with high IPF benefit more at the application-level from increased network throughput than those with low IPF, and 3) throttling applications when they have low IPF is more effective at reducing congestion, because these applications are more network-intensive.

Basic Idea: We propose an interval-based congestion control algorithm that periodically (every $100,000$ cycles, at least 10x shorter than typical application phases – see Fig. 6): 1) detects congestion based on starvation rates, 2) determines IPF of applications, 3) if the network is congested, throttles *only* the nodes on which certain applications are running (chosen based on their IPF). Our algorithm, described here, is summarized in Algos. 1, 2, and 3.

Mechanism: A *key difference* of this mechanism to the majority of currently existing congestion control mechanisms in traditional networks [35, 38] is that it is a *centrally-coordinated* algorithm. This is possible in an on-chip network, and in fact is cheaper in our case (Central vs. Distributed Comparison: §6.6).

Since the on-chip network exists within a CMP that usually runs a single operating system (i.e., there is no hardware partitioning), the system software can be aware of all hardware in the system and communicate with each router in some hardware-specific way. As our algorithm requires some computation that would be impractical to embed in dedicated hardware in the NoC, we find that a hardware/software combination is likely the most efficient approach. Because the mechanism is periodic with a relatively long period, this does not place burden on the system's CPUs. As described in detail in §6.5, the pieces that integrate tightly with the router are implemented in hardware for practicality and speed.

There are several components of the mechanism's periodic update: first, it must determine when to throttle, maintaining appropriate responsiveness without becoming too aggressive; second, it must determine whom to throttle, by estimating the IPF of each node; and third, it must determine how much to throttle in order to optimize system throughput without hurting individual applications. We address these elements and present a complete algorithm.

When to Throttle: As described in §3, *starvation rate* is a superlinear function of network congestion (Fig. 2(b)). We use starvation rate (σ) as a per-node indicator of congestion in the network. Node *i* is *congested* if:

$$\sigma_i > \min(\beta_{starve} + \alpha_{starve}/IPF_i, \gamma_{starve}) \quad (1)$$

where α is a scale factor, and β and γ are lower and upper bounds, respectively, on the threshold (we use $\alpha_{starve} = 0.4$, $\beta_{starve} = 0.0$ and $\gamma_{starve} = 0.7$ in our evaluation, determined empirically; sensitivity results and discussion can be found in §6.4). It is important to factor in IPF since network-intensive applications will naturally have higher starvation due to higher injection rates. Throttling is *active* if at least one node is congested.

Whom to Throttle: When throttling is active, a node is throttled if its intensity is above average (not all nodes are throttled). In most cases, the congested cores are not the ones throttled; only the heavily-injecting cores are throttled. The cores to throttle are chosen by observing IPF: lower IPF indicates greater network intensity, and so nodes with IPF below average are throttled. Since we use central coordination, computing the mean IPF is possible without distributed averaging or estimation. The *Throttling Criterion* is:

If *throttling is active* AND $IPF_i < mean(IPF)$.

The simplicity of this rule can be justified by our observation that IPF in most workloads tend to be widely distributed: there are memory-intensive applications and CPU-bound applications. We find the separation between application classes is clean for our workloads, so a more intelligent and complex rule is not justified.

Finally, we observe that this throttling rule results in relatively stable behavior: the decision to throttle depends only on the instructions per flit (IPF), which is *independent* of the network service provided to a given node and depends only on that node's program characteristics (e.g., cache miss rate). Hence, this throttling criterion makes a throttling decision that is robust and stable.

Determining Throttling Rate: We throttle the chosen applications proportional to their application intensity. We compute *throttling rate*, the fraction of cycles in which a node cannot inject, as:

$$R \Leftarrow \min(\beta_{throt} + \alpha_{throt}/IPF, \gamma_{throt}) \quad (2)$$

Algorithm 1 Main Control Algorithm (in software)

```
Every T cycles:
collect IPF[i], σ[i] from each node i
/* determine congestion state */
congested ⇐ false
for i = 0 to N_nodes − 1 do
    starve_thresh = min(β_starve + α_starve/IPF[i], γ_starve)
    if σ[i] > starve_thresh then
        congested ⇐ true
    end if
end for
/* set throttling rates */
throt_thresh = mean(IPF)
for i = 0 to N_nodes − 1 do
    if congested AND IPF[i] < throt_thresh then
        throttle_rate[i] = min(β_throt + α_throt/IPF[i], γ_throt)
    else
        throttle_rate[i] ⇐ 0
    end if
end for
```

Algorithm 2 Computing Starvation Rate (in hardware)

```
At node i:
σ[i] ⇐ Σ_{k=0}^{W} starved(current_cycle − k)/W
```

Algorithm 3 Simple Injection Throttling (in hardware)

```
At node i:
if trying to inject in this cycle and an output link is free then
    inj_count[i] ⇐ (inj_count[i] + 1) mod MAX_COUNT
    if inj_count[i] ≥ throttle_rate[i] * MAX_COUNT then
        allow injection
        starved(current_cycle) ⇐ false
    else
        block injection
        starved(current_cycle) ⇐ true
    end if
end if
```

Note: this is one possible way to implement throttling with simple, deterministic hardware. Randomized algorithms can also be used.

where IPF is used as a measure of application intensity, and α, β and γ set the scaling factor, lower bound and upper bound respectively, as in the starvation threshold formula. Empirically, we find $\alpha_{throt} = 0.90$, $\beta_{throt} = 0.20$ and $\gamma_{throt} = 0.75$ to work well. Sensitivity results and discussion of these parameters are in §6.4.

How to Throttle: When throttling a node, only its data requests are throttled. Responses to service requests from other nodes are *not* throttled; this could further impede a starved node's progress.

6. EVALUATION

In this section, we present an evaluation of the effectiveness of our congestion-control mechanism to address high load in small NoCs (4x4 and 8x8) and scalability in large NoCs (up to 64x64).

6.1 Methodology

We obtain results using a cycle-level simulator that models the target system. This simulator models the network routers and links, the full cache hierarchy, and the processor cores at a sufficient level of detail. For each application, we capture an instruction trace of a representative execution slice (chosen using PinPoints [56]) and replay each trace in its respective CPU core model during simulation. Importantly, the simulator is a *closed-loop* model: the backpressure of the NoC and its effect on presented load are accurately captured. Results obtained using this simulator have been published in past NoC studies [20, 22, 50]. Full parameters for the simulated system are given in Table 2). The simulation is run for 10 million cycles, meaning that our control algorithm runs 100 times per-workload.

Network topology	2D mesh, 4x4 or 8x8 size
Routing algorithm	FLIT-BLESS [50] (example in §2)
Router (Link) latency	2 (1) cycles
Core model	Out-of-order
Issue width	3 insns/cycle, 1 mem insn/cycle
Instruction window size	128 instructions
Cache block	32 bytes
L1 cache	private 128KB, 4-way
L2 cache	shared, distributed, perfect cache
L2 address mapping	Per-block interleaving, XOR mapping; randomized exponential for locality evaluations

Table 2: System parameters for evaluation.

Workloads and Their Characteristics: We evaluate 875 multiprogrammed workloads (700 16-core, 175 64-core). Each consists of independent applications executing on each core. The applications do not coordinate with each other (i.e., each makes progress independently and has its own working set), and each application is fixed to one core. Such a configuration is expected to be a common use-case for large CMPs, for example in cloud computing systems which aggregate many workloads onto one substrate [34].

Our workloads consist of applications from SPEC CPU2006 [61], a standard benchmark suite in the architecture community, as well as various desktop, workstation, and server applications. Together, these applications are representative of a wide variety of network access intensities and patterns that are present in many realistic scenarios. We classify the applications (Table 1) into three intensity levels based on their average instructions per flit (IPF), i.e., network intensity: H (Heavy) for less than 2 IPF, M (Medium) for 2 – 100 IPF, and L (Light) for > 100 IPF. We construct balanced workloads by selecting applications in seven different workload categories, each of which draws applications from the specified intensity levels: *{H,M,L,HML,HM,HL,ML}*. For a given workload category, the application at each node is chosen randomly from all applications in the given intensity levels. For example, an **H**-category workload is constructed by choosing the application at each node from among the high-network-intensity applications, while an **HL**-category workload is constructed by choosing the application at each node from among all high- and low-intensity applications.

Congestion Control Parameters: We set the following algorithm parameters based on empirical parameter optimization: the update period $T = 100K$ cycles and the starvation computation window $W = 128$. The minimum and maximum starvation rate thresholds are $\beta_{starve} = 0.0$ and $\gamma_{starve} = 0.70$ with a scaling factor of $\alpha_{starve} = 0.40$. We set the throttling minimum and maximum to $\beta_{throttle} = 0.20$ and $\gamma_{throttle} = 0.75$ with scaling factor $\alpha_{throttle} = 0.9$. Sensitivity to these parameters is evaluated in §6.4.

6.2 Application Throughput in Small NoCs

System Throughput Results: We first present the effect of our mechanism on overall system/instruction throughput (average IPC, or instructions per cycle, per node, as defined in §3.1) for both 4x4 and 8x8 systems. To present a clear view of the improvements at various levels of network load, we evaluate gains in overall system throughput plotted against the average network utilization (measured without throttling enabled). Fig. 7 presents a scatter plot that shows the percentage gain in overall system throughput with our mechanism in each of the 875 workloads on the 4x4 and 8x8 system. The maximum performance improvement under congestion (e.g., load >0.7) is 27.6% with an average improvement of 14.7%.

Fig. 8 shows the maximum, average, and minimum system throughput gains on each of the workload categories. The highest average and maximum improvements are seen when all applications in the workload have High or High/Medium intensity. As expected, our mechanism provides little improvement when all applications in the workload have Low or Medium/Low intensity, because the network is adequately provisioned for the demanded load.

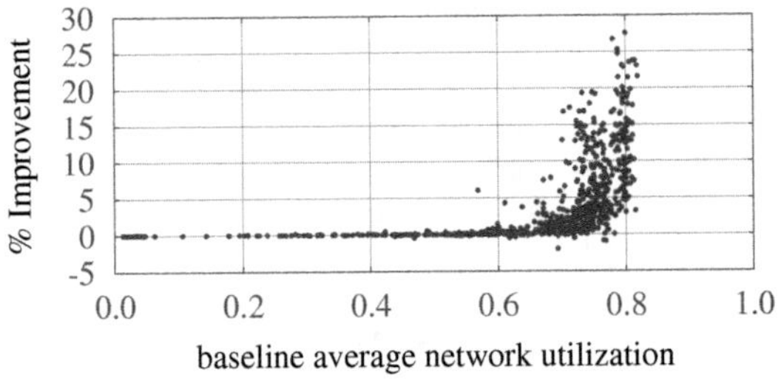

Figure 7: Improvements in overall system throughput (4x4 and 8x8).

Improvement in Network-level Admission: Fig. 9 shows the CDF of the 4x4 workloads' average starvation rate when the baseline average network utilization is greater than 60%, to provide insight into the effect of our mechanism on starvation when the network is likely to be congested. Using our mechanism, only 36% of the congested 4x4 workloads have an average starvation rate greater than 30% (0.3), whereas without our mechanism 61% have a starvation rate greater than 30%.

Effect on Weighted Speedup (Fairness): In addition to instruction throughput, a common metric for evaluation is *weighted speedup* [19,59], defined as $WS = \sum_i^N \frac{IPC_{i,shared}}{IPC_{i,alone}}$, where $IPC_{i,shared}$ and $IPC_{i,alone}$ are the *instructions per cycle* measurements for application i when run together with other applications and when run alone, respectively. WS is N in an ideal N-node system with no interference, and drops as application performance is degraded due to network contention. This metric takes into account that different applications have different "natural" execution speeds; maximizing it requires maximizing the rate of progress – compared to this natural execution speed – across *all* applications in the entire workload. In contrast, a mechanism can maximize instruction throughput by unfairly slowing down low-IPC applications. We evaluate with weighted speedup to ensure our mechanism does not penalize in this manner.

Figure 10 shows weighted speedup improvements by up to 17.2% (18.2%) in the 4x4 and 8x8 workloads respectively.

Fairness In Throttling: We further illustrate that our mechanism does not unfairly throttle applications (i.e., that the mechanism is not biased toward high-IPF applications at the expense of low-IPF applications). To do so, we evaluate the performance of applications in pairs with IPF values *IPF1* and *IPF2* when put together in a 4x4 mesh (8 instances of each application) in a checkerboard layout. We then calculate the percentage change in throughput for both applications when congestion control is applied.

Figure 11 shows the resulting performance improvement for the applications, given the IPF values of both applications. Accompanying the graph is the average baseline (un-throttled) network utilization shown in Figure 12. Clearly, when both IPF values are high, there is no change in performance since both applications are CPU bound (network utilization is low). When application 2's IPF value (IPF2) is high and application 1's IPF value (IPF1) is low (right corner of both figures), throttling shows performance improvements to application 2 since the network is congested. Importantly, however, application 1 is *not* unfairly throttled (left corner), and in fact shows some improvements using our mechanism. For example, when $IPF1 = 1000$ and $IPF2 < 1$, application 1 still shows benefits (e.g., 5-10%) by reducing overall congestion.

Key Findings: *When evaluated in 4x4 and 8x8 networks, our mechanism improves performance up to 27.6%, reduces starvation, improves weighted speedup, and does not unfairly throttle.*

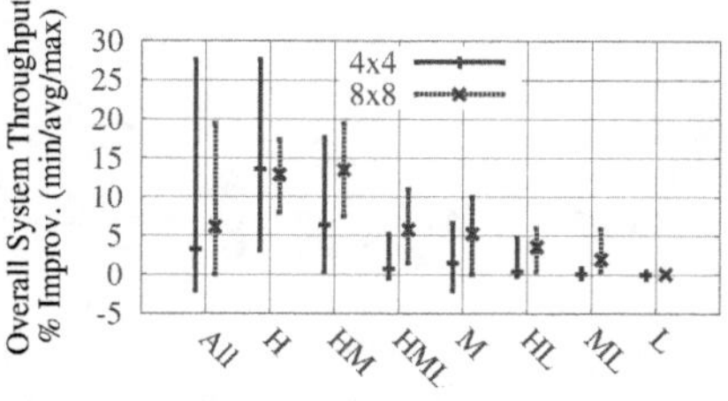

Figure 8: Improvement breakdown by category.

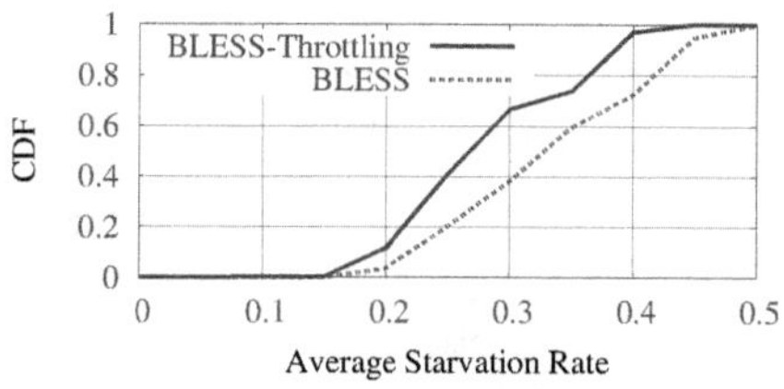

Figure 9: CDF of average starvation rates.

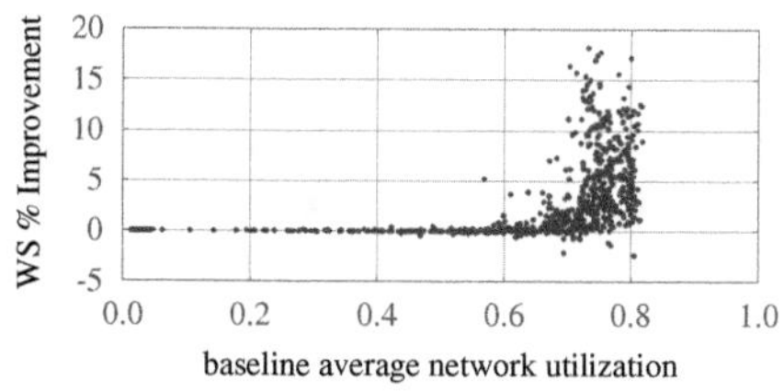

Figure 10: Improvements in weighted speedup.

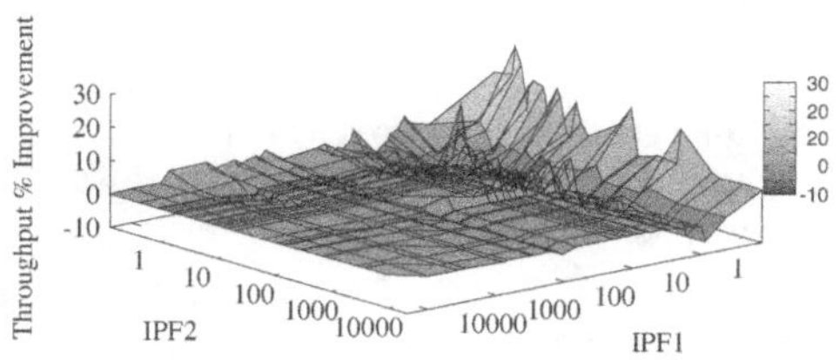

Figure 11: Percentage improvements in throughput when shared.

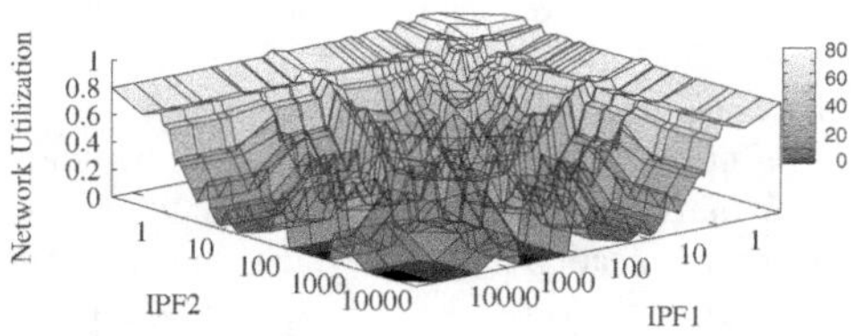

Figure 12: Average baseline network utilization when shared.

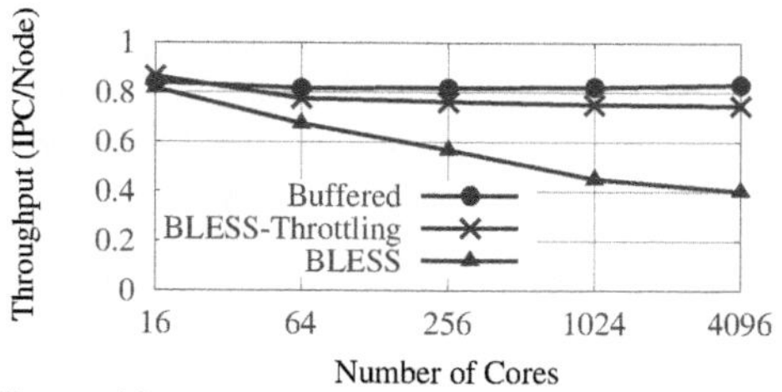

Figure 13: Per-node system throughput with scale.

6.3 Scalability in Large NoCs

In §3.2, we showed that even with fixed data locality, increases in network sizes lead to increased congestion and decreased per-node throughput. We evaluate congestion control's ability to restore scalability. Ideally, per-node throughput remains fixed with scale.

We model network scalability *with data locality* using fixed exponential distributions for each node's request destinations, as in §3.2.[4] This introduces a degree of locality achieved by compiler and hardware optimizations for data mapping. Real application traces are still executing in the processor/cache model to generate the request timing; the destinations for each data request are simply mapped according to the distribution. This allows us to study scalability independent of the effects and interactions of more complex data distributions. Mechanisms to distribute data among multiple private caches in a multicore chip have been proposed [46, 57], including one which is aware of interconnect distance/cost [46].

Note that we also include a NoC based on *virtual-channel buffered routers* [14] in our scalability comparison.[5] A buffered router can attain higher performance, but as we motivated in §1, buffers carry an area and power penalty as well. We run the same workloads on the buffered network for direct comparison with a baseline bufferless network (BLESS), and with our mechanism (BLESS-Throttling).

Figures 13, 14, 15, and 16 show the trends in network latency, network utilization, system throughput, and NoC power as network size increases with all three architectures for comparison. The baseline case mirrors what is shown in §3.2: congestion becomes a scalability bottleneck as size increases. However, congestion control successfully throttles the network back to a more efficient operating point, achieving essentially flat lines. We observed the same scalability trends in a torus topology (however, note that the torus topology yields a 10% throughput improvement for all networks).

[4]We also performed experiments with powerlaw distributions, not shown here, which resulted in similar conclusions.

[5]The network has the same topology as the baseline bufferless network, and routers have 4 VCs/input and 4 flits of buffering per VC.

We particularly note the NoC power results in Figure 16. This data comes from the BLESS router power model [20], and includes router and link power. As described in §2, a unique property of on-chip networks is a global power budget. Reducing power consumption as much as possible is therefore desirable. As our results show, through congestion control, we reduce power consumption in the bufferless network by up to 15%, and improve upon the power consumption of a buffered network by up to 19%.

6.4 Sensitivity to Algorithm Parameters

Starvation Parameters: The $\alpha_{starve} \in (0,\infty)$ parameter scales the congestion detection threshold with application network intensity, so that network-intensive applications are allowed to experience more starvation before they are considered congested. In our evaluations, $\alpha_{starve} = 0.4$; when $\alpha_{starve} > 0.6$ (which increases the threshold and hence under-throttles the network), performance decreases 25% relative to $\alpha_{starve} = 0.4$. When $\alpha_{starve} < 0.3$ (which decreases the threshold and hence over-throttles the network), performance decreases by 12% on average.

$\beta_{starve} \in (0,1)$ controls the minimum starvation rate required for a node to be considered as starved. We find that $\beta_{starve} = 0.0$ performs best. Values ranging from 0.05 to 0.2 degrade performance by 10% to 15% on average (24% maximum) with respect to $\beta_{starve} = 0.0$ because throttling is not activated as frequently.

The upper bound on the detection threshold, $\gamma_{starve} \in (0,1)$, ensures that even network-intensive applications can still trigger throttling when congested. We found that performance was not sensitive to γ_{starve} because throttling will be triggered anyway by the less network-intensive applications in a workload. We use $\gamma_{starve} = 0.7$.

Throttling Rate Parameters: $\alpha_{throt} \in (0,\infty)$ scales throttling rate with network intensity. Performance is sensitive to this parameter, with an optimal in our workloads at $\alpha_{throt} = 0.9$. When $\alpha_{throt} > 1.0$, lower-intensity applications are over throttled: more than three times as many workloads experience performance loss with our mechanism (relative to not throttling) than with $\alpha_{throt} = 0.9$. Values below 0.7 under-throttles congested workloads.

The $\beta_{throt} \in (0,1)$ parameter ensures that throttling has some effect when it is active for a given node by providing a minimum throttling rate value. We find, however, that performance is not sensitive to this value when it is small because network-intensive applications will already have high throttling rates because of their network intensity. However, a large β_{throt}, e.g. 0.25, over-throttles sensitive applications. We use $\beta_{throt} = 0.20$.

The $\gamma_{throt} \in (0,1)$ parameter provides an upper bound on the

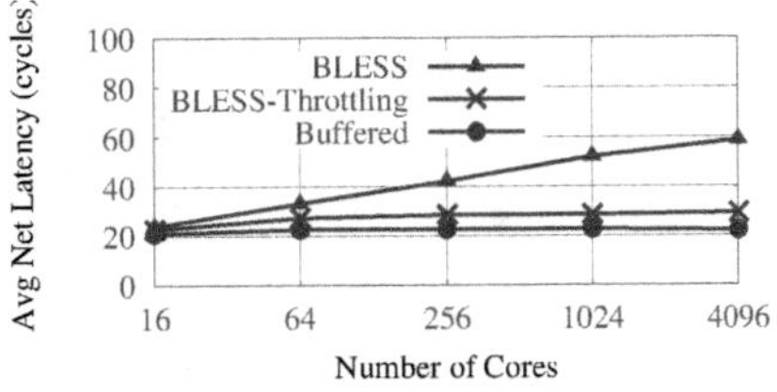

Figure 14: Network latency with scale.

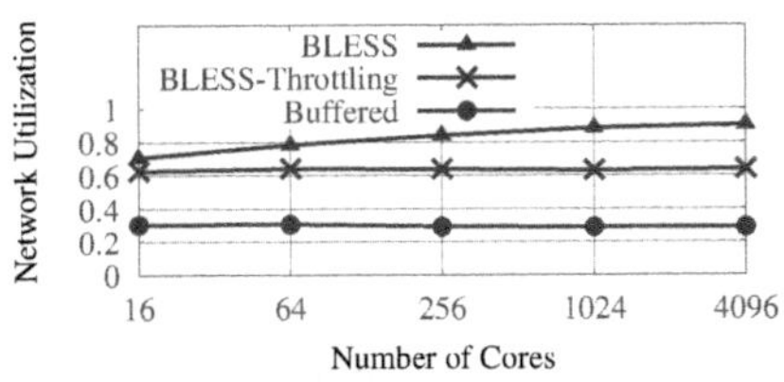

Figure 15: Network utilization with scale.

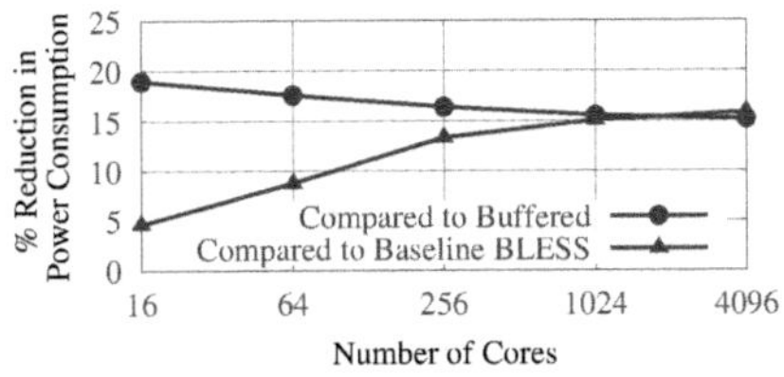

Figure 16: Reduction in power with scale.

throttling rate, ensuring that network-intensive applications will not be completely starved. Performance suffers for this reason if γ_{throt} is too large. We find $\gamma_{throt} = 0.75$ provides the best performance; if increased to 0.85, performance suffers by 30%. Reducing γ_{throt} below 0.65 hinders throttling from effectively reducing congestion.

Throttling Epoch: Throttling is re-evaluated once every 100k cycles. A shorter epoch (e.g., 1k cycles) yields a small gain (3–5%) but has significantly higher overhead. A longer epoch (e.g., 1M cycles) reduces performance dramatically because throttling is no longer sufficiently responsive to application phase changes.

6.5 Hardware Cost

Hardware is required to measure the starvation rate σ at each node, and to throttle injection. Our windowed-average starvation rate over W cycles requires a W-bit shift register and an up-down counter: in our configuration, $W = 128$. To throttle a node with a rate of r, we disallow injection for N cycles every M, such that $N/M = r$. This requires a free-running 7-bit counter and a comparator. In total, only 149 bits of storage, two counters, and one comparator are required. This is a minimal cost compared to (for example) the 128KB L1 cache.

6.6 Centralized vs. Distributed Coordination

Congestion control has historically been distributed (e.g., TCP), because centralized coordination in large networks (e.g., the Internet) is not feasible. As described in §2, NoC global coordination is often less expensive because the NoC topology and size are statically known. Here we will compare and contrast these approaches.

Centralized Coordination: To implement centrally-coordinated throttling, each node measures its own IPF and starvation rate, reports these rates to a central controller by sending small control packets, and receives a throttling rate setting for the following epoch. The central coordinator recomputes throttling rates only once every 100k cycles, and the algorithm consists only of determining average IPF and evaluating the starvation threshold and throttling-rate formulae for each node, hence has negligible overhead (can be run on a single core). Only $2n$ packets are required for n nodes every 100k cycles.

Distributed Coordination: To implement a distributed algorithm, a node must send a notification when congested (as before), but must decide on its own when and by how much to throttle. In a separate evaluation, we designed a simple distributed algorithm which (i) sets a "congested" bit on every packet that passes through a node when that node's starvation rate exceeds a threshold; and (ii) self-throttles at any node when that node sees a packet with a "congested" bit. This is a "TCP-like" congestion response mechanism: a congestion notification (e.g., a dropped packet in TCP) can be caused by congestion at any node along the path, and forces the receiver to back off. We found that because this mechanism is not selective in its throttling (i.e., it does not include application-awareness), it is far less effective at reducing NoC congestion. Alternatively, nodes could approximate the central approach by periodically broadcasting their IPF and starvation rate to other nodes. However, every node would then require storage to hold other nodes' statistics, and broadcasts would waste bandwidth.

Summary: Centralized coordination allows better throttling because the throttling algorithm has explicit knowledge of every node's state. Distributed coordination can serve as the basis for throttling in a NoC, but was found to be less effective.

7. DISCUSSION

We have provided an initial case study showing that core networking problems with novel solutions appear when designing NoCs. However, congestion control is just one avenue of networking research in this area, with many other synergies.

Traffic Engineering: Although we did not study multi-threaded applications in our work, they have been shown to have heavily local/regional communication patterns, which can create "hot-spots" of high utilization in the network. In fact, we observed similar behavior after introducing locality (§3.2) in low/medium congested workloads. Although our mechanism can provide small gains by throttling applications in the congested area, traffic engineering around the hot-spot is likely to provide even greater gains.

Due to application-phase behavior (Fig. 6), hot-spots are likely to be dynamic over run-time execution, requiring a dynamic scheme such as TeXCP [37]. The challenge will be efficiently collecting information about the network, adapting in a non-complex way, and keeping routing simple in the constrained environment. Our hope is that prior work can be leveraged, showing robust traffic engineering can be performed with fairly limited knowledge [4]. A focus could be put on a certain subset of traffic that exhibits "long-lived behavior" to make NoC traffic engineering feasible [58].

Fairness: While we have shown in §6.2 that our congestion control mechanism does not unfairly throttle applications to benefit system performance, our controller has no explicit fairness target. As described in §4, however, different applications have different rates of progress for a given network bandwidth; thus, explicitly managing fairness is a challenge. Achieving network-level fairness may not provide application level fairness. We believe the bufferless NoC provides an interesting opportunity to develop a novel application-aware fairness controller (e.g., as targeted by [10]).

Metrics: While there have been many metrics adopted by the architecture community for evaluating system performance (e.g., weighted speedup and IPC), more comprehensive metrics are needed for evaluating NoC performance. The challenge, similar to what has been discussed above, is that network performance may not accurately reflect system performance due to application-layer effects. This makes it challenging to know where and how to optimize the network. Developing a set of metrics which can reflect the coupling of network and system performance will be beneficial.

Topology: Although our study focused on the 2D mesh, a variety of on-chip topologies exist [11,29,40,41,43] and have been shown to greatly impact traffic behavior, routing, and network efficiency. Designing novel and efficient topologies is an on-going challenge, and resulting topologies found to be efficient in on-chip networks can impact off-chip topologies e.g., Data Center Networks (DCNs). NoCs and DCNs have static and known topologies, and are attempting to route large amounts of information multiple hops. Showing benefits of topology in one network may imply benefits in the other. Additionally, augmentations to network topology have gained attention, such as *express channels* [5] between separated routers.

Buffers: Both the networking and architecture communities continue to explore bufferless architectures. As optical networks [69, 71] become more widely deployed, bufferless architectures are going to become more important to the network community due to challenges of buffering in optical networks. While some constraints will likely be different (e.g., bandwidth), there will likely be strong parallels in topology, routing techniques, and congestion control. Research in this area is likely to benefit both communities.

Distributed Solutions: Like our congestion control mechanism, many NoC solutions use centralized controllers [16,18,27,65]. The benefit is a reduction in complexity and lower hardware cost. However, designing distributed network controllers with low-overhead and low-hardware cost is becoming increasingly important with scale. This can enable new techniques, utilizing distributed information to make fine-grained decisions and network adaptations.

8. RELATED WORK

Internet Congestion Control: Traditional mechanisms look to prevent *congestion collapse* and provide fairness, first addressed by TCP [35] (subsequently in other work). Given that delay increases significantly under congestion, it has been a core metric for detecting congestion in the Internet [35, 51]. In contrast, we have shown that in NoCs, network latencies remain relatively stable in the congested state. Furthermore, there is no packet loss in on-chip networks, and hence no explicit ACK/NACK feedback. More explicit congestion notification techniques have been proposed that use coordination or feedback from the network core [23,38,62]. In doing so, the network as a whole can quickly converge to optimal efficiency and avoid constant fluctuation [38]. However, our work uses application rather than network information.

NoC Congestion Control: The majority of congestion control work in NoCs has focused on buffered NoCs, and work with packets that have already entered the network, rather than control traffic at the injection point. The problems they solve are thus different in nature. Regional Congestion Awareness [27] implements a mechanism to detect congested regions in buffered NoCs and inform the routing algorithm to avoid them if possible. Some mechanisms are designed for particular types of networks or problems that arise with certain NoC designs: e.g., Baydal et al. propose techniques to optimize wormhole routing in [7]. Duato et al. give a mechanism in [18] to avoid head-of-line (HOL) blocking in buffered NoC queues by using separate queues. Throttle and Preempt [60], solves priority inversion in buffer space allocation by allowing preemption by higher-priority packets and using throttling.

Several techniques avoid congestion by deflecting traffic selectively (BLAM [64]), re-routing traffic to random intermediate locations (the Chaos router [44]), or creating path diversity to maintain more uniform latencies (Duato et al. in [24]). Proximity Congestion Awareness [52] extends a bufferless network to avoid routing toward congested regions. However, we cannot make a detailed comparison to [52] as the paper does not provide enough algorithmic detail for this purpose.

Throttling-based NoC Congestion Control: Prediction-based Flow Control [54] builds a state-space model for a buffered router in order to predict its free buffer space, and then uses this model to refrain from sending traffic when there would be no downstream space. Self-Tuned Congestion Control [65] includes a feedback-based mechanism that attempts to find the optimum throughput point dynamically. The solution is not directly applicable to our bufferless NoC problem, however, since the congestion behavior is different in a bufferless network. Furthermore, both of these prior works are application-unaware, in contrast to ours.

Adaptive Cluster Throttling [10], a recent source-throttling mechanism developed concurrently to our mechanism, is also targeted for bufferless NoCs. Unlike our mechanism, ACT operates by measuring application cache miss rates (MPKI) and performing a clustering algorithm to group applications into "clusters" which are alternately throttled in short timeslices. ACT is shown to perform well on small (4x4 and 8x8) mesh networks; we evaluate our mechanism on small networks as well as large (up to 4096-node) networks in order to address the scalability problem.

Application Awareness: Some work handles packets in an application aware manner in order to provide certain QoS guarantees or perform other traffic shaping. Several proposals, e.g., Globally Synchronized Frames [47] and Preemptive Virtual Clocks [30], explicitly address quality-of-service with in-network prioritization. Das et al. [16] propose ranking applications by their intensities and prioritizing packets in the network accordingly, defining the notion of "stall time criticality" to understand the sensitivity of each application to network behavior. Our use of the IPF metric is similar to L1 miss rate ranking. However, this work does not attempt to solve the congestion problem, instead simply scheduling packets to improve performance. In a later work, Aérgia [17] defines packet "slack" and prioritizes requests differently based on criticality.

Scalability Studies: We are aware of relatively few existing studies of large-scale 2D mesh NoCs: most NoC work in the architecture community focuses on smaller design points, e.g., 16 to 100 nodes, and the BLESS architecture in particular has been evaluated up to 64 nodes [50]. Kim et al. [42] examine scalability of ring and 2D mesh networks up to 128 nodes. Grot et al. [28] evaluate 1000-core meshes, but in a buffered network. That proposal, Kilo-NoC, addresses scalability of QoS mechanisms in particular. In contrast, our study examines congestion in a deflection network, and finds that reducing this congestion is a key enabler to scaling.

Off-Chip Similarities: The Manhattan Street Network (MSN) [48], an off-chip mesh network designed for packet communication in local and metropolitan areas, resembles the bufferless NoC in some of its properties and challenges. MSN uses drop-less deflection routing in a small-buffer design. Due to the routing and injection similarities, the MSN also suffers from starvation. Although similar in these ways, routing in the NoC is still designed for minimal complexity whereas the authors in [48] suggested more complex routing techniques which are undesirable for the NoC. Global coordination in MSNs were not feasible, yet often less complex and more efficient in the NoC (§2). Finally, link failure in the MSN was a major concern whereas in the NoC links are considered reliable.

Bufferless NoCs: In this study, we focus on bufferless NoCs, which have been the subject of significant recent work [10, 20–22, 26, 31, 49, 50, 67]. We describe some of these works in detail in §2.

9. CONCLUSIONS & FUTURE WORK

This paper studies congestion control in on-chip bufferless networks and has shown such congestion to be fundamentally different from that of other networks, for several reasons (e.g., lack of congestion collapse). We examine both network performance in moderately-sized networks and scalability in very large (4K-node) networks, and we find congestion to be a fundamental bottleneck. We develop an application-aware congestion control algorithm and show significant improvement in application-level system throughput on a wide variety of real workloads for NoCs.

More generally, NoCs are bound to become a critical system resource in many-core processors, shared by diverse applications.

Techniques from the networking research community can play a critical role to address research issues in NoCs. While we focus on congestion, we are already seeing other ties between these two fields. For example, data-center networks in which machines route packets, aggregate data, and can perform computation while forwarding (e.g., CamCube [1]) can be seen as similar to CMP NoCs. XORs as a packet coding technique, used in wireless meshes [39], are also being applied to the NoC for performance improvements [32]. We believe the proposed techniques in this paper are a starting point that can catalyze more research cross-over from the networking community to solve important NoC problems.

Acknowledgements We gratefully acknowledge the support of our industrial sponsors, AMD, Intel, Oracle, Samsung, and the Gigascale Systems Research Center. We also thank our shepherd David Wetherall, our anonymous reviewers, and Michael Papamichael for their extremely valuable feedback and insight. This research was partially supported by an NSF CAREER Award, CCF-0953246, and an NSF EAGER Grant, CCF-1147397. Chris Fallin is supported by an NSF Graduate Research Fellowship.

10. REFERENCES

[1] H. Abu-Libdeh, P. Costa, A. Rowstron, G. O'Shea, and A. Donnelly. Symbiotic routing in future data centers. SIGCOMM, 2010.
[2] M. Alizadeh et al. Data center TCP (DCTCP). SIGCOMM 2010, pages 63–74, New York, NY, USA, 2010. ACM.
[3] Appenzeller et al. Sizing router buffers. *SIGCOMM*, 2004.
[4] D. Applegate and E. Cohen. Making intra-domain routing robust to changing and uncertain traffic demands: understanding fundamental tradeoffs. SIGCOMM 2003.
[5] J. Balfour and W. J. Dally. Design tradeoffs for tiled cmp on-chip networks. Proceedings of the international conference on Supercomputing (ICS) 2006.
[6] P. Baran. On distributed communications networks. *IEEE Trans. Comm.*, 1964.
[7] E. Baydal et al. A family of mechanisms for congestion control in wormhole networks. *IEEE Trans. on Dist. Systems*, 16, 2005.
[8] L. Benini and G. D. Micheli. Networks on chips: A new SoC paradigm. *Computer*, 35:70–78, Jan 2002.
[9] S. Borkar.Thousand core chips: a technology perspective.*DAC-44*, 2007.
[10] K. Chang et al. Adaptive cluster throttling: Improving high-load performance in bufferless on-chip networks. SAFARI TR-2011-005.
[11] M. Coppola et al. Spidergon: a novel on-chip communication network. *Proc. Int'l Symposium on System on Chip*, Nov 2004.
[12] D. E. Culler et al. *Parallel Computer Architecture: A Hardware/Software Approach*. Morgan Kaufmann, 1999.
[13] W. Dally. Virtual-channel flow control. *IEEE Par. and Dist. Sys.*, '92.
[14] W. Dally and B. Towles. *Principles and Practices of Interconnection Networks*. Morgan Kaufmann, 2004.
[15] W. J. Dally and B. Towles. Route packets, not wires: On-chip interconnection networks. *DAC-38*, 2001.
[16] R. Das et al. Application-aware prioritization mechanisms for on-chip networks. *MICRO-42*, 2009.
[17] R. Das et al. Aérgia: exploiting packet latency slack in on-chip networks. *International Symposium on Computer Architecture (ISCA)*, 2010.
[18] J. Duato et al. A new scalable and cost-effective congestion management strategy for lossless multistage interconnection networks. *HPCA-11*, 2005.
[19] S. Eyerman and L. Eeckhout. System-level performance metrics for multiprogram workloads. *IEEE Micro*, 28:42–53, May 2008.
[20] C. Fallin, C. Craik, and O. Mutlu. CHIPPER: A low-complexity bufferless deflection router. *HPCA-17*, 2011.
[21] C. Fallin et al. A high-performance hierarchical ring on-chip interconnect with low-cost routers. SAFARI TR-2011-006.
[22] C. Fallin et al. MinBD: Minimally-buffered deflection routing for energy-efficient interconnect. *NOCS*, 2012.
[23] S. Floyd. Tcp and explicit congestion notification. *ACM Comm. Comm. Review, V. 24 N. 5, October 1994, p. 10-23.*
[24] D. Franco et al. A new method to make communication latency uniform: distributed routing balancing. *ICS-13*, 1999.
[25] C. Gómez, M. Gómez, P. López, and J. Duato. Reducing packet dropping in a bufferless noc. *EuroPar-14*, 2008.
[26] C. Gómez, M. E. Gómez, P. López, and J. Duato. Reducing packet dropping in a bufferless noc. *Euro-Par-14*, 2008.
[27] P. Gratz et al. Regional congestion awareness for load balance in networks-on-chip. *HPCA-14*, 2008.
[28] B. Grot et al. Kilo-NOC: A heterogeneous network-on-chip architecture for scalability and service guarantees. *ISCA-38*, 2011.
[29] B. Grot, J. Hestness, S. Keckler, and O. Mutlu. Express cube topologies for on-chip interconnects. *HPCA-15*, 2009.
[30] B. Grot, S. Keckler, and O. Mutlu. Preemptive virtual clock: A flexible, efficient, and cost-effective qos scheme for networks-on-chip. *MICRO-42*, 2009.
[31] M. Hayenga et al. Scarab: A single cycle adaptive routing and bufferless network. *MICRO-42*, 2009.
[32] M. Hayenga and M. Lipasti. The NoX router. MICRO-44, 2011.
[33] Y. Hoskote et al. A 5-ghz mesh interconnect for a teraflops processor. *In the proceedings of IEEE Micro*, 2007.
[34] Intel. Single-chip cloud computer. `http://goo.gl/SfgfN`.
[35] V. Jacobson. Congestion avoidance and control. *SIGCOMM*, 1988.
[36] S. A. R. Jafri et al. Adaptive flow control for robust performance and energy. *MICRO-43*, 2010.
[37] S. Kandula et al. Walking the Tightrope: Responsive Yet Stable Traffic Engineering. SIGCOMM 2005.
[38] D. Katabi et al. Internet congestion control for future high bandwidth-delay product environments. *SIGCOMM*, 02.
[39] S. Katti, H. Rahul, W. Hu, D. Katabi, M. Médard, and J. Crowcroft. Xors in the air: practical wireless network coding. SIGCOMM, 2006.
[40] J. Kim, W. Dally, S. Scott, and D. Abts. Technology-driven, highly-scalable dragonfly topology. *ISCA-35*, 2008.
[41] J. Kim et al. Flattened butterfly topology for on-chip networks. *IEEE Computer Architecture Letters*, 2007.
[42] J. Kim and H. Kim. Router microarchitecture and scalability of ring topology in on-chip networks. *NoCArc*, 2009.
[43] M. Kim, J. Davis, M. Oskin, and T. Austin. Polymorphic on-chip networks. *In the proceedings of ISCA-35*, 2008.
[44] S. Konstantinidou and L. Snyder. Chaos router: architecture and performance. *In the proceedings of ISCA-18*, 1991.
[45] J. Laudon and D. Lenoski. The SGI Origin: a ccNUMA Highly Scalable Server. *In the proceedings of ISCA-24*, 1997.
[46] H. Lee et al. CloudCache: Expanding and shrinking private caches. *In the proceedings of HPCA-17*, 2011.
[47] J. Lee et al. Globally-synchronized frames for guaranteed quality-of-service in on-chip networks. *ISCA-35*, 2008.
[48] N. Maxemchuk. Routing in the manhattan street network. *Communications, IEEE Transactions on*, 35(5):503 – 512, may 1987.
[49] G. Michelogiannakis et al. Evaluating bufferless flow control for on-chip networks. *NOCS-4*, 2010.
[50] T. Moscibroda and O. Mutlu. A case for bufferless routing in on-chip networks. *In the proceedings of ISCA-36*, 2009.
[51] J. Nagle. RFC 896: Congestion control in IP/TCP internetworks.
[52] E. Nilsson et al. Load distribution with the proximity congestion awareness in a network on chip. *DATE*, 2003.
[53] G. Nychis et al. Next generation on-chip networks: What kind of congestion control do we need? *Hotnets-IX*, 2010.
[54] U. Y. Ogras and R. Marculescu. Prediction-based flow control for network-on-chip traffic. *DAC-43*, 2006.
[55] J. Owens et al. Research challenges for on-chip interconnection networks. *In the proceedings of IEEE Micro*, 2007.
[56] H. Patil et al. Pinpointing representative portions of large Intel Itanium programs with dynamic instrumentation. *MICRO-37*, 2004.
[57] M. Qureshi. Adaptive spill-receive for robust high-performance caching in CMPs. *HPCA-15*, 2009.
[58] A. Shaikh, J. Rexford, and K. G. Shin. Load-sensitive routing of long-lived ip flows. SIGCOMM 2009.
[59] A. Snavely and D. M. Tullsen. Symbiotic jobscheduling for a simultaneous multithreaded processor. *ASPLOS-9*, 2000.
[60] H. Song et al. Throttle and preempt: A new flow control for real-time communications in wormhole networks. *ICPP*, 1997.
[61] Standard Performance Evaluation Corporation. SPEC CPU2006. `http://www.spec.org/cpu2006`.
[62] I. Stoica et al. Core-stateless fair queueing: A scalable architecture to approximate fair bandwidth allocations in high speed networks. *In the proceedings of SIGCOMM*, 1998.
[63] M. Taylor et al. The Raw microprocessor: A computational fabric for software circuits and general-purpose programs. *IEEE Micro*, 2002.
[64] M. Thottethodi et al. Blam: a high-performance routing algorithm for virtual cut-through networks. *ISPDP-17*, 2003.
[65] M. Thottethodi, A. Lebeck, and S. Mukherjee. Self-tuned congestion control for multiprocessor networks. *HPCA-7*, 2001.
[66] Tilera. Announces the world's first 100-core processor with the new tile-gx family. `http://goo.gl/K9c85`.
[67] S. Tota, M. Casu, and L. Macchiarulo. Implementation analysis of noc: a mpsoc trace-driven approach. *GLSVLSI-16*, 2006.
[68] University of Glasgow. Scientists squeeze more than 1,000 cores on to computer chip. `http://goo.gl/KdBbW`.
[69] A. Vishwanath et al. Enabling a Bufferless Core Network Using Edge-to-Edge Packet-Level FEC. INFOCOM 2010.
[70] D. Wentzlaff et al. On-chip interconnection architecture of the tile processor. *IEEE Micro*, 27(5):15–31, 2007.
[71] E. Wong et al. Towards a bufferless optical internet. *Journal of Lightwave Technology*, 2009.

NetPilot: Automating Datacenter Network Failure Mitigation

Xin Wu
Duke University
xinwu@cs.duke.edu

Daniel Turner
U.C. San Diego
djturner@cs.ucsd.edu

Chao-Chih Chen
Microsoft
gechen@microsoft.com

David A. Maltz
Microsoft
dmaltz@microsoft.com

Xiaowei Yang
Duke University
xwy@cs.duke.edu

Lihua Yuan
Microsoft
lyuan@microsoft.com

Ming Zhang
Microsoft
mzh@microsoft.com

ABSTRACT

The soaring demands for always-on and fast-response online services have driven modern datacenter networks to undergo tremendous growth. These networks often rely on scale-out designs with large numbers of commodity switches to reach immense capacity while keeping capital expenses under check. The downside is more devices means more failures, raising a formidable challenge for network operators to promptly handle these failures with minimal disruptions to the hosted services.

Recent research efforts have focused on automatic failure localization. Yet, resolving failures still requires significant human interventions, resulting in prolonged failure recovery time. Unlike previous work, NetPilot aims to quickly *mitigate* rather than resolve failures. NetPilot mitigates failures in much the same way operators do – by deactivating or restarting suspected offending components. NetPilot circumvents the need for knowing the exact root cause of a failure by taking an intelligent trial-and-error approach. The core of NetPilot is comprised of an Impact Estimator that helps guard against overly disruptive mitigation actions and a failure-specific mitigation planner that minimizes the number of trials. We demonstrate that NetPilot can effectively mitigate several types of critical failures commonly encountered in production datacenter networks.

Categories and Subject Descriptors

C.2.3 [**Network Operations**]: Network management

Keywords

Datacenter networks, failure mitigation

1. INTRODUCTION

The growing demand for always-on and rapid-response online services has driven datacenter networks (DCNs) to an enormous size, often comprising tens of thousands of servers, links, switches, and routers. To reduce capital expenses and increase overall system reliability, datacenter (DC) designers are increasingly building their networks using broad layers of many inexpensive commodity hardware instead of large chassis switches. However, as the number of devices grows, the failure of network devices becomes the norm rather than the exception.

Diagnosing and repairing DCN failures in a timely manner has become one of the most challenging DC management tasks. Traditionally, network operators follow a three-step procedure to react to network failures: 1) detection; 2) diagnosis; and 3) repair. Diagnosis and repair are often time-consuming, because the sources of failures vary widely, from faulty hardware components to software bugs to configuration errors. Operators must sift through many possibilities just to narrow down potential root causes. Even though automated tools exist to help localize a failure to a set of suspected components [5, 19], operators still have to manually diagnose the root cause and repair the failure. These diagnosis and repair sometimes require third-party device vendors' assistance, further lengthening the failure recovery time. Because of the above challenges, it can take a long time to recover from disruptive failures even in well-managed networks. For instance, in April 2011, a failure in Amazon's AWS service impaired the operations of many cloud services for hours [29].

Realizing the problem above, we take a fundamentally different approach to tackle the failure recovery problem in large-scale DCNs. Specifically, we advocate a four-step process to react to failures: 1) detection; 2) *mitigation*; 3) diagnosis; and 4) repair. We argue that it is more important to mitigate failures than to fix them in real-time. Here "mitigate" means taking action(s) that alleviate the symptoms of a failure, possibly at the cost of temporarily reducing spare bandwidth or redundancy. Timely and effective failure mitigation enables a DCN to operate continuously even in the presence of failures, and allows operators to dive directly into failure diagnosis and repair.

This paper presents **NetPilot**, an *automated system* that adopts our four-step process to *quickly* mitigate failures in a large-scale DCN before operators diagnose and repair the root cause. NetPilot can significantly shorten the failure disruption time by mitigating failures without human intervention. It can also improve online user experience and lower potential revenue losses that stem from service downtime. Moreover, it can lower a DC's operational costs, as it reduces the number of emergent failures and the number of midnight calls to on-call operators.

A key observation that motivates NetPilot's design is that simple actions such as deactivation or restart, coupled with the redundancy that exists in a DCN (§ 2), can effectively mitigate most types of failures in a DCN. DCNs often have extra links and switches to accommodate peak traffic load and device failures. In many cases, simple actions such as deactivating or restarting an offending com-

SIGCOMM'12, August 13–17, 2012, Helsinki, Finland.

ponent can mitigate failures with little impact on the network's normal functions. This observation differentiates NetPilot's design from conventional failure diagnosis and repair approaches, which require detailed failure-specific knowledge.

Since NetPilot has mitigation as its end goal, it can operate without human intervention and without knowing the precise failure root cause. NetPilot automatically mitigates failures in DCNs through a trial-and-error approach. First, it detects a failure and identifies the set of suspected faulty components. Then, it intelligently iterates through the suspected devices and applies mitigation actions on them one by one, until it mitigates the failure or has exhausted all possible actions.

Realizing NetPilot requires overcoming two key technical challenges. First, when mitigation actions are applied blindly, they can actually further compromise the health of the network, *e.g.*, deactivating a switch may overload other switches during peak hours. NetPilot avoids this problem by employing an Impact Estimator to accurately predict the impact of mitigation actions and executing the actions within a pre-defined safety margin.

Second, although NetPilot can safely try numerous mitigation actions before successfully mitigating a failure, an excessive number of trials will unnecessarily lengthen the failure mitigation process. NetPilot addresses this challenge with an optimized mitigation planner that uses failure-specific information to localize a failure to the most likely faulty components, and orders the sequence of mitigation actions according to their potential benefits.

To the best of our knowledge, NetPilot is the first automated failure mitigation system for DCNs. Our contributions are:

- We study and classify the high-impact failures in production DCNs over a six-month period, and find that we can mitigate most of those failures by simple actions such as restart or deactivation. We also find that there is sufficient redundancy in a DCN to accommodate the impact of mitigation actions (§ 2,3).
- We design and implement NetPilot (§ 4,5) and deploy it in a testbed that resembles a real DCN topology. We also conduct simulations using data from a production DCN to evaluate NetPilot at a large scale. We experimentally validate the accuracy of the Impact Estimator, and find that it offers an error rate of less than 8%
- We use NetPilot to automatically mitigate three types of high-impact failures that operators often encounter in production DCNs. Besides reducing operational overhead, NetPilot decreases the median mitigation time from 2 hours to 20 minutes compared to current operational practice, significantly shortening a failure's impact on online services.
- We justify the design choices made in NetPilot. Compared to simple heuristics, NetPilot can succeed with fewer trials while maintaining safe operating conditions in the network.

2. REDUNDANCY IN DATACENTER NETWORKS

In this section, we motivate NetPilot's design with the observation that today's DCNs have plenty of redundancy at the device level, protocol level, and application level. In the next section, we discuss how NetPilot takes advantage of these redundancies to automatically mitigate failures.

A DCN's must balance between scale and cost to support tens of thousands of servers with high bandwidth and at low cost. Solutions have increasingly converged to one design paradigm: using many inexpensive commodity devices to scale up capacity and to reduce cost while deploying various types of redundancy to combat unreliability [11]. We describe three redundancy types below.

2.1 Device-Level Redundancy

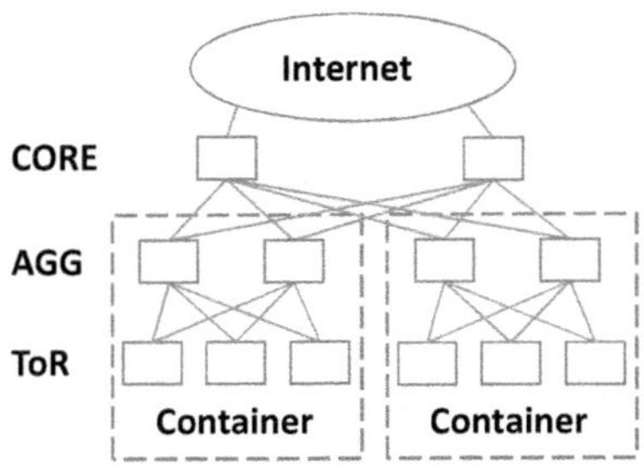

Figure 1: An example scale-out DCN topology.

A modern DCN typically uses a scale-out topology to create many redundant paths between two servers. Scale-out topologies such as a Fat-Tree [4] and Clos [8] can achieve full bi-section bandwidth using commodity switches that often have low port density.

Figure 1 shows an example scale-out topology. This topology has multiple layers: the top-of-rack (*ToR*) layer, the aggregation layer (*AGG*), and the core layer (*CORE*). A *container* is a conceptual management unit as well as a replicable building block sharing the same power and management infrastructure. A *CORE* router connects to multiple containers. For ease of exposition, in the rest of the paper, we use the terms *ToR*, *AGG*, and *CORE* to refer to a switch/router at the ToR, aggregation, or core layer, respectively.

This scale-out topology provides many paths, sometimes in the hundreds, between any two servers. Such path diversity makes the network resilient to single link, switch, or router failure. For example, in Figure 1, we find that deactivating a single link or device, with the exception of a *ToR*, will not partition the network. Even when a failed *ToR* causes network partition, it will only isolate the small number of servers connected to it.

2.2 Protocol-Level Redundancy

DCNs also use various protocol level technologies to meet traffic demands even when some devices fail. We briefly introduce three practical technologies that provide load balancing and fast failover at the link, switch, and path level. These technologies are widely available in off-the-shelf devices from major switch vendors.

Link Aggregation Control Protocol (LACP) abstracts multiple physical links into one logical link and transparently provides high aggregate bandwidth and fast failover at the link level. The resulting logical link is known as a Link Aggregation Group (LAG). LACP provides load balancing by multiplexing packets to physical links by hashing packet headers. Some LACP implementations allow a LAG to initiate from one physical switch but to terminate at multiple physical switches. A LAG can only load balance outgoing traffic, and it has no control over the incoming traffic.

Virtual switch is a logical switch composed of multiple physical switches. A network can use a virtual switch at the link or the IP layer to mask the failures of physical switches.

A virtual switch tolerates faults at the IP layer through an active/standby configuration. One switch is designated as the primary while the standby switch remains silent until it detects that the primary has failed. Two common implementations of IP layer virtual switches are the virtual redundancy router protocol (VRRP) [24] and hot standby router protocol (HSRP) [27]. Both VRRP and HSRP can be configured to provide load balancing.

A virtual switch at the link layer differs from its IP layer counterpart by allowing the physical switches to simultaneously forward traffic. Generically called Multi-Chassis LAG (MC-LAG), Virtual Port Channel (VPC) [2] and Split Multi-link Trunking [13] are two common implementations.

Full-mesh *COREs* refer to the full-mesh interconnections between *CORE*s and containers, *i.e.*, every container connects to every core switch [10,21]. The ECMP [12] routing protocols in full-mesh-*CORE*s topologies provide load balancing and fast failover for traffic between containers.

2.3 Application-Level Redundancy

Modern DCNs also deploy application-level redundancy for fault tolerance. Given that a *ToR* is a single point of failure for the servers connected to it (unless they are multi-homed), a common technique to increase failure resilience at the application level is to distribute and replicate applications under multiple *ToR*s. Therefore, stopping or restarting any switch including a *ToR* is unlikely to have more than an ephemeral impact on the applications.

3. REDUNDANCY WARRANTS AUTOMATED FAILURE MITIGATION

In this section, we analyze and classify the failure records in a six-month period from several production DCNs ($DCNs_p$). We then show that most failures are easy to detect, but difficult to diagnose or repair. However, we can mitigate them using simple actions such as deactivation or restart.

We obtained six months of failure records for $DCNs_p$, all of which were manually created by operators and contain detailed descriptions of critical failures. Network operators consider a failure critical if it is either visible to users or impacts revenue. Therefore each record represents a failure that required immediate investigation and response. The fields of interest from these records are: the data sources used to detect the failure, the techniques used to mitigate the failure, the final actions taken to repair the failure, and the start and end times.

Table 1 classifies $DCNs_p$'s critical failures. We find that the single largest source of failures are misconfigurations. Prior research made similar findings in other contexts, such as ISP networks [25]. Misconfigurations are common in DCNs due to the inherent complexity in managing configuration files in large-scale networks. Many of these misconfigurations, *e.g.*, incorrect ACL (Access Control List) rules, lead to lack of connectivity between certain hosts. Host-to-host pings can detect these misconfigurations, but fixing them is challenging and usually requires operators to manually debug the problems.

The next common type of failure is device software failures. One such example is a malfunctioning hash function that results in uneven link utilization among the physical links in a LAG. In some situations, the uneven utilization is so severe that one of the physical links becomes overloaded and discards packets. We can detect this type of failure by comparing the utilization of each link in the LAG. While we can detect software failures, diagnosing their root causes is generally nontrivial since operators know little about the inner workings of device binary code. Even for device vendors who have the source code, debugging the software is still challenging because failures cannot be easily reproduced in lab environments.

The third category of failures is hardware failures. Such failures occur frequently due to the large number of devices used in $DCNs_p$. For example, 13% of the failures are caused by a single problem, frame checksum (FCS) errors, which often significantly elevates host-to-host latencies. Like other failures, FCS errors can be detected by checking switch SNMP counters, *i.e.*, checking the number of corrupted frames received on each port. However, pinpointing the sources of FCS errors takes time because corrupted frames propagate throughout the network due to cut-through switching (§4.3.1). Even after identifying the correct source, operators have to manually replace the corrupted cable.

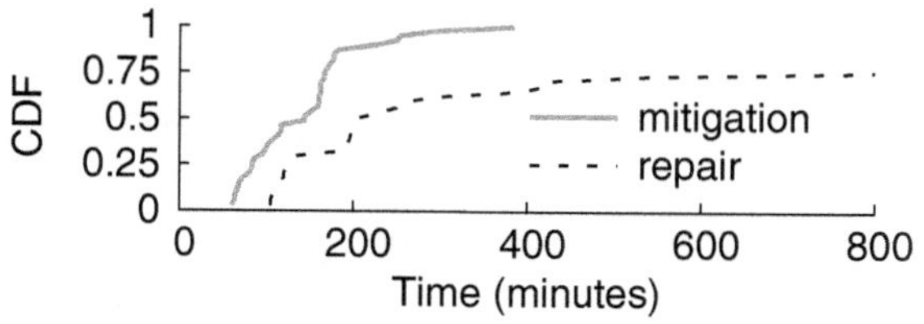

Figure 2: This figure shows the CDFs of how long it takes for $DCNs_p$'s operators to mitigate and to repair critical failures.

The last category is failures whose sources are unknown. In some failures, switches have high CPU utilization while their traffic load is low. In other cases, switches may suddenly cease to forward traffic. These failures often cause abnormally high latency or packet losses. Due to the intermittent and unpredictable nature of these failures, they are difficult to reproduce, diagnose, and repair.

3.1 Time-Consuming Failure Recovery

From studying the types of failures in $DCNs_p$, we see that completely repairing a failure may require debugging of code, software update, or hardware replacement. Further, this process may also involve multiple parties: network operators, software developers, hardware engineers, and external vendors. It is therefore difficult to automate. Figure 2 shows the CDFs of failure mitigation and repair times in $DCNs_p$. As can be seen, the failure repair times can exceed several days or even weeks.

3.2 Simple Mitigation Actions are Effective

Table 1 shows that simple actions such as deactivating or restarting a switch or port can *mitigate* most types of failures in $DCNs_p$ before the root cause can be repaired. For example, in the case of FCS errors, operators reduce latency by deactivating the corrupted links. In the case of overload triggered by load imbalance, operators restore load balance by restarting the offending switches.

In fact, operators of $DCNs_p$ already take *manual* mitigation actions to restore a network to a functioning state while diagnosing and repairing the failure. NetPilot's design goal of automating failure mitigation is partly motivated by the difficulty of manual mitigation. We compare the manual failure mitigation time with the repair time in Figure 2. As can be seen, the time it takes to mitigate failures (even manually) is much shorter than that to repair them. The median failure mitigation time is about two hours.

We note that not all failures can be mitigated by simple actions. Certain failures, *e.g.*, a global configuration error such as a misconfigured ACL, would require a network-wide reconfiguration to fix, and thus cannot be mitigated by restarting or deactivating a few offending devices. How to automatically mitigate those failures is beyond the scope of this paper.

3.3 Spare Capacity for Mitigation Actions

From the analysis above, we find that simple actions are highly effective in mitigating failures and also lead themselves to automation. An automated failure mitigation system can significantly reduce failure mitigation time, as well as the burden on operators.

However, one might be concerned that these simple mitigation actions may overload the network. To find out whether a DCN would have sufficient capacity for failure mitigation, we use the

Category	Detection	Mitigation	Repair	Percentage
software 21%	link layer loop	deactivate port	update software	19%
	imbalance triggered overload	restart switch		2%
hardware 18%	FCS error	deactivate port	replace cable	13%
	unstable power	deactivate switch	repair power	5%
unknown 23%	switch stops forwarding	restart switch	n/a	9%
	imbalance triggered overload	restart switch		7%
	lost configuration	restart switch		5%
	high CPU utilization	restart switch		2%
configuration 38%	errors on multiple switches	n/a	update configuration	32%
	errors on one switch	deactivate switch	update configuration	6%

Table 1: This table categorizes the high-impact failures in several production DCNs over a six-month period. All failures listed here either are visible to users or impact revenue.

maximum link utilization after deactivating a component to measure the amount of spare network capacity. We use the Impact Estimator to carry out this computation. (We will describe Impact Estimator in detail in §4.2.)

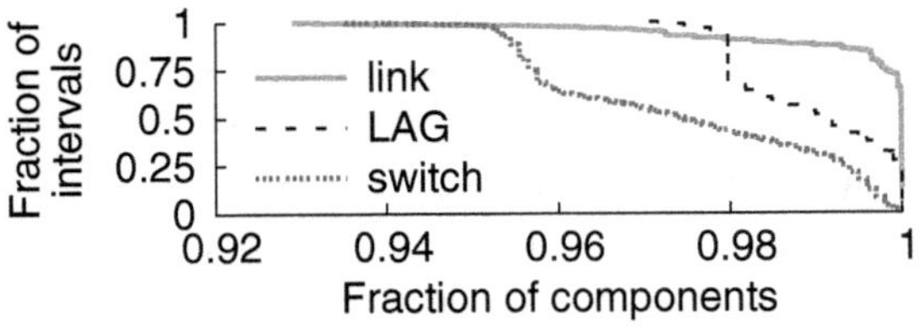

Figure 3: This figure shows the fraction of time intervals during which deactivating a link, a LAG, or a switch will not cause the maximum link utilization to exceed a target threshold (90%) versus the fraction of components for which this result holds.

We perform this computation using the traffic matrices aggregated over 10-minute intervals in one month for DCN$_p$, a single DCN in DCNs$_p$. Figure 3 shows the fraction of time intervals during which deactivating a link, a LAG, or a switch will not cause the maximum link utilization to exceed a target threshold (which is 90%) set by DCNs$_p$'s operators and the fraction of components for which this holds. As can be seen, 99% of the links can be deactivated in over 90% of the intervals without overshooting the target threshold. These numbers are slightly lower for LAGs and switches, because a LAG or a switch contains multiple links. Overall, there is sufficient redundancy in DCN$_p$ to tolerate a component deactivation most of the time.

4. NetPilot DESIGN

NetPilot quickly mitigates failures without knowing their actual root causes in four steps. The first step (S_1) is *failure detection*, in which it constantly monitors the network and detects any potential failure. The second step (S_2) is *mitigation planning*. When NetPilot detects a failure, it will propose a set of suspected components, determine the appropriate mitigation actions, and order these actions based on the likelihood of success or potential impact. The third step (S_3) is *impact estimation*. To avoid taking any action that would further degrade network health, NetPilot estimates the impact of each action and discards the actions that are considered unsafe. The last step (S_4) is *plan execution*. NetPilot will successively execute each mitigation action. If an action successfully mitigates the failure, NetPilot marks the failure as mitigated. Otherwise, NetPilot will roll back the action and try the next action.

In this section, we focus on two main technical challenges: impact estimation (S_3) and mitigation planning (S_2). We postpone the discussion of failure detection (S_1) and plan execution (S_4) to the next section. Note that impact estimation *must* be accurate in order for NetPilot to avoid actions that could further degrade network health. However, NetPilot can work properly even *without* precisely localizing a failure or ordering the mitigation actions.

4.1 Impact Metrics

A chief design goal for NetPilot is to avoid taking any mitigation action that could further degrade a DCN's health. Typically, for a given traffic matrix over a time interval T, we can assess a DCN's health via three metrics: *availability*, *packet losses* and *end-to-end latency*. The availability and packet losses of a DCN can be quantified by the fraction of servers with network connectivity to the Internet ($online\_server\_ratio$) and the total number of lost packets ($total\_lost\_pkt$) during the interval T respectively. Quantifying latency is tricky because it is difficult to predict how intra-DC network latency would change after a mitigation action. Given this problem, we use the maximum link utilization (max_link_util) across all links during the interval T as an indirect measure of network latency. Because the propagation delay is small in a DCN (no more than a few milliseconds), low link utilization implies small queuing delay and thus low network latency. Next, we will explain how to predict these metrics after a mitigation action.

4.2 Estimating Impact

The Impact Estimator aims to estimate a mitigation action's impact on a DCN. Answering this question is crucial for ensuring the safety of mitigation actions. The Impact Estimator takes an action A and a traffic matrix TM as two input variables and computes the expected impact of A under TM. Since it is straightforward to compute $online\_server\_ratio$ given a network topology, we focus on estimating max_link_util and $total\_lost\_pkt$ in the rest of the discussion.

We can get the max_link_util and $total\_lost\_pkt$ by collecting SNMP counters in a DCN. However, predicting these two metrics after a mitigation action is nontrivial because the action could change the traffic distribution in the network. In a DCN with no centralized routing control, we cannot precisely predict how packets are routed to their destinations, unless we know all the packet headers, forwarding tables, and load balancing hash functions.

Our approach to address this challenge is based on two important facts that shape the traffic distribution in DCNs. First, there are far more flows than the diversity of paths in DCNs [10, 17]. Second, hash-based flow-level load balancing is widely used at the link level, switch level, and path level in production DCNs [10, 12].

These two facts make packet headers and load balancing hash functions, which are difficult to obtain in real-time, unnecessary in predicting traffic distribution. Our intuition is that hashing many flows onto a relatively small number of paths leads to even load balancing [10]. As a result, a coarse-grained TM plus forwarding tables should enable us to estimate the real traffic distribution with reasonably high accuracy.

We choose to represent a TM at the granularity of *ToR*-to-*ToR* traffic demands instead of server-to-server, because this representation dramatically reduces the size of TM while not affecting the computation of traffic distribution at the *AGG* or *CORE* layers.

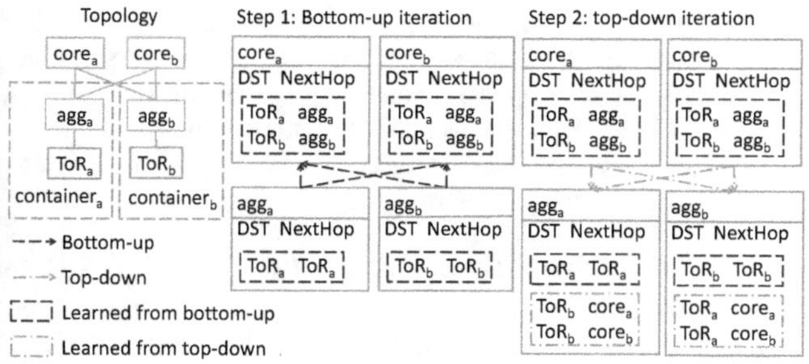

Figure 4: A switch learns the equal cost next hops to any *ToR* after two iterations in a hierarchical DCN topology.

Besides TMs, we also need the forwarding tables to know the next hops to any given destination. As explained in §2, a DCN typically follows a hierarchical structure with traffic traversing valley-free paths. This inspires us to infer the forwarding tables in a similar manner, as illustrated in Figure 4. In the first bottom-up iteration, every switch learns the routes to its descendant *ToR*s from its direct children. In the second top-down iteration, every switch learns the routes to the non-descendant *ToR*s. After these two iterations, every switch builds up the full forwarding table to any *ToR*s in the network.

Algorithm $node.Forward(load)$

1: **if** $load.dst == node$
2:     return; // reach the destination;
3: $nxtHops = node.Lookup(load.dst)$
4: **for** each node n in $nxtHops$
5:     **for** each link l between $node$ and n
6:         $subload.dst = load.dst$;
7:         $subload.volume = \frac{load.volume}{|nxtHops|} \times \frac{1}{|\text{links between } node \text{ and } n|}$;
8:         $n.Forward(subload)$;

We use the term $load$ to refer to the traffic demand between two *ToR*s. Algorithm $node.Forward$ presents how a $node$ forwards a $load$ in detail. Line 3 returns all the next hops (*nxthops*) to a destination. Assuming even load balancing for traffic crossing adjacent levels in the network hierarchy, Lines 4-8 first evenly split $load$ among the *nxthops*, and then for each next hop, the traffic is evenly split among the physical links. The second traffic split is necessary due to the presence of LAGs (described in §2). By running this algorithm on each $load$ in TM and aggregating the contribution of each $load$ on each link, we predict all the link utilizations.

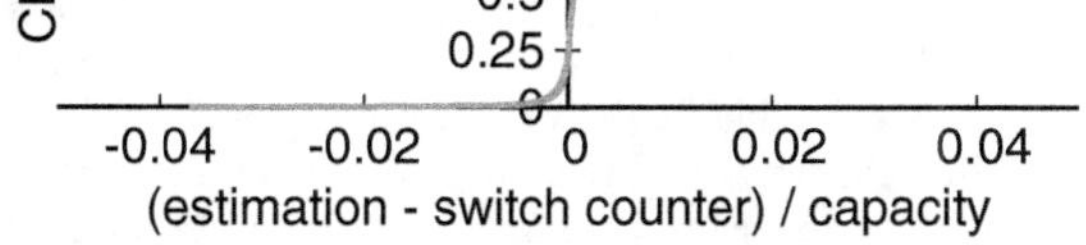

Figure 5: The differences between the link utilizations read from SNMP counters and those computed by Impact Estimator are within 4%.

We evaluate Algorithm $node.Forward$ in an 8000-server production DCN to see whether the Impact Estimator is accurate. We log one-month's socket events on all the servers and aggregate the logs into *ToR*-to-*ToR* traffic matrices at a 10-minute granularity. We also collect the link utilizations during the same month via SNMP at a 10-minute granularity. Figure 5 shows the CDF of the relative differences between the estimated link utilizations and the measured ground truth. As can be seen, the Impact Estimator works very well, and has a maximum error rate of $\pm 4\%$.

So far, we have explained how the Impact Estimator works under a known network topology and TM. To predict the impact of an action, we need to know the new topology and TM after the action is committed. Although inferring the new topology is straightforward, predicting the new TM can be tricky because a mitigation action might affect the traffic demand from minutes up to days. For a restart action which takes only several minutes, we use the TM in the most recent time interval (*e.g.*, 10 minutes) to predict the action's impact during the restart period, assuming the TM is unlikely to change dramatically in such a short time. For a deactivation action that may last days, due to a faulty component needing to be replaced, we ideally desire to use the TM's in the future days to predict the impact during the deactivation period. However, traffic prediction is a research topic by itself, and is beyond the scope of this paper. Instead, we use the TMs in the most recent n days before a deactivation action to predict the impact in the future n days, assuming that the traffic demands are stable over $2n$ days when n is small. In our evaluation (§6.5), we find that this simple heuristic works reasonably well.

4.3 Planning Mitigation

Given that NetPilot takes a trial-and-error approach toward failure mitigation, it needs a mitigation planner to localize suspected components and prioritize mitigation actions to minimize the number of trials. One simple solution is to use existing work [5, 15, 16] to localize failures and then iteratively try deactivating or restarting the suspected components. Although this simple, failure-agnostic solution might work, we choose to develop a mitigation planner that uses failure-specific knowledge to achieve finer-grained localization and more meaningful ordering of mitigation actions (*i.e.*, based on success likelihood). This in turn leads to fewer trials and shorter mitigation times. The downside is that NetPilot needs a planning module for each type of failure. However, we consider this tradeoff worthwhile since there are relatively few types of critical failures in DCNs (as shown in Table 1).

In this section, we first describe in detail mitigation planning for three types of failures: *FCS errors*, *link-down*, and *uneven-split*. We will then discuss the other failure types (listed in Table 1), which are easier to handle compared to these three types.

4.3.1 Frame Checksum Errors

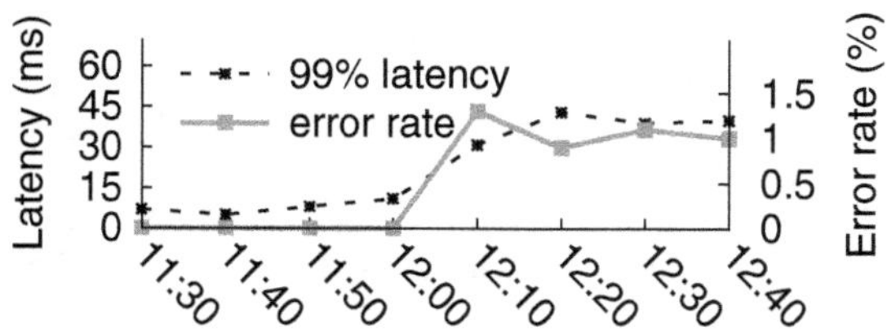

Figure 6: The 99-percentile application-level latency increases significantly due to one corrupted link in DCN$_p$.

Many of the links in a DCN are optical. When foreign material such as dust gets between an optical cable and its connector, the packets traversing the optical link can suffer bit flips. This causes a frame to mismatch its checksum. As shown in Table 1, FCS errors occur frequently in DCNs$_p$ and can significantly degrade performance. Figure 6 shows how a corrupted link impacts the application-level latency in DCN$_p$. Around 12:00pm, a link connecting a *ToR* and an *AGG* switch began to corrupt 1% of all packets. This 1% corruption rate leads to a 4.5 times increase in the 99th percentile latency for applications running under that *ToR*.

Although replacing the faulty cable is the ultimate solution, this could take days depending on staff availability. Operators can mitigate the failure by disabling the faulty link before it is replaced. However, identifying the faulty link is challenging due to the wide use of cut-through switching [1] in DCNs. Because cut-through switches start forwarding a frame before they can verify its checksum, switches can distribute corrupted packets across the network before the corrupted packets are detected locally.

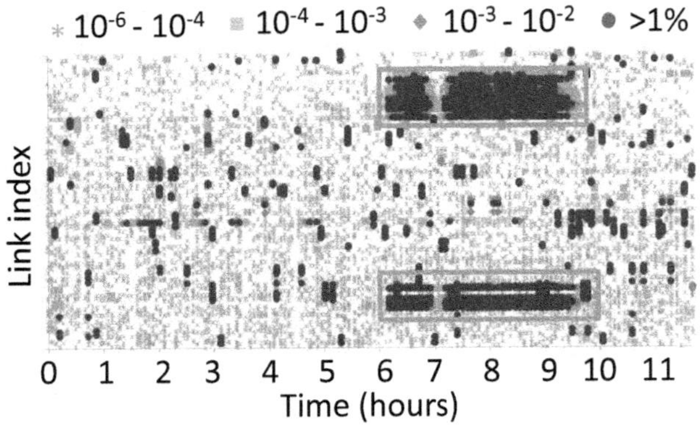

Figure 7: Each point shows a link's error rate. Darker dots indicate higher error rates. We highlight two areas that have the highest error rates (>1%) in 28 links. They are caused by two corrupted links in DCN$_p$, and it took the operators 3.5 hours and 11 trials to deactivate the two offending links.

Figure 7 exemplifies how cut-through switching affects FCS errors in DCN$_p$. The operators began to observe many corrupted packets around hour 6.5, when 28 ports' error rates exceed 1%. Over the next 3.5 hours, operators were busy deactivating the suspected ports one-by-one to determine the faulty links. Finally around hour 10, they found and deactivated the two offending ports and the link error rates returned to normal afterward.

Mitigating FCS errors. Our solution is based on two observations. First, *errors are conserved* on cut-through switches that have no faulty links, *i.e.*, the number of incoming corrupted packets should match the number of outgoing corrupted packets. This observation holds because packet losses are uncommon and broadcast/multicast packets account for only a tiny fraction of the total traffic in DCNs$_p$. Second, the error rate of each faulty link is small and the number of simultaneous faculty links is small. Therefore, it is unlikely that multiple faulty links contribute to the corruption of one packet.

Based on these two observations, we design an FCS error propagation model to localize faulty links. We use x_l to denote link l's corruption rate, p_l and e_l for the total number of packets and the number of corrupted packets traversing l respectively, and m_{kl} for the fraction of packets coming from link k that also traverse link l. Note that the number of corrupted packets coming from link l is equal to the number of packets corrupted by l plus the number of packets corrupted by other links that traverse l. By ignoring the packets corrupted by multiple links, We have:

$$e_l = \sum_{k \neq l} p_k \, x_k \, m_{kl} + p_l \, x_l \qquad (1)$$

We use the same technique as that of the Impact Estimator to compute m_{kl}. e_l, p_k and p_l can be obtained from SNMP counters. Thus, the linear equations (1) provide the same number of constraints as the number of variables (x_l's). If we get a unique solution, the faulty links are those with non-zero x_ls. If the solutions are not unique, we simply pick one with the smallest number of non-zero x_ls based on the fact that the number of simultaneous faulty links is usually small. Our evaluation shows that this approach works well in practice with very few false positives (§6.3).

4.3.2 *Link-down and Uneven-split*

Even when the network has the capacity to handle the offered load, link overloading may still occur due to load imbalance or link failure, leading to packet losses and high latencies in DCNs.

Link-down: When one link in a LAG_x is down, the LAG_x will redistribute the traffic to the remaining links. Since this process is transparent to higher layer protocols, traffic demands remain the same over LAG_x. Thus, LAG_x can become overloaded. One mitigation strategy is to deactivate the entire LAG_x and have the traffic re-routed via other LAGs to the *nxthops* (defined in §4.2). Another strategy is to deactivate all the LAGs (including LAG_x) to the *nxthops* and re-route the traffic to other switches.

Uneven-split: Due to software or hardware bugs, a switch may unevenly split traffic among the *nxthops* or the links in a LAG. In DCNs$_p$, we sometimes observe extreme traffic imbalance such as when one link in a lag carries 5Gb/s more traffic than any of the other links in the LAG. While the exact root causes might be unknown, operators have found that restarting the LAG or switches on either end rebalances the traffic (at least for some period of time).

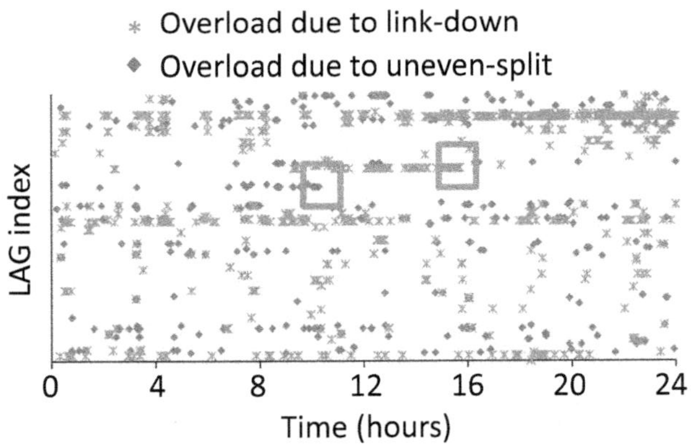

Figure 8: Each point indicates a link overloading incident in a LAG caused by load imbalance in a production DCN. We highlight two incidents. The first occurred around hour 10 was due to uneven-split and was mitigated by restarting a switch. The other occurred around hour 16 was due to link-down and was mitigated by deactivating a LAG.

Figure 8 illustrates that both types of failures are common in DCN$_p$. We collect all the link utilizations every 10 minutes for one day in DCN$_p$. In this figure, each dot represents a LAG that meets the following two conditions in a 10-minute interval: 1) at least one link is overloaded (utilization > 90%); and 2) at least one link is broken (*link-down*) or the difference between the maximum and mean link utilizations exceeds 5% (*uneven-split*). We choose

5% as the load imbalance threshold because Figure 5 not only suggests that accurate impact estimation is feasible but also suggests that large load variance (> 5%) within a LAG is a strong indication of the traffic imbalance problem. We highlight two incidents in Figure 8: one *uneven-split* failure mitigated by a switch restart around hour 10 and another *link-down* failure mitigated by a LAG deactivation around hour 16.

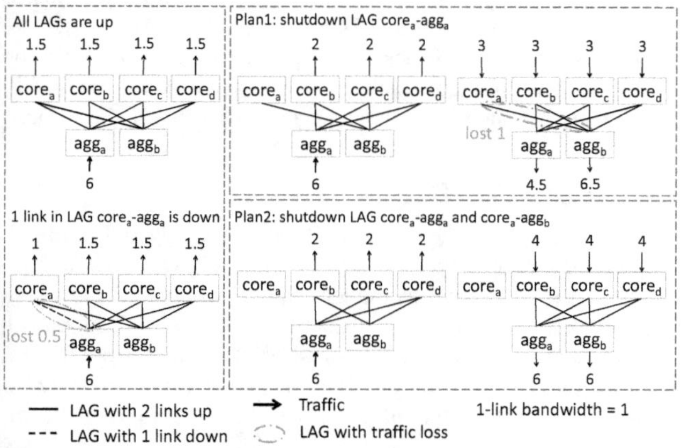

Figure 9: The first plan deactivates one LAG but still causes downward traffic loss. The second plan deactivates two LAGs without causing any traffic loss.

Mitigating a *link-down* or *uneven-split* requires some care due to the complexity of the traffic matrix and topology, as exemplified in Figure 9. Each pair of switches is connected by a LAG consisting of two physical links with a combined capacity of two units. There are six units of upward traffic from agg_a to the $cores$ and twelve units of downward traffic from $cores$ to $aggs$. Suppose one link between agg_a and $core_a$ is down, halving the corresponding LAG capacity, resulting in 0.5 unit of upward traffic loss. One obvious mitigation strategy (Plan 1) is to deactivate the entire LAG between agg_a and $core_a$. Although this prevents the upward traffic loss, it causes one unit of downward traffic loss between $core_a$ and agg_b. The correct strategy is to deactivate the LAG between $core_a$ and agg_b as well (Plan 2). This will shift the downward traffic via $core_a$ to the other $cores$ and prevent traffic loss in both directions.

Mitigating *link-down*. NetPilot mitigates *link-down* failures by estimating the impact of all possible deactivation actions and carrying out the ones with the least impact, *i.e.*, minimizing maximum link utilization. Because a link could be down for n days, NetPilot needs to estimate an action's impact during the downtime. To do so, NetPilot uses the traffic matrices of the most recent n days (§4.2) as an approximation. Such a computation is difficult for human operators to perform because the number of mitigation actions and traffic matrices to consider in concert could be quite large.

Mitigating *uneven-split*. NetPilot mitigates *uneven-split* failures by restarting LAGs or switches. To limit the temporal interruptions during restarts, NetPilot prioritizes the restart sequence based on a restart's estimated impact, while also assuming a component cannot carry any traffic during restart. Since restarting one component usually takes only a few minutes, NetPilot uses the traffic matrix in the most recent time interval (*e.g.*, 10 minutes) as an approximation of the traffic matrix during the restart. After exhaustively calculating the impact for every possible restart, the planner will first carry out the action with the least estimated impact. If this action does not mitigate the failure, the planner will re-prioritize the remaining options based on the latest traffic matrix.

4.3.3 Other Types of Failures

FCS error, *link-down*, and *uneven-split* are by no means all the failures that NetPilot can mitigate. We carefully review all the critical failures in Table 1 and find 62% of them can be localized via available data sources (such as SNMP counters and syslogs) and can be mitigated via deactivation or restart. The only exceptions are the failures due to configuration errors (38%). Although configuration errors on a single switch can be mitigated by deactivating the misconfigured switch, identifying if a configuration error involves one or multiple switches still requires human intervention. We briefly discuss how to apply NetPilot to mitigate other failures:

Link layer loop: Due to switch software bugs, link layer protocols sometimes never converge and cause severe broadcast storms. This failure can be localized by identifying the switches which become suddenly overloaded but experience little traffic demand increase. The mitigation strategy is to deactivate one of the afflicted ports or switches to restore a loop-free physical topology.

Unstable power: Failures due to unstable power are localized by searching syslogs for unexpected switch-down events. They can be mitigated by deactivating the switches impacted by unstable power.

Failures due to unknown reasons: Such failures account for 23% of all critical failures. Even if their root causes are unknown, they can be easily localized to a single switch and mitigated by a restart. For example, a switch that stops forwarding can be identified once the difference between its received and delivered bytes exceeds a threshold. It is also straightforward to identify a switch that loses its configuration or suffers from high CPU utilization.

5. NetPilot IMPLEMENTATION

NetPilot's primary implementation challenge is reliability. As a failure mitigation system, NetPilot itself must be robust to failures. We build NetPilot as a pipeline of five independent processes, as shown in Figure 10. These processes include a failure detector (§ 5.1), failure aggregator (§ 5.2), planner (§ 5.3), impact estimator (§ 5.4), and plan executor (§ 5.5). Each process records its relevant state to a replicated database so that the state can survive server crashes. Operators can also use the recorded state to determine ex post facto why NetPilot took specific actions.

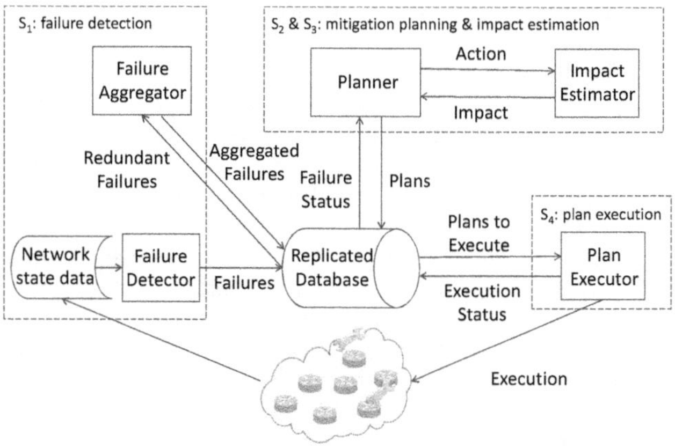

Figure 10: NetPilot implementation overview.

5.1 Failure Detector

The failure detector uses three data sources to detect failures: SNMP traps [22], switch and port counters, and syslogs [20]. The detector then applies failure-specific criteria to evaluate whether a failure has occurred. For example, the failure detector looks at the *bytes-in* and *dropped-packets* counters of a port to determine if a link is overloaded. In our implementation, values from above data sources are processed every five minutes,

When the failure detector detects a failure, it updates the database with the following information: the type of detected failure, data sources used to detect the failure, and the components that exhibit abnormal behaviors. Note that these components are not necessarily the faulty components, because the failure effects may propagate to healthy components, *e.g.*, a broken link may cause overload and hence packet losses at other links.

5.2 Failure Aggregator

Because the failure detector runs continuously and NetPilot takes a trial-and-error approach, we expect that the same failure will be detected multiple times before it is mitigated. NetPilot therefore needs a mechanism to decide whether a detected failure instance is a new or ongoing failure.

The failure aggregator compares a newly reported failure instance against all the ongoing failures recorded in the database. If it determines that the newly reported instance has not been mitigated before – determined by the failure type and components involved – it updates the database and marks the failure as ready for mitigation. If it has seen the failure and the planner is taking a mitigation action, it marks the instance as requiring no further action. If it has seen the failure and the planner has taken a mitigation action for the failure, it flags the failure as unsuccessfully mitigated. The planner may then try the next mitigation action if there is one available.

The failure aggregator does not remove the failure instance created by the failure detector, but simply marks that it has been processed so that an operator can examine the initial failure detection as well as the choices made by the failure aggregator later on.

5.3 Planner

The planner takes three steps to choose a mitigation action. First, it employs failure-specific modules to localize a failure to a set of suspected components. Second, it generates the appropriate mitigation actions against all suspected components. Third, it uses the impact estimator to estimate the impact of these actions, ranks them based on their impact or success likelihood, and then executes the best one. At the end of each step, the planner updates the database with its computation results for post-analysis.

5.4 Impact Estimator

The impact estimator implements the algorithm presented in §4.2. It uses the run-time DCN topology and historical TMs to compute $online\_server\_ratio$, max_link_util, and $total\_lost\_pkt$. We extract the run-time topology from device configurations and running state (*i.e.*, up/down). It includes both the physical and logical device connections such as a LAG that comprises multiple physical links and a virtual switch that comprises multiple physical switches. The traffic matrices are continuously collected via socket event logs on each server and are aggregated to ToR-to-ToR traffic matrices at a 10-minute granularity.

5.5 Plan Executor

Once the planner chooses a mitigation action, the *plan executor* is engaged to take the action. The executor translates the action into a series of commands recognized by switches. As the commands are vendor-specific, we create a vendor-specific *configlet* file that includes the commands for each mitigation action. A *configlet* file parameterizes configuration arguments such as port number, so it can be reused to take the same action on different switches or ports. We also implement a library that allows the executor to send commands to switches via both in-band and out-of-band channels.

After an action is taken, the executor updates the database to record the time when the action was taken and whether the action was successfully applied to the switch.

5.6 Interactions with Operators

NetPilot is fully capable of mitigating failures without human intervention. Nonetheless, NetPilot is explicitly designed to record the inputs and outputs of each mitigation step in a manner that is readily accessible to operators. Operators can later examine the decisions at each step. This design helps them debug and understand counterintuitive mitigation actions. Moreover, it helps reveal failures that are repeatedly mitigated for only a short period of time.

6. EVALUATION

In this section, we show that NetPilot can quickly and automatically mitigate several types of critical failures in DCNs. After presenting our experimental methodology, we first conduct three end-to-end experiments to highlight that, compared with the approach used by today's operators in DCNs$_p$, NetPilot can mitigate failures faster and with less disruption. Then we conduct more detailed experiments to illustrate the three reasons why NetPilot outperforms the status quo: effective failure localization, accurate impact estimation and action prioritization that minimizes disruption.

6.1 Experimental Methodology

We conduct four types of experiments to study various key aspects of NetPilot: 1) *failure-replay,* for failures with operation logs, we feed the device counters and traffic matrices when failures occurred in DCN$_p$ into NetPilot and compare NetPilot's actions with the actions that were actually taken by the operators; 2) *heuristic-replay,* for failures without operation logs, we feed the device counters and traffic matrices when failures occurred in DCN$_p$ into NetPilot and an *operator's approach* and compare their actions; 3) *testbed,* we inject traffic and failures into a testbed and compare the actions taken by NetPilot with those taken by operators; 4) *simulation,* we use large scale simulations to compare NetPilot and the operator's approach in the face of multiple simultaneous failures.

6.1.1 Operator's Approach

For failures with operation logs created by operators, we can directly compare NetPilot's actions with the actual operator's actions. For failures without operation logs, we build a model of how operators would mitigate the failures based on discussions with DCNs$_p$'s operators and our observations. We first carefully review the failures with operation logs and enumerate the typical action sequences for the types of failures discussed in §4.3. We then have operators explain the reasoning behind each sequence of actions. We summarize the operator's heuristics for mitigating each type of failure and estimating impact below.

Estimating impact: When deactivating a link or switch, all of its traffic will fail over evenly among its redundant components. Components not in the same redundancy group have no change in traffic. Because manual computation is slow and error-prone, operators cannot afford to compute the traffic changes that are more than one hop away from the deactivated component.

Mitigating FCS errors: Identify the switches having significantly more corrupted packets of outgoing than incoming, and then deactivate their ports in the descending order of error rate.

Mitigating *uneven-split*: Try restarting LAGs first and then switches, because operators believe restarting a switch is likely to have greater impact.

Mitigating *link-down*: Compute the impact of each possible action under the latest traffic loads using the impact estimation heuristic above, and then execute the action with the least impact.

6.1.2 Experimental Setup

Our data for the failure-replay and heuristic-replay experiments come from DCNs$_p$, several large production DCNs using the scale-out topology for which operators log critical failures with details and mitigation actions. We collected six months of device counters via SNMP at 5-minute intervals. We also logged the socket events on all servers during the same six months and aggregated the logs into ToR-to-ToR traffic matrices at a 10-minute granularity.

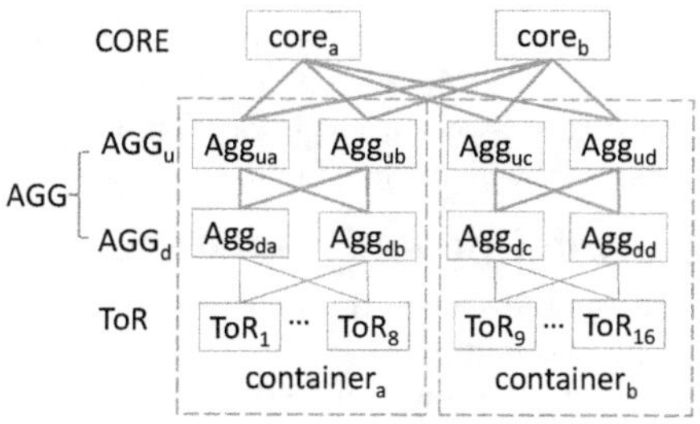

Figure 11: The testbed topology.

As shown in Figure 11, our testbed is a mini version of DCN$_p$'s scale-out topology with all the important characteristics preserved. It has a hierarchical structure with two containers. The connections between containers and *CORE*s form a full mesh running BGP ECMP for load balancing. We use multi-chassis LAG to virtualize the AGG_d switch pairs and VRRP to virtualize the AGG_u switch pairs in the same container. Each connection above AGG_d is a LAG with four 10Gb/s physical links and each connection between a *ToR* and an AGG_d is one 10Gb/s physical link. Unlike the topology in Figure 1, the testbed topology has 1) two *AGG* levels: AGG_u and AGG_d, and 2) one traffic generator instead of multiple servers under each ToR. The traffic generators can either inject arbitrary traffic or replay real server traffic traces captured in DCN$_p$.

Our simulator for the simulation experiments use the same topology and traffic matrix data as the failure-replay experiments. However, we inject hypothetical failures and compare the actions taken by NetPilot and the operator's approach.

6.2 End-to-End Failure Mitigation

In this section, we compare the state-of-the-art manual failure mitigation by operators with NetPilot for three failure incidents in DCNs$_p$. We are limited by the number of available operation logs to conduct a large-scale comparison. Nevertheless, from the failure incidents we examined, we expect that NetPilot's improvement shown in these examples is highly representative.

FCS error. Figure 7 shows the timeline of operators mitigating an FCS error incident. The operators iteratively tried deactivating links that had a significant error rate. Without an Impact Estimator, they performed manual calculations to ensure the remaining links would not become overloaded when they deactivated a link. It took a team of experienced operators nearly 3.5 hours and 10 unsuccessful trials to deactivate the two faulty links. In contrast, a failure-replay experiment shows that NetPilot can pinpoint and deactivate the two troublesome links in less than 15 minutes without any human intervention.

Overload due to *link-down*. Figure 8 depicts an incident of overloaded link caused by *link-down*. This incident persisted for almost 6 hours before the operators mitigated it by deactivating a LAG. The mitigation action was repeatedly delayed because the operator's manual impact estimation was inaccurate and informed them that deactivating the LAG would be worse than taking no action at all. After six hours of continually re-running their impact estimation, the operators deactivated the LAG around hour 16.

In contrast, a failure-replay experiment shows that NetPilot finds an alternative action that would have mitigated the incident soon after the failure was detected, which is almost 5.5 hours ahead of the operator's action. This is possible only because NetPilot's Impact Estimator is faster and more accurate and thus can exhaustively explore all options. This improved accuracy allows NetPilot to take actions that operators would have erroneously excluded.

Overload due to *uneven-split*. A known bug in the software that runs on the AGG_us causes them to occasionally stop generating routing updates. This in turn causes the *CORE*s to cease forwarding traffic to the afflicted AGG_us even though the links are up. Because it is difficult to consistently reproduce the exact same scenario in a testbed experiment, we emulate it by misconfiguring an AGG_u's ACL to block the TCP connections to the *CORE*s. Under this setting, *CORE*s' routing table entries with the afflicted AGG_u as the next hop will expire and most of the traffic will be sent to the other AGG_u in the same afflicted container.

NetPilot detects this problem as an *uneven-split* incident and attempts to mitigate it by restarting switches. Because we only install the ACL rules in the running configuration and not the startup configuration, the problem will be mitigated upon switch restart.

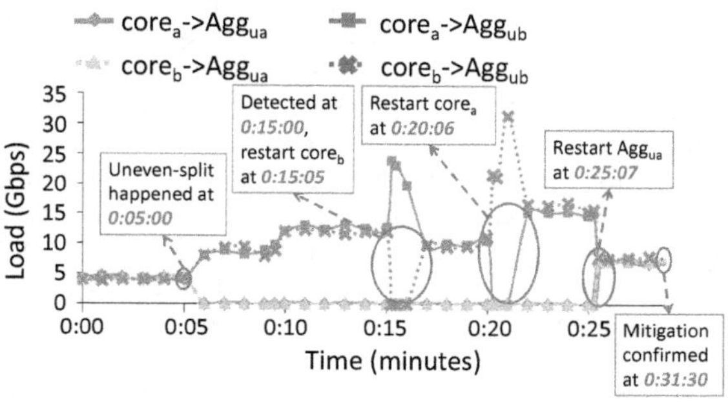

Figure 12: NetPilot mitigates link overloading due to *uneven-split* after restarting three switches.

Figure 12 shows the end-to-end mitigation process. We inject the *uneven-split* failure at minute 5 of the trace. Because switch counters are pushed to NetPilot once every 5 minutes, NetPilot responds at a granularity of 5 minutes. The first problem is identified at minute 15 because the failure aggregator waits for two consecutive counter values before declaring a failure. Approximately 5 seconds later, the planner finishes generating a list of possible actions, estimating the impact of each action via the Impact Estimator, and executing the one with the least impact. When the next set of counters gets pushed to NetPilot, the planner notices that the problem has not been mitigated and starts the next round of planning, excluding the actions that have already been taken. Five minutes later, the planner again notices the failure persists. As a result, it starts a third round of planning and mitigates the failure after restarting the third switch. Even though the first two actions are incorrect, NetPilot can mitigate the failure in approximately 20 minutes, which is still much faster than engaging the operators and manually restarting the suspected switches.

6.3 Fine-grained Failure Localization

In this section, we show that NetPilot can expedite the mitigation process using its fine-grained failure localization. We compare NetPilot's FCS model described in §4.3.1 against three alternative approaches: 1) *Greedy deactivation*, in which links are deactivated in the order of decreasing error rate; 2) *Operator's approach*, which is described in §6.1.1; and 3) *Exhaustive search*, in which we try deactivating every possible combination of the links on the switches that violate the error conservation constraint, from one up to three links in each combination, in the order of decreasing combined error rate. If deactivating one combination fails to mitigate the failure, we roll back the ineffective link deactivations and try deactivating the next combination.

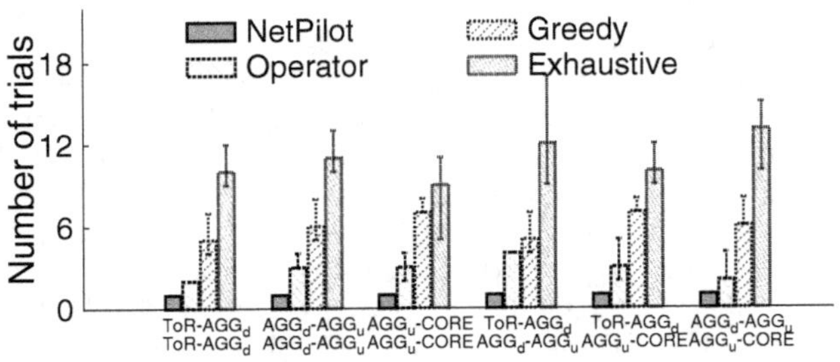

Figure 13: This figure compares the number of trials needed by different approaches to localize two simultaneously corrupted links in our testbed.

We design the first set of testbed experiments to show the number of attempts needed by each of the four algorithms to mitigate FCS errors. In each of the six experiments, we pick two links in the testbed and use a commercial FCS error injector to corrupt 1% of all packets that traverse the two links. Two simultaneous faulty links in a large-scale DCN is common, as there are tens or hundreds of thousands of links. We repeat each experiment ten times using different traffic matrices. The histogram in Figure 13 shows the median number of trials for each experiment, while the error bars mark the maximum and minimum number of trials.

In almost all cases, NetPilot can accurately locate the two corrupted links in one trial. In two out of the sixty experiments, NetPilot identifies three links, the two malfunctioning links and one false positive link. This is far more effective than the operator's approach that will cumulatively deactivate two to five links before disabling the corrupted links. On the other hand, because the exhaustive search will roll back the ineffective link deactivations, it will eventually mitigate the FCS errors without having any healthy link deactivated. However, the disruption caused to the network by the tens of trials is significant.

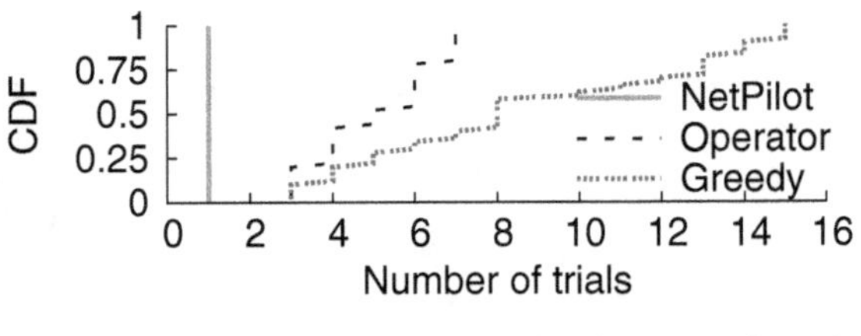

Figure 14: This figure shows the CDFs of the number of trials needed by different approaches to localize three simultaneously corrupted links. NetPilot significantly outperforms the others.

In the testbed experiments, we can only corrupt two links simultaneously due to the limitations of the FCS error injector. We use simulation to simulate more than two corrupted links in a much larger network, *i.e.*, DCN$_p$. We perform fifty simulations. In each simulation we randomly pick three links and set their corruption rates to be uniformly distributed between 1% and 5% (typically observed in DCN$_p$). NetPilot's FCS model can uniquely localize the three corrupted links in 96% of the cases. For the remaining 4%, NetPilot localizes the three corrupted links plus one false positive link. Figure 14 shows that compared to NetPilot, other approaches require far more trials to mitigate just three simultaneously corrupted links. We omit the results from exhaustive search because its search space is too large and the simulations cannot finish in a reasonable amount of time.

To study the localization accuracy of NetPilot's FCS model in the real world, we replay the traffic matrices and switch counters from 78 DCNs$_p$'s FCS error instances on NetPilot. NetPilot can generate unique solutions and accurately localize the corrupted links in over 90% of the instances. In the remaining 10% of the instances, the solutions are not unique because the failure-time traffic matrices do not provide sufficient constraints for the linear equations, *e.g.*, we cannot tell if a link is corrupted when there is no traffic on it. For these cases, we pick the solution with the smallest number of non-zero corruption rates, as discussed in Section 4.3.1. We find this approach works well with the maximum one-link false positives.

6.4 Accurate Impact Estimation

In §4.2, we showed that NetPilot can accurately estimate link utilizations when no mitigation action is taken. We now conduct testbed experiments to show that NetPilot can accurately predict link utilizations as well as packet losses after device deactivations. We compare NetPilot with the operator's approach under five types of component deactivation: a randomly selected physical link, a LAG between an AGG_d and an AGG_u, a LAG between an AGG_u and a *CORE* switch, an AGG_d switch, and an AGG_u switch. For each deactivation type, we repeat the experiments under 144 different traffic matrices generated as follows. First, we collect the ToR-to-ToR traffic matrices from DCN$_p$ at a 10-minute granularity for one day. Then we use a modulo-16 hash function (16 is the number of *ToR*s in the testbed) to map the *ToR*s in DCN$_p$ to the *ToR*s in the testbed. Finally, we map the ToR-to-ToR traffic matrices from DCN$_p$ to the testbed and scale down the traffic volume by a scaling factor. We use a scaling factor that leads to no packet loss in the testbed to study how well NetPilot estimates link utilizations. We use a different scaling factor that leads to packet losses to study how well NetPilot estimates packet losses.

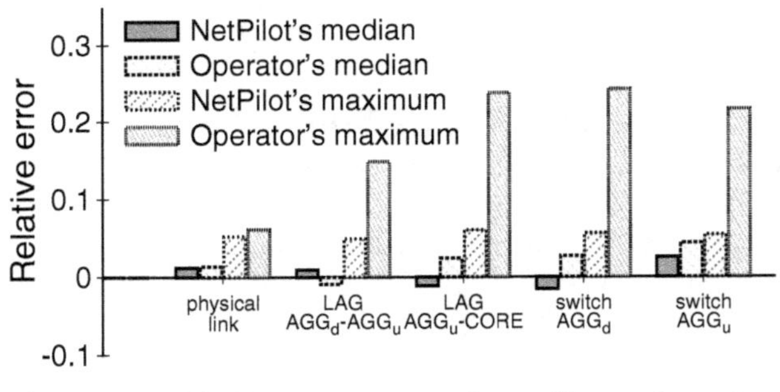

Figure 15: This figure compares the median and maximum relative errors of NetPilot's estimations of the maximum link utilization after deactivating various network components with those of the operator's manual approach. NetPilot has low estimation errors.

Figure 15 compares the relative errors of NetPilot's link utilization prediction with that of the operator's approach when there is no loss in the testbed. The relative error is defined as the difference between a predicted value and the value read from a switch counter normalized by a link's capacity: $\frac{prediction - switch\_counter}{link\_capacity}$.

NetPilot' median and maximum relative errors are less than 2.5% and 5% respectively. Although the median relative errors of the operator's approach are only slightly higher, its maximum relative errors can often exceed 20%. The main reason is that the operator's approach cannot predict the significant traffic load increase on links that are multiple hops away from a deactivated component.

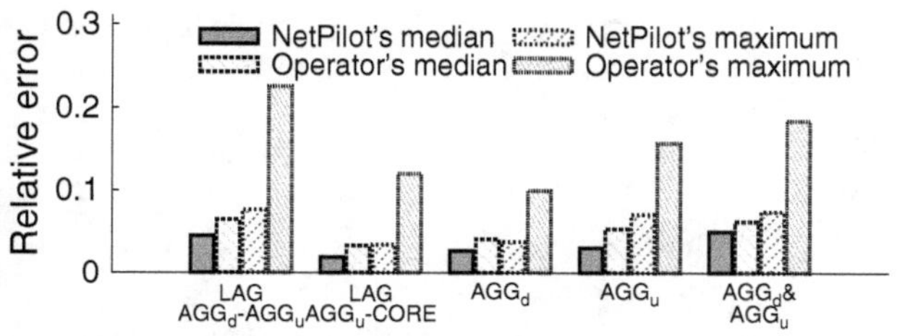

Figure 16: This figure compares the median and maximum relative errors of NetPilot's packet loss estimations with those of the operator's approach when deactivating various network components. Again NetPilot achieves low estimation errors.

Figure 16 is similar to Figure 15 except that we scale the traffic matrices to lead to packet losses in the network. We also replace the case of deactivating a single physical link with the case of deactivating both an AGG_d and AGG_u, since the former rarely leads to packet losses. Again NetPilot's loss prediction has median and maximum relative errors below 5% and 8% respectively. Yet, the maximum relative error of the operator's approach is always more than double NetPilot's.

6.5 Effective Action Planning

When mitigating a failure, NetPilot must carefully choose the order of actions to minimize network disruption. In this section, we show that correctly ordering mitigation actions is challenging and often contradicts the operator's intuition.

We first compare NetPilot with the operator's approach when mitigating *uneven-split* failures. We collect all the link utilizations at a 10-minute granularity from DCN_p for one year and identify 151 *uneven-split* incidents by applying two criteria: 1) the difference between the maximum and mean link utilizations in the same LAG exceeds 5% (the same threshold used in §4.3.2); 2) the difference above lasts at least thirty minutes. Because some of these incidents did not cause link overload and thus were not investigated by operators, we do not know which component was responsible. Therefore, we randomly assign one "responsible" component to each incident and conduct heuristic-replay experiments.

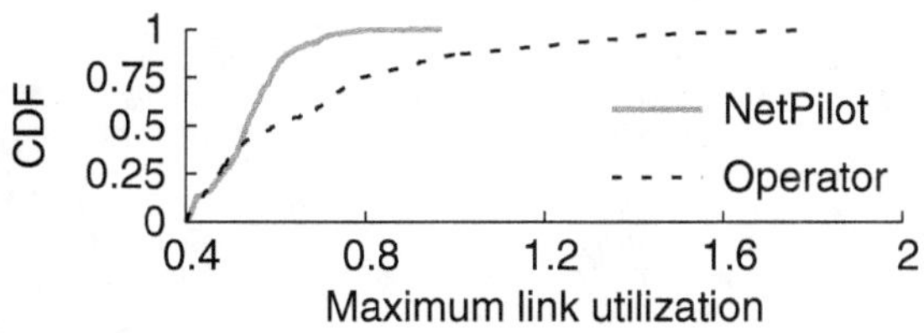

Figure 17: This figure compares the CDFs of the maximum link utilizations when NetPilot and operators restart components to mitigate the *uneven-split* failures. NetPilot causes lower maximum link utilizations during the failure mitigation period.

In these experiments, we emulate the operator's approach by first restarting LAGs and then switches (§6.1.1), whereas NetPilot exhaustively compares all possible actions under the latest TM and picks the one with the least impact (§4.3.2).

We use the CDFs of the maximum link utilizations under both approaches during the failure mitigation periods to compare NetPilot with the operator's approach, as shown in Figure 17. Because we randomly assign faulty components, it takes NetPilot and the operator's approach on average the same number of trials to successfully mitigate an *uneven-split* failure. Therefore, the failure mitigation periods under both approaches are roughly the same. The approach with lower maximum link utilizations is a better approach. For ease of presentation, we represent packet losses as link utilizations greater than 100%. As can be seen, for 60% of the incidents, NetPilot is far less disruptive than the operator's approach. NetPilot never overloads links while the operator's approach would lead to traffic losses in 20% of the incidents.

Unlike *uneven-split* failures, *link-down* failures are mitigated by deactivations in which the deactivated components may remain down for days. The operator's approach uses the latest TM right before the actions to estimate the impact of deactivations, and to choose the actions with the least negative impact. NetPilot has two main advantages over this approach: 1) it can use multiple TMs that better approximate future TMs in the deactivation periods to estimate impact; and 2) its impact estimation is more accurate (§ 6.4).

Although we have shown NetPilot's impact estimation is more accurate in the previous subsection, we further use a failure-replay experiment to show that more accurate impact estimation can lead to less disruptive mitigation actions. We replay 97 *link-down* incidents observed in DCNs_p during one year. For each incident, NetPilot uses the same TM as the operator does to estimate the impact of a deactivation action. We then compare the resulting maximum link utilizations right after NetPilot's deactivation actions with those from the real failure traces.

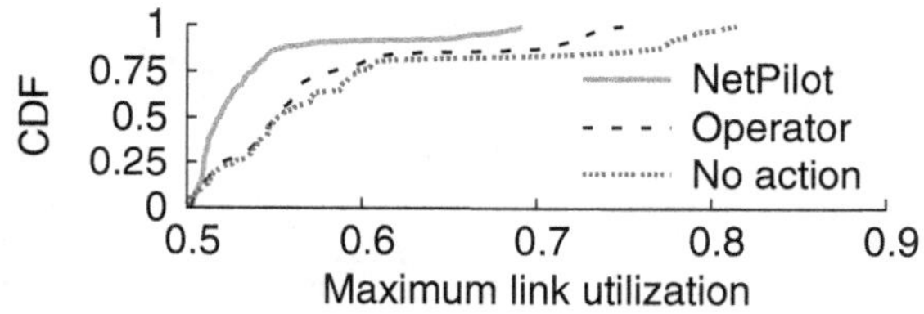

Figure 18: This figure shows the CDFs of the maximum link utilizations right after each method takes its deactivation action to mitigate *link-down* failures in DCN_p. NetPilot outperforms other methods because its impact estimation is more accurate.

Figure 18 shows the comparison results. As can be seen, NetPilot's actions are noticeably better than the actual actions taken by the operators or taking no action at all. These results suggest that NetPilot's higher estimation accuracy can lead to less disruptive mitigation actions.

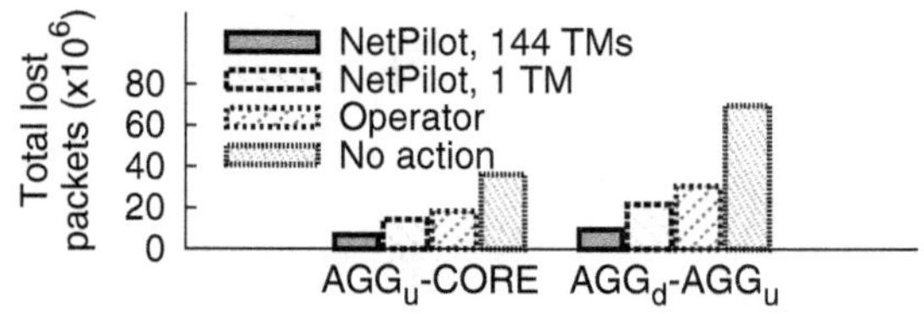

Figure 19: This figure compares the accumulated packet losses in a 24-hour period right after a deactivation action is taken in our testbed. When using multiple historical TMs to approximate future TMs, NetPilot leads to the fewest packet losses.

Next, we run testbed experiments to show that using multiple TMs to approximate future TMs during a device's deactivation period can also lead to less disruptive mitigation actions. In these experiments, we inject *link-down* failures in the testbed by shutting down half of the physical links in a LAG either between an

AGG_d and an AGG_u or between an AGG_u and a *CORE*. We then use the 144 TMs from the preceding 24 hours, rather than just the TM right before a deactivation action, to plan mitigation actions. We then measure the packet losses during the following 24 hours using the TMs in those hours.

Figure 19 presents the packet losses under each method. As can be seen, NetPilot has the fewest packet losses when using multiple historical TMs to approximate future TMs.

7. RELATED WORK

The design and implementation of NetPilot parallel the ideas from various fields. We group the related work into three categories: failure diagnosis, failure recovery, and what-if analysis.

Failure diagnosis: Automated failure diagnosis is a well-studied topic. Some systems identify failures in IP networks with active probes [6]. Others take measurement data from both end hosts and the network and build probabilistic models to localize the most likely components that are responsible for the observed failure data [15, 19]. Recent work has focused on developing systems that can pinpoint any failure that decreases application performance, whether it be hardware-related or software-related [5, 16]. There is also a large body of work on distributed system diagnosis [3,7,23]. NetPilot distinguishes itself by not attempting to find the exact root cause of a failure.

Failure recovery: The idea of automated failure recovery in DCs is not new. Isard [14] proposed an automated server management system called Autopilot based on the concept of recovery oriented computing [9]. When Autopilot detects a server is misbehaving, it takes one of three recovery actions: restart, reimage, or RMA (return merchandise authorization).

R3 [31] is a recovery service that can quickly mitigate link failures by pre-computing forwarding table updates for the loss of each link. Outside the networking domain, Total Recall [18] is a distributed storage system that adapts the amount of redundancy to compensate for host availability changes. Saxons is a peer-to-peer overlay service that can heal itself in the event of a partition [26]. NetPilot has different requirements from the systems above: failures that span multiple devices and constantly changing traffic loads make pre-computation prohibitively expensive. Also, NetPilot must minimize the adverse impact caused by mitigation.

What-if analysis: In order to make an informed decision, it is crucial for NetPilot to reason about the impact of possible mitigation actions – in essence a what-if analysis on the network. Recent work has explored the subject of what-if analysis for content distribution networks, with Tariq *et al.* [28] proposing a statistical approach and Wang *et al.* [30] proposing an empirical approach. Unlike the work above, NetPilot uses a what-if analysis technique that takes advantage of the unique properties of DCNs.

8. CONCLUSION

NetPilot is a system that automatically mitigates DCN failures. It is a departure from the status quo that relies heavily on human intervention. We believe that our work is critical to managing modern DCNs given the ballooning number of devices in these DCNs and the trend towards commodity hardware. NetPilot works by identifying a candidate set of afflicted components that are likely to cause a problem and iteratively taking mitigation actions targeting each one until the problem is alleviated. A key insight that makes this approach viable is the redundancy presented in modern DCN topologies. This redundancy reduces the potential for any single deactivated or rebooted component to disrupt a network. Our experiments show that NetPilot can successfully detect and mitigate several common types of failures both in a testbed and in a real production DCN.

9. ACKNOWLEDGMENTS

The authors would like to thank our shepherd Brad Karp and anonymous reviewers for their help in shaping this paper into its final form. We would also like to thank the infrastructure team at Microsoft Bing whose help was invaluable.

This work was largely done at Microsoft. Part of Xin Wu and Xiaowei Yang's work was supported by NSF awards CNS-0845858 and CNS-1040043.

10. REFERENCES

[1] Cut-Through and Store-and-Forward Ethernet Switching for Low-Latency Environments. `http://www.cisco.com/en/US/prod/collateral/switches/ps9441/ps9670/white_paper_c11-465436.pdf`.

[2] Virtual PortChannels: Building Networks without Spanning Tree Protocol. `http://www.cisco.com/en/US/prod/collateral/switches/ps9441/ps9402/white_paper_c11-516396.html`.

[3] M. K. Aguilera, J. C. Mogul, J. L. Wiener, P. Reynolds, and A. Muthitacharoen. Performance debugging for distributed systems of black boxes. SOSP '03.

[4] M. Al-Fares, A. Loukissas, and A. Vahdat. A scalable, commodity data center network architecture. In *SIGCOMM '08*.

[5] P. Bahl, R. Chandra, A. Greenberg, S. Kandula, D. A. Maltz, and M. Zhang. Towards highly reliable enterprise network services via inference of multi-level dependencies. In *SIGCOMM '07*.

[6] D. Banerjee, V. Madduri, and M. Srivatsa. A Framework for Distributed Monitoring and Root Cause Analysis for Large IP Networks. In *SRDS '09*.

[7] M. Y. Chen, A. Accardi, E. Kiciman, J. Lloyd, D. Patterson, A. Fox, and E. Brewer. Path-based failure and evolution management. In *NSDI '04*.

[8] W. Dally and B. Towles. *Principles and Practices of Interconnection Networks*. Morgan Kaufmann Publishers Inc., San Francisco, CA, USA, 2003.

[9] D. P. et. al. Recovery Oriented Computing: Motivation, Definition, Techniques, and Case Studies. Technical report, Berkeley Computer Science, 2002.

[10] A. Greenberg, J. R. Hamilton, N. Jain, S. Kandula, C. Kim, P. Lahiri, D. A. Maltz, P. Patel, and S. Sengupta. VL2: a scalable and flexible data center network. In *SIGCOMM '09*.

[11] U. Hoelzle and L. A. Barroso. *The Datacenter as a Computer: An Introduction to the Design of Warehouse-Scale Machines*. Morgan and Claypool Publishers.

[12] C. Hopps. Analysis of an Equal-Cost Multi-Path Algorithm. RFC 2992, 2000.

[13] IEEE. 802.3ad Link Aggregation Standard.

[14] M. Isard. Autopilot: automatic data center management. *SIGOPS Oper. Syst. Rev.*, 41:60–67, April 2007.

[15] S. Kandula, D. Katabi, and J.-P. Vasseur. Shrink: a tool for failure diagnosis in IP networks. In *MineNet '05*.

[16] S. Kandula, R. Mahajan, P. Verkaik, S. Agarwal, J. Padhye, and P. Bahl. Detailed diagnosis in enterprise networks. In *SIGCOMM '09*.

[17] S. Kandula, S. Sengupta, A. Greenberg, P. Patel, and R. Chaiken. The nature of data center traffic: measurements & analysis. In *IMC '09*.

[18] R. B. Kiran, K. Tati, Y. chung Cheng, S. Savage, and G. M. Voelker. Total Recall: System Support for Automated Availability Management. In *NSDI '04*.

[19] Kompella, R.R and Yates, Jennifer, and Greenberg, Albert and Snoeren, Alex. IP Fault Localization Via Risk Modeling. In *NSDI '05*.

[20] C. Lonvick. The BSD Syslog Protocol. RFC 3164, 2001.

[21] R. Niranjan Mysore, A. Pamboris, N. Farrington, N. Huang, P. Miri, S. Radhakrishnan, V. Subramanya, and A. Vahdat. PortLand: a scalable fault-tolerant layer 2 data center network fabric. In *SIGCOMM '09*.

[22] R. Presuhn. Management Information Base (MIB) for the Simple Network Management Protocol (SNMP). RFC 3418, 2002.

[23] P. Reynolds, J. L. Wiener, J. C. Mogul, M. K. Aguilera, and A. Vahdat. WAP5: black-box performance debugging for wide-area systems. In *WWW '06*.

[24] S. Nadas, Ericsson. Virtual Router Redundancy Protocol (VRRP) Version 3 for IPv4 and IPv6. *Internet RFC 5798*, 2010.

[25] A. Shaikh, C. Isett, A. Greenberg, M. Roughan, and J. Gottlieb. A case study of OSPF behavior in a large enterprise network. In *IMW '02*.

[26] K. Shen. Structure management for scalable overlay service construction. In *NSDI '04*.

[27] T. Li, B. Cole, P. Morton, D. Li. Cisco Hot Standby Router Protocol (HSRP). RFC 2281, 1998.

[28] M. Tariq, A. Zeitoun, V. Valancius, N. Feamster, and M. Ammar. Answering what-if deployment and configuration questions with wise. In *SIGCOMM '08*.

[29] A. A. Team. Summary of the Amazon EC2 and Amazon RDS Service Disruption in the US East Region. `http://aws.amazon.com/message/65648/`.

[30] Y. Wang, C. Huang, J. Li, and K. Ross. Estimating the performance of hypothetical cloud service deployments: A measurement-based approach. In *INFOCOM '11*.

[31] Y. Wang, H. Wang, A. Mahimkar, R. Alimi, Y. Zhang, L. Qiu, and Y. R. Yang. R3: resilient routing reconfiguration. In *SIGCOMM '10*.

Surviving Failures in Bandwidth-Constrained Datacenters

Peter Bodík
Microsoft Research
peterb@microsoft.com

Ishai Menache
Microsoft Research
ishai@microsoft.com

Mosharaf Chowdhury
UC Berkeley
mosharaf@cs.berkeley.edu

Pradeepkumar Mani
Microsoft
prmani@microsoft.com

David A. Maltz
Microsoft
dmaltz@microsoft.com

Ion Stoica
UC Berkeley
istoica@cs.berkeley.edu

Abstract— Datacenter networks have been designed to tolerate failures of network equipment and provide sufficient bandwidth. In practice, however, failures and maintenance of networking and power equipment often make tens to thousands of servers unavailable, and network congestion can increase service latency. Unfortunately, there exists an inherent tradeoff between achieving high fault tolerance and reducing bandwidth usage in network core; spreading servers across fault domains improves fault tolerance, but requires additional bandwidth, while deploying servers together reduces bandwidth usage, but also decreases fault tolerance. We present a detailed analysis of a large-scale Web application and its communication patterns. Based on that, we propose and evaluate a novel optimization framework that achieves both high fault tolerance and significantly reduces bandwidth usage in the network core by exploiting the skewness in the observed communication patterns.

Categories and Subject Descriptors

C.2.3 [**Computer-Communication Networks**]: Network Operations

Keywords

datacenter networks, fault tolerance, bandwidth

1. INTRODUCTION

Users increasingly rely on online services to manage their computation, storage, and communication requirements. Downtimes hurt them dearly. In 2010, North American businesses collectively lost an estimated $26.5 billion in revenue due to partial or complete outage of services [1]. According to [2], unplanned outages cost $5,000 per minute, on average.

Previous work on designing network topologies [16, 29] and resource allocation mechanisms [6, 17] focused on tolerating failures of network components by providing multiple paths between server pairs and on ensuring predictable network behavior through better bisection bandwidth and reservation. However, in practice, despite redundancy, failures and maintenance of network and power components cause sections of datacenters with tens to thousands of servers to become unavailable [4, 12, 14, 15]. At the same time, datacenter networks can get congested, causing spikes in network latencies. In both cases, Internet services become unavailable and/or unresponsive. Although fault tolerance and bandwidth could be improved by investing more in power and network infrastructure, such actions can be performed only at a slow time-scale, typically during datacenter construction.

In this paper, we present practical solutions that consider the requirements of the underlying applications and help right away in any network. Given a fixed network topology, our goal is to improve the fault tolerance of the deployed applications *and* reduce the bandwidth usage in the core simply by optimizing the allocation of applications to physical machines.

However, there is conflict between simultaneously improving the fault tolerance and reducing the bandwidth usage of cloud-based services. We can improve the fault tolerance by spreading machines of a particular application across many fault domains, thus reducing the impact of any single failure on the application. However, this allocation requires more bandwidth in the core of the network and thus could be prone to congestion, as illustrated in Figure 1(b). On the other hand, to reduce communication through oversubscribed network elements, we can allocate all machines of the application on the same rack or under one aggregation switch, as proposed in [6, 17]. This allocation, however, reduces the fault tolerance of the application; a failure of the top-of-rack (TOR) or aggregation switch, can make the application unavailable, as illustrated in Figure 1(a).

Figure 2 demonstrates the perils of trying to independently opti-

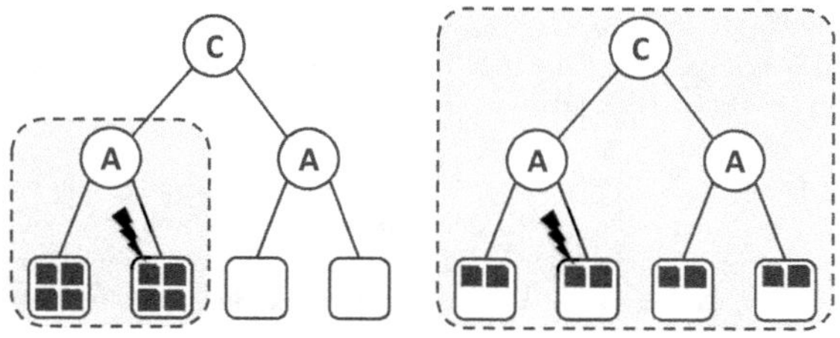

(a) Allocation optimized for bandwidth, low fault tolerance (b) High fault tolerance, but high bandwidth requirement

Figure 1: Simple network topology with two aggregation switches and four racks illustrating the tradeoff between bandwidth usage (pack servers together, (a)) and fault-tolerance (spread servers across racks, (b)). Grayed boxes indicate parts of the cluster network each allocation is using. Assuming only racks as fault domains, (a) has worst-case survival of 0.5 (four of the eight servers survive a failure of a rack), while (b) has worst-case survival of 0.75 (six of the eight servers survive).

SIGCOMM'12, August 13–17, 2012, Helsinki, Finland.

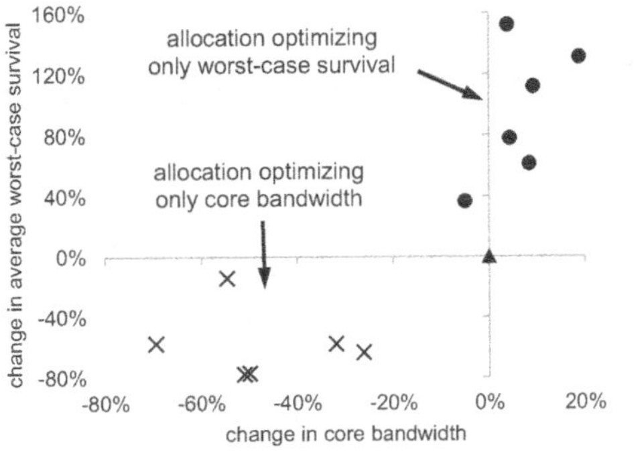

Figure 2: This plot shows the changes in core bandwidth and average worst-case survival for six different datacenters, after applying BW-only optimization (minimum k-way cut, crosses) and FT-only optimization (spreading servers, circles). For both optimizations, one of the metrics improves significantly, but the other one actually degrades.

mize for either fault tolerance or bandwidth usage. Using data from six large-scale production datacenters, we applied both bandwidth-only optimization (using minimum k-way cut of the communication graph, similar to [28][1]) and also fault tolerance-only optimization (described in Section 4). For each service in a datacenter, we measure fault tolerance using *worst-case survival* – fraction of machines of the service that remain available during a single, worst-case failure. Both algorithms significantly improve the metric they optimize for, but actually degrade the other metric. Thus, our goal is to *design algorithms that produce server allocations which achieve both high fault tolerance and reduce bandwidth requirements on the core of the network.*

Our work is motivated and evaluated on actual empirical data from large-scale production datacenters running Bing.com, each running thousands to tens of thousands of servers. The following are crucial observations from our study of the communication patterns of software services in these datacenters:

- The communication matrix between different services is very sparse – only 2% of all pairs of services in the datacenter communicate. In addition, the communication pattern is extremely skewed – the top 1% of all services generate 64% of all the network traffic.
- The fault domains of power equipment do not necessarily match the fault domains of networking equipment, thus complicating the fault-tolerant server allocation.

These observations motivate the formulation of our optimization framework that provides a principled way to explore the tradeoff between reducing bandwidth usage and improving fault tolerance. The problem of improving fault tolerance and the problem of reducing bandwidth usage are both NP-hard and hard to approximate (see Section 4 for details). With that in mind, we formulate a related convex optimization problem that incentivizes spreading machines of individual services across fault domains. In addition, we add a penalty term for machine reallocations that increase bandwidth usage. Intuitively, the combination of these two components exploits the communication skewness, e.g., spreading machines of low-communicating services improves their fault tolerance without significantly affecting bandwidth. Appealingly, the resulting algorithm is easy to integrate and computationally efficient.

Our algorithm achieves $20\% - 50\%$ reduction in bandwidth usage in the core of the network, while at the same time improving

[1] We note that [28] does not consider the joint bandwidth-fault tolerance optimization.

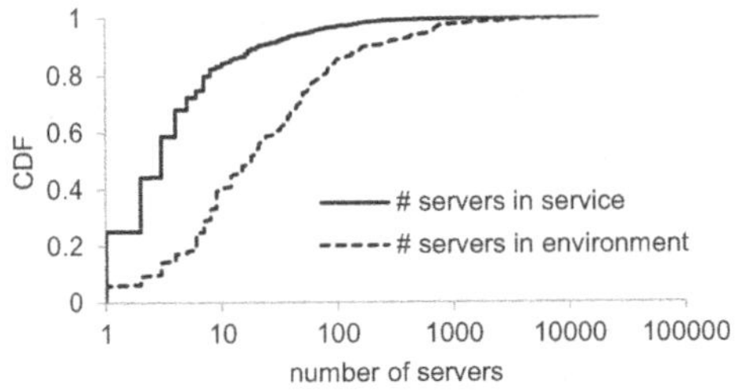

Figure 3: CDF of number of servers in individual services and number of servers in environments.

the average worst-case survival by $40\% - 120\%$. Moreover, because this algorithm minimizes the number of machine moves, it can achieve significant improvements by moving only 20% of machines in the datacenter. Our simulations show that these improvement in fault tolerance can reduce the fraction of services affected by potential hardware failures by up to a factor of 14. In addition, we present a solution based on a minimum k-way cut algorithm, which in many cases achieves reduction in bandwidth by additional $10 - 20$ percentage points.

The contributions of this paper are three-fold:

- **Measurement study.** We present detailed analysis of the structure of a modern, large-scale Web application and the observed communication patterns between the individual services comprising this application (Section 2). While there exist studies on datacenter communication patterns [8, 20], our study concentrates on the traffic at the application level.
- **Algorithms.** To the best of our knowledge, the joint fault tolerance and bandwidth optimization has not been considered in previous studies. Although fault tolerance and bandwidth are typically contradicting objectives, our optimization framework (Sections 3–4) achieves both high fault tolerance and significant reduction in bandwidth usage in the core. Our algorithms exploit the observed skewed communication pattern, scale to large datacenters, and can accommodate arbitrary fault domain structure.
- **Methodology.** We present methodology for stressing server placement algorithms and evaluating their effectiveness. In particular, we introduce a novel framework which utilizes empirical measurements for solving the joint optimization of worst-case survival of software services, machines moves, and bandwidth usage.

2. MOTIVATION AND BACKGROUND

In this section we analyze Bing.com, a large-scale Web application running in multiple datacenters around the world and use it to illustrate the issues described in the Introduction. This analysis is based on data from production clusters with thousands to tens of thousands of servers and also from a large-scale pre-production cluster consisting of thousands of machines. While this analysis is based on a single application, we note that Bing is not simply a "Web index", as it contains many components very similar to typical Web applications. We also highlight the insights we learned and we extract the key parameters, like *skew*, that impact the effectiveness of our algorithms. This enables others to apply our observations and the proposed solutions to other large scale datacenter and cloud computing environments.

We describe the application characteristics in Section 2.2, highlight the sparseness and skewness of communication patterns in 2.3, and characterize the complexities of datacenter fault domains in 2.4. Finally, we describe additional scenarios and implications of this study on our optimization framework.

Term	Definition
Logical machine	Smallest logical component of a web application
Service	Service consists of many logical machines executing the same code
Environment	Environment consists of many services
Physical machine	Physical server that can run a single logical machine
Fault domain	Set of physical machines that share a single point of failure; one machine can belong to multiple domains

Table 1: Definitions

2.1 Definitions

For the purpose of this paper, we assume a datacenter with a known network topology running one or many independent Web applications. A general web application is composed of many *environments* (or frameworks), such as front-end, back-end, or an offline, data-processing framework. Each environment is composed of one or more *services*; for example, the front-end environment might consist of services such as the web servers, caches and other logic that is necessary when processing user requests. Another example of an environment is a MapReduce-like system, with services for job execution or job submission UI. Each service requires a certain number of *logical machines* to support the workload and each logical machine is deployed to a *physical machine* (or a *server*) in the datacenter. We assume only one logical machine per physical machine and describe an extension to a more general scenario in Section 6. Reassigning a logical machine between physical machines is expensive for various reasons. It could take tens of minutes for machines to warm up their caches of in-memory data needed to answer production requests; furthermore, reimaging and making data sets available to the new logical machine takes time and consumes network resources. We assume the network can be modeled as a hierarchical tree[2]; see Section 5 for details. Table 1 summarizes the terms used in this paper.

A *fault domain* is a set of physical machines that share a single point of failure, such as a top-of-rack switch or a circuit breaker. Upon a failure, the servers in the corresponding fault domain become either unavailable or their capacity is reduced. The fault domains can overlap in non-trivial patterns as described in more detail in Section 2.4.

2.2 Service Characteristics of Bing

Bing is a large-scale Web application deployed in many datacenters around the world, serving mostly interactive user traffic with a significant offline, data processing workload. The number of services in Bing's datacenters is in the order of a thousand. Figure 3 shows the distributions of the number of servers for individual services and entire environments. The services are mostly small – 80% of them have less than ten logical machines, while a few services are very large. Similarly, most environments contain less than ten services, but some environments contain a few tens of services.

2.3 Communication Patterns

In this section, we present the analysis of communication patterns of Bing running in a preproduction environment consisting of thousands of servers. We trace communication between all pairs of servers and for each pair of services i and j, we compute the total number of bytes transmitted between this service pair, $R_{i,j}$, and also total traffic generated by individual services, T_i. We refer to (the collection of) $R_{i,j}$ as the *service communication matrix*.

[2]We note that even fat-tree-like networks can be modeled as a logical hierarchical tree with higher bandwidth at the root.

link utilization	>50%	>60%	>70%	>80%
aggregate months above utilization	115.7	47.5	18.3	6.2

Table 2: The aggregate time (during a single month) that all core links spent above certain utilization in one of our production datacenters. We cannot reveal the actual utilization of the links.

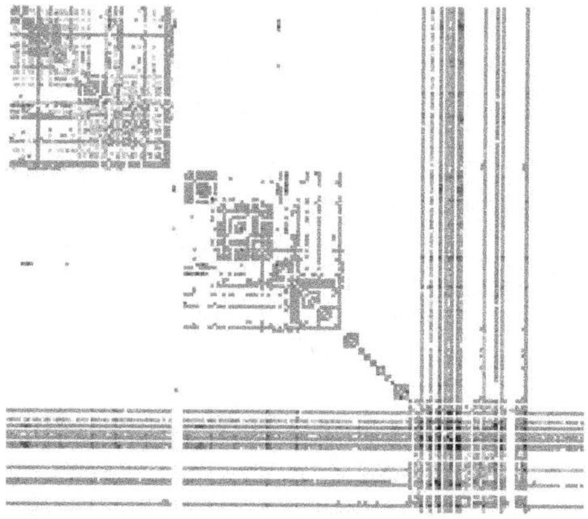

Figure 4: Visualization of communication matrix for a small subset of services using a heat map.

Datacenter network core is highly utilized. As shown in [8], the links in the network core are the most utilized in a datacenter (compared to aggregation and TOR switches) and will thus benefit the most from reduction in usage. We have experienced several incidents in our datacenters where a single application was not aware of its placement in the network topology and started a large data transfer. This overloaded the core of the network and thus impacted the rest of the datacenter. Table 2 shows the amount of time the core links in one of our production datacenters spent at various utilizations during one month. While these statistics are not normalized by the number of links, there are many links that are highly utilized for long time intervals. By reducing the bandwidth usage in the core of the network, our algorithms thus reduce the probability of congestion and packet loss.

Traffic matrix is very sparse. A section of the communication matrix is visualized as a heatmap in Figure 4, where each row and column correspond to a service and color (light to dark) that encodes the amount of communication (white meaning no communication). Notice that the communication matrix $R_{i,j}$ is extremely sparse. Our pre-production environment runs on the order of thousand different services, however, out of all the possible service pairs, only 2% service pairs communicate at all. The remaining 98% of service pairs do not exchange any data.

Communication pattern is very skewed. We observe that the communication pattern of the 2% service pairs that do communicate is extremely skewed. The dashed line in Figure 5 shows the cumulative traffic generated by certain fraction of service pairs. 0.1% of the service pairs *that communicate* (including services talking to themselves) generate 60% of all traffic and 4.8% of service pairs generate 99% of all traffic. The solid line in Figure 5 shows the cumulative traffic generated by individual services. 1% of services generate 64% of traffic and 18% of services generate 99% of traffic. Services that do not require lot of bandwidth can be spread out across the datacenter, thus improving their fault tolerance.

Most bytes stay inside environments. In Figure 4, the services are ordered alphabetically by (environment,service) so that services in

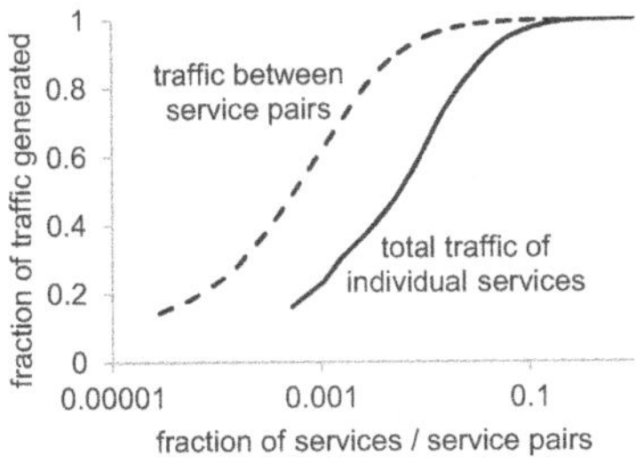

Figure 5: Cumulative fraction of total traffic (y-axis) generated by the top X% of services (x-axis, solid line) and top X% of service pairs (dashed line). The service and service pairs are in decreasing order of T_i, or $R_{i,j}$, respectively.

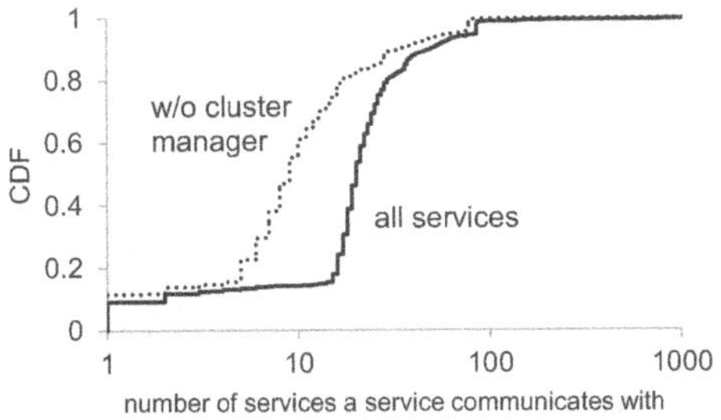

Figure 6: CDF of number of services an individual service communicates with (*i.e.,* degree of vertex in communication graph) for the full graph (solid line) and with services of cluster manager environment excluded (dotted line).

same environment appear next to each other. Most of the communication is along the diagonal (service talking to itself or other services in same environment); the two bigger squares on the diagonal represent two different environments. The vertical and horizontal lines correspond to cluster manager communicating with all machines. The majority of the traffic, 45%, stays within the same service, while 23% leaves the service but stays within the same environment, and 32% crosses environments.

Median service talks to nine other services. The communication matrix can be also represented as a graph, where vertices represent the services and edges represent the communication between service pairs. Figure 6 shows the CDF of the vertex degrees for the full communication graph (solid line). Approximately 15 services of the cluster manager environment communicate with almost all other services (see Figure 4), but this is not application traffic per-se. We therefore also show the CDF of degrees with cluster manager services excluded (dotted line); median degree is 9.

Communicating services form small and large components. The complete communication graph has a single connected component, because the services are linked by common management, logging, and data analysis systems. However, after keeping only the edges that together generate 90% of the traffic (representing the most "chatty" service pairs), this graph falls apart into many disconnected components; see Figure 7 which highlights the major communication patterns. The circles correspond to individual services, and the edges to communication between service pairs. Notice that there are two large connected components – one with star-like topology, another with longer dependencies, and many other smaller components.

2.4 Failure Characteristics

In this paper we consider failures of datacenter hardware equipment that can cause unavailability of many physical servers. Examples of devices we consider are server enclosures (containing

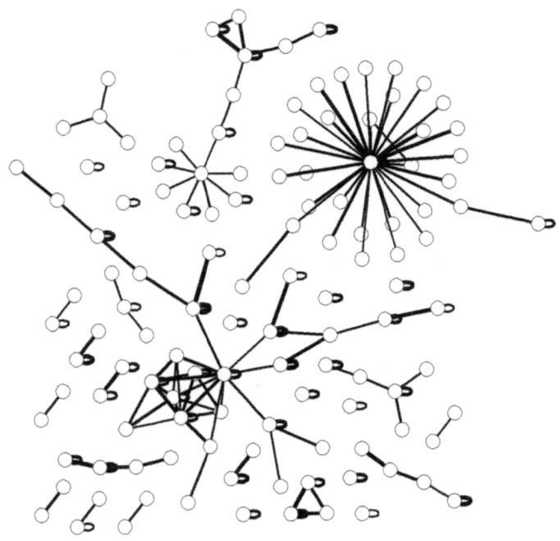

Figure 7: Communication pattern between individual services (represented by circles) after keeping only edges that generate 90% of the traffic. The width of the edges is proportional to the total bytes sent between the corresponding two components.

several servers), TOR and aggregation switches, containers, and power equipment. However, our approach is not limited to these failures and can be extended to any additional fault domains[3]. As demonstrated above, large-scale web applications are composed of many connected services and a failure of a critical service can cause failure of the whole website or significantly impact its functionality and degrade user experience.

Networking hardware failures can cause significant outages. The authors of [15] describe a large study of network failures in several Microsoft datacenters, many of them running Bing. We summarize their findings that motivate the need to improve application fault tolerance. While TOR switches and aggregation switches are fairly reliable (only 5% and 10% failure rates per year, respectively), the failures of TORs were responsible for majority of the observed downtime. While redundancy is often employed in datacenters for aggregation switches or access routers, [15] observes that redundancy reduces impact of failures on lost bytes by only 40% and thus does not prevent impact on applications. Even if these failures are rare, when large fraction of a service is deployed under the same aggregation switch, that service might become unavailable during a failure or have reduced throughput, with potentially high impact on user experience. Notice that the impact of maintenance is similar to a large-scale failure, as many servers might become unavailable. Nonetheless, since maintenance is scheduled in advance, our algorithms can be executed ahead of maintenance in order to automatically move applications out of the affected servers.

Power fault domains create non-trivial patterns. On top of fault domains created by network equipments, servers are also subject to failures in power distribution. While some of the power fault domains match the fault domains of the networking equipment, in general, the wiring of power to servers can create non-trivial patterns with power fault domains that do not match the domains created by switches and routers. For example, servers in multiple racks (but not all racks under an aggregate switch) can be connected to the same circuit, effectively putting them in the same fault domain. Therefore, distributing a service across these racks would not improve its fault tolerance. In some cases, a server and the corre-

[3]While correlated failures happen in practice, their probability is much lower than individual failures. We do not consider them in this paper.

Term	Definition
BW	Aggregate bandwidth usage on core links
WCS	Smallest fraction of machines that remain functional during a single failure in a datacenter
FT	Average WCS across all services
FTC	Fault tolerance cost function
NM	# of server moves to reach target allocation
Cell	Set of physical machines belonging to the same set of fault domains

Table 3: Definitions for the optimization framework.

sponding TOR and aggregate switches might be connected to the same power device, while in other cases they might be connected to three different power devices, again increasing the complexity of the fault domain mapping. These complexities arise from variation and evolution of power chassis architecture, container architecture and datacenter architecture as well as changes in the vendors. Due to their opaqueness, power fault domains are hard to consider for application developers and they have to be handled automatically by the server allocation algorithm.

2.5 Other Practical Considerations

Our core algorithms propose server allocations that reduce bandwidth usage in the datacenter and improve application fault tolerance. As described in Section 2.1, moving a logical machine to a new server is an expensive operation. We therefore consider machine moves as a crucial metric in our optimization framework.

We point out that our solution is orthogonal to previously mentioned solutions, such as a different network topology or bandwidth reservations. In fact, by reducing bandwidth usage we can make bandwidth available to new services that actually need it and could avoid costly and disruptive upgrades of network equipment.

There are several additional issues and scenarios that one may need to consider in practice. For example, services might have additional hard constraints on minimum worst-case survival or hardware configuration of the physical servers. We do not evaluate all such scenarios in this paper, however, our algorithms can be easily extended to handle such cases, as described in Section 6.

2.6 Implications for Optimization Framework

Our analysis of the communication patterns shows that even for a single large-scale website, only a small fraction of services communicate and the communication pattern of the ones that do is very skewed. Most of the traffic is generated by a small number of services and service pairs. *It is therefore feasible to spread a large fraction of services across the datacenter to improve their fault tolerance, without affecting much the bandwidth usage in the core of the network.*

While failures of networking and power equipment are relatively rare, given a bad allocation of services in a datacenter, even a single failure can have significant impact on a website. This motivates the use of the "worst-case survival" metric for fault tolerance (cf. Section 3) that captures the impact of the worst single failure on a service. *Our optimization framework has to consider the complex patterns of the power and networking fault domains, instead of simply spreading the services across several racks to achieve good fault tolerance.*

3. PROBLEM STATEMENT

In this section we provide the description of the problem. See formal mathematical details in Section 4.

3.1 Metrics

In this paper we consider the following three metrics.

Bandwidth. Core links are links from the root of the network topology tree to the second-level nodes. We use the sum of rates on the core links as the overall measure of the bandwidth usage at the core of network; this measure is denoted by **BW**.

Fault Tolerance. For every service, we define the Worst-Case Survival (WCS) to be smallest fraction of machines that remain functional during any *single* failure in the datacenter. For example, if the service is uniformly deployed across 3 racks, and we only consider TOR switch failures, its WCS is $2/3$. While more general survivability models have been proposed [26], we prefer this simpler and more practical metric. We use the average WCS across services as the measure for fault tolerance in the entire datacenter, and denote this average by **FT**. We note that the worst-case survival is directly related to application-level metrics such as throughput or capacity; *e.g.*, a service with WCS of 0.6 will lose at most 40% of its capacity during any single failure.

Number of Moves. The number of servers that have to be re-imaged to get from initial datacenter allocation to the proposed allocation. We denote this number by **NM**. Table 3 contains a summary of our terminology.

Remark. Reducing the bandwidth usage in the core might actually lead to reduction of the WCS of a high-communication service by allocating its machines in the same rack. To prevent such situations for important services, we may extend our basic formulation, putting hard constraints on WCS for such services, as described in Section 6. Alternatively, FT of *small*, high-communication services can be improved by fully replicating them. The FT of *large*, high-communication services is inherently high since they are deployed across multiple fault domains anyway.

3.2 Optimization

The input for our optimization framework is a network topology with initial (or current) assignment of machines to services. Unless otherwise specified, in the remainder of the paper, we consider the following optimization problem

$$\begin{aligned} \text{Maximize} \quad & FT - \tilde{\alpha} BW, \\ \text{Subject to} \quad & NM \leq N_0, \end{aligned} \qquad (1)$$

where $\tilde{\alpha}$ is a tunable positive parameter[4], and N_0 is an upper limit on the number of moves. This formulation accommodates a flexible tradeoff between BW and FT, manifested through the choice of $\tilde{\alpha}$. While we focus on (1) for concreteness, we note that our solution approach could be adapted to other formulations as well, such as minimizing NM under the constraint of certain improvements of BW and FT. See Section 6 for more details.

4. ALGORITHMIC SOLUTIONS

4.1 Problem Hardness

Even when considering the optimization of FT or BW in isolation, the resulting optimization problems are NP-hard and hard-to-approximate. Variants of the BW optimization problem have been broadly considered, see, e.g., [22] and references therein. These works obtain approximation algorithms with logarithmic guarantees, yet they restrict attention to *balanced* cuts (partitions of approximately equal size). Unfortunately, it is an open problem to find good approximation algorithms for the unbalanced case, which represents our setup (network partitions of unequal size). Already

[4]in inverse units of BW so that the objective function is unit-less.

a simple instance of the FT problem reduces to the maximum independent set problem. The known hardness bounds for that problem imply that no reasonable approximation factors are achievable [19]. On top of that, restricting the number of moves adds another level of complexity. Therefore, our focus in this paper is on designing reasonable heuristics that can consider all three metrics simultaneously. We note that we are not able to compare performance against the optimal solution, because the scale of the problem makes exhaustive search infeasible. Instead, we use other heuristics that do not limit the number of moves as performance benchmarks.

4.2 Overview of Solution Approach

In view of the hardness of our optimization problem, our algorithmic approach relies first on designing methods that individually optimize either BW or FT. We then combine the methods into algorithms that incorporate both objectives. Our solution roadmap is the following:

- We first introduce the notion of *cells*, which serve as the basic physical entities in our optimization framework, and allow for significant reduction in the size of the optimization problem.
- Instead of maximizing the intractable FT objective, we define an alternative cost function in Section 4.4, the *Fault Tolerance Cost* (FTC), which has "nice" convex structure. We show that FTC is in practice negatively correlated to FT, hence the minimization of FTC improves FT. The convexity of FTC allows us to iteratively improve the FTC with greedy moves that reduce the current value of the cost function, until reaching its global minimum (or very close to it)[5].
- Our basic method for optimizing BW is to perform a minimum k-way cut on the communication graph of the logical machines.
- CUT+FT+BW is a hybrid algorithm consisting of two-phases. First, we use a minimum k-way cut algorithm to compute an initial assignment that minimizes bandwidth at the network core. Second, we iteratively move machines to improve the FT. The drawback of this hybrid algorithm is that the graph cut procedure ignores the number of moves, and thus generates allocations that require moving almost all machines in the datacenter, which is undesirable during regular datacenter operation. Therefore, this algorithm serves mainly as a performance benchmark for the actual algorithm that we use.
- Finally, FT+BW also takes into account the NM metric. The algorithm does not perform the graph cut, but rather starts from the current allocation and improves performance through greedy moves that reduce a weighted sum of BW and FTC.

4.3 Formal Definitions

Cells. To reduce the complexity of our optimization problem due to overlapping and hierarchical fault domains, we partition the set of physical machines to *cells* (i.e., each machine belongs to exactly a single cell). A cell is a subset of physical machines that belong to exactly the same fault domains (therefore different cells do *not* overlap). For example, the datacenter in Figure 8 has four different cells, capturing all combinations of power sources and containers. Since all machines within a given cell are indistinguishable in terms of faults, it suffices to describe an assignment of machines by the set of variables $\{x_{n,k}\}$; the variable $x_{n,k}$ indicates the number of machines within cell n allocated to service k. We next define the BW and FT objectives with $\{x_{n,k}\}$ as the optimization variables. We emphasize that our cell-based formulation is general in the sense that it does not require any assumptions on the fault-domain topology.

Formal definition of BW and FT. To formally define BW, let $I(\cdot,\cdot)$ be the indicator function, whose inputs are cell pairs. For each such pair (n_1, n_2), $I(n_1, n_2) = 1$ if traffic from n_1 to n_2 (and vice-versa) traverses through a core link, and $I(n_1, n_2) = 0$ otherwise. Then BW, which is the total bandwidth consumption at the core is given by

$$BW(\mathbf{x}) = \sum_{n_1,n_2} \sum_{k_1,k_2} I(n_1,n_2) x_{n_1,k_1} x_{n_2,k_2} r_{k_1,k_2}, \qquad (2)$$

where $r_{k_1,k_2} = \frac{R_{k_1,k_2}}{s_{k_1} s_{k_2}}$ is the required bandwidth between a pair of machines from services k_1 and k_2 (s_k is the number of machines required by service k), and $\sum_{\cdot,\cdot}$ sums each pair only once.

To formally define FT, let $z_{k,j}(\mathbf{x_k}) \triangleq \sum_{n \in j} x_{n,k}$ be the total number of machines allocated to service k affected by fault j ($\mathbf{x_k}$ is the vector of allocations $\{x_{n,k}\}$ for service k). Our fault tolerance objective, FT, is formally given by

$$FT(\mathbf{x}) = \frac{1}{K} \sum_{k=1}^{K} \frac{s_k - \max_j z_{k,j}(\mathbf{x_k})}{s_k}, \qquad (3)$$

where K is the total number of services.

4.4 Basic Building Blocks

FTC and steepest descent moves for FT. As mentioned earlier, instead of maximizing (3), our fault tolerance optimization is achieved via the *minimization* of a cost function, termed FTC. This function is defined as follows:

$$FTC(\mathbf{x}) = \sum_j w_j \sum_k b_k (z_{k,j}(\mathbf{x_k}))^2, \qquad (4)$$

where $z_{k,j}$ and $\mathbf{x_k}$ are defined above, and b_k and w_j are positive weights that can be assigned to services and faults, respectively. For example, one could give higher weights to small services, as faults might have a relatively higher impact on them. More generally, these weights can be used to prioritize service placement according to its importance. The advantage of optimizing over (4) instead of optimizing directly over FT is that the former is a *convex* objective function. Convex functions are appealing, as "local" variable changes (e.g., swapping the physical machine allocation of two logical machines) that improve current cost, also get us closer to the *optimal* value (as opposed to arbitrary cost functions, in which such moves would converge to a *local* minimum). Intuitively, a decrease in FTC should lead to an increase in FT, as squaring the $z_{k,j}$ variables incentivizes keeping their values small, which is obtained by spreading the machine assignment across multiple fault domains; see Figure 8 for a simple example. Figure 9 shows the negative correlation between FTC and FT. While other convex functions could be used instead of (4), we use the sum-of-squares function, as it provides efficient and fast calculation of the delta-cost of machine reallocations, and performs very well in practice.

Our basic method for improving the FTC uses machine *swaps*, which simply switch the allocation of two logical machines, thereby preserving the number of machines assigned to each service (while increasing the NM count by two). To improve FT, we choose the swap that most improves FTC (details below). Accordingly, the swaps can informally be regarded as moves in the "steepest descent" direction.

[5]Convergence to global optimum is guaranteed for the continuous optimization problem provided that the step size is not too large. Here we deal with discrete variables, hence theoretically there is no guarantee to converge to the global optimum. Nevertheless, given the large scale of the problem, the "integrality gap" is negligible and the greedy moves are expected to converge to costs which are nearly optimal.

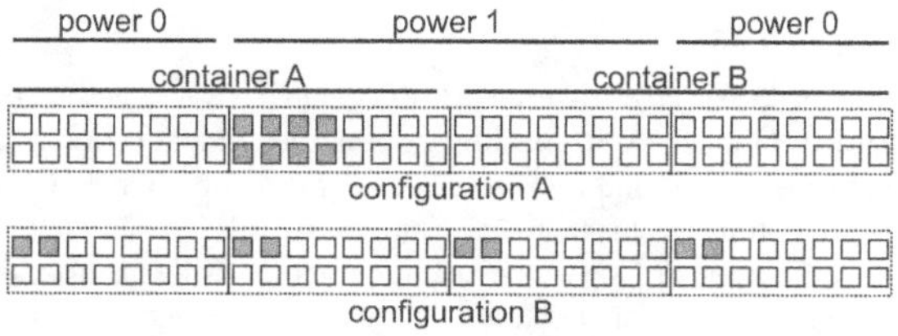

Figure 8: A simple example of a "datacenter" with two containers and two different power sources. In this example, the power fault domains overlap with the network (container) fault domains. The dotted rectangles represent the resulting cells and gray squares represent the allocated servers. Configuration A has a FTC of 128 and a FT of 0; configuration B spreads the machines across the fault domains, hence has a lower FTC of 64 and a higher FT of 1/2. Note that FTC and FT are negatively correlated.

Minimum k-way cut for BW. The basic procedure that we apply for minimizing the BW is based on minimum k-way cut, which partitions the logical machines into a given number of clusters. We often refer to this procedure as *min cut* or just *cut* for simplicity. In our setup, each of the clusters corresponds to a VLAN (we note that the traffic between different VLANs passes through core links). To utilize the cut for our purposes, we construct the datacenter communication graph as follows. Each node in the graph corresponds to a logical machine, and the weight of each edge in the graph is set as the average traffic rate between the two machines. We use Metis' graph partitioning function (gpmetis, see [21]), which allows us to define the size of each cluster within the partition. This feature is important, as the physical clusters need not be of equal size. Gpmetis outputs the set of logical machines which should belong to each cluster, while minimizing the total weight of the minimum k-way cut. Notice that the total weight is exactly BW, the bandwidth usage in the core under the suggested partition.

4.5 Algorithms to Improve both BW and FT

Based on the above building blocks, we first describe a set of algorithms that are not constrained by the number of moves. All the algorithms consist of two phases: in the first phase they minimize BW, followed by minimizing FTC with the option of penalizing swaps that increase the core bandwidth.

- **CUT+FT** : Apply CUT in the first phase, then minimize FTC in the second phase using machine swaps (no penalty for BW increase).
- **CUT+FT+BW** : As above, with the difference that in the second phase we add a penalty term for the bandwidth. More precisely, the delta cost of a swap is $\Delta FTC + \alpha \Delta BW$, where α is a weighting factor. By incorporating the penalty term we lose the convexity property, nevertheless the actual performance is not affected much.

Ignoring the number of moves for now, the algorithm CUT+FT+BW can be viewed as a plausible approach for solving our optimization problem (1): the first phase provides a low-bandwidth allocation, which is the starting allocation for the second phase. The second phase aims at reducing the FTC as much as possible, while being aware of costly bandwidth moves. The weight α specifies the required tradeoff between FT and BW[6]. Since the steepest descent direction is chosen, the cost FTC+αBW is reduced using a relatively small number of moves, which is yet another feature that prevents significant bandwidth increase.

The above algorithms can be used for the initial allocation of machines in a datacenter. However, during regular datacenter operation, these algorithms might not be applicable. This is because cut is not aware of the current machine assignment, and might therefore reassign almost all machines (e.g., if there are 100 containers, the probability that a logical machine will remain in the same container is 1%). To address this issue, we consider the following NM-aware algorithm:

- **FT+BW** : Starting from the initial allocation, do only the second phase of CUT+FT+BW.

Finally, we introduce an algorithm that directly exploits the skewness of the communication matrix.

- **CUT+RANDLOW** : Apply cut in the first phase. In the second phase, determine the subset of services whose aggregate bandwidth requirements are lower than others (the size of the subset is a parameter to the algorithm), then randomly permute the machine allocation of all the services belonging to the subset.

This algorithm takes the advantage of having services that do not consume much bandwidth. Consequently, even random "spreading" of these services could lead to significant FT improvements, without significantly affecting the bandwidth consumption.

4.6 Scaling to Large Datacenters

To scale the algorithms to large datacenters, we describe additional features of the two building blocks. Because there are many possible machine swaps, finding the actual best one would take a long time. Instead, we sample a large number of candidate swaps and choose the one that most improves FTC.[7] Due to the separable structure of FTC, we compute the improvement in FTC incrementally, allowing us to indeed examine a large number of swaps.

When performing the graph cut, the full graph representation, where each node represents a logical machine, often leads to an intractable optimization problem for our minimum k-way cut solver. To address that, we employ a *coarsening* technique, in which we group logical machines of the same service into a smaller number of representative nodes. Accordingly, the edge weight between each pair of such nodes becomes the sum of inter traffic rates between the logical machines of the two nodes.

5. EVALUATION

In this section we evaluate the algorithms described in Section 4. We first evaluate algorithms that optimize for bandwidth or fault tolerance (Section 5.3), then evaluate algorithms that optimize for both bandwidth and fault tolerance, but ignore the number of server moves required to get to the target server configuration (Section 5.4). In Section 5.5, we evaluate algorithms that also consider the number of server moves. See Section 6 for discussion on handling additional constraints within our framework.

5.1 Overview of Results

CUT+FT+BW performs the best. When ignoring the number of server moves, CUT+FT+BW achieves the best performance (see Figure 10). This algorithm achieves 30% – 60% reduction in bandwidth usage in the core of the network, while at the same time improving FT by 40% – 120%. The graph cut significantly reduces the bandwidth usage and provides a good starting point for

[6]We note, however, that since we are minimizing FTC instead of maximizing FT, the α specified in the algorithm need not be equivalent to $\tilde{\alpha}$ in the objective (1). However, since α would in any case be adapted online by the perceived performance, this is not an issue.

[7]Although this means that the optimization step is perhaps not the steepest one, it is enough to move in any descent direction in order to converge to the global optimum (see, e.g., [9]).

the steepest-descent moves that improve fault tolerance. The results of this algorithm serve as a benchmark for the other algorithms.

FT+BW is close to CUT+FT+BW. When executing FT+BW until convergence, it achieves results close to CUT+FT+BW, even without performing the graph cut (falling behind by only $10 - 20$ percentage points in BW reduction). This is perhaps surprising because FT+BW performs only steepest-descent moves, without the advantage of global optimization through graph cut. In scenarios where the number of concurrent server moves is limited, we can use FT+BW to perform incremental improvements (*e.g.*, move 10 servers every hour). This result shows that we can perform incremental updates without getting stuck at bad local minimum.

FT+BW achieves most improvement within few moves. FT+BW performs the "best" machines swaps right at the beginning and can thus achieve large improvement with few moves. For example, when limiting the number of moves to 20% of the servers, it reaches more than half of the potential benefits.

Random allocation in CUT+RANDLOW works well. Since many services transfer relatively little data, they can be spread randomly across the datacenter to achieve high fault tolerance without hurting bandwidth.

5.2 Methodology

We need the following information to perform the evaluation: a) network topology of a cluster, b) services running in a cluster and number of machines required for each service, c) list of fault domains in the cluster and list of machines belonging to each fault domain, and d) traffic matrix for services in the cluster. We describe each below.

Network Topology of a Cluster

We model the network as a hierarchical tree, each level of which *may* have a different oversubscription ratio, branching factor, and redundancy count. This model can represent a wide range of topologies from a Clos network/fat-tree (the Clos becomes a node with high branching factor, high redundancy count, and low oversubscription) to a hub-and spoke (high branching factor, redundancy count = 1, and high oversubscription). The topologies vary among the datacenters we study, including many hybrids such as a Clos network connecting a large number of sub-topologies, each of which is a 2-redundant aggregation tree. Given a hierarchical tree representing the network topology, *core links* are links from the root of the tree to the second-level nodes, which in our datacenters typically represent the VLANs. Accordingly, we use the network topology to determine whether machine pairs are located in different VLANs, which indicates that their traffic traverses through the core.

Services in a Cluster

We use four production clusters with thousands to tens of thousands of servers to obtain the list of services and the number of required machines in each service. Some statistics on the number of services and their sizes are presented in Section 2.

Fault Domains

Based on the network topology and physical wiring of the production clusters, we identified four types of fault domains: server containers (containing on the order of thousand servers), top-of-rack switches, server enclosures (containing three or four servers) and power domains (containing hundreds of machines).

Traffic Matrix

We obtain the service-to-service traffic matrix by collecting network traces of machine-to-machine communication that contain the amount of data sent and received. These traces are collected from a pre-production cluster that runs a full copy of Bing. This cluster contains thousands of servers, executes fraction of live traffic, and is used for testing before deploying code to the production datacenters. We do not have such traces for production clusters. Because machine moves are potentially expensive, we do not respond to changes in traffic patterns that happen on the order of minutes or even hours. Instead, we use the traces to compute the long-term, or steady-state, bandwidth requirements for each service pair.

Because we extract the data for network topology, fault domains and services from production clusters, we need to map the services from the pre-production cluster to services in production clusters. We directly match services with identical names and can match about $1/3$ of the total traffic in the pre-production cluster. We match the remaining production services (in descending order by number of servers) with remaining pre-production services (in descending order by required bandwidth). Such mapping increases bandwidth requirements for large services and thus makes the bandwidth optimization more difficult than other mappings.

Comparing Different Algorithms

All evaluated algorithms produce different server allocations for different input parameters, and therefore explore the tradeoff space between improving bandwidth and fault tolerance. The input parameters are the α (for CUT+FT+BW and FT+BW), number of server swaps to perform (for CUT+FT and FT+BW), and fraction of services to allocate randomly (for CUT+RANDLOW). Instead of comparing the performance of the algorithm on single configurations of problem parameters, we compare the entire achievable tradeoff *boundaries* for these algorithms. In other words, we run the algorithm with different values of the parameters and plot the BW and FT achieved (see Figure 10). The solid line in the figure clearly represents the best algorithm, since its performance curve "dominates" the respective curves of the other two algorithms.

We show the changes in BW and FT relative to the current server allocation in the datacenters. The current allocation algorithm considers communication patterns of only a few services (because this data is typically not available) and does not systematically consider all fault domains. In the following figures, three solid circles represent the FT and BW at starting allocation (at origin), after BW-only optimization (bottom-left corner), and after FT-only optimization (top-right corner) to provide context for results of the other algorithms. For each service, the relative improvement in FT corresponds to an increase in the number of available servers during a single worst-case failure; for example, a service with 20% improvement in FT will have 20% more servers available during such failure. In our graphs, we depict the average FT improvement across services.

5.3 Optimizing for Either BW or FT

Here we describe the results of applying algorithms that optimize either for bandwidth or for fault tolerance, but not both. BW-only algorithm applies the minimum k-way cut to the communication graph. The results for this algorithm to six different datacenters of various sizes (see Figure 2) imply that we can reduce the bandwidth requirement at the network core by 45% on average (relative to the current cluster allocation). However, since this algorithm clusters services in the same partitions (containers, VLANs, ...), it also significantly reduces fault tolerance (by 66%, on average).

FT-only optimization uses the FT+BW algorithm with α set to

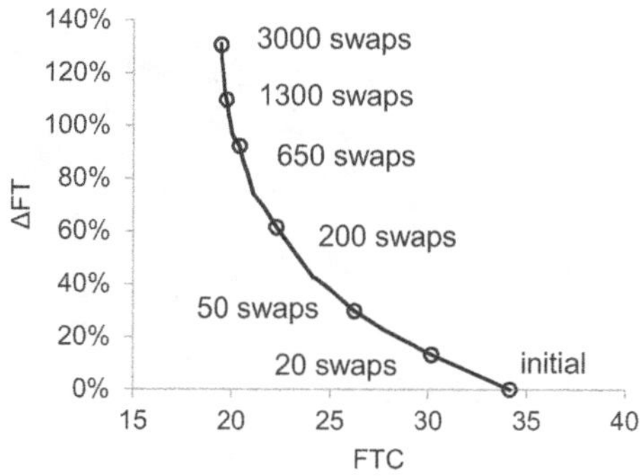

Figure 9: Convergence of FTC (the fault tolerance cost function, Equation 4) and the corresponding average worst-case survival for a smaller cluster of several thousand machines. The FT algorithm improves (reduces) FTC while indirectly improving (increasing) the fault tolerance of the services. Notice that even small changes in cost (*e.g.*, from 1,300 to 3,000 swaps) result in significant improvement in fault tolerance. Finding a swap takes from seconds to a minute, depending on the size of the cluster and the number of candidate swaps that are explored (this number is a tunable parameter in our algorithm).

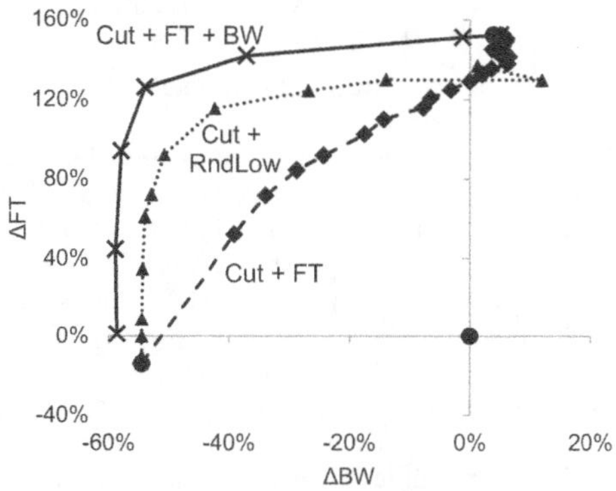

Figure 10: The achievable trade-off boundary for the CUT+FT, CUT+FT+BW and CUT+RANDLOW algorithms. For CUT+FT, each marker on the line represents the state after moving approximately additional 2% of servers in the cluster. Note that CUT+RANDLOW outperforms CUT+FT simply by randomly allocating the low-talking services; however, it does not improve the bandwidth usage of these services nor fault tolerance of the high-talkers.

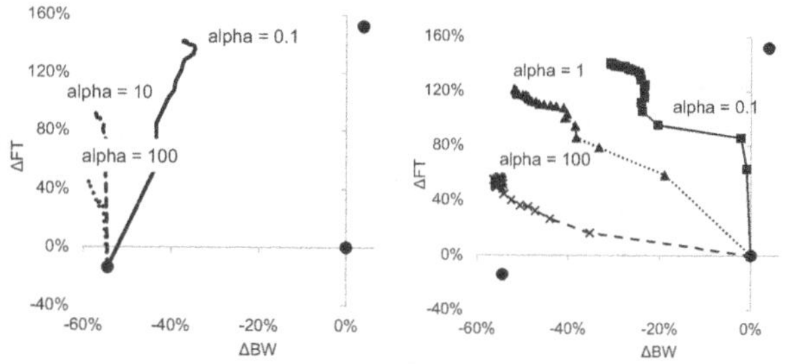

Figure 11: Change in fault tolerance and core bandwidth over time for CUT+FT+BW (left) and FT+BW (right) for three different values of α. Using larger values of α puts more weight on improving bandwidth at the expense of fault-tolerance.

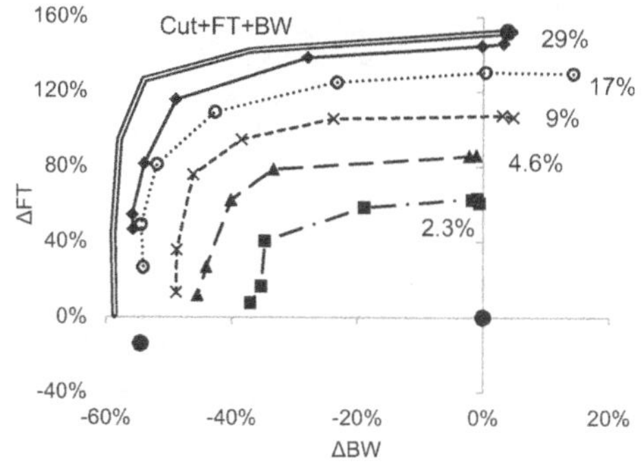

Figure 12: The achievable trade-off boundary for the FT+BW algorithm. Values next to the lines represent the fraction of the cluster that was moved. The outer-most line represents the result of the CUT+FT+BW.

zero. Applying this algorithm improves FT by 83% on average, but also increases the bandwidth requirement at the core by 7%. Although we report these changes in relation to the current server allocation at the clusters, we note that the absolute values for BW and FT metrics do not depend on the starting allocation. The graph cut approach completely ignores the starting allocation, and while the steepest descent in FT starts from a particular allocation, it converges to the globally optimal FT of the cluster.

As explained in Section 4, the FT-only algorithm uses optimization over convex FTC to indirectly improve FT. Figure 9 illustrates the negative correlation of these two metrics – as FTC decreases, the fault tolerance improves.

5.4 Optimizing for Both BW and FT

In this section we evaluate algorithms that optimize bandwidth and fault tolerance, but ignore the number of server moves required to reach the target service allocation (CUT+FT, CUT+FT+BW, and CUT+RANDLOW). All these algorithms first use the minimum k-way cut to reduce the bandwidth at the core, followed by performing gradient descent that improves fault tolerance, gradient descent that improves fault tolerance and bandwidth, and randomizing the low-talking services, respectively. We show results for one of the datacenters in Figures 10 and 11(left), and results for the remaining three datacenters in Figure 13a. We note that we have examined ways to incorporate FT optimization directly within the cut procedure. First, we tried to artificially partition each service to several subgroups, however this did not lead to satisfactory performance. We also tried to augment the cut procedure with "spreading" requirements for services. Unfortunately, this approach does not scale to large applications with hundreds of services.

CUT+FT

CUT+FT performs the minimum k-way cut (reaching the lower-left point in Figure 11(left)), followed by the steepest descent algorithm that only considers improvement in fault tolerance. We executed the algorithm many times, each time allowing it to swap an increasing number of servers. The resulting BW and FT metrics of the obtained server allocations are shown in Figure 10 (in particular, the CUT+FT curve). The diagonal line in this figure represents the achievable tradeoff boundary for this algorithm; by changing the total number of performed swaps, we can control the tradeoff between BW and FT. This formulation of the problem is convex, so performing steepest descent until convergence leads to the global minimum with respect to fault tolerance.

CUT+FT+BW

CUT+FT+BW is similar to CUT+FT, but gradient descent also considers improvement in bandwidth when selecting a machine swap. The results of this algorithm depend on the value of α;

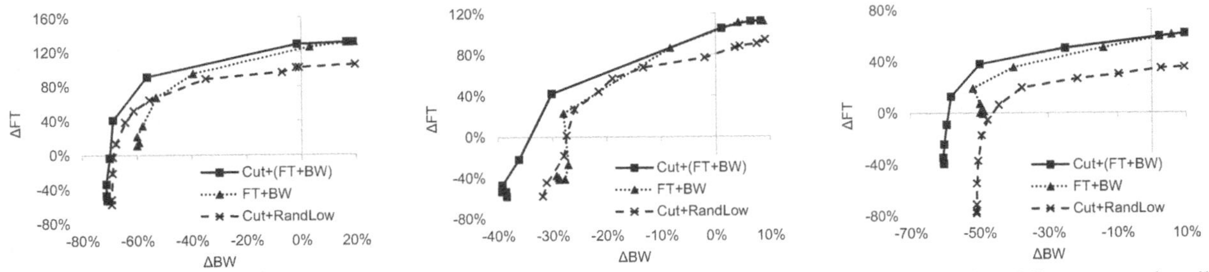

(a) Tradeoff boundary for CUT+FT+BW (solid line), CUT+RANDLOW (dashed line), and FT+BW (dotted line, executed until convergence).

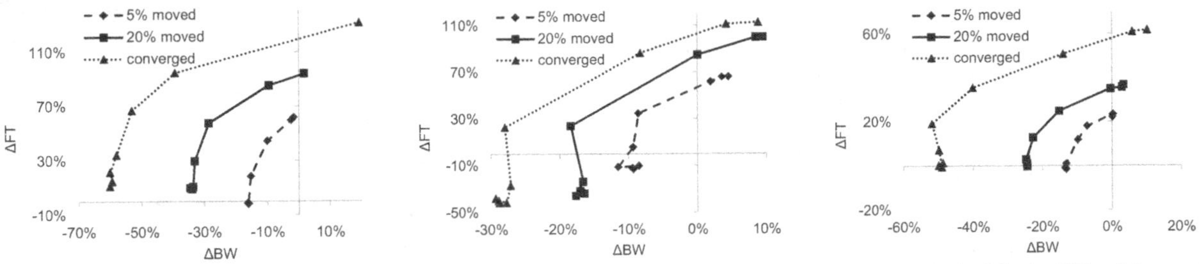

(b) Tradeoff boundary for FT+BW when limiting the number of moves to 5% of the total servers (dashed line), 20% of the total servers (solid line), or running FT+BW until convergence (dotted line, same as dotted line in (a) above).

Figure 13: Evaluation of CUT+FT+BW, CUT+RANDLOW, and FT+BW on three additional datacenters.

higher values of α put more weight on improvement of bandwidth at the cost of not improving fault tolerance as much. Figure 11 shows the progress of this algorithm for three different values of α. By running the algorithm until convergence with several different values of α, we obtain the "benchmark boundary" to which other algorithms can be compared (see the solid line in Figure 10). Because this algorithm is not optimizing over a convex function, it is not guaranteed to reach the global optimum. However, it still significantly improves both the bandwidth and fault tolerance of the cluster.

CUT+RANDLOW

CUT+RANDLOW first performs the minimum k-way cut, followed by randomizing the allocation of the least-communicating services responsible for total of $y\%$ of the total traffic in the cluster. The achievable tradeoff boundary for this algorithm[8] is in Figure 10. This algorithm achieves performance close to the CUT+FT+BW algorithm, but it does not optimize the bandwidth of the low-talking services nor the fault tolerance of the high-talking ones, which explains the gap between these two algorithms.

5.5 Optimizing for BW, FT, and NM

In this section we evaluate FT+BW, which also considers the number of server moves. This algorithm starts from the current server allocation and performs steepest descent moves on the cost function that considers the fault tolerance and bandwidth. The progress of this algorithm for different values of α is shown in Figure 11(right); as in CUT+FT+BW, using larger α skews the optimization towards optimizing bandwidth. In this figure, each marker corresponds to moving approximately additional 2% of servers. Notice that improvement is significant at the beginning and then slows down.

Figure 12 shows the achievable tradeoff boundaries of FT+BW for different fractions of the cluster that are required to move. For example, notice that we obtain significant improvements by moving just 5% of the cluster. Moving 29% of the cluster achieves results similar to moving most of machines using the CUT+FT+BW algorithm (see the outer double line in Figure 12). Results for three additional datacenters are presented in Figure 13(b).

Finally, notice that when running FT+BW until convergence (see Figure 13a), it achieves results close to CUT+FT+BW even without the global optimization of graph cut. This is significant, because it means we can use FT+BW incrementally (*e.g.,* move 2% of the servers every day) and still reach similar performance as CUT+FT+BW that reshuffles the whole datacenter at once.

5.6 Understanding the Improvements

Here we explain the actual allocation changes performed by the FT+BW algorithm for α values of 1.0 and 0.1 and how they lead to the improvements in bandwidth and fault tolerance. For α = 1.0, FT+BW reduced the core bandwidth usage by 47% and improved the average fault tolerance by 121%. Overall, the contribution of every single service on the core bandwidth consumption was reduced. This happened at the expense of fault tolerance of some services as they were packed closer to conserve bandwidth. The fault tolerance was reduced, stayed the same, and was improved for 7%, 35%, and 58% of services, respectively. Finally, Figure 14(left) shows the changes of bandwidth and fault tolerance for all services with *reduced* fault tolerance. Again, a few services contributed significantly to the 47% drop in bandwidth, but paid for it by being spread across fewer fault domains.

For α = 0.1, FT+BW achieved reduction of bandwidth usage by 26%, but improved the fault tolerance by 140%. In this case, fault tolerance was reduced only for 2.7% of the services (see the right plot on Figure 14) and the magnitude of the reduction was much smaller than for α = 1.0. This demonstrates how the value of α controls the tradeoff between fault tolerance and bandwidth usage.

To understand the impact of improved fault tolerance on the services, we compute the number of services that are affected by a potential hardware failure. We say that a service is affected by a potential hardware failure if its worst-case survival is less than a certain threshold H. We use $H = 30\%$ that is used in the alert sys-

[8]Using $y = 0, 25, 50, 60, 70, 80, 85, 90, 95, 98, 99, 99.9, 100$.

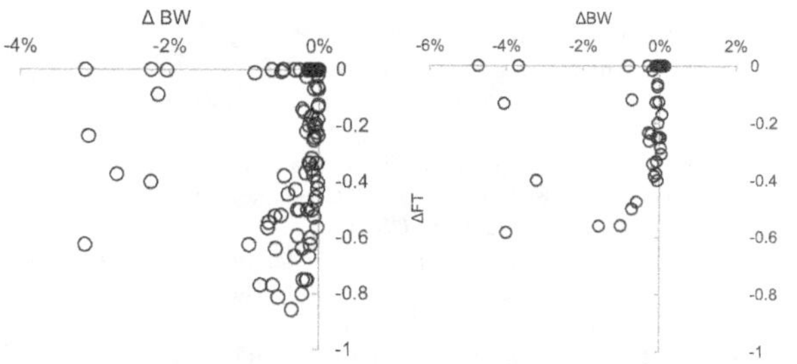

Figure 14: The relative change in core bandwidth (x-axis) and fault tolerance (y-axis) for all services (circles) that actually reduced their fault tolerance for α=1 (left) and α=0.1 (right).

tem for Bing. For FT+BW with $\alpha = 1.0$ discussed above, the fault tolerance increased by 121%. Based on simulations, the number of services potentially affected by hardware failures would drop by a factor of 14 when using the allocation proposed by FT+BW.

We believe that our approach is general, nevertheless the achieved improvements depend on communication patterns of the deployed applications, as we demonstrate below. We stress-test our algorithms using synthetically modified communication matrix to capture different extremes of communication types. In the first two experiments, we a) removed all communication between pairs of services that belong to different environments and, b) removed all communication between different services (kept only communication inside the service). The resulting matrices are close to communication patterns one might observe in a cloud computing environment with many independent parties. In both cases, we achieved results very close to using the original traffic matrix. These experiments highlight the robustness of our approach, as it can handle significantly different communication patterns.

Finally, we experimented with a traffic matrix with uniform communication pattern; in particular, for every pair of logical machines that belong to services that communicate (based on the original traffic matrix), we set their bandwidth requirement to a constant C. Thus, every pair of servers requires bandwidth of either 0 or C. Notice that the traffic in this matrix is not skewed as the original matrix. Using this matrix, we were able to achieve the same improvement in fault tolerance, since it does not depend on the traffic matrix. However, none of the algorithms were able to reduce the bandwidth usage in the core of the network; the best improvement was on the order of -10^{-7}. This demonstrates that the skewness of the communication patterns is crucial for our approach.

6. ADDITIONAL SCENARIOS

Besides the main objectives of improving fault tolerance and reducing bandwidth consumption in the core of the network, the deployment of our algorithms may need to consider further practical issues. We describe some of them here and discuss how our algorithms can be extended to support them.

Hard constraints on fault tolerance and placement. Certain important services might have hard constraints on the minimum required fault tolerance. Increasing the weight of a service monotonously improves its FT. Thus, we could exploit this monotonicity to efficiently find the weight that should be assigned to the service (through offline simulations). Further, when evaluating a machine swap, we could check the FT of the involved services and never perform a swap that would reduce FT below a specified limit. Similarly, if a logical machine has certain hardware requirements, we could also check them during swap evaluation.

Optimization of bandwidth across multiple layers. The focus in this paper is on reducing the bandwidth at the core, but the same algorithms could be applied at the lower levels of the network, such as reducing the bandwidth usage on the aggregation switch. Our optimization framework can address this extension by performing the cuts *hierarchically*.

Preparing for maintenance and online recovery. A large number of machines might become unavailable during maintenance or during long-lasting failures. Setting high weight w_j for the affected fault domain j will incentivize our algorithm to move machines away from that domain, thereby improving FT of applications.

Adapting to changes in traffic patterns. Since services might be added and removed and their traffic matrix can change over time, the FT+BW algorithm might be executed every time such a change is detected. When the change in the algorithm inputs is small, FT+BW would propose only a small number of machine swaps.

Multiple logical machines on a server. Our optimization framework could be extended to support single physical machine hosting multiple logical machines. Indeed, the basic idea of preserving a feasible allocation by swaps may carry over to the more general case where services could choose between "small", "medium" or "large" logical machines. Under such extensions, we might restrict the set of feasible swaps to avoid complicated bin packing issues. We leave this extension for future work.

In addition to the above scenarios, there are interesting extensions for future work. These include fault tolerance with correlated failures, non-uniform and time-varying communication patters, and additional performance metrics for fault tolerance and bandwidth (such as worst-case core link utilization).

7. RELATED WORK

Datacenter traffic analysis. Datacenter traffic has been studied in the past. In [8], the authors analyze the communication patterns in datacenters at the flow- and packet-level, while in [20] they concentrate on a particular, MapReduce-like application. In this paper, we study a complex Web application consisting of a large number of interconnected services and provide a detailed analysis of their communication patterns at the application level.

Datacenter resource allocation. Current approaches to resource allocation in multi-tenant datacenters can be divided into two major categories. On the one hand, Seawall [33] and NetShare [24] share the network bandwidth among coexisting tenants based on their weights. Node allocation is a non-goal for these algorithms. On the other hand, Oktopus [6] and SecondNet [17] provide guaranteed bandwidth to tenant requests. Oktopus uses greedy heuristics to allocate nodes and bandwidth for two specific types of topologies. SecondNet provides pairwise bandwidth guarantees between tenant nodes. Solutions in both categories do not consider application availability; since they are tailored for satisfying bandwidth constraints, it is nontrivial to extend them to include availability.

Virtual network embedding. Allocation of arbitrary virtual network topologies on top of physical networks is a well-studied problem [10]. Existing solutions include greedy heuristics [38], simulated annealing [27], multi-commodity flow models [34, 37], and mixed-integer formulations [11]. However, these algorithms do not consider availability constraints.

Survivable embedding. Survivability from node and link failures has been extensively investigated for optical and MPLS networks [23, 25, 35]. The key objective, however, is to ensure IP connectivity in presence of failures. Survivable virtual network embedding [31] deals with a similar problem in the context of network virtualization. These solutions do not consider node allocation, nor can they work at datacenter scale.

High availability in distributed systems. Resource allocation to achieve high availability in distributed systems has been studied before [3,5,7,36]. These projects divide the servers into fault domains with different failure probabilities, and allocate data and application components to maximize some measure of application availability. However, they do not consider bandwidth usage.

VPN and network testbed allocation. Algorithms for allocating virtual private networks (VPNs) in a shared provider topology [13, 18,30] involves finding paths for a collection of source/destination pairs given a small, pre-defined set of customer nodes. The authors of [32] consider bandwidth constraints when allocating nodes in Emulab. Neither of these considers application availability.

8. CONCLUSION

Fault tolerance and reduction of bandwidth usage are often contradictory objectives – one requires spreading machines across the datacenter, the other placing them together. Indeed, simulations on large-scale Web application demonstrate that optimizing for one of these metrics independently improves it significantly, but it actually degrades the other metric. In this paper, we propose an optimization framework that provides a principled way to explore the tradeoff between improving fault tolerance and reducing bandwidth usage. The essentials of this framework are motivated by a detailed analysis of the application's communication patterns. In particular, our analysis shows that the communication volume between pairs of services has long tail, with the majority of traffic being generated by small fraction of service pairs. This allows our optimization algorithms to spread most of the services across fault domains without significantly increasing the bandwidth usage in the core.

Acknowledgements

We thank Peter Winkler, Seffi Naor, Mohit Singh, Sharon Goldberg and Srikanth Kandula for useful discussions. We are grateful to Michael Ma, Greg Foxman, Alexander Kolomeets and other members of the Bing's Autopilot team for providing important input during the project.

9. REFERENCES

[1] The avoidable cost of downtime. ca Technologies, `http://goo.gl/SC6Ve`, 2010.

[2] Calculating the cost of data center outages. Ponemon Institute, `http://goo.gl/ijLoV`, 2011.

[3] S. Agarwal, J. Dunagan, N. Jain, S. Saroiu, and A. Wolman. Volley: Automated data placement for geo-distributed cloud services. In *NSDI*, pages 17–32, 2010.

[4] Amazon.com. Summary of the Amazon EC2, Amazon EBS, and Amazon RDS service event in the EU west region. `http://aws.amazon.com/message/2329B7/`, 2011.

[5] K. Amiri, D. Petrou, G. R. Ganger, and G. A. Gibson. Dynamic function placement for data-intensive cluster computing. In *ATC*, 2000.

[6] H. Ballani, P. Costa, T. Karagiannis, and A. Rowstron. Towards predictable datacenter networks. In *ACM SIGCOMM*, 2011.

[7] N. Bansal, R. Bhagwan, N. Jain, Y. Park, D. S. Turaga, and C. Venkatramani. Towards optimal resource allocation in partial-fault tolerant applications. In *INFOCOM*, pages 1319–1327, 2008.

[8] T. Benson, A. Akella, and D. A. Maltz. Network traffic characteristics of data centers in the wild. In *IMC*, pages 267–280, 2010.

[9] D. Bertsekas. *Nonlinear programming*. Athena Scientific, 1999.

[10] M. Chowdhury and R. Boutaba. A survey of network virtualization. *Computer Networks*, 54(5):862–876, 2010.

[11] M. Chowdhury, M. R. Rahman, and R. Boutaba. ViNEYard: Virtual network embedding algorithms with coordinated node and link mapping. *ACM/IEEE ToN*, 2011.

[12] J. Dean. Designs, lessons and advice from building large distributed systems. LADIS '09 keynote talk, 2009.

[13] N. G. Duffield, P. Goyal, A. Greenberg, P. Mishra, K. K. Ramakrishnan, and J. E. van der Merive. A flexible model for resource management in virtual private networks. In *ACM SIGCOMM*, 1999.

[14] D. Ford, F. Labelle, F. I. Popovici, M. Stokely, V.-A. Truong, L. Barroso, C. Grimes, and S. Quinlan. Availability in globally distributed storage systems. In *OSDI*, 2010.

[15] P. Gill, N. Jain, and N. Nagappan. Understanding network failures in data centers: measurement, analysis, and implications. In *SIGCOMM*, pages 350–361, 2011.

[16] A. Greenberg, J. R. Hamilton, N. Jain, S. Kandula, C. Kim, P. Lahiri, D. A. Maltz, P. Patel, and S. Sengupta. VL2: a scalable and flexible data center network. In *ACM SIGCOMM*, pages 51–62, 2009.

[17] C. Guo, G. Lu, H. J. Wang, S. Yang, C. Kong, P. Sun, W. Wu, and Y. Zhang. SecondNet: A data center network virtualization architecture with bandwidth guarantees. In *ACM CoNext*, 2010.

[18] A. Gupta, J. M. Kleinberg, A. Kumar, R. Rastogi, and B. Yener. Provisioning a virtual private network: A network design problem for multicommodity flow. In *ACM STOC*, pages 389–398, 2001.

[19] J. Håstad. Clique is hard to approximate withinn 1- ε. *Acta Mathematica*, 182(1):105–142, 1999.

[20] S. Kandula, S. Sengupta, A. Greenberg, P. Patel, and R. Chaiken. The nature of data center traffic: measurements & analysis. In *IMC*, 2009.

[21] G. Karypis and V. Kumar. A fast and high quality multilevel scheme for partitioning irregular graphs. *SIAM J. Sci. Comput.*, 20:359–392, December 1998.

[22] R. Krauthgamer, J. Naor, and R. Schwartz. Partitioning graphs into balanced components. In *SODA*, pages 942–949, 2009.

[23] M. Kurant and P. Thiran. Survivable routing of mesh topologies in IP-over-WDM networks by recursive graph contraction. *IEEE JSAC*, 25(5):922–933, 2007.

[24] T. Lam, S. Radhakrishnan, A. Vahdat, and G. Varghese. NetShare: Virtualizing data center networks across services. Technical Report CS2010-0957, UC San Diego, 2010.

[25] W. Lau and S. Jha. Failure-oriented path restoration algorithm for survivable networks. *IEEE TNSM*, 1(1):11–20, 2008.

[26] Y. Liu and K. S. Trivedi. A general framework for network survivability quantification. In *MMB*, pages 369–378, 2004.

[27] J. Lu and J. Turner. Efficient mapping of virtual networks onto a shared substrate. Technical Report WUCSE-2006-35, Washington University, 2006.

[28] X. Meng, V. Pappas, and L. Zhang. Improving the scalability of data center networks with traffic-aware virtual machine placement. In *INFOCOM*, 2010.

[29] R. N. Mysore, A. Pamboris, N. Farrington, N. Huang, P. Miri, S. Radhakrishnan, V. Subramanya, and A. Vahdat. PortLand: A scalable fault-tolerant layer 2 data center network fabric. In *ACM SIGCOMM*, pages 39–50, 2009.

[30] S. Raghunath, K. K. Ramakrishnan, S. Kalyanaraman, and C. Chase. Measurement based characterization and provisioning of IP VPNs. In *ACM IMC*, pages 342–355, 2004.

[31] M. R. Rahman, I. Aib, and R. Boutaba. Survivable virtual network embedding. In *IFIP Networking*, pages 40–52, 2010.

[32] R. Ricci, C. Alfeld, and J. Lepreau. A solver for the network testbed mapping problem. *ACM SIGCOMM CCR*, 33(2):65–81, April 2003.

[33] A. Shieh, S. Kandula, A. Greenberg, and C. Kim. Sharing the data center network. In *USENIX NSDI*, 2011.

[34] W. Szeto, Y. Iraqi, and R. Boutaba. A multi-commodity flow based approach to virtual network resource allocation. In *IEEE GLOBECOM*, pages 3004–3008, 2003.

[35] K. Thulasiraman, M. S. Javed, and G. L. Xue. Circuits/cutsets duality and a unified algorithmic framework for survivable logical topology design in IP-over-WDM optical networks. In *IEEE INFOCOM*, 2009.

[36] H. Yu, P. B. Gibbons, and S. Nath. Availability of multi-object operations. In *NSDI*, 2006.

[37] M. Yu, Y. Yi, J. Rexford, and M. Chiang. Rethinking virtual network embedding: Substrate support for path splitting and migration. *ACM SIGCOMM CCR*, 38(2):17–29, April 2008.

[38] Y. Zhu and M. Ammar. Algorithms for assigning substrate network resources to virtual network components. In *IEEE INFOCOM*, 2006.

Mirror Mirror on the Ceiling: Flexible Wireless Links for Data Centers

Xia Zhou, Zengbin Zhang, Yibo Zhu, Yubo Li[§], Saipriya Kumar,
Amin Vahdat[†], Ben Y. Zhao and Haitao Zheng
Department of Computer Science, U. C. Santa Barbara, USA
[§]Xi'an Jiaotong University, P. R. China [†]Google and U. C. San Diego, USA
{xiazhou,zengbin,yibo,saipriya,ravenben,htzheng}@cs.ucsb.edu, heatonlyb@gmail.com, vahdat@cs.ucsd.edu

ABSTRACT

Modern data centers are massive, and support a range of distributed applications across potentially hundreds of server racks. As their utilization and bandwidth needs continue to grow, traditional methods of augmenting bandwidth have proven complex and costly in time and resources. Recent measurements show that data center traffic is often limited by congestion loss caused by short traffic bursts. Thus an attractive alternative to adding physical bandwidth is to augment wired links with wireless links in the 60 GHz band.

We address two limitations with current 60 GHz wireless proposals. First, 60 GHz wireless links are limited by line-of-sight, and can be blocked by even small obstacles. Second, even beamforming links leak power, and potential interference will severely limit concurrent transmissions in dense data centers. We propose and evaluate a new wireless primitive for data centers, *3D beamforming*, where 60 GHz signals bounce off data center ceilings, thus establishing indirect line-of-sight between any two racks in a data center. We build a small 3D beamforming testbed to demonstrate its ability to address both link blockage and link interference, thus improving link range and number of concurrent transmissions in the data center. In addition, we propose a simple link scheduler and use traffic simulations to show that these 3D links significantly expand wireless capacity compared to their 2D counterparts.

Categories and Subject Descriptors

C.2.1 [**Network Architecture and Design**]: Wireless communication

Keywords

Data centers, 60 GHz wireless, wireless beamforming

1. INTRODUCTION

Modern distributed applications running in clusters and data centers can run at massive scale, with potentially tens of thousands of servers spread across hundreds of racks. The bandwidth requirements of these applications can range from the relatively modest (*e.g.*, hundreds of Mb/s per server [44]) to substantial (*e.g.*, tens of Gb/s per server for high-end scientific computing [24]). Delivering such bandwidth comes at substantial cost for the requisite switching infrastructure. As a result, a number of recent efforts have investigated techniques for deploying more efficient data center network topologies [11, 14, 20, 21, 22].

While these alternate topologies offer a range of benefits over the current state of the art, we argue that there are a number of inherent challenges with the deployment of *any* wired network technology. First, any large-scale network consists of multiple "stages," meaning a multiplicative factor in the number of fibers/wires required for every server in the cluster. The process of planning, routing, deploying, testing, and then repairing tens of thousands of fibers at the scale of a building incurs substantial cost (both in capital and operational expenditures). Based on our experience, this manual overhead often delays the operational time of large clusters by weeks or months. Lost machine depreciation and lost opportunity costs for leveraging additional capacity can produce millions of dollars in hidden cluster costs.

Second, wired deployments typically cannot anticipate bandwidth requirements to every rack, and thus must distribute a fixed amount of fiber to every rack spot. This means that we must overdeploy fiber for the "worst case" of communication requirements rather than the average case, increasing the cost and maintenance of networking infrastructure.

Third, deployed networks are extremely costly and complex to modify, partially due to the characteristics of multistage network topologies. For example, doubling the number of top of rack servers in a Clos topology network [14, 21] either requires rewiring half the existing fiber (impossible since fibers are often bundled together) or pre-deploying twice the higher-stage network switches than otherwise required (an expensive proposition). Similarly, moving a rack later based on computation needs requires running additional fiber to match. Our experience has shown that adding fiber to a pre-existing cluster is a complex process that incurs substantial delays, and sometimes simply intractable.

Of course, the key benefit of a wired network, whether electrical or optical, is the tremendous amount of bandwidth it can deliver. While the highest end of data center computing is likely to continue to require wired network deployments, in this paper we focus on the subset of applications, and perhaps the majority of applications as borne out by recent measurement studies [15, 16, 25], with more modest or

SIGCOMM'12, August 13–17, 2012, Helsinki, Finland.

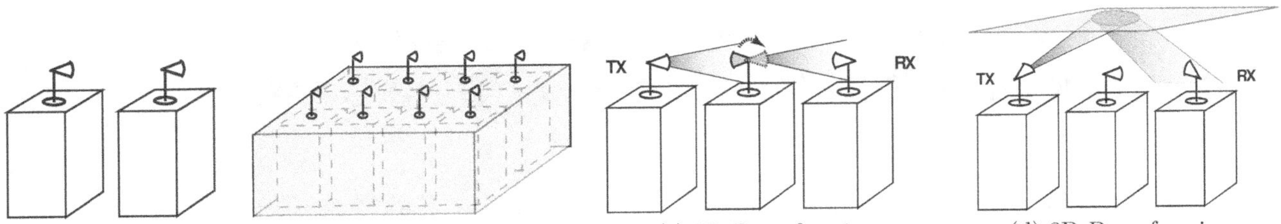

(a) Rack-based DC (b) Container-based DC (c) 2D Beamforming (d) 3D Beamforming

Figure 1: Radio transceivers are placed on top of each rack (a) or container (b). Using 2D beamforming (c), transceivers communicate with neighboring ones directly, but forward traffic in multiple hops to non-neighboring racks. Using 3D beamforming (d), the ceiling reflects the signals from each sender to its receiver, avoiding multi-hop relays.

more localized/bursty bandwidth requirements. That is, we focus on the subset that do not require (near) non-blocking all-to-all communication at data center scale.

In particular, we focus on high-throughput, beamforming wireless links in the 60 GHz band. The unlicensed 60 GHz band provides multi-Gbps data rates and can be implemented with relatively low-cost hardware. Because 60 GHz signals attenuate quickly with distance, multiple wireless links can be deployed in a single data center. In our efforts to expand the effective bandwidth of 60 GHz links, we hope to create a new primitive that can be used to either augment existing networks with *on-demand* network links, or potentially replace wired links in data centers with modest bandwidth requirements. We build on pioneering efforts of earlier work that proposed 60 GHz links to alleviate hot spots in the data center [23, 26].

However, earlier efforts face a number of limitations. First, even beamforming directional links will experience *signal leakage*, and produce a cone of interference to receivers near or behind the intended target receiver. This limits the number of links that can be active concurrently in densely occupied data centers, and reduces the aggregate throughput offered by these wireless links.

Second, these links *require direct line-of-sight* (LOS) between sender and receiver, and can be blocked by even small objects in the path. This limits the effective range of 60 GHz links to neighboring top-of-rack radios. Since hotspots occur regularly at both edge and core links [15], augmenting core links would require multiple hops through a line-of-sight 60 GHz network. Half-duplex, directional antennas mean that these multi-hop links will suffer at least a 50% throughput drop, higher-levels of potential congestion, and additional delays required to frequently adjust antenna orientation.

To address these issues, we investigate the feasibility of 60 GHz 3D beamforming as a flexible wireless primitive in data centers. In 3D beamforming, a top-of-rack directional antenna forms a wireless link by reflecting a focused beam off the ceiling towards the receiver. This reduces its interference footprint, avoids blocking obstacles, and provides an indirect line-of-sight path for reliable communication. Such a system requires only beamforming radios readily available today, and near perfect reflection can be provided by simple flat metal plates mounted on the ceiling of a data center.

3D beamforming has several distinctive advantages over prior "2D" approaches. First, bouncing the beam off the ceiling allows links to extend the reach of radio signals by avoiding blocking obstacles. Second, the 3D direction of the beam significantly reduces its interference range, allowing more nearby flows to transmit concurrently. Third, the reduced interference extends the effective range of each link, allowing our system to connect any two racks using a single hop, and mitigating the need for multihop links.

In this paper, we propose a 3D beamforming system for 60 GHz wireless transmissions in data centers. The 3D beamforming idea was first introduced by Zhang et al. in [46]. In this paper, we greatly extend the prior work, and use measurements of a local 60 GHz testbed to quantify and compare the performance of 3D and 2D beamforming links. We find that 3D wireless beamforming works well in practice, and experiences zero loss in signal or throughput from reflection. We also describe a link scheduler for 3D beamforming systems that maximizes concurrent links while also taking into account accumulative interference and antenna alignment delays. Finally, we use a detailed simulation of data center traffic hotspots to quantify the performance of 3D beamforming systems. Our results show that while 2D links can only support a small portion of hotspot traffic links, 3D beamforming can connect all rack pairs in a single hop, and can significantly reduce overall data completion time for wired networks across a range of bisection bandwidths.

While wired networks will likely remain the vehicle of choice for the high-end of distributed computing, we believe that efforts such as 3D beamforming can expand the applicability and benefits of wireless networking to a broader range of data center deployments.

2. 60 GHZ: LIMITATIONS AND SOLUTIONS

While modifying the topology of wired data centers is costly, complex, and sometimes intractable, administrators can introduce flexible point-to-point network links with the addition of wireless radios. Prior work has proposed the use of 60 GHz links to augment data center capacity [23, 26, 35, 38]. Figures 1(a)-(b) show a common deployment scenario, where wireless radios are placed on top of each rack or container to connect pairs of top-of-rack (ToR) switches.

In practice, however, data center managers remain skeptical on deploying wireless links despite their potential benefits [1]. In this section, we summarize prior work in this space, and use detailed experiments on a 60 GHz testbed to identify and quantify key limitations of current proposals.

2.1 60 GHz Links in Data Centers

Existing designs [23, 26, 27, 38] adopt 60 GHz wireless technologies for several reasons. *First*, the 7GHz spectrum

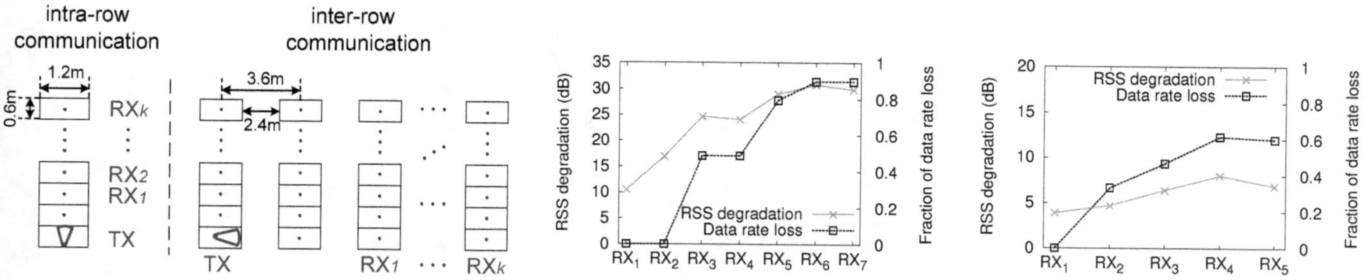

(a) Experiment Setup (b) Intra-row Blockage (c) Inter-row Blockage

Figure 2: Impact of link blockage using 2D beamforming. (a) Our experiment setup to examine antenna blockage for intra- and inter-row links. TX represents the transmitter, and RX_k is the target receiver which is blocked by k antennas. (b-c) The RSS degradation and data rate loss for intra- and inter-row scenarios.

available in this band can deliver the multi-Gbps data rates required by data centers. *Second,* 60 GHz links operate at a high carrier frequency, which limits the interference they generate [45], and is highly beneficial to data centers with dense rack deployments. *Third,* 60 GHz links can use beamforming to enhance link rate and further suppress interference [23, 27]. Beamforming is a physical layer technique to concentrate transmission energy in a specific direction. Today, 60 GHz beamforming radios are readily available and affordable, either as directional (horn) antennas [23] or antenna arrays [8]. They use either mechanical or electronic mechanisms to achieve fine-grain directional control [23, 40].

A 60GHz Testbed. As an experimental platform for 60 GHz wireless projects, we have built a small 60 GHz wireless testbed. To build a bidirectional 60 GHz link, we set two Gigalink 6451 radios from HXI Millimeter Wave Products [5] to operate using frequency division on two frequency bands near 60 GHz, centered at 58.2 GHz and 61.9 GHz. A simple On-Off-Keying modulation method delivers up to 1.25Gbps of bandwidth,but actual throughput is limited to 1 Gbps by the radio's fiber-optic Gigabit Ethernet interface.

These radios were originally designed for long distance wireless links, and each is equipped with a 60 GHz transceiver and a 41dBi directional antenna transmitting at 10dBm power. To modify them for short distance data center links in our experiments, HXI engineers modified the radio amplifier to reduce output power to 0dBm. We also replaced 41dBi antennas with a 25dBi standard gain horn antenna from Flann Microwave [3], with 3dB beamwidth of 10°. For our measurements, we read RSS values from an SNMP-based software monitor running on the radio. Since currently available 60 GHz radios all transmit at a single fixed data rate, we derive link throughput by converting RSS values to data rates according to the 802.11ad standard [6].[1]

2.2 Current Limitations

Despite the many tangible benefits of adding 60 GHz links to a data center, there are two notable limitations with current designs. Here we discuss each in detail and use data from our testbed to quantify its impact.

Link Blockage. Link blockage is a limiting factor for 60 GHz links. The 5mm wavelength of these links means that any object larger than 2.5 mm can effectively block signals or reflect them, producing multipath fading and degrading transmission rates [37]. In today's data centers, this is problematic because racks are organized in a grid, and transceivers and antennas on one rack can easily block transmissions on another rack. This has led to current designs limiting themselves to connecting neighboring racks [23, 27] (see Figure 1(c)).

We measure the severity of the problem by placing multiple antennas between two 60 GHz endpoints, and observing performance degradation on the link. We examine cases for both intra- and inter-row rack communications (see Figure 2(a)). In our intra-row case, antennas are separated by a fixed distance of 0.6m (roughly one rack width). Thus, a link blocked by k antennas, *i.e.* $TX \to RX_k$, has a distance of $(k+1)\cdot 0.6$m. We vary the number of blocking antennas in the experiment, and show the difference in performances with and without blocking antennas. Figure 2(b) shows the RSS degradation and data rate loss. Clearly, even when there is only one blocking antenna (link distance = 1.2m), RSS degradation can be as high as 10dB. Since the signal strength is still high, RSS degradation does not reduce data rates. As the number of antennas increases, the RSS degradation becomes more severe and data rates drop quickly: 3 antennas (link distance = 2.4m) can cause a 25dB RSS degradation and 50% data rate loss, while 6 blocking antennas (link distance = 4.2m) can cause a 30dB RSS degradation and nearly 90% data rate loss.

For inter-row communications, antennas are separated by a fixed distance of 3.6m (one rack length of 1.2m plus one row separation of 2.4m) [23]. Results in Figure 2(c) show similar trends as the intra-row case, except that the impact of blockage is slightly lower. This is because the RF beam emitted by the horn antenna propagates in a cone-shape. The closer the first blocking antenna is to the transmitter, the more the signal it blocks from the receiver. Overall, our measurement results clearly demonstrate that link blockages cause severe problems for 60 GHz transmissions.

To reduce link blockage, one option is to intelligently place radios on each racks, which might be effective for some rack pair connections. To connect racks dynamically, however, the radios must tune to different directions and still block

[1]The 802.11ad standard defines a set of data rates (385Mbps-6.76Gbps). Following its receiver sensitivity table, which guarantees 1% packet error rate and assumes 10dB noise figure and 5dB additional implementation loss, we derive the required SNR for each data rate. We then compute the measured SNR from each measured RSS and noise, and use the above mapping to derive the data rate supported by each measured RSS value.

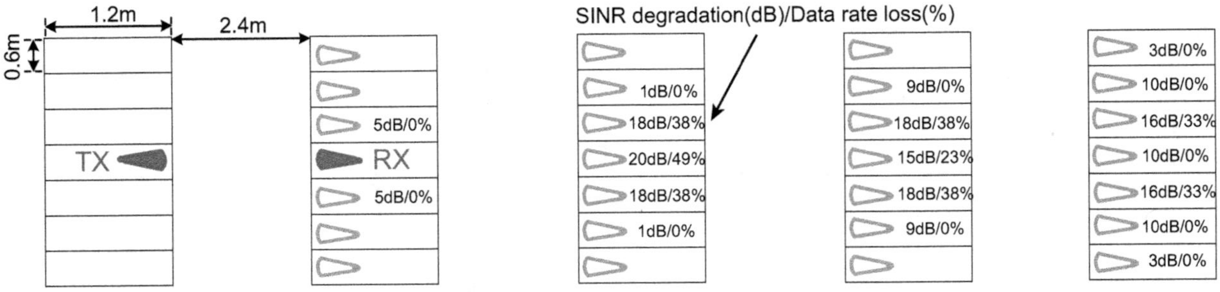

Figure 3: Our experiment to examine the impact of link interference using 2D beamforming. We place a transmitter TX in the middle of a rack row as the interfering transmitter, a receiver RX in the next rack row as its receiver, and four additional rack rows as victims. Each victim radio, in red, points its antenna to the rack of its neighboring row to the left, representing the receiver of an inter-row link of link distance 3.6m. In each rack we list the SINR degradation and data rate loss due to the interference from TX. We omit the results at racks whose SINR degradation is less than 1dB.

transmissions. Another option is to place racks in a hexagonal configuration [41]. While this approach leads to inefficient space use, it also does not solve the fundamental link blockage problem. Finally, multi-hop connections between non-neighboring racks would increase end-to-end delay, dramatically reduce throughput, and produce potential bottlenecks at racks congested by forwarding traffic from multiple links.

Radio Interference. Despite the use of beamforming to bound the transmission energy in a "narrow" direction, radio interference remains an issue for these systems. Radio design artifacts will still produce signal leaks outside of the intended direction [29, 40]. When placed in a dense rack formation, leakage produces harmful interference between nearby links and limits the density of concurrent links.

Using our testbed, we measure the impact of interference produced by a single transmission, in the presence of antenna blockage. As shown in Figure 3, we place a 60GHz transmitter (TX) in the middle of the data center, and measure the RSS at 27 racks (in red) located in four neighboring rows. At each of these 27 racks, the radio points its antenna to the rack of its immediately left row, representing the receiver of an inter-row link with link distance of 3.6m and SNR of 31dB. This experiment allows us to measure the interference experienced by each of the 27 inter-row links when TX is transmitting, from which we compute the SINR degradation and data rate loss due to interference. Results in Figure 3 show that despite the fact that TX's interference signal is blocked by various antennas, 15 inter-row links behind the destination still observe 5-20dB degradation in their SINR. 8 of these links suffer 20-49% data loss.

The spread of radio interference significantly limits the number of concurrent wireless links in a data center. One option is to separate the links in the frequency domain. But this reduces the per-link capacity, since the total available bandwidth is fixed across the frequency range. Alternatively, data center managers can increase the spacing between racks to reduce interference. But this leads to inefficient space and power usage, and weakens long-distance links.

2.3 Solution: 3D Beamforming

To address these limitations, we propose *3D beamforming*, a new beamforming approach that leverages ceiling reflections to connect racks wirelessly. An example is shown in Figure 1(d), where a transmitter bounces its signal off of the ceiling to the receiver. This creates an indirect line-of-sight path between the sender and receiver, bypassing obstacles[2] and reducing interference footprint.

To align its antenna for a transmission, the sender only needs to know the physical location of the receiver rack, and point to a position on the ceiling directly between the two racks. This is because all racks (and their 60 GHz radio antennas) are of the same height.

3D beamforming requires three hardware components:

- *Beamforming Radios*: We reuse beamforming radios [8, 23] and adjust beam directions in both azimuth and elevation by placing the horn antennas on rotators. Existing rotators can achieve an accuracy of 0.006°-0.09° [4, 7].
- *Ceiling Reflectors*: Reflectors on the ceiling act as *specular mirrors* to reflect signals. Our experiments confirm prior work [12, 36] showing that flat metal plates offer perfect specular reflection without degrading energy or changing path loss characteristics.
- *Electromagnetic Absorbers*: We place electromagnetic absorbers [12] near each antenna to prevent any local reflection and scattering. These inexpensive absorbers require no maintenance.

3D beamforming largely addresses both of the main limitations with existing 2D 60 GHz proposals. First, by bouncing beams off a reflective ceiling, it dramatically reduces the interference region for wireless links and allows deployment in densely packed data centers. Second, the reflective path avoids obstacles and creates effective line-of-sight paths between most or all rack pairs in a common 250-rack data center. Addressing these issues means we can connect most or all rack pairs using single-hop 60 GHz links, thus maximizing bandwidth and eliminating forwarding delays. It also means a large number of links can be active in a small area without causing mutual interference and limiting performance.

In this paper, we present first steps in building flexible wireless links using 3D beamforming. We identify and address practical issues in the physical and link layers, and describe experience and experimental results from a local 60

[2]Here we assume that there are no obstacles between top of racks (or containers) and the ceiling. For instance, this might require mounting the radios above cable trays, though we leave a detailed study of physical deployment to future work.

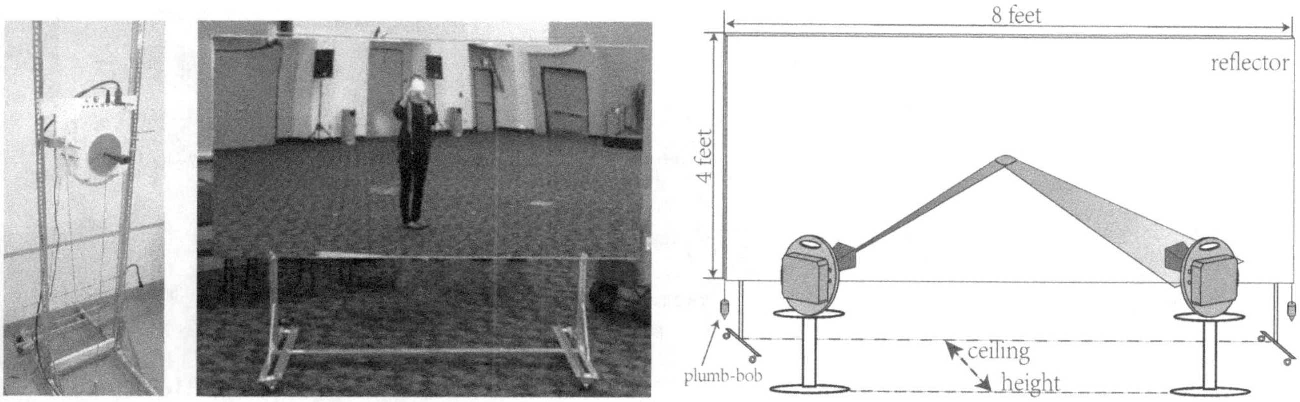

(a) Radio (b) Metal Reflector (c) Testbed Overview

Figure 4: Our 3D beamforming testbed. (a) The 60GHz radio with horn antenna, mounted on top of a mobile platform with adjustable height. (b) A 4ft×8ft, mirror-quality stainless steel reflector mounted vertically on a mobile platform. (c) An illustration of the experiment configuration.

GHz 3D beamforming testbed. We limit our discussion of 3D beamforming as a general link-layer primitive, and leave for future work other issues such as routing, traffic management, and wired/wireless co-scheduling. We believe this work addresses a few of the key concerns associated with large-scale wireless data center deployments, principally enabling substantially more bandwidth to be delivered more flexibly in the data center. However, many open questions remain before we expect to see large-scale deployments.

3. MICROBENCHMARK RESULTS

Using detailed hardware experiments, we now examine the key properties of 3D beamforming, and compare them to 2D systems. We focus specifically on physical performance characteristics of our approach, and its sensitivity to factors such as radio density, rotator accuracy, and reflection material.

3D Beamforming Testbed. As shown by Figure 4, our local testbed consists of two 60GHz beamforming radios from HXI (described in Section 2), a 4ft×8ft metal reflector, and RF absorbers from ETS-Lindgren [2]. We test two types of reflectors: commercial-grade mirror-quality stainless steel plates and off-the-shelf cheap galvanized steel sheets from our local home improvement store. To assist with rapid experimentation, we mount the reflector vertically on a mobile platform that stands in parallel to a line connecting the center of the two radio transceivers. We vertically align platform using multiple hanging plumb-bobs. The corresponding ceiling height h is the perpendicular distance between the reflector and the line. To prevent reflected signals from producing more reflections at the receiver side, we place RF absorbers under the antenna. The absorber is a surface tiled with small pyramids 7.5cm thick. It does not block 3D transmit/reflection paths, but eliminates additional reflections. Finally, instead of using rotators, we manually calibrate the orientations of the horn antennas, using high precision laser pointers for guidance. We also manually introduce calibration errors to emulate the use of rotators of different precision (Section 3.3).

We performed detailed experiments in two indoor environments: a 10m×10m conference room and a 20m×26m pavilion room.

3.1 Validating Physical Properties

Our first question is a basic one: "does 3D beamforming work, and what is the impact of reflection on signal strength and interference characteristics?"

Property 1: Extended Link Connectivity. Our first experiment looks at link connectivity. Intuitively, using ceiling reflection, 3D beamforming will bypass obstacles in the horizontal plane, eliminating the antenna blockage problem of its 2D counterpart. More importantly, since ceiling reflectors should produce no loss [12, 36], it should produce an indirect LOS path following the free-space propagation model [23, 32]:

$$P_r = \frac{P_t G_t G_r \lambda^2}{(4\pi)^2 (L^2 + 4h^2)} \quad (1)$$

where P_t and P_r are the transmit and receive power, G_t and G_r are the transmit and receive beamforming antenna gains, λ is the radio wavelength, L is the distance between the sender and receiver, and h is the distance from the antenna to the ceiling. To verify our hypothesis, we measure RSS at different link distances for both 2D (with no obstacles) and 3D beamforming. We also vary the ceiling height h between 2m and 3m.

The results confirm our hypothesis. Figure 5(a) plots the measured RSS as the function of the propagation path length, *i.e.* L for 2D and $\sqrt{L^2 + 4h^2}$ for 3D beamforming. As a point of reference, we also plot the free-space model in (1). We make three key observations. *First*, our measurement results match the model, confirming that both beamforming methods follow the free-space propagation model, and that the reflector introduces no energy loss. *Second*, a mirror-quality stainless steel plate and a cheap galvanized steel sheet both offer perfect reflection. *Third*, we found no visible difference between the results collected in the two rooms.

We also verify 3D beamforming's ability of bypassing obstacles in the 2D plane by placing absorbers along the line connecting the two radios. The height of the absorber is similar to that of the horn antenna. We observe no difference in RSS even for the longest link distance achievable in the two rooms (30m).

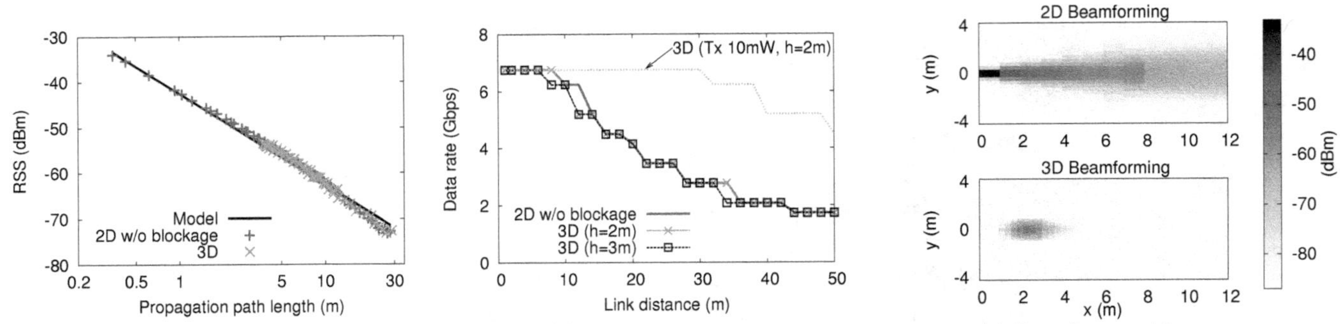

(a) RSS (b) Link Data Rate (c) Interference Map

Figure 5: Measured propagation characteristics of 3D beamforming, transmit power at 0dBm. (a) Ceiling reflection introduces no energy loss; (b) Despite having a longer propagation path, 3D beamforming can still maintain similar data rate as 2D beamforming without blockage; (c) 3D beamforming can significantly reduce the interference footprint. The sender and receiver are at (0, 0), (2.4, 0) respectively.

We also examine the link throughput of the two beamforming methods. Because the HXI radios transmit at a single data rate (1Gbps), we examine a wider range of data rates using the measured RSS and the 802.11ad's receiver sensitivity table. Figure 5(b) shows the resulting link throughput as a function of the link distance L. Because the room where we performed experiments was only 30m long, we derived the data rates for longer links using the RSS values generated by the propagation model. We see that even at a very low transmit power (0dBm), 3D beamforming can reach 6+Gbps when two endpoints are separated by 10m or less. At a link distance of 50m, it still offers nearly 2Gbps of throughput. If we set the transmit power to the standard level of 10dBm, link throughput, shown as the dotted line in the same figure, increases to 6.76Gbps at 30m and 4.5Gbps at 50m. Furthermore, compared to 2D beamforming, 3D achieves nearly the same data rate despite having a longer propagation path (see Eq. (1)).

Property 2: Reduced Radio Interference. Our second experiment examines the interference footprint of both 2D and 3D beamforming. For both methods, we first set up a target transmission link X, then keep the transmitter intact and move the receiver around to measure link X's power emission map. We divide the measurement space into 0.3m×0.15m grids. In each grid, we rotate the receiver antenna to locate the direction with the maximum signal strength, subtract this strength by the receiver antenna gain, and use the result as the maximum interference that link X produces to this location.

Figure 5(c) shows the measured interference footprint for both 2D (w/o blockage) and 3D beamforming, when the ceiling height h=2m. The sender and the receiver of the target link X are placed at position (0m, 0m) and (2.4m, 0m) on the map, respectively. For 2D beamforming, the directional wave still propagates freely in its beam direction, affecting other receivers along the path. The signal leakage also contributes to the level of interference. In contrast, 3D beamforming bounds the interference region to a much smaller area, and limits the impact of signal leakage. We also verified that the measured interference footprint aligns with the propagation model and the antenna pattern of the 10° horn antenna [28, 29]. We omit those results due to space limitations.

3.2 Multiple Radios per Rack

In practice, a single rack can host many servers (e.g., 20-80), and is likely to request multiple simultaneous data connections with other racks. With a single 60 GHz radio, these transmissions will be performed in order, with the antenna re-orienting between transmissions. A much more desirable scenario is to put multiple radios on each rack to support parallel transmissions and reduce head-of-line blocking. For today's standard racks with size (4ft×2ft) and 60GHz radio size (1ft×1ft), we can place up to 8 radios per rack.

We quantify these benefits by using simulations to compute the number of concurrent wireless links supported for two data center configurations. The first configuration has size 15m×42m and contains 250 racks, similar to the layout used in [23]. Racks are grouped into 5×5 clusters, and each cluster is a row of 10 racks with no inter-spacing. Aisles separating the clusters are 3m (between columns) and 2.4m (between rows). The second deployment uses shipping containers [10]. It consists of 2×2 container clusters. Each cluster has 8 containers in a row with inter-spacing of 0.61m. Overall, the data center has size 15m×50m, and contains 256 racks.

We configure wireless links as follows. We assign m radio transceivers per rack and allow each transceiver to associate with one link. Given the size of our deployments, we use 60GHz radios with 10dBm transmit power and standard $10°$ horn antenna, so that every rack pair connects in 1-hop at 5+Gbps in both directions. We build bi-directional links by randomly selecting rack pairs, forming arbitrary rack to rack communication. We determine the number of concurrent links as follows. We admit links one by one in a random order, compute their cumulative interference to each other, and only admit a link if all links after admission achieve their stand-alone data rates as if there were no interference. In other words, these concurrent links do not interfere with each other. We consider two cases: when all the links operate on a single 2.16GHz channel, and when three 2.16GHz channels (for the US 60GHz band) are available. Our simulator uses the free-space propagation model (defined by Eq.(1)), which we verified via experiments in Section 3.1. We compute interference as the total energy accumulated from all concurrent transmissions, accounting for the impact of both antenna orientation and radiation pattern [28, 29].

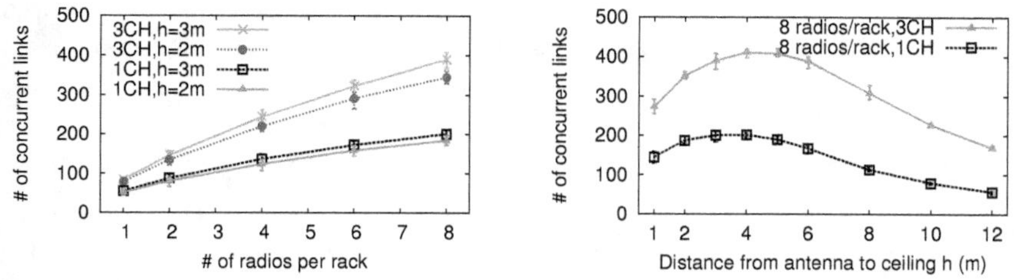

(a) Impact of # of Radios per Rack

(b) Impact of Ceiling Height h

Figure 6: The number of concurrent links using 3D beamforming, varying the number of radios per rack and ceiling height h, using one (1CH) or three channels (3CH).

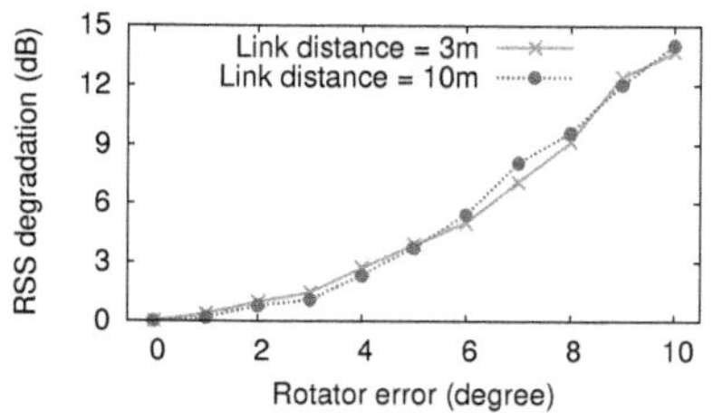

Figure 7: RSS degradation caused by antenna rotator error (measured by antenna mis-alignment angle), using 3D beamforming.

Figure 6(a) plots the number of concurrent links supported as a function of the number of radios per rack. The two topologies lead to similar results and thus we only show the result for the first. We make two key observations. *First*, with a single radio, an average of 55 randomly formed links can operate simultaneously on a single channel. When using three channels, about 88 links (70% of the total links) can operate simultaneously[3]. This result shows that we can simultaneously connect the majority of rack pairs wirelessly using a wire-like connection with 5+Gbps of bandwidth.

Second, the number of concurrent links grows linearly with the number of radios per rack. With eight radios and three channels, an average of 390 randomly formed bi-directional links can operate simultaneously, a 440% improvement over the single radio scenario. This also means that on average, each rack can communicate with four other racks simultaneously, while each bi-directional link achieves at least 5Gbps. This type of flexible and extended connectivity is particularly useful for popular data center jobs such as the "shuffle phase" of MapReduce and local multicast.

Impact of Ceiling Height h. From Figure 6(a), we also observe that increasing the ceiling height h from $2m$ to $3m$ leads to more concurrent links. This is because a larger h makes each beam arrive at its receiver at a larger elevation angle, effectively reducing the interference region. Increasing h beyond $4m$, however, leads to performance loss (Figure 6(b)). This is because increasing h also lengthens the signal propagation path and hence degrades the received signal strength. This loss starts to dominate when h exceeds $4m$, creating a sweet spot of h between 3-$4m$.

3.3 Sensitivity to Hardware

Finally, we examine the sensitivity of performance to different types of hardware and materials.

Sensitivity to Rotator Accuracy. The first question is whether the performance of 3D beamforming will degrade significantly if antenna directions are not calibrated accurately, *e.g.* due to rotator error. We verify this sensitivity using testbed experiments. To produce rotation errors, we first set up a link with accurately calibrated antennas and measure its received signal strength. We then rotate the receive antenna at 1° intervals while recording signal strengths. We repeat this experiment at various link distances and for the case of rotating the transmit antenna. Results in Figure 7 show that a misalignment within 1° leads to negligible impact on the signal strength, and an error of 5° only leads to a RSS degradation of 3.6dB. These results also closely match prior work on Kelleher's universal horn pattern [28]. Thus many existing rotators [7, 4] are sufficiently precise for our needs. Finally, we also verified that 2D and 3D beamforming have the same sensitivity to rotator errors, *i.e.* they receive the same level of performance degradation from rotator misalignments.

Impact of Reflector Materials. Throughout our experiments, our results show that both the cheap, lightweight steel plate and the mirror-quality stainless steel plate offer perfect reflection. This means that 3D beamforming does not require specialized polished metal surfaces, and can be deployed using low-cost metal sheets. Finally, we also tested the suitability of other building materials as reflectors, including standard smooth concrete and plaster walls. The results confirm results from prior studies [30]. For concrete walls, we observe a small signal strength degradation compared to the metal reflectors (roughly 3dB). For plaster walls, the degradation increases to 5dB. While more detailed study is necessary, these initial findings raise the possibility that we may be able to deploy 3D beamforming links without modifying data center ceiling materials.

4. SCHEDULING 60 GHZ LINKS

By forming high-throughput wireless interconnects on-demand, 3D beamforming can deliver additional burst bandwidth to data center applications without pre-provisioning wired capacity among all rack pairs. Prior 60 GHz data center proposals constrained links to neighboring racks, greatly limiting the distance and number of wireless links [23]. In contrast, 3D beamforming connects pairs of racks in large data centers in a single hop using indirect line-of-sight paths.

But to fully utilize the benefits of 3D beamforming, we must carefully schedule transmission links to maximize efficiency and minimize wireless interference. In this section, we identify the key challenges of scheduling 3D beamforming links in data centers, and present a centralized link scheduler to support flexible bandwidth allocation.

4.1 Challenges

Our link scheduler must address three key challenges:

First, designing our scheduler requires an accurate interference model for 3D beamforming links. Given the reflective nature of our beamforming links, the interference a receiver

[3]The number of concurrent links does not grow proportionally with the number of channels because interference patterns are not uniform across links. Such negative effects should gradually diminish in larger data center topologies.

experiences is no longer dominated by energy leakage from the nearest transmitter. Instead, because of the dense deployment of these links in data centers, the main source of interference is the accumulation of signals from the many transmitting neighbors. This *accumulative interference* effect is significant and must be accounted for. Our initial simulations show that if we use conventional pairwise interference models that ignore accumulative interference [34], up to 30% of our scheduled links will fail.

Second, our scheduler needs to handle short-lived traffic bursts [15, 25], and thus must be online. To maximize the number of concurrent transmissions and minimize job execution time, our scheduler must be efficient and lightweight, *i.e.* introduce minimal overhead in control traffic and scheduling delay.

Third, scheduling account for antenna rotation delay. This is particularly important when using horn antennas. Using today's rotators [4, 7], rotation delay ranges between 0.01 and 1 second, which is likely in the range of (and longer than) the full transmission times of some links. Furthermore, since the amount of rotation carried by a mechanical rotator directly affects its lifetime and reliability, we need to minimize such overhead.

Assumptions. We further assume that the centralized scheduler has full knowledge of the rack traffic demands to be carried by the 60GHz network. It receives link requests, generates link schedules periodically, and notifies the scheduled racks with the channel, the radio, and the beam direction they should use. We assume that control messages are sent via a separate control channel independent of the 60GHz wireless network. In practice, the scheduler can either use a provisioned wired network, or use a dedicated local WiFi network for control signaling. We leave the detailed design of control channels as future work.

4.2 Scheduler Design

We propose a greedy scheduling algorithm that addresses these challenges. Our scheduler's primary goal is to schedule as many concurrent links as possible, thus maximizing channel usage and minimizing transmission time. To do so, it derives a "conflict degree" for each link from its accumulative interference, described below. Finally, within the framework of this scheduler, we seek to reduce rotational delay by considering radio orientation in assigning links to radios. We will now describe the components of the scheduler in detail, starting with the estimation of link conflicts.

Conflict Estimation. At the core of the algorithm, we must estimate conflicts to not only calculate the conflict degrees of each link, but also to determine whether a specific radio link can be added to the existing active links without affecting their current link rates. In a nutshell, conflict estimation is based on the calculation of Signal-to-Interference-Noise-Ratio (SINR) for each link.

The scheduler derives the signal and interference using model prediction since the 3D beamforming's propagation environment in data center is very predictable. This avoids additional measurement overhead. The calculation is based on the free-space propagation model (Eq. (1)) verified by our experiments, and also the antenna orientations. We apply Kelleher's universal horn pattern [28, 29] to model the radiation pattern of the horn antenna.

To compute conflict degrees, the scheduler considers each possible pair of link requests assuming they operate on the same channel. Let D_i, D_j denote two link requests, we then calculate D_i's SINR value SINR_{ij} in the presence of D_j as:

$$SINR_{ij} = \frac{S_i}{N + I_{ji}}, \tag{2}$$

where S_i is the signal strength received at D_i's receiver, N is the noise level, and I_{ji} is the interference D_j's transmitter creates on D_i's receiver. D_i and D_j conflict if either has a SINR below the threshold of its required data rate. Thus, the conflict degree of D_i is the number of link requests that conflict with D_i when considered in this way.

To determine if a candidate radio link L_i can be admitted given the presence of scheduled links $\mathbb{L}$ on the same channel, the scheduler calculates $\text{SINR}_{k,\mathbb{L}'}$ for each $L_k \in \mathbb{L}' = \mathbb{L} \cup \{L_i\}$ as the following:

$$SINR_{k,\mathbb{L}'} = \frac{S_k}{N + \sum_{j:L_j \in \mathbb{L}' \setminus \{L_k\}} I_{jk}}, \tag{3}$$

where the notation is identical to those in Eq. (2). If each $\text{SINR}_{k,\mathbb{L}'}$ satisfies the corresponding data rate requirement, then L_i can be scheduled on this channel. This accounts for the interference accumulated by multiple links, and ensures that scheduled links can be active simultaneously without conflict.

Conflict-Degree based Greedy Scheduling. With the goal of minimizing the job completion time, the scheduling problem can be mapped to a traditional graph coloring problem that aims to use the minimal number of colors to color all nodes. In our case, the colors map to 60 GHz frequency channels and time slots. We employ techniques from the graph coloring literature [19] in a greedy fashion, where we schedule rack-level requests in an order based on their *conflict degrees*. The conflict degree d_i of an unscheduled request D_i is defined as the number of other unscheduled requests denoted by set D^C, such that if D_i and any request in D^C are on the same channel, at least one cannot achieve the required data rate.

There is an issue of link preemption in the scheduler. Given long-lived links that provide less than ideal link usage, should the scheduler preempt them, *i.e.* pause them, in order to schedule competing links? In a non-preemptive model, the scheduler keeps the unfinished links untouched, and checks which new ones can be added. This policy ensures that scheduled links will not be disturbed until they complete, thus minimizing the antenna rotation delay and control overhead by interrupting an ongoing link transmission. In a preemptive model, the scheduler pauses ongoing links, treats them as new requests with the remaining unsent traffic, and schedules them together with new link requests. Since previous link requests must compete with new requests, this policy could lead to interruptions to ongoing active links. The benefit, however, is that such a policy could increase the number of concurrent, active links. While we evaluate both policies in Section 5, our default scheduler is non-preemptive.

We note that more complex policies can be added to our scheduler, such as alternative ranking metrics that prioritize job by their deadlines in deadline-driven data centers [43]. In such priority-based scheduling policies, care must be taken to avoid link starvation by gradually increasing the priority of jobs as their waiting time increases. We leave the design of those metric-based schedulers as future work.

Assigning Radios to Links. To minimize the antenna rotation overhead, the scheduler assigns scheduled links to unassigned radios during scheduling. It applies a simple policy. First, if an idle radio on the rack is already pointing to the desired destination rack, the scheduler assigns the link to this radio. Second, if multiple candidate/idle radios exist, the scheduler checks the existing orientations of these antennas and selects the one that is closest in angle to the desired angle for the new link's destination rack.

5. ADDRESSING TRAFFIC HOTSPOTS

In this section, we use network simulations to quantify 3D beamforming's ability to deliver additional bandwidth to data center environments, and its advantages over its 2D counterpart. We consider the case of using wireless links to cover traffic hotspots on top of an existing wired network in data centers. Specifically, we seek to answer three key questions:

1) *Does adding 3D beamforming links to existing wired networks significantly increase available bandwidth for hotspots?*

2) *How significant are the benefits of 3D beamforming over 2D beamforming, and where are they most visible?*

3) *Will antenna rotation delay of today's rotators be a performance bottleneck for 3D beamforming?*

While we answer these questions using results from synthetic traffic traces, we hope trends identified by our study will serve as useful guidelines for practical deployment of 3D beamforming systems.

5.1 Simulation Setup

For our traffic hotspot simulations, we use the data center and radio configurations described in Section 3.2. It consists of 250 racks and a total of 5000 servers. The distance between the antenna and the ceiling (h) is 2m. We use 60GHz radios with 10dBm transmit power and 10° horn antennas. There are three channels of width 2.16GHz, and each radio can operate on one channel at a time. We derive data rates following the specifications of the IEEE 802.11ad standard. For this data center size, every possible pair of racks is able to form a 1-hop link at 5+Gbps using 3D beamforming.

There are 8 radios positioned at the top of each rack. To account for the rotation delay, we assume each antenna uses a rotator from FLIR [4]. The pan speed is 300°/second and the tilt speed is 60°/second. We also examine the case where the rotation is instantaneous, which represents the best possible (ideal) performance for 3D beamforming links.

Traffic Generation. Existing traces [15, 16, 23, 25] do not map data sources to rack locations. Hence, we used synthetic traffic generated based on popular workloads [18] to produce hotspots. We simulate a simple scenario, where 400 of the 5000 total servers in the data center each sends a fixed data payload to 200 other servers. The 400 servers are chosen randomly from any rack in the data center. Each data payload is 128MBytes, and each of the 400 transmitters chooses a set of 200 servers to receive its data. We refer to a complete cycle where each server sends a single payload to each of its 200 destinations as a single round.

To produce controlled traffic hotspots, we introduce a slight bias in server selection. This might emulate a slight preference for certain machines based on their properties such as compute power, uptime, memory size, or network proximity to storage servers. We identify 50 random servers in the data center as "preferred" servers. In each round, each of the 400 transmitters chooses 200 receivers to receive their data, randomly, but with a small bias. As we choose each of the 200, there is a 10% chance that the receiver server is one of the 50 preferred servers. Once we have chosen 200 unique receivers for each of the 400 transmitters, we aggregate these server-pair traffic loads based on their server locations to produce traffic demands at the rack level. We repeat the above procedure to generate 10 rounds of rack-level workloads.

We assume an underlying wired network offering 1Gbps network bisection bandwidth. We also explore larger values ranging from 2 to 6Gbps, which are typical given the oversubscription of today's data centers [14]. Since our goal is to understand how 3D beamforming addresses traffic hotspots, one issue that arises is how traffic demands are split across wired and wireless network links. Without advocating any particular allocation policy, we assume that the traffic is split between the two networks by setting a fixed "deadline" for wired links to finish their portion of the transmission. This allows us to define the portion of traffic sent over the wired network *a priori*, thus deriving the amount of "overflow" traffic allocated to the 60 GHz network. To compute job completion time, we assume that wired and wireless links send data in parallel.

5.2 Impact of Adding 2D/3D Beamforming

Coverage of 2D Links. One of the primary limitations with 2D beamforming is potential blockage issues that result in lower signal strength and loss in data throughput. Prior proposals limited 2D beamforming links to connecting neighboring racks that have no potential blockage issues. Our first experiment looks considers portion of the total overflow traffic can be sent over 2D links, where overflow traffic is the traffic that cannot be sent by the wired network by the specified deadline. Figure 8(a) plots this as a ratio of total traffic across neighboring links, *i.e.* traffic that can be sent across 2D links, over total overflow traffic. In all of our graphs, we plot error bars covering the 90% confidence interval.

Less than 3% of overflow traffic can be addressed using 2D links, regardless of how much traffic is sent across the wired network. We note that this figure might increase, depending on how well the data center managers scheduled jobs to increase rack affinity and limit hotspots to neighboring racks, but this would introduce an additional constraint in job scheduling. Note that we do not plot similar values for 3D links, because 3D links can connect all possible rack pairs in our scenario with a single hop, thus coverage is 100%.

Impact of Antenna Rotation Delay. Next, we look at 3D performance, and try to understand the impact on end-to-end latency by antenna rotation delays. In Figure 8(b), we plot the wireless completion time, *i.e.* time required to send overflow traffic over the 3D links, against the wired completion deadline. A longer wired deadline means more traffic will go over the wired network and less overflow traffic will be left for wireless links. We also draw the line "$y = x$" to show the minimum time to complete transmissions if we used the ideal traffic allocation between wired and wireless. By plotting 3D links with and without rotational delay, we see that all transmissions are completed in 9 seconds in a realistic system, but can be improved to 8 seconds if we

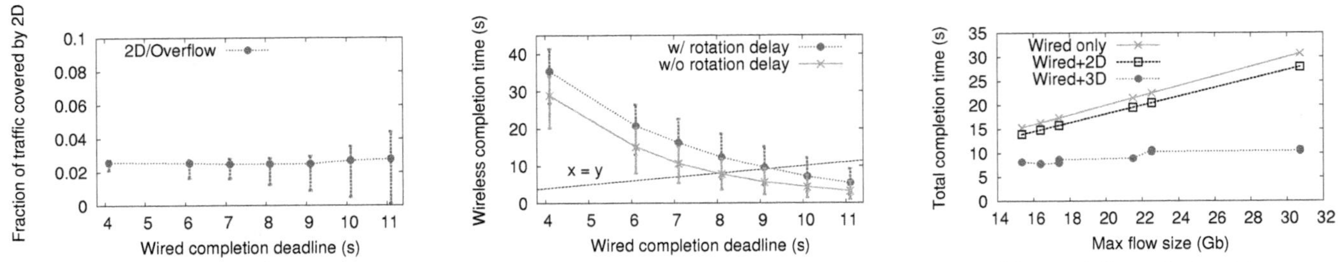

(a) Portion of Traffic Covered by 2D Links

(b) Impact of Antenna Rotation Delay on 3D Performance

(c) Total Completion Time

Figure 8: Performance of 2D and 3D beamforming links in conjunction with a wired network with 1Gbps bisection bandwidth. (a) Less than 3% of overflow traffic can be addressed using 2D links, while 3D links can cover 100%. (b) The time required for 3D to send all the overflow traffic vs. the wired completion deadline, using today's rotators or ideal rotators with zero delay. (c) Adding 3D beamforming cuts the total job completion time by half.

completely eliminate antenna rotation delays. This is the effective upper bound on how well we can perform given perfect radio assignment, and shows that antenna rotation delay is only a small component in end-to-end performance.

Total Completion Time. Next, in Figure 8(c), we plot the effective completion time (if both wired and wireless networks transmitted simultaneously with optimal traffic split) for different rounds of our simulation, assuming a 1Gbps bisection on the wired network. Regardless of the size of the biggest flow, 3D wireless is generally able to reduce total transmission time by half. Since 2D links can only address a small portion of the overflow traffic, its impact on completion time is limited.

Next, we ask the question, "will 3D beamforming links become less useful for wired networks with higher bisection bandwidth?" We vary the underlying bisection bandwidth of the wired network between 1 Gbps and 6 Gbps, and in each case, compute the minimal completion time if we performed the ideal traffic split between the wired and wireless networks. Figure 9 shows that even as the wired network grows in bisection bandwidth, adding 3D beamforming can still reduce total transmission time significantly (ranging from more than 50% to slightly less than 40% as wired bandwidth ranges from 1 Gbps to 6 Gbps). This is not surprising. Since the demand in our hotspot scenario is fixed, a wired network with larger bisection bandwidth consumes more traffic, leaving less overflow traffic for the 60 GHz links.

5.3 Impact of Scheduler Policies

We now evaluate the net impact of different design choices in our link scheduler.

Preemption vs. No Preemption We discussed the issue of preempting existing links in Section 4. To understand which policy is more preferable, we compare the job completion time when applying these two policies with the same workload as above. Non-preemptive is the default policy used in prior experiments, where once a link is scheduled, it utilizes the radio until it finishes. The preemptive policy considers all radios on a rack when scheduling requests, and where deemed appropriate by the scheduler, will pause an existing link to give its radio to a higher priority link request.

We plot the completion time of the wireless traffic load in Figure 10(a). The results match our expectations. A non-preemptive policy reduces the completion time by up

Wired completion time (s)	4	6	7	8
Rotator usage reduction via intelligent radio assignment	27%	22%	22%	21%

Table 1: Intelligent radio assignment leads to moderate reduction in rotator usage.

to 25% compared to the preemptive policy. This is because rescheduling the paused link introduces significant additional overhead (and more antenna rotation), thus reducing overall efficiency.

Intelligent Radio Assignment. Finally, we evaluate the impact of using intelligent radio assignment during link scheduling. Figure 10(b) plots the completion time for the wireless traffic load for our default scheduler (including intelligent radio assignment) and a basic scheduler, which uses random choice to choose between available radios. Neither scheduler uses link preemption.

The results are varied, depending on the traffic load on the wireless network. When the large majority of traffic is sent through the wired network, only the strongest of the hotspots remain for the wireless network. In this case, there will be high contention for radios at a small number of racks, and very little choice in terms of radio assignment. Thus the optimization shows small benefit. When a larger portion of the traffic is delegated to the wireless network, more racks carry moderate traffic that leaves some number of radios free for assignment. In this case, optimizing radio assignment provides some moderate benefit, which is ultimately bounded by the overhead of antenna rotation delay, shown in Figure 8(b).

We also compare the amount of rotations performed by each rotator using the two schedulers. Results in Table 1 show that depending on the traffic load given to the wireless network, intelligent radio assignment can reduce the rotator usage by up to 27%.

Summary. Our findings in this section can be summarized as follows. *First*, we find that in data centers with random traffic patterns, 2D beamforming restricted to neighboring racks can only address a very limited ($\sim$ 3%) portion of traffic hotspots, compared to 100% for single hop 3D beamforming links. *Second*, we find that for many scenarios involving bursty traffic hotspots, using 3D beamforming links in conjunction with the existing wired network can generally reduce completion time by half or more. *Finally*, we

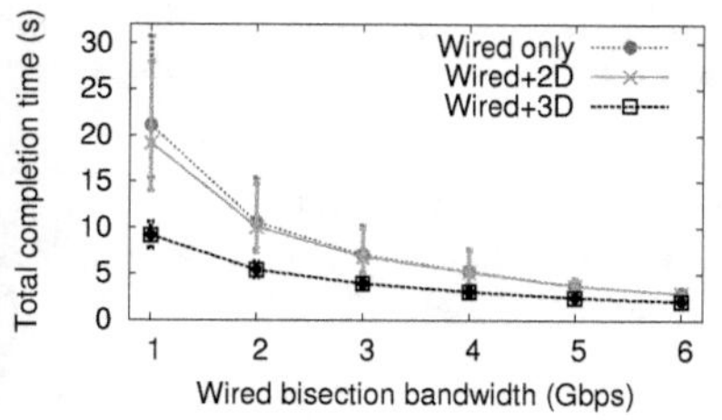

Figure 9: Overall impact of adding beamforming links.

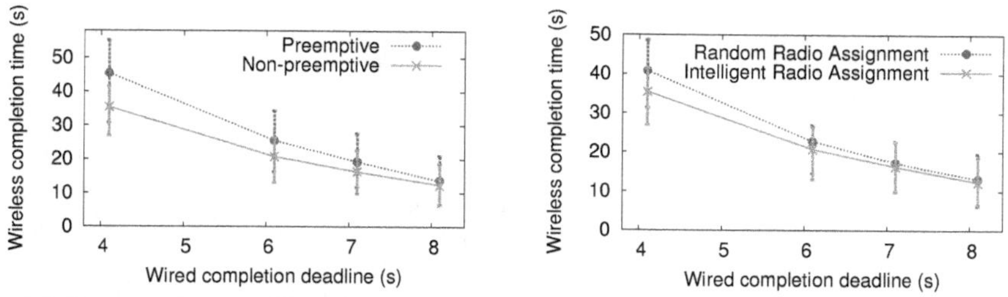

(a) Preemption vs. Non-preemption

(b) Intelligent Radio Assignment

Figure 10: Impact of different scheduling policies.

find that when sizable payloads are involved, *e.g.* 128MB, antenna rotation delays only contribute a small portion of the overall completion time, and much of that can be recovered using simple heuristics such as choosing radios that are closer to the desired transmission angle.

6. DEPLOYMENT CHALLENGES

In this section, we briefly discuss some key challenges of deploying 3D beamforming in data centers.

Physical Rack/Reflector Placement. 3D beamforming performs the best when there are no obstacles between the top of rack/container and the ceiling. When physically arranging racks/containers as well as ceiling reflectors, data center managers should avoid obstacles such as cables and cooling pipes. This is not an issue for container-based data centers [10]. For other types of data centers, raised floors can be used to house cables and pipes in the ground. Any suspended cable trays can be concealed within aluminum-plated ducts, essentially lowering the reflection point from the ceiling. When unavoidable, one can also plan multi-hop transmissions or reflect off walls to route around obstacles. An open question is whether physical rack and reflector placement can be jointly optimized with network communication patterns.

Reflector Curvature. We observe in our experiments that the performance of 3D beamforming is sensitive to the curvature of the reflector. Reflected by a flat surface, the beam propagates following the free-space model. When the surface becomes slightly concave, we observe a visible increase in the measured RSS value. Similarly, we observe a drop when the surface becomes convex. This could be the result of reflection creating multipath signals to the receiver, which could degrade the link performance. Ideally, the reflector should be kept as flat as possible. Yet an open question is whether one can manipulate reflector curvature to further improve 3D beamforming performance.

Rotator Reliability. To communicate with different racks, each transceiver must adjust its beam direction in both azimuth and elevation. For radios with horn antennas, this requires a mechanical rotator to rotate the antenna, leading to extra rotation delay and the issue of rotator reliability. Today's off-the-shelf pan-tilt rotators can provide roughly 3-5 million cycle durability [4] or 15,000-hour lifetime [9]. Clearly, the failure rates of these components must be weighed against their performance when choosing between horn antennas and antenna arrays.

Antenna Arrays. The above concerns with rotator delay and reliability can be addressed by replacing horn antennas with switched beam smart antennas or antenna arrays. Antenna arrays use electronic beam rotation, with delay as low as 50ns in existing 16-element arrays [40]. However, antenna arrays still do not eliminate issues of link blockage or interference and produce more signal leakage than horn antennas, leading to weaker link signal strength and stronger interference to neighboring flows [46]. One potential solution is to configure beam patterns to steer away from neighboring links or to nullify interference [33]. We leave exploration of these issues for future study.

7. RELATED WORK

Data Center Networks. Most prior work addresses traffic congestion through network architecture design and traffic scheduling [11, 13, 14, 18, 21, 22], or modeling network traffic characteristics [15, 16, 25, 17]. 60 GHz wireless was first proposed to data center networking in [35] as a solution to reduce the cabling complexity. Recent proposals use wireless links to augment [23, 26] or replace [38] wired links. In this work, we identify two practical issues of existing approaches, and propose a new beamforming paradigm to effectively address these issues.

While prior work has applied the principle of radio signal reflection to connect non line-of-sight links [35] or to reduce WiFi interference [31], we generalize it to 60GHz links. Our key contribution is to use ceiling reflection in the data center to extend connectivity and suppress interference. We also used detailed hardware experiments to validate our design.

Optical circuit switching [20, 39, 42] is an alternative for adding burst bandwidth to data centers. Optical circuit switching promises tremendous bandwidth but the technology incurs relatively substantial cost and does not offer some of the benefits of wireless augmentation of data center bandwidth, namely reduced cost and deployment complexity.

60GHz Wireless Technology. Prior work mainly focuses on radio and antenna design issues [8, 40], and signal propagation and reflection modeling [37, 45, 12, 30, 36]. Today, there is a wide selection of commercial 60GHz radio products [5]. One recent effort [40] developed a 16-element on-chip antenna array, allowing each radio to switch a beam to any of the 32 predefined directions within 50ns. Our work leverages readily available hardware, and focuses on designing new wireless interconnects explicitly for data centers.

8. CONCLUSION

Traffic in today's data centers is unpredictable, often producing traffic hotspots that result in congestion and delay. Instead of overprovisioning the wired network for bursts by rewiring a data center network at scale, we advocate the use of 60 GHz wireless beamforming links to alleviate traffic hotspots as they occur.

Our work addresses limitations of 60 GHz beamforming that arise from signal blockage and interference caused by signal leakage. Our insight is that by aiming 60 GHz beamforming links at a reflective ceiling, we can achieve indirect line-of-sight between most or all rack pairs in a data center, while minimizing interference. The net effect is that 3D beamforming greatly expands the reach and capacity of 60 GHz links, making them feasible as flexible and reconfigurable alternatives to wired cabling. Our testbed measurements confirm that 3D beamforming links suffer zero energy loss from reflection, and effectively avoid blocking obstacle and reduce interference footprint.

While wired networks will continue to serve high-end data center needs, we believe that efforts such as 3D beamforming can provide significant benefits to a broad range of data center deployments, by potentially reducing deployment complexity and reducing cost compared to a fully provisioned wired network.

9. ACKNOWLEDGMENTS

The authors thank Romit Roy Choudhury and the reviewers for their feedback, and Daniel Halperin and Lei Yang for their insights on 60 GHz hardware. This work is supported in part by NSF grant CNS-0905667.

10. REFERENCES

[1] Data center managers not fixed on wireless. `http://searchdatacenter.techtarget.com/tip/Data-center-managers-not-fixed-on-wireless`.

[2] ETS-Lindgren. `http://ets-lindgren.com/Absorbers`.

[3] Flann Microwave. `http://www.flann.com/`.

[4] FLIR. `http://www.flir.com/mcs/products/`.

[5] HXI Milimeter Wave Products. `http://www.hxi.com/`.

[6] IEEE P802.11ad/D0.1: Enhancements for very high throughput in the 60 GHz band. Draft 0.1, June 2010.

[7] NSI. `http://www.nearfield.com/`.

[8] SiBeam. `http://sibeam.com/whitepapers/`.

[9] Sidus. `http://www.sidus-solutions.com/`.

[10] Sun modular datacenter s20/d20 overview. `http://download.oracle.com/docs/cd/E19115-01/mod.dc.s20/index.html`.

[11] Abu-Libdeh, H., et al. Symbiotic routing in future data centers. In *Proc. of SIGCOMM* (2010).

[12] Ahmadi-Shokouh, J., et al. Reflection coefficient measurement for house flooring materials at 57-64 GHz. In *Proc. of GLOBECOM* (2009).

[13] Al-fares, M., et al. Hedera: Dynamic flow scheduling for data center networks. In *Proc. of NSDI* (2010).

[14] Al-Fares, M., Loukissas, A., and Vahdat, A. A scalable, commodity data center network architecture. In *Proc. of SIGCOMM* (2008).

[15] Benson, T., Akella, A., and Maltz, D. A. Network traffic characteristics of data centers in the wild. In *Proc. of IMC* (2010).

[16] Benson, T., et al. Understanding data center traffic characteristics. *ACM CCR 40* (2010), 92–99.

[17] Chen, Y., Ganapathi, A., Griffith, R., and Katz, R. The case for evaluating MapReduce performance using workload suites. In *Proc. of MASCOTS* (2011).

[18] Curtis, A., Kim, W., and Yalagandula, P. Mahout: Low-overhead datacenter traffic management using end-host-based elephant detection. In *INFOCOM* (2011).

[19] Diestel, R. *Graph Theory*, 4th ed. Springer-Verlag, Heidelberg, 2010.

[20] Farrington, N., et al. Helios: A hybrid electrical/optical switch architecture for modular data centers. In *Proc. of SIGCOMM* (2010).

[21] Greenberg, A., et al. VL2: A scalable and flexible data center network. In *Proc. of SIGCOMM* (2009).

[22] Guo, C., et al. BCube: a high performance, server-centric network architecture for modular data centers. In *Proc. of SIGCOMM* (2009).

[23] Halperin, D., et al. Augmenting data center networks with multi-gigabit wireless links. In *Proc. of SIGCOMM* (2011).

[24] Kamil, S., et al. Communication requirements and interconnect optimization for high-end scientific applications. *IEEE TPDS 21* (2009), 188–202.

[25] Kandula, S., et al. The nature of data center traffic: measurements & analysis. In *Proc. of IMC* (2009).

[26] Kandula, S., Padhye, J., and Bahl, V. Flyways to de-congest data center networks. In *Proc. of HotNets* (2009).

[27] Katayama, Y., et al. Wireless data center networking with steered-beam mmwave links. In *WCNC* (2011).

[28] Kelleher, K. *The Microwave Engineers' Handbook and Buyers' Guide*, 5th ed. New York: Horizon Press, 1964.

[29] Knop, C., and Wiesenfarth, H. On the radiation from an open-ended corrugated pipe carrying the HE11 mode. *IEEE Trans. on Antennas and Propagation* (1972), 644 – 648.

[30] Langen, B., Lober, G., and Herzig, W. Reflection and transmission behavior of building materials at 60 GHz. In *Proc. of PIMRC* (1994).

[31] Liu, X., et al. DIRC: Increasing indoor wireless capacity using directional antennas. In *Proc. of SIGCOMM* (2009).

[32] Manabe, T., Miura, Y., and Ihara, T. Effects of antenna directivity on indoor multipath propagation characteristics at 60GHz. In *Proc. of PIMRC* (1995).

[33] Nikolaidis, G., et al. Cone of silence: adaptively nulling interferers in wireless networks. *SIGCOMM CCR* (2010).

[34] Qiu, L., et al. A general model of wireless interference. In *Proc. of MobiCom* (2007).

[35] Ranachandran, K., et al. 60GHz data-center networking: wireless => worryless? *NEC Technical Report* (2008).

[36] Sato, K., et al. Measurements of reflection and transmission characteristics of interior structures of office building in the 60-GHz band. *IEEE Trans. on Antennas and Propagation 45*, 12 (1997), 1783 –1792.

[37] Schönthier, J. WP3-study "the 60 GHz channel and its modelling". *Tech. Report IST-2001-32686* (2003).

[38] Shin, J.-Y., et al. On the feasibility of completely wireless data centers. *Technical Reports, Cornell University* (2011).

[39] Singla, A., et al. Proteus: a topology malleable data center network. In *Proc. of HotNets* (2010).

[40] Valdes-garcia, A., et al. Single-element and phased-array transceiver chipsets for 60-GHz Gb/s communications. *IEEE Communications Magazine* (2011).

[41] Vardhan, H., et al. Wireless data center with millimeter wave network. In *Proc. of GLOBECOM* (2010).

[42] Wang, G., et al. c-through: part-time optics in data centers. In *Proc. of SIGCOMM* (2010).

[43] Wilson, C., Ballani, H., Karagiannis, T., and Rowstron, A. Better never than late: Meeting deadlines in datacenter networks. In *Proc. of SIGCOMM* (2011).

[44] Yang, F., Shanmugasundaram, J., and Yerneni, R. A scalable data platform for a large number of small applications. In *Proc. of CIDR* (2009).

[45] Yong, S.-K., Xia, P., and Valdes-Garcia, A. *60 GHz Technology for Gbps WLAN and WPAN: From Theory to Practice.* WILEY, 2011.

[46] Zhang, W., Zhou, X., Yang, L., Zhang, Z., Zhao, B. Y., and Zheng, H. 3D beamforming for wireless data centers. In *Proc. of HotNets* (2011).

Author Index

www.ingramcontent.com/pod-product-compliance
Lightning Source LLC
LaVergne TN
LVHW080005210926
848481LV00011B/2947
* 9 7 8 1 4 5 0 3 1 4 1 9 0 *